T0310657

FIELD THEORIES OF CONDENSED MATTER PHYSICS

Presenting the physics of the most challenging problems in condensed matter using the conceptual framework of quantum field theory, this book is of great interest to physicists in condensed matter and high-energy and string theorists, as well as to mathematicians. Revised and updated, this second edition features new chapters on the renormalization group, the Luttinger liquid, gauge theory, topological fluids, topological insulators, and quantum entanglement.

The book begins with the basic concepts and tools, developing them gradually to bring readers to the issues currently faced at the frontiers of research, such as topological phases of matter, quantum and classical critical phenomena, quantum Hall effects, and superconductors. Other topics covered include one-dimensional strongly correlated systems, quantum ordered and disordered phases, topological structures in condensed matter and in field theory and fractional statistics.

EDUARDO FRADKIN is a Professor in the Department of Physics, University of Illinois at Urbana-Champaign. His research interests are in condensed matter physics; disordered systems, high-temperature superconductors, and electronic liquid-crystal phases of strongly correlated systems; quantum Hall fluids and other topological phases of matter; and quantum field theory in condensed matter.

FIELD THEORIES OF CONDENSED MATTER PHYSICS

SECOND EDITION

EDUARDO FRADKIN
University of Illinois at Urbana-Champaign

CAMBRIDGE
UNIVERSITY PRESS

University Printing House, Cambridge CB2 8BS, United Kingdom

Cambridge University Press is part of the University of Cambridge.

It furthers the University's mission by disseminating knowledge in the pursuit of education, learning and research at the highest international levels of excellence.

www.cambridge.org
Information on this title: www.cambridge.org/9780521764445

First edition published by Addison Wesley, 1991
Second edition published by Cambridge University Press, 2013
3rd printing 2015

A catalogue record for this publication is available from the British Library

Library of Congress Cataloguing in Publication data
Fradkin, Eduardo.
Field theories of condensed matter physics / Eduardo Fradkin. – Second edition.
pages cm
Includes bibliographical references and index.
ISBN 978-0-521-76444-5
1. High temperature superconductivity. 2. Hubbard model. 3. Antiferromagnetism. I. Title.
QC611.98.H54F73 2013
537.6′23–dc23
2012039026

ISBN 978-0-521-76444-5 Hardback

Contents

Preface to the second edition

I am extremely happy to, at long last, be able to present the second edition of this book. In spite of what I stated in the preface of the 1991 edition, I ended up not only writing a second edition but, in a sense, a new book. So one can say, once again, that we have met the enemy and it is us. I have been pleased that the 1991 edition of this book was appreciated by many people who found it useful and stimulating. I am really happy that my effort was not in vain.

My motivation for writing this book, in 1991 and now, was to present quantum field theory as a conceptual framework to understand problems in condensed matter physics that cannot be described perturbatively, and hence do not admit a straightforward reduction to some non-interacting problem. In essence, almost all interesting problems in condensed matter physics have this character. Two prime examples of problems of this type in condensed matter physics that developed in the late 1980s, and even more so in the 1990s, are the understanding of high-temperature superconductors and the quantum Hall effects. In both areas field theory played (and plays) a central role. If anything, the use of these ideas has become widespread and increasingly plays a key role. It was lucky that the first edition of this book appeared at just about the right time, even though this meant that I had to miss out on research that was and still is important. This was probably the only time that I was on time, as people who know me can relate. Much has happened since the first edition appeared in print. The problem of the quantum Hall effects has developed into a full-fledged framework to understand topological phases of matter. Although it is still an unsolved problem, the research in high-temperature superconductors (and similar problems) has motivated theorists to look for new ways to think of these problems, and the ideas of quantum field theory have played a central role. The concepts, and subtleties, of gauge theory have come to play a key role in many areas, particularly in frustrated quantum magnetism. The interactions between condensed matter and other areas of physics, particularly high-energy physics and string theory, have become more important. Concepts in topology and other areas of mathematics rarely frequented by condensed matter physicists have also entered the field with full force. More recent developments have seen the incorporation of ideas of general relativity and quantum entanglement into the field.

These developments motivated me to work on a second edition of this book. I have to thank Simon Capelin, my editor from Cambridge University Press, who took the time to persuade me that this was not a foolish project. So, some time in 2007 (I think) I finally agreed to do it. Of course, this was a more complex project than I had expected (nothing new there!). For this reason it took until now, the Spring of 2012, for me to finish what I thought would take just one year (or so). I wish to thank Simon Capelin and the people at Cambridge University Press for working with me throughout this project.

This second edition contains essentially all that was included in the ten chapters of the first edition, with a substantial editing of misprints and "misprints." However, it has grown to have seven more chapters to incorporate some important material that I left out in 1991 and to add new material to reflect some of the new developments. The result is that this is essentially a new book. I hope that in the process of writing this second edition I have not ruined what was good in the first one, and that the new material will be useful to a wide spectrum of people, not only in condensed matter. Although the book is significantly larger than its first edition, I had to leave out some really important material. In particular, I incorporated hardly any discussion of Fermi liquids, non-Fermi liquids (except for Luttinger liquids), and superconductors, among many important problems that are also of interest to me.

Several notable books that cover some parts of the material I cover have appeared in print since 1991, such as Xiao-Gang Wen's *Quantum Field Theory of Many Body Systems* (published in 2003) and Subir Sachdev's *Quantum Phase Transitions* (published in 1999). Other books that cover some aspects of the material are Assa Auerbach's *Interacting Electrons and Magnetism* (published in 1994) and the book by A. Gogolin, A. Nersesyan, and A. Tsvelik, *Bosonization and Strongly Correlated Systems* (published in 2004), as well as the superb *Principles on Condensed Matter Physics* by Paul Chaikin and Tom Lubensky (published in 1995) and John Cardy's *Scaling and Renormalization in Statistical Physics* (published in 1996).

I am deeply indebted to many people whose work has influenced my views. I have to particularly thank Steve Kivelson for his long-term friendship and collaboration, which has had a strong impact on my work, as reflected here. I also thank my collaborators in many projects, some of which are reflected here, Chetan Nayak, Claudio Chamon, Paul Fendley, Shivaji Sondhi, Joel Moore, and Fidel Schaposnik. I am also indebted to Lenny Susskind and Steve Shenker, who played a great role during my formative years as a theorist and whose outlook has strongly influenced these pages. I also thank my former students Ana López, Christopher Mudry,

Antonio Castro Neto, Eun-Ah Kim, Michael Lawler, Kai Sun, and Benjamin Hsu, whose work is also reflected here. I am also indebted to my colleagues Mike Stone and Rob Leigh, with whom I collaborated in several projects and had countless stimulating discussions. Their work has strongly influenced my own. I also wish to thank Taylor Hughes and Shinsei Ryu for explaining their work (and others) on topological insulators, and motivating me to think on these problems. I am also grateful to Pouyan Ghaemi for reading the chapter on topological insulators and catching several misprints, and to Rodrigo Soto Garrido and to Ponnuraj Krishnakumar for proofreading the entire book and for their great help in generating the skyrmion figures for the cover.

I must also acknowledge the constant and permanent support of the Department of Physics of the University of Illinois, and my colleagues in our department. Some of the material presented here was also used in several special-topics courses I taught in Urbana over the years. I am particularly grateful to Professor Dale van Harlingen, our Department Head, for his constant support. I also wish to thank the many people who over the years have pointed out to me several conceptual issues present in the first edition as well as numerous misprints. I hope the editing of the second edition is substantially better than that of the first. I also wish to thank the National Science Foundation, which supported my research for many years.

This second edition, much like the first, could not have existed without the emotional support and love of Claudia, my wife and lifetime companion. Our children have fortunately (for them) been spared this second edition, which also could not have existed without my father constantly asking when I was going to be done with it.

Eduardo Fradkin
Urbana, Illinois, USA

Preface to the first edition

This volume is an outgrowth of the course "Physics of Strongly Correlated Systems" which I taught at the University of Illinois at Urbana-Champaign during the Fall of 1989. The goal of my course was to present the field-theoretic picture of the most interesting problems in Condensed Matter Physics, in particular those relevant to high-temperature superconductors. The content of the first six chapters is roughly what I covered in that class. The remaining four chapters were developed after January 1, 1990. Thus, that material is largely the culprit for this book being one year late! During 1990 I had to constantly struggle between finalizing the book and doing research that I just could not pass on. The result is that the book is one year late and I was late on every single paper that I thought was important! Thus, I have to agree with the opinion voiced so many times by other people who made the same mistake I did and say, don't ever write a book! Nevertheless, although the experience had its moments of satisfaction, none was like today's when I am finally done with it.

This book exists because of the physics I learned from so many people, but it is only a pale reflection of what I learned from them. I must thank my colleague Michael Stone, from whom I have learned so much. I am also indebted to Steven Kivelson, Fidel Schaposnik, and Xiao-Gang Wen, who not only informed me on many of the subjects which are discussed here but, also, more importantly, did not get too angry with me for not writing the papers I still owe them.

This book would not have existed either without the extraordinary help of Christopher Mudry, Carlos Cassanello, and Ana López, who took time off their research to help me with this crazy project. They have done an incredible job in reading the manuscript, finding my many mistakes (not just typos!), making very useful comments, and helping me with the editing of the final version. I am particularly indebted to Christopher, who made very important remarks and comments concerning the presentation of very many subjects discussed here. He also generated the figures. Mrs. Phyllis Shelton-Ball typeset the first six chapters. My wife,

Claudia, made this project possible by learning LaTeX at great speed and typesetting the last four chapters, correcting some of my very boring and awkward writing style.

This book was also made possible by the love and help of my children Ana, Andrés, and Alejandro, who had to live with a father who became a ghost for a while. Ana and Andrés helped in the proofreading, and took care of their little brother, who helped by keeping everybody happy.

Finally, I must acknowledge the support of the Department of Physics and the Center for Advanced Study of the University of Illinois. The help and understanding of the staff at Addison Wesley is also gratefully acknowledged.

Eduardo Fradkin
Urbana, Illinois, USA

1

Introduction

1.1 Field theory and condensed matter physics

Condensed matter physics is a very rich and diverse field. If we are to define it as being "whatever gets published in the condensed matter section of a physics journal," we would conclude that it ranges from problems typical of material science to subjects as fundamental as particle physics and cosmology. Because of its diversity, it is sometimes hard to figure out where the field is going, particularly if you do not work in this field. Unfortunately, this is the case for people who have to make decisions about funding, grants, tenure, and other unpleasant aspects in the life of a physicist. They have a hard time figuring out where to put this subject which is neither applied science nor dealing with the smallest length scales or the highest energies. However, the richness of the field comes precisely from its diversity.

The past few decades have witnessed the development of two areas of condensed matter physics that best illustrate the strengths of this field: critical phenomena and the quantum Hall effect. In both cases, it was the ability to produce extremely pure samples which allowed the discovery and experimental study of the phenomenon. Their physical explanation required the use of new concepts and the development of new theoretical tools, such as the renormalization group, conformal invariance, and fractional statistics.

While the concept of conformal invariance was well known in field theory before critical phenomena became recognized as a field, its importance to the complete structure of the field theory was not understood. The situation changed with the development of the renormalization group (RG). For condensed matter physics, the RG is the main tool for the interpretation of the experimental data, providing the conceptual framework and the computational algorithm which has allowed the theory to make powerful predictions. In particle physics, the RG is also a tool for the interpretation of the data. But, more importantly, the concept of an infrared-unstable fixed point has become the *definition* of the field theory itself.

Similarly, the Chern–Simons theories, which are field theories that describe systems exhibiting fractional statistics, were known before the quantum Hall effect (QHE) was discovered (actually they were discovered at about the same time), but were regarded as a curiosity of field theories below four dimensions: in other words, a beautiful piece of mathematical physics but without relevance to "the world." We have come to recognize that Chern–Simons theories are the natural theoretical framework to describe the quantum Hall effect.

Another case relevant to this point is superconductivity. Viable mechanisms for superconductivity have been known for the fifty-some years that have passed since the theory of Bardeen, Cooper, and Schrieffer (BCS). This theory has successfully explained superconductivity, and a variety of related phenomena, in very diverse areas of physics. This theory has been applied to diverse areas of physics, ranging from superconductivity in metals and superfluidity of liquid ^{3}He in condensed matter physics to neutron stars and nuclear matter in nuclear physics, and dynamical symmetry breaking and grand unification mechanisms (such as technicolor) in elementary-particle physics.

The origin of this constant interplay between field theory and condensed matter (or statistical) physics is that, despite their superficial differences, both fields deal with problems that involve a large (macroscopic) number of degrees of freedom that interact with each other. Thus, it should be no surprise that the *same techniques* can be used in both fields. The traditional trend was that field theory provided the tools (and the "sexy" terms) which were later adapted to a condensed matter problem. In turn, condensed matter models were used as "toy models" in which to try new techniques. Although this is still the case, more recent developments in condensed matter physics have allowed us to investigate new fundamental conceptual problems in quantum field theory. However, as the examples of the RG and the QHE show, the "toy models" can provide a framework for the understanding of much more general phenomenon. The *experimental accessibility* of condensed matter systems is just as important. The MOSFETs and heterostructures in which the QHE is studied have given us the surprisingly exact quantization of the Hall conductance whose understanding has required the use of topology and fiber bundles.

The importance of condensed matter physics to field theory, and vice versa, has been recognized at least since the 1950s. Landau and Feynman are perhaps the two theorists who best understood this deep connection. They worked in both fields and used their ideas and experience from one field in the other and then the converse.

1.2 What has been included in this book (first edition)

This volume is an outgrowth of the course "Physics of Strongly Correlated Systems" which I taught at the University of Illinois at Urbana-Champaign during the

Fall of 1989. Much of the material covered here has been the subject of intense research by a lot of people during the past four years. Most of what I discuss here has never been presented in a book, with the possible exception of some reprint volumes. While the *choice* of the material is motivated by current work on high-temperature superconductors, the methods and ideas have a wide range of applicability.

This book is not a textbook. Many of the problems, ideas, and methods which are discussed here have become essential to our current understanding of condensed matter physics. I have made a considerable effort to make the material largely self-contained. Many powerful methods, which are necessary for the study of condensed matter systems in the strong-fluctuation limit, are discussed and explained in some detail within the context of the applications. Thus, although the theoretical apparatus is not developed systematically and in its full glory, this material may be useful to many graduate students, in order for them to learn both the subject and the methods. For the most part I have refrained from just quoting results without explaining where they come from. So, if a particular method happens to be appropriate to the study of a particular subject, I present a more or less detailed description of the method itself. Thus, various essential theoretical tools are discussed and explained. Unfortunately, I was able to cover only part of the material I wanted to include. Perhaps the biggest omission is a description of conformal field theory. This will have to wait for a second edition, if and when I ever become crazy enough to come back to this nightmare.

The material discussed here includes path-integral methods applied to several problems in condensed matter such as the Hubbard model, quantum spin systems and the fractional quantum Hall effect; $1/N$, $1/S$, and other semi-classical expansions; coherent states; the Bethe ansatz; Jordan–Wigner transformations and bosonization; gauge invariance; topological invariants in antiferromagnets; the Hall effect; and the Chern–Simons theory of fractional statistics. The material is always developed within the context of a particular application. While there is the danger that the application may go "out of fashion," I find that it is easier to motivate and to understand this material within the framework of a concrete problem. Perhaps what this book may be good for is not so much for learning the *techniques* but as a place to find the conceptual framework of field theory in a condensed matter setting.

1.3 What was left out of the first edition

The course that I taught had as its subtitle "High Temperature Superconductors and Quantum Antiferromagnets." As the reader will soon find out, in the material that I have covered there is plenty of quantum antiferromagnetism but little superconductivity. This is not an oversight on my part. Rather, it is a reflection of what we understand today on this subject which is still a wide open field. Thus I chose not to include *the very latest fashion* on the subject but only what appears to be rather well established. This is a field that has produced a large number of very exciting ideas. However, the *gedanken theories* still dominate. To an extent, this book reflects my own efforts in transforming several fascinating *gedanken theories* into something more or less concrete.

Still, the tantalizing properties of the high-temperature superconductors seem to demand from us novel mechanisms such as Phil Anderson's RVB. But, of course, this is far from being universally accepted. After all, with a theory like BCS being around, with so many successes in its bag, it seems strange that anybody would look for any other mechanism to explain the superconductivity of a set of rather complex materials. After all, who would believe that understanding the superconductivity produced by stuff made with copper and oxygen, mixed and cooked just right, would require the development of fundamentally new ideas? Right? Well, maybe yes, maybe not.

1.4 What has been included in the second edition

I have not changed at all the content of what I wrote in the first section of this chapter back in 1991. If anything, these words are even more pertinent today.

Over the years I have often decided that it had been a mistake to include certain topics in the first edition, since they no longer seemed relevant, and regretted not having included others for similar reasons. However, there is a conservation law of good ideas in physics. So it is often the case that a theory that was proposed at a certain time in a certain setting acquires new life and meaning in a different setting. A case in point is the material on spin liquids, both chiral and non-chiral, which was discussed in Chapters 6 and 7 of the first edition. Shortly after the book appeared it became clear that the chiral spin liquid, and the anyon superconductor, do not play a role in the physics of high-temperature superconductors. This was possible since these are examples of internally consistent theories, as opposed to *gedanken* ones, which make clear predictions and hence can be tested in experiment. Nevertheless, the chapters on spin liquids and quantum dimers regained their relevance in the late 1990s and in much of the following decade as evidence for the internal consistency of topological phases in frustrated quantum magnets and quantum dimer models became more established, even though so far they have eluded experimental confirmation.

This second edition is in many ways a new book. Here is a summary of what has been included. Except for correcting a few misprints and typos, Chapter 2, The Hubbard model, is the same as in the first edition. Chapter 3, The magnetic instability of the Fermi system, has been edited to remove typos and misprints and the last section has had its mistakes purged. Chapter 4, The renormalization group, is new. In the first edition the discussion on the renormalization group was scattered throughout the text. In Chapter 4 I present a succinct but modern presentation of the subject, which sets the stage for its use in other chapters. This chapter was strongly influenced by John Cardy's beautiful textbook (Cardy, 1996). Chapter 5, One-dimensional quantum antiferromagnets, was edited and revamped. It now has three sections discussing the important subject of duality in spin systems, and another one on the one-dimensional quantum Ising model, including the exact solution. The section on Abelian bosonization was updated, particularly the notation. Chapter 6, The Luttinger liquid, is entirely new. Although some of this material also appears in Chapter 5, here I give what I think is a thorough presentation of this important problem in condensed matter. Some of the material used here is strongly inspired by reviews written by Kivelson and Emery on this problem (Emery, 1979; Carlson *et al.*, 2004). Chapter 7, Sigma models and topological terms, is a vastly revised version of what was Chapter 5 in the first edition. The main changes in this chapter are the new sections on the Wess–Zumino–Witten model and non-abelian bosonization, and another section giving a brief presentation of the main ideas of conformal field theory (a subject that has acquired widespread use in many areas of condensed matter) and their application to the Wess–Zumino–Witten model and to quantum spin chains. For the sake of brevity I chose not to include a discussion of the Kondo problem here.

In the first edition Chapters 6 and 7 dealt with spin liquids and chiral spin states, respectively. These two chapters have been completely revised, expanded and split into three chapters, Chapters 8, 9, and 10. Chapter 8, Spin-liquid states, contains much of the discussion of the old Chapter 6 on spin liquids, valence-bond states, and the gauge-theory description of antiferromagnets, but significantly edited and updated to account for the many developments. The content of the new Chapter 9, Gauge theory, dimer models, and topological phases, is completely new. Here I include an in-depth discussion of the phases and observables of gauge theories, paying special attention to their relation to time-reversal-invariant topological phases, the $\mathbb{Z}_2$ spin liquid, the Kitaev toric code, and quantum loop models. I also include a theory of quantum criticality in quantum dimer models and the quantum Lifshitz model.

In the new Chapter 10, Chiral spin states and anyons, I have merged all the discussions on the chiral spin liquid. I also expand the treatment of the Chern–Simons gauge theory and its role as a theory of fractional statistics. I also corrected some errors on the lattice version of Chern–Simons that were present in the first edition. Chapter 11, Anyon superconductivity, is a compressed version of Chapter 8 of the first edition. It is now clear that an anyon superconductor, a state resulting from the condensation of electrically charged anyons (abelian) is not essentially different from a superconductor with a spontaneously broken time-reversal symmetry, e.g. a

$p_x + ip_y$ or $d_{x^2-y^2} + id_{xy}$ superconductor. Nevertheless, I am not fond of rewriting history and for this reason I kept this chapter, after excising some results that were wrong.

Chapter 12, Topology and the quantum Hall effect, is almost the same as Chapter 9 of the first edition. I only made minor editing changes. Similarly, the new Chapter 13, The fractional quantum Hall effect, is almost the same as Chapter 10 in the first edition. The only important change here, aside from editing, was that the section on edge states is no longer in this chapter. The bulk of this chapter is devoted to a presentation of the bosonic and fermionic Chern–Simons theory of the fractional quantum Hall states.

The remaining four chapters of the new edition are new and are devoted to, respectively, Topological fluids (Chapter 14), Physics at the edge (Chapter 15), Topological insulators (Chapter 16), and Quantum entanglement (Chapter 17). Chapter 14 is devoted to the theory of topological fluids presented here as a theory of fractional quantum Hall fluids. Here I include a description of the hydrodynamic theory (of Wen and Zee), its extensions to general abelian multi-component fluids, non-abelian quantum Hall fluids, superconductors as topological fluids, and topological superconductors, and a brief presentation of the concepts of braiding and fusion. Chapter 15, Physics at the edge, is an in-depth presentation of the theory of edge states in integer and fractional quantum Hall fluids, both abelian and non-abelian, Wen's theory of bulk–edge correspondence, and the effective field theories of the non-abelian fractional quantum Hall states. I devote special sections to discussions of tunneling conductance at quantum point contacts, noise and the measurement of fractional charge, and the theory of abelian and non-abelian quantum interferometers, and there is a brief sales pitch for topological quantum computing. Chapter 16 is devoted to a brief presentation of the exciting new field of topological insulators. Here I discuss the basic concepts, band topological invariants, the anomalous quantum Hall effect, and the spin quantum Hall effect and its experimental discovery. I also discuss the extensions of these ideas to three-dimensional $\mathbb{Z}_2$ topological insulators, their relations to fractional charge (and polyacetylene) in one dimension, the Callan–Harvey effect in three dimensions, surface Weyl fermions, Majorana modes, and possible new topological insulators resulting from spontaneous symmetry breaking. Chapter 17 is devoted to the role of quantum entanglement in field theory, quantum critical systems and topological phases, and large-scale entanglement and the scaling of the entanglement entropy, as well as the relation of this problem to the modern ideas of holography and the CFT/gravity duality.

Several important subjects are not in this book. In particular, except for some cursory discussion in Chapter 2, Fermi liquids are not discussed. For this reason I have also not discussed the Bardeen–Cooper–Schrieffer theory of superconductivity and other mechanisms. I have also not discussed what happens when a Fermi liquid fails. This is an area to which I have devoted a great deal of effort, including the formulation of a new class of electronic liquid-crystal phases of strongly correlated systems. Experimentally these phases arise often in conjunction or in competition with high-temperature superconductivity. I have also not discussed the important problem of fermionic quantum criticality and non-Fermi-liquid behavior, with many interesting connections with the concept of holography, as well as extensions of bosonization to dimensionalities higher than two, which is a natural framework to describe these open problems. Absent from this book is also the discussion of disordered systems, a fascinating problem in which there are very few well-established results.

Finally, all the figures in this book are new because I had lost the source files of the figures that were used in the first edition. This new edition includes an extensive set of references at the end of the book, and a detailed index, which I hope will be useful to the reader.

2
The Hubbard model

2.1 Introduction

All theories of strongly correlated electron systems begin with the Hubbard model because of its simplicity. This is a model in which *band electrons* interact via a two-body *repulsive* Coulomb interaction. No phonons are present, and in general no explicitly attractive interactions are included. For this reason, the Hubbard model has traditionally been associated with magnetism. Superconductivity, on the other hand, has traditionally (i.e. after BCS) been interpreted as an instability of the ground state resulting from *effectively attractive* interactions (say, electron–phonon as in BCS). A novel situation has arisen with Anderson's suggestion (Anderson, 1987) that the superconductivity of the new high-T_c materials may arise from purely repulsive interactions. This suggestion was motivated by the fact that the superconductivity seems to originate from doping (i.e. extracting or adding charges) an otherwise insulating state.

The Hubbard model is a very simple model in which one imagines that, out of the many different bands which may exist in a solid, only very few states per unit cell contribute significantly to the ground-state properties. Thus, if a Bloch state of energy ϵ_p, momentum $\vec{p}$, and index α has a wavefunction $\Psi_{\vec{p},\alpha}$, one can construct Wannier states

$$\Psi_\alpha(\vec{r}_i) = \frac{1}{\sqrt{N}} \sum_{\vec{p}\in \mathrm{BZ}} e^{i\vec{p}\cdot\vec{r}_i} \Psi_{\vec{p},\alpha}(\vec{r}_i) \tag{2.1}$$

where $\vec{r}_i$ is the location of the ith atom and BZ is the Brillouin zone. The assumption here will be that only one (or a few) band indices matter, so I will drop the index α. The Coulomb interaction matrix elements are

$$U_{ij,i'j'} = \int d^3r_1\, d^3r_2\, \Psi_i^*(\vec{r}_1)\Psi_j^*(\vec{r}_2)\tilde{V}(\vec{r}_1 - \vec{r}_2)\Psi_{i'}(\vec{r}_1)\Psi_{j'}(\vec{r}_2) \tag{2.2}$$

(in three dimensions), where $\tilde{V}$ is the (screened) Coulomb interaction. Since $\tilde{V}$ is expected to decay as the separation increases, the largest term will be the "on-site" term: $U_{ii,ii} \equiv U$. Next will come nearest neighbors, etc. Moreover, since the Wannier functions have exponentially decreasing overlaps, $U_{ij,i'j'}$ is expected to decrease rather rapidly with the separation $|i - j|$.

The second quantized Hamiltonian tight binding (in the Wannier-functions basis) is

$$\begin{aligned} H = &- \sum_{\substack{\vec{r}_i,\vec{r}_j \\ \sigma=\uparrow,\downarrow}} \left(c^{\dagger}_{\sigma}(\vec{r}_i) t_{ij} c_{\sigma}(\vec{r}_j) + c^{\dagger}_{\sigma}(\vec{r}_j) t_{ij} c_{\sigma}(\vec{r}_i) \right) \\ &+ \frac{1}{2} \sum_{\substack{i,j,i',j' \\ \sigma,\sigma'=\uparrow,\downarrow}} U_{ij,i'j'} c^{\dagger}_{\sigma}(\vec{r}_i) c^{\dagger}_{\sigma'}(\vec{r}_j) c_{\sigma'}(\vec{r}_{j'}) c_{\sigma}(\vec{r}_{i'}) \end{aligned} \tag{2.3}$$

where $c^{\dagger}_{\sigma}(\vec{r})$ creates an electron at site $\vec{r}$ with spin σ (or more precisely, at the unit cell $\vec{r}$ in the band responsible for the Fermi surface) and satisfies

$$\begin{aligned} \{c_{\sigma}(\vec{r}), c^{\dagger}_{\sigma'}(\vec{r}\,')\} &= \delta_{\sigma,\sigma'} \delta_{\vec{r},\vec{r}\,'} \\ \{c_{\sigma}(\vec{r}), c_{\sigma'}(\vec{r}\,')\} &= 0 \end{aligned} \tag{2.4}$$

The Hubbard model is an approximation to the more general Hamiltonian, Eq. (2.3), in which the hopping is restricted to nearest neighboring sites:

$$t_{ij} = \begin{cases} t & \text{if } i,j \text{ are nearest neighbors} \\ 0 & \text{otherwise} \end{cases} \tag{2.5}$$

and the Coulomb interaction is assumed to be screened. If just the "on-site" term is kept,

$$U_{ij,i'j'} = U \delta_{ij} \delta_{i'j'} \delta_{ii'} \tag{2.6}$$

the resulting model Hamiltonian

$$H = -t \sum_{\substack{\langle \vec{r},\vec{r}\,' \rangle \\ \sigma=\uparrow,\downarrow}} \left(c^{\dagger}_{\sigma}(\vec{r}) c_{\sigma}(\vec{r}\,') + \text{h.c.} \right) + U \sum_{\vec{r}} n_{\uparrow}(\vec{r}) n_{\downarrow}(\vec{r}) \tag{2.7}$$

is known as the one-band Hubbard model. In Eq. (2.7), we have dropped the lattice site labels and $\langle\,,\rangle$ means nearest-neighboring sites. This is the tight-binding approximation and represents the one-band Hubbard model. We have introduced

$$n_{\sigma}(\vec{r}) = c^{\dagger}_{\sigma}(\vec{r}) c_{\sigma}(\vec{r}) \tag{2.8}$$

From the Pauli principle we get $n_{\sigma} = 0, 1$ or $n^2_{\sigma} = n_{\sigma}$ at every site.

The Hilbert space of this system is the tensor product of only *four* states per site, representing $|0\rangle$ as nothing, $|\uparrow\rangle$ as an electron with spin up, $|\downarrow\rangle$ as an electron with spin down, and $|\uparrow\downarrow\rangle$ as an up–down pair. The states $|0\rangle$ and $|\uparrow\downarrow\rangle$ are spin singlets (i.e. $S = 0$).

It is convenient to define the following operators. The spin operator $\vec{S}(\vec{r})$ is defined by (the summation convention is assumed)

$$\vec{S}(\vec{r}) = \frac{\hbar}{2} c^{\dagger}_{\sigma}(\vec{r}) \vec{\tau}_{\sigma\sigma'} c_{\sigma'}(\vec{r}) \tag{2.9}$$

where $\vec{\tau}$ are the (three) Pauli matrices

$$\tau_1 = \begin{pmatrix} 0 & 1 \\ 1 & 0 \end{pmatrix}, \quad \tau_2 = \begin{pmatrix} 0 & -i \\ i & 0 \end{pmatrix}, \quad \tau_3 = \begin{pmatrix} 1 & 0 \\ 0 & -1 \end{pmatrix} \tag{2.10}$$

The particle number operator at site $\vec{r}$ (or *charge*) is

$$n(\vec{r}) = \sum_{\sigma} n_{\sigma}(\vec{r}) = \sum_{\sigma} c^{\dagger}_{\sigma}(\vec{r}) c_{\sigma}(\vec{r}) \equiv c^{\dagger}_{\sigma}(\vec{r}) 1_{\sigma\sigma'} c_{\sigma'}(\vec{r}) \tag{2.11}$$

and the associated total charge Q is given by

$$Q = e \sum_{\vec{r}} n(\vec{r}) \equiv e N_{\mathrm{e}} \tag{2.12}$$

2.2 Symmetries of the Hubbard model

2.2.1 SU(2) spin

Suppose we rotate the local spin basis (i.e. the quantization axis)

$$c'_{\sigma}(\vec{r}) = U_{\sigma\sigma'} c_{\sigma'}(\vec{r}) \tag{2.13}$$

where U is a 2×2 SU(2) matrix. Namely, given four complex numbers a, b, c, and d, the matrix U given by

$$U = \begin{pmatrix} a & c \\ d & b \end{pmatrix} \tag{2.14}$$

must satisfy

$$U^{-1} = U^{\dagger} \equiv \left(U^{\mathrm{T}}\right)^{*} \tag{2.15}$$

together with the condition

$$\det U = 1 \tag{2.16}$$

We will parametrize the matrix $U(\vec{\theta})$ as follows:

$$U(\vec{\theta}) = e^{i\vec{\theta}\cdot\vec{\tau}} = \mathbf{1} \cos|\vec{\theta}| + i \sin|\vec{\theta}| \frac{\vec{\theta} \cdot \vec{\tau}}{|\vec{\theta}|} \tag{2.17}$$

where **1** represents the 2×2 identity matrix, $\vec{\tau}$ are the Pauli matrices, and the Euler angles $\vec{\theta} = (\theta_1, \theta_2, \theta_3)$ parametrize the SU(2) group.

Under such a unitary transformation, the spin $\vec{S}$ transforms as follows:

$$\begin{aligned} S'^a(\vec{r}) &= R^{ab} S^b(\vec{r}) \\ &= \frac{\hbar}{2} c'^\dagger(\vec{r}) \tau^a c'(\vec{r}) \\ &= \frac{\hbar}{2} c^\dagger(\vec{r}) \left(U^{-1} \tau^a U\right) c(\vec{r}) \end{aligned} \tag{2.18}$$

where R^{ab} is a rotation matrix induced by the SU(2) transformation of the fermions:

$$U^{-1} \tau^a U = R^{ab} \tau^b \tag{2.19}$$

In other words, we have a rotation of the quantization axis.

The axis of quantization can be chosen arbitrarily. Thus, the Hubbard model Hamiltonian should not change its form under a rotation of the spin quantization axis. This is not apparent in the standard form of the interaction

$$H_1 = U \sum_{\vec{r}} n_\uparrow(\vec{r}) n_\downarrow(\vec{r}) \tag{2.20}$$

But we can write this in a somewhat different form in which the SU(2) symmetry becomes explicit. Consider the operator

$$\sum_{\vec{r}} \left(\vec{S}(\vec{r})\right)^2 = \sum_{\substack{\vec{r} \\ a=1,2,3}} S^a(\vec{r}) S^a(\vec{r}) \tag{2.21}$$

By expanding the components and making use of the SU(2) identity

$$\sum_{a=1,2,3} \tau^a_{\alpha\beta} \tau^a_{\gamma\delta} = 2\delta_{\alpha\delta}\delta_{\beta\gamma} - \delta_{\alpha\beta}\delta_{\gamma\delta} \tag{2.22}$$

one gets

$$\sum_{\vec{r}} \left(\vec{S}(\vec{r})\right)^2 = \sum_{\vec{r}} \left(\frac{3}{4} n(\vec{r}) - \frac{3}{2} n_\uparrow(\vec{r}) n_\downarrow(\vec{r})\right) \tag{2.23}$$

Thus, we can write

$$H_1 = U \sum_{\vec{r}} n_\uparrow(\vec{r}) n_\downarrow(\vec{r}) = -\frac{2U}{3} \vec{S}^2(\vec{r}) + \frac{N_e U}{2} \tag{2.24}$$

The last term is a constant, which can be dropped. The Hamiltonian now has the form

$$H = -t \sum_{\substack{\langle \vec{r}, \vec{r}\,' \rangle \\ \sigma=\uparrow,\downarrow}} \left(c^\dagger_\sigma(\vec{r}) c_\sigma(\vec{r}\,') + \text{h.c.} \right) - \frac{2U}{3} \sum_{\vec{r}} \left(\vec{S}(\vec{r}) \right)^2 + \frac{N_e U}{2} \tag{2.25}$$

which is manifestly SU(2)-invariant.

For $U > 0$, the interaction energy is lowered if *the total spin at each site is maximized*. Thus, one should expect some sort of magnetic ground state, at least if each site has one particle (on average). This state requires that the system somehow should pick a global (i.e. the same for all sites) quantization axis. In other words, the global SU(2) spin symmetry may be spontaneously broken. This has important consequences, which we will discuss later.

2.2.2 U(1) charge

We are free to change the phase of the one-particle wavefunction

$$c'_\sigma(\vec{r}) = e^{i\theta} c_\sigma(\vec{r}) \tag{2.26}$$

Here, $e^{i\theta}$ is an element of the group U(1), and group elements satisfy

$$e^{i\theta} e^{i\theta'} = e^{i(\theta+\theta')} \tag{2.27}$$

The Hamiltonian is invariant under this U(1) transformation. This is nothing but charge conservation. For example, if we had terms that would not conserve charge, like

$$c^\dagger_\uparrow(\vec{r}) c^\dagger_\downarrow(\vec{r}\,') \to e^{i2\theta} c^\dagger_\uparrow(\vec{r}) c^\dagger_\downarrow(\vec{r}\,') \tag{2.28}$$

we would not have this invariance.

Suppose now that we couple this system to the electromagnetic field $(A_0, \vec{A})$. We expect three effects.

1. A *Zeeman coupling* given by

$$H_{\text{Zeeman}} = g \sum_{\vec{r}} \vec{S}(\vec{r}) \cdot \vec{B}(\vec{r}) \tag{2.29}$$

   which couples the spin $\vec{S}(\vec{r})$ with the local magnetic field $\vec{B}(\vec{r})$ so as to align it along the $\vec{B}(\vec{r})$ direction.
2. An *orbital coupling* for electrons in a crystal with one-particle Hamiltonian

$$H(\vec{p}) = \frac{1}{2m_e} \left(\vec{p} - \frac{e}{c}\vec{A} \right)^2 + V(\vec{r}) \tag{2.30}$$

   where $V(\vec{r})$ is the periodic potential imposed by the crystal. In the tight-binding approximation, we must therefore modify the kinetic-energy term according to

$$
\begin{aligned}
H_0 &\equiv -t \sum_{\substack{\langle \vec{r},\vec{r}\,' \rangle \\ \sigma=\uparrow,\downarrow}} \left(c_\sigma^\dagger(\vec{r})c_\sigma(\vec{r}\,') + \text{h.c.}\right) \\
&\to -t \sum_{\substack{\langle \vec{r},\vec{r}\,' \rangle \\ \sigma=\uparrow,\downarrow}} \left(c_\sigma^\dagger(\vec{r})e^{\frac{ie}{\hbar c}\int_{\vec{r}}^{\vec{r}\,'} d\vec{x}\cdot\vec{A}(\vec{x})}c_\sigma(\vec{r}\,') + c_\sigma^\dagger(\vec{r}\,')e^{-\frac{ie}{\hbar c}\int_{\vec{r}}^{\vec{r}\,'} d\vec{x}\cdot\vec{A}(\vec{x})}c_\sigma(\vec{r})\right)
\end{aligned}
\tag{2.31}
$$

We should now check the gauge invariance under the transformation

$$\vec{A}' = \vec{A} + \vec{\nabla}\Lambda \tag{2.32}$$

where Λ is an arbitrary function of space and time. We get the change

$$
\begin{aligned}
A'(\vec{r},\vec{r}\,') &\equiv \int_{\vec{r}}^{\vec{r}\,'} d\vec{x}\cdot\vec{A}'(\vec{x}) \\
&= A(\vec{r},\vec{r}\,') + \Lambda(\vec{r}\,') - \Lambda(\vec{r})
\end{aligned}
\tag{2.33}
$$

Thus the kinetic-energy term is gauge-invariant,

$$
\begin{aligned}
H_0' &\equiv -t \sum_{\substack{\langle \vec{r},\vec{r}\,' \rangle \\ \sigma=\uparrow,\downarrow}} \left(c_\sigma'^\dagger(\vec{r})e^{\frac{ie}{\hbar c}A'(\vec{r},\vec{r}\,')}c_\sigma'(\vec{r}\,') + \text{h.c.}\right) \\
&= -t \sum_{\substack{\langle \vec{r},\vec{r}\,' \rangle \\ \sigma=\uparrow,\downarrow}} \left(c_\sigma^\dagger(\vec{r})e^{-i\theta(\vec{r})}e^{\frac{ie}{\hbar c}\left(A(\vec{r},\vec{r}\,')+\Lambda(\vec{r}\,')-\Lambda(\vec{r})\right)}e^{+i\theta(\vec{r}\,')}c_\sigma(\vec{r}\,') + \text{h.c.}\right)
\end{aligned}
\tag{2.34}
$$

provided that the *local* change of phase is given by

$$\theta(\vec{r}) \equiv -\frac{e}{\hbar c}\Lambda(\vec{r}) \tag{2.35}$$

3. An *electrostatic coupling* given by

$$H_{\text{electrostatic}} = \sum_{\vec{r},\sigma} eA_0(\vec{r})c_\sigma^\dagger(\vec{r})c_\sigma(\vec{r}) \tag{2.36}$$

which couples the particle density to $A_0(\vec{r})$.

2.2.3 Particle–hole transformations

In the case of a *bipartite lattice* (i.e. a lattice that is the union of two interpenetrating sublattices A and B) we get additional symmetries.

1. First, the *sign of t* can be changed. Consider the transformation

$$
\begin{aligned}
c_\sigma(\vec{r}) &\to +c_\sigma(\vec{r}) \quad \text{if} \quad \vec{r}\in A \\
c_\sigma(\vec{r}) &\to -c_\sigma(\vec{r}) \quad \text{if} \quad \vec{r}\in B
\end{aligned}
\tag{2.37}
$$

under which the kinetic energy changes sign:

$$t c_\sigma^\dagger(\vec{r}) c_\sigma(\vec{r}\,') \to -t c_\sigma^\dagger(\vec{r}) c_\sigma(\vec{r}\,'), \quad \vec{r} \in A, \quad \vec{r}\,' \in B \tag{2.38}$$

while the potential energy is left unchanged. This transformation leaves the canonical commutation relations unchanged and therefore leaves the *spectrum unchanged.*

2. Now consider the *particle–hole* transformation

$$\begin{aligned} c_\uparrow(\vec{r}) &= d_\uparrow(\vec{r}) \\ c_\downarrow(\vec{r}) &= \begin{cases} +d_\downarrow^\dagger(\vec{r}), & \vec{r} \in A \\ -d_\downarrow^\dagger(\vec{r}), & \vec{r} \in B \end{cases} \end{aligned} \tag{2.39}$$

The Hamiltonian $H(t, U)$, Eq. (2.7), changes into $H(t, -U) + U N_\uparrow$, where $N_\uparrow$ is the total number of up spins (which is conserved), since under this transformation we get

$$n_\uparrow + n_\downarrow = c_\uparrow^\dagger c_\uparrow + c_\downarrow^\dagger c_\downarrow = d_\uparrow^\dagger d_\uparrow + d_\downarrow d_\downarrow^\dagger = d_\uparrow^\dagger d_\uparrow - d_\downarrow^\dagger d_\downarrow + 1 \tag{2.40}$$

and

$$n_\uparrow - n_\downarrow = c_\uparrow^\dagger c_\uparrow - c_\downarrow^\dagger c_\downarrow = d_\uparrow^\dagger d_\uparrow - d_\downarrow d_\downarrow^\dagger = d_\uparrow^\dagger d_\uparrow + d_\downarrow^\dagger d_\downarrow - 1 \tag{2.41}$$

Similarly, the charge Q and the component S_z of the total spin transform as

$$Q \to S_z + 1, \qquad S_z \to Q - 1 \tag{2.42}$$

Thus the attractive and the repulsive cases map into each other and, at the same time, spin maps into charge and vice versa. Note that for *negative* $U < 0$ the Hamiltonian favors local spin-singlet states ($S = 0$), i.e. empty and doubly occupied sites.

2.3 The strong-coupling limit

We consider now the strong-coupling limit of the Hubbard model, i.e. $U \to \infty$. The half-filled case is special but important. We will consider it first.

2.3.1 The half-filled system (U > 0)

Recall that the interaction term

$$H_{\text{int}} \equiv -\frac{2}{3} U \sum_{\vec{r}} \left(\vec{S}(\vec{r})\right)^2 \tag{2.43}$$

forces the spin $\vec{S}$ to be largest if U becomes infinitely large, i.e. doubly occupied sites are forbidden. Only $|\uparrow\rangle$ and $|\downarrow\rangle$ states are kept in this large-U limit at

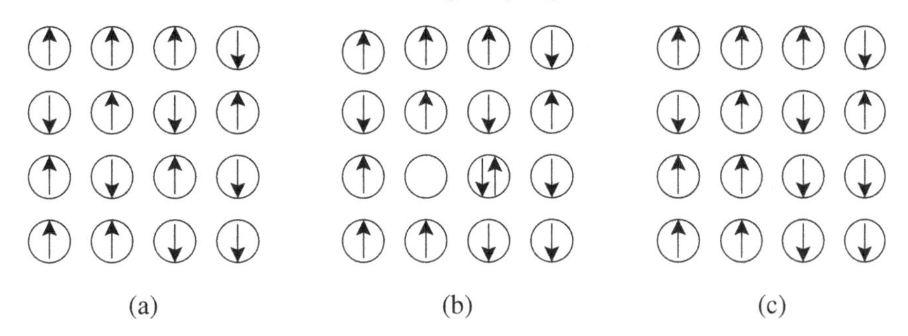

Figure 2.1 Configurations (a) and (c) are two configurations of spins corresponding to orthogonal ground states of H_0. They differ by the exchange of two neighboring spins. Configuration (b) corresponds to a virtual state. The circles represent the unit cells of the crystal and the arrows are the components of the spin along the quantization axis.

half-filling. The interaction part of the Hamiltonian has infinitely many eigenstates. Any spin configuration is an eigenstate. In order to lift this massive degeneracy we will keep the effects of fluctuations induced by the kinetic-energy term to leading order in an expansion in t/U. We have to solve a problem in degenerate perturbation theory, a strong-coupling expansion.

Suppose we begin with any configuration that can be labeled by the local z-component of the spin $|\{\sigma(\vec{r})\}\rangle$ (see Fig. 2.1(a)). In an expansion in powers of t/U, we have intermediate states in which one site will become doubly occupied and, at the same time, another site becomes empty (see Fig. 2.1(b)). This state has an energy U above that of the degenerate ground states. The matrix element (squared) is t^2. There is also a multiplicity factor of 2 since this process can occur in two different ways. Hence we expect that the relevant parameter of the effective Hamiltonian should be $2t^2/U$. Also, the final state has to be either the same one as the initial state or it can differ at most by a spin exchange (see Fig. 2.1(c)). The natural candidate for the effective Hamiltonian is the quantum Heisenberg antiferromagnet.

We can obtain this result by carrying out this expansion explicitly (Emery, 1979). Let H_0 and H_1 denote the kinetic and interaction terms of the Hubbard Hamiltonian H, Eq. (2.7),

$$\begin{aligned} H_0 &= -t \sum_{\substack{\langle \vec{r}, \vec{r}\,' \rangle \\ \sigma=\uparrow,\downarrow}} \left(c_\sigma^\dagger(\vec{r}) c_\sigma(\vec{r}\,') + \text{h.c.} \right) \\ H_1 &= U \sum_{\vec{r}} n_\uparrow(\vec{r}) n_\downarrow(\vec{r}) \end{aligned} \tag{2.44}$$

Let $|\alpha\rangle$ be any of the 2^N states with every site occupied by a spin either up or down. Here, $|\alpha\rangle$ is an eigenstate of H_1 with eigenvalue $E_1 = 0$.

We will use Brillouin–Wigner perturbation theory (Baym, 1974) Consider Schrödinger's equation

$$H|\Psi\rangle = E|\Psi\rangle \tag{2.45}$$

where $|\Psi\rangle$ is any eigenstate. We can write

$$(E - H_1)|\Psi\rangle = H_0|\Psi\rangle \tag{2.46}$$

Formally, we get

$$\begin{aligned} |\Psi\rangle &= \frac{1}{E - H_1} H_0|\Psi\rangle \\ &= \frac{\hat{P}}{E - H_1} H_0|\Psi\rangle + \sum_\alpha |\alpha\rangle \frac{\langle\alpha|H_0|\Psi\rangle}{E - E_1} \end{aligned} \tag{2.47}$$

where

$$H_1|\alpha\rangle = E_1|\alpha\rangle \tag{2.48}$$

and

$$\hat{P} = 1 - \sum_\alpha |\alpha\rangle\langle\alpha| \tag{2.49}$$

projects out of the unperturbed states. Clearly $\hat{P}$ commutes with H_1. Define $|\Psi_\alpha\rangle$ as the solution of the equation

$$|\Psi_\alpha\rangle = |\alpha\rangle + \frac{\hat{P}}{E - H_1} H_0|\Psi_\alpha\rangle \tag{2.50}$$

Let a_α be given by

$$a_\alpha = \frac{\langle\alpha|H_0|\Psi\rangle}{E - E_1} \tag{2.51}$$

Then we can write

$$|\Psi\rangle = \sum_\alpha a_\alpha|\Psi_\alpha\rangle \tag{2.52}$$

If we iterate Eq. (2.50) to first order in powers of $[\hat{P}/(E - H_1)]H_0$, we find

$$|\Psi_\alpha\rangle \approx |\alpha\rangle + \frac{\hat{P}}{E - H_1} H_0|\alpha\rangle \approx |\alpha\rangle - \frac{1}{U} H_0|\alpha\rangle \tag{2.53}$$

since $\langle\beta|H_0|\alpha\rangle = 0$ at half-filling. Thus, if we insert Eq. (2.53) into Eq. (2.52), and in turn insert this into Eq. (2.51), we get

$$(E - E_1)a_\alpha = \frac{1}{|U|} \sum_{\alpha'} \langle\alpha|H_0^2|\alpha'\rangle a_{\alpha'} \tag{2.54}$$

This is the same as the Schrödinger equation for the Hamiltonian $H_0' = H_0^2/|U|$, where H_0', at half-filling, is given by

$$H_0' = \frac{2t^2}{|U|} \sum_{\langle \vec{r}, \vec{r}\,' \rangle} \vec{S}(\vec{r}) \cdot \vec{S}(\vec{r}\,') \tag{2.55}$$

In other words, we find the spin one-half quantum Heisenberg antiferromagnet with the exchange coupling $J = 2t^2/|U|$. This result is valid for the half-filled system in any dimension and for any lattice.

2.3.2 Away from half-filling

Clearly other processes are now allowed. If $U \gg t$, doubly occupied sites are energetically very expensive. Thus the restricted Hilbert space now consists of configurations made of empty sites (holes), and up and down spins.

The kinetic-energy term will allow charge motion since empty sites (holes) will be able to move. These holes carry electric charge but have no spin. The effective Hamiltonian now has the form of the t–J model:

$$H = -t \sum_{\substack{\langle \vec{r}, \vec{r}\,' \rangle \\ \sigma = \uparrow, \downarrow}} \left(c_\sigma^\dagger(\vec{r}) c_\sigma(\vec{r}\,') + \text{h.c.} \right) + J \sum_{\langle \vec{r}, \vec{r}\,' \rangle} \vec{S}(\vec{r}) \cdot \vec{S}(\vec{r}\,') \tag{2.56}$$

(where $J = 2t^2/|U|$), with the *constraint*

$$\sum_{\sigma = \uparrow, \downarrow} c_\sigma^\dagger(\vec{r}) c_\sigma(\vec{r}) = n(\vec{r}) = 0, 1 \tag{2.57}$$

which eliminates doubly occupied sites. Now we have *two* separately conserved quantities: the charge Q, which equals the number of holes, and the spin component $S_z = \sum_{\vec{r}} S_z(\vec{r})$. It is clear that as the holes move they can induce spin-flip processes. The spin configurations get disrupted by the motion of holes and the long-range order (antiferromagnetic) may be destroyed. Presumably it should take a finite density of holes to destroy the long-range order. We will discuss this strong-coupling limit at great length. Let us first consider the opposite case.

2.4 The weak-coupling limit

In the weak-coupling limit $U \ll t$ we may think of the interaction as a weak perturbation. One therefore expects that the states of a *weakly coupled* electron gas may be qualitatively similar to the states of a free-electron gas. This picture is usually called a *Fermi liquid* (Pines and Nozières, 1966). The main *assumption* is that there is a one-to-one correspondence between the states of a free-fermion system and those in a weakly interacting one.

For free fermions, the Hamiltonian reduces to the kinetic-energy term. For the Hubbard model we have

$$H_0 = -\sum_{\substack{\vec{r},\vec{r}\,' \\ \sigma=\uparrow,\downarrow}} \left(c_\sigma^\dagger(\vec{r}) t(\vec{r}-\vec{r}\,') c_\sigma(\vec{r}\,') + c_\sigma^\dagger(\vec{r}\,') t(\vec{r}-\vec{r}\,') c_\sigma(\vec{r}) \right) \tag{2.58}$$

It is convenient to go to Fourier space (momentum). Assume that we are in d space dimensions and that the lattice has N^d sites with N even (for simplicity). With $V \equiv N^d$, we define

$$c_\sigma(\vec{r}) = \frac{1}{V}\sum_{\vec{k}} e^{i\vec{k}\cdot\vec{r}} c_\sigma(\vec{k}) \tag{2.59}$$

where

$$\vec{k} = \frac{2\pi}{N}(n_1, \ldots, n_d) - (\pi, \ldots, \pi) \tag{2.60}$$

and $1 \le n_i \le N$. Thus the momenta k_i vary over the range $-\pi + 2\pi/N \le k_i \le \pi$. In the thermodynamic limit $N \to \infty$, $2\pi/N \to 0$ and the ks become uniformly distributed in the interval $-\pi \le k_i \le \pi$, the Brillouin zone.

Remember the following properties of Fourier transforms. Let $k = (2\pi/N)\, n - \pi, n = 1, \ldots, N$ and let $f(k)$ be some function of k. We have the Riemann sums

$$\begin{aligned} \sum_k f(k) &= \frac{1}{\Delta k}\sum_k \Delta k\, f(k) \\ &\to \frac{N}{2\pi}\int_{-\pi}^{\pi} dk\, f(k) \qquad \text{as} \quad N \to \infty \end{aligned} \tag{2.61}$$

where $\Delta k = 2\pi/N$. The extension to the d-dimensional case is

$$\sum_{\vec{k}} f(\vec{k}) \to N^d \int_{-\pi}^{\pi} \frac{d^d k}{(2\pi)^d} f(\vec{k}) \quad \text{as} \quad N \to \infty \tag{2.62}$$

In particular, as $N \to \infty$

$$\begin{aligned} \frac{1}{N^d}\sum_{\vec{k}} e^{i\vec{k}\cdot(\vec{r}-\vec{r}\,')} = \delta_{\vec{r},\vec{r}\,'} &\to \int_{-\pi}^{\pi} \frac{d^d k}{(2\pi)^d} e^{i\vec{k}\cdot(\vec{r}-\vec{r}\,')} = \delta_{\vec{r},\vec{r}\,'} \\ \sum_{\vec{r}} e^{i(\vec{k}-\vec{k}')\cdot\vec{r}} = N^d \delta_{\vec{k},\vec{k}'} &\to (2\pi)^d \delta^{(d)}(\vec{k}-\vec{k}') \\ c_\sigma(\vec{r}) &\to \int_{-\pi}^{\pi} \frac{d^d k}{(2\pi)^d} e^{i\vec{k}\cdot\vec{r}} c_{\sigma(\vec{k})} \end{aligned} \tag{2.63}$$

The canonical (anti)commutation relations

$$\{c_\sigma^\dagger(\vec{r}), c_{\sigma'}(\vec{r}\,')\} = \delta_{\sigma,\sigma'}\delta_{\vec{r},\vec{r}\,'} \tag{2.64}$$

become, in the same limit,

$$\{c_\sigma^\dagger(\vec{k}), c_{\sigma'}(\vec{k}')\} = \delta_{\sigma,\sigma'}\delta_{\vec{k},\vec{k}'} \to (2\pi)^d \delta_{\sigma,\sigma'}\delta^{(d)}(\vec{k} - \vec{k}') \tag{2.65}$$

The kinetic energy then takes the form

$$\begin{aligned} H_0 &= -\sum_{\substack{\vec{r},\vec{r}\,' \\ \sigma=\uparrow,\downarrow}} \left(c_\sigma^\dagger(\vec{r})t(\vec{r}-\vec{r}\,')c_\sigma(\vec{r}\,') + c_\sigma^\dagger(\vec{r}\,')t(\vec{r}-\vec{r}\,')c_\sigma(\vec{r})\right) \\ &= -\sum_{\substack{\vec{r},\vec{r}\,' \\ \sigma=\uparrow,\downarrow}} t(\vec{r}-\vec{r}\,') \int \frac{d^d k}{(2\pi)^d} \int \frac{d^d k'}{(2\pi)^d} \left(e^{-i\vec{k}\cdot\vec{r}+i\vec{k}'\cdot\vec{r}\,'} c_\sigma^\dagger(\vec{k})c_\sigma(\vec{k}') + \text{h.c.}\right) \end{aligned} \tag{2.66}$$

If by $t(\vec{k})$ we denote the Fourier transform of $t(\vec{l})$,

$$t(\vec{k}) = \sum_{\vec{l}} t(\vec{l})e^{-i\vec{k}\cdot\vec{l}} \tag{2.67}$$

we can write

$$\sum_{\vec{r},\vec{r}'} t(\vec{r}-\vec{r}\,')e^{-i\vec{k}\cdot\vec{r}+i\vec{k}'\cdot\vec{r}\,'} = t(\vec{k})(2\pi)^d\delta^{(d)}\left(\vec{k}-\vec{k}'\right) \tag{2.68}$$

For the case

$$t(\vec{r}-\vec{r}\,') \equiv t(\vec{l}) = \begin{cases} t & \text{for nearest neighbors} \\ 0 & \text{otherwise} \end{cases} \tag{2.69}$$

we get

$$t(\vec{k}) = 2t\sum_{j=1}^{d} \cos k_j \tag{2.70}$$

and a free Hamiltonian of the form

$$H_0 = \sum_{\sigma=\uparrow,\downarrow} \int \frac{d^d k}{(2\pi)^d} \epsilon(\vec{k})c_\sigma^\dagger(\vec{k})c_\sigma(\vec{k}) \tag{2.71}$$

with

$$\epsilon(\vec{k}) = -t(\vec{k}) = -2t\sum_{j=1}^{d} \cos k_j \tag{2.72}$$

The ground state is found by filling up the Fermi sea. Thus, if we have N particles, the total number of momentum states with energy smaller than E is (assuming that $\epsilon_{\vec{k}}$ has its minimum at $\vec{k} = (0\ldots 0)$) determined by the constant-energy curves

$\epsilon(\vec{k}) \equiv \epsilon$ (see Fig. 2.2). For instance, in the one-dimensional case we find (see Fig. 2.3)

$$\epsilon(k) = -2t \cos k \tag{2.73}$$

If $\mathcal{N}$ is the number of particles and N the number of sites, we get

$$\mathcal{N} = 2N \int_{-k_F}^{k_F} \frac{dk}{2\pi} = \frac{2Nk_F}{\pi} \tag{2.74}$$

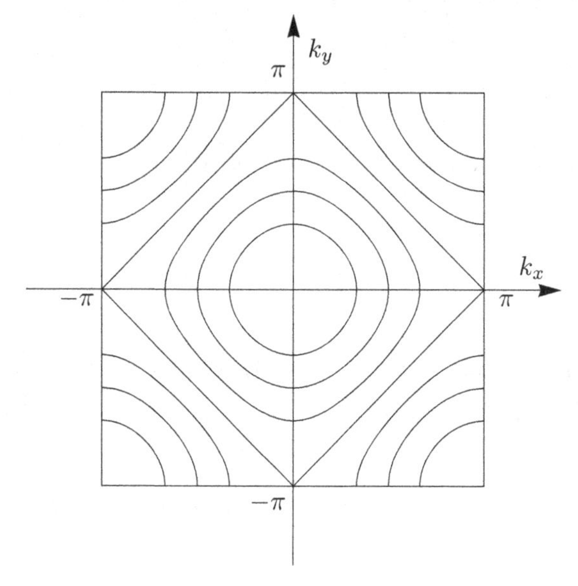

Figure 2.2 Constant-energy curves of H_0 in the first Brillouin zone of the square lattice.

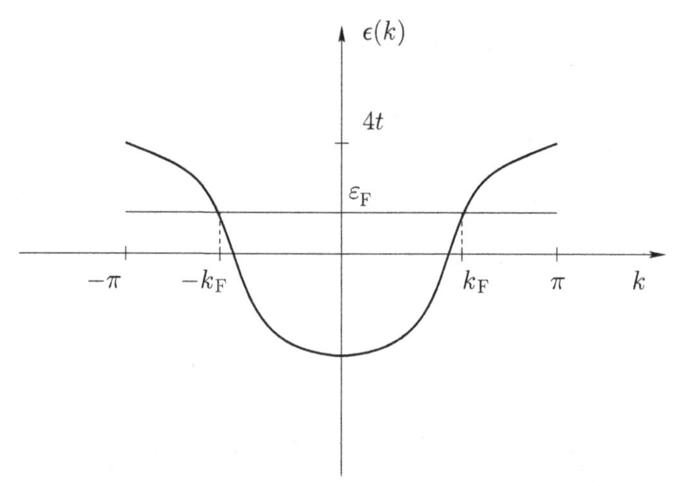

Figure 2.3 One-particle spectrum of H_0 in one dimension.

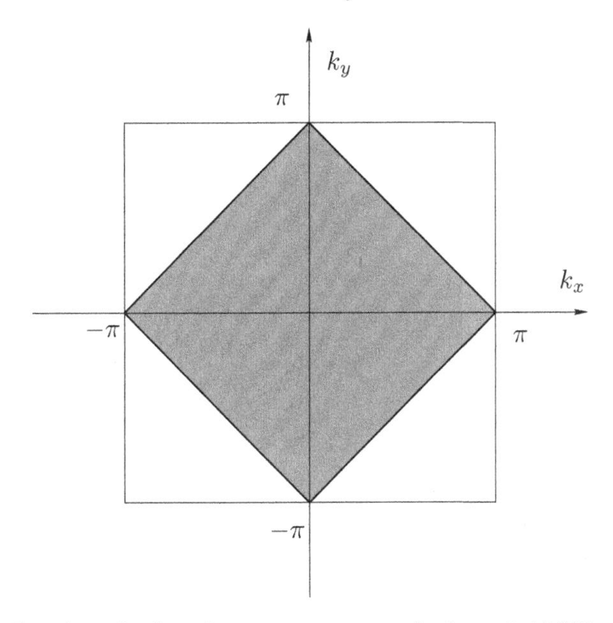

Figure 2.4 Fermi sea for free electrons on a square lattice at half-filling (the lattice spacing is unity).

and

$$k_{\rm F} = \frac{\pi \mathcal{N}}{2N} \equiv \frac{\pi}{2}\rho \tag{2.75}$$

where ρ is the linear density. At half-filling $k_{\rm F} = \pi/2$ and $\epsilon(k_{\rm F})$ vanishes. In higher dimensions we determine the constant-energy curves in the same way. For a half-filled system we just fill up the negative-energy states to obtain the Fermi sea. This is so because this band has $E \leftrightarrow -E$ symmetry ("particle–hole" symmetry) and there are as many states with positive energy as there are with negative energy. The *Fermi surface* is defined by $\epsilon_{\rm F} = 0$ and for a square lattice is rectangular (square) (see Fig. 2.4).

The expectation value of the occupation number

$$n_{\vec{k}} = \sum_{\sigma=\uparrow,\downarrow} c_\sigma^\dagger(\vec{k})c_\sigma(\vec{k}) \tag{2.76}$$

has a jump at the Fermi surface both in the free case and in the case with interaction (see Fig. 2.5).

2.5 Correlation functions

The fermion Green function (or propagator) plays an important role in the theory. We can define it in terms of field operators in the Heisenberg representation

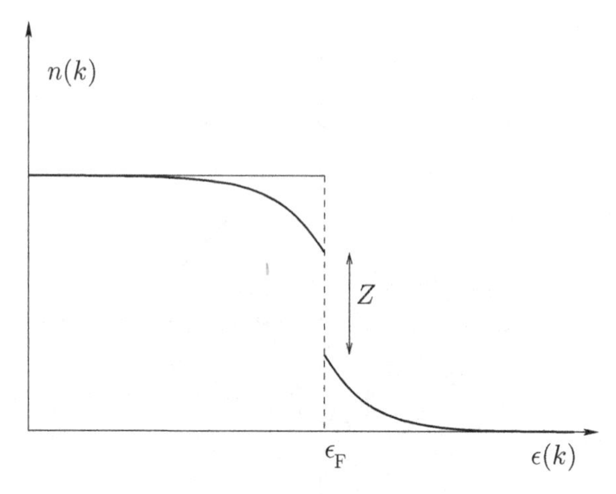

Figure 2.5 Occupation number of the energy levels labeled by k in the non-interacting case (straight line) and interacting (curved line) case; Z is the quasiparticle residue of Eq. (2.84).

$$c_\sigma(\vec{r}, t) = e^{iHt} c_\sigma(\vec{r}\,) e^{-iHt} \tag{2.77}$$

The fermion propagator is defined by

$$G_{\sigma\sigma'}(\vec{r}, t; \vec{r}\,', t') = -i\langle \mathrm{Gnd}|T c_\sigma(\vec{r}, t) c^\dagger_{\sigma'}(\vec{r}\,', t')|\mathrm{Gnd}\rangle \tag{2.78}$$

where $|\mathrm{Gnd}\rangle$ stands for the ground state of the system and T means a time-ordered product of operators,

$$T A(t) B(t') = A(t) B(t') \theta(t - t') \pm B(t') A(t) \theta(t' - t) \tag{2.79}$$

with a + (–) sign for bosons (fermions) and

$$\theta(t) = \begin{cases} 1 & \text{if } t > 0 \\ 0 & \text{if } t < 0 \end{cases} \tag{2.80}$$

For a translationally invariant and time-independent system, we can write $G_{\sigma\sigma'}(\vec{r}, t; \vec{r}\,', t')$ in terms of its Fourier transform (Abrikosov *et al.*, 1963; Fetter and Walecka, 1971; Doniach and Sondheimer, 1974)

$$G_{\sigma\sigma'}(\vec{r}, t; \vec{r}\,', t') = \int \frac{d^d k}{(2\pi)^d} \int \frac{d\omega}{2\pi} e^{i(\vec{k}\cdot(\vec{r}-\vec{r}\,') - \omega(t-t'))} G_{\sigma\sigma'}(\vec{k}, \omega) \tag{2.81}$$

In principle $G_{\sigma\sigma'}(\vec{k}, \omega)$ is a 2×2 spin matrix. In the case of a non-interacting system (and for any spin-isotropic ground state, for that matter), $G^{(0)}_{\sigma\sigma'}(\vec{k}, \omega)$ is very simple to compute (Fetter and Walecka, 1971). The result is

$$G_{\sigma\sigma'}(\vec{k}, \omega) = \delta_{\sigma\sigma'} \lim_{\nu \to 0^+} \left(\frac{\theta\left(\epsilon(\vec{k}) - \epsilon_F\right)}{\omega - \epsilon(\vec{k}) + i\nu} + \frac{\theta\left(\epsilon_F - \epsilon(\vec{k})\right)}{\omega - \epsilon(\vec{k}) - i\nu} \right) \tag{2.82}$$

The poles of $G^0_{\alpha\beta}(\vec{k}, \omega)$ exhibit the physical one-particle excitation spectrum

$$\omega = \epsilon(\vec{k}) \tag{2.83}$$

A *weakly interacting system* (a *Fermi liquid*) resembles a non-interacting one in the sense that the physical low-energy excitations look like weakly interacting fermions. Thus the fermion propagator retains its pole structure, albeit with a renormalized dispersion relation $\omega = \epsilon_{\text{ren}}(\vec{k})$ and a non-trivial residue for energies close to the Fermi energy. In other words, for $\omega \sim \epsilon_F$ the propagator should look like

$$\lim_{\substack{\omega \to \epsilon_F \\ \nu \to 0^+}} G(\vec{k}, \omega) \approx \lim_{\substack{\omega \to \epsilon_F \\ \nu \to 0^+}} \left(\frac{Z}{\omega - \epsilon_{\text{ren}}(\vec{k}) + i\nu} + G_{\text{reg}}(\vec{k}, \omega) \right) \tag{2.84}$$

where Z is the residue and $G_{\text{reg}}(\vec{k}, \omega)$ does not have any singularities close to ϵ_F. The *wave-function renormalization* Z measures the strength of the jump of the fermion occupation number $n_{\vec{k}}$ at the Fermi surface, see Fig. 2.5. These excitations are the fermion quasiparticles, the *dressed electrons*, of the Fermi liquid. These quasiparticles are assumed to be stable in the sense that the poles lie on the real energy axis. Any imaginary part would imply a decay rate of the quasiparticles ("damping"). The existence of such fermionic quasiparticles in the low-energy spectrum of a system of interacting fermions is the central content (and assumption) of the Landau theory of the Fermi liquid (Baym and Pethick, 1991).

In addition to one-particle states, a Fermi liquid has a large class of many-particle excitations. These include particle–hole excitations (i.e. density fluctuations), spin excitations, and paramagnons. These *collective modes* are *bound states* that exist only in an interacting system. Many of these modes are damped. Others are not. We can study the collective modes by means of the many-particle Green functions. Several *correlation functions* are going to be important to us. They are outlined below.

(1) The *density correlation function*

$$K_{00}(\vec{r}, t; \vec{r}\,', t') = \langle \text{Gnd} | T \hat{n}(\vec{r}, t) \hat{n}(\vec{r}\,', t') | \text{Gnd} \rangle \tag{2.85}$$

where $\hat{n}(\vec{r}, t)$ is the local normal-ordered density operator

$$\hat{n}(\vec{r}, t) = \sum_{\sigma = \uparrow, \downarrow} c^\dagger_\sigma(\vec{r}, t) c_\sigma(\vec{r}, t) - \rho \tag{2.86}$$

and ρ is the average density.

The density correlation function measures the strength of the density fluctuations in a physical system. As we will see in other chapters, it plays a key role in understanding the dielectric properties of a system. If the Hamiltonian system is translationally invariant (or, equivalently, invariant under lattice translations, as the Hubbard model is) and time-independent, then the correlation function can only be a function of the separation $\vec{r} - \vec{r}\,'$ and of the time difference $t - t'$,

$$K_{00}(\vec{r}, t; \vec{r}\,', t') \equiv K_{00}(\vec{r} - \vec{r}\,', t - t') \tag{2.87}$$

whose Fourier transform has the form $K_{00}(\vec{k}, \omega)$.

In many electronic systems, such as $NbSe_2$ (and other chalcogenides), and in strongly correlated systems, such as the high-temperature superconductor $La_{2-x}Ba_xCuO_4$ for some doping range $x \sim 1/8$, the electronic ground state is a *charged stripe* or (charge) density wave (CDW) that breaks spontaneously the lattice translation symmetry. The *CDW order parameter* $\langle \text{Gnd}|n(\vec{Q})|\text{Gnd}\rangle$ is the ground-state expectation value of the Fourier transform of the density operator at the *ordering wave vector* $\vec{Q}$ of the CDW. In a CDW state, the Fourier transform of the density correlation function takes the form

$$K_{00}(\vec{k}, \omega) = |\langle n(\vec{Q})\rangle|^2 \delta(\vec{k} - \vec{Q})\delta(\omega) + K_{00}^{\text{conn}}(\vec{k}, \omega) \tag{2.88}$$

where the connected correlation function $K_{00}^{\text{conn}}(\vec{k}, \omega)$ contains the information on the spectrum of density fluctuations of this state. This correlator can be measured by a number of experimental techniques such as light and X-ray scattering.

(2) The *current correlation function*

$$K_{ii'}(\vec{r}, t; \vec{r}\,', t') = \langle \text{Gnd}|T J_i(\vec{r}, t) J_{i'}(\vec{r}\,', t')|\text{Gnd}\rangle \tag{2.89}$$

where $J_i(\vec{r}, t)$ is the current operator, which, in the case of the Hubbard model in the absence of external electromagnetic fields, is

$$J_i(\vec{r}, t) = -it \sum_{\sigma=\uparrow,\downarrow} \left(c_\sigma^\dagger(\vec{r}, t) c_\sigma(\vec{r} + \vec{e}_i, t) - \text{h.c.} \right) \tag{2.90}$$

There is no need to normal-order this operator since the ground state is not expected to spontaneously carry a non-zero current. The current correlation function has the most direct information on the conductivity and can also be measured by light scattering and similar experimental techniques.

(3) The *spin correlation function*

$$K^{aa'}(\vec{r}, t; \vec{r}\,', t') = \langle \text{Gnd}|T S^a(\vec{r}, t) S^{a'}(\vec{r}\,', t')|\text{Gnd}\rangle \tag{2.91}$$

We will see below that, if the spin symmetry is broken spontaneously (i.e. magnetism), $K^{aa'}$ can be non-zero as $(\vec{r}, t)$ and $(\vec{r}\,', t')$ become infinitely separated from each other. In fact, the limit (at equal time!)

$$\lim_{|\vec{r}-\vec{r}\,'|\to\infty} \frac{1}{3}\sum_{a=1}^{3} K^{aa}(\vec{r},t;\vec{r}\,',t) = M^2 \tag{2.92}$$

represents the amplitude of the *ferromagnetic* order parameter M. A non-zero order parameter M signals the presence of a spontaneously broken symmetry. If the magnetic ordering has a spatial dependence with an ordering wave vector $\vec{Q}$ it is a *spin-density wave* (SDW) whose order parameter is the Fourier transform at wave vector $\vec{Q}$ of the ground-state expectation value of the spin-density operator $\langle \text{Gnd}|\vec{S}(\vec{r})|\text{Gnd}\rangle$, the local magnetization. In the case of antiferromagnetism, the ground state is a Néel state, in which case the ordering wave vector is $\vec{Q} = (\pi, \pi)$ (in two dimensions), and the order parameter is the *staggered magnetization*, which takes opposite values on the two sublattices of the square lattice.

(4) The *Cooper-pair correlation function*

All of these two-particle operators have the common feature that they conserve particle numbers locally and the excitations are electrically neutral. In the Bardeen–Cooper–Schrieffer (BCS) theory of superconductivity particle-number conservation is lost locally (but not globally), since one could break a Cooper pair at one location and form it again somewhere else. In BCS theory (Schrieffer, 1964), a Cooper pair is a bound state of an *electron* with momentum $\vec{k}$ and spin up and another *electron* with momentum $-\vec{k}$ and spin down, with $\vec{k}$ on the Fermi surface. This state has charge $2e$ and is a spin singlet. The order parameter for a superconducting state is

$$\Delta_{\sigma\sigma'}(\vec{k}) = \langle \text{Gnd}|c_{\sigma}^{\dagger}(k)c_{\sigma'}^{\dagger}(-\vec{k})|\text{Gnd}\rangle \tag{2.93}$$

In a BCS state $\vec{k}$ and $-\vec{k}$ are two vectors on the Fermi surface and $\Delta(\vec{k})$ is the Cooper-pair amplitude. If $\Delta_{\sigma\sigma'}(\vec{k})$ does not depend on the *direction* of $\vec{k}$, as in most low-temperature superconductors, such as niobium, the superconductor is in an s-wave state. Other superconductor order-parameter symmetries, i.e. other angular-momentum channels, are also possible. In general, the fermionic nature of the electrons dictates that states which pair in an *even*-angular-momentum channel (such as s-wave and d-wave states) must be spin singlets, while those which pair in an *odd*-angular-momentum channel (such as p-wave states) must be spin triplets. Thus, ^{3}He A has an order parameter that is a spin triplet and a p-wave orbital state. The order parameter of the copper-oxide high-temperature superconductors is a spin singlet and has d-wave orbital symmetry, specifically $d_{x^2-y^2}$, and is invariant under time-reversal invariance. However, superconducting states that break time-reversal invariance are also possible. An example of a time-reversal-invariance-breaking superconducting state is the case of the p-wave, specifically $p_x + ip_y$, spin-triplet state that appears to describe the superconductivity of the ruthenate Sr_2RuO_4. We can now define a Cooper-pair correlation function

$$C(\vec{k}, t; \vec{k}', t') = \langle \text{Gnd}| T \left(c_\uparrow^\dagger(\vec{k}, t) c_\downarrow^\dagger(-\vec{k}, t) c_\uparrow(\vec{k}', t') c_\downarrow(-\vec{k}', t') \right) |\text{Gnd}\rangle \quad (2.94)$$

From an experimental point of view, what one can measure are *susceptibilities*. In other words, one couples the system to a weak external field. From linear-response theory (Fetter and Walecka, 1971; Doniach and Sondheimer, 1974) we know how to relate the *response functions* (i.e. causal propagators) to the time-ordered functions. The susceptibilities are the Fourier-transformed causal propagators. For instance, the magnetic susceptibility $\chi^{aa'}(\vec{k}, \omega)$ is defined as follows:

$$\chi^{aa'}(\vec{k}, \omega) = \sum_{\vec{r}} \int dt \, e^{i(\vec{k}\cdot\vec{r} - \omega t)} \chi^{aa'}(0, 0; \vec{r}, t) \quad (2.95)$$

where $\chi^{aa'}(\vec{r}, t; \vec{r}\,', t')$ is the *causal (or retarded) propagator*

$$\chi^{aa'}(\vec{r}, t; \vec{r}\,', t') = \theta(t' - t) \langle \text{Gnd}| S^a(\vec{r}, t) S^{a'}(\vec{r}\,', t') |\text{Gnd}\rangle \quad (2.96)$$

In particular the *static susceptibility* $\chi^{aa'}(\vec{k}, 0)$ measures the response of the system to a weak external magnetic field. In the uniform limit $\vec{k} \to 0$, we are measuring the response to a weak uniform magnetic field. Thus this is the static ferromagnetic susceptibility. If we want to probe a Néel state we must couple to the *staggered magnetization* and hence use a staggered field. This is difficult to achieve. However, using neutron scattering we can measure $\chi(\vec{k}, \omega)$ for a wide range of wave vectors $\vec{k}$, in particular the case $\vec{k} = (\pi, \pi, \pi)$, which is the staggered or Néel susceptibility.

3
The magnetic instability of the Fermi system

The Hubbard model was originally introduced as the simplest system which may exhibit an insulating (Mott) state. This state is the result of strong electron–electron interactions. In this chapter we consider the Hubbard model at half-filling. The main goal here is the study of the magnetic properties of its ground state. Apart from an exact solution in one dimension, no exact results are available for this problem. This leads to the use of several approximations. The most popular one, and the oldest, is the mean-field theory (MFT). In the MFT one has the *bias* that the ground state does have some sort of magnetic order (i.e. ferromagnetic, Néel antiferromagnetic, etc.). The problem is then usually solved by means of a variational ansatz. However, one is usually interested in more than just the ground-state energy, which, after all, is not directly measurable and depends very sensitively on the properties at short distances. Most often we wish to evaluate the long-distance, low-frequency, properties of the correlation and response functions of this theory. Moreover, in some cases, such as in one dimension, the fluctuations overwhelm the MFT predictions.

In this chapter we will consider the standard MFT (i.e. Hartree–Fock), which is expected to become accurate at *weak* coupling. We will consider both ferromagnetic and antiferromagnetic states. We will also rederive these results using path integrals. As a byproduct, we will also have a theory of the fluctuations: the non-linear sigma model.

3.1 Mean-field theory

Let us consider now the effects of interactions on the unperturbed ground state. It is convenient to consider the Fourier transform of the interaction term of the Hubbard Hamiltonian, Eq. (2.7)

$$\begin{aligned} H_1 &= U \sum_{\vec{r}} n_\uparrow(\vec{r}) n_\downarrow(\vec{r}) \\ &= U \int_{\vec{k}_1 \ldots \vec{k}_4} (2\pi)^d \delta^d(-\vec{k}_1 + \vec{k}_2 - \vec{k}_3 + \vec{k}_4) c_\uparrow^\dagger(\vec{k}_1) c_\uparrow(\vec{k}_2) c_\downarrow^\dagger(\vec{k}_3) c_\downarrow(\vec{k}_4) \end{aligned} \tag{3.1}$$

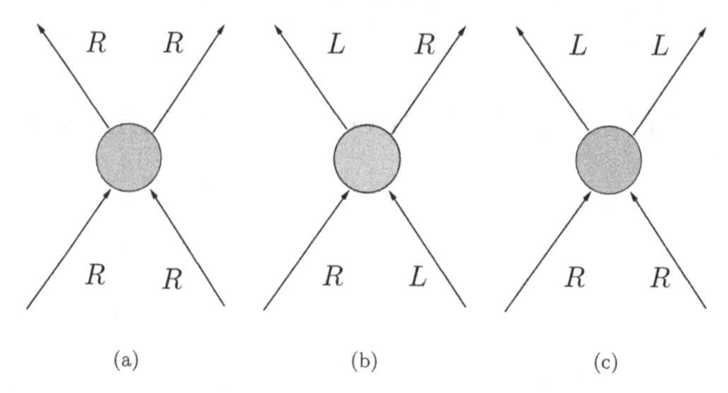

Figure 3.1 Scattering processes between right (R) and left (L) movers: (a) forward, (b) backward and (c) Umklapp.

where $\int_{\vec{k}_i}$ is a shorthand for $\int_{\mathrm{BZ}} d^d k_i/(2\pi)^d$. Notice that on a lattice momentum is conserved, modulo a reciprocal-lattice vector $\vec{G}$.

Let us discuss first the simpler one-dimensional case. There are two Fermi points (at $\pm k_{\mathrm{F}}$). Thus we can classify the excitations as left- or right-moving particles and holes with either spin orientation. We have the following scattering processes (see Fig. 3.1): (a) forward scattering, (b) backward scattering, and (c) Umklapp scattering.

In case (a) two particles scatter with a small momentum transfer and do not change the direction of their individual motion. In case (b) a right mover becomes a left mover and vice versa. In case (c) two right movers become left movers. This process violates momentum conservation but, if the total momentum violation equals a reciprocal-lattice vector, the (Umklapp) process is allowed. This occurs for $k_{\mathrm{F}} = \pi/2$, which is the half-filled case (see Fig. 3.1(c)).

Case (b), backward scattering, implies a scattering process involving two degenerate states: exchanging a right mover with a left mover and vice versa (see Fig. 3.1(b)). Since the energy denominator is zero we may have an instability of the perturbation theory. This is an antiferromagnetic instability since the momentum transfer is π. Conversely, instability in the forward-scattering channel is a symptom of ferromagnetism.

In dimensions higher than one, the situation is more complex due to the intricacies of the Fermi surface. For instance, in the case of a half-filled square lattice the Fermi surface is a square (see Fig. 2.4). A scattering process involving the (nesting) wave vectors shown in Fig. 3.2 may induce an antiferromagnetic instability. That is, we *exchange* particles with opposite spin from opposite sides of the Fermi surface. Once again this involves a momentum exchange of (π, π) or $(\pi, -\pi)$, depending the case.

However, for a nearly empty band, only quasi-forward scattering should matter, and the relevant momentum exchange should be zero (ferromagnetism). Other

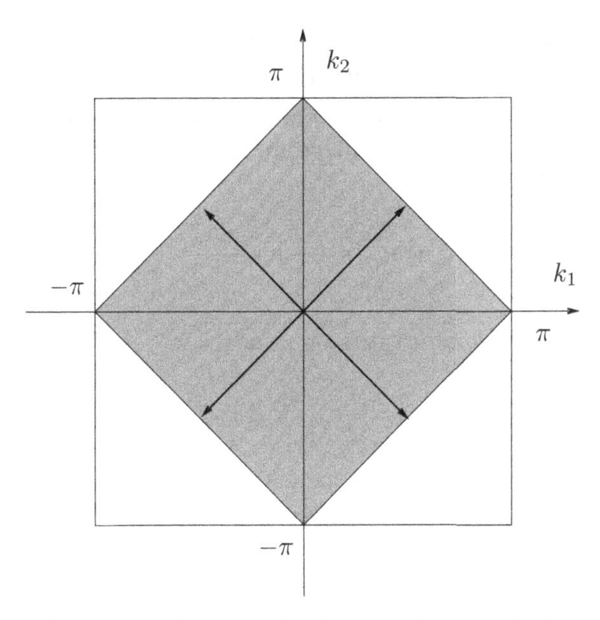

Figure 3.2 Nesting vectors for the Brillouin zone of a square lattice at half-filling. This nesting property is responsible for an antiferromagnetic instability.

cases, involving other momentum exchanges, are possible. These instabilities generally give rise to a *spin-density wave* of wave vector $\vec{k}$. The ferromagnetic state occurs when $\vec{k} = 0$ and the Néel antiferromagnetic state occurs when $\vec{k} = (\pi, \pi)$.

We want to develop a theory of these instabilities. As we see, we need to find bound states of a certain wave vector $\vec{k}$ and the ground state will have to be rebuilt in the form of a coherent superposition of these bound states. Since we do not know how to solve this problem exactly, some sort of mean-field theory is necessary. There are several ways of achieving this goal, using (a) the Hartree–Fock and random-phase approximation (RPA), (b) variational wave functions, and (c) $1/N$ expansions.

These three approaches are, to some extent, physically equivalent. While (a) and (b) (and mostly (a)) are commonly discussed in textbooks, the $1/N$ expansion is a rather novel technique and, for that reason, is not usually available to students (although it has become pervasive throughout the current literature).

The Hartree–Fock–RPA approach involves choosing a particular set of the Feynman diagrams, which one can argue gives the "most important" contributions. While in one dimension it is possible to select diagrams according to their degree of divergence in the infrared, the situation is far less clear in two or more dimensions. Typically, one has to choose a particular process and sum all the leading contributions which contribute to the process and, at the same time, do not violate any conservation laws, a "conserving approximation," in the terminology of Baym and Kadanoff (Kadanoff and Baym, 1962). Such is the spirit of Hartree–Fock–RPA theories. Similarly, in the variational-wave-function approach, one chooses variational states which are essentially inspired by RPA-like calculations.

Let us first discuss a simple form of mean-field theory. We start from the Hubbard Hamiltonian

$$H = -t \sum_{\langle \vec{r}, \vec{r}' \rangle} c^{\dagger}_{\sigma}(\vec{r}) c_{\sigma}(\vec{r}\,') + \text{h.c.} - \frac{2}{3} U \sum_{\vec{r}} \left(\vec{S}(\vec{r}) \right)^2 \tag{3.2}$$

which is quartic in the fermionic operators, since

$$\vec{S}(\vec{r}) = \frac{1}{2} c^{\dagger}_{\sigma}(\vec{r}) \vec{\tau}_{\sigma\sigma'} c_{\sigma'}(\vec{r}) \tag{3.3}$$

(From now on, I will be using the summation convention on repeated spin indices.)

The interaction term of the Hubbard Hamiltonian, Eq. (3.2), is quartic in fermion operators. In general a non-linear problem of this sort is not solvable except in some very special cases, such as one-dimensional systems. A standard approach is the mean-field approximation (or *Hartree–Fock approximation*) in which the quartic term is factorized in terms of a fermion *bilinear* times a Bose field, which is usually treated classically. In other words one simply ignores the dynamics of the Bose field. Consider, for instance, the Hamiltonian H',

$$H' = -t \sum_{\langle \vec{r}, \vec{r}\,' \rangle} c^{\dagger}_{\sigma}(\vec{r}) c_{\sigma}(\vec{r}\,') + \text{h.c.} + \frac{3}{8U} \sum_{\vec{r}} \vec{M}^2(\vec{r}) + \sum_{\vec{r}} \vec{M}(\vec{r}) \cdot \vec{S}(\vec{r}) \tag{3.4}$$

which can be regarded as a *linearized* version of H in terms of a Bose field $\vec{M}(\vec{r})$, which, as we will see below, represents the local magnetization.

However, there is something in this expression that is not quite right, since the field $\vec{M}(\vec{r})$ does not have any dynamics. It looks like a variational parameter. Indeed, in the *Hartree–Fock* approximation, one assumes that a certain operator, say $\vec{S}(\vec{r})$, picks up an expectation value. One then has to shift the operator by its expectation value and neglect fluctuations (this is the mean-field approximation). Therefore, one writes

$$\vec{S}(\vec{r}) = \langle \vec{S}(\vec{r}) \rangle + \left(\vec{S}(\vec{r}) - \langle \vec{S}(\vec{r}) \rangle \right) \tag{3.5}$$

The term in brackets clearly represents fluctuations. Thus

$$\vec{S}^2 = \langle \vec{S} \rangle^2 + \left(\vec{S} - \langle \vec{S} \rangle \right)^2 + 2 \langle \vec{S} \rangle \cdot \left(\vec{S} - \langle \vec{S} \rangle \right) \tag{3.6}$$

Neglecting fluctuations means that we drop the second term. Thus we write

$$H = H_{\text{MF}} + H_{\text{fl}} \tag{3.7}$$

where

$$H_{\text{MF}} = -t \sum_{\langle \vec{r}, \vec{r}\,' \rangle} c_\sigma^\dagger(\vec{r}) c_\sigma(\vec{r}\,') + \text{h.c.} - \frac{2}{3} U \sum_{\vec{r}} \langle \vec{S}(\vec{r}) \rangle^2 \\ - \frac{4}{3} U \sum_{\vec{r}} \langle \vec{S}(\vec{r}) \rangle \cdot \left(\vec{S}(\vec{r}) - \langle \vec{S}(\vec{r}) \rangle \right) \tag{3.8}$$

and the fluctuation part H_{fl} is the rest. We can also write

$$H_{\text{MF}} = -t \sum_{\langle \vec{r}, \vec{r}\,' \rangle} c_\sigma^\dagger(\vec{r}) c_\sigma(\vec{r}\,') + \text{h.c.} + \frac{2U}{3} \sum_{\vec{r}} \langle \vec{S}(\vec{r}) \rangle^2 - \frac{4U}{3} \sum_{\vec{r}} \langle \vec{S}(\vec{r}) \rangle \cdot \vec{S}(\vec{r}) \tag{3.9}$$

which is just H', Eq. (3.4), if we make the identification $\vec{M}(\vec{r}) \equiv -(4U/3)\langle \vec{S}(\vec{r}) \rangle$.

We can give dynamics to $\vec{M}\,(\vec{r})$ by using the following device. Consider first the simple classical oscillator problem with a degree of freedom $\vec{M}$ and Lagrangian

$$L = \frac{1}{2} m \dot{\vec{M}}^2 - \frac{g}{2} \vec{M}^2 - \vec{M} \cdot \vec{S} \tag{3.10}$$

The equations of motion of this oscillator are

$$\frac{d}{dt} \frac{\partial L}{\partial \dot{\vec{M}}} = \frac{\partial L}{\partial \vec{M}} \tag{3.11}$$

which imply

$$m \ddot{\vec{M}} = -g \vec{M} - \vec{S} \tag{3.12}$$

At the quantum level these equations become the equation of motion of the *operator* $\vec{M}(t)$ in the Heisenberg representation.

Consider now the limit $m \to 0$. The *only* smooth trajectories, i.e. with $\ddot{\vec{M}}$ finite, which are possible in this limit satisfy

$$g \vec{M} + \vec{S} = 0 \tag{3.13}$$

The Hamiltonian H' has to be regarded in precisely the same way. We have to add a kinetic-energy term at each site of the form $\sum_{\vec{r}} \vec{P}^2(\vec{r})/(2m)$, where $\vec{P}$ and $\vec{M}$ obey canonical commutation relations, and consider the limit $m \to 0$. One should not panic at the apparent divergence in the kinetic-energy term: the equations of motion are taking care of it. We are going to come back to this later on, when we discuss the path-integral form. There everything is simpler.

Thus we see that the Lagrange multiplier field $\vec{M}$ is dynamical in the sense that it follows the configurations of fermions in detail. In mean-field theory (i.e. the Hartree–Fock approximation), one replaces $\vec{M}$ by some *static* (time-independent) configuration, which, in turn, is determined by the condition that the ground-state energy should be the lowest possible energy.

Let us now look at H_{MF} in Fourier space. The Lagrange multiplier field $\vec{M}(\vec{r})$ has the Fourier transform

$$\vec{M}(\vec{r}) = \frac{1}{N^d} \sum_{\vec{k}} e^{i\vec{k}\cdot\vec{r}} \vec{M}(\vec{k}) \to \int \frac{d^d k}{(2\pi)^d} e^{i\vec{k}\cdot\vec{r}} \vec{M}(\vec{k}) \quad \text{for} \quad N \to \infty \tag{3.14}$$

We can now write H_{MF} in the form

$$H_{\text{MF}} = \int_{\vec{k}} \left(\sum_{\sigma} \epsilon(\vec{k}) n_{\sigma}(\vec{k}) + \frac{3}{8U} |\vec{M}(\vec{k})|^2 + \vec{M}^*(\vec{k}) \cdot \vec{S}(\vec{k}) \right) \tag{3.15}$$

since

$$\vec{M}(\vec{k}) = \vec{M}^*(-\vec{k}) \tag{3.16}$$

and where

$$\vec{S}(\vec{k}) = \int_{\vec{k}'} c_{\alpha}^{\dagger}(\vec{k}') \frac{\vec{\tau}_{\alpha\beta}}{2} c_{\beta}(\vec{k}' + \vec{k}) \tag{3.17}$$

with $\epsilon(\vec{k}) = -2t \sum_{j=1}^{d} \cos k_j$.

The second term on the right-hand side of Eq. (3.15) implies that a configuration with the Fourier component $\vec{M}(\vec{k})$ induces scattering processes which mix one-particle states differing by $\vec{k}$.

3.1.1 The ferromagnetic state

Let us consider first the *ferromagnetic* solution in which $\vec{M}(\vec{r})$ is a constant $\vec{M}_0$. In Fourier space, we have

$$\vec{M}(\vec{k}) = \vec{M}_0 (2\pi)^d \delta^d(\vec{k}) \tag{3.18}$$

Then H_{MF} is (with the volume $V = N^d$)

$$H_{\text{MF}} = \frac{3}{8U} V \vec{M}_0^2 + \int_{\vec{k}} \left(\epsilon(\vec{k}) c_{\sigma}^{\dagger}(\vec{k}) c_{\sigma}(\vec{k}) + c_{\sigma}^{\dagger}(\vec{k}) \frac{\vec{\tau}_{\sigma\sigma'}}{2} c_{\sigma'}(\vec{k}) \cdot \vec{M}_0 \right) \tag{3.19}$$

Since the direction of $\vec{M}_0$ is arbitrary, one can choose the z axis (i.e. the quantization axis) to be parallel to $\vec{M}_0$ without any loss of generality. One then finds

$$H_{\text{MF}} = \frac{3}{8U} V \vec{M}_0^2 + \int_{\vec{k}} \left[\left(\epsilon(\vec{k}) + \frac{1}{2} |\vec{M}_0| \right) n_{\uparrow}(\vec{k}) + \left(\epsilon(\vec{k}) - \frac{1}{2} |\vec{M}_0| \right) n_{\downarrow}(\vec{k}) \right] \tag{3.20}$$

The result is that, if $|\vec{M}_0|$ is non-zero, we can lower the *electronic* energy by filling up a number of down spin states and, at the same time, emptying the same number of up spin states. Since the first term penalizes a non-zero value of $|\vec{M}_0|$ we must search for a balance. We also need to keep track of the fact that there is a *total* of

N electrons (both those with up and those with down spins). As usual, this is taken care of by shifting H_{MF} by $\mu \hat{N} = \mu \sum_{\vec{r}} c_\sigma^\dagger(\vec{r}) c_\sigma(\vec{r})$.

Consider now a state with $N_\uparrow$ electrons with spin up and $N_\downarrow$ with spin down. Let $\epsilon_\uparrow$ ($\epsilon_\downarrow$) be the one-particle energy of the top of the filled up (down) states. The total energy of such a state E is a function of $|\vec{M}_0|$, μ, $\epsilon_\uparrow$, and $\epsilon_\downarrow$ (or, equivalently $N_\uparrow$ and $N_\downarrow$).

The energy is $\left(|\vec{M}_0| \equiv M_0\right)$

$$
\begin{aligned}
E_0(M_0, \mu, \epsilon_\uparrow, \epsilon_\downarrow) = {} & \frac{3}{8U} V M_0^2 + V \int_{\vec{k}} \left(\epsilon(\vec{k}) + \frac{1}{2} M_0\right) \theta\left(\epsilon_\uparrow - \epsilon(\vec{k})\right) \\
& + V \int_{\vec{k}} \left(\epsilon(\vec{k}) - \frac{1}{2} M_0\right) \theta\left(\epsilon_\downarrow - \epsilon(\vec{k})\right) \\
& + \mu V \int_{\vec{k}} \left(\theta\left(\epsilon_\uparrow - \epsilon(\vec{k})\right) + \theta\left(\epsilon_\downarrow - \epsilon(\vec{k})\right)\right)
\end{aligned}
\tag{3.21}
$$

By introducing the one-particle *band* density of states (DOS) (i.e. the DOS of the unperturbed system without spin), $\rho(\epsilon)$, we get

$$
\begin{aligned}
\mathcal{E} & \equiv \frac{E_0}{V} \\
& = \frac{3}{8U} M_0^2 + \int_{\epsilon_0}^{\epsilon_\uparrow} d\epsilon \left(\epsilon + \frac{1}{2} M_0\right) \rho(\epsilon) + \int_{\epsilon_0}^{\epsilon_\downarrow} d\epsilon \left(\epsilon - \frac{1}{2} M_0\right) \rho(\epsilon) \\
& \quad + \mu \left(\int_{\epsilon_0}^{\epsilon_\uparrow} d\epsilon\, \rho(\epsilon) + \int_{\epsilon_0}^{\epsilon_\downarrow} d\epsilon\, \rho(\epsilon)\right)
\end{aligned}
\tag{3.22}
$$

where ϵ_0 is the energy of the bottom of the band.

Since the ground-state energy must be an extremum (actually a minimum) we have to find the values of μ, M_0, $\epsilon_\uparrow$, and $\epsilon_\downarrow$ which make the energy density have a minimum *at fixed density*. That is

$$
\frac{\partial \mathcal{E}}{\partial \mu} = \frac{N}{V}, \qquad \frac{\partial \mathcal{E}}{\partial |\vec{M}_0|} = 0, \qquad \frac{\partial \mathcal{E}}{\partial \epsilon_\uparrow} = 0, \qquad \frac{\partial \mathcal{E}}{\partial \epsilon_\downarrow} = 0
\tag{3.23}
$$

An explicit calculation gives

$$
\begin{aligned}
\frac{N}{V} & = \int_{\epsilon_0}^{\epsilon_\uparrow} d\epsilon\, \rho(\epsilon) + \int_{\epsilon_0}^{\epsilon_\downarrow} d\epsilon\, \rho(\epsilon) \\
0 & = \frac{3}{4U} M_0 + \frac{1}{2} \int_{\epsilon_0}^{\epsilon_\uparrow} d\epsilon\, \rho(\epsilon) - \frac{1}{2} \int_{\epsilon_0}^{\epsilon_\downarrow} d\epsilon\, \rho(\epsilon) \\
0 & = \left(\epsilon_\uparrow + \frac{1}{2} M_0\right) \rho(\epsilon_\uparrow) + \mu \rho(\epsilon_\uparrow) \\
0 & = \left(\epsilon_\downarrow - \frac{1}{2} M_0\right) \rho(\epsilon_\downarrow) + \mu \rho(\epsilon_\downarrow)
\end{aligned}
\tag{3.24}
$$

Provided that $\rho(\epsilon_{\uparrow,\downarrow}) \neq 0$, we see that the polarization M_0 is given by

$$M_0 = \epsilon_\downarrow - \epsilon_\uparrow \tag{3.25}$$

and the chemical potential μ is equal to

$$\mu = -\frac{1}{2}(\epsilon_\downarrow + \epsilon_\uparrow) \tag{3.26}$$

Clearly, since M_0 is positive, $\epsilon_\uparrow < \epsilon_\downarrow$ and there are more occupied down spin states than up spin states. We can also write

$$\epsilon_\downarrow - \epsilon_\uparrow = \frac{2U}{3} \int_{\epsilon_\uparrow}^{\epsilon_\downarrow} d\epsilon \, \rho(\epsilon) \tag{3.27}$$

and

$$\frac{N}{V} = 2 \int_{\epsilon_0}^{\epsilon_\uparrow} d\epsilon \, \rho(\epsilon) + \int_{\epsilon_\uparrow}^{\epsilon_\downarrow} d\epsilon \, \rho(\epsilon) \tag{3.28}$$

Equations (3.27) and (3.28) determine $\epsilon_\uparrow$ and $\epsilon_\downarrow$ and, thus, the solution to the problem. In general these equations need to be solved numerically.

Equation (3.27) has two solutions: $\epsilon_\uparrow = \epsilon_\downarrow$ (i.e. $M_0 = 0$, the *paramagnetic* state) and $\epsilon_\uparrow \neq \epsilon_\downarrow$ ($M_0 \neq 0$, the *ferromagnetic* state). The analysis of these equations follows closely the solution of the Curie–Weiss equation in the theory of phase transitions. We can write Eq. (3.27) in the form

$$x = \frac{2U}{3} \int_0^x d\epsilon \, \rho(\epsilon + \epsilon_\uparrow) \tag{3.29}$$

where $x = \epsilon_\downarrow - \epsilon_\uparrow$. Also, we get

$$\frac{N}{V} = 2 \int_{\epsilon_0}^{\epsilon_\uparrow} d\epsilon \, \rho(\epsilon) + \frac{3}{2U} x \tag{3.30}$$

For $\epsilon_\uparrow$ given, the integral in Eq. (3.29) is a monotonically increasing function of x (see Fig. 3.3). For values of $U > U_c$ there are two solutions $x_0 = 0$ and $x_0 \neq 0$, whereas for $U < U_c$ there is only one solution, $x_0 = 0$. The critical Hubbard coupling U_c is determined by the condition

$$1 = \frac{2U_c}{3} \rho(\bar{\epsilon}_\uparrow) \tag{3.31}$$

where $\rho(\bar{\epsilon}_\uparrow)$ is determined by Eq. (3.30) at $x = 0$

$$\frac{N}{V} = 2 \int_{\epsilon_0}^{\bar{\epsilon}_\uparrow} d\epsilon \, \rho(\epsilon) \tag{3.32}$$

Equation (3.31) is known as the *Stoner criterion*. The statement is that for $U > U_c$ the *ferromagnetic* solution appears and has a lower energy than the *paramagnetic* state, $|\vec{M}_0| = 0$.

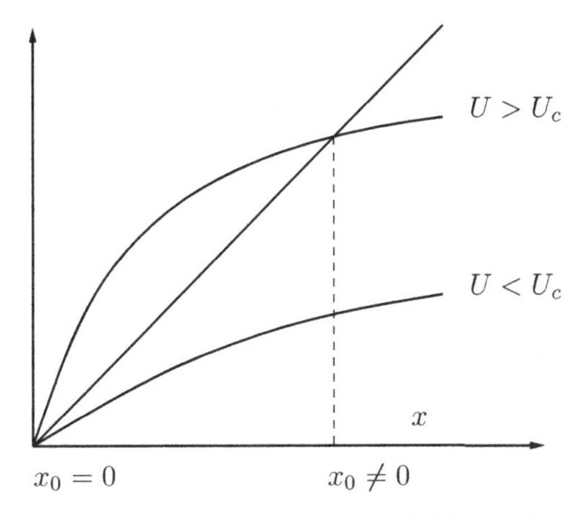

Figure 3.3 Solution of the mean-field equation.

If the DOS is a *smooth* function near the Fermi energy of the paramagnetic state, we can find the solution close to U_c by using a power-series expansion. The result, to leading order in $(U_c - U)/U_c$, is

$$x_0 = \epsilon_\downarrow - \bar{\epsilon}_\uparrow \approx \frac{2\bar{\rho}}{\bar{\rho}'}\left(\frac{U_c - U}{U_c}\right) + \cdots \tag{3.33}$$

and

$$\delta = \epsilon_\uparrow - \bar{\epsilon}_\uparrow \approx -\frac{3}{2U_c\bar{\rho}'}\left(\frac{U_c - U}{U_c}\right) + \cdots \tag{3.34}$$

where $\bar{\epsilon}_\uparrow$ satisfies Eq. (3.32) and $\bar{\rho}$ and $\bar{\rho}'$ are the DOS and its derivative at $\bar{\epsilon}_\uparrow$. There are important cases in which the DOS $\rho(\epsilon)$ has singularities at certain energies known as the van Hove singularities. This happens at half-filling for systems like a square lattice, for which the Fermi surface has the property of nesting.

3.1.2 The Néel state

We now will look for solutions of the mean-field equations in which $\vec{M}(\vec{r})$ is not a constant. Ultimately the problem boils down to a comparison of the energies for different solutions. However, for situations in which nesting takes place we can argue that a Néel state, or, more generally, a spin-density wave (SDW), is the ground state.

Let us consider the mean-field Hamiltonian of Eq. (3.15) and assume that $M(\vec{r})$ has the form

$$\vec{M}(\vec{r}) = \vec{M}_0 \cos(\vec{Q} \cdot \vec{r}) \tag{3.35}$$

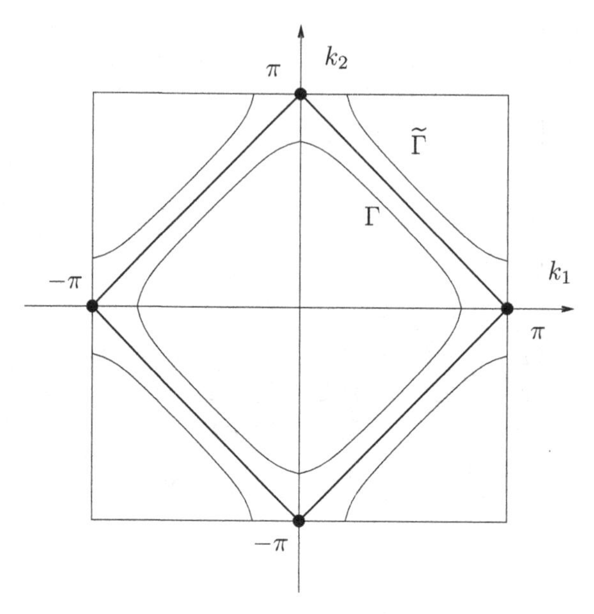

Figure 3.4 The first Brillouin zone for the square lattice for a nearest-neighbor hopping band structure. The diamond-shaped full curve is the Fermi surface (FS) at half-filling with Fermi energy $E_\mathrm{F} = 0$. Γ is an FS with $E_\mathrm{F} < 0$, and $\widetilde{\Gamma}$ is an FS with $E_\mathrm{F} > 0$. The black dots are the van Hove singularities at the saddle-points of the dispersion $E(\vec{k})$.

where $\vec{Q} = (\pi, \pi)$. We saw before that, at half-filling, the Fermi surface (FS) has the shape shown in Fig. 2.4; see here Fig. 3.4. The states across the FS differ by a wave vector $\vec{Q} = (\pi, \pi)$, which is at the Brillouin-zone (BZ) edge (this is the nesting property). Furthermore, for a square lattice, we have $\epsilon(\vec{k}) = -2t(\cos k_1 + \cos k_2)$. Thus we get

$$\epsilon(\vec{k}) = -\epsilon(\vec{k} + \vec{Q}) \tag{3.36}$$

The mean-field Hamiltonian can now be written in the form

$$\begin{aligned} H_\mathrm{MF} = {} & \frac{3}{8U}|\vec{M}_0|^2 V + \int_{\vec{k}} \epsilon(\vec{k}) c_\sigma^\dagger(\vec{k}) c_\sigma(\vec{k}) \\ & + \int_{\vec{k}} c_\alpha^\dagger(\vec{k}) \frac{1}{2} \vec{\tau}_{\alpha\beta} c_\beta(\vec{k} + \vec{Q}) \cdot \frac{1}{2}\vec{M}_0 \\ & + \int_{\vec{k}} c_\alpha^\dagger(\vec{k}) \frac{1}{2} \vec{\tau}_{\alpha\beta} c_\beta(\vec{k} - \vec{Q}) \cdot \frac{1}{2}\vec{M}_0 \end{aligned} \tag{3.37}$$

Consider the spinor $\Psi_\sigma(\vec{k})$

$$\Psi_\sigma(\vec{k}) = \begin{pmatrix} c_\sigma(\vec{k}) \\ c_\sigma(\vec{k} + \vec{Q}) \end{pmatrix} \tag{3.38}$$

If we restrict ourselves to the two-dimensional case we can now write

$$H_{\text{MF}} = \frac{3}{8U}|\vec{M}_0|^2 V + \int_{\vec{k}}' \Psi^{\dagger}_{a\sigma}(\vec{k})\mathcal{H}_{a\sigma,a'\sigma'}(\vec{k})\Psi_{a'\sigma'}(\vec{k}) \tag{3.39}$$

where the integral now ranges over the upper half of the BZ (it has been "folded") and $a = 1, 2$ and $\sigma = \uparrow, \downarrow$, indicating the upper and lower components of Ψ with either spin. The one-particle Hamiltonian $\mathcal{H}$ is a 4×4 matrix that has the block form

$$\mathcal{H} = \begin{pmatrix} \epsilon(\vec{k}) & \frac{1}{2}\vec{\tau}\cdot\vec{M}_0 \\ \frac{1}{2}\vec{\tau}\cdot\vec{M}_0 & \epsilon(\vec{k}+\vec{Q}) \end{pmatrix} = \begin{pmatrix} \epsilon(\vec{k}) & \frac{1}{2}\vec{\tau}\cdot\vec{M}_0 \\ \frac{1}{2}\vec{\tau}\cdot\vec{M}_0 & -\epsilon(\vec{k}) \end{pmatrix} \tag{3.40}$$

This matrix can be diagonalized very easily (see, for instance, the diagonalization of the Dirac Hamiltonian). It has two doubly degenerate (spin) eigenvalues $E_{\pm}(\vec{k})$ with

$$E_{\pm} = \pm\sqrt{\epsilon^2(\vec{k}) + \frac{1}{4}|\vec{M}_0|^2} \tag{3.41}$$

Thus, we see that the system has now acquired a gap Δ with

$$\Delta = |\vec{M}_0| \tag{3.42}$$

at the "Fermi surface."

The ground state is obtained by filling up the negative-energy single-particle fermionic states with spins both up and down. This state has a vanishing z-component of the total spin: $S_z = 0$. The energy density is

$$\mathcal{E} = \frac{3}{8U}|\vec{M}_0|^2 - \int_{\vec{k}}' E_{+}(\vec{k}) \tag{3.43}$$

where the integral ranges over the occupied states in the upper half of the first BZ. We can now use the symmetries of $E(\vec{k}) \equiv E_{+}(\vec{k})$ to write this expression in terms of an integral over the upper-right quadrant of the BZ

$$\mathcal{E} = \frac{3}{8U}|\vec{M}_0|^2 - 2\int_{\Omega} E(\vec{k}) \tag{3.44}$$

where the integration range is the set of points of the first BZ $\Omega = \{\vec{k}, 0 \leq k_i \leq \pi, k_1 + k_2 \leq \pi\}$.

We must now determine the value of M_0 for which $\mathcal{E}$ is lowest. The condition for an extremum is

$$\frac{\partial\mathcal{E}}{\partial M_0} = \frac{3}{4U}|\vec{M}_0| - 2\int_{\Omega} \frac{d^2k}{(2\pi)^2}\frac{\partial E(\vec{k})}{\partial M_0} = 0 \tag{3.45}$$

Clearly we get

$$\frac{3}{4U}|\vec{M}_0| - \frac{1}{2}\int_{\Omega} \frac{d^2k}{(2\pi)^2} \frac{|\vec{M}_0|}{E(\vec{k})} = 0 \tag{3.46}$$

One solution is $|\vec{M}_0| = 0$ (i.e. no long-range order). This is the paramagnetic state. The other solution obeys the gap equation

$$\frac{3}{2U} = \int_{\Omega} \frac{d^2k}{(2\pi)^2} \frac{1}{\sqrt{\epsilon^2(\vec{k}) + \frac{1}{4}|\vec{M}_0|^2}} \tag{3.47}$$

In the case of the square lattice (although similar results hold for other cases with nesting) the integral on the right-hand side of Eq. (3.47) is *divergent* (for $M_0 \to 0$). In fact, the integral is dominated by contributions with momenta close to the FS, i.e. near $k_1 + k_2 = \pi$, or for single-particle states with energy close to the Fermi energy ($E_{\rm F} = 0$ in this case). The simplest way to evaluate Eq. (3.47) is to write the momentum integral as an integral over the energies ε of the occupied single-particle states:

$$\frac{3}{2U} = \int_{E_{\rm min}}^{E_{\rm F}} d\varepsilon \, \frac{\rho(\varepsilon)}{\sqrt{\varepsilon^2 + \Delta^2}} \tag{3.48}$$

where $\rho(\varepsilon)$ is the one-particle DOS (per unit volume) for the single-particle dispersion $\epsilon(\vec{k}) = -2t(\cos k_1 + \cos k_2)$, and $\Delta = \frac{1}{2}M_0$ is the *single-particle gap*. Here, $E_{\rm F} = 0$ and $E_{\rm min} = -4t$.

The single-particle DOS can always be written as a line integral over a closed contour $\Gamma(\varepsilon)$ (the set of points with energy ε)

$$\rho(\varepsilon) = \frac{2}{(2\pi)^2} \oint_{\Gamma(\varepsilon)} \frac{d\vec{l} \cdot \hat{n}}{|\vec{v}|} = \frac{1}{2\pi^2} \oint_{\Gamma(\varepsilon)} \frac{d\vec{k} \cdot \vec{\nabla}_k \varepsilon(\vec{k})}{|\nabla_k \varepsilon(\vec{k})|^2} \tag{3.49}$$

where $\vec{v} = \vec{\nabla}_k \varepsilon(\vec{k})$ and $\hat{n}$ is the unit tangent vector, $\hat{n} = \vec{v}/|\vec{v}|$. If the DOS $\rho(\varepsilon)$ is a smooth function of the energy ε near the Fermi energy $E_{\rm F}$, the integral of Eq. (3.48) would be logarithmically divergent as the gap $\Delta \to 0$. However, for the case of the square lattice the DOS is singular for $\varepsilon \to 0$.

The FS of the half-filled square lattice has $E_{\rm F} = 0$ and hence it separates electron-like states with $E > 0$ from hole-like states with $E < 0$. The point $(0, 0)$ of the BZ is the minimum energy (for single-particle states) whereas the point (π, π) (and others related by reciprocal lattice vectors) are maxima of the dispersion. Of particular interest are the points $(\pi, 0)$ and $(0, \pi)$ (and their images under reciprocal-lattice translations) that are *saddle-points* of the dispersion. It turns out that the one-particle DOS is logarithmically divergent at these saddle-points, which are examples of van Hove singularities. An explicit evaluation of the integral of

Eq. (3.49) shows that the most singular contributions are due to the vicinity of the van Hove singularities, yielding the logarithmically divergent result

$$\rho(\varepsilon) = \frac{1}{4\pi^2 t} \ln\left(\frac{t}{\varepsilon}\right) + \cdots \qquad \text{as } \varepsilon \to 0 \tag{3.50}$$

The mean-field (gap) equation, Eq. (3.47), becomes

$$\frac{3}{2U} \simeq \int_{\Delta}^{E_{\max}} \frac{d\varepsilon}{\varepsilon} \frac{1}{4\pi^2 t} \ln\left(\frac{t}{\varepsilon}\right) \tag{3.51}$$

Upon evaluating this integral we find that the gap Δ for single-particle states in a Néel antiferromagnetic state is

$$\Delta = 2\pi^2 t e^{-\sqrt{\frac{12t}{U}}} \tag{3.52}$$

We conclude that the order parameter $\langle \vec{S}(\vec{r}) \rangle$ is also non-zero: $\langle \vec{S}(\vec{r}) \rangle = [3/(8U)]\vec{M}_0 \cos(\vec{Q} \cdot \vec{r})$. Thus an antiferromagnetic solution is found even for arbitrarily weak Hubbard coupling U. Please keep in mind that the ferromagnetic solution requires a finite value of U to exist. It is also easy to see that this *Néel state* has less energy than that of the paramagnetic state $|\vec{M}_0| = 0$. Thus, at least at half-filling, the ground state appears to be a Néel antiferromagnet.

This solution is also remarkable for other reasons. First, the dependence of $|\vec{M}_0|$ on U is highly non-analytic: we get an essential singularity. This is exactly analogous (albeit with a stronger singularity) to what one finds in BCS theory and in the case of the Peierls instability of one-dimensional electron–phonon systems (see Section 16.7). Secondly, the electronic spectrum has a *gap* Δ, which is equal to $|\vec{M}_0|$. Thus the gap also has an essential singularity in the coupling constant. But is the spectrum truly massive? Are there any gapless (or massless) excitations present? What this calculation says is that the one-particle spectrum is massive. What about the two-particle spectrum? We will see below that there are massless spin waves, in agreement with Goldstone's theorem.

3.2 Path-integral representation of the Hubbard model

So far we have discussed some features of these systems within a mean-field theory that is based on the canonical Hamiltonian formalism. It is possible to gain further insight by going to the path-integral form. These two representations are equivalent and certainly whatever one can do in one form can be reproduced in the other picture. However, certain aspects of the problem can be dealt with in a more natural and concise way in the path-integral picture. The questions relating to the symmetry, and its breaking, effective theories for the low-lying modes, etc., are more answerable in path-integral form. Also, of course, the semi-classical treatment, including non-perturbative features such as solitons and the like, is very simple to picture in terms of path integrals.

Typically we are interested in studying both zero-temperature and finite-temperature properties of the system. At finite temperature, the equilibrium properties are determined by the partition function

$$Z = \operatorname{tr} e^{-\beta H} \tag{3.53}$$

where $\beta = 1/T$ (in units in which $k_{\mathrm{B}} = 1$). Usually, one would also like to know the behavior of the correlation functions.

At zero temperature one is interested in the "vacuum persistence amplitude" (Coleman, 1985)

$$Z = \lim_{t \to \infty} \operatorname{tr} e^{iHt} \tag{3.54}$$

which is just the trace of the evolution operator at long times. Feynman showed (Feynman and Hibbs, 1965) that Eq. (3.54) can be written as a sum over histories. Also, it is apparent that Eq. (3.53) is related to Eq. (3.54) by an analytic continuation procedure known as "Wick rotation,"

$$it = -\tau \tag{3.55}$$

which amounts to going to imaginary time (Abrikosov *et al.*, 1963). I will use both forms more or less simultaneously.

I do not intend to give a thorough description of the path-integral method. Qualitatively, the path integral is derived as follows. Let H be the Hamiltonian of the system and $\{|\alpha\rangle\}$ be a set of states. In most cases we will demand that $\{|\alpha\rangle\}$ be a complete set of states. However, it will also be convenient to work with a system of *coherent states* that is over-complete. In either case, what matters is the existence of an identity of the form ("resolution of unity")

$$1 = N \int d\alpha |\alpha\rangle\langle\alpha| \tag{3.56}$$

where N is a normalization constant and $d\alpha$ is an integration measure. It is worthwhile to comment that the states $\{|\alpha\rangle\}$ need not be position eigenstates. In the usual derivation ("sum over histories") the position-space (or coordinate) representation is used. On the other hand, in many problems, such as the quantization of spin systems, there isn't a natural separation between canonical coordinates and momenta. Thus the space of states $\{|\alpha\rangle\}$ can be quite general and abstract. In fact the coherent-state representation is in a sense more primitive (or fundamental).

The standard strategy that one employs is the following. First one defines the states $\{|\alpha\rangle\}$. In the case of a many-body system such as the Hubbard model, these states should be antisymmetrized many-fermion states. I will work in the grand canonical ensemble and, hence, use second quantization. The need to antisymmetrize the states will bring some complications, which will be taken care of by using Grassmann variables.

The second step is to split up the time interval t into N_t segments of infinitesimal length Δ_t such that $N_t \Delta_t = t$. The same prescription applies to the imaginary time (Euclidean) formalism. The vacuum persistence amplitude is

$$Z = \text{tr}\, \hat{T} e^{i \int_{-\infty}^{+\infty} dt\, H(t)} \tag{3.57}$$

where $\hat{T}$ is the time-ordering operator. For a time-independent Hamiltonian, Eq. (3.57) reduces to Eq. (3.54). For infinitesimal intervals $\Delta_t \to 0$ we can write

$$Z = \text{tr}\, \hat{T} e^{i \sum_{j=1}^{N_t} \Delta_t H(t_j)} \approx \text{tr}\, \hat{T} \prod_{j=1}^{N_t} e^{i \Delta_t H(t_j)} \tag{3.58}$$

Now we proceed by inserting the resolution of unity, Eq. (3.56), at each intermediate time t_j. Let $\{|\alpha_j\rangle\}$ be a set of states at each time t_j. We get

$$Z = \sum_{\{\alpha_j\}} \prod_{j=1}^{N_t} \langle \alpha_j | e^{-i \Delta_t H(t_j)} | \alpha_{j+1} \rangle \tag{3.59}$$

with $|\alpha_{N+1}\rangle = |\alpha_1\rangle$ and where $|\alpha_j\rangle \equiv |\alpha(t_j)\rangle$ are the states at time t_j.

We can regard the $\alpha(t_j)$ as a set of parameters spanning a manifold defining the states $|\alpha_j\rangle$. Thus what we actually have is a sum over configurations $\{\alpha(t)\}$. Notice that this procedure is absolutely general. We are supposed to take the limit $\Delta_t \to 0$, $N_t \to \infty$ at the end. That this limit exists is a highly non-trivial issue and certainly not a formal matter. This procedure applies both for single-particle problems and for states of a many-body second-quantized system, i.e. a field theory.

Let us first review the simple particle in a potential problem. In this case the Hamiltonian is

$$H = \frac{\hat{p}^2}{2m} + V(\hat{q}) \tag{3.60}$$

where $\hat{p}$ and $\hat{q}$ obey canonical commutation relations ($\hbar = 1$)

$$[\hat{q}, \hat{p}] = i \tag{3.61}$$

Thus the states $|\alpha\rangle$ can be the complete set of position eigenstates $|q\rangle$. The resolution of the identity is

$$1 = \int dq |q\rangle\langle q| \tag{3.62}$$

Conversely, we could also use momentum eigenstates. The momentum operator $\hat{p}$ is not diagonal in this basis. Thus the amplitude

$$\langle q(t_j) | e^{i \Delta_t H} | q(t_{j+1}) \rangle \tag{3.63}$$

can be written in the form

$$\langle q(t_j)|e^{i\Delta_t \frac{\hat{p}^2}{2m}}|q(t_{j+1})\rangle e^{i\Delta_t V(q(t_{j+1}))} \tag{3.64}$$

where we have used the fact that $\hat{V}$ is diagonal in the coordinate representation. Now we use a complete set of momentum eigenstates $\{|p(t_j)\rangle\}$ and write

$$e^{i\Delta_t \frac{\hat{p}^2}{2m}} = \int_{-\infty}^{+\infty} \frac{dp}{2\pi} e^{i\Delta_t \frac{p^2(t_j)}{2m}} |p(t_j)\rangle\langle p(t_j)| \tag{3.65}$$

On collecting these various contributions we get

$$Z = \int \mathcal{D}p\,\mathcal{D}q \prod_{j=1}^{N_t} e^{i\Delta_t \frac{p^2(t_j)}{2m} + i\Delta_t V(q(t_{j+1}))} \langle p(t_j)|q(t_{j+1})\rangle\langle p(t_j)|q(t_j)\rangle^* \tag{3.66}$$

where I used the definition of the measure

$$\mathcal{D}p\,\mathcal{D}q = \prod_{j=1}^{N_t} \frac{dp(t_j)dq(t_j)}{2\pi} \tag{3.67}$$

Thus, we can write Eq. (3.66) in the form

$$Z = \int \mathcal{D}p\,\mathcal{D}q\; e^{-i\int dt[p\dot{q} - H(p,q)]} \tag{3.68}$$

by making use of the fact that

$$\langle p|q\rangle = e^{-i\vec{p}\cdot\vec{q}} \tag{3.69}$$

Equation (3.68) is nothing but a sum over the configurations in the *phase space* of the action S,

$$S = \int dt(p\dot{q} - H) \tag{3.70}$$

of each configuration. Since we are computing a *trace* the *field* $q(t)$ obeys periodic boundary conditions in time. Note that $\hat{p}$ and $\hat{q}$ do not commute. The phase-space integral is actually a coherent-state path integral (Faddeev, 1976). Equation (3.68) is generally valid even for Hamiltonians for which it is not possible to clearly separate coordinates and momenta. I will adopt the phase-space (or coherent-state) path integral as the definition.

This procedure can be trivially generalized to second-quantized systems. In the case of bosons we have second-quantized field operators $\hat{\Psi}(\vec{r})$ and $\hat{\Psi}^\dagger(\vec{r})$ and a Hamiltonian H. The field operators obey the equal-time commutation relations

$$\left[\hat{\Psi}(\vec{r}), \hat{\Psi}^\dagger(\vec{r}\,')\right] = \delta(\vec{r} - \vec{r}\,') \tag{3.71}$$

Consider the classical Lagrangian L

$$L = \sum_{\vec{r}} \Psi^* i\, \partial_t \Psi - H \tag{3.72}$$

The commutation relations in Eq. (3.71) follow from canonically quantizing L. The canonical momentum $\hat{\Pi}(\vec{r})$ is given by

$$\hat{\Pi}(\vec{r}) \equiv \frac{\delta L}{\delta \partial_t \hat{\Psi}(\vec{r})} \equiv i \hat{\Psi}^\dagger(\vec{r}) \tag{3.73}$$

Thus the canonical commutation relations

$$\left[\hat{\Psi}(\vec{r}), \hat{\Pi}(\vec{r}\,')\right] = i\delta(\vec{r} - \vec{r}\,') \tag{3.74}$$

are equivalent to Eq. (3.71) after $\hat{\Pi}$ has been identified with $i\hat{\Psi}^\dagger$.

A discussion analogous to what we did for the particle case yields a phase-space path integral

$$Z = \int \mathcal{D}\Psi^* \mathcal{D}\Psi\, e^{i \int dt [\sum_{\vec{r}} \Psi^* i\, \partial_t \Psi - H(\Psi^*, \Psi)]} \tag{3.75}$$

where Ψ and Ψ^* are complex c-number fields that parametrize the coherent states (Faddeev, 1976). Since we are dealing with bosons, the fields Ψ are c-numbers and commute. The boundary conditions in turn are periodic. The case of fermions can also be dealt with, provided that one takes care of the anti-commuting nature of fermion operators.

It is convenient to introduce coherent states for fermions. Let $\hat{\Psi}^\dagger$ and $\hat{\Psi}$ be Fermi creation and annihilation operators that satisfy

$$\{\hat{\Psi}, \hat{\Psi}^\dagger\} = 1 \tag{3.76}$$

and

$$\hat{\Psi}^2 = (\hat{\Psi}^\dagger)^2 = 0 \tag{3.77}$$

In the occupation-number representation we have two states, $|0\rangle$ and $|1\rangle$, with the properties

$$\begin{aligned} \hat{\Psi}|0\rangle &= 0, \quad & \hat{\Psi}^\dagger|0\rangle &= |1\rangle, \quad & \hat{\Psi}^\dagger\hat{\Psi}|0\rangle &= 0, \\ \hat{\Psi}^\dagger|1\rangle &= 0, \quad & \hat{\Psi}|1\rangle &= |0\rangle, \quad & \hat{\Psi}^\dagger\hat{\Psi}|1\rangle &= |1\rangle \end{aligned} \tag{3.78}$$

We introduce the two Grassmann numbers Ψ and $\bar{\Psi}$ that we will associate with the Fermi operators $\hat{\Psi}$ and $\hat{\Psi}^\dagger$. Their defining property is the (Grassmann) algebra

$$\{\Psi, \Psi\} = \{\bar{\Psi}, \bar{\Psi}\} = \{\Psi, \bar{\Psi}\} = 0 \tag{3.79}$$

that they satisfy. It is natural to extend these anticommutation relations by imposing (Negele and Orland, 1988)

$$\{\Psi, \hat{\Psi}\} = 0 \qquad \text{and} \qquad (\Psi\hat{\Psi})^\dagger = \hat{\Psi}^\dagger\bar{\Psi} \tag{3.80}$$

We define the coherent state $|\Psi\rangle$ in terms of the Grassmann number Ψ

$$|\Psi\rangle \equiv |0\rangle - \Psi|1\rangle = |0\rangle - \Psi\hat{\Psi}^\dagger|0\rangle \tag{3.81}$$

With the help of Eqs. (3.79) and (3.80) we obtain

$$|\Psi\rangle = \left(1 - \Psi\hat{\Psi}^\dagger\right)|0\rangle \equiv e^{-\Psi\hat{\Psi}^\dagger}|0\rangle \tag{3.82}$$

$$\hat{\Psi}|\Psi\rangle = \hat{\Psi}|0\rangle - \hat{\Psi}\Psi\hat{\Psi}^\dagger|0\rangle = \Psi|0\rangle = \Psi|\Psi\rangle \tag{3.83}$$

and

$$\hat{\Psi}^\dagger|\Psi\rangle = \hat{\Psi}^\dagger|0\rangle - \hat{\Psi}^\dagger\Psi\hat{\Psi}^\dagger|0\rangle = |1\rangle \equiv -\frac{\delta}{\delta\Psi}|\Psi\rangle \tag{3.84}$$

As usual the exponential on the right-hand side of Eq. (3.82) is defined as a power expansion in $\Psi\hat{\Psi}^\dagger$ and Eq. (3.84) *defines* operationally the "left" Grassmann derivative.

The adjoint coherent state $\langle\Psi|$ is defined in terms of the Grassmann number Ψ

$$\langle\Psi| \equiv \langle 0| - \langle 1|\bar{\Psi} = \langle 0| - \langle 0|\hat{\Psi}\bar{\Psi} \tag{3.85}$$

As before

$$\langle\Psi| = \langle 0|\left(1 - \hat{\Psi}\bar{\Psi}\right) \equiv \langle 0|e^{-\hat{\Psi}\bar{\Psi}} \tag{3.86}$$

$$\langle\Psi|\hat{\Psi}^\dagger = \langle 0|\hat{\Psi}^\dagger - \langle 0|\hat{\Psi}\bar{\Psi}\hat{\Psi}^\dagger = \langle 0|\bar{\Psi} = \langle\Psi|\bar{\Psi} \tag{3.87}$$

and

$$\langle\Psi|\hat{\Psi} = \langle 0|\hat{\Psi} - \langle 0|\hat{\Psi}\bar{\Psi}\hat{\Psi} = \langle 1| \equiv -\langle\Psi|\frac{\delta}{\delta\bar{\Psi}} \tag{3.88}$$

The right-hand side of Eq. (3.88) defines the "right" Grassmann derivative in complete analogy to Eq. (3.84). It is natural to require that the "left" ("right") derivative $\delta/\delta\Psi$ anti-commutes with Ψ so that $(\delta/\delta\Psi)\Psi = -\Psi(\delta/\delta\Psi)$.

From Eqs. (3.86), (3.82), and (3.80) the inner product $\langle\Psi|\Psi'\rangle$ is equal to

$$\langle\Psi|\Psi'\rangle = e^{\bar{\Psi}\Psi'} \tag{3.89}$$

This, together with Eqs. (3.83) and (3.87), gives for the matrix elements of a normal-ordered operator $:\mathcal{U}\left(\hat{\Psi}^\dagger, \hat{\Psi}\right):$

$$\langle\Psi| :\mathcal{U}\left(\hat{\Psi}^\dagger, \hat{\Psi}\right): |\Psi'\rangle = \mathcal{U}(\bar{\Psi}, \Psi')e^{\bar{\Psi}\Psi'} \tag{3.90}$$

where $\mathcal{U}(\bar{\Psi}, \Psi')$ is obtained by carrying out the replacements $\hat{\Psi} \to \Psi$ and $\hat{\Psi}^\dagger \to \bar{\Psi}$ inside the normal-ordered operator $: \mathcal{U} :$.

The resolution of unity in this representation is just

$$1 = \int d\bar{\Psi}\, d\Psi\; e^{-\bar{\Psi}\Psi} |\Psi\rangle\langle\Psi| \tag{3.91}$$

This identity can be checked by computing the inner product $\langle\Psi'|\Psi''\rangle$, where Ψ' and Ψ'' are the Grassmann variables Ψ or $\bar{\Psi}$. The integrals in Eq. (3.91) are understood to be linear functionals on the space of functions of the Grassmann variables with

$$\int d\Psi\, \Psi \equiv 1, \qquad \int d\Psi \equiv 0 \tag{3.92}$$

We can now repeat the procedure outlined at the beginning of this section for a second-quantized system of fermions except that now we will use fermion coherent states $|\{\Psi_\sigma(\vec{r})\}\rangle$ at each site and for each spin degree of freedom

$$|\{\Psi_\sigma(\vec{r})\}\rangle = \exp\left(-\sum_{\vec{r},\sigma} \Psi_\sigma(\vec{r})\hat{\Psi}_\sigma^\dagger(\vec{r}) \right) |0\rangle \tag{3.93}$$

where $|0\rangle$ is the empty state (not the "vacuum," as we will see in Chapter 5). Following our noses, we find

$$Z = \lim_{\substack{N_t\to\infty \\ \Delta_t\to 0}} \left(\prod_{j=1}^{N_t} \int d\bar{\Psi}(t_j) d\Psi(t_j) \right) \prod_{j=1}^{N_t} e^{-\bar{\Psi}(t_j)\Psi(t_j)} \langle\Psi(t_j)|\, (1 - i\Delta_t H)\, |\Psi(t_{j+1})\rangle \tag{3.94}$$

where, for the sake of simplicity, I have dropped the space and spin labels. In the limit $N_t \to \infty$ and $\Delta_t \to 0$, one finds

$$Z = \int \mathcal{D}\bar{\Psi}\, \mathcal{D}\Psi\; e^{i\int dt L} \tag{3.95}$$

where L is given by

$$L = \sum_{\vec{r}} \bar{\Psi}_\sigma(\vec{r}) i\, \partial_t \Psi_\sigma(\vec{r}) - H(\bar{\Psi}_\sigma(\vec{r}), \Psi_\sigma(\vec{r})) \tag{3.96}$$

For the case of the Hubbard model, or any other model, for that matter, and in the presence of a non-zero chemical potential μ, we get (see Eq. (2.25))

$$L = \sum_{\vec{r}} \bar{\Psi}_\sigma(\vec{r})(i\,\partial_t + \mu)\Psi_\sigma(\vec{r}) + t \sum_{\langle\vec{r},\vec{r}\,'\rangle} \bar{\Psi}_\sigma(\vec{r})\Psi_\sigma(\vec{r}\,') - H_{\text{int}}(\bar{\Psi}, \Psi) \tag{3.97}$$

From Eq. (2.24) the interaction term of the Hubbard model is

$$H_{\text{int}} = -\frac{U}{6} \sum_{\vec{r}} c_{\alpha}^{\dagger}(\vec{r}) \tau_{\alpha\beta}^{a} c_{\beta}(\vec{r}) c_{\gamma}^{\dagger}(\vec{r}) \tau_{\gamma\delta}^{a} c_{\delta}(\vec{r}) \tag{3.98}$$

Normal ordering relative to the empty state $|0\rangle$ gives

$$: H_{\text{int}} := -\frac{U}{6} \sum_{\vec{r}} c_{\alpha}^{\dagger}(\vec{r}) c_{\gamma}^{\dagger}(\vec{r}) c_{\delta}(\vec{r}) c_{\beta}(\vec{r}) \tau_{\alpha\beta}^{a} \tau_{\gamma\delta}^{a} - \frac{U}{2} \sum_{\vec{r}} c_{\alpha}^{\dagger}(\vec{r}) c_{\alpha}(\vec{r}) \tag{3.99}$$

Thus $H_{\text{int}}(\bar{\Psi}, \Psi)$ is given by

$$H_{\text{int}}(\bar{\Psi}, \Psi) = -\frac{U}{6} \sum_{\vec{r}} \bar{\Psi}_{\alpha}(\vec{r}) \bar{\Psi}_{\gamma}(\vec{r}) \Psi_{\delta}(\vec{r}) \Psi_{\beta}(\vec{r}) \tau_{\alpha\beta}^{a} \tau_{\gamma\delta}^{a} - \frac{U}{2} \sum_{\vec{r}} \bar{\Psi}_{\alpha}(\vec{r}) \Psi_{\alpha}(\vec{r}) \tag{3.100}$$

The last term on the right-hand side can obviously be cancelled out by means of a shift of the chemical potential μ.

The final property of Grassmann integrals which will be useful for us is the integral for actions that are quadratic in the fields,

$$S = \sum_{\vec{r}, \vec{r}\,'} \bar{\Psi}(\vec{r}) M(\vec{r}, \vec{r}\,') \Psi(\vec{r}\,') \tag{3.101}$$

where $M(\vec{r}, \vec{r}\,')$ is an antisymmetric matrix (operator). We get the Gaussian integral

$$Z = \int \mathcal{D}\bar{\Psi}\, \mathcal{D}\Psi\, e^{-\int \bar{\Psi} M \Psi} = \det M \tag{3.102}$$

This expression should be contrasted with the analogous result for bosonic fields ϕ

$$Z = \int \mathcal{D}\phi^{*}\, \mathcal{D}\phi\, e^{-\int \phi^{*} M \phi} = (\det M)^{-1} \tag{3.103}$$

Both results can be derived quite easily by expanding Ψ and $\bar{\Psi}$ in a basis of eigenstates of M, see Faddeev (1976) or Negele and Orland (1988).

3.3 Path integrals and mean-field theory

We now turn to the mean-field theory for the Hubbard model in path-integral form. The advantage of this description is that we will be able to extract an effective-field theory for the low-lying modes in the Néel state: spin waves.

The Lagrangian density for the Hubbard model in two dimensions, in real time and at zero temperature is, from Eqs. (3.97) and (3.100),

$$\begin{aligned}\mathcal{L} = & \bar{\Psi}_\alpha(\vec{r}, t)(i\,\partial_t + \mu)\Psi_\alpha(\vec{r}, t) \\ & + t \sum_{j=1,2} \left(\bar{\Psi}_\alpha(\vec{r}, t)\Psi_\alpha(\vec{r} + \vec{e}_j, t) + \bar{\Psi}_\alpha(\vec{r}, t)\Psi_\alpha(\vec{r} - \vec{e}_j, t) + \text{c.c.}\right) \\ & + \frac{U}{6}(\bar{\Psi}_\alpha(\vec{r}, t)\vec{\tau}_{\alpha\beta}\Psi_\beta(\vec{r}, t))^2\end{aligned} \tag{3.104}$$

The associated path integral contains quartic terms, the interaction, and hence we do not know how to compute the partition function. The strategy is to write another theory, which is quadratic in Grassmann fields and is equivalent to Eq. (3.104). We will make extensive use of the Gaussian identity for bosonic fields (or Hubbard–Stratonovich transformation)

$$\int d\vec{\phi}\; e^{-i\left(\frac{1}{2}\vec{\phi}^2 + \lambda\vec{\phi}\cdot\bar{\Psi}\vec{\tau}\Psi\right)} = \text{constant} \times e^{i\frac{1}{2}\lambda^2(\bar{\Psi}\vec{\tau}\Psi)^2} \tag{3.105}$$

Thus at any point in space time $(\vec{r}, t)$, we introduce a three-component real Bose field $\vec{\phi}(\vec{r}, t)$ coupled bilinearly to the fermions as in Eq. (3.105). If one chooses the coupling constant λ to be equal to $\sqrt{U/3}$, one finds the interaction term of Eq. (3.100). Thus the Lagrangian density $\mathcal{L}'$,

$$\begin{aligned}\mathcal{L}' = & \bar{\Psi}_\alpha(\vec{r}, t)(i\,\partial_t + \mu)\Psi_\alpha(\vec{r}, t) \\ & + t \sum_{j=1,2} [\bar{\Psi}_\alpha(\vec{r}, t)\Psi_\alpha(\vec{r} + \vec{e}_j, t) + \bar{\Psi}_\alpha(\vec{r}, t)\Psi_\alpha(\vec{r} - \vec{e}_j, t) + \text{c.c.}] \\ & - \sqrt{\frac{U}{3}}\vec{\phi}(\vec{r}, t) \cdot \bar{\Psi}_\alpha(\vec{r}, t)\vec{\tau}_{\alpha\beta}\Psi_\beta(\vec{r}, t) - \frac{1}{2}\vec{\phi}^2(\vec{r}, t)\end{aligned} \tag{3.106}$$

is equivalent to the Lagrangian of the Hubbard model. Equation (3.106) has the advantage of being bilinear in Fermi fields (compare Eqs. (3.106) and (3.3)). Thus, using Eq. (3.102), we can now integrate out the fermions. The result is an *effective action* for the Bose fields $\vec{\phi}$. We will see that the $\vec{\phi}$ fields represent the collective modes associated with spin fluctuations. The result is

$$Z = \int \mathcal{D}\vec{\phi}\; e^{iS_{\text{eff}}(\vec{\phi})} \tag{3.107}$$

where the effective action $S_{\text{eff}}(\vec{\phi})$ is given by

$$S_{\text{eff}}(\vec{\phi}) = -\int dt \sum_{\vec{r}} \frac{1}{2}\vec{\phi}^{\,2}(\vec{r}, t) - i \ln\det\left(i\,\partial_t + \mu - \mathcal{M}(\vec{\phi})\right) \tag{3.108}$$

The operator $\mathcal{M}(\vec{\phi})$ in Eq. (3.108) has the matrix elements

$$\begin{aligned}\langle\vec{r}t\alpha|\mathcal{M}(\vec{\phi})|\vec{r}\,'t'\beta\rangle = & -\delta_{\alpha\beta}\delta(t - t')t \sum_{j=1,2}\left(\delta_{\vec{r}\,',\vec{r}+\vec{e}_j} + \delta_{\vec{r}\,',\vec{r}-\vec{e}_j}\right) \\ & + \sqrt{\frac{U}{3}}\delta(t - t')\delta_{\vec{r},\vec{r}\,'}\vec{\phi}(\vec{r}, t) \cdot \vec{\tau}_{\alpha\beta}\end{aligned} \tag{3.109}$$

The mean-field theory for this problem is just the evaluation of the path integral Eq. (3.107) by means of the saddle-point (or stationary-phase) approximation. For this problem, this approximation is equivalent to a Hartree–Fock decoupling. The stationary condition is

$$0 = \frac{\delta S_{\text{eff}}}{\delta\phi^a(\vec{r},t)} = -\phi^a(\vec{r},t) - i\frac{\delta}{\delta\phi^a(\vec{r},t)} \ln\det(i\,\partial_t + \mu - \mathcal{M}(\vec{\phi})) \tag{3.110}$$

Using the identity

$$\ln\det A = \text{tr}\,\ln\,A \tag{3.111}$$

and Eq. (3.109), one finds

$$\begin{aligned}\phi^a(\vec{r},t) &= -i\frac{\delta}{\delta\phi^a(\vec{r},t)}\,\text{tr}\left(\ln[i\,\partial_t + \mu - \mathcal{M}(\vec{\phi})]\right)\\ &= +i\,\text{tr}\left(\frac{1}{i\,\partial_t + \mu - \mathcal{M}(\vec{\phi})}\frac{\delta M(\vec{\phi})}{\delta\phi^a(\vec{r},t)}\right)\\ &= i\sqrt{\frac{U}{3}}\langle\vec{r}t\alpha|\frac{1}{i\,\partial_t + \mu - \mathcal{M}(\vec{\phi})}|\vec{r}t\beta\rangle\tau^a_{\beta\alpha}\end{aligned} \tag{3.112}$$

The expression in angular brackets is just the (in space and time diagonal) matrix element of the fermion one-particle Green function in a background field $\vec{\phi}(\vec{r},t)$,

$$G_{\alpha\beta}(\vec{r}t;\vec{r}\,'t';\phi) \equiv -i\langle\vec{r}t\alpha|\frac{1}{i\,\partial_t + \mu - \mathcal{M}(\phi)}|\vec{r}\,'t'\beta\rangle \tag{3.113}$$

Hence we can write Eq. (3.112) in the form

$$\phi^a(\vec{r},t) = -\sqrt{\frac{U}{3}}G_{\alpha\beta}(\vec{r}t;\vec{r}t;\phi)\tau^a_{\beta\alpha} \tag{3.114}$$

On the other hand, the local magnetic moment $\langle S^a(\vec{r}t)\rangle$ is equal to

$$\langle S^a(\vec{r}t)\rangle = +G_{\alpha\beta}(\vec{r}t;\vec{r}t;\phi)\frac{\tau^a_{\beta\alpha}}{2} \tag{3.115}$$

Thus, the saddle-point approximation, Eq. (3.110), is the same as the Hartree–Fock condition

$$\phi^a(\vec{r},t) = -\sqrt{\frac{4U}{3}}\langle S^a(\vec{r},t)\rangle \tag{3.116}$$

At the level of a one-band Hubbard model, there is no quantitative justification for the validity of this approach, since there is no small parameter other than $\hbar$ to control this expansion. Thus, this is essentially a semi-classical approximation.

We also know that an angular-momentum degree of freedom, such as spin itself, becomes semi-classical if the angular momentum becomes large. The main assumption of the mean-field theory is that the order parameter thus obtained, in this case the staggered magnetization, is close to its saturation value and hence is large.

We can formally introduce a small parameter to control this expansion by means of the following device. Let us imagine that the band electrons have an orbital degeneracy labeled by an index $a = 1, \ldots, N_b$, where N_b is the number of degenerate bands. The total band spin at a given site $\vec{r}$ is now given by ($i = 1, 2, 3$)

$$S^i(\vec{r}) = \sum_{\alpha,\beta=\uparrow,\downarrow} \sum_{a=1}^{N_b} \Psi^\dagger_{\alpha,a}(\vec{r}) \tau^i_{\alpha\beta} \Psi_{\beta,a}(\vec{r}) \tag{3.117}$$

The generalized Hubbard model is then given by the Hamiltonian

$$H = -\sum_{\substack{\langle \vec{r},\vec{r}\,' \rangle \\ \alpha,a}} \Psi^\dagger_{\alpha,a}(\vec{r}) t(\vec{r},\vec{r}\,') \Psi_{\alpha,a}(\vec{r}\,') + \text{c.c.} - \frac{2}{3} U \sum_{\vec{r}} \left(\vec{S}(\vec{r})\right)^2 \tag{3.118}$$

where $\vec{S}(\vec{r})$ is the total band spin at $\vec{r}$. This system still has the global SU(2) invariance of spin rotations. For large values of U, i.e. $U/t \to \infty$, the local spin becomes as large as possible. The equivalent Heisenberg model has a total spin quantum number s at each site equal to $s = N_b/2$ or, equivalently, $N_b = 2s$. The limit $N_b \to \infty$ is then the same as the semi-classical limit $s \to \infty$. This limit is usually treated by spin-wave theory (Bloch, 1930; Holstein and Primakoff, 1940; Dyson, 1956a, b; Maleev, 1957) (for a review see the book by Mattis (1965)).

The path-integral approach is particularly well suited to deal with this limit. As a matter of fact, all the formulas derived above carry over to this case. The Hubbard–Stratonovich transformation works, with the only change being that $\vec{\phi}$ couples now to the total band spin. Since all N_b orbital species couple exactly in the same way to the Hubbard–Stratonovich field $\vec{\phi}$, the only change that occurs is that the fermion determinant *factorizes* and is given by the N_bth power of the determinant of a single species. After a trivial rescaling of the field $\vec{\phi}$ by $\sqrt{N_b}$, the effective action $S_{\text{eff}}^{N_b}$ for the theory with orbital degeneracy is simply given by

$$S_{\text{eff}}^{N_b}(\vec{\phi}) = N_b S_{\text{eff}}(\vec{\phi}) \tag{3.119}$$

In the large-N_b limit (i.e. large-s limit), the saddle-point approximation becomes exact. For the rest of this section, we will carry on with this expansion assuming that it is valid. We should keep in mind, however, that the results will become accurate only in the $s \to \infty$ limit.

It is apparent that if $\phi^a(\vec{r}, t)$ is a solution, any *uniform* rotation of it is also a solution,

$$\phi'_a = R_{ab} \phi_b \tag{3.120}$$

where R_{ab} is a constant rotation matrix. This implies that the global spin symmetry has been preserved. We will see now that this implies the existence of Goldstone modes, spin waves, if this symmetry is spontaneously broken.

Let us consider the half-filled case. Here, we expect an antiferromagnetic state. The classical solution is (for the case of a square lattice)

$$\phi^a(\vec{r}, t) = |\vec{\phi}| n^a (-1)^{x_1 + x_2} \tag{3.121}$$

This solution is (a) static, and (b) staggered (i.e. a Néel state). It really represents an infinite number of solutions parametrized by the unit vector $\vec{n}$ (in spin space). The *amplitude* $|\vec{\phi}|$ is determined by solving the saddle-point equation, Eq. (3.116). In the notation of Eq. (3.113), we can write

$$\vec{M} = \sqrt{\frac{4U}{3}} \vec{\phi} \tag{3.122}$$

In Section 3.1 we determined that the Néel state was energetically preferred both to a paramagnetic state and to a ferromagnetic state. Notice that this argument does not rule out other solutions. However, the existing numerical evidence seems to indicate that a Néel state is the ground state at half-filling except for one-dimensional systems.

In Section 3.1 we showed that (a) the amplitude $|\vec{\phi}|$ is always non-zero at zero temperature and (b) the single-particle excitation spectrum has a gap $\Delta = |\vec{M}_0|$. This last result can be seen to follow by computing the one-particle Green function $G_{\alpha\beta}(\vec{r}t; \vec{r}\,'t'; \phi)$ and writing it as a 4×4 matrix in spin and sublattice space.

Equation (3.113) is equivalent to

$$\begin{aligned} -i\delta(t - t')\delta_{\vec{r},\vec{r}\,'}\delta_{\alpha\beta} = {} & (i\,\partial_t + \mu) G_{\alpha\beta}\left(\vec{r}t; \vec{r}\,'t'; \vec{\phi}\right) \\ & + t \sum_{j=1,2} [G_{\alpha\beta}(\vec{r} + \vec{e}_j t; \vec{r}\,'t'; \vec{\phi}) + G_{\alpha\beta}(\vec{r} - \vec{e}_j t; \vec{r}\,'t'; \vec{\phi})] \\ & - \sqrt{\frac{U}{3}} |\vec{\phi}| \vec{n} \cdot \vec{\tau}_{\alpha\gamma} (-1)^{x_1 + x_2} G_{\gamma\beta}(\vec{r}t; \vec{r}\,'t'; \vec{\phi}) \end{aligned} \tag{3.123}$$

If we Fourier transform Eq. (3.123), we find

$$-i\delta_{\alpha\beta} = (\omega - \epsilon(\vec{k})) G_{\alpha\beta}(\vec{k}, \omega) - \sqrt{\frac{U}{3}} |\vec{\phi}| \vec{n} \cdot \vec{\tau}_{\alpha\gamma} G_{\gamma\beta}(\vec{k} - \vec{Q}, \omega) \tag{3.124}$$

where $\vec{Q} = (\pi, \pi)$ is the ordering wave vector (for the Néel state, the corner of the first Brillouin zone). This equation can be solved by writing (the Fourier transform of) the fermion operators in a spinor notation $\Psi_\alpha(\vec{k}, \omega)$ defined by

$$\Psi_\alpha(\vec{k}, \omega) \equiv \begin{pmatrix} \Psi_\alpha(\vec{k}, \omega) \\ \Psi_\alpha(\vec{k} - \vec{Q}, \omega) \end{pmatrix} \tag{3.125}$$

where $\vec{k} \in \mathrm{BZ}^+$ (the upper half of the first Brillouin zone). Equation (3.124) now takes the matrix form

$$-i\delta_{\alpha\beta} = \begin{pmatrix} (\omega - \epsilon(\vec{k}))\delta_{\alpha\gamma} & -\sqrt{U/3}|\vec{\phi}|\vec{n}\cdot\vec{\tau}_{\alpha\gamma} \\ -\sqrt{U/3}|\vec{\phi}|\vec{n}\cdot\vec{\tau}_{\alpha\gamma} & (\omega + \epsilon(\vec{k}))\delta_{\alpha\gamma} \end{pmatrix} \mathcal{G}_{\gamma\beta}(\vec{k}, \omega) \tag{3.126}$$

The solution of this equation is

$$\mathcal{G}(\vec{k}, \omega) = \frac{-i}{\omega^2 - \left(\epsilon^2(\vec{k}) + (U/3)|\vec{\phi}|^2\right) + i\eta} \begin{pmatrix} (\omega + \epsilon(\vec{k})) & \sqrt{U/3}|\vec{\phi}|\vec{n}\cdot\vec{\tau} \\ \sqrt{U/3}|\vec{\phi}|\vec{n}\cdot\vec{\tau} & (\omega - \epsilon(\vec{k})) \end{pmatrix} \tag{3.127}$$

with $\eta \to 0^+$. Here, the diagonal components correspond to the Fourier transform of the Green function for sites on the same sublattice, and the off-diagonal terms correspond to sites on different sublattices.

This solution clearly shows that the single-particle fermionic spectrum for the Néel state is

$$E(\vec{k}) = \sqrt{\epsilon^2(\vec{k}) + \frac{U}{3}|\vec{\phi}|^2} = \sqrt{\epsilon^2(\vec{k}) + \Delta^2} \tag{3.128}$$

and we recover Eq. (3.42), with an energy gap $\Delta = \sqrt{U/3}|\vec{\phi}|$.

3.4 Fluctuations: the non-linear sigma model

In the previous section we obtained an effective action for the order-parameter field $\vec{\phi}$ and solved the saddle-point equations. We now wish to estimate the role and size of the quantum-mechanical fluctuations about this classical Néel state.

When solving the saddle-point equations, we observed that, if a non-trivial solution $\vec{\phi}_{\mathrm{c}}$ with broken symmetry can be found, then any configuration obtained by means of a *rigid rotation* in spin space from $\vec{\phi}_{\mathrm{c}}$ is also a solution. This reflects the fact that the spin sector has a continuous symmetry group, in this case O(3).

Imagine now not a solution of the saddle-point equation but a slowly varying configuration $\vec{\phi}(\vec{r}, t)$ not far from a solution. The fluctuation part $\delta\vec{\phi}(\vec{r}, t)$ is small and slowly varying. By slowly varying, I mean slow on time scales compared with $\tau = 1/\Delta$ and smooth on length scales long compared with $\xi = v_{\mathrm{F}}/\Delta$ (where v_{F} is the Fermi velocity of the unperturbed system). This last length ξ is the (mean-field) correlation length of the system. It will turn out that ξ and τ determine the scales on which the *magnitude* $|\vec{\phi}|$ of the order parameter fluctuates, at least in mean-field theory.

The existence of an infinite number of solutions of the saddle-point equation indicates that there are configurations $\delta\vec{\phi}$ with arbitrarily low action. These are the

Goldstone bosons of this problem and are spin waves. We wish to find an effective theory for these spin waves.

Our first step will be to study the (Gaussian) fluctuations around the mean-field solution. Thus, we will expand the effective action $S_{\text{eff}}(\phi)$ in powers of $\delta\vec{\phi}(\vec{r}, t)$. Since $\mathcal{M}(\vec{\phi})$ of Eq. (3.109) is linear in $\vec{\phi}$, we can write

$$\mathcal{M}(\vec{\phi}) = \mathcal{M}(\vec{\phi}_{\text{c}}) + \mathcal{M}(\delta\vec{\phi}) \tag{3.129}$$

where the matrix elements of $\mathcal{M}(\delta\vec{\phi})$ are

$$\langle \vec{r} t \alpha | \mathcal{M}(\delta\vec{\phi}) | \vec{r}\,' t' \alpha' \rangle = \sqrt{\frac{U}{3}}\, \delta\vec{\phi}(\vec{r}, t) \cdot \vec{\tau}_{\alpha\alpha'} \delta_{\vec{r},\vec{r}\,'} \delta(t - t') \tag{3.130}$$

By expanding in powers of $\mathcal{M}(\delta\vec{\phi})$, we find that the effective action

$$S_{\text{eff}}(\vec{\phi}) = -\int dt \sum_{\vec{r}} \frac{\vec{\phi}^2(\vec{r}, t)}{2} - i \operatorname{tr} \ln\Big(i\, \partial_t + \mu - \mathcal{M}(\vec{\phi})\Big) \tag{3.131}$$

can be written in the form

$$\begin{aligned} S_{\text{eff}}(\vec{\phi}) = &-\int dt \sum_{\vec{r}} \frac{\vec{\phi}^2(\vec{r}, t)}{2} - i \operatorname{tr} \ln\Big(i\, \partial_t + \mu - \mathcal{M}(\vec{\phi}_{\text{c}})\Big) \\ &- i \operatorname{tr} \ln\Big(1 - i\mathcal{G}(\vec{\phi}_{\text{c}})\mathcal{M}(\delta\vec{\phi})\Big) \end{aligned} \tag{3.132}$$

with the *mean field* Green function $\mathcal{G}(\vec{\phi}_{\text{c}})$

$$\mathcal{G}_{\alpha\alpha'}(\vec{r}t; \vec{r}\,'t'; \vec{\phi}_{\text{c}}) = -i \left\langle \vec{r} t \alpha \left| \frac{1}{i\, \partial_t + \mu - \mathcal{M}(\vec{\phi}_{\text{c}})} \right| \vec{r}\,' t' \alpha' \right\rangle \tag{3.133}$$

Recall that the Green function $\mathcal{G}$ has a matrix structure involving both sublattices of the Néel state, cf. Eq. (3.127).

Since the Néel state breaks translation invariance by one lattice spacing (i.e. it breaks the sublattice symmetry of the square lattice), the Green function also breaks translation invariance and it is invariant only under shifts on the same sublattice. For the same reason the fluctuations $\delta\phi(\vec{r}, t)$ on the two sublattices behave differently. Hence we will also represent $\delta\phi(\vec{r}, t)$ by a two-component object denoting the two sublattices.

By expanding the logarithm in powers of the fluctuations $\delta\vec{\phi}$, we get

$$\begin{aligned} S_{\text{eff}}(\vec{\phi}) = &-\int dt \sum_{\vec{r}} \frac{\vec{\phi}^2(\vec{r}, t)}{2} - i \operatorname{tr} \ln\Big(i\, \partial_t + \mu - \mathcal{M}(\vec{\phi}_{\text{c}})\Big) \\ &+ \sum_{n=1}^{\infty} \frac{i^{n+1}}{n} \operatorname{tr}\Big(\mathcal{G}(\vec{\phi}_{\text{c}})\mathcal{M}(\delta\vec{\phi})\Big)^n \end{aligned} \tag{3.134}$$

Equation (3.134) can be organized as follows:

$$S_{\text{eff}}(\vec{\phi}) = \sum_{n=0}^{\infty} S^{(n)}(\vec{\phi}_{\text{c}}, \delta\vec{\phi}) \tag{3.135}$$

where $S^{(0)}(\vec{\phi}_{\text{c}})$ is the classical action (i.e. the action of the mean-field solution)

$$S^{(0)}(\vec{\phi}_{\text{c}}) = -\int dt \sum_{\vec{r}} \frac{\vec{\phi}_{\text{c}}^2(\vec{r},t)}{2} - i \operatorname{tr} \ln[i\,\partial_t + \mu - \mathcal{M}(\vec{\phi}_{\text{c}})] \tag{3.136}$$

Since the mean-field solution is *static* (i.e. time-independent), we can write

$$S^{(0)}(\vec{\phi}_{\text{c}}) = T\,E^{(0)}_{\text{Gnd}}(\vec{\phi}_{\text{c}}) \tag{3.137}$$

where T is the time span (not the temperature!) and $E^{(0)}_{\text{Gnd}}(\vec{\phi}_{\text{c}})$ is the ground-state energy (at the level of the mean-field theory). The first-order term in $\delta\vec{\phi}$ cancels out since $\vec{\phi}_{\text{c}}$ is a solution of the saddle-point equation

$$\frac{\delta S_{\text{eff}}}{\delta\vec{\phi}} = 0 \tag{3.138}$$

Thus we can write S_{eff} in the form

$$S_{\text{eff}}(\vec{\phi}) = T\,E^{(0)}_{\text{Gnd}}(\vec{\phi}_{\text{c}}) + \sum_{n=2}^{\infty} S^{(n)}(\vec{\phi}_{\text{c}}, \delta\vec{\phi}) \tag{3.139}$$

The Gaussian theory, i.e. the quadratic terms in Eq. (3.139), has the action $S^{(2)}(\delta\vec{\phi})$, where

$$S^{(2)}(\delta\vec{\phi}) = -\int dt \sum_{\vec{r}} \frac{\delta\vec{\phi}^2(\vec{r},t)}{2} - \frac{i}{2} \operatorname{tr}\left(\mathcal{G}(\vec{\phi}_{\text{c}})\mathcal{M}(\delta\vec{\phi})\right)^2 \tag{3.140}$$

This expression can be expanded out in components to yield

$$\begin{aligned} S^{(2)}(\delta\vec{\phi}) = & -\int dt \sum_{\vec{r}} \frac{\delta\vec{\phi}^2(\vec{r},t)}{2} \\ & - i\frac{U}{6}\int dt\,dt' \sum_{\vec{r},\vec{r}\,'} \mathcal{G}_{\alpha,\alpha'}(\vec{r}t;\vec{r}\,'t';\vec{\phi}_{\text{c}})\tau^{a'}_{\alpha'\beta}\mathcal{G}_{\beta\beta'}(\vec{r}\,'t';\vec{r}t;\vec{\phi}_{\text{c}})\tau^{a}_{\beta'\alpha} \\ & \times \delta\phi^{a'}(\vec{r}\,',t')\delta\phi^{a}(\vec{r},t) \end{aligned} \tag{3.141}$$

We will see now that fluctuations $\delta\vec{\phi}(\vec{r},t)$ with wave vector $\vec{p}$ *close* to the antiferromagnetic (Néel) ordering wave vector $\vec{Q} = (\pi,\pi)$ are gapless, i.e. have vanishingly small energy. These are the antiferromagnetic spin waves. Conversely, the excitations with $\vec{p}$ close to the zone center ($\vec{p} \approx 0$), which describe *uniform* ferromagnetic fluctuations, have *large* energies. Formally this amounts to splitting the fluctuations into a *ferromagnetic* component $\delta\phi_{\rm F}$, whose Fourier transform has wave vectors close to $(0,0)$, and an *antiferromagnetic* component $\delta\phi_{\rm AF}$, whose Fourier components have wave vectors close to the ordering wave vector $\vec{Q} = (\pi, \pi)$. From now on we will consider only the antiferromagnetic fluctuations, which we will denote by $\delta\phi_{\rm AF} = \delta\phi$.

In Fourier components, the effective action for the antiferromagnetic fluctuations $\delta\phi$ has the form

$$S^{(2)}(\delta\vec{\phi}) = \int_{\vec{p},\Omega} \frac{1}{2}\, \delta\phi_a(\vec{p},\Omega)\delta\phi_b^*(\vec{p},\Omega) K^{ab}(\vec{p},\Omega) \tag{3.142}$$

where the kernel $K^{ab}(\vec{p},\Omega)$ is given by

$$K^{ab}(\vec{p},\Omega) = -\delta^{ab} - i\frac{U}{3}\int_{\vec{k},\omega} \mathrm{tr}\left(\mathcal{G}(\vec{k},\omega)\tau^a\, T\, \mathcal{G}(\vec{k}+\vec{p},\omega+\Omega)\tau^b\, T\right) \tag{3.143}$$

where $\mathcal{G}(\vec{k},\omega)$ is given by Eq. (3.127) and T is the 2×2 matrix

$$T = \begin{pmatrix} 1 & 0 \\ 0 & 1 \end{pmatrix} \tag{3.144}$$

The following trace identities for the Pauli matrices will be useful to compute the kernel:

$$\begin{aligned} &\mathrm{tr}\,1 = 2, \qquad \mathrm{tr}\,\tau^a = 0, \qquad \mathrm{tr}(\tau^a\tau^b) = 2\delta^{ab}, \qquad \mathrm{tr}(\tau^a\tau^b\tau^c) = 2i\epsilon^{abc}, \\ &\mathrm{tr}(\tau^a\tau^b\tau^c\tau^d) = 2\left(\delta^{ab}\delta^{cd} - \delta^{ac}\delta^{bd} + \delta^{ad}\delta^{bc}\right) \end{aligned} \tag{3.145}$$

We can now write the kernel $K^{ab}(\vec{p},\Omega)$ in the form

$$K^{ab}(\vec{p},\Omega) = K_0(\vec{p},\Omega)\delta^{ab} + K_2(\vec{p},\Omega)n^a n^b \tag{3.146}$$

after inserting Eq. (3.127) into Eq. (3.143) and using the trace identities. One obtains for K_0 and K_2

$$\begin{aligned} K_0(\vec{p},\Omega) &= -1 + i\frac{4U}{3}\int_{\vec{k},\omega} \frac{\omega(\omega+\Omega) + \epsilon(\vec{k})\epsilon(\vec{k}+\vec{p}) - \Delta^2}{\left(\omega^2 - E^2(\vec{k}) + i\eta\right)\left((\omega+\Omega)^2 - E^2(\vec{k}+\vec{p}) + i\eta\right)} \\ K_2(\vec{p},\Omega) &= +i\frac{8U}{3}\Delta^2 \int_{\vec{k},\omega} \frac{1}{\left(\omega^2 - E^2(\vec{k}) + i\eta\right)\left((\omega+\Omega)^2 - E^2(\vec{k}+\vec{p}) + i\eta\right)} \end{aligned} \tag{3.147}$$

where the integrals range over the full Brillouin zone and $\Delta = \sqrt{U/3}|\vec{\phi}|$ (see Eq. (3.128)). The structure of the effective kernel, Eqs. (3.147), is a consequence of the symmetry of the spectrum, which, in turn, reflects the fact that the Néel state is invariant under the combined effects of time reversal and displacement by one lattice spacing, which amounts to the exchange of the two antiferromagnetic sublattices.

We are interested in studying the low-energy limit ($\Omega\Delta \ll 1$) of this system. It will be convenient to decompose the fluctuations $\delta\vec{\phi}(\vec{p}, \Omega)$ into a longitudinal component $\sigma(\vec{p}, \Omega)$ parallel to $\vec{n}$ and two transverse components $\pi_i(\vec{p}, \Omega)$ perpendicular to $\vec{n}$. In terms of σ and $\vec{\pi}$, the Gaussian action $S^{(2)}$, Eq. (3.142), has the form

$$S^{(2)}(\delta\vec{\phi}) = \int_{\vec{p},\Omega} \{(K_0(\vec{p}, \Omega) + K_2(\vec{p}, \Omega))|\sigma(\vec{p}, \Omega)|^2 + K_0(\vec{p}, \Omega)\vec{\pi}(\vec{p}, \Omega) \cdot \vec{\pi}^*(\vec{p}, \Omega)\} \tag{3.148}$$

I will now show that for wave vectors $\vec{p}$ close to the ordering wavevector $\vec{Q} = (\pi, \pi)$, the transverse components (represented by the fields $\vec{\pi}$) become gapless. In contrast, the longitudinal component, σ, remains massive (i.e. with a non-zero gap) over the entire Brillouin zone. Thus, in a Néel state, the Hubbard model has *two* gapless transverse spin waves $\vec{\pi}$ and a massive (gapped) longitudinal amplitude mode σ.

We can check this statement by considering the limit $\Omega \to 0$ and $\vec{p} = \vec{Q} - \vec{q}$ where $|\vec{q}|$ is small, i.e. $|\vec{q}|\xi \ll 1$. Thus, the relevant limit is

$$\Omega\Delta \ll 1, \qquad |\vec{q}|\frac{v_F}{\Delta} \ll 1 \tag{3.149}$$

We wish to expand the kernels of Eq. (3.147) in powers of $\vec{q}$ and Ω, i.e. we are performing a gradient expansion.

Let us first compute K_0 and K_2 for $\Omega = 0$ *and* $\vec{p} = \vec{Q}$. The result, for $K_0(\vec{Q}, 0)$, is zero:

$$K_0(\vec{Q}, 0) = -1 + i\frac{4U}{3}\int_{\vec{k},\omega} \frac{1}{\omega^2 - E^2(\vec{k}) + i\eta} \equiv 0 \tag{3.150}$$

This result follows from the saddle-point (gap) equation, Eq. (3.114) (or, equivalently, Eq. (3.47)). Similarly, the other kernel K_2 has the limit, after integrating over ω,

$$\begin{aligned} K_2(\vec{Q}, 0) &= i\frac{8U}{3}\Delta^2 \int_{\vec{k}}\int \frac{d\omega}{2\pi} \frac{1}{\left(\omega^2 - E^2(\vec{k}) + i\eta\right)^2} \\ &= -\frac{2U}{3}\Delta^2 \int_0^\infty d\epsilon\, \frac{\rho(\epsilon)}{(\epsilon^2 + \Delta^2)^{3/2}} \\ &\simeq -\frac{1}{12\pi^2}\left(\frac{2U}{t}\right)\ln\left(\frac{2t}{\Delta}\right) \end{aligned} \tag{3.151}$$

where $\rho(\epsilon)$ is the one-particle density of states of Eq. (3.49) and Eq. (3.50).

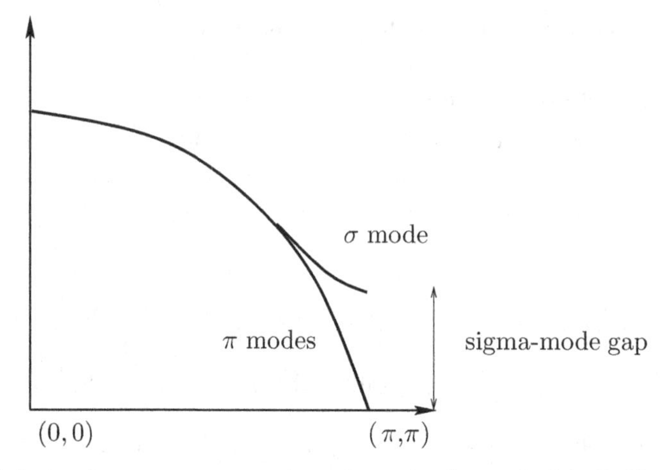

Figure 3.5 A spin-wave spectrum along the main diagonal of the Brillouin zone.

These results imply that the *longitudinal* σ mode for $\vec{p} \approx \vec{Q}$ and $\Omega \to 0$ has a finite mass (or energy gap), and imply the *absence* of such a *mass* term for both *transverse* $\vec{\pi}$ modes (see Fig. 3.5). The mass of the σ mode can be used to define a spin-correlation length ξ_{spin}. (We should note, however, that in two space dimensions the σ mode has a finite lifetime since it can decay into two $\vec{\pi}$ modes.)

We can also estimate the spin-wave stiffness from these results. However, we need to keep the leading order in both Ω and $\vec{q}$ to get these results. By expanding in powers of both $\Omega \ll \Delta$ and $\vec{q}$ (with $|\vec{q}| \ll |\vec{Q}|$), we find for $K_0(\vec{Q}-\vec{q},\Omega)$

$$K_0(\vec{Q}-\vec{q},\Omega) = a\Omega^2 - b\vec{q}^{\,2} + \text{h.o.t.} \tag{3.152}$$

The coefficients a and b are most easily found by doing first an analytic continuation of the frequency integration, $\omega \to i\omega$, together with an analytic continuation of the external frequency, $\Omega \to i\Omega$; at the end of the calculation we must analytically continue back to the real frequency axis. We obtain

$$\begin{aligned} a &= \lim_{\Omega\to 0} \frac{K_0(\vec{Q},i\Omega)}{-\Omega^2} = \frac{2U}{3}\int_{\text{BZ}} \frac{d^2k}{(2\pi)^2}\frac{1}{E^3(\vec{k})} \\ &= \frac{2U}{3}\int_0^\infty d\epsilon\, \frac{\rho(\epsilon)}{(\epsilon^2+\Delta^2)^{3/2}} \simeq \frac{U}{3\Delta^2}\rho(\Delta) \end{aligned} \tag{3.153}$$

where the last expression follows in the weak-coupling limit, $U \ll t$, where Δ is small. Similarly, we also obtain

$$
\begin{aligned}
b &= \lim_{\vec{p}\to 0} \frac{K_0(\vec{Q}-\vec{p},0)}{\vec{p}^{\,2}} \\
&= \frac{U}{3} \lim_{\vec{p}\to 0} \frac{1}{\vec{p}^{\,2}} \int_{\rm BZ} \frac{d^2k}{(2\pi)^2} \frac{\epsilon(\vec{k})(\epsilon(\vec{k}+\vec{p})-\epsilon(\vec{k}))}{E(\vec{k})E(\vec{k}+\vec{p})(E(\vec{k}+\vec{p})+E(\vec{k}))} \\
&= \frac{U}{24} \int_0^\infty d\epsilon\, \rho(\epsilon) \left[\frac{\epsilon}{\epsilon^2+\Delta^2} + \frac{6t\epsilon}{(\epsilon^2+\Delta^2)^{3/2}} \left(1-\frac{\epsilon^2}{4t^2}\right)\right] \simeq \frac{U}{16}\frac{D}{\Delta}\rho(\Delta)
\end{aligned}
\tag{3.154}
$$

where we have used that the band width is $2D = 8t$. In Eqs. (3.153) and (3.154), $\rho(\Delta)$ is the density of states of the free-fermion problem at the energy Δ.

Thus, the transverse $\vec{\pi}$ modes have the effective action for fluctuations with momenta close to the ordering wave vector $\vec{Q}$ and frequencies low with respect to the gap Δ of

$$
S_{\vec{\pi}}^{(2)} = \frac{1}{2} \int_{\vec{q},\Omega} \rho_{\rm s} \left(\frac{1}{v_{\rm s}^2}\Omega^2 - \vec{q}^{\,2}\right) |\vec{\pi}(\vec{q},\Omega)|^2 \tag{3.155}
$$

where we have defined $\rho_{\rm s}$, the spin-wave stiffness, and $v_{\rm s}$, the spin-wave velocity, respectively, by

$$
\rho_{\rm s} = b \simeq \frac{U}{16}\frac{D}{\Delta}\rho(\Delta), \quad v_{\rm s}^2 = \frac{b}{a} \simeq \frac{\sqrt{3}}{4}\sqrt{D\Delta} \tag{3.156}
$$

These results show that the transverse $\vec{\pi}$ modes are massless and have a linear dispersion relation

$$
\Omega = v_{\rm s}|\vec{q}\,| \tag{3.157}
$$

3.5 The Néel state and the non-linear sigma model

I will now show that these results can be embodied in a very simple effective Lagrangian that includes all the relevant non-linear effects. The key to the argument is the observation that the Néel state breaks, on each sublattice, the SU(2) spin rotation symmetry. This is a continuous symmetry. The transverse spin waves are gapless excitations because they involve merely tilting the spins relative to the classical mean-field state. But they do not change the amplitude of the field. Thus, it is natural to ask for the effective Lagrangian in which the amplitude fluctuations are frozen but the transverse ones are not. In mathematical terms this means that, in position space, the *staggered order parameter* $\vec{n}(\vec{r},t)$ will be slowly varying and its length will be constrained to be equal to a fixed value $|\vec{\phi}_{\rm c}|$, the classical minimum. Thus, the fields $\sigma(\vec{k},\omega)$ and $\vec{\pi}(\vec{k},\omega)$ need to be scaled by a factor of $|\vec{\phi}_{\rm c}|^{-1}$. The net effect is that the spin-wave stiffness $\rho_{\rm s}$, Eq. (3.155), gets multiplied by $|\vec{\phi}_{\rm c}|^2$, the solution of Eq. (3.110).

This calculation can in fact be carried out because the transverse spin waves remain massless to *all* orders in an expansion around the mean-field solution. This is guaranteed by a Ward identity, which is a consequence of the symmetry. We shall derive this identity below.

Let us consider the Hubbard model coupled to an external magnetic field $\vec{H}(\vec{r}, t)$,

$$H_{\text{Zeeman}} = \sum_{\vec{r}} \vec{H}(\vec{r}, t) \cdot c_\alpha^\dagger(\vec{r}, t) \vec{\tau}_{\alpha\beta} c_\beta(\vec{r}, t) \tag{3.158}$$

If we retrace the steps that led to the path integral for the Hubbard–Stratonovich field $\vec{\phi}(\vec{r}, t)$, we find a new effective action of the form

$$S_{\text{eff}}\left(\vec{\phi}, \vec{H}\right) = -\int dt \sum_{\vec{r}} \frac{\vec{\phi}^2(\vec{r}, t)}{2} - i \operatorname{tr} \ln\left(i\,\partial_t + \mu - M(\vec{\phi}) - \vec{H}(\vec{r}, t) \cdot \vec{\tau}\right) \tag{3.159}$$

where the magnetic field $\vec{H}(\vec{r}, t)$ is a c-number operator with matrix elements

$$\langle \vec{r} t \alpha | \vec{H}(\vec{r}, t) | \vec{r}\,' t' \alpha' \rangle = \delta_{\alpha, \alpha'} \delta_{\vec{r}, \vec{r}\,'} \delta(t - t') \vec{H}(\vec{r}, t) \tag{3.160}$$

In principle, we will want to study the Néel state. Hence we will choose $\vec{H}(\vec{r}, t)$ to be staggered and time-independent, i.e.

$$\vec{H}_s(\vec{r}, t) = \vec{H}_s e^{i\vec{Q}\cdot\vec{r}} \tag{3.161}$$

with

$$\vec{Q} = (\pi, \pi) \tag{3.162}$$

We can prove the existence of gapless excitations, to all orders in perturbation theory, by deriving a Ward identity. This identity can be derived by standard methods (Amit, 1980). Let us first shift the $\vec{\phi}$ field

$$\vec{\phi}\,'(\vec{r}, t) \equiv \vec{\phi}(\vec{r}, t) + \sqrt{\frac{3}{U}}\, \vec{H}(\vec{r}, t) \tag{3.163}$$

We can write

$$S_{\text{eff}}\left(\vec{\phi}, \vec{H}\right) = S_{\text{eff}}(\vec{\phi}\,', 0) + \sqrt{\frac{3}{U}} \int dt \sum_{\vec{r}} \vec{\phi}\,'(\vec{r}, t) \cdot \vec{H}(\vec{r}, t) - \frac{3}{U} \int dt \sum_{\vec{r}} \frac{\vec{H}^2(\vec{r}, t)}{2} \tag{3.164}$$

Next, we can make use of the invariance of the integration measure $\mathcal{D}\vec{\phi}$ under the rotation $\phi_a'' = R_{ab}\phi_b'$ (where R_{ab} is a rotation matrix), $\mathcal{D}\phi'' = \mathcal{D}\phi'$, to shift the coordinates of the functional integral. The rotation matrix R_{ab} can be written in terms of Euler angles θ^c and rotation generators L^c in the form

$$R_{ab} = \left(e^{-iL^c\theta^c}\right)_{ab} \tag{3.165}$$

where

$$(L^c)_{ab} = -i\epsilon_{abc} \tag{3.166}$$

Thus, for an infinitesimal rotation ($\theta^c \ll 1$)

$$\phi''_a = \phi'_a - \epsilon_{abc}\phi'_b\theta_c \tag{3.167}$$

Using the invariance of the measure, we can now write for the vacuum persistence amplitude in the presence of the "source" $\vec{H}$

$$\begin{aligned}
Z[\vec{H}] &\equiv \exp\left(iF[\vec{H}] - i\frac{3}{U}\int dt \sum_{\vec{r}} \frac{\vec{H}^2}{2}\right) \\
&= \int \mathcal{D}\vec{\phi}' \exp\left[iS_{\text{eff}}(\vec{\phi}', 0) + i\int dt \sum_{\vec{r}}\left(\sqrt{\frac{3}{U}}\vec{\phi}' \cdot \vec{H} - \frac{3}{2U}\vec{H}^2\right)\right] \\
&= \int \mathcal{D}\vec{\phi}'' \exp\left[iS_{\text{eff}}(\vec{\phi}'', 0) + i\int dt \sum_{\vec{r}}\left(\sqrt{\frac{3}{U}}\vec{\phi}'' \cdot \vec{H} - \frac{3}{2U}\vec{H}^2\right)\right] \\
&= \int \mathcal{D}\vec{\phi}' \exp\left[iS_{\text{eff}}(\vec{\phi}', 0) + i\int dt \sum_{\vec{r}}\left(\sqrt{\frac{3}{U}}\vec{\phi}' \cdot \vec{H} - \frac{3}{2U}\vec{H}^2\right)\right. \\
&\qquad \left. + i\,\delta S_{\text{eff}}(\vec{\phi}', \vec{H}, \vec{\theta})\right]
\end{aligned} \tag{3.168}$$

where

$$\delta S_{\text{eff}}(\vec{\phi}', \vec{H}, \vec{\theta}) = -\sqrt{\frac{3}{U}}\int dt \sum_{\vec{r}} \epsilon^{abc} H^a(\vec{r}, t)\phi'^b(\vec{r}, t)\theta^c \tag{3.169}$$

By expanding Eq. (3.168) in powers of θ we obtain, to leading order,

$$Z[\vec{H}] = Z[\vec{H}]\left(1 + \delta\bar{S}_{\text{eff}} + \cdots\right) \tag{3.170}$$

Thus, since the variables θ^c are arbitrary, we obtain the identity

$$0 = \int dt \sum_{\vec{r}} \epsilon^{abc} H^a(\vec{r}, t)\bar{\phi}'^b(\vec{r}, t) \tag{3.171}$$

Here $\bar{\phi}'^b$ is the exact expectation value of ϕ'^b in the presence of $\vec{H}$.

Let us define now the generating functional of vertex functions, $\Gamma[\bar{\phi}']$, by means of the Legendre transform

$$\Gamma[\bar{\phi}'] = \sqrt{\frac{3}{U}} \int dt \sum_{\vec{r}} \bar{\phi}'_a(\vec{r}, t) H_a(\vec{r}, t) - F[\vec{H}] \tag{3.172}$$

with

$$\frac{\delta F}{\delta H^b(\vec{r}, t)} = \sqrt{\frac{3}{U}} \bar{\phi}'^b(\vec{r}, t) \tag{3.173}$$

It follows (Amit, 1980) that

$$\frac{\delta \Gamma}{\delta \bar{\phi}'^a(\vec{r}, t)} = \sqrt{\frac{3}{U}} H^a(\vec{r}, t) \tag{3.174}$$

The one-particle irreducible vertex functions can be defined in terms of functional derivatives of Γ relative to $\bar{\phi}'_d$. For instance, the two-point irreducible function

$$\Gamma^{(2)}_{ab}(\vec{r}, t, \vec{r}\,', t') \equiv \frac{\delta^2 \Gamma}{\delta \bar{\phi}'_a(\vec{r}, t) \delta \bar{\phi}'_b(\vec{r}\,', t')} \tag{3.175}$$

is the inverse of the $\vec{\phi}'$ two-point (Green) function

$$\sum_{\vec{r}\,', b} \frac{\delta^2 \Gamma}{\delta \bar{\phi}'^a(\vec{r}, t) \delta \bar{\phi}'^b(\vec{r}\,', t')} \frac{\delta^2 F}{\delta H^b(\vec{r}\,', t') \delta H^c(\vec{r}\,'', t'')} = \delta_{ac} \delta_{\vec{r}, \vec{r}\,''} \delta(t - t'') \tag{3.176}$$

With those definitions, Eq. (3.171) can be brought to the form

$$0 = \int dt \sum_{\vec{r}} \left(\epsilon^{abc} \frac{\delta \Gamma}{\delta \bar{\phi}'^a(\vec{r}, t)} \bar{\phi}'^b(\vec{r}, t) \right) \tag{3.177}$$

To avoid cumbersome notation we denote from now on $(\vec{r}, t) \equiv (x_1, x_2, x_0) \equiv x$, and $\delta_{\vec{r}, \vec{r}\,'} \delta(t - t') \equiv \delta(x - x')$. By taking a further derivative with respect to $\bar{\phi}'^d(x')$, one gets the general Ward identity

$$0 = \int dt \sum_{\vec{r}} \epsilon^{abc} \left(\frac{\delta^2 \Gamma}{\delta \bar{\phi}'^d(x') \delta \bar{\phi}'^a(x)} \bar{\phi}'^b(x) + \frac{\delta \Gamma}{\delta \bar{\phi}'^a(x)} \delta^{bd} \delta(x - x') \right) \tag{3.178}$$

In particular, for the Néel state

$$\bar{\phi}'^b(x) = |\vec{\phi}| n^b (-1)^{x_1 + x_2} \tag{3.179}$$

and the corresponding staggered field of Eq. (3.161), we get the Ward identity

$$\epsilon^{adc} \sqrt{\frac{3}{U}} H^a_s(x') = -\epsilon^{abc} |\vec{\phi}| n^b \int dt \sum_{\vec{r}} \frac{\delta^2 \Gamma}{\delta \bar{\phi}'^d(x') \delta \bar{\phi}'^a(x)} (-1)^{x_1 + x_2} \tag{3.180}$$

In momentum space, Eq. (3.180) simply becomes

$$\epsilon^{acd}\sqrt{\frac{3}{U}}H_s^a(\vec{Q}) = \lim_{\substack{\omega\to 0\\ \vec{p}\to 0}} \epsilon^{abc}|\vec{\phi}|n^b\Gamma_{da}^{(2)}(\vec{Q}+\vec{p},\omega) \tag{3.181}$$

The spontaneous breaking of symmetry means that, when the external field is switched off, the order parameter remains finite. In this case, for the right-hand side of Eq. (3.181) to vanish, the contraction of the vertex two-point function with the Levi-Civita tensor and with $\vec{n}$ must vanish in this limit. Thus, if $|\vec{\phi}| \neq 0$, the *transverse* components of $\vec{\phi}$ must have a *pole* at $\omega = 0$ and $\vec{p} = \vec{Q}$ in their correlation function (see Eq. (3.175)). We found before that this was indeed the case in the leading order of an expansion around mean-field theory.

Since we know now that this leading-order pole must persist to *all orders*, we can look for an effective Lagrangian with the following properties:

(1) the amplitude fluctuations are suppressed,
(2) it has a massless pole for each transverse component, and
(3) it has a minimum number of derivatives in the order-parameter field.

The simplest expression satisfying these properties is the non-linear sigma model with a Lagrangian density given by

$$\mathcal{L}_{\text{eff}} = \frac{\rho}{2}\left((\partial_t\vec{n})^2 - v_s^2(\nabla\vec{n})^2\right) + \cdots \tag{3.182}$$

where $\vec{n}$ satisfies the constraint

$$|\vec{n}|^2 = 1 \tag{3.183}$$

Here $\vec{n}$ represents the slow fluctuations of the order-parameter field. The spin-wave stiffness ρ_s and velocity v_s appearing in Eq. (3.182) are not generally identical to the values we calculated above (see Eq. (3.156)). The reason for that is that the non-linear σ-model is the effective theory at low frequencies and for wave vectors close to $\vec{Q}$. It is the result of integrating out all fluctuations at high energies. This process significantly renormalizes the values of ρ_s and v_s. In addition, I have ignored the possible existence of topological terms. We will see in Chapter 7 that these terms are generated in one dimension, but not in two or higher dimensions.

It is a simple matter to see that, if one solves for the constraint $|\vec{n}| = 1$ in terms of a σ and a $\vec{\pi}$ field

$$\vec{n} = \begin{pmatrix}\sigma\\ \vec{\pi}\end{pmatrix} \tag{3.184}$$

and expands in powers of $\vec{\pi}$, one finds, to the leading quadratic order in $\vec{\pi}$, the effective action Eq. (3.155).

We conclude that the quantum fluctuations around a Néel state are described by a non-linear σ-model. One key assumption that we have made here is that the stiffness ρ_s was assumed to be large. Otherwise this expansion does not make sense. In fact for ρ_s sufficiently small, the non-linear σ-model has wild fluctuations which destroy the Néel state. The system becomes a (quantum) paramagnet and the spin symmetry is unbroken. We will see below that *frustrating* interactions will generally produce this effect. On the other hand, given the large renormalizations of ρ_s and v_s, it is not possible to be sure whether the half-filled Hubbard model is in a Néel state (i.e. whether the ground state is an antiferromagnet) or in a disordered state. In practice only numerical calculations, i.e. fermion Monte Carlo or finite-size exact diagonalizations, can yield more reliable answers for a specific Hubbard model. So far the evidence strongly favors a Néel state.

Finally we should consider the connection between the collective excitations $\vec{\phi}(\vec{x}, t)$ and the susceptibilities of the Hubbard model. From Eq. (3.158) we see that the field $\vec{H}(x)$ couples to the local moment $c_\alpha^\dagger(x)\vec{\tau}_{\alpha\beta}c_\beta(x)$. Thus by functionally differentiating with respect to $\vec{H}(x)$ we should be able to compute expectation values related to the spin degrees of freedom. Indeed, the spin correlation function

$$K^{aa'}(x, x') = \langle \text{Gnd}|\hat{T} S^a(x')S^{a'}(x')|\text{Gnd}\rangle \tag{3.185}$$

is equivalent to

$$K^{aa'}(x, x') = -g(x, x')\frac{\delta^2 F}{\delta H_s^a(x)\delta H_s^{a'}(x')} \equiv G_{aa'}^{(2)}(x, x') \tag{3.186}$$

where $g(x, x')$ is a sign function

$$g(x, x') = (-1)^{x_1+x_2+x_1'+x_2'} \tag{3.187}$$

Since $G_{aa'}^{(2)}(x, x')$ is the inverse of $\Gamma_{aa'}^{(2)}(x, x')$, we conclude that a *zero* in $\Gamma_{aa'}^{(2)}(\vec{p}, \omega)$ at $(\vec{Q}, 0)$ implies a divergence of $G_{aa'}^{(2)}(x, x')$ also at $(\vec{Q}, 0)$. Thus the staggered static susceptibility $\chi_\perp(\vec{Q}, 0)$, which is the Fourier transform of $K_{aa'}(x, x')$, must have a delta-function peak at $(\vec{Q}, 0)$ if the ground state is a Néel state $(\vec{Q}, 0)$. This peak appears only in the transverse components of $K_{aa'}$ since the longitudinal components are connected to excitations with a non-zero gap.

4

The renormalization group and scaling

4.1 Scale invariance

The renormalization group is a central conceptual framework for understanding the behavior of strongly coupled and critical systems. It was originally formulated in the context of perturbative quantum field theory (particularly in relation to quantum electrodynamics), and found its crisper and most powerful realization in the explanation of critical phenomena in statistical physics. The most important ideas derived from the renormalization group are the concepts of a fixed point and universality. These ideas, due primarily to Wilson and Kadanoff, in turn provided a definition of a quantum field theory outside the framework of perturbation theory. In this chapter we will present a brief exposition of the main ideas and tools of the renormalization group and their application to problems of interest in strongly correlated and quantum critical systems (Cardy, 1996).

Many problems in condensed matter physics can be described in terms of a Hamiltonian that is a sum of two terms, $H = H^* + H'$. Equivalently, in terms of the action, we can write $S = S^* + S'$, where the action S is a function of some macroscopic number of degrees of freedom, which we will represent in terms of a field $\phi(x)$ and its derivatives (in space and time). The field ϕ may obey Fermi or Bose statistics, in which case it will be represented by a set of Grassmann or scalar (or vector) fields, depending on the case. Both in classical statistical mechanics and in a quantum field theory we can formally represent the system in terms of a path integral

$$\mathcal{Z} = \int \mathcal{D}\phi \, e^{-S(\phi)} \tag{4.1}$$

In what follows we will work in imaginary time. In the following chapters we will discuss many systems that admit a representation of this form. The degrees of freedom may be spins (classical or quantum) or fermions, as well as gauge fields.

In its simplest representation the renormalization group is a transformation that maps a system with a set of coupling constants and a scale (representing the short-distance or high-energy cutoff) to another equivalent system with a different set of ("renormalized") coupling constants and a different scale. This is done by a procedure known as a "block-spin" transformation (in the language of classical statistical mechanics), by which some of the degrees of freedom, representing the short-distance physics, are integrated out and a subsequent scale transformation is performed to restore the original scale (or units).

The field $\phi(x,t) \equiv \phi(x)$ has many Fourier components (in D-dimensional space-time)

$$\phi(x) = \int \frac{d^D k}{(2\pi)^D} e^{ik\cdot x} \phi(k) \tag{4.2}$$

The large-momentum (-frequency) components correspond to configurations with larger action than those with lower momentum $k \equiv (\omega, \vec{k})$. If the high-$(k, \omega)$ components, represented by $\phi_>(x)$, are integrated out, then we can find an effective action (or Hamiltonian) for the configurations $\phi_<(x)$ with lower (k, ω). In real space-time we are eliminating degrees of freedom at short distances, or rather we are defining configurations *averaged* over a short distance scale ba (a is the lattice constant) or with wave vectors $|k| < \Lambda/b$ (where Λ is a high-energy cutoff).

In most cases $S(\phi)$ will change upon a process of this sort

$$\mathcal{Z} = \int \mathcal{D}\phi\, e^{-S(\phi)} = \int \mathcal{D}\phi_<\, \mathcal{D}\phi_>\, e^{-S(\phi_< + \phi_>)} \tag{4.3}$$

where

$$\phi = \phi_< + \phi_> \tag{4.4}$$

Hence,

$$\mathcal{Z} \equiv \int \mathcal{D}\phi_<\, e^{-S_{\text{eff}}(\phi_<)} \tag{4.5}$$

where we have defined an *effective action* S_{eff}

$$e^{-S_{\text{eff}}(\phi_<)} \equiv \int \mathcal{D}\phi_>\, e^{-S(\phi_< + \phi_>)} \tag{4.6}$$

This is, formally, our "block-spin" transformation.

However, now we have a new short-distance cutoff $a' \equiv ba$ (with $b > 1$) or, equivalently, a high-energy cutoff $\Lambda' \equiv \Lambda/b < \Lambda$. Thus, we need to rescale the lengths (and time) in order to restore the units. In order to compensate for the change in the definition of units, we rescale lengths and momenta as follows:

$$x' = \frac{x}{b} \quad \text{or} \quad k' = kb \tag{4.7}$$

such that $|k'| < \Lambda$ or $|x|' > a$ back again.

Clearly many transformations of this type can be defined and the form of the effective action will depend on this definition. Nevertheless, these transformations must obey some basic principles. The most important one is that this procedure should be compatible with the underlying symmetries of the physical system. Thus, if the system has a tendency to become anisotropic (either in space or in space-time) the rescaling will have to be consistent with this fact. For simplicity, in what follows we will assume that the rescaling is isotropic both in space and in space-time. Thus, we are assuming that there will be an effective Lorentz invariance in the system of interest.

Suppose that we are able to find an action S^* such that it remains *invariant* under the renormalization-group (RG) transformation:

$$S^*_{\text{eff}}(\phi_<) = S^*(\phi) \tag{4.8}$$

We will call this action a *fixed point* of the RG.

At a fixed point the action and the Hamiltonian reproduce themselves under an RG transformation. Thus a system at a fixed point has a new symmetry: *scale invariance*. This means that there can be no scales left in the problem. Therefore, a fixed-point action describes either

(1) a system with a vanishing correlation length, $\xi \to 0$, and hence a divergent energy gap, $E_{\text{G}} \to \infty$, or
(2) a system with a divergent correlation length, $\xi \to \infty$, and hence a vanishing energy gap, $E_{\text{G}} \to 0$.

Fixed points with vanishing correlation lengths (and divergent energy gaps) describe *stable phases of matter*. Conversely, fixed points with diverging correlation lengths (and vanishing energy gaps) describe *systems at criticality* (quantum or thermal), and thus correspond to *phase transitions*. We will see that fixed points with vanishing correlation lengths are stable under the action of all (local) perturbations, and are hence known as stable fixed points. In contrast, fixed points with divergent correlation lengths are unstable at least with respect to one local perturbation (or more), and are known as unstable fixed points.

Let us consider now a system close to a fixed point. Quite generally we can write the action of the system in the form

$$S(\phi) = S^*(\phi) + \int dx^D \sum_n \lambda_n \phi_n(x)) \tag{4.9}$$

where $\{\phi_n(x)\}$ is a complete set of operators defined in the fixed-point theory. Here $\{\lambda_n\}$ are the associated coupling constants. Under a renormalization-group transformation consisting of

(1) integrating out high-energy modes $\Lambda \to b\Lambda$ $(b < 1)$ and
(2) rescaling lengths $x \to b^{-1}x$

the action $S^*(\phi)$ remains *invariant* since it is a fixed point. The operators $\{\phi_n(x)\}$ transform irreducibly under rescalings as

$$\phi_n(xb^{-1}) = b^{\Delta_n}\phi_n(x) \tag{4.10}$$

where Δ_n is the *scaling dimension* of this operator. Operators that transform irreducibly (i.e. homogeneously) under rescalings are called primary (scaling) fields. Then, under the action of the RG the perturbation transforms as

$$\int dx^D \sum_n \lambda_n\phi_n(x) \to \int dx^D \sum_n b^{-D+\Delta_n}\lambda_n\phi_n(x) \tag{4.11}$$

Thus, the RG transformation is equivalent to a rescaling of the coupling constants

$$\lambda_n' = \lambda_n b^{\Delta_n - D} \equiv \lambda_n(b) \tag{4.12}$$

Since $b \to 1^-$, we can write $b = e^{-\delta\ell}$ and turn the transformation into a differential change of the coupling constants of terms of the *beta function*

$$\beta(\lambda_n) = \frac{\partial\lambda_n}{\partial\ell} = (D - \Delta_n)\lambda_n + \cdots \tag{4.13}$$

where the ellipsis denotes higher-order contributions to the beta function that we will discuss shortly. This result, Eq. (4.13), is usually called the *tree-level* beta function.

Alternatively, we can define *dimensionless* coupling constants $\bar{\lambda}_n = \lambda_n a^{D-\Delta_n}$. We now ask how we have to change the dimensionless coupling constants as the UV cutoff changes, $\Lambda \to b\Lambda$ or $a \to b^{-1}a$, while keeping the dimensional couplings λ_n fixed. For an infinitesimal change $a \to a + da$ this can be expressed as the condition

$$a\frac{\partial\lambda_n}{\partial a} = 0 \;\Rightarrow\; \beta(\bar{\lambda}_n) = a\left.\frac{\partial\bar{\lambda}_n}{\partial a}\right|_{\lambda_n} = (D - \Delta_n)\bar{\lambda}_n + O(\bar{\lambda}_n^2) \tag{4.14}$$

From now on λ_n will always denote dimensionless coupling constants, and we will always express the renormalization-group flows in terms of the changes of the dimensionless coupling constants. The physical reason for this is that dimensional parameters depend on the (arbitrary) choice of units.

We now see the following.

1. If $D - \Delta_n > 0$, the dimensionless coupling λ_n grows as the momentum scale decreases. In this case $\phi_n(x)$ is a *relevant* operator ($\Delta_n < D$) and drives the system away from the fixed-point action S^*, into a phase characterized by a new (stable) fixed point.

2. If $D - \Delta_n < 0$, the dimensionless coupling λ_n shrinks as the momentum scale decreases. In this case $\phi_n(x)$ is an *irrelevant* operator ($\Delta_n > D$), whose effects become asymptotically negligible at long distances and low energies.
3. If $D - \Delta_n = 0$, to leading order, the dimensionless coupling λ_n does not change as the momentum scale decreases. In this case $\phi_n(x)$ is a *marginal* operator ($\Delta_n = D$). To assess its effects, a higher-order calculation needs to be done. If this operator remains exactly marginal, then it should be included in the fixed-point action.

We will now see how this structure works in some simple examples of physical interest.

4.2 Examples of fixed points

4.2.1 The free-scalar-field-theory fixed point

A simple example of a fixed-point action is a free scalar field in D dimensions. The case $D = 2$, which represents either a classical statistical-mechanical system in two space dimensions or a quantum field theory in $(1 + 1)$ (Euclidean) dimensions, will be of special interest. In the next chapters we will see that this theory plays a key role in the theory of quantum antiferromagnets in one dimension and Luttinger liquids, which we discuss in Chapters 5 and 6. In classical statistical mechanics this is the theory of critical phenomena originally proposed by Landau (and Ginzburg) and eventually developed by Wilson and Fisher and by Kadanoff.

The action of a scalar field $\phi(x)$ in D dimensions is

$$S(\phi) = \frac{1}{2} \int d^D x (\partial \phi)^2 = \int^{\Lambda} \frac{d^D k}{(2\pi)^D} \frac{k^2}{2} |\phi(k)|^2 \tag{4.15}$$

where we have also introduced the expression in terms of the Fourier components of the field. Here Λ is the large-momentum (or short-distance) cutoff.

Let us now proceed to split the field into its slow and fast components,

$$\phi(x) = \phi_<(x) + \phi_>(x) \tag{4.16}$$

where (in terms of a slicing parameter $b < 1$)

$$\phi_<(x) \equiv \int^{b\Lambda} \frac{d^D k}{(2\pi)^D} \phi(k) e^{ik\cdot x} \tag{4.17}$$

represents the slow components, and

$$\phi_>(x) \equiv \int_{b\Lambda}^{\Lambda} \frac{d^D k}{(2\pi)^D} \phi(k) e^{ik\cdot x} \tag{4.18}$$

represents the fast components. In terms of the Fourier transform of the slow and fast fields the action takes the form

$$S_\Lambda = S_< + S_> \\ = \int^{b\Lambda} \frac{d^D k}{(2\pi)^D} \frac{k^2}{2} |\phi_<(k)|^2 + \int_{b\Lambda}^{\Lambda} \frac{d^D k}{(2\pi)^D} \frac{k^2}{2} |\phi_>(k)|^2 \tag{4.19}$$

where we have labeled the action with the cutoff Λ. Thus, for a free field (and only for a free field), the total action is the sum of the actions for the slow and fast modes independently. In this simple case we can integrate out the fast fields and obtain an effective action for the slow fields:

$$\int \mathcal{D}\phi_> \, e^{-S_\Lambda(\phi)} = \text{constant} \times e^{-S_{b\Lambda}(\phi_<)} \tag{4.20}$$

We now must make a scale transformation to restore the value of the cutoff to its original value:

$$x' = b^{-1}x, \quad k' = bk, \quad \phi(x') = b^\Delta \phi(x) \tag{4.21}$$

where Δ is the *scaling dimension* of the field ϕ. Thus, under a scale transformation,

$$\int d^D x \, \frac{1}{2} \left(\frac{\partial \phi}{\partial \vec{x}}\right)^2 = \int d^D x' \, b^{2-D} \frac{1}{2} \left(\frac{\partial \phi(\vec{x}\,' b^{-1})}{\partial \vec{x}\,'}\right)^2 \\ = \int d^D x' \, \frac{1}{2} \left(\frac{\partial \phi(\vec{x}\,')}{\partial \vec{x}\,'}\right)^2 \tag{4.22}$$

Thus, the action is invariant under scale transformation, provided that the scaling dimension of the field ϕ is $\Delta = (D-2)/2$. The case $D = 2$ is special in that ϕ has scaling dimension zero (i.e. it is dimensionless, and so are all powers of the field ϕ^n).

Let us examine now simple perturbations at this fixed point. The mass term is the operator $\int d^D x (m^2/2)\phi^2$, which is not invariant under rescalings, since the volume element scales as $\int d^2 x \to d^D x' \, b^{-D}$ and $\phi^2 \to b^{2\Delta}\phi^2$. Thus, the mass scales as $m'^2 = b^{2\Delta - D} m^2 = b^{-2} m^2$, and as b decreases the "coupling" m^2 "flows" to larger values under renormalization.

Similarly, the couplings to higher-derivative operators decrease as b decreases,

$$\int d^D x \; g(\partial^2 \phi)^2 = \int d^D x' \, b^2 g(\partial'^2 \phi')^2 \tag{4.23}$$

Hence $g' = gb^2$, and we conclude that under the RG g decreases as b decreases. This is an irrelevant operator. Obviously, operators with higher derivatives are even more irrelevant. This is consistent with the intuition that the effective-field theory at long distances must contain only operators with the smallest possible number of derivatives compatible with the symmetries of the system.

The same conclusions also follow simply from naive dimensional analysis. Thus, by demanding that the free-field action be dimensionless, we see that the units of the scalar field are $[\phi] = L^{-\Delta}$, where $\Delta = (D-2)/2$ is the scaling dimension. Similarly, the mass m has units of L^{-1} (as it should). The same line of reasoning determines the scaling of interaction terms. Thus, demanding once again that this term also be dimensionless, we find $L^D[\lambda_r][\phi]^r = 1$ Thus, the scaling dimension of the operator ϕ^r is $\Delta_r = r\Delta = r(D-2)/2$. Hence, the units of its coupling constant are $[\lambda_r] = L^{-D+r\Delta} = L^{(r/2)(D-2)-D}$, e.g. $[\lambda_4] = L^{D-4}$, and it is dimensionless at $D = 4$. We can also write down the (tree-level) beta function for the (dimensionless) coupling λ_r:

$$\beta(\lambda_r) = (D - \Delta_r)\lambda_r + O(\lambda_r^2) \tag{4.24}$$

Therefore the free-massless-scalar-field fixed point is stable provided that $D - \Delta_r < 0$. However, since

$$D - \Delta_r = \frac{r-2}{2}\left(-D + \frac{2r}{r-2}\right) \tag{4.25}$$

the fixed point is stable only if $D > 2r/(r-2)$. In particular, the operator ϕ^4 is irrelevant only for $D > 4$ and the free-field fixed point is unstable for fewer than four dimensions.

4.2.2 The non-linear sigma model

In Chapter 7 we will discuss the case of a non-linear sigma model as the field theory of a quantum antiferromagnet, i.e. an N-component scalar field that satisfies the constraint $\vec{n}^{\,2} = 1$. Owing to the constraint, this theory is no longer a free field. Also, the constraint forces the field $\vec{n}$ to be dimensionless. The action of the non-linear sigma model is

$$S = \frac{1}{2g}\int d^D x\, \left(\partial_\mu \vec{n}\right)^2 \tag{4.26}$$

with $\mu = 1, \ldots, D$. The only operator in this action has scaling dimension $\Delta = 2$. Notice that, due to the symmetry, only derivative operators are allowed in this case, and operators with higher derivatives will correspondingly have a larger scaling dimension. By dimensional analysis we see that the coupling constant has units of $[g] = L^{D-2}$, and hence it is dimensionless in $D = 2$. The tree-level beta function for the coupling constant g is $\beta(g) = -(D-2)g + \cdots$. Hence, dimensional analysis predicts that the fixed point at $g \to 0$, the classical theory, describes a stable phase for $D > 2$. We will discuss this problem in some detail in Chapter 7.

4.2.3 Anisotropic scaling

As we noted above, there are many circumstances in which space and time scale differently. One such example is the theory of a quantum ferromagnet that will be discussed in Chapter 7. There we will see that the effective action contains Berry-phase-like terms (that we refer to as Wess–Zumino terms) that are of first order in time derivatives, whereas the spatial dependence comes from terms with two space derivatives. Thus, in the case of a quantum ferromagnet time scales as two powers of length, $T \sim L^2$. This system has a dynamical critical exponent $z = 2$. This behavior is manifest in the structure of the Landau–Lifshitz equation and in the dispersion of the magnons, i.e. Bloch waves.

Yet another system in which anisotropic scaling arises is quantum dimer models at criticality. In this problem, which will be discussed in Chapter 9, the effective-field theory also has dynamical critical exponent $z = 2$. However, in contrast to the case of the ferromagnet, this theory is time-reversal-invariant. Consequently the dynamics is manifest through operators that are quadratic in time derivatives. However, since these systems have spatially anisotropic phases, at quantum criticality their spatial dependence comes from operators that are quartic in space derivatives, which is why $z = 2$. More generally, there are many systems in which spatial and temporal fluctuations scale with a non-trivial exponent $z \neq 1$. One example of this type of behavior is Fermi fluids at a nematic quantum critical point, which has $z = 3$ (Oganesyan *et al.*, 2001).

4.2.4 Scaling in Fermi liquids

A Fermi liquid is a system of fermions at finite density whose interactions, in many circumstances, with the notable exception of the case of one dimension, become very weak at low energies due to the kinematical constraints imposed by the existence of a Fermi surface. In this limit, the interacting system is smoothly connected to the physics of non-interacting fermionic quasiparticles. This is reflected in the simple scaling behavior that these systems exhibit and it is the essence of the Landau theory of the Fermi liquid (Baym and Pethick, 1991).

A system of non-interacting fermions is described by an action for the Fermi field whose kinetic energy is defined by the single-particle spectrum $\varepsilon(\vec{p})$. The Fermi surface is the locus of points $\{\vec{p}\}$ where $\varepsilon(\vec{p}) = E_{\mathrm{F}}$, the Fermi energy. At low enough energies, i.e. close to the Fermi energy, the allowed momenta of the fermion states are restricted to being close to the Fermi surface and the energy of these excitations differs little from E_F. In this regime it is legitimate to approximate the single-particle spectrum by $\varepsilon(\vec{p}) = v_F(|\vec{p}| - p_F) + \cdots$, where v_F is the Fermi velocity and p_F is the Fermi momentum (assuming, for simplicity, an isotropic Fermi surface). Thus, in this limit the energy and the momentum scale in the same way, and hence space and time scale in the same way as well (Polchinski, 1993; Shankar, 1994), $T \sim L$, and the dynamic critical exponent is $z = 1$. Notice, however, that only the normal component of the momentum enters in the scaling, since the tangential components do not change the energy. In this regime the interactions between the quasiparticles become actually irrelevant (with the important exception of the Cooper channel, the BCS interaction). The generic irrelevance of essentially all interactions is the reason for the robustness of the Landau theory of the Fermi liquid.

4.2.5 The free-relativistic-fermion theory

Relativistic free-fermionic systems obey the Dirac equation as their equation of motion. In condensed matter physics interacting Dirac fermions appear naturally in the theory of one-dimensional Fermi systems such as Luttinger liquids. They also appear as topological excitations (or solitons) of one-dimensional antiferromagnets. Dirac fermions give a natural description of the low-energy electronic states in two-dimensional semi-metals made of carbon layers (graphene), and of the fermionic quasiparticles of d-wave superconductors and in flux phases (which will be discussed in subsequent chapters).

Dirac fermions are spinor fields. Dirac fermions in one and two space dimensions are two-component complex Fermi fields, whereas in three space dimensions they are four-component fields. Let $\psi_\alpha(x)$, with $\alpha = 1, 2$, denote a two-component Fermi field in $D = 1 + 1$ and $D = 2 + 1$ space dimensions. In $D = 1 + 1$ dimensions, the two components correspond to right- and left-moving fermions. The Hamiltonian density for massless Dirac fermions in $D = 1 + 1$ is

$$\mathcal{H} = v_F \psi_\alpha^\dagger(x) i \sigma_3^{\alpha\beta}\, \partial_x \psi_\beta(x) \tag{4.27}$$

where σ_3 is the (diagonal) 2×2 Pauli matrix. The single-particle spectrum has the relativistic form $E(p) = v_F p$. In the massless (gapless) limit, the spectrum is linear. Hence, time and space scale the same way, $T \sim L$, as required by relativistic invariance: Dirac fermions have dynamic exponent $z = 1$.

The action of a free Dirac field is

$$S = \int d^2x\, \bar{\psi}_\alpha(x) i \gamma^\mu_{\alpha\beta}\, \partial_\mu \psi_\beta(x) \tag{4.28}$$

where we have used the Dirac γ-matrices, $\gamma_0 = \sigma_1$ and $\gamma_1 = -i\sigma_2$, and the notation $\bar{\psi} = \psi^\dagger \gamma^0$ (see Chapters 5 and 6). We have also used the convention that repeated indices are summed. In this relativistic notation the Fermi velocity v_F (the "speed of light") is absorbed in a redefinition of the time coordinate as $x_0 = v_F t$ or, equivalently, set to 1 by an appropriate choice of units.

The theory of free massless Dirac fermions is scale-invariant. The only quantity with units in Eq. (4.27) is the velocity v_F, which, as we said, can be set to 1. It is straightforward to determine the scaling behavior of various local operators at this fixed point. By first generalizing the action of Eq. (4.28) to D space-time dimensions, which now involve a set of D anticommuting *gamma* matrices γ_μ (whose rank depends on the dimension), we see that the Dirac field must scale as $[\psi] = L^{-(D-1)/2}$. It is easy to determine the scaling of simple local operators. The mass term, $\bar{\psi}\psi$, has scaling dimension $\Delta = D - 1$, and the Dirac mass scales as $[m] = L^{-1}$, as expected. This scaling analysis shows that the Dirac current $j_\mu = \bar{\psi}\gamma_\mu\psi$ has scaling dimension $D - 1$.

Local interactions of Dirac fermions have the form of four-fermion operators, such as $(\bar{\psi}\gamma_\mu\psi)^2$ and $(\bar{\psi}\psi)^2$, which have scaling dimension $\Delta = 2(D - 1)$. In particular, in $D = 1 + 1$ dimensions, the current has dimension 1 and the four-Fermi operators have dimension 2. Therefore, in $D = 2$ space dimensions the four-fermion interactions are *marginal* and are *irrelevant* for $D > 2$. The associated coupling constant g scales as $[g] = L^{D-2}$. Hence, the tree-level beta function is

$$\beta(g) = -(D - 2)g + O(g^2) \tag{4.29}$$

and behaves similarly to the non-linear sigma model.

Thus, a theory of free massless Dirac fermions is a fixed point of the renormalization group. At this level this fixed point is marginally stable in $D = 1 + 1$. Much as in the case of the non-linear sigma model, we will see that determining the actual stability of this fixed point requires us to take into account the effects of fluctuations which appear at order $O(g^2)$ in the beta function. On the other hand, for $D > 2$ the free-massless-Dirac-fermion fixed point is stable under small local perturbations and defines a stable phase. Thus, for $D > 2$ space-time dimensions a theory of relativistic fermions with local four-Fermi couplings is perturbatively stable (in the infrared) and has a phase transition to a phase with a non-vanishing mass gap at a critical value of the coupling constant (Wilson, 1973). An example of this is the case of graphene. For $D < 4$ the free Dirac fixed point is unstable with respect to the coupling to a dynamical gauge field. The nature of the fixed point which controls the infrared behavior of this theory is not well understood.

4.3 Scaling behavior of physical observables

One powerful consequence of the renormalization group is that, provided the structure of fixed points is understood, it will enable us to understand how physical observables behave in a physical system. This is the basis of the theory of critical phenomena, both in classical and in quantum-mechanical systems.

4.3.1 The correlation length

We will begin by considering first the behavior of the correlation length ξ for a theory close to a fixed point. The distance to the fixed point is controlled by the (dimensionless) coupling constant λ of a local operator of the theory. By dimensional analysis we can write ξ in the form

$$\xi = af(\lambda) \tag{4.30}$$

where $f(\lambda)$ is a so-far-undetermined dimensionless function of the coupling constant; a is the ultraviolet (UV) cutoff.

By demanding that physical dimensional quantities, such as the correlation length ξ, remain fixed (invariant) under the action of the renormalization group, we can derive a *flow equation* that will allow us to determine, in principle, the function $f(\lambda)$. Thus, the condition

$$\frac{\partial \xi}{\partial \ln a} = 0 = \frac{\partial}{\partial \ln a}(af(\lambda)) \tag{4.31}$$

leads to the differential flow equation for $f(\lambda)$

$$0 = af(\lambda) + a\frac{\partial f}{\partial \lambda}\frac{\partial \lambda}{\partial \ln a} \tag{4.32}$$

where we recognize the presence of the (Gell-Mann–Low) beta function, $\beta(\lambda) = a\partial\lambda/\partial a$ in the last term of Eq. (4.32). We can now write the flow equation as

$$0 = f(\lambda) + \frac{\partial f}{\partial \lambda}\beta(\lambda) \tag{4.33}$$

Flow equations of this type are known as Callan–Symanzik, or renormalization-group, equations.

It is trivial to solve these flow equations, if the beta function is known, since they are first-order partial differential equations. In this simple case we easily find

$$\frac{\partial \ln f}{\partial \lambda} = -\frac{1}{\beta(\lambda)} \tag{4.34}$$

which implies that $f(\lambda)$ must be such that

$$\ln f(\lambda) = \text{constant} - \int \frac{d\lambda'}{\beta(\lambda')} \tag{4.35}$$

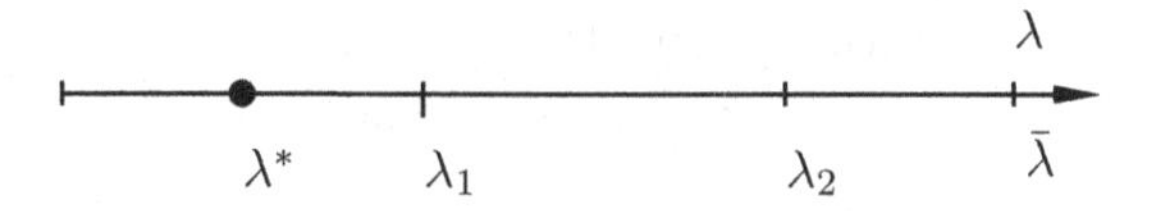

Figure 4.1 Schematic RG flow with a single relevant coupling constant. Here λ^* is the fixed point. See the text for details.

By integrating between two values λ and λ_0 of the coupling constant we obtain the following solution:

$$\ln\left(\frac{f(\lambda)}{f(\lambda_0)}\right) = -\int_{\lambda_0}^{\lambda} \frac{d\lambda'}{\beta(\lambda')} \tag{4.36}$$

We will now consider a theory with a *fixed point* at a particular value λ^* of the dimensionless coupling constant. A fixed point is defined as the value λ^* at which the beta function vanishes, $\beta(\lambda^*) = 0$. See Figure 4.1.

Let us consider the system at two values of the dimensionless coupling constant λ_1 and λ_2, where the correlation length takes the values $\xi(\lambda_1) = af(\lambda_1)$ and $\xi(\lambda_2) = af(\lambda_2)$, respectively. Then, $f(\lambda_1)$ and $f(\lambda_2)$ have the form

$$f(\lambda_1) = f(\tilde{\lambda})\exp\left(\int_{\lambda_1}^{\tilde{\lambda}} \frac{d\lambda}{\beta(\lambda)}\right) \tag{4.37}$$

and

$$f(\lambda_2) = f(\tilde{\lambda})\exp\left(\int_{\lambda_2}^{\tilde{\lambda}} \frac{d\lambda}{\beta(\lambda)}\right) \tag{4.38}$$

where $\tilde{\lambda}$ is some other value of the coupling constant. Thus, the ratio of the correlation lengths $\xi(\lambda_1)$ and $\xi(\lambda_2)$ is determined by the change of the coupling constant:

$$\frac{\xi(\lambda_1)}{\xi(\lambda_2)} = \exp\left(\int_{\lambda_1}^{\tilde{\lambda}} \frac{d\lambda}{\beta(\lambda)} - \int_{\lambda_2}^{\tilde{\lambda}} \frac{d\lambda}{\beta(\lambda)}\right) = \exp\left(\int_{\lambda_1}^{\lambda_2} \frac{d\lambda}{\beta(\lambda)}\right) \tag{4.39}$$

Therefore we find that

$$\xi(\lambda_1) = \xi(\lambda_2)\exp\left(\int_{\lambda_1}^{\lambda_2} \frac{d\lambda}{\beta(\lambda)}\right) \tag{4.40}$$

Close enough to the fixed point at λ^*, we can approximate the beta function as

$$\beta(\lambda) = \beta'(\lambda^*)(\lambda - \lambda^*) + O((\lambda - \lambda^*)^2) \tag{4.41}$$

where $\beta'(\lambda^*) = (d\beta/d\lambda)(\lambda^*)$ is the slope of the beta function at the fixed point. If we take $|\lambda_1 - \lambda^*| \ll \lambda^*$ to be very close to the fixed point and $|\lambda_2 - \lambda^*| \approx \lambda^*$ at some finite distance from it, Eq. (4.39) becomes

$$
\begin{aligned}
\xi(\lambda_1) &= \xi(\lambda_2)\exp\left(\int_{\lambda_1}^{\lambda_2} \frac{d\lambda}{\beta'(\lambda^*)(\lambda-\lambda^*)+\cdots}\right) \\
&= \xi(\lambda_2)\exp\left[\frac{1}{\beta'(\lambda^*)}\ \ln\left(\frac{\lambda_2-\lambda^*}{\lambda_1-\lambda^*}\right)+\cdots\right] \\
&= \xi(\lambda_2)\left(\frac{\lambda_2-\lambda^*}{\lambda_1-\lambda^*}\right)^{1/\beta'(\lambda^*)}(1+\cdots)
\end{aligned}
\tag{4.42}
$$

If the operator with coupling constant λ is a *relevant operator*, then the slope of the beta function must be positive, $\beta'(\lambda^*) > 0$. In this case, provided that this relevant operator drives the system to a phase with a finite correlation length, which we can take to be $\xi(\lambda_2) \approx a$, Eq. (4.42) implies that the correlation length $\xi(\lambda)$ must diverge with a power law as the fixed point is approached, $\lambda \to \lambda^*$,

$$
\xi(\lambda) \approx a\left|\frac{\lambda_2-\lambda^*}{\lambda-\lambda^*}\right|^{\nu}
\tag{4.43}
$$

where

$$
\nu = \frac{1}{\beta'(\lambda^*)}
\tag{4.44}
$$

is the *universal* critical exponent of the correlation length. By universal we mean that ν is independent of the microscopic physics of the systems and, in particular, of the choice of the short-distance cutoff a. In conclusion, the correlation length must diverge $\xi(\lambda) \to \infty$ as the fixed point is approached $\lambda \to \lambda^*$ with a universal critical exponent ν.

4.4 General consequences of scale invariance

We will now discuss some general and important properties of a theory at a fixed point. We noted before that at a fixed point the system becomes scale-invariant, a new symmetry that holds precisely only at the fixed point. We will also assume that the system is spatially homogeneous and isotropic. In what follows we will assume that at this fixed point the correlation length is infinite, $\lim_{\lambda\to\lambda^*}\xi(\lambda) = \infty$ and hence is much bigger than the short-distance cutoff a (the lattice spacing). In this regime we must be able to replace the lattice system with an effective-field theory without a lattice, i.e. we have effectively a continuum system. In quantum field theory, a system with these properties is said to be a conformal field theory.

In this discussion we will adopt a heuristic approach and we will not give a formal proof of a number of (important!) properties that will prove to be very useful. Many of these proofs (and arguments) can be found in the literature, see e.g. Polyakov (1987) and Cardy (1996). However, we will need these properties in order to show the generality of the approach that is used.

Let $\{\phi_n(\vec{r})\}$ be a set of operators of a system at a fixed point with a *scale-invariant action* S^*. Here n is a set of labels that depends on the system. We will assume that under scale transformations these operators transform as $\phi_n(b\vec{r}) = b^{-\Delta_n}\phi_n(\vec{r})$. Operators that transform irreducibly under scale transformations are called primary operators (or fields). The transformation property of the operators under scale transformations dictates the transformation properties of their correlation functions. Thus, scale invariance of the action (and of the partition function) demands that under a global change of scale $\vec{r} \to b\vec{r}$ the correlation functions of all operators will transform simply under scale transformations. This means that all expectation values at the fixed point must be homogeneous functions of the coordinates. A function $F(x)$ is homogeneous of degree k if, under a scale transformation of its variables $x \to bx$, it transforms as $F(bx) = b^k F(x)$.

4.4.1 Scale invariance and correlation functions

Let us now apply this concept to the correlation function of the operator $\phi_n(\vec{r})$ at a fixed point S^*. Below we will denote all expectation values at the fixed point as $\langle \ldots \rangle_*$. We will further assume that the operator is *normal ordered* at the fixed point, $\langle \phi_n(\vec{r}) \rangle_* = 0$, and that the (connected) correlation function *decays* at long separations $|\vec{r} - \vec{r}\,'|$. Then, the assumption of translation, rotation, and scale invariance means that the correlator must have a power-law dependence of the distance $|\vec{r} - \vec{r}\,'|$

$$\langle \phi_n(\vec{r})\phi_n(\vec{r}\,')\rangle_* = \frac{1}{|\vec{r} - \vec{r}\,'|^{2\Delta_n}} \tag{4.45}$$

where Δ_n is the *scaling dimension* of operator $\phi_n(\vec{r})$ at the fixed point, i.e. $[\phi_n] = L^{-\Delta_n}$. Here $-2\Delta_n$ is the degree of the correlator as a homogeneous function. We have normalized the operator so that the possible constant factor in the numerator is set to 1.

The scaling dimension of a local operator can be regarded as a quantum number. Indeed, in treating scale invariance as a *symmetry* we are implicitly assuming that the operators can be chosen to transform *irreducibly* under these transformations, much as angular momentum labels the representations of the group of rotations, and describe how physical observables transform. More generally, scale transformations are a special case of the group of global conformal transformations. Systems that are homogeneous, isotropic, and scale-invariant are also invariant under general global conformal transformations. The operators of a system at a fixed point are then classified by the way they transform under all of these symmetries.

Much as in the case of the group of rotations, operators in a scale-invariant system obey a generalized orthogonality property. Namely, if we consider two operators $\phi_n(\vec{r})$ and $\phi_m(\vec{r}\,')$ with scaling dimensions Δ_n and Δ_m, respectively, then the correlator at the fixed point must have the following form:

$$\langle \phi_n(\vec{x}_1)\phi_m(\vec{x}_2)\rangle_* = \frac{\delta_{\Delta_n,\Delta_m}}{|\vec{x}_1 - \vec{x}_2|^{2\Delta_n}} \tag{4.46}$$

i.e. the correlation function of operators with different scaling dimensions vanishes. This orthogonality condition follows from the following simple observations. Translation and rotational invariance require that the correlator of Eq. (4.46) be a function only of the distance $|\vec{x}_1 - \vec{x}_2|$. Scale invariance now requires that it be a homogeneous function of a distance, i.e. a power law. Thus it cannot be a separate function of each scaling dimension; the only solution is that the two scaling dimensions must be the same and that, otherwise, the correlation function must vanish. From a quantum-mechanical (or field-theory) perspective this relation states that the states created by the action of the operator $\phi_n(\vec{r})$ on the ground state are orthogonal to the states created by the action of $\phi_m(\vec{r}\,')$ on the same ground state if these operators transform differently under scale transformations, i.e. if their scaling dimensions are not equal.

Global scale invariance also constrains the form of three-point correlators. Let $\phi_n(\vec{x}_1)$, $\phi_m(\vec{x}_2)$, and $\phi_k(\vec{x}_3)$ be three operators with scaling dimensions Δ_n, Δ_m, and Δ_k, respectively. It can be shown that scale and conformal invariance require (Belavin *et al.*, 1984; Polyakov, 1987) that the three-point function of these operators must have the form

$$\langle \phi_n(\vec{x}_1)\phi_m(\vec{x}_2)\phi_k(\vec{x}_3)\rangle_* = \frac{C_{nmk}}{|\vec{x}_1 - \vec{x}_2|^{\Delta_{nm}}|\vec{x}_2 - \vec{x}_3|^{\Delta_{mk}}|\vec{x}_3 - \vec{x}_1|^{\Delta_{kn}}} \tag{4.47}$$

where the exponents are

$$\begin{aligned} \Delta_{nm} &= \Delta_n + \Delta_m - \Delta_k \\ \Delta_{mk} &= \Delta_m + \Delta_k - \Delta_n \\ \Delta_{kn} &= \Delta_k + \Delta_n - \Delta_m \end{aligned} \tag{4.48}$$

With the normalization we chose for the two-point functions, the constant C_{nmk} in Eq. (4.47) is a *universal amplitude*.

4.4.2 The operator-product expansion

Let us consider now a general correlation function at a fixed point of the form $\langle \dots \phi_n(\vec{x}_n) \dots \phi_m(\vec{x}_m) \dots\rangle^*$. We will say that the set of operators $\{\phi_n(\vec{x})\}$ is "complete" if inside a general expectation value of this type we can replace the product of a pair of these operators by a series of terms involving operators of the same set, i.e.

$$\lim_{\vec{x}_m \to \vec{x}_n} \phi_m(\vec{x}_m)\phi_n(\vec{x}_n) \equiv \lim_{\vec{x}_m \to \vec{x}_n} \sum_k \frac{C_{nmk}}{|\vec{x}_n - \vec{x}_m|^{\Delta_m+\Delta_n-\Delta_k}} \phi_k\left(\frac{\vec{x}_n + \vec{x}_m}{2}\right) \tag{4.49}$$

Here the limit is understood to mean that the distance $|\vec{x}_n - \vec{x}_m|$ between the operators ϕ_n and ϕ_m is much smaller than their distances from all other operators in the general expectation value. This identity is known as the *operator product-expansion* (or OPE). In particular, the terms of the OPE tell us how different operators *fuse* with another. Thus, the OPE can be understood as a set of *fusion rules* dictated by the scaling dimensions of the operators, $\{\Delta_n\}$, and by the coefficients $\{C_{nmk}\}$, which are known as the *structure constants* of the OPE, the universal amplitudes of the three-point function of the operators ϕ_n, ϕ_m, and ϕ_k. The fusion of two fields ϕ_n and ϕ_m, denoted by the operation $\star$, is summarized in the expression

$$\phi_n \star \phi_m = \sum_k C_{nmk}\phi_k \tag{4.50}$$

4.5 Perturbative renormalization group about a fixed point

We will now see that the scaling dimensions of the operators and the structure constants of the OPE completely determine the form of the renormalization-group beta functions in the vicinity of this fixed point. Let S be the action of a system close to a fixed point with action S^*:

$$S = S^* + \delta S = S^* + \int d^D x \sum_n g_n a^{\Delta_n - D} \phi_n(\vec{x}) \tag{4.51}$$

where, as before, $\{\Delta_n\}$ are the scaling dimensions of the operators $\{\phi_n\}$, $\{g_n\}$ are dimensionless coupling constants, and $a \sim \Lambda^{-1}$ is the short-distance cutoff. We will assume that the perturbing operators $\{\phi_n(\vec{r})\}$ are primary and obey the properties we listed above.

We will now consider the effects of the perturbations contained in δS on the partition function of the system close to the fixed point S^*:

$$Z = \text{tr}\, e^{-S(\phi)} = \text{tr}\left[e^{-S^*(\phi)} e^{-\sum_n \int d^D x\, g_n a^{\Delta_n - D}\phi_n(\vec{x})}\right] \tag{4.52}$$

where the trace indicates a sum (in the sense of a path integral) over the configurations (or histories) of the fields of the system. By expanding in powers of the couplings to low orders, we find

$$
\begin{aligned}
\frac{Z}{Z^*} = 1 &+ \sum_n \int \frac{d^D x}{a^{D-\Delta_n}} g_n \langle \phi_n(\vec{x}) \rangle^* \\
&+ \frac{1}{2!} \sum_{n,m} \int \frac{d^D x_1}{a^{D-\Delta_n}} \int \frac{d^D x_2}{a^{D-\Delta_m}} g_n g_m \langle \phi_n(\vec{x}_1) \phi_m(\vec{x}_2) \rangle^* \\
&- \frac{1}{3!} \sum_{nmk} \int \frac{d^D x_1}{a^{D-\Delta_n}} \int \frac{d^D x_2}{a^{D-\Delta_m}} \int \frac{d^D x_3}{a^{D-\Delta_k}} \\
&\times g_n g_m g_k \langle \phi_n(\vec{x}_1) \phi_m(\vec{x}_2) \phi_k(\vec{x}_3) \rangle^* + \cdots
\end{aligned}
\tag{4.53}
$$

where

$$
Z^* \equiv \operatorname{tr} e^{-S^*(\phi)} \tag{4.54}
$$

is the partition function at a fixed point S^*.

At the fixed point the correlators of the operators $\{\phi_n(x)\}$ acquire the scaling form of Eq. (4.46) and Eq. (4.47) and, as a result, the integrals that appear in each term of this expansion are typically singular. Two types of singularities are present: (a) long-distance or infrared (IR) divergences, which are cut off by a finite size L, and (b) short-distance or ultraviolet (UV) divergences, which are regulated by the short-distance cutoff a. In this form, the expansion formally looks like a partition function of a "gas" of "particles" located at coordinates $\{\vec{x}_k\}$, with the index of the operator n labeling the different "species" of particles. This picture goes back to the formulation of the RG by Kosterlitz. We will work in the regime where the coordinates of these "particles" are separated from each other by distances large compared with the short-distance cutoff and small compared with the linear size of the system, $L \gg |x_1 - x_2| \gg a$.

An RG transformation consists of (a) a change in the short-distance cutoff $a \to ba$ (with $b > 1$) and (b) a rescaling of all (dimensionless!) coupling constants to compensate for this change. In the process we will keep the partition function and the linear size of the system L (the IR cutoff) fixed.

We first rescale the cutoff $a \to ba$ $(b > 1)$ and parametrize the change as $b = e^{\delta l}$ or, equivalently, $\delta l = \ln b$. (See Fig. 4.2.) Hence, $\delta l \to 0$ as $b \to 1$. We will change the UV cutoff a by an amount b in the integrals and we will compensate by changing the couplings while keeping Z fixed. How do we change the coupling constants g_n with Z fixed? To proceed we note that the UV cutoff a appears

(1) in the factors $a^{D-\Delta_n}$ in the action,
(2) as the cutoff in the integrals, and
(3) in the L dependence (L/a) of the integrals.

The effect of the change of the cutoff a in the factors of the action is readily compensated for by a rescaling of the coupling constants g_n. Indeed, under a change

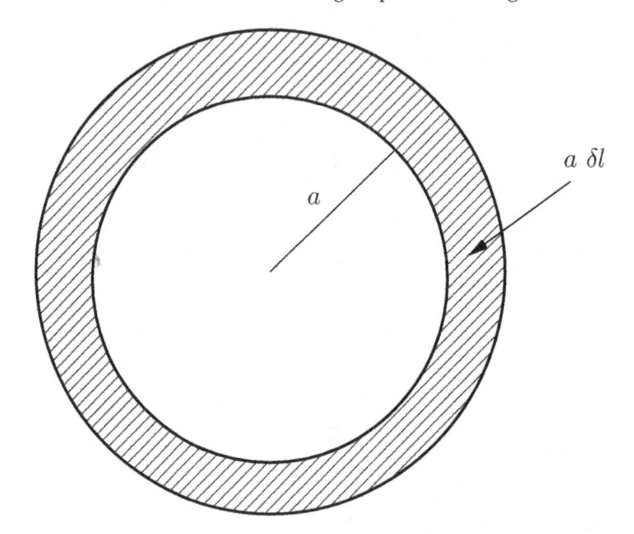

Figure 4.2 Changing the cutoff a by the annular region (shaded) of radial thickness $a\,\delta l$.

$a \to ba$ the factor involving the coupling constant g_n changes as

$$\frac{g_n}{a^{D-\Delta_n}} \to \frac{g_n}{a^{D-\Delta_n} b^{D-\Delta_n}} \tag{4.55}$$

For an infinitesimal change $\delta l = \delta a/a$, this change can be compensated for by a rescaling of the coupling constant g_n:

$$g_n \to b^{D-\Delta_n} g_n = g_n(1 + (\delta l)(D - \Delta_n) + \cdots) \tag{4.56}$$

Hence, at this level, the change in the coupling constant g_n

$$g_n \to g_n + (D - \Delta_n) g_n\, \delta l + \cdots \tag{4.57}$$

is dictated by the scaling dimension Δ_n and the dimensionality D. It is easy to see that this change is equivalent to the tree-level approximation to the beta function for g_n that we discussed above.

Next we look at the effect of changing the UV cutoff a in the integrals. The cutoff is brought into the integrals by restricting the integrations over the coordinates $\{\vec{x}_j\}$ to the range $|\vec{x}_j - \vec{x}_k| > a$. Thus, the integral over two of these coordinates, say $\vec{x}_j$ and $\vec{x}_k$, becomes modified as follows:

$$\begin{aligned} &\int_{|\vec{x}_j-\vec{x}_k|>a(1+\delta l)} d^D x_j\, d^D x_k\, F(\vec{x}_j, \vec{x}_k) \\ &= \int_{|\vec{x}_j-\vec{x}_k|>a} d^D x_j\, d^D x_k\, F(\vec{x}_j, \vec{x}_k) - \int_{a(1+\delta l)>|\vec{x}_j-\vec{x}_k|>a} d^D x_j\, d^D x_k\, F(\vec{x}_j, \vec{x}_k) \end{aligned} \tag{4.58}$$

The change we need to evaluate is then the integral over the annular region $a(1 + \delta l) > |\vec{x}_j - \vec{x}_k| > a$. We will now examine this in each term in the expansion of Eq. (4.53)

The first term of the series of Eq. (4.53) has no singular terms and is affected only by the change of the cutoff in the prefactors that we discussed above. Upon implementing this rescaling, we recover the original contribution to Z. In the second term, in addition to the prefactor rescaling, we get singular contributions in the two-point functions. Here, the change in the UV cutoff leads to the evaluation of the integral over an annular region. Inside each annular region we can compute the two-point function using the OPE, Eq. (4.49),

$$
\begin{aligned}
&\frac{1}{2}\sum_{n,m}\int d^D x_1\, d^D x_2\, \frac{g_n}{a^{D-\Delta_n}}\frac{g_m}{a^{D-\Delta_m}}\langle\phi_n(\vec{x}_1)\phi_m(\vec{x}_2)\rangle^* \\
&\equiv \frac{1}{2}\sum_{n,m,k}\int d^D x_1\, d^D x_2\, \frac{g_n\, g_m}{a^{2D-\Delta_n-\Delta_m}}\frac{C_{nmk}\langle\phi_k(\frac{1}{2}(\vec{x}_1+\vec{x}_2))\rangle^*}{|\vec{x}_1-\vec{x}_2|^{\Delta_n+\Delta_m-\Delta_k}+\cdots} \\
&= \sum_{nmk} C_{nmk} a^{\Delta_k-\Delta_n-\Delta_m} g_n g_m \int d^D x\langle\phi_k(\vec{x})\rangle^* \frac{S_D a^D\, \delta l}{a^{2D-\Delta_n-\Delta_m}} + \cdots \\
&\equiv \frac{1}{2}\sum_{n,m}\sum_k g_n g_m C_{nmk}\int \frac{d^D x}{a^{D-\Delta_k}}\langle\phi_k(\vec{x})\rangle^* S_D\, \delta l
\end{aligned}
\tag{4.59}
$$

where $S_D = 2\pi^{D/2}/\Gamma(D/2)$ is the area of a hypersphere in D dimensions, and $\Gamma(z)$ is the Euler gamma function.

From the form of Eq. (4.59), we see that the net effect of the second-order ("one-loop") contributions amounts to an additional rescaling (or renormalization) of the coupling constants

$$
g_k \to g_k - \frac{1}{2}S_D\sum_{n,m} g_n g_m C_{nmk}\, \delta l + \cdots \tag{4.60}
$$

It is straightforward to see that this repeats to all orders (cf. cubic terms and higher in Eq. (4.53)).

By combining the tree-level and one-loop rescalings we obtain a general form of the beta functions for all the coupling constants $\{g_k\}$:

$$
\frac{dg_k}{dl} = (D-\Delta_k)g_k - \frac{1}{2}S_D\sum_{nm} C_{nmk} g_n g_m + \cdots \tag{4.61}
$$

We can simplify this expression by absorbing the phase-space factors in a further simple redefinition of the coupling constants, $g_k \to (2/S_D)g_k$, and find the final form of the one-loop beta functions

$$\beta(g_k) = \frac{dg_k}{dl} = (D - \Delta_k)g_k - \sum_{nm} C_{nmk} g_n g_m + \cdots \tag{4.62}$$

We see that, in addition to the linear term which follows from knowledge of the scaling dimensions $\{\Delta_k\}$ (and the dimensionality D), the one-loop contribution amounts to quadratic (bilinear) terms in the coupling constants whose coefficients are given by the structure constants of the OPE, $\{C_{nmk}\}$. All one-loop beta functions of all perturbative RG transformations have this structure.

4.6 The Kosterlitz renormalization group

We will use this approach to derive the RG flow for the sine–Gordon (SG) model in $1 + 1$ dimensions. This is the Kosterlitz RG flow (Kosterlitz, 1974). As will be explained in Chapters 5 and 6, this RG appears in the theory of one-dimensional (1D) quantum antiferromagnets and Luttinger liquids. It was derived originally by Kosterlitz to explain the Kosterlitz–Thouless transition in two-dimensional (2D) classical superfluids and XY magnets (Kosterlitz and Thouless, 1973; José *et al.*, 1977). An RG with the same structure was found some years earlier by Anderson and Anderson, Yuval, and Hamann in their work on the Kondo problem (Anderson, 1970; Anderson *et al.*, 1970).

The SG model is a theory of a scalar field ϕ in $1 + 1$ dimensions whose Lagrangian density (in Euclidean space-time) is

$$\mathcal{L} = \frac{1}{2}(\partial\phi)^2 + \frac{u}{a^2}\cos(\beta\phi) \tag{4.63}$$

Here u is the dimensionless coupling constant and β is a (real) parameter. The first (free-field) term of the SG Lagrangian, Eq. (4.63), is invariant under shifts of the field, $\phi \to \phi + \alpha$, where α is arbitrary. The cosine term of the SG Lagrangian breaks this continuous symmetry to a discrete symmetry,

$$\phi \to \phi + 2\pi n R \tag{4.64}$$

where n is an integer and $R = 1/\beta$ is known as the "compactification radius," using a terminology borrowed from string theory. Alternatively, we can define a rescaled field φ, $\beta\phi = \varphi$, which has compactification radius 1. The SG (Euclidean) Lagrangian takes the form

$$\mathcal{L} = \frac{K}{2}(\partial\varphi)^2 + \frac{u}{a^2}\cos\varphi \tag{4.65}$$

where we introduced the stiffness $K = 1/\beta^2$. This form of the SG Lagrangian will be useful for our discussion of its behavior under scaling.

To construct a renormalization group transformation for the SG theory we will follow the approach discussed in the preceding sections. We will consider this problem in 2D Euclidean space, where it becomes equivalent to a problem in classical statistical mechanics. In a 2D Euclidean space-time it is convenient to express the Cartesian coordinates in terms of the complex coordinates $z = x + iy$ and the complex conjugate $\bar{z} = x - iy$.

The expansion of the partition function of the SG theory in powers of the coupling constant g has the same form as Eq. (4.53). On writing the cosine as a sum of two vertex operators, $\cos\varphi = \frac{1}{2}\left(e^{i\varphi} + e^{-i\varphi}\right)$, we see that the equivalent gas of particles has now two "species" or charges, ± 1, one for each type of vertex operator. On the other hand, since we are expanding about the fixed "point" (actually a fixed line, as we will see below) of the free massless scalar field in two Euclidean space-time dimensions, the propagator is a logarithmic function of the distance, and the effective interaction between the charges is also logarithmic. Therefore, the equivalent system is a 2D two-component Coulomb gas at total charge neutrality with the partition function

$$Z_{\text{CG}}^{(2)} = \sum_{\{m(\vec{r})=\pm 1\}}{}' z^{\sum_{\vec{r}} m^2(\vec{r})} \exp\left(-\frac{1}{2T_{\text{eff}}} \sum_{\vec{r},\vec{r}\,'} m(\vec{r}) U_{\text{eff}}(|\vec{r} - \vec{r}\,'|) m(\vec{r}\,')\right) \tag{4.66}$$

The effective interaction of the 2D Coulomb gas is

$$U_{\text{eff}}(r) = \frac{1}{4\pi} \ln\left(\frac{r}{a}\right) \tag{4.67}$$

with an effective "temperature" T_{eff} and "fugacity" z

$$T_{\text{eff}} = K = \frac{1}{\beta^2}, \qquad z = \frac{u}{2a^2} \tag{4.68}$$

Since the effective interaction diverges logarithmically at large distances, the weight of the configurations $\{m(\vec{r})\}$ that do not obey the charge-neutrality condition, $\sum_{\vec{r}} m(\vec{r}) = 0$, vanishes identically. The prime label on the sum of Eq. (4.66) indicates that the sum runs only over overall charge-neutral configurations. On the other hand, the effective interaction also diverges (logarithmically) at short distances and hence needs to be regularized. This is done by imposing a short-distance cutoff in the integrals, which, as in the preceding section, amounts to a "hard-core" condition at some short distance a. Thus we will cut off the integrals at that distance and absorb the short-distance behavior of the interaction in a redefinition of the fugacity.

This problem was first solved by Kosterlitz and Thouless (1973) in their theory of the thermal superfluid phase transition in two dimensions. In the superfluid case (as in the physically equivalent classical 2D XY model), the Coulomb charges represent topologically non-trivial configurations known as *vortices*. In the SG theory (and in 1D antiferromagnets and Luttinger liquids) the Coulomb charges are *instantons*. This is the Kosterlitz–Thouless transition (which also describes the roughening transition of crystal surfaces). The general problem of the D-dimensional two-component Coulomb gas was solved (using a perturbative RG) by Kosterlitz (1977).

This system has two phases separated by a (Kosterlitz–Thouless) phase transition. For low $T_{\text{eff}} \leq T_{\text{c}}$ (where T_{c} is an effective critical temperature), the logarithmic interaction forces the Coulomb charges to be rare and to be bound in neutral dipole configurations. This is the dielectric phase of the 2D neutral Coulomb gas, i.e. the "2D superfluid" phase. In contrast, for $T_{\text{eff}} > T_{\text{c}}$ the Coulomb charges *proliferate* and the long-range Coulomb interaction becomes (Debye) screened. This is the *plasma* phase of the 2D Coulomb gas. This is the normal phase, with a finite correlation length, namely the screening length of the Coulomb gas.

Here we will use the OPE approach to the perturbative RG discussed in the preceding section. It requires us (as in Kosterlitz's analysis) to expand about a specific value of the stiffness K where the cosine operator is marginal. We will see below how this physically intuitive picture works out in this approach.

The correlation function of the free field φ can be expressed as the sum of a holomorphic function $G_{\text{R}}(z)$ and an anti-holomorphic function $G_{\text{L}}(\bar{z})$,

$$\langle \varphi(0,0)\varphi(\vec{x})\rangle \equiv G(\vec{x}) = G_{\text{R}}(z) + G_{\text{L}}(\bar{z}) \tag{4.69}$$

where

$$G_{\text{R}}(z) = -\frac{1}{4\pi K} \ln z \tag{4.70}$$

and likewise for $G_{\text{L}}(\bar{z})$. Similarly, the field $\varphi(x, y)$ can be decomposed into a sum of holomorphic φ_{R} and anti-holomorphic φ_{L} components,

$$\varphi(x, y) \equiv \varphi_{\text{R}}(z) + \varphi_{\text{L}}(\bar{z}) \tag{4.71}$$

which satisfy

$$\partial_{\bar{z}}\varphi_{\text{R}} = \partial_{z}\varphi_{\text{L}} = 0 \tag{4.72}$$

and whose correlators are given by

$$\langle \varphi_{\text{R}}(0)\varphi_{\text{R}}(z)\rangle = G_{\text{R}}(z), \qquad \langle \varphi_{\text{L}}(0)\varphi_{\text{L}}(z)\rangle = G_{\text{L}}(\bar{z}) \tag{4.73}$$

Upon analytic continuation from 2D Euclidean space to $(1 + 1)$-dimensional Minkowski space-time, i.e. real time $iy \to -vt$ (with velocity v), we can identify the holomorphic component of the field with its *right-moving* component, which is a function of $x - vt$, and the anti-holomorphic component with its *left-moving* component, which is a function only of $x + vt$. The holomorphic and anti-holomorphic components, $\varphi_{\rm R}$ and $\varphi_{\rm L}$, are known as chiral bosons and will play an important role in our discussion of several systems.

In order to construct an RG transformation we first need to determine the scaling dimensions of the operators of interest. We begin with the *vertex operator* $V_n(x) = \exp(in\varphi(x))$, whose correlation functions are

$$\left\langle e^{in\varphi(0)} e^{-in\varphi(\vec{x})} \right\rangle = \exp\left(-\frac{n^2}{2} \left\langle (\varphi(0) - \varphi(|\vec{x}|))^2 \right\rangle \right) \tag{4.74}$$

This expression has a formal singularity in the $G(0)$ and requires a regularization (or cutoff). We will introduce the regularized correlation function

$$G_{\rm reg}(x) \equiv -\frac{1}{4\pi K} \ln\left(\frac{|\vec{x}|^2 + a^2}{a^2} \right) \tag{4.75}$$

where a is the short-distance cutoff, and by virtue of which $G_{\rm reg}(0) = 0$. The correlation function of the vertex operators becomes

$$\begin{aligned} \left\langle e^{in\varphi(0)} e^{-in\varphi(\vec{x})} \right\rangle &= e^{n^2 (G_{\rm reg}(|\vec{x}|) - G_{\rm reg}(0))} \\ &= \exp\left[-\frac{n^2}{4\pi K} \ln\left(\frac{|\vec{x}|^2 + a^2}{a^2} \right) \right] \\ &= \left(\frac{1}{|\vec{x}|} \right)^{n^2/(2\pi K)} = \left(\frac{1}{z} \right)^{n^2/(4\pi K)} \left(\frac{1}{\bar{z}} \right)^{n^2/(4\pi K)} \end{aligned} \tag{4.76}$$

where we have rescaled the operator to absorb the power of the cutoff a.

From this result we can read off the scaling dimension Δ_n of the vertex operator $V_n(\vec{x})$ to be

$$\Delta_n = \frac{n^2}{4\pi K} = \frac{n^2 \beta^2}{4\pi} \tag{4.77}$$

According to our general rules, the cosine operator $\frac{1}{2}(V_1 + V_{-1})$ is *irrelevant* if $\Delta_1 > 2$, *relevant* if $\Delta_1 < 2$, and marginal if $\Delta_1 = 2$. Hence, there is a critical value of the stiffness $K_{\rm c} = 1/(8\pi)$ or, equivalently, $\beta_{\rm c} = 1/R_{\rm c} = \sqrt{8\pi}$, at which the cosine operator is marginal. Since we are constructing a perturbative RG, we will have to assume that the stiffness K is *close* to its critical value $K_{\rm c}$. This is the Kosterlitz–Thouless (KT) transition.

This theory has an exactly marginal operator for all values of K. To see this, let us compute the scaling dimension Δ_0 of the (normal-ordered) operator, $: (\partial\varphi)^2 : \equiv (\partial\varphi)^2 - \langle (\partial\varphi)^2 \rangle$. We need to compute the correlation function

$$\left\langle : (\partial\varphi)^2(\vec{x}) :: (\partial\varphi)^2(\vec{y}) : \right\rangle = \left(\frac{1}{\pi K} \right)^2 \frac{1}{|\vec{x} - \vec{y}|^4} \tag{4.78}$$

which tells us that its scaling dimension is $\Delta_0 = 2$. Hence, this is a *marginal* operator.

Since the free-field action $S_* = \int (K/2)(\partial\varphi)^2$ is scale-invariant *for all values of* K, and since the scaling dimension of the vertex operator O_n varies *continuously* with K, we conclude that this theory has a *line of fixed points* rather than an isolated fixed point. Notice that the scaling dimension of $(\partial\varphi)^2$ *does not vary* and hence it remains marginal.

Our next task is to compute the coefficients of the OPE. We begin with the marginal operator O_0. Since the fixed-point action is a free field, the three-point function of the operator $:(\partial\varphi)^2:$ with itself vanishes. It then follows that the OPE coefficient of this operator with itself also vanishes,

$$C_{0,0,0} = 0 \tag{4.79}$$

On the other hand, the OPE of the vertex operators $V_n(\vec{x})$ is

$$\lim_{\vec{y}\to\vec{x}} :V_n(\vec{x})::V_{-n}(\vec{y}): = \frac{1}{|\vec{x}|^{2\Delta_n}} - \frac{1}{|\vec{x}-\vec{y}|^{2\Delta_n-2}}\frac{n^2}{4K} :(\partial\varphi(\vec{x}))^2: + \cdots \tag{4.80}$$

$$\lim_{\vec{y}\to\vec{x}} :V_n(\vec{x})::V_m(\vec{y}): = \frac{1}{|\vec{x}-\vec{y}|^{\Delta_n+\Delta_m-\Delta_{n+m}}} :V_{n+m}(\vec{x}): + \cdots \tag{4.81}$$

From Eq. (4.80) and Eq. (4.81), we see that the non-vanishing OPE coefficients $C_{n,-n,0}$ and $C_{n,m,-(n+m)}$ are

$$C_{n,-n,0} = -\frac{n^2}{4K} = -\pi\Delta_n, \qquad C_{n,m,-(n+m)} = 1 \tag{4.82}$$

In particular, Eq. (4.80) implies that there is a finite renormalization of the stiffness. On the other hand, since the scaling dimension of the vertex operators is a continuous function of the stiffness, there is also a non-linear feedback on their scaling behavior as well.

We can now derive our perturbative RG. We will consider a general perturbation of the form

$$\mathcal{L}_{\text{int}} = u\cos(n\varphi(x)) = \frac{u}{2}(V_n(x) + V_{-n}(x)) \tag{4.83}$$

although in this chapter we are interested only in the case $n = 1$. Since our procedure is perturbative, we will expand about the critical value of the stiffness $K_n = n^2/(8\pi)$ at which the operators $V_{\pm n}$ are marginal. Let us parametrize the distance to this critical value by $x(K) = 2 - \Delta_n(K) = 2 - n^2/(4\pi K)$, which satisfies $x(K_n) = 0$. The perturbative RG equations are

$$a\frac{du}{da} = (2 - \Delta_n(K))u + \cdots \tag{4.84}$$

The OPE of the vertex operators $V_{\pm n}$ generates a renormalization of the stiffness

$$a\,\frac{dK}{da} = -4\pi C_{n,-n,0}(K_n)\frac{u^2}{4} = \frac{n^2\pi}{4K} \tag{4.85}$$

where the factor of 4π arises from two contributions, a space phase factor and the definition of the stiffness. On the other hand, from the definition of the variable x, we have

$$a\,\frac{dx}{da} = \left(\frac{n^2}{4\pi K^2}\right) a\,\frac{dK}{da} = \frac{n^4}{16K^3}u^2 + \cdots \tag{4.86}$$

Near marginality, where these results are reliable, the RG beta functions, Eq. (4.84) and Eq. (4.86), can be recast in the more compact form

$$a\,\frac{du}{da} = xu + \cdots \tag{4.87}$$

$$a\,\frac{dx}{da} = A(n)^2u^2 + \cdots \tag{4.88}$$

where $A(n)^2 = 32\pi^2/n^2$. Up to a redefinition of the coupling constant, these RG equations are known as the Kosterlitz renormalization group.

The RG flows of the SG theory are the solutions of the Kosterlitz RG equations and are shown in Fig. 4.3. With the sign conventions we are using the direction of

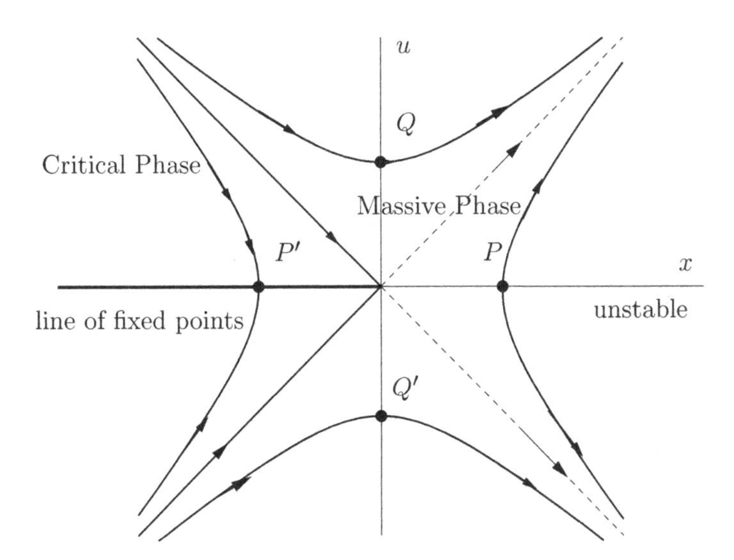

Figure 4.3 The Kosterlitz RG flow and phase diagram; u is the SG dimensionless coupling constant and $x = 2 - n^2/(4\pi K)$, where K is the stiffness. The phase boundaries are the asymptotes $u = \pm x/A(n)$ for $x < 0$. Here $P = A(n)\sqrt{|C|}$, $P' = -A(n)\sqrt{|C|}$, $Q = \sqrt{|C|}$, and $Q' = -\sqrt{|C|}$.

the flows is towards the infrared (long distances). It is straightforward to show that the RG trajectories are the hyperbolas

$$u^2 - \frac{x^2}{A(n)^2} = C \tag{4.89}$$

For $C > 0$ these hyperbolas intersect the u axis, whereas for $C < 0$ they intersect the x axis. The solutions of the RG equations are

$$x(\ell) = A(n)\sqrt{C}\,\tan(A(n)\sqrt{C}\ell), \quad u(\ell) = \pm\frac{\sqrt{C}}{\cos(A(n)\sqrt{C}\ell)}, \quad C > 0 \tag{4.90}$$

$$x(\ell) = A(n)\sqrt{|C|}\coth(A(n)\sqrt{|C|}\ell), \quad u(\ell) = \pm\frac{\sqrt{|C|}}{\sinh(A(n)\sqrt{|C|}\ell)}, \quad C < 0 \tag{4.91}$$

whereas for $C = 0$ they degenerate into the asymptotes $u = \pm x/A(n)$.

We will now discuss the case $n = 1$, although (with some minor changes) the conclusions are general. The RG flow is clearly symmetric with respect to the x axis. This is natural since the changing of the sign of the coupling constant $g \to -g$ can be compensated for by a shift of the field $\varphi \to \varphi + \pi$. Hence, it is sufficient to consider only the upper half-plane, $u \geq 0$. The most important feature of these flows is that for values of u and x located below or at the asymptote $u = -x/A(1)$ the cosine term in the Lagrangian is irrelevant and the system flows towards the *stable line of fixed points* at $u = 0$ and $x \leq 0$. For values outside this region (including the *unstable line* $u = 0$ and $x > 0$) the cosine term is *relevant* and the system flows to large (and positive) values of u and x, converging on the other asymptote, $u = x/A(1)$. Thus the SG theory describes two *phases* (see Fig. 4.3): (a) a *critical phase* and (b) a *massive* phase.

1. In the *critical phase* all correlation functions exhibit power-law behavior with a non-universal exponent equal to twice the scaling dimension of the operator. The exponent is non-universal and varies continuously with the stiffness, reaching a universal value at the end of the fixed line at $x = 0$ (i.e. at $K_c = 1/(8\pi)$).
2. In the massive phase the cosine operator is relevant and the system flows to strong coupling. In this regime the continuous shift symmetry of the field is broken and (qualitatively) the field is pinned at values $2\pi n$ ($n \in \mathbb{Z}$). In this phase the fluctuations are massive and the correlation length is finite. In this system the way the correlation length ξ scales is more complex than what we discussed in the preceding sections. Thus, along the unstable trajectory on the asymptote $u = x/A(1)$, the correlation length diverges as $x \to 0^+$ with an essential singularity $\xi(x) \sim \exp(1/x)$, whereas away from the asymptote it has a different behavior

(still with an essential singularity!) given by $\xi(x) \sim \exp(A/\sqrt{|\delta x|})$, where δx is the distance to the stable asymptote (the phase boundary of Fig. 4.3) and $A > 0$ is a non-universal constant. These results follow from the solutions of the RG flow.

3. How do we know that the correlation length is finite in the regime in which the cosine operator becomes relevant? Or, which amounts to the same thing, that the 2D Coulomb gas has a plasma phase with a finite screening length? At the level of the perturbative RG we are implementing this is an assumption whose justification is beyond the reaches of perturbation theory. We know that this is correct for several reasons. One is that the (quantum) SG field theory is integrable and its spectrum (and scattering matrix) have been computed explicitly using a Bethe-ansatz approach similar to the one we will discuss in the next chapter. From the Bethe-ansatz results in SG theory (Zamolodchikov and Zamolodchikov, 1979; Faddeev, 1984) and in the (equivalent) massive Thirring model (Bergknoff and Thacker, 1979) and earlier semi-classical results (Dashen *et al.*, 1975; Rajaraman, 1985), it is known that, in the regime in which the cosine operators become relevant, $\beta^2 < 8\pi$, the spectrum of the SG theory is massive. It consists of a boson of (renormalized) mass M (i.e. the fluctuations about any one of the classical ground states of the SG theory), which, as discussed above, generally vanishes as $M \sim \exp\left(-A/\sqrt{|\delta x|}\right)$ as $\beta^2 \to 8\pi$ from below. In addition to these boson states, for $\beta^2 \leq 4\pi$ there is also a set of soliton states (and soliton bound states) with a finite mass M_n given by

$$M_n = M \sin\left(n \frac{\pi\beta^2/2}{\sqrt{1-\beta^2/8\pi}}\right), \qquad n = 1, \ldots < \frac{8\pi}{\beta^2}\sqrt{1-\frac{\beta^2}{8\pi}} \tag{4.92}$$

We will see in Chapter 5 that abelian bosonization methods show that at $\beta^2 = 4\pi$ the SG theory is equivalent to a theory of massive free Dirac fermions.

This analysis implies that there is a finite mass (energy) gap in the spectrum and hence that the correlation length is finite. Notice that, as expected, the mass gap vanishes as $\beta^2 \to 8\pi$ (from below) Similarly, from the behavior of the lattice models of interest we know that in this regime the system flows to a phase with a finite energy gap and a finite correlation length.

We will see in Chapters 5 and 6 that the set of trajectories that are important (and therefore the scaling behavior) depends on the problem of interest.

5

One-dimensional quantum antiferromagnets

In this chapter we will discuss the physical behavior of 1D quantum antiferromagnets. It is worthwhile to study them for several reasons: (a) in many cases we have exact solutions (which are lacking in higher dimensions), (b) they exhibit a wealth of ground states, including disordered phases, and (c) they are a natural testing ground for methods and approximations. We shall first discuss the spin one-half Heisenberg chain and later discuss its generalization to (a) higher spin-S and (b) other symmetry groups.

5.1 The spin-1/2 Heisenberg chain

Consider the Heisenberg model on a one-dimensional chain of N sites. The Hamiltonian is

$$H = J \sum_{n=1}^{N} \vec{S}(n) \cdot \vec{S}(n+1) \tag{5.1}$$

where $J > 0$. I will assume that N is an even integer and that we have periodic boundary conditions. Much of what we know about this system comes from (a) the Bethe-ansatz solution for the ground state (Bethe, 1931) and excitation spectrum (Yang and Yang, 1969), (b) mapping to the sine–Gordon theory (Luther and Peschel, 1975), (c) non-abelian bosonization (Affleck, 1985), and (d) mapping to the sigma model (Haldane, 1983a, c).

The exact solution via Bethe ansatz is very peculiar to one-dimensional integrable systems and hence is not generalizable. The other methods are also very specific to one dimension but they are more generally applicable, and higher-dimensional versions of them are currently being developed. Thus we shall concentrate mainly on them. The mapping to the sine–Gordon system is based on the abelian bosonization transformation (Lieb and Mattis, 1965; Coleman, 1975; Luther and Peschel, 1975; Mandelstam, 1975). In a deep sense, it is a particular case of the non-abelian bosonization developed by Witten (1984) and by Polyakov and Wiegmann (1984), and applied to spin systems by Affleck (1985). The main advantage of all these approaches is that they are non-perturbative: they yield the exact behavior of the ground-state properties at long distances, and, in principle, one can find the low-energy spectrum. One important feature is the existence, in addition to spin waves, of soliton states. These states are highly extended configurations of spins that cannot be created locally and that comprise the lowest portion of the spectrum of these systems.

5.1.1 The Bethe-ansatz solution

I will not attempt to give a detailed description of the Bethe-ansatz solution, which is fairly technical. A good summary can be found in the Les Houches Lectures of 1982, in particular the articles by Faddeev (1984) and Lowenstein (1984).

Here I will review very quickly the method as given by Lowenstein. The main idea is to consider the wave function for a pure state of N spins one-half, each labeled by an index $s(n) = \pm\frac{1}{2}$ $(n = 1, \ldots, N)$. The total spin of the system is $\vec{S} = \sum_{n=1}^{N} \vec{S}(n)$. We will consider states $\Psi(s(1), \ldots, s(n))$ in which $(N - M)$ spins are up $(+\frac{1}{2})$ and M are down $(-\frac{1}{2})$. Thus, the total z-component of the spin is

$$S_z \Psi(s(1), \ldots, s(N)) = \left(\sum_{n=1}^{N} s(n) \right) \Psi(s(1), \ldots, s(N)) \tag{5.2}$$

with

$$\sum_{n=1}^{N} s(n) = \frac{N}{2} - M \tag{5.3}$$

We shall denote by

$$\Psi(s(1), \ldots, s(N)) \equiv \phi(x_1, \ldots, x_M) \tag{5.4}$$

a state with the jth down spin located at the site $x_j (1 \leq x_1 \leq \ldots \leq x_j \leq \ldots \leq x_M \leq N)$. Thus, if Ψ_0 is the *ferromagnetic* state

$$\Psi_0 = |\uparrow \ldots \uparrow\rangle \tag{5.5}$$

the most general state with M spins down has the form

$$\Psi = \sum_{\{x_j\}} \phi(x_1, \ldots, x_M) S^-(x_1) \ldots S^-(x_M) \Psi_0 \tag{5.6}$$

where $S^-(n)$ is the lowering operator at site n.

The Heisenberg model is translationally invariant and on a chain with periodic boundary conditions has the translation symmetry in which the nth site is identified with the $(N+n)$th site. Thus we can look for a basis in which not only are $\vec{S}^2$ and S_z diagonal but also the cyclic permutation operator $\tilde{P}$, given by

$$\tilde{P}\Psi(s(1),\ldots,s(N)) = \Psi(s(N),s(1),\ldots,s(N-1)) \tag{5.7}$$

is diagonal.

5.1.2 The basis functions

Bethe's method begins by first writing the Hamiltonian in terms of a spin-exchange operator $P_{n,m}$, where

$$P_{n,m}\Psi(s(1),\ldots,s(n),\ldots,s(m),\ldots,s(N)) \\ = \Psi(s(1),\ldots,s(m),\ldots,s(n),\ldots,s(N)) \tag{5.8}$$

in the form

$$H = J\sum_{n=1}^{N}\left(P_{n,n+1}-1\right) \tag{5.9}$$

with periodic boundary conditions. Consider first a state with one spin down

$$\Psi(s_1,\ldots,s_N) = \sum_{p=1}^{N}\phi(p)|\uparrow\ldots\downarrow\ldots\uparrow\rangle \tag{5.10}$$

where the spin at site p is down.

By using the cyclic translation operator $\tilde{P}$ we see that its main effect is just to shift the location of the down spin by one. Thus an eigenstate of $\tilde{P}$ with eigenvalue μ should satisfy

$$\tilde{P}\phi(p) = \phi(p+1) = \mu\phi(p) \quad \text{for} \quad p = 1,\ldots,N-1 \tag{5.11}$$

and

$$\tilde{P}\phi(N) = \phi(1) = \mu\phi(N) \tag{5.12}$$

Hence

$$\phi(p) = \mu^{p-1}\phi(1) \tag{5.13}$$

and, if we set $\phi(1) = 1$, we get

$$\phi(p) = \mu^{p-1} \tag{5.14}$$

and in particular μ must satisfy

$$1 = \phi(1) = \phi(N+1) = \mu^N \tag{5.15}$$

i.e. it is an Nth root of unity.

Now, a state with one spin down can be a member either of the multiplet with highest total spin $N/2$ or of the multiplet with total spin $N/2 - 1$. In the latter case the state is the highest-weight state in the multiplet and satisfies

$$S^+\Psi = 0 \tag{5.16}$$

where

$$S^+ = \sum_{n=1}^{N} S^+(n) \tag{5.17}$$

Thus we get

$$\begin{aligned} S^+\Psi &= \sum_{p=1}^{N} \phi(p) S^+ |\uparrow \dots \downarrow \dots \uparrow\rangle \\ &= \sum_{p=1}^{N} \phi(p) S^+(p) |\uparrow \dots \downarrow \dots \uparrow\rangle \\ &= \left(\sum_{p=1}^{N} \phi(p) \right) \Psi_0 \end{aligned} \tag{5.18}$$

Using Eqs. (5.14) and (5.16), we obtain

$$0 = \sum_{p=1}^{N} \phi(p) = \sum_{p=1}^{N} \mu^{p-1} = \frac{1-\mu^N}{1-\mu} \tag{5.19}$$

Thus $\mu^N = 1$ and $\mu \neq 1$. Hence we found $N - 1$ members of the spin-($N/2 - 1$) *multiplet* (the other members of the multiplet can be found by applying S^-). The case $\mu = 1$ represents the state $S = N/2$ which belongs to the multiplet of the ferromagnetic state.

If we now consider the case of M spins down, we can still find states with $S = S_z$. They satisfy

$$\tilde{P}\phi(p_1, \dots, p_M) = \phi(p_1 + 1, \dots, p_M + 1) = \lambda\phi(p_1, \dots, p_M) \qquad (p_M < N) \tag{5.20}$$

and

$$\tilde{P}\phi(p_1, \dots, p_{M-1}, N) = \phi(1, p_1 + 1, \dots, p_{M-1} + 1) = \lambda\phi(p_1, \dots, p_{M-1}, N) \tag{5.21}$$

We look for wave functions that are *products* of *single* down-spin wave functions

$$\phi(p_1, \dots, p_M) = \mu_1^{p_1-1} \dots \mu_M^{p_M-1} \tag{5.22}$$

By choosing

$$\lambda = \prod_{j=1}^{M} \mu_j \tag{5.23}$$

we can satisfy Eq. (5.20). However, as it stands, this ansatz does *not* satisfy Eq. (5.21), but if we permute the *order* of the parameters $\mu_1, \ldots, \mu_M$ we can find a solution. Thus, Bethe introduced the *Bethe-ansatz* solution

$$\phi(p_1, \ldots, p_M) = \sum_{P \in \mathcal{S}_M} A_P \mu_{P1}^{n_1} \ldots \mu_{PM}^{n_M} \tag{5.24}$$

where $n_j = p_j - 1$, and P belongs to the permutation group $\mathcal{S}_M$ (i.e. $(P1, \ldots, PM)$ is a permutation of $(1, \ldots, M)$). Now everything is consistent, provided that the identity

$$A_{P\tilde{Q}^{-1}} \equiv A_{PM,P1,\ldots,P(M-1)} = A_P \mu_{PM}^N \tag{5.25}$$

where $(\tilde{Q}1, \tilde{Q}2, \tilde{Q}3, \ldots, \tilde{Q}M) = (2, 3, 4, \ldots, M, 1)$ holds. The demand that ϕ be a highest-weight state with $S = N/2 - M \equiv S_z$ yields the constraint (Lowenstein, 1984)

$$\frac{A_{P'}}{A_P} = -\frac{2\mu_{P(k+1)} - \mu_{Pk}\mu_{P(k+1)} - 1}{2\mu_{Pk} - \mu_{Pk}\mu_{P(k+1)} - 1} \tag{5.26}$$

for all k and for all pairs of permutations P and P' such that $(P'1 \ldots P'kP'(k+1) \ldots P'M) \equiv (P1 \ldots P(k+1)Pk \ldots PM)$.

Define χ_j,

$$\mu_j = \frac{\chi_j + i\pi/2}{\chi_j - i\pi/2} \tag{5.27}$$

Then one finds

$$\frac{A_{P'}}{A_P} = \frac{\chi_{P'j} - \chi_{Pj} + i\pi}{\chi_{P'j} - \chi_{Pj} - i\pi} \tag{5.28}$$

By combining Eqs. (5.27) and (5.25) we get

$$\left(\frac{\chi_{PM} + i\pi/2}{\chi_{PM} - i\pi/2}\right)^N = \frac{A_{PMP1\ldots P(M-1)}}{A_{P1P2\ldots PM}} \tag{5.29}$$

By using Eq. (5.28) repeatedly we obtain

$$\left(\frac{\chi_{PM} + i\pi/2}{\chi_{PM} - i\pi/2}\right)^N = \prod_{j=1}^{M-1} \left(\frac{\chi_{PM} - \chi_{Pj} + i\pi}{\chi_{PM} - \chi_{Pj} - i\pi}\right) \tag{5.30}$$

Since this equation should be valid for all permutations P, we get the *Bethe-ansatz equations*

$$\left(\frac{\chi_j + i\pi/2}{\chi_j - i\pi/2}\right)^N = -\prod_{l=1}^{M}\left(\frac{\chi_j - \chi_l + i\pi}{\chi_j - \chi_l - i\pi}\right) \tag{5.31}$$

Also we see that

$$\lambda^N \equiv \prod_j \left(\frac{\chi_j + i\pi/2}{\chi_j - i\pi/2}\right)^N = 1 \tag{5.32}$$

Thus, for all M, the eigenvalue of $\tilde{P}$ is an Nth root of unity. For a given value of $S = N/2 - M$, the Hilbert space with given S and $\tilde{P}$ has a huge size. It is thus generally unlikely that this basis will diagonalize a randomly chosen Hamiltonian. It is now known that systems that can be diagonalized in this basis, such as the nearest-neighbor Heisenberg chain, have this property because they are completely integrable, i.e. obey an infinite number of conservation laws (Faddeev, 1984).

5.1.3 The spectrum

Let us now act with the Heisenberg Hamiltonian on a Bethe-ansatz wave function. The result is

$$\begin{aligned} H\phi(p_1, \ldots, p_M) = J & \sum_{\substack{j=1 \\ p_j \neq p_{j+1}-1}}^{M} \phi(p_1, \ldots, p_j + 1, \ldots, p_M) \\ & + J \sum_{\substack{j=1 \\ p_j \neq p_{j-1}+1}}^{M} \phi(p_1, \ldots, p_j - 1, \ldots, p_M) \\ & + (N - 2M) J \phi(p_1, \ldots, p_M) \\ & + 2J \sum_{\substack{j=1 \\ p_j \neq p_{j+1}-1}}^{M} \phi(p_1, \ldots, p_j, \ldots, p_M) - NJ\phi(p_1, \ldots, p_M) \end{aligned} \tag{5.33}$$

The first and second terms come from acting with $\sum_n P_{n,n+1}$ on $\uparrow\downarrow$ and $\downarrow\uparrow$ pairs. The third and fourth terms come from acting with $\sum_n P_{n,n+1}$ on $\uparrow\uparrow$ and $\downarrow\downarrow$ pairs.

Using the Bethe ansatz, Eq. (5.24) and Eq. (5.31), we can put Eq. (5.33) into the form

$$H\phi(p_1, \ldots, p_M) = J \sum_{j=1}^{M} [\mu(\chi_j) + \mu^{-1}(\chi_j) - 2]\phi(p_1, \ldots, p_M)$$

$$- J \sum_{j=1}^{M} [\phi(\ldots p_j + 1, p_j + 1 \ldots) + \phi(\ldots p_j, p_j \ldots)$$
$$- 2\phi(\ldots p_j, p_{j+1} \ldots)] \tag{5.34}$$

The last term (in brackets) is found to vanish. Thus the Bethe-ansatz state, Eq. (5.24), is an eigenstate of the Heisenberg model with eigenvalue E given by

$$E = J \sum_{j=1}^{M} [\mu(\chi_j) + \mu^{-1}(\chi_j) - 2] = -J \sum_{j=1}^{M} \frac{\pi^2}{\chi_j^2 + (\pi/2)^2} \tag{5.35}$$

We must now find solutions to the Bethe-ansatz equation, Eq. (5.31). Intuitively, if $J > 0$ (an antiferromagnet), we expect the ground state to have $S_z = 0$ ("Néel") and thus $M/N = \frac{1}{2}$. Let us assume that the solutions of the Bethe-ansatz equations are *real roots* χ_j. By taking logarithms we can write the Bethe-ansatz equations in the form

$$2N \tan^{-1}\left(\frac{\chi_j}{\pi/2}\right) - 2\sum_{i=1}^{M} \tan^{-1}\left(\frac{\chi_j - \chi_i}{\pi}\right) = 2\pi I_j \tag{5.36}$$

for $j = 1, \ldots, M$ and where I_j are integers (half-integers) for $N - M$ odd (even).

Let us now assume that $\{\chi_j\}$ is a set of real roots with $N - M$ odd. The *function* $J(\chi)$

$$J(\chi) = \frac{1}{2\pi}\left(2N \tan^{-1}\left(\frac{\chi}{\pi/2}\right) - 2\sum_{i=1}^{M} \tan^{-1}\left(\frac{\chi - \chi_i}{\pi}\right)\right) \tag{5.37}$$

is a monotonically increasing function of χ. If J happens to take the value of one of the integers I_i, $J^{-1}(I_i) = \chi$ will be equal to the corresponding root χ_i. However, it may happen that for some integers the value of χ might not be in the set $\{\chi_j\}$. Such a χ is called a hole (not to be confused with the "holes" of a more general context). If the roots are closely spaced (i.e. their separation vanishes in the thermodynamic $N \to \infty$ limit), we should be able to define a distribution of roots and holes $\rho(\chi)$

$$\rho(\chi) = \frac{dJ(\chi)}{d\chi} \tag{5.38}$$

or, equivalently,

$$J(\chi) = J(-\infty) + \int_{-\infty}^{\chi} d\chi' \, \rho(\chi') \tag{5.39}$$

Now $dJ/d\chi$ is given by differentiating Eq. (5.37),

$$\frac{dJ}{d\chi} = \rho(\chi) = \frac{N/2}{\chi^2 + (\pi/2)^2} - \sum_{j=1}^{M} \frac{1}{(\chi - \chi_i)^2 + \pi^2} \tag{5.40}$$

Let $\{\theta_j\}_{j=1\ldots n}$ denote the positions of the holes. In the $N \to \infty$ limit the following approximation is valid:

$$\sum_{i=1}^{M} f(\chi_i) = \int_{-\infty}^{+\infty} d\chi\, \rho(\chi) f(\chi) - \sum_{i=1}^{n} f(\theta_i) \tag{5.41}$$

where n is the number of holes. By using these results we find the integral equation

$$\rho(\chi) + \int_{-\infty}^{+\infty} d\chi' \frac{\rho(\chi')}{(\chi - \chi')^2 + \pi^2} = \frac{N/2}{\chi^2 + (\pi/2)^2} + \sum_{j=1}^{n} \frac{1}{(\chi - \theta_j)^2 + \pi^2} \tag{5.42}$$

Consider now the set $\{\chi_1, \ldots, \chi_M, \theta_1, \ldots, \theta_n\}$ of roots and holes and let ξ_k denote the kth element in this set, counting from left to right on the χ axis. This element is defined by

$$\int_{-\infty}^{\xi_k} \rho(\chi) d\chi = J(\xi_k) - J(-\infty) = I_k - \frac{M - N}{2} \tag{5.43}$$

The integral equation is solved by taking a Fourier transform:

$$\rho(\chi) = \int_{-\infty}^{+\infty} \frac{dp}{2\pi} e^{ip\chi} \tilde{\rho}(p) \tag{5.44}$$

One finds the solution

$$\tilde{\rho}(p) = \tilde{\rho}_0(p) + \sum_{j=1}^{n} \frac{e^{-ip\theta_j - \frac{\pi|p|}{2}}}{2\cosh(\pi p/2)} \tag{5.45}$$

with

$$\tilde{\rho}_0(p) = \frac{N/2}{2\cosh(\pi p/2)} \tag{5.46}$$

Thus

$$\rho(\chi) = \rho_0(\chi) + \sum_{j} \rho_{\text{hole}}(\chi - \theta_j) \tag{5.47}$$

and

$$\rho_0(\chi) = \frac{N}{2\cosh\chi} \tag{5.48}$$

The total number of roots M in a state with n holes is

$$M = \int_{-\infty}^{+\infty} d\chi\, \rho(\chi) - n = \tilde{\rho}(0) - n = \frac{N - n}{2} \tag{5.49}$$

Since M is an integer, n must be even (odd) for N even (odd). This state has the energy eigenvalue

$$E = -J\pi^2 \int d\chi \, \frac{\sigma(\chi)}{\chi^2 + (\pi/2)^2} \tag{5.50}$$

Here I introduced the density of roots for the Bethe-ansatz equations

$$\sigma(\chi) = \rho(\chi) - \sum_{i=1}^{n} \delta(\chi - \theta_i) \tag{5.51}$$

In Fourier space, we get

$$E = -J\pi \int dp \, \tilde{\sigma}(-p) e^{-\frac{\pi|p|}{2}} \tag{5.52}$$

with

$$\tilde{\sigma}(p) = \tilde{\rho}(p) - \sum_{i=1}^{n} e^{-ip\theta_i} \tag{5.53}$$

We find the result

$$E = E_0 + \sum_{i=1}^{n} E_h(\theta_i) \tag{5.54}$$

where $E_0 = -2NJ \ln 2$ is the ground energy state, and the "excitation energy" (i.e. that for "holes") is

$$E_h(\theta) = \frac{\pi J}{\cosh \theta} \tag{5.55}$$

Thus, we can minimize the energy by choosing the solution with real roots and no holes (complex roots are irrelevant to this issue (Lowenstein, 1984)). The total spin S for this state when N is even is obtained from Eqs. (5.3) and (5.49)

$$S = \frac{N}{2} - M = 0 \tag{5.56}$$

Thus the ground state is a *singlet* ($S = 0$). The excitations are "holes" with energy $\pi J / \cosh \theta$. For a lattice with N sites, N even (odd), there is an even (odd) number of holes. A state with one hole constructed in this manner carries $S_z = +\frac{1}{2}$. The spin-reversed hole is found by acting with S^- on this state. These states are degenerate, as required by the SU(2) symmetry.

The momentum of these states can be calculated by noting that the operator $\tilde{P}$ that translates the wave function by one lattice spacing is related to the total momentum $\bar{P}$ of the state by

$$\tilde{P}\phi(p_1, \ldots, p_M) = e^{i\bar{P}} \phi(p_1, \ldots, p_M) \tag{5.57}$$

Before we found that the eigenvalue of $\tilde{P}$ was λ. Hence

$$\bar{P} = -i \ln \lambda = -i \sum_{j=1}^{M} \ln \mu_j = -i \sum_{j=1}^{M} \ln \left(\frac{\chi_j + i\pi/2}{\chi_j - i\pi/2} \right) \tag{5.58}$$

We can also write

$$\bar{P} = -2 \sum_{j=1}^{M} \tan^{-1}\left(\frac{2\chi_j}{\pi}\right) + M\pi \tag{5.59}$$

In terms of "holes" θ_i and the distribution $\rho(\chi)$ we can write $\bar{P}$ in the form

$$\bar{P} = \bar{P}_0 + \sum_{i=1}^{n} \bar{P}_i \tag{5.60}$$

where $\bar{P}_0$ is the total momentum of the ground state,

$$\bar{P}_0 = -\int_{-\infty}^{+\infty} d\chi\, \rho_0(\chi) 2 \tan^{-1}\left(\frac{2\chi}{\pi}\right) + M\pi \tag{5.61}$$

and $\bar{P}_i$ is the contribution from the ith "hole" (see Eq. (5.45)),

$$\bar{P}_i = \int_{-\infty}^{+\infty} d\chi \int \frac{dp}{2\pi} 2\tan^{-1}\left(\frac{2\chi}{\pi}\right) \frac{e^{ip(\chi-\theta_i)}}{1+e^{-\pi|p|}} \tag{5.62}$$

Since $\rho_0(\chi)$ is even (see Eq. (5.48)), the total momentum of the ground state is (mod 2π)

$$\bar{P}_0 = M\pi \tag{5.63}$$

as predicted by Marshall's theorem (Marshall, 1955).

What is the momentum of the first excited state? From mean-field theory, which yields a Néel state, we expect that the lowest excited state should be a spin wave with wave vector $Q = \pi$ (i.e. momentum $\bar{P} = \pi$) and vanishing energy. From the excitation energy, Eq. (5.54), we learn that there are massless excitations (i.e. $E \to E_0$) if $\theta \to \pm\infty$. But, in this limit, $\bar{P}_i$ has the value

$$\begin{aligned}
\lim_{\theta_i \to \pm\infty} \bar{P}_i &= + \lim_{\theta_i \to \pm\infty} \int d\chi \int \frac{dp}{2\pi} 2\tan^{-1}\left(\frac{2\chi}{\pi}\right) \frac{e^{ip(\chi-\theta_i)}}{1+e^{-\pi|p|}} \\
&= + \lim_{\theta_i \to \pm\infty} \int d\chi \int \frac{dp}{2\pi} 2\tan^{-1}\left(\frac{2}{\pi}(\chi+\theta_i)\right) \frac{e^{ip\chi}}{1+e^{-\pi|p|}} \\
&= \pm\pi \int d\chi \int \frac{dp}{2\pi} \frac{e^{ip\chi}}{1+e^{-\pi|p|}} \\
&= \pm\pi \int dp \frac{\delta(p)}{1+e^{-\pi|p|}}
\end{aligned} \tag{5.64}$$

Thus we get

$$\lim_{\theta_i \to \pm\infty} \bar{P}_i = \pm\frac{\pi}{2} \tag{5.65}$$

This result means that the lowest excited state of a chain with N even, which has *two* "holes," has total momentum equal either to zero or to π (mod 2π). In fact we can view this state as the sum of two "single" particle states (i.e. "holes"), each with momenta $\pm\pi/2$. In other words, this state is *not* a spin wave with momentum π. Rather, the system behaves as if its elementary excitations had momenta close to $\pm\pi/2$. This resembles the physics of one-dimensional *fermions* on a half-filled chain. The Fermi "surface" is just two points, $k_{\mathrm{F}} = \pm\pi/2$. The elementary excitations are particle–hole pairs with momenta close to the Fermi points. We will see below that this system, with purely bosonic degrees of freedom, does indeed have fermions in its spectrum.

5.2 Fermions and the Heisenberg model

5.2.1 The Jordan–Wigner transformation

At first sight it may appear to be obvious that there should be fermions in the spectrum of the Heisenberg model. After all, we derived the Heisenberg model as the strong-coupling limit of a purely fermionic system, namely the half-filled Hubbard model. However, the fermions found in the last section are *not* the "constituent" band (Hubbard) fermions. For one thing, these states carry no electric charge. The spin-up and spin-down species are only degenerate precisely at the Heisenberg isotropic point. Furthermore, it is not possible to write the spin operators $S^{\pm}$ as local bilinears in those fermions.

One may also argue that the states of the spin system can be viewed as a collection of bosons with hard cores: a spin can be flipped once only. The algebra of the Pauli matrices, on the other hand, seems to have mixed properties: they *commute* on different sites and they *anti-commute* on the same sites. The anticommutativity of the Pauli matrices guarantees that the bosons do indeed have hard cores.

More formally, let us imagine that we are going to use a set of basis vectors in which $S_z \equiv S_3$ is diagonal. We can also consider the raising and lowering operators, at each site n, $S^{\pm}(n)$

$$S^{\pm}(n) = S_1(n) \pm i S_2(n) \tag{5.66}$$

where I am using the notation

$$S_i \equiv \frac{1}{2}\sigma_i, \qquad i = 1, 2, 3 \tag{5.67}$$

and the σ_is are the three Pauli matrices

$$\sigma_1 = \begin{pmatrix} 0 & 1 \\ 1 & 0 \end{pmatrix}, \qquad \sigma_2 = \begin{pmatrix} 0 & -i \\ i & 0 \end{pmatrix}, \qquad \sigma_3 = \begin{pmatrix} 1 & 0 \\ 0 & -1 \end{pmatrix} \tag{5.68}$$

The operators $S^{\pm}(n)$ *commute* on different sites

$$\left[S^+(n), S^+(m)\right] = \left[S^-(n), S^-(m)\right] = \left[S^+(n), S^-(m)\right] = 0 \tag{5.69}$$

for $m \neq n$. But on the same sites they anti-commute

$$\{S^+(n), S^-(n)\} = 1 \tag{5.70}$$

$$\{S^+(n), S^+(n)\} = \{S^-(n), S^-(n)\} = 0 \tag{5.71}$$

This last condition implies that, if $|F\rangle$ is an arbitrary state *not annihilated* by $S^+(n)$, then it is annihilated by $S^+(n)^2$

$$S^+(n)\left[S^+(n)|F\rangle\right] = 0 \tag{5.72}$$

In other words, $S^+(n)$ creates bosonic excitation at the nth site but it is not possible to have two such excitations at the same site. This is the hard-core condition.

Consider now the *kink* or *soliton* operators $K(n)$

$$K(n) = \exp\left(i\pi \sum_{j=1}^{n-1} S^+(j)S^-(j)\right) \tag{5.73}$$

In terms of $S_3(n)$ we can write

$$K(n) = \exp\left(i\pi \sum_{j=1}^{n-1}\left(S_3(j) + \frac{1}{2}\right)\right) \equiv i^{n-1} \exp\left(i\pi \sum_{j=1}^{n-1} S_3(j)\right) \tag{5.74}$$

Thus $K(n)$ is a unitary operator which, up to a phase factor, rotates the spin configurations by π around the z axis on all sites to the left of the nth site. Thus the state $|\frac{1}{2}\ldots\frac{1}{2}\rangle$, an eigenstate of S_1 on all sites, becomes

$$K(n)|\tfrac{1}{2}\ldots\tfrac{1}{2}\rangle = i^{n-1}|-\tfrac{1}{2}\ldots-\tfrac{1}{2}, \tfrac{1}{2}\ldots\tfrac{1}{2}\rangle \tag{5.75}$$

where the last flipped spin is at the site $n-1$. The operator $K(n)$ is said to create a *kink* in the spin configuration. Clearly this operator cannot have a non-vanishing expectation value in any state exhibiting long-range order. On the other hand, it may have an expectation value on states without long-range order. For this reason these operators are usually called disorder operators (Kadanoff and Ceva, 1971;

Fradkin and Susskind, 1978). Consider now the operators $c^\dagger(n)$ and $c(n)$ obtained by flipping a spin and creating a kink at the same place (Jordan and Wigner, 1928):

$$\begin{aligned} c(n) &\equiv K(n)S^-(n) = e^{i\pi\sum_{j=1}^{n-1} S^+(j)S^-(j)} S^-(n)r \\ c^\dagger(n) &\equiv S^+(n)K^\dagger(n) = S^+(n)e^{-i\pi\sum_{j=1}^{n-1} S^+(j)S^-(j)} \end{aligned} \tag{5.76}$$

The following results are easy to prove (Lieb *et al.*, 1961).

First of all,

$$\begin{aligned} c^\dagger(n)c(n) &= S^+(n)K^\dagger(n)K(n)S^-(n) \\ c(n)c^\dagger(n) &= K(n)S^-(n)S^+(n)K^\dagger(n) \end{aligned} \tag{5.77}$$

But the kink operator is unitary,

$$K^\dagger(n)K(n) = K(n)K^\dagger(n) = 1 \tag{5.78}$$

and, because $S^\pm(n)$ and $K(n)$ commute, one finds

$$\begin{aligned} c^\dagger(n)c(n) &= S^+(n)S^-(n) = \frac{1}{2} + S_3(n) \\ c(n)c^\dagger(n) &= S^-(n)S^+(n) = \frac{1}{2} - S_3(n) \end{aligned} \tag{5.79}$$

Moreover, the hard-core condition $\left(S^\pm\right)^2 = 0$ implies that the same property holds for the fermion operators,

$$\left(c^\dagger(n)\right)^2 = (c(n))^2 = 0 \tag{5.80}$$

What are the commutation relations obeyed by the operators $c^\dagger(n)$ and $c(m)$? Let us compute the products $c(n)c(m)$ and $c(m)c(n)$, say for $m > n$. Clearly $S^-(n)$ commutes with all the operators in $K(m)$ except for those at the site $j = n$, and therefore

$$S^-(n)K(m) = \prod_{j=1, j\neq n}^{m-1} e^{i\pi S^+(j)S^-(j)} S^-(n)e^{i\pi S^+(n)S^-(n)} \tag{5.81}$$

By making use of the identity

$$e^{\pm i\pi S^+(n)S^-(n)} = e^{\pm i\pi\left(\frac{1}{2}+S_3(n)\right)} = -2S_3(n) \tag{5.82}$$

we get

$$S^-(n)K(m) = -K(m)S^-(n) \tag{5.83}$$

since $\{S^-(n), S_3(n)\} = 0$ on the same site. Thus

$$\begin{aligned} c(n)c(m) &= K(n)S^-(n)K(m)S^-(m) \\ &= -K(n)K(m)S^-(n)S^-(m) \\ &= -K(m)S^-(m)K(n)S^-(n) \\ &= -c(m)c(n) \end{aligned} \tag{5.84}$$

Similarly, we can also prove $(n \neq m)$

$$\begin{aligned} c^\dagger(n)c(m) &= S^+(n)K^\dagger(n)K(m)S^-(m) \\ &= -K(m)S^-(m)S^+(n)K^\dagger(n) \\ &= -c(m)c^\dagger(n) \end{aligned} \tag{5.85}$$

In summary, the operators $c^\dagger(n)$ and $c(n)$ obey canonical anticommutation relations

$$\{c(n), c(m)\} = \{c^\dagger(n), c^\dagger(m)\} = 0 \tag{5.86}$$

and

$$\{c(n), c^\dagger(m)\} = \delta_{n,m} \tag{5.87}$$

Thus the operator $c^\dagger(n)$ $(c(n))$ creates (destroys) a fermion at site n. These operators are highly non-local. The states created by $c^\dagger(n)$ are fermions. Conversely, we can also write the inverse of the Jordan–Wigner transformation:

$$\begin{aligned} S^-(n) &= e^{-i\pi \sum_{j=1}^{n-1} c^\dagger(j)c(j)} c(n) \\ S^+(n) &= c^\dagger(n) e^{i\pi \sum_{j=1}^{n-1} c^\dagger(j)c(j)} \end{aligned} \tag{5.88}$$

5.2.2 The Heisenberg chain: fermion representation

Let us apply these results to the Heisenberg model. In terms of S^+ and S^-, the Heisenberg Hamiltonian (with anisotropy γ) is

$$\begin{aligned} H = &\frac{1}{2} J \sum_{j=1}^{N} \left(S^+(j)S^-(j+1) + S^-(j)S^+(j+1)\right) \\ &+ \gamma J \sum_{j=1}^{N} \left(S^+(j)S^-(j) - \frac{1}{2}\right)\left(S^+(j+1)S^-(j+1) - \frac{1}{2}\right) \end{aligned} \tag{5.89}$$

For $\gamma = 1$ we recover the isotropic Heisenberg model. The case $\gamma = 0$ is known as the spin one-half XY model.

We can now use the Jordan–Wigner transformation, Eq. (5.88), to get

$$\begin{aligned} S^+(j)S^-(j+1) &= c^\dagger(j)e^{-i\pi c^\dagger(j)c(j)}c(j+1) \\ &= c^\dagger(j)\left(1-2c^\dagger(j)c(j)\right)c(j+1) \\ &= c^\dagger(j)c(j+1) \end{aligned} \tag{5.90}$$

and

$$\begin{aligned} S^-(j)S^+(j+1) &= c(j)e^{+i\pi c^\dagger(j)c(j)}c^\dagger(j+1) \\ &= c(j)\left(1-2c^\dagger(j)c(j)\right)c^\dagger(j+1) \\ &= c(j)c^\dagger(j+1) - 2c(j)c^\dagger(j)c(j)c^\dagger(j+1) \\ &= c(j+1)c(j) \end{aligned} \tag{5.91}$$

The Heisenberg Hamiltonian takes the simple form (Luther and Peschel, 1975)

$$H = \frac{J}{2}\sum_{j=1}^{N}\left(c^\dagger(j)c(j+1) + \text{h.c.}\right) + \gamma J\sum_{j=1}^{N}\left(n(j)-\frac{1}{2}\right)\left(n(j+1)-\frac{1}{2}\right) \tag{5.92}$$

where $n(j)$ is the density (or occupation number) for spinless fermions

$$n(j) = c^\dagger(j)c(j) \tag{5.93}$$

What boundary conditions do the $c(j)$ operators obey? Suppose that the *spin* problem has periodic boundary conditions, i.e.

$$S_i(N+1) = S_i(1) \quad \text{for } i = 1, 2, 3 \tag{5.94}$$

In the fermion case, the periodic boundary conditions on the spin degrees of freedom imply

$$\begin{aligned} c(N+1) &= \exp\left(i\pi\sum_{j=1}^{N}S^+(j)S^-(j)\right)S^-(N+1) \\ &= \exp\left[i\pi\sum_{j=1}^{N}\left(\frac{1}{2}+S_3(j)\right)\right]S^-(1) \end{aligned} \tag{5.95}$$

where

$$c(1) \equiv S^-(1) \tag{5.96}$$

Thus, the boundary condition on the fermionic degrees of freedom is

$$c(N+1) = i^N e^{i\pi S_3}c(1) \tag{5.97}$$

where S_3 is the total z-component of the spin. But $\sum_{j=1}^{N} S^+(j)S^-(j)$ is just the total fermion number N_F, so S_3 and N_F are related by

$$S_3 = \sum_{j=1}^{N} c^\dagger(j)c(j) - \frac{N}{2} = N_\mathrm{F} - \frac{N}{2} \tag{5.98}$$

Hence, the $S_3 = 0$ sector maps into the half-filled sector for the fermions under the Jordan–Wigner transformation:

$$S_3 = 0 \Rightarrow N_\mathrm{F} = \frac{N}{2} \tag{5.99}$$

provided that N is even. Conversely, the state with $S_3 = \frac{1}{2}$ has $N_\mathrm{F} = (N+1)/2$ provided that N is odd. The boundary condition, Eq. (5.97), depends on the z-component of the total spin S_3 or, alternatively, on the total number of fermions N_F

$$c(N+1) = e^{i\pi N_\mathrm{F}} c(1) \tag{5.100}$$

For a lattice with N even and $S_3 = 0$ (i.e. $N_\mathrm{F} = N/2$) we get *periodic (anti-periodic) boundary conditions* if $N/2$ is even (odd). Thus the many-body fermion wave functions obey different boundary conditions depending on whether N_F is even or odd.

The Hamiltonian, Eq. (5.92), has quartic terms and is not readily solvable except, of course, by Bethe's method. We can gain some insight by considering the case $\gamma = 0$, the XY model.

For $\gamma = 0$, the Hamiltonian is simply

$$H_0 = \frac{J}{2} \sum_{j=1}^{N} \left(c^\dagger(j)c(j+1) + \mathrm{h.c.} \right) \tag{5.101}$$

This is a trivial problem. *The fermions are free*. As we saw before, this problem can be solved by taking the Fourier transform. Let $c(k)$ denote the Fourier modes, with $|k| \leq \pi$. The eigenvalues for a system with periodic boundary conditions are

$$H_0 = \int_{-\pi}^{\pi} \frac{dk}{2\pi} \epsilon(k) c^\dagger(k) c(k) \tag{5.102}$$

where

$$\epsilon(k) = J \cos k \tag{5.103}$$

The ground state is found by filling up the negative-energy modes. In the case of $N_\mathrm{F} = N/2$, we get two Fermi points, $k_\mathrm{F} = \pm\pi/2$. The negative-energy states have k in the interval $\pi > |k| \geq \pi/2$.

This system is *gapless*. In fact, there are no massive excitations in the one-dimensional spin one-half spin chain. This system is *critical* in the sense that *all* its correlation functions fall off as a power of the distance. We will discuss this issue below. Also, there is no long-range order in the sense that (at equal times)

$$\lim_{|m-n|\to\infty} \langle S^+(n)S^-(m)\rangle \approx (-1)^{m-n} \frac{\text{constant}}{|m-n|^{\eta}} \longrightarrow 0 \tag{5.104}$$

with an exponent η that will be computed below. Thus there is no Néel order for the chain. (Kennedy, Lieb, and Shastri (Kennedy *et al.*, 1988) have shown that for the square lattice the spin one-half XY model *does have* long-range order $\langle S^+\rangle \neq 0$.)

5.2.3 The continuum limit of the one-dimensional quantum Heisenberg antiferromagnet

We are interested in the physics at large distances compared with the lattice constant and at frequencies much lower than, say, J. In this limit some sort of continuum theory should emerge. We will see now that the continuum theory associated with this 1D system of fermions looks like a theory of "relativistic" fermions moving at the speed of "light" (with $c = v_{\mathrm{F}} = Ja_0$, the Fermi velocity, with a_0 being the lattice spacing). These results apply not only to the Hamiltonian of Eq. (5.92) but, in fact, to *all* 1D Fermi systems with local hopping Hamiltonians. A similar situation develops for fermions in a flux phase in two dimensions, as we will see in Chapter 8.

Consider first the non-interacting problem

$$H_0 = \frac{J}{2}\sum_{n=1}^{N}\left(c^\dagger(n)c(n+1) + \text{h.c.}\right) \tag{5.105}$$

which is equivalent to the XY model. We are assuming periodic boundary conditions. The dispersion law for this system is

$$\epsilon(k) = J\cos k \tag{5.106}$$

with Fermi points at $k_{\mathrm{F}} = \pm\pi/2$. The elementary excitations will have a characteristic momentum of $\pm k_{\mathrm{F}}$ and we should expect that the correlation functions of the fermions should have a rapid variation of the type $e^{ik_{\mathrm{F}}n} = i^n$ with a *slow* variation on top. It is then natural to define new fermionic variables $a(n)$ that should exhibit only a slow variation in n and hence should have a simple continuum limit. Define

$$a(n) = i^{-n}c(n) \tag{5.107}$$

The Hamiltonian H_0 now reads

$$\begin{aligned}H_0 &= \frac{J}{2}\sum_{n=1}^{N}\left(i^{-n}a^\dagger(n)i^{(n+1)}a(n+1)+\text{h.c.}\right)\\ &= \frac{J}{2}\sum_{n=1}^{N}\left(ia^\dagger(n)a(n+1)+\text{h.c.}\right)\\ &= \frac{J}{2}\sum_{n=1}^{N} ia^\dagger(n)\Big(a(n+1)-a(n-1)\Big)\end{aligned} \tag{5.108}$$

where we have used the periodic boundary conditions in the last step. By separating the sum into even and odd sites, one finds for N even

$$H_0 = \frac{J}{2}\sum_{s=1}^{N/2} i\Big\{a^\dagger(2s)\Big(a(2s+1)-a(2s-1)\Big)+a^\dagger(2s+1)\Big(a(2s+2)-a(2s)\Big)\Big\} \tag{5.109}$$

We see that even sites couple to odd sites (and vice versa) but there is no even–even or odd–odd coupling.

Define now the *spinor* field $\phi_\alpha (\alpha = 1, 2)$, by

$$\phi_\alpha(n) = \begin{cases}\phi_1(n) = a(2s) & n \text{ even}\\ \phi_2(n) = a(2s+1) & n \text{ odd}\end{cases} \tag{5.110}$$

Thus we can write

$$H_0 = i\frac{J}{2}\sum_{s=1}^{N/2}\{\phi_1^\dagger(2s)[\phi_2(2s+1)-\phi_2(2s-1)]+\phi_2^\dagger(2s+1)[\phi_1(2s+2)-\phi_1(2s)]\} \tag{5.111}$$

A Fermi field $\psi_\alpha(x)$ in the continuum is expected to obey the equal-time canonical anticommutation relations

$$\{\psi_\alpha^\dagger(x), \psi_{\alpha'}(x')\} = \delta_{\alpha\alpha'}\delta(x-x') \tag{5.112}$$

The $\phi_\alpha(n)$ fields obey

$$\{\phi_\alpha^\dagger(n), \phi_{\alpha'}(n')\} = \delta_{\alpha\alpha'}\delta_{n,n'} \tag{5.113}$$

since they are defined on a lattice. We can make these relations compatible by defining

$$\psi_\alpha(x) = \frac{1}{\sqrt{2a_0}}\phi_\alpha(n) \tag{5.114}$$

for $x = 2sa_0$ and a_0 the lattice spacing, which will be the unit of length. Thus ψ_α has dimensions of $[\text{length}]^{-1/2}$, whereas ϕ_α is dimensionless. We have assumed that the *distribution* $\delta(x - x')$ is defined by the limit

$$\delta(x - x') = \lim_{a_0 \to 0} \frac{\delta_{n,n'}}{2a_0} \tag{5.115}$$

which, of course, makes sense only as a limit.

By expanding ϕ, in Eq. (5.111), in a Taylor-series expansion,

$$\begin{aligned} \phi_2(2s+1) - \phi_2(2s-1) &\approx 2a_0(2a_0)^{1/2}\, \partial_x \psi_2(x) \\ \phi_1(2s+2) - \phi_1(2s) &\approx 2a_0(2a_0)^{1/2}\, \partial_x \psi_1(x) \end{aligned} \tag{5.116}$$

and using the fact that

$$\lim_{a_0 \to 0} \sum_s 2a_0 f(s) = \int dx\, f(x) \tag{5.117}$$

one finds the effective Hamiltonian in the continuum $\tilde{H}_0$ to be given by

$$\tilde{H}_0 = \int dx\, \psi^\dagger(x) \alpha i\, \partial_x \psi(x) \tag{5.118}$$

where

$$\tilde{H}_0 = \frac{H_0}{J a_0} \tag{5.119}$$

and the matrix

$$\alpha \equiv \sigma_1 = \begin{pmatrix} 0 & 1 \\ 1 & 0 \end{pmatrix} \tag{5.120}$$

This is just the Hamiltonian for a Dirac spinor field $\psi_\alpha(x)$ in units in which $\hbar$ and the Fermi velocity v_F are set to unity. We will see below that interactions normally lead to finite, non-universal, renormalizations of the Fermi velocity.

The upper (lower) component of ψ_α represents the amplitude on even (odd) sites. Alternatively we could have used a basis in which σ_1 is diagonal. In this basis, the upper (lower) component R (L) represents fermions moving towards the right (left) with speed $v_F = 1$. It will be, in fact, more convenient to work in the *chiral* basis

$$\begin{aligned} \psi_1(x) &= \frac{1}{\sqrt{2}} (-R(x) + L(x)) \\ \psi_2(x) &= \frac{1}{\sqrt{2}} (R(x) + L(x)) \end{aligned} \tag{5.121}$$

We get

$$\psi_1^\dagger i\, \partial_x \psi_2 + \psi_2^\dagger i\, \partial_x \psi_1 = -(R^\dagger i\, \partial_x R - L^\dagger i\, \partial_x L) \tag{5.122}$$

In the Dirac theory in $(1+1)$ dimensions one defines the γ-matrices γ_0, γ_1, and γ_5 by requiring that they satisfy

$$\{\gamma_\mu, \gamma_\nu\} = 2g_{\mu\nu}, \qquad \gamma_5 = i\gamma_0\gamma_1 \tag{5.123}$$

We can choose the chiral representation, in which

$$\begin{aligned} \gamma_5 &= \gamma_0\gamma_1 = \sigma_3 \\ \gamma_0 &= \sigma_1 \\ \gamma_1 &= -i\sigma_2 \end{aligned} \tag{5.124}$$

It is convenient to define a field $\bar{\psi}$ by

$$\bar{\psi} = \psi^\dagger \gamma_0 \tag{5.125}$$

The Hamiltonian $\tilde{H}_0$ now is

$$\tilde{H}_0 = \int dx\, \bar{\psi}(x) i\gamma_1\, \partial_x \psi(x) \tag{5.126}$$

Let us write the interaction terms of Eq. (5.92) in this formalism.

First, we note that we can rewrite

$$H_{\text{int}} = \gamma J \sum_{j=1}^{N} \left(c^\dagger(j)c(j) - \frac{1}{2}\right)\left(c^\dagger(j+1)c(j+1) - \frac{1}{2}\right) \tag{5.127}$$

in the form

$$H_{\text{int}} = -\frac{\gamma J}{2} \sum_{j=1}^{N} \left(c^\dagger(j)c(j) - c^\dagger(j+1)c(j+1)\right)^2 + \frac{1}{4}\gamma J N \tag{5.128}$$

Following the same steps as those which led to Eq. (5.126), we find that $\tilde{H}_{\text{int}}$, defined by

$$\tilde{H}_{\text{int}} = \frac{H_{\text{int}}}{J a_0} \tag{5.129}$$

has the form, up to the irrelevant additive constant $\gamma N/(4a_0)$,

$$\tilde{H}_{\text{int}} = -2\gamma \int dx \left(\bar{\psi}(x)\psi(x)\right)^2 \tag{5.130}$$

which is the interaction term of the $(1+1)$-dimensional Gross–Neveu model. The expression $\bar{\psi}\psi$ is the continuum limit of

$$\begin{aligned} \frac{1}{2a_0}(n(2s+1) - n(2s)) &\approx -\left(\psi_1^\dagger(x)\psi_1(x) - \psi_2^\dagger(x)\psi_2(x)\right) \\ &= \left(R^\dagger L + L^\dagger R\right) \equiv \bar{\psi}\psi \end{aligned} \tag{5.131}$$

Thus a non-zero average for $\bar{\psi}\psi$ breaks chiral (i.e. left–right) symmetry down to its $\mathbb{Z}_2$ (Ising) invariance. We see that this is equivalent to the development of a periodic density modulation of the lattice fermion system. Tracing our steps backwards, we interpret this state as a Néel antiferromagnet. In particular, Eq. (5.131) shows that in the continuum limit the z-component of the Néel order parameter, the z-component of the staggered magnetization N_z, is essentially the fermion mass term $\bar{\psi}\psi$.

Equation (5.130) can also be written, up to an additive constant, in the form

$$\tilde{H}_{\text{int}} = \gamma \int dx\, j_\mu j^\mu - 2\gamma \int dx \left((R^\dagger L)^2 + (L^\dagger R)^2 \right) \tag{5.132}$$

where we have used the fermionic current j_μ,

$$j_\mu = \bar{\psi}\gamma_\mu \psi \tag{5.133}$$

which, in the chiral basis, has components

$$j_0 = R^\dagger R + L^\dagger L \tag{5.134}$$

and

$$j_1 = R^\dagger R - L^\dagger L \tag{5.135}$$

Thus j_0 measures the total number of fermions, i.e. the total density, and j_1 is the difference in number of left and right movers. A system with the first term of Eq. (5.132) as its only interaction is known as the (massless) Thirring or Luttinger model.

The last term in Eq. (5.132) is peculiar. On the one hand, it appears to be superficially zero, since it is a sum of squares of Fermi fields and Fermi statistics may seem to imply that it is zero. However, all these expressions, written in the continuum, are to be interpreted as a product of operators at short distances. Furthermore, when inserted into the calculation of any expectation value, there should be singular contributions due to the presence of this operator. We are supposed to keep the leading singular term in the product. Thus, expressions such as $(\bar{\psi}\psi)^2$ and the like are to be taken in the sense of an *operator product expansion* (Kadanoff, 1969; Wilson, 1969) in which only the leading singularity is kept.

What is more important the operators $(R^\dagger L)^2$ and $(L^\dagger R)^2$ break the continuous left–right (chiral) symmetry down to a discrete subgroup. Terms of this sort arise from Umklapp scattering processes (Emery, 1979; Haldane, 1982). In the language of Feynman diagrams, these terms give contributions of the type shown in Fig. 5.1. Such processes violate momentum conservation by $4k_{\text{F}}$, which equals 2π for a half-filled system. Thus $4k_{\text{F}}$ is a reciprocal-lattice vector and hence the process is allowed, since on a lattice momentum is conserved mod 2π.

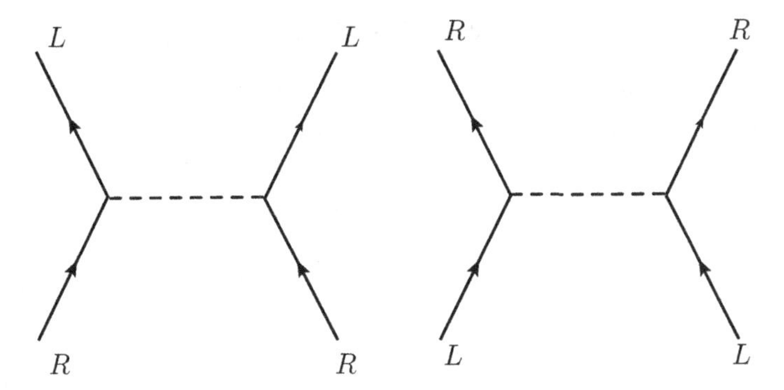

Figure 5.1 Umklapp processes.

There is a *continuous* chiral symmetry

$$\psi_\alpha = \left(e^{i\gamma_5\theta}\right)_{\alpha\beta} \psi'_\beta \tag{5.136}$$

where γ_5 is given by Eq. (5.124), and θ is an arbitrary *constant* angle.

It is easy to check that operators, such as the current $\bar{\psi}\gamma_\mu\psi$ and $\bar{\psi}i\gamma^\mu\,\partial_\mu\psi$, are invariant. Indeed, upon a chiral transformation, $\bar{\psi}$ transforms like

$$\bar{\psi} = \psi^\dagger\gamma_0 = \psi'^\dagger e^{-i\gamma_5\theta}\gamma_0 = \bar{\psi}' e^{+i\gamma_5\theta} \tag{5.137}$$

since γ_0 and γ_5 anti-commute. Thus

$$\bar{\psi}\gamma_\mu\psi = \bar{\psi}'\gamma_\mu\psi' \tag{5.138}$$

and

$$\bar{\psi}i\gamma_\mu\,\partial^\mu\psi = \bar{\psi}' i\gamma_\mu\,\partial^\mu\psi' \tag{5.139}$$

again, since $\{\gamma_5, \gamma_\mu\} = 0$. However, $\bar{\psi}\psi$ is not invariant since

$$\bar{\psi}\psi = \bar{\psi}' e^{i2\gamma_5\theta}\psi' \equiv \cos(2\theta)\bar{\psi}'\psi' + i\,\sin(2\theta)\bar{\psi}'\gamma_5\psi' \tag{5.140}$$

In particular $(\bar{\psi}\psi)^2$ has only the discrete invariance

$$\bar{\psi}\psi = -\bar{\psi}'\psi' \tag{5.141}$$

In other words, $\theta = \pi/2$. This is not so surprising. The chiral symmetry originates from the two-sublattice structure. There is always an arbitrariness in how we choose a given sublattice. Thus the discrete symmetry is genuine, but the continuous symmetry is a consequence of a carelessly taken continuum limit.

5.3 The quantum Ising chain

In this section we will discuss a very simple spin chain of great conceptual (and historical) significance: the 1D quantum Ising model, also known as the Ising model in a transverse field. We will see that the 1D version of this problem (the chain) is closely related (actually equivalent) to the 2D Ising model of classical statistical mechanics.

Let us consider again a 1D chain of N_s sites. We will take N_s to be an even number. As in the Heisenberg case, on each site n of the chain we define a spin one-half degree of freedom. Thus at each site n we have two states, $|\uparrow\rangle_n$ and $|\downarrow\rangle_n$, which I will take to be eigenstates of the Pauli matrix $\sigma_3(n)$ such that $\sigma_3(n)|\sigma(n)\rangle_n = \sigma(n)|\sigma(n)\rangle_n$. Here $\sigma = \pm 1$, for the $|\uparrow\rangle$ and $|\downarrow\rangle$, respectively. Thus, the dimensionality of the Hilbert space is 2^{N_s}. For convenience we will let the label n run from $-N_s/2 + 1$ to $N_s/2$. We will work with periodic boundary conditions (PBCs). Here this means that the states (and the operators) on the site following the N_sth site, "$N_s/2 + 1$," are identified with the states (and operators) on the first site, $-N_s/2 + 1$, e.g.

$$\sigma_3\left(\frac{N_s}{2} + 1\right) \equiv \sigma_3\left(-\frac{N_s}{2} + 1\right) \tag{5.142}$$

Thus, the chain is actually a circle of circumference $L = N_s a$, where a is the lattice spacing. For now we will set $a = 1$.

The Hamiltonian of the quantum Ising chain (with PBCs) is

$$H = -\sum_{n=-\frac{N_s}{2}+1}^{\frac{N_s}{2}} \sigma_1(n) - \lambda \sum_{n=-\frac{N_s}{2}+1}^{\frac{N_s}{2}} \sigma_3(n)\sigma_3(n+1) - h \sum_{n=-\frac{N_s}{2}+1}^{\frac{N_s}{2}} \sigma_3(n) \tag{5.143}$$

with coupling constant λ and symmetry-breaking (longitudinal) field h. The Ising model in a transverse field is defined by setting the longitudinal field $h = 0$. As before, σ_1 is the off-diagonal real Pauli matrix. In one dimension this model is exactly solvable by means of a Jordan–Wigner transformation to a system of Majorana fermions, which we will explain below.

Let us first discuss the connection between Ising models in transverse fields, for simplicity on a 1D lattice (a chain), and the *classical* Ising model on a square lattice. Similar relations exist for other lattices in all dimensions. Thus, on each site x of the 2D square lattice we have an Ising spin that can take two values, $\sigma(x) = \pm 1$, which we regard as the two z projections of the spin in a magnet with uniaxial anisotropy. Here $\{x\}$ are a set of two-component vectors that label the sites of the lattice. The classical energy functional for a spin configuration $[\sigma]$ is

$$E[\sigma] = -\sum_{\langle x,x'\rangle} J\sigma(x)\sigma(x') - \sum_{x} H\sigma(x) \tag{5.144}$$

where x and x' are (for simplicity) nearest-neighboring sites of the 2D square lattice. We have assumed that all the exchange coupling constants J and the external magnetic field H are equal and that the system is translationally invariant. We will restrict our discussion to the case of a system with ferromagnetic interactions, $J > 0$.

The partition function for a system at temperature T is the sum over all configurations of the Ising spins with a Gibbs weight for each state $[\sigma]$

$$Z = \sum_{[\sigma]} e^{-\frac{E[\sigma]}{T}} \tag{5.145}$$

The square lattice is isotropic and hence invariant under the action of the symmetry group of the square, known as C_4. We will now use this invariance to arbitrarily choose one direction, which will be denoted by $x_2 = \tau$. In what follows we will further assume that we have imposed PBCs along this direction of the lattice. We will call this the "direction of transfer" or "evolution" and assume that we have a square lattice with N_τ rows, $\tau = 1, \ldots, N_\tau$, and that each row has N_s sites, $x_1 = n = 1, \ldots, N_s$. Let us write the position vectors of each site as $x = (n, \tau)$, where we will refer to n as the "space coordinate" and to τ as the "time coordinate." In this representation we can think of a configuration $[\sigma]$ as a "history" (or evolution) of a spatial spin configuration (a *state*) from one row (labeled by τ) to the next (labeled by $\tau + 1$). We will denote the state of a row by $|[\sigma]\rangle$, where here $[\sigma]$ are the σ_3 eigenvalues at each site n of the row.

In other words, in this language the partition function of the classical system is viewed as the path integral (in discretized imaginary time) of a quantum system in one dimension, which we will see is the quantum Ising model. Using the assumption of PBCs, we will now show that the partition function can be written as the trace of a matrix called the *transfer matrix* (Schultz *et al.*, 1964), denoted by $\mathcal{T}$. For a system in D dimensions, the transfer matrix $\mathcal{T}$ will play the role of the evolution operator of a quantum problem in $d = D - 1$ space dimensions (Fradkin and Susskind, 1978). The transfer matrix $\mathcal{T}$ can be defined to be hermitian for any system in classical statistical mechanics with local and positive Gibbs weights. This feature, hermiticity of the transfer matrix, turns out to be the same as the statement that the equivalent theory in one dimension fewer has a hermitian Hamiltonian and hence a unitary time evolution.

For a system with nearest-neighbor interactions (this can be made more general) we can classify the terms of the classical energy functional $E[\sigma]$, Eq. (5.144), into terms describing the interactions between spins on the same row and others describing interactions between two (neighboring) rows. It will be convenient to break the isotropy of the lattice and to have the intra-row exchange (ferromagnetic) interactions $J_s > 0$ be different than the inter-row (also ferromagnetic) interactions $J_\tau > 0$ (above we had set $J = J_s = J_\tau$).

Let $\mathcal{T}_s$ and $\mathcal{T}_\tau$ be two matrices each of rank $2^{N_s} \times 2^{N_s}$. We will define $\mathcal{T}_s$ to be diagonal in the row states $|\sigma\rangle$ and $\mathcal{T}_\tau$ to be off-diagonal in that basis:

$$\mathcal{T}_s = \exp\left(\beta_s \sum_n \sigma_3(n)\sigma_3(n+1) + \beta_h \sum_n \sigma_3(n)\right) \tag{5.146}$$

$$\mathcal{T}_s = \left[2\sinh\left(\frac{2J_\tau}{T}\right)\right]^{N_s/2} \exp\left(\tilde{\beta}_\tau \sum_n \sigma_1(n)\right) \tag{5.147}$$

where $n = -N_s/2+1, \ldots, N_s/2$, and

$$\tanh\tilde{\beta}_\tau = e^{-2\beta_\tau}, \qquad \beta_\tau \equiv \frac{J_\tau}{T}, \qquad \beta_s \equiv \frac{J_s}{T}, \qquad \beta_h \equiv \frac{H}{T} \tag{5.148}$$

The transfer matrix $\mathcal{T}$ is defined to be

$$\mathcal{T} = \mathcal{T}_\tau^{1/2}\mathcal{T}_s\mathcal{T}_\tau^{1/2} \tag{5.149}$$

Since the matrices $\mathcal{T}_\tau$ and $\mathcal{T}_s$ are separately hermitian, the transfer matrix $\mathcal{T}$ defined above is hermitian as well.

With these definitions it is straightforward to see that, for a system with PBCs along the direction of transfer τ, the partition function is

$$Z_{\text{PBC}} = Z = \text{tr}\, \mathcal{T}^{N_\tau} \tag{5.150}$$

regardless of the boundary conditions along the "spatial" direction. For a system with fixed boundary conditions along the τ direction, with initial and final states $|[\sigma]_{\tau=1}\rangle$ and $|[\sigma]_{\tau=N_\tau}\rangle$, the partition function is

$$Z_{\text{fixed}} = \langle [\sigma]_{\tau=1}|\mathcal{T}^{N_\tau}|[\sigma]_{\tau=N_\tau}\rangle \tag{5.151}$$

It is straightforward to show that the correlation functions of the classical system

$$\langle \sigma(n,\tau)\sigma(n',\tau')\rangle = \frac{1}{Z}\sum_{[\sigma]} \sigma(n,\tau)\sigma(n',\tau')\exp\left(-\frac{E[\sigma]}{T}\right) \tag{5.152}$$

are equal to

$$\langle \hat{T}_\tau\left(\sigma_3(n,\tau)\sigma_3(n',\tau')\right)\rangle = \frac{1}{Z}\,\text{tr}\left[\hat{T}\left(\sigma_3(n,\tau)\sigma_3(n',\tau')\right)\mathcal{T}^{N_\tau}\right] \tag{5.153}$$

where we used the generalized Heisenberg representation

$$\sigma_3(n,\tau) = \mathcal{T}^{N_\tau}\sigma_3(n)\mathcal{T}_\tau^{-N_\tau} \tag{5.154}$$

and the "time-ordering" symbol $\hat{T}$.

The transcription of the classical statistical-mechanical system to this "algebraic" language suggests a mapping to a quantum system in one dimension fewer. Indeed, the formalism we just sketched parallels the relation between the

Hamiltonian (operator) and path-integral formulations of quantum mechanics (and of quantum field theory). Given this observation, we are thus tempted to write the transfer matrix $\mathcal{T}$ as an exponential of (the negative of) a quantum Hamiltonian. The problem is that, since the factor matrices $\mathcal{T}_s$ and $\mathcal{T}_\tau$ do not commute with each other, the resulting effective quantum Hamiltonian is not local (it is a sum of nested commutators of increasing order).

However, there is a procedure that will allow us to define a local quantum Hamiltonian (Fradkin and Susskind, 1978). Let us imagine stretching the lattice spacing along the horizontal direction (the rows) and compressing it by a related amount along the vertical direction (the columns). As a result the coupling J_s will become smaller while J_τ will grow bigger. Given the symmetries of this system, it is clear that the partition function $Z[J_s/T, J_\tau/T]$ must be the same (invariant) on some curves defined by a relation of the form $f(J_s/T, J_\tau/T) = \text{constant}$. As we carry on with this procedure, the curves on which the correlation functions are constant become increasingly deformed, also being stretched along the horizontal axis ("space") and squashed along the vertical axis ("time"). However, we can now also imagine increasing the number of rows by the precise amount required so that (in lattice units) we restore isotropy. If we continue indefinitely with this procedure, the spacing along the time direction becomes very small but the associated coupling is very large. However, as we can see from the expressions for the matrices $\mathcal{T}_s$ and $\mathcal{T}_\tau$, in this limit

$$\frac{J_s}{T} \to 0, \qquad \tilde{\beta}_\tau \simeq \exp\left(-\frac{2J_\tau}{T}\right) \to 0 \tag{5.155}$$

as well. Therefore, in this asymptotic regime we can write the transfer matrix in the simpler form

$$\mathcal{T} = \left[2\sinh\left(\frac{2J_\tau}{T}\right)\right]^{N_s/2} e^{-\epsilon H + O(\epsilon^2)} \tag{5.156}$$

where H is the Hamiltonian of the quantum Ising model with a symmetry-breaking field. The coupling constant λ and the symmetry-breaking field h are given by

$$\frac{J_s}{T} = \epsilon\lambda, \qquad \frac{H}{T} = \epsilon h, \qquad \epsilon = \exp\left(-\frac{2J_\tau}{T}\right) \tag{5.157}$$

In the thermodynamic limit, $N_s \to \infty$ and $N_\tau \to \infty$ (with N_s/N_τ fixed), the partition function becomes

$$Z = \lim_{\beta\to\infty} \operatorname{tr} e^{-\beta H} \tag{5.158}$$

where we keep $\beta = N_\tau \exp(-2J_\tau/T)$ fixed as $N_\tau \to \infty$, and $J_\tau/T \to \infty$.

5.4 Duality

From now on we will consider the quantum Ising chain of Eq. (5.143) at zero external field, $h = 0$. This quantum spin chain has a discrete, $\mathbb{Z}_2$ global symmetry. Let $\mathcal{I}$ be the identity transformation, $\mathcal{I}|[\sigma]\rangle \to |[\sigma]\rangle$, and let $\mathcal{R}$ be a global spin-flip transformation, $\mathcal{R}|[\sigma]\rangle \to |[-\sigma]\rangle$, where $[\sigma]$ is a configuration of the spin degrees of freedom in the σ_3 basis. Under the composition operation (sequential action), which we will denote by $\star$, of these transformations they form a group with the properties

$$\mathcal{I} \star \mathcal{I} = \mathcal{R} \star \mathcal{R} = \mathcal{I}, \qquad \mathcal{I} \star \mathcal{R} = \mathcal{R} \star \mathcal{I} = \mathcal{R} \tag{5.159}$$

Thus, this is the group $\mathbb{Z}_2$, the permutation group of two elements, $\mathcal{I}$ and $\mathcal{R}$.

If we denote by $I(n)$ the identity operator on the nth site, we can construct a representation for the symmetry operators $\mathcal{I}$ and $\mathcal{R}$ in terms of the states of the quantum Ising spin chain,

$$\mathcal{I} = \mathcal{I}^{-1} = \prod_n \otimes I(n), \qquad \mathcal{R} = \mathcal{R}^{-1} = \prod_n \otimes \sigma_1(n) \tag{5.160}$$

These operators act on the spin operators as follows:

$$\begin{aligned} \mathcal{I}\sigma_1(n)\mathcal{I}^{-1} &= \sigma_1(n), \qquad & \mathcal{I}\sigma_3(n)\mathcal{I}^{-1} &= \sigma_3(n) \\ \mathcal{R}\sigma_1(n)\mathcal{R}^{-1} &= \sigma_1(n), \qquad & \mathcal{R}\sigma_3(n)\mathcal{R}^{-1} &= -\sigma_3(n) \end{aligned} \tag{5.161}$$

Since, at $h = 0$, both operators commute with the Hamiltonian of Eq. (5.143), $[\mathcal{I}, H] = [\mathcal{R}, H] = 0$, we conclude that the Ising model in a transverse field is invariant under a global $\mathbb{Z}_2$ transformation.

As a function of the coupling constant λ, the quantum Ising chain has two phases, separated by a quantum phase transition at $\lambda_c = 1$.

1. The $\lambda < \lambda_c$ phase. This phase is best described in terms of the eigenstates of $\sigma_1(n)$, $|\pm, n\rangle$. At $\lambda = 0$, the ground state is $|\Psi_0\rangle_{\lambda=0} = \prod_n |+, n\rangle$. In this phase $\langle\Psi_0|\sigma_3(n)|\Psi_0\rangle = 0$ to all orders in perturbation theory (in λ) and all correlators decay exponentially with distance with a finite correlation length $\xi(\lambda)$.
2. The $\lambda > \lambda_c$ phase. This phase is best described in terms of the eigenstates of $\sigma_3(n)$. For all $\lambda > \lambda_c$ this phase has two degenerate ground states, $|\Psi_\pm\rangle$, related to each other by a global spin flip, $\mathcal{R}|\Psi_\pm\rangle = |\Psi_\mp\rangle$. Thus in this phase we find the phenomenon of spontaneous symmetry breaking with $\langle\Psi_\pm|\sigma_3(n)|\Psi_\pm\rangle \equiv \mathcal{M}(\lambda) \neq 0$ which plays the role of the order-parameter field.

Let us introduce the concept of a duality transformation. This transformation was first introduced by Kramers and Wannier (1941) as mapping between the low- and high-temperature phases of the 2D classical Ising model. Here we will use an equivalent mapping of the 1D quantum Hamiltonian (Fradkin and Susskind, 1978).

To this end let us introduce the dual lattice as the set of midpoint lattice sites of a 1D chain (with PBCs). We will denote by $\tilde{n}$ the site of the dual lattice midpoint between the sites n and $n+1$ of the chain. We will now define on each dual lattice site the operators $\tau_1(\tilde{n})$ and $\tau_3(\tilde{n})$,

$$\tau_1(\tilde{n}) = \sigma_3(n)\sigma_3(n+1), \qquad \tau_3(\tilde{n}) = \prod_{-\frac{N_s}{2}+1\leq p\leq n} \sigma_1(p) \tag{5.162}$$

The operators $\tau_1(\tilde{n})$, $\tau_3(\tilde{n})$, and $\tau_2(\tilde{n}) = i\tau_1(\tilde{n})\tau_3(\tilde{n})$ form a representation of the algebra of Pauli matrices.

It is trivial to see that

$$\sigma_1(n) = \tau_3(\tilde{n})\tau_3(\tilde{n}+1), \qquad \sigma_3(n)\sigma_3(n+1) = \tau_1(\tilde{n}) \tag{5.163}$$

We recognize that the dual operator $\tau_3(n)$ is essentially equivalent to the *kink-creation operator* defined by Eq. (5.74) (up to a factor of i^n and a rotation of basis). We will refer to an operator which is the dual of an order parameter as a *disorder operator.*

Using the identities of Eq. (5.163) it is apparent that the two terms of the Hamiltonian map into each other. Hence, under a duality transformation the Hamiltonian of the quantum Ising chain (at $h = 0$) at coupling constant λ, $H(\lambda)$, transforms into its dual $\widetilde{H}(\lambda)$,

$$\widetilde{H}(\lambda) = \lambda H\left(\frac{1}{\lambda}\right) \tag{5.164}$$

Thus, the strong-coupling and weak-coupling phases map into each other: this Hamiltonian is *self-dual.* If one further assumes (as Kramers and Wannier did and Onsager proved) that the transition is unique, then it must occur at the critical coupling $\lambda_c = 1$ which is invariant under duality.

Similarly, one immediately finds that the dual of the spin–spin correlation function at coupling constant λ is the same as the correlation function of two disorder operators in the dual theory at coupling constant $1/\lambda$,

$$\left\langle T\left[\sigma_3(n,\tau)\sigma_3(n',\tau')\right]\right\rangle_{\lambda} = \left\langle T\left[\tau_3(\tilde{n},\tau)\tau_3(\tilde{n}',\tau')\right]\right\rangle_{1/\lambda} \tag{5.165}$$

This result also implies that the disorder operator has an expectation value in the *disordered phase* of the Ising model (Kadanoff and Ceva, 1971; Fradkin and Susskind, 1978).

Duality also tells us how to relate seemingly dissimilar systems. Let us consider a system of two *decoupled* quantum Ising models with the same coupling constant λ and on chains of the same length. We will depict this system as a *single* quantum Ising model with only next-nearest-neighbor interactions, i.e. a system with twice as many sites but with interactions only between neighboring even sites for one chain and between neighboring odd sites for the other. Hence, this is the same as a system of two interpenetrating Ising models. Thus the Hamiltonian is the same as in Eq. (5.143) with the proviso that the interaction acts only at distance $2a$. This system ostensibly has a $\mathbb{Z}_2 \times \mathbb{Z}_2$ global symmetry, instead of a single $\mathbb{Z}_2$.

Let us now look for the dual of the Hamiltonian of this system of two decoupled Ising models using the duality transformation of Eq. (5.163) for the combined system. We find that the Hamiltonian of two decoupled Ising models, H_2, is the dual of an anisotropic XZ (equivalent to the spin one-half XY) model

$$H_2 = -\sum_n \sigma_1(n) - \lambda \sum_n \sigma_3(n)\sigma_3(n+2) \tag{5.166}$$

$$= -\sum_{\tilde{n}} \tau_3(\tilde{n})\tau_3(\tilde{n}+1) - \lambda \sum_{\tilde{n}} \tau_1(\tilde{n})\tau_1(\tilde{n}+1) \tag{5.167}$$

In this language the dual system still has a $\mathbb{Z}_2 \times \mathbb{Z}_2$ global symmetry, and duality reduces to a rotation of the basis. At the self-dual point of each Ising model, $\lambda = 1$, the quantum dual XY model is isotropic and has a global U(1) symmetry whose infinitesimal generator is $\sum_{\tilde{n}} \tau_2(\tilde{n})$. We will shortly see that at $\lambda = 1$ the system is critical. This is an example of the enhancement of a symmetry at a critical point.

Finally, we will discuss briefly the transcription of the Ising correlators in this case. However, instead of considering the usual two-point function of a single Ising chain we will consider the following four-point function with two spins on each chain:

$$\left\langle T\left[\sigma_3(n,\tau)\sigma_3(n+1,\tau)\sigma_3(n',\tau')\sigma_3(n'+1,\tau')\right]\right\rangle = \left\langle T\left[\tau_1(\tilde{n},\tau)\tau_1(\tilde{n}',\tau')\right]\right\rangle \tag{5.168}$$

Hence the correlator of two τ_1 spin operators in the dual theory is equal to the product of the correlation functions of the individual Ising chains, i.e. to the square of the spin–spin correlation function of the quantum Ising chain. Later in this chapter we will compute this correlator at $\lambda = 1$ using bosonization methods (Bander and Itzykson, 1977; Zuber and Itzykson, 1977).

5.5 The quantum Ising chain as a free-Majorana-fermion system

In a seminal paper Schultz, Mattis, and Lieb (Schultz *et al.*, 1964) calculated the partition function of the 2D classical Ising model on an anisotropic square lattice and reproduced, mapping the problem to a theory of fermions, the celebrated Onsager solution (Onsager, 1944). To this effect they used a Jordan–Wigner transformation, such as the one we used earlier in this chapter for the quantum Heisenberg antiferromagnet.

Here we will use a Jordan–Wigner transformation, similar to the one defined in Eq. (5.76), to solve the quantum Ising chain at zero external field, $h = 0$. On every site n we define the operators $\chi_1(n)$ and $\chi_2(n)$,

$$
\begin{aligned}
\chi_1(n) &= \sigma_3(n) \prod_{j<n} \sigma_1(j) \\
\chi_2(n) &= i\sigma_3(n) \prod_{j \le n} \sigma_1(j)
\end{aligned}
\tag{5.169}
$$

with the additional definitions $\chi_1(-N_s/2+1) = \sigma_1(-N_s/2+1)$ and $\chi_2(-N_s/2+1) = -\sigma_2(-N_s/2+1)$. These operators are self-adjoint, $\chi_1^\dagger(n) = \chi_1(n)$ and $\chi_2^\dagger(n) = \chi_2(n)$, anti-commute with each other and square to the identity. Hence they obey the algebra

$$
\begin{aligned}
&\{\chi_1(j), \chi_1(j')\} = \{\chi_1(j), \chi_1(j')\} = \delta_{j,j'} \\
&\{\chi_1(j), \chi_2(j')\} = 0
\end{aligned}
\tag{5.170}
$$

Operators that obey this algebra are known as Majorana fermions.

In terms of the Majorana operators, $\chi_1(n)$ and $\chi_2(n)$, the Hamiltonian of the quantum Ising chain is

$$
H = -\sum_n i\chi_1(n)\chi_2(n) - \lambda \sum_n i\chi_2(n)\chi_1(n+1)
\tag{5.171}
$$

The spin-flip operator $\mathcal{R}$, which generates the global $\mathbb{Z}_2$ symmetry, takes the form

$$
\mathcal{R} = \prod_n \sigma_1(n) = i^{N_s} \prod_n (\chi_1(n)\chi_2(n))
\tag{5.172}
$$

which (as it should do) commutes with the Hamiltonian.

5.5.1 The Majorana-fermion universality class

It is instructive to derive the (Heisenberg) equations of motion of the Majorana operators:

$$
\begin{aligned}
i\,\partial_t \chi_1(n) &= i\chi_2(n) - i\lambda\chi_2(n-1) \\
i\,\partial_t \chi_2(n) &= -i\chi_1(n) + i\lambda\chi_1(n+1)
\end{aligned}
\tag{5.173}
$$

These equations of motion are linear. Hence, the Majorana fields χ_1 and χ_2 are *free*! In contrast, it is simple to see that the equations of motion of the spin operators $\sigma_1(n)$ and $\sigma_3(n)$ are *not* linear, and hence these fields are not free.

We will now restore a lattice constant $a \neq 1$ and set $x_n = na_0$ to take the continuum limit of the equations of motion of the Majorana fields, Eq. (5.173),

$$
\begin{aligned}
\chi_1(n+1) &\approx \chi_1(x_n) + a_0\,\partial_x \chi_1(x_n) + O(a_0^2) \\
\chi_2(n-1) &\approx \chi_2(x_n) - a_0\,\partial_x \chi_2(x_n) + O(a_0^2)
\end{aligned}
\tag{5.174}
$$

Hence we can rewrite Eqs. (5.173) in the form

$$\begin{aligned} \frac{1}{a_0\lambda} i\, \partial_t \chi_1 &\simeq i \left(\frac{1-\lambda}{a_0\lambda}\right) \chi_2 + i\, \partial_x \chi_2 \\ \frac{1}{a_0\lambda} i\, \partial_t \chi_2 &\simeq -i \left(\frac{1-\lambda}{a_0\lambda}\right) \chi_1 + i\, \partial_x \chi_1 \end{aligned} \tag{5.175}$$

We now rescale the time coordinate $t \to (a_0\lambda)x_0$, relabel the space coordinate as $x \to x_1$, and rescale the Majorana fields $\chi_i(x_n) \to (1/\sqrt{2a_0})\chi_i(x_1)$ $(i = 1, 2)$. In this notation the equations of motion of the Majorana fields (in the continuum limit) become

$$\begin{aligned} i\, \partial_0 \chi_1 - i\, \partial_1 \chi_2 + im\chi_2 &= 0 \\ i\, \partial_0 \chi_2 - i\, \partial_1 \chi_1 - im\chi_1 &= 0 \end{aligned} \tag{5.176}$$

where we introduced the Majorana mass $m(\lambda)$ defined by the scaling limit

$$m = \lim_{\substack{a_0 \to 0 \\ \lambda \to 1}} \left(\frac{1-\lambda}{a_0\lambda}\right) \tag{5.177}$$

which vanishes right at $\lambda = 1$.

The (continuum) Majorana fields satisfy the equal-time anticommutation relations (with $i, j = 1, 2$)

$$\left\{\chi_i(x), \chi_j(x')\right\} = \delta_{ij}\delta(x - x') \tag{5.178}$$

Equations (5.176) are (in components) the Dirac equation for Majorana fields in $1+1$ dimensions. Indeed, if we define the two-component Majorana spinor field χ,

$$\chi = \begin{pmatrix} \chi_1 \\ \chi_2 \end{pmatrix} \tag{5.179}$$

which satisfies the Majorana condition

$$\chi^\dagger = \chi^{\mathrm{T}} = \sigma_1 \chi \tag{5.180}$$

we can write the equations of motion in the (Dirac–Majorana) form

$$i\partial\!\!\!/ \chi - im\chi^{\mathrm{T}} = 0 \tag{5.181}$$

In other terms, we have shown that the quantum Ising chain is equivalent, in the continuum limit, to a theory of free Majorana fermions in $(1+1)$ dimensions with a (Majorana) mass that tunes the distance to the self-dual point $\lambda_c = 1$. Since at $\lambda_c = 1$ the mass $m \to 0$, we conclude that the Majorana fermions are massless at λ_c. The mass m also defines a length scale that we will identify with the correlation length,

$$\xi = \frac{1}{|m|} \sim \frac{1}{|\lambda - \lambda_c|} \tag{5.182}$$

from which we conclude that the correlation-length exponent is $\nu = 1$. Hence the universality class of the Ising quantum chain (and of the 2D classical Ising model) is a theory of massless Majorana fermions.

5.5.2 Diagonalization of the Hamiltonian

We will now reexamine these results by diagonalizing the Hamiltonian of the quantum Ising chain explicitly. To do this let us return to the operators $\chi_1(n)$ and $\chi_2(n)$ of Eq. (5.169). Let us define the canonical (Dirac) fermion operator $\psi(n)$ and its adjoint $\psi^\dagger(n)$,

$$\psi(n) = \frac{1}{\sqrt{2}}\left(\chi_1(n) + i\chi_2(n)\right), \qquad \psi(n)^\dagger = \frac{1}{\sqrt{2}}\left(\chi_1(n) - i\chi_2(n)\right) \tag{5.183}$$

which satisfy the canonical anticommutator algebra

$$\left\{\psi(n), \psi(n')^\dagger\right\} = \delta_{n,n'}, \qquad \left\{\psi(n), \psi(n')\right\} = 0 \tag{5.184}$$

In terms of these fermions we obtain a Jordan–Wigner transformation

$$\begin{aligned} \sigma_1(n) &= 2\psi(n)^\dagger\psi(n) - 1 \\ \sigma_3(n) &= \left(\psi(n)^\dagger + \psi(n)\right)\exp\left(i\pi\sum_{j<n}\psi(j)^\dagger\psi(j)\right) \end{aligned} \tag{5.185}$$

We will now be more careful with the choice of boundary conditions. We will denote by $\eta = 1$ periodic and by $\eta = -1$ anti-periodic boundary conditions of the spins,

$$\sigma_3\left(\frac{N_s}{2} + 1\right) = \eta\sigma_3\left(-\frac{N_s}{2} + 1\right) \tag{5.186}$$

It follows that the fermions satisfy the boundary conditions

$$\psi\left(\frac{N_s}{2} + 1\right) = \mathcal{R}\eta\psi\left(-\frac{N_s}{2} + 1\right) \tag{5.187}$$

where $\mathcal{R}$ is the spin-flip operator, which now looks like

$$\mathcal{R} = e^{i\pi\mathcal{N}} \tag{5.188}$$

where

$$\mathcal{N} = \sum_n \psi(n)^\dagger\psi(n) \tag{5.189}$$

is the total fermion number. However, in the Ising chain

$$[\mathcal{N}, H] \neq 0, \qquad \text{but} \qquad [\mathcal{R}, H] = 0 \tag{5.190}$$

In other terms, the number of fermions $\mathcal{N}$ is not a conserved observable but conserved modulo two. Thus, *fermion parity*, measured by $\mathcal{R}$, is conserved. This is a natural consequence of the fact that this is actually a theory of real (and hence not complex) fermions, which do not have a conserved charge. Since the choice of periodic (or anti-periodic) boundary conditions for the spins does not completely specify the boundary conditions for the fermions (since their parity needs to be fixed), we conclude that there is a two-to-one relation between the system of fermions and the system of spins.

In terms of the fermion operators the Hamiltonian is

$$\begin{aligned} H = - \sum_{n=-\frac{N_s}{2}+1}^{\frac{N_s}{2}} & \left[2\psi^\dagger(n)\psi(n) + \lambda(\psi^\dagger(n) - \psi(n))(\psi^\dagger(n+1) + \psi(n+1))\right] \\ & + N_s + H_b \end{aligned} \tag{5.191}$$

where N_s is the number of sites and H_b is a boundary term for the coupling connecting the last with the first site:

$$H_b = -\lambda\eta\mathcal{R}(\psi^\dagger(N_s/2) - \psi(N_s/2))(\psi^\dagger(-N_s/2+1) + \psi(-N_s/2+1)) \tag{5.192}$$

If we compare the fermion Hamiltonian of Eq. (5.191) with the analogous fermion Hamiltonian for the fermionized version of the (anisotropic) quantum Heisenberg antiferromagnet of Eq. (5.92), we see two important differences: (a) in the Heisenberg case there is a fermion density-interaction term, which makes the system interacting except at the XY-model point; and (b) in the Heisenberg case the fermion number is conserved, whereas in the Ising case only the parity is conserved. In the Heisenberg case fermion-*number* conservation is due to the existence of the unbroken U(1) global symmetry of the anisotropic Heisenberg antiferromagnet, whereas in the Ising case the global symmetry is $\mathbb{Z}_2$, which leads to the conservation of fermion parity.

Since the flip operator $\mathcal{R}$ commutes with the Hamiltonian H, the energy eigenstates can be chosen also to be fermion-parity eigenstates. In a system with PBCs, $\eta = +1$, the ground state of the even-fermion-parity sector has lower energy than the ground state of the odd-parity sector. This is so since, in a system with PBCs, the odd-fermion-parity sector has at least one fermion and hence has an odd number of *domain walls*. Thus, we will work with PBCs for the spin system and within the even-parity fermion sector. This forces also the fermions to obey PBCs. In this sector the Hamiltonian is translationally invariant (and hence defect-free). Since $\{\sigma_3(n), \mathcal{R}\} = 0$ (for all n), the spin operator $\sigma_3(n)$ changes the boundary conditions for the fermions from periodic to anti-periodic.

Using the translation invariance of the fermion Hamiltonian, Eq. (5.191), with $\eta = +1$ and $\mathcal{R} = +1$, we will attempt to find its spectrum by means of a Fourier transform,

$$\psi(n) = \frac{1}{N_s} \sum_{k=-\frac{N_s}{2}+1}^{\frac{N_s}{2}} e^{i2\pi kn/N_s} a(k) \tag{5.193}$$

such that the operators $a(k)$ obey the same usual anticommutator algebra

$$\left\{a(k), a^\dagger(k')\right\} = \delta_{k,k'}, \qquad \left\{a(k), a(k')\right\} = \left\{a^\dagger(k), a^\dagger(k')\right\} = 0 \tag{5.194}$$

In the thermodynamic limit, $N_s \to \infty$, the integer variable k is replaced by the (lattice) momentum variable $k \equiv 2\pi k/N_s$, which takes values in the interval $-\pi \leq k < \pi$. In turn, the sums in Eq. (5.193) become momentum integrals

$$\psi(n) = \int_{-\pi}^{\pi} \frac{dk}{2\pi} e^{ikn} a(k) \tag{5.195}$$

and the non-vanishing anticommutators now are

$$\left\{a(k), a^\dagger(k')\right\} = 2\pi\,\delta(k - k') \tag{5.196}$$

where $\delta(k)$ is the periodic delta function,

$$2\pi\,\delta(k) = \lim_{N_s \to \infty} \sum_{n=-\frac{N_s}{2}+1}^{\frac{N_s}{2}} e^{ikn} \tag{5.197}$$

After some straightforward algebra we find that the Hamiltonian becomes

$$\begin{aligned} H = N_s &- \int_{-\pi}^{\pi} \frac{dk}{2\pi} 2(1 + \lambda \cos k) a^\dagger(k) a(k) \\ &- \int_{-\pi}^{\pi} \frac{dk}{2\pi} \lambda \left(e^{ik} a^\dagger(k) a^\dagger(-k) - e^{-ik} a(k) a(-k)\right) \end{aligned} \tag{5.198}$$

This Hamiltonian violates fermion-number conservation but, as it should, conserves fermion parity as fermions are created and destroyed in pairs with equal and opposite momentum. In this language this system is reminiscent of the *pairing Hamiltonian* of the Bardeen–Cooper–Schrieffer (BCS) theory of superconductivity (Schrieffer, 1964) (at the mean-field level).

It will be convenient to "fold" the momentum interval to $0 \leq k < \pi$ and to rewrite the Hamiltonian in the equivalent form

$$\begin{aligned} H = N_s &- \int_{0}^{\pi} \frac{dk}{2\pi} 2(1 + \lambda \cos k) \left(a^\dagger(k) a(k) + a^\dagger(-k) a(-k)\right) \\ &- \int_{-\pi}^{\pi} \frac{dk}{2\pi} 2i\lambda \sin k \left(a^\dagger(k) a^\dagger(-k) + a(k) a(-k)\right) \end{aligned} \tag{5.199}$$

Let us define the spinor field $\Psi(k)$ and its adjoint $\Psi^\dagger(k)$

$$\Psi(k) = \begin{pmatrix} \psi^\dagger(k) \\ \psi(-k) \end{pmatrix}, \qquad \Psi^\dagger(k) = \big(a(k), a^\dagger(-k)\big) \tag{5.200}$$

Notice that the two components of the spinor field $\Psi(k)$ are not independent. Indeed, we find that the spinor field obeys the Majorana condition

$$\Psi^\dagger(k) = [\sigma_1 \Psi(-k)]^{\mathrm{T}} = \Psi^{\mathrm{T}}(-k)\sigma_1 \tag{5.201}$$

where A^{T} is the transpose of the operator (or matrix) A, and σ_1 is the real and symmetric Pauli matrix. We have thus rederived in this language the condition that the fermions of the Ising model (and of superconductors!) are Majorana fermions.

As in the BCS theory of superconductivity, we will diagonalize the (pairing) Hamiltonian of Eq. (5.199) by means of a Bogoliubov transformation to a new set of fermions $\eta(k)$,

$$\begin{aligned} a(k) &= u(k)\eta(k) - iv(k)\eta^\dagger(-k) \\ a(-k) &= u(k)\eta(k) + iv(k)\eta^\dagger(k) \end{aligned} \tag{5.202}$$

where the amplitudes $u(k)$ and $v(k)$ are chosen to be real functions of k. The inverse transformation is

$$\begin{aligned} \eta(k) &= u(k)a(k) + iv(k)a^\dagger(-k) \\ \eta(-k) &= u(k)a(-k) - iv(k)a^\dagger(k) \end{aligned} \tag{5.203}$$

We will choose the amplitudes $u(k)$ so that the transformation is canonical, i.e. so that it preserves the (anti)commutation relations

$$\{a(k), a^\dagger(q)\} = 2\pi\delta(k-q) \Rightarrow \{\eta(k), \eta^\dagger(q)\} = 2\pi\delta(k-q) \tag{5.204}$$

This condition requires that

$$u^2(k) + v^2(k) = 1 \tag{5.205}$$

This condition is met by writing $u(k)$ and $v(k)$ in terms of a phase angle $\theta(k)$,

$$u(k) = \cos\theta(k), \qquad v(k) = \sin\theta(k) \tag{5.206}$$

We will choose the phase $\theta(k)$ in such a way that the Hamiltonian for the fermions $\eta(k)$ and $\eta(-k)$ does not contain fermion-non-conserving terms, i.e. we have that in the transformed Hamiltonian the coefficient of terms of the form $i(\eta^\dagger(k)\eta^\dagger(-k) + \eta(k)\eta(-k))$ vanishes identically. This leads to the condition

$$\tan(2\theta(k)) = \frac{\lambda \sin k}{1 + \lambda \cos k} \tag{5.207}$$

With these choices the Hamiltonian becomes

$$H = \varepsilon_0(\lambda)N_s + \int_0^{\pi} \frac{dk}{2\pi}\omega(k) \left(\eta^\dagger(k)\eta(k) + \eta^\dagger(-k)\eta(-k)\right) \tag{5.208}$$

where $\omega(k)$ is found to be

$$\omega(k) = 2\sqrt{(1+\lambda\cos k)^2 + \lambda^2\sin^2 k} \tag{5.209}$$

The ground state of this system, $|0\rangle$, has no excitations and satisfies

$$\eta(k)|0\rangle = 0, \quad \eta(-k)|0\rangle = 0 \tag{5.210}$$

The ground-state energy density, $\varepsilon_0(\lambda)$, is found to be

$$\varepsilon_0(\lambda) = -\int_0^{\pi} \frac{dk}{2\pi}\omega(k) < 0 \tag{5.211}$$

which is negative. After some algebra we can write the ground-state energy density as

$$\varepsilon_0(\lambda) = -\frac{2}{\pi}(1+\lambda)E\left(\frac{\pi}{2}, \sqrt{1-\gamma^2}\right), \qquad \gamma = \left|\frac{1-\lambda}{1+\lambda}\right| \tag{5.212}$$

where $E(\pi/2, k)$ (with $k = \sqrt{1-\gamma^2}$) is the complete elliptic integral of the second kind

$$E\left(\frac{\pi}{2}, k\right) = \int_0^{\pi/2} d\theta\sqrt{1-k^2\sin^2\theta} \tag{5.213}$$

where k is known as the modulus of the elliptic integral.

The excited states of this system are created by the fermion-creation operators $\eta^\dagger(k)$ and $\eta^\dagger(-k)$. Thus the lowest-energy excited state is $\eta^\dagger(k)|0\rangle$, which has an *excitation energy* $\omega(k)$. The lowest excited state has $\omega(k)$ smallest, which occurs at $k = \pi$, with excitation energy $E_{\text{gap}}(\lambda) = 2|1-\lambda|$. Therefore, as $\lambda \to 1$ the excitation gap vanishes with a critical exponent $\nu = 1$. As anticipated we will identify $\lambda = 1$ with the (quantum) critical point of the quantum Ising chain. The state $\eta^\dagger(k)|0\rangle$ has an odd number of fermions, and hence is not in this sector of the Hilbert space. In this sector, the lowest-energy state is a two-fermion state, $|k, p\rangle$, each with both momenta at π.

In classical statistical mechanics we know that as the critical temperature is approached the specific heat $c(T)$ diverges as

$$c(T) \simeq \text{constant} \times |T - T_c|^{-\alpha} \tag{5.214}$$

where α is a universal critical exponent. The ground-state energy density of the quantum Ising model is the same as the *free-energy* density of the classical problem. Thus, we can determine the singular behavior of the specific heat of the 2D classical Ising model by looking at the behavior of the ground-state energy density $\varepsilon_0(\lambda)$. The quantity related to the singular part of the specific heat of the classical problem is

$$c_{\text{sing}}\left(\frac{T - T_{\text{c}}}{T_{\text{c}}}\right) = -\frac{\partial^2 \varepsilon_0^{\text{sing}}(\lambda)}{\partial \lambda^2} \tag{5.215}$$

We can determine the singular part of the energy density by looking at its behavior as $\lambda \to 1$ (with $t = |\lambda - 1|$)

$$\varepsilon_0^{\text{sing}}(t) = -\frac{4}{\pi}\left[1 + \frac{t^2}{8}\left(\ln\left(\frac{8}{|t|}\right) - \frac{1}{2}\right) + \cdots\right] \tag{5.216}$$

which tells us that the specific heat of the classical 2D problem has a logarithmic divergence as $T \to T_{\text{c}}$,

$$c_{\text{sing}}(t) = \frac{1}{\pi} \ln\left(\frac{8}{|t|}\right) \tag{5.217}$$

This is the Onsager result. Hence the exponent is $\alpha = 0$, as expected since the correlation length exponent is $\nu = 1$.

The computation of the correlation functions of the Ising model is more subtle and technically more demanding than what we have done here (see e.g. McCoy and Wu (1973)). Nevertheless at the end of this chapter we will use bosonization results to compute the square of the spin–spin correlation function at the critical point.

5.6 Abelian bosonization

We now return to the fermion representation of the quantum Heisenberg antiferromagnetic chain. We are now going to discuss some subtle but very important properties of 1D Fermi systems. To date, these properties are not known to generalize to higher dimensions.

A very important tool for the understanding of 1D Fermi systems is the *bosonization transformation*. In its abelian form this transformation was first discussed by Bloch (1933) and Tomonaga (1950). It was rediscovered (and better understood) by Lieb and Mattis (1965) in the 1960s, and by Coleman (1975), Luther and Peschel (1975), and Mandelstam (1975) in the 1970s. Witten (1984) solved the non-abelian version of bosonization in 1984. In this section we will consider only the abelian case. Non-abelian bosonization will be discussed in Chapter 7.

Let us consider first a theory of non-interacting (spinless) fermions with Hamiltonian H_0 given (in units in which the Fermi velocity is $v_F = 1$) by

$$H_0 = \int dx\, \psi^\dagger i\alpha\, \partial_x \psi \tag{5.218}$$

where $\alpha = \gamma_5$ (defined in Section 5.2.3), with canonically quantized Fermi fields, i.e.

$$\begin{aligned} \{\psi_\alpha^\dagger(x), \psi_{\alpha'}(x')\} &= \delta_{\alpha\alpha'}\delta(x - x') \\ \{\psi_\alpha(x), \psi_{\alpha'}(x')\} &= \{\psi_\alpha^\dagger(x), \psi_{\alpha'}^\dagger(x')\} = 0 \end{aligned} \tag{5.219}$$

at equal times. The Hamiltonian H_0 and the canonical anticommutation relations follow from canonical quantization (for fermions!) of the system with Lagrangian density

$$\mathcal{L}_0 = \bar\psi i\gamma^\mu\, \partial_\mu \psi = \bar\psi i\gamma^0\, \partial_0\psi - \bar\psi i\gamma^1\, \partial_1\psi \tag{5.220}$$

which has the form of the relativistic Dirac Lagrangian density in $(1 + 1)$ dimensions. All along I have assumed that the metric tensor $g_{\mu\nu}$ is

$$g_{\mu\nu} = \begin{pmatrix} 1 & 0 \\ 0 & -1 \end{pmatrix} \tag{5.221}$$

This Lagrangian density is clearly invariant under global continuous chiral transformations. In fact, the Hamiltonian density, in the chiral basis, is

$$H_0 = -\int dx (R^\dagger i\, \partial_x R - L^\dagger i\, \partial_x L) \tag{5.222}$$

which implies that the right (left)-moving component R (L) moves towards the right (left) at speed 1 (in units in which $v_F = 1$).

5.6.1 Anomalous commutators

Consider now the "vacuum states" $|0\rangle$ and $|G\rangle$, where $|0\rangle$ is the *empty* state and $|G\rangle$ is the *filled Fermi sea* obtained by having occupied all the negative-energy one-particle eigenstates of the Hamiltonian Eq. (5.222). The Hamiltonian H_0 relative to both vacua differs by normal-ordering terms. Indeed, for any eigenstate $|F\rangle$ of H_0 one can write

$$H_0 =: H_0 : + E_F |F\rangle\langle F| \tag{5.223}$$

where $: H_0 :$ is the Hamiltonian normal ordered with respect to $|F\rangle$, i.e.

$$: H_0 : |F\rangle = \langle F| : H_0 := 0 \tag{5.224}$$

and E_F is the energy of $|F\rangle$,

$$H_0|F\rangle = E_F|F\rangle \tag{5.225}$$

Clearly, if we choose $|0\rangle$ or $|G\rangle$ as the reference state, E_F will be different.

The currents and densities also need to be normal-ordered. This is equivalent to the subtraction of the (infinite) background charge of the reference state, say of the filled Fermi sea. We will see that these apparently "formal" manipulations have a profound effect on the physics.

Let us compute the commutator of the charge density and current operators at equal times $[j_0(x), j_1(x')]$. Relative to the empty state $|0\rangle$, both operators are already normal-ordered since a state with no fermions has neither charge nor current, i.e.

$$j_0(x)|0\rangle = 0, \qquad j_1(x)|0\rangle = 0 \tag{5.226}$$

It will be useful to consider the right and left components of the current $j_\pm$ defined by

$$j_\pm = \frac{1}{2}(j_0 \pm j_1) \tag{5.227}$$

Clearly, we get that

$$j_+ = R^\dagger R \tag{5.228}$$

is the right-moving current, and

$$j_- = L^\dagger L \tag{5.229}$$

is the left-moving current. In Fourier components, we find

$$j_+(p) = \frac{1}{\sqrt{L_0}} \sum_k R^\dagger(k) R(k+p) \tag{5.230}$$

which annihilates the empty state $|0\rangle$. In fact, for any state $|\phi\rangle$ with a *finite* number of particles, the result is

$$[j_\pm(p), j_\pm(p')]|\phi\rangle = 0 \tag{5.231}$$

Consider now the filled Fermi sea, $|G\rangle$. Explicitly we can write

$$|G\rangle = \prod_{p<0} R^\dagger(p) \prod_{q>0} L^\dagger(q)|0\rangle \tag{5.232}$$

In other words, in $|G\rangle$ all right-moving states with negative momentum and all left-moving states with positive momentum are filled (see Fig. 5.2).

Let us compute the commutator $[j_+(x), j_+(x')]$ at equal times (see, for instance, Affleck (1986a)). The operator $j_+(x)$ is formally equal to a product of fermion

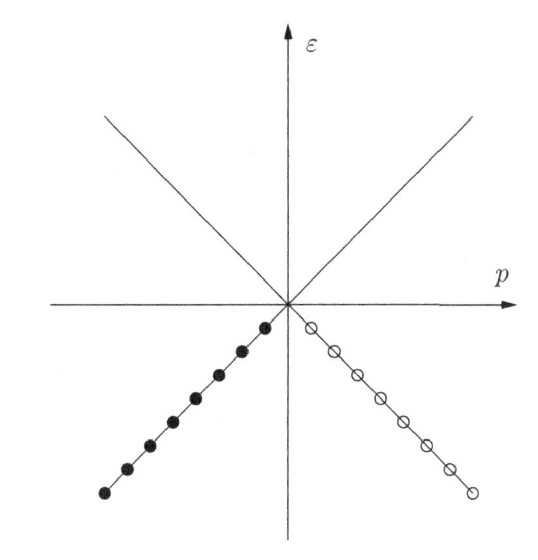

Figure 5.2 Vacuum $|G\rangle$ is obtained by filling the right-moving states with negative momentum (filled circles) and filling the left-moving states with positive momentum (empty circles).

operators at the same point. Since we anticipate divergences, we should "point-split" the product

$$j_+(x) = R^\dagger(x)R(x) = \lim_{\epsilon\to 0} R^\dagger(x+\epsilon)R(x-\epsilon) \tag{5.233}$$

and write j_+ in terms of a normal-ordered operator $:j_+:$ and a vacuum expectation value

$$j_+(x) =: j_+(x) : + \lim_{\epsilon\to 0}\langle G|R^\dagger(x+\epsilon)R(x-\epsilon)|G\rangle \tag{5.234}$$

The singularities are absorbed into the expectation value.

Consider a system on a segment of length L_0 with PBCs and expand $R(x)$ in Fourier series

$$R(x) = \frac{1}{\sqrt{L_0}}\sum_{p=-\infty}^{+\infty} R_p e^{i\frac{2\pi x p}{L_0}} \tag{5.235}$$

The vacuum expectation value to be computed is

$$\langle G|R^\dagger(x+\epsilon)R(x-\epsilon)|G\rangle = \frac{1}{L_0}\sum_{p,p'=-\infty}^{+\infty} e^{i\frac{2\pi}{L_0}[(x-\epsilon)p'-(x+\epsilon)p]}\langle G|R_p^\dagger R_p|G\rangle \tag{5.236}$$

Using the definition of the filled Fermi sea, we get

$$\langle G|R_p^\dagger R_{p'}|G\rangle = \delta_{p,p'}\theta(-p) \tag{5.237}$$

$$\langle G|L_p^\dagger L_{p'}|G\rangle = \delta_{p,p'}\theta(+p) \tag{5.238}$$

Hence

$$\langle G|R^\dagger(x+\epsilon)R(x-\epsilon)|G\rangle = \frac{1}{L_0}\sum_{p=-\infty}^{0} e^{-i\frac{2\pi p}{L_0}(2\epsilon)} \tag{5.239}$$

This is a conditionally convergent series. In order to make it convergent, we will regulate this series by damping out the contributions due to states deep below the Fermi energy. We can achieve this if we analytically continue ϵ to the upper half of the complex plane (i.e. $\epsilon \to \epsilon + i\eta$) to get the convergent expression

$$\begin{aligned}
\langle G|R^\dagger(x+\epsilon)R(x-\epsilon)|G\rangle &= \lim_{\eta\to 0}\frac{1}{L_0}\sum_{p=0}^{\infty} e^{i\frac{4\pi p}{L_0}(\epsilon+i\eta)} \\
&= \lim_{\eta\to 0}\frac{1}{L_0\left(1-e^{i\frac{4\pi}{L_0}(\epsilon+i\eta)}\right)} \\
&= \lim_{\eta\to 0}\frac{1}{L_0\left(-i(4\pi/L_0)(\epsilon+i\eta)\right)} \\
&= \frac{i}{4\pi\epsilon}
\end{aligned} \tag{5.240}$$

Thus, the result is

$$\langle G|R^\dagger(x+\epsilon)R(x-\epsilon)|G\rangle = \frac{i}{4\pi\epsilon} \tag{5.241}$$

Similarly, the expectation value $\langle G|L^\dagger(x+\epsilon)L(x-\epsilon)|G\rangle$ is found to be given by

$$\langle G|L^\dagger(x+\epsilon)L(x-\epsilon)|G\rangle = -\frac{i}{4\pi\epsilon} \tag{5.242}$$

The current commutator can now be readily evaluated:

$$\begin{aligned}
[j_+(x), j_+(x')] &= \lim_{\epsilon,\epsilon'\to 0}\left[R^\dagger(x+\epsilon)R(x-\epsilon), R^\dagger(x'-\epsilon')R(x'+\epsilon')\right] \\
&= \lim_{\epsilon,\epsilon'\to 0}\Big\{\delta(x'-x+\epsilon'+\epsilon)R^\dagger(x+\epsilon)R(x'-\epsilon') \\
&\qquad - \delta(x-x'+\epsilon'+\epsilon)R^\dagger(x'+\epsilon')R(x-\epsilon)\Big\}
\end{aligned} \tag{5.243}$$

The contributions from normal-ordered products cancel out (since they are regular). The only non-zero terms are, using Eq. (5.239),

$$[j_+(x), j_+(x')] = \lim_{\epsilon,\epsilon'\to 0}\left(\frac{i\delta(x'-x+\epsilon'+\epsilon)}{2\pi(x-x'+\epsilon+\epsilon')} - \frac{i\delta(x'-x+\epsilon+\epsilon')}{2\pi(x'-x+\epsilon+\epsilon')}\right) \tag{5.244}$$

Thus, in this limit we find

$$[j_+(x), j_+(x')] = -\frac{i}{2\pi}\,\partial_x\delta(x-x') \tag{5.245}$$

and

$$[j_-(x), j_-(x')] = +\frac{i}{2\pi}\,\partial_x\delta(x-x') \tag{5.246}$$

In terms of Lorentz components, we get

$$[j_0(x), j_1(x')] = -\frac{i}{\pi}\,\partial_x\delta(x-x') \tag{5.247}$$

whereas

$$[j_0(x), j_0(x')] = [j_1(x), j_1(x')] = 0 \tag{5.248}$$

The commutator $[j_0(x), j_1(x')]$ has a non-vanishing right-hand side, which is a c-number. These terms are generally known as *Schwinger terms*. They are pervasive in theories of relativistic fermions. But terms of this sort are also found in non-relativistic systems of fermions at finite densities. In fact, these terms are the key to the derivation of the f-sum rule (Pines and Nozières, 1966; Martin, 1967).

5.6.2 The bosonization rules

We thus notice that the equal-time current commutator $[j_0(x), j_1(x')]$ acquires a Schwinger term if the currents and densities are normal-ordered relative to the filled Fermi sea. The identity of Eq. (5.247) suggests that there should be a connection between a canonical Fermi field ψ with a filled Fermi sea and a canonical Bose field ϕ. Let $\Pi(x)$ be the canonical momentum conjugate to ϕ, i.e. at equal times

$$[\phi(x), \Pi(x')] = i\delta(x-x') \tag{5.249}$$

If we identify the normal-ordered operators

$$j_0(x) = \frac{1}{\sqrt{\pi}}\,\partial_x\phi(x) \tag{5.250}$$

and

$$j_1(x) = -\frac{1}{\sqrt{\pi}}\,\partial_t\phi(x) \equiv -\frac{1}{\sqrt{\pi}}\Pi(x) \tag{5.251}$$

we see that Eq. (5.249) implies

$$\frac{1}{\pi}\left[\partial_x\phi(x), \Pi(x')\right] = \frac{i}{\pi}\delta'(x-x') \tag{5.252}$$

which is consistent with the Schwinger term. These equations can be written in the more compact form

$$j_\mu = \frac{1}{\sqrt{\pi}} \epsilon_{\mu\nu} \, \partial^\nu \phi \tag{5.253}$$

where $\epsilon_{\mu\nu}$ is the (antisymmetric) Levi-Civita tensor and we are using *from now on* the notation $t \to x_0$, $x \to x_1$, and $x \equiv (x_0, x_1)$. We then arrive at the conclusion that the current commutator with a Schwinger term, Eq. (5.247), is equivalent to the statement that there exists a canonical Bose field ϕ whose *topological current*, Eq. (5.253), coincides with the normal-ordered fermion current.

The fermion current j_μ is conserved, i.e.

$$\partial_\mu j^\mu = 0 \tag{5.254}$$

which is automatically satisfied by Eq. (5.253). In the case of the *free* theory, the numbers of left and right movers are separately conserved. This means that not only should j_μ be conserved, but also j_μ^5, defined by

$$j_\mu^5 = \bar{\psi} \gamma_\mu \gamma^5 \psi \tag{5.255}$$

should be conserved. Using the identity

$$\gamma_\mu \gamma_5 = \epsilon_{\mu\nu} \gamma^\nu \tag{5.256}$$

we see that j_μ and j_μ^5 are in fact related by

$$j_\mu^5 = \epsilon_{\mu\nu} j^\nu \tag{5.257}$$

The divergence of j_μ^5 can be computed in terms of the Bose field ϕ as follows:

$$\partial_\mu j^{5\mu} = \epsilon^{\mu\nu} \, \partial_\mu j_\nu = \frac{1}{\sqrt{\pi}} \epsilon^{\mu\nu} \epsilon_{\nu\lambda} \, \partial_\mu \partial^\lambda \phi = \frac{1}{\sqrt{\pi}} \partial^2 \phi \tag{5.258}$$

Thus, the conservation of the *axial* current j_μ^5 implies that ϕ should be a free canonical Bose field

$$\partial_\mu j^{5\mu} = 0 \Rightarrow \partial^2 \phi = 0 \tag{5.259}$$

where

$$\partial^2 \equiv \partial_0^2 - \partial_1^2 \tag{5.260}$$

The Lagrangian for these bosons is simply given by

$$\mathcal{L}_{\rm B} = \frac{1}{2} \left(\partial_\mu \phi \right)^2 \tag{5.261}$$

Conversely, if ϕ is not free j_μ^5 should not be conserved. We will see below that this is indeed what happens in the Thirring–Luttinger model.

Before doing that, let us consider a set of identities originally derived by Mandelstam (1975). By analogy with the Jordan–Wigner transformation of Section 5.2.1, we should expect that these identities should be highly non-local, although they should have local anticommutation relations. These identities, like all others derived within the bosonization approach, only make sense within the operator-product expansion (OPE): the operators so identified give rise to the same leading singular behavior when arbitrary matrix elements are computed. Also, from the Jordan–Wigner analogy, we should expect that the fermion operators, as seen from their representation in terms of bosons, should act like operators that create *solitons*.

The free Bose field ϕ can be written in terms of creation and annihilation operators. Let $\phi^+(x)$ ($\phi^-(x)$) denote the piece of $\phi(x)$ which depends on the creation (annihilation) operators only,

$$\phi(x) = \phi^+(x) + \phi^-(x) \tag{5.262}$$

where $\phi(x)$ is a Heisenberg operator ($x \equiv (x_0, x_1)$, see Eq. (5.253)). Obviously, ϕ^- annihilates the vacuum of the Bose theory. The operators ϕ^+ and ϕ^- obey the commutation relations

$$[\phi^+(x_0, x_1), \phi^-(x_0', x_1')] = \lim_{\epsilon \to 0} \Delta_+(x_0 - x_0', x_1 - x_1') \tag{5.263}$$

where Δ_+ is given by

$$\Delta_+(x_0 - x_0', x_1 - x_1') = -\frac{1}{4\pi} \ln \left[\frac{(x_1 - x_1')^2 - (x_0 - x_0' + i\epsilon)^2}{a^2} \right] \tag{5.264}$$

where a is a short-distance cutoff, and it is necessary to make the argument of the logarithm dimensionless.

Consider now the operators $\mathcal{O}_\alpha(x)$ and $\mathcal{Q}_\beta(x)$ defined by

$$\mathcal{O}_\alpha(x) = e^{i\alpha\phi(x)} \tag{5.265}$$

and

$$\mathcal{Q}_\beta(x) = e^{i\beta \int_{-\infty}^{x_1} dx_1' \, \partial_0 \phi(x_0, x_1')} \equiv e^{i\beta \int_{-\infty}^{x_1} dx_1' \, \Pi(x_0, x_1')} \tag{5.266}$$

When acting on a state $|\{\phi(x')\}\rangle$, $\mathcal{O}_\alpha(x)$ simply multiplies the state by $e^{i\alpha\phi(x)}$. The operator $\mathcal{Q}_\beta(x)$ has quite a different effect. Since $\Pi(x)$ and $\phi(x)$ are conjugate pairs, $\mathcal{Q}_\beta(x)$ will shift the value of $\phi(x_0, x_1')$ to $\phi(x_0, x_1') + \beta$ for all $x_1' < x_1$. Thus, $\mathcal{Q}_\beta(x)$ creates a coherent state, which we can call a *soliton*:

$$\mathcal{Q}_\beta(x)|\{\phi(x_0, x_1')\}\rangle = |\{\phi(x_0, x_1') + \beta\theta(x_1 - x_1')\}\rangle \tag{5.267}$$

Consider now the operator $\psi_{\alpha,\beta}(x)$ of the form

$$\psi_{\alpha,\beta}(x) = \mathcal{O}_\alpha(x)\mathcal{Q}_\beta(x) = e^{i\alpha\phi(x) + i\beta \int_{-\infty}^{x_1} dx_1' \, \partial_0 \phi(x_0, x_1')} \tag{5.268}$$

and compute the product $\psi_{\alpha,\beta}(x)\psi_{\alpha,\beta}(x')$ at equal times ($x'_0 = x_0$). Using the Baker–Hausdorff formula

$$e^{\hat{A}}e^{\hat{B}} = e^{\hat{B}}e^{\hat{A}}e^{-[\hat{A},\hat{B}]} = e^{\hat{A}+\hat{B}-\frac{1}{2}[\hat{A},\hat{B}]} \tag{5.269}$$

where $[\hat{A}, \hat{B}]$ is a complex-valued distribution, we get

$$\psi_{\alpha,\beta}(x)\psi_{\alpha,\beta}(x') = \psi_{\alpha,\beta}(x')\psi_{\alpha,\beta}(x)e^{-i\Phi(x,x')} \tag{5.270}$$

where $\Phi(x, x')$ is given (all the commutators are understood to be at equal times and $x_0 = x'_0$ but $x'_1 \neq x_1$) by

$$\begin{aligned} i\Phi(x, x') = & -\alpha^2[\phi(x), \phi(x')] - \beta^2 \int_{-\infty}^{x_1} dy_1 \int_{-\infty}^{x'_1} dy'_1[\Pi(y), \Pi(y')] \\ & - \alpha\beta \int_{-\infty}^{x'_1} dy'_1[\phi(x), \Pi(y')] - \alpha\beta \int_{-\infty}^{x_1} dy_1[\Pi(y), \phi(x')] \\ = & -i\alpha\beta \end{aligned} \tag{5.271}$$

For the operators $\psi_{\alpha,\beta}(x)$ to have *fermion commutation relations* we need to choose $\alpha\beta = \pm\pi$. It is useful to write left and right components of the Fermi field in the form (Mandelstam, 1975)

$$R(x) = \frac{1}{\sqrt{2\pi a}} : e^{-i\frac{2\pi}{\beta}\int_{-\infty}^{x_1} dx'_1\, \Pi(x_0,x'_1)+i\frac{\beta}{2}\phi(x)} : \tag{5.272}$$

$$L(x) = \frac{1}{\sqrt{2\pi a}} : e^{-i\frac{2\pi}{\beta}\int_{-\infty}^{x_1} dx'_1\, \Pi(x_0,x'_1)-\frac{i\beta}{2}\phi(x)} : \tag{5.273}$$

The constant β is arbitrary and it can be chosen by demanding that the currents satisfy the operator identity

$$j_\mu = \frac{1}{\sqrt{\pi}}\epsilon_{\mu\nu}\, \partial^\nu \phi \tag{5.274}$$

From Eqs. (5.272) and (5.273), it follows that the free fermionic current is identified with the bosonic operator (Mandelstam, 1975)

$$j_\mu = \frac{\beta}{2\pi}\epsilon_{\mu\nu}\, \partial^\nu \phi \tag{5.275}$$

Thus, we must choose $\beta = \sqrt{4\pi}$ for the *free*-fermion problem.

The free-scalar-field operator $\phi(x)$ and the canonical momentum $\Pi(x)$ have the mode expansions

$$\phi(x) = \int_{-\infty}^{\infty} \frac{dk}{2\pi} \frac{1}{2|k|} \left(a(k) e^{i(|k|x_0 - kx_1)} + a^\dagger(k) e^{-i(|k|x_0 - kx_1)} \right)$$
$$\Pi(x) = \int_{-\infty}^{\infty} \frac{dk}{2\pi} \frac{1}{2|k|} \left(i|k| a(k) e^{i(|k|x_0 - kx_1)} - i|k| a^\dagger(k) e^{-i(|k|x_0 - kx_1)} \right) \tag{5.276}$$

where the creation and annihilation operators obey standard commutation relations, i.e. $[a(k), a^\dagger(k')] = (2\pi) 2|k| \delta(k - k')$.

The field operator $\phi(x)$ and the canonical momentum $\Pi(x)$ admit a decomposition in terms of right- and left-moving chiral bosonic fields, $\phi_R(x) \equiv \phi_R(x_0 - x_1)$ and $\phi_L(x) \equiv \phi_L(x_0 + x_1)$, which are given by

$$\phi_R(x_0 - x_1) = \int_0^{\infty} \frac{dk}{2\pi} \frac{1}{2k} \left(a(k) e^{ik(x_0 - x_1)} + a^\dagger(k) e^{-ik(x_0 - x_1)} \right) \tag{5.277a}$$

$$\phi_L(x_0 + x_1) = \int_{-\infty}^{0} \frac{dk}{2\pi} \frac{1}{2k} \left(-a(k) e^{-ik(x_0 + x_1)} + a^\dagger(k) e^{ik(x_0 + x_1)} \right) \tag{5.277b}$$

It is convenient to introduce the *dual* field $\vartheta(x)$, defined by

$$\Pi(x) = \partial_1 \vartheta(x) \tag{5.278}$$

or, equivalently (up to a suitably defined boundary condition),

$$\vartheta(x) \equiv \int_{-\infty}^{x_1} dx_1' \; \Pi(x_0, x_1') \tag{5.279}$$

The field operator $\phi(x)$ and the dual field operator $\vartheta(x)$ obey the *Cauchy–Riemann* equations

$$\partial_0 \phi = \partial_1 \vartheta, \qquad \partial_1 \phi = -\partial_0 \vartheta \tag{5.280}$$

as operator identities. The chiral decomposition reads

$$\begin{aligned} \phi(x_0, x_1) &= \phi_R(x_0 - x_1) + \phi_L(x_0 + x_1) \\ \vartheta(x_0, x_1) &= -\phi_R(x_0 - x_1) + \phi_L(x_0 + x_1) \end{aligned} \tag{5.281}$$

In this subsection we will work primarily with the free-fermion problem. In this case the Mandelstam identities, Eq. (5.272) and Eq. (5.273), take the simpler form

$$\begin{aligned} R(x) &= \frac{1}{\sqrt{2\pi a}} : e^{i2\sqrt{\pi}\phi_R(x)} : \\ L(x) &= \frac{1}{\sqrt{2\pi a}} : e^{-i2\sqrt{\pi}\phi_L(x)} : \end{aligned} \tag{5.282}$$

It is interesting to consider products of the form $\lim_{y_1 \to x_1} R^\dagger(x) L(y)$ and $\lim_{y_1 \to x_1} L^\dagger(y) R(x)$ at equal times. We will use Mandelstam's formulas,

Eqs. (5.282), to derive an operator product expansion for $R^\dagger L$ and $L^\dagger R$, both to leading order. We find

$$\lim_{y_1 \to x_1} R^\dagger(x) L(y) = \frac{1}{2\pi a} : e^{i2\sqrt{\pi}\phi_{\mathrm{R}}(x)} :: e^{-i2\sqrt{\pi}\phi_{\mathrm{L}}(y)} : \tag{5.283}$$

We can make use of the Baker–Hausdorff formula once again, now in the form

$$: e^{\hat{A}} :: e^{\hat{B}} := e^{[\hat{A}^+, \hat{B}^-]} : e^{\hat{A}+\hat{B}} : \tag{5.284}$$

and write down a *bosonic* expression for $R^\dagger L$. The normal-ordered operator is, by definition, regular. Thus we can take the limit readily to find

$$\lim_{y \to x} : e^{\hat{A}+\hat{B}} : = : e^{-i\beta\phi(x)} : \tag{5.285}$$

This operator is multiplied by a singular coefficient that compensates for the fact that $R^\dagger L$ and $e^{-i\beta\phi}$ have superficially different scaling dimensions. An explicit calculation gives the operator identity

$$\lim_{y_1 \to x_1} R^\dagger(x) L(y) = \frac{1}{2\pi a} : e^{-i2\sqrt{\pi}\phi(x)} : \tag{5.286}$$

Similarly, one finds the identification

$$\lim_{y_1 \to x_1} L^\dagger(x_0, y_1) R(x_0, x_1) = \frac{1}{2\pi a} : e^{+i2\sqrt{\pi}\phi(x)} : \tag{5.287}$$

To sum up, the Dirac mass bilinear operator $\bar{\psi}\psi$ at $\beta = \sqrt{4\pi}$ is given by

$$\bar{\psi}(x)\psi(x) \equiv \lim_{y_1 \to x_1} \bar{\psi}(x_0, x_1)\psi(x_0, y_1) = \frac{1}{\pi a} : \cos(\sqrt{4\pi}\phi(x)) : \tag{5.288}$$

In the Ising regime of the Heisenberg model, we expect $\langle\bar{\psi}\psi\rangle$ to be different from zero and therefore the bosonic theory should have a ground state such that the expectation value $\langle\cos(\sqrt{4\pi}\phi)\rangle$ is not zero. Under a chiral transformation by $\theta = \pi/2$, $\bar{\psi}\psi$ transforms as

$$\bar{\psi}\psi \to -\bar{\psi}\psi \tag{5.289}$$

which is equivalent to a sublattice exchange. In bosonic language, this transformation amounts to

$$\phi \to \phi + \frac{\pi}{\sqrt{4\pi}} \tag{5.290}$$

The Umklapp operators play a crucial role here (Emery, 1979; den Nijs, 1981; Haldane, 1982). These operators enter the interaction Hamiltonian through terms of the form (see Eq. (5.132))

$$\int dx_1 \left\{ (R^\dagger L)^2 + (L^\dagger R)^2 \right\} \tag{5.291}$$

These terms can be bosonized using the Mandelstam identities Eqs. (5.282). Indeed, we get the (equal-time) operator expansion

$$\begin{aligned}
\lim_{y_1 \to x_1} (R^\dagger(x)L(y))^2 &= \left(\frac{1}{2\pi a}\right)^2 : e^{-i\sqrt{4\pi}\phi(x)} :: e^{-i\sqrt{4\pi}\phi(y)} : \\
&= \left(\frac{1}{2\pi a}\right)^2 : e^{-4\pi[\phi^+(x),\phi^-(y)]} :: e^{-i2\sqrt{4\pi}\phi(x)} : \\
&= \left(\frac{1}{2\pi a}\right)^2 e^{-4\pi\Delta_+(0^+,x_1-y_1)} : e^{-i2\sqrt{4\pi}\phi(x)} :
\end{aligned} \tag{5.292}$$

where Eqs. (5.269) and (5.263) have been used. In short, the bosonized version of the Umklapp terms is (at $\beta = \sqrt{4\pi}$)

$$\lim_{y_1 \to x_1} (R^\dagger(x)L(y))^2 = \left(\frac{1}{2\pi a}\right)^2 : e^{-i4\sqrt{\pi}\phi(x)} : \tag{5.293}$$

and likewise

$$\lim_{y_1 \to x_1} (L^\dagger(y)R(x))^2 = \left(\frac{1}{2\pi a}\right)^2 : e^{+i4\sqrt{\pi}\phi(x)} : \tag{5.294}$$

5.6.3 The sine–Gordon theory

Now that we have done all the hard work and derived the necessary identities, we are in position to write down the bosonized form of the Lagrangian. The *fermionic* Lagrangian density (see Eqs. (5.220) and (5.132))

$$\mathcal{L}_\text{F} = \bar{\psi} i \gamma^\mu \partial_\mu \psi - \gamma(\bar{\psi}\gamma_\mu\psi)^2 + 2\gamma\left((R^\dagger L)^2 + (L^\dagger R)^2\right) \tag{5.295}$$

which we showed was equivalent to the Heisenberg model (in the continuum limit), is thus equivalent to a *bosonic* theory with Lagrangian density (see Eqs. (5.261) and (5.274))

$$\mathcal{L}_\text{B} = \frac{1}{2}\left(\partial_\mu\phi\right)^2 - \frac{\gamma}{\pi}\epsilon_{\mu\nu}\,\partial^\nu\phi\,\epsilon^{\mu\lambda}\,\partial_\lambda\phi + \frac{\gamma}{\pi^2 a^2} : \cos(4\sqrt{\pi}\phi) : \tag{5.296}$$

Lorentz invariance is kept in this form of the bosonized Lagrangian. In Chapter 6 we will do a somewhat different analysis in which in addition to a renormalization of the stiffness (or compactification radius) there is a finite non-universal renormalization of the speed.

Using the identity

$$\epsilon_{\mu\nu}\epsilon^{\mu\lambda} = -\delta^\lambda_\nu \tag{5.297}$$

we can write

$$\mathcal{L}_\text{B} = \frac{1}{2}\left(\partial_\mu\phi\right)^2 + \frac{\gamma}{\pi}\left(\partial_\mu\phi\right)^2 + \frac{\gamma}{\pi^2 a^2} : \cos(4\sqrt{\pi}\phi) : \tag{5.298}$$

Thus, the interactions in the fermions give rise to (a) a rescaling of the Bose field ϕ and (b) a non-linear term.

This Lagrangian density can be brought into the canonical form by a simple rescaling of the field $\phi(x)$

$$\left(1+\frac{2\gamma}{\pi}\right)^{1/2}\phi(x)\equiv\varphi(x) \tag{5.299}$$

If we define β by the expression

$$\beta^2=\frac{4\pi}{1+2\gamma/\pi} \tag{5.300}$$

we can write the Lagrangian in the sine–Gordon form

$$\mathcal{L}_{\mathrm{B}}=\frac{1}{2}\left(\partial_\mu\varphi\right)^2+g:\cos(2\beta\varphi): \tag{5.301}$$

where g, the sine–Gordon coupling constant, is given by

$$g\approx\frac{\gamma}{\pi^2a_0^2} \tag{5.302}$$

up to a finite non-universal multiplicative constant determined by the short-distance cutoff (i.e. we have arbitrarily set $c\mu a_0=1$). Thus, the effective bosonized theory has the sine–Gordon form, a problem that we discussed in detail in Chapter 4.

The rescaling of ϕ implies that the canonical momentum Π should also be rescaled so as to keep the form of the canonical commutation relations. Thus Π is scaled as

$$\Pi=\left(1+\frac{2\gamma}{\pi}\right)^{-1/2}\partial_0\varphi \tag{5.303}$$

The Mandelstam operators now read (see Eqs. (5.282))

$$\begin{aligned}R(x)&=\frac{1}{\sqrt{2\pi a}}:e^{-i\frac{2\pi}{\beta}\vartheta(x_0,x_1)+i\frac{\beta}{2}\varphi(x)}:\\L(x)&=\frac{1}{\sqrt{2\pi a}}:e^{-i\frac{2\pi}{\beta}\vartheta(x_0,x_1)-i\frac{\beta}{2}\varphi(x)}:\end{aligned} \tag{5.304}$$

with β given by Eq. (5.300), and ϑ is the field dual to the field φ.

Similarly, the order parameter field $\bar{\psi}\psi$ now becomes (see Eqs. (5.285), (5.299), and (5.300))

$$\bar{\psi}(x)\psi(x)=\frac{1}{\pi a}:\cos\left(\beta\varphi\right): \tag{5.305}$$

This formula will help us to determine the correlation function of the staggered longitudinal order parameter at long distances. We can also find bosonized expressions for the *transverse* components of the order parameter, i.e. $S^\pm(2s+1)-S^\pm(2s)$.

The same procedure as that which led to the relation between the (longitudinal) staggered magnetization $S_z(2s+1) - S_z(2s)$ and $\bar{\psi}\psi$, Eq. (5.131) (up to singular prefactors), now yields an operator correspondence for the transverse staggered magnetization

$$\begin{aligned} N^+(x) &= S^+(2s+1) - S^+(2s) \\ &\sim e^{-i\pi \int_{-\infty}^{x_1} dx_1' \,:\, \psi^\dagger(x_0,x_1')\psi(x_0,x_1') \,:} \left(\psi_1^\dagger(x) - \psi_2^\dagger(x)\right) \end{aligned} \tag{5.306}$$

which, in the chiral basis, has the form

$$N^+(x) \sim e^{-i\pi \int_{-\infty}^{x_1} dx_1' : j_0(x_0,x_1') :} R^\dagger(x) \tag{5.307}$$

The other transverse component, N^-, is just the hermitian conjugate of $N^+ = (N^-)^\dagger$.

We can use the bosonization identities to find an expression for $N^\pm$ in terms of the Bose field φ. The result is (up to singular coefficients, which we will not make explicit)

$$N^\pm(x) \sim \, : e^{\pm i \frac{2\pi}{\beta} \vartheta(x)} : + \cdots \tag{5.308}$$

A similar analysis yields the following operator identifications for the three components of the *magnetization*:

$$\begin{aligned} M_z &\sim \frac{1}{2}(S_z(2n) + S_z(2n+1)) \sim j_0 = \frac{\beta}{2\pi} \partial_x \varphi \\ M^\pm &\sim \frac{1}{2}(S^\pm(2n) + S^\pm(2n)) \sim L^\dagger e^{i\pi \int_{-\infty}^{x} dx' : j_0(x') :} \sim e^{i \frac{2\pi}{\beta} \vartheta + i\beta\varphi} \end{aligned} \tag{5.309}$$

The sine–Gordon potential $\cos(2\beta\varphi)$ does not affect the behavior at long distances unless the operator is *relevant*, in the sense of the renormalization group. This means that the (scaling) dimension Δ of this operator should be less than or equal to 2, the dimension of space-time. The dimension Δ_A of an operator $A(x)$ is found by considering the correlation function, say at equal times,

$$\langle A(x)A(x') \rangle \sim \frac{1}{|x_1 - x_1'|^{\eta_A}} \tag{5.310}$$

The critical exponent η_A and the dimension Δ_A are related by

$$\eta_A \equiv 2\Delta_A \tag{5.311}$$

Thus, adding the operator $A(x)$ to the Lagrangian density of the free theory, $\mathcal{L}_0 = \frac{1}{2}(\partial_\mu \varphi)^2$, does not alter the infrared behavior unless $\Delta_A \leq 2$. For $\Delta_A \leq 2$, the infrared divergences grow more and more severe with the order of perturbation theory in g_A, the coupling constant for the operator $A(x)$. Conversely, for $\Delta_A > 2$ the infrared behavior is, at every order of perturbation theory in g_A, the same as that of a theory with $g_A = 0$.

In addition to the fermions themselves, two operators $\mathcal{O}_a(x)$ and $\mathcal{Q}_b(x)$ are of importance to us:

$$\begin{aligned}\mathcal{O}_a(x) &= e^{ia\varphi(x)}\\ \mathcal{Q}_b(x) &= e^{ib\vartheta(x_0,x_1)}\end{aligned} \tag{5.312}$$

The equal-time correlation functions for $\mathcal{O}_a$ and $\mathcal{Q}_b$ are

$$\langle G| : \mathcal{O}_a(x) :: \mathcal{O}_a^\dagger(y) : |G\rangle = \text{constant} \times e^{a^2[\varphi^+(x_0,x_1),\varphi^-(x_0,y_1)]} \tag{5.313}$$

Similarly, we get

$$\langle G| : \mathcal{Q}_b(x) :: \mathcal{Q}_b^\dagger(y) : |G\rangle = \text{constant} \times e^{b^2[\vartheta^+(x_0,x_1),\vartheta^-(x_0,y_1)]} \tag{5.314}$$

After a short computation, we get for the equal-time correlation functions

$$\langle G| : \mathcal{O}_a(x) :: \mathcal{O}_a^\dagger(y) : |G\rangle = \frac{\text{constant}}{|x_1 - y_1|^{a^2/(2\pi)}} \tag{5.315}$$

and

$$\langle G| : \mathcal{Q}_b(x) :: \mathcal{Q}_b^\dagger(y) : |G\rangle = \frac{\text{constant}}{|x_1 - y_1|^{b^2/(2\pi)}} \tag{5.316}$$

Thus, the scaling dimension Δ of the operator $: \cos(2\beta\varphi) :$ is equal to

$$\Delta = \frac{\beta^2}{\pi} \tag{5.317}$$

For $\Delta \leq 2$ (i.e. $\beta^2 \leq 2\pi$) this interaction is relevant in the infrared and for $\beta^2 \geq 2\pi$ it is infrared-trivial. Thus, for values of the anisotropy γ greater than a critical value $\gamma_c \simeq \pi/2$, we expect the non-linear term to be dominant. In this regime, the field φ has small fluctuations around the classical value, which are determined by its equations of motion. The order-parameter field $\bar{\psi}\psi$ has a non-zero expectation value and the ground state is two-fold degenerate. This is the Ising regime of the Heisenberg model.

For the lattice theory one expects, and this is confirmed by a Bethe-ansatz calculation, that γ_c should be equal to unity (Luther and Peschel, 1975). In other words, the quantum Heisenberg antiferromagnet should be at this critical point. For $\gamma < \gamma_c$, XY anisotropy should dominate and the Mermin–Wagner theorem would prohibit the spontaneous breaking of the continuous symmetry of the XY model. The domain $\gamma < \gamma_c$ is a line (or segment) of critical points. A detailed theory of this phase transition in connection with the Kosterlitz–Thouless transition can be found in the work of Amit, Goldschmidt, and Grinstein (Amit *et al.*, 1980).

5.7 Phase diagrams and scaling behavior

We have shown that the one-dimensional (1D) quantum antiferromagnet is equivalent to a sine–Gordon model in $(1+1)$ dimensions. We will now apply the methods and results we derived in Chapter 4 to determine the phase diagram and the scaling behavior of the antiferromagnet.

The RG flows of the sine–Gordon theory were discussed in Chapter 4. There we saw that there is a (Kosterlitz–Thouless) transition when the value of the stiffness K is such that the cosine operator is marginal and its scaling dimension equals 2. In the bosonized treatment of the 1D quantum Heisenberg antiferromagnet this happens at $\gamma_{\rm c} \simeq \pi/2$. In the case of the Heisenberg antiferromagnet the bare values of the sine–Gordon coupling constant g and the stiffness $K = 1/\sqrt{2\beta}$ (notice the factor of 2) are not independent since both of them depend on the coupling constant γ of the Luttinger–Thirring model. This relation is not universal and depends on the cutoff scheme used. It is also affected by irrelevant operators, which have been neglected in our analysis. Nevertheless, it is useful to carry on with the analysis taking the parameters at face value.

We also saw that the RG flow, see Eq. (4.88) and Fig. 4.3, has a simple structure when expressed in terms of the SG coupling g and the parameter $x = 2 - \Delta$, where Δ is the scaling dimension of the cosine operator, which here depends also on the parameter γ. Hence the initial values of the RG flow describing the quantum Heisenberg chain lie on a curve $g = g(x)$,

$$g = \frac{1}{2\pi}\frac{2+x}{2-x} \tag{5.318}$$

which is easily obtained by eliminating the dependence of g and K (and β) on γ. The only part of this curve that matters to our analysis lies in the neighborhood of $x = 0$. In that neighborhood, $g = g(x)$ is a positive and monotonically increasing function of x that crosses the stable asymptote (see Fig. 4.3) $g = -x/A(1)$ (here $A(1) = \sqrt{32\pi^3}$) at a value $x_{\rm c}$ close to the origin, $x = 0$. Thus, for $x < x_{\rm c} < 0$ (i.e. $\gamma < \gamma_{\rm c}$) the RG flows converge on the fixed line: this is the anisotropic Heisenberg model (with XY anisotropy). In this regime the power-law behaviors we obtained are exact (up to contributions from irrelevant operators). Precisely at $x = x_{\rm c}$ the system is on the stable asymptote and it flows towards the end of the fixed line. We will see below that there is a special behavior associated with this point. Finally, for $x > x_{\rm c}$ ($\gamma > \gamma_{\rm c}$) the cosine operator is relevant and the RG flows to strong coupling. This is the Ising regime. In this regime the discrete $\mathbb{Z}_2$ Ising symmetry is spontaneously broken, there is long-range antiferromagnetism, and the energy spectrum is massive (gapped).

The correlation functions for all interesting operators on the domain $\gamma \leq \gamma_{\rm c}$ can be calculated. All the expressions listed below acquire logarithmic corrections to

Table 5.1 *Scaling dimensions at the isotropic Heisenberg antiferromagnetic point ($\beta^2 = 2\pi$) and at the XY-model point ($\beta^2 = 4\pi$).*

	$\Delta(\psi)$	$\Delta(N_z)$	$\Delta(N^\pm)$	$\Delta(M_z)$	$\Delta(M^\pm)$
Heisenberg ($\beta^2 = 2\pi$)	5/8	1/2	1/2	1	1
XY ($\beta^2 = 4\pi$)	1/2	1	1/4	1	5/4

the scaling at $\gamma = \gamma_c$. The dimensions of the fermion $\Delta(\psi)$, longitudinal $\Delta(N_z)$ (i.e. of the fermion mass term $\bar{\psi}\psi$), and transverse $\Delta(N^\pm)$ components of the staggered (Néel) order parameter, and of the components of the uniform magnetization ($\Delta(M_z)$ and $\Delta(M^\pm)$), are found to be

$$\Delta(\psi) = \frac{\pi}{\beta^2} + \frac{\beta^2}{16\pi} \tag{5.319}$$

$$\Delta(\bar{\psi}\psi) = \Delta(N_z) = \frac{\beta^2}{4\pi}, \qquad \Delta(N^\pm) = \frac{\pi}{\beta^2} \tag{5.320}$$

$$\Delta(M_z) = 1, \qquad \Delta(M^\pm) = \frac{\beta^2}{4\pi} + \frac{\pi}{\beta^2} \tag{5.321}$$

where we have kept only the contributions with smallest dimension (which are the most relevant). The scaling dimensions of these operators for the isotropic Heisenberg antiferromagnet and for the XY model (the free-fermion point) are given in Table 5.1.

At $\beta^2 = 4\pi$, the free-fermion limit, we obtain results for the quantum XY model which, as we also saw, is equivalent to two decoupled quantum Ising chains. In particular the correlation function

$$\langle T(S^+(n,\tau)S^-(n',\tau'))\rangle_{XY} = 2\langle T(\sigma_3(n,\tau)\sigma_3(n',\tau'))\rangle^2_{\text{Ising}} \tag{5.322}$$

The results of Table 5.1 show that at $\beta^2 = 4\pi$ the scaling dimension of the Néel order parameter, $N^\pm$, which in the XY model is the same as the spin operators $S^\pm$, is $\Delta = 1/4$. This result also tells us that the Ising correlator decays with an exponent $\eta = 1/4$ and therefore that the Ising spin operator σ has scaling dimension 1/8 (Bander and Itzykson, 1977; Zuber and Itzykson, 1977),

$$\langle T(\sigma_3(n,\tau)\sigma_3(n',\tau'))\rangle \sim \frac{1}{R^{1/4}} \tag{5.323}$$

where $R^2 = (n-n')^2 + (\tau-\tau')^2$.

From these results we conclude that the anisotropy disappears at $\gamma = \gamma_c$ since the *longitudinal* and *transverse* components of the Néel order parameter, the staggered magnetizations, have the *same* correlations functions at the critical point $\gamma = \gamma_c$, where they behave, up to logarithmic corrections, like

$$\langle G|N^+(x)N^-(y)|G\rangle|_{\gamma_c} \sim \langle G|\bar{\psi}(x)\psi(x)\bar{\psi}(y)\psi(y)|G\rangle|_{\gamma_c} \sim \frac{\text{constant}}{|x_1 - y_1|} \tag{5.324}$$

Similarly, the scaling dimensions of the three components of the *uniform magnetization* M_z and $M^\pm$ are also equal to each other (and to 1) for $\beta^2 = 2\pi$. This result is very significant since, as we will see in Chapter 7, these three dimension-1 operators generate a (chiral) SU(2) current algebra. However, the actual scaling behavior is a little more subtle than what our analysis shows. For instance, the three components of the Néel order parameter N^z and $N^\pm$ have dimension 1/2 on the fixed line but their correlators acquire a (multiplicative) logarithmic correction and behave as (similarly for the transverse components)

$$\langle N^z(x)N^z(y)\rangle \sim \frac{|\ln(|x-y|)|^{1/2}}{|x-y|} \tag{5.325}$$

right at γ_c. This *correction to scaling* is due to the marginally irrelevant flow along the stable asymptote towards the end of the fixed line (Affleck, 1998). However, the three components of the local magnetization densities M^z and $M^\pm$ do not acquire such corrections to scaling. This different behavior is due to the fact that the magnetization density is part of a locally conserved SU(2) current. Thus, although our formalism does not keep track of the global SU(2) symmetry of the Heisenberg model at the isotropic point, it recovers it as an effective ("dynamical") symmetry of the critical point.

For $\gamma < \gamma_c$ the correlation functions are different, although both exhibit an algebraic decay (i.e. power-law behavior) with exponents η_z and $\eta_\pm$ satisfying $\eta_z > \eta_\pm$. These exponents are universal in the sense that their numerical values are independent of the short-distance cutoff. However, the coupling constant itself does depend on the precise definition of the cutoff. Thus the value of γ_c, which is equivalent to unity in the lattice system, turns out to be close to $\pi/2$ for the continuum model. Nevertheless, it is possible to find a relationship between the continuum and lattice coupling constants (Luther and Peschel, 1975).

The fact that the correlation functions exhibit a power-law behavior means that the system, for $\gamma < \gamma_c$, is critical. It has been argued (den Nijs, 1981) that this is a line of critical points ending at γ_c, the Heisenberg point. That the system is critical means that there are no energy gaps; that is, *all* the excitations are gapless. For $\gamma > \gamma_c$ an energy gap $m(\gamma)$ develops (den Nijs, 1981). The RG analysis we discussed tells us that the energy gap exhibits the Kosterlitz–Thouless behavior

$$m(\gamma) \sim \text{constant} \times \exp\left(-\frac{\text{constant}}{\sqrt{\gamma - \gamma_c}}\right) \tag{5.326}$$

This is the regime with $\beta^2 < 2\pi$ in the sine–Gordon theory. Renormalization-group arguments imply that the operator $e^{i\beta\phi}$ exhibits long-range order and, consequently, that $\langle \bar{\psi}\psi \rangle \neq 0$.

It is natural to ask whether the fact that the spin-$1/2$ Heisenberg chain is at a critical point with gapless (neutral) fermions in the spectrum does generalize to other situations such as higher spin or higher dimensions. We will see below that in general the behavior of the half-integer-spin chains is analogous to that of the spin-$1/2$ chain, and that, in contrast, the *integer*-spin chains are not critical.

6

The Luttinger liquid

6.1 One-dimensional Fermi systems

We will now consider the case of one-dimensional (1D) Fermi systems for which the Landau theory fails. The way it fails is quite instructive since it reveals that in one dimension these systems are generally at a (quantum) critical point, and it will also teach us valuable lessons on quantum criticality. It will also turn out that the problem of 1D Fermi systems is closely related to the problem of quantum spin chains. This is a problem that has been discussed extensively by many authors, and there are several excellent reviews on the subject (Emery, 1979; Haldane, 1981; Gogolin *et al.*, 1998). Here I follow in some detail the discussion and notation of Carlson and coworkers (Carlson *et al.*, 2004).

One-dimensional (and quasi-1D) systems of fermions occur in several experimentally accessible systems. The simplest one to visualize is a *quantum wire*. A quantum wire is a system of electrons in a semiconductor, typically a GaAs–AlAs heterostructure built by molecular-beam epitaxy (MBE), in which the electronic motion is laterally confined along two directions, but not along the third. An example of such a channel of length L and width d (here shown as a two-dimensional (2D) system) is seen in Fig. 6.1. Systems of this type can be made with a high degree of purity with very long (elastic) mean free paths, often tens of micrometers or even longer. The resulting electronic system is a 1D electron gas (1DEG). In addition to quantum wires, 1DEGs also arise naturally in carbon nanotubes, where they are typically multi-component (with the number of components being determined by the diameter of the nanotube).

Other 1D Fermi systems include the edge states of *two-dimensional* electron gases (2DEGs) in large magnetic fields in the regime in which *quantum Hall effects* are seen. (We will discuss this problem later on.) This case is rather special since these edge states can propagate in only one direction, determined by the sign of the perpendicular magnetic field.

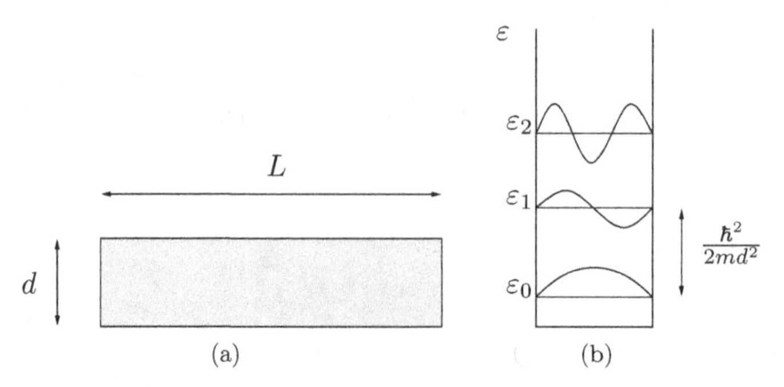

Figure 6.1 (a) A long quantum wire of length L and transverse width d ($L \gg d$): a channel for the electron fluid. Only the 2D case is shown for simplicity. (b) The square-well spectrum of the transverse quantum single-particle states confined by the finite width d of the wire.

Other (quasi-) 1D systems occur in organic compounds, such as TTFTCNQ, some Bechgaard salts (commonly called the "BEDTs" and the "ETs"), and TMSFs. There are also quasi-1D chalcogenide materials, e.g. $NbSe_3$, as well as complex oxides. There are some oxides, e.g. $Sr_{14-x}Ca_xCu_{24}O_{41}$, that can be regarded as a set of weakly coupled ladders (instead of chains). Quasi-1D Fermi systems are often used to describe complex ordered states in 2D strongly correlated systems. Typical examples are the *stripe phases* of the copper oxide high-T_c superconductors, such as $La_{2-x}Sr_xCuO_4$ and $La_{2-x}Ba_xCuO_4$.

We will consider first the conceptually simpler example of the quantum wire. We will assume that the electron density is such that the Fermi energy lies below the energy of the first excited state. The result is that the single-particle states with momenta in the range $-p_F < p < p_F$ are occupied and the states outside this range are empty. Thus the Fermi "surface" of this system reduces to two Fermi points at $\pm p_F$. We will assume that the wire is long enough, $L \gg d$, so that the single-particle states fill up densely the momentum axis, and that the density is high enough that $\Delta p = 2\pi\hbar/L \ll p_F$. On the other hand, we will assume that the wire is narrow enough that the next band of (excited) states can effectively be neglected, $\varepsilon_F \ll \hbar^2/(2md^2)$. At higher electronic densities, more than one band can intersect the Fermi energy. Each new partially occupied band is labeled by a pair of Fermi points. In practice we will work in a regime in which the following inequality holds:

$$\frac{L}{d} \gg 1 \gg \frac{d}{\lambda_F} \tag{6.1}$$

where $\lambda_F = \hbar/p_F$ is the Fermi wavelength, and we have only two Fermi points (see Fig. 6.2).

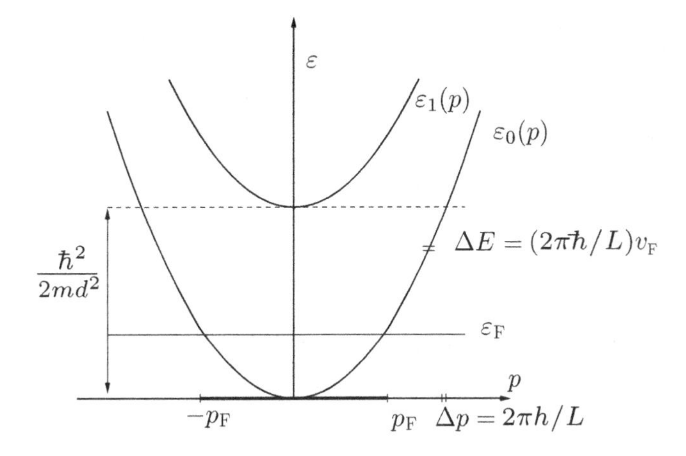

Figure 6.2 Energy–momentum relation of the two lowest bands of propagating non-relativistic free fermions along the length of the quantum wire; ε_F is the Fermi energy, $\pm p_F$ are the two Fermi points, and $v_F = p_F/m$ is the Fermi velocity. The filled Fermi sea (the occupied states) is shown as a dark segment; ΔE and Δp are the level spacings in a finite wire of length L. We have shifted the minimum of the energy of the lowest band to be at zero.

The Hamiltonian for the 1DEG is $H = H_0 + H_{\text{int}}$, where

$$H_0 = \sum_{\sigma=\uparrow,\downarrow} \int_0^L dx\, \psi_\sigma^\dagger(x)\Big(-\frac{\hbar^2}{2m}\frac{\partial^2}{\partial x^2} - \mu\Big)\psi_\sigma(x) \tag{6.2}$$

is the free-fermion Hamiltonian and

$$H_{\text{int}} = \sum_{\sigma,\sigma'=\uparrow,\downarrow} \int_0^L dx \int_0^L dx'\, \psi_\sigma^\dagger(x)\psi_\sigma(x)U(x-x')\psi_{\sigma'}^\dagger(x')\psi_{\sigma'}(x') \tag{6.3}$$

where $U(x-x')$ is the interaction potential, which can be Coulomb or short-ranged, depending on the physical situation. In what follows, for simplicity we will use periodic boundary conditions, requiring

$$\psi_\sigma(x+L) = \psi_\sigma(x) \tag{6.4}$$

which amounts to wrapping the system ("compactification") onto a circle. Sometimes we may want to use the more physical open boundary conditions.

In many cases we will be interested in lattice systems. So, consider a 1D chain of N sites (atoms) and lattice spacing a, and total length $L = Na$. A typical system of interest is the Hubbard model, whose lattice Hamiltonian is

$$H = \sum_{j=1}^{N}\sum_{\sigma=\uparrow,\downarrow} t\Big(\psi_\sigma^\dagger(j)\psi_\sigma(j+1) + \text{h.c.}\Big) + \sum_{j=1}^{N} U n_\uparrow(j) n_\downarrow(j) \tag{6.5}$$

where $n_\sigma(j) = \psi_\sigma^\dagger(j)\psi_\sigma(j)$ is the fermion occupation number with spin σ at site j and $n(j) = \sum_\sigma n_\sigma(j)$ is the total occupation number (i.e. the charge) at site j. Here U is the on-site (Hubbard) interaction. This model describes a system of electrons with hopping only between nearest neighboring sites; t, the hopping amplitude, is the local kinetic energy. This system has only one band of single-particle states with the dispersion relation

$$\varepsilon(p) = 2t\cos(pa) \tag{6.6}$$

In the thermodynamic limit, $N \to \infty$, the momenta p lie in the first Brillouin zone, $-\pi/a < p \leq \pi/a$. In general we will be interested in a system either at fixed chemical potential μ or at fixed density $n = N_e/N$. The effective model for interacting systems that we will discuss will describe equally well (with minor changes) the low-energy physics of both continuum and lattice systems.

What is special about one dimension?

1. In the Landau theory of the Fermi liquid we considered the low-energy states and we saw that they can be described in terms of *particle–hole pairs*. In dimensions $D > 1$ the momentum $\delta\vec{q}$ of the pair is not necessarily parallel to the Fermi wave vector $\mathbf{p}_F$ of the location of the Fermi surface (FS) where the pair is excited. However, in 1D δq *must* be either parallel or anti-parallel to the Fermi momentum p_F since the FS has collapsed to just two (or more) points (see Fig. 6.3).
2. This kinematic restriction implies that particle–hole pairs effectively form long-lived bound states, the collective modes, since the particle and the hole move with the same speed (the Fermi velocity). We will see that this implies that the low-energy effective theory is a theory of *bosons*. This is the main reason why the non-perturbative theory of 1D fermions, bosonization, works.
3. Another insight can be gleaned by looking at the density correlators, whose singularities are the collective modes. In $D > 1$ the retarded density–density correlation function $D^R(\mathbf{q}, \omega)$ of a free fermion is

$$D^R(\mathbf{q}, \omega) = \int \frac{d^D p}{(2\pi)^D} \frac{n_\mathbf{p} - n_{\mathbf{p}+\mathbf{q}}}{\omega - \varepsilon(\mathbf{p}+\mathbf{q}) + \varepsilon(\mathbf{p}) + i\eta} \tag{6.7}$$

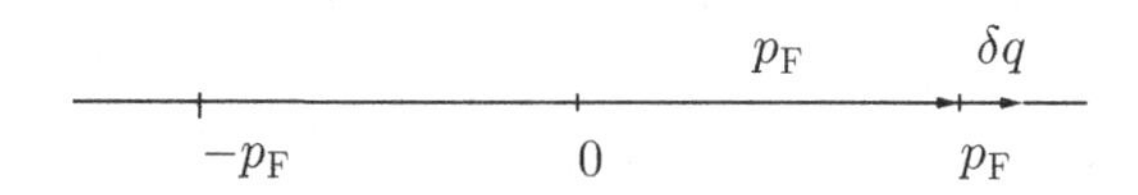

Figure 6.3 One-dimensional kinematics: the momentum of a particle–hole pair of momentum q is always parallel (or anti-parallel) to the Fermi wave vector p_F.

For low momenta $|\mathbf{q}| \ll p_F$ and at low energies $\omega \ll E_F$, $D^R(\mathbf{q}, \omega)$ can be written as an integral on the Fermi surface

$$D^R(\mathbf{q}, \omega) \simeq \int \frac{d^D p}{(2\pi)^D} \frac{\mathbf{q} \cdot \widehat{\mathbf{p}}_F}{\omega - \left(\mathbf{q} \cdot \widehat{\mathbf{p}}_F\right) v_F + i\eta} \, \delta\left(|\mathbf{p}| - p_F\right)$$
$$= \frac{p_F^{D-1}}{(2\pi)^D} \oint_{FS} d\widehat{\mathbf{p}}_F \frac{\mathbf{q} \cdot \widehat{\mathbf{p}}_F}{\omega - \left(\mathbf{q} \cdot \widehat{\mathbf{p}}_F\right) v_F + i\eta} \tag{6.8}$$

For $D > 1$ the angular integration is a function of $\mathbf{q}$ and ω that has branch cuts. For instance, in 3D the result is

$$D^R(\mathbf{q}, \omega) \sim 1 + \frac{\omega}{2qv_F} \ln \left| \frac{\omega - qv_F}{\omega + qv_F} \right| + \cdots \tag{6.9}$$

The branch cuts mean that the collective modes (zero sound) eventually become Landau damped (Abrikosov *et al.*, 1963; Baym and Pethick, 1991).

4. However, in 1D there is no such angular integration (the FS is just two points!) and the result is

$$D^R(q, \omega) \sim \frac{q}{2\pi} \left(\frac{1}{\omega - qv_F + i\eta} - \frac{1}{\omega + qv_F + i\eta} \right) \tag{6.10}$$

This expression contains two singularities, two *poles*, representing bosonic states that move to the right (the first term) or to the left (the second term). It is easy to check that this result is consistent with the f-sum rule.

Furthermore, these results suggests that a theory of free fermions in 1D must be, in some sense, equivalent to a theory of a Bose field whose excitations obey the dispersion relation $\omega = pv_F$. In other terms, the bosons are density fluctuations, which in this case are just sound waves.

6.2 Dirac fermions and the Luttinger model

We will now proceed to construct an effective low-energy theory by following a procedure similar to what led to the Landau theory of the Fermi liquid. The result, however, will be quite different in 1D.

To this end we will first look at the free-fermion system and focus on the low-energy excitations. We have already encountered this problem in Section 5.2.3. We will follow a similar line of argument. In 1D, instead of a Fermi surface we have (at least) two Fermi points at $\pm p_F$. The low-energy fermionic states thus have momenta $p \sim \pm p_F$ and a single-particle energy close to ε_F:

$$\varepsilon(p) \simeq \varepsilon_F + (|p| - p_F)v_F + \cdots \tag{6.11}$$

We are interested in the electronic states near the Fermi energy. Thus, consider the fermion operator $\psi_\sigma(x)$, whose Fourier expansion is (we will set $\hbar = 1$ from now on), in the thermodynamic limit ($L \to \infty$),

$$\psi_\sigma(x) = \int \frac{dp}{2\pi} \psi_\sigma(p) e^{ipx} \tag{6.12}$$

Only its Fourier components near $\pm p_{\mathrm{F}}$ describe low-energy states. This suggests that we restrict ourselves to the modes of the momentum expansion in a neighborhood of $\pm p_{\mathrm{F}}$ of width 2Λ, and that we write

$$\psi_\sigma(x) \simeq \int_{-\Lambda}^{\Lambda} \frac{dp}{2\pi} e^{i(p+p_{\mathrm{F}})x} \psi_\sigma(p + p_{\mathrm{F}}) + \int_{-\Lambda}^{\Lambda} \frac{dp}{2\pi} e^{i(p-p_{\mathrm{F}})x} \psi_\sigma(p - p_{\mathrm{F}}) \tag{6.13}$$

and that we define right- and left-moving fields $\psi_{\sigma,\mathrm{R}}(x)$ and $\psi_{\sigma,\mathrm{L}}(x)$ such that

$$\psi_\sigma(x) \simeq e^{ip_{\mathrm{F}}x} \psi_{\sigma,\mathrm{R}}(x) + e^{-ip_{\mathrm{F}}x} \psi_{\sigma,\mathrm{L}}(x) \tag{6.14}$$

Thus we have split off the rapidly oscillating piece of the field and we focus on the slowly varying parts, $\psi_{\sigma,\mathrm{R}}(x)$ and $\psi_{\sigma,\mathrm{L}}(x)$, whose Fourier transforms are

$$\psi_{\sigma,\mathrm{R}}(p) = \psi_\sigma(p + p_{\mathrm{F}}) \qquad \text{and} \qquad \psi_{\sigma,\mathrm{L}}(p) = \psi_\sigma(p - p_{\mathrm{F}}) \tag{6.15}$$

respectively.

The free-fermion Hamiltonian

$$H_0 = \sum_\sigma \int \frac{dp}{2\pi} \varepsilon(p) \psi_\sigma^\dagger(p) \psi_\sigma(p) \tag{6.16}$$

becomes

$$H_0 = \sum_\sigma \int_{-\Lambda}^{\Lambda} \frac{dp}{2\pi} p v_{\mathrm{F}} \left(\psi_{\sigma,\mathrm{R}}^\dagger(p) \psi_{\sigma,\mathrm{R}}(p) - \psi_{\sigma,\mathrm{L}}^\dagger(p) \psi_{\sigma,\mathrm{L}}(p) \right) \tag{6.17}$$

where we have linearized the dispersion $\varepsilon(p)$ near the Fermi momenta $\pm p_{\mathrm{F}}$. Let us define the two-component spinor

$$\psi_\sigma(x) = \begin{pmatrix} \psi_{\sigma,\mathrm{R}}(x) \\ \psi_{\sigma,\mathrm{L}}(x) \end{pmatrix} \tag{6.18}$$

in terms of which the free-fermion Hamiltonian is

$$H_0 = \sum_\sigma \int \frac{dp}{2\pi} \psi_\sigma^\dagger(p) \sigma_3 p v_{\mathrm{F}} \psi_\sigma(p) = \sum_\sigma \int dx \; \psi_\sigma^\dagger(x) \sigma_3 i v_{\mathrm{F}} \, \partial_x \psi_\sigma(x) \tag{6.19}$$

where

$$\sigma_3 = \begin{pmatrix} 1 & 0 \\ 0 & -1 \end{pmatrix} \tag{6.20}$$

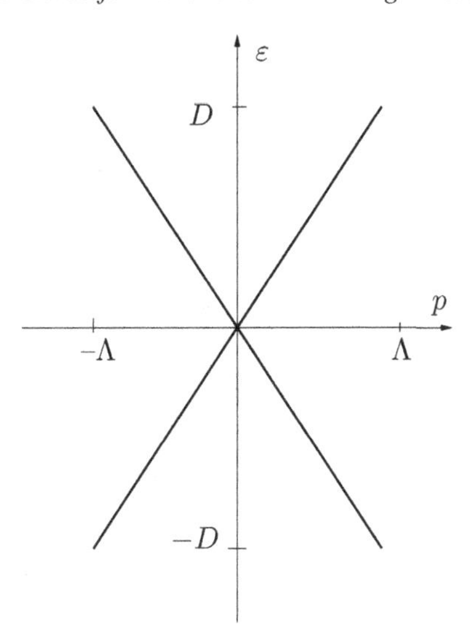

Figure 6.4 The Dirac dispersion; the slope is the Fermi velocity v_F. The momentum cutoff is Λ and the energy cutoff is $D = v_F\Lambda$.

In this form the effective low-energy Hamiltonian reduces to the (massless) Dirac Hamiltonian in 1D. In Fig. 6.4 we show the dispersion in the spinor notation.

Most of the interaction terms we discussed above can be expressed in terms of the local densities of right- and left-moving fermions

$$j_{R,\sigma}(x) = \psi^{\dagger}_{R,\sigma}(x)\psi_{R,\sigma}(x), \qquad j_{L,\sigma}(x) = \psi^{\dagger}_{L,\sigma}(x)\psi_{L,\sigma}(x) \tag{6.21}$$

from which we can write the slowly varying part of the charge-density operator $j_0(x)$ (i.e. with Fourier component with wave vectors close to $p = 0$) and the charge-current density $j_1(x)$ as

$$j_0(x) = j_R(x) + j_L(x), \qquad j_1(x) = j_R(x) - j_L(x) \tag{6.22}$$

which is a 2-vector of the form

$$j_\mu(x) = (j_0, j_1) \tag{6.23}$$

with $\mu = 0, 1$ (not to be confused with the chemical potential!). Thus, the coupling to a slowly varying external electromagnetic field $A_\mu(x) = (A_0, A_1)$ is represented by a term of the form

$$H_{\text{em}} = \int dx \left(-eA_0(x)j_0(x) + \frac{e}{c}A_1(x)j_1(x) \right) \tag{6.24}$$

The actual particle-density operator of the microscopic system,

$$\rho(x) = \sum_{\sigma} \psi^{\dagger}_{\sigma}(x)\psi_{\sigma}(x) \tag{6.25}$$

can be written in the form (using the decomposition of the Fermi field into right and left movers)

$$\rho(x) = \rho_0 + j_0(x) + \sum_\sigma \left(e^{2ip_F x} \psi^\dagger_{R,\sigma}(x)\psi_{L,\sigma}(x) + e^{-2ip_F x} \psi^\dagger_{L,\sigma}(x)\psi_{R,\sigma}(x) \right) + \cdots \tag{6.26}$$

where $\rho_0 = N_e/L = 2p_F/\pi$ is the average total density of electrons (including spin), and where $\cdots$ represents terms that oscillate more rapidly for large p_F.

The significance of the oscillatory terms can be seen by adding a coupling to a periodic potential $V(x)$ (with wave vector $2p_F$). For simplicity here we consider only potentials with wave vectors commensurate with the fermion density, $Q = 2p_F$. More general cases can also be considered and lead to interesting physical effects. We will take the potential to have the simple form

$$V(x) = V_0 \cos(2p_F x) \tag{6.27}$$

leading to a new term in the Hamiltonian of the form

$$\begin{aligned} H_{\text{pot}} &= \int dx\, V(x)\rho(x) \\ &= -e \int \frac{dp}{2\pi} \int \frac{dq}{2\pi} V(q)\psi^\dagger(p+q)\psi(p) \\ &= \int \frac{dp}{2\pi} (-eV_0)\left(\psi^\dagger_R(p)\psi_L(p) + \psi^\dagger_L(p)\psi_R(p)\right) \\ &= \int dx (-eV_0)\left(\psi^\dagger_R(x)\psi_L(x) + \psi^\dagger_L(x)\psi_R(x)\right) \end{aligned} \tag{6.28}$$

In other terms, a periodic potential of wave vector $Q = 2p_F$ causes backscattering: it scatters a right-moving fermion into a left-moving fermion (and vice versa). Similarly, a periodic potential of wave vector $Q \ll 2p_F$ scatters right movers into right movers (and left movers into left movers). From now on we will drop the spin indices, unless we state the contrary.

Thus, in the case of a free fermion coupled to a periodic potential $V(x)$ with wave vector $Q = 2p_F$, the potential induces backscattering that mixes the two Fermi points at $\pm p_F$. This leads to the existence of an energy gap at the Fermi energy. In terms of the Dirac Hamiltonian, the periodic potential $V(x)$ leads to the Hamiltonian

$$H = \int dx\, \psi^\dagger(x)\left(i v_F \sigma_3\, \partial_x + eV_0\sigma_1\right)\psi(x) \tag{6.29}$$

where

$$\sigma_1 = \begin{pmatrix} 0 & 1 \\ 1 & 0 \end{pmatrix} \tag{6.30}$$

In Eq. (6.29) $eV_0 = \Delta$ is the energy gap, and is usually denoted by $\Delta = mv_F^2$.

In the Dirac theory it is useful to define the matrices $\alpha = \sigma_3$ (in 1D) and $\beta = \sigma_1$, such that the Dirac Hamiltonian reads

$$H = \int dx\, \psi^\dagger(x)\Big(\alpha i v_{\rm F}\, \partial_x + \beta\Delta\Big)\psi(x) \tag{6.31}$$

The single-particle spectrum consists of particles and holes with energy $\varepsilon(p) = \sqrt{v_{\rm F}^2 p^2 + \Delta^2}$. In the Dirac theory it is customary to define Dirac's γ-matrices. In this 1D case there are just two of them, $\gamma_0 = \beta = \sigma_1$ and $\gamma_1 = \beta\alpha = i\sigma_2$. They satisfy the algebra

$$\{\gamma_0, \gamma_1\} = 0, \qquad \gamma_0^2 = 1, \qquad \gamma_1^2 = -1 \tag{6.32}$$

If we define $\bar{\psi} = \psi^\dagger\gamma_0$, the fermion mass term is

$$\psi_{\rm R}^\dagger\psi_{\rm L} + \psi_{\rm L}^\dagger\psi_{\rm R} = \psi^\dagger\gamma_0\psi = \bar{\psi}\psi \tag{6.33}$$

6.3 Order parameters of the one-dimensional electron gas

Similarly we can also define the matrix $\gamma^5 = \gamma_0\gamma_1 = \sigma_3$, and the bilinear $\bar{\psi}\gamma^5\psi$:

$$\bar{\psi}\gamma_5\psi = \psi_{\rm R}^\dagger\psi_{\rm L} - \psi_{\rm L}^\dagger\psi_{\rm R} \tag{6.34}$$

It is straightforward to see that the density $\rho(x)$ can be written as

$$\rho(x) = \rho_0 + j_0(x) + \cos(2p_{\rm F}x)\bar{\psi}(x)\psi(x) + i\sin(2p_{\rm F}x)\bar{\psi}(x)\gamma^5\psi(x) \tag{6.35}$$

From here we see that if $\langle\bar{\psi}(x)\psi(x)\rangle \neq 0$ (or $\langle\bar{\psi}(x)\gamma_5\psi(x)\rangle \neq 0$), then the expectation value of the charge density $\langle\rho(x)\rangle$ has a modulated component (over the background ρ_0). If this were to occur *spontaneously* (i.e. in the absence of an external periodic potential) then the ground state of the system would be a *charge-density wave* (CDW). Hence, $\langle\bar{\psi}\psi\rangle$ and $\langle i\bar{\psi}\gamma^5\psi\rangle$ play the role of the *order parameters* of the CDW state. We can also see that the expectation value of the density will be *even* (invariant) under inversion, $x \to -x$ (i.e. parity), if $\langle\bar{\psi}(x)\gamma^5\psi(x)\rangle = 0$; conversely, if this expectation value is not zero, the density will not be even under parity, which amounts to a phase shift.

In the absence of the periodic potential the original system is translationally invariant. The periodic potential breaks translation invariance. To see how that works, we define the transformation

$$\psi(x) \to e^{i\theta\gamma^5}\psi(x) = \begin{pmatrix} e^{i\theta}\psi_{\rm R}(x) \\ e^{-i\theta}\psi_{\rm L}(x) \end{pmatrix} \tag{6.36}$$

which is known as a *chiral transformation*. Under this transformation the two-component vector

$$\begin{pmatrix} \bar{\psi}\psi \\ i\bar{\psi}\gamma^5\psi \end{pmatrix} \tag{6.37}$$

transforms as a rotation

$$\begin{pmatrix} \bar{\psi}\psi \\ i\bar{\psi}\gamma^5\psi \end{pmatrix} \rightarrow \begin{pmatrix} \cos(2\theta) & -\sin(2\theta) \\ \sin(2\theta) & \cos(2\theta) \end{pmatrix} \begin{pmatrix} \bar{\psi}\psi \\ i\bar{\psi}\gamma^5\psi \end{pmatrix} \tag{6.38}$$

under which the density operator becomes

$$\begin{aligned} \rho(x) \rightarrow \rho_0 + j_0(x) + e^{i2(p_F x - \theta)}\psi_R^\dagger(x)\psi_L(x) + e^{-i2(p_F x-\theta)}\psi_L^\dagger(x)\psi_R(x) \\ = \rho\left(x - \frac{\theta}{p_F}\right) \end{aligned} \tag{6.39}$$

Therefore, a chiral transformation by an angle θ is equivalent to a translation of the charge density by a distance $d = \theta/p_F$. Notice that transformations by $\theta = n\pi$ have no physical effect since they amount to translations by a distance $n\pi/p_F = 2n\pi/Q = n\ell$, i.e. an integer number of periods $\ell = 2\pi/Q$ of the CDW. Thus only chiral transformations modulo π are observable.

In a similar fashion we can define an operator corresponding to a *spin-density wave* (SDW). Indeed, the local magnetization (or spin-polarization) density is

$$m^a(x) = \psi_\sigma^\dagger(x)\tau^a_{\sigma,\sigma'}\psi_{\sigma'}(x) \tag{6.40}$$

(where τ^a are the three Pauli matrices, acting only on the spin indices σ and σ'), which can be expressed as

$$m^a(x) = j_0^a(x) + e^{2ip_F x}\psi^\dagger_{R,\sigma}(x)\tau^a_{\sigma,\sigma'}\psi_{L,\sigma'}(x) + \text{h.c.} \tag{6.41}$$

where $j_0^a(x)$ is the slowly varying spin density

$$j_0^a(x) = \psi^\dagger_{R,\sigma}(x)\tau^a_{\sigma,\sigma'}\psi_{R,\sigma'}(x) + \psi^\dagger_{L,\sigma}(x)\tau^a_{\sigma,\sigma'}\psi_{L,\sigma'}(x) \tag{6.42}$$

and, similarly, the spin current is

$$j_1^a(x) = \psi^\dagger_{R,\sigma}(x)\tau^a_{\sigma,\sigma'}\psi_{R,\sigma'}(x) - \psi^\dagger_{L,\sigma}(x)\tau^a_{\sigma,\sigma'}\psi_{L,\sigma'}(x) \tag{6.43}$$

The SDW order parameters are

$$N^a(x) = \bar{\psi}_{s,\sigma}(x)\tau^a_{\sigma,\sigma'}\psi_{s,\sigma'}(x), \qquad N^a_c(x) = i\bar{\psi}_{s,\sigma}(x)\tau^a_{\sigma,\sigma'}\gamma^5_{s,s'}\psi_{s',\sigma'}(x) \tag{6.44}$$

(where $s, s' = \text{R}, \text{L}$) and describe modulations of the local spin polarization with wave vector $Q = 2p_F$.

Finally let us discuss *pairing* operators, which are associated with *superconductivity*. Here we will be interested in pairing operators associated with *uniform* ground states (although modulated states are also possible). Pairing operators that create a pair of quasiparticles with total momentum (close to) zero have the form

$$O_{SP}(x) = \langle \psi^\dagger_{R,\sigma}(x)\psi^\dagger_{L,-\sigma}(x)\rangle, \qquad O_{TP}(x) = \langle \psi^\dagger_{R,\sigma}(x)\psi^\dagger_{L,\sigma}(x)\rangle \tag{6.45}$$

where $O_{\mathrm{SP}}(x)$ corresponds to (spin) *single* pairing, and $O_{\mathrm{TP}}(x)$ to (spin) *triplet* pairing. Differently from all the operators we discussed so far, the pairing operators do not conserve particle number.

We will see below that all of these order parameters break some (generally continuous) symmetry of the system: translation invariance for the CDW, spin rotations and translation invariance for the SDW, and (global) gauge invariance (associated with particle-number conservation) for the superconducting case. There is a theorem, known as the *Mermin–Wagner theorem*, that states that in a 1D quantum system continuous symmetries cannot be spontaneously broken. More precisely, this theorem states that correlation functions of order parameters that transform under a continuous global symmetry *cannot decay more slowly than as a power-law function of distance (or time)*. We will now see that in the case of the Luttinger model the behavior is exactly a power law. We will interpret this as saying that the system is at a (quantum) critical point. (In high-energy physics this theorem is often attributed to S. Coleman.)

6.4 The Luttinger model: bosonization

We will now consider the Luttinger (Tomonaga) model (also known as the massless Thirring model in high-energy physics). We will consider first the case of spinless fermions. The Hamiltonian density $\mathcal{H}$ of the Luttinger model is

$$\mathcal{H} = \psi^\dagger(x)\Big(\alpha i v_{\mathrm{F}}\,\partial_x + \beta\Delta\Big)\psi(x) + 2g_2\rho_{\mathrm{R}}(x)\rho_{\mathrm{L}}(x) + g_4\Big(\rho_{\mathrm{R}}(x)^2 + \rho_{\mathrm{L}}(x)^2\Big) \tag{6.46}$$

where $\rho_{\mathrm{R}}(x) \equiv \psi_{\mathrm{R}}^\dagger(x)\psi_{\mathrm{R}}(x)$ and $\rho_{\mathrm{L}}(x) \equiv \psi_{\mathrm{L}}^\dagger(x)\psi_{\mathrm{L}}(x)$ denote the densities of right and left movers, respectively. Here $g_2 = \widetilde{V}(0) - \widetilde{V}(2p_{\mathrm{F}})$ and $g_4 = \widetilde{V}(0)/2$, where $\widetilde{V}(q)$ is the Fourier transform of the interaction potential. Hence, g_2 measures the strength of the backscattering interactions and g_4 that of the forward-scattering interactions. For a model of spinless fermions on a lattice near half-filling with nearest-neighbor interactions with coupling constant V, the coupling constants become $g_2 = 2V$ and $g_4 = V$.

Notice that the Hamiltonian of the Luttinger model has the same form as that in the Landau theory of the Fermi liquid in which the quasiparticles have only forward-scattering interactions, here represented by g_4. Here we have also included backscattering processes labeled by g_2, with a wave vector $2p_{\mathrm{F}}$ (i.e. across the "Fermi surface"). Owing to the kinematical restrictions of a *curved* Fermi surface, in the Landau theory backscattering processes have negligible effects. We will see that in 1D (where there is no curvature) they play a key role.

Precisely at half-filling, in addition to the backscattering and forward-scattering interactions (Figs. 3.1(a) and (b)), an Umklapp interaction must also be considered:

this is a scattering process in which momentum conservation is conserved up to a reciprocal-lattice vector $G = 2\pi$ (Fig. 3.1(c)). An Umklapp process has the form

$$\mathcal{H}_{\text{Umklapp}} = g_{\text{u}} \lim_{y \to x} \left(\psi_{\text{R}}^{\dagger}(x)\psi_{\text{R}}^{\dagger}(y)\psi_{\text{L}}(x)\psi_{\text{L}}(y) + R \leftrightarrow L \right) \tag{6.47}$$

where $g_{\text{u}} = \widetilde{V}(4p_{\text{F}} \simeq 2\pi)$. This coupling cannot be expressed in terms of densities of right and left movers. Since the Hamiltonian of the Luttinger model is written in terms of $\rho_{\text{R}}(x)$ and $\rho_{\text{L}}(x)$, it is invariant under a continuous chiral transformation, i.e. it is invariant under an arbitrary continuous translation. An Umklapp term reduces this continuous symmetry to the (discrete) symmetry of lattice displacements.

We will now see that the Luttinger model can be solved exactly by abelian bosonization (see Section 5.6). We will now use the identities we derived in Chapter 5 to find the bosonized form of the Hamiltonian of the Luttinger model and of the observables. As we saw, the free-fermion system maps onto the free-boson system (with the same velocity v_{F}). Hence the free-fermion Hamiltonian density (the Dirac Hamiltonian density) becomes

$$\mathcal{H}_0 = \frac{v_{\text{F}}}{2}\left(\Pi^2 + (\partial_x \phi)^2\right) \tag{6.48}$$

which, in terms of the field ϕ and the dual ϑ, has the symmetric (*self-dual*) form

$$\mathcal{H}_0 = \frac{v_{\text{F}}}{2}\left((\partial_x \vartheta)^2 + (\partial_x \phi)^2\right) \tag{6.49}$$

where ϑ is the dual field defined in Eqs. (5.278), (5.279), and (5.281). The right- and left-moving (fermion) densities ρ_{R} and ρ_{L} map onto

$$\rho_{\text{R}} = \frac{1}{2\sqrt{\pi}}\left(\partial_x \phi - \Pi\right) \equiv \frac{1}{2\sqrt{\pi}}\,\partial_x(\phi - \vartheta) \tag{6.50}$$

$$\rho_{\text{L}} = \frac{1}{2\sqrt{\pi}}\left(\partial_x \phi + \Pi\right) \equiv \frac{1}{2\sqrt{\pi}}\,\partial_x(\phi + \vartheta) \tag{6.51}$$

In terms of the right- and left-moving densities the Hamiltonian takes the Sugawara form

$$\mathcal{H} = (\pi v_{\text{F}} + g_4)\left(\rho_{\text{R}}^2 + \rho_{\text{L}}^2\right) + 2g_2\rho_{\text{R}}\rho_{\text{L}} \tag{6.52}$$

Hence, the *forward-scattering* term of the Luttinger Hamiltonian becomes

$$g_4\left(\rho_{\text{R}}^2 + \rho_{\text{L}}^2\right) \to \frac{g_4}{2\pi}\left(\Pi^2 + (\partial_x \phi)^2\right) \tag{6.53}$$

Similarly, the backscattering term becomes

$$2g_2\rho_{\text{R}}\rho_{\text{L}} \to \frac{g_2}{2\pi}\left((\partial_x \phi)^2 - \Pi^2\right) \tag{6.54}$$

Thus, we see that the Hamiltonian of the Luttinger model can be represented by an effective bosonized theory, which includes the total effects of forward-scattering and backscattering interactions, and which has the (seemingly) free-bosonic Hamiltonian of the form

$$\mathcal{H} \equiv \frac{v}{2}\left(\frac{1}{K}\Pi^2 + K(\partial_x\phi)^2\right) \tag{6.55}$$

with an effective velocity v and stiffness K (also known as the *Luttinger parameter*) given by

$$v = \sqrt{\left(v_{\text{F}} + \frac{g_4}{\pi}\right)^2 - \left(\frac{g_2}{\pi}\right)^2} \tag{6.56}$$

$$K = \sqrt{\frac{v_{\text{F}} + g_4/\pi + g_2/\pi}{v_{\text{F}} + g_4/\pi - g_2/\pi}} \tag{6.57}$$

In terms of the field ϕ and its dual field ϑ the bosonized Luttinger Hamiltonian has the symmetric form

$$\mathcal{H} = \frac{v}{2}\left(\frac{1}{K}\left(\partial_x\vartheta\right)^2 + K\left(\partial_x\phi\right)^2\right) \tag{6.58}$$

which is manifestly invariant (*self-dual*) under the duality transformation

$$\phi \leftrightarrow \vartheta, \qquad K \leftrightarrow \frac{1}{K} \tag{6.59}$$

In string theory this transformation is known as *T-duality* and the Luttinger parameter is known as the *compactification radius* (see e.g. Polchinski (1998) and Di Francesco *et al.* (1997)).

Thus, we see that the Luttinger model, which describes the *density fluctuations* of a 1D interacting fermion system, is effectively equivalent to a free Bose field with (in addition to the renormalized stiffness K) an effective speed v for the propagation of the bosons (the density fluctuations). We see immediately several effects.

1. The only effect of the forward-scattering interactions, parametrized by the coupling g_4, is to renormalize the velocity.
2. The backscattering interactions, with coupling g_2, renormalize the velocity and the stiffness. Furthermore, for *repulsive* interactions $g_2 > 0$, the stiffness is renormalized upwards, $K > 1$, whereas for *attractive* interactions, $g_2 < 0$, it is renormalized downwards. We will see that these effects are very important.
3. The bosonized form of the Luttinger model has the obvious invariance under $\phi \to \phi + \theta$, where θ is arbitrary. This is the bosonized version of the continuous chiral symmetry of the Luttinger model or, equivalently, the invariance of

the original fermionic system under a rigid displacement of the density profile. Owing to this invariance the system has long-lived long-wavelength density (particle–hole) fluctuations that propagate with speed v. In other words, the system has long-lived (undamped) sound modes (i.e. phonons) much as a 1D quantum elastic solid would.

4. We saw that in higher dimensions there are similar collective modes, *zero sound*, which eventually become (Landau) damped. In 1D for a system with a strictly linearized dispersion these modes are never damped.
5. This feature of the Luttinger model is, naturally, spoiled by microscopic effects we have ignored, such as band curvature that can be shown to contribute non-quadratic terms to the bosonized Hamiltonian of the form $(\partial_x\phi)^3$ and similar. These non-linear terms have two main effects: (a) they break the inherent particle–hole symmetry of the Luttinger model, and (b) they cause the boson (the sound modes) to interact with each other and decay, which leads to damping.

At half-filling (obviously on a lattice) we have to consider also the Umklapp term, which becomes

$$\mathcal{H}_{\rm u} \sim g_{\rm u} \cos(4\sqrt{\pi}\phi) \tag{6.60}$$

This term formally breaks the continuous U(1) chiral symmetry $\phi \to \phi + \theta$ to a discrete symmetry subgroup $\phi \to \phi + n\sqrt{\pi}/4$, where $n \in \mathbb{Z}$. We will see that when the effects of this operator are important ("relevant") there is a density modulation (a CDW) which is commensurate with the underlying lattice and there is a gap in the fermionic spectrum. In its absence, the fermions remain gapless and the CDW correlations are incommensurate.

The local electron density in bosonized form becomes

$$\rho(x) = \rho_0 + \frac{1}{\sqrt{\pi}}\partial_x\phi + \frac{1}{2\pi a}\left\{e^{i2(p_{\rm F}x-\theta)}e^{i\sqrt{4\pi}\phi(x)} + e^{-i2(p_{\rm F}x-\theta)}e^{-i\sqrt{4\pi}\phi(x)}\right\} \tag{6.61}$$

and the total charge of the system is

$$Q = -e\int dx\, j_0(x) = -\frac{e}{\sqrt{\pi}}\int dx\, \partial_x\phi(x) = -\frac{e}{\sqrt{\pi}}\Delta\phi \tag{6.62}$$

where $\Delta\phi = \phi(+\infty) - \phi(-\infty)$. Hence, in the *charge-neutral* sector the system must obey *periodic boundary conditions*, $\Delta\phi = 0$. Conversely, boundary conditions involving the winding of the boson by $\Delta\phi = N\sqrt{\pi}$, where $N \in \mathbb{Z}$, amount to the sector with charge $Q = -Ne$.

We now summarize our main operator identifications:

$$j_0 \to \frac{1}{\sqrt{\pi}} \partial_x \phi, \qquad j_1 \to -\frac{1}{\sqrt{\pi}} \partial_x \vartheta \tag{6.63}$$

$$\psi_R \to \frac{1}{\sqrt{2\pi a}} e^{i2\sqrt{\pi}\phi_R}, \qquad \psi_L \to \frac{1}{\sqrt{2\pi a}} e^{-i2\sqrt{\pi}\phi_L} \tag{6.64}$$

$$\bar{\psi}\psi \to \frac{1}{\pi a} \cos(2\sqrt{\pi}\phi), \qquad i\bar{\psi}\gamma^5\psi \to \frac{1}{\pi a} \sin(2\sqrt{\pi}\phi) \tag{6.65}$$

$$\psi_R^\dagger \psi_L^\dagger \to \frac{1}{\pi a} e^{i2\sqrt{\pi}\vartheta} \qquad \psi_R^\dagger \psi_L^\dagger \psi_R \psi_L \to \frac{1}{\pi a} e^{i4\sqrt{\pi}\phi} \tag{6.66}$$

6.5 Spin and the Luttinger model

We will now consider the case of the Luttinger model for spin-1/2 fermions, and use the same bosonization approach as before. In this context it is known as *abelian bosonization* since the SU(2) symmetry of spin is not treated in full. A more correct (and more sophisticated) approach that involves *non-abelian bosonization* (Witten, 1984) will be discussed in Section 7.10.

The Hamiltonian density for the Luttinger model for spin-1/2 fermions with both chiralities, denoted below by $s = +1$ (for R) and $s = -1$ (for L), is

$$\begin{aligned} \mathcal{H} = & -i v_F \sum_{\sigma=\uparrow,\downarrow} \sum_{s=\pm 1} s \psi_{s,\sigma}^\dagger \partial_x \psi_{s,\sigma} \\ & + g_4 \sum_{\sigma,s} \psi_{s,\sigma}^\dagger \psi_{s,-\sigma}^\dagger \psi_{s,-\sigma} \psi_{s,\sigma} \\ & + g_2 \sum_{\sigma,\sigma'} \psi_{1,\sigma}^\dagger \psi_{-1,\sigma'}^\dagger \psi_{-1,\sigma'} \psi_{1,\sigma} \\ & + g_{1,\parallel} \sum_{\sigma} \psi_{1,\sigma}^\dagger \psi_{-1,\sigma}^\dagger \psi_{1,\sigma} \psi_{-1,\sigma} \\ & + g_{1,\perp} \sum_{\sigma} \psi_{1,\sigma}^\dagger \psi_{-1,-\sigma}^\dagger \psi_{1,-\sigma} \psi_{-1,\sigma} \end{aligned} \tag{6.67}$$

Here g_4 represents forward-scattering processes of fermions of the same branch (and opposite spin), g_2 forward-scattering processes on opposite branches, $g_{1,\parallel}$ backscattering processes without spin flip, and $g_{1,\perp}$ scattering processes on opposite branches with spin flip. There is also a possible Umklapp scattering term whose form is

$$\mathcal{H}_u = g_3 e^{i(4p_F - G)x} \psi_{-1,\uparrow}^\dagger \psi_{-1,\downarrow}^\dagger \psi_{1,\downarrow} \psi_{1,\uparrow} + \text{h.c.} \tag{6.68}$$

where G is a reciprocal-lattice vector. As before, we will ignore Umklapp processes unless we are at half-filling. For the special case of the 1D Hubbard model, Eq. (6.5), the relations between the coupling constants of the spin-1/2 Luttinger model and the (Hubbard) coupling constant U are $g_2 = g_4 = g_{1,\perp} = g_{1,\parallel} = U$, and $g_3 = U$ (if the Umklapp process is allowed).

For the system to be manifestly invariant under SU(2) spin rotations (as the Hubbard model is) it must be possible to rewrite the Hamiltonian in an explicitly SU(2)-invariant form. To see that this is true we introduce the right- and left-moving SU(2) (chiral) spin currents $J_{\mathrm{R}}^a(x)$ and $J_{\mathrm{L}}^a(x)$ ($a = 1, 2, 3$)

$$J_{\mathrm{R}}^a(x) = \frac{1}{2}\psi_{\mathrm{R},\sigma}^\dagger(x)\tau_{\sigma,\sigma'}^a\psi_{\mathrm{R},\sigma'}(x), \qquad J_{\mathrm{L}}^a(x) = \frac{1}{2}\psi_{\mathrm{L},\sigma}^\dagger(x)\tau_{\sigma,\sigma'}^a\psi_{\mathrm{L},\sigma'}(x) \tag{6.69}$$

where the τ^a (again with $a = 1, 2, 3$) are the three Pauli matrices and the factor of $\frac{1}{2}$ is included in order to use the standard normalization of the SU(2) generators. The right- and left-moving U(1) (chiral) charge currents $J_{\mathrm{R}}(x)$ and $J_{\mathrm{L}}(x)$ are given by

$$J_{\mathrm{R}}(x) = \psi_{\mathrm{R},\sigma}^\dagger(x)\psi_{\mathrm{R},\sigma}(x), \qquad J_{\mathrm{L}}(x) = \psi_{\mathrm{L},\sigma}^\dagger(x)\psi_{\mathrm{L},\sigma}(x) \tag{6.70}$$

In Chapter 5 we derived the algebra of the chiral currents for the U(1) case, the chiral U(1) Kac–Moody algebra of Eqs. (5.245) and (5.246). In the SU(2) × U(1) (i.e. U(2)) case we are interested in here we find, instead, that the two chiral SU(2) and U(1) currents obey the Kac–Moody algebras (Witten, 1984)

$$\begin{aligned} \left[J_{\mathrm{R}}^a(x), J_{\mathrm{R}}^b(y)\right] &= i\epsilon^{abc}J_{\mathrm{R}}^c(x)\delta(x-y) + i\frac{k}{4\pi}\delta^{ab}\delta'(x-y) \\ \left[J_{\mathrm{L}}^a(x), J_{\mathrm{L}}^b(y)\right] &= i\epsilon^{abc}J_{\mathrm{L}}^c(x)\delta(x-y) - i\frac{k}{4\pi}\delta^{ab}\delta'(x-y) \end{aligned} \tag{6.71}$$

and

$$\begin{aligned} \left[J_{\mathrm{R}}(x), J_{\mathrm{R}}(y)\right] &= \frac{i}{\pi}\delta'(x-y) \\ \left[J_{\mathrm{L}}(x), J_{\mathrm{L}}(y)\right] &= -\frac{i}{\pi}\delta'(x-y) \end{aligned} \tag{6.72}$$

where, once again, the primes in Eqs. (6.71) and (6.72) denote derivatives of the delta functions. The integer k (which in this case is $k = 1$) is called the level of the $\mathrm{SU}(2)_k$ Kac–Moody algebra. The current algebras defined by Eqs. (6.71) are the basis of non-abelian bosonization (Witten, 1984), which we will discuss in Section 7.10. Please note that there is a factor of 2 difference between Eqs. (6.72) and Eqs. (5.245) and (5.246) due to the fact that, for SU(2) × U(1), the U(1) chiral anomaly (the Schwinger term) is doubled.

The Hamiltonian of Eq. (6.67) is SU(2)-invariant if the couplings satisfy $g_{1,\parallel} = g_{1,\perp} \equiv g_1$. Indeed, if this condition holds it is possible to rewrite the effective Hamiltonian of Eq. (6.67) (i.e. the effective Hamiltonian without Umklapp terms) in the more compact and SU(2) × U(1)-invariant form

$$\begin{aligned}\mathcal{H} = & \frac{\pi}{2} v_{\mathrm{F}} \Big(J_{\mathrm{R}}(x) J_{\mathrm{R}}(x) + J_{\mathrm{L}}(x) J_{\mathrm{L}}(x) \Big) \\ & + \frac{2\pi}{3} v_{\mathrm{F}} \Big(\vec{J}_{\mathrm{R}}(x) \cdot \vec{J}_{\mathrm{R}}(x) + \vec{J}_{\mathrm{L}}(x) \cdot \vec{J}_{\mathrm{L}}(x) \Big) \\ & + \frac{g_4}{4} \Big(J_{\mathrm{R}}(x) J_{\mathrm{R}}(x) + J_{\mathrm{L}}(x) J_{\mathrm{L}}(x) \Big) \\ & - g_4 \Big(\vec{J}_{\mathrm{R}}(x) \cdot \vec{J}_{\mathrm{R}}(x) + \vec{J}_{\mathrm{L}}(x) \cdot \vec{J}_{\mathrm{L}}(x) \Big) \\ & + \frac{1}{2} (2g_2 - g_1) J_{\mathrm{R}}(x) J_{\mathrm{L}}(x) \\ & - 2g_1 \vec{J}_{\mathrm{R}}(x) \cdot \vec{J}_{\mathrm{L}}(x) \end{aligned} \tag{6.73}$$

where we expressed the (free) kinetic-energy term as a quadratic form in the currents (Dashen and Frishman, 1975; Affleck, 1986a). This is the Sugawara form of the Hamiltonian.

It is useful to write this effective-field theory also in the form of a Lagrangian (density) for the two Dirac spinor fields, $\psi_{\mathrm{R},\sigma}$ and $\psi_{\mathrm{L},\sigma}$ (with $\sigma = \uparrow, \downarrow$). Ignoring the forward-scattering terms (with coupling constant g_4), the effective Lagrangian $\mathcal{L}$ is

$$\mathcal{L} = \bar{\psi} i \gamma^\mu \partial_\mu \psi - \frac{1}{8}(2g_2 - g_1) \left(\bar{\psi}\gamma^\mu \psi\right)^2 + \frac{g_1}{8} \left(\bar{\psi}\gamma_\mu \vec{\sigma} \psi\right)^2 \tag{6.74}$$

where we have dropped the spin indices for clarity. Forward-scattering terms amount to finite renormalizations of the velocity of the collective modes of the charge and spin sectors. Since spin and charge degrees of freedom are effectively split this amounts to a separate rescaling of the dependence on the time (or space) coordinates of these degrees of freedom.

The Lagrangian of Eq. (6.74) is known as the SU(2) × U(1) Thirring model. Using the (Fierz) identity

$$\left(\bar{\psi}\gamma^\mu \vec{\sigma} \psi\right)^2 + \left(\bar{\psi}\gamma_\mu \psi\right)^2 = -2\left[\left(\bar{\psi}\psi\right)^2 - \left(\bar{\psi}\gamma_5\psi\right)^2\right] \tag{6.75}$$

the effective Lagrangian of Eq. (6.74) can be written in the equivalent form (again we are dropping the spin indices)

$$\mathcal{L} = \bar{\psi} i \gamma^\mu \partial_\mu \psi - \frac{g_2}{4} \left(\bar{\psi}\gamma^\mu \psi\right)^2 + \frac{g_1}{4} \left[\left(\bar{\psi}\psi\right)^2 - \left(\bar{\psi}\gamma_5\psi\right)^2\right] \tag{6.76}$$

which is known as the chiral Gross–Neveu model.

The effective Lagrangians of Eq. (6.74) and Eq. (6.76) are invariant under the continuous chiral transformation $\psi \to \exp(i\theta\gamma_5)\psi$, which, as we saw earlier in this chapter, amounts to a rigid translation in space of the electronic charge density $\rho(x)$ by an amount proportional to the chiral angle θ. As we saw, away from half-filling, this is a symmetry of the Luttinger liquid. At half-filling this global continuous chiral symmetry is broken down to a discrete symmetry subgroup by the Umklapp term in the Hamiltonian, which, in this notation, has the manifestly SU(2) × U(1)-invariant form

$$\mathcal{L}_{\text{Umklapp}} \sim \epsilon_{\alpha\beta}\psi^{\dagger}_{\text{L},\alpha}\psi^{\dagger}_{\text{L},\beta}\epsilon_{\gamma\delta}\psi_{\text{R},\gamma}\psi_{\text{R},\delta} + \text{h.c.} \tag{6.77}$$

This interaction term represents a process in which a spin-singlet pair of right movers is destroyed and a spin-singlet pair of left movers is created (and vice versa).

6.5.1 Abelian bosonization of the Luttinger liquid

We are now ready to proceed with the (abelian) bosonization of the spin-1/2 Luttinger model. Once again we begin with the free fermion. We then introduce two Bose fields, $\phi_\uparrow$ and $\phi_\downarrow$, and their respective canonical momenta, $\Pi_\uparrow$ and $\Pi_\downarrow$. The corresponding free-boson Hamiltonian is

$$\mathcal{H}_0 = \frac{v_{\text{F}}}{2}\sum_{\sigma}\left(\Pi_\sigma^2 + (\partial_x\phi_\sigma)^2\right) \tag{6.78}$$

We now define the charge and spin Bose fields ϕ_{c} and ϕ_{s},

$$\phi_{\text{c}} = \frac{1}{\sqrt{2}}\left(\phi_\uparrow + \phi_\downarrow\right) \tag{6.79}$$

$$\phi_{\text{s}} = \frac{1}{\sqrt{2}}\left(\phi_\uparrow - \phi_\downarrow\right) \tag{6.80}$$

in terms of which $\mathcal{H}_0$ becomes a sum over the charge and spin sectors

$$\mathcal{H}_0 = \frac{v_{\text{F}}}{2}\left(\Pi_{\text{c}}^2 + (\partial_x\phi_{\text{c}})^2\right) + \frac{v_{\text{F}}}{2}\left(\Pi_{\text{s}}^2 + (\partial_x\phi_{\text{s}})^2\right) \tag{6.81}$$

where Π_{c} and Π_{s} are the momenta canonically conjugate to ϕ_{c} and ϕ_{s}. By analogy with the spinless case we now define for the charge and spin fields ϕ_{c} and ϕ_{s} their respective dual fields ϑ_{c} and ϑ_{s} (cf. Eq. (5.278) and Eq. (5.279)).

We now see that the interactions will lead to a finite renormalization of these parameters, leading to the introduction of a charge and a spin velocity, v_{c} and v_{s}, and of the charge and spin Luttinger parameters K_{c} and K_{s}.

The charge and spin densities and currents are

$$j_0^{\text{c}} = j_0^{\uparrow} + j_0^{\downarrow} = \frac{1}{\sqrt{\pi}}\partial_x(\phi_\uparrow + \phi_\downarrow) = \sqrt{\frac{2}{\pi}}\,\partial_x\phi_{\text{c}} \tag{6.82}$$

$$j_0^{\text{s}} = \frac{1}{2}\left(j_0^{\uparrow} - j_0^{\downarrow}\right) = \frac{1}{\sqrt{\pi}}\partial_x(\phi_\uparrow - \phi_\downarrow) = \sqrt{\frac{2}{\pi}}\,\partial_x\phi_{\text{s}} \tag{6.83}$$

Using the bosonization identities, we can write the Luttinger Hamiltonian in the form

$$\mathcal{H} = \frac{v_c}{2}\left(\frac{1}{K_c}\Pi_c^2 + K_c(\partial_x\phi_c)^2\right) + \frac{v_s}{2}\left(\frac{1}{K_s}\Pi_s^2 + K_s(\partial_x\phi_s)^2\right) + V_c\cos(2\sqrt{2\pi}\phi_c) + V_s\cos(2\sqrt{2\pi}\phi_s) \tag{6.84}$$

where v_c and v_s are the charge and spin velocities,

$$v_c = \sqrt{\left(v_F + \frac{g_4}{2\pi}\right)^2 - \left(\frac{g_{1,\parallel}}{2\pi} - \frac{g_2}{\pi}\right)^2}$$
$$v_s = \sqrt{\left(v_F - \frac{g_4}{2\pi}\right)^2 - \left(\frac{g_{1,\parallel}}{2\pi}\right)^2} \tag{6.85}$$

K_c and K_s are the charge and spin Luttinger parameters,

$$K_c = \sqrt{\frac{2\pi v_F + g_4 + 2g_2 - g_{1,\parallel}}{2\pi v_F + g_4 - 2g_2 + g_{1,\parallel}}}, \qquad K_s = \sqrt{\frac{2\pi v_F - g_4 - g_{1,\parallel}}{2\pi v_F - g_4 + g_{1,\parallel}}} \tag{6.86}$$

The couplings V_c and V_s, due to Umklapp and backscattering with spin flip, respectively, are given by

$$V_c = \frac{g_3}{2(\pi a)^2}, \qquad V_s = \frac{g_{1,\perp}}{2(\pi a)^2} \tag{6.87}$$

In what follows we will neglect Umklapp processes and hence set $V_c = 0$. In the absence of backscattering, $g_{1,\parallel} = g_{1,\perp} = 0$ (and hence $V_s = 0$); this model is known as the Tomonaga–Luttinger model. Notice that in this case $K_s = 1$ automatically. This is a consequence of the SU(2) symmetry of spin.

1. We now see that this model describes a system with charge and spin bosons, the charge and spin collective modes of the fermionic system. In general the charge and spin velocities are different.
2. There is no mixing between charge and spin bosons: *spin–charge separation*.
3. We also see that for repulsive interactions the charge mode propagates faster than the spin mode, $v_c > v_s$.
4. In the same regime, $K_c > 1$ while $K_s < 1$. This will have important consequences.
5. The fermion operators, with chirality $\eta = \pm$ (for right and left) and spin σ can now be expressed in terms of the right- and left-moving charge and spin Bose fields $\phi_{\eta,\sigma}$ (by using Eq. (5.282)) as

$$\psi_{\eta,\sigma} = \frac{1}{\sqrt{2\pi a}} F_{\eta,\sigma} e^{-i\eta\sqrt{2\pi}(\phi_{\eta,c} + \sigma\phi_{\eta,s})} \tag{6.88}$$

where $F_{\eta,\sigma}$ are Klein factors that ensure that fermions with different labels anticommute with each other,

$$\{F_{\eta,\sigma}, F_{\eta'\sigma'}\} = \delta_{\eta,\eta'}\delta_{\sigma,\sigma'} \tag{6.89}$$

and a is a short-distance cutoff.

We can now express all the operators in which we are interested in terms of charge and spin bosons. In the following subsections we will use these expressions to compute their correlation functions and several observables of physical interest.

1. *SU(2) spin currents*. The SU(2) chiral currents are given by

$$J_{\mathrm{R}}^3 = \frac{1}{\sqrt{2\pi}}\,\partial_x\phi_{\mathrm{R},s}, \qquad J_{\mathrm{R}}^{\pm} = \frac{1}{2\pi a}e^{\mp i2\sqrt{2\pi}\phi_{\mathrm{R},s}} \tag{6.90}$$

$$J_{\mathrm{L}}^3 = \frac{1}{\sqrt{2\pi}}\,\partial_x\phi_{\mathrm{L},s}, \qquad J_{\mathrm{L}}^{\pm} = \frac{1}{2\pi a}e^{\pm i2\sqrt{2\pi}\phi_{\mathrm{L},s}} \tag{6.91}$$

It is straightforward to check that at the free-fermion point all these operators have dimension 1 (as they should), are conserved, and obey the $\mathrm{SU}(2)_1$ Kac–Moody algebra.

2. *Charge-density wave*. The CDW order parameter has the bosonized expression

$$\mathcal{O}_{\mathrm{CDW}} = e^{-i2p_{\mathrm{F}}x}\sum_{\sigma}\psi_{1,\sigma}^{\dagger}(x)\psi_{-1,\sigma} \to \frac{1}{\pi a}e^{-i2p_{\mathrm{F}}x}\cos\left(\sqrt{2\pi}\phi_{\mathrm{s}}\right)e^{-i\sqrt{2\pi}\phi_{\mathrm{c}}(x)} \tag{6.92}$$

3. *Spin-density wave*. The bosonized forms of the three components of the SDW order parameter are

$$\begin{aligned}\mathcal{O}_{\mathrm{SDW}}^{(3)} &= e^{-i2p_{\mathrm{F}}x}\sum_{\sigma,\sigma'}\psi_{1,\sigma}^{\dagger}(x)\tau_{\sigma,\sigma'}^{3}\psi_{-1,\sigma'}\\ &\to -\frac{1}{\pi a}e^{-i2p_{\mathrm{F}}x}\,2i\sin\left(\sqrt{2\pi}\phi_{\mathrm{s}}\right)e^{-i\sqrt{2\pi}\phi_{\mathrm{c}}(x)}\end{aligned} \tag{6.93}$$

$$\begin{aligned}\mathcal{O}_{\mathrm{SDW}}^{(\pm)} &= e^{-i2p_{\mathrm{F}}x}\sum_{\sigma,\sigma'}\psi_{1,\sigma}^{\dagger}(x)\tau_{\sigma,\sigma'}^{\pm}\psi_{-1,\sigma'}\\ &\to \frac{1}{\pi a}e^{-i2p_{\mathrm{F}}x}e^{-i\sqrt{2\pi}\phi_{\mathrm{c}}}e^{\pm i\sqrt{2\pi}\vartheta_{\mathrm{s}}(x)}\end{aligned} \tag{6.94}$$

4. *Singlet superconductivity*. The singlet superconducting order parameter, i.e. the singlet Cooper-pair amplitude, has the bosonized expression

$$\mathcal{O}_{\mathrm{SS}} = \psi_{\mathrm{R},\uparrow}^{\dagger}\psi_{\mathrm{L},\downarrow}^{\dagger} \to e^{i\sqrt{2\pi}\vartheta_{\mathrm{c}}}e^{-i\sqrt{2\pi}\phi_{\mathrm{s}}} \tag{6.95}$$

5. *Triplet superconductivity*. The triplet Cooper-pair operator is

$$\mathcal{O}_{\mathrm{TS}}^{(1)} = \psi_{\mathrm{R},\uparrow}^{\dagger}\psi_{\mathrm{L},\uparrow}^{\dagger} \to e^{i\sqrt{2\pi}\vartheta_{\mathrm{c}}}e^{i\sqrt{2\pi}\vartheta_{\mathrm{s}}} \tag{6.96}$$

$$\mathcal{O}_{\mathrm{TS}}^{(-1)} = \psi_{\mathrm{R},\downarrow}^{\dagger}\psi_{\mathrm{L},\downarrow}^{\dagger} \to e^{i\sqrt{2\pi}\vartheta_{\mathrm{c}}}e^{-i\sqrt{2\pi}\vartheta_{\mathrm{s}}} \tag{6.97}$$

6.6 Scaling and renormalization in the Luttinger model

We will now discuss the scaling behavior of the Luttinger model. We saw before that it generally exhibits the phenomenon of spin–charge separation which, at the level of the effective low-energy Hamiltonian density $\mathcal{H}$, means that it is a sum of two decoupled terms

$$\mathcal{H} = \mathcal{H}_c + \mathcal{H}_s \tag{6.98}$$

where both the charge sector and the spin sector are represented at low energies by sine–Gordon Hamiltonians. Indeed, the bosonized form of the Hamiltonian density for the charge sector $\mathcal{H}_c$ is

$$\mathcal{H}_c = \frac{v_c}{2}\left[\frac{1}{K_s}\Pi_c^2 + K_s(\partial_x\phi_c)^2\right] + V_c \cos\left(2\sqrt{2\pi}\phi_c\right) \tag{6.99}$$

As we saw above, the last term is present only at half-filling, and is due to an Umklapp process. In its presence the system develops a (Mott) charge gap and it is an insulator.

The bosonized form of the sine–Gordon Hamiltonian density for the spin sector $\mathcal{H}_s$ is

$$\mathcal{H}_s = \frac{v_s}{2}\left(\frac{1}{K_s}\Pi_s^2 + K_s(\partial_r\phi_s)^2\right) + V_s \cos(2\sqrt{2\pi}\phi_s) \tag{6.100}$$

In contrast to the cosine operator of the charge sector, the last term of the Hamiltonian density of the spin sector $\mathcal{H}_s$ is always present since it represents backscattering processes with spin flip.

Thus, we can use the analysis of the sine–Gordon model to derive scaling laws for the Luttinger model. In the preceding sections we discussed the behavior of several correlation functions and susceptibilities of interest in the absence of the sine–Gordon operators. Following the same analysis as in Chapter 4, we begin by computing the scaling dimensions of the cosine operators both in the charge sector and in the spin sector.

6.6.1 The charge sector

The scaling dimension of the cosine operator (charge Umklapp processes) is

$$\Delta_c(2\sqrt{2\pi}) = \frac{(2\sqrt{2\pi})^2}{4\pi K_c} = \frac{2}{K_c} \tag{6.101}$$

Since it is an Umklapp scattering process it is present only at half-filling. Away from half-filling this process has zero amplitude. For free fermions the charge Luttinger parameter is $K_c = 1$, the scaling dimension is $\Delta_c = 2$, and Umklapp processes are marginal. For a Luttinger model with repulsive interactions, the charge Luttinger parameter obeys $K_c > 1$, the scaling dimension is $\Delta_c = 2/K_c < 2$, and Umklapp processes are *relevant*. This is the case of the 1D Hubbard model at half-filling. Our analysis of the scaling behavior of sine–Gordon theory in Chapter 4 in terms of the Kosterlitz RG flow tells us that in this case the charge sector flows to strong coupling where the charge boson ϕ_c is pinned and hence acquires an expectation value. The charge fluctuations become massive and there is an energy gap in the charge spectrum. This is a Mott insulating state. Conversely, for a system with attractive interactions, the charge Luttinger parameter now has $K_c < 1$, the scaling dimension is $\Delta_c > 2$, and Umklapp processes are not allowed by symmetry, and the charge sector remains gapless.

6.6.2 The spin sector

In the spin sector we have to look at the relevance or irrelevance of the cosine operator of the bosonized theory representing backscattering processes with spin flip. The scaling dimension is

$$\Delta_s(2\sqrt{2\pi}) = \frac{2}{K_s} \tag{6.102}$$

Once again, for free fermions the spin Luttinger parameter is $K_s = 1$, the scaling dimension is $\Delta_s = 2$, and the operator is marginal. In the case of an interacting theory, i.e. in the Luttinger model, for repulsive interactions the spin Luttinger parameter satisfies $K_s < 1$, which implies that the scaling dimension is $\Delta_s = 2/K_s > 2$ and, thus, these processes are irrelevant. Thus, for repulsive interactions we expect the spin sector to remain gapless, and hence critical. With some caveats (which we address below), this is what happens in the case of the 1D Hubbard model in the repulsive regime. Conversely, for attractive interactions the spin Luttinger parameter now obeys $K_s > 1$, the scaling dimension is $\Delta_s < 2$, and backscattering processes with spin flip are relevant. In this regime the spin boson ϕ_s becomes pinned and its fluctuations are massive. Hence, for attractive interactions we generally expect that the spin sector becomes gapped.

6.6.3 Scaling analysis of the one-dimensional Hubbard model

In the case of the 1D Hubbard model all the Luttinger couplings are equal. As we saw above, this condition in part is a consequence of the spin SU(2) symmetry. However, this symmetry alone does not require that the other couplings be equal. An examination of the charge and spin Luttinger velocities and parameters, Eq. (6.85) and Eq. (6.86), reveals that the Hubbard model has an additional symmetry:

$$K_c = \sqrt{1 + \frac{U}{\pi v_F}}, \qquad K_s = \sqrt{1 - \frac{U}{\pi v_F}} \tag{6.103}$$

$$v_c = \sqrt{\left(v_F + \frac{U}{2\pi}\right)^2 - \left(\frac{U}{2\pi}\right)^2}, \qquad v_s = \sqrt{\left(v_F - \frac{U}{2\pi}\right)^2 - \left(\frac{U}{2\pi}\right)^2} \tag{6.104}$$

$$g_c = \frac{U}{2\pi^2 v_c}, \qquad g_s = \frac{U}{2\pi^2 v_s} \tag{6.105}$$

where we introduced the dimensionless (sine–Gordon) couplings g_c and g_s. There is a clear symmetry $U \leftrightarrow -U$, which amounts to exchanging the spin and charge sectors.

However, since the spin sector has an SU(2) symmetry (and its associated $\mathrm{SU}(2)_1$ Kac–Moody current algebra) we guess that there must be a "hidden" SU(2) symmetry (and a current algebra) in the charge sector as well. Indeed, let us decompose the charged (Dirac) Fermi fields $\psi_{\eta,\sigma}(x)$ into their (real and imaginary) Majorana fermion components,

$$\psi_{\eta,\sigma} = \xi_{1,\eta,\sigma} + i\xi_{2,\eta,\sigma} \tag{6.106}$$

The Majorana fermions $\xi_{i,\eta,\sigma}(x)$ satisfy the (Majorana) anticommutation relations

$$\{\xi_{i,\eta,\sigma}(x), \xi_{i',\eta',\sigma'}(x')\} = \delta_{i,i'}\delta(x - x') \tag{6.107}$$

In terms of these Majorana fermions, the $\mathrm{SU}(2) \times \mathrm{U}(1) \simeq \mathrm{U}(2)$ symmetry actually becomes an SO(4) symmetry. In fact it is easy to construct an additional set of SU(2) currents in terms of the Majorana fields $\xi_{i,\eta,\sigma}$. The three generators are the chiral charge current $\psi^\dagger_{\mathrm{R},\sigma}\psi_{\mathrm{R},\sigma} = 2i\xi_{1,\eta,\sigma}\xi_{2,\eta,\sigma}$ and the chiral pair fields $\psi^\dagger_{\mathrm{R},\uparrow}\psi^\dagger_{\mathrm{R},\downarrow} = -2i\xi_{1,\eta,\uparrow}\xi_{2,\eta,\downarrow}$. They also obey an $\mathrm{SU}(2)_1$ Kac–Moody algebra (and the same applies to their left-moving counterparts).

This SO(4) symmetry is not an accident of the continuum Luttinger model, since it is also present in the Hubbard model on general lattices. In the case of the 1D chain, the SO(4) symmetry plays a key role in its exact solution (see Essler *et al.* (2005)). In the 1D lattice model, the pairing operator is known as "eta pairing" and is given by

$$\eta_n^\dagger = \sum_{n=1}^{N} (-1)^n c^\dagger_{n,\uparrow} c^\dagger_{n,\downarrow} \tag{6.108}$$

which creates a spin-singlet pair on sites with momentum π.

Let us now apply our RG results (derived in Chapter 4) for the case of the 1D Hubbard model (Fig. 6.5). Insofar as the charge sector is concerned there are two cases: at half-filling (for which the Umklapp operator is present and we expect a Mott (charge) gap) and away from half-filling (for which there is no Umklapp term and there is no charge gap). In the latter case the charge sector remains strictly marginal. However, even in this case, the spin sector still flows because there is always a potentially (marginally) relevant cosine operator.

We consider first the half-filled case. Since the Kosterlitz RG is perturbative, we will need to know only these relations in the weak-coupling regime of small g_s (or, which amounts to the same thing, small Hubbard U). Clearly the SG couplings,

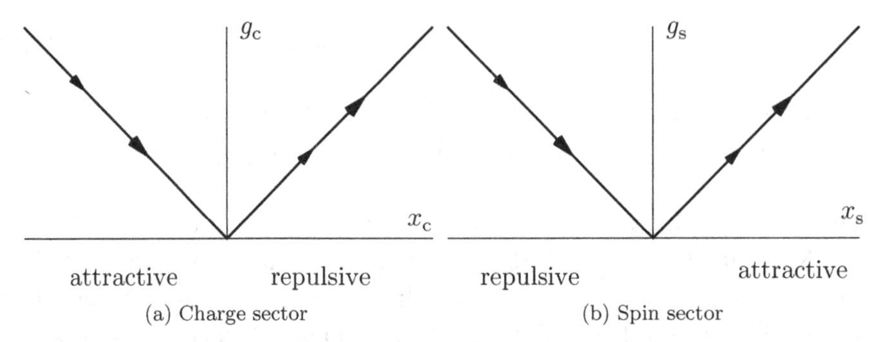

Figure 6.5 Schematic RG flows for the (a) charge and (b) spin sectors of the 1D Hubbard model at half-filling. Here g_c and g_s are the effective sine–Gordon coupling constants in the charge and spin sectors, $x_c = 2 - 2/K_c$, and $x_s = 2 - 2/K_s$. Attractive means $U < 0$ and repulsive $U > 0$. Away from half-filling, the charge sector remains marginal and does not flow, and the spin sector has the same flow as in (b).

g_c (which is non-vanishing at half-filling) and g_s (which is always non-zero), and the stiffnesses, K_c and K_s, are related to each other. These relations set the allowed initial values of the RG flow and only those points in the (x_c, g_c) and (x_s, g_s) planes to which the RG flow has access are physically relevant (here we set $x_c = 2-2/K_c$ and $x_s = 2 - 2/K_s$, respectively). For small g_c and g_s the Luttinger parameters become $K_c = 1 + \pi g_c + O(g_c^2)$, $K_s = 1 - \pi g_s + O(g_s^2)$, $x_c = 2\pi g_c + O(g_c^2)$, and $x_s = -2\pi g_s + O(g_s^2)$. Thus, we find that the initial values of the RG flows must be such that $x = \pm 2\pi g$ (see Eq. (4.88) and Figs. 4.3 and 6.5), which are just the asymptotes of the Kosterlitz RG flows.

Thus, as expected, the charge and spin flows are not identical but are symmetric. For repulsive interactions, $U > 0$, the charge flow is on the unstable trajectory on one of the asymptotes, and the spin RG flow is on the stable trajectory, on the opposite asymptote. For attractive interactions, $U < 0$, the two sectors switch roles. It is straightforward to see that *along the asymptotes* the RG flows are given by the beta functions

$$a\frac{dg_c}{da} = 2\pi g_c^2, \qquad a\frac{dg_s}{da} = -2\pi g_s^2 \tag{6.109}$$

Thus, for $U > 0$, the RG flow of the charge sector is *marginally relevant* while the RG flow spin sector is *marginally irrelevant*, and conversely for $U < 0$. In the next chapter we will examine the predictions of asymptotically free beta functions, Eqs. (6.109), and find that in the marginally relevant case there is a gap in the spectrum scaling as

$$M = \text{constant} \times \exp\left(-\frac{1}{2\pi g}\right) \tag{6.110}$$

In the opposite case, for a marginally irrelevant operator, the system flows to $g \to 0$ but very slowly, with logarithmic corrections to scaling. Thus, for a half-filled repulsive Hubbard model, the charge sector is gapped while the spin sector is gapless, and conversely for attractive interactions.

On the other hand, away from half-filling there is no Umklapp operator and the charge sector remains gapless and marginal, and there is no Mott gap. The spin sector is also gapless for repulsive interactions. In this regime the system is a Luttinger liquid. In contrast, for attractive interactions there is a spin gap while the charge sector is gapless. This regime is known as the Luther–Emery liquid.

We close this section by discussing briefly the consequences of breaking the accidental symmetry of the Hubbard model explicitly. We noticed earlier that the hidden SU(2) symmetry of the charge sector of the Hubbard model is a consequence of the special form of the interaction. Indeed, the addition of a simple interaction V between nearest-neighboring charge densities (i.e. a "Coulomb" interaction) causes this accidental symmetry to become broken explicitly. In this case only the U(1) charge symmetry remains, in addition to the SU(2) symmetry of the spin sector. This slightly modified system is known as the extended Hubbard model. In this case the RG flows of the charge sector become the generic Kosterlitz flows and no longer lie precisely on the unstable asymptote. However, unless the coupling g_1 becomes large enough, the RG flows of the charge sector still approach the unstable asymptote, and the charge sector remains gapped. On the other hand, if the global (but not accidental) SU(2) symmetry of spin were to be explicitly broken down to a $\mathbb{Z}_2$ (easy-axis, Ising symmetry) by a magnetic anisotropy of the material the RG flows of the spin sector would no longer be on the stable asymptote and would, in fact, converge to the stable asymptote, leading to a system with a finite spin gap. Conversely, for an easy-plane (XY) symmetry, the RG flows in the spin sector converge on the line of fixed points and there is no spin gap.

6.7 Correlation functions of the Luttinger model

We will now compute the correlation functions of the Luttinger model. We will first consider the spinless case.

6.7.1 The spinless case

The bosonized Luttinger Hamiltonian density for spinless fermions is

$$\mathcal{H} = (\pi v_F + g_4)\left(\rho_L^2 + \rho_R^2\right) + 2g_2\rho_R\rho_L \tag{6.111}$$

We will diagonalize this Hamiltonian by means of a Bogoliubov transformation (which is canonical):

$$\rho_R = \cosh\lambda\ \tilde{\rho}_R + \sinh\lambda\ \tilde{\rho}_L \tag{6.112}$$

$$\rho_{\mathrm{L}} = \sinh\lambda\ \tilde{\rho}_{\mathrm{R}} + \cosh\lambda\ \tilde{\rho}_{\mathrm{L}} \tag{6.113}$$

where

$$\tilde{\rho}_{\mathrm{R}} = \frac{1}{\sqrt{\pi}}\,\partial_x\tilde{\phi}_{\mathrm{R}}, \qquad \tilde{\rho}_{\mathrm{L}} = \frac{1}{\sqrt{\pi}}\,\partial_x\tilde{\phi}_{\mathrm{L}} \tag{6.114}$$

With the choice

$$\tanh(2\lambda) = -\frac{g_2}{\pi v_{\mathrm{F}} + g_4} \tag{6.115}$$

the Hamiltonian becomes

$$\mathcal{H} = \pi v\left(\tilde{\rho}_{\mathrm{R}}^2 + \tilde{\rho}_{\mathrm{L}}^2\right) = \frac{v}{2}\left((\partial_x\tilde{\vartheta})^2 + (\partial_x\tilde{\phi})^2\right) \tag{6.116}$$

where, as before,

$$\pi v = \sqrt{(\pi v_{\mathrm{F}} + g_4)^2 - g_2^2} \tag{6.117}$$

and

$$\cosh\lambda = \frac{K+1}{2\sqrt{K}}, \quad \sinh\lambda = \frac{K-1}{2\sqrt{K}} \tag{6.118}$$

and

$$K = \sqrt{\frac{\pi v_{\mathrm{F}} + g_4 + g_2}{\pi v_{\mathrm{F}} + g_4 - g_2}} \tag{6.119}$$

The propagator of the field $\tilde{\phi} = \tilde{\phi}_{\mathrm{R}} + \tilde{\phi}_{\mathrm{L}}$, using a regularization in which it vanishes as $x' \to x$ and $t' \to t$, is given by

$$\left\langle T\left(\tilde{\phi}(x,t)\tilde{\phi}(x',t')\right)\right\rangle = -\frac{1}{4\pi}\,\ln\left(\frac{(x-x')^2 - v^2(t-t')^2 + a_0^2 + i\epsilon}{a_0^2}\right) \tag{6.120}$$

from which we get

$$\begin{aligned}
\left\langle T\left(\tilde{\phi}_{\mathrm{R}}(x,t)\tilde{\phi}_{\mathrm{R}}(x',t')\right)\right\rangle &= -\frac{1}{4\pi}\,\ln\left(\frac{(x-x') - v(t-t') + i\epsilon}{a_0}\right)\\
\left\langle T\left(\tilde{\phi}_{\mathrm{L}}(x,t)\tilde{\phi}_{\mathrm{L}}(x',t')\right)\right\rangle &= -\frac{1}{4\pi}\,\ln\left(\frac{(x-x') + v(t-t') + i\epsilon}{a_0}\right)
\end{aligned} \tag{6.121}$$

Using these expressions, we get

$$\begin{aligned}
\langle T(\phi_{\mathrm{R}}(x,t)\phi_{\mathrm{R}}(x',t'))\rangle &= \alpha\langle T(\tilde{\phi}_{\mathrm{R}}(x,t)\tilde{\phi}_{\mathrm{R}}(x',t'))\rangle + \beta\langle T(\tilde{\phi}_{\mathrm{L}}(x,t)\tilde{\phi}_{\mathrm{L}}(x',t'))\rangle\\
\langle T(\phi_{\mathrm{L}}(x,t)\phi_{\mathrm{L}}(x',t'))\rangle &= \beta\langle T(\tilde{\phi}_{\mathrm{R}}(x,t)\tilde{\phi}_{\mathrm{R}}(x',t'))\rangle + \alpha\langle T(\tilde{\phi}_{\mathrm{L}}(x,t)\tilde{\phi}_{\mathrm{L}}(x',t'))\rangle
\end{aligned} \tag{6.122}$$

with

$$\alpha = \frac{(K+1)^2}{4K}, \qquad \beta = \frac{(K-1)^2}{4K} \tag{6.123}$$

The fermion propagator

The propagator for right-moving fermions is

$$\begin{aligned}
\langle T(\psi_{\mathrm{R}}(x,t)\psi_{\mathrm{R}}^{\dagger}(x',t'))\rangle &\sim \frac{1}{2\pi a_0}\langle T(e^{i2\sqrt{\pi}\phi_{\mathrm{R}}(x,t)}e^{-i2\sqrt{\pi}\phi_{\mathrm{R}}(x',t')})\rangle \\
&= \frac{1}{2\pi a_0}e^{2\pi\langle T(\phi_{\mathrm{R}}(x,t)\phi_{\mathrm{R}}(x',t'))\rangle} \\
&= \frac{1}{2\pi a_0}\left(\frac{a_0}{(x-x')-v(t-t')+i\epsilon}\right)^{(K+1)^2/(4K)} \\
&\quad\times\left(\frac{a_0}{(x-x')+v(t-t')+i\epsilon}\right)^{(K-1)^2/(4K)}
\end{aligned} \tag{6.124}$$

and that for left-moving fermions is

$$\begin{aligned}
\langle T(\psi_{\mathrm{L}}(x,t)\psi_{\mathrm{L}}^{\dagger}(x',t'))\rangle &\sim \frac{1}{2\pi a_0}\langle T(e^{-i2\sqrt{\pi}\phi_{\mathrm{L}}(x,t)}e^{i2\sqrt{\pi}\phi_{\mathrm{L}}(x',t')})\rangle \\
&= \frac{1}{2\pi a_0}e^{4\pi\langle T(\phi_{\mathrm{L}}(x,t)-\phi_{\mathrm{L}}(x',t'))\rangle} \\
&= \frac{1}{2\pi a_0}\left(\frac{a_0}{(x-x')+v(t-t')+i\epsilon}\right)^{(K+1)^2/(4K)} \\
&\quad\times\left(\frac{a_0}{(x-x')-v(t-t')+i\epsilon}\right)^{(K-1)^2/(4K)}
\end{aligned} \tag{6.125}$$

In the free-fermion case, $K = 1$, for right-moving fermions this propagator just becomes

$$\langle T(\psi_{\mathrm{R}}(x,t)\psi_{\mathrm{R}}^{\dagger}(x',t'))\rangle = \frac{1}{2\pi a_0}\frac{a_0}{(x-x')-v(t-t')+i\epsilon} \tag{6.126}$$

(and a similar expression applies for left-moving fermions). We see that while in the free-fermion case the propagator has a simple pole, and hence a finite fermion residue $Z = 1$, as soon as the interactions are turned on the pole disappears and is replaced by a branch cut.

We also see that the fermion propagators given by Eqs. (6.124) and (6.125) factorize into right- and left-moving contributions. This happens since the interactions mix the right- and left-moving sectors, which leads to the electron operator acquiring an anomalous dimension. From Eqs. (6.124) and (6.125) we see that the scaling dimension of both right- and left-moving fermions now is

$$\Delta_{\text{fermion}} = \frac{1}{4}\left(K + \frac{1}{K}\right) \tag{6.127}$$

and it no longer scales as a free fermion. Notice that these fermion propagators still describe fermions since they change sign under a permutation (understood as a rotation by π in complex coordinates), i.e. the "spin" s is

$$s = \frac{(K+1)^2}{8K} - \frac{(K-1)^2}{8K} = \frac{1}{2} \tag{6.128}$$

However, this also has the interpretation that the fermion factorizes (or *fractionalizes*) into a right-moving soliton, with scaling dimension $\Delta_{\rm R}$ and "conformal spin" $s_{\rm R}$ (Di Francesco *et al.*, 1997),

$$(\Delta_{\rm R}, s_{\rm R}) = \left(\frac{(K+1)^2}{2K}, \frac{(K+1)^2}{2K} \right) \tag{6.129}$$

and a left-moving soliton with scaling dimension and spin

$$(\Delta_{\rm L}, s_{\rm L}) = \left(\frac{(K-1)^2}{2K}, \frac{(K-1)^2}{2K} \right) \tag{6.130}$$

and vice versa for right-moving fermions. In this regime there are no states in the spectrum with the quantum numbers of the electron, and instead the spectrum is described by gapless solitons. Since the quasiparticle residue Z measures the overlap between an asymptotic state with the quantum numbers of a free electron and the actual eigenstates of the interacting system, the vanishing of the quasiparticle residue means that the exact eigenstates are orthogonal to the asymptotic ("incoming") electron. This feature has been dubbed (by P. W. Anderson) the "orthogonality catastrophe."

Correlators of order parameters

The order parameters also exhibit anomalous dimensions, which can be read off directly from their correlation functions. The correlator of the CDW order parameter is found to be

$$\frac{1}{(2\pi a_0)^2} \left\langle T \left(e^{i2\sqrt{\pi}\phi(x)} e^{-i2\sqrt{\pi}\phi(x)} \right) \right\rangle \sim \frac{1}{(2\pi a_0)^2} \left(\frac{a_0^2}{(x-x')^2 - v^2(t-t')^2 + i\epsilon} \right)^{1/K} \tag{6.131}$$

Hence the scaling dimension of the CDW order parameter is $\Delta_{\rm CDW} = 1/K$ and only takes the naive dimension 1 for free fermions ($K = 1$). The correlator for the superconducting order parameter is, instead,

$$\frac{1}{(2\pi a_0)^2} \left\langle T \left(e^{i2\sqrt{\pi}\vartheta(x)} e^{-i2\sqrt{\pi}\vartheta(x)} \right) \right\rangle \sim \frac{1}{(2\pi a_0)^2} \left(\frac{a_0^2}{(x-x')^2 - v^2(t-t')^2 + i\epsilon} \right)^{K} \tag{6.132}$$

Hence, the scaling dimension of the superconducting order parameter is $\Delta_{\mathrm{SC}} = K$. This order parameter only has its naive scaling dimension, 1, for free fermions.

6.7.2 The spin-1/2 case

The behavior of the correlation functions for the case of spin-1/2 fermions can be computed similarly. Since the Hamiltonian of the Luttinger model decomposes into a sum of terms for the charge and spin sectors, respectively, we will find that the correlation functions *factorize* into a contribution from the charge sector and a contribution from the spin sector. We will not examine all possible cases, just the most interesting ones.

Since $\mathcal{H} = \mathcal{H}_{\mathrm{c}} + \mathcal{H}_{\mathrm{s}}$, the propagators factorize. In other terms, the system behaves as if the electrons have *fractionalized* into two independent excitations: (a) a spinless *holon* with charge $-e$ and (b) a spin-1/2 charge-neutral *spinon*. This feature is known as *spin–charge separation*. It is a robust feature of these 1D systems in the low-energy limit.

We will follow the same approach as in the spinless case, although we will implement it less explicitly. Here too we define the densities of right- and left-moving fermions with either spin polarization,

$$\rho_{\mathrm{R},\sigma} = \frac{1}{\sqrt{\pi}}\,\partial_x \phi_{\mathrm{R},\sigma}, \qquad \rho_{\mathrm{L},\sigma} = \frac{1}{\sqrt{\pi}}\,\partial_x \phi_{\mathrm{L},\sigma} \tag{6.133}$$

and write the Luttinger Hamiltonian in terms of these densities. It reduces to

$$\mathcal{H} = \mathcal{H}_{\mathrm{c}} + \mathcal{H}_{\mathrm{s}} \tag{6.134}$$

where

$$\mathcal{H}_{\mathrm{c}} = \frac{1}{2}\left(\pi v_{\mathrm{F}} + g_4\right)\left(\rho_{\mathrm{c,R}}^2 + \rho_{\mathrm{c,L}}^2\right) + \frac{1}{2}\left(2g_2 - g_{1,\parallel}\right)\rho_{\mathrm{c,R}}\rho_{\mathrm{c,L}} \tag{6.135}$$

$$\mathcal{H}_{\mathrm{s}} = \frac{1}{2}\left(\pi v_{\mathrm{F}} - g_4\right)\left(\rho_{\mathrm{s,R}}^2 + \rho_{\mathrm{s,L}}^2\right) - \frac{1}{2}\, g_{1,\parallel}\rho_{\mathrm{s,R}}\rho_{\mathrm{s,L}} \tag{6.136}$$

We now perform Bogoliubov transformations (separately) for charge and spin, whose parameters λ_{c} and λ_{s} are

$$\tanh(2\lambda_{\mathrm{c}}) = -\frac{2g_2 - g_{1,\parallel}}{\pi v_{\mathrm{F}} + g_4} \tag{6.137}$$

$$\tanh(2\lambda_{\mathrm{s}}) = +\frac{g_{1,\parallel}}{\pi v_{\mathrm{F}} - g_4} \tag{6.138}$$

The Luttinger parameters K_{c} and K_{s} are

$$K_{\mathrm{c}} = e^{2\lambda_{\mathrm{c}}} = \sqrt{\frac{\pi v_{\mathrm{F}} - g_4 + 2g_2 - g_{1,\parallel}}{\pi v_{\mathrm{F}} + g_4 - 2g_2 + g_{1,\parallel}}} \tag{6.139}$$

$$K_{\mathrm{s}} = e^{2\lambda_{\mathrm{s}}} = \sqrt{\frac{\pi v_{\mathrm{F}} - g_4 - g_{1,\parallel}}{\pi v_{\mathrm{F}} - g_4 - g_{1,\parallel}}} \tag{6.140}$$

and

$$\pi v_{\mathrm{c}} = \sqrt{(\pi v_{\mathrm{F}} + g_4)^2 - \left(2g_2 - g_{1,\parallel}\right)^2} \tag{6.141}$$

$$\pi v_{\mathrm{s}} = \sqrt{(\pi v_{\mathrm{F}} - g_4)^2 - g_{-1,\parallel}^2} \tag{6.142}$$

The transformed densities and bosons are denoted by

$$\tilde{\rho}_{\mathrm{c,R}} = \frac{1}{\sqrt{\pi}}\,\partial_x \tilde{\phi}_{\mathrm{c,R}}, \qquad \tilde{\rho}_{\mathrm{c,L}} = \frac{1}{\sqrt{\pi}}\,\partial_x \tilde{\phi}_{\mathrm{c,L}} \tag{6.143}$$

$$\tilde{\rho}_{\mathrm{s,R}} = \frac{1}{\sqrt{\pi}}\,\partial_x \tilde{\phi}_{\mathrm{s,R}}, \qquad \tilde{\rho}_{\mathrm{s,L}} = \frac{1}{\sqrt{\pi}}\,\partial_x \tilde{\phi}_{\mathrm{s,L}} \tag{6.144}$$

The fermion propagator

The operators for right- and left-moving fermions with spin σ now take the form

$$\psi_{\mathrm{R},\sigma} \sim \frac{1}{\sqrt{2\pi a_0}} e^{i\sqrt{2\pi}\phi_{\mathrm{R,c}}} e^{i\sigma\sqrt{2\pi}\phi_{\mathrm{R,s}}} \tag{6.145}$$

$$\psi_{\mathrm{L},\sigma} \sim \frac{1}{\sqrt{2\pi a_0}} e^{-i\sqrt{2\pi}\phi_{\mathrm{L,c}}} e^{-i\sigma\sqrt{2\pi}\phi_{\mathrm{L,s}}} \tag{6.146}$$

After some algebra we find

$$\begin{aligned}\langle T\psi_{\mathrm{R},\uparrow}(x,t)\psi^\dagger_{\mathrm{R},\uparrow}(0,0)\rangle &= \langle T\psi_{\mathrm{R},\downarrow}(x,t)\psi^\dagger_{\mathrm{R},\downarrow}(0,0)\rangle \\ &= \frac{a_0^{\gamma_{\mathrm{c}}+\gamma_{\mathrm{s}}}}{2\pi}\Big(a_0 + i(v_{\mathrm{c}}t - x)\Big)^{-1/2}\Big(a_0 + i(v_{\mathrm{s}}t - x)\Big)^{-1/2} \\ &\quad\times\Big(x^2 + (a_0 + i v_{\mathrm{c}}t)^2\Big)^{-\gamma_{\mathrm{c}}/2} \\ &\quad\times\Big(x^2 + (a_0 + i v_{\mathrm{s}}t)^2\Big)^{-\gamma_{\mathrm{s}}/2}\end{aligned} \tag{6.147}$$

and

$$\begin{aligned}\langle T\psi_{\mathrm{L},\uparrow}(x,t)\psi^\dagger_{\mathrm{L},\uparrow}(0,0)\rangle &= \langle T\psi_{\mathrm{L},\downarrow}(x,t)\psi^\dagger_{\mathrm{L},\downarrow}(0,0)\rangle \\ &= \frac{a_0^{\gamma_{\mathrm{c}}+\gamma_{\mathrm{s}}}}{2\pi}\Big(a_0 + i(v_{\mathrm{c}}t + x)\Big)^{-1/2}\Big(a_0 + i(v_{\mathrm{s}}t + x)\Big)^{-1/2} \\ &\quad\times\Big(x^2 + (a_0 + i v_{\mathrm{c}}t)^2\Big)^{-\gamma_{\mathrm{c}}/2} \\ &\quad\times\Big(x^2 + (a_0 + i v_{\mathrm{s}}t)^2\Big)^{-\gamma_{\mathrm{s}}/2}\end{aligned} \tag{6.148}$$

where

$$\gamma_{c,s} = \frac{1}{4}\left(K_{c,s} + \frac{1}{K_{c,s}}\right) - \frac{1}{2} \tag{6.149}$$

Just as in the spinless case, we see that the electron becomes fractionalized and acquires an anomalous dimension, which we can read off immediately as

$$\Delta_{\text{fermion}} = \frac{1}{8}\left(K_c + \frac{1}{K_c}\right) + \frac{1}{4} \tag{6.150}$$

where we have set the spin Luttinger parameter $K_s = 1$ for the SU(2)-invariant system. We have also neglected the logarithmic correction to scaling arising from the spin sector as K_s flows to 1.

Correlators of the order parameters

It is now straightforward to find the correlators of the order parameters. The CDW correlator is

$$\begin{aligned}
&\langle T \mathcal{O}_{\text{CDW}}(x,t)\mathcal{O}^\dagger_{\text{CDW}}(0,0)\rangle \\
&= \frac{1}{(\pi a_0)^2}\langle T\cos(\sqrt{2\pi}\phi_s(x,t))\cos(\sqrt{2\pi}\phi_s(0,0))\rangle \\
&\quad\times \langle T e^{-i\sqrt{2\pi}\phi_c(x,t)} e^{i\sqrt{2\pi}\phi_c(0,0)}\rangle \\
&= \frac{2}{(\pi a_0)^2}\left(\frac{a_0^2}{x^2 - v_c^2 t^2 + a_0^2 + i\epsilon}\right)^{1/(2K_c)}\left(\frac{a_0^2}{x^2 - v_s^2 t^2 + a_0^2 + i\epsilon}\right)^{1/(2K_s)}
\end{aligned} \tag{6.151}$$

The scaling dimension of the CDW order parameter is

$$\Delta_{\text{CDW}} = \frac{1}{2}\left(\frac{1}{K_c} + 1\right) \tag{6.152}$$

where we set $K_s = 1$ for SU(2)-invariance (and also neglected the logarithmic correction).

The (transverse) SDW correlator is

$$\begin{aligned}
&\langle T \mathcal{O}^{(\pm)}_{\text{SDW}}(x,t)\mathcal{O}^{(\pm)\,\dagger}_{\text{SDW}}(0,0)\rangle \\
&= \frac{1}{(2\pi a_0)^2}\langle T e^{\pm i\sqrt{2\pi}\vartheta_s(x,t)} e^{\mp i\sqrt{2\pi}\vartheta_s(0,0)}\rangle\langle T e^{-i\sqrt{2\pi}\phi_c(x,t)} e^{i\sqrt{2\pi}\phi_c(0,0)}\rangle \\
&= \frac{1}{(2\pi a_0)^2}\left(\frac{a_0^2}{x^2 - v_c^2 t^2 + a_0^2 + i\epsilon}\right)^{1/(2K_c)}\left(\frac{a_0^2}{x^2 - v_s^2 t^2 + a_0^2 + i\epsilon}\right)^{K_s/2}
\end{aligned} \tag{6.153}$$

which, for an SU(2)-invariant system, has the same scaling dimension as the CDW order parameter, although the logarithmic correction to scaling is different.

The singlet superconductor correlator is

$$
\begin{aligned}
&\langle T \mathcal{O}_{\rm SS}(x,t) \mathcal{O}_{\rm SS}^{\dagger}(0,0) \rangle \\
&= \frac{1}{(2\pi a_0)^2} \langle T e^{i\sqrt{2\pi}\vartheta_{\rm c}(x,t)} e^{-i\sqrt{2\pi}\vartheta_{\rm c}(0,0)} \rangle \langle T e^{-i\sqrt{2\pi}\phi_{\rm s}(x,t)} e^{i\sqrt{2\pi}\phi_{\rm s}(0,0)} \rangle \\
&= \frac{2}{(\pi a_0)^2} \left(\frac{a_0^2}{x^2 - v_{\rm c}^2 t^2 + a_0^2 + i\epsilon} \right)^{K_{\rm c}/2} \left(\frac{a_0^2}{x^2 - v_{\rm s}^2 t^2 + a_0^2 + i\epsilon} \right)^{1/(2K_{\rm s})}
\end{aligned}
\tag{6.154}
$$

with scaling dimension

$$
\Delta_{\rm SC} = \frac{1}{2}(K_{\rm c} + 1) \tag{6.155}
$$

6.8 Susceptibilities of the Luttinger model

6.8.1 The fermion spectral function

The fermion (electron) spectral function $\mathcal{A}_{s,\sigma}(p,\omega)$ (where $s = {\rm R, L}$ and $\sigma = \uparrow, \downarrow$) is defined by

$$
\begin{aligned}
\mathcal{A}_{{\rm s},\sigma}(p,\omega) &\equiv -\frac{1}{\pi} \, {\rm Im} \, G_{s,\sigma}^{\rm ret}(p,\omega) \\
&= \frac{1}{2\pi} \int_{-\infty}^{\infty} dx \int_{-\infty}^{\infty} dt \, e^{-i(px-\omega t)} \Big(G(x,t) + G(-x,-t) \Big)
\end{aligned}
\tag{6.156}
$$

where

$$
G_{s,\sigma}^{\rm ret}(x,t) = -i\theta(t) \Big\langle \Big\{ \psi_{s,\sigma}(x,t), \psi_{s,\sigma}(0,0) \Big\} \Big\rangle \tag{6.157}
$$

is the fermion retarded Green function, $G(x,t) = G_{{\rm R},\uparrow}(x,t)$ (since the system is invariant under parity and spin-reversal) is the time-ordered propagator we derived before, and p is measured from the Fermi point at $p_{\rm F}$. The detailed form of the spectral function for the general case is complicated. Explicit expressions are given in Chapter 19 of the book by Gogolin, Nersesyan, and Tsvelik (Gogolin *et al.*, 1998), where they use the notation $\rho_{s,\sigma}(p,\omega)$ for the spectral function. Here we will just quote the main results and analyze its consequences.

For a free-fermion system the Luttinger parameters $K_{\rm c} = K_{\rm s} = 1$ and hence $\gamma_{\rm c} = \gamma_{\rm s} = 0$. Similarly, the charge and spin velocities are equal in that case, $v_{\rm c} = v_{\rm s} = v_{\rm F}$. Hence, in the free-fermion case, we see that the spectral function $\mathcal{A}_{s,\sigma}(p,\omega)$ reduces to the sum of two poles (resulting from the poles in the propagator), for right- and left-moving fermions, respectively, each with a quasiparticle residue $Z = 1$.

The situation changes dramatically for the interacting case no matter how weak the interactions are. For simplicity we will discuss only the case in which the system has a full SU(2) spin invariance, in which case $K_{\mathrm{s}} = 1$ and $\gamma_{\mathrm{s}} = 0$. We see that, instead of poles, the fermion propagator has branch cuts, whose tips are located at $\omega = \pm p v_{\mathrm{c,s}}$ ($\pm$ here stands for right- and left-moving fermions). An analysis of the integral shows that close to these singularities the spectral function has the behavior

$$\mathcal{A}(p, \omega \simeq p v_{\mathrm{c}}) \sim \theta(\omega - p v_{\mathrm{c}})(\omega - p v_{\mathrm{c}})^{(\gamma_{\mathrm{c}}-1)/2} \tag{6.158}$$

$$\mathcal{A}(p, \omega \simeq -p v_{\mathrm{c}}) \sim \theta(-\omega - p v_{\mathrm{c}})(-\omega - p v_{\mathrm{c}})^{\gamma_{\mathrm{c}}} \tag{6.159}$$

$$\mathcal{A}(p, \omega \simeq p v_{\mathrm{s}}) \sim \theta(\omega - p v_{\mathrm{s}})(\omega - p v_{\mathrm{s}})^{\gamma_{\mathrm{c}}-1/2} \tag{6.160}$$

where p is the momentum of the incoming fermion measured from p_{F}. For the SU(2)-symmetric case, $\gamma_{\mathrm{s}} = 0$ and there is no singularity at $\omega \sim -p v_{\mathrm{s}}$.

Thus, the free-fermion poles are replaced in the interaction system by power-law singularities. These results show clearly the spin–charge separation: an injected electron has decomposed into (soliton-like) excitations, namely holons and spinons, that disperse at characteristic (and different) speeds.

In angle-resolved photoemission spectroscopy (ARPES) high-energy photons impinge on the surface of a system. If the photons' energy is high enough (typically the photons are X-rays from synchrotron radiation of a particle accelerator), there is a finite amplitude for an electron to be ejected from the system (a photo-electron), leaving a hole behind. In an ARPES experiment the energy and momentum (including the direction) of the photo-electron are measured. It turns out that the intensity of the emitted photo-electrons is proportional to the *spectral function* of the hole left behind at a known momentum and energy. Although it is not technically possible to do an ARPES experiment in a literally 1D system, it is possible to do such experiments in quasi-1D systems, namely arrays of weakly coupled 1DEGs. Experiments of this type have been done in systems of this type, such as the blue bronzes, although their degree of quasi-one-dimensionality is not strong enough for one to see the effects we discuss here.

The data from ARPES experiments are usually presented in terms of cuts of the spectral function: (a) as *energy-distribution curves* (EDCs), in which case the spectral function at fixed momentum is plotted as a function of energy; and (b) as *momentum-distribution curves* (MDCs), in which case the spectral function at fixed energy is plotted as a function of momentum (see Figs. 6.6(a)–(d)). Even if an ARPES experiment could be done in a Luttinger liquid, it is important to include the effects of thermal fluctuations since all experiments are done at finite temperature. One important effect is that the singularities of the spectral functions will be rounded at finite temperature. For example, the singularity of the EDC near the charge right-moving branch for $p = 0$ (near p_{F}), which diverges as $\omega^{(\gamma_{\mathrm{c}}-1)/2}$

as $\omega \to 0$ at $T = 0$, saturates at finite temperature T with a maximum $\sim T^{(\gamma_c - 1)/2}$ (which will increase as T is lowered). The same holds for the EDC at $\omega = 0$ as a function of momentum p (at p_F), which will saturate at a value $\sim (T/v_c)^{(\gamma_c - 1)/2}$. A more detailed study of the spectral function at finite temperature T (which must be done numerically) shows that the EDCs are much broader than the MDCs and look like what is shown in Fig. 6.6. Similar behaviors are seen in ARPES experiments in high-temperature superconductors.

6.8.2 The tunneling density of states

In a scanning tunneling microscopy (STM) experiment (Fig. 6.7), a (very sharp) metallic tip (typically made of a simple metal such as gold) is placed near a very flat (and clean) surface of an electronic system. There a finite voltage difference V is applied between the tip and the system, and, depending on its sign, electrons will tunnel from the tip to the system or vice versa. An STM instrument is operated by scanning the system (i.e. by displacing the tip) while keeping the tip at a fixed

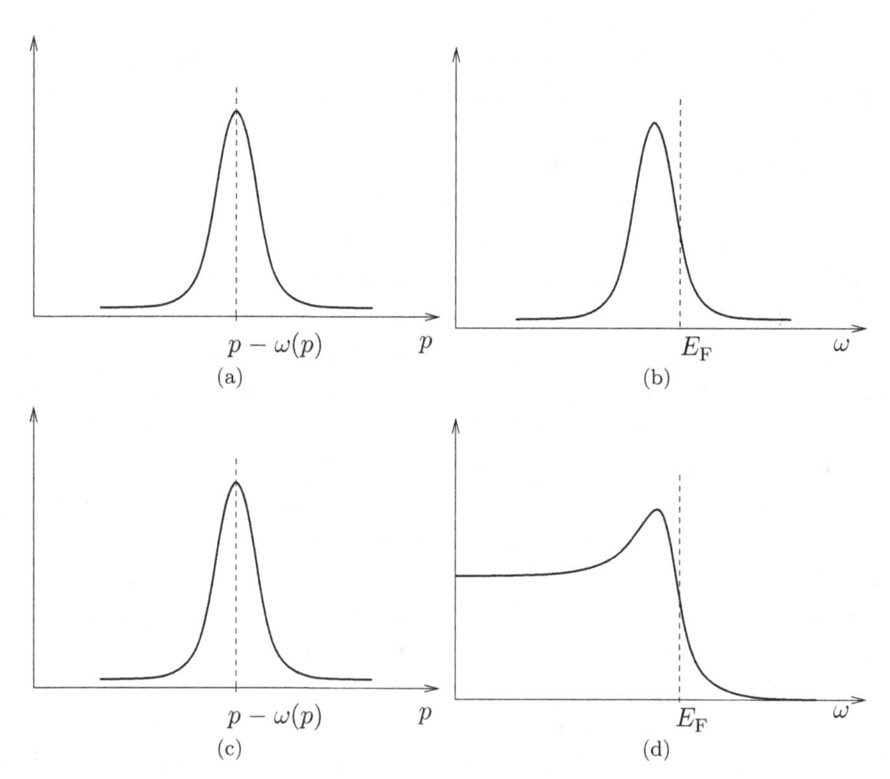

Figure 6.6 ARPES spectra: (a) MDC in a Fermi liquid, (b) EDC in a Fermi liquid, (c) MDC in a Luttinger liquid, and (d) EDC in a Luttinger liquid.

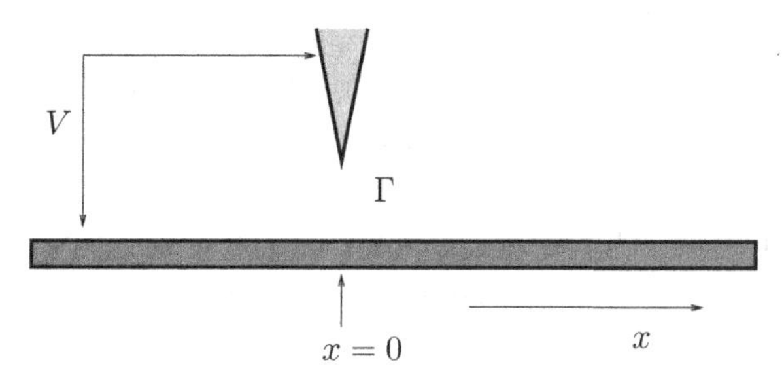

Figure 6.7 A sketch of an STM setup.

distance from the surface and at a fixed voltage difference. If the tip is sharp enough the intensity of the measured tunneling current (of electrons), which reflects the local changes of the electronic structure, can be used to map the local environment with atomic precision.

We will now see that the local differential conductance measured in STM contains direct information on the local density of states. To see how this works, let us consider a simple model of the operation of the STM. Let H_{tip} be the Hamiltonian which describes the electronic states in the tip. Let us denote by $\psi_{\text{tip}}(t)$ the fermionic operator that removes an electron from the tip in some one-particle state (which we will not need to know) close to the tip Fermi energy.

The tunneling process from the tip to the 1DEG at a point $x = 0$ is described by a term in the Hamiltonian of the form

$$\begin{aligned} \mathcal{H}_{\text{tunnel}} &= \delta(x)\Gamma \sum_{\sigma} \psi_{\sigma}^{\dagger}(0)\psi_{\text{tip}} + \text{h.c.} \\ &\equiv \delta(x)\Gamma \sum_{\sigma} \left(\psi_{\text{R},\sigma}^{\dagger}(0) + \psi_{\text{L},\sigma}^{\dagger}(0)\right)\psi_{\text{tip}} + \text{h.c.} \end{aligned} \tag{6.161}$$

To generate a tunneling current across the contact we need to couple the tip and the system to an infinitesimal external electromagnetic field. Let $A(t)$ be the (line integral of the) component of the electromagnetic time-dependent vector potential connecting the tip and the system at the point contact. By virtue of gauge invariance the electromagnetic coupling at the point contact amounts to modifying the tunneling Hamiltonian by changing the amplitude $\Gamma \to \Gamma \exp(ieA(t)/\hbar)$. The gauge-invariant tunneling-current operator J at the point contact is given by

$$J = \hbar \left.\frac{\delta \mathcal{H}_{\text{tunnel}}}{\delta A(t)}\right|_{A(t)=0} = ie\Gamma \sum_{\sigma} \left[\left(\psi_{\text{R},\sigma}^{\dagger}(0) + \psi_{\text{L},\sigma}^{\dagger}(0)\right)\psi_{\text{tip}} - \text{h.c.}\right] \tag{6.162}$$

We will assume that the energy of this state is higher than the Fermi energy in the Luttinger liquid by an amount equal to eV, where V is the voltage difference. We will assume that this is a rather uninteresting metal well described by a Fermi liquid with a density of one-particle states $\rho_{\text{tip}}(E)$ that is essentially constant for the range of voltages V used. Hence we can make the approximation that the density of states of the tip is constant, $\rho_{\text{tip}}(E_{\text{F}} + eV) \simeq \rho_{\text{tip}}(E_{\text{F}})$.

A fixed voltage V is equivalent to a difference of the chemical potentials of eV between the tip and the 1DEG. The same physics can be described by assigning to the tunneling matrix element Γ the phase factor

$$\Gamma \to \Gamma e^{i\frac{e}{\hbar}Vt} \tag{6.163}$$

both in the tunneling Hamiltonian and in the definition of a tunneling current. We recognize that the phase factor plays the same role as the vector potential we invoked just above. This is equivalent to a time-dependent gauge transformation of one of the Fermi fields, say that of the tip. The simplest way to see that this is true is to write down the Lagrangian density involving the tip degrees of freedom,

$$\mathcal{L}_{\text{tip}} = \psi_{\text{tip}}^{\dagger}(i\hbar\,\partial_t - eV)\psi_{\text{tip}} - \mathcal{H}_{\text{tip}} - \mathcal{H}_{\text{tunnel}} \tag{6.164}$$

and perform the time-dependent transformation

$$\psi_{\text{tip}} \to e^{i\frac{e}{\hbar}Vt}\psi_{\text{tip}} \tag{6.165}$$

The only terms in $\mathcal{L}_{\text{tip}}$ affected by this transformation are the first (which involves a time derivative) and the tunneling term, $\mathcal{H}_{\text{tunnel}}$. The change of the first term amounts to an extra term, which precisely cancels out the second term, which is where the voltage difference is specified. The change in the tunneling term is equivalent to the substitution given in Eq. (6.163). Notice that the tunneling-current operator, Eq. (6.162), is similarly affected by this transformation.

We now use perturbation theory in powers of the tunneling matrix element Γ to find the expectation value of the current operator, which will be denoted by I. To the lowest possible order in Γ, I is given by

$$I = 2\pi\frac{e}{\hbar}|\Gamma|^2 \int_{-eV}^{0} dE\ \rho_{\text{LL}}(E,T)\rho_{\text{tip}}(E + eV, T) \tag{6.166}$$

which follows from the well-known Fermi golden rule. The differential tunneling conductance $G(V,T)$ is found by differentiation:

$$G(V,T) = \frac{dI}{dV} \simeq \frac{2\pi e}{\hbar}|\Gamma|^2 \rho_{\text{tip}}(0)\rho_{\text{LL}}(E,T) \tag{6.167}$$

Here $\rho_{\text{LL}}(E,T)$ is the one-particle local density of states of the Luttinger liquid

$$\begin{aligned}\rho_{\text{LL}}(E,T) &= -\frac{1}{\pi}\,\text{Im}\,G_{\text{LL}}^{\text{ret}}(x=0,\omega=E,T)\\ &= -\frac{4}{\pi}\,\text{Im}\,G_{\text{R},\uparrow}^{\text{ret}}(x=0,E,T) = 4\int_{-\infty}^{\infty}\frac{dp}{2\pi}\mathcal{A}_{\text{R},\uparrow}(p,E,T)\end{aligned} \tag{6.168}$$

where $\mathcal{A}_{R,\uparrow}(p, E, T)$ is the spectral function defined above, and the factor of 4 arises since right and left movers (with both spin orientations) contribute equally at equal positions (denoted by $x = 0$). Alternatively,

$$\rho_{LL}(E, T) = -\frac{4}{\pi} \operatorname{Im} \int_{-\infty}^{\infty} dt\, e^{-i\frac{E}{\hbar}t} G^{\text{ret}}_{R,\uparrow}(x = 0, t, T) \tag{6.169}$$

At $T = 0$, by computing this Fourier transform one finds that $\rho_{LL}(E)$ has a power-law behavior,

$$\rho_{LL}(E) \propto E^{2(\gamma_c+\gamma_s)} \tag{6.170}$$

Hence, the differential tunneling conductance essentially measures the local density of states of the Luttinger liquid. Therefore, at $T = 0$, the differential tunneling conductance behaves as

$$G_{LL}(V) \propto V^{2(\gamma_c+\gamma_s)} \tag{6.171}$$

whereas for $T > 0$ one finds a saturation for $V \ll T$:

$$G_{LL}(V, T) \propto T^{2(\gamma_c+\gamma_s)} \tag{6.172}$$

The crossover between the $T > 0$, $V \to 0$ Ohmic behavior and the $T \to 0$, $V > 0$ Luttinger behavior occurs for $eV \sim k_B T$.

In contrast, for a free fermion (and for a Landau Fermi liquid)

$$G_{FL}(V) = \text{constant} \tag{6.173}$$

since in this case $\gamma_c = \gamma_s = 0$.

Therefore, for a system of free fermions we find that the point contact is Ohmic, $I \propto V$. For a Luttinger liquid there is instead a *power-law suppression* of the tunneling differential conductance for $T \ll V$ (see Fig. 6.8), and Ohmic behavior for $T \gg V$ (with a conductance that scales as a power of T). These behaviors reflect the fact that there are no stable electron-like quasiparticles in the Luttinger liquid: the electron states are *orthogonal* to the states in the spectrum of the Luttinger liquid, leading to a vanishing of the quasiparticle residue and to characteristic power-law behaviors in many quantities. This fact is known as the *orthogonality catastrophe*.

6.8.3 The fermion momentum distribution function

We will now discuss the fermion momentum distribution functions at zero temperature, $T = 0$. Since we have right and left movers, with both spin orientations, in principle we have four such functions. However, the Luttinger liquid state is invariant under global spin flips, $\uparrow \leftrightarrow \downarrow$, and under parity, $R \leftrightarrow L$. Thus all four

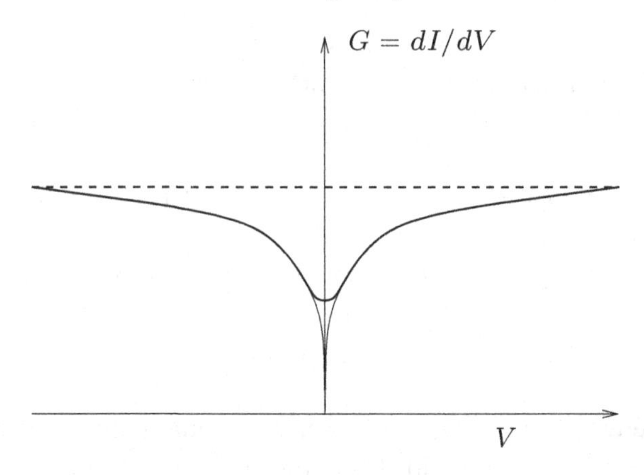

Figure 6.8 The differential tunneling conductance $G = dI/dV$ as a function of the bias voltage V, in a Fermi liquid (dashed line), and in a Luttinger liquid at $T = 0$ (thick line) and at $T > 0$ (thin line).

momentum distributions are equal to each other. Let us compute, say, $n_{\mathrm{R},\uparrow}(p)$, which is given by the *equal-time* correlator

$$\begin{aligned} n_{\mathrm{R},\uparrow}(p) &= \lim_{t' \to t+0^+} \langle \psi^\dagger_{\mathrm{R},\uparrow}(p,t)\psi_{\mathrm{R},\uparrow}(p,t')\rangle \\ &= \lim_{L\to\infty} \frac{1}{L}\int_{-L/2}^{+L/2} dx \int_{-L/2}^{+L/2} dx'\, e^{-ip(x-x')} \\ &\quad \times \lim_{t'\to t+0^+} \langle T\psi^\dagger_{\mathrm{R},\uparrow}(x,t)\psi_{\mathrm{R},\uparrow}(x',t')\rangle \\ &= \int_{-\infty}^{\infty} \frac{d\omega}{2\pi} A_{\mathrm{R},\uparrow}(p,\omega) \end{aligned} \tag{6.174}$$

(here T means time-ordering!).

A lengthy computation of the Fourier transforms leads to the result at $T = 0$ (here p is measured from p_{F}):

$$n_{\mathrm{R},\uparrow}(p) \sim \text{constant} + A|p|^{2(\gamma_c+\gamma_s)}\,\mathrm{sign}(p) \tag{6.175}$$

where A is a positive non-universal constant. At finite temperature $T > 0$ this singularity is rounded by thermal fluctuations, which dominate for momenta $|p| \lesssim k_{\mathrm{B}}T/v_c$, which lead to a smooth momentum dependence in this regime. This is why Luttinger behavior is difficult to detect in the momentum distribution function.

Thus, instead of a jump (or discontinuity) of Z (the quasiparticle residue) at p_{F} (the Fermi-liquid result), in a Luttinger liquid there is no jump (since $Z = 0$!). Instead we find that the momentum distribution function has a weak singularity at p_{F} (Fig. 6.9). This is what replaces the "Fermi surface" in a Luttinger liquid. We will show below that this happens since the Luttinger liquid is a (quantum) critical system and the fermions have an *anomalous dimension* given by $2(\gamma_c + \gamma_s)$.

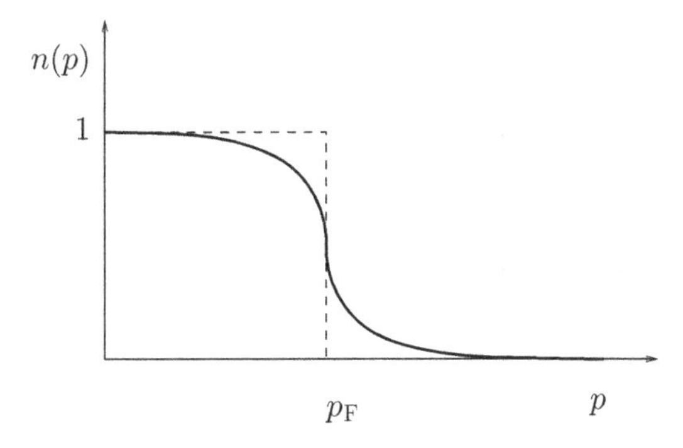

Figure 6.9 The fermion momentum distribution function in a Luttinger liquid.

6.8.4 Dynamical susceptibilities at finite temperature

Finally, we will discuss the behavior of dynamical susceptibilities at finite $T > 0$. In the preceding sections we gave explicit expressions for the correlators (time-ordered) of various physical quantities (order parameters and currents) at $T = 0$ in real space and time. Here we will need the dynamical susceptibilities, which are the associated retarded (instead of time-ordered) correlators at finite $T > 0$ in real momentum and frequency.

We saw earlier that we can determine all of these properties from the temperature correlators, i.e. in imaginary time τ, restricted to the interval $0 \leq \tau < 1/T$ (with $k_B = 1$). We accomplish this by first implementing the analytic continuation

$$vt \to -ivt \tag{6.176}$$

which implies introducing the complex coordinates

$$x - vt \to z = x + ivt \quad \text{and} \quad x + vt \to \bar{z} = x - ivt \tag{6.177}$$

Next we perform the *conformal mapping* from the complex plane labeled by the coordinates z to the cylinder, labeled by the coordinates $w = x + iv\tau$ (see Fig. 6.10)

$$x + ivt \to e^{2\pi \frac{T}{v}(x+iv\tau)} \tag{6.178}$$

Thus, the long axis of this cylinder is space (labeled by $-\infty \leq x \leq \infty$), and the circumference is the imaginary time τ, $0 \leq \tau \leq 1/T$. Under the conformal mapping the boson propagator (in imaginary time t) tuns out to transform as

$$\langle \phi(x,t)\phi(x',t')\rangle \to \frac{1}{2\pi} \ln \left| \frac{\pi T}{\sinh\left(\pi (T/v)(w - w')\right)} \right| \tag{6.179}$$

where $w = x + iv\tau$. This is discussed in more detail in Section 7.11.

The computation of the correlators of the observables we are interested in is the same as at $T = 0$, except that the boson propagator changes as shown above.

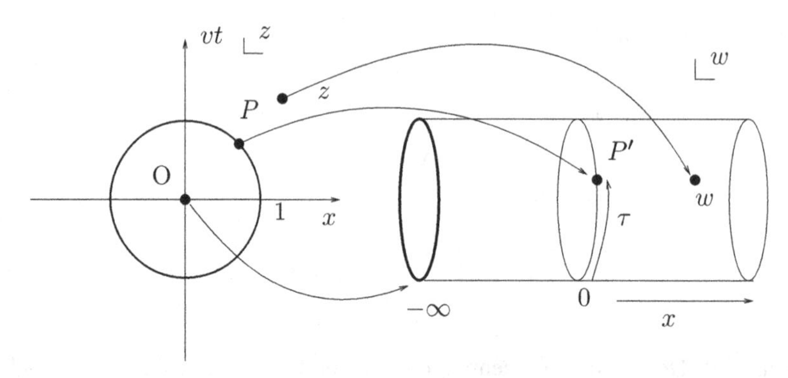

Figure 6.10 The conformal mapping $z = e^{2\pi T w/v}$ which maps the complex plane $z = x + ivt$ to the cylinder $w = x + iv\tau$. Under this mapping the origin O on the plane maps onto $-\infty$ on the cylinder.

The form of the boson propagator on the cylinder insures that the correlators are translation-invariant and periodic (or anti-periodic for fermions). This result can also be derived by an explicit calculation of the propagator (without using conformal mappings).

Thus, to compute the temperature propagators we perform (a) the analytic continuation followed by (b) the conformal mapping. This leads to the following identification for the power-law factors in the correlators (to restore proper units we must set $T/v \to k_B T/(\hbar v)$):

$$\left(\frac{1}{(x-x') \mp v(t-t') + i\epsilon}\right)^{\gamma} \to \left(\frac{1}{(x-x') \pm iv(t-t')}\right)^{\gamma}$$
$$\to \left(\frac{\pi T/v}{\sinh\Big((\pi T/v)\big[(x-x') \pm iv(\tau-\tau')\big]\Big)}\right)^{\gamma} \tag{6.180}$$

where γ is an exponent. Here we use $k_B = \hbar = 1$. Notice that the temperature changes the behavior of the boson propagator on distance scales that are long compared with the thermal wavelength v/T to

$$\langle \phi(x,t)\phi(x',t')\rangle \to \ln\left(\frac{\pi T}{2}\right) - \frac{\pi T}{v}|x-x'| + \cdots \tag{6.181}$$

The long-distance behavior of the boson propagator at finite T changes the behavior of the other correlators as well. In this regime they exhibit exponential decay of correlations over distances that are long compared with the thermal wavelength.

In contrast to what we did in perturbation theory, where the correlators are given in momentum and frequency space, the bosonization approach yields the exact correlators in real space and time. Thus, to compute spectral functions and other quantities of interest we now must perform Fourier transforms on the analytic continuations of these expressions (some of which have a somewhat involved analytic structure). But the expressions we have are not perturbative, they are exact!

The CDW susceptibility

The thermal CDW correlation function, i.e. the CDW propagator in imaginary time at finite temperature T, is

$$\begin{aligned} D_{\mathrm{CDW}}(x,\tau;T) &= \left\langle T_\tau \left(\mathcal{O}_{\mathrm{CDW}}(x,\tau)\mathcal{O}^\dagger_{\mathrm{CDW}}(0,0)\right)\right\rangle \\ &\sim \left[\frac{(\pi T/v_{\mathrm{c}})^2}{\sinh\left[(\pi T/v_{\mathrm{c}})(x+iv_{\mathrm{c}}\tau)\right]\sinh\left[(\pi T/v_{\mathrm{c}})(x-iv_{\mathrm{c}}\tau)\right]}\right]^{1/(2K_{\mathrm{c}})} \\ &\quad\times \left[\frac{(\pi T/v_{\mathrm{s}})^2}{\sinh\left[(\pi T/v_{\mathrm{s}})(x+iv_{\mathrm{s}}\tau)\right]\sinh\left[(\pi T/v_{\mathrm{s}})(x-iv_{\mathrm{s}}\tau)\right]}\right]^{1/(2K_{\mathrm{s}})} \end{aligned} \tag{6.182}$$

The CDW dynamical susceptibility at finite temperature $\chi_{\mathrm{CDW}}(p,\omega;T)$ is the Fourier transform of this expression in x and τ (after an analytic continuation to real time t). The Fourier transform has a complex analytic structure due to the branch cuts and to the difference in the charge and spin velocities. Of direct physical interest is the imaginary time of the dynamical susceptibility, $\chi''_{\mathrm{CDW}}(p,\omega;T)$ at finite temperature, which is measured by inelastic X-ray scattering (up to a Bose factor $\chi''_{\mathrm{CDW}}(p,\omega)$ is proportional to the inelastic cross section).

Although the general form of $\chi''_{\mathrm{CDW}}(p,\omega;T)$ can be determined numerically, a simple expression (which captures the main physics) can be obtained by setting $v_{\mathrm{c}} = v_{\mathrm{s}} = v$:

$$\chi''_{\mathrm{CDW}}(\omega, p>0;T) \sim -\frac{\sin(\pi\gamma)}{T^{2(1-\gamma)}}\,\mathrm{Im}\left\{ f\left(\frac{\omega - pv}{4\pi T}\right) f\left(\frac{\omega + pv}{4\pi T}\right)\right\} \tag{6.183}$$

Here p is measured from $2p_{\mathrm{F}}$, and

$$\gamma = \frac{1}{2}\left(\frac{1}{K_{\mathrm{c}}} + \frac{1}{K_{\mathrm{s}}}\right) \tag{6.184}$$

The complex function $f(x)$ is given by

$$f(x) = \frac{\Gamma(\gamma/2 - ix)}{\Gamma(1-\gamma/2-ix)} \tag{6.185}$$

where $\Gamma(z)$ is the Euler gamma function. At very low temperatures, such that $|\omega \pm pv| \gg T$, $\chi''_{\text{CDW}}(p, \omega; T)$ converges to the $T = 0$ result:

$$\chi''_{\text{CDW}}(p, \omega; T = 0) \propto \left| \frac{\omega^2 - p^2 v^2}{4\pi^2} \right|^{-\frac{1}{2}(1-1/K_c)} \left(\theta(\omega - pv) + \theta(-(\omega + pv)) \right) \tag{6.186}$$

where I have set $K_s = 1$ (we will see below that this is required by the SU(2) spin rotational invariance). Since $K_c > 1$ (for repulsive interactions) we see that the spectral function is largest near $\omega = \pm pv$ ("on-shell"), where it diverges as a power law. Notice that this is a one-sided singularity since the spectral function vanishes on the other side of the mass-shell condition. This divergence is cut off at finite temperature T, where it takes the maximum value determined by T.

The same behavior is found for the full *static susceptibility* at the ordering wave vector (which here means $p = 0$). Although it can be determined from the general expression for $\chi''_{\text{CDW}}(p, \omega; T)$, it is instructive to determine it more directly by the following simple argument. At finite temperature, the long-distance behavior of the correlators in real space for distances long compared with the thermal wave length v/T is an exponential decay. This behavior effectively cuts off the infrared singularities in many quantities such as the static susceptibility at the ordering wavevector $Q = 2p_{\text{F}}$, $\chi_{\text{CDW}}(\omega = 0, p = 0)$ (again, the momentum p is measured from the ordering wave vector). This quantity can be computed directly from a Fourier transform of the thermal CDW correlation function in imaginary time at $T = 0$ with a long-distance cutoff of v/T:

$$\begin{aligned} \chi_{\text{CDW}}(0, 0; T) &\sim \int_{|x|<v/T} dx \int_{|t|<1/T} dt \; D_{\text{CDW}}(x, t; T = 0) \\ &\sim 2\pi \times \text{constant} \times \int_a^{v/T} \frac{r\,dr}{r^{2\gamma}} \propto T^{-(1-1/K_c)} \end{aligned} \tag{6.187}$$

where we have set $K_s = 1$. Thus, for $K_c > 1$ (repulsive interactions) the CDW susceptibility at the ordering wave vector $Q = 2p_{\text{F}}$ diverges as $T \to 0$. This means that the $T = 0$ system is *almost ordered*. We will see below that this behavior means that it is actually *critical at* $T = 0$, and we will identify the exponent γ with the *scaling dimension* (or *dimension* in short) of the CDW order parameter.

A similar analysis can be used to find the *structure factor* $S(p; T)$ (once again measured from $Q = 2p_{\text{F}}$) at temperature T, the Fourier transform *in space* of the *equal-time* correlation function,

$$S(p; T) = \int_{-\infty}^{\infty} dx \; e^{-ipx} D(x, t = 0; T) \tag{6.188}$$

$S(p; T)$ is measured by X-ray-diffraction experiments. An analysis of this Fourier transform (similar to what we did above) leads to the following result:

$$S(p; T) \propto 2K_c a^{-1/K_c} - \frac{K_c^2 \Gamma(1/K_c)}{K_c - 1} \times \begin{cases} \cos[\pi/(2K_c)]|p|^{1/K_c}, & \text{for } T/v \ll |p| \\ (T/v)^{1/K_c}, & \text{for } T/v \gg |p| \end{cases} \tag{6.189}$$

Hence, in contrast with the static susceptibility, the structure factor at $T = 0$, $S(p, T = 0)$, does not diverge as $T \to 0$, and instead has a weak singularity (a *cusp*) at the ordering wave vector.

Finally, we note that for spin-rotational invariant systems (for which $K_s \to 1$) there are logarithmic corrections to these results due to corrections to scaling effects (we will not give a derivation of these corrections here). We can put all of this together in terms of the equal-time density correlation function (including the logarithmic correction):

$$\langle \rho(x)\rho(0)\rangle = \frac{1}{K_c(\pi x)^2} + \text{constant} \times \frac{\cos(2p_F x)}{|x|^{1+1/K_c}} |\ln|x||^{-3/2} + \cdots \tag{6.190}$$

and the $2p_F$ static CDW susceptibility (including the logarithmic correction) is

$$\chi_{\text{CDW}}(T) \sim \frac{|\ln T|^{-3/2}}{T^{1-1/K_c}} \tag{6.191}$$

The SDW susceptibility

The correlation function of the SDW order parameter can be analyzed in a similar fashion. The dynamical susceptibility is measured by inelastic neutron scattering. Here I will quote only the main results.

The equal-time transverse spin-correlation function turns out to be (setting $K_s = 1$)

$$\langle \vec{S}(x) \cdot \vec{S}(0)\rangle = \frac{1}{(\pi x)^2} + \text{constant} \times \frac{\cos(2p_F x)}{|x|^{1+1/K_c}} \sqrt{|\ln|x||} + \cdots \tag{6.192}$$

and the SDW transverse susceptibility (at the ordering wavevector $2p_F$) is

$$\chi_{\text{SDW}}^{\perp}(T) \sim \frac{\sqrt{|\ln T|}}{T^{1-1/K_c}} \tag{6.193}$$

So, up to logarithmic corrections, it has the same behavior as the CDW correlators and susceptibilities. This is a special property for the spin-rotational invariant system.

The superconducting susceptibility

Finally, we quote the results for the singlet superconductor equal-time correlation function

$$\langle \mathcal{O}_{\rm SS}^{\dagger}(x)\mathcal{O}_{\rm SS}(0)\rangle \sim \frac{\text{constant}}{|x|^{1+K_{\rm c}}}|\ln|x||^{-3/2} \tag{6.194}$$

and finite-temperature susceptibility

$$\chi_{\rm SS}(T) \sim T^{K_{\rm c}-1}|\ln\, T|^{-3/2} \tag{6.195}$$

Hence, the superconducting static susceptibility does not diverge as $T \to 0$ for $K_{\rm c} > 1$ (repulsive interactions) but it does for $K_{\rm c} < 1$ (attractive interactions). The different behavior of the superconducting and CDW susceptibilities follows directly from duality.

7

Sigma models and topological terms

7.1 Generalized spin chains: the Haldane conjecture

The phenomenology which emerges from the spin one-half Heisenberg antiferromagnetic chain is quite striking: there is no long-range order, and there are gapless states, in particular, gapless spinless fermions (which, in the Heisenberg picture, are solitons). From the point of view of the Hubbard model, the Heisenberg model occurs at infinite coupling, where the charge-bearing degrees of freedom acquire a gap that is infinitely large. Thus spin and charge degrees of freedom are separated and the spin sector is at a critical point. This phenomenology inspired Anderson (1987) to propose a similar picture for the two-dimensional systems, the resonating-valence-bond (RVB) picture. However, most of this picture surely should not generalize. Critical points are not generic and, in general, it is not possible to have gapless states without the spontaneous breaking of a continuous symmetry except in one dimension due to the Mermin–Wagner theorem. In higher dimensions gapless states without a broken symmetry may be possible in a Coulomb phase of a gauge theory with a continuous gauge group. Thus, the 1D spin one-half case may be more the exception than the rule. For instance, it may be possible that the system is in a state without long-range order, which is likely to be massive. For this reason, it is important to consider generalizations of the Heisenberg model. This problem has been studied extensively. Two different approaches have been considered in one dimension: (a) enlarging the representation (higher spin, same symmetry group SU(2)) and (b) higher symmetry groups (SU(N), for instance).

Haldane considered the generalization to higher spin but keeping the symmetry group SU(2) (Haldane, 1983a, 1983c, 1985b). He first considered the *large-spin limit*, which should have semi-classical character. He showed that in this limit the effective Lagrangian was *almost* the Lagrangian of the quantum non-linear sigma model. That the non-linear sigma model should appear in a *semi-classical*

($S \to \infty$) limit should be of no surprise: one finds the same answer in mean-field theory. But there is something wrong with this picture. The non-linear sigma model is known to have no long-range order and, in fact, it has a finite correlation length (Polyakov, 1975). Thus, if the sigma model truly was the infrared limit of the Heisenberg model, it could not possibly be a critical system, at least for S sufficiently large. Haldane found that this is indeed the case for spin systems in which S is an *integer*. For *half-integer* spins, he found that, in addition to the sigma model, there is an extra term that changes the physics drastically. The extra term turned out to be proportional to a topological invariant, namely the winding number or Pontryagin index of the (smooth) spin configuration. Thus it would appear that integer and half-integer *spin chains* behave rather differently.

Generalized spin systems with other symmetry groups have also been considered. These include SU(N) generalizations of the (SU(2)) Heisenberg model for various representations of the group. Affleck studied *a large-N limit* in which he was able to show that the ground state does not have long-range order and that there are no gapless states (Affleck, 1985). However, other SU(N) generalizations of the Heisenberg model have been considered. For special choices of parameters, these systems are integrable (in the Bethe-ansatz sense), and they are also at a critical point (Babujian and Tsvelik, 1986). Their critical behavior is, however, different from the one we discussed in the Heisenberg case. Thus, it appears that, at least in one dimension, these systems are either critical or in a *disordered state*, i.e. a state without long-range order and with only short-range spin correlations.

Let us first discuss the spin-S quantum Heisenberg chain. I will do so by introducing a path-integral method for spin systems that *does* generalize to higher dimensions, groups, representations, etc.

7.2 Path integrals for spin systems: the single-spin problem

In Section 3.2 we developed a path-integral method for Fermi systems of the Hubbard type (i.e. with local interactions). Using a Hubbard–Stratonovich transformation we were able to derive an effective action for the low-energy degrees of freedom, the spin fluctuations. The result was a path-integral representation of the long-range spin fluctuations, the quantum-mechanical non-linear sigma model.

We also showed that, in the strong-coupling limit, the half-filled Hubbard model maps onto the quantum Heisenberg model. In this limit the "band" fermions are tightly bound into localized spins. There is no motion of the fermionic degrees of freedom since, in this limit, the energy gap for charge fluctuations is infinitely large. It is natural to ask for an alternative derivation of the effective action for the spin fluctuations that should not be based on the weak-coupling mean-field theory, as we did in Chapter 3. Also we will now be careful enough to keep terms of topological significance, something we did not do in Chapter 3, and to assess their importance.

We begin with the discussion of an extremely simple system: a spin-S degree of freedom coupled to an external field through a Zeeman term. From the standard treatment in elementary quantum mechanics (Baym, 1974) we know that the $(2S+1)$-fold degeneracy is lifted by the Zeeman interaction, resulting in $2S+1$ non-degenerate levels. The path integral will enable us to study the evolution operator between arbitrary initial and final states.

There are several published path-integral treatments of spin degrees of freedom. They all share the feature that they deal with coherent states rather than the more familiar complete states (Schulman, 1981). The method of coherent states has been extensively reviewed by A. Perelomov (1986). We will use a special version of the method of coherent states that keeps the spin symmetry intact, which was first introduced by Wiegmann (1988) and by Fradkin and Stone (1988).

Let us begin by describing the Hilbert space. It is very simple. We have $2S+1$ states that transform like a spin-S representation of SU(2). Let $|0\rangle$ denote the *highest-weight state* in this representation,

$$|0\rangle = |S, S\rangle \tag{7.1}$$

This state is an eigenstate both of S_3, the (only) diagonal generator of SU(2), and of the quadratic Casimir invariant $\vec{S}^2$:

$$S_3|0\rangle = S|0\rangle \tag{7.2}$$
$$\vec{S}^2|0\rangle = S(S+1)|0\rangle \tag{7.3}$$

Consider now the state $|\vec{n}\rangle$ labeled by the unit vector $\vec{n}$ which is obtained by the rotation (see Fig. 7.1)

$$|\vec{n}\rangle = e^{i\theta(\vec{n}_0\times\vec{n})\cdot\vec{S}}|S, S\rangle \tag{7.4}$$

where $\vec{n}_0$ is a unit vector along the quantization axis, θ is the co-latitude

$$\vec{n}\cdot\vec{n}_0 = \cos\theta \tag{7.5}$$

and S_i ($i = 1, 2, 3$) are the (three) generators of SU(2) in the spin-S representation. For a review of SU(2) and its representations, see, for instance, Georgi (1982).

The state $|\vec{n}\rangle$ can be expanded in a complete basis of the spin-S irreducible representation $\{|S, M\rangle\}$, where M labels the eigenvalue of S_3,

$$S_3|S, M\rangle = m|S, M\rangle \tag{7.6}$$
$$\vec{S}^2|S, M\rangle = S(S+1)|S, M\rangle \tag{7.7}$$

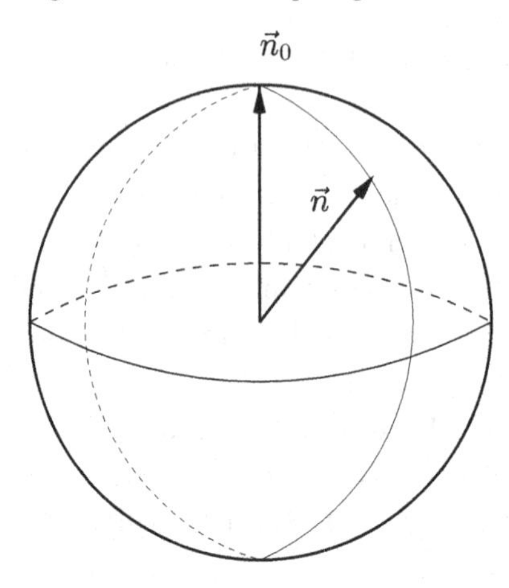

Figure 7.1 The unit sphere S_2 and the unit vectors $\vec{n}_0$ and $\vec{n}$.

and $-S \leq M \leq S$, in integer steps. The coefficients of the expansion are the representation matrices $D^{(S)}(\vec{n})_{MS}$

$$|\vec{n}\rangle = \sum_{M=-S}^{S} D^{(S)}(\vec{n})_{MS}|S, M\rangle \tag{7.8}$$

Clearly, there are many other rotations, differing from one another by multiplication on the right by rotations about the z axis. This will give rise to the same state, except for an overall phase. In more formal terms, the observable states are in a one-to-one correspondence with the right cosets $\mathrm{SU}(2)/\mathrm{U}(1)$, where $\mathrm{U}(1)$ represents phase transformations generated by the *diagonal* generator of $\mathrm{SU}(2)$. Clearly the coset is isomorphic to the 2-sphere: $\mathrm{SU}(2)/\mathrm{U}(1) \simeq S_2$. In the language of differential geometry, the coherent states form a hermitian line bundle associated with the Hopf, or monopole, principal bundle.

The matrices $D^{(S)}$ do not form a group but rather satisfy the algebra

$$D^{(S)}(\vec{n}_1)D^{(S)}(\vec{n}_2) = D^{(S)}(\vec{n}_3)e^{i\Phi(\vec{n}_1,\vec{n}_2,\vec{n}_3)S_3} \tag{7.9}$$

where $\vec{n}_1, \vec{n}_2$, and $\vec{n}_3$ are three arbitrary unit vectors on the unit sphere S_2, and $\Phi(\vec{n}_1, \vec{n}_2, \vec{n}_3)$ is the area of the spherical triangle with vertices at $\vec{n}_1$, $\vec{n}_2$, and $\vec{n}_3$ (see Fig. 7.2). Equation (7.9) is simply saying that the $D^{(S)}$ matrices form a group up to an element generated by the diagonal generators, the Cartan subalgebra. Since the sphere S_2 is a closed manifold (and hence without boundaries), the area of a spherical triangle is not uniquely defined. The indicated areas of the sphere in Figs. 7.2(a) and (b) are equally good definitions of the area. The difference of

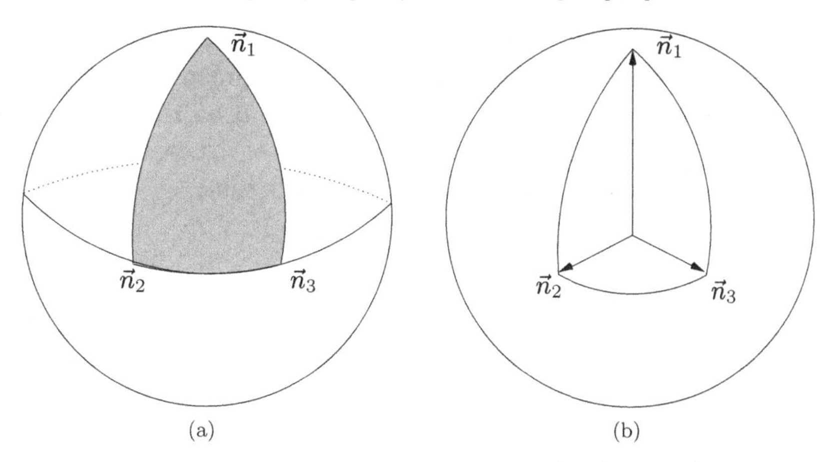

Figure 7.2 The spherical triangle with vertices at $\vec{n}_1$, $\vec{n}_2$, and $\vec{n}_3$. Its area is not unambiguously defined. The "inner" area is shown in (a) and the "outer" area in (b).

the *oriented areas* is 4π. Since S_3 has eigenvalues equal to M, which is either an integer or a half-integer, this ambiguity has no physical manifestation since

$$e^{i4\pi M} = 1 \tag{7.10}$$

We can regard the requirement that the ambiguity in the definition of the area should lead to no physical consequences as the origin of the quantization of spin.

Other useful properties of the spin coherent states $|\vec{n}\rangle$ are the inner product $\langle \vec{n}_1|\vec{n}_2\rangle$,

$$\langle \vec{n}_1|\vec{n}_2\rangle = \langle 0|D^{(S)\dagger}(\vec{n}_1)D^{(S)}(\vec{n}_2)|0\rangle \tag{7.11}$$

$$= e^{i\Phi(\vec{n}_1,\vec{n}_2,\vec{n}_0)S}\left(\frac{1+\vec{n}_1\cdot\vec{n}_2}{2}\right)^S \tag{7.12}$$

the diagonal matrix elements of the SU(2) generators $\vec{S}$,

$$\langle \vec{n}|\vec{S}|\vec{n}\rangle = S\vec{n} \tag{7.13}$$

and the "resolution of the identity," which is an expression of the identity operator $\hat{I}$ in terms of the coherent-state operators $|\vec{n}\rangle\langle\vec{n}|$,

$$\hat{I} = \int d\mu(\vec{n})|\vec{n}\rangle\langle\vec{n}| \tag{7.14}$$

The integration measure $d\mu(\vec{n})$ is given by the invariant measure

$$d\mu(\vec{n}) = \left(\frac{2s+1}{4\pi}\right) d^3n\, \delta(\vec{n}^2 - 1) \tag{7.15}$$

We are now in a position to write down an expression for the path integral in this coherent-state representation. Its generalization to other groups is straightforward and has been given by Wiegmann (1989). Let $H(\vec{S}) = \vec{B} \cdot \vec{S}$ be the Zeeman-like Hamiltonian for a spin system with one spin-S degree of freedom. I will consider the representation of the evolution operator in imaginary time:

$$Z = \operatorname{tr} e^{iHT} = \operatorname{tr} e^{-\beta H} \tag{7.16}$$

In other words, we are assuming that the initial and final states are identified. Let us split the imaginary-time interval into N_t steps each of length δt and consider the limit $N_t \to \infty$ and $\delta t \to 0$ while keeping $N_t\,\delta t = \beta$ constant. As usual we make use of the Trotter formula

$$Z = \operatorname{tr} e^{-\beta H} = \lim_{\substack{N_t \to \infty \\ \delta t \to 0}} \left(e^{-\delta t\, H}\right)^{N_t} \tag{7.17}$$

and insert the "resolution of identity," Eq. (7.14), at every intermediate time t_j,

$$Z = \lim_{\substack{N_t \to \infty \\ \delta t \to 0}} \left(\prod_{j=1}^{N_t} \int d\mu(\vec{n}_j)\right) \left(\prod_{j=1}^{N_t} \langle \vec{n}(t_j)| e^{-\delta t\, H} |\vec{n}(t_{j+1})\rangle\right) \tag{7.18}$$

with periodic boundary conditions. Here $\{t_j\}$ is a set of intermediate times in the imaginary-time interval $[0, \beta]$. Since δt is small we can approximate Eq. (7.18) as

$$Z = \lim_{\substack{N_t \to \infty \\ \delta t \to 0}} \left(\prod_{j=1}^{N_t} \int d\mu(\vec{n}_j)\right) \left(\prod_{j=1}^{N_t} \left[\langle \vec{n}(t_j)|\vec{n}(t_{j+1})\rangle - \delta t \langle \vec{n}(t_j)| H |\vec{n}(t_{j+1})\rangle\right]\right) \tag{7.19}$$

Within the same approximation we can write

$$\frac{\langle \vec{n}(t_j)| H |\vec{n}(t_{j+1})\rangle}{\langle \vec{n}(t_j)|\vec{n}(t_{j+1})\rangle} \simeq \langle \vec{n}(t_j)| H |\vec{n}(t_j)\rangle + O(\delta t) \tag{7.20}$$

Using the inner-product formula, Eq. (7.12), we get

$$\langle \vec{n}(t_j)|\vec{n}(t_{j+1})\rangle = e^{i\Phi(\vec{n}(t_j),\vec{n}(t_{j+1}),\vec{n}_0)S} \left(\frac{1 + \vec{n}(t_j) \cdot \vec{n}(t_{j+1})}{2}\right)^S \tag{7.21}$$

We now insert Eqs. (7.20) and (7.21) into Eq. (7.19) to find the (formal) expression for the path integral

$$Z = \lim_{\substack{N_t \to \infty \\ \delta t \to 0}} \int \mathcal{D}\vec{n}\; e^{-S_{\mathrm{E}}[\vec{n}]} \tag{7.22}$$

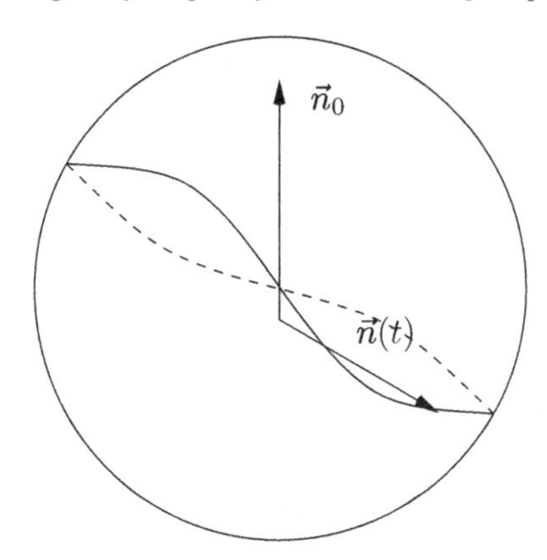

Figure 7.3 Closed smooth trajectories on S_2.

where the measure $\mathcal{D}\vec{n}$ is given by

$$\mathcal{D}\vec{n} = \prod_{j=1}^{N_t} d\mu(\vec{n}(t_j)) \tag{7.23}$$

and the Euclidean action $\mathcal{S}_{\text{E}}[\vec{n}]$ is given by

$$-\mathcal{S}_{\text{E}}[\vec{n}] = iS\sum_{j=1}^{N_t} \Phi(\vec{n}(t_j), n(t_{j+1}), \vec{n}_0) + S\sum_{j=1}^{N_t} \ln\left(\frac{1+\vec{n}(t_j)\cdot\vec{n}(t_{j+1})}{2}\right)$$
$$-\sum_{j=1}^{N_t} \langle\vec{n}(t_j)|H|\vec{n}(t_j)\rangle \tag{7.24}$$

In this derivation, we have assumed that the unit vectors $\{\vec{n}(t_j)\}$ are closed trajectories (because $\vec{n}(t_0) = \vec{n}(t_{N+1})$) on the sphere S_2 which are sufficiently smooth that all the approximations of Eq. (7.20) make sense (see Fig. 7.3). This is not quite the case, as emphasized by Klauder (1979). But these technicalities, as well as operator-ordering problems, can be taken care of without affecting the physics. We will ignore these difficulties from now on. Our path integral will be as good a mathematical object as any other path integral.

The first term of the effective Euclidean action is complex. It leads to a sum over trajectories weighted by phases (even though we are working in imaginary time!) of the form

$$e^{iS\mathcal{A}[\vec{n}]} \tag{7.25}$$

where $\mathcal{A}[\vec{n}]$ is the limit

$$\mathcal{A}[\vec{n}] = \lim_{\substack{N_t\to\infty \\ \delta t\to 0}} \sum_{j=1}^{N_t} \Phi(\vec{n}(t_j), \vec{n}(t_{j+1}), \vec{n}_0) \tag{7.26}$$

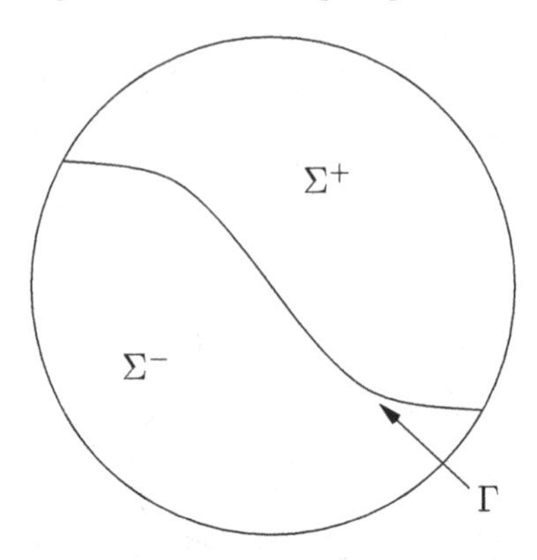

Figure 7.4 The trajectory Γ and the caps Σ^+ and Σ^-.

Since each term of this sum is the area of the spherical triangle with vertices at $\vec{n}(t_j)$, $\vec{n}(t_{j+1})$, and $\vec{n}_0$, the sum, i.e. the sum of these areas, is just equal to the total area of the cap Σ^+ bounded by the trajectory Γ parametrized by $\vec{n}(t)$ (see Fig. 7.4). Once again, since S_2 has no boundaries, there are two caps, Σ^+ and Σ^-. The oriented areas of Σ^+ and Σ^- also differ by 4π,

$$\mathcal{A}(\Sigma^+) + \mathcal{A}(\Sigma^-) = 4\pi \tag{7.27}$$

This is the same ambiguity as we encountered before. It does not lead us to any observable effects since S is restricted to being an integer or a half-integer. The area of the cap Σ, say Σ^+, is given by (in the limit $N_t \to \infty$, $\delta t \to 0$)

$$\mathcal{A}(\Sigma^+) = \int_0^1 d\tau \int_0^\beta dt\, \vec{n}(t,\tau) \cdot (\partial_t \vec{n}(t,\tau) \times \partial_\tau \vec{n}(t,\tau)) \equiv \mathcal{S}_{\mathrm{WZ}}[\vec{n}] \tag{7.28}$$

where $\vec{n}(t,\tau)$ is an arbitrary, smooth parametrization of the cap Σ^+ bounded by Γ that satisfies the boundary conditions

$$\vec{n}(t,0) \equiv \vec{n}(t), \quad \vec{n}(t,1) \equiv \vec{n}_0, \quad \vec{n}(0,\tau) = \vec{n}(\beta,\tau) \tag{7.29}$$

where $t \in [0,\beta]$ and $\tau \in [0,1]$. Terms of this sort are generically called Wess–Zumino terms, although, for reasons that will be explained later, sometimes they are also referred to as Chern–Simons terms.

We now proceed to take a naive continuum limit ($N_t \to \infty$, $\delta t \to 0$) and find from Eqs. (7.28) and (7.24) the Euclidean action

$$\mathcal{S}_{\mathrm{E}}[\vec{n}] = -iS\mathcal{S}_{\mathrm{WZ}}[\vec{n}] + \frac{S\,\delta t}{4}\int_0^\beta dt (\partial_t \vec{n}(t))^2 + S\int_0^\beta dt\, \vec{B} \cdot \vec{n}(t) \tag{7.30}$$

where $\vec{B}$ is an external magnetic field.

We can get back to real time x_0, with

$$t = ix_0, \qquad \beta = iT \tag{7.31}$$

where T is the (imaginary) time span, by writing

$$Z = \int \mathcal{D}\vec{n}\ e^{i\mathcal{S}_{\rm M}[\vec{n}]} \tag{7.32}$$

where $\mathcal{S}_{\rm M}[\vec{n}]$ is given by

$$\mathcal{S}_{\rm M}[\vec{n}] = S\mathcal{S}_{\rm WZ}[\vec{n}] + \frac{S\,\delta t}{4}\int_0^T dx_0(\partial_0\vec{n}(x_0))^2 - S\int_0^T dx_0\,\vec{B}\cdot\vec{n}(x_0) \tag{7.33}$$

This expression has a simple mechanical analogy. Let us imagine that $\vec{n}(x_0)$ is the position vector of a charged particle at time x_0. The particle has a small mass $m = S\,\delta t/2$ (with $m \to 0$) and is constrained to move on the surface of the unit sphere, S_2. A magnetic monopole with magnetic charge S is placed at the center of the sphere. The usual minimal electromagnetic coupling gives a contribution to the action of the form (Landau and Lifshitz, 1975b)

$$\mathcal{S}_{\rm em} = \oint dx_0\,\vec{A}\cdot\frac{\partial\vec{n}}{\partial x_0} \tag{7.34}$$

where $\vec{A}$ is the vector potential at position $\vec{n}(x_0)$. In order to represent a monopole, the vector potential has to have a singular piece that describes the Dirac string. We can use Stokes' theorem to write $\mathcal{S}_{\rm em}$ in terms of a two-form instead of the one-form $\vec{A}$. Stokes' theorem simply says that $\mathcal{S}_{\rm em}$ is given by the flux of the magnetic monopole through the area of S_2 bounded by the trajectory Γ (see Fig. 7.5). This is nothing but the magnetic charge S of the monopole multiplied by the area of S_2 bounded by Γ, in other words, the cap Σ of Fig. 7.4. This is precisely identical to the first term in the action Eq. (7.33). Ideas of this sort were first popularized by Witten (1983) in his discussion of Wess–Zumino terms; see also Stone (1986).

The magnetic monopole gives rise to a uniform radial magnetic field on the surface of the sphere with total flux equal to the magnetic charge S. It is well known that the eigenstates of such a particle are monopole spherical harmonics. The ground state is $(2S + 1)$-fold degenerate and it is separated from the higher-angular-momentum states (i.e. Landau "orbits") by an energy gap that scales with the mass of the particle like $1/m$. Thus, in the small-mass limit ($m \to 0$) the system is projected onto the ground state. In this way the subspace of the "lowest-Landau orbit" on a spherical geometry becomes identical to the space of the spin-S representation of SU(2). In retrospect, it would have been possible to describe spin in terms of the path integral with Eq. (7.33) for its action directly, without reference to coherent states.

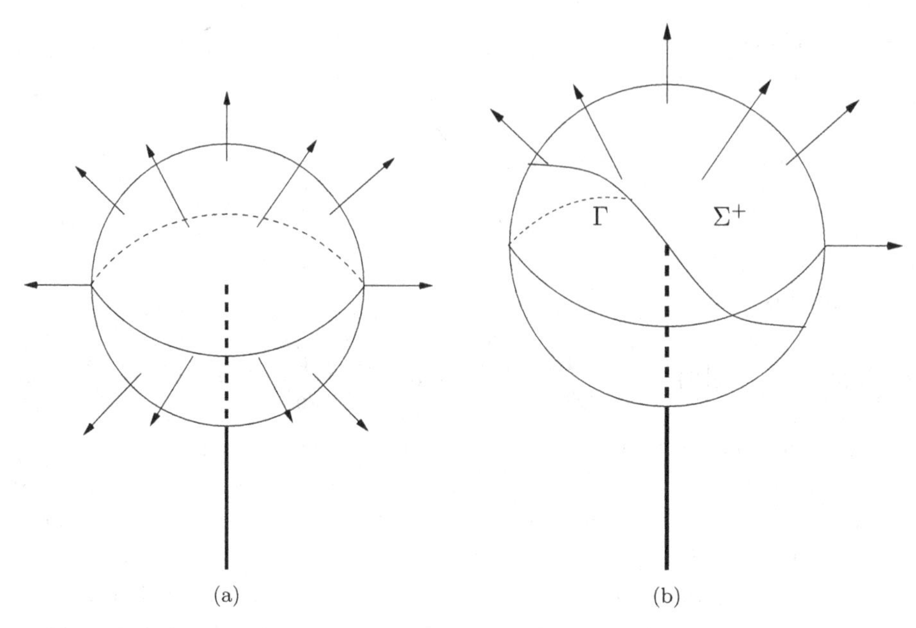

Figure 7.5 A magnetic monopole at the center of the unit sphere in (a) and the flux through the cap Σ^+ bounded by the trajectory Γ in (b). The thick line represents an infinitely long solenoid of infinitesimal thickness (a Dirac string).

Returning now to Eq. (7.34), we may consider the physical meaning of the vector potential $\vec{A}(x_0)$ in terms of the states of the spin. It is straightforward to show that the circulation of the vector field $\vec{A}(x_0)$ is just the accumulated change in the phase of the spin state under an adiabatic time evolution, i.e.

$$\oint d\vec{n} \cdot \vec{A}[\vec{n}(t)] = \int_0^T dt \langle \vec{n}(t) | \partial_t \vec{n}(t) \rangle \tag{7.35}$$

(with $|\vec{n}(0)\rangle = |\vec{n}(T)\rangle$), which is known as the *Berry phase* (Avron *et al.*, 1983; Simon, 1983; Berry, 1984), and the vector field $\vec{A}[\vec{n}(t)]$ is known as the Berry connection. In other chapters of this book we will encounter many manifestations of the Berry phase (and of the Berry connection).

7.3 The path integral for many-spin systems

It is trivial to generalize the one-spin problem to a many (or infinitely many!)-spin system. Once again, I will follow the treatment of Fradkin and Stone (1988).

The Hilbert space of a many-spin system is just the tensor product of the Hilbert space of the individual spins. Let H be the (Heisenberg) Hamiltonian for a spin-S system on an *arbitrary lattice*,

$$H = J \sum_{(\vec{r},\vec{r}')} \vec{S}(\vec{r}) \cdot \vec{S}(\vec{r}') \tag{7.36}$$

where $(\vec{r}, \vec{r}')$ are pairs of sites on that lattice. We can now use the identity $\langle \vec{n}|\vec{S}|\vec{n}\rangle = S\vec{n}$ to write down the imaginary-time action for the many-spin system

$$\begin{aligned} \mathcal{S}_{\mathrm{E}}[\vec{n}] = -iS \sum_{\vec{r}} \mathcal{S}_{\mathrm{WZ}}[\vec{n}(\vec{r})] + \frac{m}{2} \int_0^\beta dt \sum_{\vec{r}} (\partial_t \vec{n}(\vec{r}, t))^2 \\ + \int_0^\beta dt \sum_{(\vec{r},\vec{r}')} JS^2 \vec{n}(\vec{r}, t) \cdot \vec{n}(\vec{r}', t) \end{aligned} \tag{7.37}$$

where we are supposed to take the limit $m \to 0$ (it will be dropped from now on). The sums in Eq. (7.37) run over all the sites of the lattice. The first term is just the sum of the Wess–Zumino terms of the individual spins. Note that the only real-time dependence enters through the Wess–Zumino terms.

We can Wick-rotate back to real time, $t = ix_0$, $\beta = iT$, and write the corresponding real-time action, $\mathcal{S}_{\mathrm{M}}$, as

$$\mathcal{S}_{\mathrm{M}}[\vec{n}] = S \sum_{\vec{r}} \mathcal{S}_{\mathrm{WZ}}[\vec{n}(\vec{r})] - \int_0^T dx_0 \sum_{(\vec{r},\vec{r}')} JS^2 \vec{n}(\vec{r}, x_0) \cdot \vec{n}(\vec{r}', x_0) \tag{7.38}$$

The effective action $\mathcal{S}_{\mathrm{M}}[\vec{n}]$ scales like S, the spin representation. Thus, in the large-spin limit $S \to \infty$, the path integral

$$Z = \int \mathcal{D}\vec{n}\, e^{i\mathcal{S}_{\mathrm{M}}[\vec{n}]} \tag{7.39}$$

should be dominated by the stationary points of the action $\mathcal{S}_{\mathrm{M}}[\vec{n}]$. This is the semi-classical limit. Corrections to the large-S limit can be arranged in an expansion in powers of $1/S$. This is the content of the Holstein–Primakoff expansion (Holstein and Primakoff, 1940). Note, however, that we did not make use of the semi-classical limit in order to derive the path integral. Let us consider a number of cases of interest.

7.4 Quantum ferromagnets

In this case we set $J = -|J|$. I will consider the case of a hypercubic lattice and restrict the sum over pairs of sites to nearest neighbors. The results can be generalized very easily to any other lattice.

I first make use of the constraint $\vec{n}^2 = 1$ to write the action in the form

$$\mathcal{S}_{\mathrm{M}}[\vec{n}] = S \sum_{\vec{r}} \mathcal{S}_{\mathrm{WZ}}[\vec{n}(\vec{r})] - \frac{|J|S^2}{2} \sum_{\langle\vec{r},\vec{r}'\rangle} \int_0^T dx_0 [\vec{n}(\vec{r}, x_0) - \vec{n}(\vec{r}', x_0)]^2 \tag{7.40}$$

up to an additive constant. Consider now the long-wavelength limit, in which $\vec{n}(\vec{r}, x_0)$ is a smooth function of the spatial coordinates. If we denote by a_0 a short-distance cutoff (i.e. the lattice spacing) we can write an effective continuum action for the long-wavelength fluctuations

$$\mathcal{S}_{\text{M}}[\vec{n}] = \int d^d x \, \frac{S}{a_0^d} \mathcal{S}_{\text{WZ}}[\vec{n}] - \frac{|J|S^2}{2a_0^{d-2}} \int d^d x \int_0^T dx_0 (\nabla_i \cdot \vec{n}(\vec{x}, x_0))^2 \tag{7.41}$$

It is important to stress that the effective continuum action for the *quantum ferromagnet* does not have the standard non-linear sigma-model form which is of second order in time derivatives, which leads to Goldstone bosons with a *linear* dispersion law. As we will see (and as we already saw in Chapter 3), this is the correct result for *antiferromagnetic magnons* but not for ferromagnetic ones. It is well known that ferromagnetic magnons have a quadratic dispersion relation (Bloch, 1930). Thus, in the ferromagnetic case we expect the effective action to have twice as many spatial derivatives as temporal derivatives of the fields. In other terms, the dynamic critical exponent for a quantum ferromagnet is $z = 2$, whereas for an antiferromagnet it is $z = 1$.

To see how all this comes about, we will derive the classical equations of motion for the effective action of the quantum ferromagnet, Eq. (7.41). We take care of the *local constraint*

$$\vec{n}^2(\vec{x}, x_0) = 1 \tag{7.42}$$

by introducing the Lagrangian multiplier field $\lambda(\vec{x}, x_0)$ which enforces the constraint in the path integral through an extra term in the action

$$\mathcal{S}_{\text{extra}}[\vec{n}, \lambda] = \int d^d x \int_0^T dx_0 \, \frac{\lambda(\vec{x}, x_0)}{2} \left(\vec{n}^2(\vec{x}, x_0) - 1\right) \tag{7.43}$$

The classical equations of motion result from demanding that the total action

$$\mathcal{S}_{\text{tot}}[\vec{n}, \lambda] = \mathcal{S}_{\text{M}}[\vec{n}] + \mathcal{S}_{\text{extra}}[\vec{n}, \lambda] \tag{7.44}$$

be stationary,

$$\delta \mathcal{S}_{\text{tot}} = 0 \tag{7.45}$$

The variation of the local Wess–Zumino action is very simple. Indeed, $\mathcal{S}_{\text{WZ}}$ is essentially the area of the sphere bounded by the trajectory $\vec{n}(\vec{x}, x_0)$ (at each point $\vec{x}$) on the "target manifold" (the 2-sphere S_2). Thus the variation $\delta\mathcal{S}_{\text{WZ}}$ due to a small change of the trajectory $\delta\vec{n}$ is simply equal to

$$\delta \mathcal{S}_{\text{WZ}} = \delta\vec{n} \cdot (\vec{n} \times \partial_0 \vec{n}) \tag{7.46}$$

Hence, we get the classical equations of motion

$$\frac{\delta S_{\text{tot}}}{\delta \vec{n}} = \nabla_i \left(\frac{\delta S_{\text{tot}}}{\delta \nabla_i \vec{n}} \right) \tag{7.47}$$

supplemented by the constraint Eq. (7.42). More explicitly, we get

$$\frac{S}{a_0^d} \vec{n} \times \partial_0 \vec{n} + \lambda \vec{n} = -\frac{|J| S^2}{a_0^{d-2}} \nabla^2 \vec{n} \tag{7.48}$$

The classical value of the Lagrange multiplier field λ can be evaluated by computing the scalar product of Eq. (7.48) with $\vec{n}$. The result is

$$\lambda = -\frac{|J| s^2}{a_0^{d-2}} (\vec{n} \cdot \nabla^2 \vec{n}) \tag{7.49}$$

On substituting Eq. (7.49) back into Eq. (7.48) we get the equation of motion for the quantum ferromagnet:

$$\frac{S}{a_0^d} \vec{n} \times \partial_0 \vec{n} + \frac{|J| S^2}{a_0^{d-2}} \left(\nabla^2 - (\vec{n} \cdot \nabla^2 \vec{n}) \right) \vec{n} = 0 \tag{7.50}$$

By using elementary algebra as well as Eq. (7.42), this equation can be brought to the form

$$\partial_0 \vec{n} = |J| S a_0^2 \vec{n} \times \nabla^2 \vec{n} \tag{7.51}$$

This equation is known as the Landau–Lifshitz equation. The derivation shown here is due to M. Stone.

The Landau–Lifshitz equation has several interesting properties. It is a non-linear equation with first-order time derivatives and second-order space derivatives. Thus the solutions of Eq. (7.51) have a quadratic dispersion law, as they should. The spins move in a precessional fashion with an angular velocity $\vec{\Omega}$ given by

$$\vec{\Omega} = -|J| S a_0^2 \nabla^2 \vec{n} \tag{7.52}$$

The Landau–Lifshitz equations can be solved in the linear regime. Let us parametrize $\vec{n}$ by the components

$$\vec{n} = \begin{pmatrix} \sigma \\ \vec{\pi} \end{pmatrix} \tag{7.53}$$

where σ and π_i $(i = 1, 2)$ satisfy the constraint

$$\sigma^2 + \vec{\pi}^2 = 1 \tag{7.54}$$

The (linearized) Landau–Lifshitz equations are

$$\begin{aligned} \partial_0 \pi_1 &\approx -|J| S a_0^2 \nabla^2 \pi_2 \\ \partial_0 \pi_2 &\approx +|J| S a_0^2 \nabla^2 \pi_1 \end{aligned} \tag{7.55}$$

to leading order in $\vec{\pi}$. From Eqs. (7.55) we find the dispersion relation for ferromagnetic spin waves

$$|p_0| \approx |J|Sa_0^2|\vec{p}|^2 \tag{7.56}$$

which is known as Bloch's law (Bloch, 1930). As expected, we find that the frequency of the low-energy excitations of a quantum ferromagnet scales as the square of the momentum.

7.5 The effective action for one-dimensional quantum antiferromagnets

We will not consider here frustrated systems. Thus, and for the sake of simplicity, we will consider the case of quantum antiferromagnets on *bipartite* lattices, such as the hypercubic lattice. We will see that, unlike in the case of the ferromagnets, the effective low-energy action is *different* for 1D systems and for higher-dimensional cases such as the square and cubic lattices. In all cases we will find a non-linear sigma model, in agreement with our previous discussion (see Chapter 3) that was based on a mean-field weak-coupling treatment of the Hubbard model. But we will get more. For the spin-chain case we will find that the action has an extra term, a topological term.

The starting point will be, once again, the real-time action of Eq. (7.39) with a nearest-neighbor *antiferromagnetic* coupling constant $J > 0$. Since we expect that *at least* the short-range order should have Néel character, it is natural to consider the *staggered* and *uniform* components of the spin field $\vec{n}$. This construction, as is, works only for two-sublattice systems close to a Néel state, although it is possible to generalize it to other cases.

Consider a spin chain with an *even* number of sites N occupied by spin-S degrees of freedom. The sites of the lattice are labeled by an integer $j = 1, \dots, N$. The real-time action is

$$\mathcal{S}_{\rm M}[\vec{n}] = S\sum_{j=1}^{N} \mathcal{S}_{\rm WZ}[\vec{n}(j)] - \int_0^T dx_0 \sum_{j=1}^{N} JS^2\vec{n}(j,x_0)\cdot\vec{n}(j+1,x_0) \tag{7.57}$$

where we have assumed periodic boundary conditions. Since we expect to be close to a Néel state, we will stagger the configuration

$$\vec{n}(j) \to (-1)^j\vec{n}(j) \tag{7.58}$$

On a bipartite lattice, the substitution of Eq. (7.58) into Eq. (7.57) will change the sign of the *exchange term* of the action to a *ferromagnetic* one. The Wess–Zumino terms are *odd* under the replacement of Eq. (7.58) and thus become *staggered*. Thus, it is the Wess–Zumino term, a purely quantum-mechanical effect, which will distinguish ferromagnets from antiferromagnets. After staggering the spins we get, up to an additive constant,

$$\mathcal{S}_{\rm M}[\vec{n}] = S\sum_{j=1}^{N}(-1)^j \mathcal{S}_{\rm WZ}[\vec{n}(j)] - \frac{JS^2}{2}\int_0^T dx_0 \sum_{j=1}^{N}(\vec{n}(j,x_0)-\vec{n}(j+1,x_0))^2 \tag{7.59}$$

We now split the (staggered) spin field $\vec{n}$ into a slowly varying piece $\vec{m}(j)$, the order parameter field, and a small rapidly varying part, $\vec{l}(j)$, which roughly represents the average spin (Affleck, 1990). Hence, we write

$$\vec{n}(j) = \vec{m}(j) + (-1)^j a_0 \vec{l}(j) \tag{7.60}$$

The constraint $\vec{n}^2 = 1$ and the requirement that the order-parameter field $\vec{m}$ should obey the same constraint, $\vec{m}^2 = 1$, demand that $\vec{m}$ and $\vec{l}$ be orthogonal vectors:

$$\vec{m}\cdot\vec{l} = 0 \tag{7.61}$$

The Wess–Zumino terms are rewritten as

$$S\sum_{j=1}^{N}(-1)^j \mathcal{S}_{\rm WZ}[\vec{n}(j)] = S\sum_{r=1}^{N/2}(\mathcal{S}_{\rm WZ}[\vec{n}(2r)] - \mathcal{S}_{\rm WZ}[\vec{n}(2r-1)]) \tag{7.62}$$

which, by making use of the approximation

$$\begin{aligned}\vec{n}(2r)-\vec{n}(2r-1) &= \vec{m}(2r)-\vec{m}(2r-1)+a_0(\vec{l}(2r)+\vec{l}(2r-1))\\ &= a_0\left(\partial_1\vec{m}(2r)+2\vec{l}(2r)\right)+O(a_0^2)\end{aligned} \tag{7.63}$$

becomes

$$\begin{aligned}S\sum_{j=1}^{N}(-1)^j \mathcal{S}_{\rm WZ}[\vec{n}(j)] &\approx S\sum_{r=1}^{N/2}\int_0^T dx_0\, \delta\vec{n}(2r,x_0)\cdot(\vec{n}(2r,x_0)\times\partial_0\vec{n}(2r,x_0))\\ &\approx S\sum_{r=1}^{N/2}\int_0^T dx_0\left(a_0\,\partial_1\vec{m}(2r,x_0)+2a_0\vec{l}(2r,x_0)\right)\\ &\quad\times(\vec{m}(2r,x_0)\times\partial_0\vec{m}(2r,x_0))\end{aligned} \tag{7.64}$$

Thus, in the continuum limit, one finds

$$\begin{aligned}\lim_{a_0\to 0} S\sum_{j=1}^{N}(-1)^j \mathcal{S}_{\rm WZ}[\vec{n}(j)] \approx \frac{S}{2}\int d^2x\, \vec{m}\cdot(\partial_0\vec{m}\times\partial_1\vec{m})\\ + S\int d^2x\, \vec{l}\cdot(\vec{m}\times\partial_0\vec{m})\end{aligned} \tag{7.65}$$

Similarly, the continuum limit of the potential-energy terms can also be found to be given by

$$\lim_{a_0\to 0} \frac{JS^2}{2} \sum_{j=1}^{N} \int_0^T dx_0 (\vec{n}(j, x_0) - \vec{n}(j+1, x_0))^2$$
$$\simeq \frac{a_0 JS^2}{2} \int d^2x \left((\partial_1 \vec{m})^2 + 4\vec{l}^2 \right) \tag{7.66}$$

On collecting terms we find a Lagrangian density involving both the order-parameter field $\vec{m}$ and the local spin density $\vec{l}$,

$$\mathcal{L}_{\text{M}}(\vec{m}, \vec{l}\,) = -2a_0 J S^2 \vec{l}^2 + s\vec{l} \cdot (\vec{m} \times \partial_0 \vec{m}) - \frac{a_0 J S^2}{2} (\partial_1 \vec{m})^2$$
$$+ \frac{S}{2} \vec{m} \cdot (\partial_0 \vec{m} \times \partial_1 \vec{m}) \tag{7.67}$$

The fluctuations in the average spin density $\vec{l}$ can be integrated out. The result is the Lagrangian density of the non-linear sigma model,

$$\mathcal{L}_{\text{M}}(\vec{m}) = \frac{1}{2g} \left(\frac{1}{v_{\text{s}}} (\partial_0 \vec{m})^2 - v_{\text{s}} (\partial_1 \vec{m})^2 \right) + \frac{\theta}{8\pi} \epsilon_{\mu\nu} \vec{m} \cdot \left(\partial_\mu \vec{m} \times \partial_\nu \vec{m} \right) \tag{7.68}$$

where g and v_{s} are, respectively, the coupling constant and spin-wave velocity:

$$g = \frac{2}{S} \tag{7.69}$$

$$v_{\text{s}} = 2a_0 JS \tag{7.70}$$

The last term in Eq. (7.68) has topological significance. We have chosen the normalization so that the coupling constant θ is given by

$$\theta = 2\pi S \tag{7.71}$$

The tensor $\epsilon_{\mu\nu}$ is the usual Levi-Civita antisymmetric tensor in two dimensions.

Thus, apart from an anisotropy determined by the spin-wave velocity v_{s} and apart from the topological term, we find that the effective action for the low-frequency, long-wavelength fluctuation about a state with *short-range* Néel order is given by the non-linear sigma model. We reached the same results within the weak-coupling mean-field theory of the half-filled Hubbard model of Chapter 3. Indeed, using that approach, it is also possible to get the topological term (Wen and Zee, 1988).

7.6 The role of topology

In the past section we reached the conclusion that the low-energy excitations of a 1D quantum antiferromagnet with short-range Néel order can be described by the path integral of a non-linear sigma model with a topological term

$$Z = \int \mathcal{D}\vec{m} \prod_x \delta(\vec{m}^2(x) - 1) e^{i S_{\text{eff}}[\vec{m}(x)]} \tag{7.72}$$

with the effective action obtained from Eq. (7.68). Before considering the role of local quantum fluctuations, which are of fundamental importance here, we look at the role of the last term in the action, the topological term $\mathcal{S}_{\text{topo}}$:

$$\mathcal{S}_{\text{topo}} = \frac{\theta}{8\pi} \int d^2x \, \epsilon_{\mu\nu} \vec{m} \cdot \left(\partial_\mu \vec{m} \times \partial_\nu \vec{m}\right) \tag{7.73}$$

Let us consider first the Euclidean sector of the theory (i.e. we are back to imaginary time $x_2 = i x_0$) with the Lagrangian density $\mathcal{L}_{\text{E}}$,

$$\mathcal{L}_{\text{E}} = \frac{1}{2g} \left(v_{\text{s}} (\partial_1 \vec{m})^2 + \frac{1}{v_{\text{s}}} (\partial_2 \vec{m})^2 \right) + i \frac{\theta}{8\pi} \epsilon_{ij} \vec{m} \cdot \left(\partial_i \vec{m} \times \partial_j \vec{m}\right) \tag{7.74}$$

We now define the *Pontryagin index* or *topological charge* (or *winding number*) $\mathcal{Q}$ of the Euclidean-space spin configuration $\{\vec{m}(x)\}$ by the expression

$$\mathcal{Q} = \frac{1}{8\pi} \int d^2x \, \epsilon_{ij} \vec{m} \cdot \left(\partial_i \vec{m} \times \partial_j \vec{m}\right) \tag{7.75}$$

We impose the boundary condition that the Euclidean action $\int d^2x \, \mathcal{L}_{\text{E}}[\vec{m}]$ be finite. This is equivalent to the requirement that asymptotically $\vec{m}$ becomes a *constant* (but arbitrary) vector $\vec{m}_0$ at spatial-time infinity,

$$\lim_{|\vec{x}| \to \infty} \vec{m}(\vec{x}) = \vec{m}_0 \tag{7.76}$$

Thus, topologically, 2D Euclidean space-time is isomorphic to a sphere S_2 since the fields are identified with $\vec{m}_0$ at the point of infinity (Fig. 7.6). However, the order-parameter manifold (the "target space") is also isomorphic to a sphere S_2, since the constraint $\vec{m}^2 = 1$ has to be satisfied everywhere. Therefore, a field configuration $\vec{m}(x)$ with *finite Euclidean action* is thus a smooth (differentiable) mapping from the S_2 of Euclidean space-time to the S_2 of the order-parameter manifold (the target space) (Fig. 7.7).

The Pontryagin index $\mathcal{Q}[\vec{m}]$ is the topological charge (or winding number) in the sense that it counts how many times the spin configuration $\vec{m}$ has wrapped around the sphere S_2, as can be checked by comparing the definition of $\mathcal{Q}$, Eq. (7.75), with

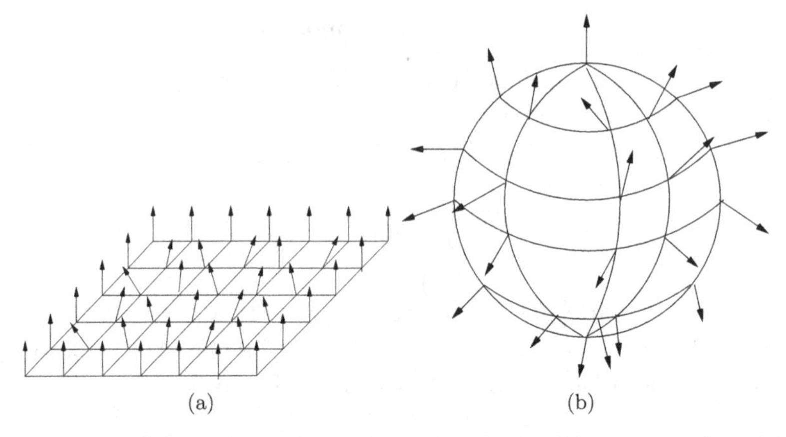

Figure 7.6 A finite-action spin configuration in Euclidean space-time (a) is isomorphic to one on the sphere S_2 (b).

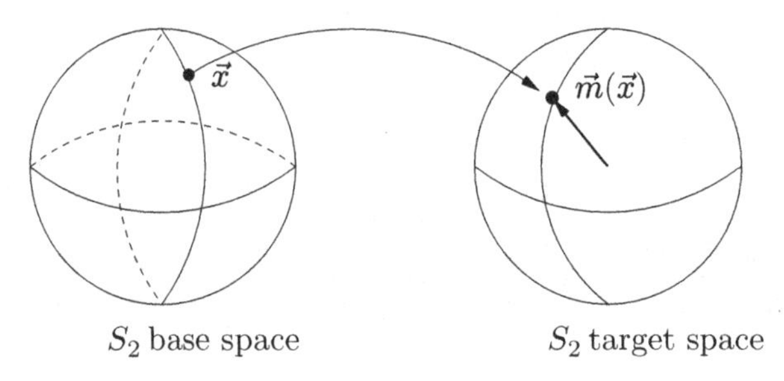

Figure 7.7 The mapping $\vec{m}(\vec{x}): S_2 \to S_2$ of the 2-sphere base space to the 2-sphere target space.

the area formula of Eq. (7.28). We can make these ideas more concrete by considering a configuration $\vec{m}(x)$ representing an *instanton* (Fig. 7.8(a)). Let the field at infinity point parallel to $\vec{m}_0$, the north pole of S_2. In the case of an instanton, the field near the origin points *oppositely* to $\vec{m}_0$, i.e. in the direction of the south pole. Alternatively, we can look at the configuration on S_2. Here it looks like a magnetic monopole or a hairy ball (Fig. 7.8(b)). The winding number $\mathcal{Q}$ of this configuration is determined by the area of the sphere divided by 4π (i.e. the "magnetic flux"),

$$\mathcal{Q} = \left(\frac{1}{4\pi}\right) 4\pi = +1 \tag{7.77}$$

Thus, an instanton has winding number (or topological charge) $\mathcal{Q} = +1$. An anti-instanton has $\mathcal{Q} = -1$. It is also possible to find multi-instanton configurations with arbitrary *integral* winding number $\mathcal{Q}$.

We conclude that the smooth configurations $\vec{m}(x)$ can be classified according to their winding number or topological charge $\mathcal{Q}$: configurations that can be

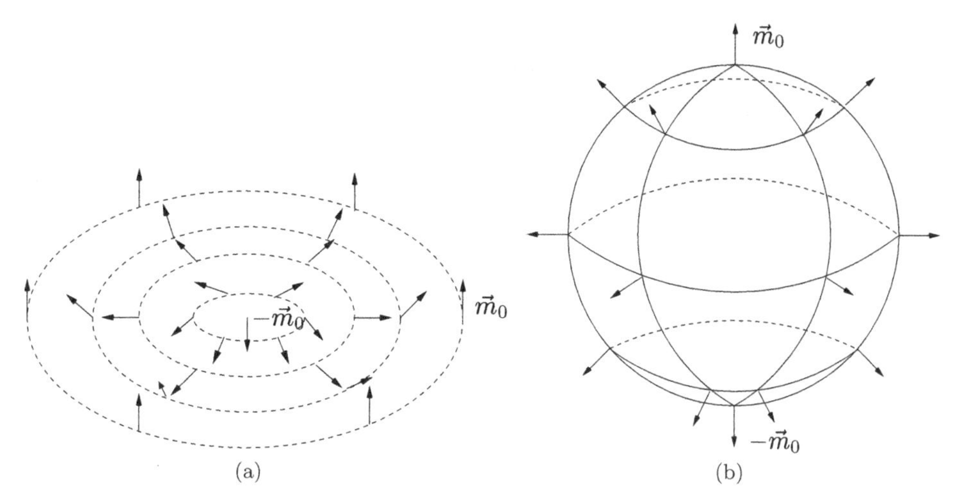

Figure 7.8 (a) An instanton configuration in 2D Euclidean space-time, with topological charge $\mathcal{Q} = 1$. In $(2 + 1)$ dimensions this configuration is known as a skyrmion or hedgehog (see the more artistic version of the skyrmion on the cover of this book). (b) An instanton on S_2 has the same topology as a monopole, that of a "hairy ball."

smoothly deformed into each other have the same topological charge $\mathcal{Q}$. Thus smooth configurations $\{\vec{m}(x)\}$ can be classified into a discrete set of equivalence classes, each labeled by an integer, their topological charge. Such smooth configurations, known as *homotopies*, form a group, known as a *homotopy group*, since the *composition* of two configurations (two smooth mappings) yields another smooth configuration whose topological charge is the sum (with their signs) of the individual topological charges. In other words, the configurations $\vec{m}(x)$ are mappings of S_2 into S_2 with *homotopy* classes classified by a *topological invariant*, namely the Pontryagin index $\mathcal{Q}$, which can take only integer values (positive or negative). In mathematical terms this homotopy group is represented by the expression

$$\pi_2(S_2) = \mathbb{Z} \tag{7.78}$$

A clear and detailed explanation of homotopy theory in condensed matter physics is given in Mermin (1979).

Similarly, the vortices of a 2D superfluid (or, equivalently, of a classical XY model), as discussed in Chapter 4, are topological excitations classified by a topological invariant $n \in \mathbb{Z}$, the winding number of the vortex. This winding number classifies the maps of the phase of the order-parameter field on a large circumference S_1 onto the target space of the complex order-parameter field itself, which is another S_1. Hence the homotopy classes for 2D vortices are

$$\pi_1(S_1) \simeq \mathbb{Z} \tag{7.79}$$

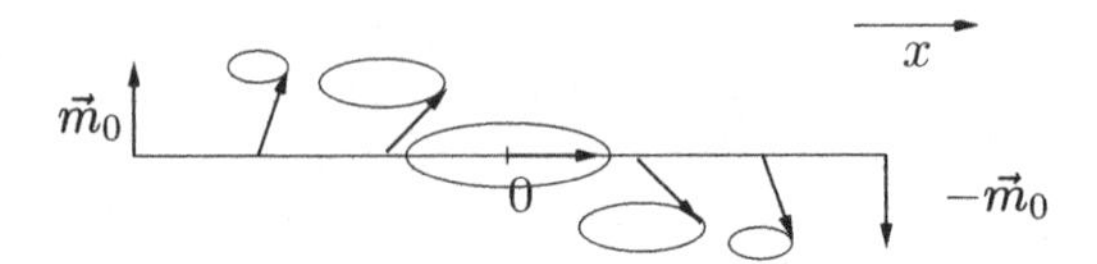

Figure 7.9 A half-twist soliton: the circles represent the precession of the spins.

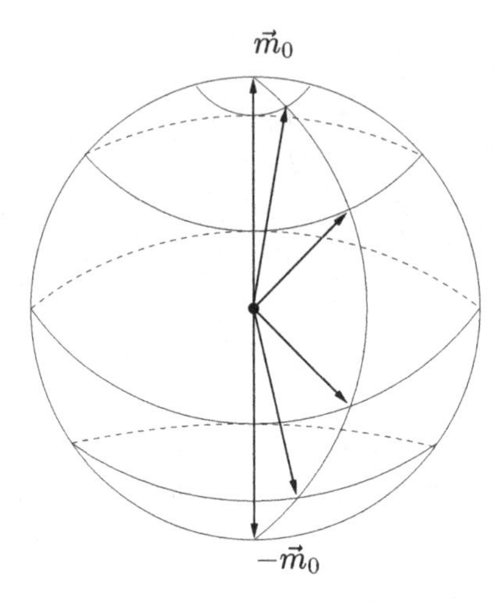

Figure 7.10 The history of a half-twist soliton: the circles ("parallels") represent the precession of the spins.

Back in real time we can consider *soliton* configurations, such as the half-twist of Fig. 7.9. As time goes by, each spin traces a closed path on the target sphere S_2 and hence it sweeps an area bound by the path. If we define that the area swept by a spin at $-\infty$ is equal to zero, we see that as we move from left to right the spins sweep an increasingly large area. At $+\infty$ the area swept is that of a full sphere, 4π. It is easy to see that $\mathcal{Q}$ is also equal to 1 for the half-twist. At each point in space, the spins are coherently precessing and keeping their relative angles constant. In other words, the spins trace lines of longitude on a sphere. The global configuration still looks like a monopole and hence also has winding number $\mathcal{Q} = +1$ (Fig. 7.10).

The final conclusion is that the topological term, Eq. (7.73), is proportional to an integer $\mathcal{Q}$. The action in the path integral of Eq. (7.72) has a contribution equal to $2\pi S\mathcal{Q}$, which should be added to the standard sigma-model term. Since S is an integer or a half-integer, we find that the extra, topological, term gives a contribution to the weight of a configuration in the path integral of

$$e^{i2\pi S\mathcal{Q}} = (-1)^{2S\mathcal{Q}} \tag{7.80}$$

Thus, if S is an integer, the spin chain is described at low energies by the standard non-linear sigma model, *without* a topological term. On the other hand, for half-integral S, each topological *class* contributes to the weight of the path integral with a sign that is positive (negative) if the winding number Q is even (odd). Note that the sign does not depend on the actual value of the spin S, but only on whether it is an integer or a half-integer. This means that the physics of this problem is *not* analytic in S: the integer- and half-integer-spin chains fall in different universality classes. We will now see that this property implies a very important result, known as Haldane's conjecture, that states that the integer-spin chains are massive (i.e. have an energy gap), whereas the half-integer-spin chains are massless as in the spin one-half case.

7.7 Quantum fluctuations and the renormalization group

In the previous section we saw that the configuration space of the non-linear sigma model can be partitioned into classes classified by their winding numbers

$$\mathcal{Q} = \frac{1}{8\pi}\int d^2x\, \epsilon_{abc}\epsilon_{ij} m_a\, \partial_i m_b\, \partial_j m_c \tag{7.81}$$

which is a topological invariant. Thus, the partition function can be represented as a sum over distinct topological sectors, labeled by the topological charge Q,

$$Z = \int \mathcal{D}\vec{m}\, e^{-\mathcal{S}_{\rm E}[\vec{m}]} = \sum_{\mathcal{Q}=0}^{\infty}\int_{\mathcal{Q}} \mathcal{D}\vec{m}\, e^{-\mathcal{S}_0^{\rm E}[\vec{m}]} e^{i2\pi S\mathcal{Q}} \tag{7.82}$$

where the subindex $\mathcal{Q}$ indicates that the path integral is to be taken over configurations with a fixed winding number $\mathcal{Q}$ and $\mathcal{S}_0^{\rm E}[\vec{m}]$ is the standard action of the non-linear sigma model

$$\mathcal{S}_0^{\rm E}[\vec{m}] = \int d^2x\, \frac{1}{2g}(\nabla_i\vec{m})^2 \tag{7.83}$$

where space and time have been rescaled so as to have $v_{\rm s} = 1$.

In this section we will consider the role of quantum fluctuations. We can do so by considering each topological class separately since these quantum fluctuations are local and do not alter the winding number. In other words, the winding number of a class of configurations cannot be changed by local fluctuations, since the former is a global property, whereas the latter are purely local. Naturally, for this picture to hold it is necessary that the short distance (ultraviolet) and the long distance (infrared) of the theory remain separate. We will see that this is not the case in one space dimension. The behavior of the non-linear sigma model is dominated by *infrared* fluctuations. Thus the actual role, in detail, of topological sectors is unclear.

We will pretend that the fluctuations are local and reasonably small. This assumption amounts to a semi-classical treatment of the path integral. Formally, this can be achieved only if the coupling constant g is small, i.e. in the limit $S \to \infty$. The standard perturbative treatment of the non-linear sigma model is thus equivalent, at low energies, to the $1/S$ expansion of the Heisenberg antiferromagnet (Haldane, 1983a, 1983c). The classical action of the non-linear sigma model, Eq. (7.83), has a very important property: it is *scale-invariant*. In other words the scale transformation

$$(x, t) \to \lambda(x, t), \quad \vec{m} \to \vec{m} \tag{7.84}$$

leaves the *action* invariant. Recall that $\vec{m}$ is dimensionless, and that the coupling constant g is also *dimensionless* in $(1+1)$ dimensions. In higher dimensions, g is *dimension-full*. Let us define the *dimensionless coupling constant* u,

$$u = g a_0^{2-d} \tag{7.85}$$

where d is the dimension of space-time. Thus the action now reads

$$\mathcal{S}_0^{\mathrm{E}}[\vec{m}] = \frac{1}{2 u a_0^{d-2}} \int d^d x (\nabla_i \vec{m})^2 \tag{7.86}$$

where $i = 1, \ldots, d$. For the sake of simplicity the discussion will be carried out in Euclidean space (i.e. imaginary time).

In renormalization-group theory (Wilson and Kogut, 1974), which was discussed in detail in Chapter 4, the fact that the classical action is scale-invariant means that $g = 0$ is a fixed point of the renormalization group (RG). I will define *a renormalization-group transformation* for the non-linear sigma model by progressively integrating out the faster modes and obtaining an effective theory for the slower modes. This procedure involves only *local* degrees of freedom. Topological invariants, such as a θ term in the Lagrangian of the non-linear sigma model, do not get renormalized under the effects of integrating out local fluctuations. In addition, as we saw before, the value of $\theta = n\pi$ (with $n \in \mathbb{Z}$) is fixed by the requirement of time-reversal invariance, a symmetry that the antiferromagnet has (in combination with a unit lattice translation).

In general, the field $\vec{m}$ will have Fourier components with momenta $\vec{p}$ ranging from the infrared ($|\vec{p}| \approx 0$) to the ultraviolet ($|\vec{p}| \approx 1/a_0$). We can also use the constraint $\vec{m}^2 = 1$ to demand that one of the components of the field $\vec{m}$, say m_3, has only fast components and that it be small (Kogut, 1979). Let m_1 and m_2 be parametrized by m_3 and ϕ $(0 \leq \phi \leq 2\pi)$,

$$m_1 = \sqrt{1 - m_3^2} \cos\phi, \quad m_2 = \sqrt{1 - m_3^2} \sin\phi \tag{7.87}$$

so as to solve the constraint $\vec{m}^2 = 1$. The Euclidean Lagrangian density now reads

$$\mathcal{L}_0^{\mathrm{E}} = \frac{1}{2ua_0^{d-2}} (\nabla_i \vec{m})^2$$
$$= \frac{1}{2ua_0^{d-2}} \left[(\nabla_i m_3)^2 + (1 - m_3^2)(\nabla_i \phi)^2 + \frac{(m_3 \nabla_i m_3)^2}{1 - m_3^2} \right] \tag{7.88}$$

Let us rescale the field m_3,

$$m_3 = \sqrt{ua_0^{d-2}}\varphi \tag{7.89}$$

and write

$$\mathcal{L}_0^{\mathrm{E}} = \frac{1}{2}(\nabla_i \varphi)^2 + \frac{1}{2ua_0^{d-2}}(1 - ua_0^{d-2}\varphi^2)(\nabla_i \phi)^2$$
$$+ \frac{1}{2}\left(\frac{ua_0^{d-2}}{1 - ua_0^{d-2}\varphi^2} \right) (\varphi \nabla_i \varphi)^2 \tag{7.90}$$

We will be interested in the behavior for small g (i.e. small u). In this limit we can approximate $\mathcal{L}_0^{\mathrm{E}}$ by the expression

$$\mathcal{L}_0^{\mathrm{E}} = \frac{1}{2} (\nabla_i \varphi)^2 + \frac{1}{2ua_0^{d-2}}(\nabla_i \phi)^2 - \frac{1}{2}\varphi^2 (\nabla_i \phi)^2$$
$$+ \frac{1}{2}ua_0^{d-2} (\varphi \nabla_i \varphi)^2 + \frac{1}{2}u^2 a_0^{2(d-2)} \varphi^2 (\varphi \nabla_i \varphi)^2 + O(u^3) \tag{7.91}$$

Both φ and θ have Fourier components all the way from zero momentum up to the cutoff $\Lambda \sim 1/a_0$. The behavior at large momenta $|\vec{p}| \sim \Lambda$ should not affect very strongly phenomena taking place for small values of $\vec{p}$. It is then natural to integrate out such fluctuations.

Consider the momentum shell $b\Lambda < |\vec{p}| < \Lambda$ with $b < 1$ and the fluctuations with momenta inside that shell (fast modes). We now will carry out the functional integral

$$\int_{b\Lambda < |\vec{p}| < \Lambda} \mathcal{D}\varphi \; e^{-S_0^{\mathrm{E}}[\varphi, \phi]}$$
$$= \int_{b\Lambda < |\vec{p}| < \Lambda} \mathcal{D}\varphi(\vec{p}) \exp\left[-\frac{1}{2} \int d^d x \left((\nabla_i \varphi)^2 + \frac{1}{ua_0^{d-2}} (\nabla_i \phi)^2 - \varphi^2 (\nabla_i \phi)^2 + O(u) \right) \right] \tag{7.92}$$

I will assume that ϕ is slowly varying and, hence, that $(\nabla_i \phi)^2$ is small and does not have Fourier components in the shell $b\Lambda < |\vec{p}| < \Lambda$ provided that $b \to 1$. Thus,

$$\int_{b\Lambda<|\vec{p}|<\Lambda} \mathcal{D}\varphi(\vec{p})\ \exp\left(-\frac{1}{2}\int \frac{d^dp}{(2\pi)^d}\vec{p}^2|\varphi(\vec{p})|^2 + \frac{1}{2}\int \frac{d^dp}{(2\pi)^d}|\varphi(\vec{p})|^2(\nabla_i\phi)^2\right)$$
$$\approx \prod_{b\Lambda<|\vec{p}|<\Lambda}\left[\frac{2\pi}{\vec{p}^2-(\nabla\phi)^2}\right]^{1/2} \tag{7.93}$$

The right-hand side of Eq. (7.93) can be exponentiated and approximated by the expression

$$\exp\left[\frac{1}{2}\int_{b\Lambda<|\vec{p}|<\Lambda}\frac{d^dp}{(2\pi)^d}\ln\left(\frac{2\pi}{\vec{p}^2}\right)+\frac{1}{2}(\nabla_i\phi)^2\int_{b\Lambda<|\vec{p}|<\Lambda}\frac{d^dp}{(2\pi)^d}\frac{1}{\vec{p}^2}\right] \tag{7.94}$$

To lowest order in u, the main effects of integrating out the fast modes in φ are twofold: (a) a shift of energy and (b) a shift, or renormalization, of the coupling constant u. Indeed, we can recast Eqs. (7.92)–(7.94) into the effective Lagrangian density

$$\begin{aligned}\mathcal{L}^{\mathrm{E}}_{\mathrm{eff}}[\phi] = &-\frac{1}{2}\int_{b\Lambda<|\vec{p}|<\Lambda}\frac{d^dp}{(2\pi)^d}\ln\left(\frac{2\pi}{\vec{p}^2}\right)\\ &+\frac{1}{2}(\nabla_i\varphi)^2+\frac{1}{2}\left(\frac{1}{ua_0^{d-2}}-\int_{b\Lambda<|\vec{p}|<\Lambda}\frac{d^dp}{(2\pi)^d}\frac{1}{\vec{p}^2}\right)(\nabla_i\phi)^2\\ &-\frac{1}{2}\varphi^2(\nabla_i\phi)^2+O(u),\end{aligned} \tag{7.95}$$

with a momentum cutoff Λ' which has been *reduced* by b. Equivalently, the spatial cutoff a_0' has been *increased* by $1/b$:

$$\Lambda' = b\Lambda, \quad a_0' = \frac{a_0}{b} \tag{7.96}$$

The effective Lagrangian density for the slow modes $\mathcal{L}^{\mathrm{E}}_{\mathrm{eff}}[\phi]$ has the same form as the old Lagrangian density except for a constant shift (of the energy density), a new rescaled cutoff a_0' ($a_0' > a_0$), and a new renormalized coupling constant u' defined by

$$\frac{1}{u'{a_0'}^{d-2}} = \frac{1}{u{a_0}^{d-2}} - \int_{b\Lambda<|\vec{p}|<\Lambda}\frac{d^dp}{(2\pi)^d}\frac{1}{\vec{p}^2} \tag{7.97}$$

After evaluating the integral, we get

$$\frac{1}{u'{a_0'}^{d-2}} = \frac{1}{ua_0^{d-2}} - \frac{S_d}{(2\pi)^d}\left(\frac{1-b^{d-2}}{d-2}\right)\Lambda^{d-2} \tag{7.98}$$

where S_d is the area of the d-dimensional unit sphere. Since $b\to 1$ and $a_0' = a_0/b$, we can write

$$-\ln b = \frac{da_0}{a_0} \tag{7.99}$$

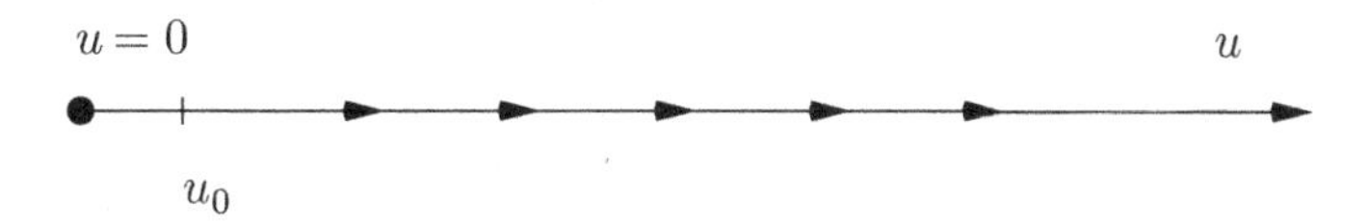

Figure 7.11 Renormalization-group infrared flow for 1D quantum spin chains; u_0 is the bare coupling constant ($u_0 = (2/S)a_0^{2-d}$).

and find the (one-loop) RG β-function

$$\beta(u) = a_0 \frac{du}{da_0} \tag{7.100}$$

to be given by

$$\beta(u) = -\epsilon u + \frac{u^2}{2\pi} + O(u^3) \tag{7.101}$$

for $\epsilon = d - 2$ *small.*

In particular in $(1 + 1)$ dimensions ($d = 2$) we find a *positive* β-function (Polyakov, 1975).

$$\beta(u) = \frac{u^2}{2\pi} + O(u^2) \tag{7.102}$$

This result means that as the cutoff a_0 is increased, and we look at longer and longer distances, the fluctuations *increase* the effective value of the coupling constant at such scales (see Fig. 7.11). Thus, even though the *bare* coupling constant $u_0 \propto 1/S$ may be initially small, as we consider the effective theory at lower energies we find that the effective coupling ("effective S") increases (decreases). From classical statistical mechanics we know that the sigma model at strong coupling (i.e. the classical Heisenberg ferromagnet at high temperatures) is disordered and has a finite correlation length. Thus, in the language of the quantum spin chains, we get that as the "effective S" decreases the semi-classical behavior gets wiped out. Instead we find a state *without* spontaneous symmetry breaking and with short-range correlations.

7.8 Asymptotic freedom and Haldane's conjecture

In the last section we found the result that the effective coupling constant of the non-linear sigma model in $(1 + 1)$ dimensions increases with the length scale. We have chosen to present this result in the form of a β-function, Eq. (7.102), which measures the change of the coupling constant u as the cutoff a_0 (the lattice constant) is increased and the fast degrees of freedom of the system are progressively integrated out. Alternatively, we could have kept the cutoff fixed and varied

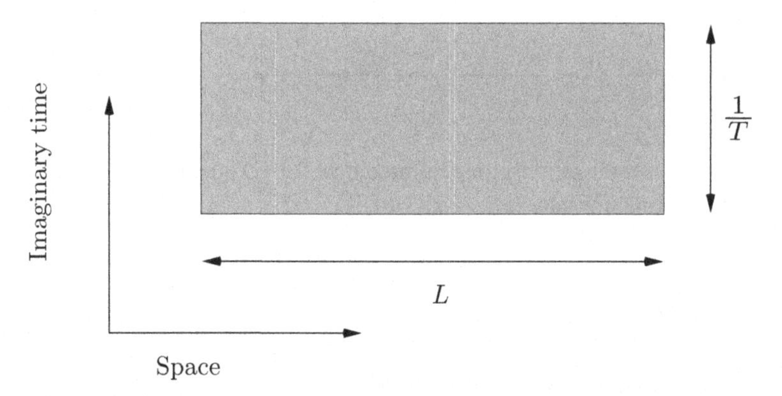

Figure 7.12 Euclidean space-time for a system of length L at temperature T. With periodic boundary conditions in imaginary time, the time axis becomes a circle of length $1/T$ and Euclidean space-time is a cylinder.

a physical scale such as the length L of the chain or an energy scale such as the temperature T.

At finite temperature T, the system can be viewed (in imaginary time) as a non-linear sigma model on a strip of length L (the linear size of the chain) and width $1/T$ with periodic boundary conditions in imaginary time. This is the standard statement that the partition function of a quantum system, with a global symmetry, in d space dimensions is equivalent to a classical mechanics problem in $(d+1)$ dimensions with imaginary time being the extra dimension (Fradkin and Susskind, 1978) (Fig. 7.12). The renormalization group of the last section can easily be generalized to an anisotropic system with spin-wave velocity $v_s \neq 1$, which is kept fixed in the RG process.

We begin our RG process with some lattice constant a_0, bare coupling $u_0 \propto 1/S$, and spin-wave velocity v_s. As we integrate out degrees of freedom the effective coupling grows and the spatial cutoff increases. At some point, the cutoff a becomes of the order of v_s/T. At this point the quantum fluctuations are negligible since the cutoff is as large as the width of the strip and we have effectively a *non-linear* sigma model at finite temperature T. In turn the non-linear sigma model, in imaginary time, is identical to the *classical* Heisenberg model in d space-time Euclidean dimensions. It can be easily proven that a classical Heisenberg model (or non-linear sigma model) in one dimension, like all 1D classical systems with short-range interactions, has a finite correlation length ξ_c at all temperatures (Landau and Lifshitz, 1975a).

We can now ask how much the effective coupling u differs from the bare coupling u_0 if the cutoff is changed from a_0 to $\bar{a}_0 \sim v_s/T$. The β-function tells us the dependence of u on the cutoff, at least for small enough u. The result of integrating the differential equation

$$\beta(u) \equiv a_0 \frac{du}{da_0} = \frac{u^2}{2\pi} \tag{7.103}$$

is

$$\frac{1}{u(\bar{a}_0)} = \frac{1}{u(a_0)} + \frac{1}{2\pi} \ln\left(\frac{a_0}{\bar{a}_0}\right) \tag{7.104}$$

By choosing $\bar{a}_0$ to be of the order of $1/T$,

$$\bar{a}_0 = \frac{v_s}{T} \tag{7.105}$$

we find the temperature dependence of the coupling constant u to be

$$\frac{1}{u(T)} = \frac{1}{u_0} + \frac{1}{2\pi} \ln\left(\frac{a_0 T}{v_s}\right) \tag{7.106}$$

Equivalently, we can write

$$u(T) = \frac{u_0}{1 + (u_0/(2\pi))\ln(a_0 T/v_s)} \tag{7.107}$$

Thus, at high temperatures, $T \gg v_s/a_0$, we find that the effective coupling $u(T)$ becomes small,

$$u(T) \approx \frac{2\pi}{\ln(a_0 T/v_s)} \to 0 \quad \text{for} \quad T \to \infty \tag{7.108}$$

In other words, the effective coupling at short distances or at high temperatures is *small*. This result is known as *asymptotic freedom* and, in this context, was first discussed by Polyakov (1975).

Conversely, as the temperature is lowered, the effective coupling u becomes large (Fig. 7.13). Equation (7.107) exhibits an apparent divergence at a temperature T_0, where

$$T_0 \approx \frac{v_s}{a_0} e^{-\frac{2\pi}{u_0}} = \frac{v_s}{a_0} e^{-\pi S} \tag{7.109}$$

The meaning of T_0 is that of the temperature at which the weak-coupling (i.e. $1/S$) expansion breaks down. To continue down to lower temperatures, we must take into account the fact that for $T \leq T_0$ the sigma model has a large effective coupling. At this point we notice that, at large values of the coupling constant, the sigma model is disordered no matter what the dimensionality of space-time is. Thus we expect a finite, and short, correlation length ξ and a finite mass (or energy) gap $\Delta = v_s/\xi$. The effective coupling should saturate due to lattice effects and the constraint $\vec{m}^2 = 1$. These ideas have been confirmed by Monte Carlo RG studies (Shenker and Tobochnik, 1980).

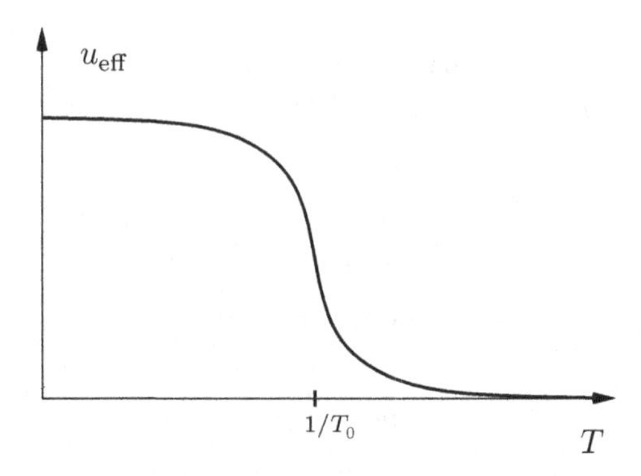

Figure 7.13 Crossover of the effective (or *running*) coupling $u_{\rm eff}(T)$. Here T is the temperature.

We can also use the RG to estimate the dependence of the correlation length ξ on the bare coupling constant $u_0 = 2/S$. Under the RG, the correlation length ξ, like all other physical observables, remains invariant. From dimensional analysis we expect ξ, which is a length scale, to have the form

$$\xi(u) = a_0 f(u) \tag{7.110}$$

where f is a function of u, the coupling constant *at the scale* a_0. Being an RG invariant, the correlation length ξ must obey

$$a_0 \frac{d\xi}{da_0} = 0 \tag{7.111}$$

which implies that $f(u)$ satisfies the differential equation

$$\beta(u)\frac{df}{du} + f(u) = 0 \tag{7.112}$$

The solution to Eq. (7.112) is

$$f(u) = f(u')\exp\left(-\int_{u'}^{u} \frac{dz}{\beta(z)}\right) = f(u')\exp\left[2\pi\left(\frac{1}{u} - \frac{1}{u'}\right)\right] \tag{7.113}$$

where u and u' are connected by the RG flow.

Consider now the correlation length ξ at *two* different values of u, namely u_1 and u_2, for the *same* value of the lattice constant a_0. Let u^* be a large reference value of the coupling u. From Eqs. (7.110) and (7.113), we find that the correlation length ξ obeys

$$\xi(u_i) = a_0 f(u_i) = a_0 f(u^*)\exp\left(-\int_{u^*}^{u_1} \frac{du}{\beta(u)}\right) \tag{7.114}$$

for $i = 1, 2$. Thus the *ratio* of two values of ξ for two different couplings and the same lattice spacing, a_0, is given by

$$\frac{\xi(u_1)}{\xi(u_2)} = \frac{a_0 f(u^*) \exp\left(-\int_{u^*}^{u_1} \frac{dz}{\beta(z)}\right)}{a_0 f(u^*) \exp\left(-\int_{u^*}^{u_2} \frac{dz}{\beta(z)}\right)} \tag{7.115}$$

Thus, we get

$$\frac{\xi(u_1)}{\xi(u_2)} = \exp\left(-\int_{u_2}^{u_1} \frac{dz}{\beta(z)}\right) \tag{7.116}$$

The integral can be easily evaluated to find

$$\frac{\xi(u_1)}{\xi(u_2)} = \exp\left(\frac{2\pi}{u_1} - \frac{2\pi}{u_2}\right) \tag{7.117}$$

For the case in which $u_1 = u_0 = 2/S$ and u_2 is large, we find that

$$\xi(u_0) \approx \xi(u_2) e^{\pi S} \tag{7.118}$$

What value should we assign to $\lim_{u_2 \to \infty} \xi(u_2)$? The answer depends on whether the spin is integer or half-integer.

Integer spin. In this case we do not get a topological term. As was emphasized above, the sigma model is always disordered at strong coupling. Thus, we expect $\xi(u_2) \approx a_0$ and we find a *finite* correlation length

$$\xi_0 = \xi(u_0) \approx a_0 e^{\pi S} \tag{7.119}$$

There is no long-range order (i.e. no Néel state). The spectrum has a gap

$$\Delta = \frac{v_s}{\xi_0} \tag{7.120}$$

and the ground state is unique. Equation (7.119) shows that the correlation length is non-perturbative in the $1/S$ expansion.

Half-integer spin. The sigma-model coupling constant u still scales to strong coupling but the topological term remains unchanged at the value $\theta = 2\pi S$ (mod 2π). However, the coupling constant $g \propto u$ is related to the spin through $S = 2/g$. Thus strong coupling is equivalent to low spin. Hence the behavior of *all* half-integral-spin chains is qualitatively identical to the spin one-half case for which $u_0 \propto 4$. The spin one-half case is *gapless*, as we saw from the Bethe-ansatz and other approaches. Thus, $\xi(\infty)$ is still infinite. All half-integral-spin chains are at a critical point with infinite correlation length. At first sight, this result seems to be paradoxical. We started with smooth configurations with well-defined winding numbers and a weak coupling g. As the energy scale was lowered the effective coupling of the sigma model grew *but* the topological coupling remained unaffected.

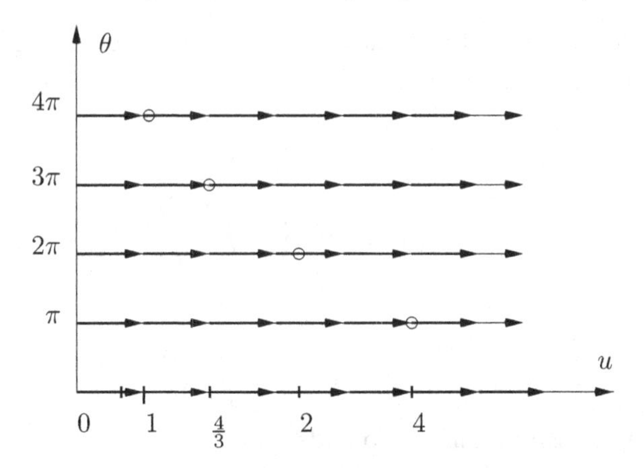

Figure 7.14 Schematic RG flows of quantum Heisenberg antiferromagnetic chains for $S = \frac{1}{2}, 1, \frac{3}{2}, 2$. The open circles represent their sigma-model bare coupling constant $u \approx 2/S$. They all iterate to $u^* = \infty$. The difference in their behavior is a consequence of the presence of a $\theta = \pi$ term in the sigma model for the half-integer-spin chains. Notice that the value of θ *does not flow under the RG*.

Thus, at low energies, the configurations become *rough* and the actual meaning of the topological term in this situation is unclear. This poses no problems for the *integer*-spin chains since the topological term does not contribute in this case ($\theta = 2\pi S$). In contrast, for half-integer-spin chains, this result simply means that, for *all* half-integer values of $S > \frac{1}{2}$, the systems behave *qualitatively* in the same way as in the $S = \frac{1}{2}$ case.

This result, namely S integer is disordered and S half-integer is critical, is known as Haldane's conjecture (Haldane, 1983a, 1983c). It has also been checked by accurate numerical calculations using exact diagonalization on finite (but large) chains (Moreo, 1987; Ziman and Schulz, 1987) and by Green-function Monte Carlo simulations (Liang, 1990b). Affleck and Haldane (1987) have also found the same result using non-abelian bosonization (Witten, 1984). The RG flows are shown in Fig. 7.14.

7.9 Hopf term or no Hopf term?

The 1D spin chains have a very unusual behavior: disorder (integer spin) or critical (half-integer spin) ground states, neutral fermions that are massless for the half-integer case and massive for S integer, etc. There is nothing in this picture that is remotely close to the physics that emerges from the mean-field theory of Chapter 3. It is then natural to ask whether or not this picture is peculiar to 1D systems or whether there is a natural generalization to higher dimensions. It is a trivial matter to generalize the 1D formalism to the case of a square lattice. The lattice action is a simple generalization of Eq. (7.57). Let $\vec{r}$ span a square lattice of size $N \times N : \vec{r} = (x_1, x_2)$, where $x_1, x_2 = 1, \ldots, N$. I will assume that N is even. The action is

$$\mathcal{S}_{\rm M}[\vec{n}] = S \sum_{\vec{r}} \mathcal{S}_{\rm WZ}[\vec{n}(\vec{r})] - \int_0^T dx_0 \sum_{\langle \vec{r}, \vec{r}\,' \rangle} JS^2 \vec{n}(\vec{r}, x_0) \vec{n}(\vec{r}\,', x_0) \tag{7.121}$$

where $\vec{r}$ and $\vec{r}\,'$ are nearest-neighboring sites on the square lattice. Since the square lattice is bipartite and we expect short-range Néel order, we will once again stagger the field configurations and find

$$\mathcal{S}_{\rm M}[\vec{n}] = S \sum_{\vec{r}} (-1)^{x_1+x_2} \mathcal{S}_{\rm WZ}[\vec{n}(\vec{r})] + \int_0^T dx_0 \sum_{\langle \vec{r}, \vec{r}\,' \rangle} JS^2 \vec{n}(\vec{r}, x_0) \vec{n}(\vec{r}\,', x_0) \tag{7.122}$$

It is straightforward, but tedious, to derive the effective action for the slowly varying fields. Once again, on the basis of symmetry, we expect a non-linear sigma model. The issue is whether or not there is a topological term in the effective action.

Before deriving the effective action by an explicit calculation, let us consider what topological terms are possible. In the $(1+1)$-dimensional case we saw that the configurations were classified in terms of an index, the topological charge, which labels the homotopy class of the configuration. The existence of such an index was guaranteed by the fact that the configurations fall into homotopy classes that form the group $\pi_2(S_2)$ of smooth maps of the 2D Euclidean space-time S_2 into the S_2 of the order-parameter manifold. This homotopy group $\pi_2(S_2)$ is isomorphic to $\mathbb{Z}$, the group of integers, i.e. the winding numbers of the topological classes. In $(2+1)$ dimensions the situation is rather different. Once again, the Euclidean space-time can be regarded as a sphere S_3 and the configurations are maps of $S_3 \to S_2$.

However, there are no smooth solutions of the classical Euclidean equations of motion with non-trivial winding numbers. There are *singular* solutions known as hedgehogs (see Fig. 7.8(b)), which have a linearly divergent action. Haldane has argued that these hedgehogs may become relevant if the sigma model becomes disordered by some mechanism (Haldane, 1988b). In the next chapter, we will see that a next-nearest-neighbor antiferromagnetic interaction can trigger a quantum phase transition to a dimerized state, a state with a spin gap that does not break the SU(2) symmetry of spin and breaks translation invariance and the point-group symmetry of the lattice. A scenario known as deconfined quantum criticality (Senthil *et al.*, 2004a), in which hedgehog configurations may play a central role in this quantum phase transition, has been proposed.

On the other hand, there are non-trivial configurations in Minkowski space-time (i.e. in real time). Consider at some time $t = t_0$ a configuration of sigma-model fields identical to one of the instantons of Section 7.6, see Fig. 7.8(a). Now it represents the snapshot of an eigenstate, a soliton known as a *skyrmion*. Thus the *configuration space* of a 2D quantum non-linear sigma model is also a sphere S_2 and is usually denoted by $\Omega_2 S_2$. Consider now the real-time evolution of such a state with periodic boundary conditions in time, i.e. consider histories in which the initial state is the same as the final state. Thus, a history is a closed curve in the configuration space $\Omega_2 S_2$. In quantum mechanics we are told to sum over all histories and to assign a phase to each history, i.e. to each curve in $\Omega_2 S_2$. Since a phase is an element of S_1 (the unit circle) we have constructed the set of maps $\pi_1(\Omega_2 S_2)$. However, we know that the configurations at any given time are maps of S_2 (space) into S_2 (field), i.e. homotopy classes of $\pi_2(S_2)$, which we saw was isomorphic to the group of integers $\mathbb{Z}$. Hence the configuration space $\Omega_2 S_2$ is decomposed into a disjoint union of path-connected pieces, each characterized by the winding number or soliton number Q. Thus each disconnected piece of the Hilbert space will have a separate time evolution and will have to be summed with separate phases. Since the classical paths are continuous curves in $\Omega_2 S_2$ classified by $\pi_3(S_2) = \mathbb{Z}$, the relevant issue is now what topological invariant is associated with such histories.

Consider a history of the order-parameter field $\vec{m}(\vec{x}, t)$ in $(2+1)$ dimensions. We can define a *topological current* J_μ by

$$J_\mu = \frac{1}{8\pi} \epsilon_{\mu\nu\lambda} \epsilon_{abc} m_a \, \partial^\nu m_b \, \partial^\lambda m_c \tag{7.123}$$

with $\mu = 0, 1, 2$ and $a, b, c = 1, 2, 3$. The topological current J_μ is clearly conserved,

$$\partial^\mu J_\mu = 0 \tag{7.124}$$

Therefore the total *topological charge* $Q = \int d^2x \, J^0(\vec{x}, t)$ is constant in time,

$$Q = \int d^2x \; J^0(\vec{x}, t) = \int d^2x \; \frac{1}{8\pi} \epsilon_{0ij} \epsilon_{abc} m_a \, \partial^i m_b \, \partial^j m_c \tag{7.125}$$

Clearly Q is identical to the winding number Q of Eq. (7.75).

Consider now a soliton state with $Q = 1$ (Fig. 7.8(a)). Imagine a time evolution in which the soliton skyrmion rotates slowly around its center and executes exactly n turns during its lifespan. Each point on the equator of the soliton traces a curve ("worldline") which wraps n times around the other curves traced by the other points, a "world-tube" (Fig. 7.15). An easy way to compute the winding number of this history is to imagine that each worldline is a wire carrying a unit of current. As the soliton rotates, the worldlines ("wires") are *braided*. The natural topological invariant is the *linking number* of these worldlines (Fig. 7.16). If we denote by $\vec{j}$ the current carried by the wires and by $\vec{B}$ the magnetostatic field they create, the linking number is simply given by Ampère's law

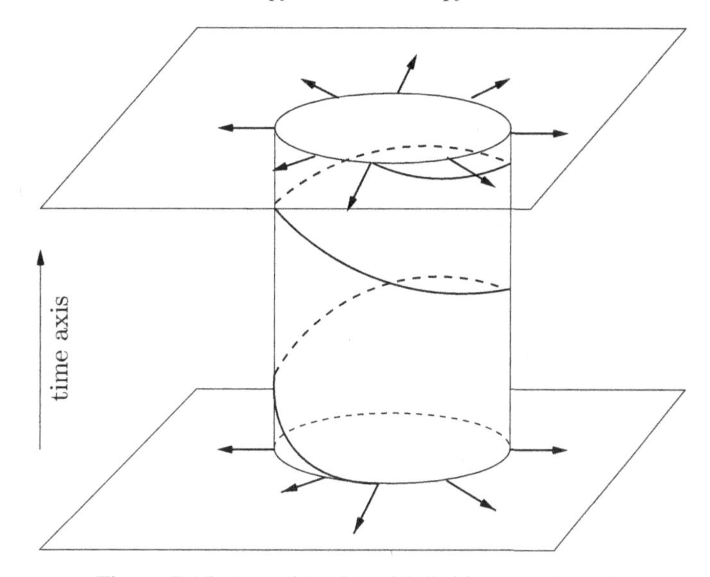

Figure 7.15 A world-tube with linking number +2.

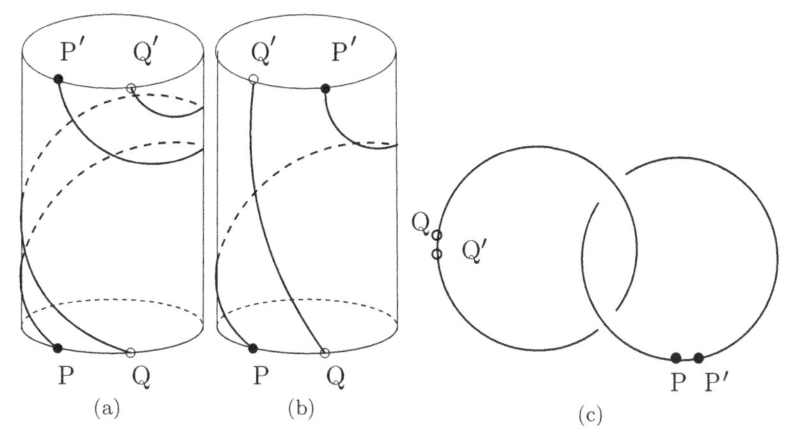

Figure 7.16 (a) Two worldlines PP′ and QQ′ with linking number +1. (b) Smooth deformation of the worldline PP′ and QQ′. (c) Periodic boundary conditions in time are enforced, P and P′ as well as Q and Q′ are identified and the worldlines form a braid.

$$\int d^3x \, \vec{j} \cdot \vec{B} = 2\pi n \tag{7.126}$$

where n is an integer that counts the number of turns.

We can make this analogy more precise by using the *Hopf map*, which maps the 3-sphere S_3 of Euclidean space-time onto the 2-sphere S_2 of the target manifold of the O(3) non-linear sigma model. Let z_1 and z_2 be two complex numbers satisfying

$$|z_1|^2 + |z_2|^2 = 1 \tag{7.127}$$

Clearly (z_1, z_2) span the 3-sphere S_3. Define now the *spinor* z_α ($\alpha = 1, 2$). The order-parameter field $\vec{m}$ is related to the complex spinor z_α through the map

$$m^a = z_\alpha^* \sigma_{\alpha\beta}^a z_\beta \tag{7.128}$$

where $\{\sigma^a\}_{a=1,2,3}$ are the Pauli matrices. The order parameter $\vec{m}$ also satisfies $\vec{m}^2 = 1$. This is the Hopf map.

It is clear that (z_1, z_2) has three independent parameters, whereas $\vec{m}$ has only two. But one of these parameters, or degrees of freedom, is unobservable since a global charge of phase of the spinor z_α

$$\begin{pmatrix} z_1 \\ z_2 \end{pmatrix} \to e^{i\phi} \begin{pmatrix} z_1 \\ z_2 \end{pmatrix} \tag{7.129}$$

does not lead to any observable effects because $\vec{m}$ is *invariant* under such *gauge transformations*.

Furthermore, the action of the non-linear sigma model itself can be written in terms of the spinor field z_α. This is the CP^1 model. To simplify matters I will consider the problem with spin-wave velocity $v_s = 1$. Let z_α be a CP^1 field and A_μ an unconstrained gauge field, with Lagrangian density

$$\mathcal{L}_{CP^1} = \frac{1}{2g} |D_\mu z|^2 \tag{7.130}$$

where g is a coupling constant and D_μ is the covariant derivative

$$D_\mu = \partial_\mu - i A_\mu \tag{7.131}$$

The functional integral is

$$Z = \int \mathcal{D}\bar{z} \mathcal{D}z \, \delta(|z|^2 - 1) \int \mathcal{D}A \, e^{iS_{CP^1}[z,A]} \tag{7.132}$$

Since $\mathcal{L}_{CP^1}$ is quadratic in the gauge field A_μ, it can be integrated out exactly by a saddle-point calculation. The saddle-point condition

$$\frac{\delta \mathcal{L}_{CP^1}}{\delta A_\mu} = 0 \tag{7.133}$$

yields the gauge field as a function of the CP^1 field:

$$A_\mu = \frac{i}{2}(z_\alpha^* \, \partial_\mu z_\alpha - z_\alpha \, \partial_\mu z_\alpha^*) \equiv -\frac{i}{2} z_\alpha^* \, \overleftrightarrow{\partial_\mu} \, z_\alpha \tag{7.134}$$

By substituting Eq. (7.134) into the Lagrangian density, Eq. (7.130), one finds

$$\frac{1}{2g}(\partial_\mu \vec{m})^2 = \frac{1}{g}|D_\mu z|^2 \tag{7.135}$$

In other words, the CP^1 model and the O(3) non-linear sigma model are equivalent.

The topological invariant, or *Hopf invariant*, has a very simple and natural form in terms of the vector potentials A_μ. Consider a term in the Lagrangian density of the form

$$\mathcal{L}_{\rm CS} = \frac{\theta}{4\pi}\epsilon_{\mu\nu\lambda}A^\mu F^{\nu\lambda} \tag{7.136}$$

which is known as a Chern–Simons term. The gauge field A_μ is *constrained* to be given by Eq. (7.134), and its field strength $F_{\mu\nu}$ can be related back to the sigma-model field $\vec{m}$ by

$$F_{\mu\nu} = \partial_\mu A_\nu - \partial_\nu A_\mu \equiv \vec{m}\cdot(\partial_\mu \vec{m}\times \partial_\nu \vec{m}) \tag{7.137}$$

Thus, the flux associated with the gauge field A_μ is simply related to the topological current. The *Hopf invariant* H is simply

$$H = \frac{\theta}{8\pi}\int d^3x\, \epsilon_{\mu\nu\lambda}A^\mu F^{\nu\lambda} \tag{7.138}$$

with A_μ and $F_{\nu\lambda}$ given by Eqs. (7.134) and (7.137). We will see in Chapter 9 that a non-zero value of θ will change the statistics of the solitons (skyrmions).

But is there a Hopf term in the effective action of the quantum Heisenberg antiferromagnet in a 2D square lattice? The only way to determine that is to compute the effective action carefully. Dzyaloshinskii, Polyakov and Wiegmann (Dzyaloshinskii *et al.*, 1988; Wiegmann, 1988) have conjectured that the effective action of the quantum antiferromagnet is a non-linear sigma model with a Hopf term with $\theta = 2\pi S$. This is a subtle business since Wu and Zee have shown that, in its CP^1 form, the Hopf term is a total derivative that does not alter the equations of motion but changes the spin and statistics of the topological excitations (Wu and Zee, 1984).

To see whether a Hopf term does (or does not) arise, let us first derive the effective action, following the methods of Fradkin and Stone (1988). The result will be a sigma model *without* a topological term (Dombre and Read, 1988; Fradkin and Stone, 1988; Haldane, 1988b; Ioffe and Larkin, 1988; Wen and Zee, 1988).

First, we need to integrate out the fast degrees of freedom. We write

$$\vec{n}(\vec{r}) = m(\vec{r}) + (-1)^{x_1+x_2}a_0\vec{l}(\vec{r}) \tag{7.139}$$

Following the *same* procedure as that which in the 1D case led to a sigma model *with* a topological term (see Eq. (7.67)), we find

$$\mathcal{L}_{\text{eff}}^{\text{M}}(\vec{m}, \vec{l}) = -\frac{JS^2}{2}\left((\partial_i \vec{m})^2 + 8\vec{l}^{\,2}\right) + \frac{S}{a_0}\vec{l} \cdot (\vec{m} \times \partial_0 \vec{m}) \tag{7.140}$$

If we now proceed to integrate out the fast modes, the $\vec{l}$ field, we find a non-linear sigma model *without* a topological term. The (bare) coupling constant and (bare) spin-wave velocity are given by (see Eqs. (7.69) and (7.70))

$$\begin{aligned} g &= \sqrt{2}a_0 \frac{2}{S} \\ v_s &= \sqrt{2}2a_0 JS \end{aligned} \tag{7.141}$$

The terms which in the 1D case gave rise to the topological term now have cancelled each other out (at least for smooth configurations). The reason for this cancellation can be traced back to the staggered character of the Néel state. Naively, we expect that each row will make a contribution similar to the 1D result. But neighboring rows are staggered in the opposite way. The result is that the terms originating from each pair of neighboring rows now effectively cancel out. We are assuming a lattice with an even number of rows and columns. In the case of an *odd* number of rows, we may get a non-zero contribution from the last row. However, this is a boundary-condition effect which, incidentally, was not needed in the case of the chains. But we do expect to see changes in the spectrum of elementary excitations if we change the boundary conditions.

The argument which led to the cancellation is a bit too naive and maybe it is dangerous. We know from the work of Wu and Zee that, at least in the CP^1 representation, the Hopf term is a total derivative. Thus, a local cancellation is not a sufficient argument for the study of a *global* effect. Slowly varying configurations may have an accumulated effect near the boundaries and yield a non-zero answer. We can check this by computing the alternating sum Φ,

$$\Phi = s \sum_{\vec{r}} (-1)^{x_1 + x_2} S_{\text{WZ}}[\vec{n}(\vec{r})] \tag{7.142}$$

for a configuration that, in the continuum limit, has soliton number $Q = 1$. If we let this soliton configuration rotate slowly around its center such that it turns exactly once during its history, the history of this configuration should have Hopf number or linking number $+1$. We should choose a lattice configuration that, in the limit of soliton radius r_s *large* compared with the lattice spacing a_0, should go smoothly over to the continuum soliton. Any soliton profile should do the job. For instance, we can imagine a configuration obtained by a stereographic configuration

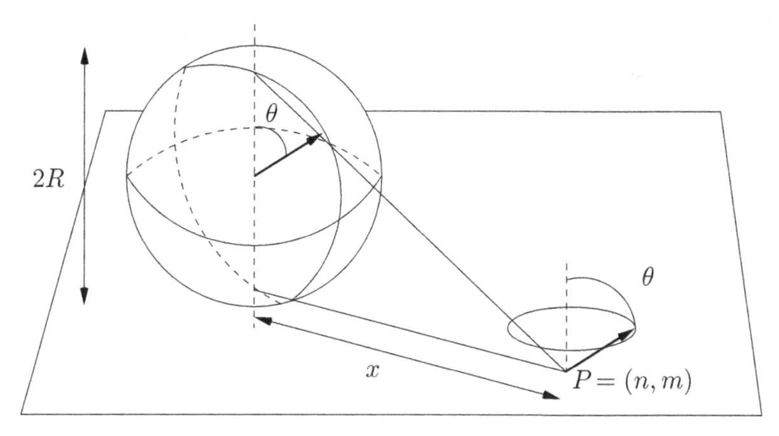

Figure 7.17 A soliton configuration can be generated using a stereographic projection. The spin is parallel-transported from the sphere to point $P = (n, m)$. Its history is pictured as an ellipse.

(Fig. 7.17). The area swept by each spin is $a = 2\pi R^2(1 - \cos\theta)$. Thus the sum $\Phi(N)$, for a system of size $2N \times 2N$, is given by

$$\Phi(N) = \sum_{n,m=-N+1}^{N} 2\pi s R^2(1 - \cos\theta(n, m))(-1)^{n+m} \tag{7.143}$$

The sphere has radius R and its south pole, which has coordinates (α_1, α_2), is in the first unit cell. The radius of the soliton r_s is equal to the diameter $2R$ of the sphere, if we define the radius as the location in which the spins are orthogonal to the asymptotic configuration at spatial infinity. Hence we find

$$\Phi(N) = \sum_{n,m=-N+1}^{N} \frac{16\pi s R^4 e^{i\pi(n+m)}}{4R^2 + (n - \alpha_1)^2 + (m - \alpha_2)^2} \tag{7.144}$$

In the thermodynamic limit, $N \to \infty$, and by making use of the Poisson summation formula

$$\sum_{n=-\infty}^{+\infty} f(n) = \sum_{n=-\infty}^{+\infty} \int_{-\infty}^{+\infty} \frac{dk}{2\pi} e^{i2\pi kn} f(k) \tag{7.145}$$

we get for $\Phi \equiv \lim_{N\to\infty} \Phi(N)$

$$\Phi = \int \frac{d^2k}{(2\pi)^2} \sum_{\vec{n}} \frac{16\pi s R^4 e^{i2\pi(\vec{k}+\vec{\alpha})\cdot(\vec{n}+\vec{G})}}{4R^2 + \vec{k}^2} \tag{7.146}$$

where $\vec{G} = (\frac{1}{2}, \frac{1}{2})$. In the limit $R \to \infty$ it is easy to see that Φ is exponentially small since we can write

$$\Phi = \sum_{n_1,n_2=-\infty}^{+\infty} 8s R^4 e^{i2\pi\vec{\alpha}\cdot(\vec{n}+\vec{G})} K_0(\pi r_s|\vec{n} + \vec{G}|) \tag{7.147}$$

where $K_0(x)$ is the modified Bessel function. Thus, for $r_s \gg 1$, we may keep just the leading terms:

$$\Phi \approx 4s\left(\frac{r_s}{\sqrt{2}}\right)^{5/2} e^{-\frac{2\pi}{\sqrt{2}}r_s + i2\pi(\alpha_1+\alpha_2)} \cos(\pi\alpha_1)\cos(\pi\alpha_2) \tag{7.148}$$

This expression vanishes exponentially fast for solitons with radius $r_s \gg 1/(\pi\sqrt{2})$. Notice that even fairly small solitons with radius $r_s \approx 1$ are "large" according to this criterion. We must conclude that if we expect to see Néel order (even if this were true only at short distances!) the effective theory at long distances is given by a non-linear sigma model with renormalized coupling constant and spin-wave velocity. Phenomenologically this is what the experiments in La_2CuO_4 indicate (Chakravarty *et al.*, 1988). In Section 7.7 we calculated the one-loop β-function for the non-linear sigma model in $(2 + \epsilon)$ dimensions (here 2 means $(1 + 1)$). We found the result (see Eq. (7.101))

$$\beta(u) = -\epsilon u + \frac{u^2}{2\pi} + O(u^3) \tag{7.149}$$

For space-time dimensions $D > 2$, the fixed point at the origin, $u^* = 0$, is *infrared-stable*. This means that, if the bare dimensionless coupling constant u is sufficiently small, the effective coupling flows toward the $u = 0$ fixed point and we have a Néel state with weakly coupled spin waves. However, for space-time dimensions $D > 2$ $(\epsilon > 0)$, Eq. (7.101) has another fixed point at $u^* \approx 2\pi\epsilon$, which is *infrared-unstable*. This fixed point is the location of a second-order (continuous) quantum phase transition as a function of the coupling constant. Beyond this fixed point, i.e. for $u > u^*$, the effective coupling flows toward the $u = \infty$ fixed point just as in the $(1+1)$-dimensional case. However, now we no longer have a topological term. Thus, we must conclude that, for $u > u^*$, the system is disordered (and hence has a finite energy gap) at distances longer than some correlation length $\xi \sim |u - u^*|^{-\nu}$, where $\nu = 1/(D-2) + O(D-2)$, and Néel-like order at scales between the lattice constant a_0 and the correlation length ξ. Such a state is a *zero-temperature quantum paramagnet* (QP), i.e. a paramagnetic state driven purely by quantum fluctuations, with the absence of thermal fluctuations. A finite correlation length without long-range order means that the ground state is unique, and there is an energy gap $\Delta = v_s/\xi$ for the elementary excitations ("spin waves").

The theory described here, which is based on the $(2+\epsilon)$ expansion, is too crude to reliably predict the value of u^*, and hence of the gap. Since we saw that our approximations were equivalent to (a resummation of) the $1/S$ expansion, we must also conclude that u^* cannot be calculated with confidence from the $1/S$ expansion either. Qualitatively, we should still expect a non-trivial fixed point for $\epsilon = 1$. The perturbative β-function predicts that for $\epsilon \approx 1$ even $S = \frac{1}{2}$ on a square lattice is on the Néel side of the phase transition, although not far from it. This result appears to be consistent with existent experimental data on quasi-2D systems believed to be reasonably well described by the $S = \frac{1}{2}$ quantum Heisenberg antiferromagnet such as La_2CuO_4. Experimentally (Shirane *et al.*, 1987) one sees a Néel state but with a magnetic moment about 50% of its classical value. The dynamical structure factor predicted by the non-linear sigma model (Chakravarty *et al.*, 1988) is also consistent with these experiments. Numerical calculations on 2D quantum Heisenberg models also exhibit a similar behavior (Liang *et al.*, 1988; Liang, 1990a; Manousakis, 1991).

7.10 The Wess–Zumino–Witten model

We will now go back to the problem of the quantum Heisenberg antiferromagnet in one dimension. In Section 7.5 we found that the effective action for a model with spin degrees of freedom in the spin-S representation is described by a non-linear sigma model with a topological θ term, Eq. (7.68). Here $\theta = 2\pi S$, which is an even multiple of π for S integer and an odd multiple of π for S a half-integer. We saw that the role of the topological term was to weight by $\exp(i\theta Q)$ the contribution to the path integral of configurations with topological charge Q. Since the topological charge is an integer the weight is equal to 1 for integer-spin chains and $(-1)^Q$ for half-integer-spin chains.

A subsequent RG analysis showed that the non-linear sigma model is asymptotically free and hence that the effective coupling constant always flows to strong coupling. For integer-spin chains this implies that there is a finite mass (or energy) gap in the spectrum and that there is no long-range antiferromagnetic order. On the other hand, for half-integer-spin chains the coupling constant still flows to strong coupling under the RG flow, but the topological term does not, since (being topological) it is unaffected by local fluctuations. Hence all half-integer-spin chains flow to the same strong-coupling fixed point, which we identified as that of the spin-$1/2$ chain, which is gapless (and hence critical) according to the Bethe ansatz. This is the essence of Haldane's result.

This result is very elegant and clearly shows why the two classes of spin chains must behave differently, but it does not identify the non-trivial fixed point for the half-integer-spin chains. In this section we will show (Affleck, 1986a; Affleck and

Haldane, 1987) that this fixed point is the $SU(2)_1$ ("level 1") Wess–Zumino–Witten (WZW) model (Witten, 1984), which is a scale-invariant field theory in $(1 + 1)$ dimensions that is exactly solvable using methods of conformal field theory (CFT) (Knizhnik and Zamolodchikov, 1984). We will not go over the full structure of CFT, a subject for which there are extensive reviews (Belavin *et al.*, 1984; Di Francesco *et al.*, 1997), but instead we will discuss the basic results and their physical consequences. Since all half-integer-spin chains flow to the same fixed point, it will be sufficient to study this question for the spin-1/2 chain. The key tool in the construction of the fixed-point theory of the spin-1/2 chain is the concept of non-abelian bosonization (Witten, 1984).

We begin with a brief description of non-abelian bosonization. As in the abelian case, discussed in detail in Section 5.6, non-abelian bosonization expresses the algebra of the currents and densities, in this case the spin currents and densities. We will seek a bosonic theory with the same global symmetries as the fermionic theory and whose currents satisfy the same algebra as that satisfied by their fermionic cousins. Thus we consider a system with two chiral fermionic fields $\psi_{\mathrm{R},\sigma}$ and $\psi_{\mathrm{L},\sigma}$ (with $\sigma = \uparrow, \downarrow$) and the associated spin currents $J_{\mathrm{R}}^a(x) = \frac{1}{2}\psi_{\mathrm{R},\sigma}^\dagger(x)\tau_{\sigma,\sigma'}^a\psi_{\mathrm{R},\sigma'}(x)$ and $J_{\mathrm{L}}^a(x) = \frac{1}{2}\psi_{\mathrm{L},\sigma}^\dagger(x)\tau_{\sigma,\sigma'}^a\psi_{\mathrm{L},\sigma'}(x)$ (with $a = 1, 2, 3$) (Eq. (6.69)), which satisfy the $SU(2)_1$ Kac–Moody (current) algebra of Eq. (6.71),

$$\begin{aligned}\left[J_{\mathrm{R}}^a(x), J_{\mathrm{R}}^b(y)\right] &= i\epsilon^{abc}J_{\mathrm{R}}^c(x)\delta(x-y) + i\frac{k}{4\pi}\delta^{ab}\delta'(x-y)\\ \left[J_{\mathrm{L}}^a(x), J_{\mathrm{L}}^b(y)\right] &= i\epsilon^{abc}J_{\mathrm{L}}^c(x)\delta(x-y) - i\frac{k}{4\pi}\delta^{ab}\delta'(x-y)\end{aligned} \tag{7.150}$$

with level $k = 1$. Since we are dealing with charged (Dirac) fermions, there are also two chiral currents associated with the global $U(1)$ symmetry, which we associate with charge and hence with gauge invariance. In the more general case in which there are N chiral fermions (instead of two for the case of spin), the symmetry becomes $SU(N)$. In this case the ϵ^{abc} tensor in Eq. (7.150) (the same as Eq. (6.71)) is replaced by the structure constants f^{abc} of $SU(N)$.

Since both up and down fermions contribute to the $U(1)$ current, the associated $U(1)$ Kac–Moody algebra has level $k = 2$. The total symmetry of the free fermions with spin is thus $U(1)_2\times SU(2)_1$. Equivalently, we can represent each Dirac fermion in terms of a pair of (neutral) Majorana fermions, $\psi_{\mathrm{R},\sigma}(x) = \chi_{\mathrm{R},\sigma}(x) + i\eta_{\mathrm{R},\sigma}(x)$ (and similarly for the left-moving fields). We can regard our set of $2N$ ($N = 2$ for $SU(2)$) chiral Majorana fermions as forming a vector of $2N$ components. The symmetry is now (in general) $O(2N) \simeq U(1) \times SU(N)$ and the fermions transform as the $2N$ (fundamental) representation of $O(2N)$.

Following Witten (1984) we seek a representation of this current algebra in terms of a set of bosonic fields. (An alternative derivation of non-abelian bosonization using an analysis of field-theoretic anomalies in the path-integral formulation was developed by Polyakov and Wiegmann (1983, 1984).) However, unlike in the abelian case, the bosonic fields are not free. Rather the bosonized theory is a special type of non-linear sigma model whose fields will be denoted by $g(x)$, which, in the case of electrons with spin 1/2, are 2×2 matrices that take values on the elements of the group SU(2). In the general case the matrix field $g(x)$ takes values on a compact Lie group G. Thus, the non-linear sigma model will in general be given in terms of a matrix-valued field, i.e. an element of a compact Lie group at every point of space-time. In quantum field theory, a non-linear sigma model taking values on a group manifold is known as the principal chiral field (Polyakov, 1987). On the other hand, a matrix-valued field has a larger $G_{\rm L} \times G_{\rm R}$ symmetry $g(x) \to h_{\rm L} g(x) h_{\rm R}^{-1}$ since the field can be multiplied by two independent constant matrices $h_{\rm R}^{-1}$ (multiplying on the right), with $h_{\rm R} \in G_{\rm R}$, and $h_{\rm L}$ (multiplying on the left), with $h_{\rm L} \in G_{\rm L}$. We will see that these two symmetries are associated with the two chiral currents.

In what follows we will discuss a theory of N free Dirac (charged) Fermi fields $\psi_{\rm R}^i(x)$ and $\psi_{\rm L}^i(x)$, with $i = 1, \ldots, N$, in terms of which we can construct a U(1) current

$$J_{\rm R} = \psi_{\rm R}^{i\dagger} \psi_{\rm R}^i, \qquad J_{\rm L} = \psi_{\rm L}^{i\dagger} \psi_{\rm L}^i \tag{7.151}$$

and the SU(N) currents

$$J_{\rm R}^a = \psi_{\rm R}^{i\dagger} t_{ij}^a \psi_{\rm R}^j, \qquad J_{\rm L}^a = \psi_{\rm L}^{i\dagger} t_{ij}^a \psi_{\rm L}^j \tag{7.152}$$

where $\{t_{ij}^a\}$ are $N \times N$ (hermitian) matrices, the $N^2 - 1$ generators of SU(N) (in the fundamental representation). (We use the normalization $\operatorname{tr}(t^a t^b) = \frac{1}{2}\delta_{ab}$.)

The Lagrangian for free massless Dirac fermions in $(1+1)$ dimensions is

$$\mathcal{L} = \bar{\psi}_i^\dagger(x) i \gamma_\mu \partial^\mu \psi_i(x) \tag{7.153}$$

where $\bar{\psi}_i = \psi_i^\dagger \gamma_0$. The free massless Dirac fermion Lagrangian of Eq. (7.153) is a fixed point under the renormalization group and it represents a scale-invariant system.

For free (massless) fermions the chiral currents $J_{\rm R}^{ij}$ and $J_{\rm L}^{ij}$ obey the conservation laws

$$\partial_- J_{\rm R}^{ij} = 0, \qquad \partial_+ J_{\rm L}^{ij} = 0 \tag{7.154}$$

where I use so-called light-cone components with $x_\pm = (1/\sqrt{2})(x_0 \pm x_1)$ and $\partial_\pm = \partial/\partial x_\pm$. These conservation laws can be satisfied by writing (Witten, 1984)

$$J_{\rm R}(x) = \frac{i}{2\pi} g^{-1}(x) \partial_+ g(x), \qquad J_{\rm L}(x) = -\frac{i}{2\pi} (\partial_- g(x)) g^{-1}(x) \tag{7.155}$$

where I have suppressed the indices i, j. The two conservation laws are now $\partial_-(g^{-1}\partial_+ g) = 0$ and $\partial_+((\partial_- g) g^{-1}) = 0$ (which are equivalent to each other).

Now we ask which action the scalar (bosonic) field $g(x)$ should obey. Witten showed that the simple guess that it should be the action of the principal chiral-field non-linear sigma model (with coupling constant λ)

$$\mathcal{L} = \frac{1}{4\lambda^2} \int d^2x \, \text{tr}\left(\partial_\mu g \, \partial^\mu g^{-1}\right) \tag{7.156}$$

(where tr is a matrix trace) does not work, for several reasons. One important reason is that the RG beta function is in general non-vanishing (unless the (Ricci) curvature of the group manifold G vanishes and the manifold is then said to be "Ricci flat") (Friedan, 1985) and hence it is not scale-invariant. For $G = \text{SU}(N)$ the beta function is positive, $\beta(\lambda) > 0$, and hence this theory is asymptotically free. Thus, for any $\lambda > 0$, the theory of Eq. (7.156) represents a theory with a non-vanishing mass gap while the theory of free fermions is massless. Hence, they cannot be equivalent to each other.

The equivalent bosonized theory turns out to be more subtle. Witten showed that the action of the bosonized theory is the action of the non-linear sigma model of Eq. (7.156) with an additional term, a Wess–Zumino term $\Gamma[g]$, which we will now define. Let us consider for the moment the theory in Euclidean space-time and work with boundary conditions in which the fields take an arbitrary but constant value at infinity. In this case, as we did before in our discussion of the skyrmion and of the instanton, the Euclidean space-time becomes isomorphic to a 2-sphere S^2. For any compact Lie group G, such as $\text{O}(N)$ or $\text{SU}(N)$, the field configurations are maps $S^2 \to G$, which are topologically trivial and have a trivial homotopy group, $\pi_2(G) = 0$. Hence, it is possible to extend smoothly the field configuration $g(x)$ from S^2 to a function $\bar{g}(y)$ defined in the ball B, the interior of S^2 (here y_i are the coordinates in B). The Wess–Zumino action is given by

$$\Gamma[g] = \frac{1}{24\pi} \int_B d^3y \, \epsilon^{ijk} \, \text{tr}\left(\bar{g}^{-1} \, \partial_i \bar{g} \bar{g}^{-1} \, \partial_j \bar{g} \bar{g}^{-1} \, \partial_k \bar{g}\right) \tag{7.157}$$

However, this action is not single-valued as there are topologically inequivalent ways to extend a configuration $g(x)$ from S^2 to the ball B. To see why this is so, let us consider that, if we work on a compactified 3D space, isomorphic to S^3, there is an ambiguity in what we regard as the "interior" of the 2-sphere S^2, and hence on how to extend the configuration. The difference of two extensions (defined on the "interior" of S^2 and on the "exterior" of S^2) is a map from $S^3 \to G$. For $G = \text{SU}(N)$, such maps are topologically non-trivial and are classified by the homotopy group

$$\pi_3(\text{SU}(N)) \simeq \mathbb{Z} \tag{7.158}$$

It follows from this result that the Wess–Zumino action is multivalued and that for two inequivalent extensions $\Gamma[g]$ is defined only modulo 2π, $\Gamma \to \Gamma + 2\pi r$, with

$r \in \mathbb{Z}$. The reader should by now have seen the close analogy between this line of argument and the one we used in the path integral for spin at the beginning of this chapter.

These considerations led Witten to conjecture that the correct non-linear sigma model has the following action:

$$S[g] = \frac{1}{4\lambda^2} \int d^2x \, \mathrm{tr}\left(\partial_\mu g \, \partial^\mu g^{-1}\right) + k\Gamma[g] \tag{7.159}$$

where $k \in \mathbb{Z}$ for the weight of the path integral to be single-valued. The action of Eq. (7.159) is known as the Wess–Zumino–Witten (WZW) model. The integer-valued coupling constant k is known as the level of the WZW model and of the associated Kac–Moody algebra, Eq. (7.150).

For the WZW model to be equivalent to a theory of free fermions (or to some other fixed-point theory) it must have a fixed point for some value of the coupling constant λ and of the level k. At small λ the non-linear sigma-model term dominates and we know that in this regime this theory has a positive beta function and it is asymptotically free. Thus, the fixed point we are looking for must occur at some finite value of the coupling constant. A one-loop computation of the beta function (Witten, 1984) yields the result for $G = \mathrm{SU}(N)$:

$$\beta(\lambda, k) = \left(\frac{N}{4\pi}\right) \lambda^2 \left[1 - \left(\frac{\lambda^2 k}{4\pi}\right)^2\right] + \cdots \tag{7.160}$$

The constraint that the path integral be single-valued fixes the level k to be an integer. As such, it does not flow under the local fluctuations, and it is invariant under the action of the RG. Nevertheless, the Wess–Zumino term does contribute to the beta function of the coupling constant λ, as shown in Eq. (7.160).

Thus, provided that this one-loop result turns out to be exact, this result predicts the existence of a *stable* fixed point at a critical value of the coupling constant $\lambda_c^2 = 4\pi/k$. Furthermore, Witten also showed that at the classical level the WZW model is compatible with the chiral conservation laws of Eq. (7.154) for the bosonized currents of Eq. (7.155) only if the coupling constant λ and the level k satisfy $\lambda^2 = 4\pi/k$. In particular, the level k of the WZW model and that of the Kac–Moody algebra are the same. This led to the conjecture that the WZW model must indeed have an exact fixed point (i.e. an exact zero of the beta function) at this value of the coupling constant even beyond perturbation theory. This was proven to be correct by Knizhnik and Zamolodchikov (1984) using methods of CFT.

Returning to the relation between the free Dirac fermion theory (with $G = \mathrm{SU}(N)$ global symmetry) and the WZW theory, we are led to conclude that this identification should hold for level $k = 1$. Thus, a theory of N free Dirac fermions is equivalent to an $\mathrm{SU}(N)_1$ WZW model at its fixed point and a U(1) free boson.

Thus, the conjectured equivalent bosonized action is a sum of two terms, one representing the U(1) currents and another a WZW model with SU(N) symmetry

$$S = \int d^2x \, \frac{1}{2}\left(\partial_\mu \varphi\right)^2 + S_{\text{WZW}}^{k=1}[g] \tag{7.161}$$

where $S_{\text{WZW}}^{k=1}[g]$ is the action of the WZW model at its fixed point $\lambda^2 = 4\pi/k$ at level $k = 1$.

At the operator level this implies that the non-abelian chiral currents of the fermionic theory, Eq. (7.152), must be identified with the chiral currents of the WZW model, Eq. (7.155):

$$\begin{aligned} J_{\text{R}}^a &= \psi_{\text{R}}^{i\dagger} t_{ij}^a \psi_{\text{R}}^j = \frac{1}{2\pi} \operatorname{tr}\left(t^a g^{-1}(\partial_+ g)\right) \\ J_{\text{L}}^a &= \psi_{\text{L}}^{i\dagger} t_{ij}^a \psi_{\text{R}}^j = \frac{1}{2\pi} \operatorname{tr}\left(t^a (\partial_- g) g^{-1}\right) \end{aligned} \tag{7.162}$$

The abelian U(1) currents are bosonized as before,

$$J_\mu = \sqrt{\frac{N}{\pi}} \epsilon_{\mu\nu} \partial^\nu \varphi \tag{7.163}$$

A similar identification can be made between the currents of the WZW model and the fermionic currents. Indeed, the operators of Eq. (7.162) have *the same correlation functions* as their fermionic counterparts,

$$\langle J_{\text{R}}^a(z) J_{\text{R}}^b(0)\rangle = \frac{\delta_{ab}}{z^2}, \qquad \langle J_{\text{L}}^a(\bar{z}) J_{\text{L}}^b(0)\rangle = \frac{\delta_{ab}}{\bar{z}^2} \tag{7.164}$$

where (in Euclidean space-time) $z = x_1 + ix_0$ and $\bar{z} = x_1 - ix_0$. Hence, the respective operators have the same scaling dimension $\Delta_J = 1$. They also have the same three-point functions (and hence they also have the same OPEs) because they satisfy the same Kac–Moody algebra (as it turns out, this holds for all values of the level k). Since the Kac–Moody currents generate the spectrum, their spectra are also the same.

It remains to find an identification (a bosonization identity) for the Dirac fermion bilinears which mix right- and left-moving sectors, i.e. mass terms, of the form $Q^a = \operatorname{tr}\left(\psi_{\text{R}}^{i\dagger} t_{ij}^a \psi_{\text{L}}^j\right)$. In the free-fermion theory this operator has scaling dimension 1 and it transforms *as a group element* under the SU(N) $\times$ U(1) symmetry (both right and left). The natural candidate identification is

$$\operatorname{tr}\left(\psi_{\text{R}}^{i\dagger} t_{ij}^a \psi_{\text{L}}^j\right) \propto \operatorname{tr}\left(t^a g(x)\right) e^{i\sqrt{\frac{4\pi}{N}}\varphi(x)} \tag{7.165}$$

The last factor is due to the U(1) sector and has scaling dimension $1/N$. On the other hand, since the WZW field $g(x)$ is classically dimensionless, for this identity to hold it is required that at the quantum level it should have an *anomalous dimension* $(N-1)/N$ so that the free Dirac bilinear has scaling dimension 1. We will see that this is indeed the case.

7.11 A (brief) introduction to conformal field theory

To prove that the operator identifications we just discussed are correct it is necessary to have an exact solution of the WZW at its fixed point, $\lambda_c^2 = 4\pi/k$. This is a non-trivial problem whose solution was found by Knizhnik and Zamolodchikov (1984) using methods of conformal field theory (CFT). We will not give a derivation of these results, but instead give a brief description suitable for our purposes (Affleck, 1986a, 1990). An in-depth treatment of CFT can be found in several texts, see e.g. Di Francesco *et al.* (1997) and Polchinski (1998).

In Eq. (6.73) we showed that the free (Dirac)-fermion Hamiltonian density can be written as a Sugawara form, a quadratic form of the SU(2) and U(1) currents in which the chiral components are decoupled. For a system with $SU(N) \times U(1)$ symmetry the free-Dirac-fermion Hamiltonian density is

$$\mathcal{H} = T_{\rm R} + T_{\rm L} \tag{7.166}$$

where $T_{\rm R}$ and $T_{\rm L}$ are the right- and left-moving components of the energy–momentum tensor, which in this system can be written in terms of the currents (here we have set the velocities to 1 for simplicity)

$$\begin{aligned} T_{\rm R} &= \frac{\pi}{N} J_{\rm R} J_{\rm R} + \frac{2\pi}{N+1} J_{\rm R}^a J_{\rm R}^a \\ T_{\rm L} &= \frac{\pi}{N} J_{\rm L} J_{\rm L} + \frac{2\pi}{N+1} J_{\rm L}^a J_{\rm L}^a \end{aligned} \tag{7.167}$$

which implies that the spectrum is generated by the Kac–Moody currents.

In a theory with translation and Lorentz invariance, the energy–momentum tensor $T_{\mu\nu}$ is locally conserved and symmetric,

$$\partial^\mu T_{\mu\nu} = 0, \qquad T_{\mu\nu} = T_{\nu\mu} \tag{7.168}$$

If the system is also scale-invariant, $T_{\mu\nu}$ is also traceless, $T_\mu^\mu = 0$. Since $T^{00} =$ and T^{01} are the Hamiltonian and linear momentum densities, $\mathcal{H}$ and $\mathcal{P}$, we have

$$T^{00} = -T^{11} = \mathcal{H} = T_{\rm R} + T_{\rm L}, \qquad T^{01} = T^{10} = \mathcal{P} = T_{\rm R} - T_{\rm L} \tag{7.169}$$

On the other hand, the chiral components of the energy–momentum tensor, $T_{\rm R}$ and $T_{\rm L}$, are the local generators of the (infinite-dimensional) group of local conformal transformations in $(1+1)$ dimensions (Di Francesco *et al.*, 1997). They obey the equal-time commutation relations

$$\left[T_{\mathrm{R}}(x), T_{\mathrm{R}}(x')\right] = i\delta(x-x')T'_{\mathrm{R}}(x) + i2\delta'(x-x')T_{\mathrm{R}}(x) + i\frac{c}{24\pi}\delta'''(x-x')$$
$$\left[T_{\mathrm{L}}(x), T_{\mathrm{L}}(x')\right] = i\delta(x-x')T'_{\mathrm{L}}(x) + i2\delta'(x-x')T_{\mathrm{L}}(x) - i\frac{c}{24\pi}\delta'''(x-x') \tag{7.170}$$

These operator identities are known as the Virasoro algebra. The last term is known as the conformal anomaly of the Virasoro algebra, and the (positive) real number c is the central charge of this algebra.

Conformal field theory is a symmetry-based theory of the classification of fixed points in $(1+1)$ dimensions (Belavin *et al.*, 1984; Cardy, 1984; Friedan *et al.*, 1984). It works much in the same way as group theory allows for a classification of wave functions and quantum numbers of states in quantum mechanics without reference to a specific Hamiltonian. A key CFT result is the statement that the fixed-point theories in $(1+1)$-dimensional quantum systems and, equivalently, 2D classical critical statistical-mechanical systems, are representations of the (infinite-dimensional) conformal group generated by the energy–momentum tensor (which satisfies the Virasoro algebra). A given fixed point is characterized by a complete set of operators $\{\phi_\ell(x)\}$, called primary fields, which transform simply (irreducibly) under the action of the conformal group, i.e. under a local scale transformation $x \to \lambda x$,

$$\phi_\ell(\lambda x) \to \lambda^{-\Delta_\ell}\phi_\ell(x) \tag{7.171}$$

where we recognize $\{\Delta_\ell\}$ as the set of the scaling dimensions which play the role of the quantum numbers. Similarly, under a Lorentz transformation (or a rotation in the Euclidean metric), the primary fields transform with a well-defined (conformal) spin, $\{s_\ell\}$. In CFT the chiral components of the energy–momentum tensor, T_{R} and T_{L}, act independently. This leads to two separate conformal dimensions (or weights), usually denoted by h_ℓ and $\bar{h}_\ell$. For a chirally invariant (symmetric) system, the scaling dimension and spin are expressed as $\Delta_\ell = h_\ell + \bar{h}_\ell$ and $s_\ell = h_\ell - \bar{h}_\ell$. A CFT is given by a specific value of the central charge c, a complete set of primary fields, their scaling dimensions (and conformal spins, not to be confused with the conventional spin), and the full set of universal coefficients of their OPE.

The central charge c of the Virasoro algebra, Eq. (7.170), plays a key role in CFT. As we noted above, in a CFT the energy–momentum tensor $T_{\mu\nu}$ is conserved, symmetric, and traceless. These symmetry properties follow from the definition of the energy–momentum tensor as the response to a change of the metric $g_{\mu\nu}$,

$$T_{\mu\nu}(x) = \frac{\delta S}{\delta g^{\mu\nu}(x)} \tag{7.172}$$

and from the Lorentz invariance of the action (rotational invariance in the Euclidean metric, where it is interpreted as a stress tensor).

We will now consider a general $(1 + 1)$-dimensional CFT in (2D) Euclidean space-time (i.e. in imaginary time) and consider the system to be defined on a cylinder: an infinitely long strip of width β (with periodic boundary conditions along this direction). The path integral of the CFT can be regarded as the partition function of a system in classical statistical mechanics on a finite-sized region or, equivalently, as the quantum partition function of a conformally invariant quantum field theory at temperature $T = \beta^{-1}$. An operator ϕ (a primary field) with scaling dimension Δ_ϕ has the (connected) correlation function in the infinite Euclidean plane

$$\langle \phi(z_1)\phi(z_2)\rangle_c = \frac{1}{|z_1 - z_2|^{2\Delta_\phi}} \tag{7.173}$$

The correlator on the strip can be found by the conformal mapping

$$w = \frac{\beta}{2\pi} \ln z \tag{7.174}$$

Under a conformal mapping the correlator of a primary field obeys the transformation law (Belavin *et al.*, 1984)

$$\langle \phi(z_1)\phi(z_2)\rangle = |w'(z_1)|^{\Delta_\phi} |w'(z_2)|^{\Delta_\phi} \langle \phi(w(z_1))\phi(w(z_2))\rangle \tag{7.175}$$

Thus, at "equal times" (the same value of the periodic or compactified coordinate) the correlator becomes

$$\langle \phi(0)\phi(w)\rangle = \frac{1}{[(\beta/\pi)\sinh(\pi w/\beta)]^{2\Delta_\phi}} \tag{7.176}$$

This shows that on the cylinder (i.e. at finite temperature T) the correlation function decays exponentially at long separations as

$$\langle \phi(0)\phi(w)\rangle \sim e^{-|w|/\ell_\phi} \tag{7.177}$$

with

$$\ell_\phi = (2\pi T \Delta_\phi)^{-1} \tag{7.178}$$

If we now regard the cylinder as running along the imaginary-time direction and having circumference L, this result can be interpreted as the statement that the autocorrelation function (i.e. at equal position in space) decays exponentially in (imaginary) time, with a characteristic energy gap $\Delta(L)$ for the excitation created by the primary field ϕ which scales to zero as $L \to \infty$ (since the theory in the thermodynamic limit must be gapless to be conformal) with the law

$$\Delta(L) = \frac{2\pi \Delta_\phi}{L} \tag{7.179}$$

These results offer a direct and practical way to compute scaling dimensions and this method is used extensively in numerical simulations.

The energy–momentum tensor is traceless only if the action, and hence the partition function, is scale-invariant. However, if the metric $g_{\mu\nu}$ has a non-vanishing scalar curvature $R(x)$ (i.e. in the Euclidean case the system is placed on a curved surface), which now supplies a scale, the energy–momentum tensor can have a non-vanishing trace. A key result from CFT is an identity, known as the trace (or *conformal*) anomaly, that relates the expectation value of the trace of the energy–momentum tensor, $\langle T_\mu^\mu(x)\rangle$, to the central charge c of the CFT (Polyakov, 1981; Friedan, 1984):

$$-g^{\mu\nu}(x)\frac{\delta Z}{\delta g^{\mu\nu}(x)} = \langle T_\mu^\mu(x)\rangle = \frac{c}{48\pi}\left(R(x)+\mu^2\right) \tag{7.180}$$

where $R(x)$ is the scalar curvature and μ^2 is a non-universal constant (of dimension L^{-2}). This general result has many important consequences.

On an infinitely long strip geometry, the free energy $F = -\ln Z$ for large values of the strip width β (i.e. low temperatures in the quantum version) is expected to obey the finite-size scaling behavior

$$F = f\beta + f^* + \frac{A}{\beta} + O(\beta^{-2}) \tag{7.181}$$

where f is the free energy per unit length, f^* is a constant term due to edge contributions (and hence absent for systems with periodic boundary conditions such as a cylinder), and A is a dimensionless constant that is (presumably) universal (though dependent on the choice of boundary conditions).

We will see now that CFT predicts a specific relation between the constant A and the central charge c: $A = -\pi c/6$ for periodic boundary conditions ($A = -\pi c/24$ for fixed and free boundary conditions) (Affleck, 1986b; Blöte *et al.*, 1986). We will follow Blöte *et al.* (1986) and consider a coordinate transformation $x^\mu \to x^\mu + \alpha^\mu$ (of the strip), which is not a conformal transformation (and hence it is not a symmetry transformation), e.g. a shear distortion. Such a coordinate transformation causes a change in the action S of the system of the form

$$\delta S = -\int \frac{\partial\alpha^\mu}{\partial x_\nu} T_{\mu\nu}(x) d^2x \tag{7.182}$$

If we denote the coordinates on the strip by (u, v) we can denote an infinitesimal non-conformal transformation as $u' = u(1-\lambda)$ and $v' = v(1+\lambda)$ with $\lambda \ll 1$. Indeed, the change in the expectation value of the action (or the internal energy in the classical statistical-mechanical version) is (for a system with periodic boundary conditions)

$$\delta\langle S\rangle = 2\beta\lambda\int_{-\infty}^{\infty}\left(\langle T\rangle + \langle\bar{T}\rangle\right)du \tag{7.183}$$

The invariance of the partition function implies that there is a compensating change in the free energy $\delta F = -2\lambda A/\beta$. Therefore, $A = 2\beta^2\langle T\rangle$ (since $\langle T\rangle = \langle\bar{T}\rangle$). On the other hand, the (response) change in $\langle T\rangle$ to a change δS in the action, to lowest order ("linear response"), is

$$\delta\langle T(0)\rangle = -2\lambda\int_{-\infty}^{\infty}du\int_0^\beta dv\langle T(0,0)T(u,v)\rangle_{\rm c} \tag{7.184}$$

We now note that the (connected) correlators of the energy–momentum tensor on the strip also obey a scaling law

$$\langle T(0)T(w)\rangle_{\rm c} = \frac{c/2}{[(\beta/\pi)\sinh(\pi w/\beta)]^4} \tag{7.185}$$

which follows from the fact that the scaling dimension of the energy–momentum tensor is 2. This is an exact property of the energy–momentum tensor in all CFTs, which is protected by the fact that it is a conserved current (and hence cannot have an anomalous dimension). The prefactor of $c/2$ arises from the conformal anomaly of the Virasoro algebra. Here we work in the Euclidean metric and denote $T_{\rm R} = T$ and $T_{\rm L} = \bar{T}$, which are holomorphic and antiholomorphic functions of $z = x_1 + ix_2$ and $\bar{z} = x_1 - ix_2$, respectively. Even if $\langle T\rangle = 0$ on the infinite plane, it is generally non-vanishing on the strip geometry. It is precisely this change that we will need in order to derive Eq. (7.181) and to evaluate the constant A. By evaluating the integrals in Eq. (7.184) we find $\delta\langle T\rangle = \pi^2 c/(6\beta^2)$, which must equal $\delta\langle T\rangle = -A\lambda/\beta^2$. Hence, $A = -(\pi/6)c$.

Two conclusions can be drawn from this result. The first is that the ground-state energy density $\varepsilon(L)$ at $T = 0$ of a critical system with length L with periodic boundary conditions in space obeys the finite-size scaling behavior (Affleck, 1986b; Blöte *et al.*, 1986)

$$\varepsilon(L) = \varepsilon_0 - \frac{\pi c}{6vL^2} + \cdots \tag{7.186}$$

where ε_0 is the (non-universal) ground-state energy density, c is the central charge, and v is the speed of the excitations (which above we have set to 1). The second term in Eq. (7.186) is the Casimir energy. It represents the leading finite-size correction, and also has the interpretation of an effective interaction between the edges due to the quantum vacuum fluctuations.

For the second case, we now consider an infinite system at finite temperature $T > 0$, whereupon this result becomes an asymptotic expansion for the free-energy density $f(T)$ at low temperatures,

$$f(T) = \varepsilon_0 - \frac{\pi c}{6v}T^2 + \cdots \tag{7.187}$$

From this result it follows that the low-temperature heat capacity $C(T)$ is

$$\frac{C(T)}{L} = \frac{\pi c k_B^2 T}{3\hbar v} + \cdots \tag{7.188}$$

(where we restored standard units). Therefore, the specific heat of a critical (Lorentz-invariant) 1D quantum system has the universal form of Eq. (7.188), and is proportional to the central charge c of the CFT. This result motivates the interpretation of the central charge c as counting the number of degrees of freedom in a physical system.

7.12 The Wess–Zumino–Witten conformal field theory

In the case of systems with a locally conserved current (associated with a global continuous symmetry), such as free Dirac fermions (among others), the representations of the Kac–Moody algebras are automatically representations of the Virasoro algebra since the energy–momentum tensor can (in that case) be expressed in terms of the Kac–Moody currents which generate the spectrum. Thus, in this case the central charge c, the scaling dimensions of the primary fields, and their OPE coefficients are fully determined in terms of the level k of the Kac–Moody algebra and by the transformation laws of the states under the symmetry generated by these currents. The level k is the central charge (or extension) of the Kac–Moody algebra of the currents.

For example, this is the case for free Dirac fermions with a U(1) symmetry. In the free-Dirac-fermion system the primary fields are the Fermi fields themselves and the composite operators such as fermion bilinears, i.e. the U(1) currents and the order parameters that we discussed in Section 5.6. In that section we introduced abelian bosonization and saw that there is an equivalent bosonic theory in which the operators of the fermionic theory are represented by vertex operators with the form of exponentials of the bosonic field. Both theories, the free Dirac fermion and the free bosonic (scalar) field, are fixed-point theories and are conformally invariant. However, not all possible vertex operators of the bosonic system have a counterpart in the fermionic system. In fact the number of allowed vertex operators is certainly smaller than the set of all possible ones. The way we constructed the set of allowed vertex operators was to first find a mapping of the currents and then a mapping of the fermionic operators themselves in their Mandelstam representation. The other operators were then obtained using the OPE.

We can reverse the logic of this construction and ask which vertex operators of the bosonic theory should be allowed in the first place. The key property obeyed by all the allowed vertex operators we constructed is that they are *local* with respect to the Dirac field (or, more properly, with respect to its bosonized version). This property means that at equal times the allowed operators commute (or anti-commute) with the Dirac fermion. The vertex operators which obey this property are those of a bosonic field $\phi(x)$ with compactification radius $R = 1/\sqrt{4\pi}$ (in the normalization used in Section 5.6), presented in Eqs. (6.63)–(6.66). Recall that the free bosonic field is a scalar field with the global U(1) shift symmetry $\phi(x) \to \phi(x) + \alpha$ (with α real and arbitrary).

In the Dirac theory this U(1) (shift) transformation corresponds to a continuous chiral transformation, $\psi(x) \to \exp(i\alpha\gamma_5)\psi(x)$, which, as we saw in our discussion of the Luttinger liquid, corresponds to a rigid translation of the charge-density profile. Thus, the quantum numbers of the vertex operators of the bosonic theory are the charges of these operators under the U(1) chiral symmetry. More importantly, it can be shown (Di Francesco *et al.*, 1997) that the bosonic theory (with compactification radius $R = 1/\sqrt{4\pi}$) and that of the free Dirac fermion have the same partition function and therefore that their spectrum is the same.

Conformal field theory is a theory of fixed points (in the RG sense). Thus, we are interested in looking at the operators that survive under the action of the RG flow. As we know, the only operators that survive the RG flow are either relevant (those whose weight in the effective action grows indefinitely under the action of the RG) or marginal (those whose weight in the action either remains unchanged or changes by a finite amount under the RG). In contrast, the weight of irrelevant operators to the effective action flows to zero under the RG. The counterpart of these observations is that in the OPE of the primary fields irrelevant operators enter in the form of analytic (non-singular) terms that vanish in the asymptotic limit.

These observations motivate the consideration of CFTs with a finite number of primary fields, which are either relevant or marginal, i.e. with scaling dimension less than or equal to 2, and hence their OPEs are singular. In such a theory each primary field has an associated set of states, called a Verma module, created by the so-called descendants (irrelevant operators) of the primary field. A CFT with a finite number of primary fields is called a rational CFT (RCFT) (Ginsparg, 1989). The bosonized theory of the Dirac fermion is an example of an RCFT.

Perhaps the best-known (and most famous) and simplest example of an RCFT is the fixed-point theory of the classical 2D Ising model (which was solved originally by Onsager in 1944) and its quantum counterpart, the quantum Ising chain in a transverse field (see Chapter 5). The transfer matrix of the classical 2D model, or alternatively the Hamiltonian of the transverse Ising model chain, can be mapped using a Jordan–Wigner transformation to a system of 1D lattice fermions, qualitatively representing domain walls of the spin system (Schultz *et al.*, 1964). However, due to the $\mathbb{Z}_2$ symmetry of the Ising model, the quantum Hamiltonian of the equivalent system does not conserve fermion number but only fermionic parity. In the critical regime, in which the gap of the fermionic spectrum becomes vanishingly small, this system is represented by an effective field theory of Majorana fermions, instead of Dirac fermions, whose mass is tuned to zero at the (quantum) critical point. Since a theory of Majorana fermions has half the number of degrees of freedom of the Dirac field, the central charge of the Ising CFT is $c = 1/2$ (instead of $c = 1$ for Dirac fermions). In addition to the identity operator I, the CFT 2D Ising model has only three primary fields: (a) the spin field σ, the order-parameter field of the Ising model, with scaling dimension $1/8$; (b) the Majorana fermion ψ with scaling dimension $1/2$; and (c) the energy density ε of the Ising model (the mass term of the Majorana field) with scaling dimension 1. The spin field is also known as the twist field since it changes the boundary conditions of the Majorana fermion from periodic to anti-periodic. Other examples of RCFTs are the fixed points of the three-state Potts model, the Ising antiferromagnet in an uniform field, and many other so-called "minimal models" (Belavin *et al.*, 1984; Friedan *et al.*, 1984; Ginsparg, 1989; Di Francesco *et al.*, 1997).

The Wess–Zumino–Witten model is another important example of an RCFT (Knizhnik and Zamolodchikov, 1984), which is the main focus of our interest in this chapter. For simplicity we will discuss WZW models with symmetry group $G = \mathrm{SU}(2)$ at general level k. We will not derive these results here because this is fairly technical and it is done in several excellent and standard texts devoted to CFT (Di Francesco *et al.*, 1997).

The central charge of $\mathrm{SU}(N)_k$ WZW models was found by Knizhnik and Zamolodchikov,

$$c(\mathrm{SU}(N)_k) = \frac{k \dim G}{k + \mathfrak{g}} = \frac{k(N^2-1)}{k+N} \tag{7.189}$$

where $\dim G$ is the rank of the algebra of G and $\mathfrak{g}$ is the "dual Coxeter number" of G. For $\mathrm{SU}(N)$ they are given by $\dim G = N^2 - 1$ and $\mathfrak{g} = N$, respectively. For the special case of $\mathrm{SU}(N)_1$ ($k = 1$), the central charge reduces to

$$c(\mathrm{SU}(N)_1) = N - 1 \tag{7.190}$$

which is an integer. Thus, one expects the $\mathrm{SU}(N)_1$ WZW models must also be describable as $N - 1$ free fields. We will see below that this is indeed the case and that it is the content of non-abelian bosonization.

The one-loop beta function of the WZW model (accurate in the limit $k \to \infty$) was given in Eq. (7.160) and predicted the existence of an infrared-stable fixed point at the value of the coupling constant $\lambda^2 = 4\pi/k$. This result was (essentially) confirmed by the CFT of the WZW model, which yields the slope of the beta function at the fixed point $\lambda^2 = 4\pi/k$ as

$$\left.\frac{d\beta}{d\lambda^2}\right|_{\lambda^2=4\pi/k} = -\frac{2N}{N+k} \tag{7.191}$$

which agrees with the one-loop result for k large and confirms that $\lambda^2 = 4\pi/k$ is indeed an infrared-stable fixed point (and hence a CFT).

Let us now discuss the primary fields of the WZW CFT (only for SU(2)) and their scaling dimensions. Since the global symmetry of WZW models is SU(2) $\times$ SU(2) the primary fields carry the labels (j, m) of the representations of the Lie group SU(2), where j is a positive integer or half-integer and $|m| \leq j$ (in integer steps). Thus, the WZW field $g(x)$ is a 2×2 unitary matrix whose rows (and columns) transform as the fundamental (spinor) representation of each SU(2) and hence carries both quantum numbers,

$$g(x) = \begin{pmatrix} g_{\frac{1}{2},\frac{1}{2}} & g_{\frac{1}{2},-\frac{1}{2}} \\ g_{\frac{1}{2},-\frac{1}{2}} & g_{-\frac{1}{2},-\frac{1}{2}} \end{pmatrix} \tag{7.192}$$

However, while for the Lie group SU(2) there is no upper bound to the value of j, for an $\text{SU}(2)_k$ WZW model (or equivalently for an $\text{SU}(2)_k$ Kac–Moody algebra) the tower of "angular-momentum" states is truncated at the upper bound $j \leq j_{\text{max}} = k/2$. In other words, there is a finite number of allowed primary fields labeled by $0 \leq j \leq k/2$. This is a general feature of WZW models.

Thus, the $\text{SU}(2)_1$ WZW model has only two primary fields: (a) the identity I and (b) the spinor representation given by the WZW field $g(x)$ itself. This is also the spectrum of primary fields of the $\text{SU}(N)_1$ WZW model. The drastic effects of truncation can be seen in the OPE of the field $g(x)$ with itself: the OPE of g with itself reduces to the identity field. Symbolically we denote this as $1/2 \otimes 1/2 = 0$ and we say that two WZW fields *fuse* into the identity (only for $\text{SU}(2)_1$!). Notice that the (expected) fusion into the spin-1 representation is absent since this is not allowed by the truncation of the spectrum of primary fields.

More generally, $\text{SU}(2)_k$ has the primary fields with quantum numbers (j, m), labeled by $\Phi_{(j,m)}$, where j is an integer or half-integer in the range $0 \leq j \leq k/2$ for k even (or up to $(k-1)/2$ if k is odd), m are integers or half-integers in the range $-j \leq m \leq j$, and the identifications $(j, m) \cong (k/2 - j, m + k/2)$, $(j, m) \cong (j, m + k)$ hold. The fields $\Phi_{(j,m)}$ have scaling dimensions

$$\Delta_{(j,m)} = \frac{j(j+1)}{k+2} \tag{7.193}$$

In the case of $\text{SU}(2)_2$ there are three primary fields: (a) the identity I, (b) the WZW field (again the spinor representation) with scaling dimension $\Delta_{(1/2,\pm 1/2)} = 3/16$, and (c) the spin-1, or adjoint, representation, with scaling dimension $\Delta_{(1,m)} = \frac{1}{2}$. Thus, at level $k = 2$ two WZW fields can now fuse either into the identity I or into the adjoint primary field. However, due to the truncation, two adjoint primary fields can fuse only into the identity.

The two-point function of the WZW field g is (in Euclidean space-time with $z = x_1 + ix_2$ and $\bar{z}$ being the complex conjugate)

$$\langle g_{\alpha_1}^{\beta_1}(z,\bar{z}) g^{-1}{}^{\alpha_2}_{\beta_2}(0,0)\rangle = M^{-2\Delta_g} \frac{\delta_{\alpha_1}^{\alpha_2}\delta_{\beta_1}^{\beta_2}}{(z\bar{z})^{\Delta_g}} \tag{7.194}$$

where Δ_g is the scaling dimension of the field g, and M is an ultraviolet cutoff with units of mass (or length^{-1}). For an $\mathrm{SU}(N)_k$ WZW theory Δ_g is given by

$$\Delta_g(\mathrm{SU}(N)_k) = \frac{N^2-1}{N(N+k)} \tag{7.195}$$

The WZW field g (just like the other fields such as the adjoint primary) is classically dimensionless and hence has scaling dimension 0 at the trivial (and unstable) $\lambda \to 0$ fixed point. Thus, the non-trivial scaling dimension of Eq. (7.195) is the *anomalous dimension* of the field g at the non-trivial (infrared-stable) fixed point of the WZW model.

From Eq. (7.195) we see that, for $\mathrm{SU}(N)_1$, the scaling dimension of the WZW field g is indeed

$$\Delta_g(\mathrm{SU}(N)_1) = \frac{N-1}{N} \tag{7.196}$$

as we deduced it should be the case for non-abelian bosonization to work; see the discussion below Eq. (7.165). Hence, at least at the level of matching the scaling dimensions, this result justifies this operator identification of non-abelian bosonization for fermionic bilinears.

However, to prove that this is an operator identity it is necessary to show not only that the scaling dimensions match but also that their correlation functions are the same. Knizhnik and Zamolodchikov (1984) showed that the correlators of Eq. (7.165) are identical for $\mathrm{SU}(N)_1$. In the derivation of this result the important fact that the OPE of the WZW field contains only the identity field (and that no other representations appear) was used. Hence the truncation of the spectrum of primaries is essential in order for the non-abelian bosonization identity to hold.

In general the four-point function of the WZW field has the form ($i = 1, \ldots, 4$)

$$\begin{aligned} G[z_i,\bar{z}_i] &= \langle g(z_1,\bar{z}_1)g^{-1}(z_2,\bar{z}_2)g^{-1}(z_3,\bar{z}_3)g(z_4,\bar{z}_4)\rangle \\ &= [(z_1-z_4)(z_2-z_3)(\bar{z}_1-\bar{z}_4)(\bar{z}_2-\bar{z}_3)]^{-\Delta_g} G(x,\bar{x}) \end{aligned} \tag{7.197}$$

where $G(x,\bar{x})$ is a function that depends only on the cross ratio x (and its complex conjugate $\bar{x}$),

$$x = \frac{(z_1-z_2)(z_3-z_4)}{(z_1-z_4)(z_2-z_3)} \tag{7.198}$$

For general level k the structure of the function $G(x, \bar{x})$ is complex and reflects the fusion channels available to the primary field. We will return to this problem in Chapters 14 and 15, where we discuss non-abelian quantum Hall states. Here we will focus on the case of $\mathrm{SU}(N)_1$, where the structure of $G(x, \bar{x})$ is simple (reflecting the fact that for $k = 1$ the WZW field can only fuse into the identity):

$$G(x, \bar{x}) = [x\bar{x}(1-x)(1-\bar{x})]^{1/N} \left(\frac{\delta^{\alpha_2}_{\alpha_1}\delta^{\alpha_4}_{\alpha_3}}{x} + \frac{\delta^{\alpha_4}_{\alpha_1}\delta^{\alpha_2}_{\alpha_3}}{1-x} \right) \left(\frac{\delta^{\beta_1}_{\beta_2}\delta^{\beta_4}_{\beta_3}}{\bar{x}} + \frac{\delta^{\beta_4}_{\beta_2}\delta^{\beta_1}_{\beta_3}}{1-\bar{x}} \right) \tag{7.199}$$

7.13 Applications of non-abelian bosonization

We will now discuss a few applications of non-abelian bosonization. Our principal aim is to consider quantum spin chains.

7.13.1 *Free fermions*

We are now ready to discuss the bosonization of free fermions. Again we consider a theory of N free Dirac fields in $(1+1)$ dimensions with both chiralities, i.e. $\psi_{\mathrm{R},\alpha}$ and $\psi_{\mathrm{L},\alpha}$ with $\alpha = 1, \ldots, N$. The free Dirac theory has a $\mathrm{U}(N)$ symmetry that can be regarded as $\mathrm{U}(N) \simeq \mathrm{SU}(N) \times \mathrm{U}(1)$, where $\mathrm{U}(1)$ is the charge sector and $\mathrm{SU}(N)$ is the "spin" sector. We return to the conjecture that the fermionic bilinears are given by a factorized operator of the form (which is the same as Eq. (7.165))

$$: \psi_{\mathrm{R},\alpha}(z)\psi^{\dagger\beta}_{\mathrm{L}}(\bar{z}) :\sim M e^{i\sqrt{\frac{4\pi}{N}}\phi(z,\bar{z})} g^{\beta}_{\alpha}(z, \bar{z}) \tag{7.200}$$

The correlation functions of the right-hand side (r.h.s.) factorize into a contribution of the scalar field ϕ and a contribution of the WZW field g since these two fields are decoupled from each other. Using the results of the previous section and similar results for the vertex operator of the ϕ-field with the lowest charge, $\exp(i\sqrt{4\pi/N}\phi)$, with scaling dimension $1/N$, we find that the four-point function of the r.h.s. of Eq. (7.200) has the same form as Eq. (7.197) but with a shifted scaling dimension $\tilde{\Delta}$, which is the sum of the scaling dimensions of the WZW field g and of the vertex operator,

$$\tilde{\Delta} = \Delta_g(\mathrm{SU}(N)_1) + \frac{1}{N} = \frac{N-1}{N} + \frac{1}{N} = 1 \tag{7.201}$$

Hence, the composite operator of the r.h.s. of Eq. (7.200) has scaling dimension 1, as it should be for a free-fermion bilinear. However, notice that the individual factors separately have non-trivial scaling dimensions. Similarly, the central charge of the theory is the sum of the central charge of the ϕ-field and the central charge of the WZW field. For a level $k = 1$ WZW theory we find that the central charge is

$$c = c(\mathrm{U}(1)) + c(\mathrm{SU}(N)_1) = 1 + N - 1 = N \tag{7.202}$$

which is the central charge of N free Dirac fields.

These arguments prove that the composite operator of the r.h.s. of Eq. (7.200) is indeed equivalent to the fermionic operator (the l.h.s.), including the SU(N) group-theoretic tensors given by the function $G(x, \bar{x})$. This completes the proof of non-abelian bosonization.

These results have been generalized to the case of a theory of Dirac fermions with N_{c} colors and N_{f} flavors (Affleck, 1986a), $\psi_{\mathrm{R},i,f}$ and $\psi_{\mathrm{L},a,\alpha}$ with $i = 1, \ldots, N_{\mathrm{c}}$ and $f = 1, \ldots, N_{\mathrm{f}}$. This theory has a U($N_{\mathrm{c}}N_{\mathrm{f}}$) symmetry that can be decomposed into a U(1) charge sector, an SU(N_{c}) color sector and an SU(N_{f}) flavor sector. The energy–momentum tensor (and the Hamiltonian) decompose again into a sum of right- and left-moving terms, each of which has again a Sugawara form and is expressed as a bilinear of the charge, color and spin currents J, J^A (with $A = 1, \ldots, N_{\mathrm{c}}^2$), and J^a (with $a = 1, \ldots, N_{\mathrm{f}}$) respectively. The currents obey a level-$N_{\mathrm{c}}N_{\mathrm{f}}$ U(1) Kac–Moody algebra (charge sector), a level-N_{f} SU(N_{c}) Kac–Moody algebra (color sector), and a level-N_{c} SU(N_{f}) algebra (flavor sector). Thus we have decomposed a level-1 U($N_{\mathrm{c}}N_{\mathrm{f}}$) theory as

$$\mathrm{U}(N_{\mathrm{c}}N_{\mathrm{f}})_1 \simeq \mathrm{U}(1)_{N_{\mathrm{c}}N_{\mathrm{f}}} \times \mathrm{SU}(N_{\mathrm{c}})_{N_{\mathrm{f}}} \times \mathrm{SU}(N_{\mathrm{f}})_{N_{\mathrm{c}}} \tag{7.203}$$

The Hamiltonian for the right-moving fields is

$$\mathcal{H} = \frac{\pi}{N_{\mathrm{c}}N_{\mathrm{f}}} J J + \frac{2\pi}{N_{\mathrm{c}} + N_{\mathrm{f}}} J^A J^A + \frac{2\pi}{N_{\mathrm{c}} + N_{\mathrm{f}}} J^a J^a \tag{7.204}$$

and similarly for the left-moving fields. Each non-abelian sector has a bosonized effective action in terms of $\mathrm{SU}(N_{\mathrm{c}})_{N_{\mathrm{f}}}$ and $\mathrm{SU}(N_{\mathrm{f}})_{N_{\mathrm{c}}}$ WZW theories (of each chirality). It is easy to see that the (Virasoro) central charges add up to the right value

$$\begin{aligned} c(\mathrm{U}(N_{\mathrm{c}}N_{\mathrm{f}})_1) &= c(\mathrm{U}(1)) + c(\mathrm{SU}(N_{\mathrm{c}})_{N_{\mathrm{f}}}) + c(\mathrm{SU}(N_{\mathrm{f}})_{N_{\mathrm{c}}}) \\ &= 1 + \frac{N_{\mathrm{f}}(N_{\mathrm{c}}^2 - 1)}{N_{\mathrm{c}} + N_{\mathrm{f}}} + \frac{N_{\mathrm{c}}(N_{\mathrm{f}}^2 - 1)}{N_{\mathrm{f}} + N_{\mathrm{c}}} = N_{\mathrm{f}}N_{\mathrm{c}} \end{aligned} \tag{7.205}$$

In particular the generalization of Eq. (7.200) for the fermion bilinears is now

$$: \psi_{\mathrm{R},i,f}(z)\psi^{\dagger\, j,l}_{\mathrm{L}}(\bar{z}) : \sim M e^{i\sqrt{\frac{4\pi}{N_{\mathrm{c}}N_{\mathrm{f}}}}\phi(z,\bar{z})} g_i^j(z,\bar{z}) h^l_f(z,\bar{z}) \tag{7.206}$$

where $g \in \mathrm{SU}(N_c)$ and $h \in \mathrm{SU}(N_f)$. The scaling (trivial!) dimension of the fermion bilinear is the sum of the (non-trivial!) scaling dimensions of the factors.

$$\Delta = \frac{1}{N_c N_f} + \frac{N_c^2 - 1}{N_c(N_c + N_f)} + \frac{N_f^2 - 1}{N_f(N_c + N_f)} = 1 \tag{7.207}$$

7.13.2 Fermions with repulsive interactions: gapping the charge sector

One physical system in which the fermions acquire a charge gap but not a spin gap is the 1D Hubbard model (see Chapter 6). In the weak-coupling regime the Hubbard model is equivalent to a theory of $N = 2$ Dirac fermions with various interactions, cf. Eq. (6.67) and Eq. (6.73). In that framework the gap in the charge sector (with U(1) symmetry) arises due to an Umklapp process. A key feature of the Hubbard model is spin–charge separation. Thus, the effective interacting fermionic field theory involves two Dirac fermions (one for each spin component) with current–current interactions (as well as an Umklapp term at half-filling), all of which exhibit the phenomenon of spin–charge separation.

As we also saw in Chapter 6, spin–charge separation is also apparent in the abelian bosonization form of the theory in which the effective Lagrangian decouples into a sum of two terms, one for the charge sector and one for the spin sector, with the charge sector becoming gapped at half-filling due to the effects of the marginally relevant Umklapp interaction.

Here we will consider again the same system using the non-abelian bosonization discussed in detail by Affleck and Haldane (1987). To simplify the discussion I will ignore the purely forward-scattering interaction g_4, which merely renormalizes (in opposite ways) the velocities of the charge and spin modes. (I will also choose "relativistic" units in which the Fermi velocity is set to unity, $v_F = 1$.) As we saw earlier in this section, the free-fermion piece of the Hamiltonian of Eq. (6.73) maps into a free scalar ϕ_c for the charge sector and an SU(2) level $k = 1$ WZW field g for the spin sector, with the following action:

$$S_0 = \int d^2x \, \frac{1}{2}\left(\partial_\mu \phi_c\right)^2 + S_{\mathrm{WZW}}^{k=1}[g] \tag{7.208}$$

The charge sector of the theory has an effective Lagrangian that includes the effects both of the backscattering interaction term of the Lagrangian of the U(1) currents,

$$\mathcal{L}_{\mathrm{int}}^{\mathrm{charge}} = -\frac{1}{2}(2g_2 - g_1) J_R J_L \tag{7.209}$$

a marginal perturbation whose effect is to renormalize the Luttinger charge parameter away from the free-fermion value $K_c = 1$, resulting again in $K_c > 1$ for repulsive interactions, and of the Umklapp interaction term. Here we have kept the theory

Lorentz-invariant and set the velocity of the charge and spin modes to 1. Because of this, the resulting expression for $K_c = [1 + (2g_2 - g_1)/(2\pi)]^{1/2}$ differs from that given in Eq. (6.139).

After bosonization the Lagrangian for the charge sector changes to

$$\mathcal{L}_{\text{charge}}[\phi_c] = \frac{K_c}{2} \left(\partial_\mu \phi_c\right)^2 + \text{constant} \times \cos(2\sqrt{2\pi}\,\phi_c) \tag{7.210}$$

Here we have used the fact that the Umklapp term describes processes in which a spin-singlet pair of right movers becomes a spin-singlet pair of left movers (and vice versa). Since the operators involved are separately spin singlets, they are independent of the WZW field g (more precisely, they involve $\det g$, which is a spin singlet and is equal to 1 for a unitary group). Moreover, as before, the Umklapp term is marginally relevant at the free-fermion fixed point. Thus, the charge sector flows to a strong-coupling fixed point corresponding to the massive phase of a sine–Gordon field theory in which the operator $\cos(2\sqrt{2\pi}\,\phi_c)$ has a non-vanishing expectation value. The mass gap of the sine–Gordon theory is the charge gap of the Hubbard model (in the scaling regime).

Classically this phase corresponds to pinning of the charge boson to the classical ground states $\phi_c = \sqrt{\pi/2} n_c$, where n_c is an arbitrary integer. Semi-classically, one can qualitatively describe this phase using a harmonic approximation, which amounts to expanding the fluctuations of the charge boson ϕ_c about these classical ground states, leading to a finite effective mass of the field ϕ_c. A more accurate description uses the mass gap of the quantum sine–Gordon theory, a quantum integrable system which is also solvable by a generalization of the Bethe ansatz (Faddeev, 1984; Rajaraman, 1985). Nevertheless, in spite of this "classical" behavior, the expectation value of the (spin-singlet) CDW order-parameter operator, $\langle \psi^\dagger_{R,\alpha} \psi_{L,\alpha}\rangle + \text{c.c.} \propto \langle \cos(\sqrt{2\pi}\phi_c)\rangle$, vanishes in this phase since this operator is odd under the exact remaining symmetry of the Hamiltonian, $n_c \to n_c + 1$. Hence, in this phase with a gapped charge sector the charge density remains uniform.

Similarly, the backscattering coupling term of the chiral SU(2) spin currents leads to a term in the (non-abelian) bosonized action, which now becomes

$$S^{\text{spin}}_{\text{int}} = S^{k=1}_{\text{WZW}}[g] + \int d^2x\, 2g_1 \vec{J}_R \cdot \vec{J}_L \tag{7.211}$$

where the currents are given by their bosonized expressions in Eq. (7.155). We will now compute the one-loop beta function for the coupling constant g_1 using the perturbative RG method of Section 4.5. As we saw there, we need to know the scaling dimension of the perturbation and the coefficients of the OPE, cf. Eq. (4.61). The fermionic spin currents $\vec{J}_R$ and $\vec{J}_L$ obey the $SU(2)_1$ Kac–Moody algebra of Eq. (7.150). It is easy to show that as a result they also obey the following OPEs (Knizhnik and Zamolodchikov, 1984; Di Francesco *et al.*, 1997):

$$J_{\rm R}^a(z) J_{\rm R}^b(w) \sim \frac{1}{2\pi(z-w)^2}\delta_{ab} + i\frac{1}{2\pi(z-w)}\epsilon^{abc} J_{\rm R}^c(w) \tag{7.212}$$

and similarly for the left movers. Here z and w are complex coordinates and $\bar{z}$ and $\bar{w}$ are their complex conjugates.

To compute the one-loop beta function we need the OPE of the operator $\vec{J}_{\rm R} \cdot \vec{J}_{\rm L}$ with itself fusing into itself,

$$\left[\vec{J}_{\rm R}(z) \cdot \vec{J}_{\rm L}(\bar{z})\right]\left[\vec{J}_{\rm R}(w) \cdot \vec{J}_{\rm L}(\bar{w})\right] \sim \frac{C}{|z-w|^2}\vec{J}_{\rm R}(w) \cdot \vec{J}_{\rm L}(\bar{w}) \tag{7.213}$$

where we have kept only the term we are interested in of this OPE. For $\mathrm{SU}(N)_1$ the OPE coefficient C is

$$C = \frac{N}{2\pi^2} \tag{7.214}$$

where we used the fact that for the group $\mathrm{SU}(N)$ the structure constants f^{abc} satisfy

$$\sum_{a,b} f^{abc} f^{abd} = 2N\delta_{cd} \tag{7.215}$$

and that the $\mathrm{SU}(N)$ generators $\{t^a\}$ satisfy

$$\sum_a t_{ij}^a t_{kl}^a = N\delta_{il}\delta_{jk} - \delta_{ij}\delta_{kl} \tag{7.216}$$

For SU(2), the structure constants are given by the Levi-Civita tensor, $f^{abc} = \epsilon^{abc}$.

On the other hand, the scaling dimension of the chiral current backscattering interaction is $\Delta = 2$, which is to say that this is a marginal operator. Hence, Eq. (4.61) tells us that the one-loop beta function for the coupling g_1 is given by (using that $S_D = 2\pi$ for $D = 2$)

$$\beta(g_1) = -\frac{N}{\pi} g_1^2 + O(g_1^3) \tag{7.217}$$

This result implies that for repulsive interactions, for which $g_1 > 0$, the backscattering coupling of the chiral spin currents is a *marginally irrelevant* operator. Therefore, the effective backscattering coupling in the spin channel flows to zero, although very slowly. The main effect of this slow flow to zero coupling is that there are logarithmic corrections to scaling in the correlators of the (Néel) order parameter.

So, tentatively, we will identify the WZW CFT with the fixed point for the spin sector. There is a possible pitfall in this argument. The WZW theory has a relevant operator, the WZW field g, which, as we saw for SU(2), has scaling dimension $\frac{1}{2}$ (cf. Eq. (7.196)). However, the WZW field g breaks the symmetry $g \to -g$. This amounts to a change in the sign of the trace of the order parameter, represented here by the mass term $\psi^\dagger_{\mathrm{R},\alpha}\psi_{\mathrm{L},\beta}$; see Eq. (7.165). Since this operator involves the mixing of right and left movers, it carries (lattice) momentum π and breaks translation invariance. Therefore this operator is not allowed in the Hamiltonian unless translation invariance is explicitly broken in the system, say by a spin-Peierls period-2 modulation of the effective exchange interaction, which is a spin-singlet operator that breaks translation invariance of displacements by one lattice spacing (but not two), or by a staggered magnetic field, which couples linearly to the Néel order parameter, the staggered magnetization, and breaks translation invariance (again of displacements by one lattice spacing) and global SU(2) symmetry. Thus, if translation and global SU(2) invariance are global symmetries of the Hamiltonian, the WZW field g cannot appear linearly in the effective field theory. The same argument applies to models with $\mathrm{SU}(N)_1$ symmetry.

7.13.3 Back to spin chains

We end this discussion by analyzing the implications of these results for the case of the spin-$\frac{1}{2}$ quantum Heisenberg antiferromagnetic chains. From the above analysis we conclude, again up to logarithmic corrections, that the spin sector of the interacting fermionic system is described at low energies by the $\mathrm{SU}(N)_1$ WZW fixed point. Since the charge sector is massive (gapped) for $\mathrm{SU}(2)_1$, the charge degrees of freedom effectively decouple at low energies and, effectively, are projected out. Thus the WZW CFT is the actual description of the long-distance physics of the 1D Hubbard model at half-filling. Since the charge gap grows monotonically as the interaction increases, this result should apply all the way to the strong-coupling limit where the half-filled Hubbard model reduces to the 1D spin-$\frac{1}{2}$ quantum antiferromagnet, albeit with a finite non-universal renormalization of the spin-wave velocity.

In Section 5.7 we discussed the scaling behavior of the quantum Heisenberg antiferromagnetic chain using abelian bosonization and computed the scaling dimensions of the operators (see Table 5.1). We can now see that the results from non-abelian bosonization are consistent with this analysis. Indeed, the Néel order-parameter operator N^a of the interacting fermionic system is given by

$$N^a(x) \sim \psi^\dagger_{\mathrm{R},\alpha}(x) t^a_{\alpha\beta}\psi_{\mathrm{L},\beta}(x) + \mathrm{h.c.} \sim \cos(\sqrt{2\pi}\phi)\mathrm{tr}(t^a g(x)) \tag{7.218}$$

Since the charge field ϕ is effectively massive (due to the charge gap), we can set the factor corresponding to the vertex operator of the charge boson field ϕ to its non-vanishing expectation value, $\langle\cos(\sqrt{2\pi}\phi(x))\rangle = \text{constant}$. Hence, the Néel order-parameter field is effectively the WZW field $g(x)$, $N^a(x) \sim \mathrm{tr}(t^a g(x))$.

This identification allows us to read off the scaling dimension of the Néel order-parameter field given that the correlator of the WZW g field is known from

Eq. (7.194). Thus, we conclude that the scaling dimension of the Néel order-parameter operator is $\Delta = \frac{1}{2}$, which agrees with what we have already found using abelian bosonization. This result was also confirmed numerically (including the effects of the logarithmic corrections to scaling) by Moreo (1987), as well as from a scaling analysis (Essler *et al.*, 2005) of the exact Bethe-ansatz solution of the 1D Hubbard model (Lieb and Wu, 1968). This operator identification also allows us to draw the non-trivial conclusion that the four-point function of the Néel order parameter is given (up to a multiplication constant) by the four-point function of the WZW field, Eq. (7.197) and Eq. (7.199).

7.13.4 Fermions with attractive interactions: gapping the spin sector

We can also apply these ideas to the case of a system with generic interactions g_1 and g_2. This corresponds to a generalized (or extended) Hubbard model. Before we considered the case of repulsive interactions for which $g_1 > 0$ and saw that, for a half-filled system, the charge sector flows to a gapped state while the spin sector flows to the fixed point of an $SU(2)_1$ WZW model and remains gapless.

Let us consider now the case in which $g_1 < 0$, which for the lattice model corresponds to an attractive Hubbard interaction $U < 0$. The RG flow of Eq. (7.217) still applies. However, for $g_1 < 0$ the beta function now has the opposite sign,

$$\beta(|g_1|) = +\frac{N}{\pi} g_1^2 + O(g_1^3) \tag{7.219}$$

which now has the same form as in the 2D non-linear sigma model. Hence we conclude that for attractive interactions the backscattering interaction of the chiral spin currents is marginally relevant and flows to strong coupling. Therefore, in this regime we expect the spin sector to be massive. Indeed, if $g_1 < 0$ flows to strong (attractive) coupling, the SU(2) currents, $J_\mu(x)^a \sim \frac{1}{2}\bar{\psi}_\alpha(x)\gamma^\mu t^a_{\alpha\beta}\psi_\beta(x)$, as *local operators* must vanish when acting on the low-energy Hilbert space of states,

$$J^a_\mu(x)|\text{Phys}\rangle = 0 \tag{7.220}$$

Thus, in this phase the ground state is an SU(2) singlet and all excited states with non-trivial SU(2) quantum numbers are massive. In the strong-coupling limit, all states with SU(2) (spin) quantum numbers are effectively projected out, or gauged away, much in the same way as the charge degrees of freedom become projected out in the repulsive case. The effective field theory of the strong-coupling fixed point is a gauged WZW model. In this case it is an SU(2) subgroup that has been gauged.

On the other hand, the fate of the charge sector at half-filling depends now on the sign of the combination of coupling constants $g_c = 2g_2 - g_1$ for the chiral charge currents. If the microscopic interactions are such that this effective coupling is still repulsive, $g_c > 0$, then the Umklapp term will also be marginally relevant and also flows to strong coupling. Thus, in this case both the spin and the charge sectors are massive. An analysis of the lattice model reveals that this phase has a broken symmetry of translations by one lattice spacing, i.e. a period-2 CDW phase such as one with a dimerized ground state. On the other hand, if the effective coupling constant for backscattering interaction of the chiral charge currents also changes sign, $g_c < 0$, as in the case of the 1D Hubbard model, the phase diagram for the charge sector now involves the full Kosterlitz–Thouless RG flow, resulting either in a massive phase (dimerized) or in a line of fixed points and a massless charge phase. In the case of the Hubbard model the charge sector is gapless. A uniform phase with gapless charge excitations and gapped spin excitations is a Luther–Emery liquid and is effectively a 1D superconductor.

8

Spin-liquid states

In the previous chapters we discussed mostly ordered Néel-like ground states of spin systems. The sole exception was the case of the spin chains in which the ordered state is *always* destroyed by quantum fluctuations. In this chapter we begin a discussion of the ground states of quantum magnets, which, as a result of strong fluctuations, lose the long-range order of their spin degrees of freedom. The key driving force behind this quantum disorder is *frustration*.

8.1 Frustration and disordered spin states

It is possible to drive a Heisenberg model toward a disordered state. One way to do that is to add extra interactions, which, if they are strong enough, may destroy the Néel behavior. A popular choice is to consider next-nearest-neighbor interactions with strength J_2 (Fig. 8.1). These interactions *frustrate* the system in the sense that, for nearest-neighbor interaction J_1 close to J_2, the *classical* Néel state becomes degenerate in energy with other classical configurations that differ from it by *local* spin flips. Quantum mechanically, one may expect a substantial increase of fluctuations, which should further decrease the value of the moment.

By following the steps that led to the non-linear sigma model (see Sections 7.5 and 7.9) and to the bare coupling constant g and spin-wave velocity v_s (Eq. (7.141)), we can compute the new values of g and v_s if we assume that at least the short-range order has the Néel structure of $J_2 = 0$. Clearly, this assumption is correct only for small J_2 and should break down for $J_2 \approx J_1$. We find

$$g' = \frac{g}{\sqrt{1 - 2J_2/J_1}} \equiv u' a_0 \tag{8.1}$$

$$v_s' = v_s \sqrt{1 - 2J_2/J_1} \tag{8.2}$$

Thus, the main effects of frustrating interactions, in the neighborhood of a Néel ordered state, are the increase of the bare coupling g and the decrease of the

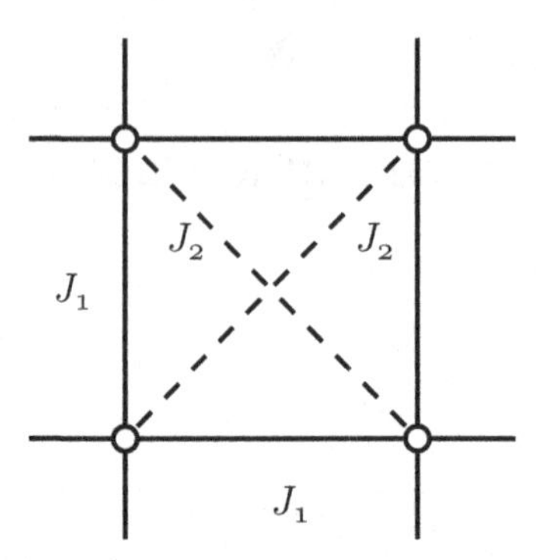

Figure 8.1 A square lattice with nearest-neighbor J_1 and next-nearest-neighbor J_2 interactions.

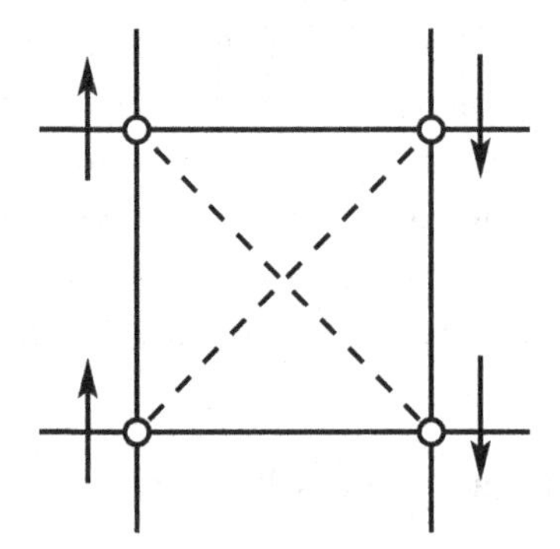

Figure 8.2 A $\vec{Q} = (\pi, 0)$ Néel state.

spin-wave velocity v_s. It is also clear that, for values of J_2 that are sufficiently large, the bare dimensionless coupling constant u' will become larger than the critical value u^*. Consequently, there should be a critical value of the next-nearest-neighbor coupling strength J_{2c} beyond which the long-range Néel order is destroyed. This theory would then predict that for $J_2 \geq J_{2c}$ the system becomes a quantum paramagnet.

It is also clear that if J_2 becomes large enough a new form of long-range order should be found. Indeed, if $J_2 \gg J_1$ a Néel-like state but with wave vector $\vec{Q} = (\pi, 0)$ or $(0, \pi)$ is favored, instead of the usual $\vec{Q} = (\pi, \pi)$ ordered state (Fig. 8.2). This Néel-like state is antiferromagnetic along the x axis but ferromagnetic along the y axis. This form of antiferromagnetism occurs, for instance, in the iron pnictide materials, which are also high-tempertature superconductors.

The low-energy effective action for this state should be a mixture of a sigma model that describes antiferromagnetism and a ferromagnetic Lagrangian of the form of Eq. (7.41). As a matter of fact, the Wess–Zumino terms of the individual spins do not completely cancel out in this case. A term of the form

$$\gamma \int d^3x \, \vec{m} \cdot (\partial_0 \vec{m} \times \partial_1 \vec{m}) \tag{8.3}$$

is found, where γ is a parameter. However, this is *not* a topological (Hopf) term. It merely states that nearby chains exhibit the *same* antiferromagnetic order and that the spins on one chain precess in the average field of the neighboring chains. In reality the effective-field theory of this state is somewhat more complex.

The $\vec{Q} = (\pi, 0)$ state can also be described as two Néel states on two interpenetrating square lattices (rotated by 45°) with order parameters $\vec{n}_1$ and $\vec{n}_2$. The effective action is that of a non-linear sigma model with two coupled fields with an extra coupling $\sim\lambda(\vec{n}_1 \cdot \vec{n}_2)^2$. For $\lambda < 0$ this coupling favors a state in which $\vec{n}_1 \cdot \vec{n}_2 = \pm 1$. Thus suggests that there is a possible state in which the expectation values $\langle \vec{n}_1 \rangle = \langle \vec{n}_2 \rangle = 0$ but $\langle \vec{n}_1 \cdot \vec{n}_2 \rangle \neq 0$, an (Ising) nematic spin state (Chandra *et al.*, 1990).

These states should also become *unstable* for values lower than $J_2 \approx J_1$. Thus, near the classically frustrated limit, $J_1 = 2J_2$, new phases should appear. There are several possibilities. One possible phase is a state without long-range magnetic order, with a gap for spin excitations and a unique ground state. This is the usual paramagnetic state in the quantum zero-temperature limit (QP). We can think of other possible states by considering that when the spin-correlation length becomes very short (i.e. of the order of the lattice constant), the ground state is more naturally described in terms of pairs of spins forming $S = 0$ singlet states over fairly short distances. These states are dubbed *valence-bond* (VB) states. Various disordered states that are based on the VB picture have been proposed. They include VB crystals and resonating-valence-bond (RVB) states, of both long- (Anderson, 1987) and short-range (Kivelson *et al.*, 1987) varieties. Yet other proposals entertain the idea of ground states with *broken time-reversal invariance*. Such is the case of the Kalmeyer–Laughlin (KL) state for the triangular lattice (Kalmeyer and Laughlin, 1987), the chiral spin states for frustrated square lattices of Wen, Wilczek, and Zee (WWZ) (Wen *et al.*, 1989), and states with non-collinear long-range order such as the multi-sublattice Néel states including spirals of Shraiman and Siggia (1989) and of Kane *et al.* (1990). In this chapter we will deal with the disordered phases. Affleck and collaborators (Affleck *et al.*, 1988b) found a class of lattice models whose exact ground states are disordered.

8.2 Valence bonds and disordered spin states

Imagine for the moment a microscopic spin system with interactions which are so strong that the Néel state is destroyed. If the local coupling between the spins is

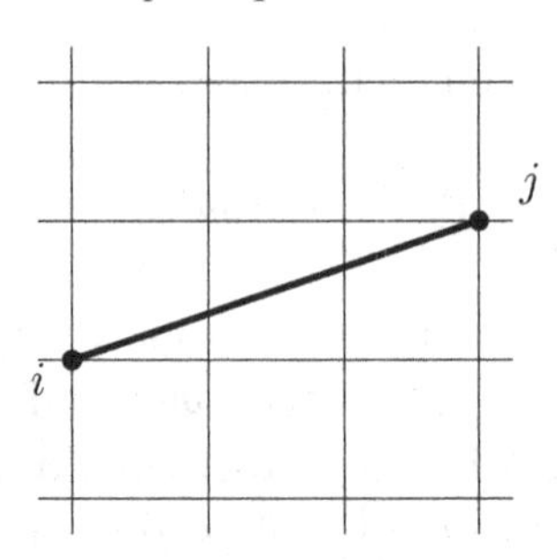

Figure 8.3 A valence bond $|(ij)\rangle$ on a 4×4 square lattice.

very strong, we should expect that a picture based on spin waves, even massive ones, will not work very well. An alternative is to *pair up* the spins into singlet pairs or valence bonds (Anderson, 1973).

Our basic building block will be a singlet pair (valence bond) of two spins at sites i and j of the lattice, which are not necessarily nearest neighbors. Let $|(ij)\rangle$ denote a valence bond pairing up sites i and j (Fig. 8.3). The state $|(ij)\rangle$ is the antisymmetric combination of up and down spins on sites i and j:

$$|(ij)\rangle = \frac{1}{\sqrt{2}}\big(|\uparrow_i\downarrow_j\rangle - |\downarrow_i\uparrow_j\rangle\big) \tag{8.4}$$

This is a spin-singlet state with respect to the total spin operators $\vec{S}^2$ and S_3,

$$\vec{S}^2|(ij)\rangle = 0 \tag{8.5}$$

$$S_3|(ij)\rangle = 0 \tag{8.6}$$

with

$$\vec{S} = \vec{S}_i + \vec{S}_j \tag{8.7}$$

Next, we proceed to partition the sites of a lattice (with an even number of sites) into sets of all possible pairs of sites. If we assign a valence bond to each pair of a given partition, we can define a VB state for the partition as a tensor product of the valence bonds for each pair of sites (Fig. 8.4):

$$|\mathrm{VB}\rangle = \prod_{\mathrm{pairs}} |(i_k j_k)\rangle \tag{8.8}$$

Since each valence bond is odd under the exchange of sites, the overall sign of the VB state is defined only up to a convention regarding how one labels the sites. I will assume that a *fixed* convention has been chosen. Since each pair is a spin singlet, the total spin of the system is necessarily equal to zero. However, zero total spin is not a good definition of a disordered spin state, as we will see below.

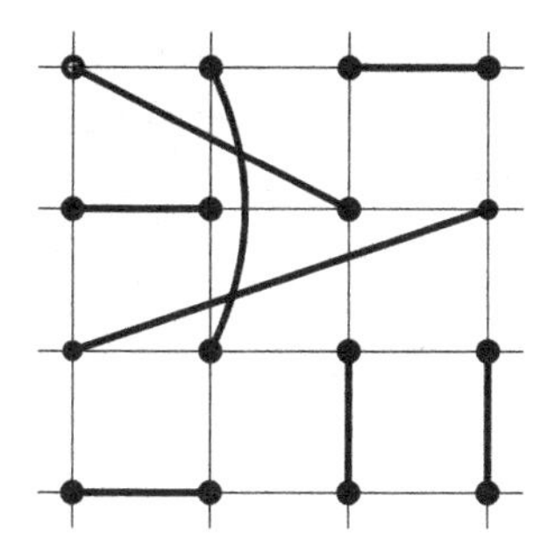

Figure 8.4 A VB state $|\text{VB}\rangle$ on a 4×4 square lattice is the product of eight valence bonds $|(i, j)\rangle$.

A priori we are tempted to consider an *arbitrary* spin-singlet state as a linear superposition of VB states

$$|\Psi\rangle = \sum_P A(P) \prod_{\text{pairs}} |(i_k j_k)\rangle \tag{8.9}$$

which is a sum over all partitions $P = \{(i_k j_k)\}$ with amplitude $A(P)$. However, we run into a difficulty here. The VB states are not orthogonal and, which is more important, in general they cannot all be linearly independent at the same time. The set of VB states is, in general, an over-complete set of states. Therefore, they are not good states for expanding a general wave function. On the other hand, if one is interested in just constructing a variational wave function, it may be convenient to write expressions of the type of Eq. (8.9) with variational parameters. One popular wave function has a *factorized* amplitude. In other words, $A(P)$ is written in the form

$$A(P) = \prod_{\text{pairs}} a(i_k, j_k) \tag{8.10}$$

and the total wave function looks like

$$|\Psi\rangle = \sum_P \prod_{\text{pairs}} a(i_k, j_k)|(i_k j_k)\rangle \tag{8.11}$$

If we further assume that $a(i_k, j_k)$ is only a function of the distance between the paired sites i_k and j_k

$$a(i_k, j_k) = a(|i_k - j_k|) \tag{8.12}$$

we have a resonating-valence-bond (RVB) state (Anderson, 1973). This state has "resonances" in the sense that all valence bonds with sites at the same relative distance enter with the same amplitude. The optimal function $a(|\vec{x}|)$ can be determined by a variational calculation.

The most extensive study of the Heisenberg model using states of this sort was carried out by Liang, Douçot, and Anderson (Liang *et al.*, 1988). The physical properties of a system depend on how fast the function $a(|\vec{x}|)$ decays at infinity. For a power-law ansatz

$$a(|\vec{x}|) \sim \frac{\text{constant}}{|\vec{x}|^{\sigma}} \qquad \text{for large } |\vec{x}| \tag{8.13}$$

They found that for $\sigma < 5$ there is Néel long-range order, even though the wave function is a *global* spin singlet. Conversely, for $\sigma \geq 5$ they do not find Néel order beyond a scale ξ, the correlation length, which is finite.

An extreme case of an RVB state is the short-range RVB state, which is defined as follows. Consider the VB states in which the paired sites are nearest neighbors to each other. There is a one-to-one correspondence between the underlying configurations of valence bonds and the configurations of classical *dimers* (Fig. 8.5) which occupy the bonds. The short-range RVB state, or nearest-neighbor RVB (NNRVB) state, is simply the linear superposition of all such configurations with *equal* amplitude (Kivelson *et al.*, 1987). Thus, states that differ by a local change in the dimer covering have exactly the same amplitude (resonance).

The NNRVB states have one important useful property: they are linearly independent. However, they are not orthogonal. To see this, consider two dimer coverings (a dimer covering is when every lattice site is connected to exactly one of its nearest neighbors by a dimer) that differ only by a local rearrangement of a few nearby spins, such as the example of Fig. 8.6. I will pick the following convention for the signs of the VB states. We will discuss here the case of a bipartite lattice (square). Later in this chapter, and in the next, we will consider the case of the triangular lattice, which is non-bipartite and frustrated. Since the lattice is bipartite, it can be partitioned into two interpenetrating sublattices called R (red) and B (black). A valence bond, or dimer, always joins a red site to a black site. The sign

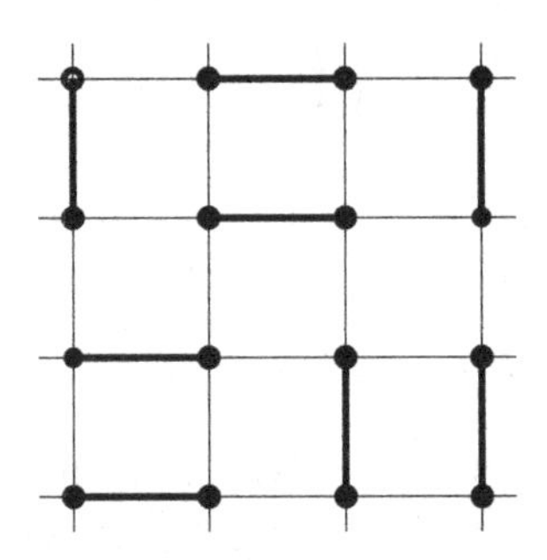

Figure 8.5 A short-range VB state on a 4×4 square lattice. The dark links ("dimers") are valence bonds.

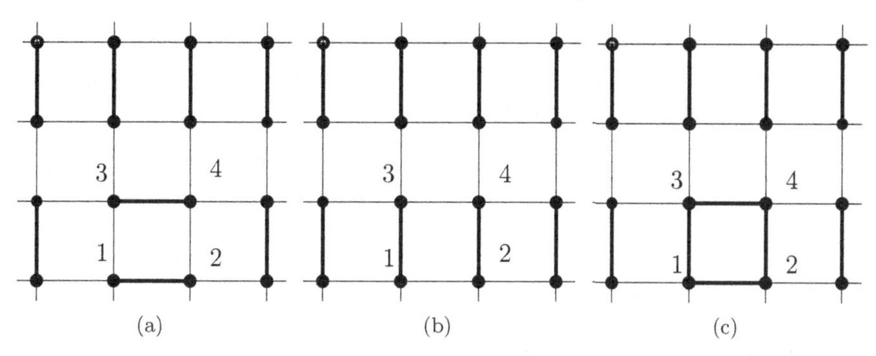

Figure 8.6 In (a) and (b) are shown two configurations of dimers that differ only in the local arrangement of the dimers at the sites 1, 2, 3 and 4. In (c) is shown the superposition of (a) and (b). The closed loop with non-vanishing area connects the sites 1, 2, 3, and 4 with four dimers, and represents the overlap of the non-orthogonal VB states $|a\rangle$ and $|b\rangle$ associated with the dimer covering in (a) and (b).

convention I pick assigns a positive amplitude for every VB state, provided that the red site appears (in the wave function) to the left of the black site. Equivalently, we can give an orientation to the valence bonds: positive for red $\to$ black and negative for black $\to$ red (Kivelson *et al.*, 1987). We can picture this either by assigning an arrow to each VB or by coloring the sites, i.e. the endpoints of the bonds.

Once we have picked a sign convention, we can unambiguously compute overlaps. The overlap between the short-range VB states shown in Figs. 8.6(a) and (b), call them $|a\rangle$ and $|b\rangle$, reduces to the overlap between the product of the two valence bonds which have been rearranged, since the other valence bonds have norm 1 by definition. Let sites 1 and 4 (2 and 3) belong to the red (black) sublattice. The overlap $\langle a|b\rangle$ is equal to

$$\langle a|b\rangle = \langle \overline{12}, \overline{43} | \overline{13}, \overline{42}\rangle \tag{8.14}$$

where $|\overline{12}\rangle$, for instance, denotes the VB

$$|\overline{12}\rangle = \frac{1}{\sqrt{2}}(|\uparrow_1\downarrow_2\rangle - |\downarrow_1\uparrow_2\rangle) \tag{8.15}$$

Thus, $\langle a|b\rangle$ is simply given by

$$\langle a|b\rangle = \frac{1}{4}(\langle\uparrow_1\downarrow_2\uparrow_4\downarrow_3||\uparrow_1\downarrow_2\uparrow_4\downarrow_3\rangle + \langle\downarrow_1\uparrow_2\downarrow_4\uparrow_3||\downarrow_1\uparrow_2\downarrow_4\uparrow_3\rangle) = \frac{1}{2} \tag{8.16}$$

More generally, overlaps between two arbitrary short-range VB states, say $|\Psi_a\rangle$ and $|\Psi_b\rangle$, will not be zero. These overlaps can be represented, and calculated, as a sum over all the closed loops on the square lattice obtained by superposing the dimer coverings associated with $|\Psi_a\rangle$ and $|\Psi_b\rangle$. The length of a loop Γ in units of the lattice spacing is $2L(\Gamma)$, where $L(\Gamma) = 1, 2, \ldots$ Its contribution to the overlap is equal to $2 \times 2^{-L(\Gamma)}$ (the factor $2^{-L(\Gamma)}$ comes from the choice of normalization, Eq. (8.4), while the factor 2 counts the number of ways to antiferromagnetically assign the spins on the sites of a loop) and therefore

$$\langle \Psi_a | \Psi_b \rangle = \prod_{\Gamma} 2 \times 2^{-L(\Gamma)} = 2^{\sum_\Gamma} \times 2^{-\frac{1}{2} \sum_\Gamma 2L(\Gamma)} \tag{8.17}$$

$$= 2^{P(a,b)} \times 2^{-N/2} \tag{8.18}$$

where $P(a,b)$ ($P_{2L}(a,b)$) is the total number of loops (of loops of length $2L$) in the *loop covering* (a,b) and N is the (even) number of sites. For example, the loop covering of the 4×4 square lattice shown in Fig. 8.6(c) has seven loops: six of length 2, which, with our normalization, give factors of 1; and one of length 4, which gives a factor of $\frac{1}{2}$. Thus, the NNRVB state $|\Psi\rangle = \sum_a |\Psi_a\rangle$ has a wave-function normalization $\langle \Psi | \Psi \rangle$ that can be written as a sum of contributions from loops (Sutherland, 1988) of the form

$$\langle \Psi | \Psi \rangle = \sum_{a,b} \langle \Psi_a | \Psi_b \rangle \tag{8.19}$$

$$= 2^{-N/2} \sum_{a,b} 2^{P(a,b)} \times 2^{P(a,b) - P_2(a,b)} \tag{8.20}$$

$$\equiv 2^{-N/2} \sum_{a,b} x^{P_2(a,b)} y^{P(a,b) - P_2(a,b)} \tag{8.21}$$

with $x = 2$ and $y = 4$. Here, the factor $2^{P(a,b) - P_2(a,b)}$ accounts for the fact that there are two ways to have a loop of length $2L > 2$ with a given antiferromagnetic spin assignment on the sites of the loop.

Not only can $\langle \Psi | \Psi \rangle$ be written as a statistical sum such as Eq. (8.21), but also the staggered spin–spin correlator can be written in a similar form. Let $G(\vec{x})$ denote the staggered correlation function

$$G(\vec{x}) = 4(-1)^{x_1 + x_2} \frac{\langle \Psi | \sigma_z(\vec{0}) \sigma_z(\vec{x}) | \Psi \rangle}{\langle \Psi | \Psi \rangle} \tag{8.22}$$

For any loop covering (a,b), there are two possibilities (Kohmoto and Shapir, 1988): (i) the two points $\vec{0}$ and $\vec{x}$ are on the same loop, in which case, due to the antiferromagnetic ordering on the loop, the contribution to the *staggered* correlation function is independent of their relative position; and (ii) the two points belong to different loops and the loop covering does not contribute to the correlation function. In other words,

$$G(\vec{x}) = \frac{\sum_{a,b} \chi(\vec{x}) x^{P_2(a,b)} y^{P(a,b) - P_2(a,b)}}{\sum_{a,b} x^{P_2(a,b)} y^{P(a,b) - P_2(a,b)}} \tag{8.23}$$

where

$$\chi(\vec{x}) = \begin{cases} 1 & \text{if } \vec{0} \text{ and } \vec{x} \text{ are on the same loop} \\ 0 & \text{otherwise} \end{cases} \tag{8.24}$$

We can recast Eq. (8.23) in terms of sums over loops of non-vanishing area. If $L(a, b)$ is the total length of all loops with non-vanishing area for the loop covering (a, b), then $2P_2(a, b) + L(a, b) = N$. Now,

$$G(\vec{x}) = \frac{\sum \chi(\vec{x}) x^{-L/2} y^{P-P_2} d(P_2)}{\sum x^{-L/2} y^{P-P_2} d(P_2)} \tag{8.25}$$

where the summations are only for configurations of loops with non-vanishing areas and $d(P_2)$ is the number of configurations of loops of length 2. Thus the staggered correlation function gives us the probability that the two sites belong to the same loop in a "gas" of loops. Since x and y are fairly small, the loop gas is reasonably dilute. A "quick and dirty" argument shows that the leading contribution to $G(\vec{x})$ should come from the smallest loop that contains both $\vec{0}$ and $\vec{x}$

$$G(\vec{x}) = \frac{x^{\frac{1}{2}(N-2(|\vec{x}|/a_0+1))} y^1 + \cdots}{x^{N/2} y^0 + \cdots} \propto e^{-(|\vec{x}|/a_0) \ln 2} \tag{8.26}$$

Kohmoto and Shapir (1988) have given a more refined argument, which shows that $G(\vec{x})$ is bounded from above by an exponentially decreasing function with correlation length $\xi \approx a_0 e^{+1/\sqrt{2}}$. Thus, short-range RVB *wave functions* represent states with total spin equal to zero and exponentially decreasing correlation functions.

But are any of these RVB states, of either short or long range, good approximations to the ground-state wavefunction of a quantum Heisenberg model? The numerical evidence (Liang *et al.*, 1988) indicates that for the unfrustrated model an RVB-like wave function with fairly long range is a good approximation to the ground state, but it is a Néel state! The short-range RVB is not a good approximation for this system. In fact the short-range spin correlations of the (short-range) RVB state indicate that it may describe the ground state of a Hamiltonian with a finite gap to all spin excitations, a spin gap.

The overlaps and the norm of the RVB wave function thus map onto a problem associated with the classical statistical mechanics of loop models on a given lattice. Classical loop models in two dimensions have a rich phase diagram that depends on the lattice on which the loops are defined. The arguments given above imply that the spin-correlation functions are short-ranged. However, the loops themselves have a more complex behavior. Quite generally a loop model will assign to a loop configuration a weight that will depend on the number of loops and on their length (Nienhuis, 1987; Kondev and Henley, 1996; Kondev, 1997; Fendley *et al.*, 2006). We will see in the next chapter that quantum-dimer models, which have a short-range RVB wave function, are critical on bipartite 2D lattices but not on non-bipartite lattices, such as the triangular lattice. The short-range RVB state on a triangular lattice is gapped and is a $\mathbb{Z}_2$ topological fluid (Moessner and Sondhi, 2001b).

In the case of a *frustrated* system, such as the Heisenberg antiferromagnet on a *triangular* lattice, the situation is less clear. The best available numerical calculations yield a non-collinear magnetically ordered state in which the moments on the vertices of each triangle are rotated by an angle of 120° relative to each other (depicted in Fig. 8.7), with a much smaller magnetic moment than that for the square lattice (Singh and Huse, 1992; Elstner *et al.*, 1993). In contrast, the spin-1/2 quantum Heisenberg antiferromagnet on a kagome lattice (the medial lattice of the honeycomb lattice), which is a frustrated system, appears to be quantum disordered and has a spin gap. In this case there is no long-range magnetic order, and it is a good candidate for either a spin liquid phase or a valence-bond solid phase (Elstner *et al.*, 1993; Leung and Elser, 1993).

On the other hand, density-matrix renormalization-group (Jiang *et al.*, 2011) and tensor-product state (Wang *et al.*, 2011) simulations of the J_1-J_2 spin-1/2 quantum antiferromagnet on a square (frustrated) lattice find strong evidence for a gapped $\mathbb{Z}_2$ spin-liquid phase in the region $0.41 \leq J_2/J_1 \leq 0.62$, separating conventional Néel and striped antiferromagnetic states for smaller and larger J_2/J_1, respectively. In 1991 Wen (Wen, 1991c) had proposed using mean-field-theory arguments of the type described later in this chapter (which were later extended by Mudry and Fradkin (1994)) that the J_1-J_2 quantum antiferromagnet may have a $\mathbb{Z}_2$ topological (spin) liquid phase qualitatively similar to the short-range RVB state.

Cano and Fendley found an SU(2)-invariant Hamiltonian with local interactions (involving local clusters of eight spins on the square lattice!) for which the short-range RVB state is the exact ground state (Cano and Fendley, 2010). Although this

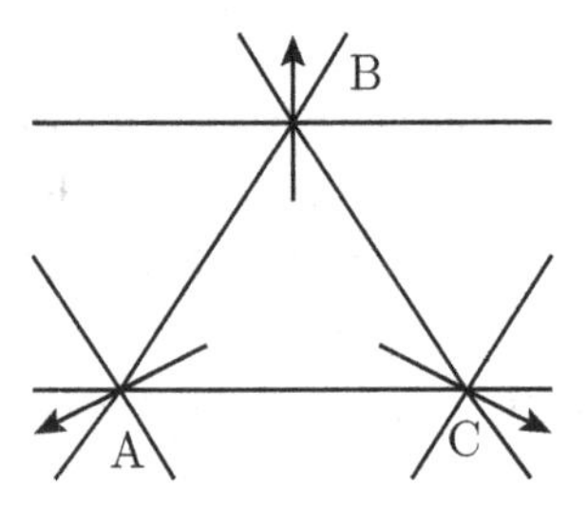

Figure 8.7 The 120° non-collinear three-sublattice magnetic order on the triangular lattice.

state has a finite spin gap, in the case of the square lattice Monte Carlo simulations of this wave function have found that it has long-range (power-law) correlations of valence-bond operators (Albuquerque and Alet, 2010; Tang *et al.*, 2011b) similar to what it is found in the simpler quantum-dimer models (which will be discussed in the next chapter).

8.3 Spinons, holons, and valence-bond states

We will now turn to other states that have been proposed. Since there is good evidence that the Heisenberg antiferromagnet may be in a Néel state, I will take the point of view that these phases may be realized by relatively small modifications of this Hamiltonian. Thus, I will carry out most of the discussion with the Heisenberg (or Hubbard) model in mind as a rather generic example.

At this point, it is convenient to go back to a representation of the spins in terms of either fermion operators or Bose operators. For the most part we have been using a fermion representation of the spins,

$$\vec{S}(\vec{x}) = \frac{1}{2} c^\dagger_\alpha(\vec{x}) \vec{\sigma}_{\alpha\beta} c_\beta(\vec{x}) \tag{8.27}$$

The main motivation for this choice is that the fermion operators $c^\dagger_\alpha(\vec{x})$ are the fermion operators of the Hubbard model. Equation (8.27) reproduces the angular-momentum algebra for spin $S = \frac{1}{2}$ only if the Hilbert space is restricted by the condition

$$n(\vec{x}) = c^\dagger_\alpha(\vec{x}) c_\alpha(\vec{x}) = 1 \tag{8.28}$$

which implies that each site is occupied by a *single* fermion with either up or down spin.

Alternatively we may use bosons to represent spin. Let $a_\alpha(\vec{x})$ be a set of boson-destruction operators. The boson bilinears

$$\vec{S}(\vec{x}) = \frac{1}{2} a^\dagger_\alpha(\vec{x}) \vec{\sigma}_{\alpha\beta} a_\beta(\vec{x}) \tag{8.29}$$

obey the angular-momentum algebra for $S = \frac{1}{2}$ only if the bosons obey the hard-core constraint

$$a^\dagger_\alpha(\vec{x}) a_\alpha(\vec{x}) = 1 \tag{8.30}$$

These formulas, known as the Schwinger boson representation of the spin-1/2 algebra, are reminiscent of the CP^1 representation of the non-linear sigma model of Section 7.5. Indeed, it is possible to derive the CP^1 model using bosons as a starting point. There is an extensive literature on this approach (see Arovas and Auerbach (1988) and Auerbach (1994)), so we will not do this here. This *boson* representation is closely related to standard spin-wave theory (Holstein and Primakoff, 1940).

Let us begin by looking for a representation of the valence bonds in terms of fermions. Let $|0\rangle$ represent the *empty* state. The valence bond on a pair of sites i and j is simply given by

$$|(ij)\rangle \equiv \epsilon_{\alpha\beta} c_\alpha^\dagger(i) c_\beta^\dagger(j)|0\rangle \equiv \left(c_\uparrow^\dagger(i) c_\downarrow^\dagger(j) - c_\downarrow^\dagger(i) c_\uparrow^\dagger(j)\right)|0\rangle \tag{8.31}$$

We will be interested, for the moment, in the half-filled system. Thus the average number of particles per site is one and, because of the constraint, no doubly occupied sites are allowed. For *finite* Hubbard U some doubly occupied sites, as well as empty sites, will occur. We may try to solve the constraint of there being no doubly occupied sites by using a "slave-boson" construction (Coleman, 1984; Read and Newns, 1983). This leads to the RVB theories of Baskaran, Zou, and Anderson (BZA) (Baskaran *et al.*, 1987) and Ruckenstein, Hirschfeld, and Appel (Ruckenstein *et al.*, 1987). In principle, there are several ways of implementing the slave-boson approach. Let us consider the fermion operators to be normal ordered with respect to the half-filled state. In other words, we will assume that we are not too far from half-filling. Let us now *define* a set of Bose and Fermi operators at each site, $b(\vec{x})$ and $f_\alpha(\vec{x})$, respectively, satisfying the constraint (at each site)

$$b^\dagger(\vec{x})b(\vec{x}) + f_\alpha^\dagger(\vec{x})f_\alpha(\vec{x}) = 1 \tag{8.32}$$

Let $|\bar{0}\rangle$ be the reference state for these operators and define the states $|h\rangle$, $|\uparrow\rangle$, and $|\downarrow\rangle$ representing a "hole" (or *holon*) with charge $+e$ and *spin zero* and a *spinon* $|\uparrow\rangle$ ($|\downarrow\rangle$) with *spin up (down)* and *no charge*:

$$\begin{aligned} |h\rangle &\equiv |e, 0\rangle = b^\dagger|\bar{0}\rangle \\ |\uparrow\rangle &\equiv |0, \uparrow\rangle = f_\uparrow^\dagger|\bar{0}\rangle \\ |\downarrow\rangle &\equiv |0, \downarrow\rangle = f_\downarrow^\dagger|\bar{0}\rangle \end{aligned} \tag{8.33}$$

Thus, the only possible states are a holon and a spinon of either orientation. More formally, we can write the operator $c_\sigma^\dagger(\vec{x})$ which creates a band fermion of charge e and spin σ at site $\vec{x}$ in the form

$$c_\sigma^\dagger(\vec{x}) = b(\vec{x}) f_\sigma^\dagger(\vec{x}) \tag{8.34}$$

Alternatively, we can also write $c_\sigma^\dagger(\vec{x})$ in the form

$$c_\sigma^\dagger(\vec{x}) = a(\vec{x}) z_\sigma^\dagger(\vec{x}) \tag{8.35}$$

where a is a spinless charged fermion and the z_σ are Schwinger bosons satisfying the constraint

$$z^\dagger_\alpha(\vec{x}) z_\alpha(\vec{x}) = 1 \tag{8.36}$$

In this representation, the hole (or holon) is a fermion and the spinon is a boson. In either representation, at half-filling, there are no holons. Away from half-filling a number of holons will be present. In the boson-holon version, the holons will superficially appear to undergo a condensation transition, which originally was mistakenly confused with "high-T_c."

8.4 The gauge-field picture of the disordered spin states

I will consider now a particular form of mean-field theory for the Heisenberg antiferromagnet, which was first proposed by Affleck and Marston (1988) and by Kotliar (1988). In this mean-field theory, one focuses on the valence-bond operator of Eq. (8.31). The spin-exchange interaction term, $\vec{S}(\vec{x}) \cdot \vec{S}(\vec{y})$, can be written in the form

$$\vec{S}(\vec{x}) \cdot \vec{S}(\vec{y}) = \frac{1}{2} c^\dagger_\alpha(\vec{x}) c_\beta(\vec{x}) c^\dagger_\beta(\vec{y}) c_\alpha(\vec{y}) - \frac{1}{4} n(\vec{x}) n(\vec{y}) \tag{8.37}$$

Thus, up to an additive constant, we have the fermion problem with the Hamiltonian

$$H = \frac{J}{2} \sum_{\vec{x}, j=1,2} c^\dagger_\alpha(\vec{x}) c_\beta(\vec{x}) c^\dagger_\beta(\vec{x}+e_j) c_\alpha(\vec{x}+e_j) \tag{8.38}$$

which has to be supplemented by the local constraint

$$n(\vec{x}) \equiv c^\dagger_\alpha(\vec{x}) c_\alpha(\vec{x}) = 1 \tag{8.39}$$

In Eq. (8.37), an underlying square lattice has been assumed and $j = 1, 2$ represents the x_1 and x_2 directions, with e_1 and e_2 being the corresponding unit vectors. This approach can be easily generalized to other lattices as well.

The path-integral picture of this system involves the use of the Lagrangian

$$L = \sum_{\vec{x}} c^\dagger_\alpha(\vec{x}, t)(i\,\partial_t + \mu) c_\alpha(\vec{x}, t) + \sum_{\vec{x}} \varphi(\vec{x}, t)(c^\dagger_\alpha(\vec{x}, t) c_\alpha(\vec{x}, t) - 1) - H \tag{8.40}$$

The second term in Eq. (8.40) contains the Lagrange multiplier field $\varphi(\vec{x}, t)$ which enforces the constraint of single occupancy, Eq. (8.39), at all times.

The Affleck–Marston mean-field theory involves a Hubbard–Stratonovich factorization in terms of the link variables $\chi_j(\vec{x})$, which are complex Bose (c-number) fields. The Lagrangian L' is given by

$$
\begin{aligned}
L' =& \sum_{\vec{x}} c^\dagger_\alpha(x)(i\,\partial_t + \mu)c_\alpha(x) + \sum_x \varphi(x)(c^\dagger_\alpha(x)c_\alpha(x) - 1) - \frac{2}{J}\sum_{\vec{x},j} |\chi_j(x)|^2 \\
&+ \sum_{\vec{x},j} \left(c^\dagger_\alpha(\vec{x},t)\chi_j(\vec{x},t)c_\alpha(\vec{x}+e_j,t) + c^\dagger_\alpha(\vec{x}+e_j,t)\chi^*_j(\vec{x},t)c_\alpha(\vec{x},t)\right)
\end{aligned}
\tag{8.41}
$$

where $x \equiv (\vec{x}, t)$. This Lagrangian is equivalent to L upon a Gaussian integration of the Hubbard–Stratonovich fields $\chi_j(x)$. Here, the link variables $\chi_j(x)$ satisfy the relations $\chi_j(\vec{x}, t) = \chi^*_{-j}(\vec{x} + e_j, t)$ since the current operator associated with an electron hopping from $\vec{x}$ to $\vec{x} + e_j$ is the adjoint of the operator associated with the (reverse) hopping from $\vec{x} + e_j$ back to $\vec{x}$.

The mean-field theory (MFT) consists, as usual, in integrating out the fermions, *at a fixed density*, and treating the Bose (c-number) fields $\chi_j(x)$ within a saddle-point expansion. The fields $\chi_j(x)$, being complex, can be parametrized in terms of two real fields $\rho_j(x)$ and $\mathcal{A}_j(x)$ representing the amplitude and phase of $\chi_j(x)$, respectively. Before carrying out the MFT, it is important to consider the symmetries of this Lagrangian. Consider the *local time-dependent gauge transformations*

$$
\begin{aligned}
\mathcal{A}_j(\vec{x}, t) &= \mathcal{A}'_j(\vec{x}, t) + \Delta_j\phi(\vec{x}, t) \\
\varphi(\vec{x}, t) &= \varphi'(\vec{x}, t) + \partial_t\phi(\vec{x}, t) \\
c_\alpha(x) &= e^{i\phi(x)}c'_\alpha(x)
\end{aligned}
\tag{8.42}
$$

These transformations leave the Lagrangian unchanged up to a total time derivative, the term $\sum_{\vec{x}} \partial_t\phi$. Thus, the Lagrange multiplier field, φ, transforms like the time component $\mathcal{A}_0$ of a U(1) gauge field. We must then conclude that this system has a "secret" gauge (local) symmetry.

The effective Lagrangian Eq. (8.41) is reminiscent of the Lagrangians of lattice gauge theories (Kogut, 1984). There are a few significant differences: (a) here the amplitude field $|\chi_j(x)| = \rho_j(x)$ fluctuates; (b) there is no *explicit* kinetic-energy term for the gauge fields $\mathcal{A}_\mu$ (i.e. an $F^2_{\mu\nu}$); and (c) there is an extra term in the Lagrangian that is proportional to φ, i.e. to $\mathcal{A}_0$. This last term may seem to break gauge invariance, since, according to Eq. (8.42), φ transforms like $\varphi \to \varphi' + \partial_t\phi$. However, we must keep in mind that what matters is not the Lagrangian but the *action*, $\mathcal{S}$,

$$
\mathcal{S} = \int dt\, L \tag{8.43}
$$

Under a gauge transformation, the extra term will transform the action by

$$\begin{aligned} \mathcal{S} \to \mathcal{S} &- \sum_{\vec{x}} \int dt\, \partial_t \phi(\vec{x}, t) \\ &= \mathcal{S} - \sum_{\vec{x}} (\phi(\vec{x}, t \to +\infty) - \phi(\vec{x}, t \to -\infty)) \end{aligned} \tag{8.44}$$

If we impose periodic boundary conditions (in time) on the gauge fields, as we must when computing a *trace* over Bose (or Fermi) fields, we must allow only for local gauge transformations that respect the boundary conditions. Thus, the field $\phi(\vec{x}, t)$ must obey periodic boundary conditions in time, $\phi(\vec{x}, t \to +\infty) = \phi(\vec{x}, t \to -\infty)$, which leave the action unchanged. We can relax this condition to a small extent. Let us notice that the "extra term" can be extracted from the action and written into the integrand in the form of a product of operators of the form

$$e^{-i \int dt \sum_{\vec{x}} \varphi(\vec{x}, t)} \equiv \prod_{\vec{x}} e^{-i \int dt\, \varphi(\vec{x}, t)} \tag{8.45}$$

Since φ can be identified with $\mathcal{A}_0$, the time component of a vector potential $\mathcal{A}_\mu$ that obeys periodic boundary conditions, we can write the extra terms in the form of time-ordered exponentials of line integrals over loops $\Gamma(\vec{x})$ that close around the time direction (see Fig. 8.8). These operators are generally called *Wilson loops*:

$$e^{-i \sum_{\vec{x}} \int dt\, \varphi(\vec{x}, t)} \equiv \prod_{\vec{x}} e^{-i \oint dt\, \mathcal{A}_0(\vec{x}, t)} \equiv \prod_{\vec{x}} e^{-i \oint_{\Gamma(\vec{x})} dx_\mu \mathcal{A}^\mu} \tag{8.46}$$

For the Wilson loops to be gauge-invariant operators,

$$\oint_{\Gamma(\vec{x})} dx_\mu\, \mathcal{A}^\mu = \oint_{\Gamma(\vec{x})} dx_\mu\, \mathcal{A}'^\mu + \oint_{\Gamma(\vec{x})} dx_\mu\, \partial^\mu \phi = \oint_{\Gamma(\vec{x})} dx_\mu\, \mathcal{A}'^\mu \tag{8.47}$$

it is *sufficient* that $d\phi$ is exact, i.e. the gauge transformation is *non-singular everywhere*. Recall that these Wilson loops appeared in our problem since we had to enforce the *constraint* of single occupancy at *every* site and at *all* times.

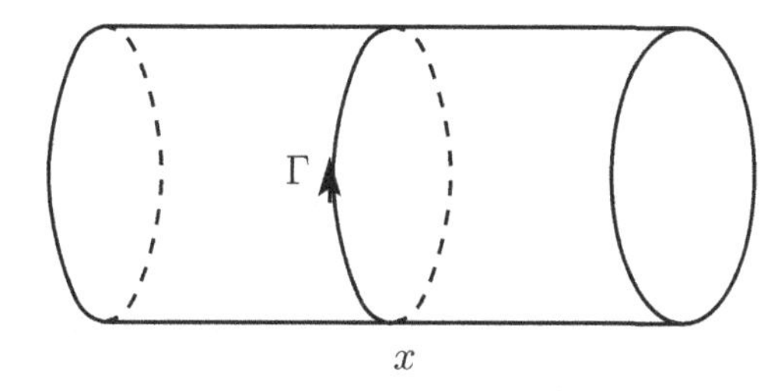

Figure 8.8 A Wilson loop along the closed curve $\Gamma(\vec{x})$ in the time direction.

Because of the gauge invariance, we need to impose the constraint of single occupancy, Eq. (8.39), only on the configuration space at some initial time surface, $t = t_0$. The local gauge invariance implies that the spin configurations at an arbitrary later time t must still obey the same constraint; i.e. they are smooth *deformations* of the initial configuration. For instance, we cannot try to fix the gauge $\mathcal{A}_0 = 0$ if only non-singular gauge transformations are allowed. This gauge is not consistent with the constraint of single occupancy since a configuration with $\mathcal{A}_0 = 0$ has $\oint dt\, \mathcal{A}_0 = 0$ and, because of gauge invariance, it cannot evolve into configurations with $\oint dt\, \mathcal{A}_0 \neq 0$. At best we can fix $\mathcal{A}_0(\vec{x}, t)$ to be a time-independent arbitrary function $\mathcal{A}_0(\vec{x})$ through

$$\oint dt\, \mathcal{A}_0(\vec{x}, t) \equiv T\mathcal{A}_0(\vec{x}) \equiv \bar{\mathcal{A}}_0(\vec{x}) \tag{8.48}$$

where T is the time span. Alternatively, we may also choose the gauge

$$\mathcal{A}_0'(\vec{x}, t) \equiv \bar{\mathcal{A}}_0(\vec{x})\delta(t - t_0) \tag{8.49}$$

which yields the same value of the line integral. This choice means that, at $t = t_0$, we restrict the space of configurations to obey the constraint $n(\vec{x}) = 1$ at all points $\vec{x}$. Gauge invariance then takes care of choosing only the time-evolving configurations which satisfy this property.

However, it is worth noticing that, from the point of view of quantum mechanics, what matters is not the invariance of the action $\mathcal{S}$ but the invariance of the amplitude $e^{i\mathcal{S}}$ assigned to a given history (Feynman and Hibbs, 1965). Thus gauge transformations that change during the time span T by $\Delta\phi(\vec{x}) = 2\pi m(\vec{x})$ (an arbitrary integer modulo 2π at each point $\vec{x}$) *are allowed,* since they do not change the amplitude, although they do change the action. These are the so-called *large* gauge transformations. These transformations change the time-like Wilson loops accordingly:

$$\oint dt\, \mathcal{A}_0 = \oint dt\, \mathcal{A}_0' + 2\pi m(\vec{x}) \tag{8.50}$$

and thus are *singular* or *large* gauge transformations. A correct description of these systems, particularly at non-zero temperatures, requires a careful treatment of these large gauge transformations.

We wish to evaluate the functional integral for a system with a Lagrangian of the form of Eq. (8.41). We will attempt a semi-classical treatment of this theory. One difficulty that we will encounter will be that there is no small parameter to organize this semi-classical expansion. Thus we should have every reason to suspect that the results might not be quite reliable. Indeed, using this approach, it is quite hard to reproduce a Néel state. This is so because the approximations that we will make will be accurate for systems that can be described in terms of valence bonds. In this representation we deal with local spin singlets and the spins fluctuate very fast. Conversely, in a Néel state, the spins are *slow* variables but the VBs are *fast* ones. These are complementary descriptions.

Several systematic procedures have been devised in order to control the fluctuations in this problem. Affleck and Marston (Affleck and Marston, 1988; Marston and Affleck, 1989) proposed studying generalizations of the quantum Heisenberg model to a system with an SU(N) symmetry by attaching a "color" index $\alpha = 1, \ldots, N$ to the fermionic degrees of freedom. The spin-1/2 model was obtained by considering the $N = 2$ (SU(2)) case. The Affleck–Marston Lagrangian has, after an RVB decoupling by means of a link variable $\chi_j(\vec{x}, t)$, the same form as the Lagrangian of Eq. (8.41) except that (a) $\alpha = 1, \ldots, N$ (not just 1 and 2, or $\uparrow$ and $\downarrow$) and (b) the local occupancy is equal not to 1 but to a suitably chosen function $n(\vec{x})$,

$$\sum_{\alpha=1}^{N} c_\alpha^\dagger(\vec{x}) c_\alpha(\vec{x}) = n(\vec{x}) \tag{8.51}$$

which they proposed could take one of two forms on a system with two interpenetrating sublattices, A and B (suitable for bipartite lattices such as the square of the honeycomb lattice):

$$n(\vec{x}) = \begin{cases} 1 & \vec{x} \in A \\ N-1 & \vec{x} \in B \end{cases} \tag{8.52}$$

or

$$n(\vec{x}) = \frac{N}{2} \qquad \vec{x} \in A \text{ or } \vec{x} \in B \tag{8.53}$$

Read and Sachdev (1989) further generalized this model and considered an SU(N) "Heisenberg antiferromagnet" of the form

$$H = \frac{J}{N} \sum_{(\vec{x}, \vec{x}')} \sum_{\alpha,\beta=1}^{N} S_\alpha^\beta(\vec{x}) S_\beta^\alpha(\vec{x}') \tag{8.54}$$

where $\vec{x} \in A$ and $\vec{x}' \in B$. The operators $S_\beta^\alpha(\vec{x})$ are generators of the Lie group SU(N). If we choose a representation of SU(N) with a Young tableau with m rows and n_c columns ($0 < m < N$) on sublattice A and $N - m$ rows and n_c columns on sublattice B (i.e. the *conjugate* of the representation on sublattice A) (Fig. 8.9) we can write $S_\beta^\alpha(\vec{x})$ in terms of fermions as follows:

$$S_\alpha^\beta(\vec{x}) = \sum_{a=1}^{n_c} c_{\alpha a}^\dagger(\vec{x}) c^{\beta a}(\vec{x}) - \delta_\alpha^\beta \frac{n_c}{2} \tag{8.55}$$

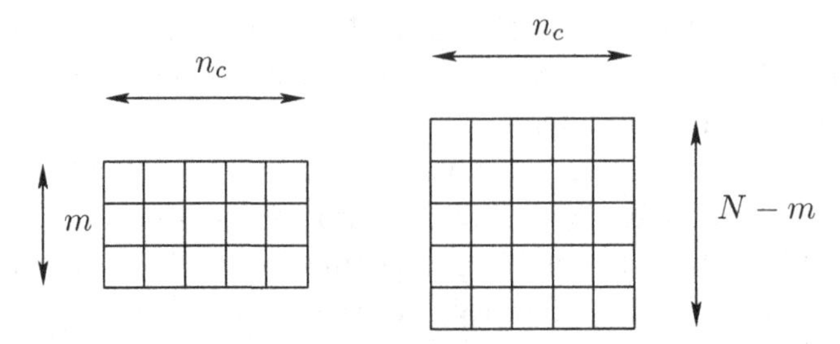

Figure 8.9 Conjugate representations of SU(N) on a bipartite lattice. The case shown here is the $(m, n_c) = (3, 5)$ representation of SU(8) and its conjugate (5, 5).

at the price of introducing an extra ("flavor") index $a = 1, \ldots, n_c$. The representation is fixed by the constraint (Read and Sachdev, 1989)

$$\sum_{\alpha=1}^{N} c^{\dagger}_{\alpha a}(\vec{x}) c^{\alpha b}(\vec{x}) = \begin{cases} \delta^b_a m & \vec{x} \in A \\ \delta^b_a (N-m) & \vec{x} \in B \end{cases} \tag{8.56}$$

Hence, there are mn_c fermions on sublattice A and $(N-m)n_c$ fermions on sublattice B. For example, for $N = 2$ (SU(2)), the only available value of m is 1 and n_c is arbitrary. It is easy to show that this representation has n_c spins one-half. The constraint means that the allowed states are *symmetric* under a permutation of the spins. This is the representation (or multiplet) with spin $s = n_c/2$. Thus, all the representations of SU(2) have been included. This is important since the limit $n_c \to \infty$, $N = 2$, is the spin-wave theory. The $1/S$ expansion discussed in Chapter 7 is simply the $1/n_c$ expansion here (since $S = n_c/2$).

A general difficulty of using this approach is that it breaks translation invariance if non-self-conjugate representations are placed on the two sublattices, and, even in that case, it can be implemented only for bipartite lattices. To sidestep this problem Read and Sachdev (1991) introduced a different generalization of the Heisenberg model for general, including non-bipartite, lattices, such as the triangular and kagome lattices. This approach amounts to generalizing the SU(2) spin-1/2 degrees of freedom to the symplectic group Sp(N) (instead of the unitary group SU(N)): the group of $2N \times 2N$ unitary matrices U that leave the (fermion or boson) bilinear, the valence-bond operator on lattice sites i and j,

$$\mathcal{J}^{aa'}_{\sigma\sigma'} c^{\dagger}_{ia\sigma} c^{\dagger}_{ja'\sigma'} \tag{8.57}$$

(where $\mathcal{J}^{aa'}_{\sigma\sigma'} = \delta_{aa'}\epsilon_{\sigma\sigma'}$, $\sigma, \sigma' = 1, 2$, and $a, a' = 1, \ldots, N$) under the transformation $c \to Uc$. For $N = 1$ this reduces to SU(2). As in the other generalizations, the number of particles, fermions n_f or bosons n_b, must be constrained at every given site to specify the chosen representation of Sp(N).

Here we will follow the analysis of Affleck and Marston and consider only the case of self-conjugate representations of SU(N) (i.e. the Young tableaux have the same number of rows $m = N/2$ for both sublattices). This is possible only for N even. We will consider only the fundamental representation, which has $n_c = 1$. The limit n_c large is more conveniently described in terms of Schwinger bosons (Arovas and Auerbach, 1988) or in terms of a coherent-state path integral (Read and Sachdev, 1989). Both representations lead to a generalization of the non-linear sigma model of Chapter 7. We will not pursue this approach here.

The Lagrangian density of Eq. (8.41) now has the form

$$\begin{aligned}\mathcal{L}' = c^\dagger_{\alpha a}(\vec{x}, t)(i\,\partial_t + \mu)c_{\alpha a}(\vec{x}, t) + \varphi_{ab}(\vec{x}, t)\left(c^\dagger_{\alpha a}(\vec{x}, t)c_{\alpha b}(\vec{x}, t) - \delta_{ab}\frac{N}{2}\right)\\ - \frac{N}{J}|\chi^{ab}_j(\vec{x}, t)|^2 + c^\dagger_{\alpha a}(\vec{x}, t)\chi^{ab}_j(\vec{x}, t)c_{\alpha b}(\vec{x} + e_j, t)\\ + c^\dagger_{\alpha b}(\vec{x} + e_j, t)\chi^{ab}_j(\vec{x}, t)^* c_{\alpha a}(\vec{x}, t)\end{aligned} \tag{8.58}$$

where $\chi^{ab}_j(\vec{x}, t)$ is an $n_c \times n_c$ complex matrix field on each link of the square lattice (labeled by the site $\vec{x}$ and the direction $j = 1, 2$) satisfying

$$\chi^{ab}_j(\vec{x}, t) = \chi^{ba}_{-j}(\vec{x} + e_j, t)^* \tag{8.59}$$

The field $\chi^{ab}_j(\vec{x}, t)$ is a generalization of $\chi_j(\vec{x}, t)$ in Eq. (8.41). This Lagrangian density has a non-abelian gauge invariance that is a generalization of Eq. (8.42). The functional integral is

$$Z = \int \mathcal{D}\chi\, \mathcal{D}\varphi\, \mathcal{D}c^\dagger\, \mathcal{D}c\; e^{i\mathcal{S}} \prod_{\vec{x}} e^{-i\frac{N}{2}\oint dt \varphi_{aa}(\vec{x}, t)} \tag{8.60}$$

The action $\mathcal{S}$ is a *bilinear* form in fermions. Hence, once again, they can be integrated out at the expense of a determinant. The effective action $\mathcal{S}_{\text{eff}}$, resulting from integrating out the fermions, is

$$\mathcal{S}_{\text{eff}}[\varphi, \chi_j] = N\bar{\mathcal{S}}[\varphi, \chi_j] \tag{8.61}$$

where

$$\begin{aligned}\bar{\mathcal{S}}[\varphi, \chi_j] = -i\,\mathrm{tr}\ln[((i\,\partial_t + \mu)\delta_{ab} + \varphi_{ab}(\vec{x}, t))\,\delta_{\vec{x},\vec{x}'}\delta_{t,t'}\\ + (\chi^{ab}_j(\vec{x}, t)\delta_{\vec{x}',\vec{x}+e_j} + \chi^{ba}_j(\vec{x} - e_j, t)^*\delta_{\vec{x}',\vec{x}-e_j})\delta_{t,t'}]\\ - \int dt \sum_{\vec{x}} \frac{1}{J}|\chi^{ab}_j(\vec{x}, t)|^2\end{aligned} \tag{8.62}$$

We can also decompose $\chi^{ab}_j(\vec{x}, t)$ into an amplitude and a phase,

$$\chi^{ab}_j(\vec{x}, t) = \rho^{ab}_j(\vec{x}, t)e^{i\mathcal{A}^{ab}_j(\vec{x}, t)} \tag{8.63}$$

where $\rho_j^{ab}(\vec{x},t)$ is a positive-definite real symmetric matrix and $\mathcal{A}_j^{ab}(\vec{x},t)$ is in the Lie algebra of SU(N) (i.e. $e^{i\mathcal{A}}$ is a group element). Clearly, $\varphi^{ab}(\vec{x},t)$ can be regarded as the time component $\mathcal{A}_0^{ab}(\vec{x},t)$ of the non-abelian vector potential $\mathcal{A}_\mu^{ab}(\vec{x},t)$, while $\mathcal{A}_j^{ab}(\vec{x},t)$ are its space components. The saddle-point approximation is justified if we take the limit $N \to \infty$ keeping $n_c < \infty$. In the Bose representation, on the other hand, the limit one is forced to consider has $N < \infty$ and $n_c \to \infty$. Thus, although the theories should be equivalent, their saddle-point approximations have quite different physics. The limit $n_c \to \infty$ means high representations and Néel-like behavior. The opposite limit, $N \to \infty$, n_c fixed, has VB states and flux phases but no Néel states.

8.5 Flux phases, valence-bond crystals, and spin liquids

For the most part I will consider only the case $n_c = 1$, which is simplest. However, there are some important new features that arise for $n_c > 1$, which I will mention in passing. For $n_c = 1$, the symmetry is abelian.

The saddle-point approximation implies considering configurations of $\bar{\rho}_j(\vec{x},t)$ and $\bar{\mathcal{A}}_\mu(\vec{x},t)$ such that

$$\frac{\delta \mathcal{S}_{\text{tot}}}{\delta \bar{\rho}_j(\vec{x},t)} = 0 \tag{8.64}$$

and

$$\frac{\delta \mathcal{S}_{\text{tot}}}{\delta \bar{\mathcal{A}}_\mu(\vec{x},t)} = 0 \tag{8.65}$$

where $\mathcal{S}_{\text{tot}}$ is given from Eqs. (8.61) and (8.60) by

$$\mathcal{S}_{\text{tot}} = \mathcal{S}_{\text{eff}} - \frac{1}{2}\sum_{\vec{x}} \oint dt\, \mathcal{A}_0 = \mathcal{S}_{\text{eff}} - \sum_{\vec{x}} \oint dt\, J_\mu \mathcal{A}^\mu \tag{8.66}$$

with $J_\mu = \frac{1}{2}\delta_{\mu 0}$. Equation (8.64) determines the value (or configuration) of $\rho(\vec{x},t)$ which extremizes the action. Similarly, Eq. (8.65) implies the absence of fermion currents j_μ^{F} in the ground state

$$\frac{\delta \mathcal{S}_{\text{tot}}}{\delta \bar{\mathcal{A}}_\mu(x)} = \frac{\delta \mathcal{S}_{\text{eff}}}{\delta \bar{\mathcal{A}}_\mu(x)} - J_\mu(x) \equiv j_\mu^{\text{F}}(x) - J_\mu(x) = 0 \tag{8.67}$$

In other words, the average fermion density is equal to unity, as required by the constraint, and the average current vanishes.

Two types of solutions have been proposed to solve the saddle-point equations: (i) flux phases and (ii) valence-bond-crystal (or Peierls) phases.

Let us look first for solutions of the saddle-point equations with maximal symmetry. For instance, we want solutions of Eq. (8.64) that are independent of $(\vec{x}, t)$ and of j:

$$\bar{\rho}_j(\vec{x}, t) = \bar{\rho} \tag{8.68}$$

We may also ask for a possible solution with non-zero value of $\bar{\mathcal{A}}_j(\vec{x}, t)$ but with $\bar{\mathcal{A}}_0 = 0$. The value of $\bar{\mathcal{A}}_j(\vec{x}, t)$ may be chosen to be time-independent but not constant in space since, in that case, it would be gauge equivalent to zero. Thus we require that the *circulation* of $\bar{\mathcal{A}}_j(\vec{x}, t)$, or *flux* $\bar{\mathcal{B}}$, around *any* elementary plaquette be constant,

$$\sum_{\text{plaquette}} \bar{\mathcal{A}}_j(\vec{x}, t) = \bar{\mathcal{B}} \tag{8.69}$$

In general, a non-zero flux $\bar{\mathcal{B}}$ violates time-reversal invariance since the time-reversal transformation maps $\bar{\mathcal{B}} \to -\bar{\mathcal{B}}$. But this system is periodic in $\mathcal{A}_j$, i.e. $\mathcal{A}_j$ and $\mathcal{A}'_j = \mathcal{A}_j + 2\pi n_j$ (here n_j is an arbitrary integer) cannot be distinguished. Thus $\bar{\mathcal{B}}$ is defined up to an integer multiple of 2π. There are two values of $\bar{\mathcal{B}}$ compatible with time-reversal invariance: $\bar{\mathcal{B}} = 0, \pi$. Any other value of $\bar{\mathcal{B}}$ represents a state with broken time-reversal symmetry, and the state is chiral. We will see below that phases of this type can arise in *frustrated* quantum antiferromagnets. These phases are called *chiral spin liquids*, and will be discussed in Chapter 10.

On the other hand, there are also solutions that break translation and/or rotation invariance, namely *valence-bond crystals*. In these VB states, the field $\bar{\chi}_j(\vec{x}, t)$ has an amplitude $\bar{\rho}_j(\vec{x}, t)$ that takes non-zero values only on dimer configurations: $\bar{\rho}_j(\vec{x}, t) = \bar{\rho}$ on those links covered by dimers and zero elsewhere. In Chapter 9 we will see that the quantum fluctuations of dimer configurations are described by quantum-dimer models. These models have crystalline phases. They also have phases in which translation and rotation invariance are restored. These are $\mathbb{Z}_2$ spin-liquid phases.

Let us consider the saddle-point equations for $n_c = 1$ in more detail. We look for solutions that are time-independent and have $\bar{\mathcal{A}}_0 = 0$. Thus, $\bar{\rho}_j$ and $\bar{\mathcal{B}}$ are constant in time. From Eq. (8.58) we infer that the dynamics of the fermions, the *spinons* of this system, is governed by the effective Hamiltonian

$$\begin{aligned} H_{\text{MF}} = &- \sum_{\vec{x}, j} \bar{\rho}_j(\vec{x}) \left(c_\alpha^\dagger(\vec{x}) e^{i\bar{\mathcal{A}}_j(\vec{x})} c_\alpha(\vec{x} + e_j) + c_\alpha^\dagger(\vec{x} + e_j) e^{-i\bar{\mathcal{A}}_j(\vec{x})} c_\alpha(\vec{x}) \right) \\ &+ \frac{N}{J} \sum_{\vec{x}, j} \bar{\rho}_j^2(\vec{x}) \end{aligned} \tag{8.70}$$

in the background $\{\bar{\rho}_j(\vec{x}), \bar{\mathcal{B}}(\vec{x})\}$. Here, we have $\frac{1}{2} N L^2$ fermions in a system with the linear dimension L.

The BZA phases. Let us consider first the uniform solutions which have $\bar{\rho}_j(\vec{x}) = \bar{\rho}$ (constant). We saw above that there are only two allowed values of $\bar{\mathcal{B}}$ consistent with time-reversal invariance. For $\bar{\mathcal{B}} = 0$, the spinons have a square Fermi surface (see Fig. 2.2). This is the state found by Baskaran, Zou, and Anderson (BZA). The total energy of the BZA state is

$$E_{\text{BZA}} = \frac{2NL^2}{J}\bar{\rho}^2 - \frac{8}{\pi^2}NL^2\bar{\rho} \tag{8.71}$$

The minimum is attained for $\bar{\rho} = 2J/\pi^2$ and $E_{\text{BZA}} = -8NL^2J/\pi^4$. Superficially, this state looks like a Fermi liquid of spinons. However, the fluctuations are likely to destroy this state. There are, naturally, amplitude fluctuations, $\tilde{\rho}_j(x) = \rho_j(x) - \bar{\rho}$. These fluctuations are essentially local in character and may trigger an instability towards a *Peierls* state in which $\bar{\rho}$ may have a periodic component in space. More importantly, the gauge fields are completely unconstrained. The result is a state in which the constraint of single occupancy is enforced and in which there is no current flow.

Flux phases. The state with $\bar{\rho}_j(\vec{x}) = \bar{\rho}$ (constant) and $\bar{\mathcal{B}} = \pi$, everywhere, is called the flux phase. In the flux phase, the dynamics of the spinons is also governed by a mean-field Hamiltonian H_{flux} of the form of Eq. (8.70),

$$\begin{aligned} H_{\text{flux}} = &-\bar{\rho}\sum_{\vec{x},j}\left(c_\alpha^\dagger(\vec{x})e^{i\bar{A}_j(\vec{x})}c_\alpha(\vec{x}+e_j) + c_\alpha^\dagger(\vec{x}+e_j)e^{-i\bar{A}_j(\vec{x})}c_\alpha(\vec{x})\right) \\ &+ \frac{2NL^2}{J}\bar{\rho}^2 \end{aligned} \tag{8.72}$$

The vector potentials $\bar{A}_j(\vec{x})$ should have circulation equal to π around every elementary plaquette,

$$\sum_{\text{plaquette}} \bar{A}_j(\vec{x}) = \pi \tag{8.73}$$

We can solve this requirement by the (gauge-dependent) *choice*

$$\begin{aligned} \bar{A}_1(\vec{x}) &= +\frac{\pi}{2} \\ \bar{A}_2(\vec{x}) &= -\frac{\pi}{2}(-1)^{x_1} \end{aligned} \tag{8.74}$$

In this phase, the (spinon) Fermi fields $c_\alpha(\vec{x}, t)$ satisfy the equation of motion

$$\begin{aligned} i\,\partial_t c_\alpha(\vec{x}, t) &= \left[c_\alpha(\vec{x}, t), H_{\text{flux}}\right] \\ &= -\bar{\rho}\sum_{j=1,2}\left(e^{i\bar{A}_j(\vec{x})}c_\alpha(\vec{x}+e_j, t) + e^{-i\bar{A}_j(\vec{x}-e_j)}c_\alpha(\vec{x}-e_j, t)\right) \end{aligned} \tag{8.75}$$

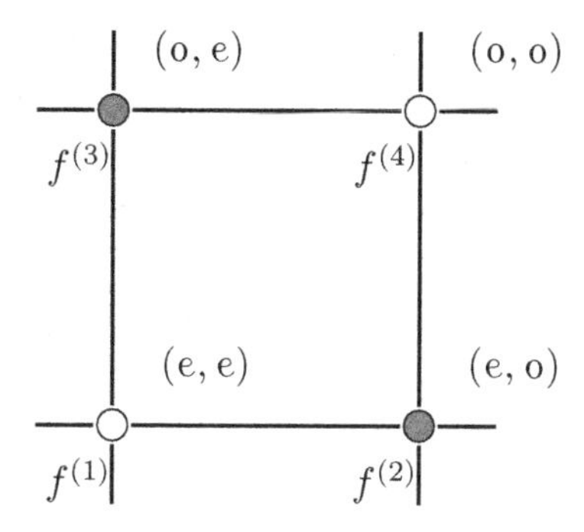

Figure 8.10 The four sublattices associated with a flux phase; (e, e), (e, o), (o, e) and (o, o) denote four sublattices with x_1 even (e) or odd (o), and x_2 even or odd, respectively.

It is convenient to split the square lattice into four sublattices, as shown in Fig. 8.10, and to introduce a separate amplitude f_α (with $\alpha = 1, \ldots, 4$) for each sublattice

$$
\begin{aligned}
i\,\partial_0 f_\alpha^{(1)}(\vec{x}) &= -i\bar{\rho}\left[f_\alpha^{(2)}(\vec{x}+e_1) - f_\alpha^{(2)}(\vec{x}-e_1)\right] \\
&\quad + i\bar{\rho}\left[f_\alpha^{(3)}(\vec{x}+e_2) - f_\alpha^{(3)}(\vec{x}-e_2)\right] \\
i\,\partial_0 f_\alpha^{(2)}(\vec{x}+e_1) &= -i\bar{\rho}\left[f_\alpha^{(1)}(\vec{x}+2e_1) - f_\alpha^{(1)}(\vec{x})\right] \\
&\quad - i\bar{\rho}\left[f_\alpha^{(4)}(\vec{x}+e_1+e_2) - f_\alpha^{(4)}(\vec{x}+e_1-e_2)\right] \\
i\,\partial_0 f_\alpha^{(3)}(\vec{x}+e_2) &= -i\bar{\rho}\left[f_\alpha^{(4)}(\vec{x}+e_1+e_2) - f_\alpha^{(4)}(\vec{x}-e_1+e_2)\right] \\
&\quad + i\bar{\rho}\left[f_\alpha^{(1)}(\vec{x}+2e_2) - f_\alpha^{(1)}(\vec{x})\right] \\
i\,\partial_0 f_\alpha^{(4)}(\vec{x}+e_1+e_2) &= -i\bar{\rho}\left[f_\alpha^{(3)}(\vec{x}+2e_1+e_2) - f_\alpha^{(3)}(\vec{x}+e_2)\right] \\
&\quad - i\bar{\rho}\left[f_\alpha^{(2)}(\vec{x}+e_1+2e_2) - f_\alpha^{(2)}(\vec{x}+e_1)\right]
\end{aligned}
\tag{8.76}
$$

If we denote by $\Delta_j\phi(\vec{x}, t)$ the finite symmetric difference

$$
\Delta_j\phi(\vec{x}, t) = \phi(\vec{x}+e_j, t) - \phi(\vec{x}-e_j, t)
\tag{8.77}
$$

we can write the equation of motion, Eq. (8.76), in vector form (with $a = 1, 2, 3, 4$),

$$
i\,\partial_t f_\alpha^{(a)}(\vec{x}, t) = -i\bar{\rho}M^{ab} f_\alpha^{(b)}(\vec{x}, t)
\tag{8.78}
$$

provided that $\vec{x}$ stands for an (e, e) site and the $f^{(1)}$, $f^{(2)}$, $f^{(3)}$, and $f^{(4)}$ components have the coordinates shown in Fig. 8.10. The matrix M^{ab} is given in terms of the symmetric difference operators Δ_j ($j = 1, 2$) by

$$M^{ab} = \begin{pmatrix} 0 & \Delta_1 & -\Delta_2 & 0 \\ \Delta_1 & 0 & 0 & \Delta_2 \\ -\Delta_2 & 0 & 0 & \Delta_1 \\ 0 & \Delta_2 & \Delta_1 & 0 \end{pmatrix} \tag{8.79}$$

Consider now the linear combinations, the two-component spinor $u^{(a)}$

$$\begin{aligned} u_\alpha^{(1)}(\vec{x}, t) &= f_\alpha^{(1)}(\vec{x}, t) + f_\alpha^{(2)}(\vec{x} + e_1, t) \\ u_\alpha^{(2)}(\vec{x}, t) &= f_\alpha^{(3)}(\vec{x} + e_2, t) - f_\alpha^{(4)}(\vec{x} + e_1 + e_2, t) \end{aligned} \tag{8.80}$$

and $v^{(a)}$

$$\begin{aligned} v_\alpha^{(1)}(\vec{x}, t) &= f_\alpha^{(3)}(\vec{x} + e_2, t) + f_\alpha^{(4)}(\vec{x} + e_1 + e_2, t) \\ v_\alpha^{(2)}(\vec{x}, t) &= f_\alpha^{(1)}(\vec{x}, t) - f_\alpha^{(2)}(\vec{x} + e_1, t) \end{aligned} \tag{8.81}$$

In terms of the spinors $u_\alpha^{(a)}$ and $v_\alpha^{(a)}$ $(a = 1, 2)$ we can write the equation of motion in the standard (two-component) Dirac form

$$i\,\partial_0 u_\alpha^{(a)}(\vec{x}, t) = -i\bar{\rho}(\sigma_3)_{ab}\Delta_1 u_\alpha^{(b)}(\vec{x}, t) + i\bar{\rho}(\sigma_1)_{ab}\Delta_2 u_\alpha^{(b)}(\vec{x}, t) \tag{8.82}$$

and the same equation for $v_\alpha^{(a)}(\vec{x}, t)$.

Let us now define the 2×2 Dirac matrices γ_0, γ_1, and γ_2:

$$\gamma_0 = -\sigma_2, \quad \gamma_1 = -i\sigma_1, \quad \gamma_2 = -i\sigma_3 \tag{8.83}$$

In this notation Eq. (8.76) has the simpler form

$$\begin{aligned} i\left(\gamma_0\,\partial_0 - v_F\vec{\gamma}\cdot\vec{\nabla}\right)_{ab} u_\alpha^{(b)} &= 0 \\ i\left(\gamma_0\,\partial_0 - v_F\vec{\gamma}\cdot\vec{\nabla}\right)_{ab} v_\alpha^{(b)} &= 0 \end{aligned} \tag{8.84}$$

where I have taken the continuum limit and the Fermi velocity v_F is

$$v_F = 2a_0\bar{\rho} \tag{8.85}$$

The eigenvalues of these Dirac operators are, in momentum space,

$$\epsilon(\vec{p}) = \pm 2\bar{\rho}\sqrt{\sin^2 p_1 + \sin^2 p_2} \tag{8.86}$$

with $|p_i| \le \pi/2$. These dispersion relations form conical surfaces near $\vec{p} = 0$ that are characteristic of a continuum relativistic system (shown in Fig. 8.11). Such "Dirac cones" are also found in the band structure of materials such as graphene, a 2D material of carbon atoms arranged on a honeycomb lattice, and in the quasiparticle spectrum of d-wave superconductors. In other words, the spinon quasiparticles of the mean-field flux phases are Dirac fermions.

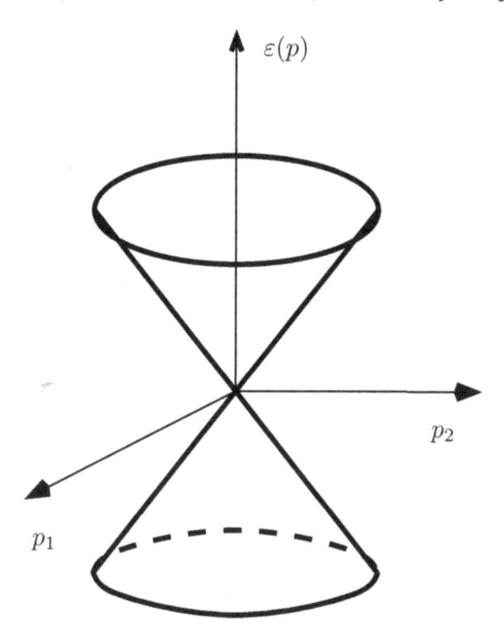

Figure 8.11 The Dirac cone: the dispersion law for spinons in a flux phase.

The ground-state energy in the flux phase is given by

$$E_{\text{flux}} = \frac{2NL^2}{J}\bar{\rho}^2 - 2 \times 2NL^2\bar{\rho} \int_{|p_i| \leq \frac{\pi}{2}} \frac{d^2p}{(2\pi)^2}\sqrt{\sin^2 p_1 + \sin^2 p_2}$$
$$\equiv \frac{2NL^2}{J}\bar{\rho}^2 - NL^2\alpha\bar{\rho} \tag{8.87}$$

where the factor of 2 is due to the contribution of both u and v spinon branches. The minimum is attained at $\bar{\rho} = \frac{1}{4}\alpha J$ and the total energy of the flux phase is

$$E_{\text{flux}} = -\frac{\alpha^2}{8}NL^2J \approx -0.115NL^2J \tag{8.88}$$

which is *lower* than that of the BZA state, namely $E_{\text{BZA}} = -8NL^2J/\pi^4 \approx -0.082NL^2J$.

Quantum dimer phases. Let us now turn our attention to a different set of solutions of the saddle-point equations, which is based on valence-bond states. Consider a configuration of $\bar{\rho}_j(\vec{x})$ that equals $\bar{\rho}$ on a set of links occupied by dimers such as in Figs. 8.12(a) and (b),

$$\bar{\rho}_j(\vec{x}) = \begin{cases} \bar{\rho} & \text{if the link } (\vec{x}, \vec{x} + e_j) \text{ is occupied by a dimer} \\ 0 & \text{otherwise} \end{cases} \tag{8.89}$$

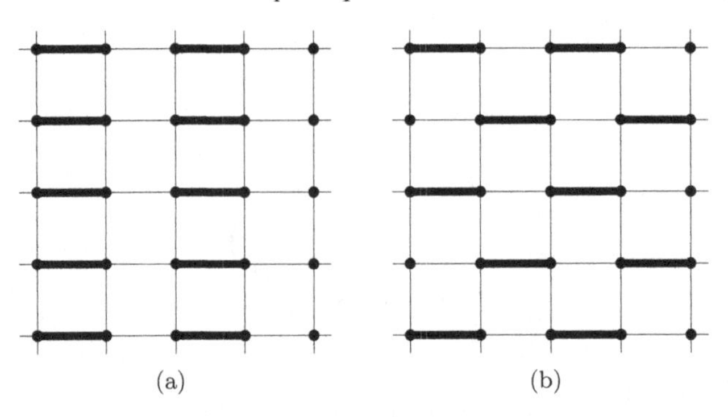

Figure 8.12 Valence-bond crystalline states: (a) one of the four columnar (or Peierls) states and (b) a staggered state.

The mean-field Hamiltonian, Eq. (8.70), with $\bar{\rho}_j(\vec{x})$ as given in Eq. (8.89), describes a set of spinons tightly *confined* (bound) inside the links, the VB states. Thus, we do not have spinon states propagating beyond the size of a dimer (one lattice spacing) in this dimer limit. Fluctuations will enable the effective size of a VB state to grow from the lattice-spacing scale up to some finite length scale ξ. This scale *is* the spin-correlation length for this system in this VB crystal phase. It is also clear that, at the level of mean-field theory, the average flux is not determined. This is simply reflecting the fact that the fluctuations of the gauge field are so strong that the average flux is wiped out. We will see later that, if the effects of the dynamics of holes are taken into account, a flux phase may also develop. The energy of a VB crystal state is

$$E_{\mathrm{VB}} = \frac{2NL^2}{J}\bar{\rho}^2 - NL^2\bar{\rho} \tag{8.90}$$

which is minimized by the choice

$$\bar{\rho} = \frac{J}{4} \tag{8.91}$$

and has the ground-state energy

$$E_{\mathrm{dimer}} = -\frac{J}{8}NL^2 \tag{8.92}$$

for *all* dimer configurations. These states clearly have less energy, $-0.125JN$ per site, than both BZA and flux states.

However, now we no longer get a unique ground state at $N = \infty$. This degeneracy is lifted by fluctuations in the amplitude that appear at order $1/N$. Several possible phases can result from the effects of these fluctuations. For example, Dombre and Kotliar (1989) as well as Read and Sachdev (1989) found that, for the case $n_c = 1$, the *four* columnar or Peierls states are chosen (shown in Fig. 8.12(a)).

However, another possible state is a spin liquid, which in this language appears as a condensate of valence bonds on next-nearest-neighbor bonds. These states are possible in the case of a frustrated antiferromagnet, and formally break the U(1) local symmetry down to a local $\mathbb{Z}_2$ gauge symmetry that can lead to a $\mathbb{Z}_2$ spin liquid (Read and Sachdev, 1991; Wen, 1991c; Mudry and Fradkin, 1994). This case will be discussed in Chapter 9.

8.6 Is the large-N mean-field theory reliable?

Both BZA and flux solutions have gapless excitations that carry a spin-$\frac{1}{2}$ degree of freedom (for SU(2)) or, more generally, SU(N) color quantum numbers. While this spectrum appears to be stable at the level of mean-field theory, we will find problems once fluctuations are taken into account. First of all, we will find that a set of *dimer* states has lower energy both than the BZA states and than the flux states. It is plausible, however, that reasonable generalizations of this Hamiltonian such that the flux state may be preferred do exist. Affleck and Marston have indeed found such generalizations.

But what is more serious about these mean-field theories is the fact that they violate the *local* gauge invariance present in the full theory. In fact, we find spin-non-singlet excitations that are not gauge-invariant: the spinon states themselves. In lattice gauge theories, there is a *theorem*, known as *Elitzur's theorem*, which states that, in a theory with local interactions and with local gauge invariance, *only* locally gauge-invariant operators can have non-zero vacuum expectation values. In other words, the only states present in the spectrum are *local* gauge singlets.

This result may appear to be puzzling at first glance. After all, even in theories with a global symmetry, such as the Ising model, the low-temperature magnetization is zero if the averages are computed over the entire configuration space. The procedure to remedy this problem is well known, and it is crucial to a correct understanding of spontaneous breaking of *global* symmetries. First one considers a *finite* system of linear size L and the allowed space of configurations is reduced either by choosing a boundary condition (that fixes the asymptotic behavior of the spins at spatial infinity) or by turning on a weak external symmetry-breaking field. Next, the thermodynamic limit $L \to \infty$ is considered in the presence of a fixed symmetry-breaking procedure, which is removed *after* the thermodynamic limit has been taken. This procedure yields a non-zero magnetization because *in the thermodynamic limit* it takes an infinite order in the low-temperature expansion, i.e. the expansion around the state with broken symmetry, to mix the two degenerate classes of configurations. Hence, there is no mixing and the magnetization is non-zero if the expansion has a finite radius of convergence.

However, if the symmetry is local, the situation is radically different. It always takes a finite order (of the order of the coordination number) in perturbation theory to mix states related through *local* gauge transformations. The behavior of the system at the boundaries has little effect on the behavior near its center. The expressions for local expectation values are analytic functions of the coupling fields, *even in the thermodynamic limit*, $L \to \infty$. Thus, in the absence of external fields or gauge-fixing conditions, expectation values of locally gauge-non-invariant operators must be zero. This is the content of Elitzur's theorem (see also Chapter 9).

However, a gauge theory may be in a non-confined phase in which a gauge-invariant operator creates a *quark* (*spinon* in the terminology of magnetism) and an antiquark (anti-spinon) at distances R, which can be separated all the way to infinity and still yield a non-zero amplitude. But, for that to happen, the fluctuations of the gauge fields, or rather of their field strengths, need to be controlled. This is *not* the case for the "RVB-type" mean-field theories since there is no term to control the fluctuations of the gauge fields here. The gauge fields fluctuate so strongly that (a) they are able to enforce the local constraint and (b) they project out all current-carrying states. The conclusion is that the BZA and flux states need not only a Gutzwiller projection but *also* an additional procedure to eliminate all processes involving transport of spin over any significant distance. In conclusion, the physical stability (and significance) of gauge-symmetry-breaking mean-field states, such as the flux phases and their generalizations, the projective-symmetry group states (PSG) (Wen, 2002), must be assessed by a non-perturbative procedure that satisfies the requirements of Elitzur's theorem.

There is a possible loophole in this argument. Although the gauge theory alone may be in a confined phase, it may possibly become confined when it is coupled to a matter field. In Chapter 9 we will see that this scenario is possible only if the matter fields "condense" in such a way that they spontaneously break the U(1) gauge symmetry down to a discrete subgroup such as $\mathbb{Z}_2$. In this case the theory effectively becomes a $\mathbb{Z}_2$ gauge theory, which in $(2+1)$ dimensions has a deconfined (topological) phase, namely, the $\mathbb{Z}_2$ spin liquid, which is a fully gapped phase. Another mechanism that will be discussed extensively in Chapter 10 is to have a phase that breaks time-reversal invariance dynamically. We will see that in this case the monopoles are suppressed and the flux phase is stable, even though it becomes massive.

On the other hand, unless some mechanism is found to suppress the monopole configurations, it is much more difficult to stabilize the gapless phases such as the BZA or flux states. One way to accomplish this goal is to consider a theory with flavor degrees of freedom, which so far we have set to unity. Indeed, if the number of flavors is large enough, it is possible to suppress the monopoles by making them irrelevant. This has been shown to hold in the limit in which the number of flavors is also taken to infinity (Metlitski *et al.*, 2008). This result is similar, and closely related, to the well-known fact that a gauge theory with a compact gauge group coupled to a large number of fermionic flavors becomes deconfining since it has an infrared-stable perturbative expansion. However, in the case of the BZA state, the suppression of the monopoles is not sufficient, since there remains the coupling between the spinons close to the Fermi surface and long-wavelength gauge fields. This coupling is known to lead to infrared singularities in the fermion propagators (in all gauges). This problem also arises in the context of quantum chromodynamics at finite density (such as in heavy-ion collisions) and in the compressible phases of 2D electron gases in large magnetic fields (i.e. when there are no fractional quantum Hall states allowed). Although many solutions have been proposed (e.g. Kwon *et al.*, 1994; Polchinski, 1994; Lee, 2008; Metlitski and Sachdev, 2010) this is still an unsolved problem. However, the available quantum Monte Carlo data do not favor these scenarios for the case of systems with SU(2) symmetry such as the quantum Heisenberg antiferromagnets (Assaad, 2005; Armour *et al.*, 2011).

On the other hand, the valence-bond states are *manifestly local singlets*, are locally gauge-invariant, and are thus free from these problems. Thus gauge-field fluctuations will play a rather small role in this case. We should expect states that are based on a VB description to be more stable. The problem of finding a "true" spin-liquid state, i.e. a state without broken symmetries *and* with spinon states in its spectrum, is a subtle one and for the most part remains essentially an open issue. We will return to this problem in Chapter 9.

8.7 SU(2) gauge invariance and Heisenberg models

There is something peculiar in the way we have treated the spin degrees of freedom. For the most part, the spin degrees of freedom are either "swallowed" by dimers or appear in an almost trivial factor as in the large-N limit (N being the number of spin degrees of freedom!). Nowhere in our discussion do we see even a hint of the fact that the spins, say for $S = \frac{1}{2}$, have an SU(2) symmetry. The reason for this can be traced back to the way we decoupled the quartic interaction in terms of an *abelian* field χ_{ij} living on the links. In the past section, we showed that, for $N = 2$, there are two types of spinors, up and down, coupled to amplitudes and gauge fields. (In reality, there are four because of the doubling.) It may seem that, if there are spinons in the excitation spectrum, then even without doubling there should be *four* elementary excitations bearing spin: spinon particles and holes of either spin orientation. The gauge fields, however, make sure that the constraint of single occupancy is strictly enforced. Thus, at each site, only two, not four, degrees of freedom are allowed, each allowed by the orientation of the spin. We must conclude that the particle and hole excitations of the spinons cannot possibly be independent degrees of freedom. We also know that, in the absence of holes, particle–hole symmetry is strictly respected. Hence, the natural conclusion is that the spinon hole with, say, spin down must be the same physical excitation as the spinon particle with spin up, and vice versa. It is clear, then, that a combination of particle–hole and spin symmetries is playing a fundamental role in these systems. The gauge symmetry must then be larger than the local U(1) symmetry implied by the $1/N$ expansion or, for that matter, by any RVB-like abelian decoupling of the Heisenberg interaction.

We will show now that a spin-$\frac{1}{2}$ Heisenberg antiferromagnet, on any lattice and in any dimension, is equivalent to the strong-coupling limit of an SU(2) gauge theory coupled to fermions (Affleck *et al.*, 1988a; Dagotto *et al.*, 1988). Let $\vec{x}$ and $\vec{x}'$ be two sites of a lattice. The term in the Heisenberg Hamiltonian which describes the antiferromagnetic coupling between spins at points $\vec{x}$ and $\vec{x}'$ (not necessarily nearest neighbors) is

$$J\vec{S}(\vec{x}) \cdot \vec{S}(\vec{x}') \tag{8.93}$$

Once again, we will use a fermion description of the spins,

$$\vec{S}(\vec{x}) = c_\alpha^\dagger(\vec{x})\vec{\tau}_{\alpha\beta}c_\beta(\vec{x}) \tag{8.94}$$

where $\vec{\tau}$ is the set of 2×2 Pauli matrices and we require single occupancy at $\vec{x}$ and $\vec{x}'$:

$$1 = c_\alpha^\dagger(\vec{x})c_\alpha(\vec{x}) = c_\alpha^\dagger(\vec{x}')c_\alpha(\vec{x}') \tag{8.95}$$

Let us perform a particle–hole transformation at every site so as to ensure that the reference state satisfies Eq. (8.95). We define new fermion operators $\psi_1(\vec{x})$ and $\psi_2(\vec{x})$ given by the relationships

$$\begin{aligned} c_\uparrow(\vec{x}) &= \psi_1(\vec{x}), \quad & c_\uparrow^\dagger(\vec{x}) &= \psi_1^\dagger(\vec{x}) \\ c_\downarrow(\vec{x}) &= \psi_2^\dagger(\vec{x}), \quad & c_\downarrow^\dagger(\vec{x}) &= \psi_2(\vec{x}) \end{aligned} \tag{8.96}$$

This canonical transformation amounts to an "exchange" of charge and spin operators since

$$\begin{aligned} c_\uparrow^\dagger(\vec{x})c_\uparrow(\vec{x}) + c_\downarrow^\dagger(\vec{x})c_\downarrow(\vec{x}) &= \psi_1^\dagger(\vec{x})\psi_1(\vec{x}) - \psi_2^\dagger(\vec{x})\psi_2(\vec{x}) + 1 \\ c_\uparrow^\dagger(\vec{x})c_\uparrow(\vec{x}) - c_\downarrow^\dagger(\vec{x})c_\downarrow(\vec{x}) &= \psi_1^\dagger(\vec{x})\psi_1(\vec{x}) + \psi_2^\dagger(\vec{x})\psi_2(\vec{x}) - 1 \end{aligned} \tag{8.97}$$

Hence, the constraint

$$c_\uparrow^\dagger(\vec{x})c_\uparrow(\vec{x}) + c_\downarrow^\dagger(\vec{x})c_\downarrow(\vec{x}) = 1 \tag{8.98}$$

is equivalent to

$$\psi_1^\dagger(\vec{x})\psi_1(\vec{x}) - \psi_2^\dagger(\vec{x})\psi_2(\vec{x}) = 0 \tag{8.99}$$

In other words, we are projecting onto the subspace with an equal number of quantum numbers 1 and 2 per site. Such states are denoted by $|\text{Phys}\rangle$. The constraint, Eq. (8.99), has the equivalent form

$$\psi^\dagger(\vec{x})\tau_3\psi(\vec{x})|\text{Phys}\rangle = 0 \tag{8.100}$$

However, Eq. (8.99) implies that the following identities must also hold:

$$\begin{aligned}\psi^\dagger(\vec{x})\tau_1\psi(\vec{x})|\text{Phys}\rangle &= \left(\psi_1^\dagger(\vec{x})\psi_2(\vec{x}) + \psi_2^\dagger(\vec{x})\psi_1(\vec{x})\right)|\text{Phys}\rangle = 0 \\ \psi^\dagger(\vec{x})\tau_2\psi(\vec{x})|\text{Phys}\rangle &= i\left(\psi_1^\dagger(\vec{x})\psi_2(\vec{x}) - \psi_2^\dagger(\vec{x})\psi_1(\vec{x})\right)|\text{Phys}\rangle = 0\end{aligned} \tag{8.101}$$

Indeed, Eq. (8.101) is equivalent to the statements

$$\begin{aligned}\left(c_\uparrow^\dagger(\vec{x})c_\downarrow^\dagger(\vec{x}) + c_\downarrow(\vec{x})c_\uparrow(\vec{x})\right)|\text{Phys}\rangle &= 0 \\ i\left(c_\uparrow^\dagger(\vec{x})c_\downarrow^\dagger(\vec{x}) - c_\downarrow(\vec{x})c_\uparrow(\vec{x})\right)|\text{Phys}\rangle &= 0\end{aligned} \tag{8.102}$$

which are true since the states $|\text{Phys}\rangle$ are singly occupied. Therefore, we have the *local* constraint on the space of *allowed* states

$$\psi^\dagger(\vec{x})\vec{\tau}\psi(\vec{x})|\text{Phys}\rangle = 0 \tag{8.103}$$

at each site of the lattice. Note, however, that $\psi^\dagger(\vec{x})\vec{\tau}\psi(\vec{x})$ is *not* a spin operator. Rather, the spin operators $S_a(\vec{x})$, $a = 1, 2, 3$, are now given by

$$\begin{aligned}S_1(\vec{x}) &\equiv c_\uparrow^\dagger(\vec{x})c_\downarrow(\vec{x}) + c_\downarrow^\dagger(\vec{x})c_\uparrow(\vec{x}) = \psi_1^\dagger(\vec{x})\psi_2^\dagger(\vec{x}) + \psi_2(\vec{x})\psi_1(\vec{x}) \\ S_2(\vec{x}) &\equiv i\left(c_\uparrow^\dagger(\vec{x})c_\downarrow(\vec{x}) - c_\downarrow^\dagger(\vec{x})c_\uparrow(\vec{x})\right) = i\left(\psi_1^\dagger(\vec{x})\psi_2^\dagger(\vec{x}) - \psi_2(\vec{x})\psi_1(\vec{x})\right) \\ S_3(\vec{x}) &\equiv c_\uparrow^\dagger(\vec{x})c_\uparrow(\vec{x}) - c_\downarrow^\dagger(\vec{x})c_\downarrow(\vec{x}) = \psi_1^\dagger(\vec{x})\psi_1(\vec{x}) + \psi_2^\dagger(\vec{x})\psi_2(\vec{x}) - 1\end{aligned} \tag{8.104}$$

This set of operators has a remarkable local symmetry. Let $\psi'(\vec{x})$ be a new spinor related to $\psi(\vec{x})$ by means of an SU(2) transformation $U(\vec{x})$:

$$\psi'_\alpha(\vec{x}) = U_{\alpha\beta}(\vec{x})\psi_\beta(\vec{x}) \tag{8.105}$$

Clearly, under such a transformation, we have

$$\begin{aligned}\psi'^\dagger_\alpha(\vec{x})\tau^a_{\alpha\beta}\psi'_\beta(\vec{x}) &= \psi_\alpha^\dagger(\vec{x})\left(U^{-1}(\vec{x})\tau^a U(\vec{x})\right)_{\alpha\beta}\psi_\beta(\vec{x}) \\ &\equiv R^{ab}(\vec{x})\psi_\alpha^\dagger(\vec{x})\tau^b_{\alpha\beta}\psi_\beta(\vec{x})\end{aligned} \tag{8.106}$$

where $R(\vec{x})$ is the SO(3) rotation associated with the SU(2) transformation $U(\vec{x})$.

The spin operators $S_a(\vec{x})$, $a = 1, 2, 3$, are *invariant* under this SU(2) transformation. First, $S_3(\vec{x})$ is clearly invariant:

$$S_3(\vec{x}) = \psi_\alpha^\dagger(\vec{x})\psi_\alpha(\vec{x}) - 1 = \psi'^\dagger_\alpha(\vec{x})\psi'_\alpha(\vec{x}) - 1 \tag{8.107}$$

Secondly, the invariance of $S_1(\vec{x})$ and $S_2(\vec{x})$ follows from the fact that the operators $\chi(\vec{x})$ and $\chi^\dagger(\vec{x})$, defined by

$$\chi(\vec{x}) \equiv \frac{1}{2}\epsilon_{ij}\psi_i(\vec{x})\psi_j(\vec{x}) \tag{8.108}$$

are also invariant under SU(2).

It is convenient to introduce the SU(2)-invariant operator $M(\vec{x})$,

$$M(\vec{x}) \equiv \psi_\alpha^\dagger(\vec{x})\psi_\alpha(\vec{x}) \tag{8.109}$$

It is easy to show now that the Heisenberg Hamiltonian on *any* lattice and in *any* dimension with a translationally invariant interaction $J(\vec{l})$ ($\vec{l}$ is the relative position vector of a pair of spins) is equivalent to the following Hamiltonian:

$$\begin{aligned} H = & -N_{\rm s}\left(\sum_{\vec{l}} J(\vec{l})\right)(1+2m_3) \\ & +\sum_{\vec{x},\vec{l}} J(\vec{l})\left(M(\vec{x})M(\vec{x}+\vec{l})+2\left(\chi^\dagger(\vec{x})\chi(\vec{x}+\vec{l})+\chi^\dagger(\vec{x}+\vec{l})\chi(\vec{x})\right)\right) \end{aligned} \tag{8.110}$$

where $N_{\rm s}$ is the total number of sites on the lattice and m_3 is the *total* polarization of the allowed Hilbert space,

$$\frac{1}{N_{\rm s}}\sum_{\vec{x}} S_3(\vec{x})|\text{Phys}\rangle = m_3|\text{Phys}\rangle \tag{8.111}$$

The Heisenberg Hamiltonian in the form given by Eq. (8.110) is manifestly invariant under the *local* SU(2) transformations of Eq. (8.105) since it is written in terms of $M(\vec{x})$, $\chi(\vec{x})$, and m_3, which are locally invariant.

It is important to stress that this local SU(2) symmetry, which involves *both* spin rotations *and* a particle–hole transformation, is unrelated to the *global* SU(2) invariance

$$c_\alpha(\vec{x}) \to c'_\alpha(\vec{x}) = V_{\alpha\beta}c_\beta(\vec{x}) \tag{8.112}$$

which induces *global* rotations of the spin polarization,

$$S_a(\vec{x}) \to S'_a(\vec{x}) = R^{ab}S_b(\vec{x}) \tag{8.113}$$

In Section 2.3.1, we showed that the Heisenberg antiferromagnet is the $U \to \infty$ limit of a half-filled Hubbard model. I will now show that it is also the strong-coupling limit of an SU(2) lattice gauge theory. Consider a system of fermions, with creation and annihilation operators $\psi_\alpha^\dagger(\vec{x})$ and $\psi_\alpha(\vec{x})$, respectively, coupled to a set of SU(2) gauge degrees of freedom $U(\vec{x},\vec{x}')$ on the bonds $(\vec{x},\vec{x}')$ of a lattice.

The Hilbert space of this system is a tensor product of fermionic states on the sites of the lattice multiplied by states on the links associated with gauge degrees of freedom. Let $A^a(\vec{x}, \vec{x}')$ be an operator that transforms like a *vector* under SU(2), i.e. $\vec{A}(\vec{x}, \vec{x}') \cdot \vec{\tau}$ is an element of the Lie algebra. Let us label the states on the links by the (real) eigenvalues of $A^a(\vec{x}, \vec{x}')$, e.g. $|\{A^a(\vec{x}, \vec{x}')\}\rangle$. The operators $U(\vec{x}, \vec{x}')$ are 2×2 matrices defined by

$$U(\vec{x}, \vec{x}') = e^{i\tau^a A^a(\vec{x}, \vec{x}')} \tag{8.114}$$

where the τ^a are the generators of SU(2) in the fundamental (spinor) representation. Moreover, we demand

$$A^a(\vec{x}, \vec{x}') = -A^a(\vec{x}', \vec{x}) \tag{8.115}$$

Equivalently, the $U(\vec{x}, \vec{x}')$ operators must satisfy the condition

$$U(\vec{x}, \vec{x}') = U^{\dagger}(\vec{x}', \vec{x}) \tag{8.116}$$

Let $E^a(\vec{x}, \vec{x}')$ be a set of operators acting on this Hilbert space. We will require that these operators be canonically conjugate to the $A^a(\vec{x}, \vec{x}')$, i.e.

$$[A^a(\vec{x}, \vec{x}'), E^b(\vec{y}, \vec{y}\,')] = i\delta^{ab} \cdot \delta_{\vec{x},\vec{y}}\delta_{\vec{x}',\vec{y}'} \tag{8.117}$$

In addition, the operators $E^a(\vec{x}, \vec{x}')$ satisfy the SU(2) (angular-momentum) algebra

$$[E^a(\vec{x}, \vec{x}'), E^b(\vec{y}, \vec{y}')] = i\epsilon^{abc} E^c(\vec{x}, \vec{x}') \cdot \delta_{\vec{x},\vec{y}}\delta_{\vec{x}',\vec{y}'} \tag{8.118}$$

In other words, the operators $E^a(\vec{x}, \vec{x}')$ transform like group generators. Clearly, the operators $E^a(\vec{x}, \vec{x}')$ and the SU(2) matrices $U(\vec{x}, \vec{x}')$ satisfy the commutation relations

$$[E^a(\vec{x}, \vec{x}'), U(\vec{y}, \vec{y}\,')] = \tau^a U(\vec{x}, \vec{x}') \cdot \delta_{\vec{x},\vec{y}}\delta_{\vec{x}',\vec{y}'} \tag{8.119}$$

All the commutators so defined (Eqs. (8.117)–(8.119)) vanish if the operators act on the Hilbert spaces associated with different links.

Consider now the Hamiltonian $\tilde{H}$ acting on the Hilbert space of gauge-invariant states:

$$\tilde{H} = \frac{G}{2} \sum_{(\vec{x},\vec{x}')a} E^a(\vec{x}, \vec{x}')E^a(\vec{x}, \vec{x}') + \frac{i}{2} \sum_{(\vec{x},\vec{x}')\alpha\beta} \left(\psi_\alpha^\dagger(\vec{x})U_{\alpha\beta}(\vec{x}, \vec{x}')\psi_\beta(\vec{x}') - \text{h.c.}\right) \tag{8.120}$$

where G is a coupling constant and $(\vec{x}, \vec{x}')$ are pairs of sites on an arbitrary lattice.

On a given lattice, the equivalence between the system described by the Hamiltonian $\tilde{H}$ and the Heisenberg model holds in the limit $G \to \infty$. The argument goes as follows. First we note that $\tilde{H}$ is invariant under time-independent local SU(2) gauge transformations,

$$\begin{aligned} U_{\alpha\beta}(\vec{x},\vec{x}') &= W^{-1}_{\alpha\gamma}(\vec{x})U'_{\gamma\delta}(\vec{x},\vec{x}')W_{\delta\beta}(\vec{x}') \\ \psi_\alpha(\vec{x}) &= W^{-1}_{\alpha\beta}(\vec{x})\psi'_\beta(\vec{x}) \end{aligned} \tag{8.121}$$

In the limit $G \to \infty$, the ground state of the system has a huge degeneracy. In fact, to leading order in an expansion in powers of $1/G$, the low-lying states are the gauge singlets which satisfy

$$E^a(\vec{x},\vec{x}')E^a(\vec{x},\vec{x}')|\Psi\rangle = 0 \tag{8.122}$$

(on all links) and obey the constraint

$$\mathcal{Q}^a(\vec{x})|\Psi\rangle \equiv \psi^\dagger_\alpha(\vec{x})\tau^a_{\alpha\beta}\psi_\beta(\vec{x})|\Psi\rangle = 0 \tag{8.123}$$

The last condition implies that at each site $\vec{x}$ we have either a state with no fermion, $|0\rangle$, or a "baryon" state, $|\chi\rangle = \chi^\dagger|0\rangle$. We now can apply a degenerate perturbation theory exactly identical to the one we used to derive the Heisenberg model from the Hubbard model. The first available excited state, $|\Psi_{\text{exc}}\rangle$, has a link excited to a state with angular-momentum quantum number $\frac{1}{2}$,

$$E^a(\vec{x},\vec{x}')E^a(\vec{x},\vec{x}')|\Psi_{\text{exc}}\rangle = \frac{3}{4}|\Psi_{\text{exc}}\rangle \tag{8.124}$$

only on that link. The effective Heisenberg exchange interaction thus obtained is equal to $J = 2/(3G)$.

What is the physical meaning of this symmetry? What we have actually shown is that the strong-correlation limit of the Hubbard model at half-filling has an effective local SU(2) gauge invariance. This gauge invariance, which is a mixture of a local particle–hole transformation and a spin rotation, merely reflects the fact that in the strong-correlation limit the only excitations left do not violate the local constraint. Hence no charge motion is possible and the system is an insulator. The charge-carrying states are either holes or doubly occupied sites, both of which violate the constraint and pay a large energy penalty of order U, the Hubbard coupling constant. The remaining states are charge-neutral states, which may, or might not, carry spin. It is thus no surprise that the gauge theory satisfies not only the constraint

$$\mathcal{Q}^a(\vec{x})|\text{Phys}\rangle = 0 \tag{8.125}$$

but also the related condition for the current,

$$J^a_i(\vec{x})|\text{Phys}\rangle = 0 \tag{8.126}$$

In other words, the current must also be zero.

In a sense, we can think of the Heisenberg model as a "free-particle" problem with its large Hilbert space projected onto a subspace of states with zero current and zero charge, at the scale of the lattice spacing. The insulating phase of the Hubbard model, on the other hand, satisfies the same condition at length scales larger than the inverse of the charge gap. Thus, the low-energy behavior of the Hubbard insulator is also described by a system with a gauge symmetry. This property is clearly violated once one considers states with non-zero charge. Indeed, the chemical potential, which couples to the charge density $c_\alpha^\dagger(\vec{x})c_\alpha(\vec{x})$, yields a term in the Hamiltonian $\tilde{H}$ of the form

$$\tilde{H}_{\text{charge}} = \mu \sum_{\vec{x}} \psi^\dagger(\vec{x})\tau_3\psi(\vec{x}) \tag{8.127}$$

which clearly violates the SU(2) symmetry. Similarly, the fermion-hopping term becomes

$$\tilde{H}_{\text{hop}} = t \sum_{\langle\vec{x},\vec{x}'\rangle} c_\sigma^\dagger(\vec{x})c_\sigma(\vec{x}') = t \sum_{\langle\vec{x},\vec{x}'\rangle} \psi_\alpha^\dagger(\vec{x})\tau_3^{\alpha\beta}\psi_\beta(\vec{x}') \tag{8.128}$$

which also violates the local SU(2) gauge invariance. We will come back to these issues later on. Let us point out now that the symmetry does imply that the spinon particle (hole) state with spin up is the same state as a spinon hole (particle) with spin down. Thus local SU(2) tells us that there are only two spinon states, which is as it should be.

9

Gauge theory, dimer models, and topological phases

In the last chapter we introduced the concept of valence-bond states and discussed several quantum disordered phases in this language. Here we will see that the quantum fluctuations of valence-bond systems are best captured in terms of a much simpler effective theory, the quantum-dimer models. An understanding of these types of phases is best accomplished in terms of gauge theories. The phases of gauge theories and their topological properties will allow us to introduce the concept of a topological phase of matter in a precise way.

9.1 Fluctuations of valence bonds: quantum-dimer models

The valence-bond crystal of Section 8.5 has a spin-correlation length of the order of one lattice constant. It represents a quantum paramagnet. However, it is *not* a translationally invariant state, unlike the equal-amplitude short-range RVB state. It has crystalline order of its valence bonds and it is a four-fold degenerate state.

Alternatively we can imagine that the amplitude fluctuations, which represent transitions to states with broken valence bonds, are suppressed. The only way the system has to minimize its energy is by finding a coherent rearrangement of valence bonds. If the amplitude fluctuations are frozen out, the system has states labeled by quantum numbers that describe the covering of the lattice by dimers. For the rest of our discussion we will ignore the SU(N) structure. In this approximation the space of states is identified with the set of configurations $\{\mathcal{C}\}$ of dimer coverings of the lattice. In particular we will take this basis to be orthonormal,

$$\langle\mathcal{C}|\mathcal{C}'\rangle = \delta_{\mathcal{C},\mathcal{C}'} \tag{9.1}$$

and complete (although the valence-bond singlet states are over-complete).

Quantum-dimer models also arise in certain limits of *frustrated* Ising models in transverse fields (Moessner *et al.*, 2000; Moessner and Sondhi, 2001b). Two examples are the Ising antiferromagnet in a transverse field on a triangular lattice and the fully frustrated Ising model (also in a transverse field). In these cases, as in their classical counterparts, dimer coverings of the lattice represent the configurations of "unsatisfied bonds." These are classically degenerate configurations (states). The quantum-dimer model is the effective quantum Hamiltonian acting on this degenerate manifold of states, much in the same way as the quantum Heisenberg antiferromagnet arises as the strong-coupling limit of the Hubbard model (at half-filling).

The structure of quantum-dimer models and of their phase diagrams depends on whether the lattice is bipartite or not, and, if it is, on the coordination number. We will focus primarily on the cases of the square and triangular lattices. Let us consider first the case of a square lattice and let $l_j(\vec{x})$ be an integer-valued variable associated with the bond $(\vec{x}, \vec{x}+e_j)$. The Hilbert space is the space of states of the form $\{|\{l_j(\vec{x})\}\rangle\}$, where the integer l_j is either equal to zero (no dimer) or one (dimer). Every site has to belong to one and only one dimer. This requirement leads to the local constraint

$$l_1(\vec{x}) + l_2(\vec{x}) + l_1(\vec{x}-e_1) + l_2(\vec{x}-e_2) = 1 \tag{9.2}$$

For the case of the triangular and honeycomb lattices there is an analogous construction with a space of states labeled by an integer l, the dimer occupation number of the link, taking the values $l = 0, 1$. The dimer occupation numbers are subject to the same constraint that their sum on links sharing a given site is fixed to be 1, indicating that a lattice site belongs to one and only one dimer at a time. Each lattice is composed of sites, nearest-neighboring pairs of which denote the links. Planar lattices are uniformly tiled by plaquettes (squares, triangles, hexagons, etc.).

The Hamiltonian of the quantum-dimer model (QDM) is in all cases a sum of a resonance term (the kinetic energy) and a diagonal term (the potential energy) with a structure that is different for each type of lattice,

$$H_{\text{QDM}} = H_{\text{res}} + H_{\text{diag}} \tag{9.3}$$

which acts on a space of states subject to the local constraint of Eq. (9.2).

The "resonance" process of Fig. 9.1 is represented by an off-diagonal matrix element in which the integer degrees of freedom l_j for parallel bonds of a plaquette are raised from zero to one if the values for the other two bonds are lowered from one to zero. This process can be described by a term in the effective Hamiltonian of the form H_{res}, which for the case of the square lattice is (Rokhsar and Kivelson, 1988)

$$H_{\text{res}} = \bar{J} \sum_{\text{plaquettes}} \left(|\square\rangle\langle\square| + |\square\rangle\langle\square| \right) \tag{9.4}$$

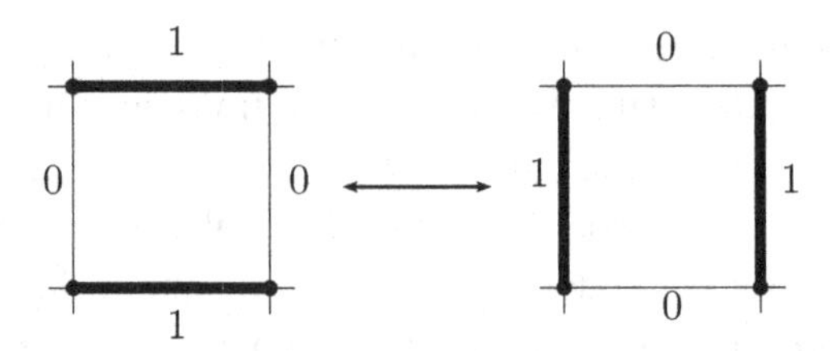

Figure 9.1 The resonance process. The integers $l = 0, 1$ represent the bond occupation by a dimer (or valence bond).

where $\bar{J}$ is an effective coupling constant, $\bar{J} \propto J$. When acting on a given plaquette, this operator annihilates states with no dimer or only one dimer on that plaquette because these states are orthogonal. For the case of the triangular and hexagonal lattices there is also a resonance term describing resonant processes for dimer configurations covering every other link of the smallest possible plaquettes of the lattice (hexagons for the honeycomb lattice and parallelograms for the triangular lattice) (Moessner and Sondhi, 2001a, 2001b).

The diagonal matrix elements are described by a term in the Hamiltonian H_{diag}, which gives an energy V to a pair of neighboring parallel dimers. For the square lattice it has the form

$$H_{\text{diag}} = V \sum_{\text{plaquettes}} \left(\left| \,=\, \right\rangle \left\langle \,=\, \right| + \left| \,\|\,\| \, \right\rangle \left\langle \,\|\,\|\, \right| \right) \tag{9.5}$$

This operator assigns a plaquette energy 0 to all states with no dimer or only one dimer on that plaquette. For the case of triangular and hexagonal lattices the structure of H_{diag} is analogous and it is also associated with the smallest possible plaquettes.

In spite of its apparent simplicity, the QDM is not easy to solve for arbitrary values of the parameters. A partial exact solution exists for a particular set of values of the parameters (which will be discussed below). We will now describe its phases and phase diagrams, which turn out to depend on whether the lattice is bipartite or not. In most (but not all) cases the ground state is a VB crystal. Thus, even the QDM, which was originally proposed by Rokhsar and Kivelson (1988) as a model with a short-range RVB state as its ground state, has, in general, a crystalline ground state.

This is easy to see in the "classical limit" in which $|V| \gg |J|$. Indeed, if V is large and negative, the diagonal term tells us that dimers on parallel links of a plaquette attract each other. Thus in this case we need to maximize the number of parallel dimers on each plaquette (regardless of the lattice structure). For the case of the square lattice the result is the columnar ordered state shown in Fig. 8.12(a). Similar "ideal" ordered phases also occur for the triangular and honeycomb lattices. In the opposite limit, with V large and positive, dimers on parallel links repel each other. In this limit too the ground state is a VB crystal, which for the case of the square lattice is the staggered crystal shown in Fig. 8.12(b). Again, analogous VB crystalline phases exist for other lattices as well.

From a symmetry point of view, a state with columnar order can be regarded as a phase with a unidirectional modulation of the dimer density, a "VB density wave." Such a phase breaks (spontaneously) translational invariance as well as rotational invariance or, more properly, the point-group symmetry of the lattice. In the cases we discuss here the VB density wave is commensurate with the underlying lattice, and in the particular case of columnar order the period is 2. Valence-bond crystalline states with more complex orders can also exist (Fradkin *et al.*, 2004; Vishwanath *et al.*, 2004; Papanikolaou *et al.*, 2007a). Crystalline phases that respect the point-group symmetry can also exist. One such example is a state with "plaquette" order (Jalabert and Sachdev, 1991), which can be regarded as a state with a bidirectional VB density wave (or as a state in which dimers resonate on a subset of plaquettes).

For general values (and signs) of the coupling constants V and J the QDM Hamiltonian must be diagonalized numerically. This is a numerically hard problem even for this very simple system. Quantum Monte Carlo simulations (Jalabert and Sachdev, 1991) and finite-size exact diagonalizations (Leung *et al.*, 1996) have been performed and confirm the statement that the "generic" ground state is a crystal rather than a liquid. For $V > |\bar{J}|$, the staggered valence-bond crystal of Fig. 8.12(b) is the exact ground state and it has zero energy for all $V > |J|$.

The exact ground state of the QDM is known for a particular value of $\bar{J}/V$, namely $\bar{J}/V = -1$ (with $V > 0$), which is known as the Rokhsar–Kivelson (RK) point. At this value of $\bar{J}/V$ Rokhsar and Kivelson (1988) found that the short-range RVB wavefunction is the exact ground-state wave function and that it has zero energy. The reason for this behavior is that precisely at $\bar{J} = -V$ the QDM Hamiltonian (in *all planar lattices*) can be written as a *sum of projection operators* that locally project out the linear superposition of the two parallel configurations of dimers on each plaquette. Thus, at the RK point and for *all lattices* the Hamiltonian is a positive semi-definite hermitian operator whose eigenvalues are non-negative. Hence, all zero-energy states are exact (zero energy!) ground states.

Up to the effects of boundary conditions (which we will discuss below) the ground state of the QDM Hamiltonian at the RK point is the short-range RVB state $|\Psi_{\text{sRVB}}\rangle$, the equal-amplitude superposition of dimer configurations $\{C\}$,

$$|\Psi_{\text{sRVB}}\rangle = \sum_{\{C\}} |C\rangle \tag{9.6}$$

which, *at the RK point*, clearly obeys

$$H_{\text{QDM}}|\Psi_{\text{sRVB}}\rangle = 0 \tag{9.7}$$

This state has a number of simple and remarkable properties. Owing to the orthonormality of the dimer configurations,

$$\langle C|C'\rangle = \delta_{C,C'} \tag{9.8}$$

the norm of this state, $||\Psi_{\text{sRVB}}||^2$, is simply the sum of the dimer configurations on the given lattice. In other terms,

$$||\Psi_{\text{sRVB}}||^2 = Z_{\text{dimer}} \tag{9.9}$$

where Z_{dimer} is the *classical* dimer partition function for that lattice. In addition, the equal-time correlation functions of local operators of the QDM that are diagonal in the dimer basis, e.g. the local dimer-density operator, is *equal* to the correlation function of the *same* physical observable in the classical dimer model about which much is known!

We saw before that the correlation length of the spins in short-ranged valence-bond states is short-ranged (Kohmoto and Shapir, 1988). In the dimer limit which we are considering here, the spins are permanently bound inside dimers and the spectrum of spin excitations has effectively an infinite energy gap for all spin-carrying excitations. In particular, in this limit the spin-correlation length cannot exceed the size of a dimer and it is effectively zero. However, this does not imply that all other correlation functions must also necessarily be short-ranged. The actual behavior turns out to depend on whether the lattice is bipartite or not.

9.2 Bipartite lattices: valence-bond order and quantum criticality

For the square and honeycomb lattices, both of which are bipartite, the RK point dimer-density correlation function is not short-ranged. This correlation function, which measures the probability on finding two dimers separated by some distance R (say on parallel links) on the lattice is, in this state, equal to the correlation function for finding two parallel dimers in a random distribution of classical dimers covering the lattice. For the case of the square lattice, Fisher and Stephenson (1963) solved this problem exactly (using Pfaffian methods), and found that this correlation function, $G(R)$, obeys a power law

$$G(R) \propto \frac{1}{R^2} \tag{9.10}$$

The precise form of the correlation function is actually more complicated. In addition to the power law shown in Eq. (9.10), it depends also on the relative sub-lattices on which the dimers reside as well as on their relative orientation (Fisher and Stephenson, 1963; Youngblood *et al.*, 1980). The same result essentially also holds for the honeycomb lattice. This result implies that the ground state at the

RK point of the QDM on the square lattice, the short-range RVB state, does not have long-range dimer order and it is not a VB crystal. However, it is not a liquid either, since the connected correlator does not decay exponentially with distance as it should in a liquid state. In fact, both on the square and on the honeycomb lattices the short-range RVB state is at a *critical point* between two VB crystals, which are dimer solids. Fisher and Stephenson also calculated the correlation function $C(R)$ of two *holes* (monomers) separated by a distance R and found the result (again up to a dependence on the relative sublattice)

$$C(R) = \frac{\text{constant}}{R^{\eta}} \tag{9.11}$$

with the anomalous dimension $\eta = \frac{1}{2}$. In the QDM this corresponds to the equal-time correlation function of an operator that creates two holes (or holons) separated by the same distance R. In the classical dimer model this corresponds to the correlator of two monomers (Rokhsar and Kivelson, 1988) (a detailed description was given by Papanikolaou and coworkers (Papanikolaou *et al.*, 2007b)).

If the equal-time correlation functions of local operators have a power-law behavior (at long distances) we expect the excitation spectrum to be gapless, with the energy gap scaling as some power z of the momentum of the excitation. In the case of QDMs at the RK point on the square lattice, where the dimer correlation has the power-law decay of Eq. (9.10), a variational calculation by Rokhsar and Kivelson (1988) (using a standard argument due to Feynman (1972)) of the excited state created by the local dimer-density operator predicts the value of the dynamical quantum critical exponent to be $z = 2$, i.e. that the "resonon" excitation energy, $\omega_{\text{resonon}}(\vec{p})$, of a state with small momentum $\vec{p}$ (measured from (π, π)) scales as

$$\omega_{\text{resonon}}(\vec{p}) = \text{constant} \times |\vec{p}|^2 \tag{9.12}$$

We will see below why this result is natural and probably exact. However, we expect that the existence of this gapless state is a special feature of the RK point and that this excitation should be unstable with respect to perturbations away from the RK point. We will also see below that the RK point (if it is actually accessible) is a quantum critical point with special properties.

9.3 Non-bipartite lattices: topological phases

The behavior is markedly different on a triangular lattice. Numerical simulations (Moessner and Sondhi, 2001b) and exact results (Fendley *et al.*, 2002) (also obtained using Pfaffian methods) for the classical dimer model on the triangular lattice indicate instead that the (connected) dimer correlation function also does not exhibit long-range order but instead is now *short-ranged*, i.e. it decays *exponentially* with distance with a correlation length $\xi \sim a$, where a is the lattice spacing. Thus, the short-range RVB state describes a *dimer liquid*, a uniform state without any type of long-range order. As we will discuss below, this state is a *topological fluid* and hence a true spin-liquid state. Similarly, the (connected) monomer (hole) correlation function also decays exponentially at long distances with a finite (and quite short) correlation length.

In the same work, Fendley, Moessner, and Sondhi also investigated the behavior for an anisotropic lattice, namely a square lattice with extra links running along one of the directions of its diagonal bonds (say SW–NW), and assigned a separate fugacity t for dimers on the diagonal bond. For $t = 1$ this system is equivalent to the isotropic triangular lattice, while for $t = 0$ it reduces to the square lattice. Using the same methods, they found that $t = 0$ is indeed a critical point, and that for all $t > 0$ the dimer system is in a liquid phase with a correlation length that diverges as the square lattice limit is approached ($t \to 0$) as $\xi(t) \sim t^{-1}$.

On the other hand, since the RVB state on the triangular (and, hence, non-bipartite) lattice has a finite correlation length and the Hamiltonian is local, it is natural to expect that the excitation spectrum be gapped. This *expectation* is known to be correct in systems with a relativistic spectrum in which energy scales like the momentum (and hence these systems have a dynamical exponent $z = 1$). It is believed to hold more generally, and it is believed to be rigorously correct for generic systems with local interactions. A theorem by Hastings and Koma (2006) proves that algebraically decaying correlators imply a gapless spectrum for local Hamiltonians. Since the energy gap is finite at the RK point, we expect that it will remain finite at least for some neighborhood of the RK point. Thus, the liquid state should describe a *phase* of the QDM on the triangular lattice.

The ground state on a bipartite lattice turns out to depend in subtle ways on the boundary conditions. For a system with open boundary conditions (essentially a disk or, equivalently, a sphere), the ground state is unique. However, for a system with periodic boundary conditions (a surface that has the topology of a torus) the ground state is four-fold degenerate. This degeneracy is not the result of any broken symmetry, since this state is translationally invariant and the correlation functions of all local operators are short-ranged. More importantly, the degeneracy depends only on the topology of the surface. In Section 9.6 we show that states with these properties are *topological phases*. This topological phase is known as a $\mathbb{Z}_2$ topological fluid.

9.4 Generalized quantum-dimer models

It is possible to construct generalized QDMs whose Hamiltonians are also the sum of local projection operators at their respective RK points (Ardonne *et al.*, 2004; Castelnovo *et al.*, 2004; Papanikolaou *et al.*, 2007b). The configuration spaces (the

Hilbert spaces) of these models in some cases are dimer coverings of the lattice, and in others are arrows defined on links. The amplitudes described by these states always have the form of a product of local weights for a given configuration C:

$$|\Psi\rangle = \sum_{\{C\}} w[C]|C\rangle \tag{9.13}$$

where $w[C]$ is a *product* of amplitudes assigned to sites, links, plaquettes, etc., of the lattice. For example Papanikolaou *et al.* (2007b) and Castelnovo *et al.* (2004) considered a generalized QDM with an exact ground state at its "RK point" of the form of Eq. (9.13) with

$$w[C] = \prod_{p} e^{-\frac{u}{2}\Phi_p[C]} \tag{9.14}$$

where p spans the plaquettes of the square lattice, and u is a parameter. Here $\Phi_p[C] = 1$ if configuration C has a pair of parallel dimers (vertical or horizontal) on plaquette p, and $\Phi_p[C] = 0$ otherwise. Since this state depends on the arbitrary parameter u, the RK point is actually a line. Once again, the norm of this state is a classical partition function, a sum over dimer configurations with a Gibbs weight $|w[C]|^2$,

$$Z = \sum_{\{C\}} |w[C]|^2 \sim \sum_{\{C\}} e^{-u\sum_p \Phi_p[C]} = \sum_{\{C\}} e^{-uN_\parallel[C]} \tag{9.15}$$

where $N_\parallel$ is the number of plaquettes in configuration C with parallel dimers. Formally, this is a classical dimer problem at "finite temperature" $T = u^{-1}$. This classical system remains critical from the dimer-model limit at $u = 1$ up to a critical value u_c, where it has a Kosterlitz–Thouless transition. For $u > u_c$ this system has columnar order (Alet *et al.*, 2005; Papanikolaou *et al.*, 2007b). Generalizations of the QDMs whose configurations are loops and nets defined on various lattices have been discussed in the context of topological phases (Freedman *et al.*, 2004; Fendley and Fradkin, 2005; Levin and Wen, 2005; Fidkowski *et al.*, 2009) and will be discussed in a later section.

How much do these results change if instead of dimer configurations we consider a short-range RVB state? This is important since, as we noted, valence-bond states are over-complete and hence do not constitute an orthonormal basis. This problem has been investigated numerically by Monte Carlo simulations (Albuquerque and Alet, 2010; Tang *et al.*, 2011b) for the case of the square lattice. The results of the Monte Carlo simulations show that the correlation function $G_{\text{RVB}}(R)$ of valence-bond densities obeys (on the square lattice) power-law correlations as does $G(R)$, but with an exponent $\alpha_{\text{RVB}} \simeq 1.15$ instead of $\alpha = 2$ for the dimer states (see Eq. (9.10)). In a later section we will see that $\alpha = 2$ is a consequence of the conservation law of dimer models. Thus, although the RK-type dimer states and the short-range RVB state have similar power-law correlations, they describe somewhat different physics.

9.5 Quantum dimers and gauge theories

We wish to consider the full quantum dynamics of the QDM. We will find it most profitable to map this problem into a lattice gauge theory (Kogut, 1979). In a sense this mapping is suggested by the RVB mean-field decoupling that we have been using all along. Baskaran and Anderson (1988) first introduced a mapping of the static interactions of the RVB mean-field theory to a gauge theory. Here I am following the work by Kivelson and me (Fradkin, 1990b; Fradkin and Kivelson, 1990).

Let us begin by defining an enlarged Hilbert space on the links of the lattice. Let $\{l_j(\vec{x})\}$ be a set of integer-valued variables defined on the links $\{(\vec{x}, \vec{x}+e_j)\}$ of the lattice. The states $|\{l_j(\vec{x})\}\rangle$ span the unrestricted Hilbert space. The angular-momentum operators $L_j(\vec{x})$ have the integers $l_j(\vec{x})$ as their eigenvalues and $|\{l_j(\vec{x})\}\rangle$ as their eigenstates. If we wish to restrict this Hilbert space to the subspace in which $l_j = 0, 1$, we can do so by assigning an infinite energy to all unwanted states. Thus, let us define a dimer contribution, or kinetic-energy term, which enforces the restriction and is nothing other than a hard-core condition. We can write H_{dimer} in the form

$$H_{\text{dimer}} = \frac{1}{2k}\sum_{\vec{x},j}\left[\left(L_j(\vec{x}) - \frac{1}{2}\right)^2 - \frac{1}{4}\right] \tag{9.16}$$

For any value of the coupling constant k, the configurations with $l_j = 0, 1$ have exactly zero energy, while *any* other state will have energy growing like $1/k$ as $k \to 0$.

We need two terms: one for resonance and the other for the diagonal terms. In order to discuss resonance we need to introduce the variable $a_j(\vec{x})$ at each link, which should be the eigenvalue of the operator $a_j(\vec{x})$ canonically conjugate to $L_j(\vec{x})$, i.e.

$$[a_j(\vec{x}), L_{j'}(\vec{x}')] = i\delta_{jj'}\delta_{\vec{x},\vec{x}'} \tag{9.17}$$

Since the spectrum of $L_j(\vec{x})$ is the integers $l_j(\vec{x})$, $a_j(\vec{x})$ should be an angle

$$0 \le a_j(\vec{x}) < 2\pi \tag{9.18}$$

and the Hilbert space is the space of the periodic functions of $a_j(\vec{x})$ with period 2π, independently at each link. Using the commutation relations Eq. (9.17), we see that the operator $e^{im_j a_j}$ acts like a *ladder* operator with step size m_j, where m_j is an integer. Indeed, we can write for any site

$$L_j e^{im_j a_j} |l_j\rangle = e^{im_j a_j} \left(e^{-im_j a_j} L_j e^{im_j a_j}\right) |l_j\rangle \tag{9.19}$$

The commutation relations tell us that the operator within brackets in Eq. (9.19) is the shifted operator

$$e^{-im_j a_j} L_j e^{im_j a_j} = L_j + m_j \tag{9.20}$$

Thus, we get

$$L_j e^{im_j a_j} |l_j\rangle = e^{im_j a_j} (l_j + m_j)|l_j\rangle = (l_j + m_j) e^{im_j a_j} |l_j\rangle \tag{9.21}$$

and we can identify

$$e^{im_j a_j} |l_j\rangle = |l_j + m_j\rangle \tag{9.22}$$

The resonance term should remove from a plaquette two parallel dimers and replace them by another pair of parallel dimers but in the orthogonal direction (Fig. 9.1). We can accomplish this by writing, in terms of raising and lowering operators, the term

$$\begin{aligned} H_{\text{res}} =& \bar{J} \sum_{\vec{x}} (e^{i[a_1(\vec{x})+a_1(\vec{x}+e_2)-a_2(\vec{x})-a_2(\vec{x}+e_1)]} \\ &+ e^{i[a_2(\vec{x})+a_2(\vec{x}+e_1)-a_1(\vec{x})-a_1(\vec{x}+e_2)]}) \end{aligned} \tag{9.23}$$

The diagonal terms are now

$$H_{\text{diag}} = V \sum_{\vec{x}} (L_1(\vec{x}) L_1(\vec{x}+e_2) + L_2(\vec{x}) L_2(\vec{x}+e_1)) \tag{9.24}$$

and the constraint is

$$Q(\vec{x}) = L_1(\vec{x}) + L_1(\vec{x}-e_1) + L_2(\vec{x}) + L_2(\vec{x}-e_2) - 1 = 0 \tag{9.25}$$

This equation looks peculiar since the left-hand side is an operator and the right-hand side is a number. The meaning of this equation is that the allowed states of the Hilbert states, which I will call $|\text{Phys}\rangle$, satisfy

$$Q(\vec{x})|\text{Phys}\rangle = 0 \tag{9.26}$$

For this condition to be consistent, $Q(\vec{x})$ should be diagonalizable simultaneously with the total Hamiltonian H, i.e.

$$[Q(\vec{x}), H] = 0 \tag{9.27}$$

where

$$H = H_{\text{dimer}} + H_{\text{res}} + H_{\text{diag}} \tag{9.28}$$

This is indeed the case, since $Q(\vec{x})$ simply counts all the dimers touching a given site and this number is a constant of motion.

The operator $Q(\vec{x})$ generates a set of local time-independent transformations that leave the physical states invariant,

$$e^{i\sum_{\vec{x}}\alpha(\vec{x})Q(\vec{x})}|\text{Phys}\rangle = |\text{Phys}\rangle \tag{9.29}$$

which therefore leave H unchanged. Since the spectra of the operators $\{Q(\vec{x})\}$ are the integers, the gauge transformations are parametrized by phases, periodic variables $\{\alpha(\vec{x})\}$ defined on the interval $[0, 2\pi)$.

Thus we discover that H has a *local gauge symmetry* and Q is the generator of local gauge transformations. The constraint equation is simply a version of Gauss's law. This local symmetry simply reflects the fact that we are free to change the phases of the valence bonds on each site independently. In this language, the wave functions which are being considered must have the form

$$|\Psi\rangle = \sum_{\{c\}} A(c)e^{i\Phi(c)}|c\rangle \tag{9.30}$$

where $\{c\}$ is a set of (linearly independent) VB states (i.e. dimer coverings), $A(c)$ is a real amplitude for configuration c, and $\Phi(c)$ is the phase. The phase $\Phi(c)$ depends on the configuration and we have chosen to write $\Phi(c)$ in the form of a sum over links,

$$\Phi(c) = \sum_{\vec{x},j} a_j(\vec{x}) \tag{9.31}$$

States of the form of Eq. (9.30) are coherent states parametrized by the variables $a_j(\vec{x})$.

We can write these formulas in a much more transparent and familiar way, by staggering the configuration $\{a_j(\vec{x})\}$. Clearly this can be done consistently only for a bipartite lattice. Let us define the staggered gauge field $A_j(\vec{x})$ and "electric fields" $E_j(\vec{x})$ by

$$A_j(\vec{x}) = e^{i\vec{Q}_0\cdot\vec{x}}a_j(\vec{x}) \tag{9.32}$$

$$E_j(\vec{x}) = e^{i\vec{Q}_0\cdot\vec{x}}L_j(\vec{x}) \tag{9.33}$$

with $\vec{Q}_0 = (\pi, \pi)$. It should be stressed that these fields do *not* represent the electromagnetic fields. With these definitions, we can rewrite the constraints of Eq. (9.26) in the form

$$[\Delta_j E_j(\vec{x}) - \rho(\vec{x})]|\text{Phys}\rangle = 0 \tag{9.34}$$

where Δ_j is the lattice divergence

$$\Delta_j E_j(\vec{x}) \equiv E_1(\vec{x}) - E_1(\vec{x} - e_1) + E_2(\vec{x}) - E_2(\vec{x} - e_2) \tag{9.35}$$

and the density $\rho(\vec{x})$ is

$$\rho(\vec{x}) = e^{i\vec{Q}_0 \cdot \vec{x}} \tag{9.36}$$

Equation (9.34) now has the standard form of Gauss's law. Note that $\rho(\vec{x})$ represents a background staggered charge density that equals $+1$ (-1) on red (black) sites, and enforces the condition that each site should belong to one and only one dimer. In the presence of holes, $\rho(\vec{x})$ will vanish on sites occupied by holes.

In this formulation the Hamiltonian reads

$$\begin{aligned} H = & \frac{1}{2k} \sum_{\vec{x},j} \left([E_j(\vec{x}) - \alpha_j(\vec{x})]^2 - \alpha_j^2(\vec{x}) \right) + 2\bar{J} \sum_{\vec{x}} \cos \left(\sum_{\text{plaquette}} A_j(\vec{x}) \right) \\ & - V \sum_{\vec{x}} (E_1(\vec{x}) E_1(\vec{x} + e_2) + E_2(\vec{x}) E_2(\vec{x} + e_2)) \end{aligned} \tag{9.37}$$

where $\sum_{\text{plaquette}} A_j(\vec{x})$ stands for the oriented sum of staggered vector potentials $A_j(\vec{x})$ around the elementary plaquette labeled by $\vec{x}$ (its southwest corner):

$$\begin{aligned} \sum_{\text{plaquette}} A_j(\vec{x}) \equiv & \; e^{i\vec{Q}_0 \cdot \vec{x}} \left(a_1(\vec{x}) + a_1(\vec{x} + e_2) - a_2(\vec{x}) - a_2(\vec{x} + e_1) \right) \\ & = A_1(\vec{x}) - A_1(\vec{x} + e_2) - A_2(\vec{x}) + A_2(\vec{x} + e_1) \\ & = \Delta_2 A_1(\vec{x}) - \Delta_1 A_2(\vec{x}) \end{aligned} \tag{9.38}$$

and is interpreted as a plaquette flux, and $\alpha_j(\vec{x})$ is

$$\alpha_j(\vec{x}) = \frac{1}{2} e^{i\vec{Q}_0 \cdot \vec{x}} \tag{9.39}$$

By expanding the square in the first term in Eq. (9.37), and using Eq. (9.39), we can write the first term of the Hamiltonian in the form

$$\frac{1}{2k} \left(\sum_{\vec{x},j} E_j^2(\vec{x}) - \frac{L^2}{2} \right) \tag{9.40}$$

where L is the linear size of the square lattice.

We can take all these considerations into account by writing the full Hamiltonian in the form

$$\begin{aligned} H = & \frac{1}{2k} \left(\sum_{\vec{x},j} E_j^2(\vec{x}) - \frac{L^2}{2} \right) + 2\bar{J} \sum_{\vec{x}} \cos \left(\sum_{\text{plaquette}} A_j(\vec{x}) \right) \\ & + \frac{V}{2} \sum_{\vec{x}} \left((\Delta_1 E_2(\vec{x}))^2 + (\Delta_2 E_1(\vec{x}))^2 \right) - \frac{V}{2} L^2 \end{aligned} \tag{9.41}$$

and considering the limit $k \to 0$. The states are restricted by demanding that Gauss's law, Eq. (9.34), be exactly satisfied.

9.6 The Ising gauge theory

In this section we will discuss results from lattice gauge theory that are relevant to understanding the problem at hand. We will be interested in the spectrum of states and of allowed observables in the different phases that these theories have. Although the problems we are interested in are, as we saw, gauge theories, we will typically be interested not in the vacuum sector but, as we saw in the last section, in sectors with lots of background charges. This sector, which is sometimes called the "odd" gauge-theory sector (Moessner *et al.*, 2001), has distinct properties. Here we will focus on the standard vacuum sector.

The simplest example of a gauge theory is the Ising gauge theory, which, as we will see, is relevant to the problems we discuss here. This is a gauge theory with a discrete gauge group $\mathbb{Z}_2$. In this theory the degrees of freedom are Ising variables, the diagonal Pauli matrices $\sigma_j^z(\vec{x})$, with $j = 1, 2$. The Hamiltonian for the Ising gauge theory (on a square lattice) is written in terms of the "vector potentials" $\{\sigma_j^z(\vec{x})\}$ (defined on the links of the square lattice) and of the "electric fields," the off-diagonal Pauli matrices $\{\sigma_j^x(\vec{x})\}$. The Hamiltonian is (Fradkin and Susskind, 1978)

$$H = -g \sum_{\vec{x},j} \sigma_j^x(\vec{x}) - \frac{1}{g} \sum_{\vec{x}} \sigma_1^z(\vec{x})\sigma_2^z(\vec{x} + e_1)\sigma_1^z(\vec{x} + e_2)\sigma_2^z(\vec{x}) \tag{9.42}$$

In what follows we will refer to the first term in the Hamiltonian as the kinetic energy and to the second term as the potential energy. We will also refer to the first term as the "electric-field" term and to the second as the magnetic (or flip) term.

The gauge-invariant states of this theory satisfy the "Gauss-law" condition, which here takes the form

$$\sigma_1^x(\vec{x})\sigma_1^x(\vec{x} - e_1)\sigma_2^x(\vec{x})\sigma_2^x(\vec{x} - e_2)|\text{Phys}\rangle = |\text{Phys}\rangle \tag{9.43}$$

Indeed, the Hamiltonian of Eq. (9.42) is invariant under the local gauge transformations generated by the operators

$$Q(\vec{x}) = \sigma_1^x(\vec{x})\sigma_1^x(\vec{x} - e_1)\sigma_2^x(\vec{x})\sigma_2^x(\vec{x} - e_2) \tag{9.44}$$

For all $\vec{x}$ these local operators commute with each other, $[Q(\vec{x}), Q(\vec{y})] = 0$, and with the Hamiltonian, $[Q(\vec{x}), H] = 0$. Hence, for all $\vec{x}$, the eigenstates of the Hamiltonian are also eigenstates of the generators $\{Q(\vec{x})\}$ and satisfy the local constraint of Eq. (9.43).

At every point $\vec{x}$ in space we can define two operators, the identity I and Q, where we see that $Q^2 = I$. Hence the local symmetry group of this problem is the discrete group $\mathbb{Z}_2$, the permutation group of two elements. The main (and important) difference between this theory and the standard Ising model in a transverse field is that this theory has a local $\mathbb{Z}_2$ symmetry, whereas the Ising model has a global $\mathbb{Z}_2$ symmetry.

The most important and central result that we will need is known as Elitzur's theorem (already discussed in Section 8.5), which states that local (gauge) symmetries cannot be spontaneously broken (Elitzur, 1975). A key consequence of this theorem is the fact that only locally gauge-invariant operators can have a non-vanishing expectation value. Thus, in contrast to what happens in systems with global symmetries (such as the antiferromagnets discussed in earlier chapters), the phase diagrams of gauge theories are classified in terms of the behavior of their gauge-invariant operators and the associated spectrum of gauge-invariant states. In particular, except in the trivial case of Maxwell's electrodynamics, a local order parameter does not generally exist (Fradkin and Shenker, 1979).

The gauge-invariant observables of this theory are as follows.

1. The Wilson loop operator on the closed loop Γ,

$$W_\Gamma = \prod_{(\vec{x},j)\in\Gamma} \sigma_j^z(\vec{x}) \tag{9.45}$$

   where $\{(\vec{x}, j)\}$ (with $j = 1, 2$) are the links of the loop Γ. The plaquette operator of the Hamiltonian is a particular case of a Wilson loop.
2. The electric-field operator on link $(\vec{x}, j)$, $\sigma_j^x(\vec{x})$.
3. An ("electric") charge created at point $\vec{x}$ amounts to requiring that the physical states have $Q(\vec{x}) = -1$, where $Q(\vec{x})$ is defined by Eq . (9.44). Owing to the $\mathbb{Z}_2$ symmetry, only the *parity* of the charge is well defined.
4. In a system with periodic boundary conditions (i.e. on a torus) the magnetic 't Hooft loop operator ('t Hooft, 1979) $\tilde{W}_{\tilde{\Gamma}}$ along a non-contractible loop $\tilde{\Gamma}$ on the dual lattice of the torus (shown in Fig. 9.2(a)) is

$$\tilde{W}_{\tilde{\Gamma}} = \prod_{(\vec{x},j)\in\tilde{\Gamma}} \sigma_j^x(\vec{x}) \tag{9.46}$$

   This operator represents the product of σ^x operators of the links of the lattice threaded by the loop $\tilde{\Gamma}$. By gauge invariance, $Q(\vec{x}) = 1$ everywhere, the actual path is unimportant; only the fact that it is globally non-contractible matters.
5. The magnetic charge operator $\tau^z(\vec{r})$

$$\tau^z(\vec{r}) = \prod_{(\vec{x},j)\ \text{pierced by}\ \tilde{\gamma}(\vec{r})} \sigma_j^x(\vec{x}) \tag{9.47}$$

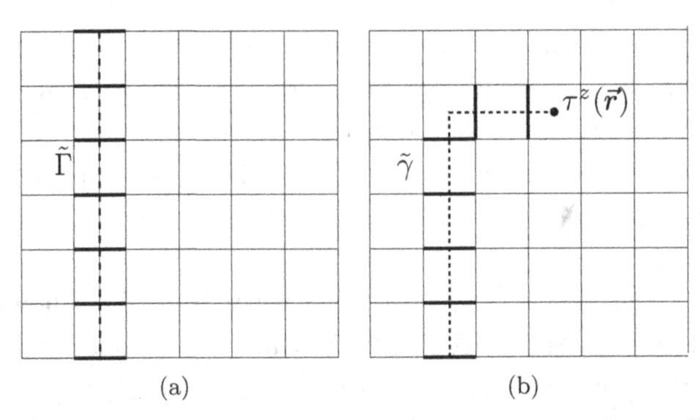

Figure 9.2 (a) A magnetic ('t Hooft) loop on a non-contractible loop $\tilde{\Gamma}$ of the dual lattice; the dark links pierced by the loop represent a product of σ^x operators on each link. (b) A Dirac string of σ^x operators on the open path $\tilde{\gamma}$ of the dual lattice creates magnetic charge (a "vison") at the dual site $\vec{r}$.

Here too, in the gauge-invariant sector, with $Q(\vec{x}) = 1$ everywhere, this operator depends only on the location $\vec{r}$ of the dual lattice and it is independent of the shape of the Dirac string, the rest of the path $\tilde{\gamma}(\vec{r})$ with its endpoint at $\vec{r}$.

It is straightforward to see that the magnetic-charge operator $\tau^z(\vec{r})$ on the dual site $\vec{r}$ (the center of a plaquette) *anti-commutes* with the plaquette Wilson loop operator $W_p(\vec{r})$,

$$\{W_p(\vec{r}), \tau^z(\vec{r})\} = 0 \tag{9.48}$$

Since $W_p(\vec{r})^2 = 1$ and $\tau_z(\vec{r})^2 = 1$, we can identify the plaquette Wilson loop operators with the Pauli matrices $W_p(\vec{r}) = \tau^x(\vec{r})$ defined on the dual sites. Similarly, the operator $\sigma_j^x(\vec{x})$ on a link of the direct lattice is easily seen to be given by

$$\sigma_j^x(\vec{x}) = \tau^z(\vec{r})\tau^z(\vec{r} + e_j) \tag{9.49}$$

which automatically satisfies the condition $Q(\vec{x}) = 1$ everywhere.

It is now easy to see that, in terms of the operators $\{\tau^z(\vec{r})\}$ and $\{\tau^x(\vec{r})\}$ defined on the dual lattice, the Hamiltonian of the gauge theory becomes

$$H = -g \sum_{\vec{r},j} \tau^z(\vec{r})\tau^z(\vec{r} + e_j) - \frac{1}{g} \sum_{\vec{r}} \tau^x(\vec{r}) \tag{9.50}$$

which we recognize as the Hamiltonian of the Ising model in a transverse field (or quantum Ising model) on a 2D square lattice. What we have done is to prove that the transverse-field Ising model and the Ising gauge theory in (2+1) dimensions are dual to each other. In particular, the operator τ^z plays the role of the order parameter of the transverse-field Ising model and of the magnetic-charge (or *monopole*) operator in the gauge theory (Fradkin and Susskind, 1978; Kogut, 1979).

As discussed in Chapter 5, the Ising model in a transverse field in d space dimensions is equivalent to to the classical Ising model in $(d+1)$ dimensions (Fradkin and Susskind, 1978). This relationship follows from the fact that the partition function of the classical model in $(d+1)$ dimensions (with periodic boundary conditions) can be written as

$$Z = \text{tr}\, T^N \tag{9.51}$$

where T is the transfer matrix and N is the number of rows (or hyperplanes) along the discrete "imaginary-time" direction. For the case of the Ising model (in all dimensions) the transfer matrix T has the form of a product of two matrices, each involving the kinetic- and potential-energy terms of the Hamiltonian of Eq. (9.50). A well-defined sequence of approximations (equivalent to taking the time continuum limit) maps the classical problem in $(d+1)$ dimensions to the quantum Hamiltonian in d dimensions (Fradkin and Susskind, 1978).

For a system that satisfies the property of reflection positivity, that is that its correlation functions are real, positive, and invariant under reflection across a hyperplane, the transfer matrix can always be constructed to be a hermitian matrix. This relation is, of course, the same as that between the path-integral and Hamiltonian formulations of quantum field theory, with reflection positivity being the Euclidean version of unitarity. It holds for many problems of interest, not just the Ising model, and it holds in all dimensions. It also holds for the gauge theory. In particular, the dual of the classical Ising model in three dimensions is a 3D theory with a local $\mathbb{Z}_2$ invariance, the Ising gauge theory (Wegner, 1971; Balian *et al.*, 1975). In systems that are isotropic, as classical systems in $(d+1)$ dimensions (as are the Ising model and gauge theory), the direction chosen to be the "imaginary-time" direction (i.e. the direction of transfer) is arbitrary. From this it follows that at their critical points these systems are equivalent to Lorentz-invariant field theories. Hence, they have an associated quantum-dynamical critical exponent of $z = 1$.

The $\mathbb{Z}_2$ gauge theory has two phases: (a) a weak-coupling, $g < g_c$, deconfined phase; and (b) a strong-coupling, $g > g_c$, confined phase.

9.7 The $\mathbb{Z}_2$ confining phase

Let us now turn to the strong-coupling phase, which we will find is confining. Although we will focus our discussion on the case of two space dimensions, the results apply to higher dimensions as well. In the strong-coupling regime, the spectrum can be determined using the strong-coupling expansion. In this expansion, which is conceptually a Brillouin–Wigner expansion similar to the one we used to derive the Heisenberg Hamiltonian from the Hubbard model in Chapter 2, the ground state is approximately an eigenstate of the electric-field operators $\sigma_j^x(\vec{x})$. To leading order in an expansion in powers in $1/g$, the ground state $|G\rangle$ is an eigenstate of the kinetic-energy term, and hence of the link "electric-field" operators $\sigma_j^x(x)$:

$$|G\rangle_{g\to\infty} = \prod_{(\vec{x},j)} |\sigma_j^x(\vec{x}) = 1\rangle \tag{9.52}$$

Thus, in this state

$$\sigma_j^x(\vec{x})|G\rangle_{g\to\infty} = +|G\rangle_{g\to\infty} \tag{9.53}$$

It is also easy to see that there is a finite energy gap. Indeed, due to the Gauss-law condition, Eq. (9.43), the allowed states must have an even number of links sharing site $\vec{x}$ with $\sigma^x = -1$ on those links. The allowed states which obey this constraint are closed loops on the lattice, i.e. the set of links on which $\sigma^x = -1$. Hence, the spectrum of states in the strong-coupling regime consists of electric loops. Since the energy cost over the ground state of each excited link is $2g$, the total energy of an allowed excited state consisting of loops of length ℓ is $\Delta E_{\text{loop}} = 2g\ell$. Thus, the lowest excited state is the elementary loop or plaquette state, a state created by the magnetic (flip) term, and consists of electric fields being excited on the perimeter of the elementary plaquette. The energy of the plaquette state is $\Delta E_{\text{plaquette}} = 8g$.

Hence, the spectrum of excited states has a finite (and large) energy gap in the strong-coupling limit. It is easy to show that for finite but large g this state is stable in the strong-coupling expansion, and that it is separated from the first excited state by a finite energy gap up to some critical coupling g_c. Indeed, the general form of the ground state in the strong-coupling phase is a superposition of states with loops of varying length, with the contribution of long loops to the amplitude becoming exponentially small as the length of loop increases. Using this line of argument it is possible, with a moderate amount of work, to show that the strong-coupling expansion has a finite radius of convergence, with the critical coupling g_c limiting the convergence of the expansion (Kogut, 1984). Clearly, as the quantum phase transition at g_c is approached, the loops contributing to the ground state progressively grow in length (and number). If the transition at g_c is continuous, their size diverges as $g \to g_c$, and the electric loops proliferate. At the same time, the energy gap $\Delta E_{\text{plaquette}}(g)$ to the lowest excited state, which as we saw is large and of the order of $8g$ in the strong-coupling regime, becomes smaller as $g \to g_c$ and vanishes with a universal critical exponent

$$\Delta E_{\text{plaquette}}(g) = \text{constant} \times |g - g_c|^{\Delta} \tag{9.54}$$

The critical coupling and the gap exponent have been calculated numerically using quantum Monte Carlo methods (Rieger and Kawashima, 1999) and RG methods (Evenbly and Vidal, 2009) (for the dual Ising model in a transverse field) with

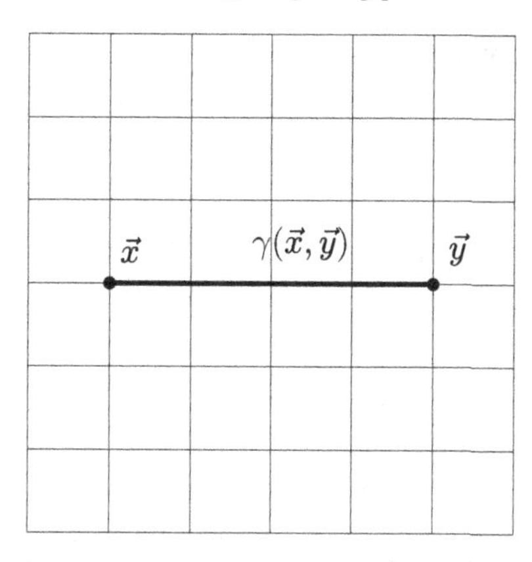

Figure 9.3 The ground state with two charges at $\vec{x}$ and $\vec{y}$ in the confinement phase in which the energy grows linearly with separation; $\sigma^x = -1$ on each link of the path $\gamma(\vec{x}, \vec{y})$.

the estimates $1/g_c^2 = 3.044$ and $\Delta = 0.622$, which are consistent with a quantum critical point with a Lorentz-invariant value of the dynamic critical exponent $z = 1$, and a correlation-length exponent $\nu = \Delta$.

To see that this phase is confining, we will simply compare the ground-state energy of the vacuum state (without electric charges) with the ground state in the sector with a charge located at $\vec{x}$ and another at $\vec{y}$, $|G; \vec{x}, \vec{y}\rangle$, i.e. $Q = -1$ at $\vec{x}$ and at $\vec{y}$, and $Q = +1$ everywhere else. The ground state of this sector is defined, to leading order, by $\sigma_j^x = 1$ on every link of the lattice *except* along the path $\gamma(\vec{x}, \vec{y})$ (shown in Fig. 9.3) stretching from $\vec{x}$ to $\vec{y}$ on whose links $\sigma_j^x = -1$. Thus, in this sector the ground state has a *string* along the shortest path $\gamma(\vec{x}, \vec{y})$ between sites $\vec{x}$ and $\vec{y}$. The energy difference between the ground state with the two charges and the vacuum state (the ground state without charges) is $\Delta E = 2gR + O(1/g)$, where R is the distance (in lattice units) between $\vec{x}$ and $\vec{y}$, which *grows linearly* with separation. From the convergence of the strong-coupling expansion we expect the same behavior throughout this phase,

$$\Delta E(R) = \sigma R \tag{9.55}$$

where $\sigma = 2g + O(1/g)$ is the "string tension." Hence in this phase the energy needed to separate the sources at infinite distance is infinite. Hence, this phase is said to be confining. The string tension has units of energy per unit length. The characteristic energy scale of this system is the energy gap, which scales as $(g - g_c)^\Delta$, and the characteristic length scale is the correlation length (or confinement scale) $\xi(g)$, which at the quantum phase transition scales as

$$\xi(g) \sim (g - g_c)^{-\nu} \tag{9.56}$$

Therefore, the string tension $\sigma(g)$ must scale as

$$\sigma(g) \sim (g - g_c)^{2\nu} \tag{9.57}$$

(since $\Delta = z\nu = \nu$ in this case) near the quantum phase transition. Hence, as the quantum critical point at g_c is approached the string tension vanishes.

Similarly, in the strong-coupling phase the Wilson loop operator W_Γ obeys an area law. Indeed, the action of the Wilson loop on the strong-coupling state $|G\rangle_{g\to\infty}$ yields an orthogonal state, and hence $\langle G|W_\Gamma|G\rangle_{g\to\infty} = 0$. The lowest order in perturbation theory (in powers of $1/g^2$) in which these states mix is n, which is the number of plaquettes enclosed by the loop Γ, i.e. the area $A[\Gamma] = n$ of the region inside Γ. Hence the leading non-vanishing contribution to the expectation value of the Wilson loop operator is

$$\langle G|W_\Gamma|G\rangle = \text{constant} \times \left(\frac{1}{g^2}\right)^n + \cdots = \text{constant} \times e^{-\mu(g)A[\Gamma]} \tag{9.58}$$

with $\mu(g) = \ln(g^2) + O(1/g^2)$. The quantity $\mu(g)$ is related to the string tension $\sigma(g)$ and also vanishes as $g \to g_c$ with the same exponent 2ν.

The duality transformation offers an alternative and intuitive picture of the confining phase. In the dual picture, the strong-coupling phase maps onto the *ordered phase* of the Ising model, which is the *weak-coupling* phase of this model. In this phase, the ground-state expectation value of the magnetic-charge (or monopole) operator maps onto the expectation value of the *order parameter* of the (dual) Ising model, the local magnetization, which in this phase is finite. Hence, in the strong-coupling phase of the gauge theory we find that the magnetic-charge operator has an expectation value of order unity and vanishes as the critical coupling g_c is approached with an exponent

$$\langle G|\tau^z(\vec{r})|G\rangle = \text{constant} \times (g - g_c)^\beta \tag{9.59}$$

with $\beta = 0.326$ (Rieger and Kawashima, 1999; Evenbly and Vidal, 2009).

We can then picture the confining phase as a *condensate of magnetic charges* and regard the magnetic-charge operator as a *disorder operator*, an operator that has an expectation value in the disordered (strong-coupling) phase of the theory. This picture of a quantum disordered phase as a condensate of a disorder operator goes back to the work of Kadanoff and Ceva in the 2D classical Ising model (Kadanoff and Ceva, 1971) and to the work of Susskind and myself (Fradkin and Susskind, 1978) in gauge theory. We have already encountered an analogous disorder operator in the kink operator of the quantum disordered phase of the 1D Ising model in a transverse field, as discussed in Chapter 5. Moreover, in the dual-Ising-model picture, the string tension corresponds to changing the sign of the Ising coupling constant from ferromagnetic to antiferromagnetic on all the bonds of the dual theory pierced by the path of Fig. 9.3. In the ordered phase of the dual Ising model this is equivalent to a defect favoring a fractional domain wall (of length R) along that path. In the ordered phase of the Ising model this defect does indeed have an energy cost that is linear in the length of the wall, as in Eq. (9.55).

9.8 The Ising deconfining phase: the $\mathbb{Z}_2$ topological fluid

Let us now turn to the weak-coupling phase. As with the strong-coupling phase, we will begin with the extreme weak-coupling limit, $g \to 0$ in this case, and construct the ground state and the spectrum in this regime. After that we will show (or rather argue) that there is a convergent weak-coupling expansion, which implies that the ground state and spectrum found in the $g \to 0$ limit are stable.

At $g = 0$ the eigenstates of the Hamiltonian are eigenstates of the magnetic (flip) operator. The ground state at $g = 0$ is in the sector in which all the plaquette operators are equal to one on all the plaquettes of the lattice (i.e. there is no flux). Excited states are created by the monopole (or vison)-creation operator $\tau^z(\vec{r})$, which flips the state of the plaquette centered at $\vec{r}$.

In this limit, we can choose the states to be in the representation of the eigenstates of the link operators $\sigma_j^z(\vec{x})$. This is the analog of the standard representation of the quantum states of quantum electrodynamics in terms of configurations of vector potentials. Since the vector potentials, here represented by the operators $\{\sigma_j^z(\vec{x})\}$, are not gauge-invariant (they do not commute with the generators of gauge transformations $\{Q(\vec{x})\}$), this representation requires that a gauge be fixed. For a system on a disk (or rectangle) with free boundary conditions, a suitable (but certainly not unique) gauge-fixing condition is to impose that the states satisfy the axial gauge condition $\sigma_1^z(\vec{x}) = 1$ on all links in the x_1 direction. In this gauge, a state is fully specified by giving the states of the $\sigma_2^z(\vec{x})$ operators on the links in the x_2 direction. Thus, *in the axial gauge* the ground state at $g = 0$ is simply

$$|G\rangle_{g=0} = \prod_{\vec{x},j} |\sigma_j^z(\vec{x}) = 1\rangle \tag{9.60}$$

However, it is also possible to construct an explicitly gauge-invariant state at $g = 0$, a state which is simultaneously an eigenstate of all the plaquette (or flip) operators *and* of the gauge generators $Q(\vec{x})$ (for all $\vec{x}$), i.e. to satisfy the $\mathbb{Z}_2$ version of the Gauss law everywhere. This state was constructed by Kitaev (2003) in his work on the toric code, which is equivalent to the $g = 0$ limit of the $\mathbb{Z}_2$ gauge theory. Since the state is required to be an eigenstate of the generators of the time-independent gauge transformation, Kitaev's state is in the "electric-field" representation in which the states are eigenstates of the operators $\sigma_j^x(\vec{x})$ used above to describe the strong-coupling limit.

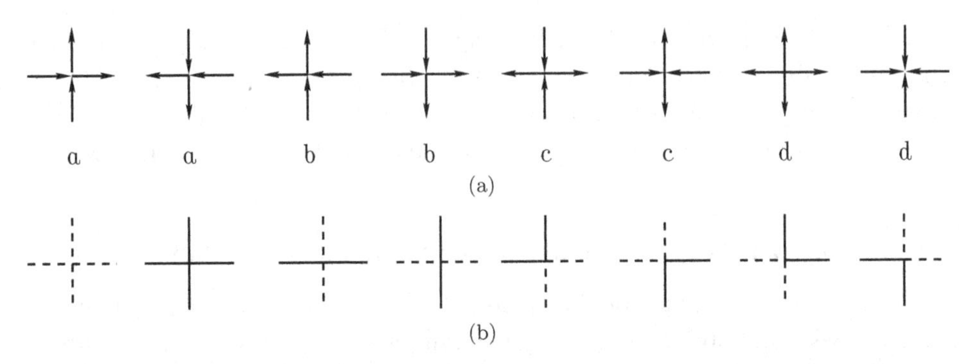

Figure 9.4 (a) The eight configurations of σ^x eigenstates on the links allowed by the $\mathbb{Z}_2$ Gauss-law condition, Eq. (9.43). (b) The corresponding loop configurations: broken lines have $\sigma^x = 1$ and full lines have $\sigma^x = -1$.

Let us denote by $\mathcal{C}$ the configurations allowed by the $\mathbb{Z}_2$ Gauss law of Eq. (9.43). We can denote a $\sigma^x = 1$ state on a horizontal link by an arrow pointing to the right, $\rightarrow$, a $\sigma^x = -1$ state by a left-pointing arrow on that link, $\leftarrow$, a $\sigma^x = 1$ state on a vertical link by an up arrow, $\uparrow$, and a $\sigma^x = -1$ state by a down arrow, $\downarrow$. The condition that $Q(\vec{x}) = 1$ at every site $\vec{x}$ implies that the allowed configurations $\mathcal{C}$ have an even number of arrows pointing in or out of the site (or vertex), as shown in Fig. 9.4(a).

We can also give an equivalent graphical representation of the allowed states by painting all links with a $\sigma^x = 1$ state (states without a $\mathbb{Z}_2$ electric field) with a broken line and all links with a $\sigma^x = -1$ state with a full line (states with a $\mathbb{Z}_2$ electric field), as shown in Fig. 9.4(b). In the latter picture the allowed configurations are the set of possible loop coverings of the square lattice. An example of such a state is the elementary loop excitation created by the magnetic plaquette operator, which we discussed in the strong-coupling phase. Thus the $\mathbb{Z}_2$ gauge theory can be viewed as a quantum-loop model.

However, the plaquette operators are not diagonal in this representation. Indeed, the action of a plaquette operator centered at dual site $\vec{r}$ (and with SW corner at the direct site $\vec{x}$) on a state in which its links have the general configuration $\{\sigma^x(i)\}$ (with $i = 1, \ldots, 4$ the four links of the plaquette) is to flip this state to the opposite configuration, $\{-\sigma^x(i)\}$. The result is that the state is the *equal-amplitude superposition* of all states of the lattice (in the σ^x representation) satisfying the Gauss-law condition. Let $\mathcal{C}$ represent the set of all such configurations. The Kitaev state is then

$$|G\rangle_{\text{Kitaev}} = \sum_{\mathcal{C}} |\mathcal{C}\rangle \tag{9.61}$$

Thus, the Kitaev representation of the ground state at $g = 0$ is the linear superposition of *all* loop configurations $\mathcal{C}$ with equal amplitude. In this phase the loops have *proliferated* (or "condensed") and have all possible lengths. In particular this state includes loops stretching between opposite boundaries of the lattice, even in the thermodynamic limit, with the same amplitude as short loops. We will see shortly that the deconfined phase is actually a *topological phase* and that this is related to the fact that it represents a state in which loops have proliferated (Freedman, 2003; Levin and Wen, 2005).

The Kitaev state and the $|G\rangle_{g=0}$ state constructed above in the σ^z representation are obviously the same state since the eigenstates of σ^z are the symmetric and antisymmetric superpositions of the eigenstates of σ^x (and vice versa). However, although the Kitaev state is formally analogous to the ground state of the QDM at the RK point, i.e. the short-range RVB state, in that they are both the equal-amplitude sum over all the allowed configurations, its properties are actually dramatically different on the square lattice but are analogous on the triangular lattice.

To see that the weak-coupling phase is deconfining, we will compute for this case the energy of two point sources separated by a distance R. It is easy to see that in the limit $g \to 0$ this energy vanishes. In this limit we must find first the ground state with two sources, i.e. a state with two points $\vec{x}$ and $\vec{y}$ on the lattice where $Q = -1$. We saw that in the strong-coupling limit the ground state in this sector has a string of smallest possible length R, the straight line between $\vec{x}$ and $\vec{y}$ if they lie on the same row or column of the lattice (other cases are similar but more complicated). We also saw that the vacuum sector at $g = 0$ (with $Q = 1$ everywhere) is the Kitaev state in which loops of all sizes (and numbers) have the same amplitude.

In the sector with two sources the ground state is also a linear superposition (with equal amplitude in the $g \to 0$ limit) of all possible configurations of closed loops but with an open string stretching from $\vec{x}$ to $\vec{y}$ *with all possible lengths*. This is so since the plaquette operator acting on a string only deforms the string, yielding another string configuration. Thus the state now is also a superposition of all possible strings (in additions to loops). However, this state has exactly the same energy as that of the vacuum sector. Therefore, the energy cost $\Delta E(R) = 0$ at $g = 0$. On the other hand, the first-order correction, in an expansion in powers of g^2 now, of both ground states yields a finite energy cost (which is independent of the distance R for large separations). In general, in this phase $\Delta E(R)$ has the long-distance behavior

$$\Delta E(R) = 2E_0(g) + V(g, R) \tag{9.62}$$

where $E_0(g) \propto g^2 + O(g^4)$ is the *self-energy* of the sources (which is vanishingly small as $g \to 0$) and

$$V(g, R) \sim A(g)e^{-R/\xi_s(g)} \tag{9.63}$$

is the *effective interaction* between the sources, which in this phase is screened (as would be expected, since the energy spectrum is gapped) with a screening length $\xi_s(g)$ that vanishes as $g \to 0$, i.e. in this phase the effective interaction is short-ranged. Therefore in the weak-coupling phase the external sources can be separated at an infinite distance with a finite self-energy and a weak (exponentially small) interaction. This is what we mean by deconfinement.

This result can also be understood by computing the expectation value of the Wilson loop operator W_Γ. Since the Wilson loop operator is a product of the σ^z operators on the perimeter of the loop Γ, this can be done more easily in the σ^z representation of the ground-state wave function. Indeed, since in the axial gauge ($\sigma_1^z = 1$) the ground state is simply $\sigma^z = 1$ on all links, the Wilson loop operator is simply

$$\langle G|W_\Gamma|G\rangle_{g=0} = 1 \tag{9.64}$$

and we see that the effective interaction vanishes at $g = 0$. It is also straightforward to see that for $g > 0$ the Wilson loop obeys a *perimeter law* in this phase,

$$\langle G|W_\Gamma|G\rangle = e^{-\rho(g)L[\Gamma]} \tag{9.65}$$

where $L[\Gamma]$ is the perimeter of the loop Γ and $\rho(g)$ is a function of g that vanishes smoothly as $g \to 0$.

We close this subsection with some comments on the behavior of the Wilson loop and of the effective interaction at the quantum critical point g_c. As $g \to g_c$ from above the string tension $\sigma(g)$ vanishes (following a power law) and the confinement scale ξ diverges (also with a power law), and as $g \to g_c$ from below the screening length $\xi_s(g)$ diverges. If the quantum phase transition at g_c is continuous, and in this case it is, in both observables we expect to obtain a behavior intermediate between confinement and deconfinement. In the case of the Wilson loop, scale invariance suggests that it should be a universal function of the aspect ratio of the loop. This assumption presumes that the loop is smooth and that it has no corners or cusps. However, on a lattice corners in closed Wilson loops are unavoidable. It is well known that in Wilson loops with corners, corners (and more generally cusps) contribute (logarithmic) singularities that we will not be concerned with here.

An elegant scaling (actually RG) argument (Peskin, 1980) shows that the effective interaction *in all dimensions* at the quantum critical point has the universal form

$$V(R) = -\frac{c}{R} \tag{9.66}$$

where c is a universal number. Hence, at g_c we expect the effective interaction to obey a universal $1/R$ ("Coulomb") law even in two dimensions, even in this theory with a discrete symmetry.

Returning to the eight possible configurations of σ_x eigenstates shown in Fig. 9.4, we can define a generalized eight-vertex wave function of the form (Ardonne *et al.*, 2004)

$$|\Psi\rangle = \sum_{\mathcal{C}} a^{N_a[\mathcal{C}]} b^{N_b[\mathcal{C}]} c^{N_c[\mathcal{C}]} d^{N_d[\mathcal{C}]} |\mathcal{C}\rangle \tag{9.67}$$

where a, b, c, and d are four real and positive amplitudes, and $N_a[\mathcal{C}]$, $N_b[\mathcal{C}]$, $N_c[\mathcal{C}]$, and $N_d[\mathcal{C}]$ are the numbers of vertices of types a, b, c and d present in configuration $\mathcal{C}$, respectively. This eight-vertex state has the property that its norm is equal to the partition function of the 2D classical Baxter eight-vertex model (Baxter, 1982). Therefore this wave function represents a generalized dimer-type model whose phase diagram is that of the classical 2D Baxter (eight-vertex) model. Ardonne and coworkers showed that it is possible to write a quantum Hamiltonian that is the sum of projection operators for which the eight-vertex wave function is the ground state (Ardonne *et al.*, 2004).

A case of special interest is the choice $a = b = 1$. The phase diagram for this case is shown in Fig. 9.5. It has two ordered phases and a disordered topological phase, separated by lines of continuous phase transitions with continuously varying exponents. The bottom line of the phase diagram is the scaling limit of the classical six-vertex model. The special point with weights $c = \sqrt{2}$ and $d = 0$ is the dimer model on the square lattice, which, as we saw, is a critical system. There is a Kosterlitz–Thouless transition at $(c^2, d^2) = (2, 0)$ to an ordered phase.

9.9 Boundary conditions and topology

Let us now consider the effects of boundary conditions on the phases of the $\mathbb{Z}_2$ gauge theory. In the preceding subsections we assumed that the system was defined to be a large disk (or rectangle) with fixed boundary conditions. We will now examine what happens when we impose periodic boundary conditions or, which amounts to the same thing, place the system on a torus.

The choice of boundary condition clearly does not affect the confinement phase in any essential way. Indeed, in our analysis of the state in the strong-coupling regime we found that it has a unique gauge-invariant ground state and, in fact, we did not even have to fix a gauge. Thus, in the strong-coupling phase we expect (and get) a unique ground state regardless of the choice of boundary conditions. However, there is a subtlety. If a link is in the state $\sigma^x = 1$, then in this state the

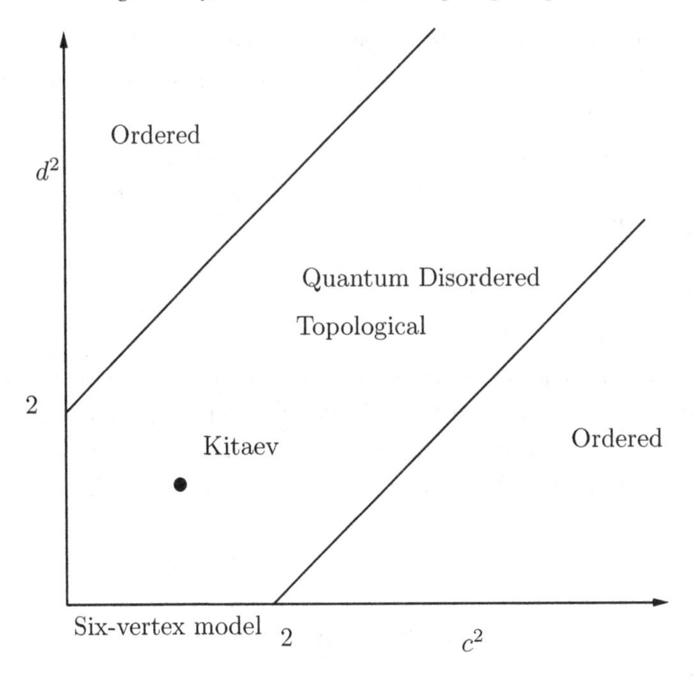

Figure 9.5 The phase diagram for the eight-vertex wave function. The phase boundaries are lines of continuous phase transitions. The topological phase is the quantum disordered phase. The Kitaev toric-code state is the special point in the middle of the disordered phase. The bottom line, $d = 0$, is the six-vertex model, and the special point $(c^2, d^2) = (\sqrt{2}, 0)$ is the dimer model on the square lattice. The effective-field theory along this line is the quantum Lifshitz model discussed at the end of this chapter (Ardonne *et al.*, 2004).

product of the τ^z operators for the two surrounding dual sites must also be equal to 1. Hence, the strong-coupling state of the gauge theory is equivalent to all the states of the dual Ising model that do not have any domain walls (as these will lead to some $\tau^z = -1$). There are two domain-wall free states in the dual Ising model: all up and all down. Thus, the duality transformation is a two-to-one mapping and does not distinguish one broken symmetry state from the other.

Another way to see this is to ask how the generator of global $\mathbb{Z}_2$ transformations of the dual Ising model, which I will denote by $\tilde{Q}$, behaves in the gauge-theory picture. The generator $\tilde{Q}$ is the product of all the $\tau^x(\vec{r})$ of the dual lattice. Under duality this operator maps to the Wilson loop acting on links of the *boundary* Γ of the entire system:

$$\tilde{Q} = \prod_{\vec{r}} \tau^x(\vec{r}) = \prod_{(\vec{x},j)\in\Gamma} \sigma_j^z(\vec{x}) \tag{9.68}$$

This operator has a vanishing expectation value in the strong-coupling phase of the gauge theory, since the latter is essentially an eigenstate of σ^x, $\langle G|\tilde{Q}|G\rangle_{g\to\infty} = 0$.

This is also the case in the dual Ising model since $\tilde{Q}$ maps one broken symmetry ground state into the other and these states are orthogonal.

In the weak-coupling phase the choice of boundary condition has an even subtler and more interesting effect. We saw that in the weak-coupling phase in the σ^z representation a gauge-fixing condition is required, and that for a system with an open boundary the axial gauge condition (such as $\sigma_1^z = 1$) completely and unambiguously fixes the gauge and defines the state. Notice that in this state the expectation value $\langle G|\tilde{Q}|G\rangle_{g=0} = 1$, which is also correct in the dual Ising model in its disordered phase.

Let us now consider the weak-coupling phase for a system on a torus. A torus is a topologically non-trivial surface (or manifold) that has the defining property that it has two non-contractible closed curves (or loops), denoted in Fig. 9.6 by γ_1 and γ_2. The Wilson loop operators on γ_1 and γ_2, W_{γ_1} and W_{γ_2}, respectively, are gauge-invariant operators whose expectation values cannot be changed by local gauge-fixing conditions. Since the Wilson loop operators are products of σ_j^z link operators, in the $g = 0$ limit $W_{\gamma_1} = \pm 1$ and $W_{\gamma_2} = \pm 1$. Hence we find four inequivalent versions of the weak-coupling state, and conclude that on a torus the axial gauge condition does not completely specify the state.

We can also define magnetic 't Hooft operators on non-contractible loops, $\tilde{W}_{\gamma_1}$ and $\tilde{W}_{\gamma_2}$ (as defined in Eq. (9.46) and Fig. 9.2), which are also gauge-invariant, and play a key role in this problem (Moessner *et al.*, 2001). Gauge-invariant observables defined on non-contractible curves of a manifold (such as the torus) define non-trivial (magnetic) *holonomies* of the system. The Wilson loop and

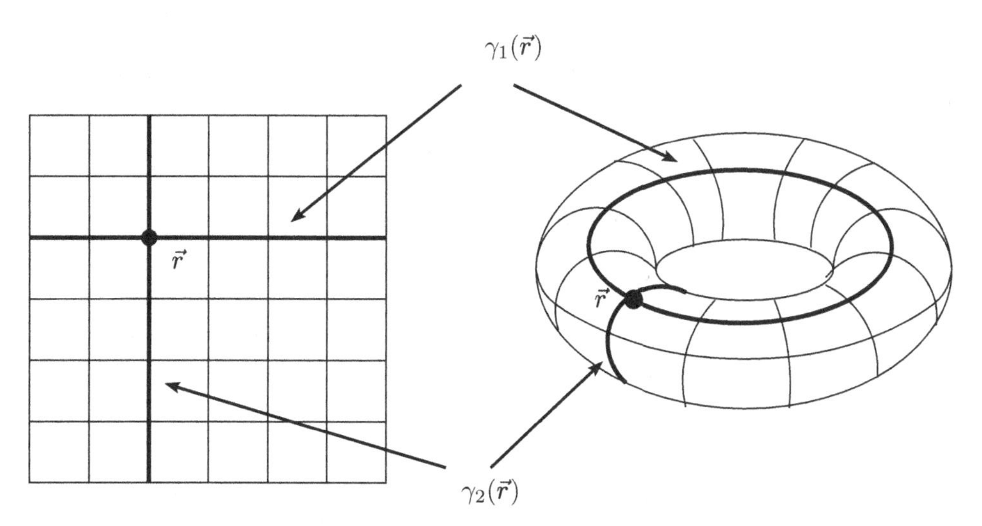

Figure 9.6 A square lattice with periodic boundary conditions is isomorphic to a torus. The two non-contractible loops $\gamma_1(\vec{r})$ and $\gamma_2(\vec{r})$ are shown.

't Hooft loop operators on non-contractible closed loops obey the commutation relations

$$\{W_{\gamma_1}, \tilde{W}_{\gamma_2}\} = 0, \qquad \{W_{\gamma_2}, \tilde{W}_{\gamma_1}\} = 0$$
$$[W_{\gamma_1}, \tilde{W}_{\gamma_1}] = 0, \qquad [W_{\gamma_2}, \tilde{W}_{\gamma_2}] = 0 \tag{9.69}$$

In addition the holonomies of the 't Hooft magnetic operators, $\tilde{W}_{\gamma_j}$, commute with the Hamiltonian of the $\mathbb{Z}_2$ gauge theory. Hence the eigenstates (and consequently the ground states as well) of the Hamiltonian are also eigenstates of the magnetic holonomies.

Precisely at $g = 0$ the Wilson loop ("electric") holonomies W_{γ_j} also commute with the Hamiltonian. Hence, at $g = 0$ all four states are exactly degenerate. For all $g > 0$ the Wilson loop holonomies no longer commute with the Hamiltonian (although the 't Hooft holonomies still do). Hence these four states are no longer degenerate. However, given their topological nature, on a torus of infinite size (the thermodynamic limit) these four states *do not mix* at any finite order in this (convergent!) perturbation theory. On a finite-sized torus, of linear size L, the mixing is a finite-size effect and it first occurs at order g^L. Hence the resulting energy splitting is exponentially small, $\sim e^{-|\ln g|L}$, and vanishes in the thermodynamic limit ($L \to \infty$). Therefore the ground-state degeneracy is an exact property of the deconfined phase.

We saw that the ground state of the deconfined theory can be written as a Kitaev state, i.e. as a linear superposition of electric-loop configurations. As a consequence of the $\mathbb{Z}_2$ Gauss law only the parity of these loops (and not their number) is conserved under the quantum evolution (by the action of the Hamiltonian). On a surface with non-trivial topology, such as the torus, one can classify the states by how many electric-loop operators *wind* around the non-contractible loops of the manifold. However, only their parity is a conserved quantity. Thus, there are four classes of states: states with an even or odd number of electric loops winding around each holonomy of the torus. Let us denote these classes by (s_1, s_2) with $s_i = \pm 1$ ($i = 1, 2$) representing states with even ($+1$) and odd (-1) numbers of winding loops.

Let us consider the class of states $(+, +)$ with an even number of loops winding on each direction. These states are eigenstates of the magnetic holonomies $\tilde{W}_{\gamma_j}$ (with $j = 1, 2$) with eigenvalue $+1$:

$$\tilde{W}_{\gamma_j}|+, +\rangle = +|+, +\rangle \tag{9.70}$$

Let us also consider the state resulting from the action of the Wilson loop operators W_{γ_j} on these states. Since these Wilson loop holonomies are products of σ^z operators on closed non-contractible paths γ_1 and γ_2, they change the parity of the winding states. Indeed, it is easy to see that, if $|+,+\rangle$ is any state in the $(+,+)$ class, then

$$\tilde{W}_{\gamma_2} W_{\gamma_1} |+,+\rangle = -W_{\gamma_1} \tilde{W}_{\gamma_2} |+,+\rangle = -|-,+\rangle \tag{9.71}$$

Thus the state resulting from the action of W_{γ_1} maps the class $(+,+)$ onto the class $(-,+)$. Similarly, W_{γ_2} maps the $(+,+)$ class onto the $(+,-)$ class, and $W_{\gamma_1} W_{\gamma_2}$ maps the $(+,+)$ class onto the $(-,-)$ class.

Therefore we can construct a total of *four* Kitaev ground states, each defined by the eigenvalues of the corresponding magnetic holonomy 't Hooft operators. Each of these equally deconfined ground states is linearly independent, and they are orthogonal to each other. We have then to conclude that on a torus the deconfined ground state of the $\mathbb{Z}_2$ gauge theory is four-fold degenerate. It is also easy to see that if we consider the theory being placed on a manifold with a more complex topology, say a closed surface with g handles (known as the *genus* of the surface), the ground-state degeneracy is 4^g. For example, on a surface with no handles (the sphere or the disk) $g = 0$ and the theory has a unique ground state.

We are very familiar with the concept of a ground-state degeneracy arising from the spontaneous breaking of a global symmetry (such as the examples discussed in Chapter 3 or in the case of the dual Ising model discussed in this section). In those cases the degenerate sectors are identified with the symmetries broken by the order-parameter field. Thus in an Ising model the two degenerate sectors are labeled by the sign of the expectation value of the order-parameter field in that state. In these systems the degeneracy is determined completely by the nature of the broken symmetry and it is independent of the topology of the surface on which the system lives. In other terms, the degeneracy does not know about topology.

What we just found in the deconfined phase is a very different situation. To begin with, the deconfined phase is not associated with the spontaneous breaking of any symmetry. More important is the fact that the degeneracy is determined by the topology of the surface and grows with its complexity. Phases of matter with these features are called *topological phases* (Wen, 1990c). However, the deconfined phase does not break *any* symmetry of the system. In particular the ground-state degeneracy we just found is not the result of spontaneous symmetry breaking. It is, however, the consequence of the topological nature of the state. Indeed, we saw that the deconfined phase is a state in which electric loops proliferate (or condense) and hence the wave functions of these states include states in which the loops wind around the non-contractible loops of the torus. In contrast, the ground state in the confined phase is unique and it is dominated by finite (and typically small) electric loops.

The deconfined phase of the $\mathbb{Z}_2$ gauge theory is in fact the simplest topological phase, known as the $\mathbb{Z}_2$ topological fluid. In Section 9.3 we saw that QDMs on non-bipartite lattices are also in topological phases with the same properties as the $\mathbb{Z}_2$ topological fluid. Spin-liquid phases, whenever they have been found to be the ground states, are actually deconfined phases and are also topological, as we will see. In later chapters we will discuss the quantum Hall phases of 2D electron gases in large magnetic fields, which are also topological phases. In that context we will see that the low-energy sector of topological phases is described by topological field theories. Density-matrix RG and tensor-product-state results for the spin-$\frac{1}{2}$ quantum Heisenberg antiferromagnet on a frustrated square lattice suggest that it has a $\mathbb{Z}_2$ topological phase (Jiang *et al.*, 2012; Wang *et al.*, 2011).

9.10 Generalized $\mathbb{Z}_2$ gauge theory: matter fields

We will now discuss a more general $\mathbb{Z}_2$ gauge theory in which we will include a dynamical matter field. Since the local symmetry is $\mathbb{Z}_2$ the matter field must also transform under this symmetry, and the simplest example is just a quantum Ising model. To this end, we define an Ising degree of freedom represented by the Pauli matrix $\tau^z(\vec{x})$ on each site $\vec{x}$ of the lattice (not to be confused with the *dual* Ising model whose degrees of freedom reside on the dual lattice!). The Hamiltonian for the $\mathbb{Z}_2$ gauge theory with (Ising) matter is

$$H = -g\sum_{\vec{x},j}\sigma_j^x(\vec{x}) - \frac{1}{g}\sum_{\vec{x}}\sigma_1^z(\vec{x})\sigma_2^z(\vec{x}+e_1)\sigma_1^z(\vec{x}+e_2)\sigma_2^z(\vec{x})$$
$$-\frac{1}{\lambda}\sum_{\vec{x}}\tau^x(\vec{x}) - \lambda\sum_{\vec{x},j}\tau^z(\vec{x})\sigma_j^z(\vec{x})\tau^z(\vec{x}+e_j) \qquad (9.72)$$

This Hamiltonian commutes with the new operators $Q(\vec{x})$ (defined on each site $\vec{x}$ of the lattice)

$$Q(\vec{x}) = \sigma_1^x(\vec{x})\sigma_1^x(\vec{x}-e_1)\sigma_2^x(\vec{x})\sigma_2^x(\vec{x}-e_2)\tau^x(\vec{x}) \qquad (9.73)$$

which square to the identity, $Q^2(\vec{x}) = 1$.

Since these operators commute with each other, $[Q(\vec{x}), Q(\vec{y})] = 0$, *and* with the Hamiltonian, they generate local $\mathbb{Z}_2$ gauge transformations. The physical states $|\text{Phys}\rangle$ are thus simultaneous eigenstates of the Hamiltonian H and of all the generators $Q(\vec{x})$. We will define the new Hilbert space of gauge-invariant states by $Q(\vec{x})|\text{Phys}\rangle = |\text{Phys}\rangle$. However, if we compare the generator $Q(\vec{x})$ as defined by Eq. (9.73) with that of the pure gauge theory, Eq. (9.44), we see that they differ by a factor of $\tau^x(\vec{x})$. Thus the new Hilbert space modifies the Gauss-law condition of the pure gauge theory of Eq. (9.43) by allowing for *dynamical* sources in the form of matter fields.

The presence of dynamical matter fields changes the spectrum in important ways. Thus, in addition to closed $\mathbb{Z}_2$ electric loops and $\mathbb{Z}_2$ magnetic charges, the spectrum now also contains bound states of matter fields, which can be regarded as open $\mathbb{Z}_2$ electric strings. States of the latter type are associated with the gauge-invariant operator

$$C_{\gamma(\vec{x},\vec{y})} = \tau^z(\vec{x}) \left(\prod_{(\vec{z},j)\in\gamma(\vec{x},\vec{y})} \sigma_j^z(\vec{z}) \right) \tau^z(\vec{y}) \tag{9.74}$$

where $\gamma(\vec{x}, \vec{y})$ is an open path on the lattice with endpoints at $\vec{x}$ and $\vec{y}$. There is still the possibility of the existence of states representing isolated $\mathbb{Z}_2$ electric charges. We will see that these states do exist in the deconfined (and topological) phase, although these states are not created by local operators. In some sense, the free $\mathbb{Z}_2$ electric charges are solitons of this theory. The existence of such "free" (in the sense of isolated) states in the deconfined phase is a form of "fractionalization" and it is one of the main interests of this theory in the context of possible spin-liquid phases.

Except at $\lambda = 0$ and at $g = 0$ the states of the system with the Hamiltonian of Eq. (9.72) can be fully specified in the unitary gauge (the analog of the London gauge in superconductivity)

$$\tau^z(\vec{x})|\text{Phys}\rangle = |\text{Phys}\rangle \tag{9.75}$$

Unlike the axial gauge, this gauge is always globally well defined. In this gauge the Ising degrees of freedom can be eliminated (provided that the generalized Gauss law implied by Eq. (9.73) is imposed), resulting in the following effective Hamiltonian involving only the gauge fields:

$$\begin{aligned} H = &-g\sum_{\vec{x},j}\sigma_j^x(\vec{x}) - \frac{1}{g}\sum_{\vec{x}}\sigma_1^z(\vec{x})\sigma_2^z(\vec{x}+e_1)\sigma_1^z(\vec{x}+e_2)\sigma_2^z(\vec{x}) \\ &-\frac{1}{\lambda}\sum_{\vec{x}}\sigma_1^x(\vec{x})\sigma_1^x(\vec{x}-e_1)\sigma_2^x(\vec{x})\sigma_2^x(\vec{x}-e_2) - \lambda\sum_{\vec{x},j}\sigma_j^z(\vec{x}) \end{aligned} \tag{9.76}$$

where the second term acts on the plaquettes and the third term acts on the sites. Since the degrees of freedom of this Hamiltonian reside on the links of the lattice, it is automatically self-dual in the sense that they also reside on the links of the dual lattice. Thus, a duality transformation effectively amounts to exchanging plaquettes with sites and rotating the basis from σ^z to σ^x. Under this transformation the Hamiltonian remains invariant up to the replacement

$$\lambda \leftrightarrow \frac{1}{g} \tag{9.77}$$

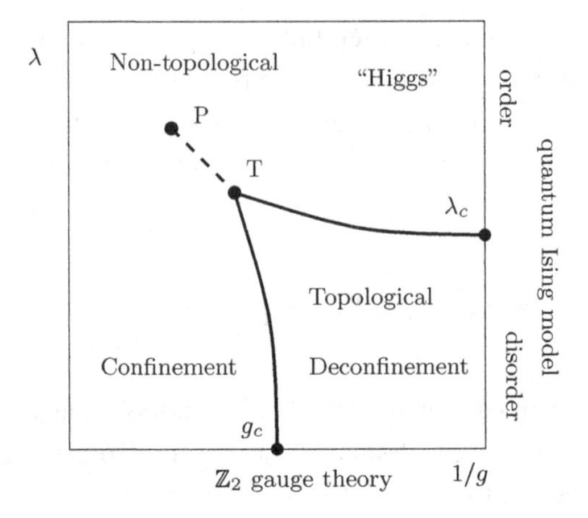

Figure 9.7 A schematic phase diagram for the $\mathbb{Z}_2$ gauge theory with Ising matter fields. Notice that the confinement phase of the gauge theory and the "Higgs" phase (the ordered phase of the Ising model) are smoothly connected and belong to the same phase. As a consequence of self-duality the phase diagram is symmetric under reflection across the anti-diagonal. P is a critical endpoint and T is a tricritical point. The smooth curves are continuous transitions and the broken line is a line of first-order transitions.

Thus the physics of this system is symmetric under a reflection of the anti-diagonal of the phase diagram of Fig. 9.7. In the limit $g \to 0$ and $\lambda \to 0$ the Hamiltonian reduces to the Hamiltonian for the Kitaev toric code (Kitaev, 2003), which is given by the second and third terms of the Hamiltonian Eq. (9.76).

Let us analyze the phase diagram of this system (Fradkin and Shenker, 1979), which is shown in Fig. 9.7. We will do this by looking at extreme regimes of the two coupling constants g and λ.

We will begin at the bottom of Fig. 9.7. This axis represents the phase diagram of the pure $\mathbb{Z}_2$ gauge theory, which, as we saw, has a quantum critical point at a critical value g_c of the coupling constant. Right on this axis the coupling constant of the Ising model vanishes, $\lambda = 0$. For λ small enough the matter fields are massive (have a large energy gap $\sim\lambda^{-1}$). Hence their fluctuations can be integrated out, resulting in a finite (and analytic) renormalization of the coupling constant of the gauge theory. Thus, no new phase transitions can possibly occur for λ small, and the only significant effect is that the critical coupling g_c becomes a smooth function λ. There are, however, some qualitative changes at small but finite λ.

Let us discuss first the confinement regime, g large and λ small. We saw before that in this regime the ground state is most easily described in the σ^x basis. At $\lambda = 0$ the ground state and the excitation spectrum can be pictured in terms of small closed electric loops on the square lattice (representing links where $\sigma^x = -1$). However, if $\lambda > 0$, no matter how small, then, in addition to closed loops, we will have states represented by open strings, and at λ small they are suppressed. However, the fact that there are open-string states does not imply deconfinement since the open strings (which carry the quantum numbers of the $\mathbb{Z}_2$ Ising matter field at their endpoints) have finite spatial extent. Thus even in this case deconfinement can take place only if the open and closed strings become macroscopic in size.

Nevertheless, the existence of open strings changes the behavior of the Wilson loop operator. Since dynamical $\mathbb{Z}_2$ Ising matter fields carry *the same charge as the external sources*, the Wilson loop operator W_Γ becomes screened by the fluctuations of the matter field. It is straightforward to see that the lowest-order contribution of the matter field to a Wilson loop of perimeter L is of order $\sim\lambda^L$ in an expansion in powers of λ, which for loops that are large enough always prevails over the area-law contribution, $\sim g^{-A}$. Hence area-law behavior of the pure gauge theory crosses over to a perimeter law for large enough loops (no matter how small λ is, provided that it is not zero). Likewise, the ground-state energy of two sources now will grow linearly with separation up to a scale of order λ^{-1}.

Let us now increase λ to arbitrary values (while keeping g large). By examining the Hamiltonian of Eq. (9.76) we see that the ground state and the excitation spectrum evolve smoothly as λ increases. Although the basis for the simplest description of the spectrum rotates smoothly from the σ^x basis at $\lambda = 0$ to the σ^z basis as $\lambda \to \infty$, this does not require the crossing of any phase boundary: the spectrum has an *analytic* evolution. In fact, the spectrum evolves with an analytic dependence on λ and g on a finite strip of the phase diagram ranging from the confinement regime (the lower-left corner of the phase diagram of Fig. 9.7), up the λ axis, turning right along the top of the phase diagram, and stretching all the way to large λ and $g \to 0$ (the upper-right corner of the phase diagram of Fig. 9.7). In particular, this implies that the ground-state energy is an analytic function of g and λ in that region. This argument was proved rigorously by Fradkin and Shenker (1979), who showed that these expansions have a finite radius of convergence inside this strip. However, analyticity means that there are no phase transitions (continuous or discontinuous) in this regime. Hence there is no global qualitative difference between the seemingly opposite regimes of λ small and g large ("confinement") and λ large and g small ("Higgs"), which are as smoothly connected to each other as liquid water is to steam.

On the other hand, on the extreme right of the diagram the gauge coupling vanishes, $g \to 0$. In this limit the plaquette term of the Hamiltonian forces the gauge fields to be a pure gauge (no flux). Up to the effects of boundary conditions, we can locally fix the gauge $\sigma_1^z = 1$ everywhere and the zero-flux condition then forces that $\sigma_2^z = 1$ also everywhere. Thus, in this limit the gauge fields are frozen out and we recover the Hamiltonian for the Ising model in a transverse field along the entire λ axis. At large λ the Ising model is ordered and has an order parameter with a non-vanishing expectation value. However, the Ising order parameter is not locally gauge-invariant, and by Elitzur's theorem its expectation value will vanish for any $g > 0$. One may wonder whether there is an operator that, in a suitable gauge, reduces to the order parameter. Such an operator does exist in Maxwell's electrodynamics, which has a *non-compact* gauge group. In the theory of superconductivity this is the conventional pair field defined in the Coulomb gauge. We will encounter an analog of this operator in the theory of quantum Hall effects. However, for a system with a *compact* gauge group (such as $\mathbb{Z}_2$) it is not possible to construct a *locally* gauge-invariant order parameter. Nevertheless, the spectrum still has an analytic ("adiabatic") evolution reaching all the way to the ordered phase. This is the Higgs phase (Fradkin and Shenker, 1979).

Let us finally discuss the stability of the deconfined phase, the lower-right corner of the phase diagram. Above we constructed the ground state at $\lambda = 0$ and showed that in this state the (electric) loops have divergent sizes and proliferate. We also showed that on a torus this phase has a four-fold degenerate ground state, labeled by the eigenvalues of the magnetic 't Hooft loop operators. The expansion in powers of g and λ is also convergent with a finite radius of convergence. This is expected since at $g = \lambda = 0$ the spectrum has a large energy gap. Thus the deconfined phase occupies a finite region of the phase diagram (as depicted in Fig. 9.7), and the deconfinement phase transition at $g = g_c$ (and $\lambda = 0$) and the Ising phase transition at $\lambda = \lambda_c$ (and $g = 0$) survive at finite λ and at finite g, respectively. It is a deconfined phase in the sense that there are finite-energy states in the spectrum that carry the $\mathbb{Z}_2$ "charge" and are free (and hence are not confined). This is possible since in this phase the electric strings have proliferated and are of macroscopic size. However, this phase is *not* characterized by a *local* order parameter. Nevertheless, it is still characterized by a finite ground-state degeneracy on a torus in the thermodynamic limit. This is true even though the 't Hooft magnetic holonomies do not commute with the Hamiltonian for any finite $\lambda > 0$.

Thus we conclude that matter fields make qualitative changes in the phase structure. The resulting phase diagram, shown in Fig. 9.7, has two phases: (a) a non-topological phase ranging from the confinement regime to the broken-symmetry regime and (b) a topological deconfined phase. The features of the theory that we described here, including the topology of the phase diagram and the existence of a critical endpoint P and of a tricritical point T, have been confirmed by several numerical Monte Carlo simulations (Jongeward *et al.*, 1980; Trebst *et al.*, 2007; Tupitsyn *et al.*, 2010). This phase diagram turns out to be *generic* for all gauge theories with a compact symmetry group and matter fields that carry the fundamental representation of the gauge group (i.e. the lowest allowed charge). Some important details of the phases do depend on the gauge group. In particular the deconfined phase is topological (i.e. with a finitely degenerate ground state on a torus) only for discrete gauge groups, but not if the symmetry group is continuous. It also holds for a theory with a compact gauge group spontaneously broken to a discrete subgroup (Krauss and Wilczek, 1989; Preskill and Krauss, 1990; Bais *et al.*, 1992).

9.11 Compact quantum electrodynamics

The Hamiltonian of the QDM of Eq. (9.41) is closely related to a problem solved by Polyakov in 1977 (Polyakov, 1977): compact quantum electrodynamics (CQED) in (2 + 1) dimensions. It is compact in the sense that its degrees of freedom, the gauge fields A_j, or rather the exponentials $e^{iA_j(\vec{x})}$, are elements of the compact Lie group U(1). The Hamiltonian of CQED has the simpler form (Kogut, 1979)

$$H_{\text{CQED}} = \frac{g}{2} \sum_{\vec{x};j} E_j^2(\vec{x}) - \frac{1}{g} \sum_{\vec{x};j,k} \cos F_{jk}(\vec{x}) \tag{9.78}$$

where the gauge variables $A_j(\vec{x})$ and the conjugate "electric" fields $E_j(\vec{x})$ satisfy canonical equal-time commutation relations, $[E_j(\vec{x}), A_k(\vec{y})] = i\delta_{jk}\delta_{\vec{x},\vec{y}}$, and $F_{jk}(\vec{x}) = \sum_{\text{plaquette}} A_j(\vec{x})$ is the gauge flux for the plaquette $(\vec{x};\, j, k)$.

As before, the local operator $Q(\vec{x}) = \Delta_j E_j(\vec{x})$ is the generator of local time-independent gauge transformations of the form

$$U[\alpha(\vec{x})] = \exp\left(i \sum_{\vec{x}} \alpha(\vec{x}) Q(\vec{x}) \right) \tag{9.79}$$

which are elements of the gauge group U(1). Since the gauge generators $\{Q(\vec{x})\}$ commute with each other and with the Hamiltonian,

$$[Q(\vec{x}), Q(\vec{y})] = 0, \qquad [Q(\vec{x}), H] = 0 \tag{9.80}$$

the states in the Hilbert space of physical states, $\{|\text{Phys}\rangle\}$, are simultaneous eigenstates of all the gauge generators $\{Q(\vec{x})\}$ and hence are gauge-invariant, i.e. they obey Gauss's law

$$\Delta_j E_j(\vec{x})|\text{Phys}\rangle = 0 \tag{9.81}$$

The Hamiltonian of Polyakov's compact QED, Eq. (9.78), differs from the Hamiltonian of the QDM, Eq. (9.41), in that (a) $\bar{J}$ has the wrong sign and (b) the constraint selects a space of states that is not the usual vacuum ($\rho = 0$) but has an array of sources, $\rho(\vec{x}) = \pm 1$. The first problem can be solved very easily (in the absence of holes) by shifting the gauge variables $A_j = A'_j + \delta A_j$ in such a way that $\sum_{\text{plaquette}} \delta A_j = \pi$. For instance, we can shift A_1 by π on every other horizontal row. Once this has been done, the first two terms of the Hamiltonian of Eq. (9.41) become essentially identical to the Hamiltonian for compact electrodynamics (Kogut, 1979). The second caveat, (b), is intrinsic and cannot be done away with by any redefinition of variables. The shift δA_j says that Eq. (9.41) represents a system that likes to have flux π per plaquette, on average. This result is reminiscent of the flux phase. Thus, in terms of shifted variables, H has exactly the same form but with $\bar{J} \leftrightarrow -\bar{J}$. Thus, although the Hamiltonian QDM can be written as a lattice gauge theory, related to compact QED in this case, the physical sector of the QDM is not the vacuum sector of CQED. We will see below that this makes a significant difference.

In $(2+1)$ dimensions Polyakov's compact electrodynamics is in a confining phase for all values of the coupling constant $g > 0$. In fact, the lowest (space-time) dimension for a deconfined phase of a pure gauge theory with a continuous compact gauge group is $D = 4$ $(3+1)$, the dimension at which the gauge coupling constant is dimensionless. In $(2+1)$ dimensions all gauge theories (again with a compact gauge group) have only one phase, confinement. This is easy to see in the strong-coupling limit (large g), where the ground state is an eigenstate of the electric fields on the links (just as we saw in the $\mathbb{Z}_2$ gauge theory). In the case of CQED the ground state has $E_j(\vec{x}) = 0$ in all links of the lattice. This state, $|\{E_j(\vec{x}) = 0\}\rangle$, obviously satisfies the Gauss-law constraint of Eq. (9.81), and hence it is gauge-invariant. In the presence of two static sources, say a source with charge $+1$ at $\vec{x}$ and another one with charge -1 at $\vec{y}$, the ground-state energy will increase by an amount $\Delta E(\vec{x}, \vec{y})$. The lowest-energy state in the strong-coupling limit has $E_j(\vec{z}) = 1$ on the links of the shortest path between $\vec{x}$ and $\vec{y}$, and $E_j(\vec{x}) = 0$ everywhere else. Thus the excess energy is once again linear in the (lattice) distance R between the two sources, $\Delta E(R) = \sigma R$, with $\sigma = g$. Thus, this is the confining phase. The strong-coupling expansion in powers of $1/g$ is once again convergent, with a finite radius of convergence. Therefore we expect that at sufficiently large coupling the theory will be in a confining phase, just as in the $\mathbb{Z}_2$ case.

What is less obvious is the fact that the confining phase extends all the way to $g = 0$. This result, which was obtained originally by Polyakov (1977) is based on an analysis of the role of instantons on the imaginary-time path integral. It is a semi-classical analysis that considers the effects of the compact nature of the U(1) group and, hence, of the periodicity requirements on the vector potentials and fluxes. The instantons of this theory are magnetic monopoles and represent tunneling events between vacua with different flux periods. A detailed analysis is given in Section 9.14, where we use the same approach to discuss the physics of the QDM. For a semi-classical description of monopoles see e.g. Rajaraman (1985). An insightful discussion of monopoles was given by Goldhaber (1998).

What matters here is Polyakov's result, which shows that the path integral for compact electrodynamics in $(2+1)$ Euclidean dimensions is equivalent (actually *dual*) to the partition function of the *3D Coulomb gas*. In contrast to the 2D Coulomb gas, a system with a Kosterlitz–Thouless transition, the 3D neutral

Coulomb gas is always in the plasma phase. This is so since the self-energy of a 3D (magnetic) charge is finite in the infrared. Hence, the entropy effects always overwhelm the energy in the partition function. Thus, the 3D neutral Coulomb gas always exhibits Debye screening of external static (magnetic!) charges. On the other hand, since this is a magnetic condensate (or more properly a phase in which monopoles and anti-monopoles proliferate) the electric Wilson loop has an area-law decay due to the violent fluctuations of unbound magnetic charges passing through its area. In the weak-coupling limit the theory is still confining, with an effective potential between two oppositely charged sources that remains a linear function of their separation, with a finite string tension that vanishes with an essential singularity as $g \to 0$, $\sigma(g) \sim e^{-A/g}$, where A is a non-universal constant (since in $(2+1)$ dimensions the gauge coupling constant is not dimensionless).

Thus, in the absence of dynamical matter fields, due to the monopole-proliferation mechanism compact electrodynamics (the U(1) gauge theory) in $(2+1)$ dimensions has only one phase, confinement. All gauge theories with a non-abelian gauge group are confining below four space-time dimensions, the lowest critical dimension for a deconfined phase to occur (Kadanoff, 1977). As is well known, even in four dimensions non-abelian gauge theories are asymptotically free, and hence the non-linearities of these theories are marginally relevant operators that lead to a confined phase at all values of the coupling constant. The U(1) gauge theory is special in that it also has a deconfined (Maxwell) phase in four dimensions with massless photons and heavy (but free) magnetic monopoles.

9.12 Deconfinement and topological phases in the U(1) gauge theory

Let us now briefly consider the more general problem of compact electrodynamics coupled to charged (bosonic) matter fields with some integer charge $q \in \mathbb{Z}$. In quantum field theory this problem is known as the abelian Higgs model. Many of the arguments we use below apply to the more general case of a theory with a compact gauge group G coupled to a matter (scalar) field that carries the quantum numbers of a representation of G (Fradkin and Shenker, 1979). At the classical level this is the typical situation of what is usually called a gauge theory that is "spontaneously broken" in a Higgs phase. As it stands, this concept is well defined only in perturbation theory. However, it violates Elitzur's theorem since, as we noted above for compact gauge groups, it is not possible to define locally gauge-invariant order parameters.

We will represent the matter field by an element of the U(1) group, defined by a set of angle-valued variables $\{\theta(\vec{x})\}$ residing at the sites to the lattice each carrying charge q and coupled minimally to the gauge fields $\{A_j(\vec{x})\}$. Since the matter fields are parametrized by the angular variables $\theta(\vec{x})$, they can be regarded as planar rigid rotors whose associated canonical conjugate variables are the angular momenta $\{L(\vec{x})\}$ and obey the commutation relations

$$[\theta(\vec{x}), L(\vec{y})] = i\delta_{\vec{x},\vec{y}} \tag{9.82}$$

Therefore the spectrum of the momenta $L(\vec{x})$ are the integers and measure the amount of charge at site $\vec{x}$. The rigid-rotor representation is often used to describe Josephson-junction arrays where $\theta(\vec{x})$ is the phase of a superconducting "grain" at $\vec{x}$ and $L(\vec{x})$ is the quantized electric charge, the number of Cooper pairs.

The Hamiltonian for the U(1) gauge theory coupled to a charge $q \in \mathbb{Z}$ matter field is

$$\begin{aligned} H = &\sum_{\vec{x}} \frac{1}{2\lambda} L^2(\vec{x}) - \sum_{\vec{x}, j=1,2} \lambda \cos\left(\Delta_j \theta(\vec{x}) - qA_j(\vec{x})\right) \\ &+ \frac{g}{2} \sum_{\vec{x};j} E_j^2(\vec{x}) - \frac{1}{g} \sum_{\vec{x};j,k} \cos F_{jk}(\vec{x}) \end{aligned} \tag{9.83}$$

where $\Delta_j \theta(\vec{x}) = \theta(\vec{x} + e_j) - \theta(\vec{x})$ is the finite difference along the direction j. The generators of time-independent gauge transformations now are

$$Q(\vec{x}) = \Delta_j E_j(\vec{x}) - L(\vec{x}) \tag{9.84}$$

and the gauge-invariant states satisfy $Q(\vec{x})|\text{Phys}\rangle = 0$.

The physics and phase diagram for the case in which the matter field carries the smallest charge, $q = 1$, which we will refer to as "the fundamental representation," are essentially the same as in the $\mathbb{Z}_2$ gauge theory shown in Fig. 9.7. In that case only the fundamental "charge" is allowed. However, for a matter field with charge $q > 1$ the phase diagram (shown in Fig. 9.8) has important differences that we will now discuss.

The pure gauge theory, obtained in the limit $\lambda = 0$, is confining for all values of the gauge coupling constant g. Hence in the U(1) case the bottom axis of the phase diagram of Fig. 9.7 does not have a phase transition. In this case too the confining phase is stable and survives at finite $\lambda > 0$. The behavior of the Wilson loops has some subtleties discussed below.

The vertical axis on the right extreme of the phase diagram now describes a quantum-rotor model with global U(1) symmetry. We will call it the XY or planar model (not to be confused with the spin-1/2 quantum XY model discussed in Chapter 5). This model is in the universality class of the 3D classical XY model, the standard model of the superfluid transition and easy-plane classical ferromagnets. It has a phase transition at a critical coupling constant λ_c. For $\lambda < \lambda_c$ the XY model has a unique ground state, roughly described by setting $L(\vec{x}) = 0$ on all sites (the exact ground state at $\lambda = 0$). The spectrum in this disordered phase

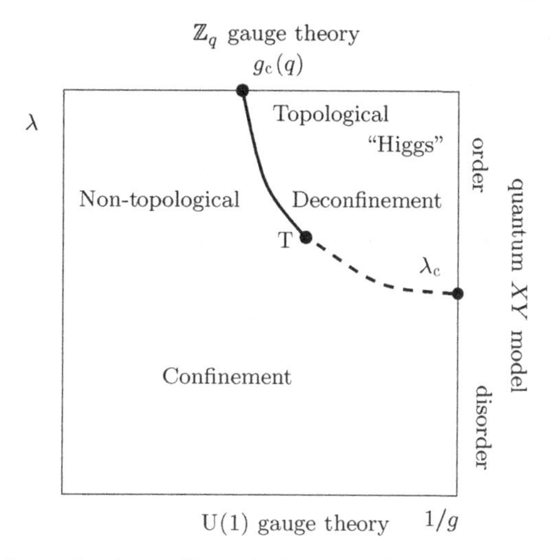

Figure 9.8 A schematic phase diagram for the U(1) gauge theory with charge-q matter fields (with $q \geq 2$). The confinement phase of the gauge theory and the "Higgs" phase (the ordered phase of the XY model) are *no longer* smoothly connected as in Fig. 9.7. The top of the phase diagram is the $\mathbb{Z}_q$ discrete gauge theory and $g_c(q)$ is its critical point. T is a tricritical point. The smooth curves are continuous transitions and the broken line is a line of first-order transitions.

has a finite energy gap $\Delta E(\lambda) > 0$. The phase transition at λ_c is the universality class of the 3D classical XY model and hence it is also continuous with universal critical exponents. This transition too is relativistic-like and has a dynamic critical exponent $z = 1$. For $\lambda > \lambda_c$ this system exhibits long-range order, with $\exp(i\theta)$ playing the role of the order parameter. For $\lambda > \lambda_c$, $M(\lambda) = \langle e^{i\theta} \rangle \neq 0$. The broken-symmetry phase (the "Higgs" phase) has a gapless Goldstone mode, which propagates with a linear dispersion $\omega(\vec{p}) = v_s|\vec{p}|$ for small $|\vec{p}|$.

The behavior at the top end of the phase diagram now depends on the value of the charge q carried by the matter field. This is most easily seen in the unitary gauge, $\theta = 0 \pmod{2\pi}$. This is literally the same as the London gauge of superconductors. In this gauge, in the limit of $\lambda \to \infty$ the gauge field on each link is constrained to take the values

$$A_j(\vec{x}) = \frac{2\pi}{q} p_j(\vec{x}) \tag{9.85}$$

where $p_j(\vec{x}) = 0, 1, \ldots, q-1$ are integers (mod q). This means that the gauge fields are now q discrete possible angles, the integer multiples of $2\pi/q$.

Hence, in this limit we obtain a gauge theory with a *discrete* gauge group $\mathbb{Z}_q$. In particular, for $q = 2$ the U(1) gauge group has been "broken" down to $\mathbb{Z}_2$, and becomes a $\mathbb{Z}_2$ gauge theory. For general q, along the top of the phase diagram the U(1) gauge theory coupled to a charge-q matter field is equivalent to a discrete gauge theory with gauge group $\mathbb{Z}_q$. For $q > 1$ this theory has a phase transition at a critical value $g_c(q)$ from a confining strong-coupling phase to a deconfining weak-coupling phase. However, for $q = 1$ the gauge fields are effectively frozen out because they have to take the value 0 (mod 2π). Hence, for $q = 1$ there is no phase transition along the top of the diagram, whereas there is a (continuous) phase transition for $q \geq 2$ between a confining phase and a deconfining phase.

The behavior of the Wilson loops depends on the charge q of the matter field and on the charge r of the Wilson loop. Let us consider here only the Wilson loops with the smallest charge $r = 1$. If the matter fields also carry charge $q = 1$ the arguments used in the $\mathbb{Z}_2$ case also apply here: the "probe" loop is "algebraically" screened by the dynamical matter field (which carries the same charge) by means of the pair-creation process we described above. Thus for $q = 1$ the Wilson loops have a perimeter law but the theory is still confining.

However, for $q \geq 2$ we can either probe the system with a Wilson loop with charge $r = 1$ or $r = 2$ (or higher). If the Wilson loop carries charge $r = 1$, the pair-creation mechanism no longer works since charge-2 particles cannot screen charge-1 particles (unless the charges condense). Thus, the charge $r = 1$ Wilson loop retains its area-law behavior but a charge $r = 2$ Wilson loop will obey a perimeter-law scaling. Thus, in U(1) gauge theories with matter fields that carry charge $q > 1$, charge $r = 1$ Wilson loops obey an area law in the confining region of the phase diagram. On the other hand, in the deconfined phase *all* Wilson loops follow a perimeter law, regardless of their charge r. The local excitations of the deconfined phase are plaquette excitations with magnetic flux $2\pi p/q$ (with $p = 1, \ldots, q-1$) and essentially free electric charges that carry charge 1. Therefore, we do have two distinct phases.

The deconfined phase is topological. It has a non-trivial behavior under the large gauge transformations of the discrete gauge symmetry, $\mathbb{Z}_q$. Indeed we can now repeat almost verbatim the arguments we used in the $\mathbb{Z}_2$ case. In this system too we can define magnetic 't Hooft holonomies along the two non-contractible paths of the torus. The elementary magnetic ('t Hooft) holonomy now is

$$\tilde{W}_j = e^{i\frac{2\pi}{q}\sum_{\vec{x}}\Theta[\gamma_j]\epsilon_{jk}L_k(\vec{x})} \tag{9.86}$$

where γ_j are the two non-contractible loops (of the torus) on the dual lattice, with $\Theta[\gamma_j] = 1$ on the links of the direct lattice crossed by the path γ_j and $\Theta[\gamma_j] = 0$ otherwise (as in the example depicted in Fig. 9.2(a)). The fundamental Wilson and 't Hooft holonomies now form an algebra that generalizes Eq. (9.69) to

$$\begin{aligned} &W_{\gamma_1}\tilde{W}_{\gamma_2} = e^{i\frac{2\pi}{q}}\tilde{W}_{\gamma_2}W_{\gamma_1}, \qquad W_{\gamma_2}\tilde{W}_{\gamma_1} = e^{i\frac{2\pi}{q}}\tilde{W}_{\gamma_1}W_{\gamma_2} \\ &[W_{\gamma_1}, \tilde{W}_{\gamma_1}] = 0, \qquad [W_{\gamma_2}, \tilde{W}_{\gamma_2}] = 0 \end{aligned} \tag{9.87}$$

For $q > 2$ this type of algebraic structure gives rise to the concept of an anyon.

Deep in the deconfined phase the low-energy states are simultaneous eigenstates of the Hamiltonian and either the electric or the magnetic holonomies (but not both since they don't commute). We can once again take the eigenstates of the Hamiltonian to be eigenstates of the magnetic holonomies. Hence they are eigenstates of $\tilde{W}_{\gamma_j}$. It is straightforward to see that their eigenvalues are simply $e^{i2\pi p/q}$, with $p = 0, 1, \ldots, q-1$. Hence on a torus we have a degeneracy of q^2 since we have two non-contractible loops. These states are created by the repeated action of the two electric Wilson loops, which act as ladder operators. For a surface of genus g, i.e., with g handles, the degeneracy is $(q^2)^g$.

In the rest of this chapter we will see how the ideas that we presented in the context of these gauge theories apply in the context of the problem of the phases of strongly correlated systems without long-range spin order. We will see that the valence-bond crystal phases result in the confining regimes of the effective gauge theories and that the spin-liquid phases occur when the gauge theory is deconfined. The condensation of objects carrying charges larger than the fundamental charge turns out to be a generic way to generate a deconfined phase. If the charge of the condensing field is such that the remaining gauge symmetry is discrete, the resulting phase is topological.

9.13 Duality transformation and dimer models

The suggestive analogy with CQED may lead us to think that the ground state of this system (after shifting) has weakly fluctuating gauge fields. In such a case one may expect that the elementary excitation should have A_j small, slowly varying, and gapless, and that there should be a "photon" excitation in the spectrum. However, we must recall that we are working with staggered variables, and hence this "photon" should have wave vectors close to $\vec{Q}_0 = (\pi, \pi)$. This is the *resonon* of Kivelson and Rokhsar, who argued that it exists for $-\bar{J} = V$.

However, this choice of couplings is very special. In fact, on the square lattice it represents a *quantum critical point*, and, away from $|\bar{J}| = V$, the resonon excitation does not exist! This is so since, as Polyakov showed, compact QED is a *confining* theory. His results, which he derived for the case $\rho(\vec{x}) = 0$ (i.e. the usual vacuum sector), imply that (i) the ground state is unique and it is a gauge singlet, (ii) the spectrum has a gap, and (iii) only gauge-invariant states are present (in particular, there is no "photon"). We will see now, by following Polyakov's ideas and using the methods of Banks, Myerson, and Kogut (Banks *et al.*, 1977), and Fradkin and Susskind (1978), how these results are modified by the presence of a non-zero $\rho(\vec{x})$. Here I have kept the description, used in the first edition of this book, of QDMs in terms of compact quantum electrodynamics. The QDM can be described equally well by an Ising gauge theory in its "odd" sector, as is is done by Moessner *et al.* (2001).

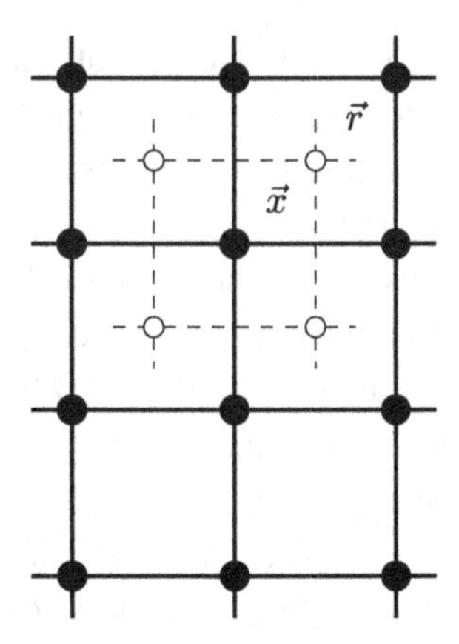

Figure 9.9 The sites of the direct lattice (filled circles) are labeled by $\vec{x}$, and the sites of the dual lattice (empty circles) are labeled by $\vec{r}$.

Since we expect, after Polyakov, that the physics of the ground state and low-lying excitations might not be accessible by means of a perturbative expansion around a state with some background classical field A_j, it is useful to identify the topological excitations of this system. If we consider the Euclidean evolution of the system (i.e. imaginary time), the field configurations which disorder the long-range properties of the classical background state look like Dirac magnetic monopoles with integer charge. Polyakov's observation was, and this will also be crucial to our problem, that fluctuations around a background configuration with monopoles induce an interaction among them that is identical to that of a (neutral) Coulomb gas in three (Euclidean) dimensions. Since the Coulomb gas has the property of screening of external charges *for all values of the coupling constant*, the ground state is unique and has a gap $\Delta \sim 1/\xi_s$, where ξ_s is the screening length of the monopole–anti-monopole plasma. Let us rederive these results and, at the same time, keep track of the sources $\rho(\vec{x})$.

The first step is a *dual transformation.* We will define this transformation in terms of the solution of the constraint equation, Eq. (9.34). Let $\vec{r}$ label the sites of the dual lattice, which is also a square lattice (Fig. 9.9). Let $N(\vec{r})$ be an operator defined on *sites* of the dual lattice with a spectrum labeled by the integers $N(\vec{r})$. Similarly, $B_j(\vec{r})$ is a *classical* background real-valued field that resides on the *links* of the dual lattice. I require that

$$E_j(\vec{x}) = \epsilon_{jk}(\Delta_k N(\vec{r}) + B_k(\vec{r})) \tag{9.88}$$

where ϵ_{jk} is the Levi-Civita tensor and $i, j = 1, 2$. If we now substitute Eq. (9.88) into the constraint Eq. (9.34), then, in the subspace of physical states, we get

$$\begin{aligned}\Delta_j E_j(\vec{x}) &= \epsilon_{jk}\big(\Delta_j \Delta_k N(\vec{r}) + \Delta_j B_k(\vec{r})\big)\\ &= \epsilon_{jk}\Delta_j B_k(\vec{r})\\ &= \rho(\vec{x})\end{aligned} \tag{9.89}$$

where I used the antisymmetry of the ϵ_{jk}. Thus, the background fields $B_k(\vec{r})$ are determined by the condition

$$\epsilon_{jk}\Delta_j B_k(\vec{r}) = \rho(\vec{x}) = (-1)^{x_1+x_2} \tag{9.90}$$

Notice that the *electrostatic-like* constraint Eq. (9.34) (i.e. Gauss's law) has become the *magnetostatic* constraint Eq. (9.90). This is the usual electric–magnetic duality.

The set of solutions of Eq. (9.90) is in one-to-one correspondence with the dimer configurations of the lattice since this equation is the dual version of the constraint, Eq. (9.26). Moreover, two different solutions $B_k(\vec{r})$ and $B'_k(\vec{r})$ are related through a gauge transformation since their difference $\bar{B}_k(\vec{r}) \equiv B_k(\vec{r}) - B'_k(\vec{r})$ must satisfy

$$\begin{aligned}\epsilon_{jk}\Delta_j \bar{B}_k(\vec{r}) &= \epsilon_{jk}\Delta_j B_k(\vec{r}) - \epsilon_{jk}\Delta_j B'_k(\vec{r})\\ &= \rho(\vec{x}) - \rho(\vec{x})\\ &= 0\end{aligned} \tag{9.91}$$

In other words, $\bar{B}_k(\vec{r})$ is curl-free. Hence, at least locally, $\bar{B}_k(\vec{r})$ must be a pure gradient

$$\bar{B}_k(\vec{r}) \equiv \Delta_k \Gamma(\vec{r}) \tag{9.92}$$

Without loss of generality, $\Gamma(\vec{r})$ is taken to be an integer-valued function on the dual lattice.

A local change in the gauge of $B_k(\vec{r})$ can thus be absorbed into an appropriate redefinition of the operators $N(\vec{r})$,

$$N(\vec{r}) = N'(\vec{r}) - \Gamma(\vec{r}) \tag{9.93}$$

There exists, however, a set of $\bar{B}_k(\vec{r})$ that cannot be done away with by a suitable redefinition of the variables $N(\vec{r})$. They correspond to *large* gauge transformations, i.e. gauge transformations that change the value of the line integral (or sum) of $\bar{B}_k(\vec{r})$ along a non-contractible loop around the torus (see Fig. 9.6).

There are two generically non-contractible loops: one along the x_1 direction, $\gamma_1(\vec{r})$, and the other along the x_2 direction, $\gamma_2(\vec{r})$; where $\gamma_1(\vec{r})$ and $\gamma_2(\vec{r})$ go through the dual site $\vec{r}$ (Fig. 9.6). Thus the line integrals $I_{\gamma_1(\vec{r})}[\vec{B}]$ and $I_{\gamma_2(\vec{r})}[\vec{B}]$, usually referred to as holonomies, defined by

$$I_{\gamma_1(\vec{r})}[\vec{B}] \equiv \sum_{\gamma_1(\vec{r})} B_1(\vec{r}) \equiv \sum_{n_1=1}^{L} B_1(\vec{r} + n_1 e_1)$$
$$I_{\gamma_2(\vec{r})}[\vec{B}] \equiv \sum_{\gamma_2(\vec{r})} B_2(\vec{r}) \equiv \sum_{n_2=1}^{L} B_2(\vec{r} + n_2 e_2 \tag{9.94}$$

are invariant under ("small") gauge transformations (which satisfy periodic boundary conditions). However, ("large") gauge transformations, which do not respect the periodic boundary conditions, do change the values of $I_{\gamma_1(\vec{r})}[\vec{B}]$ and $I_{\gamma_2(\vec{r})}[\vec{B}]$.

The constraint of Eq. (9.40)

$$\sum_{\vec{x},j} E_j^2(\vec{x}) = \frac{L^2}{2} \tag{9.95}$$

requires that there should be no bond occupied by more than one dimer. These restrictions imply that the only allowed *large* gauge transformations have to satisfy a uniformity condition. For instance, a large gauge transformation that raises $I_{\gamma_1(\vec{r})}[\vec{B}]$ by +1 everywhere has the form (see Fig. 9.10)

$$\bar{B}_k(\vec{r}) = \delta_{r_2,n_0}\delta_{k,1} \tag{9.96}$$

where n_0 is an integer $1 \leq n_0 \leq L$.

What is the meaning of these large gauge transformations? Recall that $E_j(\vec{x})$ is given by

$$E_j(\vec{x}) = \epsilon_{jk}(\Delta_k N(\vec{r}) + B_k(\vec{r})) \tag{9.97}$$

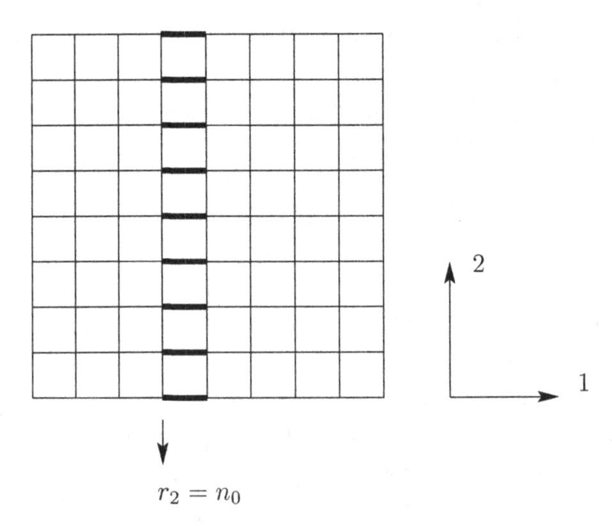

Figure 9.10 A large gauge transformation.

If we regard the operators $N(\vec{r})$ as the *quantum fluctuations* and $B_k(\vec{r})$ as a *classical background*, we see that the configurations with $N(\vec{r}) = 0$ (or constant) have $E_j(\vec{x}) = \epsilon_{jk} B_k(\vec{r})$. In other words, the classical background fields $B_k(\vec{r})$ represent a set of classical dimer configurations that can be regarded as the *parent states* for the quantum evolution of the system. Indeed, the line integral $I_{\gamma_i(\vec{r})}[\vec{B}]$ is then, from Eq. (9.33),

$$\begin{aligned} I_{\gamma_i(\vec{r})}[\vec{B}] &= \sum_{\gamma_i(\vec{r})} B_i(\vec{r}) \\ &= \sum_{\gamma_i(\vec{x})} \epsilon_{ji} E_j(\vec{x}) \\ &= \epsilon_{ji} \sum_{n_i=1}^{L} (-1)^{x_1+x_2+n_i} L_j(\vec{x} + n_i e_i) \end{aligned} \tag{9.98}$$

Thus, $I_{\gamma_i(\vec{r})}[\vec{B}]$ is the sum of the differences in the number of dimers occupying neighboring parallel links. This quantity is invariant under the dynamics of the QDM. Solutions that differ by local gauge transformations are equivalent to classical dimer configurations that differ by the "resonating" (or flipping) of a set (or sets) of plaquettes whose boundaries are contractible loops. Large gauge transformations correspond to processes in which a set of valence bonds circulate all the way around a non-contractible loop. Thus, the dimer configurations can be classified by the value of the circulation $\sum_{\gamma_i(\vec{r})} B_i(\vec{r})$ along a non-contractible loop. We can then identify $I_{\gamma_i(\vec{r})}[\vec{B}]$ with the *winding number* introduced by Rokhsar and Kivelson (1988).

Consider, for instance, configurations that belong to the class with vanishing winding numbers $\sum_{\gamma_i(\vec{r})} B_i(\vec{r}) = 0$. In the gauge $B_1(\vec{r}) = 0$, there are two possible solutions to Eq. (9.90):

$$B_1^{(1)}(\vec{r}) = 0, \qquad B_2^{(1)}(\vec{r}) = -\left(\frac{1 + (-1)^{r_1}}{2}\right)(-1)^{r_2} \tag{9.99}$$

$$B_1^{(2)}(\vec{r}) = 0, \qquad B_2^{(2)}(\vec{r}) = +\left(\frac{1 - (-1)^{r_1}}{2}\right)(-1)^{r_2} \tag{9.100}$$

In the gauge $B_2(\vec{r}) = 0$, there are also two analogous solutions. It is easy to see that these solutions are in a one-to-one correspondence with the four degenerate columnar or Peierls states (Fig. 9.11). It is clear that there should be a connection between the degeneracy of the ground state and its winding number. Indeed, the number of distinct solutions of Eq. (9.90) for a sector with a given winding number *is* equal to the degeneracy of the ground state in that sector. Since the line integrals do not change under the dynamics and the B_k terms determine the subspaces of states which are being considered, we expect that the winding number should

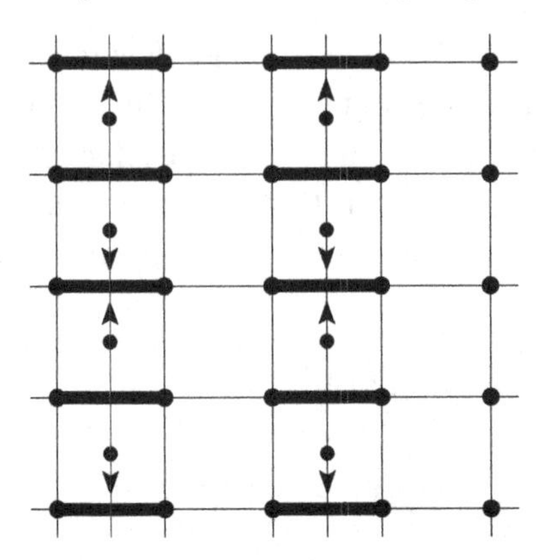

Figure 9.11 A columnar state and the background configuration of the B_k terms associated with it.

determine the ground-state degeneracy of these broken-symmetry states of the full quantum theory *unless* extra degeneracies occur, as a result of one or more modes becoming gapless. These arguments can be generalized to systems with valence bonds of finite but arbitrary length. In terms of the $1/N$ expansion, this means that this degeneracy is valid order by order in the $1/N$ expansion.

There also exist states with non-vanishing winding numbers, $I_{\gamma_i(\vec{r})}$. In this case the columns cannot reach all the way to the boundaries. For instance, a typical state with non-zero winding number I_{γ_2} can be found by taking columnar states of total height n and stacking them on top of each other after a horizontal shift of one lattice unit. States of this type are said to be *tilted* with a tilt (or slope) of $1/n$. The staggered state, shown in Fig. 8.12(b), is an example of a tilted state with $n = 1$ and has maximal winding number.

The columnar states (tilted or not) can be regarded as unidirectional dimer-density waves. Their ground-state degeneracies are specified by the period (or wave length) of the density wave. For example, the columnar state has period 2. Other types of phases with bidirectional dimer-density-wave order can also exist. More generic states of this type have non-vanishing winding numbers along the two orthogonal directions. Although in the extreme classical limit not all of these states are allowed, some of them become possible once quantum fluctuations are taken into account. An example is the plaquette state in which the valence bonds resonate on a sublattice of plaquettes with period 2. States of this type are also found in the large-N limit of the Sp(N) Hamiltonian (Read and Sachdev, 1991) and in some finite-size exact diagonalizations of the QDM (Leung *et al.*, 1996). Numerical results indicate that the QDM on the square lattice has a direct quantum phase transition from a columnar phase to a staggered phase at the Rokhsar–Kivelson (RK) point (with a possible plaquette phase intervening in between).

However, simple short-range modifications of the interactions described in the QDM Hamiltonian also allow a variety of other tilted states to become accessible, including states with asymptotically incommensurate values of the tilt that exhibit complex phase diagrams (Fradkin *et al.*, 2004; Vishwanath *et al.*, 2004; Papanikolaou *et al.*, 2007a).

Now that we have solved the constraint Eq. (9.34), we can write the dual form of the Hamiltonian. I will assume that the constraint has been solved in a sector with winding number $I_{\gamma_i(\vec{r})}[\vec{B}]$, $i = 1, 2$. We will have to find which sector yields the lowest ground-state energy. The solution of the constraint, Eq. (9.34),

$$E_j(\vec{x}) = \epsilon_{jk}[\Delta_k N(\vec{r}) + B_k(\vec{r})] \tag{9.101}$$

is one of the equations we need. We also need to define the *momentum* $P(\vec{r})$ canonically conjugate to $N(\vec{r})$ such that

$$[P(\vec{r}), N(\vec{r}\,')] = i\delta_{\vec{r},\vec{r}\,'} \tag{9.102}$$

Since the spectrum of $N(\vec{r})$ is the set of integers, the operator $P(\vec{r})$ should have eigenvalues $P(\vec{r})$ in the range $0 \leq P(\vec{r}) < 2\pi$. It is easy to see that the circulation $\sum_{\text{plaquette}} A_j(\vec{x})$, around an elementary plaquette centered at dual site $\vec{r}$, has the same effect on its Hilbert space as $P(\vec{r})$ has on the integer $N(\vec{r})$. More specifically, according to Eqs. (9.22) and (9.33), the raising operator $\exp(i \sum_{\text{plaquette}} A_j(\vec{x}))$ shifts the eigenvalues of $E_j(\vec{x})$ by $+1$ on the oriented path around the plaquette. This has exactly the effect of raising $N(\vec{r})$, on the dual lattice, also by $+1$. Thus, we identify

$$\sum_{\text{plaquette}} A_j(\vec{x}) \equiv P(\vec{r}) \tag{9.103}$$

Alternatively, it is easy to check the consistency of this identification by an explicit calculation of the commutation relations.

The Hamiltonian dual to that of Eq. (9.41) is

$$\begin{aligned} H = {} & \frac{1}{2k}\left(\sum_{\vec{r},k}(\Delta_k N(\vec{r}) + B_k(\vec{r}))^2 - \frac{L^2}{2}\right) - 2\bar{J}\sum_{\vec{r}}\cos\left(P(\vec{r})\right) \\ & + \frac{V}{2}\sum_{\vec{r}}\left[(\Delta_1(\Delta_1 N(\vec{r}) + B_1(\vec{r})))^2 + (\Delta_2(\Delta_2 N(\vec{r}) + B_2(\vec{r})))^2\right] \\ & - \frac{VL^2}{2} \end{aligned} \tag{9.104}$$

where the limit $k \to 0$ is always meant.

Also, in principle, all winding sectors have to be considered. We will keep the sector which minimizes the ground-state energy. All the inequivalent solutions of Eq. (9.90) will represent degenerate states. The manifold of degenerate states is

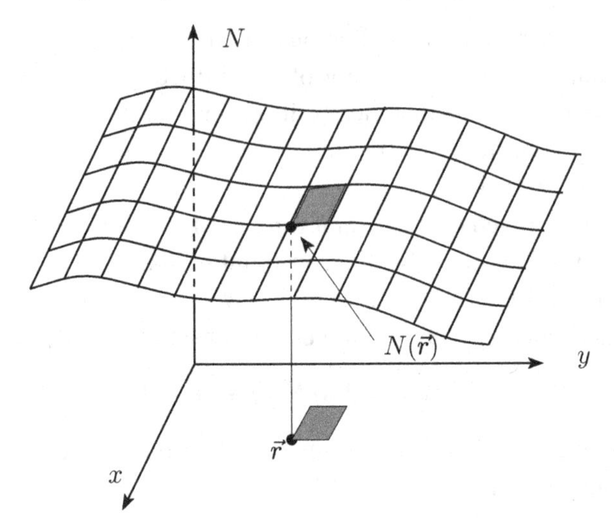

Figure 9.12 The configuration $\{N(\vec{r})\}$ parametrizes a surface in the solid-on-solid model.

closed under the group of lattice translations and rotations by $\pi/2$. From now on we will work within a given winding sector.

On comparing the QDM Hamiltonian, Eq. (9.41), and its dual, Eq. (9.104), we notice several features: (i) the kinetic- and potential-energy terms have been exchanged; (ii) the degrees of freedom in Eq. (9.41) are phases (i.e. elements of the group U(1)), whereas the degrees of freedom in Eq. (9.104) are integers; and (iii) the Hamiltonian of Eq. (9.104) has a *global* symmetry $N(\vec{r}) \to N'(\vec{r}) + n_0$ (with n_0 an arbitrary *integer*), whereas Eq. (9.41) has a local gauge symmetry. These features are present for all systems related through a duality transformation except (iii), which holds only in $(2+1)$ dimensions (Fradkin and Susskind, 1978).

A system with integer-valued degrees of freedom is usually referred to as a discrete Gaussian (DG) or solid-on-solid (SOS) model. It was originally introduced by Onsager for the study of the statistical mechanics of *classical* interfaces (Chaikin and Lubensky, 1995). In that context $N(\vec{r})$ represents the height of a column of identical atoms standing atop the lattice site $\vec{r}$. The set of values of $\{N(\vec{r})\}$ can then be regarded as the surface (or interface) of a 3D solid (Fig. 9.12). The constraint implied by the limit $k \to 0$ represents a *restriction* on this DG model. The last term in Eq. (9.104), which represents a next-nearest-neighbor interaction between atoms, has the form of a *Laplacian* coupling. The second term is responsible for the quantum dynamics of the system. There is a very large body of literature on SOS and DG models. We will not discuss it here. The most studied such system has the classical Hamiltonian

$$H_c = \frac{\gamma}{2} \sum_{\vec{r},k=1,2} (\Delta_k N(\vec{r} + e_k))^2 \tag{9.105}$$

where γ is a constant. Most studies deal with this classical problem, although the role of quantum fluctuations has also been considered.

Classically, systems such as the SOS model usually exhibit two distinct phases. At high temperatures, $T > T_R$, the interface has large transversal fluctuations and the surface is *rough*: the r.m.s. $\langle N^2 \rangle \sim \log L$ (where L is the linear size of the system) and has power-law decaying correlations. Instead, at low temperatures, $T < T_R$, the interface is *smooth*: the r.m.s. fluctuations of the surface are massive with exponentially decaying correlations. The temperature T_R is the location of a critical point at which this *roughening* transition takes place. The natural correlation functions of this problem are the height–height correlation function $G(\vec{r} - \vec{r}\,')$,

$$G(\vec{r} - \vec{r}\,') = \langle N(\vec{r}) - N(\vec{r}\,') \rangle \tag{9.106}$$

and the order-parameter correlation function $g_\alpha(\vec{r} - \vec{r}\,')$,

$$g_\alpha(\vec{r} - \vec{r}\,') = \langle e^{i\alpha N(\vec{r})} e^{-i\alpha N(\vec{r}\,')} \rangle \tag{9.107}$$

where α is an arbitrary angle.

For the *classical unrestricted* system, one finds the asymptotic behavior of $g_\alpha(R)$, where $R \equiv |\vec{r} - \vec{r}\,'| \gg a_0$ (a_0 is the lattice constant), to be

$$g_\alpha(R) \approx \begin{cases} M^2 + \text{constant} \times e^{-\frac{R}{\xi(T)}} & \text{for } T < T_R \text{ (smooth phase)} \\ \text{constant} \times R^{-\eta(\alpha,T)} & \text{for } T > T_R \text{ (rough phase)} \end{cases} \tag{9.108}$$

where $\xi(T)$ is the correlation length, M^2 is the square of the order parameter, and the exponent η is a function of α and the temperature. The corresponding behavior of $G(R)$ is

$$G(R) \approx \begin{cases} m^2 + \text{constant} \times e^{-\frac{R}{\xi(T)}} & \text{for } T < T_R \text{ (smooth phase)} \\ \text{constant} \times \ln(R/a_0) & \text{for } T < T_R \text{ (rough phase)} \end{cases} \tag{9.109}$$

where m^2 represents the square of the average height, $\langle N(\vec{r}) \rangle$.

The quantum fluctuations change this picture completely. If we ignore the restriction ($k \to 0$) and neglect the effects of *frustration* (introduced by the fields B_k), we arrive at the quantum DG model.

Let us introduce a path integral for this system. It will be convenient for us to work in imaginary time so that we can also discuss thermal fluctuations. At a non-zero temperature T, the partition function of the quantum system is

$$Z = \text{tr}\, e^{-\beta H} \tag{9.110}$$

where $\beta = 1/T$ and H is the Hamiltonian of Eq. (9.104). In order to derive a path integral we proceed in the usual fashion (Feynman and Hibbs, 1965). We first split up the imaginary-time interval $0 \le \tau \le \beta$ into N_τ time-steps, each of size Δ_τ, such that

$$N_\tau \Delta_\tau = \beta \tag{9.111}$$

The limit $\Delta_\tau \to 0$ and $N_\tau \to \infty$ is always implied. Next, we write

$$\begin{aligned} Z &= \lim_{\substack{\Delta_\tau \to 0 \\ N_\tau \to \infty}} \operatorname{tr}\left([e^{-(\Delta_\tau)H}]^{N_\tau}\right) \\ &\equiv \lim_{\substack{\Delta_\tau \to 0 \\ N_\tau \to \infty}} \operatorname{tr}\left([e^{-(\Delta_\tau)H_{\text{kin}}} e^{-(\Delta_\tau)H_{\text{pot}}}]^{N_\tau}\right) \end{aligned} \tag{9.112}$$

where we have split the Hamiltonian into a kinetic-energy term (the second term of Eq. (9.104)) H_{kin} and a potential-energy term H_{pot} (the rest). The next step is to introduce a resolution of the identity in terms of a complete set of eigenstates $|\{N(\vec{r}, t)\}\rangle$ of the operators $\{N(\vec{r}, t)\}$ between neighboring factors of $e^{-\beta H}$:

$$Z = \lim_{\substack{\Delta_\tau \to 0 \\ N_\tau \to \infty}} \sum_{\{N(\vec{r}, j)\} = -\infty}^{+\infty} \prod_{j=1}^{N_\tau} \langle \{N(\vec{r}, j)\} | e^{-\Delta_\tau H} | \{N(\vec{r}, j+1)\} \rangle \tag{9.113}$$

with periodic boundary conditions in time, i.e.

$$|\{N(\vec{r}, N_\tau + 1)\}\rangle \equiv |\{N(\vec{r}, 1)\}\rangle \tag{9.114}$$

In Eq. (9.113) the integer j represents the jth time-step and $\tau_j = \tau_0 + j\Delta_\tau$.

Let us compute the matrix elements

$$\begin{aligned} &\langle \{N(\vec{r}, j)\} | e^{-\Delta_\tau H} | \{N(\vec{r}, j+1)\} \rangle \\ &\approx \langle \{N(\vec{r}, j)\} | e^{-\Delta_\tau H_{\text{kin}}} e^{-\Delta_\tau H_{\text{pot}}} | \{N(\vec{r}, j+1)\} \rangle \\ &= \langle \{N(\vec{r}, j)\} | e^{-\Delta_\tau H_{\text{kin}}} | \{N(\vec{r}, j+1)\} \rangle e^{-\Delta_\tau H_{\text{pot}}(\{N(\vec{r}, j+1)\})} \end{aligned} \tag{9.115}$$

where I used the facts that Δ_τ is small and that H_{pot} is *diagonal* in the basis $|\{N(\vec{r}, j)\}\rangle$. In fact,

$$e^{-\Delta_\tau H_{\text{pot}}} |\{N(\vec{r}, j)\}\rangle = e^{-\Delta_\tau H_{\text{pot}}(\{N(\vec{r}, j+1)\})} |\{N(\vec{r}, j)\}\rangle \tag{9.116}$$

with an eigenvalue $H_{\text{pot}}(\{N(\vec{r}, j)\})$ given by

$$\begin{aligned} H_{\text{pot}}(\{N(\vec{r}, j)\}) &= \frac{1}{2k} \left(\sum_{\vec{r}, k=1,2} (\Delta_k N(\vec{r}, j) + B_k(\vec{r}, j))^2 - \frac{L^2}{2} \right) \\ &\quad + \frac{V}{2} \sum_{\vec{r},\, k=1,2} \left(\Delta_k^2 N(\vec{r}, j) + \Delta_k B_k(\vec{r}, j)\right)^2 - \frac{VL^2}{2} \end{aligned} \tag{9.117}$$

The off-diagonal matrix elements

$$\langle\{N(\vec{r},j)\}|e^{-\Delta_\tau H_{\rm kin}}|\{N(\vec{r},j+1)\}\rangle$$
$$= \langle\{N(\vec{r},j)\}|e^{2(\Delta_\tau)\bar{J}\sum_{\vec{r}}\cos(P(\vec{r}))}|\{N(\vec{r},j+1)\}\rangle \tag{9.118}$$

can be evaluated by repeated use of the expansion

$$e^{z\cos p} = \sum_{l=-\infty}^{\infty} I_l(z)e^{ilp} \tag{9.119}$$

where $I_l(z)$ is the Bessel function of order l of imaginary argument. The matrix elements of Eq. (9.118) are products of matrix elements of the form

$$\langle N_j|e^{2(\Delta_\tau)\bar{J}\cos(P)}|N_{j+1}\rangle \tag{9.120}$$

which we can write in the form

$$\sum_{l=-\infty}^{+\infty}\langle N_j|e^{ilP}|N_{j+1}\rangle I_l(2\bar{J}\Delta_\tau) = I_{|N_{j+1}-N_j|}(2\bar{J}\Delta_\tau) \tag{9.121}$$

In this equation we have used the orthogonality of the states $|N_j\rangle$. For convenience, and simplicity, we will use the following approximate expression for the Bessel function:

$$I_l(z) = \frac{1}{\sqrt{2\pi}}e^z e^{-\frac{l^2}{2z}}\left(1+O(z^{-1})\right) \tag{9.122}$$

On putting it all together, we can write the partition function in the suggestive form

$$Z = \lim_{\substack{\Delta_\tau\to 0\\ N_\tau\to\infty}} \sum_{\{N(\vec{r},j)\}} e^{-\mathcal{S}[N]} \tag{9.123}$$

where the Euclidean (discretized) action $\mathcal{S}[N]$ is given by $(j=1,\ldots,N_\tau)$

$$\begin{aligned}\mathcal{S}[N] = &\frac{1}{4\bar{J}\Delta_\tau}\sum_{\vec{r},j}\left[\Delta_0 N(\vec{r},j)\right]^2\\ &+\frac{\Delta_\tau}{2k}\left(\sum_{\substack{\vec{r},j\\ l=1,2}}(\Delta_l N(\vec{r},j)+B_l(\vec{r},j))^2-\frac{L^2}{2}N_\tau\right)\\ &+\frac{V\Delta_\tau}{2}\sum_{\substack{\vec{r},j\\ l=1,2}}\left(\left(\Delta_l^2 N(\vec{r},j)+\Delta_l B_l(\vec{r},j)\right)^2\right)\end{aligned} \tag{9.124}$$

I have also used the notation

$$\Delta_0 N(\vec{r},j) \equiv N(\vec{r},j) - N(\vec{r},j-1) \tag{9.125}$$

Thus the quantum partition function of the dimer model is given by the classical partition function of a *discrete Gaussian model* in *three* Euclidean dimensions on a cubic lattice of size $L^2 N_\tau$. This system looks very similar to its 2D classical counterpart Eq. (9.105), except for the fact that it is frustrated ($B_k \neq 0$), restricted ($k \to 0$), and has second-nearest-neighbor interactions ($V \neq 0$) in space.

If we work in the sector with zero winding number, the configurations with $N(\vec{r}, j) = n_0$, which is a constant, represent the columnar states. Conversely, in the sector with maximal winding number, for instance $\sum_{\gamma_1(\vec{r})} B_1(\vec{r}) = L/2$, the configuration $N(\vec{r}, j) = n_0$ is a staggered crystal. Which state dominates can be discerned only by solving the partition function Eq. (9.123). The action $\mathcal{S}[N]$, Eq. (9.124), is such that, for small Δ_τ, the fluctuations of $N(\vec{r}, t)$ in time tend to be suppressed. The columnar states have a finite degeneracy and a finite entropy, whereas the staggered states, due to the constraints, have virtually no excitations. Numerical simulations indicate that, for V small and positive, the columnar state is stable. For large V the staggered state should win, at least at low temperatures. Hence we expect that the QDGM should be in a smooth phase, albeit degenerate (see the discussion above). We will see below that at the quantum phase transition between the columnar (or plaquette) phase and the staggered phase the higher-derivative terms of Eq. (9.124) (associated with V) play an important role.

9.14 Quantum-dimer models and monopole gases

In Section 9.5 we used an intuitive argument which indicated that monopole configurations of the gauge fields play a fundamental role in this problem. We will now examine this issue more closely for the case $V = 0$.

The easiest way to relate the QDM to a gas of monopoles is to apply the Poisson summation formula

$$\sum_{n=-\infty}^{+\infty} f(n) = \sum_{m=-\infty}^{+\infty} \int d\phi \, e^{i2\pi m\phi} f(\phi) \tag{9.126}$$

to the *three-dimensional* discrete Gaussian model with action Eq. (9.124). This amounts to replacing all the integer variables $\{N(\vec{r}, j)\}$ by a continuous variable $\{\phi(r)\}$ and another set of integers $\{m(r)\}$, where now $r = (r_0, r_1, r_2)$ are 3D lattice vectors in Euclidean space-time:

$$Z = \lim_{\substack{\Delta_\tau \to 0 \\ N_\tau \to \infty}} \sum_{\{S\}} e^{-\mathcal{S}[S]}$$

$$
\begin{aligned}
&= \lim_{\substack{\Delta_\tau \to 0 \\ N_\tau \to \infty}} \sum_{\{m(r)\}} \int \mathcal{D}\phi \exp\left(2\pi i \sum_r m(r)\phi(r) - \mathcal{S}[\phi]\right) \\
&= \lim_{\substack{\Delta_\tau \to 0 \\ N_\tau \to \infty}} \sum_{\{m(r)\}} \int \mathcal{D}\phi \exp\left[\frac{\Delta_\tau}{2k}\left(\frac{N_\tau L^2}{2} - \sum_{r;l=1,2} B_l^2(r)\right)\right. \\
&\qquad\qquad \left. - \frac{V\Delta_\tau}{2} \sum_{r;l=1,2} (\Delta_l B_l(r))^2\right] \\
&\quad \times \exp\left\{-\sum_r \left[\frac{1}{4\bar{J}\Delta_\tau}(\Delta_0\phi(r))^2\right.\right. \\
&\qquad\qquad \left.\left. + \frac{\Delta_\tau}{2k}\sum_{l=1,2}(\Delta_l\phi(r))^2 + \frac{V\Delta_\tau}{2}\sum_{l=1,2}\left(\Delta_l^2\phi(r)\right)^2\right]\right\} \\
&\quad \times \exp\left[-\sum_r \phi(r)\left(2\pi i m(r) + \frac{\Delta_\tau}{k}\sum_{l=1,2}\Delta_l B_l(r)\right.\right. \\
&\qquad\qquad \left.\left. + V\Delta_\tau \sum_{l=1,2}(\Delta_l B_l(r))^2\right)\right]
\end{aligned}
\tag{9.127}
$$

We now notice the important fact that the action of Eq. (9.127) (and the partition function) is invariant under a uniform constant (in space and imaginary time) shift of the field $\phi(r) \to \phi(r) + \bar{\phi}$, provided that the integer-valued fields $m(r)$ satisfy the "neutrality" condition

$$
\sum_r m(r) = 0 \tag{9.128}
$$

We will see below that this condition translates into the requirement that the monopole gas be neutral.

Since $\mathcal{S}[\phi]$ is *quadratic* in ϕ, these fields can be integrated out. Assuming periodic boundary conditions and working in the zero-tilt sector, we obtain the result

$$
\begin{aligned}
Z = &\lim_{\substack{\Delta_\tau \to 0 \\ N_\tau \to \infty}} \exp\left[\frac{\Delta_\tau}{2k}\left(\frac{N_\tau L^2}{2} - \sum_{r,k=1,2} B_k^2(\vec{r})\right)\right]\left(\frac{\operatorname{Det} M}{2\pi}\right)^{-1/2} \\
&\times \exp\left[+\frac{1}{2}\left(\frac{\Delta_\tau}{k}\right)^2 \sum_{r,r'} \epsilon_{\alpha\mu\lambda}\Delta_\lambda^r B_\mu(r) G_0(r-r')\epsilon_{\alpha\nu\rho}\Delta_\nu^{r'} B_\rho(r')\right] Z_{\mathrm{CG}}
\end{aligned}
\tag{9.129}
$$

where Z_{CG} (defined below) is the partition function for a generalized Coulomb gas and

$$\text{Det}\, M = \text{Det}\left(\frac{1}{2\bar{J}\Delta_\tau}\Delta_0^2 + \frac{\Delta_\tau}{k}\sum_{j=1,2}\Delta_j^2 - V\Delta_\tau\left(\sum_{j=1,2}\Delta_j^2\right)^2\right) \tag{9.130}$$

The Green function associated with the operator M is $G_0(r-r')$, the 3D anisotropic lattice Green function, which is defined by

$$-\left(\frac{1}{2\bar{J}\Delta_\tau}\Delta_0^2 + \frac{\Delta_\tau}{k}\sum_{j=1,2}\Delta_j^2 - V\Delta_\tau\left(\sum_{j=1,2}\Delta_j^2\right)^2\right) G_0(r-r') = \delta_{r,r'} \tag{9.131}$$

(the minus sign comes from a "partial integration").

Then the partition function Z or Eq. (9.129) is, up to an essentially uninteresting factor, proportional to the partition function of a generalized 3D Coulomb gas Z_{CG}, which is given by

$$Z_{\text{CG}} = \sum_{\{m(r)\}}{}' \exp\left(-2\pi^2\sum_{r,r'} m(r)G_0(r-r')m(r')\right)\exp\left(2\pi i\sum_r m(r)\Psi(r)\right) \tag{9.132}$$

where $r = (\vec{r}, \tau)$ runs over the labels of the 3D cubic lattice. Just as in our discussion of the 2D Coulomb gas of Chapter 4, this partition function is constrained (indicated by the prime label in Eq. (9.132)) to configurations that obey the condition of overall charge (monopole charge in this case) neutrality, Eq. (9.128).

This partition function differs from the usual one for a Coulomb gas by the complex phase factors in Eq. (9.132), which are expressed in terms of the phase $\Psi(r)$,

$$\Psi(r) = \sum_{r'} G_0(r-r')\Delta_l^{r'} B_l(r') \tag{9.133}$$

In the thermodynamic limit $(L, N_\tau \to \infty)$ (Banks *et al.*, 1977) and at zero temperature, $G_0(r-r')$ is given by

$$G_0(r-r') = \int_{-\pi}^{\pi}\frac{d^3q}{(2\pi)^3} \times \frac{\frac{1}{4}e^{i\vec{q}\cdot(\vec{r}-\vec{r}')}}{(1/(2\bar{J}\Delta_\tau))\sin^2(q_0/2) + \sum_{j=1,2}(\Delta_\tau/k)\sin^2(q_j/2)} \tag{9.134}$$

In the time-continuum limit we find ($\omega \equiv \Delta_\tau q_0$)

$$\lim_{\Delta_\tau \to 0} G_0(\vec{r} - \vec{r}\,', \tau - \tau') = \lim_{\Delta_\tau \to 0} \int_{-\frac{\pi}{\Delta_\tau}}^{\frac{\pi}{\Delta_\tau}} \frac{d\omega}{2\pi} \int_{-\pi}^{\pi} \frac{d^2 q}{(2\pi)^2} \times \frac{e^{i\left(\omega(\tau-\tau') + \vec{q}\cdot(\vec{r}-\vec{r}\,')\right)}}{\omega^2/(2\bar{J}) + (4\Delta_\tau/k)\sum_{j=1,2} \sin^2(q_j/2)} \tag{9.135}$$

At long distances ($R = |\vec{r} - \vec{r}\,'| \gg a_0$), and at long (Euclidean) times ($\bar{\tau} = |\tau - \tau'| \gg \Delta_\tau$), $G_0(R, \bar{\tau})$ has the asymptotic behavior

$$G_0(R, \bar{\tau}) \approx \frac{k}{4\pi} \frac{1}{\sqrt{\bar{\tau}^2 + (2\bar{J}\Delta_\tau k)R^2}} \tag{9.136}$$

Except for the anisotropy ($2\bar{J}\Delta_\tau/k \neq 1$), this is just the 3D Coulomb interaction.

Thus, this problem is equivalent to a gas of monopoles (and anti-monopoles) obeying overall (magnetic) charge neutrality. The monopoles behave like a gas of charged particles (of both signs) in three dimensions, with an effective interaction V_{eff} (again regularized at short distances), which in the long-distance limit is given by

$$V_{\text{eff}}(R, \bar{\tau}) = 2\pi^2 G_0(R, \bar{\tau}) \approx \frac{\pi k}{2} \frac{1}{\sqrt{\bar{\tau}^2 + (2\bar{J}\Delta_\tau/k)R^2}} \tag{9.137}$$

The total partition function is

$$Z = \text{constant} \left(\frac{\text{Det}\, M}{2\pi}\right)^{-1/2} \times \sum_{\{m(r)\}}{}' \exp\left(-\frac{1}{2}\sum_{r,r'} m(r) V_{\text{eff}}(r, r') m(r') + 2\pi i m(r)\Psi(r)\right) \tag{9.138}$$

where Det M is given in Eq. (9.130).

The time-independent phase $\theta(r) = 2\pi\Psi(r)$ (see Eq. (9.132)) turns out to take one of four possible values, one for each sublattice:

$$\theta(r) = \begin{cases} -\pi/4 & \text{for } r_1 \text{ even, } r_2 \text{ even} \\ +3\pi/4 & \text{for } r_1 \text{ odd, } r_2 \text{ odd} \\ +\pi/4 & \text{for } r_1 \text{ odd, } r_2 \text{ even} \\ -3\pi/4 & \text{for } r_1 \text{ even, } r_2 \text{ odd} \end{cases} \tag{9.139}$$

The conclusion is that in this case, just as in Polyakov's compact electrodynamics (Polyakov, 1977), the system is also equivalent to a 3D Coulomb gas. The main difference between Polyakov's case (and the 2D case discussed in Section 4.6) and the present problem is the presence of the phases $\theta(r)$ in the weight factors of the Coulomb gas. These phases can be thought of as *Berry phases*, since they arise from non-trivial overlaps of the evolution of state of the system at nearby times. They originate from the requirement that every site of the lattice belongs to one (and only one) dimer and that the time evolution can occur only by moving dimers around in a manner compatible with this constraint. In other terms, the Berry phases reflect the fact that the QDMs are described not by the vacuum sector of the gauge theory but by the sector of the Hilbert space with alternating sources on the lattice. Read and Sachdev (1991) derived these phases, following Haldane's original suggestion, by means of an adiabatic-process calculation. It is remarkable that we find the same answer even though we started from a regime in which a non-linear sigma model cannot possibly work. However, the Berry phases make a profound difference both in terms of the nature of the ground states and in terms of the quantum phase transitions.

In Chapter 4 we showed that there is a close connection between the 2D (neutral) Coulomb gas and the sine–Gordon field theory also in two dimensions. The same relationship also exists in three dimensions and it is at the root of Polyakov's analysis. To understand the differences between compact QED and the QDM, we will revisit the derivation of this effective theory and compare the two cases.

Let us return to Eq. (9.127) and, instead of integrating out the fields $\phi(r)$, we will attempt to integrate out the integer-valued monopole fields $m(r)$. We will assume that the monopoles are dilute and we will keep only the lowest charges, $m(r) = \pm 1, 0$. This amounts to assuming that the monopole fugacity is low and, at least formally, adding to the action of Eq. (9.127) an extra term of the form

$$\mathcal{S}_{\text{core}} = u \sum_r m(r)^2 \tag{9.140}$$

where the coupling constant u can be regarded as a core energy of the monopoles and the fugacity is $z = 2e^{-u}$. This term penalizes charges with $|m| \geq 2$.

In the absence of the Berry phase terms of Eq. (9.139) (or, equivalently, if the background fields B_l are absent), the gradient terms penalize fluctuations of the field ϕ varying on short length scales while the cosine operator will penalize fluctuations of the ϕ field away from constant integer values. Thus, in this case it is possible to effectively derive (or to propose) a simple candidate continuum field theory to describe this system, namely the sine–Gordon field theory, with the effective (Euclidean) action

$$S = \int d^D x \left[\frac{K}{2} \left(\partial_\mu \phi \right)^2 - g \cos(2\pi\phi) \right] \tag{9.141}$$

where once again the sine–Gordon coupling constant $g = z/a^D$ is (essentially) the fugacity z of the Coulomb gas, and the stiffness K, determined by the parameter J, is related to the inverse temperature of the Coulomb gas.

This is Polyakov's result (Polyakov, 1977). The main difference between the 2D and 3D cases is that the 3D Coulomb gas is always in a plasma phase. This can be seen by generalizing the RG that we discussed for the sine–Gordon theory in Section 4.6 for 2D to the 3D case. The extension of the RG for $D > 2$ was done by Kosterlitz (1977), who showed that $D = 2$ is a special case, and that for $D > 2$ the Coulomb gas is always in a screening phase. In the language of the RG, for $D > 2$ the cosine term is always relevant and its coupling constant g flows to strong coupling. In this regime the discreteness of the charges of the Coulomb gas is obliterated by the strong fluctuations, leading to a phase with perfect (Debye) screening. For the same reason, the fluctuations of the field ϕ become suppressed, since in this phase the cosine operator pins the fluctuations of the coarse-grained height field ϕ to one of its minima. Thus, the field ϕ is effectively massive. In this phase monopoles with the lowest magnetic charge condense (or proliferate). As shown by Polyakov, monopole condensation implies that the Wilson loop has an area law and fundamental electric charges are confined.

The Berry phases change the structure of the effective-field theory. If we were to proceed naively, we would now integrate out the monopoles and obtain a discretized effective action for the field ϕ. In this representation (as in the representation in terms of the height fields $N(r)$) the action is *real*, and here the fields B_j couple to the field ϕ as a background static spatial gauge field whose curl represents the constraints of the quantum dimer model. In contrast, in the monopole-gas representation the action has an imaginary part, the Berry phases $\Psi(r)$, which is due to the background gauge fields B_j. However, what we want is an effective action for slowly varying fields. The background fields B_l (which are the version of the Berry phases in this representation) make the low-energy configurations vary rapidly on the scale of the lattice spacing, corresponding to the changes in the height field needed to describe dimer configurations. In this form, the height configurations are not single-valued.

Thus, in order to derive an effective-field theory for this problem it is necessary to first coarse-grain the ϕ field. We note that the Berry phases take different values on different sublattices and hence favor states that break translation invariance. However, the average of the Berry phases on the four sublattices (or equivalently the average of the electric charges in the dual gauge-theory picture) is zero. Thus the "flat" configurations (coarse-grained over blocks of size 2×2) see zero background fields and are suitably slowly varying.

On the other hand, while fluctuations due to unit-charge monopoles are strongly affected by the Berry-phase terms, monopoles with charge multiples of 4 are not affected by the Berry phases. Now, the Berry phases, and the background gauge fields B_l which they represent, reflect the fact that the square lattice has two sublattices and that, in order to represent dimer configurations, we placed equal and opposite unit electric charges on the two sublattices. The redundancy of the height representation with period 4 simply reflects the fact that on a square lattice there are four possible dimer configurations associated with each site. This also means that configurations of the dual-height model are physically equivalent if the height variables are shifted by multiples of 4. This condition restricts the structure and the allowed operators of the effective theory. In other terms, the effective-field theory we are seeking must describe fluctuations relative to some ideal state such as the columnar configurations. The theory we need must then treat all columnar states and therefore must also be able to describe the spontaneous breaking of rotational and translation invariance.

9.15 The quantum Lifshitz model

We have seen that QDMs (and their generalizations) can have different types of ordered and topological phases. We will now discuss the nature of the quantum phase transitions between them. We saw that on the square lattice the QDM at the RK point has a ground state with power-law correlations and gapless excitations. This result is surprising since the surrounding phases are either ordered VB crystals or a topological phase. This seemingly violates a general result from the theory of phase transitions, which is largely based on the analysis of Landau and Ginzburg, and was refined (and extended) by Wilson and Kadanoff (and Fisher) with the development of the RG, which states that (quantum and thermal) phase transitions between ordered phases are typically of first order. Thus we should generally expect that the phase transitions between different ordered dimer phases should also be first-order transitions.

This standard result of critical behavior in classical (Goldenfeld, 1992; Cardy, 1996) and quantum (Sachdev, 1999) critical phenomena, and in quantum field theory (Zinn-Justin, 2002), is based on the notion that the phase transitions occur primarily due to the strong fluctuations of local fields representing order parameters. Since the order parameters break spontaneously global symmetries, the associated field theories in general also have a global symmetry. Gauge theories also have phase transitions, which may be also either of first order or continuous ("second order"). Phase transitions in gauge theories, as we saw, also have universality classes and are classified not according to the behavior of local operators but according to that of their (generally non-local) observables such as Wilson loops. In fact, from an RG perspective, local quantum field theories are *defined* by the scaling behavior of theories in the vicinity of continuous phase transitions (with global or local symmetries). It is only in this regime that local field theories (without reference to a "microscopic" cutoff) can be defined as continuum field theories (Wilson, 1973, 1974, 1983; Polchinski, 1984). From this framework it seems that a continuous quantum phase transition between two dimer-ordered phases is a violation of these basic Landau rules.

A conceptually important feature of QDMs is that they have topological defects, namely "holons" and "spinons," which are forbidden to exist (and actually confined into bound states) in the ordered phases but are allowed, and hence "deconfined," as gapless excitations at these quantum critical points. These topological defects are gapped and free in the topological phase. A related problem that has been the focus of a lot of work is that of the possible quantum phase transitions in 2D spin-$1/2$ quantum Heisenberg antiferromagnets with four-spin ("ring-exchange") interactions between a Néel phase and a VB crystal (a state with columnar order) at a critical value of the ring-exchange coupling. Extensive numerical quantum Monte Carlo simulations have provided strong evidence both for a first-order transition (Kuklov *et al.*, 2008) and for a continuous phase transition (Sandvik, 2007, 2010) depending on details of the four-spin-interaction term. It has been proposed that these "Landau-forbidden" continuous quantum phase transitions have spin-$1/2$ gapless excitations, which are thus "deconfined," a form of deconfined quantum criticality (Senthil *et al.*, 2004a, 2004b).

We will now see that the quantum critical points of QDMs are actually the simplest deconfined quantum critical points (Moessner *et al.*, 2001). This interpretation is based on the gauge-theory description of QDMs. A more direct way to derive an effective-field theory is to adapt the methods used in classical dimer and loop models (Nienhuis, 1987; Kondev and Henley, 1996) to the generalized QDMs (Ardonne *et al.*, 2004; Fradkin *et al.*, 2004). We will first take a step back and change the representation of dimers in terms of height models. As before we will assign heights to plaquettes, taking into account the two sublattices (even and odd) of the square lattice. Thus, while going around a site of the even sublattice in a counterclockwise fashion, we will require that the heights on neighboring plaquettes change by $+3$ if the link they share is occupied by a dimer and by -1 if it is not. Conversely, for a site on the odd sublattice the heights change by -3 if the link is occupied and by $+1$ if it is empty (see Fig. 9.13). To avoid over-counting we identify the height h with the height $h + 4$. Notice that the assigned heights locally have jumps across a dimer. In this language, in a columnar state the average height field, defined by the average of the heights on the four plaquettes surrounding a given site of the original lattice, has a non-vanishing uniform expectation value, and in a staggered (or tilted) state the gradients of the height have an expectation value. Similarly, the action of the plaquette (flip) operator on a plaquette amounts to a shift of the height on that plaquette by one unit, $h \to h \pm 1$.

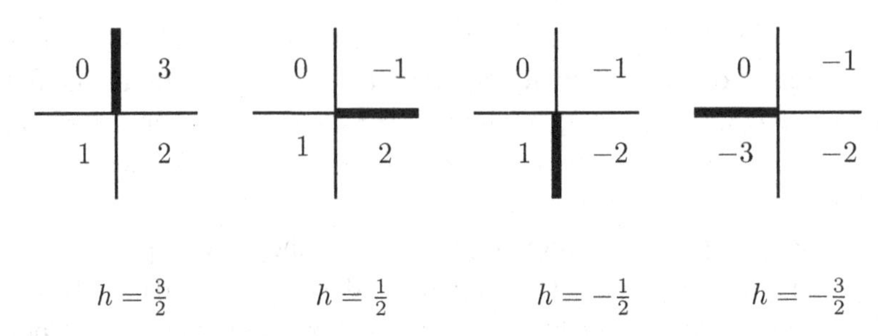

Figure 9.13 Dimer configurations around a site of the even sublattice and their associated heights on the dual lattice. The heights on the dual lattice are defined up to a uniform shift by 4, and wind around the sites of the direct lattice.

These assignments can also be represented in terms of electric fields (or currents) on the links of the direct lattice. Thus, a dimer on the x link (going from the even site to an odd site) corresponds to an electric field of 3 units leaving the even site, and to three electric fields of unit strength entering the even site on the three remaining links. Thus the allowed configurations of electric fields are ± 3 and ± 1 and satisfy a local conservation law, $\Delta \cdot \vec{E} = 0$.

We will now seek an effective continuum field to describe the ordered phases of the QDM and their quantum phase transitions. The degrees of freedom of the effective field theory are coarse-grained height variables, which we will denote by h. We will give a heuristic construction of the effective-field theory for the QDMs on the square lattice. It is based on the following requirements (Ardonne *et al.*, 2004; Fradkin *et al.*, 2004) (a similar construction holds for the hexagonal lattice).

1. The Hamiltonian for the field h must select $h \in \mathbb{Z}$ as the preferred values.
2. Field configurations h that differ by integer multiples of 4 (corresponding to the period-4 condition of the lattice heights) must be equivalent. Hence we will make the identification $h \equiv h + 4n$, with $n \in \mathbb{Z}$. This condition defines the compactification radius to be $R = 4$.
3. All the allowed operators (including the Hamiltonian) must be blind under the equivalency condition.
4. The effective-field theory must have four classical ("ideal") ground states, $h = 0, 1, 2, 3$, each corresponding to the four possible columnar states on the square lattice.
5. For a special value (or values) of the parameter(s) of the effective action, the equal-time correlation functions of the effective-field theory must be equal to the asymptotic long-distance correlation functions of the generalized classical dimer models. This defines the RK point.

In what follows it will be more convenient to work with the rescaled (angular) field $\varphi = (\pi/2)h$, which is 2π periodic and has compactification radius $R = 1$. The ideal states are now $\varphi = 0, \pi/2, \pi, 3\pi/2$. The states with $\varphi = 0, \pi$ correspond (respectively) to the two columnar states modulated along the x axis (i.e. with horizontal dimers), while those with $\varphi = \pi/2$, $3\pi/2$ correspond to the two columnar states modulated along the y axis (vertical dimers).

9.15.1 Field theory for two-dimensional classical dimers

An effective-field theory of the type we are seeking has been known for a long time in 2D classical critical phenomena, where it is known as the Gaussian model, the free "compactified" boson in the language of conformal field theory (CFT). It successfully describes the long-distance behavior of many systems of interest, including dimer models, two decoupled Ising models, Baxter and Ashkin–Teller models, and planar (XY) models (Kadanoff, 1979; Kadanoff and Brown, 1979). The derivation (or "mapping") of the Gaussian model for classical dimer models on bipartite lattices can be found in Nienhuis (1987) and Kondev and Henley (1996). The degree of freedom of the Gaussian model is a free real scalar field $\varphi(\vec{x})$ in 2D (Euclidean) space whose effective action (or "free energy") is simply given by

$$S_{2D}[\varphi] = \int d^2x \, \frac{K}{2} \left(\vec{\nabla}\varphi(\vec{x})\right)^2 \tag{9.142}$$

We have already encountered this model in Chapter 4, where we discussed its connection with the Kosterlitz–Thouless transition, and in Chapters 5 and 6, where it was discussed in the context of the 1D quantum Heisenberg model and of the Luttinger model, respectively. In all of these cases the field φ is treated as an angular variable with a compactification radius, which here we will set to be $R = 1$. This means that the observables of this theory are the electric and magnetic vertex operators $V_n(\vec{x})$ (with an "electric" charge n) and $\widetilde{V}_m(\vec{x})$ (with magnetic charge m)

$$V_n(\vec{x}) = e^{in\varphi(\vec{x})}, \qquad \widetilde{V}_m(\vec{x}) = e^{im\vartheta(\vec{x})} \tag{9.143}$$

The dual field $\vartheta(\vec{x})$ and the field $\varphi(\vec{x})$ are related by the Cauchy–Riemann relation,

$$\partial_i \vartheta = \epsilon_{ij}\, \partial_j \varphi \tag{9.144}$$

In the case of the free-dimer model on bipartite lattices its correlation functions are known explicitly (Fisher and Stephenson, 1963; Youngblood *et al.*, 1980). An identification of the dimer-model densities on the links $(\vec{x}, \vec{x} + e_x)$ and $(\vec{x}, \vec{x} + e_y)$ of the direct lattice, $n_x(\vec{x})$ and $n_y(\vec{x})$, in terms of the observables of the Gaussian model (for the case of the square lattice) is (see Fradkin *et al.* (2004))

$$n_x - \frac{1}{4} = \frac{1}{2\pi}(-1)^{x+y}\,\partial_y\varphi + \frac{1}{2}[(-1)^x e^{i\varphi} + \text{h.c.}] \tag{9.145}$$

$$n_y - \frac{1}{4} = \frac{1}{2\pi}(-1)^{x+y+1}\,\partial_x\varphi + \frac{1}{2}[(-1)^y i e^{i\varphi} + \text{h.c.}] \tag{9.146}$$

where we have used the fact that the average dimer density is $1/4$. These expansions of dimer densities are analogous to the expansion we used in Chapter 6 for the electron density in Luttinger liquids, and constitute the standard description of density-wave order (Chaikin and Lubensky, 1995).

Although the system is critical, the bipartite nature of the lattice enters into the structure of the correlation functions. In addition, on finite geometries, such as a long cylinder of finite diameter, the classical dimer model is known to have long-range columnar order (with the columns running around the circumference of the cylinder). Extensive 2D Monte Carlo numerical simulations show that, if dimer interactions are included, the classical dimer model can also exhibit columnar or staggered order below a critical temperature (Alet *et al.*, 2005, 2006; Papanikolaou *et al.*, 2007c).

As in the case of charge-density waves, we will identify the (normalized) $n = 1$ electric operator

$$V_1(\vec{x}) = e^{i\varphi(\vec{x})} \equiv O_{\rm c}(\vec{x}) \tag{9.147}$$

(taking the classical values $0, i, -1, -i$) with the columnar order parameter $O_{\rm c}(\vec{x})$. Similarly, the $n = 2$ electric operator

$$V_2(\vec{x}) = e^{i2\varphi(\vec{x})} \equiv O_{\rm o}(\vec{x}) \tag{9.148}$$

is the order parameter for *orientational order*, $O_{\rm o}(\vec{x})$. Indeed, this operator takes the values $1, -1, 1, -1$ on the respective ideal states and distinguishes the orientation of the dimers but not their displacements.

In Chapter 4 we showed that the (regularized) correlator of the classical field φ is

$$G_{\rm 2D}(|\vec{x} - \vec{y}|) = \langle\varphi(\vec{x})\varphi(\vec{y})\rangle = -\frac{1}{4\pi K}\ln(|\vec{x} - \vec{y}|^2) \tag{9.149}$$

and, as a result, the correlators of the vertex operators $V_n(\vec{x})$ and $\widetilde{V}_m(\vec{x})$ are

$$\langle V_n(\vec{x})V_{n'}(\vec{y})\rangle = \frac{\delta_{n,-n'}}{|\vec{x} - \vec{y}|^{2\Delta_n}}, \qquad \langle\widetilde{V}_m(\vec{x})\widetilde{V}_{m'}(\vec{y})\rangle = \frac{\delta_{m,-m'}}{|\vec{x} - \vec{y}|^{2\widetilde{\Delta}_m}} \tag{9.150}$$

where the scaling dimensions Δ_n and $\widetilde{\Delta}_m$ are

$$\Delta_n = \frac{n^2}{4\pi K}, \qquad \widetilde{\Delta}_m = \pi K m^2 \tag{9.151}$$

Similarly, we find that the slowly varying density operators $\delta n_j(\vec{x}) = (1/(2\pi))\partial_j\varphi(\vec{x})$ have scaling dimension 1 and hence have a power-law decay $\sim 1/|\vec{x}-\vec{y}|^2$.

The expressions for the scaling dimensions of the "electric" and "magnetic" vertex operators, $V_n(\vec{x})$ and $\tilde{V}_m(\vec{x})$, given in Eqs. (9.150) show that they transform into each other under the *duality* transformation which maps (Kadanoff, 1979)

$$K \leftrightarrow \frac{1}{K}, \qquad n \leftrightarrow m \tag{9.152}$$

This transformation is closely related to the Kramers–Wannier duality of the 1D quantum Ising model, discussed in Chapter 5, or equivalently to the classical 2D Ising model. It is formally equivalent to the duality (or T-duality) symmetry of the Luttinger liquid discussed in Chapter 6.

The exactly known correlators of dimer densities (Youngblood *et al.*, 1980) have a long-distance behavior with a single power law $1/|\vec{x}-\vec{y}|^2$ (with a sublattice structure). Our identification of the dimer-density operators of Eq. (9.145) and Eq. (9.146) predicts the same correlation functions, provided that the charge $n = 1$ vertex operator $V_1(\vec{x}) = O_c(\vec{x})$ has scaling dimension $\Delta_1(\text{free}) = 1$ at the free-dimer point. This leads us to identify, for the free-dimer model, the (non-universal) stiffness K_{free} as being given by

$$K_{\text{free}} = \frac{1}{4\pi} \tag{9.153}$$

Let us now considered the correlation function of two *monomers*, two sites of the lattice that do not belong to a dimer. In our language we will call this the *hole* operator. Fisher and Stephenson (1963) showed that in the free-dimer model (on a square lattice) the correlator of two monomers has instead a power-law decay $\sim 1/\sqrt{|\vec{x}-\vec{y}|}$. Hence the scaling dimension of a monomer (or hole) is $1/4$. We can see by inspection that this is precisely the scaling dimension of the magnetic operator $\tilde{V}_1$ at the free-dimer point. This is not an accident, since magnetic operators correspond to violations of the dimer rule.

The periodic nature of the height variable and the requirement that configurations which differ by 4 units be physically equivalent imply that the charge $n = 4$ electric operator $\sim g\cos(4\varphi)$ should also be allowed in the effective action. However, at the free-dimer point, $K = 1/(4\pi)$, the charge $n = 4$ operator has scaling dimension $\Delta_4(1/(4\pi)) = 16$. This operator is strongly irrelevant and hence in practice can be ignored.

Away from the free-dimer point, the interacting-dimer model also maps to a Gaussian CFT with the same compactification radius $R = 1$ (and hence the same set of allowed operators), but with a non-universal stiffness $K(u)$, which is now a (non-universal) function of the interaction parameter u. Thus, classical dimer models also have a line of fixed points. Much as in the case of the classical XY model, the interacting-dimer model has a KT transition at some value of the interaction parameter u_c at which $K(u_c) = 2/\pi$. At this point the charge $n = 4$ electric operator has scaling dimension $\Delta_4(2/\pi) = 2$ and is marginal. Hence, the interacting-dimer model has a KT phase transition at u_c to an ordered columnar phase (Alet *et al.*, 2005; Papanikolaou *et al.*, 2007a).

9.15.2 Field theory for two-dimensional quantum-dimer models

Let us now develop an effective-field theory for *quantum*-dimer models. We will follow the same approach as in the classical case and use the same operator identifications. However, we will need to define the quantum dynamics of the effective-field theory and to relate that to the Hamiltonians of generalized QDMs (and eight-vertex models).

The degree of freedom of the effective-field theory of the quantum case in (2+1) dimensions will also be the coarse-grained phase field $\varphi(\vec{x}, t)$. Let us define $\Pi(\vec{x})$ to be the momentum canonically conjugate to the field $\varphi(\vec{x})$, which now satisfy standard equal-time commutation relations

$$[\varphi(\vec{x}), \Pi(\vec{y})] = i\delta^{(2)}(\vec{x} - \vec{y}) \tag{9.154}$$

The simplest translation- and rotation-invariant free Hamiltonian which obeys all the requirements is

$$H_0 = \int d^2x \left[\frac{1}{2}\Pi(\vec{x})^2 + \frac{A}{2}(\vec{\nabla}\varphi(\vec{x}))^2 + \frac{\kappa^2}{2}(\nabla^2\varphi(\vec{x}))^2\right] \tag{9.155}$$

where A and κ are non-universal constants that will be determined shortly using the mapping to the QDM. We will discuss below two cases.

1. $A \neq 0$, in which case space scales the same way as time, $[T] = [L]$, and the dynamic critical exponent is $z = 1$. In this case the field has the canonical units of a relativistic field in $(2 + 1)$ dimensions, $[\varphi] = [L^{-1}]$, as does the canonical momentum, $[\Pi] = [L^{-1}]$. We will see that this case describes the ordered phases.
2. $A = 0$, in which case time scales as two powers of space, $[T] = [L^2]$, and the dynamic critical exponent is $z = 2$. In this case the field φ is dimensionless, $[\varphi] = [L^0]$, and the canonical momentum has units of $[\Pi] = [L^{-2}]$. We will see that this case describes a quantum critical point.

The associated action in (2 + 1) (Euclidean) dimensions is (with $x = (\vec{x}, \tau)$)

$$S_0 = \int d^2x\, d\tau \left[\frac{1}{2}(\partial_\tau \varphi(x))^2 + \frac{A}{2}(\vec{\nabla}\varphi(x))^2 + \frac{\kappa^2}{2}(\nabla^2\varphi(x))^2\right] \tag{9.156}$$

For the effective action (and Hamiltonian) to describe the QDM we need to add to S_0 (and to H_0) a term of the form

$$S_{\text{int}} = \int d^2x\, d\tau\, g\, \cos(2\pi h) = \int d^2x\, d\tau\, g\, \cos(4\varphi) \tag{9.157}$$

which enforces the periodicity condition. In this form the partition function is

$$\mathcal{Z} = \int \mathcal{D}\varphi\, e^{-S_0[\varphi]-S_{\text{int}}[\varphi]} \tag{9.158}$$

Let us consider first the case in which the constant $A > 0$. In this case, the $(\nabla^2\varphi)^2$ term is irrelevant compared with the $(\nabla\varphi)^2$ term, which has two fewer derivatives. The free-field action now looks like a standard free-field theory. However, the cosine operator of S_{int} is always a relevant operator that must be included. The result is the sine–Gordon theory, and we are back to Polyakov's case:

$$S_{\text{eff}} = \int d^2x\, d\tau \left[\frac{1}{2}(\partial_\tau \varphi(x))^2 + \frac{A}{2}(\vec{\nabla}\varphi(x))^2 - g\cos(4\varphi(x))\right] \tag{9.159}$$

This theory is relativistic since space scales the same way as time, and hence we have a dynamic critical exponent $z = 1$.

From Polyakov's analysis (or using the Kosterlitz RG) we saw that in this theory the cosine operator is always relevant and that the theory is controlled by a strong-coupling (large-g) fixed point. At this fixed point the field φ is pinned at the minima of the cosine operators and its fluctuations are gapped and have a finite effective mass (squared) $m_{\text{eff}}^2 \simeq 16g/\sqrt{A}$. We can determine what dimer state is described by this theory (for $A > 0$ and all g) by noting that, if the field φ is pinned, the columnar order parameter $O_c = \exp(i\varphi)$ has a non-vanishing expectation value. If we include the effects of quantum fluctuations in the pinned state to lowest (quadratic) order we find that the order parameter takes the values $1, i, -1, -i$ times a function of $\sqrt{A}$, g, and a (sharp) ultraviolet (UV) momentum cutoff Λ

$$\langle e^{i\varphi}\rangle \approx \{1, i, -1, -i\} \times \exp\left(-\frac{1}{2\pi^2}\frac{\Lambda}{\sqrt{A}}\left(\Lambda - \frac{\pi}{2}m_{\text{eff}}\right)\right) \tag{9.160}$$

Therefore, for $A > 0$ this effective-field theory describes a state with columnar order.

Let us now consider the special but important case of a theory at $A = 0$ and examine a theory with the following effective action:

$$S_{\rm QLM} = \int d^2x\, d\tau \left[\frac{1}{2}(\partial_\tau \varphi)^2 + \frac{\kappa^2}{2}(\nabla^2 \varphi)^2\right] \tag{9.161}$$

We will refer to this action as the *quantum Lifshitz model.*

On the other hand, for $A < 0$ the free-field part of the action (with terms with only two derivatives) is unstable since it has a negative stiffness ($A < 0$). In this case the momentum (or wave vector) of the ordered state shifts away from $(\pi, 0)$ or $(0, \pi)$, the ordering wave vector(s) of the columnar states. The stable state (and the shift of the ordering wave vector, $\vec{Q}$) is determined by terms not included in the quantum Lifshitz model, Eq. (9.161). The leading perturbations involve the commensurability operator $\cos(4\varphi)$ (which is already included in the action of Eq. (9.159)) and the quartic operator

$$S_4 = \int d^2x\, d\tau\, g_4 (\vec{\nabla}\varphi(\vec{x}))^4 \tag{9.162}$$

We will see shortly that this operator is marginally irrelevant at the quantum critical point described by the quantum Lifshitz model. However, despite being irrelevant, it stabilizes the ordered phase for $A < 0$. Indeed, if $A < 0$, the minimum-energy state has a wave vector shifted by an amount $\vec{Q}$,

$$\varphi(x) = \vec{Q} \cdot \vec{x} + \delta\varphi(x) \tag{9.163}$$

which we will describe as the *tilt* of the columnar state. $\vec{Q}$ is determined by minimizing the action $S_{\rm QLM} + S_4 + S_{\rm int}$. If we assume that $g_4 > 0$ and we momentarily neglect the effects of the commensurability interaction term, $S_{\rm int}$, we find that the tilt $\vec{Q}$ is

$$|\vec{Q}| = \sqrt{\frac{|A|}{4g_4}} \tag{9.164}$$

On the other hand, if $g_4 \leq 0$ the tilt grows without limit and is stabilized by lattice effects as is the case in the QDM, which for $V > J$ has a staggered phase, with wave vector (π, π). This, however, is not the full story. A tilted phase with $|\vec{Q}|$ varying continuously is an incommensurate state, which is in conflict with the commensurability interaction $S_{\rm int}$. It turns out that the resulting state is either commensurate and pinned (and hence confining), or incommensurate and not pinned (and gapless) and hence deconfined. However, the incommensurate (deconfined) gapless phases form a Cantor set (of finite measure) (Fradkin *et al.*, 2004).

Let us now discuss the special but important case of $A = 0$ which plays the role of the quantum critical point. We call this a quantum Lifshitz model by analogy with the theory of the Lifshitz point in liquid crystals and in helical magnets

(Grinstein, 1981; Chaikin and Lubensky, 1995). If we regard the imaginary-time coordinate as the z coordinate of a 3D system, the field $\varphi(\vec{x}, \tau)$ can be regarded as the spatial modulation of the height of a set of smectic layers of nematic molecules (essentially rod-shaped objects) stacked along the z axis.

The quantum Lifshitz model is dual to a gauge theory. However, since the quantum Lifshitz model has dynamic exponent $z = 2$, the dual-gauge theory cannot be the Maxwell theory since the latter has photons (with only one polarization state since we are in $(2 + 1)$ dimensions) whose energy is a linear function of the momentum. We will show that the Hamiltonian of the dual-gauge theory has the unconventional form

$$H_{\text{QLM-gauge}} = \int d^2x \left[\frac{\kappa^2}{2} \left(\vec{\nabla} \times \vec{E} \right)^2 + \frac{1}{2} B^2 \right] \tag{9.165}$$

where E_j is the electric field and $B = \epsilon_{jk}\, \partial_j A_k$ is the magnetic field (a pseudo-scalar in 2D). The electric field E_j and the vector potential A_j obey canonical commutation relations in the gauge $A_0 = 0$,

$$[E_j(\vec{x}), A_k(\vec{y})] = i\delta_{jk}\delta^{(2)}(\vec{x} - \vec{y}) \tag{9.166}$$

The physical states, $|\text{Phys}\rangle$, as usual obey Gauss's law

$$\partial_j E_j |\text{Phys}\rangle = 0 \tag{9.167}$$

In the absence of external sources (or "matter fields") the Gauss-law constraint can be solved trivially by writing

$$E_j = \epsilon_{jk}\, \partial_k \varphi \tag{9.168}$$

where φ is a scalar. Then the canonical commutation relation becomes

$$[\varphi(\vec{x}), B(\vec{y})] = i\delta^{(2)}(\vec{x} - \vec{y}) \tag{9.169}$$

and we identify the magnetic field B with Π, the momentum canonically conjugate with φ. Hence, the gauge theory and the quantum Lifshitz model are physically equivalent.

What is the gauge-theory picture of the observables of the quantum Lifshitz model? Let us consider first the operator $O_n(\vec{x})$ of the gauge theory that creates a magnetic charge (in the gauge-theory language) of charge n at location $\vec{x}$. $O_n(\vec{x})$ is given by

$$\begin{aligned} O_n(\vec{x}) &= \exp\left(in \int_{\gamma(\vec{x})} dy_j\, \epsilon_{jk} \theta(y_j - x_j) \delta(y_k - x_k) E_k(\vec{y}) \right) \\ &= \exp\left(in \int_{\gamma(\vec{x})} dx_j \partial_j \varphi(\vec{y}) \theta(y_j - x_j) \delta(y_k - x_k) \right) \\ &= e^{in\varphi(\vec{x})} \end{aligned} \tag{9.170}$$

where $\gamma(\vec{x})$ is any curve beginning somewhere on the boundary of the system and ending at $\vec{x}$; here x_j and x_k (and y_j and y_k) are the tangent and normal directions to the path γ. As we can see, it is what we call an electric operator in the quantum Lifshitz model. These vertex operators are consistent with the compactification radius $R = 1$ that we have imposed on the field φ since they are invariant under shifts of the field variable by integer multiples of 2π.

Similarly, let us consider an operator that creates an electric charge in the gauge theory, i.e. leads to the condition

$$\partial_j E_j(\vec{y}) = m\delta^{(2)}(\vec{y} - \vec{x}) \tag{9.171}$$

The solution now is

$$E_j(\vec{y}) = \epsilon_{jk}(\partial_k \varphi(\vec{y}) + \mathcal{B}_k(\vec{y})) \tag{9.172}$$

which requires that

$$\epsilon_{jk}\, \partial_j \mathcal{B}_k(\vec{y}) = m\delta^{(2)}(\vec{y} - \vec{x}) \tag{9.173}$$

In other terms, the dual-scalar field φ is coupled to a background gauge field $\mathcal{B}_k(\vec{y})$ whose magnetic charge is m: this is the magnetic-charge operator of the quantum Lifshitz model.

We now need to determine (or interpret the meaning of) the constants A and κ. In terms of the QDM we tentatively assign the RK point (which we know is critical) to the point $A = 0$ of the effective theory. Hence, close to the RK point we will write $A = c(J - V)$, where c is a constant. To determine the value of κ in Eq. (9.161) we will discuss now the behavior of this theory and find a mapping to the RK point. We will do this in two different ways.

Let us first find the *wave function* of the ground state of the quantum Lifshitz model (Ardonne *et al.*, 2004). To do this, we turn to the Schrödinger representation of the quantum Lifshitz field theory. We will work in the field representation in which the states are eigenstates of the field operator $\varphi(\vec{x})$. In this representation the wave functions (functionals) are

$$\Psi[\{\varphi(\vec{x})\}] = \langle \Psi | \{\varphi(\vec{x})\}\rangle \tag{9.174}$$

Notice that the field representation of the quantum Lifshitz model is the same as the electric-field representation of the states in the gauge theory (instead of the vector-potential representation). In this representation the canonical momentum $\Pi(\vec{x})$ is a functional differential operator

$$\Pi(\vec{x}) = i\,\frac{\delta}{\delta\varphi(\vec{x})} \tag{9.175}$$

The quantum Lifshitz Hamiltonian, Eq. (9.155), with $A = 0$, now leads to a Schrödinger equation for $\Psi[\varphi]$ of the form

$$\begin{aligned} H\Psi[\varphi] &= \int d^2x \left[-\frac{1}{2} \frac{\delta^2}{\delta\varphi(\vec{x})^2} + \frac{\kappa^2}{2} \left(\nabla^2\varphi(\vec{x})\right)^2 \right] \Psi[\varphi] = E\Psi[\varphi] \\ &= \int d^2x \, \frac{1}{2} \left\{ Q[\varphi], Q^\dagger[\varphi] \right\} \Psi[\varphi] \end{aligned} \tag{9.176}$$

where the braces denote (as usual) the anticommutator. Here we introduced the "creation" operator $Q^\dagger[\varphi]$ and its adjoint, the "annihilation" operator $Q[\varphi]$, defined by

$$Q[\varphi] = \frac{1}{\sqrt{2}} \left[-\frac{\delta}{\delta\varphi(\vec{x})} + \kappa \, \nabla^2\varphi(\vec{x}) \right] \tag{9.177}$$

As in the theory of the linear harmonic oscillator in quantum mechanics, the ground-state wave function(al) is annihilated by the "annihilation" operator $Q[\varphi]$. This leads to the simple first-order equation

$$Q\Psi_0[\varphi] = \frac{1}{\sqrt{2}} \left[-\frac{\delta}{\delta\varphi(\vec{x})} + \kappa \, \nabla^2\varphi \right] \Psi_0[\varphi] = 0 \tag{9.178}$$

whose (normalized) solution is

$$\Psi_0[\varphi] = \frac{1}{\sqrt{Z_0}} \exp\left(-\int d^2x \, \frac{\kappa}{2} \left(\vec{\nabla}\varphi(\vec{x})\right)^2 \right) \tag{9.179}$$

where Z_0, the norm (squared) of the wave function, is

$$Z_0 = \int \mathcal{D}\varphi \, \exp\left(-\int d^2x \, \kappa \left(\vec{\nabla}\varphi(\vec{x})\right)^2 \right) \tag{9.180}$$

We see that the amplitude of a field configuration $|[\varphi]\rangle$ in the ground-state wave function $\Psi_0[\varphi]$ has the form of the Gibbs weight for a 2D Gaussian model and that its norm Z_0 has the form of the partition function of the Gaussian model. Thus we find a relation between the stiffness K of the 2D classical Gaussian model (cf. Eq. (9.142)) and the parameter κ of the quantum Lifshitz model:

$$K = 2\kappa \tag{9.181}$$

We find that the ground-state wave function of the quantum Lifshitz model is *scale-invariant*! Since 2D classical scale-invariant systems are also conformally invariant (and are examples of CFTs), we will refer to this as a conformal quantum critical point (Ardonne *et al.*, 2004).

We should note that a scale-invariant wave function is not generic of quantum critical systems but rather is a peculiar feature of this theory. Although this means that it represents a quantum critical point, the converse is not true: the ground-state wave function of a quantum critical system, although it must scale, is not necessarily scale-invariant. A simple counterexample is the Luttinger model (Fradkin *et al.*, 1993).

It is also interesting to rewrite the ground-state wave function $\Psi_0[\varphi]$, Eq. (9.179), in the language of the dual-gauge theory. Since the coarse-grained height field φ is simply the curl of the dual electric field, $\vec{E} = \vec{\nabla} \times \varphi$, it is easy to show that in the gauge theory the wave function is a state in the electric-field representation:

$$\Psi_0[\vec{E}(\vec{x})] = \frac{1}{\sqrt{Z_0}} \exp\left(-\int d^2x \, \frac{\kappa}{2}\vec{E}(\vec{x})^2\right) \prod_{\vec{x}} \delta(\vec{\nabla} \cdot \vec{E}(\vec{x})) \tag{9.182}$$

In other terms, it is a simple Gaussian function of the electric-field configuration subject to the Gauss-law constraint without sources.

In this representation the equal-time correlation function of N charge operators $O_n(\vec{x})$ in the quantum Lifshitz ground state $|\Psi_0\rangle$ is given by

$$\begin{aligned}
&\langle \Psi_0 | O_{n_1}(\vec{x}_1) \ldots O_{n_N}(\vec{x}_N) | \Psi_0 \rangle_\kappa \\
&\quad = \frac{1}{Z_0} \int \mathcal{D}\varphi \; O_{n_1}(\vec{x}_1) \ldots O_{n_N}(\vec{x}_N) \exp\left(-\int d^2x \; \kappa \left(\vec{\nabla}\varphi(\vec{x})\right)^2\right) \\
&\quad = \langle O_{n_1}(\vec{x}_1) \ldots O_{n_N}(\vec{x}_N) \rangle_{K=2\kappa}
\end{aligned} \tag{9.183}$$

where the last line is the expectation value of the same operators in the *classical* Gaussian model with stiffness $K = 2\kappa$. Since we know how to relate the stiffness K of the classical model to the "microscopic" (classical) dimer model, this identity shows that the equal-time correlation functions of the quantum Lifshitz model do indeed reproduce the correlation functions of the classical dimer model, provided that we set $\kappa_{\text{free}} = 1/(8\pi)$. In particular, this mapping also tells us that the scaling dimensions *are the same* in both theories. Thus, the scaling dimensions of the charge operators $O_n[\varphi]$ of the quantum Lifshitz model are

$$\Delta_n = \frac{n^2}{8\pi\kappa} \tag{9.184}$$

We can also find a representation of the magnetic (vortex) operators. The vortex operators are

$$\tilde{O}_m(\vec{x}) = \exp\left(i \int d^2z \; \alpha(\vec{z})\Pi(\vec{z})\right) \tag{9.185}$$

where

$$\alpha(\vec{z}) = m \arg(\vec{z} - \vec{x}) \tag{9.186}$$

where $0 \leq \arg(\vec{z} - \vec{x}) \leq 2\pi$ is the argument of the vector $\vec{z} - \vec{x}$ (with a branch cut defined arbitrarily along the negative-x axis). The action of the operator $\tilde{O}_m(\vec{x})$ on an eigenstate of the field operator $|[\varphi]\rangle$ is simply a shift

$$\exp\left(i\int d^2z\,\alpha(\vec{z})\Pi(\vec{z})\right)|[\varphi]\rangle = |[\varphi(\vec{x}) - \alpha(\vec{x})]\rangle \tag{9.187}$$

In other words, it amounts to a *singular gauge transformation*. Therefore, its action is equivalent to coupling the field φ to a vector potential whose space components $\vec{A}$ satisfy

$$\oint_\gamma d\vec{z} \cdot \vec{A}[\vec{z}] = 2\pi m \tag{9.188}$$

for all closed paths γ that have the point $\vec{x}$ in their interior, and zero otherwise. In particular, the wave function of the state resulting from the action of the vortex operator on the ground state is

$$\Psi_m[\vec{x}] = \langle[\varphi]|\tilde{\mathcal{O}}_m(\vec{x})|\Psi_0\rangle = \frac{1}{\sqrt{Z_0}}\exp\left(-\frac{\kappa}{2}\int d^2z\,\left(\vec{\nabla}\varphi - \vec{A}\right)^2\right) \tag{9.189}$$

where $\vec{A}$ is any vector field that satisfies Eq. (9.188). The (equal-time) ground-state expectation value of a product of vortex operators with magnetic charges $\{m_l\}$, i.e. the *overlap* of the state with M vortices at locations $\vec{x}_l$ and magnetic charge m_l with the vortex-free ground-state wave function, is therefore

$$\begin{aligned}
&\langle\Psi_0|\widetilde{\mathcal{O}}_{m_1}(\vec{x}_1)\ldots\widetilde{\mathcal{O}}_{m_M}(\vec{x}_M)|\Psi_0\rangle_\kappa \\
&\qquad = \frac{1}{Z_0}\int \mathcal{D}\varphi\,\exp\left(-\kappa\int d^2z\left(\vec{\nabla}\varphi - \vec{A}\right)^2\right) \\
&\qquad = \langle\widetilde{\mathcal{O}}_{m_1}(\vec{x}_1)\ldots\widetilde{\mathcal{O}}_{m_k}(\vec{x}_M)\rangle_{K=2\kappa}
\end{aligned} \tag{9.190}$$

where Z_0 is given by Eq. (9.180), and the last line is an expectation value of M vortex operators in the Gaussian model, namely the 2D classical compactified boson. The vector potential in Eq. (9.190) satisfies

$$\varepsilon_{ij}\,\partial_i A_j = 2\pi\sum_{l=1}^{M} m_l\delta^2(\vec{z} - \vec{x}_l) \tag{9.191}$$

These results also show that the scaling dimensions of the vortex operators $\widetilde{\Delta}_m$ are also the same in both theories (if we set $K = 2\kappa$). Therefore the scaling dimensions of the vortex operators (or holes) in the quantum Lifshitz model are

$$\widetilde{\Delta}_m = 2\pi\kappa m^2 \tag{9.192}$$

We can also gain insight into this problem by looking at the time dependence of the correlation functions. To this end, we return to the path-integral picture and compute the propagator of the field $\varphi(\vec{x}, \tau)$. It is now easy to see that the boson propagator of this theory, in imaginary time τ, is

$$\begin{aligned} G(\vec{x}-\vec{x}', \tau-\tau') &= \langle \varphi(\vec{x},\tau)\varphi(\vec{x}',\tau')\rangle \\ &= \int \frac{d\omega}{2\pi} \int \frac{d^2q}{(2\pi)^2} \frac{e^{i\omega(\tau-\tau')-i\vec{q}\cdot(\vec{x}-\vec{x}')}}{\omega^2+\kappa^2\left(\vec{q}^{\,2}\right)^2} \end{aligned} \tag{9.193}$$

From the denominator of the integrand of Eq. (9.193) we learn that, in terms of real frequencies, the excitations of the field φ are states that propagate with an energy–momentum relation $\omega(\vec{q}) = \kappa \vec{q}^{\,2}$. This is the same as the resonon state of Rokhsar and Kivelson (1988).

This propagator has a short-distance logarithmic divergence. From now on we will use instead the regularized (subtracted) propagator

$$\begin{aligned} G_{\text{reg}}(\vec{x},\tau) &\equiv G(\vec{x},\tau) - G(a,0) \\ &= -\frac{1}{8\pi\kappa}\left[\ln\left(\frac{|\vec{x}|^2}{a^2}\right) + \Gamma\left(0, \frac{|\vec{x}|^2}{4\kappa|\tau|}\right)\right] \end{aligned} \tag{9.194}$$

where a is a short-distance cutoff and $\Gamma(0,z)$ is the incomplete Gamma function

$$\Gamma(0,z) = \int_z^\infty \frac{ds}{s} e^{-s} \tag{9.195}$$

The regularized propagator has the asymptotic behaviors

$$G_{\text{reg}}(\vec{x},\tau) = \begin{cases} -\dfrac{1}{4\pi\kappa}\ln\left(\dfrac{|\vec{x}|}{a}\right), & \text{for } |t| \to 0 \\[2ex] -\dfrac{1}{8\pi\kappa}\ln\left(\dfrac{4\kappa|\tau|}{a^2\gamma}\right), & \text{for } |\vec{x}| \to a \end{cases} \tag{9.196}$$

where $\ln\gamma = \mathbf{C} = 0.577\ldots$ is the Euler constant.

The time-dependent correlation functions of the charge operators are

$$\langle \mathcal{O}_n(\vec{x},\tau)^\dagger \mathcal{O}_n(\vec{x}\,',\tau')\rangle = e^{n^2 G_{\text{reg}}(\vec{x}-\vec{x}',\tau-\tau')} \tag{9.197}$$

At equal (imaginary) times, $|\tau-\tau'| \to 0$, it behaves like

$$\langle \mathcal{O}_n(\vec{x},0)^\dagger \mathcal{O}_n(\vec{x}\,',0)\rangle = \left(\frac{a}{|\vec{x}-\vec{x}'|}\right)^{n^2/(4\pi\kappa)} \tag{9.198}$$

and we recover the result that the operator $\mathcal{O}_n$ has (spatial) scaling dimension $\Delta_n = n^2/(8\pi\kappa)$. For $|\vec{x}-\vec{x}'| \to a$, its asymptotic behavior is instead given by

$$\langle \mathcal{O}_n(\vec{0},\tau)^\dagger \mathcal{O}_n(\vec{0},\tau')\rangle = \left(\frac{a^2\gamma}{4\kappa|\tau-\tau'|}\right)^{n^2/(8\pi\kappa)} \tag{9.199}$$

This behavior is manifestly consistent with a dynamical critical exponent $z = 2$. Similar results can be derived for the magnetic (vortex) operators.

9.15.3 Scaling at the quantum Lifshitz multicritical point

The scaling properties of this system were studied by Grinstein (1981) in the context of the theory of anisotropic scaling at Lifshitz points in helimagnets and by Grinstein and Pelcovits (1982) in the context of the theory of non-linear elasticity in smectic liquid crystals in three dimensions. As noted by Grinstein (1981), this system is in many ways a 3D analog of the Gaussian model. Hence its phase transitions are very similar to the Kosterlitz–Thouless transition of 2D statistical mechanics.

With some caveats, most of these results from classical statistical mechanics apply to this quantum critical point. Here we list the scaling properties of the main operators and how they affect the physics (Grinstein, 1981; Fradkin *et al.*, 2004; Vishwanath *et al.*, 2004). The quantum Lifshitz model is a 2D quantum critical system with dynamic critical exponent $z = 2$. Thus, its scaling properties are those of a system with total effective (Euclidean) dimension $D = z+d = 4$. This tells us that all operators with scaling dimension $\Delta > 4$ are irrelevant, whereas operators with scaling dimension $\Delta < 4$ are relevant. This system is actually a multicritical point with an exact line of fixed points parametrized by the coupling constant κ, the stiffness of the operator $(\nabla^2\varphi)^2$. This operator has scaling dimension $\Delta = 4$ and it is marginal. Since this fixed point is a free-field theory, the operator $(\nabla^2\varphi)^2$ is also exactly marginal and, in the absence of all other operators, has a vanishing beta function.

On the other hand, the operator $(\vec{\nabla}\varphi)^2$ has scaling dimension $\Delta = 2 < 4$ and it is relevant. As we saw, the sign of its coupling constant A tunes this quantum phase transition. Similarly, the operator $(\vec{\nabla}\varphi)^4$ has scaling dimension $\Delta = 4$ and is (superficially) marginal. However, Grinstein (1981) showed that the beta function for its coupling constant g_4 (defined in Eq. (9.162)) is *negative*,

$$\beta(g_4) = a\,\frac{\partial g_4}{\partial a} = -cg_4^2 + \cdots \tag{9.200}$$

where a is a length scale and c is a positive dimensionless constant. Therefore, this operator is actually marginally *irrelevant* and its coupling constant g_4 scales (logarithmically slowly) to zero at long distances. If $g_4 \neq 0$ its flow leads to logarithmic corrections to scaling in the correlation functions.

The quantum criticality of the dimer model on the square lattice at the RK point is well described by the quantum Lifshitz model (Fradkin *et al.*, 2004). In this case even the marginally irrelevant coupling g_4 is absent, so it leads to no corrections to scaling effects. However, for the case of the honeycomb lattice, the quantum Lifshitz model admits a possible cubic term in the action of the form $g_3(\partial_x\varphi)((\partial_x\varphi)^2 - 3(\partial_y\varphi)^2)$, which is invariant under a $\pi/3$ rotation and an inversion $\varphi \to -\varphi$. This operator has scaling dimension $\Delta_3 = 3$ and it is relevant. For

$g_3 \neq 0$ the coupling constant g_3 grows, and the system flows to a fixed point with a finite correlation length and a first-order phase transition. By symmetry, operators of this type cannot arise for a system on the square lattice.

The scaling dimension of the charge operators $O_n[\varphi]$, $\Delta_n = n^2/(8\pi\kappa)$, varies continuously as a function of κ. These operators are irrelevant for $n > \sqrt{32\pi\kappa}$ and relevant otherwise. For the QDM at the RK point (on the square lattice), all operators with $n > 2$ are irrelevant. In particular, at the free-dimer point, $\kappa = 1/(8\pi)$, the operator O_4 has dimension $\Delta_4 = 16$ and it is strongly irrelevant. Additional interactions in the lattice model cause dimers to attract each other, leading to an *increase* in the value of $\kappa > 1/(8\pi)$ and a decrease of the scaling dimensions. On the other hand, the combined effect of the marginally irrelevant operator $(\vec{\nabla}\varphi)^4$ and of the (commensurability) charge operator O_4 drives the system into a sequence of commensurate phases known as an (incomplete) devil's staircase (Fradkin *et al.*, 2004), thus avoiding the quantum Lifshitz critical point. This, however, is not what happens in the simple QDM on the square lattice where it is accessible.

How about magnetic (vortex) operators $\widetilde{O}_m$? As we saw, operators of this type violate the dimer constraint. For instance, we associated the operator $\widetilde{O}_1$ with the hole-creation operator. Similarly, dimers (or valence bonds) connecting two nearest-neighboring sites of the same sublattice also violate the constraint. On the square lattice operators of this type, formally operators with magnetic charge 2, cause a crossover to the QDM on the triangular lattice (Ardonne *et al.*, 2004), a system that is known to be in a topological phase (Moessner and Sondhi, 2001b) akin to the deconfined phase of the $\mathbb{Z}_2$ gauge theory. At the free-dimer point these operators have scaling dimension $\widetilde{\Delta}_m = m^2/4$ and are relevant for all $m < 4$. In particular, the operators with magnetic charge 2 have dimension 1, and are strongly relevant. When added to the action, these operators destabilize the quantum Lifshitz fixed point since they cause the magnetic excitations to proliferate. Their condensation caused the gauge symmetry to be reduced to its $\mathbb{Z}_2$ subgroup and the system is in a deconfined phase. The mechanism of proliferation (or condensation) of charge-2 operators driving the system into a topological (deconfined) $\mathbb{Z}_2$ phase was noted by Sachdev and Read (1991) and by Mudry and Fradkin (1994) on the basis of earlier work in gauge theories by Fradkin and Shenker (1979) that we have already discussed.

10

Chiral spin states and anyons

10.1 Chiral spin liquids

In Chapter 8 we considered solutions of the mean-field equations of quantum antiferromagnets, Eq. (8.64) and Eq. (8.65), that respect time-reversal invariance. We will now consider a frustrated quantum antiferromagnet and look for states for which time-reversal invariance is spontaneously broken. In terms of the mean-field theory of Section 8.4, we will consider situations in which the *phase* $\bar{\mathcal{A}}_j(\vec{x},t)$ of the link variable $\bar{\chi}_j(\vec{x},t)$ has a non-zero curl $\bar{\mathcal{B}}(\vec{x},t)$ around an elementary plaquette

$$\bar{\mathcal{B}}(\vec{x},t) = \sum_{\text{plaquette}} \bar{\mathcal{A}}_j(\vec{x},t) = \Delta_1 \bar{\mathcal{A}}_2(\vec{x},t) - \Delta_2 \bar{\mathcal{A}}_1(\vec{x},t) \tag{10.1}$$

In Section 8.5 we argued that such flux states violate time-reversal invariance unless $\bar{\mathcal{B}}(\vec{x},t) = 0, \pi$. A solution $\bar{\chi}_j(\vec{x},t)$ of the saddle-point equation applied to Eq. (8.41) satisfies

$$\frac{2}{J}\langle \bar{\chi}_j^*(\vec{x},t)\rangle = \langle c_\alpha^\dagger(\vec{x},t) c_\alpha(\vec{x}+\hat{e}_j,t)\rangle \tag{10.2}$$

For a solution with $\bar{\rho}_j(\vec{x},t) = \bar{\rho}_j$ a constant and $\bar{\mathcal{A}}_j(\vec{x},t) \neq 0$, we get

$$\frac{2}{J}\bar{\rho}_j e^{-i\bar{\mathcal{A}}_j(\vec{x},t)} = \langle c_\alpha^\dagger(\vec{x},t) c_\alpha(\vec{x}+\hat{e}_j,t)\rangle \tag{10.3}$$

Thus a flux phase implies that the *product of the band amplitudes* $\langle c_\alpha^\dagger(\vec{x},t) c_\alpha(\vec{x}+\hat{e}_j,t)\rangle$ around a closed loop γ of the lattice should have a *phase* determined by the *flux* going through the loop. Alternatively, we can consider not the product (around the loop) of expectation values $\langle c_\alpha^\dagger(\vec{x},t) c_\alpha(\vec{x}+\hat{e}_j,t)\rangle$, but the expectation value of the Wilson loop operator, the path-ordered product

$$W(\gamma) = \left\langle \prod_{(\vec{x},\vec{x}')\in\gamma} c_\alpha^\dagger(\vec{x},t) c_\alpha(\vec{x}',t)\right\rangle \tag{10.4}$$

where $(\vec{x}, \vec{x}')$ denotes a link of the lattice, with endpoints at $\vec{x}$ and $\vec{x}'$, which belongs to the *closed* path γ. The expectation value $\langle c_\alpha^\dagger(\vec{x}, t) c_\alpha(\vec{x} + \hat{e}_j, t)\rangle$ is not gauge-invariant. Accordingly, Elitzur's theorem implies that this expectation value is actually equal to zero. As a matter of fact, the solutions of the saddle-point equations are not unique. All the configurations which can be reached by means of a local gauge transformation from a given solution are solutions too. The saddle-point approximation violates this condition. The invariance is restored by fluctuations. The main effect of fluctuations is to rid the system of spurious states that violate gauge invariance. We will come back to this point shortly, when we discuss the spectrum of disordered spin states more generally.

How can we compute expectation values such as $W(\gamma)$ from a path integral written in terms of $\chi_j(\vec{x}, t)$ fields? Let us go back to the path integral for this system with the effective Lagrangian density of Eq. (8.58). I will discuss only the simpler $n_c = 1$ case. Let us shift the $\mathcal{A}_0(\vec{x}, t)$ and $\chi_j(\vec{x}, t)$ variables each by a fixed, but arbitrary, amount $\tilde{\mathcal{A}}_0(\vec{x}, t)$ and $\tilde{\chi}_j(\vec{x}, t)$. This is essentially a mathematical device to compute expectation values involving Fermi field currents. We can regard the $\tilde{\mathcal{A}}_0$ and $\tilde{\chi}_j$ as external sources in terms of which the shifted Lagrangian density, $\mathcal{L}'$, reads (for $n_c = 1$)

$$\begin{aligned} \mathcal{L}' = c_\alpha^\dagger(x)(i\,\partial_t + \mu)c_\alpha(x) + \left(\mathcal{A}_0(x) + \tilde{\mathcal{A}}_0(x)\right)\left(c_\alpha^\dagger(x)c_\alpha(x) - \frac{N}{2}\right) \\ - \frac{N}{J}|\chi_j(x)|^2 + c_\alpha^\dagger(\vec{x}, t)\big(\chi_j(\vec{x}, t) + \tilde{\chi}_j(\vec{x}, t)\big)c_\alpha(\vec{x} + \hat{e}_j, t) + \text{h.c.} \end{aligned} \tag{10.5}$$

where we recall that according to Eq. (8.59) $\tilde{\chi}_j(\vec{x}, t) = \tilde{\chi}_{-j}^*(\vec{x} + \hat{e}_j, t)$. Since $\tilde{\chi}_j(\vec{x}, t)$ couples to the term for hopping from site $\vec{x} + \hat{e}_j$ to site $\vec{x}$, it is clear that the functional differentiation of the action S by $\tilde{\chi}_j(\vec{x}, t)$ yields

$$\frac{\delta S}{\delta \tilde{\chi}_j(\vec{x}, t)} = \sum_{\alpha=1}^{N} \left(c_\alpha^\dagger(\vec{x}, t)c_\alpha(\vec{x} + \hat{e}_j, t)\right) \tag{10.6}$$

while functional differentiation with respect to $\tilde{\mathcal{A}}_0(\vec{x}, t)$ gives

$$\frac{\delta S}{\delta \tilde{\mathcal{A}}_0(\vec{x}, t)} = \sum_{\alpha=1}^{N} c_\alpha^\dagger(\vec{x}, t)c_\alpha(\vec{x}, t) - \frac{N}{2} = 0 \tag{10.7}$$

as follows from the constraint of Eq. (8.56).

Thus, by computing functional derivatives we can compute the desired expectation values. For instance,

$$\frac{\delta Z}{\delta \tilde{\chi}_j(\vec{x}, t)} = \int \mathcal{D}\chi\, \mathcal{D}\mathcal{A}_0\, \mathcal{D}c^\dagger\, \mathcal{D}c\; e^{iS}i\, \frac{\delta S}{\delta \tilde{\chi}_j(\vec{x}, t)} \tag{10.8}$$

and

$$-\frac{i}{Z}\frac{\delta Z}{\delta \tilde{\chi}_j(\vec{x},t)} = \left\langle \frac{\delta S}{\delta \tilde{\chi}_j(\vec{x},t)} \right\rangle = \left\langle \sum_{\alpha=1}^{N} \left(c_\alpha^\dagger(\vec{x},t) c_\alpha(\vec{x}+\hat{e}_j,t) \right) \right\rangle \tag{10.9}$$

In particular, the path-ordered product $W(\gamma)$ can also be computed. Let p label the pth link on the path γ and $\tilde{\chi}(p)$ the corresponding $\chi_j(\vec{x},t)$, i.e. the link $(\vec{x}, \vec{x} + \hat{e}_j, t)$ is the pth link of the path starting at some arbitrary site $\vec{x}_0$ on the path. We can write, for a closed path γ with perimeter $L(\gamma)$,

$$\begin{aligned}\frac{1}{Z}\frac{\delta^L Z}{\delta \tilde{\chi}(1)\ldots\delta\tilde{\chi}(L(\gamma))} &= i^{L(\gamma)} \left\langle \prod_{p=1}^{L(\gamma)} \left(\sum_{\alpha=1}^{N} c_\alpha^\dagger(\vec{x},t) c_\alpha(\vec{x}+\hat{e}_j,t) \right) \right\rangle \\ &\equiv i^{L(\gamma)} W(\gamma)\end{aligned} \tag{10.10}$$

On the other hand, the $\chi_j(\vec{x},t)$ degrees of freedom can be shifted without affecting the value of the partition function:

$$\begin{aligned}\mathcal{A}_0(\vec{x},t) &= \mathcal{A}_0'(\vec{x},t) - \tilde{\mathcal{A}}_0(\vec{x},t)\\ \chi_j(\vec{x},t) &= \chi_j'(\vec{x},t) - \tilde{\chi}_j(\vec{x},t)\end{aligned} \tag{10.11}$$

After this has been done, all the information about the sources is in the quadratic term of $\mathcal{L}'$,

$$\begin{aligned}\mathcal{L}' = & c_\alpha^\dagger(x)(i\,\partial_t + \mu)c_\alpha(x) + \mathcal{A}_0'(x)\left(c_\alpha^\dagger(x)c_\alpha(x) - \frac{N}{2}\right)\\ & - \frac{N}{J}\left(\chi_j'(x) - \tilde{\chi}_j(x)\right)\left(\chi_j'^*(x) - \tilde{\chi}_j^*(x)\right)\\ & + c_\alpha^\dagger(\vec{x},t)\chi_j'(\vec{x},t)c_\alpha(\vec{x}+\hat{e}_j,t) + c_\alpha^\dagger(\vec{x}+\hat{e}_j,t)\chi_j'^*(\vec{x},t)c_\alpha(\vec{x},t)\end{aligned} \tag{10.12}$$

Thus,

$$\left. \left\langle \frac{\delta S}{\delta \tilde{\chi}_j(\vec{x},t)} \right\rangle \right|_{\tilde{\chi}_j=0} = \left\langle \sum_{\alpha=1}^{N} c_\alpha^\dagger(\vec{x},t) c_\alpha(\vec{x}+\hat{e}_j,t) \right\rangle = \frac{2N}{J}\langle \chi_j'^*(\vec{x},t)\rangle \tag{10.13}$$

Similarly, $W(\gamma)$ is given by

$$W(\gamma) = \left\langle \prod_{p=1}^{L(\gamma)} \frac{2N}{J}\chi^*(p) \right\rangle \tag{10.14}$$

Notice that there is no quadratic term in the action for $\mathcal{A}_0$. Thus, all functional derivatives of Z with respect to $\tilde{\mathcal{A}}_0$ are identically equal to zero:

$$\frac{\delta Z}{\delta \tilde{\mathcal{A}}_0} = 0 \tag{10.15}$$

This merely means that the constraint

$$\sum_{\alpha=1}^{N} c_\alpha^\dagger(\vec{x},t)c_\alpha(\vec{x},t) - \frac{N}{2} = 0 \tag{10.16}$$

is strictly enforced at all times and everywhere.

The quadratic terms in $\mathcal{L}'$ express the fluctuations of the amplitude $\rho_j(\vec{x},t)$ of $\chi_j(\vec{x},t)$ but not of its phase, the gauge field $\mathcal{A}_j(\vec{x},t)$. Thus, if we imagine a state with $\bar{\rho}_j(\vec{x},t) = \bar{\rho}$, we will still have the fluctuations of the gauge fields $\mathcal{A}_j$ to deal with. The path-ordered product is, in this approximation, equal to

$$W(\gamma) \approx \left(\frac{2N}{J}\bar{\rho}\right)^{L(\gamma)} \left\langle e^{i\sum_{l\in\gamma}\bar{\mathcal{A}}(l)}\right\rangle \tag{10.17}$$

This last expectation value, $\langle \exp(i\sum_{l\in\gamma}\bar{\mathcal{A}}(l))\rangle$, is the Wilson loop operator which we have already discussed in Chapter 9. It was introduced in the context of gauge theories of strong interactions (in particle physics) as a way to measure the interaction between quarks. In the present context, it measures the interactions between ideal *static* spinons that are carried around the loop γ. The interaction is mediated by the fluctuations of the field χ_j. The relevance of Wilson loops for flux spin states was first emphasized by Wiegmann (1988).

If the saddle-point approximation were exact, the fluctuations of the gauge field $\mathcal{A}_j$ could be neglected. In this case, $W(\gamma)$ would yield the result

$$W(\gamma) \approx \left(\frac{2N}{J}\bar{\rho}\right)^{L(\gamma)} \left\langle e^{i\sum_{l\in\gamma}\bar{\mathcal{A}}(l)}\right\rangle \tag{10.18}$$

Let $a(\gamma)$ be the *area* of the lattice enclosed by the path γ. Using Stokes' theorem, we would then get

$$W(\gamma) \approx \left(\frac{2N}{J}\bar{\rho}\right)^{L(\gamma)} \left\langle e^{ia(\gamma)\bar{\mathcal{B}}}\right\rangle \tag{10.19}$$

where $\bar{\mathcal{B}}$ is the flux per plaquette. If we denote by $\delta\mathcal{A}_j(\vec{x},t)$ the fluctuating part of the gauge field $\mathcal{A}_j(\vec{x},t)$, i.e. the deviation from the saddle-point configuration, we get for $W(\gamma)$

$$W(\gamma) \approx \left(\frac{2N}{J}\bar{\rho}\right)^{L(\gamma)} \left\langle e^{ia(\gamma)\bar{\mathcal{B}}} e^{i\sum_{l\in\gamma}\delta\mathcal{A}(l)}\right\rangle \tag{10.20}$$

where the expectation value involves only the fluctuating pieces. It has been argued that flux phases can generally be defined as phases in which $\ln W(\gamma)$ has an imaginary part that scales like the area enclosed by the loop γ (Wiegmann, 1988; Wen *et al.*, 1989). It is also constructive to consider the situation in which an extra fermion, i.e. a spinon, is added at some site $\vec{x}$ and another one is removed from site $\vec{x}'$. The constraints at $\vec{x}$ and $\vec{x}'$ are

$$\sum_{\alpha=1}^{N} c_{\alpha}^{\dagger}(\vec{y}, t) c_{\alpha}(\vec{y}, t) - \frac{N}{2} = \delta_{\vec{y}, \vec{x}} - \delta_{\vec{y}, \vec{x}'} \tag{10.21}$$

This means that two extra factors enter into the partition function. They have the form $\exp(\pm i \int dt\, \mathcal{A}_0(\vec{x}, t))$. We can close the paths both in the remote past and in the remote future (assuming an adiabatic switching on and off, i.e. a smooth path) and write the extra contribution as an integral over a closed path γ_t,

$$W(\gamma_t) \propto \left\langle e^{i \sum_{l \in \gamma_t} \mathcal{A}(l)} \right\rangle \Big|_{\gamma_t} \tag{10.22}$$

where γ_t stands for a space-time closed loop (see Fig. 10.1) of time span τ and spatial extent R. Thus $W(\gamma_t)$ measures the change of the ground-state energy $\Delta E(\vec{x})$ of the system as a result of the presence of the static spinons,

$$W(\gamma_t) = e^{i \tau \Delta E(\vec{x})} \tag{10.23}$$

This expression is valid for $\tau \gg R$. Thus the effective interaction between static sources $V_{\text{eff}}(\vec{x})$ is

$$V_{\text{eff}}(\vec{x}) = \Delta E(\vec{x}) = \lim_{\tau \to \infty} \left[-\frac{i}{\tau} \ln(W(\gamma_t)) \right] \tag{10.24}$$

Notice that there is no classical flux associated with space-time loops γ. Thus $W(\gamma_t)$ does not necessarily exhibit the area law of Eq. (10.19) associated with the flux phase which we found for space loops. In fact, both $W(\gamma_t)$ and the fluctuating

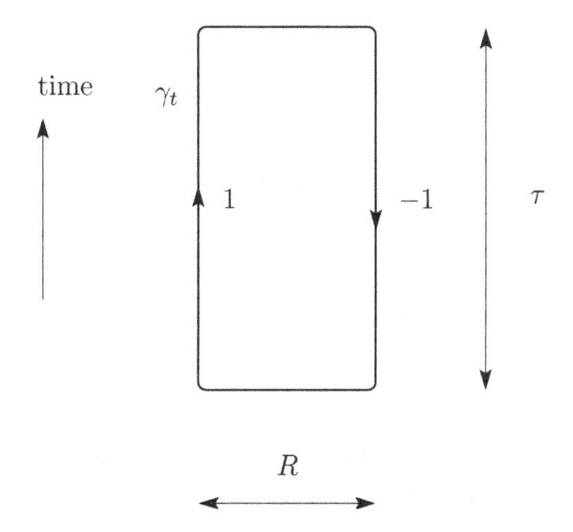

Figure 10.1 A space-time loop γ_t of size $R \times \tau$, showing a static spinon $(+1)$ and anti-spinon (-1) separated by a distance R.

components of the space-like loops have a phase that decays like the *perimeter* of the loop, not its area. This is so because, unlike the case of confining gauge theories without dynamical matter fields, we have only gauge fields associated with a dynamical matter field. The gauge fields themselves do not have any other dynamics of their own. We will return to this important point in the next section.

There is an alternative way of understanding the products over closed loops. Consider the case of three spins, $\vec{S}(1)$, $\vec{S}(2)$, and $\vec{S}(3)$. Let us form the mixed product $\hat{E}_{123}$ which Wen, Wilczek, and Zee call the chiral operator,

$$\hat{E}_{123} \equiv \vec{S}(1) \cdot \left(\vec{S}(2) \times \vec{S}(3) \right) \tag{10.25}$$

Under time reversal $\hat{T}$ we have

$$\hat{T}^{-1} \vec{S} \hat{T} = -\vec{S} \tag{10.26}$$

Thus $\hat{E}_{123}$ is *odd* under $\hat{T}$,

$$\hat{T}^{-1} \hat{E}_{123} \hat{T} = -\hat{E}_{123} \tag{10.27}$$

Similarly, under parity, $\hat{P}$, which in two space dimensions is the same as reflection through a link, we have

$$\begin{aligned} \hat{P}^{-1} \hat{E}_{123} \hat{P} &= \vec{S}(1) \cdot \left(\vec{S}(3) \times \vec{S}(2) \right) \\ &= +E_{132} = -\hat{E}_{123} \end{aligned} \tag{10.28}$$

where we have exchanged sites 2 and 3, keeping site 1 fixed. Thus, for the three spins, parity implies turning an even permutation of the three spins into an odd permutation.

Let us now write $\hat{E}_{123}$ in terms of the link operators $\hat{\chi}(i,j) \equiv c_\alpha^\dagger(i) c_\alpha(j)$. Explicitly one finds (Wen *et al.*, 1989)

$$\hat{E}_{123} = \frac{i}{4} \left(\hat{\chi}(1,2)\hat{\chi}(2,3)\hat{\chi}(3,1) - \hat{\chi}(1,3)\hat{\chi}(3,2)\hat{\chi}(2,1) \right) \tag{10.29}$$

If we consider now four spins, 1, 2, 3, and 4, we get

$$\begin{aligned} &\hat{\chi}(1,2)\hat{\chi}(2,3)\hat{\chi}(3,4)\hat{\chi}(4,1) - \hat{\chi}(1,4)\hat{\chi}(4,3)\hat{\chi}(3,2)\hat{\chi}(2,1) \\ &\quad = 2i \left(-\hat{E}_{123} - \hat{E}_{134} - \hat{E}_{124} + \hat{E}_{234} \right) \end{aligned} \tag{10.30}$$

Thus, if $\hat{E}_{123}$ acquires an expectation value, then we should expect the spatial Wilson loops implied by Eq. (10.29) and Eq. (10.30) to exhibit a now-trivial phase (which can be regarded as a Berry phase). At the level of the saddle-point approximation, we expect

$$
\begin{aligned}
\langle \hat{E}_{123} \rangle &= \frac{i}{4}\langle \hat{\chi}(1,2)\hat{\chi}(2,3)\hat{\chi}(3,1) - \hat{\chi}(1,3)\hat{\chi}(3,2)\hat{\chi}(2,1)\rangle \\
&\approx \frac{i}{4}\left(\frac{2N\bar{\rho}}{J}\right)^3 \left(e^{i\bar{\mathcal{B}}_\Delta} - e^{-i\bar{\mathcal{B}}_\Delta}\right) \\
&= -\frac{1}{2}\left(\frac{2N\bar{\rho}}{J}\right)^3 \sin(\bar{\mathcal{B}}_\Delta)
\end{aligned}
\tag{10.31}
$$

where $\bar{\mathcal{B}}_\Delta$ is the flux through the triangle with vertices at sites 1, 2, and 3. Thus, in a chiral phase, $\hat{E}_{123}$ should have a non-zero expectation value. Please notice that for the non-chiral flux phase, $\bar{\mathcal{B}}_\Delta = \pi$, $\langle \hat{E}_{123} \rangle = 0$ since in this case time-reversal invariance is not broken.

For a system with just three spins one-half we can get a very simple interpretation of this statement. For three spins one-half, the Hilbert space is $2^3 =$ 8-dimensional. The total spin is $\vec{S} = \vec{S}(1) + \vec{S}(2) + \vec{S}(3)$. The quadratic Casimir operator $\vec{S}^2$ and, say, S_3 commute with each other. What is important is that they also commute with $\hat{E}_{123}$. Thus, $\vec{S}^2$, S_3, and $\hat{E}_{123}$ can be diagonalized simultaneously. I will refer to the eigenvalues of $\hat{E}_{123}$ as the *chirality* χ of the state. The states of the three spins will thus be labeled accordingly by $|S, M; \chi\rangle$, where S is the spin quantum number, M is the total spin z-projection, and χ is the chirality. The total spin S is either 1/2 or 3/2. The spin-3/2 sector can be obtained trivially by applying the lowering operator S^- to the highest-weight state $|\uparrow\uparrow\uparrow\rangle$:

$$
\begin{aligned}
|\uparrow\uparrow\uparrow\rangle &= |\tfrac{3}{2}, \tfrac{3}{2}; 0\rangle \\
|\tfrac{3}{2}, \tfrac{3}{2} - M; 0\rangle &= (S^-)^M |\tfrac{3}{2}, \tfrac{3}{2}; 0\rangle
\end{aligned}
\tag{10.32}
$$

The state $|\uparrow\uparrow\uparrow\rangle$ has zero chirality since it is invariant under a permutation of any pair of spins. In terms of raising and lowering operators $S^\pm$ and S_3, $\hat{E}_{123}$ has the form

$$
\begin{aligned}
\hat{E}_{123} = \frac{i}{2}\big(&- S^-(1)S^+(2)S_3(3) + S^+(1)S^-(2)S_3(3) \\
&+ S^-(1)S_3(2)S^+(3) - S^+(1)S_3(2)S^-(3) \\
&- S_3(1)S^-(2)S^+(3) + S_3(1)S^+(2)S^-(3)\big)
\end{aligned}
\tag{10.33}
$$

Clearly

$$
\hat{E}_{123}|\tfrac{3}{2}, \tfrac{3}{2}; \chi\rangle = \hat{E}_{123}|\uparrow\uparrow\uparrow\rangle = 0 \tag{10.34}
$$

which proves that $\chi_{\uparrow\uparrow\uparrow} = 0$. From the form of $\hat{E}_{123}$ in Eq. (10.34) we see that all the states in the same multiplet defined by S and M have the same chirality.

There are two, orthogonal, sectors with $S = 1/2$, $M = \pm 1/2$. They differ by their chirality χ. Consider the state $|+\rangle$, defined by the linear superposition

$$|+\rangle = \frac{1}{\sqrt{3}}\left(|\uparrow\uparrow\downarrow\rangle + |\uparrow\downarrow\uparrow\rangle e^{i\frac{2\pi}{3}} + |\downarrow\uparrow\uparrow\rangle e^{-i\frac{2\pi}{3}}\right) \tag{10.35}$$

This state $|+\rangle$ is an eigenstate of $\hat{E}_{123}$ with eigenvalue χ_+ given by

$$\hat{E}_{123}|+\rangle = -\frac{1}{2}\sin\left(\frac{2\pi}{3}\right)|+\rangle \tag{10.36}$$

Thus $\chi_+ = -\frac{1}{2}\sin(2\pi/3)$. Similarly the state $|-\rangle$,

$$|-\rangle = \frac{1}{\sqrt{3}}\left(|\uparrow\uparrow\downarrow\rangle + |\uparrow\downarrow\uparrow\rangle e^{-i\frac{2\pi}{3}} + |\downarrow\uparrow\uparrow\rangle e^{i\frac{2\pi}{3}}\right) \tag{10.37}$$

has eigenvalue $\chi_- = +\frac{1}{2}\sin(2\pi/3)$. Both states, $|+\rangle$ and $|-\rangle$, have $S_3 = +\frac{1}{2}$. Thus we denote $|\pm\rangle$ as the states $|\frac{1}{2}, \frac{1}{2}; \pm\rangle$. Similarly the states with spin down can also have either chirality. These two remaining states are denoted by $|\frac{1}{2}, -\frac{1}{2}; \pm\rangle$.

The most singlet-like states, i.e. those with smallest spins, can thus be arranged to have non-zero chirality. By inspection of Eq. (10.35) and Eq. (10.37), we see that a state with non-zero chirality is a state in which a spin down moves around the triangle with a non-zero angular momentum $l = \pm 1$. Thus, a state with non-zero chirality is a state in which there is a non-zero spin current since a down spin is being transported, at a fixed rate, around the triangle.

For a macroscopic system, we can picture a situation in which $\langle \hat{E}_{123}\rangle$ is different from zero everywhere, as in a flux state, by saying that flux states are states in which there are non-vanishing *orbital* spin currents around every elementary plaquette. If we demand that the flux $\vec{B}$ be uniform throughout the system, we are in fact requiring that the state should exhibit an *orbital ferromagnetism* of some sort (Volovik, 1988).

There is one interesting analogy here with the behavior of orbital angular momentum in the A phase of ^{3}He. As is well known, ^{3}He becomes a superfluid by forming bound states of two ^{3}He atoms. The bound state has total spin $S = 1$ (triplet) and orbital angular momentum $l = 1$ (p-wave) (Leggett, 1975). In ^{3}He A, the orbital angular-momentum vector $\vec{l}$ and the spin $\vec{S}$ of the state are orthogonal to each other. In a thin-film geometry, the orbital angular momenta $\vec{l}$ are all parallel to each other and perpendicular to the surface of the film. This superfluid has orbital ferromagnetism.

10.2 Mean-field theory of chiral spin liquids

Let us consider the mean-field theory, i.e. the large-N limit, of the frustrated Heisenberg antiferromagnet on a square lattice. We have two coupling constants: J_1 (for nearest neighbors) and J_2 (for next-nearest neighbors). We considered this problem in Section 8.1, in which we discussed the effects of J_2 on the Néel state.

The effective Lagrangian Eq. (8.41) can be easily generalized in order to include the effects of the J_2 coupling. All we have to do is decouple the next-nearest-neighbor term using the same procedure as for the nearest-neighbor term (Wen *et al.*, 1989). The Lagrangian density now is, including both J_1 and J_2,

$$\begin{aligned}\mathcal{L}' = & c_\sigma^\dagger(x)(i\,\partial_t + \mu)c_\sigma(x) + \mathcal{A}_0(x)\left(c_\alpha^\dagger(x)c_\sigma(x) - \frac{N}{2}\right)\\ & - \frac{N}{J_1}|\chi_j(x)|^2 - \frac{N}{J_2}|\chi_{j'}(x)|^2\\ & + c_\sigma^\dagger(\vec{x},t)\chi_j(\vec{x},t)c_\sigma(\vec{x}+\hat{e}_j,t) + \text{h.c.}\\ & + c_\sigma^\dagger(\vec{x},t)\chi_{j'}(\vec{x},t)c_\sigma(\vec{x}+\hat{e}_1+j'\hat{e}_2,t) + \text{h.c.}\end{aligned} \tag{10.38}$$

where $j' = \pm$.

The saddle-point procedure can be carried out along very similar lines. At this level, we assume that the amplitudes $\bar{\chi}_j(\vec{x},t)$ and $\bar{\chi}_{j'}(\vec{x},t)$ are constant in time and as uniform as possible in space. If we choose the gauge of Eq. (8.74), as in our earlier discussion of the flux phase, we get (Wen *et al.*, 1989)

$$\begin{aligned}\bar{\chi}_1(\vec{x}) &= \bar{\rho}e^{+i\frac{\pi}{2}}, & \bar{\chi}_2(\vec{x}) &= \bar{\rho}e^{-\sigma i\frac{\pi}{2}}\\ \bar{\chi}_+(\vec{x}) &= \bar{\lambda}e^{+\sigma i\frac{\pi}{2}}, & \bar{\chi}_-(\vec{x}) &= \bar{\lambda}e^{+\sigma i\frac{\pi}{2}}\end{aligned} \tag{10.39}$$

with $\sigma = (-1)^{x_1}$. Notice that the flux per plaquette $\mathcal{B}_{\text{plaquette}} = \pi$, but for the triangles we have $\mathcal{B}_\Delta = +\pi/2$ for $\bar{\lambda} > 0$ and $\mathcal{B}_\Delta = -\pi/2$ for $\bar{\lambda} < 0$ (see Fig. 10.2).

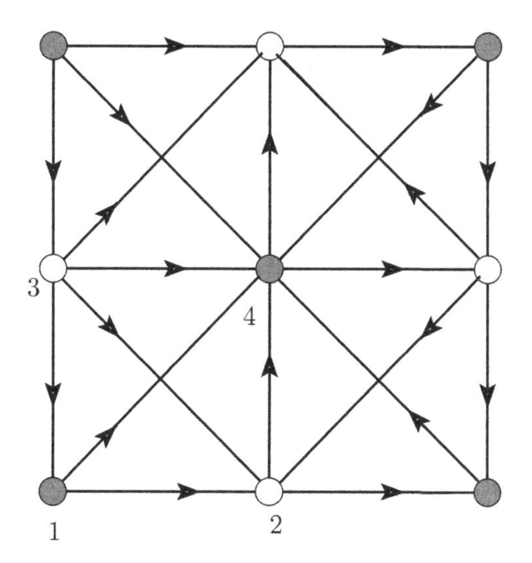

Figure 10.2 Gauge-field conventions for a chiral spin state on a frustrated square lattice. The lower-left corner is an even–even site. The arrows on the links represent a phase of $\pi/2$. The flux on a plaquette is π if the lower-left corner is on an even column. Otherwise it is $-\pi$.

Thus, this state is chiral. At this level of approximation, the spinons behave like fermions moving on a frustrated lattice with the amplitudes listed in Eq. (10.39). Since the flux on the triangles is $\pm\pi/2$, some of the amplitudes must be complex no matter what gauge we choose. Thus, the effective one-particle Hamiltonian which controls the motion of spinons is complex (but still hermitian!). This means that time-reversal invariance (and parity) are broken. Since in this system we do not have any terms that explicitly break time-reversal invariance, what we are looking for is states with *spontaneously broken time-reversal invariance* and *parity.*

Using once again the notation of Section 8.5, and the symbol Δ_i^+, with $i = 1, 2$, for the symmetric lattice-difference operator $\Delta_i^+ f(\vec{x}) \equiv f(\vec{x}+\hat{e}_i) - f(\vec{x}-\hat{e}_i)$, we can write down the equations of motion Eq. (8.76) including the effects of $\bar{\lambda}$. The new equations of motion are

$$
\begin{aligned}
i\,\partial_t f_\sigma^{(1)}(\vec{x}) =& -i\bar{\rho}\Delta_1 f_\sigma^{(2)}(\vec{x}) + i\bar{\rho}\Delta_2 f_\sigma^{(3)}(\vec{x}) \\
& - i\bar{\lambda}\Delta_2^+ f_\sigma^{(4)}(\vec{x}+\hat{e}_1) - i\bar{\lambda}\Delta_2^+ f_\sigma^{(4)}(\vec{x}-\hat{e}_1) \\
i\,\partial_t f_\sigma^{(2)}(\vec{x}+\hat{e}_1) =& -i\bar{\rho}\Delta_1 f_\sigma^{(1)}(\vec{x}+\hat{e}_1) - i\bar{\rho}\Delta_2 f_\sigma^{(4)}(\vec{x}+\hat{e}_1) \\
& + i\bar{\lambda}\Delta_2^+ f_\sigma^{(3)}(\vec{x}+2\hat{e}_1) + i\bar{\lambda}\Delta_2^+ f_\sigma^{(3)}(\vec{x}) \\
i\,\partial_t f_\sigma^{(3)}(\vec{x}+\hat{e}_2) =& -i\bar{\rho}\Delta_1 f_\sigma^{(4)}(\vec{x}+\hat{e}_2) + i\bar{\rho}\Delta_2 f_\sigma^{(1)}(\vec{x}+\hat{e}_2) \\
& - i\bar{\lambda}\Delta_2^+ f_\sigma^{(2)}(\vec{x}+\hat{e}_2+\hat{e}_1) - i\bar{\lambda}\Delta_2^+ f_\sigma^{(2)}(\vec{x}+\hat{e}_2-\hat{e}_1) \\
i\,\partial_t f_\sigma^{(4)}(\vec{x}+\hat{e}_1+\hat{e}_2) =& -i\bar{\rho}\Delta_1 f_\sigma^{(3)}(\vec{x}+\hat{e}_1+\hat{e}_2) - i\bar{\rho}\Delta_2 f_\sigma^{(2)}(\vec{x}+\hat{e}_1+\hat{e}_2) \\
& + i\bar{\lambda}\Delta_2^+ f_\sigma^{(1)}(\vec{x}+2\hat{e}_1+\hat{e}_2) + i\bar{\lambda}\Delta_2^+ f_\sigma^{(1)}(\vec{x}+\hat{e}_2)
\end{aligned} \tag{10.40}
$$

In Fourier space, Eq. (10.40) becomes

$$
\begin{aligned}
i\,\partial_t f_\sigma^{(1)}(\vec{p}) &= 2\bar{\rho}\sin p_1\, f_\sigma^{(2)}(\vec{p}) - 2\bar{\rho}\sin p_2\, f_\sigma^{(3)}(\vec{p}) - 4i\bar{\lambda}\cos p_1\cos p_2\, f_\sigma^{(4)}(\vec{p}) \\
i\,\partial_t f_\sigma^{(2)}(\vec{p}) &= 2\bar{\rho}\sin p_1\, f_\sigma^{(1)}(\vec{p}) + 2\bar{\rho}\sin p_2\, f_\sigma^{(4)}(\vec{p}) + 4i\bar{\lambda}\cos p_1\cos p_2\, f_\sigma^{(3)}(\vec{p}) \\
i\,\partial_t f_\sigma^{(3)}(\vec{p}) &= 2\bar{\rho}\sin p_1\, f_\sigma^{(4)}(\vec{p}) - 2\bar{\rho}\sin p_2\, f_\sigma^{(1)}(\vec{p}) - 4i\bar{\lambda}\cos p_1\cos p_2\, f_\sigma^{(2)}(\vec{p}) \\
i\,\partial_t f_\sigma^{(4)}(\vec{p}) &= 2\bar{\rho}\sin p_1\, f_\sigma^{(3)}(\vec{p}) + 2\bar{\rho}\sin p_2\, f_\sigma^{(2)}(\vec{p}) + 4i\bar{\lambda}\cos p_1\cos p_2\, f_\sigma^{(1)}(\vec{p})
\end{aligned} \tag{10.41}
$$

As with Eq. (8.80) and Eq. (8.81), we define the spinors $u_\sigma^{(a)}$ and $v_\sigma^{(a)}$ $(a = 1, 2)$

$$u_\sigma^{(1)}(\vec{p}) = f_\sigma^{(1)}(\vec{p}) + f_\sigma^{(2)}(\vec{p}) \tag{10.42}$$

$$u_\sigma^{(2)}(\vec{p}) = f_\sigma^{(3)}(\vec{p}) - f_\sigma^{(4)}(\vec{p}) \tag{10.43}$$

and

$$v_\sigma^{(1)}(\vec{p}) = f_\sigma^{(3)}(\vec{p}) + f_\sigma^{(4)}(\vec{p}) \tag{10.44}$$

$$v_\sigma^{(2)}(\vec{p}) = f_\sigma^{(1)}(\vec{p}) - f_\sigma^{(2)}(\vec{p}) \tag{10.45}$$

In matrix notation, we can now write $(a, b = 1, 2)$

$$i\partial_t u_\sigma^{(a)}(\vec{p}) = \left(2\bar{\rho}\sin p_1\sigma_3 - 2\bar{\rho}\sin p_2\sigma_1 - 4\bar{\lambda}\cos p_1\cos p_2\sigma_2\right)_{ab} u_\sigma^{(b)}(\vec{p}) \quad (10.46)$$

where σ_1, σ_2, and σ_3 are the three Pauli matrices. The other spinor, $v_\sigma^{(a)}(\vec{p})$, obeys the *same* equation.

We can also write Eq. (10.46) in a Dirac form by defining the α and γ matrices through

$$\alpha_1 \equiv \gamma_0\gamma_1 \equiv +\sigma_3, \quad \alpha_2 \equiv \gamma_0\gamma_2 \equiv -\sigma_1, \quad \beta \equiv \gamma_0 \equiv -\sigma_2 \quad (10.47)$$

In this notation, the equation of motion Eq. (10.46) takes the Dirac form

$$i\,\partial_t u_\sigma^{(a)}(\vec{p}) = \left(2\bar{\rho}\sum_{i=1,2}\sin p_i\alpha_i + 4\bar{\lambda}\cos p_1\cos p_2\beta\right)_{ab} u_\sigma^{(b)}(\vec{p}) \quad (10.48)$$

Thus, in the small-momentum limit $|\vec{p}| \to 0$, we obtain the equation for *two* Dirac spinors, u_σ and v_σ, in the continuum with the *same* Fermi velocity $v_{\rm F} = 2a_0\bar{\rho}$ and, more importantly, the *same effective mass* $m_{\rm c} = \bar{\lambda}/(\bar{\rho}^2 a_0^2)$. Notice that *both* species (or "valleys" in the jargon of graphene), u_σ *and* v_σ, have the same *sign* of the effective mass $m_{\rm c}$. The one-particle Hamiltonian

$$H_{\rm chiral}(\vec{p}) = 2\bar{\rho}\sum_{i=1,2}\sin p_i\;\alpha_i + 4\bar{\lambda}\cos p_1\cos p_2\;\beta \quad (10.49)$$

is *complex* (and hermitian) since all three Pauli matrices are present. This fact is, in turn, the result of the breaking of time-reversal invariance. We will see in another section of this chapter that this result gives rise to a *parity anomaly*, which greatly changes the behavior of the low-lying excitations.

The eigenvalues of $H_{\rm chiral}$ are

$$\epsilon(\vec{p}) = \pm\sqrt{4\bar{\rho}^2(\sin^2 p_1 + \sin^2 p_2) + 16\bar{\lambda}^2\cos^2 p_1\cos^2 p_2} \quad (10.50)$$

This is what we found for the flux phase, Eq. (8.86), except for a mass term proportional to the next-nearest amplitude $\bar{\lambda}$. The two branches nearly touch at $(p_1, p_2) = (0, 0)$.

Thus far, we have not discussed energetics. Wen, Wilczek, and Zee (Wen *et al.*, 1989) studied this problem in some detail. They found that, as J_2 increases, the energy of the chiral state drops below that of the flux state and becomes close to the energy of the ordered dimer state. For the square lattice, even in the classically frustrated limit $J_1 = 2J_2$, it appears that the dimer states are still preferred, although not by much. Furthermore, at least in the large-N limit, the Néel states are not favored when $J_1 \approx 2J_2$. There is numerical evidence, from the exact diagonalization of small clusters of up to 30 sites, that the Néel states are not favored for $J_1 \approx 2J_2$. In fact, at least for such small systems, the columnar states appear to be the ground states in this regime (Dagotto and Moreo, 1989). Thus, although the chiral states are locally stable, they do not appear to be the global minimum of energy. But it is quite conceivable to imagine slight modifications of the Hamiltonian that will drive the mean-field ground-state energy of the chiral states down and make them a global minimum. Results from more recent density-matrix RG studies strongly indicate that there is a time-reversal-invariant $\mathbb{Z}_2$ spin-liquid state in the strong-frustration regime (Jiang *et al.*, 2012).

What appears to be more serious is the fact that the chiral mean-field theory has low-lying excited states, namely the spinons, which are not gauge-invariant. The removal of gauge-non-invariant states is likely to raise the energy of the ground state. We will come back to these issues in the next section.

Finally, it is instructive to consider the effects of a Peierls gap, i.e. the gap which appears in the presence of a columnar state. This problem was studied by Dombre and Kotliar (1989). Consider a columnar state of the type depicted in Fig. 8.12(a). There are four such states. With the choice of gauge, Eq. (10.39), the simplest case to consider has a columnar state with the "dimers" on the y axis and the columns running along the x axis. We can represent such a state by a *modulation* of the *amplitude* $\bar{\rho}_j(\vec{x})$ such that $\bar{\rho}_j(\vec{x})$ equals $\bar{\rho} + \delta\bar{\rho}$ if there is a dimer in the bond $(\vec{x}, \vec{x} + \hat{e}_j)$ and equals $\bar{\rho} - \delta\bar{\rho}$ if there isn't a dimer in that bond. The next-nearest-neighbor hopping terms have the same form as in Eq. (10.39). Thus, we can consider the competition between the Peierls state and the chiral state. We will see that, unlike the chiral state, which breaks parity and thus leads to a complex Hamiltonian, the Peierls state does not break parity. If we assume that the selected Peierls state has the (vertical) dimers with their lower endpoints on *even* rows, the modified equations of motion are

$$
\begin{aligned}
i\,\partial_t f_\sigma^{(1)}(\vec{x}) =& -i\bar{\rho}\Delta_1 f_\sigma^{(2)}(\vec{x}) + i\bar{\rho}\Delta_2 f_\sigma^{(3)}(\vec{x}) \\
&+ i\,\delta\bar{\rho}\,\Delta_2^+ f_\sigma^{(3)}(\vec{x}) \\
&- i\bar{\lambda}\Delta_2^+ f_\sigma^{(4)}(\vec{x}+\hat{e}_1) - i\bar{\lambda}\Delta_2^+ f_\sigma^{(4)}(\vec{x}-\hat{e}_1) \\
i\,\partial_t f_\sigma^{(2)}(\vec{x}+\hat{e}_1) =& -i\bar{\rho}\Delta_1 f_\sigma^{(1)}(\vec{x}+\hat{e}_1) - i\bar{\rho}\Delta_2 f_\sigma^{(4)}(\vec{x}+\hat{e}_1) \\
&- i\,\delta\bar{\rho}\,\Delta_2^+ f_\sigma^{(4)}(\vec{x}+\hat{e}_1) \\
&+ i\bar{\lambda}\Delta_2^+ f_\sigma^{(3)}(\vec{x}+2\hat{e}_1) + i\bar{\lambda}\Delta_2^+ f_\sigma^{(3)}(\vec{x}) \\
i\,\partial_t f_\sigma^{(3)}(\vec{x}+\hat{e}_2) =& -i\bar{\rho}\Delta_1 f_\sigma^{(4)}(\vec{x}+\hat{e}_2) + i\bar{\rho}\Delta_2 f_\sigma^{(1)}(\vec{x}+\hat{e}_2) \\
&- i\,\delta\bar{\rho}\,\Delta_2^+ f_\sigma^{(1)}(\vec{x}+\hat{e}_2) \\
&- i\bar{\lambda}\Delta_2^+ f_\sigma^{(2)}(\vec{x}+\hat{e}_2+\hat{e}_1) - i\bar{\lambda}\Delta_2^+ f_\sigma^{(2)}(\vec{x}+\hat{e}_2-\hat{e}_1) \\
i\,\partial_t f_\sigma^{(4)}(\vec{x}+\hat{e}_1+\hat{e}_2) =& -i\bar{\rho}\Delta_1 f_\sigma^{(3)}(\vec{x}+\hat{e}_1+\hat{e}_2) - i\bar{\rho}\Delta_2 f_\sigma^{(2)}(\vec{x}+\hat{e}_1+\hat{e}_2) \\
&+ i\,\delta\bar{\rho}\,\Delta_2^+ f_\sigma^{(2)}(\vec{x}+\hat{e}_1+\hat{e}_2) \\
&+ i\bar{\lambda}\Delta_2^+ f_\sigma^{(1)}(\vec{x}+2\hat{e}_1+\hat{e}_2) + i\bar{\lambda}\Delta_2^+ f_\sigma^{(1)}(\vec{x}+\hat{e}_2)
\end{aligned}
\tag{10.51}
$$

In terms of the spinors u_σ and v_σ of Eq. (10.43) and Eq. (10.45), we get a modified mass term, which is *different* for u_σ and v_σ.

The one-particle Hamiltonian now is

$$H_{\text{Peierls}}(\vec{p}) = 2\bar{\rho} \sum_{i=1,2} \sin p_i \, \alpha_i + \left(4\bar{\lambda} \cos p_1 \cos p_2 \pm 2\,\delta\bar{\rho} \cos p_2\right)\beta \tag{10.52}$$

where the $+$ $(-)$ sign stands for the u_σ (v_σ) spinor. Thus, the low-energy spectrum still looks like two massive Dirac fermions that are propagating at the same speed but with different masses. What matters here is that the *sign* of the mass term depends on the relative strengths of $\delta\bar{\rho}$ and $\bar{\lambda}$. Indeed, for $|\vec{p}| \to 0$, we find that $H_{\text{Peierls}}(\vec{p})$ takes the form

$$H_{\text{Peierls}}(\vec{p}) \approx 2\bar{\rho} \left(\sum_{i=1,2} \alpha_i p_i + (m_{\text{c}} \pm \delta m)\beta \right) \tag{10.53}$$

where m_{c} is the chiral mass and δm is the splitting ($a_0 \equiv 1$)

$$m_{\text{c}} = \frac{\bar{\lambda}}{\bar{\rho}^2}, \qquad \delta m = \frac{\delta\bar{\rho}}{2\bar{\rho}^2} \tag{10.54}$$

Hence, for $\delta m < m_{\text{c}}$, the u_α and v_α have different masses m_u and m_v ($m_u > m_v$, for $\delta\bar{\rho} > 0$), both with the *same* sign. Conversely, for $\delta m > m_{\text{c}}$, m_u and m_v are not only different but also have *opposite* signs. If $\bar{\lambda}$ is set to zero (i.e. there is no chiral state), there is a perfect symmetry. Thus, the *Peierls mass* does not lead to a complex Hamiltonian and consequently it does not break parity. We will see later in this chapter that the *relative sign* of the masses of the elementary excitations has very important consequences for the overall behavior of the system in the generalized flux states. The eigenvalues of H_{Peierls} are

$$\epsilon(\vec{p}) = \pm\sqrt{4\bar{\rho}^2(\sin^2 p_1 + \sin^2 p_2) + (4\bar{\lambda} \cos p_1 \cos p_2 \pm 2\,\delta\bar{\rho} \cos p_2)^2} \tag{10.55}$$

10.3 Fluctuations and flux phases

So far we have considered only flux states at the mean-field level and fluctuations have not been taken into account. We have already pointed out that this approach is not consistent, since the fluctuations of the gauge fields, unlike the fluctuations of the amplitude, are completely out of control.

We shall consider first amplitude fluctuations around a flux phase with flux π per plaquette. The Lagrangian density of Eq. (10.38) has degrees of freedom which, in addition to inducing both chiral and non-chiral mass terms in the low-energy sector of the theory, can effectively drive the system into a highly anisotropic state, a *dimer state*. Since we are interested in understanding how these different mechanisms compete with each other, it is convenient to parametrize the fluctuations of the bond lengths in such a way that these processes are most apparent. Thus, we are led to consider configurations in which the bond amplitudes vary slowly at the scale of the lattice spacing (I will refer to these processes as being uniform or unstaggered). In addition, there are fluctuations that vary rapidly at the scale of the lattice constant. These fast fluctuations induce scattering processes that mix different sublattices very strongly. We will refer to them as *staggered* amplitude fluctuations. Hence, the bond *amplitude* for the bond $(\vec{x}, \vec{x}+\hat{e}_j)$ has the form

$$\rho_j(\vec{x}) = \rho_j^{\mathrm{u}}(\vec{x}) + \rho_j^{\mathrm{s}}(\vec{x}) \tag{10.56}$$

where $\rho_j^{\mathrm{u}}(\vec{x})$ is the unstaggered (or uniform) amplitude and $\rho_j^{\mathrm{s}}(\vec{x})$ is the staggered amplitude. While $\rho_j^{\mathrm{u}}(\vec{x})$ is slowly varying, $\rho_j^{\mathrm{s}}(\vec{x})$ changes its sign from one bond to the next. Since we anticipate that the system may choose an average uniform bond length $\bar{\rho}_j$, we write $\rho_j^{\mathrm{u}}(\vec{x})$ and $\rho_j^{\mathrm{s}}(\vec{x})$ in the form

$$\begin{aligned} \rho_j^{\mathrm{u}}(\vec{x}) &= \bar{\rho}_j \left(1 + \delta\rho_j^{\mathrm{u}}(\vec{x})\right) \\ \rho_j^{\mathrm{s}}(\vec{x}) &= \delta\rho_j^{\mathrm{s}}(\vec{x}) \end{aligned} \tag{10.57}$$

Although these amplitudes vary very slowly and over long wavelengths, they can be significantly different from each other. Thus the effective Dirac fermions may have different Fermi velocities along the x_1 and x_2 directions. More importantly, since these generalized Heisenberg models do not have any intrinsic length scale, apart from the lattice constant itself, there is an essential "softness" in the system, which favors strong anisotropy. This can be clearly seen by writing down the spinon energy of such a state, which for a non-chiral state has the form (see Eq. (10.55))

$$\epsilon(\vec{p}) = \pm\sqrt{(2\rho_1^{\mathrm{u}} \sin p_1)^2 + (2\rho_2^{\mathrm{u}} \sin p_2)^2 + (2\,\delta\rho_2^{\mathrm{s}} \cos p_2)^2} \tag{10.58}$$

This energy can be made large and negative by setting

$$\delta\rho_1^{\mathrm{u}} = -1, \quad |\delta\rho_2^{\mathrm{s}}| = |\rho_2^{\mathrm{u}}| \tag{10.59}$$

which is the dimer limit. The symmetric amplitude $\delta\rho_j^{\mathrm{u}}$ cannot grow any larger than this without driving the total amplitude into negative values. Thus, this is the saturation limit. In this limit, the spin gap is infinitely large since all spinons are in singlet bond states one lattice spacing long, namely the valence-bond states. The fluctuations of the gauge fields only cause dimer rearrangements, as in our discussion of the QDM. This phase does not break time-reversal invariance.

The tendency to a collapse towards dimers can be suppressed by a suitable local modification of the Hamiltonian (Marston and Affleck, 1989). All that is needed is to have a scale $\bar{\rho}_0$ for the average bond amplitude around which they fluctuate. In the SU(N) model, this involves an interaction that is quartic in the spins. This possibility is not available for the case of interest, the nearest-neighbor spin one-half Heisenberg model, but it may occur in further-neighbor interactions. There is strong numerical evidence that strong enough four-spin (ring-exchange) interactions do lead to a dimerized state (Sandvik, 2010).

Let us assume for the moment that dimer collapse has been avoided. Now the flux phase may be unstable against the development of both chiral and non-chiral mass terms. In turn, it is easy to write down an effective theory for the low-energy modes. The effective Lagrangian density should include the (doubled) spinon modes (u_α and v_α). It should also contain both staggered amplitudes, which, after normalization, can be denoted by two real Bose fields, ϕ_1 and ϕ_2. The chiral modes are also bosonic and real and can be denoted by χ. The effective Lagrangian density should then have the form

$$\begin{aligned}\mathcal{L} = &\left(\bar{u}_\alpha i\gamma_\mu D_\mu u_\alpha + \bar{v}_\alpha i\gamma_\mu D_\mu v_\alpha\right)\\ &- \phi_1(\bar{u}_\alpha v_\alpha + \bar{v}_\alpha u_\alpha) - \phi_2(\bar{u}_\alpha u_\alpha - \bar{v}_\alpha v_\alpha) - \frac{N}{J}U(\phi_1^2, \phi_2^2)\\ &- \chi(\bar{u}_\alpha u_\alpha + \bar{v}_\alpha v_\alpha) - \frac{N}{J'}U'(\chi^2)\end{aligned} \tag{10.60}$$

where the potentials U and U' are *even* functions of ϕ_1, ϕ_2, and χ separately. The phases of the bond amplitudes, the gauge fields, have been included through the covariant derivatives D_μ,

$$D_\mu = \partial_\mu - i\mathcal{A}_\mu \tag{10.61}$$

The potentials U and U' are assumed to have a sharp minimum at $\phi_1 = \phi_2 = \chi = 0$ and to grow rapidly as the values of their arguments increase. The latter condition is needed in order to avoid collapse towards a dimer state. The requirement that the potentials U and U' be even functions of their arguments implies a four-fold degeneracy of the ground state. In the absence of collapse the symmetric amplitude modes, which represent local fluctuations of the length scale (i.e. the Fermi velocity) and of anisotropy, do not change the qualitative physical properties of the system. The assumption that there is a well-defined, and sharp, average bond amplitude $\bar{\rho}_0$ means that local dilatations and shears are strongly suppressed. When integrated out, these fluctuations merely lead to effective interactions of the fermions that involve operators with many derivatives. In an RG sense, such terms are irrelevant. This is equivalent to saying that, if the physics of the system is correctly described by the continuum model, then operators with many derivatives may become important only if the fluctuations have large Fourier components at large values of the momentum. However, the main assumption of the continuum model is precisely that such Fourier components are small, since only smooth configurations are correctly described by this model. Under these assumptions, the effective Lagrangian density of Eq. (10.60) is a good description of the physics of the system.

The fluctuations which are described in detail by Eq. (10.60) are the fluctuations of the gauge field $\mathcal{A}_\mu$ and of the amplitudes ϕ_i and χ. The fluctuations of the amplitudes ϕ_i and χ lead to a phase transition, in which one or several amplitudes have a non-zero expectation value, only if N is not too large. This can be checked by looking for solutions of the saddle-point equations. These equations, in the absence of a dimer solution, do not have a solution with $\langle\phi_0\rangle \neq 0$ (or $\langle\chi\rangle \neq 0$) unless N is smaller than some critical value N_c. The value of N_c depends on the details of the model. This regime is still described correctly by the $1/N$ expansion. Thus, unless one happens to be interested in unphysically large values of N, one of $\langle\phi_1\rangle$, $\langle\phi_2\rangle$ or $\langle\chi\rangle$ will become non-zero. The fluctuations around this state are small and have very short correlation lengths.

From this discussion, we may conclude that, unless $N > N_\mathrm{c} \gg 1$, there are spinons in the spectrum but they have a finite gap. This result would hold if we can ignore the fact that the gauge group is compact and hence that there are magnetic-monopole configurations in its space of states. Provided that this can be done consistently, we conclude that, for $N > N_\mathrm{c}$, the spinons would be massless (i.e. there is no gap) and deconfined. Thus this model appears to predict the existence of electrically neutral spin-bearing excitations. However, this conclusion is not well founded, since the fluctuations of the gauge field $\mathcal{A}_\mu$ have been ignored altogether. A massless deconfined phase may still occur in the large-N limit since this is equivalent to increasing the number of flavors rather than the rank of the gauge group. In this limit the monopole operators are irrelevant (Metlitski *et al.*, 2008), and the compact nature of the gauge group is not important. However, monopoles play a dominant role for smaller values of N. This problem has been investigated in detail by Monte Carlo simulations. The most recent results indicate that for $N \leq 4$ the ground state is gapped and confining (Armour *et al.*, 2011). We will also see in the next sections that in states that break time-reversal invariance monopoles are suppressed and the results of the large-N theory are at least qualitatively correct even for smaller values of N.

What are the effects of the gauge fields $\mathcal{A}_\mu$? We will examine this problem now assuming that the fluctuations of the gauge fields are arbitrary but smooth enough so that we can ignore monopole configurations. A simple inspection of the effective Lagrangian density, Eq. (10.60), shows that the gauge fields appear *only* in the kinetic-energy term of the spinons, through the covariant derivatives. There is no separate term in this Lagrangian density that will control the fluctuations of the gauge field, such as $F_{\mu\nu}F^{\mu\nu}$ in electrodynamics. Since the Lagrangian density is *linear* in the gauge field $\mathcal{A}_\mu$, we can integrate the gauge fields out exactly. The integral over the gauge field $\mathcal{A}_\mu$ yields

$$\int \mathcal{D}\mathcal{A}_\mu \, e^{iS[u,v,\phi_1,\phi_2,\chi,\mathcal{A}_\mu]} = e^{iS[u,v,\phi_1,\phi_2,\chi]} \times \int \mathcal{D}\mathcal{A}_\mu \exp\left(i \int d^3x \, \mathcal{A}_\mu J^\mu\right) \tag{10.62}$$

where J^μ is the total spinon gauge current density

$$J^\mu = \bar{u}_\alpha \gamma^\mu u_\alpha + \bar{v}_\alpha \gamma^\mu v_\alpha \tag{10.63}$$

The last factor in Eq. (10.60) shows that the integral over the gauge field $\mathcal{A}_\mu$ is just a constraint

$$\int \mathcal{D}\mathcal{A}_\mu \exp\left(i \int d^3x \, \mathcal{A}_\mu J^\mu\right) = \prod_x \delta^3(J^\mu(\vec{x}, t)) \tag{10.64}$$

Hence, the only states allowed in the Hilbert space, let's call them $|\text{Phys}\rangle$, satisfy

$$J^\mu(\vec{x}, t)|\text{Phys}\rangle = 0 \tag{10.65}$$

which is a *local* condition. In components, this constraint is equivalent to the statement that the normal-ordered spinon density $j_0(\vec{x}, t)$,

$$j_0(\vec{x}, t) \equiv \rho(\vec{x}, t) - \langle \rho(\vec{x}, t) \rangle \tag{10.66}$$

with $\rho(\vec{x}, t)$ being the electron density, and the currents $j_i(\vec{x}, t)$ ($i = 1, 2$) annihilate the physical states. Thus, the condition of $N/2$ occupancy is exactly satisfied. However, this also means that the allowed states carry zero spinon current and that there are no states in the spectrum of this system carrying the spinon quantum numbers, i.e. spin one-half in the SU(2) case. As a result, these *spin-liquid phases* do not have *spinon* states in their spectra. This is not to say that the spinons do not have a role. Gauge-invariant spinon bound states do not have spinon quantum numbers and hence are allowed. In spin one-half language, these states are either spin singlets (valence bonds) or triplets. These bound states have large energy gaps, with the singlets being the states of lowest energy.

10.4 Chiral spin liquids and Chern–Simons gauge theory

In Section 10.1 we encountered a state, the chiral spin state (CSS), which spontaneously violates time reversal and parity invariance. We will see in this chapter that this feature of the CSS has far-reaching novel consequences. There are other states of condensed matter in which time-reversal invariance is broken. A ferromagnet has such a property. However, unlike the CSS, the ferromagnetic ground state does not violate parity, and its properties are quite different from what we will find in the CSS.

A system of electrons moving on a plane, in the presence of a perpendicular magnetic field, does not have time-reversal invariance. It is explicitly broken by the magnetic field. If the electrons are spin-polarized, then in some sense parity is also broken due to the orbital nature of the coupling. The results are the fascinating properties of the quantum Hall effect (QHE), in its integer and fractional forms. In this chapter and the coming ones, we will discuss the deep connections between the CSS and the QHE. We will see that, as a result of the combined effect of violation of parity and time-reversal invariance, both systems have an extra term, the so-called Chern–Simons term, in the effective Lagrangians for their low-energy degrees of freedom. These Lagrangians also provide a natural phenomenological description of the physics. In particular, both systems have low-energy excitations with fractional statistics or *anyons*. We will see in the next chapter that, if the system is *compressible*, these excitations lead to a novel form of superconductivity called *anyon superconductivity*. Deep and far-reaching connections among the CSS, the QHE, the mathematical theories of knots, and, more generally, topological field theory will be described. We begin this chapter by going beyond the discussion of the previous section on the fluctuations around a CSS.

Under what circumstances should we expect to get "free spinons," i.e. states with finite energy that carry spinon quantum numbers? The arguments at the end of the last section show that this is not possible unless the fluctuations of the gauge fields are somehow suppressed. Terms of the $F_{\mu\nu}^2$ type do not efficiently suppress fluctuations. Gauge-field mass terms are, on the other hand, very efficient at suppressing fluctuations. In (2+1) space-time dimensions two gauge-field mass terms can arise. The simplest one, A_μ^2, explicitly breaks the gauge symmetry and can arise only if the system becomes superconducting. This is a possible scenario at non-zero hole density but not at half-filling.

In $(2+1)$ dimensions, there is another possible source of mass for the gauge fields: the topological or Chern–Simons mass terms (Schonfeld, 1981; Deser *et al.*, 1982). The Chern–Simons term is a locally gauge-invariant Lagrangian that breaks parity $\mathcal{P}$ and time-reversal $\mathcal{T}$ invariance. For the case of an abelian gauge field it has the form

$$\mathcal{L}_{\mathrm{CS}} = \frac{\theta}{4}\epsilon_{\mu\nu\lambda} A^\mu F^{\nu\lambda} \tag{10.67}$$

The coupling constant θ is dimensionless and measures the strength of $\mathcal{P}$ and $\mathcal{T}$ (but not $\mathcal{C}$) violations.

We will see below that a Chern–Simons term arises in the effective action of the RVB gauge field A_μ from the fermionic fluctuations in a CSS. Since the gauge fields now have a mass, one does expect to get spinon states in the spectrum. These states are massive, i.e. have a non-vanishing mass. We will also see in the next chapter that, if holes are allowed, the system develops a novel form of superconductivity driven by excitations with fractional statistics called *anyons*. We will also see in later chapters that the Chern–Simons action plays a key role in the theory of the quantum Hall effects.

In the absence of mass terms for the fluctuations of the gauge fields the spinons disappear from the spectrum. The only low-lying excitations of the system are associated with the gauge field A_μ. It is then natural to ask for the effective Lagrangian which governs the dynamics of the gauge fields. The $1/N$ expansion provides a simple way to determine not only the effective action of the gauge field A_μ, but also that of the amplitudes ϕ_i and χ introduced in Eq. (10.60). This is done by first integrating out the spinon fields and later expanding around one of the saddle-points of the resulting action. The effective action determined in this way is

$$\begin{aligned} S_{\text{eff}}[\phi_i, \chi, A_\mu] = \int d^3x \left(-\frac{N}{J} U(\phi_1^2, \phi_2^2) - \frac{N}{J'} U'(\chi^2) \right) \\ - iN \ln \det \begin{pmatrix} i\gamma_\mu D^\mu - \chi - \phi_2 & -\phi_1 \\ -\phi_1 & i\gamma_\mu D^\mu - \chi + \phi_2 \end{pmatrix} \end{aligned} \tag{10.68}$$

where the 2×2 matrix in Eq. (10.68) occurs because of the spinon doubling in terms of u and v components of Eqs. (10.43) and Eq. (10.45). At the saddle-point level we have

$$\langle A_\mu \rangle = 0, \quad \langle \phi_i \rangle = \bar{\phi}_i, \quad \langle \chi \rangle = \bar{\chi} \tag{10.69}$$

Let us now consider the effects of fluctuations around this state. Let $\tilde{\phi}_i$ and $\tilde{\chi}$ denote the fluctuation components of the amplitude fields. The vector potential A_μ has zero average, Eq. (10.69), and hence it represents a fluctuation. The fluctuations of the amplitude fields are massive and thus do not lead to any new physics, provided, of course, that the saddle-point represents a stable state. We will not consider the effects of such fluctuations here. Qualitatively, amplitude fluctuations are important in the dimer limit. We have already considered such effects in Chapter 8.

The fluctuations of the vector potentials A_μ lead to interesting effects. Their effective action can be calculated by expanding S_{eff} of Eq. (10.68) in powers of A_μ. To second order, we get $S^{(2)}_{\text{gauge}}[A_\mu]$ given by

$$S^{(2)}_{\text{gauge}}[A_\mu] = \frac{1}{2} \int d^3x\, d^3y\, A^\mu(x) \Pi_{\mu\nu}(x, y) A^\nu(y) \tag{10.70}$$

where $\Pi_{\mu\nu}(x, y)$ is the one-particle irreducible fermion current–current correlation function (or polarization tensor)

$$\Pi_{\mu\nu} = \langle J_\mu(x) J_\nu(y) \rangle \tag{10.71}$$

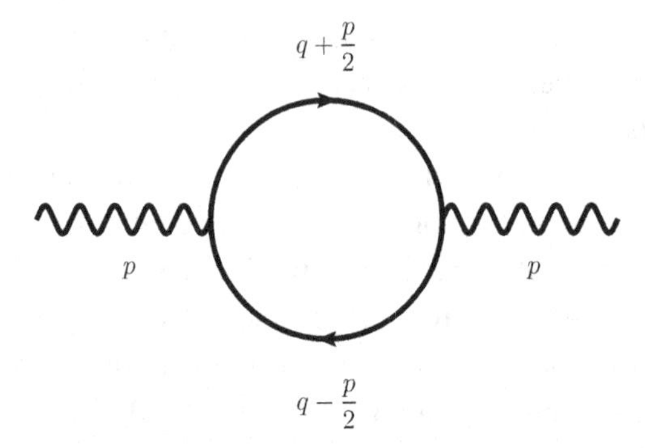

Figure 10.3 The one-loop contribution to the effective action of the gauge field.

In momentum space, we can write $S^{(2)}_{\text{gauge}}[A_\mu]$ in the form

$$S^{(2)}_{\text{gauge}}[A_\mu] = i\frac{N}{2}\int \frac{d^3p}{(2\pi)^3}\int \frac{d^3q}{(2\pi)^3}\,\text{tr}\left[S\left(\frac{p}{2}+q\right)\gamma^\mu S\left(-\frac{p}{2}+q\right)\gamma^\nu\right] \times A_\mu(p)A_\nu(-p) \tag{10.72}$$

in terms of the fermion propagator in momentum space $S(p)$,

$$S(p) = \frac{1}{p_\mu\gamma^\mu - \bar{\chi} - \bar{\phi}_i T_i} \tag{10.73}$$

where the 2×2 matrices T_1 and T_2 are given by the Pauli matrices σ_1 and σ_3, respectively. This one-loop contribution to the effective action of the gauge field can be represented by the Feynman diagram of Fig. 10.3.

An explicit computation of the operator $\Pi_{\mu\nu}(p)$ (in momentum space) yields the result

$$\begin{aligned}\Pi^{\mu\nu}(p) &= \int \frac{d^3q}{(2\pi)^3} iN\,\text{tr}\left[S\left(q+\frac{p}{2}\right)\gamma^\mu S\left(q-\frac{p}{2}\right)\gamma^\nu\right]\\ &= (p^2g^{\mu\nu} - p^\mu p^\nu)\Pi_0(p^2) - i\epsilon^{\mu\nu\lambda}p_\lambda\Pi_A(p^2)\end{aligned} \tag{10.74}$$

The kernels $\Pi_0(p^2)$ and $\Pi_A(p^2)$ have the following explicit forms:

$$\Pi_0(p^2) = -\frac{N|m_+|}{4\pi p^2} + \frac{N}{8\pi\sqrt{p^2}}\left(\frac{4m_+^2}{p^2}+1\right)\sinh^{-1}\left(\frac{1}{\sqrt{4m_+^2/p^2-1}}\right) + (m_+ \leftrightarrow m_-) \tag{10.75}$$

$$\Pi_A(p^2) = -\frac{m_+N}{2\pi\sqrt{p^2}}\sinh^{-1}\left(\frac{1}{\sqrt{4m_+^2/p^2-1}}\right) + (m_+ \leftrightarrow m_-) \tag{10.76}$$

where $m_\pm$ denotes the mass gaps (from the poles of the propagator of Eq. (10.73)) for the two species of fermions (including their *signs*),

$$m_\pm = \bar{\chi} \pm \sqrt{\bar{\phi}_1^2 + \bar{\phi}_2^2} \tag{10.77}$$

These expressions can now be used to find the effective Lagrangian $\mathcal{L}^{(2)}_{\text{gauge}}[A_\mu]$ that governs the dynamics of the "RVB" gauge field A_μ at low energies. By virtue of gauge invariance we know that only locally gauge-invariant terms can possibly occur. To lowest order in a gradient expansion, i.e. in powers of p^2/m^2, we expect a Maxwell-like term $F_{\mu\nu}F^{\mu\nu}$. However, in $(2+1)$ dimensions a Chern–Simons (CS) term, Eq. (10.67), is also allowed. The CS term, although gauge-invariant, breaks parity ($\mathcal{P}$) and time reversal ($\mathcal{T}$). Thus, it may occur in a chiral spin state. Indeed, this is what actually does happen!

By explicit calculation we find that the effective Lagrangian $\mathcal{L}^{(2)}_{\text{gauge}}[A_\mu]$ does have the low-energy form

$$\mathcal{L}^{(2)}_{\text{gauge}}[A_\mu] = -\frac{1}{4g^2}F_{\mu\nu}F^{\mu\nu} + \frac{\theta}{4}\epsilon_{\mu\nu\lambda}A^\mu F^{\nu\lambda} \tag{10.78}$$

The gauge coupling constant g^2 ("spinon charge") and the CS coupling constant θ are equal to

$$\frac{1}{g^2} = \frac{N}{\pi}\left(\frac{1}{\mid m_+ \mid} + \frac{1}{\mid m_- \mid}\right) \tag{10.79}$$

and

$$\theta = \frac{N}{4\pi}(\text{sgn}(m_+) + \text{sgn}(m_-)) \tag{10.80}$$

This result was first obtained by Redlich (1984).

Clearly, θ vanishes if $\text{sgn}(m_+) = -\text{sgn}(m_-)$. This is to be expected, since time reversal is not violated if the masses have opposite signs. This is the *non-chiral* spin-liquid state. In the *chiral* state, $\text{sgn}(m_+) = \text{sgn}(m_-)$ and either sign, plus or minus, can occur. Thus, in a *chiral spin-liquid* state we find that the CS coupling constant is $\theta = \pm N/(2\pi)$ and does not vanish. In other terms, in a massive relativistic system of Dirac spinors, the effective CS coupling θ is equal to $1/(4\pi)$ per species of Dirac fermion. A two-component Dirac spinor is known as a Weyl spinor. We will see below, in Section 11.5.1, and in more detail in Section 12.5, that the prefactor of the CS term is equal to $\sigma_{xy}/4$, where σ_{xy} is the Hall conductance. Thus this calculation predicts that each species of 2D Weyl fermions has a Hall conductance of $\frac{1}{2}e^2/h$. We will return to this question in Chapter 16, where it plays an important role.

We can gain some insight into the meaning of this result by considering the propagator of the gauge field. In particular, we want to know whether there is a massless "photon" state in the spectrum. If such a state were to be present the whole approach would be in doubt, since in our problem the vector potentials A_μ would fluctuate wildly and, as we showed in Chapter 8, the spinons would in fact be confined by the monopoles of the field A_μ. However, if the field A_μ were to become massive, the scenario would be completely different. Let us consider this question more closely. The propagator of the gauge fields $G_{\mu\nu}(x, x')$ is

$$G_{\mu\nu}(x, x') = \langle T A_\mu(x) A_\nu(x')\rangle \tag{10.81}$$

and it is *not* gauge-invariant. It only makes sense *after* a gauge has been fixed. We do this by the standard procedure (Itzykson and Zuber, 1980) of adding to the Lagrangian $\mathcal{L}^{(2)}_{\text{gauge}}[A_\mu]$ a gauge-fixing term of the form

$$\mathcal{L}^{(2)}_{\text{fixing}}[A_\mu] = \frac{\alpha}{2g^2}(\partial_\mu A^\mu)^2 \tag{10.82}$$

In particular, I will work in the Lorentz gauge in which $\alpha \to \infty$ (i.e. $\partial^\mu A_\mu = 0$). The propagator of the gauge fields, in the Lorentz gauge, is given in momentum space by

$$G^{\mu\nu}(p) = \frac{g^2}{p^2 - g^4\theta^2}\left(g^{\mu\nu} - \frac{p^\mu p^\nu}{p^2}\right) - \frac{g^4\theta i\epsilon^{\mu\nu\lambda}}{p^2(p^2 - g^4\theta^2)}p_\lambda \tag{10.83}$$

This propagator has a *pole* at $p^2 - g^4\theta^2 = 0$. This "photon" state is massive and its mass m_γ is equal to $g^2|\theta|$. Hence a chiral state implies a massive RVB gauge field. This mass term does not spoil gauge invariance, and it does *not* imply the occurrence of superconductivity. However, it is just as efficient in suppressing the fluctuations of the RVB gauge field. We have already discussed in Chapter 9 how the wild fluctuations of this gauge field, parametrized in terms of monopoles, are responsible for the confinement of excitations bearing the fundamental quantum number, the spin. Conversely, we are led to suspect that the presence of an induced CS term may signal the liberation of the *spinons* by suppressing the monopoles. We saw that monopoles were responsible for disordering the Wilson loops, leading to confinement, which, in the present context, means a valence-bond crystal. However, the presence of the induced CS term makes a significant difference.

Let us first discuss the fate of the monopoles. Consider a configuration $A^{(c)}_\mu$ that represents a set of monopoles (with their strings) and assume that they are well separated. A configuration of monopoles and anti-monopoles is generated by a set of sources in the form of infinitesimally thin solenoids joining each monopole to an anti-monopole. The issue here is the existence of a long-range monopole field in the presence of the (induced) CS term. But the CS term causes the gauge field to be massive. In such a situation, an elementary study of the Euclidean equations of motion reveals that, for instance, in the case of a simple monopole–anti-monopole pair, the RVB magnetic field does not extend beyond a distance $R \sim 1/(g^2\theta)$ away from the solenoid. Thus, the *dominant* contribution to the Euclidean action comes from this effective flux tube. If the linear size of the monopole–anti-monopole pair is R, the action of a monopole–anti-monopole pair grows linearly with their separation R. Hence, monopoles and anti-monopoles are confined and their contribution to the expectation value of gauge-invariant operators is exponentially small and can be neglected.

There is a further subtlety here. Under a local gauge transformation the CS Lagrangian density changes by a total derivative. If the space-time has boundaries (for instance, if 2D space has the topology of a disk), the CS *action* is not gauge-invariant. In that case there are degrees of freedom at the boundary, which play a key role since these are the only low-energy states left. As we will see in later chapters, this is what happens in the quantum Hall states in which there are edge states. On the other hand, we will also see below that even if the space-time manifold is closed (and hence has no boundaries) one has to carefully consider the effects of large gauge transformations, which (as we will see) force the CS coupling constant to be quantized.

The quantization of the coupling constant can be seen in the more general case of a non-abelian gauge field A_μ that takes values in the algebra of a compact Lie group G, such as $\mathrm{SU}(N)$. The general form of the CS action is (Deser *et al.*, 1982)

$$S_{\mathrm{CS}}[A_\mu] = \int_\Sigma d^3x \; \frac{\theta}{4} \epsilon_{\mu\nu\lambda} \; \mathrm{tr}\left[A^\mu F^{\nu\lambda} + \frac{2}{3} A^\mu A^\nu A^\lambda\right] \tag{10.84}$$

where Σ is a closed 3D manifold, e.g. the sphere, the torus, etc. Under a gauge transformation $U(x) \in G$ the vector potential A_μ transforms as

$$A_\mu \to U^{-1} A_\mu U + U^{-1} \partial_\mu U \tag{10.85}$$

Under a local gauge transformation $U(x)$ the CS action is invariant. However, under a large gauge transformation it transforms as

$$S_{\mathrm{CS}} \to S_{\mathrm{CS}} + 4\pi^2 w[U] \tag{10.86}$$

where $w[U]$, given by

$$w[U] = \frac{1}{24\pi^2} \int_\Sigma d^3x \; \epsilon^{\mu\nu\lambda} \; \mathrm{tr}\left[\left(U^{-1}\,\partial_\mu U\right)\left(U^{-1}\,\partial_\nu U\right)\left(U^{-1}\,\partial_\lambda U\right)\right] \tag{10.87}$$

is a *topological invariant* of the maps of the manifold Σ to the group G, known as the *winding number* (or first Chern invariant). Thus, in the case of the sphere, $\Sigma = S^3$, the winding number $w[u]$ labels the homotopy classes $\pi_3(G) \simeq \mathbb{Z}$, which we have already encountered in Section 7.10. On the other hand, on physical grounds we must require that the weight of the path integral $\exp(i S_{\mathrm{CS}})$ be gauge-invariant, which can happen only if the coupling constant is quantized, $\theta = k/(2\pi)$, with $k \in \mathbb{Z}$. The integer k is known as the level of the Chern–Simons gauge theory.

These considerations affect the behavior of this theory in the presence of monopoles (and anti-monopoles). Indeed, from the point of view of the CS action, the core of the monopoles is effectively an *edge*, and our gauge-fixing procedure fails near the core. A careful consideration of the fermion determinant in the background of gauge-field configurations shows that it vanishes identically if the monopoles are present (Affleck *et al.*, 1989; Fradkin and Schaposnik, 1991). In other terms, the fermion path integral, in a chiral spin state, loses its gauge invariance in the presence of monopoles. The result is the suppression of the monopoles and the deconfinement of the spinons. In consequence, in the chiral spin liquid there is genuine separation of spin and charge.

10.5 The statistics of spinons

What properties do liberated spinons have? The best way to address this question is to look at how spinons propagate in this system. Consider the amplitude $W^{(1)}(\vec{x}, \vec{x}; T)$ for a spinon, of any type, created in the remote past at site $\vec{x}$, to propagate throughout the system and finally return to the same place $\vec{x}$, in the same state, in the remote future, $T \to \infty$. The (imaginary-time) path integral for this amplitude is

$$\begin{aligned} \lim_{T\to\infty} W^{(1)}(\vec{x}, \vec{x}; T) &= \lim_{T\to\infty} \operatorname{tr} S_{\mathrm{F}}\left(\vec{x}, -\frac{T}{2}; \vec{x}, +\frac{T}{2}\right) \\ &= \int \mathcal{D}A_\mu \operatorname{tr}\left\langle \vec{x}, -\frac{T}{2} \left| \frac{1}{D_\mu \gamma^\mu + m} \right| \vec{x}, \frac{T}{2} \right\rangle e^{-S_{\mathrm{eff}}(A)} \end{aligned} \tag{10.88}$$

where I have used the (imaginary-time) spinon propagator in a fixed background configuration of RVB vector potentials. We can now use the Feynman picture of a sum over paths by first writing (Polyakov, 1987) the spinon propagator in the form

$$\begin{aligned} &\operatorname{tr}\left\langle \vec{x}, -\frac{T}{2} \left| \frac{1}{D_\mu \gamma^\mu + m} \right| \vec{x}, \frac{T}{2} \right\rangle \\ &\quad = \operatorname{tr}\left\langle \vec{x}, -\frac{T}{2} \left| (-D_\mu \gamma^\mu + m) \right| z \right\rangle \left\langle z \left| \frac{1}{-D^2 + m^2} \right| \vec{x}, +\frac{T}{2} \right\rangle \end{aligned} \tag{10.89}$$

where we have introduced a complete set of states $|z\rangle$, labeled by the space-time coordinate z. The sum (integral) over all values of z is left implicit here and below.

The proper-time representation of the propagator yields the following expression for the trace in Eq. (10.89):

$$\operatorname{tr}\left\langle x, -\frac{T}{2} \left| (-D_\mu \gamma^\mu + m) \right| z \right\rangle \int_0^\infty d\tau \left\langle z \left| e^{+\tau D^2} \right| \vec{x}, \frac{T}{2} \right\rangle e^{-\tau m^2} \tag{10.90}$$

The operator D_μ is the Euclidean covariant derivative, $D_\mu = \nabla_\mu + i A_\mu$. The (Euclidean) Dirac matrices are present only in the prefactors.

Notice that by taking a trace we are effectively summing over all spinon polarizations. A standard path-integral argument now yields an expression for $W^{(1)}(\vec{x}, \vec{x}; T)$ in terms of sums over paths Γ, the worldlines of the histories of the quasiparticle, of arbitrary length τ. The boundary conditions that we are using here imply that the sum over paths runs over contributions with paths which close on the imaginary-time direction (i.e. run around the cylinder). The result is the path integral

$$\left\langle z \left| e^{+\tau D^2} \right| \vec{x}, +\frac{T}{2} \right\rangle = \int \mathcal{D}x \, \exp\left\{ -\int_0^\tau dt \left[\frac{1}{2} \left(\frac{d\vec{x}}{dt} \right)^2 + i \vec{A} \cdot \frac{d\vec{x}}{dt} \right] \right\} \tag{10.91}$$

which is the sum over paths Γ of length τ satisfying the boundary condition $x(0) = z$ and $x(\tau) = (\vec{x}, T)$.

The amplitude $W^{(1)}(\vec{x}, \vec{x}; T)$ can now be written in the form

$$\begin{aligned} W^{(1)}(\vec{x}, \vec{x}; T) = & \int \mathcal{D}A_\mu \, e^{-S_{\text{eff}}(A_\mu)} \, \text{tr} \left\langle \vec{x}, -\frac{T}{2} \right| (-D_\mu \gamma^\mu + m) \left| z \right\rangle \\ & \times \int_0^\infty d\tau \, e^{-\tau m^2} \left\langle z \left| e^{+\tau D^2} \right| \vec{x}, +\frac{T}{2} \right\rangle. \end{aligned} \tag{10.92}$$

Equivalently, $W^{(1)}(\vec{x}, \vec{x}; T)$ can be written in the form

$$\begin{aligned} & \int_0^\infty d\tau \, e^{-\tau m^2} \int \mathcal{D}x_\mu \int \mathcal{D}A_\mu \, \text{tr} \left\langle \vec{x}, -\frac{T}{2} \right| (-D_\mu \gamma^\mu + m) \left| z \right\rangle \\ & \qquad \times \exp\left[-\int_0^\tau dt \, \frac{1}{2} \left(\frac{d\vec{x}}{dt} \right)^2 \right] \exp\left(i \oint_\Gamma A_\mu \, dx_\mu \right) \end{aligned} \tag{10.93}$$

where I have used the fact that the paths Γ are closed and, consequently, the term $\int_0^\tau dt \, \vec{A} \cdot d\vec{x}/dt$ is simply the circulation of $\vec{A}$ around Γ. Notice that this quantity is gauge-invariant and it arises because we are considering paths that close (wrap) around the cylinder. The path integral requires that this amplitude be averaged over all the configurations of the RVB gauge fields, for each path Γ. After doing that we get, using an obvious notation,

$$\langle W^{(1)} \rangle \sim \sum_\Gamma (\text{amplitude})_\Gamma \times \langle e^{i \oint_\Gamma dx_\mu A^\mu} \rangle \tag{10.94}$$

which involves the Wilson loop operator.

If we ignore the contribution of the monopoles, the amplitude $W^{(1)}$ can be estimated just by using the effective action of Eq. (10.78). The average in Eq. (10.94), the expectation value of the Wilson loop operator for trajectory Γ,

$$\left\langle \exp\left(i \oint_\Gamma dx_\mu \, A^\mu \right) \right\rangle \tag{10.95}$$

can now be calculated quite easily. Let $J_\mu(\Gamma)$ be a current in $(2+1)$-dimensional Euclidean space defined by

$$J_\mu(\Gamma) = \begin{cases} S_\mu(x), & x \in \Gamma \\ 0, & \text{otherwise} \end{cases} \tag{10.96}$$

where $S_\mu(x)$ is the unit vector tangent to the path Γ at x. The expectation value to be computed has the form

$$\left\langle \exp\left(i \oint_\Gamma dx_\mu \, A^\mu \right) \right\rangle = \left\langle \exp\left(i \int d^3x \; J_\mu(x) A^\mu(x) \right) \right\rangle \tag{10.97}$$

Since the effective action of Eq. (10.78) is quadratic in A_μ, the average, Eq. (10.95) is simply given by

$$\left\langle \exp\left(i \oint_\Gamma dx_\mu \, A^\mu(x) \right) \right\rangle_{\text{CS}} = \exp\left(-\frac{i}{2} \int d^3x \int d^3x' \; J_\mu(x) G^{\mu\nu}(x - x') J_\nu(x') \right) \tag{10.98}$$

where the propagator $G_{\mu\nu}(x, x')$ in the Lorentz gauge has the Fourier transform given in Eq. (10.83). In real space-time $G_{\mu\nu}(x, y)$ is given by

$$\begin{aligned} G^{\mu\nu}(x, y) &= \int \frac{d^3p}{(2\pi)^3} G^{\mu\nu}(p) e^{i\vec{p}\cdot(\vec{x}-\vec{y})} \\ &= -g^2 \left\langle x \left| \left(\frac{1}{\partial^2 + g^4\theta^2} \right) \left(g^{\mu\nu} - \frac{\partial^\mu \partial^\nu}{\partial^2} \right) \right| y \right\rangle \\ &\quad + g^4 \theta \epsilon^{\mu\nu\lambda} \left\langle x \left| \frac{1}{\partial^2(\partial^2 + g^4\theta^2)} \partial_\lambda \right| y \right\rangle \end{aligned} \tag{10.99}$$

Thus, the argument I of the exponential on the right-hand side of Eq. (10.98) reads

$$\begin{aligned} I &\equiv -\frac{i}{2} \int d^3x \int d^3y \; J_\mu(x) G^{\mu\nu}(x - y) J_\nu(y) \\ &= -\frac{ig^2}{2} \int d^3x \int d^3y \; J_\mu(x) G_0(x, y; m^2) J^\mu(y) \\ &\quad -\frac{i}{2} g^4 \theta \int d^3x \int d^3y \; J_\mu(x) \epsilon^{\mu\nu\lambda} \left\langle x \left| \frac{1}{\partial^2(\partial^2 + g^4\theta^2)} \partial_\lambda \right| y \right\rangle J_\nu(y) \end{aligned} \tag{10.100}$$

Here I used $G_0(x, y; m^2)$ as the propagator for a massive field with $m^2 = g^4\theta^2$, which obeys

$$(-\partial^2 - m^2) G_0(x, y; m^2) = \delta(x - y) \tag{10.101}$$

If we restrict our discussion to long and smooth loops, we can make the long-distance approximation in Eq. (10.101), which now becomes ultra-local

$$G_0(x, y; m^2) \approx \frac{1}{m^2}\delta^{(3)}(x-y). \tag{10.102}$$

In this limit we find the exponent I to be given by

$$I \simeq -\frac{iL(\Gamma)}{2g^2\theta^2} + \frac{i}{2\theta}\int d^3x \int d^3y \; J_\mu(x)\epsilon^{\mu\nu\lambda} G_0(x, y; 0)\partial_\lambda J_\nu(y) \tag{10.103}$$

where $L(\Gamma)$ is the perimeter of the loop Γ, and $G_0(x, y; 0)$ is the propagator of a *massless* scalar field, which, in $(2+1)$ (Euclidean) space-time dimensions, is

$$G_0(x, y; 0) = \frac{1}{4\pi|x-y|} \tag{10.104}$$

We will see below that the non-local term in Eq. (10.103) plays a crucial role in the determination of the exponent I.

The first term in Eq. (10.103) embodies the quantum corrections to the propagation amplitude of the spinon. Hence it can be interpreted as a finite renormalization of its mass. The second term in Eq. (10.103) is more interesting. Let us examine the quantity $R(\Gamma)$ given by

$$R(\Gamma) = \int d^3x \int d^3y \; J_\mu(x)\epsilon^{\mu\nu\lambda} G_0(x, y; 0)\partial^y_\lambda J_\nu(y) \tag{10.105}$$

Below we will see that $R(\Gamma)$ is in fact a *topological invariant* known as the *Hopf invariant* or *linking number*. After integration by parts and using the definition of the current $J^\mu(x)$, we can write $R(\Gamma)$ in the form

$$R(\Gamma) = \oint_\Gamma dx_\mu \oint_\Gamma dy_\nu \; \epsilon^{\mu\nu\lambda} \; \partial^x_\lambda G_0(x, y; 0) \tag{10.106}$$

We can make more sense of $R(\Gamma)$ by means of the following magnetostatic analogy. In order to make these ideas precise, it is necessary to momentarily go to Euclidean space. Now $G_0(x, y; 0)$ is just the inverse Laplacian in three dimensions:

$$G_0(x, y; 0) \equiv \left\langle \vec{x} \left| \frac{1}{-\nabla^2} \right| \vec{y} \right\rangle \tag{10.107}$$

Let us regard $J_\mu(x)$ as an electric current in 3D space. This current establishes the static magnetostatic field $B_\mu(x)$, which satisfies

$$\vec{\nabla} \times \vec{B} = \vec{J}, \qquad \vec{\nabla} \cdot \vec{B} = 0 \tag{10.108}$$

i.e. Ampère's law. This observation allows us to solve for B_μ by means of the Green function $G_0(x, y; 0)$ in the form

$$B_\mu(x) = \int d^3y \; G_0(x, y; 0)\epsilon_{\mu\nu\lambda} \; \partial_\nu J_\lambda(y) \tag{10.109}$$

Thus, $R(\Gamma)$ can be written in the more compact form

$$R(\Gamma) = \int d^3x \; J_\mu(x) B_\mu(x) \tag{10.110}$$

where B_μ is the field established by J_μ. This is a self-interaction effect. Now we can use the definition of J_μ and Stokes' theorem to get $R(\Gamma)$ in the form of a surface integral,

$$R(\Gamma) = \oint_\Gamma dx_\mu \; B_\mu = \int_\Sigma d\sigma \; n_\mu \epsilon_{\mu\nu\lambda} \; \partial_\nu B_\lambda \tag{10.111}$$

where Σ is an open surface whose boundary is Γ. By substitution of Eq. (10.108) into Eq. (10.111) we get

$$R(\Gamma) = \int_\Sigma d\vec{\sigma} \cdot \vec{J} \tag{10.112}$$

i.e. $R(\Gamma)$ is the flux of $\vec{J}$, the current, through a surface bounded by itself (see Eq. (10.98)).

Thus, at least qualitatively, $R(\Gamma)$ should be equal to the self-linking, or writhing, number of the path Γ, which measures the number of times a vector normal to Γ winds as the loop is traced. Polyakov (1988), who was the first to put these arguments forward, argued that the *writhing* $R(\Gamma)$ of the path should be interpreted as an intrinsic *spin*. On the other hand, this spin makes sense only after one has chosen a specific prescription for measuring lengths along the path (i.e. made a choice of metric) and a short-distance regularization of the integrals involved in $R(\Gamma)$. In his seminal work relating the theory of knots and CS gauge theory, Witten (1989) showed that these definitions depend on the choice of regularization at short distances ("the framing of the knot"). In the problem that we are considering here, the CS gauge theory (abelian in our case) appears as the effective theory at distances long compared with the inverse spinon gap of the mean-field theory. It is unclear what regularization one should adopt in this case. It is conceivable that the anomalous spin predicted by Polyakov may, but need not, be present depending on the size of the spinon gap.

Let us consider the properties of spinons upon exchange processes. That is to say, we want to know which statistics they obey. Microscopically, we have *defined* the spinons to be fermions. The CS term may change that. To see how that can happen, let us consider the propagation amplitude $W^{(2)}(\{\vec{x}, \vec{y}\}, \{\vec{x}, \vec{y}\}; T)$ for *two spinons*, which in the remote past were located at $\vec{x}$ and $\vec{y}$, *either* to end up at the same locations in the remote future ($T \to +\infty$) *or* to exchange their positions. Once again, we will carry out the computation in the imaginary-time formalism in which the time direction is real and periodic, i.e. the space-time has, at least, the topology of a cylinder. The two-particle amplitude will be represented as a sum over paths that close on the time direction. In principle, we will be dealing with two different paths Γ_1 and Γ_2, representing the evolution of each spinon. These paths may, or may not, be linked. In other words, the paths are equivalent to *knots* or *braids*. We will see that the path integral can be written as a sum over classes of topologically inequivalent knots. Each class will be characterized by a phase factor. These phase factors can effectively alter the statistics of the spinons. The two-spinon amplitude $W^{(2)}$ has the form

$$W^{(2)} = \pm \sum_{\nu=0}^{\infty} W_\nu^{(2)} e^{i\phi_\nu} \tag{10.113}$$

where ν is the linking number of the paths (or worldlines), to be defined below. The $\pm$ sign represents the two possible processes, direct and exchange. We will primarily be interested in the computation of the phases ϕ_ν. The amplitudes $W_\nu^{(2)}$ are renormalizations of the spinon self-energies, scattering amplitudes, etc.

In terms of a sum over paths Γ, which is the union of the individual paths of the spinons, $W^{(2)}$ has the form

$$W^{(2)} = \pm \sum_{\Gamma \equiv \Gamma_1 \cup \Gamma_2} \mathcal{A}(\Gamma) \left\langle \exp\left(i \oint_\Gamma dx_\mu \, A^\mu(x) \right) \right\rangle \tag{10.114}$$

where $\mathcal{A}(\Gamma)$ is the absolute value of the amplitude. After a little algebra we get

$$W^{(2)} = \pm \sum_\Gamma \mathcal{A}(\Gamma) \exp\left(\frac{i}{2} \int d^3x \int d^3x' \; J_\mu(x) G_{\mu\nu}(x, x') J_\nu(x') \right) \tag{10.115}$$

where J_μ is the sum of the currents which define the paths Γ_1 and Γ_2, and $G_{\mu\nu}(x, x')$ is the analytic continuation to imaginary time of the propagator of the gauge fields. We will be interested only in the behavior of very large loops in the Euclidean space. The paths for direct and exchange processes become closed on identifying their endpoints. Thus, exchange and direct processes have an extra *relative* linking number. It is this extra linking number which is responsible for the fractional statistics.

It will be sufficient for our purposes to compute just the relative linking number. Thus, we can consider a simple direct process, in which the paths Γ_1 and Γ_2 are not linked, and a simple exchange process in which the two paths are linked in such a way that they form a *single* path. Now, the *linking number* of a *single* path is its *writhing number* $R(\Gamma)$. However, there are no regularization ambiguities now, since the path winds around the cylinder exactly *once*.

The cylinder (of periodic boundary conditions in imaginary time) represents a topological obstruction and no redefinition of the metric on the path (for example by stretching it) can change this number. Thus the exchange process and the direct process have a relative linking number of ± 1. The sign depends on the process by which we define exchange. If we define exchange by a counterclockwise (clockwise) rotation of one spinon around the other by an angle of π, followed by a translation equal to their relative separation, the sign is $+1$ (-1).

In the phase of the amplitude of the path integral the relative linking number $R(\Gamma)$ enters multiplying a factor of $1/(2\theta)$. Hence, the *total* amplitude changes by a factor of $-e^{\pm \frac{i}{2\theta}}$ when two particles are exchanged, i.e.

$$W_{\rm d}^{(2)} = -e^{\pm \frac{i}{2\theta}} W_{\rm e}^{(2)} \tag{10.116}$$

Equation (10.116) implies that the spinons have *fractional statistics* with a *statistical angle* δ equal to

$$\delta = \frac{1}{2\theta} \equiv \frac{\pi}{N} \tag{10.117}$$

defined relative to the fermion sign. In particular, Eq. (10.116) and Eq. (10.117) require that the two-spinon state should have a multivalued wave function (Wilczek, 1982)

$$\psi^{(2)}(1,2) = -e^{\pm i\delta} \psi(2,1) \tag{10.118}$$

For the case of physical interest, $N = 2$, the statistical angle $\delta = \pi/2$ and the wave function is multiplied by $\pm i$ when two spinons are exchanged. Since this phase factor is exactly half-way between fermions (− sign) and bosons (+ sign), these excitations have been dubbed *semions* (Laughlin, 1988b). In general, they are *anyons*, particles with fractional statistics (Wilczek, 1982). When the anyon is characterized by a single phase factor, it constitutes a one-dimensional representation of the braid group, and it is said to be an *abelian anyon*. We will see in later chapters that there are multi-dimensional representations in which the anyon is characterized by a matrix of phase factors. Such anyons are non-abelian since the matrices do not commute. For a discussion of the braid group and fractional statistics see Section 10.7.

It remains only to compute the phases ϕ_ν in Eq. (10.113) and Eq. (10.115). Let us write the phase ϕ_ν in the form

$$\phi_\nu = \frac{1}{2\theta} R(\Gamma_1, \Gamma_2) \tag{10.119}$$

Clearly, since $R(\Gamma_1, \Gamma_2)$ is bilinear in the currents, we can write $R(\Gamma_1, \Gamma_2)$ in terms of the writhing numbers of the individual paths and of the linking number $\nu = \bar{R}(\Gamma_1, \Gamma_2)$,

$$R(\Gamma_1, \Gamma_2) = R(\Gamma_1) + R(\Gamma_2) + 2\bar{R}(\Gamma_1, \Gamma_2) \tag{10.120}$$

with

$$\bar{R}(\Gamma_1, \Gamma_2) = \frac{1}{2} \oint_{\Gamma_1} dx_\mu \oint_{\Gamma_2} dx'_\nu \, G_{\mu\nu}(x - x') \tag{10.121}$$

We can now use the magnetostatic analogy once again. Let $J_\mu^{(1)}$ and $J_\mu^{(2)}$ be the two currents which establish the static fields $B_\mu^{(1)}$ and $B_\mu^{(2)}$, respectively. We get

$$\nu = \bar{R}(\Gamma_1, \Gamma_2) = \int d^3x \; J_\mu^{(1)}(x) B_\mu^{(2)}(x) \equiv \oint_{\Gamma_1} d\vec{x} \cdot \bar{B}^{(2)}(\vec{x}) \tag{10.122}$$

as a *circulation* of the field $\vec{B}^{(2)}$ (established by Γ_2) around Γ_1. Using now Stokes' theorem, we write ν as the surface integral

$$\begin{aligned} \nu &= \oint_{\Gamma_1} d\vec{x} \cdot \vec{B}^{(2)}(\vec{x}) \\ &\equiv \int_{\Sigma_1} d\sigma \, \vec{n} \cdot \vec{\nabla} \times \vec{B}^{(2)} \\ &= \int d\sigma \, \vec{n} \cdot \vec{J} \end{aligned} \tag{10.123}$$

where Σ_1 is an arbitrary surface with boundary Γ_1. Thus, ν counts how many times the loop Γ_2 winds around Γ_1.

After putting it all together, we get a formula for the two-spinon amplitude $W^{(2)}$ of the form

$$W^{(2)} = \sum_{\Gamma_1, \Gamma_2} \left[\mathcal{A}(\Gamma) e^{\frac{i}{2\theta}((R(\Gamma_1) + R(\Gamma_2))} \right] e^{\frac{i}{\theta} \nu(\Gamma_1, \Gamma_2)} \tag{10.124}$$

which, for an exchange process, acquires an additional factor $-e^{\pm i\delta}$. The quantity in brackets in Eq. (10.124) is a renormalized amplitude including possibly an anomalous spin. It represents the total two-spinon amplitude in the topological sector with fixed linking number ν.

In the next sections we will find that the remarkable properties of the spinons in the CS theory are generically present for any system with anyons.

10.6 Fractional statistics

One of the fundamental, and most cherished, axioms of local quantum field theory is the spin-statistics theorem. In the way in which it is most commonly stated, it says that particles with integer (half-integer) spin are bosons (fermions) and that the corresponding second-quantized fields obey canonical equal-time commutation (anticommutation) relations. At the root of this theorem is the need to preserve causality in a theory with local interactions, as well as the requirement for the existence of a lowest energy state. Spin can only be integer or half-integer since the fields should transform as an irreducible representation of the Lorentz group in $(3+1)$ dimensions: SO(3, 1). Even in a non-relativistic setting, the same requirements arise since the group of rotations SO(3) is a subgroup of SO(3, 1).

Furthermore, the many-particle wave functions should be either symmetric or antisymmetric under the exchange of any pair of particles, giving rise again to bosons and fermions. Thus, it may appear that these are the only possibilities.

The situation becomes radically different if the dimension of space-time is less than four. It has been known for a very long time (Jordan and Wigner, 1928) that in one space dimension the statistics is essentially arbitrary. This is basically a kinematic effect. Fermions on a line cannot experience their statistics since they cannot get past each other, and neither can bosons with hard cores. The Jordan–Wigner transformation, which we discussed in Sections 5.2 and 5.5, gives an explicit construction of a boson operator $a^\dagger(j)$ at the jth site of a 1D lattice as a non-local function of fermion densities (see Eq. (5.88)),

$$a^\dagger(j) = c^\dagger(j) e^{i\pi \sum_{m<j} c^\dagger(m) c(m)} \tag{10.125}$$

where the operators $c^\dagger(j)$ and $c(j)$ obey canonical anticommutation relations.

In continuum quantum field theory, there exists an analogous construction known as *bosonization* (see Section 5.6), which yields a connection between a *canonical* Dirac Fermi field $\psi_\alpha(x)$ ($\alpha = 1, 2$) and a *canonical* Bose field $\phi(x)$ in $(1+1)$ dimensions given by the Mandelstam formula (see Eq. (5.272) and Eq. (5.273))

$$\psi_\alpha(x) = e^{\frac{i}{\sqrt{\pi}} \int_{-\infty}^{x} dy\, \partial_0 \phi(y) \pm i\sqrt{\pi}\phi(x)} \tag{10.126}$$

with $\alpha = 1$ (2) for $+$ $(-)$.

Both constructions are based on the idea that in order to change the statistics one has to multiply an operator that creates a particle, such as $c^\dagger(j)$, by an operator that creates a *kink*, i.e. a topological soliton. This idea, to some extent, can be generalized to higher dimensions. For instance, in $(3+1)$ dimensions a *dyon*, a bound state of a charged Bose particle and a Dirac magnetic monopole, behaves as a fermion. However, unlike the 1D cases, all the examples in $(3+1)$ dimensions are semiclassical in character. Furthermore, in one space dimension, it is also possible to get *fractional* statistics (i.e. a case intermediate between Fermi and Bose statistics). A simple way to do that is to change the exponent of the kink operator in the Jordan–Wigner formula by replacing π by an arbitrary angle δ. The resulting operators $a^\dagger(j)$ do not obey Bose commutation relations, but instead exhibit fractional statistics, i.e.

$$a(j) a^\dagger(k) = \delta_{jk} - e^{i\delta} a^\dagger(k) a(j) \tag{10.127}$$

These operators, which are also known as parafermion operators (Fradkin and Kadanoff, 1980), are generalizations of the fermion operators which are essential to the solution of the 2D classical Ising model (Kadanoff and Ceva, 1971). They occur naturally in a number of quantum theories in $(1+1)$ dimensions, such as the Gross–Neveu model, and in 2D classical statistical mechanics. These operators have been found to play an important role in the critical behavior of the clock (or $\mathbb{Z}_N$) models in two dimensions, when studied using the methods of CFT (Dotsenko, 1984).

We will consider now the construction of anyon or parafermion operators more closely (Fradkin and Kadanoff, 1980). From the point of view of our discussion, the interest of this classical construction is that it has a natural generalization to $(2+1)$ dimensions that has turned out to be quite useful. Consider a 2D *classical statistical-mechanics* model such as the $\mathbb{Z}_N$ model on a square lattice. In the $\mathbb{Z}_N$ model, one defines an angle-like variable $\theta(\vec{r})$ residing at each site of a lattice. The angle $\theta(\vec{r})$ takes the discrete values $\theta = 2\pi p/N$ at each site, where p and N are positive integers and $p = 1, \ldots, N$. The classical Hamiltonian H is chosen to be a local function of the angles $\theta(\vec{r})$ and invariant under *global* $\mathbb{Z}_N$ transformations $\theta(\vec{r}) \to \theta(\vec{r}) + 2\pi m/N$, where m is a constant integer ($1 < m \leq N$). The classical partition function is

$$\mathcal{Z} = \sum_{\{\theta(\vec{r})\}} e^{\beta \sum_{\vec{r},\mu} \cos(\Delta_\mu \theta(\vec{r}))} \tag{10.128}$$

where $\beta = 1/T$ is the inverse temperature and $\mu = 1, 2$. In this case, the parafermion consists of an order operator $\mathcal{O}_m(\vec{r}) = \exp(i(2\pi m/N)\phi(\vec{r}))$ that measures the *order* at a site $\vec{r}$ of the lattice that is the endpoint of a *defect* or *domain wall*, which flips the $\mathbb{Z}_N$ spins by a fixed angle $2\pi q/N$. This defect, which tries to create a *fractional vortex* of strength $2\pi q/N$, is most easily described by means of a gauge field $\mathcal{A}_j(\vec{x})$ defined on all the links of the square lattice. The $\mathbb{Z}_N$ spins and the gauge fields are minimally coupled through the covariant difference $\Delta_j \theta(\vec{r}) + \mathcal{A}_j(\vec{r})$. The vector potential can be chosen to have non-vanishing curl equal to $2\pi q/N$ on any arbitrary closed loop on the lattice that contains the site $\vec{R}$ on the dual lattice at the plaquette north-east of the site $\vec{r}$. A popular choice is to have $\mathcal{A}_j = 0$ except on a path on the dual lattice ending at $\vec{R}$ (a Dirac string). From this construction it is apparent that the fractional statistics of these operators results from a mechanism closely related to the Aharonov–Bohm effect.

It is now easy to check that the *correlation functions* of these operators are multivalued. Consider for instance the two-point function $\mathcal{G}_{pq}(\vec{r}, \vec{r}\,')$ that measures the correlations between operators $\mathcal{O}_p$, in the presence of defects of strengths $\pm 2\pi q/N$, at sites $\vec{r}$ and $\vec{r}\,'$, respectively. Let $\mathcal{K}_q$ be the operator which creates a defect of strength $2\pi q/N$. Imagine carrying the site $\vec{r}\,'$ around site $\vec{r}$ on a closed loop Γ. After a full round trip, the spin operators have returned to their original locations but the Dirac strings are now misplaced: if in the original situation the spin at $\vec{r}$ was north of the string, now it is located south of it. The string can be returned to its original position by means of a gauge transformation. However, the spin operator is not invariant under this operation. As a result, the correlation function acquires a phase of $(4\pi/N)pq$. Hence the *composite operator* $\Psi_{pq} = \mathcal{O}_p\mathcal{K}_q$ creates an excitation that is an anyon with statistical angle $\delta = (2\pi/N)pq$.

From the discussion outlined above, in terms of a quantum-mechanical interpretation, it is apparent that any statistics is possible in one space dimension. Furthermore, the states created by operators that obey fractional statistics are, up to a boundary condition, completely determined by the coordinates of the particles on the line. In three dimensions, on the other hand, there does not seem to be room for *particles* with exotic statistics. However, 't Hooft (1978) showed that there can be string-like states in 4D gauge theories that obey commutation relations with fractional statistics.

In two dimensions, however, one finds a very interesting situation. The Lorentz group for a 2D system is SO(2, 1). The rotation group, which is crucial to both relativistic and non-relativistic systems, is SO(2). This group has only one generator L_z, the generator of infinitesimal rotations in the plane, and hence it is abelian. Thus, all of its representations are one-dimensional and labeled by the angular-momentum quantum number ℓ. If the wave functions of the excitations are *required* to be single-valued, the angular momentum ℓ can only be an integer. However, *fractional* shifts of ℓ are also compatible with the algebra of SO(2). States with fractional angular momentum have multivalued wave functions. In the Hilbert space which represents particles that move on the plane but are not allowed to sit on top of each other (a "punctured" plane) such wave functions are indeed allowed (Leinaas and Myrheim, 1977). The plane becomes isomorphic to a Riemann surface punctured at the locations of the particles, and different points are identified up to a phase determined by the fractional angular momentum. This framework provides for a natural construction of wave functions that obey fractional statistics.

Wilczek (1982) proposed the first fully quantum-mechanical prescription regarding how to make such particles. He dubbed them *anyons*. Wilczek's model makes use of the Aharonov–Bohm effect experienced by a particle of charge q moving on the plane in the presence of a magnetic solenoid with flux ϕ perpendicular to the plane. More precisely, he assumed that each particle is *rigidly* bound to a solenoid that moves along with it. Consider now two such bound states. Let us perform the *Gedankenexperiment* of adiabatically carrying one bound state around the other along some closed curve Γ. Because of the Aharonov–Bohm effect, the wave function Ψ of the bound state changes by an overall phase factor

$$\Psi \to \exp\left(i\frac{qe}{\hbar c} \oint_\Gamma \vec{A}(\vec{x}) \cdot d\vec{x} \right) \Psi \tag{10.129}$$

where $\vec{A}$ is the vector potential associated with the magnetic flux of the solenoid. The angular momentum ℓ of the state is then equal to $(qe/(hc))\phi$. If we denote by ϕ_0 the flux quantum, $\phi_0 = hc/e$, we can write the angular momentum ℓ in the form $\ell = q\alpha$, where $\alpha = \phi/\phi_0$. The angular momentum is not an integer if the Dirac quantization condition is not satisfied. The *statistics* obeyed by the bound states can be computed by considering an *exchange* process in which one bound state goes *half* its way around the other and, afterwards, both objects are shifted rigidly in such a way that they now have exchanged their initial positions. In this process their joint wave function has picked up a phase factor exactly equal to *half* of that for a full round trip around the other particle, i.e. $e^{i\pi\alpha}$. This definition is peculiar in the sense that the statistics of a state is determined by an adiabatic transport of the bound states in such a way that they never get on top of each other. Clockwise and counterclockwise processes yield complex-conjugate phase factors. These wave functions are not representations of the permutation group. These states form representations of the braid group. These states are not defined in terms of the coordinates of the bound states alone. We have seen in Section 10.5 that the amplitude for the propagation of a pair of spinons in a chiral spin liquid has precisely these properties. In that case, the fractional statistics was a consequence of the presence of an induced Chern–Simons term in the effective action for the low-energy degrees of freedom. We will see below that the Chern–Simons term is the most general local gauge-invariant Lagrangian which binds particles and fluxes together. In Chapter 13 we will see that the quasiholes of the Laughlin ground state for the FQHE have very similar properties.

10.7 Chern–Simons gauge theory: a field theory of anyons

In order to make further progress we need a theory that will bind particles and fluxes together. Fluxes are most simply described as curls of a gauge field, which is usually called the statistical gauge field. Also, we want the particles to feel the fluxes through an Aharonov–Bohm mechanism. This means that the particles have to be minimally coupled to the statistical gauge fields through the covariant derivative. There is a problem with this approach. In most cases, a fluxoid that is electromagnetically coupled to a charged particle is not usually bound to it. The Aharonov–Bohm effect is not a bound-state problem. Rather, the amplitudes for the propagation of the particle are modified, by a phase factor, in the presence of flux. Thus, in the usual case, particles and fluxoids move quite independently of each other. In the problem that we are discussing, we want to *force* particles and fluxes to move *together*, as if they were the constituents of a bound state. There is a theory that does all of that in a simple and straightforward way, namely the Chern–Simons gauge theory.

Let us imagine that we have a set of N particles. In a path-integral picture, the motion of the particles is described in terms of a set of trajectories $\Gamma = \Gamma_1 + \cdots + \Gamma_N$ with specified initial and final conditions. Quantum mechanics tells us that we have to sum over all possible trajectories, weighting each history by the usual phase factor $\exp((i/\hbar)S(\gamma))$ in terms of the classical action of that particular history. If the particles have mass m, the classical action S_m of the particles is

$$S_m(\gamma) = \int_{t_i}^{t_f} dt \sum_{j=1}^{N} \left(\frac{1}{2} m \left(\frac{d\vec{x}_j}{dt} \right)^2 + \frac{d\vec{x}_j}{dt} \cdot \vec{\mathcal{A}}(\vec{x}_j, t) - \mathcal{A}_0(\vec{x}_j, t) \right) \tag{10.130}$$

The second term implies that the particle trajectories can also be regarded as a set of currents (and densities) $J_\mu = (J_0, \vec{J}\,)$ ($\mu = 0, 1, 2$) that are different from zero only on the trajectories of the particles and carry the unit of charge.

What should the action for the statistical gauge fields be? It cannot have the standard Maxwell form since purely electrodynamic processes do not yield bound states of particles and fluxes. What is needed is a *constraint* that will *rigidly* bind particles and fluxes. There is only one *gauge-invariant, local* expression that does the job: the Chern–Simons action. Let us consider a theory of a vector field, $\mathcal{A}_\mu$, minimally coupled to a current J_μ (representing matter fields), with the Chern–Simons action

$$S[\mathcal{A}_\mu] = \int d^3x\, \frac{\theta}{4} \epsilon_{\mu\nu\lambda} \mathcal{A}^\mu \mathcal{F}^{\nu\lambda} - \int d^3x\, \mathcal{A}_\mu J^\mu \tag{10.131}$$

The Chern–Simons term, the first term of this action, is the unique local Lagrangian which is locally gauge-invariant (up to some caveats discussed below) with the smallest number of derivatives (one!), that breaks 2D parity $\mathcal{P}$, $x \to -x$ and $y \to y$, *and* time reversal $\mathcal{T}$, while keeping the product $\mathcal{PT}$ invariant. We will see in Chapters 12 and 13 that these features of the Chern–Simons term will enable us to construct a natural effective hydrodynamic theory of the quantum Hall effects (integer and fractional).

To clarify the physical implications of this action it is useful to expand it in components. We find, up to total derivatives, that the Lagrangian density is

$$\mathcal{L}[\mathcal{A}_\mu] = \mathcal{A}_0(x)[\theta\mathcal{B}(x) - J_0(x)] - \frac{\theta}{2}\epsilon_{ij}\mathcal{A}_i(x)\partial_0\mathcal{A}_j(x) - J_i(x)\mathcal{A}_i(x) \tag{10.132}$$

where $\mathcal{B} = \epsilon^{ij}\,\partial_i\mathcal{A}_j$ is the local flux (or "magnetic" field).

The binding (or attachment) of particles and fluxes follows from the observation that in the Lagrangian of Eq. (10.132) the time component of the statistical vector potential $\mathcal{A}_0$ plays the role of a Lagrange multiplier field, which enforces the local constraint

$$J_0 = \frac{\theta}{2}\epsilon_{ij}\mathcal{F}^{ij} \equiv \theta\mathcal{B} \tag{10.133}$$

This constraint simply means that a statistical flux of strength $1/\theta$ is present wherever there is a particle. In Section 10.5 we saw that the presence of a Chern–Simons term modifies the two particle amplitudes in such a way that they exhibit fractional statistics.

However, as we can also see from Eq. (10.132), the Chern–Simons term does more than attaching particles to fluxes. It also determines the canonical structure of this system. Indeed, we can apply the formalism of canonical quantization, which in a gauge theory is most transparent in the temporal ($\mathcal{A}_0 = 0$) gauge. We see that, in this gauge, the canonical momentum conjugate to the component $\mathcal{A}_1$ of the gauge field *is not* the electric field (as is the case in Maxwell's electrodynamics) but (essentially) the other component, $\theta \mathcal{A}_2$. Thus the canonical equal-time commutation relations of Chern–Simons gauge theory are

$$\left[\mathcal{A}_1(\vec{x}), \mathcal{A}_2(\vec{y})\right] = \frac{i}{\theta}\delta^{(2)}(\vec{x} - \vec{y}) \tag{10.134}$$

In fact, the choice of $\mathcal{A}_1$ and $\mathcal{A}_2$ as canonical pairs is arbitrary, since we could have chosen other linear combinations (called *polarizations*). A convenient choice that is often used is the holomorphic polarization with canonical pairs $\mathcal{A}_z = \mathcal{A}_1 + i\mathcal{A}_2$ and $\mathcal{A}_{\bar{z}} = \mathcal{A}_1 - i\mathcal{A}_2$ (a deeper and more general discussion can be found in the work of Elitzur *et al.* (1989)).

In addition, in this gauge, the physical states $|\text{Phys}\rangle$ are required to satisfy (as usual) the Gauss-law condition, which here reduces to the constraint

$$(\theta \mathcal{B}(\vec{x}) - J_0(\vec{x}))|\text{Phys}\rangle = 0 \tag{10.135}$$

What about the Hamiltonian? From Eq. (10.132) it is easy to see that the Hamiltonian is

$$H = \int d^2x \; J_i(\vec{x})\mathcal{A}_i(\vec{x}) \tag{10.136}$$

Thus, in the absence of external currents (i.e. matter fields), the Hamiltonian of the Chern–Simons theory *vanishes* identically.

We conclude that the Chern–Simons gauge theory is equivalent to (a) a constraint relating flux to charge, (b) a set of commutation relations, and (c) a vanishing Hamiltonian. These seemingly peculiar properties follow from the fact that the Chern–Simons gauge theory is a *topological field theory*. As such it is *independent* of the choice of coordinates of space-time and thus it does not depend of the *metric* $g_{\mu\nu}$, from which it follows that the energy–momentum tensor must vanish,

$$T^{\mu\nu} = \frac{\delta S_{\text{CS}}}{\delta g_{\mu\nu}} = 0 \tag{10.137}$$

As a consequence, the Hamiltonian vanishes identically. It also follows from the independence of the metric that the choice of polarization (canonical pairs) must be to a large extent arbitrary since it requires a choice of gauge and a choice of coordinates (all of which explicitly break the topological invariance).

In Section 10.4 we showed that the Chern–Simons action is invariant under local (and smooth) gauge transformations, and that the invariance of the path integral under large gauge transformations requires that the weight of the path integral be single-valued on space-times without boundaries. As we saw, this leads to the quantization of the coupling constant of the Chern–Simons gauge theory, $\theta = k/(2\pi)$, with k being an integer. Witten showed that the expectation values of the gauge-invariant observables, the Wilson loops, are topological invariants and provided a way to compute them (Witten, 1989). On the other hand, on a space-time manifold with boundaries, the Chern–Simons action is not invariant under gauge transformations that do not vanish at the boundary.

This observation is relevant in two cases of interest. One is a system defined on a finite region of space (for example a disk D of circumference L) for all times. In this case the space-time manifold is the filled cylinder $\Sigma = D \times S_1$, where the circle S_1 represents imaginary time. In this case the only states of the Chern–Simons theory (in the absence of matter fields) have vanishing flux, $\mathcal{B} = 0$, and hence are gauge transformations, $\mathcal{A}_\mu = \partial_\mu \phi$ (for the abelian case) and $\mathcal{A}_\mu = U^{-1}\, \partial_\mu U$ for the non-abelian case. Furthermore, a set of coordinates must be specified at the boundary. The effective action for these states depends *only* on the boundary values of ϕ (or U). If we define the boundary to be tangent to the direction x_1, we can write

$$S_{\rm CS}[\phi] = \int dx_0 \int_0^L dx_1\, \frac{k}{4\pi}\left(\partial_0\phi\, \partial_1\phi - (\partial_1\phi)^2\right) \tag{10.138}$$

which is a $(1+1)$-dimensional scale-invariant (and conformal!) field theory, the chiral boson theory. In the non-abelian case the theory also projects to the boundary with an action

$$S_{\rm CS}[U] = k S_{\rm WZW}[U] \tag{10.139}$$

where $kS_{\rm WZW}[U]$ is the action of the *level-k* Wess–Zumino–Witten model (see Chapter 7), another conformal field theory in $(1+1)$ dimensions. Both systems, abelian and non-abelian, will play a key role in the theory of the edge states of quantum Hall fluids. We will discuss this problem in Chapter 13.

The other case of interest is the path integral for the *wave functional* of Chern–Simons theory. By definition, the ground-state wave function Ψ_0 is the amplitude of the evolution from some specified initial state to the final (vacuum) state. Thus in this case the space-time manifold is open and has a boundary, the initial time surface. This discussion tells us that the wave function depends on a choice of coordinates to define the initial state (which hence breaks general coordinate invariance) and also on the choice of polarization. Since the path integral is no longer gauge-invariant (it depends on the choice of gauge for the initial state) neither is the wave function. This is analogous to the gauge covariance of the wave functions of charged particles in magnetic fields. Nevertheless, the inner products of these states and the expectation values of (gauge-invariant) physical observables are gauge-invariant (Witten, 1992).

The key result from Witten's development of Chern–Simons gauge theory (Witten, 1989) is the identification of the expectation value of the Wilson loop operators with the topological invariants of the theory of knots, the Jones polynomial. Since the action is topological, the expectation values of the Wilson loops depend not on the shape and size of the loops but only on the topological character of the knots they form. The Wilson loops, on the other hand, are idealized representations of the worldlines of heavy particles in different physical processes (as we have seen earlier in this chapter).

The expectation values of the Wilson loops describe the quantum-mechanical amplitudes for adiabatic processes of these heavy excitations. These adiabatic processes are equivalent to an exchange of the particles, the braiding of their worldlines. For this reason, the fractional statistics that results, Eq. (10.117), is better called (abelian) *braiding statistics*. Braids satisfy a group property, in the sense that two consecutive braids are equivalent to a braid. This defines the *braid group*. The Hilbert space of states described by Chern–Simons gauge theory and its observables, the Wilson loops, can be classified by their braiding properties. Thus, these quantum states are representations of the braid group. The spinons of the chiral spin liquid and the quasiparticles of the Laughlin states of the FQHE (which we will discuss later) are one-dimensional representations of the braid group, described by *a single phase*, the statistical angle. Since these representations are one-dimensional, this type of fractional statistics is said to be abelian since different processes commute with each other.

However, the braid group has finite-dimensional representations. Witten showed that this happens in a Chern–Simons gauge theory with a non-abelian gauge group, such as SU(2), provided that the level $k > 1$. These representations are labeled by a *matrix* of phases. Consequently exchange processes of particles with this property do not commute. This type of fractional statistics is called non-abelian. We will see in Chapter 13 that the quasihole excitations of certain fractional quantum Hall states are examples of non-abelian fractional statistics. A key feature of these states is that the wave function for N of these particles is not fully specified by their coordinates. This results in a finite degeneracy of these states. A braiding process of two of the particles in one of these states is equivalent to a linear combination of the degenerate states. These different linear combinations define their braiding matrices. This startling property led Kitaev (2003) and Freedman *et al.* (2002b) to propose the use of these states to make a *topological* quantum computer! (Das Sarma *et al.*, 2008).

The previous discussion tells us that Chern–Simons gauge theory does not have local degrees of freedom, which is natural since it is a topological field theory. However, this does not mean that its Hilbert space is trivial. On a closed topologically non-trivial manifold, such as the torus, Chern–Simons gauge theories have a finite-dimensional Hilbert space whose dimension is determined by the quantized coupling constant, the gauge group, and the topology of the manifold (Witten, 1989).

10.8 Periodicity and families of Chern–Simons theories

The results of the last section allow us to conclude that a theory of *anyons* with statistical phase δ can be *defined* in terms of a theory of *fermions* coupled to a *Chern–Simons gauge field* with a coupling constant $\theta = 1/(2\delta)$. Likewise, the *same* theory of anyons can also be *defined* in terms of a theory of *bosons with hard cores* coupled to a Chern–Simons gauge field but with a coupling constant $\theta = 1/(2(\delta \pm \pi))$. This equivalence is the starting point of the boson approach.

However, there is an apparent discrepancy between the fermion (or boson) and anyon theories. The problem is that the anyon commutation relations are *periodic* in the statistical phase δ. Nothing changes in the anyon problem if the statistical phase is shifted by $\delta \to \delta + 2\pi n$, where n is an *arbitrary integer*, not necessarily positive. On the other hand, the only information in Chern–Simons theory about the statistics of the particles is in the coupling constant θ. It is not obvious that the Chern–Simons theory is *invariant* under the change in its coupling constant $1/\theta \to 1/\theta + 4\pi n$ as is required by the anyon commutation relations. This issue is of particular importance, since all of the approximations which are commonly made, such as the average-field approximation of Laughlin, work only in one particular period, i.e. for a choice of n. Fortunately, it is possible to show that the Chern–Simons theory is indeed invariant under shifts. Notice that a shift of δ by $2\pi n$ is equivalent to attaching an additional *even* number $2n$ of flux quanta to each one of the particles. The argument is the following.

Let us first prove that "an even number of flux quanta is the same as nothing." Consider a system of fermions coupled to a Chern–Simons gauge field with coupling constant $\theta = 1/(4\pi n)$. In first quantization, the functional integral reduces to a sum over all the histories of the particles and gauge fields. In Section 10.5 we showed that the trajectories of the fermions form braids. If we compare two histories that differ just by the relative braiding of two particles, the propagation amplitude changes just by a phase factor $\exp(i\,\Delta\nu/(2\theta))$, where $\Delta\nu$ is the change in the linking number. Thus, for $\theta = 1/(4\pi n)$, all scattering amplitudes remain unchanged since the phase change is just an integer multiple of 2π.

This suggests that, if we want to attach an additional even number of fluxes to each particle, then we have to couple the system of fermions to a *new* Chern–Simons gauge field, let us call it $\mathcal{A}'_\mu$, with a coupling constant $\theta' = 1/(4\pi n)$. Thus, the fermions end up being coupled to *two* Chern–Simons gauge fields, of which one is responsible for the fractional statistics and the other for the periodicity.

However, the resulting theory seems to be unnecessarily complicated. This problem can be remedied quite easily. Since the (abelian!) Chern–Simons action is bilinear in the fields, we can integrate out one of the two gauge fields. More precisely, let us consider a problem in which two Chern–Simons gauge fields, $\mathcal{A}_\mu$ and $\mathcal{A}'_\mu$, are both coupled to the same Fermi field ψ through the Lagrangian density $\mathcal{L}$ (I drop the subindex of the gauge fields)

$$\mathcal{L} = \mathcal{L}_{\rm F}[\psi, \mathcal{A} + \mathcal{A}'] + \theta_1 \mathcal{L}_{\rm CS}[\mathcal{A}] + \theta_2 \mathcal{L}_{\rm CS}[\mathcal{A}'] \tag{10.140}$$

where $\mathcal{L}_{\rm F}[\psi, \mathcal{A}+\mathcal{A}']$ is the fermion part of the Lagrangian. Note that the fermions are assumed to couple in the same way to both gauge fields. This is needed in order for the fluxes to be additive. We can use the invariance of the integration measure to define a new gauge field $A = \mathcal{A} + \mathcal{A}'$. The fermion couples only to the field A.

Let us now compute the functional integral over the fields $\mathcal{A}'$. After the shift the Lagrangian reads

$$\begin{aligned} \mathcal{L} &= \mathcal{L}_{\rm F}[\psi, A] + \theta_1 \mathcal{L}_{\rm CS}[A - \mathcal{A}'] + \theta_2 \mathcal{L}_{\rm CS}[\mathcal{A}'] \\ &= \mathcal{L}_{\rm F}[\psi, A] + (\theta_1 + \theta_2) \mathcal{L}_{\rm CS}[\mathcal{A}'] + \theta_1 \mathcal{L}_{\rm CS}[A] - \frac{\theta_1}{2} \epsilon_{\mu\nu\lambda} \mathcal{A}'_\mu F_{\nu\lambda} \end{aligned} \tag{10.141}$$

The functional integral over the $\mathcal{A}'_\mu$ fields can be carried out exactly. As usual, one first shifts the field $\mathcal{A}'_\mu \to \mathcal{A}'_\mu + \tilde{\mathcal{A}}_\mu$, and $\tilde{\mathcal{A}}_\mu$ is then determined from the condition that the terms linear in $\mathcal{A}'_\mu$ are exactly cancelled out. This condition yields the result

$$\tilde{\mathcal{A}}_\mu = \left(\frac{\theta_1}{\theta_1 + \theta_2} \right) A_\mu \tag{10.142}$$

The fermions are coupled to a *single* Chern–Simons gauge field A_μ with the effective Lagrangian

$$\mathcal{L}_{\rm eff} = \mathcal{L}_{\rm F}[\psi, A_\mu] + \theta_{\rm eff} \mathcal{L}_{\rm CS}[A_\mu] \tag{10.143}$$

The effective Chern–Simons coupling $\theta_{\rm eff}$ given by

$$\frac{1}{\theta_{\rm eff}} = \frac{1}{\theta_1} + \frac{1}{\theta_2} \tag{10.144}$$

If we now make the choice $\theta_2 = 1/(4\pi n)$ we get the desired result.

Thus, Chern–Simons theories with coupling constants θ of the form $1/\theta = 2\delta + 4\pi n$ have the *same physical properties*. This result is often called the flux-attachment transformation. Although this equivalence is an exact result, approximations for each member of this sequence yield quite different results. This property will be of great importance for our discussion of the FQHE in Chapter 13.

10.9 Quantization of the global degrees of freedom

In this section we consider the role and quantization of global degrees of freedom. Here I follow the results of Wen, Dagotto, and myself (Wen *et al.*, 1990), and of Section 9.9. The *global* gauge degrees of freedom $\bar{A}_j$ are completely unaffected by the Jordan–Wigner transformation, which involves only local transformations. They satisfy the *homogeneous* constraint equation

$$\epsilon_{jk}\Delta_j\bar{A}_k = 0 \tag{10.145}$$

As was discussed above, these degrees of freedom cannot be eliminated by local ("small") gauge transformations since they have non-vanishing circulation Γ_j along any large circles C_j of the torus, i.e. the holonomies of the torus. The "best" we can do, for instance, is to pick the gauge in which the fields $\bar{A}_j$ are constant in space (but not in time!),

$$\bar{A}_j = \frac{\Gamma_j(t)}{L_j} \tag{10.146}$$

(no sum over j is implied).

These relations allow us to derive an effective Lagrangian for the global degrees of freedom $\Gamma_j(t)$ and to extract from it the quantum dynamics of the global degrees of freedom. By carrying out the canonical formalism to completion, it is easy to check that the non-integrable phases obey the commutation relations

$$[\Gamma_1, \Gamma_2] = \frac{i}{\theta} \tag{10.147}$$

Hence, Γ_1 and $\theta\Gamma_2$ form a *canonical pair* and cannot be diagonalized simultaneously. This feature is not present in 1D systems, for which there is only one non-integrable phase, which is just a c-number. The global degrees of freedom in one dimension are just boundary conditions. In (2 + 1) dimensions, we discover that the global degrees of freedom acquire a life of their own. We will see now that, as a result of this feature, the states of anyon systems on a torus are not determined by the location of the particle alone.

It is now easy to check that the operators $\exp(i\Gamma_j)$ satisfy the algebra

$$e^{i\Gamma_1}e^{i\Gamma_2} = e^{-\frac{i}{\theta}}e^{i\Gamma_2}e^{i\Gamma_1} \tag{10.148}$$

Let us denote the exponential operators $\exp(i\Gamma_j)$ by T_j. These operators will give an extra phase to any state as the anyons move around each other. Furthermore, since Γ_1 and Γ_2 do not commute, the *eigenstates* of the Hamiltonian are functions of either variable but not of both at the same time. Also, both Γ_1 and Γ_2 enter only through the exponential operators T_j. Thus we can always choose, say, Γ_1 to be an angle with a range $[0, 2\pi]$. Hence $\theta\Gamma_2$ is an angular-momentum-like operator whose spectrum is the set of integers. In all cases of physical interest, the statistical angle θ can only take the restricted set of values $\theta = m/(2\pi n)$, where m and n are integers. After all *local* gauge degrees of freedom have been eliminated, we find that the effective Hamiltonian for the anyon system has the form

$$H = \sum_{\vec{x}, j=1,2} a^\dagger(\vec{x}) \exp\left[i\left(\mathcal{A}_j(\vec{x}) + \frac{\Gamma_j}{L_j}\right)\right] a(\vec{x} + e_j) + \text{h.c.} \tag{10.149}$$

where $\mathcal{A}_j$ is given by the solution of the local constraint. This Hamiltonian is *almost* identical to the "free-anyon" Hamiltonian. The only difference here is the presence of the *global* degrees of freedom Γ_j, which were not included in our original naive expression. We will adopt this generalized version as the *definition* of the anyon Hamiltonian. In other words, the global degrees of freedom are an intrinsic feature of the anyon system on a torus. Clearly, if the manifold on which the anyons move is not a torus, but some other manifold, the properties of the global degrees of freedom will be different. For instance, if the system is quantized on a manifold with a boundary, such as a disk, there are no global degrees of freedom. Instead, gauge invariance requires the existence of edge states, which have very interesting properties.

The form of the Hamiltonian suggests that its eigenstates are not functions only of the coordinates of the anyons, since H involves the global degrees of freedom as well. Let us denote by Ψ_0 an eigenstate of H. We can also choose Ψ_0 to be an eigenstate of Γ_1 with zero eigenvalue or, which is equivalent, to be an eigenstate of T_1 with unit eigenvalue

$$T_1 \Psi_0 = \Psi_0 \tag{10.150}$$

Thus $|\Psi_0\rangle$ is the "highest-weight state." Let us consider now the state Ψ_p defined by

$$\Psi_p \equiv T_2^p \Psi_0 \tag{10.151}$$

The state Ψ_p is an eigenstate of T_1,

$$T_1 \Psi_p \equiv T_1 T_2^p \Psi_0 = e^{-\frac{ip}{\theta}} T_2^p T_1 \Psi_0 = e^{-\frac{ip}{\theta}} \Psi_p \tag{10.152}$$

with the eigenvalue $e^{-\frac{ip}{\theta}}$. Thus, for all cases of physical interest ($\theta = m/(2\pi n)$), there are m distinct eigenstates, each labeled by the integer p. The states of the Hilbert space are thus labeled by the anyon coordinates *and* by the quantum number

p describing the state of the global degrees of freedom. In particular, the condensed states of the anyon system do exhibit this degeneracy. The idea that such topological degeneracies occur quite generally in spin-liquid states and other topological fluid states, which we discussed extensively in Chapter 9, was originally due to Wen (1989).

10.10 Flux phases and the fractional quantum Hall effect

In Section 8.5 we considered solutions to the saddle-point equations, Eq. (8.64) and Eq. (8.65), with a spontaneously generated *flux* of π per plaquette. The problem was shown to be equivalent, at the saddle-point level, to a system of fermions moving in a *uniform average field* with a one-half flux quantum per plaquette. In Section 10.1, we saw that a next-nearest-neighbor exchange coupling, which frustrates the system, effectively lowers the energy of the flux state. Furthermore, it drives the flux state into a chiral phase with spontaneously broken time-reversal invariance. The flux phase has two bands that become degenerate at four points of the Brillouin zone. The chiral states have gaps at those points, and the gaps grow larger as the frustrated regime $J_1 \simeq 2J_2$ is approached.

If the fluctuations around the mean field are ignored (in the first stage), a flux phase is then equivalent to two species (up and down spinons) of fermions moving in that flux. In the chiral phase we also have a gap, which grows larger as frustration increases (i.e. for increasing J_2/J_1). The one-particle spinon states can, in this limit, be approximated by the eigenstates of the lowest Landau level of a continuum problem in which the fermions move in a field with the same total flux. This approximation should be qualitatively correct, provided that no level crossings occur. However, as we stressed previously, it is not possible to ignore the fluctuations around the mean field. Nevertheless, such an analogy offers the possibility of a new sort of spin liquid: a Laughlin state.

Laughlin states (Laughlin, 1983) are condensed states of N fermions moving on a plane in the presence of an external magnetic field. These *incompressible* states, which have been shown to exhibit the *fractional Hall effect*, represent a featureless liquid. It is tempting to speculate that the spin-liquid states, which are also incompressible if there is a gap, may be described in terms of a Laughlin wave function, which we will discuss below.

Kalmeyer and Laughlin (1987) showed that, in the case of frustrated quantum spin systems, there is indeed a close analogy with the Hall-effect system except for the fact that, here, we have bosons instead of fermions. Let us discuss the Kalmeyer–Laughlin picture in more detail. Consider a frustrated quantum spin system, such as the square lattice with $J_1 = 2J_2$ or the triangular lattice. Let us assume that the Hamiltonian is still given by the usual Heisenberg exchange Hamiltonian.

Instead of representing spins in terms of constituent bands of fermions, one can use *hard-core bosons* instead. This idea goes back to Holstein and Primakoff. Let $|F\rangle$ represent the ferromagnetic state, which we will use as a *reference* state, not necessarily the ground state. Relative to $|F\rangle$, the raising operator $S^+(\vec{r})$ acts like a boson-creation operator, a *spin-flip* being the boson. Since it is not possible to flip a spin twice, the bosons should have hard cores: a site cannot be occupied by more than one boson. More formally, we can write

$$\begin{aligned} S^+(\vec{r}) &= S_1(\vec{r}) + i S_2(\vec{r}) \equiv a^\dagger(\vec{r}) \\ S^-(\vec{r}) &= S_1(\vec{r}) - i S_2(\vec{r}) \equiv a(\vec{r}) \end{aligned} \tag{10.153}$$

and

$$S_z(\vec{r}) = a^\dagger(\vec{r})a(\vec{r}) - \frac{1}{2} \tag{10.154}$$

where the operators a and $a^\dagger$ are bosons and, hence, satisfy the commutation relations

$$[a(\vec{r}), a^\dagger(\vec{r}\,')] = \delta_{\vec{r},\vec{r}'} \tag{10.155}$$

The Pauli spin algebra requires that these operators also satisfy a hard-core condition,

$$a^2 = (a^\dagger)^2 = 0 \tag{10.156}$$

Using these identities, it is now easy to write the Heisenberg Hamiltonian in terms of hard-core bosons. Notice that these identities follow just from the nature of the states at each site. Thus, they hold for any lattice and dimension.

Thus, the quantum Heisenberg antiferromagnet can be written as an equivalent model of *hard-core* bosons with a Hamiltonian of the form

$$\begin{aligned} H = & \frac{J}{2} \sum_{\langle \vec{r},\vec{r}\,' \rangle} \left(a^+(\vec{r})a(\vec{r}\,') + a^+(\vec{r}\,')a(\vec{r}) \right) \\ & - \gamma J \sum_{\vec{r}} a^+(\vec{r})a(\vec{r}) \\ & + J \sum_{\langle \vec{r},\vec{r}\,' \rangle} a^+(\vec{r})a(\vec{r})a^+(\vec{r}\,')a(\vec{r}\,') \\ & + \frac{\gamma N J}{4} + U_\infty \sum_{\vec{r}} a^+(\vec{r})a(\vec{r})(a^+(\vec{r})a(\vec{r}) - 1) \end{aligned} \tag{10.157}$$

where $\langle \vec{r}, \vec{r}\,' \rangle$ stands for the nearest-neighboring sites $\vec{r}$ and $\vec{r}\,'$ (on that lattice), γ is the coordination number, and N is the total number of sites. The last term enforces the hard-core condition since at $U_\infty \to \infty$ the only states in the Hilbert space with finite energy are occupied by at most one boson.

We are interested in studying the sector of the Heisenberg model with $S_z^{\text{tot}} = 0$. This implies that the bosons half-fill the system. Thus, if N_{B} is the number of bosons, we have

$$S_z^{\text{tot}} = \sum_{\vec{r}} \left(\frac{1}{2} - a^+(\vec{r})a(\vec{r}) \right) = \frac{N}{2} - N_{\text{B}} = 0 \tag{10.158}$$

i.e. the number of bosons equals half the number of sites, $N_{\text{B}} = N/2$.

The first term of the Hamiltonian Eq. (10.157) can be regarded as a kinetic-energy term for the bosons. However, it has the wrong sign. We can remedy this problem by means of the following trick. Let $A(\vec{r}, \vec{r}\,')$ be a fixed gauge field defined on each link. Let us write the Hamiltonian Eq. (10.157) in the form

$$\begin{aligned} H = &-\frac{J}{2} \sum_{\langle \vec{r}, \vec{r}\,' \rangle} \left(a^+(\vec{r}) e^{iA(\vec{r}, \vec{r}\,')} a(\vec{r}\,') + \text{h.c.} \right) + J \sum_{\langle \vec{r}, \vec{r}\,' \rangle} a^+(\vec{r})a(\vec{r})a^+(\vec{r}\,')a(\vec{r}\,') \\ &+ U_\infty \sum_{\vec{r}} a^+(\vec{r})a(\vec{r})a^+(\vec{r})a(\vec{r}) - \left(\frac{\gamma N J}{4} + \frac{U_\infty N}{2} \right) \end{aligned} \tag{10.159}$$

This expression is consistent with Eq. (10.157) provided that $A(\vec{r}, \vec{r}\,') = \pi$ for all bonds of the lattice. Now, the first term does have the interpretation of the kinetic-energy operator for the bosons, but there is an external fixed gauge field $A(\vec{r}, \vec{r}\,')$. This gauge field, or rather its *circulation*, represents the frustration of the spin system. For the case of a bipartite lattice, such as the square lattice, this gauge field can be removed. This is so because the circulation of $A(\vec{r}, \vec{r}\,')$ around any elementary plaquette of the square lattice is always equal to 2π, which, by periodicity, is equivalent to zero. Indeed, on the square lattice, the transformation

$$a(\vec{r}) \to (-1)^{x_1 + x_2} a(\vec{r}) \tag{10.160}$$

flips all the signs and we get a kinetic-energy operator with a proper sign.

However, for a frustrated lattice, it is not possible to do this. In the case of a triangular lattice the circulation is 3π, which is equivalent to π (mod 2π). Thus, the flux is intrinsic and it is determined by the lattice structure. Moreover, we conclude that the bosons behave like particles of charge e moving in an external magnetic field B with a flux of half of the flux quantum per triangle. This result motivates the following approximation (Kalmeyer and Laughlin, 1987).

Consider a system of hard-core bosons with an effective mass M moving on a plane in the presence of an external magnetic field B and of a periodic potential $V(\vec{r})$ that localizes the bosons on the lattice sites. The bosons also have a short-range interaction. Now one imagines varying the periodic potential from some weak value to the strong tight-binding limit, in which Eq. (10.159) holds. The magnetic field B is fine-tuned so as to always give one-half of a flux quantum per triangle. If we denote the lattice spacing by a_0 and the magnetic (or cyclotron) length by l_0, we can fulfill the requirements mentioned above by setting $B = 1/l_0^2$ and $l_0 = a_0(\sqrt{3}/(4\pi))^{1/2}$, in units in which the flux quantum ϕ_0 equals 2π.

Assume for the moment that we can make the further approximation that the tight-binding (lattice) limit and the weak-potential limit are smoothly connected. In this limit a simple physical picture can be drawn. The problem we are dealing with is that of a set of bosons with hard cores and short-range interactions, which carry the unit of charge and are moving on a plane in the presence of an external magnetic field perpendicular to the plane. Except for the fact that these particles are bosons, this situation appears to be identical to the problem of the fractional quantum Hall effect. In that case *fermions* (electrons each with charge e) move on a plane in the presence of a magnetic field B with the same geometry. The electrons have short-range interactions. This problem was solved by Laughlin (1983) who guessed a wave function for it that appears to have exceedingly good properties. It then appears that the chiral spin state and the fractional quantum Hall effect (FQHE) belong to a general class of problems that are characterized by strong correlation and broken time-reversal invariance. In the FQHE case, time reversal is broken *explicitly* by the presence of the external magnetic field. In the chiral-spin-state case time-reversal symmetry is spontaneously broken. We will see below that, at long distances and low energies, both problems have effective Lagrangians that include a Chern–Simons term. In a sense it is this Chern–Simons term which defines this problem.

10.11 Anyons at finite density

In this section we consider a simple model that describes a gas of anyons at finite density. Since we are interested in systems in their thermodynamic limit, this theory is necessarily a field theory of anyons. The model that we will discuss is a system of "free" anyons on a square lattice (in space) with the topology of a torus. We choose to work on a spatial lattice both in order to avoid regularization problems and with an eye on applications to theories of high-T_c superconductors. The time variable will remain continuous. This choice simplifies the formalism without any significant loss of generality. The model can also be defined rigorously on a space-time lattice (Fröhlich and Marchetti, 1988). The results have much wider applications than our derivation may suggest. For instance, as a byproduct, we will derive a Jordan–Wigner transformation for systems in two space dimensions. This transformation is of great use for the study of 2D quantum magnets. The theory can also be considered in the continuum, although some care has to be exercised at short distances. Chen, Wilczek, Witten, and Halperin (Chen *et al.*, 1989) considered the continuum non-relativistic theory in great detail. In this section, I discuss the problem on a 2D square lattice, following the results of Fradkin (1989), which were expanded and clarified (and corrected) by Eliezer and Semenoff (1992a, 1992b).

In the model that we consider, the anyons are free in the sense that the Hamiltonian contains only a nearest-neighbor hopping term. However, the anyons will be assumed to have *hard cores*. This requirement is essential to the whole construction, since otherwise the anyon worldlines can cross and the notion of braids falls apart.

Let us now show that the problem of a gas of $N_{\rm a}$ anyons with hard cores on a square lattice is equivalent to a gas of $N_{\rm f} = N_{\rm a}$ fermions, on the square lattice, coupled to a Chern–Simons gauge field defined on the links of that lattice. To be more precise, let $a^\dagger(\vec{x})$ and $a(\vec{x})$ be a set of anyon-creation and -annihilation operators defined on the sites $\{\vec{x}\}$ of the square lattice that satisfy the generalized equal-time commutation relations

$$a(\vec{x})a^\dagger(\vec{y}) = \delta_{\vec{x},\vec{y}} \ - \ e^{i\delta}a^\dagger(\vec{y})a(\vec{x}) \tag{10.161}$$

The angle δ indicates that we are dealing with fractional statistics. The choice of sign is such that for $\delta = 0$ we have fermions, whereas for $\delta = \pi$ we have bosons. The hard-core condition implies that, when acting on physical states, these operators obey

$$a^\dagger(\vec{x})a^\dagger(\vec{x}) = a(\vec{x})a(\vec{x}) = 0 \tag{10.162}$$

The second quantized Hamiltonian is simply given by

$$H = \sum_{\langle\vec{x},\vec{y}\rangle} a^\dagger(\vec{x})a(\vec{y}) + \text{h.c.} \tag{10.163}$$

where $\langle\vec{x}, \vec{y}\rangle$ are nearest-neighboring sites on the square lattice.

Consider now a set of *fermion*-creation and -annihilation operators $c^\dagger(\vec{x})$ and $c(\vec{x})$ on the same square lattice. Let $A_j(\vec{x})$ be a set of *boson* operators defined on the links of the lattice $\{(\vec{x}, \vec{x} + e_j)\}$ (with $j = 1, 2$) representing statistical gauge fields. A naive transcription of the commutation relations of the Chern–Simons gauge theory in the continuum to a discrete lattice would lead to the requirement that the gauge fields defined on the links of the lattice must satisfy equal-time commutation relations of the form

$$\left[A_1(\vec{x}), A_2(\vec{y})\right] = \frac{i}{\theta}\delta_{\vec{x},\vec{y}} \tag{10.164}$$

Notice that, at every point $\vec{x}$ of the lattice, the component of the vector potential along the direction x_1 is the canonical pair of the component along the direction x_2. We will see below that these commutation relations are not quite correct, and that this leads to inconsistencies.

The dynamics of the system is governed by the Hamiltonian

$$H_{\rm f} = \sum_{\vec{x},j} c^{\dagger}(\vec{x}) e^{iA_j(\vec{x})} c(\vec{x}+e_j) + \text{h.c.} \tag{10.165}$$

and the physical states $\{|\text{Phys}\rangle\}$ are required to satisfy a local constraint ("Gauss's law") between the fermion density $\rho(\vec{x})$ and the local magnetic flux $B(\vec{x})$ of the statistical gauge fields,

$$(\rho(\vec{x}) - \theta B(\vec{x}))|\text{Phys}\rangle = 0 \tag{10.166}$$

This constraint implies that a fluxoid of strength $1/\theta$ is attached to each particle at the level of the lattice scale. The local statistical flux $B(\vec{x})$ is given by the usual formula

$$B(\vec{x}) = \Delta_1 A_2(\vec{x}) - \Delta_2 A_1(\vec{x}) \tag{10.167}$$

where Δ_j is the finite-difference operator in the direction j. The flux thus defined effectively exists only on the dual lattice. This formulation has the additional advantage that the particles are not allowed to get "inside" the flux.

However, we must now check that the Gauss-law generators $G(\vec{x})$ defined by Eq. (10.166) commute with each other at all lattice points

$$\left[G(\vec{x}), G(\vec{y})\right] = 0 \tag{10.168}$$

This condition, a requirement for the consistency of a gauge theory, is not satisfied with the naive commutation relations we assumed in Eq. (10.164). There is, however, a consistent set of commutation relations compatible with the Gauss-law constraints, which can be derived from the following lattice action (in continuous time) for the Chern–Simons gauge field (Eliezer and Semenoff, 1992a, 1992b) $(i, j = 1, 2)$:

$$S_{\rm CS}[A] = \int dt \sum_{\vec{x}} A_0(\vec{x},t)\epsilon^{ij} A_j(\vec{x},t) - \frac{1}{2}\int dt \sum_{\vec{x}} A_i(\vec{x},t) L_{ij}\, \partial_t A_j(\vec{x},t) \tag{10.169}$$

where L_{ij} is the following lattice operator:

$$L_{ij} = -\frac{1}{2}\begin{pmatrix} \Delta_2^+ + \Delta_2^- & -2 - 2\Delta_1^+ + 2\Delta_2^- + \Delta_2^-\Delta_1^+ \\ 2 + 2\Delta_2^+ - 2\Delta_1^- - \Delta_1^-\Delta_2^+ & -\Delta_1^+ - \Delta_2^- \end{pmatrix} \tag{10.170}$$

where $\Delta_i^{\pm}$ are the forward (+) and backward (−) difference operators along direction i: $\Delta_i^{\pm} f(\vec{x}) = f(\vec{x} \pm e_i) - f(\vec{x})$, with e_i the lattice unit vector along direction i. The second term of this action defines the canonical commutation relations, which, although they are less local than the naive ones of Eq. (10.164), lead to compatible constraints.

The Hamiltonian H, together with the constraint and the commutation relations, follows from the canonical quantization in the gauge $A_0 = 0$ of the lattice action (again with continuous time)

$$S = \int dt \sum_{\vec{x}} c^\dagger(\vec{x}, t)(i\,\partial_0 + A_0 + \mu)c(\vec{x}, t) - \int dt\; H_{\rm f}(c^\dagger, c, \vec{A}) - S_{\rm CS}[A] \tag{10.171}$$

Here $H_{\rm f}$ is the fermion Hamiltonian of Eq. (10.165), μ is the chemical potential, $x = (\vec{x}, t)$, and $S_{\rm CS}$ is the Chern–Simons action of Eq. (10.169). This action is explicitly invariant under local, time-dependent, gauge transformations.

The equivalence between the anyon Hamiltonian and the Chern–Simons gauge theory coupled to fermions is established by solving the constraint Eq. (10.166), the flux-attachment condition that relates the local flux to the local density. This can be accomplished by fixing the remaining invariance under local time-independent gauge transformations. We will choose the *Coulomb* or *anyon* gauge $\vec{\nabla} \cdot \vec{A}(\vec{x}) = 0$. The statistical vector potential $\vec{A}(\vec{x})$ which is the solution of the constraint in this gauge is an explicit function only of the local particle density. Thus it may appear that there are no gauge degrees of freedom left. This, however, is not generally the case. Whether or not there are any gauge degrees of freedom left depends on the boundary conditions. On a torus, there are global gauge degrees of freedom that are not affected by the local fixing of the gauge.

We now have to solve the constraint, Eq. (10.166), for a square lattice with the topology of a torus. Let L_1 and L_2 be the linear dimensions of the lattice along directions 1 and 2, respectively. It is impossible to eliminate all the gauge degrees of freedom by solving the constraint equation no matter what gauge is chosen unless *large* gauge transformations that wrap around the torus along direction 1 or 2 are included. Following our discussion in Chapter 9, let us consider the circulation of the statistical vector potential on a *non-contractible* closed loop wrapping around the torus along one of its large circles $\mathcal{C}_j$ ($j = 1, 2$). Any *local* time-independent gauge transformation shifts the spatial components of the vector potential A_k by the gradient of a smooth function of the coordinates $\Lambda(\vec{x})$, i.e. $A_k(\vec{x}, t) \to A_k(\vec{x}, t) + \Delta_k \Lambda(\vec{x})$. Thus, the circulation Γ_j, with $\Gamma_j = \oint_{\mathcal{C}_j} d\vec{x} \cdot \vec{A}(\vec{x})$ (which can be defined on the square lattice in an obvious way), is unchanged since Λ is a smooth and single-valued function of $\vec{x}$. Notice that this is the case even in the absence of fermions! Thus, the circulations Γ_j, or *non-integrable phases*, are *global* degrees of freedom of the gauge field. A consistent treatment of this problem must take into account their dynamics.

There is a simple way to take care of both global and local gauge degrees of freedom. The local gauge degrees of freedom are non-local functions of the local particle density $\rho(\vec{x}, t)$ given by the solution of the local constraint equation in some particular gauge. The global degrees of freedom are the non-integrable phases Γ_j. To make any further progress it is necessary to fix the gauge. At the level of the functional integral, we first observe that the component A_0 of the statistical gauge field can always be integrated out, giving rise to the local constraint at all times. We next write the spatial components of the statistical vector potential A_j in the form

$$A_j(x) = \mathcal{A}_j(x) + \bar{A}_j(x) \tag{10.172}$$

where $\mathcal{A}_j$ is a particular solution of the constraint equation and $\bar{A}_j$ generates the non-integrable phases which are solutions to the homogeneous constraint equation (i.e. without fermions). We can completely determine all of these fields by choosing a particular gauge.

Let us consider first the local gauge degrees of freedom. In the Coulomb gauge, the inhomogeneous solution for the constraint equation is given in terms of the scalar field $\Phi(\vec{r})$:

$$\mathcal{A}_j(\vec{x}) = \epsilon_{jk}\Delta_k\Phi(\vec{r}) \tag{10.173}$$

where $\vec{r}$ are the sites of the dual lattice (Fradkin, 1989). Here the scalar field Φ is the solution to the equation (see Eq. (10.166) and Eq. (10.167))

$$\Delta^2\Phi(\vec{r}) = -\frac{1}{\theta}\rho(\vec{x}) \tag{10.174}$$

where $\vec{r}$ is the site on the dual lattice located northeast of the site $\vec{x}$ on the direct lattice and Δ^2 is the lattice Laplacian.

In this approach, fluxes are on the dual lattice while particles are on the direct lattice. Particles and fluxes never sit on top of each other and we have no ambiguities. On the other hand, we could have chosen to put the flux southeast of the particle, or some other similar prescription. These different prescriptions are related to the possible existence of a self-linking number and an anomalous spin. We will not explore these issues any further. Let us simply note that this lattice regularization provides a natural way to separate particles and fluxes while keeping all the relevant symmetries intact. Also note the close analogy with the order–disorder operator construction for 2D classical statistical-mechanical systems. This feature is also present in the 2D classical Ising model and it reflects the fact that the Onsager fermions are two-component spinors (Kadanoff and Ceva, 1971).

We now use the lattice Green function $G(\vec{r}, \vec{r}\,')$, that is, the solution of the partial-difference equation

$$\Delta_{\vec{r}}^2 G(\vec{r}, \vec{r}\,') = \delta_{\vec{r},\vec{r}\,'} - \frac{1}{L_1 L_2} \tag{10.175}$$

The last term of this equation, while unimportant in the thermodynamic limit, is necessary in order to define the Green function in a finite system without boundaries, no matter how large it is. The solution for the scalar field has the form

$$\Phi(\vec{r}) = \frac{1}{\theta} \sum_{\vec{r}\,'} G(\vec{r}, \vec{r}\,') \rho(\vec{x}') \tag{10.176}$$

Thus, by inserting Eq. (10.176) into Eq. (10.173), we can write the vector potentials $\mathcal{A}_j$ in the form

$$\mathcal{A}_j(\vec{x}) = \frac{1}{\theta} \epsilon_{jk} \Delta_k \sum_{\vec{r}'} G(\vec{r}, \vec{r}\,') \rho(\vec{x}') \tag{10.177}$$

Let us *define* the multivalued function $\Theta(\vec{x}, \vec{r}')$ as the solution for the lattice version of the Cauchy–Riemann equation

$$-\Delta_j G(\vec{r}, \vec{r}\,') = \epsilon_{jk} \Delta_k \Theta(\vec{x}, \vec{r}\,') \tag{10.178}$$

The function $\Theta(\vec{x}, \vec{r}\,')$ is found by integrating the Cauchy–Riemann equation along a path $\Gamma(\vec{x}, \vec{x}')$, on the direct lattice, going from $\vec{x}$ to $\vec{x}'$ which leaves the point $\vec{r}$ to its left. For a finite system, the function $\Theta(\vec{x}, \vec{r}\,')$ obtained by this procedure is *path-dependent.* Moreover, along a closed path Γ on the direct lattice, which has the point $\vec{r}$ of the dual lattice in its interior region, the function Θ has a discontinuity $(\Delta\Theta)_\Gamma$. We can compute this discontinuity by using the Cauchy–Riemann equation

$$(\Delta\Theta)_\Gamma = \sum_\Gamma s_j(\Gamma) \Delta_j \Theta = \sum_\Gamma s_j(\Gamma) \epsilon_{jk} \Delta_k G \tag{10.179}$$

where $s_j(\Gamma)$ is a vector field that is equal to unity on the path Γ and zero everywhere else. The last "line integral" in this equation can be computed by first using a discrete version of Gauss's theorem and then inserting Eq. (10.175) to yield

$$(\Delta\Theta)_\Gamma = \sum_{\bar{\Gamma}} \Delta^2 G = 1 - \frac{\mathcal{A}(\bar{\Gamma})}{L_1 L_2} \tag{10.180}$$

where $\bar{\Gamma}$ is the region of the dual lattice inside the closed path Γ and $\mathcal{A}(\bar{\Gamma})$ is its area. Thus, *in the thermodynamic limit*, the function Θ has a jump equal to unity as a closed path Γ is traversed. Equivalently, we can say that Θ is a *multivalued function* that has a branch cut representing a jump by one unit. Using the same line of reasoning, one can show that the following important identity holds:

$$\Theta(\vec{x}, \vec{r}\,') - \Theta(\vec{x}', \vec{r}) = \frac{1}{2} \tag{10.181}$$

This equation can be derived by using the following geometric construction. Draw a rectangle centered at $\vec{x}$ that has corners at $\vec{x} + \vec{R}$ and $\vec{x} - \vec{R}$ along a diagonal.

We now consider the paths Γ_1, $\vec{x}+\vec{R} \to \vec{x}-\vec{R}$ without crossing the cut, and Γ_2, $\vec{x}+\vec{R} \to \vec{x}-\vec{R}$ crossing the cut. By symmetry, we have

$$\left(\Theta(\vec{x}-\vec{R},\vec{r}) - \Theta(\vec{x}+\vec{R},\vec{r})\right)_{\Gamma_1} = \left(\Theta(\vec{x}+\vec{R},\vec{r}) - \Theta(\vec{x}-\vec{R},\vec{r})\right)_{\Gamma_2} \quad (10.182)$$

Since the total discontinuity of Θ is unity, $[\Delta\Theta]_{\Gamma_1+\Gamma_2} = 1$, we get just half that result for a "half-way trip."

We can now use the Cauchy–Riemann equation, Eq. (10.178), to write the vector potential $\mathcal{A}_j$ in Eq. (10.177) as the *gradient* of a scalar "function" $\phi(\vec{x})$:

$$\mathcal{A}_j(\vec{x}) = \Delta_j \phi(\vec{x}) \quad (10.183)$$

where ϕ is given by

$$\phi(\vec{x}) = \frac{1}{\theta} \sum_{\vec{x}'} \Theta(\vec{x},\vec{r}\,')\rho(\vec{x}') \quad (10.184)$$

Therefore, the vector potentials associated with the *local* gauge degrees of freedom are pure gradients, and they can be "eliminated" by means of the (singular) "gauge transformation"

$$a(\vec{x}) = e^{-i\phi(\vec{x})} c(\vec{x}) \quad (10.185)$$

However, since ϕ is a function of the local density $\rho(\vec{x})$, the phase factor $e^{-i\phi}$ is not a c-number but an operator. This operator creates a coherent state of vector potentials, which represents the flux attached to the particles. The operators $a(\vec{x})$ *so defined* satisfy the anyon commutation relations and the hard-core condition. Indeed, after some straightforward algebra we get that the operators $a(\vec{x})$ satisfy the commutation relations

$$\hat{a}(\vec{x})\hat{a}^\dagger(\vec{y}) = \delta_{\vec{x},\vec{y}} - e^{i\delta}\hat{a}^\dagger(\vec{y})\hat{a}(\vec{x}) \quad (10.186)$$

where the statistical phase δ is given by

$$\delta = \frac{1}{\theta}\left(\Theta(\vec{x},\vec{r}\,') - \Theta(\vec{x}',\vec{r})\right) = \frac{1}{2\theta} \quad (10.187)$$

The hard-core condition $a(\vec{x})^2 = 0$ is a consequence of the fact that the operator $c(\vec{x})$ is a fermion. Thus, the operators $a(\vec{x})$ and $a^\dagger(\vec{x})$ are anyon-destruction and -creation operators. The statistical angle δ and the Chern–Simons coupling constant θ are related by

$$\delta = \frac{1}{2\theta} \quad (10.188)$$

This is the same result as that we derived in Section 10.5 by considering a first-quantized path-integral approach.

It is clear that much of what was done above for a lattice theory can also be done in the continuum case. Thus, the identification of anyons with either fermions or bosons coupled to Chern–Simons gauge fields is also valid for continuum systems (Semenoff, 1988), but with one caveat. The notion of attaching fluxes to particles in the continuum is a very tricky one. We remarked in Section 10.5 that, in addition to fractional statistics, the particles may acquire a fractional induced spin depending on the definition of the problem at short distances. For example, if the particle and the charge literally "sit on top of each other," there is no relative winding and no extra phase can possibly appear. But, if the particle and the flux are separated by some distance, they can wind around each other. As a result an extra phase may appear in the propagation amplitudes. This extra phase can be interpreted as an induced fractional spin. The lattice theory that we have discussed above does separate particles from fluxes in a natural and gauge-invariant way. We then expect that lattice anyons should have an induced fractional spin.

10.12 The Jordan–Wigner transformation in two dimensions

The identity

$$\hat{a}(\vec{x})\hat{a}^\dagger(\vec{y}) = \delta_{\vec{x},\vec{y}} - e^{i\delta}\hat{a}^\dagger(\vec{y})\hat{a}(\vec{x}) \tag{10.189}$$

leads to the 2D analog of the Jordan–Wigner transformation discussed in Chapter 5. In particular, for $\theta = 1/(2\pi m)$ we get $\delta = \pi m$. Hence, for m odd the operators $a(\vec{x})$ obey equal-time *boson* commutation relations and a hard-core condition. If we recall the mapping in Section 10.6 between bosons with hard cores and spin-1/2 Pauli operators:

$$\sigma^+(\vec{x}) = a^\dagger(\vec{x}), \quad \sigma^-(\vec{x}) = a(\vec{x}), \quad \sigma_3(\vec{x}) = 2a^\dagger(\vec{x})a(\vec{x}) - 1 \tag{10.190}$$

we get from Eq. (10.185)

$$\sigma^+(\vec{x}) = c^\dagger(\vec{x})e^{+i\phi(\vec{x})}, \quad \sigma^-(\vec{x}) = e^{-i\phi(\vec{x})}c(\vec{x}), \quad \sigma_3(\vec{x}) = 2c^\dagger(\vec{x})c(\vec{x}) - 1 \tag{10.191}$$

These equations tell us that the 2D quantum Heisenberg antiferromagnet on a square lattice is *exactly equivalent* to a theory of spinless fermions *on the same lattice* coupled to a Chern–Simons gauge field. In addition, there is a direct density–density repulsive force among nearest neighbors. Thus, unlike the familiar results from one dimension, in which the fermions are *free* (see Chapter 5) in the XY limit, there is a long-range gauge interaction in two dimensions even in the XY limit. This property is due to the fact that, in one dimension, the only possible flux that the fermions can feel is a global effect determined by the boundary conditions. In two dimensions, the fermions feel both a local and a global flux.

As we will see next, even the global flux is non-trivial. Although the resulting fermion theory is not free, approximations and perturbation theory in one scheme still turn into a non-perturbative feature in the other. The approach employing the Jordan–Wigner transformation, combined with an average field approximation (and quantum corrections), has been applied with success to the case of the 2D spin-$1/2$ antiferromagnet (López *et al.*, 1994). At the mean-field (average-field) level it has also been applied to the quantum antiferromagnet on some frustrated lattices (Yang *et al.*, 1993; Misguich *et al.*, 2001).

11
Anyon superconductivity

11.1 Anyon superconductivity

In this chapter we will consider the problem of predicting the behavior of an assembly of particles obeying fractional statistics. We have already considered the problem of the quantum mechanics of systems of anyons. However, we did not consider what new phenomena may arise if the system has a macroscopic number of anyons present. At the time of writing, the physical reality of this problem is still unclear. However, this is such a fascinating problem that we will discuss it despite the lack of firm experimental support for the model.

There are two different physical situations in which the problem of anyons at finite density is important. Halperin (1984) observed that the quasiparticles of the Laughlin state for the FQHE obeyed fractional statistics (i.e. they are *anyons*). In Chapter 13 we will discuss Halperin's theory. Furthermore, Halperin and Haldane suggested that, for filling fractions of a Landau level different from the $1/m$ Laughlin sequence, the ground state of a 2D electron gas in a strong magnetic field could be understood as a Laughlin state of anyons. Shortly afterwards, Arovas, Schrieffer, Wilczek, and Zee (Arovas *et al.*, 1985) studied the high-temperature behavior of a gas of anyons and calculated the second virial coefficient.

Much of the original interest in this problem was connected to its possible relevance to high-temperature superconductors. Since anyons "interpolate" between fermions and bosons, it is natural to ask whether an assembly of anyons at finite density is more "fermion-like" or "boson-like." Fermions have non-condensed ground states with Fermi surfaces, whereas bosons undergo Bose condensation and are superfluids. In two remarkable papers, Laughlin (1988a, 1988b) argued that anyons generally form "condensates" in the sense that their ground states exhibit superfluid properties. Fetter, Hanna, and Laughlin (Fetter *et al.*, 1989) developed a mean-field theory for the free-anyon gas in the continuum that has generally confirmed these conjectures. They argued that, if one represents anyons in terms of fermions coupled to fractional fluxoids, then the fermions feel an effective average flux determined by the particle density. A quantum-Hall-effect-like picture could then be used, at least within mean-field theory. In a sense this is a very surprising result since a quantum-Hall-effect system is *incompressible* and, thus, does not have any low-energy modes. However, the flux is uniform only on average since the constraints force it to fluctuate together with the particle density. Fetter, Hanna, and Laughlin showed that this was indeed the case. They did a calculation with the flavor of a random-phase approximation (RPA) and found a Goldstone pole in the (fermion) current–current correlation function. Hence, the fluctuations restore the *compressibility* which is necessary in order for the system to behave like a condensate. They argued that this pole implies the presence of a Meissner effect for an external electromagnetic field. This picture relies on two crucial assumptions: (1) the fermions can effectively be stripped of their fluxes and (2) the Goldstone pole is robust against fluctuations.

The predictions of Fetter, Hanna, and Laughlin have, to some extent, been confirmed by extensive numerical calculations (Canright *et al.*, 1989). Chen, Wilczek, Witten, and Halperin (Chen *et al.*, 1989) offered an argument to explain why the Goldstone pole is exact that is based on the *f*-sum rule, which is a consequence of gauge invariance (see Chapter 12). For a nice derivation see, for instance, the book by Martin (1967).

Wen and Zee (1990), Lee and Fisher (1989), and Kitazawa and Murayama (1990) considered this problem from a bosonic point of view. In this language, one focuses more directly on the role of vortices, anyons binding into "Cooper bound states," etc. The emerging picture is complementary to the fermion description. Local operators in one language are non-local "disorder" operators in the other. It is worth noting that a similar picture has been developed for the FQHE (Girvin and MacDonald, 1987; Read, 1989; Zhang *et al.*, 1989), as we will see in Chapter 13.

In our discussion here, I will follow my own work, which is based on the path-integral approach for fermions coupled to the Chern–Simons theory (Fradkin, 1990a) (see also Randjbar-Daemi *et al.* (1990) and Hosotani and Chakravarty (1990)). In this approach, the exactness of the Goldstone modes follows from the topological invariance of an effective Hall conductance. In Chapters 12 and 13, we discuss these issues of topological invariance and quantization at great length within the framework of the theory of the quantum Hall effect.

11.2 The functional-integral formulation of the Chern–Simons theory

In this section I consider the functional-integral formalism for a system of $\mathcal{N}_{\mathrm{a}}$ anyons at zero temperature. I will use the fermion formalism discussed above. I will work with a chemical potential μ, which will be determined later from the requirement that the density ρ be equal to $\mathcal{N}_a/L^2$ for a system with L^2 sites (I assume a square lattice of L^2 sites with lattice constant $a_0 = 1$).

The functional-integral representation for the partition function of this system at zero temperature (in real time) with chemical potential μ and background electromagnetic fields A_μ ($\mu = 0, 1, 2$) is given by

$$Z = \int \mathcal{D}\bar{\psi}\, \mathcal{D}\psi\, \mathcal{D}\mathcal{A}\, e^{i \int dt\, L} \tag{11.1}$$

where ψ and $\bar{\psi}$ are Grassmann fields and some gauge-fixing procedure is implicitly assumed. This functional integral has to be understood as a coherent-state path integral. Let us consider the gauge-field sector for the moment. The fermion sector is already known to be a coherent-state path integral. In Section 10.9, I showed, with some caveats, that $\mathcal{A}_1$ and $\theta\mathcal{A}_2$ form a canonical pair. Notice that $\mathcal{A}_1$ resides on the link $(\vec{x}, \vec{x}+\hat{e}_1)$, whereas $\theta\mathcal{A}_2$ resides on the orthogonal link $(\vec{x}, \vec{x}+\hat{e}_2)$. In the derivation of the path integral one has to introduce complete sets of states at every intermediate time of the evolution. However, since $\mathcal{A}_1$ and $\mathcal{A}_2$ do not commute, we cannot define a complete set of states in which both are diagonal. Let us say we choose a basis in which $\mathcal{A}_1$ is diagonal and that we now insert a complete set of such states at every intermediate time. In addition, the states have to be restricted so as to satisfy the *local* constraint. This is implemented by means of a Lagrange multiplier $\mathcal{A}_0(\vec{x}, t)$ at every lattice site and at all times. The matrix elements of the time-evolution operator for an infinitesimal time δt are not easy to compute in such a basis. Thus, it is convenient to introduce also a complete set of states in which $\mathcal{A}_2$ is diagonal. It is easy to show that, in addition to a term of the form $\theta\mathcal{A}_0\mathcal{B}$ that arises from the constraint, we get an extra term of the form $\theta\sum_{\vec{x}} \mathcal{A}_2(\vec{x})\partial_0\mathcal{A}_1(\vec{x})$ in the Lagrangian. This term is generated by the overlaps of $\mathcal{A}_1$ and $\mathcal{A}_2$ states on neighboring time slices. The two sets of terms can be condensed into a single expression: the Chern–Simons Lagrangian. Hence, the functional integral is just the phase-space integral for the canonical pair $\mathcal{A}_1$ and $\theta\mathcal{A}_2$.

The anyons are coupled to the electromagnetic field via the minimal-coupling prescription. Thus, all we need to do in order to include the chemical potential μ and the electromagnetic fields A_μ is to modify the derivatives and amplitudes in the usual manner:

$$\begin{aligned} D_0 = \partial_0 - i\mathcal{A}_0 &\to \partial_0 - i(\mathcal{A}_0 + A_0 + \mu) \\ e^{i\mathcal{A}_j(x)} &\to e^{i(\mathcal{A}_j(x) + A_j(x))} \end{aligned} \tag{11.2}$$

Notice that, as usual, the chemical potential μ can be regarded as a constant shift of A_0. The integration measures are invariant measures.

11.3 Correlation functions

The response of the system to slowly varying electromagnetic fields can be studied in terms of the current correlation functions. In addition, we will be interested also in correlations that probe other features of the spectrum of the system. In particular, it is of interest to study the *gauge-invariant* fermion propagator

$$G_\Gamma(x, x') = \left\langle \bar{\psi}(x) e^{i\int_\Gamma \mathcal{A}} \psi(x') \right\rangle \tag{11.3}$$

where $\int_\Gamma \mathcal{A}$ is shorthand for the line integral of the statistical vector potentials along some path Γ. Likewise the pair correlation function can be calculated in terms of the gauge-invariant four-point function, and so on. Other probes of interest are Wilson loops for the statistical vector potential $\mathcal{A}$ along a closed path Γ:

$$W_{\mathcal{A}}[\Gamma] = \left\langle e^{i\oint_\Gamma \mathcal{A}} \right\rangle \tag{11.4}$$

In particular a space-like Wilson loop for a closed path Γ on the square lattice must represent, as a result of the constraint, the fluctuation in the number of fermions $\mathcal{N}_\text{a}(\Sigma)$ (and hence anyons) inside the region Σ bounded by the loop Γ:

$$W_\text{space}\left[\Gamma = \partial\Sigma\right] = \left\langle e^{i\oint_\Gamma \mathcal{A}} \right\rangle = \left\langle e^{\frac{i}{\theta}\int_\Sigma j_0(x)} \right\rangle \equiv \left\langle e^{\frac{i}{\theta}\mathcal{N}_a(\Sigma)} \right\rangle \tag{11.5}$$

In the case of a time-like loop Γ, the constraint implies that a static particle has been added at one point and subtracted at another point. Thus $\langle e^{i\oint_\Gamma \mathcal{A}} \rangle$ for time-like loops roughly represents the energy cost for adding a particle, say at $\vec{r}$, and removing it at $\vec{r}\,'$. This is the standard interpretation of the Wilson loop. Notice, however, that now a particle is added without adding a flux. Hence we are creating a mismatch between charge and flux.

Analogously, we can create a coherent state that represents a (static) flux piercing a given plaquette at a dual site $\vec{r}$. The operator which creates this state is

$$K(\vec{x}) \equiv e^{i\theta\sum_{\vec{x}'}\epsilon_{jk}\mathcal{A}_k(\vec{x}')\mathcal{A}_j^\text{c}(\vec{x}')} \tag{11.6}$$

where $\mathcal{A}_j^\text{c}(\vec{x}')$ is a background static vector potential with a curl equal to the flux. For $\mathcal{A}_j^\text{c}(\vec{x}')$ to represent a flux we must demand that $\mathcal{A}_j^\text{c}(\vec{x}') = (1/\theta)\Delta_j\Theta(\vec{x}', \vec{r})$. It is easy to show that this operator $K(\vec{x})$ is precisely identical to the operator $e^{i\phi}$ defined in Eq. (10.184). Indeed, using Gauss's law, Eq. (10.166),

$$\theta\epsilon_{jk}\Delta_j\mathcal{A}_k - j_0 = 0 \tag{11.7}$$

one finds (up to boundary terms)

$$\begin{aligned}\theta\sum_{\vec{x}'}\epsilon_{jk}\mathcal{A}_k(\vec{x}\,')\mathcal{A}^{\rm c}_j(\vec{x}\,') &= \sum_{\vec{x}'}\epsilon_{jk}\mathcal{A}_k(\vec{x}\,')\Delta_j\Theta(\vec{x}',\vec{r})\\ &= -\sum_{\vec{x}'}\epsilon_{jk}\Big(\Delta_j\mathcal{A}_k(\vec{x}')\Big)\Theta(\vec{x}',\vec{r})\\ &= -\frac{1}{\theta}\sum_{\vec{x}'}j_0(\vec{x}')\Theta(\vec{x}',\vec{r})\end{aligned} \tag{11.8}$$

By means of the identity (see Eq. (10.181))

$$\Theta(\vec{x},\vec{r}\,') - \Theta(\vec{x}',\vec{r}) = \frac{1}{2} \tag{11.9}$$

one finds

$$\begin{aligned}\theta\sum_{\vec{x}'}\epsilon_{jk}\mathcal{A}_k(\vec{x}')\mathcal{A}^{\rm c}_j(\vec{x}') &= \frac{1}{2\theta}\sum_{\vec{x}'}j_0(\vec{x}') - \frac{1}{\theta}\sum_{\vec{x}}j_0(\vec{x}')\Theta(\vec{x},\vec{r}\,')\\ &\equiv +\frac{1}{2\theta}\mathcal{N}_{\rm a} - \phi(\vec{x})\end{aligned} \tag{11.10}$$

Hence

$$K(\vec{x}) = e^{\frac{i}{2\theta}\mathcal{N}_{\rm a}}e^{-i\phi(\vec{x})} \tag{11.11}$$

Clearly $K(\vec{x})$ *is not invariant* under local gauge transformation of the statistical gauge field. Indeed, for a gauge transformation $\mathcal{A}_j(\vec{x}) \to \mathcal{A}_j(\vec{x}) + \Delta_j\varphi(\vec{x})$, we get

$$\theta\sum_{\vec{x}'}\epsilon_{jk}\mathcal{A}_k(\vec{x}')\mathcal{A}^{\rm c}_j(\vec{x}\,') \to \theta\sum_{\vec{x}'}\epsilon_{jk}\mathcal{A}_k(\vec{x}')\mathcal{A}^{\rm c}_j(\vec{x}') + \theta\sum_{\vec{x}'}\epsilon_{jk}\Delta_k\varphi(\vec{x})\mathcal{A}^{\rm c}_j(\vec{x}) \tag{11.12}$$

By integrating by parts and with the help of Eq. (10.178) and Eq. (10.175) one finds

$$K(\vec{x}) \to K(\vec{x})e^{+i\varphi(\vec{x})} \tag{11.13}$$

and thus the product

$$a(\vec{x}) = K(\vec{x})c(\vec{x}) \tag{11.14}$$

is gauge-invariant. Obviously the operator $a(\vec{x})$ is nothing other than the anyon operator.

11.4 The semi-classical approximation

We are interested in studying the physical properties of the partition function of a gas of anyons. In particular we want to understand the following issues: (1)

the spectrum of low-lying excitations, (2) the statistics of the quasiparticles, (3) whether it exhibits superfluidity, (4) whether there is a Meissner effect, and (5) the behavior of correlation functions.

I will study this problem by treating the functional integral within the semi-classical (saddle-point) expansion. Formally this requires the presence of a large coefficient in front of the action $S = \int dt\, L$. This system does not have such a coefficient (apart from $1/\hbar$ itself). It is plausible that at large densities the saddle-point approximation may become accurate. Such is the case for the (weakly interacting) electron gas, for which the RPA works very well. Since the statistical angle δ happens to be equal to $1/(2\theta)$ one expects that this approximation may also work *for large values of* θ (i.e. almost a fermion). This is the limit studied by Chen, Wilczek, Witten, and Halperin. In the Bose limit ($\theta = 1/(2\pi)$) the results depend crucially on the density. In fact it is well known that the hard-core Bose gas, at moderate densities, can be treated within the RPA due to the effective softening of the hard cores at such densities. At high densities on a lattice, this approximation breaks down and the hard cores cause the existence of crystalline states or off-diagonal long-range order (ODLRO) for the spin one-half XY model. However, it is conceivable that there may exist regimes of Bose systems for which the results of a *fermion* mean-field theory may still be qualitatively correct. The results of Lee and Fisher (1989) suggest that this may be the case.

The saddle-point approximation (SPA) may also be formally justified by considering a system of M *species* of anyons (each with $\mathcal{N}_a$ particles), which are "free" in the sense that there is no explicit interaction term in the Hamiltonian. The requirement of fractional statistics, of course, amounts to an interaction since it is equivalent to the statement that all M species of *fermions* interact through the *same* statistical vector potential $\mathcal{A}_\mu$. At large M, with $\theta = \theta_0 M$, the SPA is formally correct. For the sake of simplicity I will consider only $M = 1$ and assume that the approximation is, *at least*, qualitatively correct.

The SPA is now carried out in the usual fashion. One first observes that the action is a bilinear form in fermion variables. Thus the fermions can be integrated out explicitly. The result is naturally a determinant,

$$\int \mathcal{D}\bar{\psi}\, \mathcal{D}\psi\, e^{iS_{\text{F}}} = \det\Big[iD_0 - h[\mathcal{A}_j + A_j]\Big] \tag{11.15}$$

where S_{F} is the *fermion* part of the action,

$$S_{\text{F}} = \int dt \sum_{\vec{x},\vec{x}'} \bar{\psi}(x)\Big(iD_0\delta_{\vec{x},\vec{x}'} - h[\mathcal{A}_j + A_j]\Big)\psi(x') \tag{11.16}$$

and the *one-particle Hamiltonian* $h[\mathcal{A}]$ is

$$h[\mathcal{A}_j] = \tau \sum_{j=1,2} e^{i\mathcal{A}_j(x)} \delta_{\vec{x}',\vec{x}+\hat{e}_j} \tag{11.17}$$

The condition $\mathcal{A}_j(x) = -\mathcal{A}_{-j}(x + e_j)$ guarantees the hermiticity of the Hamiltonian.

Therefore, the statistical vector potentials $\mathcal{A}_\mu$ have the *effective action* S_{eff} given by

$$S_{\text{eff}}[\mathcal{A}_\mu, A_\mu] = -i \,\text{tr}\, \ln\left(i D_0 - h[\mathcal{A} + A]\right) - i\frac{\theta}{4} S_{\text{CS}}[\mathcal{A}] \tag{11.18}$$

We can use the invariance of the measure to shift the statistical vector potentials $\mathcal{A}_\mu + A_\mu \rightarrow \mathcal{A}_\mu$. The result is that the effective action now reads

$$S_{\text{eff}}[\mathcal{A}_\mu, A_\mu] = -i \,\text{tr}\, \ln(i D_0 + \mu - h[\mathcal{A}]) - i\frac{\theta}{4} S_{\text{CS}}[\mathcal{A} - A] \tag{11.19}$$

where we have pulled the chemical potential μ out of the definition of D_0. In this form, the electromagnetic fields appear only in the Chern–Simons term which is *quadratic* in the fields. We can thus write

$$S_{\text{CS}}[\mathcal{A} - A] = S_{\text{CS}}[\mathcal{A}] + S_{\text{CS}}[A] - \epsilon_{\mu\nu\lambda}\left(\mathcal{A}^\mu F^{\nu\lambda} + A^\mu \mathcal{F}^{\nu\lambda}\right) \tag{11.20}$$

We will assume that the electromagnetic field A_μ is small and has zero *average* strength. In this case we may treat A_μ as a perturbation (i.e. linear-response theory). Note that a non-zero uniform external magnetic field cannot be treated in perturbation theory. This is crucial for the correct study of the Meissner effect. Let us consider, for the moment, the SPA in the absence of external electromagnetic fields.

We demand that S_{eff} be stationary around some configuration $\mathcal{A}_\mu$, which is assumed to be time independent (i.e. zero "electrical" statistical field $\bar{\mathcal{E}}$) and with uniform statistical "magnetic" field $\bar{\mathcal{B}}$. Thus

$$\left.\frac{\delta S_{\text{eff}}}{\delta \mathcal{A}_\mu}\right|_{\bar{\mathcal{A}}_\mu} = 0 \tag{11.21}$$

yields the saddle-point equation (SPE)

$$\left\langle j_\mu^{\text{F}} \right\rangle_{\bar{\mathcal{A}}} = \frac{\theta}{2} \epsilon_{\mu\nu\lambda}\left(\bar{\mathcal{F}}^{\nu\lambda} - F^{\nu\lambda}\right) \tag{11.22}$$

where $\langle j_\mu^{\text{F}} \rangle$ is the gauge-invariant fermion current.

Since the electromagnetic field will be assumed to be small and with zero average, we will set $F^{\nu\lambda} = 0$ in the SPE for the rest of this section. In this case, and for solutions with $\bar{\mathcal{B}} = \text{constant}$, $\bar{\mathcal{E}} = 0$, we find

$$\rho = \theta \bar{\mathcal{B}} \tag{11.23}$$

where ρ is the fermion density.

The requirement that there should be $\mathcal{N}_{\text{a}}$ particles is met by requiring

$$-\frac{i}{Z}\frac{\partial Z}{\partial \mu} = \mathcal{N}_{\text{a}} \tag{11.24}$$

Since μ is nothing but a constant shift of $\mathcal{A}_0$, one finds

$$\mathcal{N}_{\text{a}} = \theta \Phi \tag{11.25}$$

where Φ is the *total flux*

$$\Phi = \bar{\mathcal{B}} L^2 \tag{11.26}$$

Thus, we find

$$\theta \bar{\mathcal{B}} = \frac{\mathcal{N}_{\text{a}}}{L^2} \tag{11.27}$$

which is Laughlin's result. Thus, at the saddle-point level, the fermions feel an effective flux $\bar{\mathcal{B}}$ per plaquette.

The spectrum of this problem was studied by Hofstadter, and its properties are summarized in Chapter 12. He found that, if the *number of particles is fixed*, then as $\bar{\mathcal{B}}$ *varies* the spectrum of the effective one-particle Hamiltonian is very rich and complex and, as a function of $\bar{\mathcal{B}}$, it has a fractal structure. However, in the problem at hand, $\bar{\mathcal{B}}$ *is determined by the number of particles*. In fact, for a system of $\mathcal{N}_{\text{a}}$ anyons on a lattice with L^2 sites, the density ρ is $\mathcal{N}_{\text{a}}/L^2$ and therefore can be written as a ratio of two relatively prime integers r and q, i.e.

$$\rho = \frac{r}{q} \tag{11.28}$$

Similarly, we can also write the statistical phase δ in the form of an irreducible fraction in terms of two relatively prime integers n and m,

$$\delta = \pi \frac{n}{m} \tag{11.29}$$

Equivalently, the Chern–Simons coupling constant θ is given by

$$\theta = \frac{m}{2\pi n} \tag{11.30}$$

The effective field $\bar{\mathcal{B}}$ is a fraction of the flux quantum, 2π,

$$\bar{\mathcal{B}} = 2\pi \frac{P}{Q} \tag{11.31}$$

where the two relatively prime integers P and Q are given from Eq. (11.27) by

$$2\pi \frac{P}{Q} = \frac{\rho}{\theta} \tag{11.32}$$

Hence, we can write

$$\frac{P}{Q} = \frac{nr}{mq} \tag{11.33}$$

The spectrum of one-particle states, the Hofstadter problem, for *rational fluxes* $\bar{\mathcal{B}} = 2\pi P/Q$, consists of q Landau bands each with L^2/Q degenerate states (see Section 12.2). In the continuum limit, these bands become the usual Landau levels. If we denote by f the fraction of occupied Landau bands, then f must be $\mathcal{N}_a \times Q/L^2$ since there is a total of $\mathcal{N}_a$ particles. The density is then f/Q. Using Eq. (11.28), we get $f = (r/q)Q$. Thus, f is an integer if and only if q is a factor of Q.

Let (a, b) denote the largest common factor of the pair of integers a and b. Let k and l be two integers defined satisfying $k = (n, q)$ and $l = (m, r)$. Hence, there exist four integers $\bar{n}$, $\bar{m}$, $\bar{r}$, and $\bar{q}$ such that

$$\begin{aligned} n &= k\bar{n}, & q &= k\bar{q} \\ m &= l\bar{m}, & r &= l\bar{r} \\ (\bar{n}, \bar{q}) &= 1, & (\bar{m}, \bar{r}) &= 1 \\ (\bar{n}, \bar{m}) &= 1, & (\bar{r}, \bar{q}) &= 1 \end{aligned} \tag{11.34}$$

Thus,

$$\frac{P}{Q} = \frac{\bar{n}\bar{r}}{\bar{m}\bar{q}} \tag{11.35}$$

and

$$P = \bar{n}\bar{r}, \qquad Q = \bar{m}\bar{q} \tag{11.36}$$

Therefore, the fraction f of occupied Landau bands is

$$f = \frac{r}{q} Q = \frac{l}{k} \bar{r}\bar{m} \tag{11.37}$$

It is easy to show that k does not have any common factors with any one of l, $\bar{r}$, and $\bar{m}$. In general, f is an irreducible fraction, unless one of the following conditions is satisfied:

$$(n, q) = 1, \qquad (n, q) = \frac{m}{(m, r)}, \qquad (n, q) = \frac{r}{(m, r)} \tag{11.38}$$

If f is not an integer, then there is no gap. Fluctuation effects should overwhelm the saddle-point results and this theory will generally be unstable. Hence, whenever possible, one must have f integer since, except for one special case, there is always a gap. In summary, for arbitrary density ρ and Chern–Simons coupling constant θ, it is not always possible to require f to be an integer. On the other hand, for the "happy fractions" listed above f is an integer and we have an integer number f of filled Landau bands. The physical behavior of the system will depend on which of the conditions listed above is realized. Thus, the physics of this problem is determined not just by the density and the statistics, but also by number-theoretic conditions, i.e. commensurability conditions.

Of particular importance will be the sequence $\theta = m/(2\pi)$, i.e. $n = 1$. In this case we have $k = (n, q) = 1$ and f is indeed an integer, $f = mr/(m, r)$. For this sequence, we have an integer number f of Landau bands filled for a system with *arbitrary* density $\rho = r/q$ and statistical parameter $\delta = \pi/m$. This is the case considered by Chen, Wilczek, Witten, and Halperin. The saddle-point approximation is expected to work for θ large (i.e. large m), which is the limit in which the anyons are *almost* fermions. The case of semions has $m = 2$.

An exceptional case occurs if $\rho = \frac{1}{2}$ (i.e. for half-filling) and $\theta = m/(2\pi n)$, with n an *odd* integer. In this case we get $P = n$, $Q = 2m$, and $f = m$. This means that all states with energy less than zero are filled and that the Fermi level is at $E = 0$. It has been known since Hofstadter's work that, in this case, there is a band crossing in the spectrum (see Chapter 12). These bands cross at $E = 0$ at Q points of the Brillouin zone. In fact the case $m = 1$ and $\rho = \frac{1}{2}$ corresponds to a "flux phase" with $\bar{\mathcal{B}} = \pi$ (i.e. half-flux quantum per plaquette). In this case the fermion spectrum is effectively relativistic. In fact, it has long been recognized that hopping in a frustrated lattice is an efficient way to set up the Dirac equation on a lattice. In general one finds Q species of Dirac fermions. Fluctuations in the statistical gauge fields may open up a gap in the spectrum. It is possible that this may be done in a manner in which time-reversal invariance is violated explicitly, or it may be spontaneously broken by fluctuations. In the field-theory language, one is asking whether a parity anomaly is present. This problem is exactly the same as the one we have already encountered in our study of the chiral spin liquid in Chapter 10. For lattice systems one has to deal with the "fermion-doubling" problem (here it is Q-fold!). In most cases one expects no anomalies unless a perturbation that breaks time reversal is explicitly introduced. We are going to see in the next section that these issues are quite relevant for our problem.

Thus, the SPA to the partition function yields Laughlin's result that the mean-field theory for the anyon system should be equivalent to a system of particles (say fermions) moving in an effective magnetic field determined by their density. It is clear that this approximation assumes that the flux subsystem is rigid in the sense that the *average field*, determined at the saddle-point level, will not be modified by the fluctuations. In this high-density-like approximation, the fluctuations around the average field $\bar{\mathcal{B}}$ should be small in order for this approximation to be stable. The local value of the field is, however, still being determined by local fluctuations of the density. In this sense, the system is *compressible*. If the local fluctuations are massive, the spectrum should have a gap and the system will truly be rigid. But, if the fluctuations have a gapless state, the system will not be rigid. Indeed, this "fluctuation-induced compressibility" is the very origin of the superfluidity.

11.5 Effective action and topological invariance

11.5.1 Effective action

In the previous section we discussed the SPA to the path integral for the anyon gas. Fetter, Hanna, and Laughlin claimed that the fluctuations around the state with average flux $\bar{\mathcal{B}} = 2\pi/Q$ induce a pole in the current–current correlation function, which, in turn, is responsible for the superfluidity. This is the Fetter–Hanna–Laughlin Goldstone boson. At first sight, this result seems to be quite surprising. In fact, fermions in a *background* magnetic field always lead to a spectrum with a gap, as in quantum-Hall-effect systems. What is different here is that the magnetic fields *do not constitute a fixed background*, since they are generated by the particles themselves. The fluctuations of the system retain this character. The SPA fixes only the average field, not its fluctuations, and one is led to study the effects of fluctuations of the statistical gauge fields about the mean field. It is natural to compute the effective action of the statistical gauge fields including the effects of fermion loops. In this sense this calculation is close to the standard RPA.

Purely on the grounds of gauge and translation invariance, we can assert that the effective action for the statistical gauge fields at low energies and long distances (i.e. to leading order in a gradient expansion) should only be a function of the fluctuating part of $\mathcal{F}_{\mu\nu}$ (with the smallest number of gradients) plus a term with the same form as the bare Chern–Simons term. Banks and Lykken (1990) argued that, if the effective action has an induced Chern–Simons term that happens to cancel out the bare one, then the Goldstone boson is found, and it is nothing other than the *massless transverse* component of the fluctuating statistical vector potential. However, it is necessary to explain why this crucial cancellation, present to leading order, survives renormalization. This is in fact far from obvious, since the coefficients of the other terms do get renormalized.

Let us now investigate how the Goldstone pole appears within this path-integral framework. We will see that the exactness of the Goldstone boson is a consequence of the *topological invariance* of the quantized Hall conductance for this system of fermions. Thus our problem is naturally related to the *integer* quantum Hall effect (IQHE) on lattices. In fact, we are going to be using many results of the theory of the quantum Hall effect. Most of these results are discussed in Chapter 12.

Let us first consider the quadratic (i.e. Gaussian) fluctuations around the mean field. The effective action for the fluctuating part of the statistical gauge fields, hereafter denoted by $\mathcal{A}_\mu$, to quadratic order, $S^{(2)}$, is given by

$$S^{(2)}[\mathcal{A}_\mu] = \sum_{x,x'} \frac{\delta^2 S_{\text{eff}}}{\delta\mathcal{A}_\mu(x)\delta\mathcal{A}_\nu(x')} \mathcal{A}_\mu(x)\mathcal{A}_\nu(x') \tag{11.39}$$

where $x \equiv (\vec{x}, t)$ and $x' \equiv (\vec{x}', t')$, $\vec{x}$ and $\vec{x}'$ take values on the square lattice, and t and t' are continuous (time) variables. Since S_{eff} is a sum of a fermionic part and a

Chern–Simons term, $S^{(2)}$ also is a sum of two terms. The first term, which comes from the fermion loops, is nothing but the *polarization operator* $\Pi_{\mu\nu}(x, x')$. The second term is just the Chern–Simons term itself,

$$S^{(2)}[\mathcal{A}_\mu] = \sum_{x,x'} \Pi_{\mu\nu}(x, x')\mathcal{A}_\mu(x)\mathcal{A}_\nu(x') - \frac{\theta}{4} S_{\text{CS}}[\mathcal{A}_\mu] \tag{11.40}$$

where $\Pi_{\mu\nu}(x, x')$ is the polarization operator for a system of *fermions* on a lattice in the presence of the background magnetic field $\bar{\mathcal{B}}$. Thus it is just the usual linear-response-theory kernel, the current–current correlation function of fermions.

However, the polarization tensor $\Pi_{\mu\nu}$ of the fermions in this mean-field state should not be confused with the actual polarization tensor of the full theory, which we will denote by $K_{\mu\nu}$. While $\Pi_{\mu\nu}$ is the current-correlation function of the *fermions* of the mean-field theory, $K_{\mu\nu}$ is the current-correlation function of the *anyon* system. These two tensors are not the same and their properties are quite different. We will discuss their different behaviors in detail in Chapter 13, where we will discuss the fermion Chern–Simons theory of the FQHE, in which context similar questions arise. For the purposes of our discussion of the anyon system, it will be sufficient to note that, while a system of fermions in a magnetic field has an explicitly broken time-reversal invariance and a finite Hall conductance, a system of anyons at finite density (in zero external field) is time-reversal invariant and should not have a "zero-field" Hall conductance (i.e. in the absence of an external magnetic field). We will postpone the discussion of the electrodynamic properties of the anyon system to Chapter 13, where it will be discussed together with the theory of the FQHE. Here we will just derive an effective action for the low-energy modes.

The polarization operator $\Pi_{\mu\nu}$ has an interesting structure, whose form is strongly constrained by gauge invariance and translation invariance. This structure will be discussed in detail in Chapter 13. Here we will use a few important features of its structure. The long-distance, low-energy behavior of $S^{(2)}$ can be obtained simply by noting that it has to satisfy the requirements of translation and gauge invariance. If there is a gap in the spectrum, $\Pi_{\mu\nu}$ is also local and it has a gradient expansion. Thus the effective action for fluctuations at distances larger than the inter-particle separation and energies less than the gap has the form

$$S^{(2)}[\mathcal{A}] \approx \int d^2x\, dt \left[\frac{\epsilon}{2}\vec{\mathcal{E}}^{\,2}(\vec{x}, t) - \frac{\chi}{2}\mathcal{B}^2(\vec{x}, t) + \frac{1}{4}(\sigma_{xy} - \theta)\epsilon_{\mu\nu\lambda}\mathcal{A}^\mu \mathcal{F}^{\nu\lambda}\right] + \text{h.o.t.} \tag{11.41}$$

where ϵ, χ, and σ_{xy} are the (long-wavelength, low-frequency) dielectric constant, diamagnetic susceptibility, and Hall conductance of the Fermi system, respectively. Note that the term which contains the Hall conductance has the *same form* as but *opposite sign* to the bare Chern–Simons term which determines the statistics of the anyons. This effective action is correct in the limit in which the frequencies of the modes are small while their momentum is held fixed, but not in the opposite limit of small momentum and fixed frequency.

The parameters (ϵ, χ, and σ_{xy}) are in principle determined by integrating out all fluctuations from the highest energies allowed in this problem down to the only physical scale this system has: the gap. One expects that these coefficients will be heavily renormalized away from their saddle-point values. For the "almost-fermion" limit of large θ, the renormalizations are expected to be small, of order $1/\theta$. Thus, although explicit expressions for these coefficients can be found (they are given by various pieces of the polarization operator $\Pi_{\mu\nu}$), their precise form is not in principle very important due to the above-mentioned renormalization effects.

While these considerations apply to ϵ and χ, as well as to the higher-order terms in the effective action which we have neglected, the value of σ_{xy} *is completely determined already at the saddle-point level.* This is so because σ_{xy} is the Hall conductance for a system of fermions on a lattice, with an integer number f of Landau bands exactly filled, which has been shown to be *quantized*.

11.5.2 Quantized Hall conductance and compressibility

The quantization of σ_{xy} has been studied extensively in the context of the quantum Hall effect. Thouless, Kohmoto, Nightingale, and den Nijs (TKNN) (Thouless *et al.*, 1982) showed that the σ_{xy} associated with the Hofstadter problem is quantized in terms of an integer t in the range $-Q/2 < t < Q/2$, which, in turn, is determined by a Diophantine equation. The theory of TKNN is discussed in Chapter 12. The following results are relevant to our problem. If j denotes the jth gap of a Hofstadter problem with $\bar{\mathcal{B}} = 2\pi P/Q$, there exist two integers t_j and s_j (with t_j in the same range as t and s_j unconstrained) such that

$$j = Qs_j + Pt_j \tag{11.42}$$

If the Fermi energy lies in the fth gap, the Hall conductance is given by

$$\sigma_{xy} = \frac{e^2}{\hbar}\sum_{j=1}^{f}(t_j - t_{j-1}) \tag{11.43}$$

with $t_0 = 0$. Thus, in units in which $e^2 = \hbar = 1$, we find

$$\sigma_{xy} = \frac{t_f}{2\pi} \tag{11.44}$$

where t_f is the solution of the Diophantine equation for the fth gap. We may now combine these results to get

$$\sigma_{xy} = \theta\left(1 - \frac{s_f}{\rho}\right) \tag{11.45}$$

The Diophantine equation has solutions in the form of a pair of integers (s_f, t_f). The solution is, in most cases, unique and, in general, both s_f *and* t_f will be different from zero. Under special circumstances, we will find families of solutions with $s_f = 0$. Also, in some special cases, the solution is not unique. The solutions with $s_f = 0$ play a special role for, as we will see, they represent the *compressible* states.

Let us first consider the sequence $\theta = m/(2\pi)$ and ρ arbitrary. The mean-field theory yields the values $P = r/(m, r)$, $Q = mq/(m, r)$ and it requires that exactly $f = mr/(m, r)$ Landau bands are filled. The Diophantine equation has, for $j = f$, the *unique* solution

$$s_f = \begin{cases} 0, & t_f = m, \quad \text{if } |m| < mq/(2(m, r)) \\ r/(m, r), & t_f = m - mq/(m, r), \quad \text{if } |m| > mq/(2(m, r)) \end{cases} \tag{11.46}$$

There are degenerate solutions whenever $|m| = mq/(2(m, r))$. In this case both solutions are possible and the value of t_f is ambiguous. It is easy to see that a degeneracy occurs whenever $m = mq/(2(m, r))$, i.e. for $q = 2(m, r)$ and q even. This includes the half-filled even-denominator case $\rho = \frac{1}{2}$. Which solution is realized depends on how this degeneracy is lifted by additional terms in the Hamiltonian. It is natural to assume that it is always possible to find terms that will remove this degeneracy. The physical properties of the system will depend on the way we choose to render the system non-degenerate.

Thus, in the absence of degeneracies, the solution is unique and one finds $t_f = m$ and $s_f = 0$ if $q > 2(m, r)$. Hence, we get

$$\sigma_{xy} = \frac{m}{2\pi} \tag{11.47}$$

which is exactly identical to θ! We then conclude that, at least at the level of the SPA and in the absence of degeneracies, $\sigma_{xy} = \theta$ and the Chern–Simons term in the effective action for the fluctuating statistical gauge fields is cancelled out provided that $q > 2(m, r)$. As Banks and Lykken observed, this is a *sufficient* condition for the existence of the Goldstone boson. Conversely, for $q < 2(m, r)$, the solution has $s_f = r/(m, r) \neq 0$, and there is no cancellation and no Goldstone boson.

For other sequences, such as $n \neq 1$, it is not possible to find a solution of the Diophantine equation with $s_f = 0$. It is easy to check that this solution exists only if n is a factor of m, which is impossible since $(n, m) = 1$ except for the case $n = 1$. Thus, the other sequences do not exhibit superfluidity. These non-superfluid states cannot be found in the continuum theory. They are the result of *diffraction* effects generated by the underlying lattice. It is clear that, in the low-density limit, these effects do not impose an overwhelming constraint, provided that the Fermi energy lies in one of the main energy gaps. In this case, there is a smooth continuum limit at low densities. However, if the Fermi energy is in one of the lower gaps, we will not get a cancellation, even in the low-density limit. Thus, the continuum limit is tricky to get. We should then expect that the properties of the ground state should depend on some details of the behavior of the system at short distances. This problem will come back when we consider the role of higher-order fluctuations.

11.5.3 Stability of the mean-field state

One might wonder about the stability of this crucial result once fluctuations about mean-field theory are considered. Two problems naturally arise. First of all, one must worry about infinite renormalizations. In continuum relativistic-field theories it is known that the Chern–Simons term does not acquire infinite renormalizations (Semenoff *et al.*, 1989). Non-relativistic theories are not expected to be any more singular. Thus divergent renormalizations of σ_{xy} are not to be expected. However, *finite* renormalizations are not excluded by such arguments. The stability of the Goldstone boson requires no renormalization at all, neither infinite nor finite.

No-renormalization theorems usually follow from symmetry considerations or as a result of topology (or both). For the case of the lattice system, Kohmoto, and Avron, Seiler, and Simon, showed that σ_{xy} is a topological invariant (see Chapter 12). The topological invariance of σ_{xy} follows from the fact that the Brillouin zone of a 2D system with periodic boundary conditions is a 2-torus. The integer t_m is the first Chern number of the fiber bundle associated with the Berry connection induced by the wave functions on the 2-torus. Small changes in the microscopic Hamiltonian will not change this number, provided that no band crossings occur as a result of such changes. Qualitatively speaking, fluctuations about a solution with a finite gap are expected to have the same effect. After all, the fluctuations, configuration by configuration, will modulate the gap. Since each configuration yields the *same* value for σ_{xy}, the final result should be the same, provided that the sum over configurations makes sense. Once again, this argument requires the existence of a non-zero energy gap. Niu, Thouless, and Wu have also given an argument for the stability of the quantization of σ_{xy} including many-body effects (i.e. fluctuations). They showed that, if the *many-body wave function* for the ground state winds by the phases α and β along the x_1 and x_2 directions, then the value of σ_{xy}^{av}, *averaged over* α *and* β, is a topological invariant and hence it is quantized in the full theory. For a system with specific boundary conditions, say periodic, they showed that σ_{xy}^{pbc} differs from σ_{xy}^{av} by terms that vanish exponentially fast in the thermodynamic limit, provided that the system has a non-zero energy gap. For more details, see the discussion in Chapter 12.

I thus conclude that the topological invariance of σ_{xy} guarantees that the Goldstone boson is stable to all orders in perturbation theory.

11.5.4 Low-energy spectrum

We must then conclude that the anyon gas can exist in only one of two possible states, each defined by a low-energy effective action of the form of a QED-type theory with possibly a Chern–Simons term with some effective coupling. For the case of the "happy fractions," $\theta = m/(2\pi)$ and $\rho = r/q$, the effective action does not have a net Chern–Simons term. The effective action has the form

$$S^{(2)}[\mathcal{A}] = \int d^2x\, dt \left[\frac{\epsilon}{2}\vec{\mathcal{E}}^2(\vec{x}, t) - \frac{\chi}{2}\mathcal{B}^2(\vec{x}, t)\right] + \text{h.o.t.} \tag{11.48}$$

which is the action of free "Maxwell" electrodynamics in $(2+1)$ dimensions. Here I have neglected terms that vanish in the infrared limit.

Let us now consider the *dual* of this theory. Here we understand duality in the statistical-mechanical sense in which a *gauge* theory in $(2+1)$ dimensions is *dual* to a theory with a *global* symmetry. Since the gauge field of this problem, the statistical gauge field, has a U(1) symmetry, its dual is a *phase* field. Let $\Lambda_{\mu\nu}(\vec{x}, t)$ be a real antisymmetric tensor field. Since we are dealing with an anisotropic theory, it is convenient to define $\Lambda_{0i} = \vec{e}_i$ and $\Lambda_{ij} = \epsilon_{ij} b$, where $\vec{e}_i$ and b are real functions of space and time.

Consider now the modified action S'

$$S' = \int dt\, d^2x \left(-\frac{1}{2\epsilon}\vec{e}^{\,2} + \frac{1}{2\chi}b^2 + \frac{1}{2}\Lambda_{\mu\nu}\mathcal{F}^{\mu\nu}\right) \tag{11.49}$$

We can identify the path integrals with actions S and S' after a shift of the Gaussian variables $\Lambda_{\mu\nu}$, except for an irrelevant constant. The fluctuating statistical gauge fields $\mathcal{A}_\mu$ can now be integrated out, yielding the constraint on the $\Lambda_{\mu\nu}$ fields

$$\partial^\mu \Lambda_{\mu\nu} = 0 \tag{11.50}$$

This constraint can be solved by means of the phase field ω defined by

$$\Lambda_{\mu\nu} = \frac{1}{2\pi}\epsilon_{\mu\nu\lambda}\,\partial^\lambda \omega \tag{11.51}$$

By substituting back into the effective action, we get the effective Lagrangian density in terms of the ω field

$$\mathcal{L} = \frac{1}{8\pi^2\chi}\,(\partial_0\omega + mA_0)^2 - \frac{1}{8\pi^2\epsilon}\,(\partial_i\omega + mA_i)^2 + \text{h.o.t.} \tag{11.52}$$

which has the same form as in a conventional superfluid. This effective Lagrangian was first obtained by Banks and Lykken. In this derivation I used the fact that $\theta = m/(2\pi)$.

This theory has only one transverse degree of freedom, the "photon." Note that this has nothing to do with the *real* electromagnetic field. It originates from the fluxes associated with the anyons. This "photon" is the only massless excitation of this theory. It is precisely the Goldstone boson. It is responsible both for the phase mode necessary for superconductivity and for a direct Coulomb-like static interaction among sources (or excitations) that couple to the *statistical* gauge field. At long distances, the $(2+1)$-dimensional Coulomb interaction goes like $\ln R$, where R is the separation between two *sources* of the field $\mathcal{A}$. Thus, the energy necessary to create a *fermion* diverges logarithmically with the size of the system. The same happens with the energy required in order to add a flux to the system.

An anyon, however, is a gauge-invariant state. As such it couples only weakly to the fluctuations of the statistical gauge field since it is neutral but not quite point-like. Thus, we expect the energy of an anyon-like excitation to be finite and its value to be determined primarily by short-distance effects. Let us consider an operator that creates an anyon at point $\vec{x}$. It is easy to compute correlation functions of these gauge-invariant operators in the Coulomb (or *anyon*) gauge. In this gauge, we can write

$$\mathcal{A}_j(\vec{x},t) = \epsilon_{jk}\Delta_k\phi(\vec{x},t) \tag{11.53}$$

where ϕ also obeys periodic boundary conditions. If we now substitute Eq. (11.53) back into Eq. (11.6) then, after an integration by parts of the argument of the exponential, we find

$$K(\vec{x},t) = e^{i\theta\sum_{\vec{y}}\phi(\vec{y},t)\Delta_j\mathcal{A}_j^{\mathrm{c}}(\vec{y})} \tag{11.54}$$

If we also choose the Coulomb gauge to describe the classical fluxes, i.e. $\Delta_j\mathcal{A}_j^{\mathrm{c}} = 0$, we see that, *in this gauge*, $K(\vec{x},t)$ *is equivalent to the identity operator*. Thus, the correlation function for anyon operators is, in the Coulomb gauge, the same as the (gauge-dependent) fermion propagator evaluated in the same gauge. In the Coulomb gauge, the fermion propagator has the following properties: (a) it is multivalued and (b) it is short-ranged. It is multivalued, since the one-particle wave functions are multivalued in this gauge. It is short-ranged, since the ground state has *filled* Landau bands and the only possible one-particle states available are in the next unfilled Landau band. These states are separated from the ground state by the energy gap between Landau bands, which is *finite*.

In contrast, the elementary fermion excitations have a logarithmically divergent self-energy. This is so because the operators that create these states are not gauge-invariant, reflecting the fact that these are not neutral states. A gauge-invariant fermion operator can be defined. This is achieved by inserting the usual exponential of the line integral, along some path Γ, of the statistical vector potential between a pair of fermion-creation and -annihilation operators some distance apart from each other:

$$c^\dagger(\vec{x}, t) \exp\left(i \int_\Gamma \mathcal{A}_\mu \, dx^\mu \right) c(\vec{x}', t') \tag{11.55}$$

The massless "photon" gives rise to a logarithmically divergent fermion self-energy. A similar treatment can be given to flux states. The operator K which creates fluxes is not gauge-invariant. A way to make it invariant is to multiply this operator by a fermion operator that represents anyons, not fermions or fluxes. However, it is still possible to multiply $K(\vec{x}, t)$ by a line integral, just as in the fermion case. The resulting operator is a boson, and it is manifestly gauge-invariant. The one-particle states created by these operators also have logarithmically divergent energy and exactly for the same reason: the exchange of massless "photons."

Let us end this section by briefly considering the state in which the effective action has a non-zero effective Chern–Simons term. I will call this phase the quantum Hall state. The effective Chern–Simons coupling constant $\bar{\theta}$ equals

$$\bar{\theta} = -s_f \frac{\theta}{\rho} \tag{11.56}$$

Thus, a non-zero s_f means $\bar{\theta} \neq 0$. A theory with a non-zero Chern–Simons coupling constant is known to contain a *massive* photon. The mass of the photon is proportional to $\bar{\theta}$ and hence it is determined by s_f. Thus, the quantum Hall state has short-range *gauge* interactions mediated by the statistical gauge field. These fluctuations are effectively suppressed and the state is effectively incompressible.

12

Topology and the quantum Hall effect

In this chapter I discuss the problem of electrons moving on a plane in the presence of an external uniform magnetic field perpendicular to the system. This is a subject of great interest from the point of view of both theory and experiment. The explanation of the remarkable quantization of the Hall conductance observed in MOSFETs and in heterostructures has demanded a great deal of theoretical sophistication. Concepts drawn from branches of mathematics, such as topology and differential geometry, have become essential to the understanding of this phenomenon. In this chapter I will consider only the quantum Hall effect in non-interacting systems. This is the theory of the *integer* Hall effect. The fractional quantum Hall effect (FQHE) is discussed in Chapter 13. The related subject of topological insulators is discussed in Chapter 16.

The chapter begins with a description of the one-electron states, both in the continuum and on a 2D lattice, followed by a summary of the observed phenomenology of the quantum Hall effect. A brief discussion of linear-response theory is also presented. The rest of the chapter is devoted to the problem of topological quantization of the Hall conductance.

12.1 Quantum mechanics of charged particles in magnetic fields

Let us review the Landau problem of the states of charged particles moving on a plane in the presence of a perpendicular uniform magnetic field B. We will consider both continuum and lattice versions of the problem.

We consider first the continuum problem. Let us think of a spinless particle of mass M and charge $-e$. The one-particle Hamiltonian which describes the dynamics of this system is

$$\mathcal{H} = \frac{1}{2M}\left[\left(-i\hbar\frac{\partial}{\partial x_1} - \frac{e}{c}A_1\right)^2 + \left(-i\hbar\frac{\partial}{\partial x_2} - \frac{e}{c}A_2\right)^2\right] \tag{12.1}$$

The vector potential $\vec{A}$ is such that its curl is equal to B, the perpendicular component of the field,

$$B = \epsilon_{ij}\, \partial_i A_j \tag{12.2}$$

If the linear size of the plane is L, the total flux Φ is

$$\Phi = BL^2 \tag{12.3}$$

In what follows, I will assume that there is an exact *integer* number N_ϕ of flux quanta ϕ_0 piercing the plane

$$\Phi = N_\phi \phi_0 \equiv N_\phi \frac{hc}{e} \tag{12.4}$$

If we choose units such that $\hbar = e = c = 1$, the flux quantum ϕ_0 is just equal to 2π. In these units we can write $\Phi = 2\pi N_\phi$. Also, we are going to measure lengths in units of the magnetic length l_0 defined to be $l_0 = B^{-1/2}$.

We will work in the isotropic gauge

$$A_i = -\frac{1}{2} B \epsilon_{ij} x_j \tag{12.5}$$

In this gauge, it is convenient to work in complex coordinates $z = x_1 + ix_2$. Let us factor an exponentially decaying function of $|z|^2$ out of the wave function. This procedure automatically introduces an apparently special point, the origin $z = 0$. Since the location of the origin must be arbitrary, there should exist an operator that will remove this arbitrariness. We will see that this is the case. As a byproduct, we will also find not only that the energy eigenvalues, the Landau levels, are degenerate but also that this degeneracy is generated by a special group of transformations, the group of magnetic translations (Zak, 1964). If we are dealing with a rotationally invariant system, such as a disk, it is convenient to write the wave functions in the form

$$\Psi(z, \bar{z}) = f(z, \bar{z}) e^{-\lambda |z|^2} \tag{12.6}$$

which decays exponentially fast at infinity. For this Hilbert space, the disk is topologically equivalent to a 2-sphere.

If we now choose for λ the value

$$\lambda = \frac{e|B|}{4\hbar c} = \left(\frac{e}{\hbar c}\right) \frac{|B|}{4} \equiv \frac{1}{4 l_0^2} \tag{12.7}$$

(where we introduced the magnetic length l_0), the function $f(z, \bar{z})$ is found to satisfy an equation that, in complex coordinates, has the form

$$-\frac{2\hbar^2}{M} \partial_z \partial_{\bar{z}} f + \frac{e|B|\hbar}{Mc} \bar{z}\, \partial_{\bar{z}} f + \frac{e|B|\hbar}{2Mc} f = Ef \tag{12.8}$$

for $B > 0$. For $B < 0$ we must replace z by $\bar{z}$. In Eq. (12.8), we have introduced the operators ∂_z and $\partial_{\bar{z}}$ defined by

$$\partial_z = \frac{1}{2}(\partial_1 - i\,\partial_2), \qquad \partial_{\bar{z}} = \frac{1}{2}(\partial_1 + i\,\partial_2) \tag{12.9}$$

It will be sufficient to discuss the case of $B > 0$.

Any *analytic* function $f(z)$ is a solution of Eq. (12.8). A complete basis $\{f_n(z)\}$ has the form

$$f_n = z^n \tag{12.10}$$

which are also eigenstates of the angular-momentum operator L_z

$$L_z = -i\hbar(x_1\,\partial_2 - x_2\,\partial_1) \equiv +\hbar(z\,\partial_z - \bar{z}\,\partial_{\bar{z}}) \tag{12.11}$$

with energy and angular-momentum eigenvalues

$$E_0 = \frac{\hbar\omega_{\rm c}}{2}, \qquad L_z = n\hbar \tag{12.12}$$

An antianalytic function, $\bar{z}^m$ is an eigenstate of the mth Landau level with energy

$$E_m = \hbar\omega_{\rm c}\left(m + \frac{1}{2}\right) \tag{12.13}$$

where

$$\omega_{\rm c} = \frac{eB\hbar}{Mc} \tag{12.14}$$

is the cyclotron frequency. The Landau levels have a huge degeneracy, which is the same for all the Landau levels and is equal to N_ϕ.

In order to make this degeneracy more apparent, let us introduce the operators of *magnetic translation* and the group of transformations induced by them. Let $\vec{a}$ and $\vec{b}$ be two vectors on the plane. For a system in a magnetic field B $(B > 0)$, the canonical momentum operator $\vec{P}$ is given by the usual minimal-coupling definition

$$\begin{aligned} \vec{P} &= -i\hbar\vec{\nabla} - \frac{e}{c}\vec{A} \\ B &= \epsilon_{ij}\,\partial_i A_j \end{aligned} \tag{12.15}$$

The generator of infinitesimal magnetic translations $\vec{k}$ (Zak, 1964) is

$$k_i = P_i - \frac{eB}{c}\epsilon_{ij}x_j \equiv P_i(-B) \tag{12.16}$$

A *finite* magnetic translation by a vector $\vec{a}$ is represented by the operator $\hat{t}(\vec{a})$,

$$\hat{t}(\vec{a}) = e^{i\vec{a}\cdot\frac{\vec{k}}{\hbar}} \tag{12.17}$$

These *magnetic translation operators* obey the so-called *magnetic algebra*

$$\hat{t}(\vec{a})\hat{t}(\vec{b}) = \exp\left(-i(\vec{a} \times \vec{b}) \cdot \frac{\hat{z}}{l_0^2}\right) \hat{t}(\vec{b})\hat{t}(\vec{a}) \tag{12.18}$$

where $\hat{z}$ is a unit vector normal to the plane.

The magnetic translations form a group in the sense that the operators $\hat{t}(\vec{a})$ obey the composition law

$$\hat{t}(\vec{a})\hat{t}(\vec{b}) = \exp\left(\frac{i}{2l_0^2}(\vec{a} \times \vec{b}) \cdot \hat{z}\right) \hat{t}(\vec{b} + \vec{a}) \tag{12.19}$$

Thus, the operators $\hat{t}(\vec{a})$ form a representation of the group of magnetic translations. Equation (12.19) has an extra phase factor that is not present in the usual group composition law. The existence of this phase, which is known in mathematics as a *cocycle*, indicates that the operators $\hat{t}(\vec{a})$ form a *ray representation* of the group of magnetic translations.

The Hamiltonian for a charged particle moving in a magnetic field can now be written in the standard form, $H = \vec{P}^2/(2M)$. The canonical momentum operators $\vec{P}$ and the generators of magnetic translations $\vec{k}$ commute with each other,

$$[k_i, P_j] = 0 \tag{12.20}$$

although the *different* components of $\vec{k}$ (and $\vec{P}$) do not commute among themselves,

$$[k_i, k_j] = -[P_i, P_j] = i\frac{e\hbar B}{c}\epsilon_{ij} \tag{12.21}$$

Thus, the two components of $\vec{k}$ commute with the Hamiltonian,

$$[k_i, H] = \frac{1}{2M}[k_i, \vec{P}^2] = 0 \tag{12.22}$$

and are constants of motion. However, since k_1 and k_2 do not commute with each other, they cannot be diagonalized simultaneously. We can then use k_1 or k_2, or some linear combination thereof, to label the degenerate states. Which combination is convenient depends on the choice of boundary conditions.

Let us assume, for the moment, that the system has the shape of a rectangle with linear dimensions L_1 and L_2 along the (orthogonal) directions $\hat{e}_1$ and $\hat{e}_2$, respectively ($\hat{e}_i \cdot \hat{e}_j = \delta_{ij}, i = 1, 2$). The total flux Φ passing through the rectangle is $\Phi = BL_1L_2$. In units of the flux quantum $\phi_0 = hc/e$, the total flux is an integer $N_\phi = \Phi/\phi_0$. Alternatively, N_ϕ can be given in terms of the magnetic length l_0 and the area of the system L_1L_2 in the equivalent form

$$\frac{L_1L_2}{l_0^2} = 2\pi N_\phi \tag{12.23}$$

Let us now consider the operators $\hat{T}_1$ and $\hat{T}_2$ which represent magnetic translations by L_1/N_ϕ and L_2/N_ϕ along the directions $\hat{e}_1$ and $\hat{e}_2$, respectively:

$$\begin{aligned}\hat{T}_1 &\equiv \hat{t}\left(\frac{L_1}{N_\phi}\hat{e}_1\right)\\ \hat{T}_2 &\equiv \hat{t}\left(\frac{L_2}{N_\phi}\hat{e}_2\right)\end{aligned} \tag{12.24}$$

The operators $\hat{T}_1$ and $\hat{T}_2$ obey the algebra

$$\hat{T}_1\hat{T}_2 = e^{-i\frac{2\pi}{N_\phi}}\hat{T}_2\hat{T}_1 \tag{12.25}$$

which is often also referred to as the *magnetic algebra*. In Chapter 10 we discussed this algebra in the context of the commutation relations for anyon operators.

Let us now assume that we have a state $\Psi_{n,0}$ that is an eigenstate of the Hamiltonian in the nth Landau level and that it is also an eigenstate of, say, $\hat{T}_1$, i.e.

$$\begin{aligned}\hat{H}\Psi_{n,0} &= E_n\Psi_{n,0}\\ \hat{T}_1\Psi_{n,0} &= e^{i\lambda_0}\Psi_{n,0}\end{aligned} \tag{12.26}$$

where E_n and λ_0 are the eigenvalues. Consider now the state $\Psi_{n,m}$,

$$\Psi_{n,m} = \hat{T}_2^m\Psi_{n,0} \tag{12.27}$$

Since both $\hat{T}_1$ and $\hat{T}_2$ commute with $\hat{H}$, it follows that all the states $\Psi_{n,m}$ have energy E_n,

$$\hat{H}\Psi_{n,m} = \hat{H}\hat{T}_2^m\Psi_{n,0} = \hat{T}_2^m\hat{H}\Psi_{n,0} = E_n\Psi_{n,m} \tag{12.28}$$

However, the states $\Psi_{n,m}$ have different eigenvalues of $\hat{T}_1$,

$$\hat{T}_1\Psi_{n,m} = e^{-i2\pi\frac{m}{N_\phi}+i\lambda_0}\Psi_{n,m} \tag{12.29}$$

Thus, there are exactly N_ϕ linearly independent degenerate eigenstates of the Hamiltonian in a given Landau level. For a system with wave functions vanishing at spatial infinity (i.e. a "disk") the operators k_1 and k_2 are replaced by their counterparts in complex coordinates, k and $\bar{k}$,

$$\begin{aligned}k &= \frac{i}{2\hbar}(k_1 - ik_2) = \partial_z - \frac{\bar{z}}{4l_0^2}\\ \bar{k} &= \frac{i}{2\hbar}(k_1 + ik_2) = \partial_{\bar{z}} + \frac{z}{4l_0^2}\end{aligned} \tag{12.30}$$

which also commute with the momenta (in complex coordinates) P and $\bar{P}$,

$$\begin{aligned} P &= \frac{i}{2\hbar}(P_1 + iP_2) = \partial_{\bar{z}} - \frac{z}{4l_0^2} \\ \bar{P} &= \frac{i}{2\hbar}(P_1 - iP_2) = \partial_z + \frac{\bar{z}}{4l_0^2} \end{aligned} \tag{12.31}$$

The complex-coordinate analogs of $\hat{T}_1$ and $\hat{T}_2$, namely T and $\bar{T}$, are defined by

$$\begin{aligned} T &= e^{\frac{2L}{N_\phi}k} \\ \bar{T} &= e^{i\frac{L}{N_\phi}\bar{k}} \end{aligned} \tag{12.32}$$

for a system with $L_1 = L_2 = L$. The operators T and $\bar{T}$ also satisfy the magnetic algebra Eq. (12.25). The operator $\bar{k}$ annihilates the wave function $\Psi_n(z, \bar{z})$:

$$\bar{k}\Psi_n = 0 \qquad \text{with} \qquad \Psi_n = c_n z^n e^{-\frac{|z|^2}{4l_0^2}} \tag{12.33}$$

Thus, Ψ_n is an eigenstate of $\bar{T}$ with unit eigenvalue

$$\bar{T}\Psi_n = e^{i\frac{L}{N_\phi}\bar{k}}\Psi_n = \Psi_n \tag{12.34}$$

A complete set of eigenstates of the nth Landau level $\{\Psi_{n,m}\}$ can now be constructed quite easily ($m = 1, \ldots, N_\phi$):

$$\Psi_{n,m}(z, \bar{z}) = T^m \Psi_n(z, \bar{z}) \equiv C_{n,m} e^{+2L\frac{m}{N_\phi}k}\Psi_n(z, \bar{z}) \tag{12.35}$$

The states in the set $\{\Psi_{n,m}(z, \bar{z})\}$ have eigenvalues

$$\begin{aligned} H\Psi_{n,m}(z, \bar{z}) &= E_n \Psi_{n,m}(z, \bar{z}) \\ \bar{T}\Psi_{n,m}(z, \bar{z}) &= e^{-i2\pi\frac{m}{N_\phi}}\Psi_{n,m}(z, \bar{z}) \end{aligned} \tag{12.36}$$

with

$$H = \frac{2\hbar^2}{M}\left[-P\bar{P} + \frac{eB}{4\hbar c}\right] \tag{12.37}$$

If instead of open (or vanishing) boundary conditions we want to consider a system on a *torus*, i.e. periodic boundary conditions along the directions $\hat{e}_1$ and $\hat{e}_2$ of a rectangle, the wave functions will have to satisfy a periodicity condition. It is customary to demand that

$$\Psi(x_1, x_2) = \Psi(x_1 + L_1, x_2) = \Psi(x_1, x_2 + L_2) \tag{12.38}$$

However, it is not possible to satisfy this condition if a non-zero magnetic field is present. The vector potential violates translation invariance. Thus, the wave functions cannot obey periodic boundary conditions (PBCs) since no flux could possibly go through the system if PBCs are to be obeyed. In such a case, the circulation of $\vec{A}$ around the boundary equals zero. In order to accommodate a non-zero external flux, the vector potentials *and* the wave functions have to change by a large gauge transformation as we traverse the system (Haldane and Rezayi, 1985),

$$\begin{aligned} A_i(x_1 + L_1, x_2) &= A_i(x_1, x_2) + \partial_i \beta_1(x_1, x_2) \\ A_i(x_1, x_2 + L_2) &= A_i(x_1, x_2) + \partial_i \beta_2(x_1, x_2) \end{aligned} \tag{12.39}$$

such that the circulation around the boundary Γ equals the flux Φ. This requirement implies that β_1 and β_2 must satisfy the condition

$$[\beta_2(x_1 + L_1, x_2) - \beta_2(x_1, x_2)] - [\beta_1(x_1, x_2 + L_2) - \beta_1(x_1, x_2)] = \Phi \tag{12.40}$$

It is sufficient to give just one solution to this equation, which we choose to be

$$\beta_i = -\frac{1}{2}\Phi\epsilon_{ij}\frac{x_j}{L_j} \tag{12.41}$$

The requirement of gauge invariance forces the wave functions $\Psi(x_1, x_2)$ to transform as

$$\Psi(x_1, x_2) \to \exp\left(-i\frac{e}{\hbar c}\Lambda(x_1, x_2)\right)\Psi(x_1, x_2) \tag{12.42}$$

under a gauge transformation in which A_i changes by $\partial_i \Lambda(x_1, x_2)$.

Thus, under the large gauge transformation of Eq. (12.39), the wave functions must change like

$$\begin{aligned} \Psi(x_1 + L_1, x_2) &= e^{i\frac{e}{\hbar c}\beta_1(x_1,x_2)}\Psi(x_1, x_2) \\ \Psi(x_1, x_2 + L_2) &= e^{i\frac{e}{\hbar c}\beta_2(x_1,x_2)}\Psi(x_1, x_2) \end{aligned} \tag{12.43}$$

The boundary conditions of Eq. (12.39) and Eq. (12.43) are consistent provided that the translations $(x_1, x_2) \to (x_1 + L_1, x_2) \to (x_1 + L_1, x_2 + L_2)$ and $(x_1, x_2) \to (x_1, x_2 + L_2) \to (x_1 + L_1, x_2 + L_2)$ lead to the same value of the wave function. It is easy to check that this consistency condition leads to the flux quantization $\Phi = N_\phi \phi_0$. This result should come as no surprise, since we are in the situation of the Aharonov–Bohm effect. In other words, the system has single-valued wave functions on the torus only if the flux is quantized. The (single-valued) wave functions constructed with this prescription are (doubly) periodic and form N_ϕ-fold-degenerate multiplets. If the flux is not quantized (e.g. a rational multiple of ϕ_0) the wave functions are multivalued and have branch cuts.

12.2 The Hofstadter wave functions

In the last section we considered the quantum-mechanical motion of charged particles moving in a plane in the presence of an external magnetic field perpendicular to the plane. There are many physical situations in which the presence of a *lattice* cannot be ignored. In most cases these effects are quite small. Magnetic fields are relativistic effects and, if we want to pass a sizable fraction of the flux quantum ϕ through a plaquette of a physical lattice (with spacing $a_0 \approx 10\,\text{Å}$), astronomically large magnetic fields are required. Thus, for problems such as electrons in a heterostructure, lattice effects are, in practice, negligible. However, when we are dealing with a chiral spin state, we discover the existence of dynamically generated gauge fields with *large* fluxes. Here, of course, lattice effects become dominant.

The problem of the quantum motion on 2D lattices in external magnetic fields was first studied by Hofstadter (1976). He considered the problem of a particle of charge e hopping on a square lattice, with hopping amplitude t, in the presence of an external uniform magnetic field B. Let $|\vec{x}\rangle$ denote the (Wannier) state localized at site $\vec{x}$ of the square lattice. The hopping (tight-binding) Hamiltonian H is

$$H = -t \sum_{\vec{x}, j=1,2} |\vec{x}\rangle e^{i\frac{e}{\hbar c}A_j(\vec{x})} \langle \vec{x} + e_j| + \text{h.c.} \tag{12.44}$$

The vector potentials $A_j(\vec{x})$ reside on the links and represent the external flux. The total flux Φ going through any individual plaquette (with lattice spacing $a_0 = 1$) is B,

$$\sum_{\square} A_j = \Delta_1 A_2 - \Delta_2 A_1 = B \tag{12.45}$$

If we demand that the system be a torus, it is customary to work in the *Landau gauge*

$$A_1 = -Bx_2, \quad A_2 = 0 \tag{12.46}$$

where x_1 and x_2 are integers ($0 \leq x_i \leq L_i,\ i = 1, 2$). From now on, I will assume that $Ba^2 = (p/q)\phi_0$, with p and q a pair of relatively prime integers. In other words, the flux going through an elementary plaquette is a finite fraction (p/q) of the flux quantum ϕ_0.

The eigenstates $|\Psi\rangle$ of the system can be expanded in terms of a set of site (or Wannier) states

$$|\Psi\rangle = \sum_{\vec{x}} \Psi(\vec{x})|\vec{x}\rangle \tag{12.47}$$

and obey the discrete Schrödinger equation

$$\begin{aligned} -t\left\{e^{-i2\pi\frac{p}{q}x_2}\Psi(x_1+1, x_2) + e^{+i2\pi\frac{p}{q}x_2}\Psi(x_1-1, x_2)\right\} \\ - t\{\Psi(x_1, x_2+1) + \Psi(x_1, x_2-1)\} = E\Psi(x_1, x_2) \end{aligned} \tag{12.48}$$

This Hamiltonian is not invariant under translations by one lattice spacing. However, in the Landau gauge, it is invariant under the translations

$$\begin{aligned}(x_1, x_2) &\to (x_1 + q, x_2)\\ (x_1, x_2) &\to (x_1, x_2 + 1)\end{aligned} \tag{12.49}$$

Hence, the unit cell has q elementary plaquettes. With the present choice of gauge, the unit cell is $1 \times q$. The total flux passing through the unit cell is

$$\Phi_{\text{cell}} = q\Phi_{\text{plaquette}} = p \tag{12.50}$$

which is an integer. Naturally, this is not an accident.

The gauge-invariant operator for translations $e^{i\hat{P}_j}$ is (in units such that $e = \hbar = c = a_0 = 1$)

$$e^{i\hat{P}_j} = \sum_{\vec{x}} |\vec{x}\rangle e^{iA_j(\vec{x})} \langle \vec{x} + \hat{e}_j| \tag{12.51}$$

These operators satisfy the algebra

$$e^{i\hat{P}_1} e^{i\hat{P}_2} = e^{i2\pi\frac{p}{q}} e^{i\hat{P}_2} e^{i\hat{P}_1} \tag{12.52}$$

and, hence, do not generally commute with each other. But $e^{in_1\hat{P}_1}$ and $e^{in_2\hat{P}_2}$ do commute with each other if

$$\frac{p}{q} n_1 n_2 \in \mathbb{Z} \tag{12.53}$$

Thus, the translations $e^{in_1\hat{P}_1}$ and $e^{in_2\hat{P}_2}$ commute if and only if the flux passing through the rectangle with edges n_1 and n_2 is an integer multiple of the flux quantum. The smallest rectangle satisfying Eq. (12.53) is known as the *magnetic unit cell.*

The hopping Hamiltonian can now be written in terms of the operators $e^{i\hat{P}_j}$ in the form

$$H = -t \sum_{j=1,2} (e^{i\hat{P}_j} + e^{-i\hat{P}_j}) \tag{12.54}$$

The eigenstates of H are eigenstates not of e^{iP_j}, but of the operators $e^{i\hat{k}_j}$ which generate finite (i.e. lattice) *magnetic translations*. The operators $e^{i\hat{k}_j}$ are defined by

$$e^{i\hat{k}_j} \equiv \sum_{\vec{x}} |\vec{x}\rangle e^{iA'_j(\vec{x})} \langle \vec{x} + \hat{e}_j| \tag{12.55}$$

where the vector potentials $A'_j(\vec{x})$ have to be chosen so that the *magnetic translation* operators $e^{i\hat{k}_j}$ *commute* with the elementary *lattice translations* $e^{i\hat{P}_j}$ and, hence, with the Hamiltonian H. These conditions are met if we choose ($j \neq k$) $\Delta_j A'_k(\vec{x}) = \Delta_k A_j(\vec{x})$.

So, once again, we find

$$\hat{k}_j = \hat{P}_j(-B) \tag{12.56}$$

but in the specific choice of gauge:

$$A_1'(\vec{x}) = 0, \qquad A_2'(\vec{x}) = -2\pi \frac{p}{q} x_1 \tag{12.57}$$

The operators $e^{i\hat{k}_j}$ do not commute with each other. Rather, they obey

$$e^{i\hat{k}_1} e^{i\hat{k}_2} = e^{i2\pi \frac{p}{q}} e^{i\hat{k}_2} e^{i\hat{k}_1} \tag{12.58}$$

Consider now the magnetic translations by n_1 steps along x_1 and n_2 steps along x_2 (no sum over j),

$$\hat{T}_j^{n_j} = e^{i n_j \hat{k}_j} \tag{12.59}$$

These operators commute with each other if n_1 and n_2 satisfy Eq. (12.53).

Thus, the eigenstates of H are also eigenstates of $\hat{T}_1^{n_1}$ and $\hat{T}_2^{n_2}$. With the choice of Eq. (12.57), we see that $\hat{T}_1$ and $\hat{T}_2^q$ satisfy

$$[\hat{T}_1, \hat{T}_2^q] = [\hat{T}_1, \hat{H}] = [\hat{T}_2^q, H] = 0 \tag{12.60}$$

and their eigenstates can be used to label the eigenstates of H. The eigenstates of $\hat{T}_1$ and $\hat{T}_2^q$ are of the form $|k_1, k_2\rangle$:

$$\begin{aligned} \hat{T}_1 |k_1, k_2\rangle &= e^{ik_1} |k_1, k_2\rangle \\ \hat{T}_2^q |k_1, k_2\rangle &= e^{iqk_2} |k_1, k_2\rangle \end{aligned} \tag{12.61}$$

and must satisfy periodic boundary conditions

$$\begin{aligned} \hat{T}_1^{L_1} |k_1, k_2\rangle &= |k_1, k_2\rangle \\ (\hat{T}_2^q)^{L_2/q} |k_1, k_2\rangle &= |k_1, k_2\rangle \end{aligned} \tag{12.62}$$

These conditions can be met only if (k_1, k_2) belongs to the *magnetic Brillouin zone* ($-\pi \le k_1 < \pi$ and $-\pi/q \le k_2 < \pi/q$). Clearly, these boundary conditions can be imposed only if L_2 is an integer multiple of q. That is to say, the total flux Φ going through the entire system has to be an integer N_ϕ multiple of the flux quantum ϕ_0, with $N_\phi = (p/q) L_1 L_2$.

The magnetic Brillouin zone labels a total of $L_1 L_2/q$ states. We will see now that this system has q Landau (or Hofstadter) bands, each with $L_1 L_2/q$ states. This is the discrete version of the degeneracy of the continuum problem.

Let us now expand the states $\Psi(\vec{x})$ in terms of magnetic-translation eigenstates:

$$\Psi(x_1, x_2) = \frac{1}{q} \sum_{r=1}^{q} \int_{-\pi}^{\pi} \frac{dk_1}{2\pi} \int_{-\frac{\pi}{q}}^{\frac{\pi}{q}} \frac{dk_2}{2\pi/q} e^{i(k_1 x_1 + k_2 x_2)} \Psi\left(k_1, k_2 + 2\pi \frac{p}{q} r\right) \tag{12.63}$$

It is now convenient to define the q-component vector, $\Psi_r(k_1, k_2)$, as

$$\Psi_r(k_1, k_2) \equiv \Psi\left(k_1, k_2 + 2\pi \frac{p}{q} r\right), \qquad r = 1, \ldots, q \tag{12.64}$$

We recognize in this vector a generalization of the spinons used to study the flux phase, where $p/q = \frac{1}{2}$. The (discrete) Schrödinger equation now reads

$$\begin{aligned} -t[e^{ik_1}\Psi_{r+1}(k_1, k_2) + e^{-ik_1}\Psi_{r-1}(k_1, k_2)] - 2t \cos\left(k_2 + 2\pi \frac{p}{q} r\right) \Psi_r(k_1, k_2) \\ = E(k_1, k_2)\Psi_r(k_1, k_2) \end{aligned} \tag{12.65}$$

This equation is also known as the Harper equation and plays an important role in the theory of the electronic structure of incommensurate systems. The amplitudes $\Psi_r(r_1, r_2)$ are periodic functions on the magnetic Brillouin zone and thus satisfy

$$\begin{aligned} \Psi_r(k_1 + 2\pi n_1, k_2) &= \Psi_r(k_1, k_2) \\ \Psi_r\left(k_1, k_2 + \frac{2\pi}{q} n_2\right) &= \Psi_{r+n_2}(k_1, k_2) \\ \Psi_{r+q}(k_1, k_2) &= \Psi_r(k_1, k_2) \end{aligned} \tag{12.66}$$

where n_1 and n_2 are integers. Equation (12.66) implies that the magnetic Brillouin zone has the topology of a 2-torus. The amplitudes $\Psi_r(k_1, k_2)$, which are solutions of Eq. (12.65), form an r-component complex vector field that is continuous on the torus.

For arbitrary values of the integers p and q (p and q relatively prime), the spectrum determined from Eq. (12.65) has a very complex structure. For instance, if p and q are chosen to belong to some infinite sequence such that, in the limit, p/q becomes arbitrarily close to an irrational number, the spectrum becomes a Cantor set (Hofstadter, 1976) and the wave functions exhibit self-similar behavior (Kohmoto, 1983). Even if the problem is restricted to commensurate flux only ($\Phi = 2\pi(p/q)\phi_0$), the spectrum has energy gaps that, as q is increased, exhibit a hierarchical structure. We will not consider these issues here. Rather, we will consider only the broad qualitative properties of the spectrum and wave functions. In general, Eq. (12.65) has to be solved numerically.

For generic values of p and q, the spectrum has q bands. For any arbitrary pair of relatively prime integers p and q, the Hamiltonian $\mathcal{H}(k_1, k_2)$ associated with the Schrödinger equation Eq. (12.65) has a number of symmetries (Wen and Zee, 1989). Let $\hat{A}$ and $\hat{B}$ be a pair of $q \times q$ matrices defined by

$$\hat{A}_{jk} = \omega^k \delta_{jk}, \qquad \hat{B}_{jk} = \delta_{j,k-1} \tag{12.67}$$

where $j, k = 1, \ldots, q$ and $\omega = e^{-i2\pi p/q}$, satisfying the algebra $AB = e^{i2\pi \frac{p}{q}} BA$. The Hamiltonian $\mathcal{H}(k_1, k_2)$ is given by

$$\mathcal{H}(k_1, k_2) = e^{-ik_2} \hat{A} + e^{+ik_1} \hat{B} + \text{h.c.} \tag{12.68}$$

Given p and q, we can always find a pair of (relatively prime) integers n and m such that $1 = np + mq$. It is easy to check that the matrices $\tilde{A} \equiv \hat{A}^n$ and $\tilde{B} \equiv \hat{B}^m$ satisfy the following identities:

$$\tilde{A}\mathcal{H}(k_1, k_2)\tilde{A}^{-1} = \mathcal{H}\left(k_1 + \frac{2\pi}{q} n, k_2\right) \tag{12.69}$$

$$\tilde{B}\mathcal{H}(k_1, k_2)\tilde{B}^{-1} = \mathcal{H}\left(k_1, k_2 + \frac{2\pi}{q}\right) \tag{12.70}$$

$$\mathcal{H}\left(k_1 + \frac{2\pi}{q} - 2\pi m, k_2\right) = \mathcal{H}\left(k_1 + \frac{2\pi}{q}, k_2\right) \tag{12.71}$$

$$\tilde{A}\tilde{B} = e^{-i2\pi p \frac{n^2}{q}} \tilde{B}\tilde{A} \tag{12.72}$$

Thus, if $\Psi(k_1, k_2)$ is an eigenstate of $\mathcal{H}(k_1, k_2)$ with energy $E(k_1, k_2)$, the state $\Psi'(k_1, k_2)$,

$$\Psi'(k_1, k_2) = \tilde{A}\Psi(k_1, k_2) \tag{12.73}$$

is an eigenstate of $\mathcal{H}'(k_1, k_2) = \mathcal{H}(k_1 + 2\pi/q, k_2)$ with the *same* eigenvalue $E(k_1, k_2)$. In other words, there is a one-to-one correspondence between the spectrum at (k_1, k_2) and that at $(k_1 + 2\pi/q, k_2)$. An analogous argument shows that the spectra at $(k_1, k_2 + 2\pi/q)$ and (k_1, k_2) are also identical to each other. In addition, under the translation $(k_1, k_2) \to (k_1 + \pi, k_2 + \pi)$, the Hamiltonian changes sign, i.e. $\mathcal{H}(k_1 + \pi, k_2 + \pi) = -\mathcal{H}(k_1, k_2)$, and $E(k_1 + \pi, k_2 + \pi) = -E(k_1, k_2)$. For *q even*, this operation is a particular case of Eq. (12.69). Thus, if q is even, then, for each eigenstate of $\mathcal{H}$ with energy E, there exists an eigenstate with energy $-E$. The operator that connects states with opposite signs of the energy, let us call it Γ, must anti-commute with $\mathcal{H}$ and be hermitian. It is easy to check that the matrix Γ_{jk},

$$\Gamma_{jk} = (-1)^j i^{q/2} \delta_{k, j+q/2} \tag{12.74}$$

has the desired properties

$$\{\mathcal{H}, \Gamma\} = 0, \qquad \Gamma^2 = I \tag{12.75}$$

and that Γ also anti-commutes with $\hat{A}$, $\hat{B}$, $\tilde{A}$, and $\tilde{B}$.

Furthermore, it is possible to show that, for q even, there are at least q eigenstates of $\mathcal{H}$ with zero energy (Wen and Zee, 1989). The argument uses the topology of the torus in an essential way. It can be regarded as a generalization of the Nielsen–Ninomiya theorem for the absence of Weyl fermions in lattice systems (Nielsen and Ninomiya, 1981; Friedan, 1984). The magnetic Brillouin zone is locally isomorphic to the complex plane ($w = k_1 + ik_2$) and globally equivalent to a torus. Let us consider a point w on the magnetic Brillouin zone and assume that the eigenstates of $\mathcal{H}$ at $w = k_1 + ik_2$ are all different from zero. It is possible to choose a basis of states in which Γ is diagonal. In this basis we can write

$$\Gamma = \begin{pmatrix} I & 0 \\ 0 & -I \end{pmatrix}, \qquad \mathcal{H} = \begin{pmatrix} 0 & h^+ \\ h & 0 \end{pmatrix} \tag{12.76}$$

where h is a $q/2 \times q/2$ matrix and h^+ is its adjoint. In other words, $\mathcal{H}$ has the same structure as the Dirac Hamiltonian in the chiral basis. If in the neighborhood of w there are no zero energy eigenvalues of $\mathcal{H}$, the determinant

$$\det \mathcal{H} = -|\det h| \neq 0 \tag{12.77}$$

is non-zero. Let us denote by D the determinant of h, $D = \det h$. D is locally an analytic function of w. Thus, it is possible to define the vector field $\mathcal{A}_i (i = 1, 2)$,

$$\mathcal{A}_i = D^{-1} \frac{\partial}{\partial k_i} D \tag{12.78}$$

that, in fiber-bundle terminology, is a 1-form, a connection. In any neighborhood of w free of zero-energy eigenvalues, the 1-form $\mathcal{A}_i$ is closed, i.e.

$$\epsilon_{ij}\, \partial_i \mathcal{A}_j = \epsilon_{ij}\, \partial_i \partial_j \ln D = 0 \tag{12.79}$$

but, in general, it is not exact. The circulation ν of $\vec{\mathcal{A}}$ on an arbitrary contour $\mathcal{C}$ of the magnetic Brillouin zone

$$\nu = \frac{1}{2\pi} \oint_{\mathcal{C}} d\vec{k} \cdot \vec{\mathcal{A}} \neq 0 \tag{12.80}$$

is in general different from zero.

If ν is not zero, the determinant $D(\vec{k}) = \det h(\vec{k})$ must have a zero at some point $\vec{k}_0$ somewhere inside $\mathcal{C}$. We now follow Wen and Zee (1989), and consider a path $\mathcal{C}$ that is a rectangle with corners at (k_1, k_2), $(k_1 + 2\pi/q, k_2)$, $(k_1 + 2\pi/q, k_2 + 2\pi/q)$, and $(k_1, k_2 + 2\pi/q)$. From the symmetries of $\mathcal{H}$, it is possible to show that $D(k_1, k_2)$ satisfies

$$\begin{aligned} D(k_1, k_2) &= -D^*\left(k_1 + \frac{2\pi}{q}, k_2\right) = -D\left(k_1 + \frac{2\pi}{q}, k_2 + \frac{2\pi}{q}\right) \\ &= D^*\left(k_1, k_2 + \frac{2\pi}{q}\right) \end{aligned} \tag{12.81}$$

Equation (12.81) implies that the *phase* of D must wind as the path $\mathcal{C}$ is traversed. In general, $D(\vec{k})$, being a complex number, will trace a closed path $\mathcal{D}$ on the complex plane as $\vec{k}$ traces the path $\mathcal{C}$. If D does not have a zero inside $\mathcal{C}$, the winding number ν will vanish and $\mathcal{C}$, and hence $\mathcal{D}$, can be smoothly shrunk to zero. If there is a zero, D will have a singularity and $\mathcal{C}$ cannot be deformed to zero. The path $\mathcal{D}$ will now wind around the origin $\mathcal{D} = 0$ a number of times before closing. The winding number ν of Eq. (12.80) is precisely this winding number. Since $D(\vec{k})$ is not a constant, we conclude that it must have zeros at certain isolated locations. However, the translation symmetries of Eq. (12.69) require that, if $\vec{k}_0$ is a zero of $\mathcal{H}$, then $\vec{k}_0 + (2\pi/q)(n_1\hat{e}_1 + n_2\hat{e}_2)$ must also be zeros of $\mathcal{H}$. This lattice of zeros of $\mathcal{H}$ must be periodic. The only values of $\vec{k}_0$ consistent with these demands are $\vec{k}_0 = (\pi/2, \pi/2)$ and its translations.

There are exactly q distinct points in this lattice. Thus, for q even, the Hamiltonian has exactly q zeros. Note that the flux phase is a particular case of this problem. The doubling of spinon species that we found there is a particular case of the q-fold multiplicity discussed in this section.

12.3 The quantum Hall effect

In this section we will discuss the most qualitative features of a very fascinating problem: the quantum Hall effect. It is not within the scope of this book to give an exhaustive review on this subject. Reviews are widely available, in particular the excellent volume by R. Prange and S. Girvin (Prange and Girvin, 1990).

However, there are very close analogies and connections between the theories of the fractional and integer quantum Hall effects and the theories of chiral spin liquids. We will devote considerable attention to these analogies.

In 1980, K. von Klitzing, G. Dorda, and M. Pepper (von Klitzing *et al.*, 1980) announced the discovery of very unusual transport properties of a 2D electron gas in a high magnetic field. They were studying the Hall conductance of 2D inversion layers or MOSFETs. In these systems, the electrons of a semiconductor move on quantum states that are localized within atomic scales of the layer. They are almost free to move inside the layer. Von Klitzing and his collaborators noticed that, when they measured the Hall conductance σ_{xy} of the layer at very low temperatures, the conductance had a stepwise dependence on the external magnetic field. At the same time, the longitudinal conductivity, σ_{xx}, appeared to be essentially zero when σ_{xy} was nearly constant, the so-called plateaus. For values of the field at which σ_{xy} varied, σ_{xx} was non-zero.

What was very unusual was the values that σ_{xy} attained at the plateaus. It appeared to be *quantized* at integer multiples of e^2/h. Furthermore, the quantization appeared to be sharper at lower temperatures and, oddly enough, for the more disordered samples. This phenomenon is known today as the integer quantum Hall effect.

In 1982, D. Tsui, H. Stormer, and A. Gossard (Tsui *et al.*, 1982), performed a similar series of experiments but on highly pure GaAs–AlAs heterojunctions. Here too, the electrons are bound to a surface and are essentially free to move inside the surface. They found a fractional quantum Hall effect. In fact, their results were very similar to what von Klitzing *et al.* had seen, except that σ_{xy} was not an integer multiple of e^2/h but a *fraction*. In particular, they were able to observe the fractions $\frac{1}{3}$, $\frac{2}{5}$, and others. It is a simple matter to argue that, if a Landau level is completely filled, the Hall conductance has to be quantized. In the case of a *translationally invariant* system a simple argument can be made. Let us imagine that we have an external magnetic field B perpendicular to the sample and that there is an external electric field $\vec{E}$ parallel to the sample. By coupling the system to a source and a sink of electrons, a current is established. In such a situation, there is a Lorentz force that pushes all the electrons sideways. Also, if some of the Landau levels are completely filled, leaving all others empty, there cannot be any component of the current parallel to $\vec{E}$ since it would require processes that are suppressed by an energy gap equal to $\hbar\omega_c$. If the electric field is small, and the system is translationally invariant, there is a reference frame moving at a velocity $\vec{v}$ relative to the laboratory such that $\vec{v}/c \times \vec{B} = -\vec{E}$. In this frame the electric field is absent. A completely filled Landau level has $N = N_\phi$ electrons. If there are n Landau levels that are filled, the total charge Q is $Q = nN_\phi$. The current $\vec{J}$ is then equal to $\vec{J} = +Qe\vec{v}$. Putting it all together, we conclude that the current density, $\vec{j} = \vec{J}/L^2$, has components

$$j_i = \frac{Qe}{L^2} v_i = \left(\frac{Qec}{BL^2}\right) \epsilon_{ij} E_j \tag{12.82}$$

From Eq. (12.82) we conclude that the Hall conductance σ_{xy}, i.e. the coefficient of E_j, is equal to Qec/BL^2. By using the fact that there are n filled Landau levels and that the flux BL^2 is equal to N_ϕ times the flux quantum hc/e, we get

$$\sigma_{xy} = \frac{Qec}{BL^2} = \frac{nN_\phi ec}{(hc/e)N_\phi} = n\frac{e^2}{h} \tag{12.83}$$

Notice that h, and hence quantum mechanics, enters only through the flux quantum hc/e.

This is an appealing argument, but it is deceptive. First of all, it does not apply to systems that are not translationally invariant. However, a detailed calculation shows that Eq. (12.83) is valid even in that case. The second and most serious problem with this argument is that it cannot predict the existence of the fractional values of σ_{xy}. In fact, the absence of the parallel, or dissipative, component of the current was argued by recalling the fact that, if an integer number of levels is exactly filled, no scattering is possible. If some level is only partially filled, there are states available for scattering and the argument seems to fall apart. Thirdly, this argument *alone* cannot explain the fact that the effect is actually observed. It cannot explain either the incredible accuracy to which the quantization is measured (one part per million for the integer steps). In the experimental setup, the charge density or the external magnetic field can be varied. In either case, the chemical potential must lie between Landau levels in order for one Landau level to be filled and the next one to be empty. As the density increases, the chemical potential (i.e. Fermi energy) jumps discontinuously from Landau level to Landau level. It remains fixed at the energy of a given level until the level is completely filled. This argument suggests that σ_{xy} should be a monotonically increasing function of the electron density. So, why do we see steps?

The resolution of all of these paradoxes has required a significant amount of theoretical effort. The explanation of the *observability* of the steps in σ_{xy} (i.e. the plateaus) involves the presence both of *impurities* and of *states at the edge* of the sample. The *accuracy* of the effect turned out to be connected to the *topological properties* of the quantum states. The *fractional* effect required the discovery of a new condensed state of matter, the *Laughlin state*.

The in-depth study of all these issues lies far from the main scope of this book. Besides, excellent reviews are widely available. I will hence not discuss the role of disorder beyond giving a very qualitative description. The role of topology and the Laughlin wave function will be discussed in the next sections.

12.4 The quantum Hall effect and disorder

Let us briefly discuss the role of disorder. In part for the sake of simplicity, but also because the problem is not fully understood, we will focus just on the non-interacting problem. It is widely suspected that disorder is as essential to the observability of the fractional effect as it is to the integer effect. So, we wish to understand why the integer quantum Hall effect is observed in the more disordered samples. We saw above that a simple model of free electrons in Landau levels does not explain the plateaus which are characteristics of the integer Hall effect. The reason behind the monotonic increase was the fact that the Fermi level *jumps* from one Landau level to the next as the level gets filled up. If there were extra states "in the gap" (i.e. "between Landau levels"), the Fermi energy would have to progress through those levels until they too became filled. However, these extra states should not contribute to the value of σ_{xy} for the plateaus to remain sharp.

Disorder offers a natural way to generate states "between Landau levels." First of all, any degree of randomness, usually represented by a *random potential* $V(\vec{x})$, will lift the degeneracy of each Landau level, making them become narrow bands. From studies of electron states in random potentials one expects that at least some states should become *localized* (Anderson, 1958). In the *absence* of a magnetic field, it is widely believed that *all electronic* states of 2D disordered systems are localized (Abrahams *et al.*, 1979). The arguments involve both scaling ideas and a mapping of the problem onto a special type of non-linear sigma model (Wegner, 1979). If the electrons move in the presence of a *weak* magnetic field, the same arguments apply. It turns out that the presence of the field has only two effects: (1) the symmetry of the non-linear sigma model is unitary (which reflects the fact that, in the presence of the field, there is no time-reversal invariance) and (2) the presence of a *topological term* in the effective action (Levine *et al.*, 1983; Pruisken, 1984). There is an excellent review by Pruisken on this subject in the book by Prange and Girvin (1990).

The non-linear sigma model represents the physics of the diffusive modes in the presence of the external field. It is a correct description if the elastic mean-free path λ is *short* compared with the magnetic length l_0, $\lambda \ll l_0$. This condition can be achieved only in the weak-field limit. The diffusive modes are represented in the replica formalism, by $2N \times 2N$ hermitian matrices $Q_{i\alpha, j\beta}(x)$, where $i, j = 1, \ldots, N$ and $\alpha, \beta = \pm$. The Latin indices i and j represent the "replicas" and the Greek indices, α and β, represent the particle and hole channels. The non-linear sigma model has the effective Lagrangian (Levine *et al.*, 1983)

$$\mathcal{L} = -\frac{\sigma_{xx}^0}{8}\,\text{tr}[\partial_\mu Q\,\partial_\mu Q] + \frac{\sigma_{xy}^0}{8}(B)\text{tr}[\epsilon_{\mu\nu} Q\,\partial_\mu Q\,\partial_\nu Q] \tag{12.84}$$

valid in the replica limit, $N \to 0$.

The coefficients in this Lagrangian σ_{xx}^0 and $\sigma_{xy}^0(B)$ represent the values of the longitudinal (σ_{xx}^0) and Hall (σ_{xy}^0) conductance at the length scales of the elastic mean-free path (i.e. their Boltzmann values). This non-linear sigma model is invariant under global unitary transformations in the coset $\text{U}(2N)/\text{U}(N)\times\text{U}(N)$. Notice that the topological term in Eq. (12.84) has the *same* structure as the topological terms that we discussed for antiferromagnets in Chapter 7. At scales l that are *long* compared with λ but *short* compared with l_0, the effective values of $\sigma_{xx}(l)$ and $\sigma_{xy}(l)$ are strongly renormalized. The non-linear sigma model of Eq. (12.84) is asymptotically free, which means that $\sigma_{xx}(l) \to 0$ for $l \ll \lambda$. In this infrared limit $\sigma_{xy}(l)$ is quantized, $\sigma_{xy}(l) \to (n/(2\pi))(e^2/\hbar)$. This quantization has the same topological origin as the quantization of spin and of the coefficients of the topological terms, which we discussed in Chapter 7.

Thus, this calculation shows that σ_{xy} is indeed quantized and that σ_{xx} is zero whenever the magnetic field B is in a plateau of the Hall conductivity σ_{xy}. However, the replica limit obscures the physical mechanism by which all of this takes place. It almost seems like magic! Moreover, the actual mechanism by which the system manages even to support a Hall current is very obscure in this picture. But it does point to the fact that it is the physics of localization that makes the effect *observable* in the first place and that the *topological* properties of the quantum states are responsible for the exact *quantization* of σ_{xy}.

If topology is to be the source both of the quantization and of the accuracy of the quantum Hall effect, it appears that the mechanism which supports the Hall current should not be linked to disorder in an essential way. Halperin proposed that the states which carry the Hall current reside at the *edge* of the system (Halperin, 1982). Roughly speaking, the electrons are kept inside the sample by a potential that rises towards the physical edge of the system. On some set of points close to the edge, the potential is equal to the Fermi energy. This set of points constitutes a closed curve. The edge states are the waves of the electron liquid spilling over this curve. The presence of disorder complicates the picture. The landscape of the potential can be quite rough. Semi-classically the ground state can be viewed as a set of equipotential curves. In the high-field limit, equipotential curves will generally be closed, and enclose regions that are quite small and are occupied by electrons. As the field is lowered, these regions will begin to merge and, at some critical value B_c of the field, a percolation phenomenon occurs (Trugman, 1983). At B_c there is at least one curve that percolates throughout the system. This curve is a "new edge," which is thus capable of carrying current. The electron states associated with these "edges" have a very special property: they are "chiral" (Wen, 1991b). What this means is that the electrons have to *drift* in the field, and hence the *direction* of their motion is determined by the sign of the magnetic field. Roughly, the electrons move at the drift velocity cE/B. Since the electrons in the edge states move in only one direction, the only possible effect of impurities on them is just a phase shift of the wave propagating forwards. There are no backward-scattering processes. Localization is due to a multiple-scattering process in which forward- and backward-scattering events interfere so much that the electron is unable to propagate. In the absence of backward scattering, there are no localized states. The edge states carry the full current.

12.5 Linear-response theory and correlation functions

In this section, we derive a set of formulas that will enable us to calculate the Hall conductance, as well as other response functions, in terms of the Green functions of the system. In the next section it will be shown that these formulas, when used to compute σ_{xy} for a system with an energy gap, have a hidden topological structure.

Let us consider the system of fermions coupled to an external electromagnetic field. We will consider cases of the fermions moving in free space and on a lattice in the tight-binding limit. In both cases, the generating functional of the fermion Green functions is a functional integral $\mathcal{Z}[A_\mu]$, which is a functional of the external electromagnetic field A_μ. Let us further assume that A_μ is a small fluctuating component of the external field. The average field $\langle A_\mu \rangle$ is absorbed into the definition of the system. Under such circumstances, it makes sense to determine $\mathcal{Z}[A]$ in perturbation theory, i.e. as a series expansion in powers of A_μ. The leading term in this expansion is known as linear-response theory (Fetter and Walecka, 1971). This series can be written in the exponentiated form

$$\mathcal{Z}[A_\mu] = \mathcal{Z}[0] \exp\left\{\frac{i}{2}\int d^D x \int d^D y\, A_\mu(x)\Pi_{\mu\nu}(x, y)A_\nu(y) + \cdots\right\} \tag{12.85}$$

where $\Pi_{\mu\nu}(x, y)$ is the *polarization tensor* and $D = d + 1$, with d being the dimension of the space. For a tight-binding model the spatial integrals are replaced by sums.

The underlying fermion system is *gauge-invariant.* Thus, upon an arbitrary local gauge transformation $\phi(x)$,

$$\begin{aligned} A_\mu &= A'_\mu + \partial_\mu \phi(x) \\ \psi(x) &= e^{i\frac{e}{\hbar c}\phi(x)}\psi'(x) \end{aligned} \tag{12.86}$$

the functional $\mathcal{Z}[A]$ is invariant. Thus, the linear-response term must also be gauge-invariant. This is possible only if the polarization tensor $\Pi_{\mu\nu}(x, y)$ is *transverse*, i.e.

$$\partial^x_\mu \Pi_{\mu\nu}(x, y) = 0 \tag{12.87}$$

To be more precise, we consider either a system without boundaries or one in which only "small" gauge transformations are allowed, i.e. those transformations which vanish at the boundaries, $\lim_{|x|\to\infty}\phi(x) = 0$. If the actual boundaries are to be taken into account, such as in cases in which the system is physically coupled to external leads of batteries or measuring instruments, then the values of the gauge transformations at the boundaries become physical degrees of freedom, i.e. the voltage of a battery. Similarly, for a system without boundaries, the circulations of the vector potential A_μ around closed loops Γ that wrap around the system are gauge-invariant operators. These gauge-invariant operators are physical degrees of freedom. An example are the loops Γ which are topologically equivalent to the large circles of a torus. The line integrals $\oint_\Gamma dx_\mu\, A_\mu$ are the so-called holonomies of the gauge fields on the torus.

The transversality condition Eq. (12.87) then follows from a simple algebraic manipulation of the exponent in Eq. (12.85):

$$\begin{aligned} I &= \frac{i}{2}\int d^D x \int d^D y\, A_\mu(x)\Pi_{\mu\nu}(x, y)A_\nu(y) \\ &= \frac{i}{2}\int d^D x \int d^D y [A'_\mu(x) + \partial_\mu\phi(x)]\Pi_{\mu\nu}(x, y)[A'_\nu(y) + \partial_\nu\phi(y)] \\ &= \frac{i}{2}\int d^D x \int d^D y\, A'_\mu(x)\Pi_{\mu\nu}(x, y)A'_\nu(y) + \delta I \end{aligned} \tag{12.88}$$

where the change δI is given by

$$\delta I = \frac{i}{2} \int d^D x \int d^D y \{\partial_\mu \phi(x) \Pi_{\mu\nu}(x, y) A'_\nu(y) + A'_\mu(x) \Pi_{\mu\nu}(x, y) \partial_\nu \phi(y) + \partial_\mu \phi(x) \Pi_{\mu\nu}(x, y) \partial_\nu \phi(y)\} \tag{12.89}$$

Then $\mathcal{Z}[A]$ is gauge-invariant if and only if $\delta I \equiv 0$. Upon integration by parts we get

$$\delta I = -\frac{i}{2} \int d^D x \int d^D y \{\phi(x) \partial^x_\mu \Pi_{\mu\nu}(x, y) A'_\nu(y) + A'_\mu(x) \partial^y_\nu \Pi_{\mu\nu}(x, y) \phi(y) - \phi(x) \partial^x_\mu \partial^y_\nu \Pi_{\mu\nu}(x, y) \phi(y)\} + \text{surface terms} \tag{12.90}$$

Since $\phi(x)$ is arbitrary, δI vanishes identically if and only if $\Pi_{\mu\nu}(x, y)$ is transverse. The surface terms are zero since either ϕ vanishes at the surface or there are no boundaries.

It is possible to relate $\Pi_{\mu\nu}$ to a fermion-current-correlation function. The gauge-invariant fermion current $J_\mu(x)$ is

$$J_\mu(x) = \frac{\delta S}{\delta A_\mu(x)} \tag{12.91}$$

where S is the total action of the system. The current J_μ is gauge-invariant because the action S itself is invariant. For the problem of fermions in free space, J_μ is just the usual fermion current with the diamagnetic term included (the spin is omitted):

$$\begin{aligned} J_0 &= e\psi^\dagger \psi \\ J_j &= \frac{e\hbar}{2imc}[\psi^\dagger (\partial_j \psi) - (\partial_j \psi^\dagger)\psi] - \frac{e^2}{mc^2} A_j \psi^\dagger \psi \end{aligned} \tag{12.92}$$

The spatial components of the current can be written in the more manifestly gauge-invariant form

$$J_j = \frac{e\hbar}{2mic}[\psi^\dagger D_j \psi - (D_j \psi)^\dagger \psi] \tag{12.93}$$

where D_j is, once again, the covariant derivative

$$D_j = \partial_j - \frac{ie}{\hbar c} A_j \tag{12.94}$$

and e is the (negative) electron charge.

For a lattice system, J_j has the form

$$J_j(\vec{x}) = \frac{t}{2i} \left[\psi^\dagger(x) \exp\left(i \frac{e}{\hbar c} \int_{\vec{x}}^{\vec{x}+\hat{e}_j} \vec{A}(\vec{z}) \cdot d\vec{z} \right) \psi(\vec{x} + \hat{e}_j) - \text{h.c.} \right] \tag{12.95}$$

where t is a hopping amplitude and $\hat{e}_j$ is the vector difference of the positions of two lattice sites along the direction j on the lattice.

Since $J_\mu = \delta S/\delta A_\mu$, we can compute expectation values of products of currents by functional differentiation of $\mathcal{Z}[A]$. The average current $\langle J_\mu(x)\rangle$ is given by

$$\langle J_\mu(x)\rangle = \frac{-i}{\hbar}\frac{1}{\mathcal{Z}[A]}\frac{\delta \mathcal{Z}[A]}{\delta A_\mu(x)} \tag{12.96}$$

The polarization tensor $\Pi_{\mu\nu}(x, y)$ can be computed from its definition. We get

$$\Pi_{\mu\nu}(x, y) = -i\hbar \frac{\delta^2}{\delta A_\mu(x)\delta A_\nu(y)} \ln \mathcal{Z}[A] \tag{12.97}$$

A straightforward algebraic manipulation yields the expression

$$\begin{aligned}\Pi_{\mu\nu}(x, y) &= -i\hbar \frac{\delta}{\delta A_\mu(x)}\left(\frac{1}{\mathcal{Z}[A]}\frac{\delta \mathcal{Z}[A]}{\delta A_\nu(y)}\right)\\ &= i\hbar\left(\frac{1}{\mathcal{Z}[A]}\frac{\delta \mathcal{Z}[A]}{\delta A_\mu(x)}\right)\left(\frac{1}{\mathcal{Z}[A]}\frac{\delta \mathcal{Z}[A]}{\delta A_\nu(y)}\right)\\ &\quad - i\hbar\frac{1}{\mathcal{Z}}\frac{\delta^2 \mathcal{Z}[A]}{\delta A_\mu(x)\delta A_\nu(y)}\end{aligned} \tag{12.98}$$

Hence, we get

$$\Pi_{\mu\nu}(x, y) = \frac{i}{\hbar}\langle J_\mu(x) J_\nu(y)\rangle_{\rm c} + \left\langle \frac{\delta J_\mu(x)}{\delta A_\nu(y)}\right\rangle \tag{12.99}$$

where $\langle J_\mu(x) J_\nu(y)\rangle_{\rm c}$ is the connected time-ordered current–current correlation function $D_{\mu\nu}(x, y)$ defined by

$$\frac{\hbar}{i} D_{\mu\nu}(x, y) \equiv \langle J_\mu(x) J_\nu(y)\rangle_{\rm c} = \langle J_\mu(x) J_\nu(y)\rangle - \langle J_\mu(x)\rangle\langle J_\nu(y)\rangle \tag{12.100}$$

The last term in Eq. (12.99) is usually called the "tadpole" term and follows from the diamagnetic piece of the current.

Since $\Pi_{\mu\nu}$ has to be transverse in order for the system to be gauge-invariant, $D_{\mu\nu}$ must obey a similar conservation law. However, $D_{\mu\nu}$ is not quite transverse because of the presence of the tadpole term in Eq. (12.99). Indeed, from the transversality of $\Pi_{\mu\nu}$ we get the equation

$$0 = \partial_\mu^x \Pi_{\mu\nu}(x, y) = \partial_\mu^x D_{\mu\nu}(x, y) + \partial_\mu^x \left\langle \frac{\delta J_\mu(x)}{\delta A_\nu(y)}\right\rangle \tag{12.101}$$

Thus the divergence of $D_{\mu\nu}$ is

$$\partial_\mu^x D_{\mu\nu}(x, y) = -\partial_\mu^x \left\langle \frac{\delta J_\mu(x)}{\delta A_\nu(y)}\right\rangle \tag{12.102}$$

Since $D_{\mu\nu}$ is time-ordered and J_ν is conserved ($\partial_\mu J_\mu = 0$), we can write the l.h.s. of Eq. (12.98) as

$$\begin{aligned}
\partial_\mu^x D_{\mu\nu}(x,y) &= \frac{i}{\hbar}\,\partial_\mu^x \langle T J_\mu(x) J_\nu(y)\rangle \\
&= \frac{i}{\hbar}\,\partial_\mu^x \left(\theta(x_0-y_0)\langle J_\mu(x)J_\nu(y)\rangle + \theta(y_0-x_0)\langle J_\nu(y)J_\mu(x)\rangle\right) \\
&= \frac{i}{\hbar}\delta(x_0-y_0)\langle [J_0(x), J_\nu(y)]\rangle + \frac{i}{\hbar}\langle T\,\partial_\mu^x J_\mu(x) J_\nu(y)\rangle \\
&= \frac{i}{\hbar}\delta(x_0-y_0)\langle [J_0(x), J_\nu(y)]\rangle
\end{aligned} \tag{12.103}$$

The r.h.s. of Eq. (12.102) is equal to

$$\begin{aligned}
&\partial_\mu^x \left\langle \frac{\delta J_\mu(x)}{\delta A_0(y)}\right\rangle = 0 \\
&\partial_\mu^x \left\langle \frac{\delta J_\mu(x)}{\delta A_l(y)}\right\rangle = \partial_k^x \left\langle \frac{\delta J_k(x)}{\delta A_l(y)}\right\rangle = -\frac{e}{mc^2}\,\partial_k^x\left[\delta(x-y)\langle J_0(x)\rangle\right]
\end{aligned} \tag{12.104}$$

On collecting terms, we get the following identities for the ground-state equal-time expectation value of the commutators:

$$\begin{aligned}
&\delta(x_0-y_0)\langle [J_0(x), J_k(y)]\rangle = \frac{ie}{\hbar mc^2}\,\partial_k^x\left[\delta(x-y)\langle J_0(x)\rangle\right] \\
&\delta(x_0-y_0)\langle [J_0(x), J_0(y)]\rangle = 0
\end{aligned} \tag{12.105}$$

which are the Ward identities for this system. These identities are the key to the derivation of the f-sum rule (Kadanoff and Martin, 1961). These identities show that, even though $D_{\mu\nu}(x,y)$ is a correlation function of conserved currents, $D_{\mu\nu}$ itself it is not conserved:

$$\begin{aligned}
&\partial_\mu^x D_{\mu 0}(x,y) = 0 \\
&\partial_\mu^x D_{\mu k}(x,y) = \frac{ie}{\hbar mc^2}\,\partial_k^x\left[\delta(x-y)\langle J_0(x)\rangle\right]
\end{aligned} \tag{12.106}$$

On the other hand, $\Pi_{\mu\nu}$ is strictly conserved, since $\partial_\mu^x \Pi_{\mu\nu} = 0$. The non-vanishing r.h.s. of Eq. (12.106) is an example of what in quantum field theory is commonly called a *Schwinger term*. We have already encountered a Schwinger term in Section 5.6.1. There, the Schwinger term resulted from the lack of chiral symmetry in a gauge-invariant theory of 1D relativistic fermions. In a sense, it is due to an effect produced by the "bottom" of the Fermi sea. In the problem discussed in the present section it follows from the definition of the current.

The results of this section are valid in the most general condensed matter systems. They hold regardless of the statistics of the charge carriers. In the derivation that is usually presented in textbooks (Pines and Nozières, 1966; Mahan, 1990), the proof is done within the framework of Fermi-liquid theory. The argument presented here is more general and follows in spirit the discussion by Kadanoff and Martin (1961). These conservation laws and sum rules are, in fact, a direct consequence of local gauge invariance. In other words, they follow from local charge conservation. It is important to stress that they also hold in phases with "spontaneously broken gauge invariance," such as superconducting states. The quotation marks are meant to stress that *local* gauge invariance *cannot* be spontaneously broken, as dictated by Elitzur's theorem (see Chapter 9). In superconducting states the *global* phase invariance (a subgroup of local gauge transformations) is spontaneously broken in the *absence* of an electromagnetic gauge field. The sum rules are a statement about the system as a whole, and they hold provided that both the normal and the superfluid contributions are taken into account.

Let us now find an explicit expression for $\Pi_{\mu\nu}(x, y)$ for a simple system. For the sake of simplicity I will discuss only the non-interacting fermion case. Interactions can be introduced in the standard fashion. Let us discuss the problem of non-interacting electrons moving in free space coupled to an external electromagnetic field A_μ. Once again, A_μ represents a small fluctuating component with vanishing average. All averages $\langle A_\mu \rangle$ are absorbed in the definition of the otherwise non-interacting fermions. The action for this system is (ignoring spin)

$$S[\psi^*, \psi, A] = \int d^d x \; \psi^*(iD_0 + \mu - h[\langle A_\mu \rangle + A_\mu])\psi \tag{12.107}$$

where $h[A_\mu]$ is a one-particle Hamiltonian that describes the dynamics of particles coupled to a gauge field. For free fermions in the continuum (*no lattice*) it is simply given by

$$h[A_\mu] = -\frac{\hbar^2 \vec{D}_j^2}{2m} \tag{12.108}$$

and D_0 and D_j are covariant derivatives. The generating functional of the current-correlation functions $\mathcal{Z}[A]$ is given by

$$\mathcal{Z}[A] = \int \mathcal{D}\psi^* \mathcal{D}\psi \; \exp\left(\frac{i}{\hbar} S[\psi^*, \psi, A]\right) \tag{12.109}$$

Since the ψ fields represent fermions, we get (after setting $\hbar = 1$)

$$\mathcal{Z}[A] = \det(iD_0 + \mu - h[A]) \tag{12.110}$$

Thus, the effective action for the gauge field A_μ due to the motion of the charged particles is

$$S_{\text{eff}}[A] = -i \; \mathrm{tr} \ln(iD_0 + \mu - h[A]) \tag{12.111}$$

We have encountered expressions of this sort several times in the previous sections of this book. We will deal with it in exactly the same way here.

If A_μ is small, $S_{\text{eff}}[A_\mu]$ can be expanded in powers of A_μ; and, if A_μ has zero average, the first non-zero term is quadratic in A_μ. A straightforward calculation yields the following expressions for $\Pi_{\mu\nu}(x, y)$ in terms of the fermion propagator (the one-particle fermion Green function) in a background gauge field $G(x, y)$. $G(x, y)$ satisfies the equation of motion

$$(iD_0 + \mu - h[\langle A \rangle])_x G(x, y) = \delta(x - y) \tag{12.112}$$

that is,

$$G(x, y) = \langle x | \frac{1}{iD_0 + \mu - h[\langle A \rangle]} | y \rangle \tag{12.113}$$

The components of the polarization tensor $\Pi_{\mu\nu}(x, y)$ can now be written as (again with $\hbar = 1$)

$$\begin{aligned}
\Pi_{00}(x, y) &= iG(x, y)G(y, x) \\
\Pi_{0j}(x, y) &= \frac{1}{2m}\{G(x, y)D_j^y G(y, x) - G(y, x)D_j^{y\dagger} G(x, y)\} \\
\Pi_{j0}(x, y) &= \frac{1}{2m}\{-G(x, y)D_j^{x\dagger} G(y, x) + G(y, x)D_j^x G(x, y)\} \\
\Pi_{jk}(x, y) &= \frac{i}{m}\delta(x - y)\delta_{jk}G(x, y) - \frac{i}{4m^2}(D_j^x G(x, y))(D_k^y G(y, x)) \\
&\quad - \frac{i}{4m^2}(D_j^{x\dagger} G(y, x))(D_k^{y\dagger} G(x, y)) \\
&\quad + \frac{i}{4m^2}G(y, x)(D_j^x D_k^{y\dagger} G(x, y)) \\
&\quad + \frac{i}{4m^2}(D_j^{x\dagger} D_k^y G(y, x))G(x, y)
\end{aligned} \tag{12.114}$$

These formulas, in addition to satisfying the requirements of gauge invariance, are also translation-invariant if the external fields are uniform.

In the next section we will make use of these formulas, particularly that for Π_{0j}, to compute the Hall conductance. Notice that all the expressions in this section hold for *time-ordered* correlation functions. In order to compute the conductivities it is necessary to go to retarded functions (Fetter and Walecka, 1971). Fortunately, the static limit of the Hall conductance can also be calculated directly from the time-ordered functions.

The tight-binding case (on a cubic or square lattice) can be treated using a similar line of argument. In fact, the polarization tensor $\Pi_{\mu\nu}$ for the lattice case can be obtained in the following manner. First the spatial integrals are replaced by sums over lattice sites $\{\vec{x}\}$. The covariant derivatives are replaced by covariant differences according to the rule

$$D_j^x G(x, y) \to \Delta_j^x G(x, y) \equiv G(x + e_j, y) \exp\left(i \int_x^{x+e_j} \vec{A} \cdot \vec{dl}\right) - G(x, y) \tag{12.115}$$

and the hopping amplitude t and the mass m are related by $1/t = ma_0^2$, where a_0 is the lattice constant. Once these identifications have been made, the continuum result becomes valid for the lattice case.

We will be interested primarily in the low-frequency, long-wave-length limit of the effective action. On the basis of gauge and translation invariance we can write the effective action $S_{\text{eff}}[A]$ in terms of an expansion in powers of the gradients of A_μ. The leading-order terms are (in two space dimensions)

$$S_{\text{eff}}[A_\mu] = \int d^2x\, dt\, \left(\frac{\epsilon}{2}\vec{E}^2 - \frac{\chi}{2}B^2 + \gamma(\vec{\nabla} \cdot \vec{E})B + \frac{\sigma_{xy}}{4}\epsilon_{\mu\nu\lambda}A_\mu F_{\nu\lambda} + \cdots\right) \tag{12.116}$$

where $\vec{E}$ and B are the fluctuating pieces of the external electromagnetic field. The coefficients ϵ, χ, γ, and σ_{xy} can be determined from $\Pi_{\mu\nu}$. In particular, ϵ and χ are the static dielectric constant and diamagnetic susceptibility of the system, and σ_{xy} is the static Hall conductance. Notice that the Hall term is precisely the Chern–Simons term that we encountered in Section 10.4. Indeed, the last term gives a contribution to the average current $\langle J_k \rangle_{xy}$,

$$\langle J_k \rangle_{xy} = \sigma_{xy}\epsilon_{kl}E_l \tag{12.117}$$

which has precisely the correct form for the Hall current. The static Hall conductance σ_{xy} can be obtained from the Fourier transform $\Pi_{\mu\nu}(Q)$ of the polarization tensor,

$$\sigma_{xy} = \lim_{Q \to 0} \frac{i}{2} \frac{\epsilon^{\mu\nu\lambda} Q_\lambda}{Q^2} \Pi_{\mu\nu}(Q) = \lim_{Q_0 \to 0} \frac{i}{Q_0} \Pi_{xy}(Q_0, \vec{Q} = 0) \tag{12.118}$$

where $Q = (Q_0, \vec{Q})$. Thus, the Hall conductivity is determined from the $\vec{Q} = 0$ limit of the xy component of the current–current correlation function.

12.6 The Hall conductance and topological invariance

The most remarkable feature of the quantum Hall effect is the quantization of the Hall conductance, i.e. the very existence of the effect itself! The arguments of the previous section show that σ_{xy} is determined from $\Pi_{\mu\nu}$. However, the coefficients of the gradient expansion of the effective action $S_{\text{eff}}[A]$ are usually renormalized away from the values predicted by a theory of weakly interacting fermions. In effect, the $\Pi_{\mu\nu}$ of the last section is just the leading-order (RPA) approximation to the full $\Pi_{\mu\nu}$. Furthermore, the higher-order terms of the gradient expansion are also expected to give contributions at lower orders. This is so since the higher-order terms are important for wave vectors $|\vec{Q}|$ larger than the inverse cyclotron length and frequencies Q_0 larger than the inverse Landau gap. The effective low-energy (hydrodynamic) theory is determined by integrating out (or summing over) the high-momentum and high-frequency modes. All these processes will contribute with effective (usually finite) renormalization of the parameters ϵ, χ, σ_{xy}, and γ. On these grounds, it is not obvious why σ_{xy} should be given exactly by some integer (or fractional) multiple of $e^2/\hbar$.

In the general case (i.e. arbitrary density and arbitrary external field) σ_{xy} does get renormalized. However, there is a special, but very important, case in which σ_{xy} does not get renormalized. This happens whenever the ground state and the lower-energy excitations of the system are separated by a non-zero energy gap.

We will show now, following the arguments due to Thouless, Kohmoto, Nightingale, and den Nijs (Thouless *et al.*, 1982), that in this case σ_{xy} is not renormalized by fluctuations. The key to the argument is the observation that σ_{xy} is determined by a *topological invariant*. We will follow the arguments first presented by M. Kohmoto (Kohmoto, 1985) and by Q. Niu, D. J. Thouless, and Y. S. Wu (Niu *et al.*, 1985). In this section I will discuss the topological invariance in terms of the more general problem of boundary conditions in a many-body system with an energy gap.

12.6.1 The Kubo formula

Let us consider a system that is in its ground state $|\Psi_0\rangle$ and for which there is a gap to all excitations. Let us assume that, in addition to a uniform magnetic field B, the system is allowed to interact with a *small, slowly varying* external electromagnetic field. In this limit, perturbation theory reduces to the *adiabatic approximation* or Born–Oppenheimer approximation. To first order in the time derivative, the perturbed eigenstates are

$$|\Psi_\alpha(t)\rangle = \exp\left(-\frac{i}{\hbar}\int_0^t dt'\,\epsilon_\alpha(t')\right)\left[|\alpha(t)\rangle + i\hbar\sum_{\beta\neq\alpha}\frac{|\beta(t)\rangle\langle\beta(t)|\partial_t|\alpha(t)\rangle}{\varepsilon_\beta(t)-\varepsilon_\alpha(t)}\right] \tag{12.119}$$

where $|\alpha(t)\rangle$ is an instantaneous eigenstate of the time-dependent Schrödinger equation

$$H(t)|\alpha(t)\rangle = \varepsilon_\alpha(t)|\alpha(t)\rangle \tag{12.120}$$

which is a parametric function of time t. If we choose the gauge $A_0 = 0$, time enters into the Hamiltonian $\hat{H}$ only through the space components of the vector potential, which have now the extra term $\delta\vec{A}$,

$$\delta\vec{A} = \vec{E}(t)t \tag{12.121}$$

where $\vec{E}$ is a very weak, slowly varying electric field. The expectation value of an arbitrary operator $\hat{M}$ in this state is

$$\langle\Psi_\alpha(t)|\hat{M}|\Psi_\alpha(t)\rangle = i\hbar\sum_{\beta\neq\alpha}\frac{\langle\alpha|\hat{M}|\beta\rangle\langle\beta|\partial/\partial t|\alpha\rangle + \langle\alpha|\partial/\partial t|\beta\rangle\langle\beta|\hat{M}|\alpha\rangle}{\varepsilon_\beta(t) - \varepsilon_\alpha(t)} \tag{12.122}$$

Let us compute the expectation value of the current operator $\hat{J}_k(x)$. We recall that the states $\{|\alpha\rangle\}$ obey the time-dependent Schrödinger equation. The expectation value $\langle\alpha|\partial_t|\beta\rangle$ is given by

$$\langle\alpha|\partial_t|\beta\rangle = \frac{\langle\alpha|\partial_t\hat{H}|\beta\rangle}{\varepsilon_\beta(t) - \varepsilon_\alpha(t)} \tag{12.123}$$

The Hamiltonian $\hat{H}$ is a slowly varying function of time. But time only enters into $\hat{H}$ through its dependence on the vector potential $\hat{A}$. From this observation, and from the definition of the current as the functional derivative of the Hamiltonian $\hat{H}$, it follows that the Hall conductance σ_{xy} can be written in the form

$$(\sigma_{xy})_\alpha = -i\hbar L_1 L_2\sum_{\beta\neq\alpha}\frac{\langle\alpha|\hat{J}_1|\beta\rangle\langle\beta|\hat{J}_2|\alpha\rangle - \langle\alpha|\hat{J}_2|\beta\rangle\langle\beta|\hat{J}_1|\alpha\rangle}{(\varepsilon_\beta(t) - \varepsilon_\alpha(t))^2} \tag{12.124}$$

This expression is known as the Kubo formula for the Hall conductance σ_{xy}. Analogous formulas can be derived for other components of the conductivity tensor as well as for other transport properties. It is important to stress that the states $\{|\alpha\rangle\}$ are the *exact* eigenstates of the full *many-body system* described by $\hat{H}$ and that $\{\varepsilon_\alpha(t)\}$ are the exact energy levels. They should not be confused with the one-particle states and levels of the non-interacting system, which are quite different.

12.6.2 Generalized toroidal boundary conditions

There is an alternative approach that yields a more suggestive and useful expression for σ_{xy}. Let us use the Schrödinger equation to write an equivalent expression for the Hall conductance. Let us imagine that the system under consideration has N particles inside a rectangle of sides L_1 and L_2. Since the external (weak) electric field is taken to be uniform in space, we can write the associated electrostatic potential $U(\vec{x})$ in the form

$$U(\vec{x}) = \vec{E}\cdot\vec{x} \tag{12.125}$$

and $\vec{E} = \vec{\nabla} U$. Thus, the extra term in the vector potential $\delta \vec{A}$ is simply

$$\delta \vec{A} = \vec{E} t = \vec{\nabla}[U(\vec{x})t] \tag{12.126}$$

Since $\delta \vec{A}$ is a pure gradient, it can be eliminated by a suitable gauge transformation of the fermion operator of the form

$$\Psi(\vec{x}) \to \exp\left(i \frac{e}{\hbar c} U(\vec{x}) t\right) \Psi(\vec{x}) \tag{12.127}$$

Notice, however, that such *local* gauge transformations cannot change the value of the circulation of the vector potential $\delta \vec{A}$ on closed non-contractible loops. More specifically, the line integrals I_j,

$$I_j = \oint_{\Gamma_j} \delta \vec{A} \cdot d\vec{l} = t \oint_{\Gamma_j} \vec{E} \cdot d\vec{l} \equiv t E_j L_j \tag{12.128}$$

on paths Γ_j which wrap around the system along the x_1 and x_2 directions, respectively, are gauge-invariant if the fermions move on the torus. Thus, although the vector potential $\delta \vec{A}$ disappears from the problem, the holonomies do not. In fact, they enter into the boundary conditions. Line integrals of a gauge field on non-contractible loops in space (or space-time) are called the *holonomies* of the gauge field.

The problem of assigning boundary conditions to quantum-mechanical systems on a closed manifold is a very subtle one. For instance, if the fermions move on a torus and no magnetic field is present, it is perfectly consistent to use periodic or twisted boundary conditions, which, for an N-particle system, are

$$\Psi(\vec{x}_1, \ldots, \vec{x}_N) = e^{-i\vec{\theta}\cdot\vec{L}} \Psi(\vec{x}_1 + \vec{L}, \ldots, \vec{x}_N + \vec{L}) \tag{12.129}$$

where $\vec{\theta}$ is an arbitrary two-component vector and $\vec{L}$ is a displacement along x_1 by a distance L_1 or along x_2 by a distance L_2. These boundary conditions are perfectly consistent since, in the absence of a magnetic field, the total momentum $\vec{P}$ is a constant of motion. The momentum of the only eigenstate compatible with the boundary conditions is $\vec{\theta}/|\vec{L}|$. But, if a magnetic field is present, the situation is somewhat different. In Section 12.1 we introduced the *magnetic translation operators*. These operators commute with the one-particle Hamiltonian. In fact, they also commute with the Hamiltonian of the full interacting system. In Section 12.1, we also found that the only consistent boundary conditions for the wave functions (generalized now to the N-particle case) for charged particles moving on a torus in the presence of a non-zero magnetic field B are

$$A_1(x_1, x_2 + L_2) = A_1(x_1, x_2) + \partial_1 \beta_2(x_1, x_2)$$
$$A_2(x_1 + L_1, x_2) = A_2(x_1, x_2) + \partial_2 \beta_1(x_1, x_2)$$

$$\Psi(\{x_1^{(j)} + L_1\}; \{x_2^{(j)}\}) = \exp\left(\frac{-ie}{\hbar c}\sum_{j=1}^{N}\beta_1(x_1^{(j)}; x_2^{(j)}) + i\theta_1\right) \times \Psi(\{x_1^{(j)}\}; \{x_2^{(j)}\})$$
$$\Psi(\{x_1^{(j)}\}; \{x_2^{(j)} + L_2\}) = \exp\left(\frac{-ie}{\hbar c}\sum_{j=1}^{N}\beta_2(x_1^{(j)}; x_2^{(j)}) + i\theta_2\right) \times \Psi(\{x_1^{(j)}\}; \{x_2^{(j)}\}) \tag{12.130}$$

where we have included the effect of the electric fields through the angles θ_1 and θ_2. The boundary phases θ_1 and θ_2 are related to the electric field by

$$\theta_j = \frac{et}{\hbar c}E_j L_j \equiv \frac{e}{\hbar c}I_j \tag{12.131}$$

Thus, in addition to the phase twist $\vec{\theta}$, the requirement that the states be eigenstates of the magnetic translation operator leads naturally to the generalized boundary conditions. We will see below that the additional phase factors arise from the impossibility of defining the phase of the wave function globally and smoothly on the torus. The wave functions for particles on a torus in the presence of a magnetic field form a fiber bundle. The conditions, or rather the requirement that the states be eigenstates of the magnetic translations, define the fiber bundle. We will see below that a similar difficulty arises when one tries to define the dependence of the phase of the wave function on the twist angles $\vec{\theta}$.

12.6.3 The Kubo formula for σ_{xy} and the first Chern number

From now on, we will assume that the vector $\vec{\theta}$ represents two constant angles. In any case, all of the time dependence of the states enters through $\vec{\theta}$. All time derivatives become derivatives relative to the phase θ_j. The Kubo formulas can now be written in the form ($\partial_j \equiv \partial_{\theta_j}$)

$$(\sigma_{xy})_\alpha = \frac{ie^2}{\hbar}\Big[\partial_1\langle\alpha|\partial_2|\alpha\rangle - \partial_2\langle\alpha|\partial_1|\alpha\rangle\Big] \tag{12.132}$$

In this form, this formula was first derived by Niu, Thouless, and Wu (Niu *et al.*, 1985), who also considered the *average* $\langle(\sigma_{xy})\rangle$ over the torus of boundary conditions

$$\begin{aligned}\langle(\sigma_{xy})_\alpha\rangle &= \int_0^{2\pi}\frac{d\theta_1}{2\pi}\int_0^{2\pi}\frac{d\theta_2}{2\pi}(\sigma_{xy})_\alpha \\ &= \frac{e^2}{i\hbar}\int_0^{2\pi}\int_0^{2\pi}\frac{d\theta_1\, d\theta_2}{2\pi\ 2\pi}\Big[\partial_2\langle\alpha|\partial_1|\alpha\rangle - \partial_1\langle\alpha|\partial_2|\alpha\rangle\Big]\end{aligned} \tag{12.133}$$

What matters to our discussion is the fact that $\langle(\sigma_{xy})_\alpha\rangle$ is proportional to a quantity known as the first Chern number, C_1, which is a topological invariant. A similar expression also appears in the tight-binding case, which will be discussed below.

Before we consider what the average conductance is, we must face the fact that in any physically relevant situation the boundary conditions are fixed. Thus it might appear that, while $\langle(\sigma_{xy})_\alpha\rangle$ may be an interesting quantity to compute, it is not directly relevant. This is true. However, we are considering a special situation in which there is a finite energy gap between the ground state and the first excited state. It is easy to argue that, if the gap is finite, then the difference of the *measured* value of $(\sigma_{xy})_\alpha^\theta$, with *fixed* boundary conditions, and $\langle(\sigma_{xy})_\alpha\rangle$ vanishes in the thermodynamic limit, at least like $1/L$.

Let us consider the differential change

$$\frac{\partial(\sigma_{xy})_\alpha^\theta}{\partial\theta_1} \tag{12.134}$$

The dependence of the conductance on the phase angles θ_1 and θ_2 enters through the Hamiltonian $\hat{H}$. But $\hat{H}$ is a function of θ_1/L_1 and θ_2/L_2 only. Thus, all changes must be of the form ($j = 1, 2$)

$$\frac{1}{L_j}\frac{\partial(\sigma_{xy})_\alpha}{\partial(\theta_j/L_j)} \tag{12.135}$$

Since there is a non-vanishing gap, all small changes in the parameters of the Hamiltonian $\hat{H}$ must lead to changes of order unity in all local quantities. This includes changes in the energies and wave functions of local excitations. Thus, the derivatives $\partial(\sigma_{xy})_\alpha/\partial(\theta_j/L_j)$ must have finite limits for thermodynamically large systems. Hence, $\partial(\sigma_{xy})_\alpha/\partial\theta_j \propto \text{constant}/L_1$ for asymptotically large systems. This justifies the use of the conductance averaged over all boundary conditions (Niu *et al.*, 1985).

12.6.4 Fiber bundles and the quantum Hall conductance

Let us now turn to the issue of the topological invariance of $\langle(\sigma_{xy})_\alpha\rangle$. The argument goes as follows. The boundary-condition angles θ_1 and θ_2, being phases, are defined modulo 2π. Each choice of a boundary condition amounts to a choice of a point $\vec{\theta}$ on the torus $S_1 \times S_1$ of boundary conditions. For each point $\vec{\theta}$ we have a unique eigenstate $\Psi_\alpha(\{\vec{x}\}; \vec{\theta})$ of the full many-body Hamiltonian $\hat{H}$. In mathematical jargon, we have a *fiber bundle*. The wave function has an amplitude and a phase that are smooth functions of $\vec{\theta}$. Now, the total phase of the wave function is not a physical observable. But *changes* of the phase are. In particular, let us imagine that, at some initial time t_0, we have defined an initial boundary condition $\vec{\theta}(t_0)$

with a phase for the state arg $[\Psi(\theta(t_0))]$. The external electromagnetic field is now allowed to couple to the system in such a way that the boundary conditions change as a function $\vec{\theta}(t)$ and return to the initial value $\vec{\theta}(t_0)$ after some very long time T. During this process the vector $\vec{\theta}(t)$ traces a closed curve Γ on the torus $S_1 \times S_1$. At the same time, the phase of the wave function changes by an amount δ_Γ,

$$\delta_\Gamma = \Delta\arg[\Psi] = \Delta\mathrm{Im}\ln[\Psi] = \arg[\Psi(\theta(t_0+T))] - \arg[\Psi(\theta(t_0))] \quad (12.136)$$

If Ψ is an *analytic non-vanishing function* of $\vec{\theta}$, the phase change, δ_Γ must be zero. This is so because, in such a case the contour can be deformed to zero. However, the only analytic function on a torus is a constant. Thus, a non-vanishing adiabatic phase change δ_Γ requires that the function $\ln\Psi$ be non-analytic on the torus of boundary conditions. In this case, closed contours that enclose singularities of $\ln[\Psi]$ are non-contractible and δ_Γ is non-zero for such contours. Non-zero adiabatic changes of the phases of wave functions of quantum-mechanical systems are known as *Berry phases* (Simon, 1983; Berry, 1984). Since the wave function $\Psi_\alpha(\vec{x}_1, \ldots, \vec{x}_N; \vec{\theta})$ is a smooth function of its arguments, a non-analyticity in $\ln\Psi$ amounts to zeros of Ψ for some values of $\vec{\theta}$. Smoothness requires that the zeros be isolated points on the torus $S_1 \times S_1$. The Berry phase δ_Γ counts the number of zeros of Ψ enclosed by the contour Γ.

How is the phase of the wave function $\Psi^{(\alpha)}(\{\vec{x}\}; \vec{\theta})$ related to the Hall conductance? In order to investigate this issue, let us introduce the following suggestive notation which was originally introduced by Kohmoto. Let $\mathcal{A}_k^{(\alpha)}(\theta_1, \theta_2)$ be on a vector field on the torus $S_1 \times S_1$ defined by

$$\mathcal{A}_k^{(\alpha)} = i\langle\alpha|\frac{\partial}{\partial\theta_k}|\alpha\rangle \equiv i\left\langle\Psi_{\vec{\theta}}^{(\alpha)}\middle|\frac{\partial}{\partial\theta_k}\middle|\Psi_{\vec{\theta}}^{(\alpha)}\right\rangle \quad (12.137)$$

With this notation, the (averaged) Hall conductance is

$$\langle(\sigma_{xy})_\alpha\rangle = \frac{e^2}{\hbar}\int_0^{2\pi}\frac{d\theta_1}{2\pi}\int_0^{2\pi}\frac{d\theta_2}{2\pi}(\partial_1\mathcal{A}_2 - \partial_2\mathcal{A}_1) \quad (12.138)$$

This is the Niu–Thouless–Wu formula (Niu *et al.*, 1985). In other words, $\langle(\sigma_{xy})_\alpha\rangle$ is the flux *through the torus* $S_1 \times S_1$ *of the vector field* $\vec{\mathcal{A}}(\vec{\theta})$.

Furthermore, the states $|\Psi^{(\alpha)}(\vec{\theta})\rangle$ are defined up to an overall phase factor. Thus, the states $|\Psi^{(\alpha)}(\vec{\theta})\rangle$ and $e^{if(\vec{\theta})}|\Psi^{(\alpha)}(\vec{\theta})\rangle$ are physically equivalent. Notice that the phase factor does not modify the boundary conditions. Under a phase change, the vector field $\vec{\mathcal{A}}(\vec{\theta})$ transforms like a gauge transformation,

$$\mathcal{A}_k(\vec{\theta}) = i\langle\alpha|\partial_k|\alpha\rangle \to i\langle\alpha|\partial_k|\alpha\rangle - \partial_k f(\vec{\theta}) \quad (12.139)$$

Thus phase factors in the wave functions translate into a gauge transformation for the vector field $\mathcal{A}_k$ defined on the torus of boundary conditions. We can now use

Stokes' theorem to write the averaged Hall conductance in the form

$$\langle(\sigma_{xy})_\alpha\rangle = \frac{e^2}{\hbar}\oint_{\bar{\Gamma}} \mathcal{A}_k(\vec{\theta})d\theta_k \tag{12.140}$$

where $\bar{\Gamma}$ is the rectangular contour with corners at (θ_1, θ_2), $(\theta_1 + 2\pi, \theta_2)$, $(\theta_1, \theta_2 + 2\pi)$, and $(\theta_1 + 2\pi, \theta_2 + 2\pi)$. A *non-zero Hall conductance* means that the vector field $\vec{\mathcal{A}}$ *cannot be a periodic function on the torus* $S_1 \times S_1$ of boundary conditions. This, in turn, implies that along non-contractible closed contours Γ_1 and Γ_2, which wrap around the torus along the θ_1 and θ_2 directions, respectively, $\mathcal{A}_k$ and the wave functions must change as follows:

$$\begin{aligned}
\mathcal{A}_k(\theta_1 + 2\pi, \theta_2) &= \mathcal{A}_k(\theta_1, \theta_2) + \partial_k f_1(\theta_1, \theta_2)\\
\mathcal{A}_k(\theta_1, \theta_2 + 2\pi) &= \mathcal{A}_k(\theta_1, \theta_2) + \partial_k f_2(\theta_1, \theta_2)\\
\Psi^{(\alpha)}(\{\vec{x}\}; \theta_1 + 2\pi, \theta_2) &= e^{if_1(\theta_1,\theta_2)}\Psi^{(\alpha)}(\{\vec{x}\}; \theta_1, \theta_2)\\
\Psi^{(\alpha)}(\{\vec{x}\}; \theta_1, \theta_2 + 2\pi) &= e^{if_2(\theta_1,\theta_2)}\Psi^{(\alpha)}(\{\vec{x}\}; \theta_1, \theta_2)
\end{aligned} \tag{12.141}$$

This topological structure is strongly reminiscent of the Wu–Yang construction for the wave functions of charged particles moving in the presence of a Dirac magnetic monopole (Eguchi *et al.*, 1980; Nash and Sen, 1983). We can make the analogy even sharper. Let us suppose that we have a wave function $\Psi_\alpha(\{\vec{x}\}, \vec{\theta})$, which satisfies boundary conditions determined by the point $\vec{\theta}$ on $S_1 \times S_1$. Now, given $\Psi_\alpha(\{\vec{x}\}, \vec{\theta})$, can we unambiguously and completely determine $\Psi_\alpha(\{\vec{x}\}, \vec{\theta}')$ for some other *arbitrary* point $\vec{\theta}'$ on $S_1 \times S_1$? The answer to this question is *no*. The phase of Ψ_α *cannot* be determined uniquely and smoothly over the boundary-condition torus *unless* the Hall conductance is equal to zero. This is so because at the zeros of Ψ_α its phase is undefined.

Let us consider the simpler case of a wave function that vanishes at just one point $\vec{\theta}_0$ on $S_1 \times S_1$. We now split the torus $T \equiv S_1 \times S_1$ into two disjoint subsets (or patches) T_{I} and T_{II} such that $\vec{\theta}_0$ is in T_{I}, as shown in Fig. 12.1. Since T_{II} does not contain point $\vec{\theta}_0$ where Ψ_α is zero, the phase of Ψ_α can be determined globally on T_{II}. For instance, we can choose to make Ψ_α real on T_{II}. However, on T_{I} there is a point $\vec{\theta}_0$ where $\Psi_\alpha(\vec{\theta}_0) = 0$. We can always define the phase of Ψ_α at $\vec{\theta} = \vec{\theta}_0$ to be some arbitrarily chosen value. Once a value has been chosen, the phase of Ψ_α can be defined by continuity in an arbitrary neighborhood of $\vec{\theta}_0$ that is not equal to the whole torus T.

Thus we have two different definitions of the phase of Ψ_α on T_{I} and T_{II}. Obviously these definitions must amount to a gauge transformation, i.e.

$$\Psi_\alpha^{\text{I}}(\{\vec{x}\}, \vec{\theta}) = e^{if(\vec{\theta})}\Psi_\alpha^{\text{II}}(\{\vec{x}\}, \vec{\theta}) \tag{12.142}$$

where $f(\vec{\theta})$ is a smooth function on the closed curve γ (the boundary between T_{I} and T_{II}), and it is known as the transition function. Similarly, the vector field $\mathcal{A}_k$

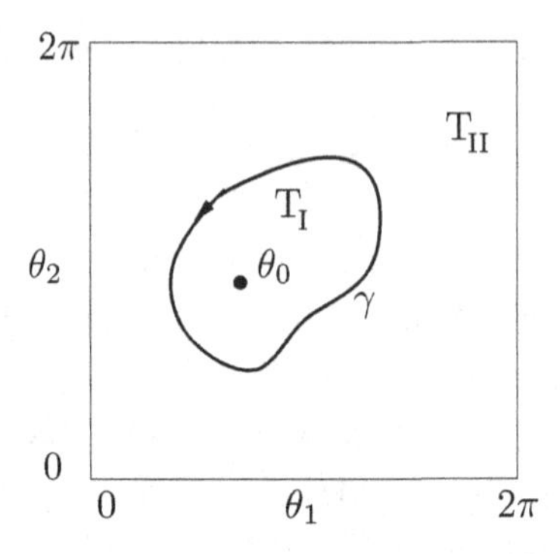

Figure 12.1 $T = S_1 \times S_2 = \{0 \leq \theta_1 < 2\pi,\ 0 \leq \theta_2 < 2\pi\}$ is the torus of boundary conditions for wave functions on a spatial torus. The wave function $\Psi_\alpha(\vec{\theta})$ with boundary condition $\vec{\theta} = (\theta_1, \theta_2)$ is well defined on the region (patch) $T_{\rm I}$ but not on the full torus, and vanishes at $\vec{\theta}_0$. $T_{\rm II}$ is the complement of $T_{\rm I}$ on the torus. The oriented path γ is their common boundary. See the text.

also has two different definitions on $T_{\rm I}$ and $T_{\rm II}$, which, again, differ by a gauge transformation

$$\mathcal{A}_k^{\rm I}(\vec{\theta}) - \mathcal{A}_k^{\rm II}(\vec{\theta}) = \partial_k f(\vec{\theta}) \tag{12.143}$$

We conclude that the Hall conductance reduces to a sum of two contributions, one from $T_{\rm I}$ and the other from $T_{\rm II}$. Since the regions $T_{\rm I}$ and $T_{\rm II}$ share a common boundary, γ, we can readily use Stokes' theorem to write

$$\begin{aligned}\langle(\sigma_{xy})_\alpha\rangle &\equiv \frac{e^2}{(2\pi)^2\hbar}\left\{\int_{T_{\rm I}} d\theta_1\, d\theta_2(\partial_1\mathcal{A}_2 - \partial_2\mathcal{A}_1)\right.\\ &\quad \left.+\int_{T_{\rm II}} d\theta_1\, d\theta_2(\partial_1\mathcal{A}_2 - \partial_2\mathcal{A}_1)\right\}\\ &= \frac{e^2}{(2\pi)^2\hbar}\left\{\int_\gamma \vec{\mathcal{A}}_{\rm I}\cdot d\vec{\theta} - \int_\gamma \vec{\mathcal{A}}_{\rm II}\cdot d\vec{\theta}\right\}\end{aligned} \tag{12.144}$$

where we have used the fact that the boundaries of $T_{\rm I}$ and $T_{\rm II}$ have opposite orientations. Thus, we find

$$\langle(\sigma_{xy})_\alpha\rangle = \frac{e^2}{(2\pi)^2\hbar}\int_\gamma(\vec{\mathcal{A}}_I - \vec{\mathcal{A}}_{II})\cdot d\vec{\theta} = \frac{e^2}{(2\pi)^2\hbar}\int_\gamma \vec{\partial} f\cdot d\vec{\theta} \tag{12.145}$$

Thus, $\langle(\sigma_{xy})_\alpha\rangle$ counts the number of times the gauge transformation $f(\vec{\theta})$ winds by 2π as $\vec{\theta}$ traces the closed loop γ. The *winding number*

$$C_1 = \frac{1}{2\pi}\int_\gamma \vec{\partial} f\cdot d\vec{\theta} \tag{12.146}$$

is a topological invariant known as the first Chern number. It is a topological invariant since it cannot change upon any smooth deformation of the contour γ. However, if, under a deformation, one or more additional zeros of Ψ_α cross the boundary into region $T_{\rm I}$, the winding number will jump by integer amounts. These processes correspond to crossings of energy levels.

The fiber bundle associated with this problem can be defined in the following way (Kohmoto, 1985). With every point $\vec{\theta}$ on T we associate a state $\Psi_\alpha(\vec{\theta})$. States $\Psi'_\alpha(\vec{\theta})$, which differ from $\Psi_\alpha(\vec{\theta})$ by a gauge transformation, $f(\vec{\theta})$, are physically equivalent. Thus, at every point $\vec{\theta} \in T$ we have associated the *ray or bundle* of states related to $\Psi_\alpha(\vec{\theta})$ by a gauge transformation. The torus T is partitioned into a union of sets $T_{\rm I}, T_{\rm II}, \ldots$ each containing *at most* one zero of Ψ_α. The phase of Ψ_α is defined for each set, which results in a set of state vectors $\Psi_\alpha^{\rm I}, \Psi_\alpha^{\rm II}, \ldots$ whose phases are smoothly defined on $T_{\rm I}, T_{\rm II}, \ldots$

These state vectors differ from each other just by gauge transformations that are smooth functions $f(\vec{\theta})$ on the overlap between two regions, say $T_{\rm I}$ and $T_{\rm II}$. The *transition function* $f(\vec{\theta})$ is a smooth map from the closed curve $\gamma \subset T_{\rm I} \bigcap T_{\rm II}$ to the group U(1) of phases $e^{if(\vec{\theta})}$. Since γ is isomorphic to U(1), the transition function is a smooth map from U(1) onto U(1). These maps can be classified into *homotopy classes*, with each class defined by the winding number C_1 of Eq. (12.146). This map is known as the principal $U(1)$ bundle over the torus T. The vector field $\mathcal{A}_k(\vec{\theta})$ defines a *connection*.

Let's define now the 1-form $d\mathcal{A} = \mathcal{A}_k \, d\theta_k$. A connection 1-form can be written as $\Omega = \mathcal{A} + d\mathcal{A}$. The transition functions act on fibers (i.e. state vectors) by multiplication. Once a connection $\mathcal{A}_k$ has been given, a curvature 2-form $F = d\mathcal{A}$ can be defined, and it is known as the first Chern form. The integral of this 2-form is the first Chern number.

Let us now note the following interesting analogy. In Section 12.1 we discussed the problem of the quantization of the motion of a charged particle in a uniform magnetic field with the particle constrained to move on the surface of a torus in *space*. There, we found how the wave functions transform under magnetic translations. In this section, we showed how to construct the wave function on different *patches* of the torus of boundary conditions. The relation between the wave functions on different patches is analogous to the way in which the wave functions transform under magnetic translations. However, here we are discussing phases of many-body wave functions on the torus of boundary conditions! At the root of this analogy is the fact that the many-body wave functions are also representations of the group of magnetic translations. Here too, if the wave functions $\Psi^{(\alpha)}(\{\vec{x}\}, \vec{\theta})$ are required to be single-valued functions on the torus $S_1 \times S_1$, the same consistency condition as that discussed in Section 12.1 implies that the total flux through the torus should be an *integer* n multiple of 2π. Otherwise different paths from

$\vec{\theta} = (0, 0)$ to, say, $\vec{\theta} = (2\pi, 2\pi)$ would lead to inequivalent phases for the wave function $\Psi^{(\alpha)}$. We conclude that, in this case, the averaged Hall conductance is quantized to be an integer multiple of e^2/h.

This argument is actually much too strong. In fact, it appears to require that $\langle(\sigma_{xy})_\alpha\rangle$ should *always* be an integer multiple of e^2/h. The experimental observation of the fractional quantum Hall effect, as well as the success of Laughlin's theory, indicates that this argument cannot be literally correct. Indeed, the observation of the FQHE, which has $\sigma_{xy} = (e^2/h)(n/m)$, requires that, for the case of toroidal boundary conditions, the wave functions $\Psi^{(\alpha)}(\{\vec{x}\}; \vec{\theta})$ must be *multivalued* functions on the *torus of boundary conditions*. This means that the eigenstates of $\hat{H}$ must have several components and behave like vectors under periodic changes of boundary conditions.

Hence, rather than requiring that $\Psi_\alpha(\{\vec{x}\}; \vec{\theta})$ be single-valued on the torus $S_1 \times S_1$, we should demand that Ψ_α must have m components (where m is some integer). The wave function returns to its initial value only after the torus has been covered m times. In this case the averaged Hall conductance is equal to $(e^2/h)(n/m)$. We saw in Chapter 9 that the ground state on a torus of a theory of a topological phase does indeed have a finite degeneracy that is characteristic of this phase. In Chapter 13 we will see that the quantum Hall fluid in a state with a *fractional* Hall conductivity is indeed an example of a topological phase: the ground-state wave function has several components or, equivalently, it has a finite ground-state degeneracy of a torus (Wen and Niu, 1990).

The integers n and m cannot be determined by topological arguments alone. They have to be calculated from some microscopic theory. In the next sections we will discuss a few examples: (a) free electrons filling up one Landau level, (b) the tight-binding Hofstadter problem, and (c) Laughlin's theory of the FQHE. In each case n and m turn out to be different. However, the importance of the topological argument is that, for the specific task of computing σ_{xy}, it *suffices* to consider just some simple limit in which the calculation can be done easily. The topological invariance of $\langle\sigma_{xy}\rangle$ insures that it cannot change under smooth deformations of the underlying Hamiltonian (unless, of course, during this process there is a level crossing).

12.6.5 How many components does the wave function have?

Let us point out that subtle, but important, differences in the behavior of the system arise depending on the choice of boundary conditions. In this section we have considered mainly the case of generalized periodic boundary conditions (GPBCs). These GPBCs require that the fermions move on a 2D torus in space. While this choice is *convenient* from the point of view of the mathematics, it is not very natural from an experimental standpoint.

Experimentally, the natural choice is a rectangle coupled to a four-point probe, which is a set of sources and sinks of charge. In practice this means taking charge from one point on the edge of the sample (the sink) and injecting it back into the system at another point (the source). Typically this process involves the use of wires, batteries, etc. In a sense the *measuring devices* implement the GPBCs. The voltage drop across the device is proportional to the boundary-condition angles θ_1 and θ_2. Yet, another physically relevant situation is a *disk* without wires. If the disk is isotropic and thermodynamically large, then the wave function vanishes exponentially fast as the difference of the particle coordinates becomes large. This can happen due to the presence of an isotropic potential that confines the particles inside some region of the disk. In this case the points on the edge of the disk are asymptotically equivalent to each other. The thermodynamic limit of this case is thus identical to that of a set of particles moving on the surface of a sphere with uniform radial magnetic field, i.e. a magnetic monopole (Haldane, 1983b).

Niu, Thouless, and Wu (Niu *et al.*, 1985) observed that GPBCs *require* multi-component wave functions. Spherical (or disk-like) boundary conditions have only one-component wave functions (Laughlin, 1983; Haldane, 1983b). This issue has caused a great deal of confusion, which was partly due to the fact that the components of the wave functions for GPBCs were originally thought of as resulting from the spontaneous breakdown of some unknown discrete symmetry. Indeed, in systems in which a global discrete symmetry is spontaneously broken, there is a finite number of degenerate ground states, which are related by a symmetry operation.

This phenomenon is quite common in magnetic systems with spontaneously broken discrete global symmetries. The most common example is the Ising model, which we discussed in Chapter 5, whose discrete global symmetry is a global spin flip. Multiple ground states are also present in commensurate charge-density-wave systems, such as polyacetylene, which will be discussed in Section 16.7, where they result from the spontaneous breaking of discrete global translation symmetries. However, these analogies are quite misleading. In the case of quantum Hall systems, the multi-component structure is a feature of the *entire* Hilbert space, not just of the ground state. The Hilbert space is split into a number of disconnected pieces that are not related by a symmetry operation. In other words, this structure is not the result of the spontaneous breakdown of any symmetry. Rather, this feature of the Hilbert space merely reflects the global non-triviality of the manifold on which the particles move.

As a matter of fact, the number of components of the wave functions is different on different manifolds (Wen, 1990c; Wen and Niu, 1990). For example, instead of a torus, let us consider a sphere. All closed loops on the surface of a sphere are contractible. Thus, all the holonomies are trivial. The wave functions for charged particles moving on the surface of the sphere in the presence of a uniform radial magnetic field (i.e. a magnetic monopole) still form a non-trivial fiber bundle, known as the monopole bundle (Wu and Yang, 1975). But the arguments given above indicate that the states are now non-degenerate.

What is the physical significance of this degeneracy? There are two schools of thought on this issue. According to one school, the degeneracy should not be regarded as being physical since it changes with the boundary conditions. According to this point of view, the degeneracy merely reflects the fact that the location of the center of mass is quantized if the system is placed on a torus. Indeed, Haldane (1985a) undertook a detailed study of the symmetries of the states on the torus and showed that the degeneracy arises from the magnetic translations of the center of mass independently of the physical properties of the system. He further showed that, in general, there are no additional degeneracies and that the states for the relative coordinates are generally non-degenerate. But, by the same token, it is clear that there are no states on a sphere that can carry a current.

Thus, if we wish to have a state with a non-vanishing current we must put the system on a torus. This statement naturally applies only to systems on closed surfaces. All experimental systems have edges, sources, and drains, and, as we well know, can have a steady current. Wen (1989, 1990c) gave a very general argument showing that if the surface on which the fermions move has g handles (a genus-g Riemann surface) the degeneracy is k^g, if k is the degeneracy on the torus. (See the discussion on topological phases in gauge theory in Chapter 9.) From this point of view, the topological degeneracy is a fundamental qualitative feature of the system.

12.7 Quantized Hall conductance of a non-interacting system

In this section we will discuss the fairly simple but interesting problem of the computation of the Hall conductance for an assembly of non-interacting electrons moving freely on a torus. We will assume that the external magnetic field and the electron density are such that there is an *integer* number of completely filled Landau levels.

Let us begin by discussing the nature of the one-particle states. Let $\vec{x}$ denote the coordinate of a particle of charge e and mass m. The magnetic field is B and the torus has linear dimensions L_1 and L_2 along its main circles. In Section 12.1 we constructed the single-particle states for the case of an isotropic disk. For simplicity we will restrict our discussion to the case of particles on the lowest Landau level. In Section 12.1 we found that the single-particle states for the lowest Landau level $\Psi(z, \bar{z})$ have the form

$$\Psi(z, \bar{z}) = f(z, \bar{z}) e^{-|z|^2/(4l_0^2)} \tag{12.147}$$

where $f(z, \bar{z})$ is an analytic function, i.e. $\partial_{\bar{z}} f = 0$.

A basis of (analytic) functions is constituted by the powers z^m. For a system with N_ϕ flux quanta there are N_ϕ linearly independent states. Thus, an arbitrary state in the lowest Landau level is a polynomial in z of degree N_ϕ times the exponential factor.

Let us consider now the case of a system with exactly $N = N_\phi$ electrons in a magnetic field B with N_ϕ flux quanta. The ground-state wave function Ψ_N for the N-particle system is the Slater determinant

$$\Psi_N(z_1, \ldots, z_N) = \begin{vmatrix} 1 & \ldots & 1 \\ z_1 & \ldots & z_N \\ \vdots & \vdots & \vdots \\ z_1^N & \ldots & z_N^N \end{vmatrix} \exp\left(-\frac{1}{4l_0^2}\sum_{j=1}^{N}|z_j|^2\right) \tag{12.148}$$

This determinant has the form of a Vandermonde determinant. By application of a standard algebraic identity, the wave function Ψ_N can be written in the form

$$\Psi_N(z_1, \ldots, z_N) = \prod_{1\leq j<k\leq N}(z_j - z_k) \times \exp\left(-\frac{1}{4l_0^2}\sum_{j=1}^{N}|z_j|^2\right) \tag{12.149}$$

We want to compute the Hall conductance for this system. We will use the Niu–Thouless–Wu formula. However, in order to use that formula we need to write down a wave function that is an explicit function of the boundary-condition angles θ_1 and θ_2. What we need is to generalize the state for a system on a torus (instead of a disk) of linear dimensions L_1 and L_2 satisfying the GPBCs of Haldane (Haldane and Rezayi, 1985), whose work we follow here.

Since toroidal boundary conditions break rotational invariance, it is more natural to work in the axial (or Landau) gauge $A_1 = -Bx_2, A_2 = 0$. It can be easily checked that the wave functions for the states in the lowest Landau level have the form

$$\Psi(x_1, x_2) = f(z)\exp\left(-\frac{x_2^2}{2l_0^2}\right) \tag{12.150}$$

where $z = x_1 + ix_2$ and $f(z)$ is an analytic function.

The GPBCs imply that $f(z)$ must satisfy the consistency conditions

$$\begin{aligned} f(z + L_1) &= e^{i\theta_1} f(z) \\ f(z + iL_2) &= \exp\left[i\theta_2 - i\pi N_\phi\left(\frac{2z}{L_1} + \tau\right)\right] f(z) \end{aligned} \tag{12.151}$$

where $\tau = iL_2/L_1$ is the modular parameter of the torus. The analytic function $f(z)$ must have zeros inside the rectangle with vertices at $(L_1/2)(\pm 1 \pm \tau)$. Thus, $f(z)$ must have N_ϕ zeros. Indeed, the integral

$$\int_\gamma dz\, \frac{f'(z)}{f(z)} = N_\phi \tag{12.152}$$

where γ is the path around the edges of the rectangle of sides L_1 and L_2, is equal to N_ϕ since the total change of the *phase* of $f(z)$ around the edges of the rectangle is $2\pi N_\phi$.

A fundamental theorem of the theory of complex functions tells us that functions $f(z)$ which are analytic inside the rectangle and satisfy the consistency conditions must be analytic functions with exactly N_ϕ zeros. The most general form that $f(z)$ can take is (Haldane and Rezayi, 1985)

$$f(z) = e^{ikz} \prod_{j=1}^{N_\phi} \vartheta_1 \left(\frac{z - z_j}{L_1} \middle| \tau \right) \tag{12.153}$$

where $\vartheta_1(u|\tau)$ is the first odd elliptic theta function (Erdélyi, 1953),

$$\vartheta_1(u|\tau) = i \sum_{n=-\infty}^{+\infty} (-1)^n \, \exp\left[i\pi\tau \left(n - \frac{1}{2}\right)^2 + i\pi(2n-1)u \right] \tag{12.154}$$

The parameter k is a real number in the range $0 \leq |k| \leq \pi N_\phi L_2/L_1^2$. The solutions are thus parametrized by the set of N_ϕ complex numbers $\{z_j\}$ that determine the locations of the zeros of the function $f(z)$ and by k. By direct substitution we find that k and $z_0 \equiv \sum_{j=1}^{N_\phi} z_j$ are the solutions to the set of equations

$$\begin{aligned} e^{i\theta_1} &= \exp\left[ikL_1(-1)^{N_\phi}\right] \\ e^{i\theta_2} &= \exp\left(-kL_2 + i\pi \frac{z_0}{L_1}\right) \end{aligned} \tag{12.155}$$

which have the unique solution

$$k = \frac{\theta_1 + \pi N_\phi}{L_1}, \qquad z_0 = \frac{\theta_2 L_1}{\pi} - ik\frac{L_1 L_2}{\pi} \tag{12.156}$$

The locations of the zeros are determined by requiring that the wave functions $f(z)$ form a complete set of orthogonal wave functions that are eigenstates of the magnetic translation operators. A simple way to construct such a set (i.e. a basis for the Hilbert space of the lowest Landau levels) is to choose a set of zeros, $\{z_j\}$, that satisfies $z_{j+i} = z_j + L_1/N_\phi$. Thus the dimension of the Hilbert space equals N_ϕ, as it should be.

The N-particle states are constructed in very much the same fashion. Here we consider the case of $N = N_\phi$ particles and, once again, we have filled up the lowest Landau level. The only difference here is that we will separate the coordinates $z = \sum_{j=1}^{N} z_j$ for the center of mass (CM) of the system from the set of relative coordinates $\{z_j - z_k\}$. The antisymmetric (fermionic!) N-particle wave function Ψ_N in the lowest Landau level has the form

$$\Psi_N = \mathcal{N}\Psi_{\mathrm{CM}}(z)\left[\prod_{1\leq j<k\leq N} f(z_j - z_k)\right] \times \exp\left(-\sum_{j=1}^{N}\frac{(x_2^j)^2}{2l_0^2}\right) \tag{12.157}$$

where x_2^j is the x_2 coordinate of the jth particle and $\mathcal{N}$ is a normalization constant. The wave functions Ψ_{CM} and $f(z)$ for the center-of-mass and relative coordinates are determined by demanding that Ψ_N satisfy the GPBCs. The "pair wave functions" $f(z_j - z_k)$ do not change if *all* particles are (magnetically) translated simultaneously. Only the center-of-mass wave function Ψ_{CM} is sensitive to a uniform translation of the system as a whole. On the other hand, if a particle (say the jth particle) is transported around the torus exactly once, then the wave function must change by a sign determined by its antisymmetry property. These conditions can be met by requiring $\Psi_{\mathrm{CM}}(z)$ and $f(z)$ to satisfy

$$\begin{aligned}
f(z+L_1) &= f(z)\\
f(z+iL_2) &= f(z)\exp\left[i\pi\left(\frac{2z}{L_1}+\tau\right)\right]\\
\Psi_{\mathrm{CM}}(z+L_1) &= e^{i\theta_1}(-1)^{N-1}\Psi_{\mathrm{CM}}(z)\\
\Psi_{\mathrm{CM}}(z+iL_2) &= e^{i\theta_2}(-1)^{N-1}\ \exp\left[-i\pi\left(\frac{2z}{L_1}+\tau\right)\right]\Psi_{\mathrm{CM}}(z)
\end{aligned} \tag{12.158}$$

The conditions of Eqs. (12.158) imply that both $f(z)$ and $\Psi_{\mathrm{CM}}(z)$ are entire (doubly) periodic functions with just one zero in the principal region. The solution is again the odd elliptic theta function

$$f(z_j - z_k) = \vartheta_1\left(\frac{z_j - z_k}{L_1}\middle|\tau\right) \tag{12.159}$$

The wave function $\Psi_{\mathrm{CM}}(z)$ can also be written in terms of a theta function

$$\Psi_{\mathrm{CM}}(z) = e^{ikz}\vartheta_1\left(\frac{z - z_0}{L_1}\middle|\tau\right) \tag{12.160}$$

This solution has three parameters (k and z_0, the coordinates of the zero of Ψ_{CM}), which are determined by the set of consistency conditions

$$\begin{aligned}
e^{ikL_1} &= (-1)^N e^{i\theta_1}\\
\exp\left(i2\pi\frac{z_0}{L_1}\right) &= (-1)^N e^{i\theta_2+kL_2}
\end{aligned} \tag{12.161}$$

which has the unique solution

$$k = \frac{\pi N}{L_1} + \frac{\theta_1}{L_1}, \qquad z_0 = L_1 \left(\frac{\theta_2}{2\pi} + \frac{N}{2} \right) - i L_2 \left(\frac{N}{2} + \frac{\theta_1}{2\pi} \right) \tag{12.162}$$

Therefore the wave function for one filled Landau level on a torus is unique. Notice that, in contrast, the single-particle states have an N-fold degeneracy.

One important feature of the wave function Ψ_N is the fact that the twist angles θ_1 and θ_2 affect the wave function of the center of mass only through $\Psi_{\rm CM}(z)$. The wave function $\Psi_{\rm CM}(z)$ can be viewed as the wave function for a single particle located at z with charge $-Ne$ moving on a torus in the presence of a uniform external magnetic field with $N_\phi = N$ units of flux. Thus, the center of mass carries the full current. The Niu–Thouless–Wu formula can now be used to yield the result

$$\begin{aligned} \langle \sigma_{xy} \rangle &= \frac{e^2}{ih} \oint \frac{d\theta_j}{2\pi} \left\langle \Psi_N \left| \frac{\partial}{\partial \theta_j} \right| \Psi_N \right\rangle \\ &\equiv \frac{e^2}{ih} \oint \frac{d\theta_j}{2\pi} \int_0^{L_1} dx_1 \int_0^{L_1} dx_2 |\Psi_N|^2 \frac{\partial}{\partial \theta_j} \ln \Psi_{\rm CM}(z, \vec{\theta}) \end{aligned} \tag{12.163}$$

Therefore, the average Hall conductance on a torus is determined by the *average* change of the phase of the wave function for the center of mass on a closed loop on the edges of the torus of boundary conditions. Since $\Psi_{\rm CM}(z, \vec{\theta})$ is an entire function with exactly one zero in the principal region of the elliptic theta function, the theory of functions of complex variables tells us that the integral has the value

$$\oint d\theta_j \frac{\partial}{\partial \theta_j} \ln \Psi_{\rm CM}(z, \vec{\theta}) = 2\pi i \tag{12.164}$$

Thus we find that the Hall conductivity σ_{xy} (averaged over all boundary conditions on the torus) is

$$\langle \sigma_{xy} \rangle = \frac{e^2}{h} \cdot 1 \tag{12.165}$$

This is the result we expected to get. As we can see, it reflects the fact that the wave function for a full Landau level on a torus is unique.

12.8 Quantized Hall conductance of Hofstadter bands

We now turn to the far less trivial question of computing the value of σ_{xy} for the problem of charged particles moving on a square lattice in the presence of a uniform commensurate magnetic field, namely the Hofstadter problem. In Section 12.2 we presented a description of its single-particle states. Let us recall that, if the flux per plaquette is $(p/q)\phi_0$, there are q single-particle Landau bands, each with $L_1 L_2/q$ states. In principle, if we solve the Schrödinger equation, we can construct all the wave functions, whence we can compute anything we wish. These equations are very complicated and yield only to numerical solution. However, the computation of σ_{xy} is considerably simplified by the fact that, here too, it is related to a topological invariant. Thus, we can calculate σ_{xy} within some approximate scheme and still get the exact answer.

Let us first derive an expression for σ_{xy} for a lattice system with periodic boundary conditions. Unlike the continuum case of the last section, the lattice problem is considerably simpler since the main effect of the magnetic field is to generate a sublattice structure. Indeed, in Section 12.2 we saw that the requirement that there should be an integer number of flux quanta piercing the lattice means that either L_1 or L_2 must be an integer multiple of q. Since the magnetic unit cell has q plaquettes, there are $L_1 L_2/q$ magnetic unit cells. We have q sublattices and a Schrödinger equation satisfied by the q sublattices. Hence, unlike in the continuum case, we can apply periodic boundary conditions directly. The reason is that for this lattice problem what matters is not the vector potential $A_j(\vec{r})$ on the link $(\vec{r}, \vec{r} + \hat{e}_j)$ but rather the phase $e^{iA_j(\vec{r})}$, which is invariant under the shifts $A_j(\vec{r}) \to A_j(\vec{r}) + 2\pi l_j(\vec{r})$, where $\{l_j(\vec{r})\}$ is a set of arbitrary integers.

Furthermore, in Section 12.2 we saw that, even though the discrete magnetic translations do not commute with each other, there is a subset of discrete magnetic translations (i.e. those generated by $\hat{T}_1$ and $\hat{T}_2^q$) that commute among themselves and with the Hamiltonian. This subset, which defines the magnetic Brillouin zone, consists of the set of translations by integer numbers of magnetic unit cells. Thus, in units of the magnetic unit cell, the Hamiltonian is translationally invariant. It is then perfectly consistent to impose conventional periodic boundary conditions since the wave functions in *real space* are globally defined. However, they are not globally defined on the *momentum-space torus*, the magnetic Brillouin zone $(-\pi \leq k_1 < \pi, -\pi/q \leq k_2 < \pi/q)$.

Let us derive a version of the Niu–Thouless–Wu formula for the case of a tight-binding system (Kohmoto, 1985; Fradkin and Kohmoto, 1987). In the case of a tight-binding system, the current operator $\hat{J}_k(\vec{r})$ flowing on the link $(\vec{r}, \vec{r} + \hat{e}_j)$ can be obtained by differentiation of the Hamiltonian

$$\hat{J}_k(\vec{r}) = \frac{\delta H}{\delta A_k(\vec{r})} \tag{12.166}$$

where H is an arbitrary (generally interacting) tight-binding Hamiltonian. We will assume here that the external vector potential $A_k(\vec{r})$ enters only in the kinetic-energy term of the Hamiltonian, which in momentum space generally has the form

$$H_{\text{kin}} = \int_{\vec{k},\vec{k}'} c^\dagger(\vec{k}) h_{\text{kin}}(\vec{k}, \vec{k}') c(\vec{k}') \tag{12.167}$$

where $c(\vec{r})$ and $c^\dagger(\vec{r})$ are fermion-destruction and -creation operators at site $\vec{r}$ (the spin indices are omitted), and $h_{\rm kin}(\vec{k}, \vec{k}')$ is the (hermitian) one-particle non-interacting Hamiltonian. This kinetic-energy term applies equally for fermions and bosons. Here we are interested in the fermionic case.

In the case of a system coupled to an external electric field $\vec{E}$, the vector potential $\vec{A}$ gets shifted by $\vec{E}t$. It is easy to show that, when $\vec{E}$ is not zero, the kinetic part of the one-particle Hamiltonian $h_{\rm kin}$ takes the form

$$h_{\rm kin}(\vec{k}, \vec{k}'; \vec{E}) \equiv h_{\rm kin}\left(\vec{k} + \frac{e}{\hbar c}\vec{E}t, \vec{k}' + \frac{e}{\hbar c}\vec{E}t\right) \tag{12.168}$$

Thus, the external uniform electric field $\vec{E}$ (or a twist $\vec{\theta} \equiv (et/(\hbar c))\vec{E}$) is equivalent to a shift of the momentum of each particle by $(et/(\hbar c))\vec{E}$.

The Kubo formula can be written in the following simple form ($j, l = 1, 2$):

$$(\sigma_{xy})_\alpha = -i\hbar L_1 L_2 \epsilon_{kl} \frac{\delta}{\delta A_j}\langle\alpha|\frac{\delta}{\delta A_l}|\alpha\rangle \tag{12.169}$$

For the case of a non-interacting system this expression reduces to a sum over all the occupied one-particle states $\{|n\rangle\}$, i.e. with single-particle energy $E_n < E_{\rm F}$, where $E_{\rm F}$ is the Fermi energy,

$$(\sigma_{xy})_\alpha = \frac{e^2}{\hbar}\sum_{\{n\}} \epsilon_{jl} \frac{\partial}{\partial k_j}\langle n|\frac{\partial}{\partial k_l}|n\rangle \tag{12.170}$$

The one-particle states $\{|n\rangle\}$ are labeled by a band index r ($1 \leq r \leq q-1$) and by a momentum label $\vec{k}$, where $\vec{k}$ lies in the magnetic Brillouin zone. $\Psi_r(\vec{k})$ are the eigenstates of the Schrödinger equation which satisfy the boundary condition $\Psi_{r+q}(\vec{k}) = \Psi_r(\vec{k})$. Let λ be a small parameter ($\lambda \to 1$) that we will use to define (formally) a perturbation theory. This parameter enters into the Schrödinger equation in the form of a Harper equation,

$$\begin{aligned} &-\lambda t[e^{ik_1}\Psi_{r+1}(k_1, k_2) + e^{-ik_1}\Psi_{r-1}(k_1, k_2)] \\ &- 2t\cos\left(k_2 + 2\pi\frac{p}{q}r\right)\Psi_r(k_1, k_2) = E(k_1, k_2)\Psi_r(k_1, k_2) \end{aligned} \tag{12.171}$$

This equation has a set of q linearly independent solutions $\{\Psi_r^{(j)}\}(\vec{k})$ ($j = 1, \ldots, q$). Each solution $\Psi_r^{(j)}(\vec{k})$ has an eigenvalue $E_j(\vec{k})$. These are the Landau–Hofstadter bands.

Let us now consider the case in which the number of particles N is such that there is an integer number r of exactly filled Landau–Hofstadter bands. This requirement defines the state $|\alpha\rangle$. The Hall conductance $(\sigma_{xy})_\alpha$ is then a sum of contributions, one from each filled band, of the form

$$(\sigma_{xy})_\alpha = \frac{e^2}{i\hbar} \sum_{n=1}^{r} \int_{-\pi}^{\pi} \frac{dk_1}{2\pi} \int_{-\frac{\pi}{q}}^{\frac{\pi}{q}} \frac{dk_2}{2\pi} \sum_{p=1}^{q} \epsilon_{jl}\, \partial_{k_j} \Psi_p^{(n)*}(\vec{k}) \partial_{k_l} \Psi_p^{(n)}(\vec{k}) \tag{12.172}$$

We can define a vector field $\mathcal{A}_j^{(n)}(\vec{k})$, for $\vec{k}$ on the magnetic Brillouin zone, to be

$$\mathcal{A}_j^{(n)}(\vec{k}) = \sum_{p=1}^{q} \Psi_p^{(n)*}(\vec{k})(-i)\partial_{k_j} \Psi_p^{(n)}(\vec{k}) \tag{12.173}$$

The Hall conductance is essentially the flux of $\mathcal{A}_j^{(n)}$ through the magnetic Brillouin zone,

$$(\sigma_{xy})_\alpha = \frac{e^2}{\hbar} \sum_{n=1}^{r} \int_{-\pi}^{\pi} \frac{dk_1}{2\pi} \int_{-\frac{\pi}{q}}^{\frac{\pi}{q}} \frac{dk_2}{2\pi} \epsilon_{jl}\, \partial_{k_j} \mathcal{A}_l^{(n)}(\vec{k}) \tag{12.174}$$

Once again, $(\sigma_{xy})_\alpha$ is identified with a Chern number, which counts the winding number of the phase of the wave functions as $\vec{k}$ traces the boundary of the magnetic Brillouin zone. Let us denote by I_n the Chern number for the nth band. I_n is given by

$$I_n = \frac{1}{2\pi} \int_{-\pi}^{\pi} dk_1 \int_{-\frac{\pi}{q}}^{\frac{\pi}{q}} dk_2\, \epsilon_{jl} \sum_{p=1}^{q} \partial_{k_j} \Psi_p^{(n)*}(\vec{k}) \partial_{k_l} \Psi_p^{(n)}(\vec{k}) \tag{12.175}$$

Since the Chern numbers I_n are topological invariants, we can compute their exact values by considering a *smooth* deformation of the Schrödinger equation. For instance, we can compute the integers I_n in the limit $\lambda \to 0$ (or rather a perturbative expansion in powers of λ). If, as λ is varied from $\lambda = 0$ to $\lambda = 1$, there are no band crossings, the integers I_n will not change.

Let us now discuss the qualitative features of a (degenerate) perturbation theory in λ. At $\lambda = 0$ the eigenstates $\Psi_p^{(n)}(\vec{k})$ are $(n = 1, \ldots, q)$

$$\Psi_p^{(n)}(\vec{k}) = \delta_{pn} \tag{12.176}$$

with eigenvalues $E_n^{(0)}(\vec{k})$

$$E_n^{(0)}(\vec{k}) = -2t \cos\left(k_2 + 2\pi \frac{p}{q} n\right) \tag{12.177}$$

The spectrum then has q generally non-degenerate bands with dispersion laws $E_n^{(0)}(\vec{k})$.

On the magnetic Brillouin zone ($-\pi \leq k_1 < \pi$, $-\pi/q \leq k_2 < \pi/q$), the unperturbed bands of Eq. (12.177) cross at $\vec{k} = (k, 0)$ and $\vec{k} = (k_1, \pi/q)$. For example, the lowest band ($n = 1$) crosses the next ($n = 2$) band at $k_2 = \pi/q$. The second band crosses the third one ($n = 3$) at $k_2 = 0$, etc. In general, the nth band (for n *even*) crosses the $(n-1)$th band at $k_2 = \pi/q$ (the bottom of the nth band) and the $(n+1)$th band at $k_2 = 0$ (the top of the nth band).

Conversely, for n *odd*, the top of the nth band is at $k_2 = 0$ (where it crosses the $(n+1)$th band), while the bottom is at $k_2 = \pi/q$ (where it crosses the $(n-1)$th band). The integer n labels the bands as well as the gaps. The top band ($n = q$) has only one crossing of the band with $n = q - 1$ at $k_2 = 0$ (q even) or $k_2 = \pi/q$ (q odd).

The integers I_n are determined by the changes of the phases of the wave function as $\vec{k}$ passes through the degeneracy points. We can determine these phases by using Brillouin–Wigner perturbation theory (see Section 2.3.1). The nth band (for p and q fixed) crosses the mth band if $m = n - l_n$, where the integer l ($|l_n| \leq q/2$) is the solution of the Diophantine equation

$$n = qs_n + pl_n \tag{12.178}$$

a result first derived by Thouless, Kohmoto, Nightingale, and den Nijs (Thouless *et al.*, 1982).

The Schrödinger equation mixes $\Psi^{(n)}$ only with $\Psi^{(n\pm1)}$. Thus, it takes l_n orders of perturbation theory to mix $\Psi^{(n)}$ and $\Psi^{(n-l)}$. For $\vec{k}$ close to the degeneracy points, the eigenstates will have almost all of their weight in $\Psi^{(n)}$ and $\Psi^{(n-l)}$. Thus, we get an effective Schrödinger equation of the form

$$\begin{aligned} E_n^{(0)}\Psi_n + V_{n,n-l}\Psi_{n-l} &= E\Psi_n \\ V_{n,n-l}\Psi_n + E_{n-l}^{(0)}\Psi_{n-l} &= E\Psi_{n-l} \end{aligned} \tag{12.179}$$

The matrix element $V_{n,n-l}$ is (approximately) equal to

$$V_{n,n-l} = V_{n-l,n}^* \simeq (-\lambda t e^{-ik_1}) \prod_{r=n-l+1}^{n-1} \left[\frac{-\lambda t e^{-ik_1}}{\frac{1}{2}(E_n^{(0)} + E_{n-l}^{(0)}) - E_r^{(0)}} \right] \tag{12.180}$$

where $E_n^{(0)}(\vec{k}) = -2t\cos(k_2 + 2\pi(p/q)n)$ are the unperturbed energy bands.

The eigenvalues of Eq. (12.179) are

$$E^{\pm}(\vec{k}) = \frac{1}{2}(E_n^{(0)} + E_{n-l}^{(0)}) \pm \sqrt{\frac{(E_n^{(0)} - E_{n-l}^{(0)})^2}{2} + |V_{n,n-l}|^2} \tag{12.181}$$

The eigenstates have amplitudes $(\Psi_n^{(\pm)}, \Psi_{n-l_n}^{\pm})$ of the form

$$\Psi_n^{(\pm)} = |\Psi_n^{(\pm)}| e^{i\theta_n^{(\pm)}} \tag{12.182}$$

with a similar expression for $\Psi_{n-l_n}^{(\pm)}$. The amplitudes $|\Psi_n^{(\pm)}|$ are

$$\begin{aligned} |\Psi_n^{(\pm)}| &= \frac{|V_{n,n_l}|}{\sqrt{|E^{(\pm)} - E_n^{(0)}|^2 + |V_{n,n-l_n}|^2}} \\ |\Psi_{n-l_n}^{(\pm)}| &= \frac{|E^{(\pm)} - E_n^{(0)}|}{\sqrt{|E^{(\pm)} - E_n^{(0)}|^2 + |V_{n,n-l_n}|^2}} \end{aligned} \tag{12.183}$$

The phases $\theta_n^{(\pm)}$ are given by

$$\begin{aligned} \theta_n^{(+)} - \theta_{n-l_n}^{(+)} &= \arg(V_{n,n-l_n}) + \pi = -k_1 l_n - (l_n - 1)\pi \\ \theta_n^{(-)} - \theta_{n-l_n}^{(-)} &= \arg(V_{n,n-l_n}) = -k_1 l_n - l_n \pi \end{aligned} \tag{12.184}$$

Let us consider the nth band with n even. The result is the same for n odd. At $k_2 = \pi/q$ it crosses the $(n+1)$th band. At this degeneracy we have to choose the solution $E^{(-)}$ for the top of the nth band. Conversely, at $k_2 = 0$, the nth band crosses the $(n-1)$th band. Thus, we have to choose the solution $E^{(+)}$ for the bottom of the nth band.

Let us compute the circulation of the vector field $\mathcal{A}_j^{(n)}(\vec{k})$ for the nth band for $\vec{k} = (k_1, k_2)$ along the closed contour γ (which encloses half of the magnetic Brillouin zone):

$$\gamma : \ (0,0) \to (\pi, 0) \to (\pi, \pi/q) \to (0, \pi/q) \to (0,0) \tag{12.185}$$

On the first and third segments of the contour γ, k_2 is constant, while k_1 changes from zero to π and from π to zero, respectively. The component $\mathcal{A}_1^{(n)}$ is then equal to

$$\begin{aligned} \mathcal{A}_1^{(n)}|_{k_2=\pi/q} &= \frac{\partial}{\partial k_1} \arg\left[V_{n,n-l_n}\right]\Big|_{k_2=\pi/q} = -l_n \\ \mathcal{A}_1^{(n)}|_{k_2=0} &= \frac{\partial}{\partial k_1} \arg\left[V_{n-1,n-1-l_{n-1}}\right]\Big|_{k_2=0} = -l_{n-1} \end{aligned} \tag{12.186}$$

For the second and fourth segments we need to compute $\mathcal{A}_2^{(n)}$. Since the phases have no essential dependence on k_2, we get

$$\mathcal{A}_2^{(n)}|_{k_1=0,\pi} = 0 \tag{12.187}$$

The results summarized by Eq. (12.184) and Eq. (12.186) show that the circulation of $\mathcal{A}_j^{(n)}$ on the contour γ of Eq. (12.185) is

$$I_n = \frac{1}{2\pi} \oint_{\gamma} \mathcal{A}_j^{(n)} \, dk_j = \int_0^{\pi} \frac{dk_1}{2\pi} \left[\mathcal{A}_1^{(n)}(k_1, 0) - \mathcal{A}_1^{(n)}\left(k_1, \frac{\pi}{q}\right) \right] \tag{12.188}$$

Thus, I_n of the nth band is

$$I_n = l_n - l_{n-1} \tag{12.189}$$

Therefore, the contribution from the nth band to the Hall conductance is

$$(\sigma_{xy})^{(n)} = \frac{e^2}{h}(l_n - l_{n-1}) \tag{12.190}$$

For a problem with r filled bands we have

$$(\sigma_{xy}) = \frac{e^2}{h} \sum_{n=1}^{r} (l_n - l_{n-1}) = \frac{e^2}{h}(l_r - l_0) \equiv \frac{e^2}{h} l_r \tag{12.191}$$

where we have used the definition $l_0 = 0$.

This result, which was originally derived by Thouless, Kohmoto, Nightingale, and den Nijs (Thouless *et al.*, 1982), shows that σ_{xy} is determined by the topological invariant I_n which characterizes the Landau–Hofstadter bands. This integer is the solution of the Diophantine equation. The integers l_n may be positive or negative and are restricted to be in the range $|l| \leq q/2$. Thus, in contrast to the continuum result, the quantized Hall conductance of a filled Landau–Hofstadter band may be positive or negative. This surprising result is a Bragg-scattering effect due to the magnetic unit cells. Let us consider an example with $p = 11$ and $q = 7$. There are seven bands. Let us use the notation (s_n, l_n) for the two integers which solve the Diophantine equation. The solutions are $(-3, 2)$, $(-6, 4)$, $(2, -1)$, $(-1, 1)$, $(7, -4)$, $(4, -2)$, and $(1, 0)$ for n ranging from $n = 1$ up to $n = 7$. Notice that the bands with $n = 3$, 5, and 6 have $l = -1, -4$, and -2, respectively, and carry *negative* Hall conductance.

The Diophantine equation has a unique solution for q odd. For q even, the band with index $n = q/2$ has two possible solutions, namely $((1 - p)/2, q/2)$ and $((1 + p)/2, -q/2)$. What happens here is that, for q even and $n = q/2$, the Landau–Hofstadter bands have a degeneracy, which we have already discussed in Section 12.2. Depending on how this degeneracy is removed, the conductance is $+q/2$, $-q/2$, or even zero. This observation is important to the physics of flux phases. Let us finally remark that the solutions of the Diophantine equation, Eq. (12.178), $\sum_{n=1}^{q} s_n$ and $\sum_{n=1}^{q} l_n$ obey the sum rules

$$\sum_{n=1}^{q} s_n = \frac{q+1}{2}, \qquad \sum_{n=1}^{q} l_n = 0 \tag{12.192}$$

for q odd, and

$$\sum_{n=1}^{q} s_n = \frac{q+1 \mp p}{2}, \qquad \sum_{n=1}^{q} l_n = \pm\frac{q}{2} \tag{12.193}$$

for q even. The ambiguity in the sum rule is due precisely to the double solution at $n = q/2$ (q even).

13

The fractional quantum Hall effect

In this chapter we discuss the theory of the fractional quantum Hall effect (FQHE). The explanation of this phenomenon has required the development of completely new ideas and methods. The concept of fractional statistics has become a crucial element of the theory.

The physical system involves fermions in strong correlation in the absence of time-reversal symmetry. The treatment of systems with these features cannot be achieved successfully within the conventional Hartree–Fock approach to correlations in condensed matter physics. A new condensed state of matter, the Laughlin state, had to be discovered.

The Chern–Simons gauge theory, which has already been discussed in Chapter 10, has come to play an essential role in the theory of the FQHE, both as a way to describe the low-energy phenomena and as a theoretical tool to explain the most important features of the problem.

We begin with a detailed description of the theory of the Laughlin wave function, which is followed by the field-theory approaches to the FQHE.

13.1 The Laughlin wave function

In the last two sections of the previous chapter we considered the problem of electrons moving on a 2D surface in the presence of a perpendicular magnetic field. We assumed that the electron density was such that an integer number of Landau levels (or bands) would be completely filled. Because the system has an energy gap, the interactions do not play a very important role. In fact, a perturbative expansion (in powers of the coupling constant) around a state with one filled Landau level (or more) is likely to be well behaved. Since all processes involve exciting one or several electrons across the gap, the energy denominators are always different from zero. The ground-state wave function for the interacting system is smoothly connected to the ground-state wave function of the non-interacting system. The arguments of the last three sections of Chapter 12 indicate that the topological properties of the wave function for the interacting and the non-interacting systems will then be the same. In other words, naive perturbation theory is a good approximation in this case.

However, if one Landau level (or band) is partially filled, perturbation theory breaks down. Consider for simplicity the case of N particles in a magnetic field B with N_ϕ quanta of flux piercing the surface. The filling fraction $\nu = N/N_\phi$ is not an integer. We will consider the simpler (and popular) case of $\nu = 1/m$, where m is an odd integer and for each electron there are m quanta of flux. We further assume that the magnetic field is sufficiently large that all the Zeeman energies are so large that the system is completely spin-polarized. This is the case for most, but not all, of the experimentally accessible systems. In this limit, the electrons behave as charged spinless fermions, each carrying an electric charge of $-e$.

In Section 12.7 we saw that, if just one Landau level is filled ($m = 1$), the ground state is non-degenerate and its wave function is a Slater determinant. For $m > 1$ only a fraction $1/m$ of the states in the first Landau level will be occupied. The remaining $(m - 1)/m$ states are empty. However, occupied and unoccupied states have exactly the same energy. The actual ground state has then to be determined through some sort of degenerate-perturbation-theory scheme. This procedure is bound to be very complex due to the macroscopic degeneracy of the Landau level. The resulting state is likely to have properties that are completely different from those of the unperturbed state.

The observed phenomenology of the FQHE also suggests the need for a completely different state. A non-interacting fractionally filled state would still exhibit a fractional Hall conductance σ_{xy} since, at least for a Galilean-invariant system, the conductance is determined by the amount of charge present. But such a state would not support the very precise plateaus which are seen in experiments, since additional particles can be added at almost no energy cost. The fact that the FQHE is seen only in the purest samples indicates that the effect is the result of electron correlations due to the Coulomb interactions. Moreover, the "quenching" of the single-particle kinetic energies by the magnetic field is telling us that the interactions play a dominant role. The FQHE is the result of the competition between degeneracy and interactions. In this sense, the FQHE is an example of strongly correlated electron systems.

The model which naturally describes the essential features of the physical system consists of an assembly of N electrons that occupy a fraction ν of the N_ϕ states of the lowest Landau level and interact with each other via Coulomb interactions. The ground state of this system must be such that it should not support any gapless excitations (otherwise the plateaux of σ_{xy} could not be so sharp) and it should be essentially insensitive to the presence of impurities. The wave function should be a complex function of the electron coordinates. This requirement follows from the fact that, if a magnetic field is present, time-reversal invariance is broken explicitly. Finally, Fermi statistics demands that the wave function $\Psi_N(\vec{r}_1, \ldots, \vec{r}_N)$ should be antisymmetric under the permutation of the positions of any pair of particles. Thus, Ψ_N vanishes as the positions of two particles approach each other.

We will now construct a wave function that satisfies all these requirements. Here we follow closely Laughlin's construction (Laughlin, 1983, 1987). Let us consider first the *low-density* limit $\nu \ll 1$ $(m \gg 1)$. In this limit, the average separation between two electrons is much larger than the single-particle magnetic length l_0 $(a_0 \gg l_0)$. The electrons do not venture very far away and interactions further restrict their motion. The natural ground state in this limit is an electron crystal, known as a Wigner crystal. The electrons are able to minimize the total energy by arranging themselves on a triangular lattice. Actually the "guiding center coordinates" form a triangular lattice. A Hartree approximation yields a Wigner crystal state $\Psi_{\rm W}(z_1, \ldots, z_N)$ of the form (Laughlin, 1987)

$$\Psi_{\rm W}(z_1, \ldots, z_N) = \sum_P (-1)^P \phi_{j_1 l_1}(z_{P_1}) \ldots \phi_{j_N l_N}(z_{P_N}) \tag{13.1}$$

where the single-particle states $\phi_{jl}(z)$ are

$$\phi_{jl}(z) \approx \exp\left(-\frac{1}{4l_0^2}|z_{jl}^{(0)}|^2 + \frac{1}{2l_0^2}\bar{z} z_{jl}^{(0)} - \frac{1}{4l_0^2}|z|^2 \right) \tag{13.2}$$

and $z_{jl}^{(0)}$ are the (complex) coordinates of the (j, l) site of a triangular lattice,

$$z_{jl}^{(0)} = l_0 \sqrt{\frac{4\pi m}{\sqrt{3}}} \left(j + \left(\frac{1}{2} + i\frac{\sqrt{3}}{2} \right) l \right) \tag{13.3}$$

The Wigner crystal state $\Psi_{\rm W}$ does satisfy a number of the requirements listed above but not all of them. First, it *does* support elementary excitations with arbitrarily low excitation energy, namely the sound waves of the Wigner crystal. Since the state is a periodic array of charges, the charge density is not uniform and it is *strongly* affected by the presence of impurities, which can, and do, pin the crystal at the impurity sites. Thus, this *pinned* state does not support any charge current unless the electric field is larger than some critical threshold value E_0 determined by the local pinning forces. This behavior is commonly seen in other charge crystals, such as incommensurate charge-density waves. The best-known examples are the quasi-1D system $NbSe_3$ and the quasi-2D system $NbSe_2$.

As the electron density increases (i.e. m grows smaller) the inter-particle separation a_0 decreases. For a triangular lattice, we have that a_0 is related to the filling fraction ν and the cyclotron length l_0 through the relation $\nu = (4\pi/\sqrt{3})(l_0/a_0)^2$. As ν approaches unity, the ratio l_0/a_0 becomes also a number of order unity. Thus, as ν grows larger, there should be a phase transition from a Wigner crystal to a state that supports a Hall current. Indeed, as ν grows larger and the cyclotron length approaches the inter-particle spacing, the quantum fluctuations should increase. The leading fluctuations should involve exchanges of a small number of nearby particles. In particular, there are processes that involve three-particle exchanges around an elementary triangle (or "ring").

Such processes spoil the long-range positional order of the Wigner crystal. If these ring exchanges are able to proliferate, the Wigner crystal melts and there is a transition to a liquid state (Kivelson *et al.*, 1986). This phase transition is most likely to be of first order but, depending on microscopic properties, it can also be of second order. The resulting *liquid* state is expected to have uniform density. What is more important, and far less trivial to see from this point of view, is that it should have a gap to all excitations. The phonon of the Wigner crystal should disappear from the physical spectrum. This phenomenon is strongly reminiscent of the Higgs mechanism in a superconductor coupled to a *dynamical* gauge field: the phase mode of the superconductor gets "eaten" by the gauge field, which, in the process, becomes massive. We will see below that the FQHE *has a hidden, dynamically generated, gauge field* that is responsible for the most striking features of this phenomenon.

The liquid state should be regarded as a new condensed state of matter. Laughlin was the first to realize that this state is fundamentally different from other known condensed states, such as magnetism or superconductivity. Drawing on intuition he gained by studying systems with small numbers of particles, Laughlin proposed the following class of wave functions (Laughlin, 1983):

$$\Psi_N(\vec{r}_1, \dots, \vec{r}_N) = \prod_{1 \le j < k \le N} f(z_j - z_k) \exp\left(-\sum_{j=1}^{N} \frac{|z_j|^2}{4l_0^2}\right) \tag{13.4}$$

where $f(z)$ is a suitably chosen *analytic* function of the complex coordinates $\{z_1, \dots, z_N\}$, i.e. single-particle states from just the lowest Landau level. Fermi statistics demands that $f(z_j - z_k)$ be an *odd* function of $z_j - z_k$ that vanishes as $z_j \to z_k$. These requirements, together with the demand that Ψ_N should be an eigenstate of the total L_z orbital angular momentum, can be met by the simple choice of $f(z) \sim z^m$, where m is an odd integer. We thus arrive at the celebrated Laughlin wave function Ψ_m,

$$\Psi_m(\vec{r}_1, \dots, \vec{r}_N) = \prod_{1 \le j < k \le N} (z_j - z_k)^m \exp\left(-\sum_{j=1}^{N} \frac{|z_j|^2}{4l_0^2}\right) \tag{13.5}$$

This wave function is remarkable in several ways. Laughlin has computed the overlap between Ψ_m and the *exact* wave function of a small cluster of electrons (with $N \leq 3$) and interaction pair potentials $u(r) = 1/r, -\ln r, \exp(-r^2/2)$. He found that in all cases the overlap was better than 99%. For a special potential, namely $u(r) = u_0 \nabla^2 \delta(\vec{r})$, Trugman and Kivelson (1985) showed that $\Psi_m^{(N)}$ is the *exact* ground-state wave function for all m. Haldane (1983b) constructed a class of Hamiltonians for which Laughlin-like states are the exact ground states (see below). Laughlin originally thought of Ψ_m as a *variational* wave function, with a Jastrow form, which is commonly used to construct variational states for superfluid liquid helium (Feenberg, 1969).

However, Ψ_m does not contain any variational parameters! The ground state is determined by just finding the values of m that minimize the energy. But m is in fact determined by the total angular momentum! It is remarkable that this guess works so well. It is an important problem for theorists to explain why this is such a good state. The Laughlin wave function also admits a number of generalizations that describe other filling fractions. These are the hierarchical wave functions of Haldane (1983b) and Halperin (1984). We will consider mostly the $1/m$ Laughlin states. In the next chapters we will see why these wave functions, and their generalizations, are good representations of fluids that exhibit the FQHE.

13.1.1 The plasma analogy

We now follow Laughlin and determine the optimal value of m, as well as the nature of the correlations present in Ψ_m, by using the *plasma analogy.* Let $\rho(z_1, \ldots, z_N)$ be the *joint probability distribution* function

$$\rho(z_1, \ldots, z_N) = |\Psi_m(z_1, \ldots, z_N)|^2 \tag{13.6}$$

which can be thought of as a classical probability distribution for a one-component plasma with N particles located at $\{z_1, \ldots, z_N\}$. Let $U(z_1, \ldots, z_N)$ be the classical potential energy, and let β be an *effective inverse temperature* ($\beta = m$). The potential U is defined by demanding that ρ should have the Gibbs form

$$|\Psi_m(z_1, \ldots, z_N)|^2 = e^{-\beta U(z_1, \ldots, z_N)} \tag{13.7}$$

The classical potential energy $U(z_1, \ldots, z_N)$ is given by

$$U(z_1, \ldots, z_N) = -2 \sum_{1 \leq j < k \leq N} \ln|z_j - z_k| + \frac{1}{2m} \sum_{j=1}^{N} |z_j|^2 \tag{13.8}$$

where we have used units of length such that $l_0 = 1$. The potential $U(\{z_j\})$ is equal to the total energy of a gas of classical particles each carrying charge $q = 1$, which interact with each other via the 2D Coulomb pair potential,

$V_C(z_j - z_k) = -\ln|z_j - z_k|$, and with a uniform neutralizing background charge of density $\rho_0 = 1/(2\pi m)$. The interaction with the background charge is represented in $U(z)$ by the last term. This can be checked by noting that $\nabla^2(1/(2m))|z|^2 = 2/m$, which agrees with the density being uniform and equal to $1/(2\pi m)$. This is the one-component classical plasma.

The plasma analogy is a very powerful tool for the investigation of the properties of the Laughlin wave function. All expectation values of local operators in the Laughlin state can be represented as an ensemble average in the plasma. There is a well-developed body of knowledge on this subject. For instance, the average electron density at point z, $\langle \rho(z) \rangle$, is

$$\langle \rho(z) \rangle = \frac{\int d^2z_1 \dots d^2z_N \ \rho(z) |\Psi_m(z_1, \dots, z_N)|^2}{\int d^2z_1 \dots d^2z_N |\Psi_m(z_1, \dots, z_N)|^2} \tag{13.9}$$

where the local charge density $\rho(z)$ is equal to

$$\rho(z) = \sum_{j=1}^{N} \delta(z - z_j) \tag{13.10}$$

In the plasma analogy, we write the average charge density $\langle \rho(z) \rangle$ in the form of a weighted average over the positions of the classical charges

$$\langle \rho(z) \rangle = \frac{1}{Z_{\text{plasma}}} \int d^2z_1 \dots d^2z_N \ \rho(z) e^{-\beta U(z_1, \dots, z_N)} \tag{13.11}$$

where Z_{plasma} is the partition function for a classical one-component plasma. The potential energy $U(z_1, \dots, z_N)$ has a simple form in terms of the density variable $\rho(z)$:

$$U[\rho(z)] = \int d^2z \int d^2z' (\rho(z) - \rho_0) V(z - z') (\rho(z') - \rho_0) \tag{13.12}$$

where $V(z - z')$ is the 2D Coulomb pair potential

$$V(z - z') = -\ln|z - z'| \tag{13.13}$$

and ρ_0 is the background charge.

If the density is low, the quantization of the charge of the individual electrons is very important. The dominant configuration in this limit is a Wigner crystal. But, as the density increases, the local density experiences larger fluctuations. As a result, the local average charge is not equal to the electron charge. In other words, at high densities, the local average density $\rho(z)$ becomes a continuous variable. In this limit, any additional local charge will be rapidly screened, and the local average density should become equal to the background charge. Conversely, at low densities, screening is very poor and the local density can deviate significantly from the value of the background charge density. Thus, the electron liquid corresponds to the (high-density) plasma phase of the one-component Coulomb gas. The approximation in which the local density becomes a continuous variable is known as the Debye–Hückel theory. It is straightforward to verify that in this limit $\langle\rho(z)\rangle = \rho_0$. This result is also seen to hold in Monte Carlo simulations, at least for $m \leq 5$. More details on how the plasma methods are applied to the theory of the Laughlin state can be found in Laughlin's article in the book edited by Prange and Girvin (Laughlin, 1987), where he uses extensively the methods described by G. Stell for classical fluids (Stell, 1964).

13.1.2 The Haldane Hamiltonians

Let us now discuss Haldane's construction of a class of Hamiltonians that have the Laughlin state as their exact ground state. Haldane begins by noticing that a system with a disk geometry with wave functions that vanish on the boundary (in the thermodynamic limit) is equivalent to a (large) sphere of radius R. A uniform magnetic field flows outwards from the sphere. The Laughlin states are then isotropic on the sphere. Since the magnetic field is normal to the sphere, it is the field of a magnetic monopole in the center of the sphere with magnetic charge equal to the total flux. Let $2S$ be the total flux, in units of the flux quantum hc/e. The single-particle states for particles of charge e moving on the surface of the sphere have to be smooth and single-valued. This demand forces the magnetic charge $2S$ of the monopole to be an integer. This is the famous Dirac quantization condition (Dirac, 1931). We have already encountered this problem in Chapter 7 when we described the path-integral formalism for spin.

The single-particle Hamiltonian H now becomes

$$H = \frac{\omega_c}{2\hbar S}\left[\vec{r} \wedge \left(\vec{p} + \frac{e}{c}\vec{A}\right)\right]^2 \tag{13.14}$$

where ω_c is the cyclotron frequency. Let $\vec{n}$ be a unit vector normal to the surface of the sphere, i.e. $\vec{n} = \vec{r}/R$. The magnetic field of the monopole is $\vec{\nabla} \wedge \vec{A} = B\vec{n}$, where $B = \hbar c S/(eR^2)$. The vector $\vec{\Lambda} = \vec{n} \wedge (\vec{p} + (e/c)\vec{A})$ satisfies the algebra

$$[\Lambda_a, \Lambda_b] = i\hbar\epsilon_{abc}(\Lambda_c - \hbar S n_c) \tag{13.15}$$

and $\vec{\Lambda} \cdot \vec{n} = \vec{n} \cdot \vec{\Lambda} = 0$. Of course, this is the same problem with the gauge-covariant momentum as that which we discussed in Section 12.1. Here too we should define another operator, which should generate the magnetic translations. For the spherical geometry, this is just rotations. The generators of rotations are $\vec{L} = \vec{\Lambda} + \hbar S\vec{n}$ and satisfy the algebra

$$\begin{aligned}
&[L_a, L_b] = i\hbar\epsilon_{abc}L_c \\
&\left[L_a, \vec{L}^2\right] = 0 \\
&[L_a, n_b] = i\hbar\epsilon_{abc}n_c \\
&[L_a, \Lambda_b] = i\hbar\epsilon_{abc}\Lambda_c
\end{aligned} \tag{13.16}$$

The last condition implies that L_a commutes with $\vec{\Lambda}^2$ and hence with H. Thus, L_a and H can be diagonalized simultaneously. The first two equations are telling us that the operators L_a satisfy the algebra of angular momentum. The eigenvalues of $\vec{L}^2$ are $\hbar^2 l(l+1)$, where $l = S + n$, n is a positive integer (or zero), and $2S$ is an integer. This is just the Dirac quantization condition. Thus, $\vec{\Lambda}^2$ is equal to $\vec{L}^2 - \hbar^2 S^2$. We conclude that the single-particle Hamiltonian has eigenstates $|m, l\rangle$ such that

$$\begin{aligned}
L_3|m, l\rangle &= \hbar m|m, l\rangle \\
\vec{L}^2|m, l\rangle &= \hbar^2 l(l+1)|m, l\rangle \\
H|m, l\rangle &= \hbar\omega_c \frac{l(l+1) - S}{2S}|m, l\rangle
\end{aligned} \tag{13.17}$$

where $|m| \leq l$. Thus, each level is $(2l+1)$-fold degenerate. In terms of n and S the degeneracy is $2n + 1 + 2S$. The lowest energy level, which corresponds to the lowest Landau level, has $n = 0$ $(l = S)$, and it is $(2S+1)$-fold degenerate. If we represent the unit vector $\vec{n}$ in terms of a two-component spinor $\vec{u} = (u, v)$, since $\vec{n} = u_\alpha^* \vec{\tau}_{\alpha\beta} u_\beta$ ($\vec{\tau}_{\alpha\beta}$ are the Pauli matrices), then the Hilbert space of the lowest Landau level is spanned by the coherent states $\Psi^{(S)}_{(\alpha,\beta)}(u, v) = (\alpha^* u + \beta^* v)^{2S}$, with $|\alpha|^2 + |\beta|^2 = 1$, which are polynomials of degree $2S$.

In this notation the Laughlin states Ψ_m are

$$\Psi_m = \prod_{1 \leq j < k \leq N} (u_j v_k - u_k v_k)^m \tag{13.18}$$

with $S = \frac{1}{2}m(N-1)$ for states with N particles. It can be readily checked that this state is also an eigenstate of $\vec{L}^2 = (\sum_{j=1}^N \vec{L}_j)^2$ with zero eigenvalue since the three operators $L^+ = \hbar\sum_{j=1}^N u_j\, \partial/\partial v_j$, $L^- = \hbar\sum_{j=1}^N v_j\, \partial/\partial u_j$, and $L_3 = \hbar\sum_{j=1}^N (u_j\, \partial/\partial u_j - v_j\, \partial/\partial v_j)$ annihilate Ψ_m. The state Ψ_m is thus rotationally and translationally invariant on the sphere.

Haldane further remarked that the states Ψ_m are *exact* eigenstates of a class of Hamiltonians constructed in the following manner. Let $P_J(L)$ be a projection operator on states with $\vec{L}^2$ eigenvalue equal to $\hbar^2 J(J+1)$, and let Π_S be the projection operator onto the Hilbert space of the lowest Landau level. Haldane proposed to write a projected Hamiltonian as

$$\Pi_S H \Pi_S = \sum_{1 \le j < k \le N} \left\{ \sum_{j > 2S - m} P_J(\vec{L}_j + \vec{L}_k) V_J \right\} \tag{13.19}$$

which, by construction, annihilates the state Ψ_m.

13.1.3 Elementary excitations of the Laughlin state

The Laughlin wave function is an accurate approximation for the ground state of the system only if the electron density ρ_0 and the magnetic field B are such that the filling fraction ν is exactly equal to $1/m$. For densities and fields for which ν is close, but not equal, to $1/m$, it is no longer a good approximation. As we will see below, the states with $\nu \approx 1/m$ are excited states of the $\nu = 1/m$ state. It is an essential feature of the Laughlin state that these states are not degenerate with the ground state even in the thermodynamic limit. The Laughlin state is found to have a non-zero gap for *all* elementary excitations. The Laughlin state thus represents a uniform *incompressible* fluid.

Several excited states are possible. We may change the magnetic field locally without changing the total number of particles. This can be achieved by inserting an infinitesimally thin solenoid, carrying exactly one flux quantum, at one point of the sample (say, the origin $z = 0$). Or, we may add (or subtract) an electron without changing the external field. Furthermore, we may imagine local density fluctuations that change neither the field nor the total particle number. Among these excitations, there are density fluctuations involving states only in the lowest Landau levels (phonons) or states in the first (or higher) excited Landau levels (plasmons). For the sake of simplicity, in this section I will consider only the state obtained by the addition of a solenoid. This state is a *Laughlin quasihole*. We will briefly discuss the collective modes in a later section in which we will discuss the field-theory picture of the Laughlin state.

The Laughlin state Ψ_m for $\nu = 1/m$ is the product of a polynomial in the particle coordinates and an exponential factor. We can expand Ψ_m in a series of the form

$$\Psi_m(z_1, \ldots, z_N) = \sum_{\{k_1, \ldots, k_N\}} C_{k_1, \ldots, k_N} z_1^{mk_1} \ldots z_N^{mk_N} \exp\left(-\sum_{j=1}^{N} \frac{|z_j|^2}{4l_0^2} \right) \tag{13.20}$$

The integers $\{k_1, \ldots, k_N\}$ run from 0 to N with the restriction

$$\sum_{j=1}^{N} k_j = \frac{1}{2} N(N-1) \tag{13.21}$$

For the wave function to describe a system of fermions, the coefficients $C_{k_1,\ldots,k_N}$ must be antisymmetric under the permutation of the indices.

Under a rigid rotation of the system as a whole by an angle θ about the origin, the coordinate z_j of each particle is multiplied by a phase factor $e^{i\theta}$. Thus, Ψ_m transforms like

$$\Psi_m(e^{i\theta}z_1,\ldots,e^{i\theta}z_n) = e^{im\frac{N}{2}(N-1)\theta}\Psi_m(z_1,\ldots,z_N) \tag{13.22}$$

which means that the total L_z angular momentum of Ψ_m is equal to

$$M_m = \frac{1}{2}mN(N-1) \tag{13.23}$$

Let us now imagine that an infinitesimally thin solenoid carrying one unit of flux is introduced adiabatically into the system and pierces the disk at the origin, $z = 0$. For flux ϕ, the single-particle state changes from $z^n e^{-|z|^2/(4l_0^2)}$ to $z^{n+\alpha}e^{-|z|^2/(4l_0^2)}$, where $\alpha = \phi/\phi_0$, with ϕ_0 being the flux quantum hc/e. Thus, if $\phi = \phi_0$, the nth state in the first Landau level becomes the $(n+1)$th state in the same Landau level.

The Laughlin state reacts very much in the same way, by shifting each $z_j^{mk_j}$ to $z_j^{mk_j+1}$ and undergoing a change in the coefficients. This process does not alter the exponential factor. If we ignore the change in the coefficients $C_{k_1,\ldots,k_N}$ the shift can be seen to be the same as a multiplication of Ψ_m by a factor of the form $\Pi_{j=1}^N z_j$. This observation, which was also made first by Laughlin (1983), motivates the choice of the following ansatz for the wave function $\Psi_m^{(+)}(z_0;\{z_j\})$ of the quasihole state created by the adiabatic insertion of a solenoid:

$$\Psi_m^{(+)}(z_0;z_1,\ldots,z_N) = \prod_{j=1}^{N}(z_j - z_0)\Psi_m(z_1,\ldots,z_N) \tag{13.24}$$

This state has angular momentum $M_m^{(+)} = M_m + N$. Furthermore, the amplitude $\Psi_m^{(+)}$ vanishes whenever the coordinate z_j of any of the N electrons approaches z_0. Thus, at z_0 the effect of the solenoid is to *deplete* the charge density. Hence, this state can be regarded as a *quasihole*. Naturally, since the total charge is the same as in the Laughlin state and since the charge density away from z_0 should be uniform, the only place where the charge missing from z_0 could have gone to is infinity – or, rather, the physical boundary of the system. Thus, the solenoid causes the electron liquid to swell and to spill over the region it had occupied before the solenoid was introduced.

The quasihole excitation energy ϵ_0 can be calculated using the plasma analogy. I will not describe this calculation here, since it demands getting into a very technical plasma calculation that is better described elsewhere. The computation is given in considerable detail by Laughlin in his excellent review on the FQHE (Laughlin,

1987). What will matter, for the purposes of our discussion, is that the excitation energy is finite and has a finite limit as $N \to \infty$. Thus, the spectrum of quasiholes has an energy gap ϵ_0.

The *charge* q_0 of the quasihole can also be determined using the plasma analogy. It turns out that q_0 is a fraction of the electron charge, namely $q_0 = +e/m$. The argument goes as follows. The normalization of the quasihole wave function is

$$|\Psi_m^{(+)}(z_0; z_1, \ldots, z_N)|^2 = \Pi_{j=1}^N |z_j - z_0|^2 |\Psi_m(z_1, \ldots, z_N)|^2 \tag{13.25}$$

We can rewrite this expression in terms of a modified classical potential energy $U(z_0; z_1, \ldots, z_N)$, which has the simple form

$$U(z_0; z_1, \ldots, z_N) = U(z_1, \ldots, z_N) - \frac{2}{m} \sum_{j=1}^{N} \ln|z_j - z_0| \tag{13.26}$$

where $U(z_1, \ldots, z_N)$ is the classical potential energy for the one-component plasma.

The potential energy $U(z_0; z_1, \ldots, z_N)$ represents a classical one-component plasma interacting with a charge $-1/m$, which is held fixed at $z = z_0$. The most important properties of a plasma are its uniform density (in the absence of external probes) and the exact screening of all external probes. Since the external probe has charge $-1/m$, it repels the charges of the plasma within a distance ξ, which is the plasma screening length. For $|z - z_0| < \xi$, the plasma density is suppressed by the repulsive force due to the probe. The amount of charge expelled from the vicinity of z_0 is equal to $-1/m$, so there is a *missing charge* of $+1/m$, which neutralizes the charge of the probe. This behavior is indeed seen in detailed calculations, such as the ones reported by Laughlin (1987). Thus, the quasihole behaves like a *positive* charge $q_0 = +e/m$. Away from the quasihole, the charge density is uniform and equal to its value in the absence of the quasihole. Where has the missing charge gone? To the boundary, of course! Indeed, if the N-particle system occupies an area of radius R in the absence of the quasihole, its presence forces the liquid to expand from R to $R + \delta R$. The extra area occupied by the deformed liquid is $\pi(R + \delta R)^2 - \pi R^2$. Since R is large, the density is uniform and equal to $1/(2\pi m l_0^2)$. The radius R has to grow just enough to accommodate the extra charge $1/m$. Thus, we get the relation

$$[\pi(R + \delta R)^2 - \pi R^2] \frac{1}{2\pi m l_0^2} = \frac{1}{m} \tag{13.27}$$

where R/l_0 is given by

$$\frac{R}{l_0} = \sqrt{2mN} \tag{13.28}$$

The total change δR of the radius is

$$\frac{\delta R}{2l_0} = \sqrt{mN+1} - \sqrt{mN} \tag{13.29}$$

By inspecting the expansions in single-particle wave functions both of the Laughlin state and of the quasihole, we see that the highest single-particle angular momentum which enters into the Laughlin state has angular momentum equal to mN. For the quasihole the highest occupied state has angular momentum $mN + 1$. Indeed, the change δR of the radius is exactly the amount necessary to include the $(mN + 1)$th state inside the region occupied by the liquid. On the other hand, had we added or extracted a whole *particle* from the liquid ($N \to \pm 1$), the change in the area would have been m times the amount we just calculated. This can be seen quite easily in the expansion of the Laughlin state in single-particle Landau states. We conclude that the quasihole has *fractional charge* $+e/m$.

The quasi-electron can be constructed (qualitatively) in a similar manner. Instead of adiabatically introducing a solenoid that increases the local magnetic field, the solenoid now carries a flux that decreases the field by exactly one flux quantum. An argument along the lines of what we did above for the quasihole shows that a solenoid carrying a negative flux decreases the angular momentum of each single-particle state by one unit. Except for the state with angular momentum zero, which gets shifted to a state on the first excited Landau level, the addition of a solenoid with negative flux is equivalent to a downwards shift of the angular momentum of all single-particle states by one unit. At the level of the Laughlin wave function, this is accomplished by a derivative operator that acts on the polynomial factor in the wave function (Laughlin, 1983),

$$\Psi_m^{(-)}\left(z_0; \{z_j\}\right) = \exp\left(-\sum_{j=1}^{N} \frac{|z_j|^2}{4l_0^2}\right) \prod_{j=1}^{N} \left(2\frac{\partial}{\partial z_j} - \frac{\bar{z}_0}{l_0^2}\right) \prod_{1\leq j<k\leq N} (z_j - z_k)^m \tag{13.30}$$

The same line of argument as that used above on the quasihole shows that the charge q_0 of the quasi-electron is also fractional, but negative, $q_0 = -e/m$.

The construction of the quasihole, as well as that of the quasi-electron, has a strong resemblance to the construction of soliton states in 1D systems in quantum field theory (Jackiw and Rebbi, 1976) and in 1D condensed matter systems (Su and Schrieffer, 1981; Heeger *et al.*, 1988) (see Chapter 16). However, these two problems are qualitatively different. In fact, the Laughlin states either are non-degenerate, as in the case of a spherical geometry, or have a degeneracy of topological origin, as in the case of a torus. In contrast, the 1D systems which have solitons have ground states that spontaneously break a (discrete) global symmetry.

The degeneracy of their ground states is a consequence of this phenomenon. Nevertheless, the operator which introduces an extra solenoid has some of the characteristic features of a soliton operator. While the short-distance details are unimportant, the topological property of the extra vector potential (i.e. the line integral on a non-contractible loop) is the only essential property of the "solenoid" or quasihole operator. In fact, the addition of the solenoid changes the value of the circulation of the vector potential around the physical boundary of the system. In turn, this change determines the amount of charge which is "spilled over the edge." This extra charge becomes an excitation of the states at the edge of the system.

The quasiholes and quasi-electrons cannot be made in isolation directly by just adding or subtracting electrons. As a matter of fact a *hole* (not a quasihole) requires the removal of a full electron, which carries integer charge. Thus, electrons and holes are equivalent to bound states of m fractionally charged quasiparticles. For certain definite electron densities, the excess electrons which cannot be accommodated into a $1/m$ Laughlin state can be placed into a generalized Laughlin state. The excess electrons can be regarded as bound states of quasiholes or quasi-electrons, which, if their number is right, can form a Laughlin state. But this is a Laughlin state for anyons, not electrons. This mechanism is known as the hierarchy scheme of Haldane and Halperin.

The construction of the quasihole also suggests a different interpretation of the Laughlin wave function as well as generalizations that are valid for other filling fractions. Let us write the Laughlin wave function Ψ_m in the following suggestive form due to J. Jain (Jain, 1989a, 1989b, 1990):

$$\Psi_m(z_1, \ldots, z_N) = \prod_{1 \le j < k \le N} (z_j - z_k)^{m-1} \Psi_1(z_1, \ldots, z_N) \tag{13.31}$$

The factor $\Psi_1(z_1, \ldots, z_N)$ is just the wave function for N particles exactly filling up the lowest Landau level. Following the construction of the quasihole, the factor in front of Ψ_1 is interpreted as the result of having attached a solenoid to each particle. The flux carried for each solenoid is equal to $(m - 1)$ flux quanta. Unlike in the quasihole construction, the solenoids are physically attached to the particles, which fill up the Landau level and move around with them. This factorization, which appears to be quite innocent, has the virtue (and the beauty) of bringing the fractional and integer Hall states together. It is also telling us that the Laughlin state can be viewed as the result of a dynamical generation of a local gauge field that generates the solenoids which partially screen the external magnetic flux. In fact, the amount of screening is sufficient to turn the fractional filling of a Landau level of the bare field into the complete filling of a Landau level of the unscreened part of the field. Later in this chapter we will see that this is the starting point of the field-theoretic description of the FQHE.

In summary, the $1/m$ Laughlin states are seen to have quasihole and quasi-electron excitations that have fractional charge $\pm e/m$ and fractional statistics $\pm\pi/m$. These quasiparticles are obtained by the *adiabatic* addition or removal of infinitesimally thin solenoids carrying one flux quantum. The adiabatic nature of this process is essential to this construction, since it is necessary to make the fluid swell enough to include one additional Landau orbit without promoting electrons to higher Landau levels or producing ripples in the fluid. All these bulk excitations have finite energy gaps. This is required by the incompressibility of the fluid, which guarantees the accuracy of the adiabatic process.

13.1.4 The statistics of quasiparticles in Laughlin's theory

In this section we will discuss the statistics of the quasiholes within the first-quantized picture of the FQHE. In the last section of this chapter we will return to this problem and derive the main results directly from the field theory. The statistics of the quasi-electron can also be discussed along very similar lines.

The quasihole wave function discussed in Section 13.1.3 is given up to a normalization factor. For a single quasihole, the amplitude of this wave function is not very important. However, at the moment we wish to construct a wave function for two or more quasiholes, the normalization begins to play a rather subtle but important role. During a process that involves dragging a single quasihole very slowly around a closed loop, the phase of the quasihole wave function undergoes very important changes. Indeed, since the quasihole carries an electric charge of $-e/m$, we should expect an Aharonov–Bohm effect $1/m$ times smaller than the value for electrons. In fact, the Aharonov–Bohm effect is perhaps the "operationally correct" way of measuring the charge of a quasiparticle.

The quasihole wave function is physically appealing, but it has several drawbacks. Consider, for example, a naively constructed wave function for two quasiholes located at $z = u$ and $z = w$, respectively,

$$\Psi^{(+)}(u, w, z_1, \ldots, z_N) = N(u, w) \prod_{j=1}^{N} (z_j - u)(z_j - w)\Psi_m(z_1, \ldots, z_N) \quad (13.32)$$

The factor $N(u, w)$ has a subtle origin. On the one hand, it can be regarded as the normalization constant for the state with two quasiholes. However, if that were indeed the case, N would have to be a function not only of u and w, but also of $\bar{u}$ and $\bar{w}$, and it would not be analytic. More importantly, this amplitude has to be determined from the requirement that it represents the *physical* process of adiabatic insertion of two thin solenoids. In Section 13.1.3 we saw that the form of the wave function for one quasihole was *suggested* by the observation that the *adiabatic* insertion of a solenoid carrying one flux quantum implied an increase of the angular momentum *relative to the location of the solenoid* by one unit per particle. We also argued that the quasihole carries charge e/m. Later in this section we will give a path-integral argument to support this picture.

But let us assume that we have already manufactured one quasihole, which is sitting at $z = u$. We now want to create another quasihole, but this time at $z = w$. The *adiabatic* addition of the extra solenoid must change the angular momentum of the particle also by one unit, but this time the angular momentum is measured *relative* to w, not to u. Furthermore, since the quasihole carries electric charge equal to e/m, as we drag one quasihole slowly around the other we should pick up an extra Aharonov–Bohm phase factor. This phase factor should correspond to an Aharonov–Bohm effect for a charge equal to e/m (Kivelson and Roček, 1985).

We are going to determine the amplitude (or "normalization constant") $N(u, w)$ by demanding that the following conditions are met: (a) the wave function should be an analytic function of the coordinates of the electrons $\{z_1, \ldots, z_N\}$ and of the quasiholes u and w up to exponential factors; and (b) the normalization of this wave function should be invariant under translations, i.e. it should be a function of differences of the coordinates $\{z_1, \ldots, z_N, u, w\}$. The analyticity condition is just the requirement that the wave function should have contributions only from the lowest Landau level. These conditions, as well as the solution, were first proposed by Halperin (1983, 1984).

The normalization of the state (or, rather, the probability density with two quasiholes at coordinates u and w) is

$$|\Psi_m^{(+)}(u, w, z_1, \ldots, z_N)|^2 = e^{-\beta U_{\text{eff}}(u,w,z_1,\ldots,z_N)} \tag{13.33}$$

where the effective potential U_{eff} is given by

$$U_{\text{eff}}(u, w, z_1, \ldots, z_N) = U(z_1, \ldots, z_N) - \frac{2}{m}\sum_{j=1}^{N}(\ln|u - z_j| + \ln|w - z_j|) + \frac{2}{m}\ln|N(u, w)| \tag{13.34}$$

The translation invariance and analyticity requirements are met by choosing the factor $N(u, w)$ to be

$$N(u, w) = N_0(u - w)^{1/m}\exp\left(-\frac{|u|^2 + |w|^2}{4l_0^2 m}\right) \tag{13.35}$$

With this choice, the Halperin wave function for two quasiholes is

$$\Psi_m^{(+)}(u, w; \{z_j\}) = N_0(u-w)^{1/m} \prod_{j=1}^{N} [(u-z_j)(w-z_j)] \\ \times \exp\left(-\frac{1}{4ml_0^2}(|u|^2+|w|^2)\right) \Psi_m(z_1, \ldots, z_N) \tag{13.36}$$

and the effective potential U_{eff} is

$$U_{\text{eff}}(u, w; \{z_j\}) = -2 \sum_{1\leq j<k\leq N} \ln|z_j - z_k| \\ - \frac{2}{m}\sum_{j=1}^{N}(\ln|z_j-u| + \ln|z_j-w|) - \frac{2}{m^2}\ln|u-w| \\ + \frac{1}{2ml_0^2}\sum_{j=1}^{N}|z_j|^2 + \frac{1}{2m^2l_0^2}(|u|^2+|w|^2) \tag{13.37}$$

In plasma language, this is the potential energy of a set of N classical particles (each carrying charge (-1)) at sites $\{z_1, \ldots, z_N\}$ interacting with two extra particles (each with charge $-1/m$) at u and w. All $N+2$ charges are coupled to a neutralizing background charge of density $1/(2\pi m l_0^2)$. The manifest translation invariance of U_{eff} takes care of the translation invariance requirement.

The wave function for two quasiholes is a multivalued function of the complex coordinates of the two quasiholes. As a result, if the quasiholes undergo a counterclockwise exchange process, defined as a counterclockwise rotation by π of one quasihole around the other followed by a translation that restores the relative position of the quasiholes, *the phase of this wave function changes by* π/m,

$$\Psi_m^{(+)}(u, w; \{z_j\}) = e^{+\frac{i\pi}{m}}\Psi_m^{(+)}(w, u; \{z_j\}) \tag{13.38}$$

Thus, the quasiholes are *anyons* with statistical angle $\delta = \pi/m$ relative to bosons or $\delta = ((m-1)/m)\pi$ relative to fermions. This remarkable result suggests that the FQHE can be described in terms of a theory of either bosons or fermions coupled to a hidden (or dynamically generated) Chern–Simons gauge field. In the next section we will describe both a "Landau–Ginzburg" approach to the FQHE and a field theory that are based on this idea.

We conclude that the quasiholes of the Laughlin state carry fractional charge $+e/m$ and have fractional statistics π/m. This is a very striking result. Arovas, Schrieffer, and Wilczek (Arovas *et al.*, 1984) have given an alternative derivation of both results using an argument based on the concept of Berry phases (Simon, 1983; Berry, 1984). Rather than following that path, we will now construct a path integral to represent the motion of the quasiholes. The key ingredient of our construction is the observation that the quasihole wave functions are *coherent states* (Kivelson *et al.*, 1987). Thus, we can adapt the formalism described in Chapter 7 to construct the path integral for spin-S particles, to treat the quantum dynamics of the quasiholes. The reader should keep in mind that the following arguments are heuristic at best. In Sections 13.8 and 13.9 I give a different derivation of the same result, which is based on the field-theory approach to the FQHE. Naturally, the results agree!

Let us begin with the wave function for a *single* quasihole. Let us define the state $|z\rangle$ as

$$|z\rangle = \exp\left(-\frac{1}{4ml_0^2}|z|^2\right)\prod_{j=1}^{N}(z_j - z)|m\rangle \tag{13.39}$$

where $|m\rangle$ is the Laughlin state. The set of states $\{|z\rangle\}$ is over-complete (Laughlin, 1987). The overlap between two states $|z\rangle$ and $|w\rangle$ is

$$\langle z|w\rangle = \exp\left(-\frac{1}{4ml_0^2}(|z|^2 + |w|^2)\right)\langle m|\prod_{j=1}^{N}[(\bar{z}_j - \bar{z})(w_j - w)]|m\rangle \tag{13.40}$$

Except for the exponential factor, $\langle z|w\rangle$ is an analytic function of $\bar{z}$ and w separately. Thus, $\langle z|w\rangle$ can be related to $\langle z|z\rangle$ by analytic continuation (Laughlin, 1987). The result is that the inner product is given by

$$\langle z|w\rangle = \exp\left(-\frac{1}{4ml_0^2}(|z|^2 + |w|^2) + \frac{1}{2ml_0^2}\bar{z}w\right)\langle z|z\rangle \tag{13.41}$$

Indeed, the translation invariance of the 2D one-component plasma guarantees that the overlap $\langle z|z\rangle$ is just a constant that is independent of z. Also, up to a normalization constant we can write the resolution of the identity

$$1 = \mathcal{N}\int |z\rangle\langle z| d^2z \tag{13.42}$$

We consider now a process in which we prepare the quasihole in a coherent state $|z_0\rangle$ at time $t = t_0$. We now ask for the quantum-mechanical amplitude $\langle z_0, t_0 + T|z_0, t_0\rangle$ for the quasihole to return to $|z_0\rangle$ after a very long time T. Upon inserting the resolution of identity at $\mathcal{N}_\tau$ intermediate times $t_n = t_0 + n\,\Delta t$ in the limit $\mathcal{N}_\tau \to \infty$ and $\Delta \to 0$ with $T = \mathcal{N}_\tau\,\Delta t$ fixed, we can write

$$\langle z_0, T + t_0|z_0, t_0\rangle = \mathcal{N}\int \prod_{n=1}^{\mathcal{N}_\tau} d^2z_n \prod_{n=1}^{\mathcal{N}_\tau}\langle z_n|z_{n+1}\rangle \tag{13.43}$$

where $z_n = z(t_0 + n\,\Delta t)$ and $z_{\mathcal{N}_\tau} = z_0$.

In the limit $\Delta t \to 0$ we can approximate the overlaps by the expression

$$\langle z_{n+1}|z_n\rangle \approx \langle z_n|z_n\rangle \exp\left[\frac{1}{4ml_0^2}\left(\bar{z}_n \frac{dz_n}{dt} - z_n \frac{d\bar{z}_n}{dt}\right)\Delta t\right] \tag{13.44}$$

Thus, the path integral for one quasihole is

$$\langle z_0, T + t_0|z_0, t_0\rangle = \mathcal{N} \int \mathcal{D}z \, \exp\left(\frac{1}{2ml_0^2}\int_0^T dt \, \bar{z}\frac{dz}{dt}\right) \tag{13.45}$$

By expanding the exponent in its real and imaginary components we get the identity

$$\frac{1}{2ml_0^2}\int_0^T dt \, \bar{z}\frac{dz}{dt} = i\frac{e/m}{\hbar c}\oint_\Gamma \vec{A}(\vec{x})\cdot d\vec{x} \tag{13.46}$$

where $\vec{A}$ is the vector potential for the field B in the isotropic gauge and Γ is the path. Thus, the amplitude is given by the path integral

$$\langle z_0, T + t_0|z_0, t_0\rangle = \mathcal{N} \int \mathcal{D}\vec{x} \, \exp\left(i\frac{e/m}{\hbar c}\oint_\Gamma d\vec{x}\cdot\vec{A}(\vec{x})\right) \tag{13.47}$$

which is just the path integral for a particle of mass M and charge e/m moving in the field $B = \vec{\nabla}\times\vec{A}\cdot\hat{z}$ in the limit $M \to 0$. This limit is just the projection onto the lowest Landau level. Notice that the normalization constants $\langle z|z\rangle$ have been absorbed into the uninteresting factor $\mathcal{N}$. The amplitude for the path Γ of this path integral is just the Aharonov–Bohm phase factor (Arovas *et al.*, 1984). At the end of this chapter we give a derivation of this result that is based on the field-theory approach, which does not require the choice of a set of wave functions with a specific form.

Let us briefly discuss the generalization of this result for the problem of two quasiholes. Let us assume that at some initial time t_0 the quasiholes are prepared in the state $|z_0, w_0\rangle$. Once again we ask for the amplitude $\langle z_0', w_0'; t_0 + T|z_0, w_0; t\rangle$ after a very long time T. The normalized two-quasihole states $|z, w\rangle$ will be taken to be of the Halperin form. The derivation for two quasiholes follows quite closely the arguments given for one quasihole. However, the two results differ in two important aspects: (a) the multivalued phase factors $(z - w)^{1/m}$ lead to an "induced" gauge interaction, and (b) the diagonal overlaps are no longer constant but functions of $|z - w|$. The final result is

$$\langle \vec{z}_0', \vec{w}_0'|\vec{z}_0, \vec{w}_0\rangle = \mathcal{N}\int \mathcal{D}\vec{z}\,\mathcal{D}\vec{w}\, \exp\left(\frac{i}{\hbar}S_{\text{eff}}^{(2)}(\vec{z}, \vec{w})\right) \tag{13.48}$$

where $S_{\text{eff}}^{(2)}(\vec{z}, \vec{w})$ is the effective action for two quasiholes.

The integration measure, denoted here by $\mathcal{D}\vec{z}\ \mathcal{D}\vec{w}$, has absorbed the diagonal overlaps $\prod_n \langle \vec{z}_n, \vec{w}_n | \vec{z}_n, \vec{w}_n \rangle$. Laughlin (1987) has shown that these factors are constant at long distances but vanish at short distances like $|\vec{z} - \vec{w}|^{2/m}$. Thus, their main effect is to remove from the path integral the paths in which the particles get to be too close to each other. This feature of the integration measure is essential, since fractional statistics cannot be defined if the paths of the particles are allowed to cross.

The effective action $S_{\text{eff}}^{(2)}$ for two holes is

$$S_{\text{eff}}^{(2)}(\vec{z}, \vec{w}) = \int_{t_0}^{t_0+T} dt \left\{ \frac{d\vec{z}}{dt} \cdot \left(\frac{e/m}{c} \vec{A}(\vec{z}) + \frac{\hbar}{m} \vec{\mathcal{A}}(\vec{z} - \vec{w}) \right) + \frac{d\vec{w}}{dt} \cdot \left(\frac{e/m}{c} \vec{A}(\vec{w}) + \frac{\hbar}{m} \vec{\mathcal{A}}(\vec{w} - \vec{z}) \right) \right\} \tag{13.49}$$

where m is the index of the Laughlin state (not to be confused with a mass!) and $\vec{A}$ is the electromagnetic vector potential. The "induced" vector potential $\vec{\mathcal{A}}$ arises from the multivalued factors. It is given by the total change of phase accumulated during the process, i.e.

$$\frac{1}{m} \int_{t_0}^{t_0+T} dt \left(\vec{\mathcal{A}}(\vec{z} - \vec{w}) \cdot \frac{d\vec{z}}{dt} + \vec{\mathcal{A}}(\vec{w} - \vec{z}) \cdot \frac{d\vec{w}}{dt} \right) = \frac{1}{m} [\arg(z_0' - w_0') - \arg(z_0 - w_0)] \tag{13.50}$$

This equation requires only that the "induced" vector potential $\vec{\mathcal{A}}$ give the correct winding number. It is clear that $\vec{\mathcal{A}}$ can be represented by an effective Chern–Simons gauge field with an appropriately chosen coupling constant. One possible choice for $\vec{\mathcal{A}}$ was given by Arovas, Schrieffer, and Wilczek (Arovas *et al.*, 1984) (in the isotropic gauge),

$$\mathcal{A}_j(\vec{z} - \vec{w}) = \frac{\epsilon_{jk}(z - w)_k}{|\vec{z} - \vec{w}|^2} \tag{13.51}$$

which has the quantized circulation

$$\oint_{C[\vec{w}]} \mathcal{A}_j(\vec{z} - \vec{w}) dz_j = 2\pi \tag{13.52}$$

for any closed path $C[\vec{w}]$ that encloses the point $\vec{w}$.

Hence, each quasihole carries a solenoid with just one flux quantum. In agreement with our discussion of Section 10.5, these *"induced" or statistical gauge fields* change the statistics of the quasiparticles. In the problem of spinons in the chiral spin state (see Section 10.4) the quasiparticles are semions or half-fermions. The quasiholes of the FQHE have statistical angle equal to π/m. This property can be seen very directly from the coherent-state path integral. Let us consider a process in which two quasiholes undergo a counterclockwise exchange, during which $\Delta\left[\arg(z_0 - w_0)\right] = \pi$. The amplitude of the path integral picks up a phase of $e^{i\pi/m}$. Below, when we derive the Laughlin theory from a field theory, we will see that these phase factors arise directly from a Chern–Simons gauge field.

13.2 Composite particles

In the past sections we discussed the first quantization approach to the FQHE. Here we will discuss an alternative approach that is based on a special form of field theory, the Chern–Simons theory, which we discussed extensively in Chapter 10 in the context of theories of anyons. Here we will show that the Chern–Simons theory is quite useful from two different points of view: (a) as a Landau–Ginzburg theory for the long-distance phenomenology and (b) as a way to derive the Laughlin state from a microscopic theory. For reasons of space and conciseness, in this chapter I will discuss only the simplest case of fully polarized (i.e. "spinless") electrons. Also I will restrict myself for now to the theory of the Laughlin sequence and to the first level of the hierarchy. Generalizations of this theory are discussed in Chapter 14.

In Section 13.1.3 we saw that the construction of the state for the quasihole suggested a different interpretation of the Laughlin wave function that was first proposed by Jain. This structure of the state for the quasihole gave rise to the picture of the FQHE as a ground state of "electrons bound to fluxes." From this point of view, all that the long-range correlations do is make it possible for the electrons to "nucleate" flux. Jain (1989a) realized that, in the Laughlin state, the electrons nucleate enough flux that the bound states *exactly* fill up an integer number of the Landau levels of the *unscreened* part of the field. In this formulation, the FQHE is an integer quantum Hall effect of the bound states. Jain proposed writing the Laughlin wave function in the suggestive factorized form

$$\Psi(z_1, \ldots, z_N) = \prod_{i<j} (z_i - z_j)^{m-1} \chi_1(z_1, \ldots, z_N) \tag{13.53}$$

where χ_1 is the wave function for a completely filled lowest Landau level

$$\chi_1(z_1, \ldots, z_N) = \prod_{i<j} (z_i - z_j) \exp\left(-\sum_{i=1}^{N} \frac{|z_i|^2}{4\ell^2}\right) \tag{13.54}$$

The phases associated with the factor multiplying χ_1 can be regarded as representing an even number $(m - 1)$ of fluxes that are attached to each coordinate z_i where an electron is present. It is a crucial feature of this picture that the electrons bind to an *even* number of flux quanta and, in this way, they retain their fermion character. We will also see below that this approach has allowed a simple description of the so-called hierarchy states in terms of wave functions that have a factorized structure.

In Chapter 10 we saw that there is a natural and local way to attach particles and fluxes together: the Chern–Simons gauge theory. Girvin and MacDonald (1987) were the first to propose that the Laughlin state had a hidden form of off-diagonal long-range order (ODLRO). They suggested an order parameter for the Laughlin state, but it turned out to be non-local. As a matter of fact, the Girvin–MacDonald order parameter is closely related to the anyon operators constructed in Chapter 10. We also saw that it is always possible to map *any* 2D fermion system into an equivalent problem with arbitrarily chosen statistics. We are going to use this mapping in two different ways: as a mapping (a) to a theory of bosons and (b) to a theory of fermions (each coupled to a Chern–Simons gauge field with a suitably chosen coupling constant).

The Girvin–MacDonald argument that the Laughlin state has a hidden form of ODLRO goes as follows. The ground-state correlation function $\rho(z, z')$ for the electron operator (also called the one-particle density matrix) in the mth Laughlin state for a system with N particles $|0_m; N\rangle$ is given by the expansion

$$\rho(z, z') \equiv \langle 0_m; N|\hat{\psi}^\dagger(z)\hat{\psi}(z')|0_m; N\rangle = \sum_{n,k} \varphi_n^*(z)\varphi_k(z')\langle 0_m; N|\hat{\psi}_n^\dagger\hat{\psi}_k|0_m; N\rangle \tag{13.55}$$

where $\{\varphi_n(z)\}$ is the set of one-particle wave functions of the lowest Landau level (see Section 12.1), and n and k run over all the occupied states.

Since the states $\{\varphi_n(z)\}$ all have different values of angular momentum, the expectation value $\langle 0_m|\hat{\psi}_n^\dagger\hat{\psi}_n|0_m; N\rangle$ in an isotropic uniform state, such as the Laughlin state, takes the very simple form

$$\langle 0_m; N|\hat{\psi}_n^\dagger\hat{\psi}_k|0_m; N\rangle = \nu\delta_{nk} \tag{13.56}$$

where ν is the filling fraction. The correlation function $\rho(z, z')$ can be shown to be given by (Girvin and Jach, 1984)

$$\rho(z, z') = \frac{\nu}{2\pi}\exp\left(-\frac{|z - z'|^2}{4\ell^2} + \frac{1}{4\ell^2}(z^*z' - z'^*z)\right) \tag{13.57}$$

This identity shows that the one-particle electron-correlation function decays exponentially fast in a Laughlin ground state.

Consider now the *composite operator* $\hat{K}(z)$, which was introduced by Read (1989) (see also Rezayi and Haldane (1988)), who refined the arguments of Girvin and MacDonald. The operator $\hat{K}(z)$, which creates one electron, together with a solenoid carrying m flux quanta, at point z is

$$\hat{K}(z) = \hat{\psi}^\dagger(z)\hat{U}^m(z) \tag{13.58}$$

where $\hat{U}(z)$ is the second-quantized operator that creates a quasihole at z.

Each quasihole has charge $1/m$ and fractional statistics π/m, and m quasiholes have charge 1 and statistics π. Thus, m holes have the same quantum numbers as a *missing electron*. Furthermore, the operator $\hat{K}(z)$ obeys *bosonic* commutation relations. This implies that the operator $\hat{K}(z)$ must have a non-vanishing expectation value in a ground state with an indefinite number of particles. This property is indeed strongly reminiscent of Bose condensation. More precisely, Read showed that the following identity holds (Read, 1989):

$$\langle 0_m; N|\hat{K}^\dagger(z)\hat{K}(z')|0_m; N\rangle = \frac{1}{\rho_0}\langle 0_m; N+1|\hat{\rho}(z)\hat{\rho}(z')|0_m; N+1\rangle \to \rho_0 \tag{13.59}$$

where $\hat{\rho}(z)$ is the density operator and its expectation value is $\rho_0 = 1/(2\pi m)$. Thus, there is ODLRO in the Laughlin state.

Since ODLRO is the hallmark of superfluidity, its existence suggested the idea that there should be a Landau–Ginzburg theory for the FQHE. However, unlike superfluids, the Laughlin state is an incompressible state and it does not have excitations with arbitrarily low energy (in the bulk!). So, whatever the Landau–Ginzburg theory happens to be, it cannot describe a system with any Goldstone modes. Now, a system with an order parameter that is complex, as the Girvin–MacDonald order parameter is, in principle should have Goldstone modes, unless the order parameter is coupled to a fluctuating gauge field. In this case, the gauge field would "eat" the Goldstone mode and, at the same time, become massive. Hence, there would not be any gapless modes left. This phenomenon, which is usually called the Higgs mechanism, does take place in charged superfluids, i.e. superconductors. This is the Meissner state of a superconductor.

The problem with this picture is that the fractional quantum Hall fluid is not a superconducting state! As we will see, although the Landau–Ginzburg theory of the fractional quantum Hall state is strongly reminiscent of (and suggested by) the physics of a superconductor, it is a theory with a dynamical gauge field that "eats" the would-be Goldstone boson, leaving behind nothing to be eaten by the electromagnetic gauge field. As a consequence, there is no flux expulsion in the Laughlin state and no Meissner effect. Furthermore, the absence of a Goldstone mode from the spectrum of the fractional quantum Hall state also implies that it does not support a Josephson effect, the physical signature of superconducting coherence. Moreover, the non-locality of the Girvin–MacDonald order parameter is clearly indicating that a naive application of the Higgs mechanism is not possible. So the gauge fields have to arise from the fluctuations about the Laughlin ground state rather than coming from "honest-to-god" electromagnetism. In other words, the gauge field in question has to be self-generated by the correlations that describe this phase of matter. Furthermore, since the Laughlin state is not a superconductor, the mechanism for generation of mass (or gaps) to all excitations should be manifestly gauge-invariant. This fact suggested to Girvin and MacDonald that the gauge field should have a Chern–Simons form.

13.3 Landau–Ginzburg theory of the fractional quantum Hall effect

The methods that we have discussed for the field-theoretic treatment of anyons can also be used to study the FQHE. Zhang, Hansson, and Kivelson (Zhang *et al.*, 1989) used a mapping to *bosons* in terms of a Chern–Simons gauge field. This procedure allowed them to derive the qualitative features of a Landau–Ginzburg theory for the FQHE. Their Landau–Ginzburg approach, which is valid at low energies and long distances, qualitatively confirmed the idea that the FQHE had a hidden form of ODLRO without Goldstone bosons. Read (1989) gave a careful derivation of the Landau–Ginzburg theory directly from the Laughlin wave function.

Let us use now the methods of Chapter 10 to derive the Landau–Ginzburg theory. Consider once again a system of N electrons moving on a plane in the presence of an external uniform magnetic field B perpendicular to the plane. The electrons will be assumed to have an inter-particle interaction governed by a pair potential $V(|\vec{r}|)$, for two electrons separated a distance $|\vec{r}|$ on the plane. The magnetic field will be assumed to be so large that the system is completely polarized and that we can ignore the spin degrees of freedom. The eigenstates $\Psi(\vec{x}_1, \ldots, \vec{x}_N)$ are eigenfunctions of the (first-quantized) Hamiltonian $\hat{H}$

$$\hat{H} = \sum_{i=1}^{N} \left\{ \frac{1}{2M} \left(\vec{p}_j - \frac{e}{c} \vec{A}_j(\vec{x}_j) \right)^2 + eA_0(\vec{x}_j) \right\} + \sum_{i<j} V(|\vec{x}_i - \vec{x}_j|) \tag{13.60}$$

where we have included the coupling both to the electromagnetic vector potential $\vec{A}$ and to the scalar potential A_0. Hence, we are dealing with N spinless fermions of charge $-e$ and mass M. In second-quantized notation, the electron operator is $\psi(x)$ and the dynamics of the system is governed by the action $\mathcal{S}$

$$\begin{aligned} \mathcal{S} = & \int d^3z \left\{ \psi^*(z)[iD_0 + \mu]\psi(z) + \frac{\hbar^2}{2M} |\vec{D}\psi(z)|^2 \right\} \\ & - \frac{1}{2} \int d^3z \int d^3z' (|\psi(z)|^2 - \rho_0) V(|\vec{z} - \vec{z}\,'|)(|\psi(z')|^2 - \rho_0) \end{aligned} \tag{13.61}$$

where ρ_0 is the average density. The quantum partition function $\mathcal{Z}$ for this system is (at zero temperature and in real time)

$$\mathcal{Z} = \int \mathcal{D}\psi^* \mathcal{D}\psi \; e^{\frac{i}{\hbar}\mathcal{S}} \tag{13.62}$$

13.3.1 Composite bosons

In Chapter 10 we showed that a system of fermions in two dimensions is equivalent to a system of ("composite") bosons coupled to a Chern–Simons gauge field $\mathcal{A}_\mu$. The action $\mathcal{S}_\text{B}$ for the Bose system is

$$\begin{aligned}\mathcal{S}_\text{B} = &\int d^3z \left\{\phi^*(z)[iD_0 + \mu]\phi(z) + \frac{1}{2M}|\vec{D}\phi(z)|^2 + \frac{\theta}{4}\epsilon_{\mu\nu\lambda}\mathcal{A}^\mu \mathcal{F}^{\nu\lambda}\right\} \\ &- \frac{1}{2}\int d^3z \int d^3z' (|\phi(z)|^2 - \rho_0) V(|\vec{z} - \vec{z}\,'|)(|\phi(z')|^2 - \rho_0)\end{aligned} \tag{13.63}$$

The covariant derivatives D_μ in this action contain both electromagnetic and Chern–Simons gauge fields, i.e.

$$D_\mu = \partial_\mu + i\frac{e}{\hbar c}A_\mu + i\mathcal{A}_\mu \tag{13.64}$$

In Eq. (13.63) $\phi(z)$ is the Bose field, $\theta = 1/(2\pi n)$, and n, for the moment, is an arbitrary *odd* integer. The reader should note that the Chern–Simons coupling constant $\theta = 1/(2\pi n)$ that we have to use for the flux-attachment transformation formally violates the requirement that it be quantized in order for the theory to be gauge-invariant on closed surfaces. The formulation that we are using here (and in the fermionic version as well) is correct only for a theory defined on a surface with the topology of a large disk. In Section 14.1.1 we will give a consistent formulation of flux attachment on a torus.

It is an implicit assumption of this theory that the bosons must have a hard core since, otherwise, the fractional-statistics transformation does not make sense. It is very difficult to keep track of this constraint in the continuum. On a lattice the hard-core constraint does not pose any serious problem. However, if we are interested *only* in the long-distance and low-energy behavior, we can replace the hard core by an effective short-distance repulsive force. This change amounts to adding an extra term $\mathcal{S}_\text{hc}$ to the action of the form

$$\mathcal{S}_\text{hc} = \int d^3z(-\lambda|\phi(z)|^4) \tag{13.65}$$

The total action is $\mathcal{S}_\text{eff} = \mathcal{S}_\text{B} + \mathcal{S}_\text{hc}$ and we have now a bosonic functional integral

$$\mathcal{Z} = \int \mathcal{D}\phi^* \mathcal{D}\phi \, \mathcal{D}\mathcal{A}_\mu \, \exp\left(\frac{i}{\hbar}\mathcal{S}_\text{eff}[\phi, \phi^*, \mathcal{A}_\mu; A_\mu]\right) \tag{13.66}$$

This functional integral can be regarded as a Landau–Ginzburg theory and was first proposed by Zhang, Hansson, and Kivelson (Zhang *et al.*, 1989). As in the Bogoliubov theory of the dilute Bose gas, the parameter λ cannot be calculated directly from this theory. Zhang, Hansson, and Kivelson dropped the repulsive-pair-potential term altogether and replaced it by the $|\phi|^4$ term. We now follow their treatment and extract the low-energy behavior.

13.3.2 Landau–Ginzburg theory

The effective theory looks like a theory of bosons coupled to a gauge field. In the absence of the gauge field, the bosons condense and spontaneously break the *global* U(1) phase symmetry

$$\phi(z) \to e^{i\alpha}\phi(z) \tag{13.67}$$

The system is then a superfluid and its spectrum has a massless excitation, namely the phase ω of ϕ, which is the Goldstone boson associated with the broken U(1) symmetry. We will see now that this Goldstone boson disappears from the spectrum once the system is coupled to the statistical gauge field.

Let us consider the behavior of the system in the semi-classical (mean-field) limit. In that limit, the fluctuations of the amplitude of the Bose field ϕ are small. Let us write ϕ in the form

$$\phi(z) = \sqrt{\rho(z)}e^{i\omega(z)} \tag{13.68}$$

The classical equations of motion of the Bose theory (i.e. the mean-field equations) are (in units such that $\hbar = c = e = 1$)

$$\begin{aligned}
(iD_0+\mu)\phi(x) - \frac{1}{2M}\vec{D}^2\phi(x) - 2\lambda|\phi(x)|^2\phi(x)& \\
-\phi(x)\int d^3x'\; V(x-x')(|\phi(x')|^2-\rho_0)^2 &= 0 \\
\theta\mathcal{B}(x) + |\phi(x)|^2 &= 0 \\
\theta\epsilon_{i\alpha\beta}\,\partial^\alpha \mathcal{A}^\beta + \frac{i}{2M}\left[\phi^*(x)D_i\phi(x) - (D_i\phi(x))^*\phi(x)\right] &= 0 \\
\int d^3x|\phi(x)|^2 &= \rho_0 L^2 T
\end{aligned} \tag{13.69}$$

where $D_0 = \partial_0 - i(A_0 + \mathcal{A}_0)$, $\vec{D} = \vec{\nabla} - i(\vec{A} + \vec{\mathcal{A}})$, L^2 is the area of the system, and T is the time span.

For a configuration ϕ with constant amplitude (the ground state) for a system with (areal) density ρ_0 these equations become

$$
\begin{aligned}
|\phi|^2 &= \rho_0 \\
\rho_0 + \theta \langle \mathcal{B} \rangle &= 0 \\
\mu - 2\lambda \rho_0 &= 0 \\
\langle \mathcal{A}_\mu \rangle + A_\mu &= 0 \\
\rho_0 - \frac{\nu}{2\pi \ell_0^2} &= 0
\end{aligned}
\tag{13.70}
$$

where $\ell_0 = 1/\sqrt{B}$ is the cyclotron length and $\nu = 1/m$ is the filling fraction.

Thus, the *average* statistical gauge field $\langle \mathcal{A}_\mu \rangle$ exactly cancels out, or *screens*, the electromagnetic field A_μ. Consequently, we get $\langle \mathcal{B} \rangle = -B$. However, the first of the equations of Eq. (13.70) requires the average statistical magnetic field to be proportional to the average particle density. Hence, the density and the field are not independent of each other but satisfy $\rho_0 = \theta B$. Recall the definition of the filling fraction $\nu = (\rho_0/B)\phi_0$, where ϕ_0 is the flux quantum $\phi_0 = hc/e$ (in standard units). Thus, the classical equations of motion have uniform solutions only if the filling fraction is $\nu = \theta/(2\pi) = 1/m$, with m an odd integer. We can then identify the odd integer m with the index of the Laughlin wave function, which is also odd. Thus, the Landau theory suggests the picture of the FQHE as a problem of bosons in an average magnetic field that is determined by the number of bosons! Notice that, with the identification of m as the index of the Laughlin wave function, the constraint implies that each boson is made of a fermion and m flux quanta. This is precisely what the arguments of Girvin and MacDonald, and Read told us.

13.3.3 Low-energy fluctuations

However, this story does not end at the level of mean-field theory. The fluctuations play a very important role in this problem. Mean-field theory told us that the average particle density and average statistical magnetic field are fixed. But the fluctuations of the phase ω appear to be completely unconstrained. In order to investigate this problem we need an effective action for the slow modes of the phase field. This effective action can be obtained by integrating out the amplitude fluctuations. Indeed, we can write the field ϕ in the form

$$
\phi(z) = \sqrt{\rho_0 + \delta\rho(z)}e^{i\omega(z)} \tag{13.71}
$$

The fluctuations of the gauge field are

$$
A_\mu + \mathcal{A}_\mu = \delta\mathcal{A}_\mu \tag{13.72}
$$

where we used that $A_\mu + \langle \mathcal{A}_\mu \rangle = 0$. We now substitute this expression back into the Landau–Ginzburg action to obtain

$$
\begin{aligned}
\mathcal{S}_{\text{eff}}[\delta\rho, \delta\mathcal{A}_\mu, \mu] = \int d^3x \Big\{ & \sqrt{\rho_0 + \delta\rho}e^{-i\omega} \left[i\,\partial_0 + \delta\mathcal{A}_0 + \mu \right] \sqrt{\rho_0 + \delta\rho}e^{i\omega} \\
& - \frac{1}{2M} \left| i\,\vec{\nabla}\left(\sqrt{\rho_0 + \delta\rho}e^{i\omega}\right) + \delta\vec{\mathcal{A}}\,\sqrt{\rho_0 + \delta\rho}e^{i\omega} \right|^2 \\
& - \lambda(\rho_0 + \delta\rho)^2 \Big\} \\
& - \frac{1}{2}\int dt \int d^2x \int d^2x'\, \delta\rho(x) V(\vec{x} - \vec{x}\,')\delta\rho(x') \\
& + \int d^3x\, \frac{\theta}{4}\epsilon_{\mu\nu\lambda}\,\delta\mathcal{A}^\mu\,\delta\mathcal{F}^{\nu\lambda}
\end{aligned}
\tag{13.73}
$$

We now expand the effective action in powers of the density fluctuation $\delta\rho(z)$ up to second order to get (using Eq. (13.70))

$$
\begin{aligned}
\mathcal{S}_{\text{eff}}[\delta\rho, \delta\mathcal{A}_\mu, \mu] \simeq \int d^3x & \left[-\frac{1}{8M\rho_0}\left(\vec{\nabla}\delta\rho\right)^2 - \lambda\,(\delta\rho)^2 \right] \\
& - \frac{1}{2}\int dt \int d^2x \int d^2x'\, \delta\rho(x) V(\vec{x} - \vec{x}\,')\delta\rho(x') \\
& + \int d^3x\, (\delta\rho)\, \left[(\delta\mathcal{A}_0 - \partial_0\omega) - \frac{1}{2M}\left(\vec{\nabla}\omega - \delta\vec{\mathcal{A}}\right)^2 \right] \\
& + \int d^3x\, \frac{\theta}{4}\epsilon_{\mu\nu\lambda}\,\delta\mathcal{A}^\mu\,\delta\mathcal{F}^{\nu\lambda}
\end{aligned}
\tag{13.74}
$$

We can now integrate out the massive density fluctuations to get the effective Lagrangian for the fluctuations of the phase and statistical gauge fields,

$$
\mathcal{L}_{\text{eff}} = \frac{\kappa}{2}\left(\partial_0\omega - \delta\mathcal{A}_0\right)^2 - \frac{\rho_{\text{s}}}{2}\left(\vec{\nabla}\omega - \delta\vec{\mathcal{A}}\right)^2 + \frac{\theta}{4}\epsilon_{\mu\nu\lambda}\mathcal{A}_\mu\mathcal{F}_{\nu\lambda}
\tag{13.75}
$$

where κ is the compressibility of this Bose gas, which, with the approximations we made, is

$$
\kappa = \frac{1}{2\lambda + \bar{V}/(4M\rho_0)}
\tag{13.76}
$$

(where we defined an effective short-range interaction $\bar{V}$), and ρ_{s} is the effective superfluid density (of the bosons),

$$
\rho_{\text{s}} = \frac{\rho_0}{M}
\tag{13.77}
$$

However, since the density $\rho_0 = \theta B$, we can also write the superfluid density ρ_{s} in terms of the filling fraction $\nu = 1/m$ and the cyclotron frequency $\omega_{\text{c}} = eB/(Mc)$:

$$
\rho_{\text{s}} = \theta\frac{B}{M} = \frac{\nu}{2\pi}\hbar\omega_{\text{c}}
\tag{13.78}
$$

This effective Lagrangian has the same form as the one we derived for the anyon superconductor in Chapter 10 except for the very important difference that the gauge field here is the statistical one, whereas there it was the electromagnetic field. Nevertheless, the phase field still disappears from the spectrum. Indeed, the phase field ω can be eliminated by a gauge transformation $\mathcal{A}_\mu = \mathcal{A}'_\mu - \partial_\mu \omega$. The resulting theory is that of a gauge field that has just two massive modes. The masses were also calculated in Chapter 10. Thus, this is an incompressible ground state. The two massive modes represent the magneto-phonon and magneto-plasmon which were derived directly from Laughlin's theory by Girvin, MacDonald, and Platzman (Girvin *et al.*, 1986).

13.3.4 Hall conductance

Let us now turn to the problem of the electromagnetic response of the Hall fluid and to the computation of the Hall conductivity. The Hall conductivity is a response perpendicular to an applied electric field and hence it does not involve dissipation. Thus, unlike the longitudinal resistivity, the Hall response can be determined in a system in the absence of impurities or phonons. We will compute the Hall conductivity by looking at the response to a weak classical (i.e. unquantized) electromagnetic perturbation in the form of a vector potential $\delta A_\mu = (\delta A_0, \delta \vec{A})$ (not to be confused with the uniform static magnetic field).

The effective Lagrangian in the presence of the electromagnetic field A_μ is dictated by electromagnetic gauge invariance, and it is given by

$$\mathcal{L}_{\text{eff}}[\omega, \mathcal{A}_\mu, \delta A_\mu] = \frac{\kappa}{2}\Big(\partial_0\omega + \mathcal{A}_0 + e\,\delta A_0\Big)^2 - \frac{\rho_{\text{s}}}{2}\left(\vec{\nabla}\omega + \vec{\mathcal{A}} + e\,\delta\vec{A}\right)^2 + \frac{\theta}{4}\epsilon_{\mu\nu\lambda}\mathcal{A}_\mu\mathcal{F}_{\nu\lambda} \tag{13.79}$$

The electromagnetic response is obtained from the effective action of the electromagnetic field $S_{\text{eff}}[A_\mu]$,

$$e^{\,iS_{\text{eff}}[\delta A_\mu]} = \int \mathcal{D}\omega\, \mathcal{D}\mathcal{A}_\mu\, \exp\left(i\int d^3x\, \mathcal{L}_{\text{eff}}[\omega, \mathcal{A}_\mu, \delta A_\mu]\right) \tag{13.80}$$

Since this theory is gauge-invariant and has a dynamical gauge field, namely the statistical field $\mathcal{A}_\mu$, we can do this computation in the London gauge, $\omega = 0$, in which the Goldstone boson of the Bose field, the phase field ω, is eaten by the dynamical gauge field. This is just the Higgs mechanism. In the London gauge the effective Lagrangian is

$$\mathcal{L}_{\text{eff}}[\mathcal{A}_\mu, \delta A_\mu] = \frac{\kappa}{2}\left(\mathcal{A}_0 + e\,\delta A_0\right)^2 - \frac{\rho_{\text{s}}}{2}\left(\vec{\mathcal{A}} + e\,\delta\vec{A}\right)^2 + \frac{\theta}{4}\epsilon_{\mu\nu\lambda}\mathcal{A}_\mu\mathcal{F}_{\nu\lambda} \tag{13.81}$$

This Lagrangian tells us that the phase mode of the Bose field (the phonon of the Bose fluid) is now absent from the spectrum and that the fluctuations of the statistical gauge field $\mathcal{A}_\mu$ are now massive (even in the absence of the Chern–Simons term) with a mass (squared) given by the superfluid density $\rho_s = \rho_0/M$ of the bosons. In other terms there is a Meissner effect of the statistical gauge field. The first term of the effective Lagrangian of Eq. (13.81) implies that there is complete screening of the electric field. For a system with Coulomb interactions, $V(r) \propto 1/r$, the screening of a 2D electron gas (2DEG) is incomplete.

Upon integrating out the statistical gauge field $\mathcal{A}_\mu$ we can now compute the effective action for the external electromagnetic perturbation, $S_{\text{eff}}[\delta A_\mu]$. By keeping only the terms with the fewest derivatives (one!) we find that the effective action of the electromagnetic field is just the Chern–Simons term:

$$S_{\text{eff}}[\delta A_\mu] = e^2 \frac{\theta}{2} \int d^3x \; \epsilon_{\mu\nu\lambda} \, \delta A^\mu \, \partial^\nu \delta A^\lambda + \cdots \tag{13.82}$$

The induced current $J_\mu(x) = (J_0(x), \vec{J}(x))$ is obtained by differentiation with respect to the electromagnetic field:

$$J_\mu(x) = \theta e^2 \epsilon_{\mu\nu\lambda} \, \partial^\nu \delta A^\lambda(x) + \cdots \tag{13.83}$$

In particular, the spatial components $J_i(x)$ of the induced current are

$$J_i(x) = \theta e^2 \epsilon_{ij} \, \delta E_j(x) \tag{13.84}$$

where $\delta \vec{E}(x)$ is the applied electric field. Thus, we see that the induced current is nothing but the Hall current. From this result we can read off the Hall conductivity σ_{xy} as being (after restoring physical units)

$$\sigma_{xy} = \theta \left(\frac{e^2}{\hbar c} \right) = \frac{1}{m} \left(\frac{e^2}{h} \right) \tag{13.85}$$

where we used the fact that $\theta = 1/(2\pi m)$. In other terms, we find that the fluid exhibits the FQHE for a 2DEG with filling fraction $\nu = 1/m$.

13.3.5 Vortices

The classical equations of motion of Eq. (13.70) admit static vortex solutions with the asymptotic behavior

$$\lim_{|\vec{x}| \to \infty} \phi(\vec{x}) = \sqrt{\rho_0} e^{\,i\varphi(\vec{x})} \tag{13.86}$$

$$\delta \mathcal{A}_0 = 0 \tag{13.87}$$

$$\lim_{|\vec{x}| \to \infty} \delta \mathcal{A}_i(\vec{x}) = \pm \vec{\nabla} \varphi(\vec{x}) = \pm \epsilon_{ij} \frac{x_j}{|\vec{x}|^2} \tag{13.88}$$

where $\varphi(\vec{x})$ is the azimuthal angle on the plane,

$$\varphi(\vec{x}) = \tan^{-1}\left(\frac{y}{x}\right) \tag{13.89}$$

This solution is called a vortex.

In a neutral superfluid (without any gauge fields) the energy of the vortex is (as we saw in Chapter 4)

$$E_{\text{vortex}} = \frac{\rho_{\text{s}}}{2}\int d^2x\left(\vec{\nabla}\omega(\vec{x})\right)^2 \approx \frac{\rho_{\text{s}}}{2}\int \frac{d^2x}{|\vec{x}|^2} \simeq \frac{\rho_{\text{s}}}{2}\ln\left(\frac{R}{a_0}\right) \tag{13.90}$$

where a_0 is a short-distance cutoff (of the order of the inter-particle spacing) and R is the linear size of the system. Hence, as we discussed in the context of the problem of the Kosterlitz–Thouless transition, isolated vortices are very expensive (energetically speaking) for a neutral superfluid and occur only (as excitations) as vortex–anti-vortex pairs.

However, in the problem at hand we have a dynamical gauge field, the statistical field $\mathcal{A}_\mu$, which affects the computation of the energy (much in the same way as in the theory of superconductivity). In this case we find finite-energy vortex solutions provided that the Bose field $\phi(x)$ and the statistical gauge field $\mathcal{A}_\mu$ obey the asymptotic behavior

$$\lim_{|\vec{x}|\to\infty}\left|\left(i\,\vec{\nabla} - \vec{\mathcal{A}}\right)\phi(\vec{x})\right|^2 = 0 \tag{13.91}$$

which is satisfied by Eq. (13.88). Thus, at very long distances we find the condition of Eq. (13.88), $\vec{\mathcal{A}} = \pm\vec{\nabla}\omega$. This solution is regular (except at the core of the vortex) and has finite energy, provided that the circulation of the statistical gauge field on any large closed contour Γ enclosing the vortex satisfies the flux quantization condition

$$\oint_\Gamma \vec{\mathcal{A}}\cdot d\vec{x} = \pm 2\pi \tag{13.92}$$

However, the Chern–Simons term tells us that a vortex has an associated electric charge, and that it is a dyonic object with an induced charge density $J_0(\vec{x})$,

$$\begin{aligned} J_0(\vec{x}) &= -e\,\frac{\delta S_{\text{eff}}}{\delta A_0(\vec{x})} = -e\,\frac{\delta S_{\text{eff}}}{\delta \mathcal{A}_0(\vec{x})} \\ &= +e\,\frac{\delta S_{\text{CS}}}{\delta \mathcal{A}_0(\vec{x})} = \theta\epsilon_{ij}\,\partial_i\mathcal{A}_j(\vec{x}) \end{aligned} \tag{13.93}$$

Therefore the total induced (or excess) charge $Q_{\rm v}$ due to a positive vortex is

$$
\begin{aligned}
Q_{\rm v} &= e\int d^2x\ J_0(\vec{x}) = \theta\int d^2x\ \epsilon_{ij}\ \partial_i \mathcal{A}_j(\vec{x}) \\
&= e\theta \oint_\Gamma d\vec{x}\cdot\vec{\mathcal{A}}(\vec{x}) \\
&= \frac{e}{m}
\end{aligned}
\tag{13.94}
$$

So we conclude that a positive vortex has a fractional charge e/m and represents a Laughlin quasihole, whereas a negative vortex is a Laughlin quasiparticle with negative fractional charge $-e/m$.

In Section 13.9 we will revisit this problem within the fermion Chern–Simons formulation, where we will show that these excitations have fractional statistics.

13.3.6 The order parameter

In hindsight, we can construct the order parameter directly in the theory of bosons, without having to rely on the Landau–Ginzburg theory. The first guess is that the order parameter is the Bose field ϕ itself. However, ϕ is not invariant under gauge transformations of the statistical gauge field. Thus, its expectation value, as well as the expectation values of any product of ϕ fields, is zero when averaged over *all* configurations of the gauge field. It may be argued that this is not much of a problem since one always has to fix the gauge. Since this gauge theory is abelian and non-compact, all small gauge transformations (i.e. those which do not wind around the system) are connected to the identity, and it is possible to fix the gauge completely. Now, the expectation value of products of ϕ fields will depend on the gauge in which it is evaluated. Thus, it does not represent a physical observable. However, all we need is *an* operator that in some convenient *gauge* reduces to a product of ϕ fields. Fortunately, it is quite easy to construct such operators. We discussed a similar question in Chapter 9.

Let us consider the case of the boson-correlation function, which is the expectation value of the product $\phi^\dagger(x)\phi(y)$, where x and y are two arbitrary points in $(2+1)$-dimensional space-time. Under a gauge transformation $\phi(x)\to\exp(i\Lambda(x))\phi(x)$, the product transforms like

$$
\phi^\dagger(x)\phi(y)\to e^{i(-\Lambda(x)+\Lambda(y))}\phi^\dagger(x)\phi(y) \tag{13.95}
$$

Thus, we need to find an operator that transforms in the opposite way and cancels out the unwanted phase factor. One possibility is the exponential of the line integral $\int_\Gamma \mathcal{A}_\mu\, dx_\mu$, where Γ is a path that goes from x to y. But this is just an

Aharonov–Bohm phase factor, which fluctuates very rapidly and does not vanish in any gauge. It can be shown that the expectation value of the product

$$\phi^\dagger(x) \exp\left(i \int_\Gamma \mathcal{A}_\mu dx_\mu\right) \phi(y) \tag{13.96}$$

decays rapidly as $|x - y| \to \infty$.

Let us consider the operator $\mathcal{O}^\dagger(x)\mathcal{O}(y)$,

$$\mathcal{O}^\dagger(x)\mathcal{O}(y) \equiv e^{i \int d^3z \, \mathcal{A}^c_\mu(z) B^c_\mu(z)} \phi^\dagger(x)\phi(y) \tag{13.97}$$

where A^c_μ is some suitably chosen fixed classical configuration with field strength $B^c_\mu(z) = \epsilon_{\mu\nu\lambda} \partial_\nu A^c_\lambda$. We will choose B^c_μ in such a way that the product $\mathcal{O}^\dagger(x)\mathcal{O}(y)$ is gauge-invariant and that in the Landau–Lorentz gauge ($\partial_\mu \mathcal{A}_\mu = 0$) it reduces to the product of local operators $\phi^\dagger(x)\phi(y)$. Under a gauge transformation that vanishes at infinity $\lim_{|x|\to\infty} \Lambda(x) = 0$,

$$\begin{aligned} \phi(x) &= e^{i\Lambda(x)} \phi'(x) \\ \mathcal{A}_\mu(x) &= \mathcal{A}'_\mu(x) - \partial_\mu \Lambda(x) \end{aligned} \tag{13.98}$$

the operator $\mathcal{O}^\dagger(x)\mathcal{O}(y)$ transforms as

$$\mathcal{O}^\dagger(x)\mathcal{O}(y) = e^{i\Phi} \mathcal{O}^\dagger(x)\mathcal{O}(y) \tag{13.99}$$

where Φ is given by

$$\Phi = \Lambda(y) - \Lambda(x) + \int d^3z \, \Lambda(z) \partial_\mu B^c_\mu(z) \tag{13.100}$$

Gauge invariance demands that $\Phi \equiv 0$ for *all* gauge transformations $\Lambda(z)$ and for *all* points x and y. The only way to meet these requirements is for $B^c_\mu(z)$ to satisfy the equation

$$\partial_\mu B^c_\mu(z) = \delta(z - x) - \delta(z - y) \tag{13.101}$$

We can think of $B^c_\mu(z)$ as being the *classical* magnetic field of two magnetic monopoles of (opposite) unit magnetic charge located at x and y. If we denote the "potential" by $U(z)$, we get

$$\begin{aligned} B^c_\mu(z) &= \partial_\mu U(z) \\ \nabla^2 U(z) &= \delta(z - x) - \delta(z - y) \end{aligned} \tag{13.102}$$

the solution of which is just the electrostatic potential for two unit and opposite charges.

Having checked that it is gauge-invariant, we now want to see what this operator is in the Landau–Lorentz gauge ($\partial_\mu \mathcal{A}_\mu = 0$). In this gauge, the argument of the exponential part of the operator vanishes identically,

$$\int d^3z\, \mathcal{A}_\mu(z) B^{\text{c}}_\mu(z) = \int d^3z\, \mathcal{A}_\mu(z)\partial_\mu U(z) = -\int d^3z\, \partial_\mu \mathcal{A}_\mu(z) U(z) = 0 \tag{13.103}$$

Thus, in the Landau–Lorentz gauge, we get

$$\mathcal{O}^\dagger(x)\mathcal{O}(y) \equiv \phi^\dagger(x)\phi(y) \tag{13.104}$$

Therefore, the operator $\mathcal{O}(x)$, defined by

$$\mathcal{O}(x) \equiv \phi(x) \exp\left(i \int d^3z\, \mathcal{A}_\mu(z) B^{\text{c}}_\mu(z)\right) \tag{13.105}$$

where $B^{\text{c}}_\mu(z) = \partial_\mu U(z)$ is the field created by a single charge at x, is the gauge-invariant order-parameter operator for this problem in the boson description since, in this gauge, it becomes identical to the field operator of the bosons. Thus, the correlation functions of this operator exhibit long-range order.

In an arbitrary gauge, this operator is highly non-local. But, in the Landau–Lorentz gauge, it becomes local and just simple. This is not a surprise since, for instance, the order parameter of an ordinary BCS-like superconductor is local only in this gauge. Indeed, it is possible to define an order parameter for a superconductor in the same way. For practical purposes, in the case of a superconductor, this is not very useful since the electromagnetic field is not usually treated as a dynamical field. In the problem of the FQHE, the gauge field is dynamically generated, and it plays an essential role.

13.4 Fermion field theory of the fractional quantum Hall effect

In this section we derive a field theory for the FQHE that is based on the fermion picture. These methods, which have been so successful in the treatment of anyon superfluidity (see Chapter 10), are also very useful for the study of the FQHE. They have a great advantage over the boson theories in that there is no difficulty in handling the short-distance behavior, unlike in the case of bosons. It is quite easy to derive an effective action for the fluctuations that explicitly involves Chern–Simons gauge fields. The Landau–Ginzburg theory can be seen to be the *dual* of the fermion theory in very much the same way as in the case of the anyon superconductor. The fermion field theory was developed by López and myself (López and Fradkin, 1991).

Let us go back to the second-quantized form of the problem of electrons in a magnetic field. In its standard form, the dynamics is governed by the action

$$\mathcal{S} = \int d^3z \left\{ \psi^*(z)[iD_0 + \mu]\psi(z) + \frac{1}{2M}|\vec{D}\psi(z)|^2 \right\}$$
$$- \frac{1}{2}\int d^3z \int d^3z' (|\psi(z)|^2 - \rho_0) V(|\vec{z} - \vec{z}\,'|)(|\psi(z')|^2 - \rho_0) \qquad (13.106)$$

Since we are dealing with a problem in which one Landau level is fractionally filled, we do not expect that the semi-classical approximation for this problem will, in general, be very reliable, unless, of course, the ground state of the system is such that there is a gap in the energy spectrum. For example, in the low-density limit, the system can lower its energy by modulating the electron density and forming a Wigner crystal. Wigner crystals can also be studied with a path integral of this section, but we will not do it here.

Let us recall Jain's interpretation of the Laughlin state as a state in which the electrons "nucleate" flux to screen enough of the external magnetic field, so that the bound states of electrons plus fluxes exactly filled an integer number of Landau levels. In this section we are going to use the *periodicity property* of theories of fermions coupled to Chern–Simons gauge fields, which was Derived in chapter 10, to make this nucleation picture more explicit.

In Chapter 10 we saw that a system of fermion could be mapped into a system of fermions coupled to Chern–Simons gauge fields if the Chern–Simons coupling constant were chosen to be equal to $\theta = 1/(2\pi n)$, where n is an *even* integer. Thus, the problem becomes equivalent to a theory with fermions and gauge fields with an action given by

$$\mathcal{S}_\theta = \int d^3z \left\{ \psi^*(z)[iD_0 + \mu]\psi(z) + \frac{1}{2M}|\vec{D}\psi(z)|^2 + \frac{\theta}{4}\epsilon_{\mu\nu\lambda}\mathcal{A}^\mu \mathcal{F}^{\nu\lambda} \right\}$$
$$- \frac{1}{2}\int d^3z \int d^3z' (|\psi(z)|^2 - \rho_0) V(|\vec{z} - \vec{z}\,'|)(|\psi(z')|^2 - \rho_0) \qquad (13.107)$$

where $\psi(z)$ is a second-quantized Fermi field, μ is the chemical potential, and D_μ is the covariant derivative which couples the fermions both to the external electromagnetic field A_μ and to the statistical gauge field $\mathcal{A}_\mu$,

$$D_\mu = \partial_\mu + i\frac{e}{c}A_\mu + i\mathcal{A}_\mu \qquad (13.108)$$

We are going to see below that the *even* integer n has to be identified with $m - 1$, where m is the index of the Laughlin state.

13.4.1 The semi-classical limit and the Laughlin state

We will show that the *semi-classical limit* of the theory described by the action $\mathcal{S}_\theta$, with $1/\theta = 2\pi(m-1)$, yields the same physics as the Laughlin state. In order to prove this statement we will develop a semi-classical approach to this problem. In principle, this formalism provides a procedure by which to compute the corrections to the Laughlin approximation. This is, to the best of my knowledge, the first formalism for which the Laughlin ansatz arises as the first of a series of approximations.

The action $\mathcal{S}_\theta$ governs the dynamics of a system of spinless *fermions* interacting through a pair-interaction potential $V(|\vec{x} - \vec{x}'|)$ coupled both to electromagnetic and to statistical gauge fields. The starting point of the semi-classical approximation maps this FQHE problem into an *equivalent* IQHE system. This mapping is made possible by the statistical or Chern–Simons gauge fields, which screen out enough of the external magnetic field, to the point that the number of flux quanta of the effective magnetic field which is left is an exact factor of the total number of particles. Naturally, this perfect screening is not possible for arbitrary values of the external magnetic field for a fixed number of electrons. The values of the filling fraction for which this perfect screening can be accomplished happen to be the same as the Laughlin sequence with filling factors $\nu = 1/m$ and the first level of the hierarchy. For all other cases, there will be some partially filled level left over. As we discussed in Section 10.5, these quasiparticles are anyons.

Consider the quantum partition function for this problem (at $T = 0$),

$$\mathcal{Z} = \int \mathcal{D}\psi^* \, \mathcal{D}\psi \, \mathcal{D}\mathcal{A}_\mu \, e^{i\mathcal{S}_\theta} \tag{13.109}$$

We will treat this path integral in the semi-classical approximation. In order to do that, we will first integrate out the fermions and treat the resulting theory within the saddle-point expansion. For this procedure to be accurate, there should be a small parameter in the theory to control this expansion. For instance, in Chapter 3 we used a similar procedure to study the magnetic instabilities of a dense Fermi system. In that case the small parameter was $1/N$, where N was the number of fermion species (orbitals). For spin-S antiferromagnets we used a similar approach, with $1/S$ being the small parameter.

However, in the case at hand there is no such small parameter. Nevertheless, we will find that we will be able to construct sequences of gapped states corresponding to fractional quantum Hall states. In the presence of an energy gap what we call here a semi-classical approximation, i.e. the average-field approximation with one-loop quantum corrections (conventionally called the random-phase approximation), will yield *exact* results for universal long-distance quantities protected by symmetries (and sum rules) and topology such as the Hall conductance, and the charge and statistics of the quasiparticles. On the other hand, with the sole exception of the cyclotron resonance which, as we will see, is protected by Galilean invariance, other quantities that are not dimensionless, such as energy gaps, have large systematic errors that are very difficult to correct. These difficulties become extreme in regimes in which the energy gap vanishes, where the theory has infrared divergences in various quantities of interest.

The root of these problems lies not so much in the lack of a formal small parameter to justify the expansion, but more in the approximations done on the theory with flux attachment. Before flux attachment, one has a theory of fermions partially filling a Landau level. As we saw, a Landau level has an extensive degeneracy, and simple perturbative approaches to dense Fermi systems (such as the Hartree–Fock approach) fail, since in a Landau level *all* interactions are strong no matter how small the nominal coupling constant is. Flux attachment, i.e. mapping to an equivalent theory of fermions coupled to a Chern–Simons gauge field, sidesteps this problem, but at the price of introducing a large amount of mixing between Landau levels, whose Hilbert spaces are now rearranged in a non-trivial way. We will find that, if the system manages to have a finite energy gap, the effects of Landau-level mixing become negligible in the long-distance and low-energy regime. This is what happens in the fractional quantum Hall states. However, if the system does not have a gap, and hence becomes compressible, the effects of Landau-level mixing cannot be disentangled.

The procedure is almost identical to the theory of anyon superconductivity discussed in Chapter 10. In the absence of electron–electron interactions the fermions can be integrated out immediately, since the action becomes quadratic in Fermi fields. In the presence of interactions, this is no longer possible, since the interaction term makes the action quartic in the Fermi fields. This problem can be sidestepped by means of a Hubbard–Stratonovich transformation by which we trade a quartic form in fermions for a quadratic action coupled to a new Bose field, the density fluctuation. This procedure will allow us to give a full description of the spectrum of collective modes of the FQHE states. Note that, since we are dealing with a gauge theory, a gauge has to be specified in order to make the functional integral well defined. We will assume that a gauge-fixing condition has been imposed, but, for the moment, we will not make any specific choice of gauge.

Before we proceed to integrate out the Fermi degrees of freedom, we need to deal with the interaction term of the action. Here we could perform the Hubbard–Stratonovich transformation in terms of a scalar Bose field $\lambda(x)$. Let F be the weight in the path-integral amplitude which contains in its exponent the terms in the action which are quartic in the Fermi field ψ,

$$F = \exp\left(-i \int d^3z \int d^3z' \, \frac{1}{2}(|\psi(z)|^2 - \rho_0) V(z - z')(|\psi(z')|^2 - \rho_0)\right) \tag{13.110}$$

The Hubbard–Stratonovich transformation allows us to write F as a Gaussian functional integral over a Bose field $\lambda(x)$. However, the Hubbard–Stratonovich λ represents density fluctuations and couples linearly to the fermion-density operator. Thus, $\lambda(x)$ enters into the time-covariant derivative in the same way as the time component of the statistical gauge field, $\mathcal{A}_0(x)$, and can be integrated out exactly. The net result is that the Hubbard–Stratonovich transformation is equivalent to the replacement of the fermion density $|\psi(x)|^2$ in the interaction term by $\theta\mathcal{B}(x)$ (where we used the Chern–Simons constraint). Therefore we can replace the interaction term by the equivalent expression

$$S_{\text{int}}[\mathcal{A}_\mu] = -\frac{1}{2}\int d^3z\, d^3z'\, [\theta\mathcal{B}(z) - \rho_0]V(z-z')\left[\theta\mathcal{B}(z') - \rho_0\right] \tag{13.111}$$

where $V(z-z')$ represents the instantaneous pair interaction, i.e.

$$V(z-z') = V(|\vec{z}-\vec{z}\,'|)\delta(t-t') \tag{13.112}$$

I will assume that the physics of the FQHE can be studied in a model system in which the pair potential is reasonably local.

The partition function $\mathcal{Z}$ can be written in the form of a functional integral involving the Fermi fields ψ and the statistical gauge fields $\mathcal{A}_\mu$. The action for the system is now given by

$$\begin{aligned} S = &\int d^3z \left\{\psi^*(z)(iD_0 + \mu)\psi(z) + \frac{1}{2M}|\vec{D}\psi(z)|^2\right\} \\ &+ \int d^3z\, \frac{\theta}{4}\epsilon_{\mu\nu\lambda}\mathcal{A}^\mu\mathcal{F}^{\nu\lambda} + S_{\text{int}}[\mathcal{A}_\mu] \end{aligned} \tag{13.113}$$

The Fermi fields can be integrated out without any difficulty, yielding a fermion determinant. The resulting partition function can thus be written in terms of an effective action S_{eff} given by

$$S_{\text{eff}} = -i \,\text{tr}\ln\left[iD_0 + \mu + \lambda + \frac{1}{2M}\vec{D}^2\right] + \theta S_{\text{CS}}(\mathcal{A}_\mu - A_\mu) + S_{\text{int}}[\mathcal{A}_\mu - A_\mu] \tag{13.114}$$

where D_0 and $\vec{D}$ are the covariant derivatives and S_{CS} is the Chern–Simons action for $\theta = 1$. The field A_μ represents a small fluctuating electromagnetic field, with vanishing average everywhere, which will be used to probe the system. The electromagnetic currents will be calculated as first derivatives of $\mathcal{Z}$ with respect to A_μ. The full electromagnetic response can be obtained in this way.

We are now ready to proceed with the semi-classical approximation. The path integral $\mathcal{Z}$ will be approximated by expanding its degrees of freedom around stationary configurations of the effective action S_{eff} in powers of the fluctuations. This is the conventional WKB approximation. The classical configurations $\bar{\mathcal{A}}_\mu(z)$ can be obtained by demanding that S_{eff} be stationary under small fluctuations. This requirement yields the classical equations of motion

$$\left.\frac{\delta S_{\text{eff}}}{\delta \mathcal{A}_\mu(z)}\right|_{\bar{\mathcal{A}}} = 0 \tag{13.115}$$

By varying S_{eff} with respect to $\mathcal{A}_\mu(z)$ we get

$$\langle j_\mu^{\text{F}}(z)\rangle + \frac{\theta}{2}\epsilon_{\mu\nu\lambda}\left[\langle \mathcal{F}^{\nu\lambda}(z)\rangle - eF^{\nu\lambda}\right] = 0 \tag{13.116}$$

In addition, we must fix the particle density to be uniform and equal to $\bar{\rho}$,

$$\langle j_0(z)\rangle = \rho_0 \tag{13.117}$$

If the external electromagnetic fluctuation is assumed to have zero average, the only time-independent uniform solutions have uniform average statistical flux $\langle \mathcal{B}\rangle$ and vanishing average statistical electric field $\langle \vec{\mathcal{E}}\rangle$ (unless there is a non-zero current in the ground state), and satisfy

$$\langle \mathcal{B}\rangle = -\frac{\rho_0}{\theta}, \qquad \langle \vec{\mathcal{E}}\rangle = 0 \tag{13.118}$$

The non-uniform solutions have $\langle \mathcal{A}_0(z)\rangle$ a periodic function that induces a periodic modulation of the electron density. These solutions are Wigner crystals and stripe phases. Notice that, in principle, the crystalline solutions have a modulation both in the charge density and in the local statistical flux. We will not discuss these states here.

The equations of motion show that, for a translationally invariant ground state, the effect of the statistical gauge fields, at the level of the saddle-point approximation, is to *reduce* the effective flux experienced by the fermions. The total effective field is thus reduced from the value of the external field B down to B_{eff}, given by

$$B_{\text{eff}} = B + \langle \mathcal{B}\rangle = B - \frac{\rho_0}{\theta} \tag{13.119}$$

Notice that, since ρ_0/θ can be either smaller or larger than B, the effective field B_{eff} can be parallel or anti-parallel to B.

Let us assume that we are trying to find the ground state of N (interacting) electrons in the presence of an external magnetic field of strength B. We will further assume that the linear size L of the sample is such that a total of N_ϕ quanta of the magnetic flux will be piercing the surface. In general, the filling fraction $\nu = N/N_\phi$ is not an integer. Thus, a perturbative approach based on a Slater-determinant wave function of the occupied single-particle states does not yield a stable answer. This is so because there is a macroscopic number of essentially degenerate states that will mix with this trial state. On the other hand, a Laughlin state is known to represent a

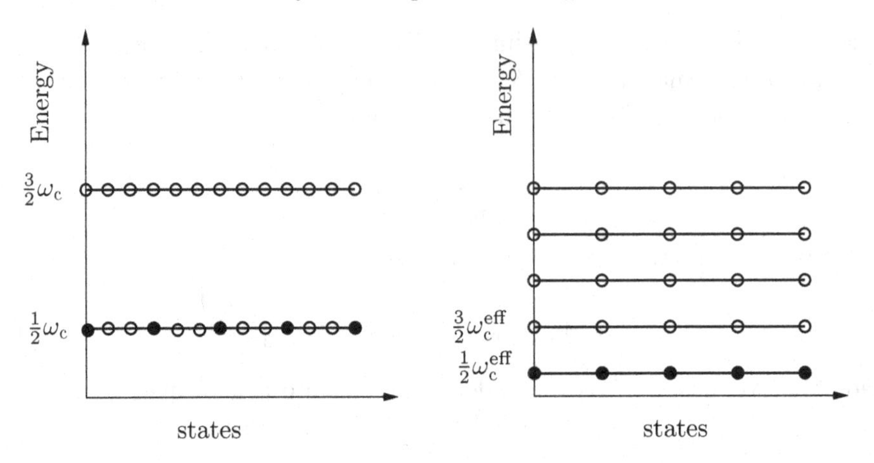

Figure 13.1 The composite-fermion mean-field picture of the Laughlin state at filling fraction $\nu = 1/3$. Left panel: electrons in a 1/3-filled lowest Landau level. Right panel: composite fermions in a fully filled, $p = 1$, lowest effective Landau level. Full circles, filled states; open circles, empty states; ω_c and ω_c^{eff} are the cyclotron frequency of the electrons and that of the composite fermions. See the text.

state with an energy gap. Thus, the correlations have removed the massive degeneracy of the free electrons. Since this gap is not equal to the Landau gap of the non-interacting electrons, we can expect our saddle-point expansion to succeed only if the *effective* theory ends up with a non-zero gap.

It is easy to check that the uniform-saddle-point state has a gap *only if* the effective field B_{eff} experienced by the N fermions is such that the *fermions fill exactly an integer number p of the effective Landau levels*, see Fig. 13.1. This is precisely the point of view advocated by Jain: the mean-field state of the 2D electron gas in a fractional quantum Hall state is an integer quantum Hall state of a system of *composite fermions*, electrons dressed by an even number of flux quanta. However, this condition cannot be met for arbitrary values of the filling fraction ν at fixed field (or at fixed density). Let N_ϕ^{eff} denote the effective number of flux quanta piercing the surface after screening. It is given by

$$2\pi N_\phi^{eff} = 2\pi N_\phi - \frac{\rho_0}{\theta} L^2 = 2\pi (N_\phi - 2sN) \tag{13.120}$$

where $2s$ is an even integer (that before we had denoted by n). The spectrum supported by this state has an energy gap if the N fermions fill exactly p of the Landau levels created by the effective field B_{eff}. In other words, the *effective* filling fraction is $\nu_{eff} \equiv N/N_\phi^{eff} = p$. Using these results, we find that the filling fraction ν and the external magnetic field B must satisfy

$$\frac{N}{p} = \frac{N}{\nu} - 2sN \tag{13.121}$$

or, equivalently,

$$\frac{1}{\nu} = \frac{1}{p} + 2s \tag{13.122}$$

Since the filling fraction ν is in general equal to the ratio of two integers, a solution exists for filling fractions on the Jain sequences:

$$\nu = \frac{p}{2sp+1} \tag{13.123}$$

On the other hand, since the effective field B_{eff} can be positive or negative, the number p of filled effective Landau levels should also take either sign. Hence, we will denote the number of effective filled Landau levels by $\pm p$ (with $p > 0$) to indicate the case in which $B_{\text{eff}} > 0$ or $B_{\text{eff}} < 0$. We can regard the case $B_{\text{eff}} < 0$ as an integer quantum Hall effect of holes.

In order to allow for both cases we will write Eq. (13.122) in the form

$$\frac{1}{\nu} = \pm\frac{1}{p} + 2s \tag{13.124}$$

or, which amounts to the same thing,

$$\nu_{\pm}(p, s) = \frac{p}{2sp \pm 1} \tag{13.125}$$

In this approximation the ground state is then interpreted as a system of N composite fermions filling up exactly p Landau levels of the effective field B_{eff},

$$|B_{\text{eff}}| = \frac{B}{2sp+1} \tag{13.126}$$

Similarly, the effective cyclotron frequency, i.e. the gap between effective Landau levels, is also reduced by the same amount,

$$\omega_{\text{c}}^{\text{eff}} = \frac{\omega_{\text{c}}}{2sp+1} \tag{13.127}$$

Hence, for a fixed number $2s$ of attached flux quanta, the splitting of the effective energy levels becomes smaller as the level p in the Jain hierarchy increases.

The states are thus parametrized by two integers, p (the number of filled Landau levels of the *effective field*) and $2s$ (the number of flux quanta carried by each fermion). The Laughlin sequence is an obvious solution since, for $p = 1$ and m an odd integer, we get the unique solution $p = 1$ and $2s = m - 1$. The effective fermions thus fill up exactly one Landau level and θ has to be chosen to be

$1/\theta = 2\pi(m-1)$. This result agrees with Jain's theory. At this mean-field level the wave function is the Slater determinant for one filled Landau level χ_1. The additional factor, $\prod_{i<j}(z_i - z_j)^{m-1}$, is due to the fluctuations of the statistical gauge fields.

Many states of the Jain hierarchy, Eq. (13.125), have been seen in experiments on 2D electron gases in large magnetic fields in quantum wells and heterostructures (see Pan *et al.* (2003, 2008) and Xia *et al.* (2004)). Of the principal Jain sequence, with $2s = 2$ (i.e. two fluxes attached to each electron) the following states have been seen experimentally (Pan *et al.*, 2008): $1/3, 2/5, 3/7, \ldots, 10/21$ ($p = 1, \ldots, 10$); and of the "reverse" (or "hole") sequence: $1, 2/3, 3/5, 4/7, \ldots,$ $10/19$ ($p = 1, \ldots, 10$). In the experiments the most prominent states have wider plateaus (when sweeping in magnetic field) in their quantum Hall conductances. It is also found that the widths of the plateaus of the fractional quantum Hall states on a given Jain sequence decrease as the order in a given sequence increases. In the experiments, it is also found that the states with the wider plateaus have larger energy gaps (measured from the temperature dependence of the longitudinal resistivity). These observations have given credence to the composite-fermion picture (Heinonen, 1998; Jain, 2007).

All the states mentioned above can also be described by the "bosonic" Haldane–Halperin hierarchy. However, empirically the stronger fractional quantum Hall states, defined by the width of the observed plateau in the Hall conductance, are naturally described by the Jain sequences. On the other hand, there are several observed fractional quantum Hall states that do not fit in the Jain sequences, such as the state at filling fraction $4/11$. Such a state can be described as a fractional quantum Hall state in the bosonic hierarchy or as a "next-generation" Jain state, a fractional quantum Hall state of the quasiparticles (vortices) of the primary Jain sequence. More interesting are the states with even denominators, such as at $\nu = 5/2$ (Willett *et al.*, 1987; Pan *et al.*, 1999), which cannot be described by either hierarchy. We will see in Chapters 14 and 15 that this is a *paired* or *non-abelian* fractional quantum Hall state.

Thus, at the level of the average-field approximation we find that this problem is equivalent to the integer Hall state of a system of composite fermions. But is this the correct answer? We will see shortly that the answer is no. The reason is that fluctuations, even at the (one-loop) Gaussian level, change the physics completely. In particular, we will see that, unlike other mean-field theories, the average-field approximation yields the incorrect quantum numbers of the excitations. The reason is the Chern–Simons constraint that relates flux to charge, which tells us that a charge fluctuation is always accompanied by a flux fluctuation. This fact, and the commutation relations between the gauge fields, lead to profound changes in the spectrum of states and the response of the system.

13.4.2 Compressible states

An interesting feature of the Jain sequences is the filling fractions of their limiting values:

$$\nu_\infty = \lim_{p\to\infty} \frac{p}{2sp \pm 1} = \frac{1}{2s} \tag{13.128}$$

As these limiting values are approached the composite fermion gaps vanish,

$$\lim_{p\to\infty} \hbar\omega_c^{\text{eff}}(p, s) = 0 \tag{13.129}$$

Thus, in this limit, and at the mean-field level, the composite fermions see a vanishing effective field, $B_{\text{eff}} \to 0$. If we momentarily ignore the (extremely important!) fact that the fermions are coupled to a Chern–Simons (statistical) gauge field, $\mathcal{A}_\mu$, we are led to the conclusion that for $\nu_\infty = 1/(2s)$ the composite fermions behave as a system of fermions at finite density ρ_0 in a vanishing magnetic field.

Thus, if this picture is correct, we expect the composite fermions to fill up a Fermi sea, with a finite Fermi momentum p_{F} determined by the density ρ_0 of composite fermions (the same as the electron density!), which is related to the filling fraction $\nu_\infty = 1/(2s)$ and to the magnetic length $\ell_0 = \sqrt{eB/(\hbar c)}$ by

$$\rho_0 = \frac{\nu_\infty}{2\pi\ell_0^2} \tag{13.130}$$

We can determine the Fermi momentum p_{F} by the standard relation

$$\int_{|\vec{p}|\leq p_{\text{F}}} \frac{d^2p}{(2\pi\hbar)^2} = \rho_0 \tag{13.131}$$

which tells us that we have filled all single composite-fermion states with momentum less than p_{F}. We then find that p_{F} is given by

$$p_{\text{F}} = \sqrt{2\nu_\infty}\,\frac{\hbar}{\ell_0} \tag{13.132}$$

and we have a Fermi energy

$$E_{\text{F}} = \frac{p_{\text{F}}^2}{2M} = \nu_\infty \frac{\hbar}{M\ell_0^2} \tag{13.133}$$

Provided that the approximations we just made can be trusted (which is a big "if" since this theory does not have a small parameter), we conclude that at the limiting fractions ν_∞ behaves as a system of composite fermions filling up a Fermi disk. As we saw in Chapter 2, a state of this type does not have an energy gap. Its low-lying excitations are composite-fermion quasiparticles as well as quasiparticle–quasihole pairs (with arbitrarily low energy).

Naturally this picture is much too naive since the composite fermions (a) interact with each other via the pair interaction and (b) are coupled to a now dynamical gauge field, the statistical gauge field. In the next subsections we will discuss in detail the role of quantum fluctuations for the incompressible states. In that case, although they play a key role, and change the physics in a qualitative way, in the long-distance and low-energy regime these corrections are free of infrared divergences and hence are controlled. Indeed, in the low-energy regime we will find results that are actually exact.

However, the situation is drastically different for the compressible states since they do not have a gap at the level of the average-field approximation. This naturally would lead us to suspect that the corrections may have infrared divergences, which would signal an instability of the mean-field state. If the only important correlations were due to the pair interactions between composite fermions, the result would be a renormalized Fermi-liquid state, a quantum liquid adiabatically connected to the free-fermion state. In this case it is known that the effect of interactions primarily results in a set of renormalized Fermi-liquid parameters that control the forward-scattering interactions of the quasiparticles (Baym and Pethick, 1991; Shankar, 1994). In a weakly coupled Fermi-liquid state, the only possible instability is with respect to a superconducting state (in this case of composite fermions) in the $p_x + ip_y$ channel (Kohn and Luttinger, 1965; Chubukov, 1993; Raghu and Kivelson, 2011). At any rate, the lack of a small parameter leads one to question the applicability of these results to this case.

A much more serious problem in this system is that the gauge-field fluctuations lead to strong infrared divergences in the forward-scattering channel. The problem of a system of fermions at finite density coupled to dynamical gauge fields is a problem that has been researched in high-energy and condensed matter physics for many years, and it is still an essentially unsolved problem. In spite of the lack of a solution, what is clear is that the main effect of the gauge-field fluctuations is to invalidate the quasiparticle picture and hence put into question the basis of the Fermi-liquid scenario (Halperin *et al.*, 1993; Kwon *et al.*, 1994; Nayak and Wilczek, 1994; Polchinski, 1994).

In spite of these questions, a theory of the compressible state which is based on a Fermi-liquid theory of composite fermions that yields a good phenomenological description of most of the experiments in this regime has been constructed by Halperin, Lee, and Read (Halperin *et al.*, 1993). In this theory it is shown that the gauge-invariant current correlators are free of infrared divergences and can be used to compute response functions that agree qualitatively with the results from experiments. Infrared divergences do appear in the (gauge-dependent) fermion propagators, leading to a divergence of the composite-fermion effective mass and the vanishing of the quasiparticle residue (as discussed in Chapter 2). On the other hand, numerical simulations with a variational wave function with the form of a Slater determinant of free composite fermions at fixed density, projected onto the lowest Landau level,

$$\Psi_{\text{FL}}[z_i] = P_{\text{LLL}}\left[\det\{e^{i\vec{k}\cdot\vec{r}_i}\}\right]\prod_{i<j}(z_i - z_j)^2 \tag{13.134}$$

yield results in good agreement with exact diagonalizations of the problem of electrons in a half-filled Landau level. Here P_{LLL} denotes the projection onto the lowest Landau level and $\{\vec{r}_i\}$ are the electron coordinates. A conceptually serious question is that one expects the quasiparticles to have vanishing charge, but not a vanishing dipole moment. A theory defined directly in the lowest Landau level (without flux attachment) that is based on physical observables, namely the non-commuting guiding center coordinates, has been proposed (Pasquier and Haldane, 1998; Read, 1998). But it has proven to be very difficult to make progress with this approach.

13.5 The semi-classical excitation spectrum

We will now consider the role of the Gaussian fluctuations around the classical solutions. This is equivalent to a WKB approximation of the functional integral. We begin by considering the effective action. We showed that the saddle-point approximation has a uniform liquid-like solution. Let $\mathcal{A}_\mu(x)$ denote the *fluctuations* of the statistical vector potential $\mathcal{A}_\mu$ (from its average value) and $A_\mu(x)$ be an external weak electromagnetic field acting as a probe (not the uniform field). The effective action can be expanded in a series in powers of the fluctuations. We will be interested only in keeping the terms up to quadratic order in the fluctuations. As usual, the linear terms are cancelled out if the saddle-point equations are satisfied. This means that the Chern–Simons piece of the action now has the form $S_{\text{CS}}(\mathcal{A}_\mu - A_\mu)$.

At the quadratic (Gaussian) level the effective action has the form

$$S^{(2)} = \frac{1}{2}\int d^3x\, d^3y\; \mathcal{A}_\mu(x)\Pi_{\text{F}}^{\mu\nu}(x,y)\mathcal{A}_\nu(y) + \theta S_{\text{CS}}(\mathcal{A}_\mu - A_\mu) + S_{\text{int}}(\mathcal{A}_\mu - A_\mu) \tag{13.135}$$

where S_{int} is the part of the effective action for the interactions. After cancelling the external uniform magnetic field with the average statistical magnetic field, S_{int} becomes a function of the fluctuations $\mathcal{A}_\mu - A_\mu$ (where A_μ denotes the external probe electromagnetic field):

$$S_{\text{int}}(\mathcal{A}_\mu - A_\mu) = -\frac{\theta^2}{2}\int d^3z \int d^3z'[\mathcal{B}_\mu(z) - B_\mu(z)]V(z-z')[\mathcal{B}_\mu(z') - B_\mu(z')] \tag{13.136}$$

The general form of the polarization tensor $\Pi_{\rm F}^{\mu\nu}$ for free fermions in an external field was derived in Section 12.5. For a system with an integer number of Landau levels, the most important properties of $\Pi_{\mu\nu}$ are that it is transverse, i.e. $\partial_\mu \Pi_{\mu\nu} = 0$, Eq. (12.87), and that it can be expanded in powers of gradients. The latter property is a consequence of the fact that the system has an energy gap. Thus, gauge invariance and locality will be sufficient to fix the form of the effective action for the low-energy fluctuations.

In momentum (and frequency) space the (transverse) polarization tensor of the composite fermions, $\Pi_{\rm F}^{\mu\nu}(Q)$ (with $Q = (Q_0, \vec{Q})$, where $Q_0 = \omega$ is the frequency), has the form of a linear combination of explicitly transverse tensors, of which two are even under parity and time reversal separately, and a third breaks both parity and time reversal (but not their product). When expanded in components, $\Pi^{\mu\nu}(Q)$ has the structure (López and Fradkin, 1991, 1993)

$$\begin{aligned}
\Pi_{00}^{\rm F} &= \vec{Q}^2 \Pi_0^{\rm F}(\omega, \vec{Q}) \\
\Pi_{0j}^{\rm F} &= \omega \vec{Q}_j \Pi_0^{\rm F}(\omega, \vec{Q}) + i\epsilon_{jk} Q_k \Pi_1^{\rm F}(\omega, \vec{Q}) \\
\Pi_{j0}^{\rm F} &= \omega \vec{Q}_j \Pi_0^{\rm F}(\omega, \vec{Q}) - i\epsilon_{jk} Q_k \Pi_1^{\rm F}(\omega, \vec{Q}) \\
\Pi_{ij}^{\rm F} &= \omega^2 \delta_{ij} \Pi_0^{\rm F}(\omega, \vec{Q}) - i\epsilon_{ij} \omega \Pi_1^{\rm F}(\omega, \vec{Q}) + (\vec{Q}^2 \delta_{ij} - Q_i Q_j) \Pi_2^{\rm F}(\omega, \vec{Q})
\end{aligned} \tag{13.137}$$

The kernels $\Pi_0^{\rm F}(Q)$, $\Pi_1^{\rm F}(Q)$, and $\Pi_2^{\rm F}(Q)$ represent charge-conserving fluctuations in the system of composite fermions in the effective magnetic field $B_{\rm eff}$. $\Pi_0^{\rm F}$ and $\Pi_2^{\rm F}$ are associated with the parity and time-reversal even processes, while $\Pi_1^{\rm F}$ is associated with the parity and time-reversal odd processes (and has the tensorial structure of the Chern–Simons term of the action).

For a Jain state at level $p \geq 1$, which has a mean-field ground state with p filled effective Landau levels, the tensors have the form of a series of terms, each representing a process with a particle–hole excitation, and have simple poles at the particle–hole excitation energies $\omega_{mn} = (m-n)\omega_{\rm c}^{\rm eff}$, with $m > p$ (particle) and $n \leq p$ (hole). Each term has a residue given in terms of powers of Q^2 (or $\vec{Q}^2$) and Laguerre polynomials of Q (given in detail in López and Fradkin (1991, 1993)).

In the limit of zero frequency and zero momentum, $\omega = 0$ and $\vec{Q} = 0$, for a Jain state at level p, they take the limiting values

$$\begin{aligned}
\Pi_0^{\rm F}(0,0) &= \frac{1}{2\pi}\frac{pM}{B_{\rm eff}} \equiv \epsilon \\
\Pi_1^{\rm F}(0,0) &= \pm\frac{p}{2\pi} \equiv \sigma_{xy}^0 \\
\Pi_1^{\rm F}(0,0) &= -\frac{1}{2\pi}\frac{p^2}{M} \equiv -\chi
\end{aligned} \tag{13.138}$$

To leading order in fluctuations and in gradients, we get the following effective action:

$$
\begin{aligned}
S_{\text{eff}} = & \int d^3z \left(\frac{\epsilon}{2} \vec{\mathcal{E}}^2 - \frac{\chi}{2} \mathcal{B}^2 \right) \\
& + (\sigma_{xy}^0 + \theta) \mathcal{S}_{\text{CS}}(\mathcal{A}_\mu) + \theta \mathcal{S}_{\text{CS}}(A_\mu) - \int d^3z \, \frac{\theta}{2} \epsilon_{\mu\nu\lambda} \mathcal{A}^\mu F^{\nu\lambda} \\
& - \int d^3z \int d^3z' \, \frac{\theta^2}{2} \left(\mathcal{B}(z) - B(z)\right) V(z - z') \left(\mathcal{B}(z') - B(z') \right)
\end{aligned} \tag{13.139}
$$

where we have expanded the Chern–Simons term of the action. Once again, we find that the effective action is parametrized in terms of the three quantities ϵ, χ, and σ_{xy}^0, which we have already discussed in Chapter 10. Following exactly the same arguments, we expect that ϵ and χ will have significant finite renormalizations, but the Hall conductance σ_{xy}^0 will remain unrenormalized at the value predicted by mean-field theory. Thus, we know that, for a state with an integer number p of filled Landau levels, $\sigma_{xy}^0 = \pm p/(2\pi)$ (in units of $e^2/\hbar$).

13.6 The electromagnetic response and collective modes

To determine the full electromagnetic response and the collective modes, we need to calculate the polarization tensor $\Pi_{\mu\nu}$ of the external electromagnetic perturbation A_μ, defined from its effective action

$$
S_{\text{eff}}[A_\mu] = \frac{1}{2} \int d^3z \, d^3z' \, A_\mu(z) \Pi^{\mu\nu}(z, z') A_\nu(z') + \cdots \tag{13.140}
$$

To compute this effective action, we return to the action for the quadratic fluctuations of the statistical field $\mathcal{A}_\mu$ given in Eq. (13.139). Upon integrating out the Gaussian fluctuations of the statistical gauge field $\mathcal{A}_\mu$, we find an effective action for the electromagnetic perturbation A_μ of the form of Eq. (13.140). In momentum and frequency space the polarization tensor $\Pi_{\mu\nu}$ has the same tensorial structure as $\Pi_{\mu\nu}^{\text{F}}$ of Eq. (13.137) (as is required by gauge invariance and charge conservation), but with a new set of kernels, $\Pi_0(\omega, \vec{Q})$, $\Pi_1(\omega, \vec{Q})$, and $\Pi_2(\omega, \vec{Q})$, given by

$$
\Pi_0(\omega, \vec{Q}) = -\theta^2 \frac{\Pi_0^{\text{F}}(\omega, \vec{Q})}{D(\omega, \vec{Q})} \tag{13.141}
$$

$$
\Pi_1(\omega, \vec{Q}) = \theta + \theta^2 \frac{\theta + \Pi_1^{\text{F}}(\omega, \vec{Q})}{D(\omega, \vec{Q})} + \theta^3 V(\vec{Q}) \vec{Q}^2 \frac{\Pi_0^{\text{F}}(\omega, \vec{Q})}{D(\omega, \vec{Q})} \tag{13.142}
$$

$$\Pi_2(\omega, \vec{Q}) = -\theta^2 \frac{\Pi_2^{\rm F}(\omega, \vec{Q})}{D(\omega, \vec{Q})}$$
$$+ \frac{V(\vec{Q})}{D(\omega, \vec{Q})} \left[\omega^2 \Pi_0^{\rm F}(\omega, \vec{Q})^2 - \Pi_1 F(\omega, \vec{Q})^2 \right.$$
$$\left. + \vec{Q}^2 \Pi_0^{\rm F}(\omega, \vec{Q}) \Pi_2^{\rm F}(\omega, \vec{Q}) \right] \tag{13.143}$$

$$D(\omega, \vec{Q}) = \omega^2 (\Pi_0^{\rm F}(\omega, \vec{Q}))^2 - \left(\theta + \Pi_1^{\rm F}(\omega, \vec{Q})\right)^2$$
$$+ \vec{Q}^{\,2} \Pi_0^{\rm F}(\omega, \vec{Q}) \left(\Pi_2^{\rm F}(\omega, \vec{Q}) - \theta^2 V(\vec{Q})\right) \tag{13.144}$$

In spite of the complexity of these formulas, some important consequences are easily extracted from them. The physical excitations (the collective modes) are the poles of the kernels $\Pi_0(\omega, \vec{Q})$, $\Pi_1(\omega, \vec{Q})$, and $\Pi_2(\omega, \vec{Q})$, which are the zeros of the function $D(\omega, \vec{Q}) = 0$ (defined in Eq. (13.144)).

The f-sum rule and Kohn's theorem

In Chapter 12 we showed that the current–current retarded correlation function of a physical system, $D^{\rm R}(x, x') = -i\theta(x_0 - x_0') \langle [J_\mu(x), J_\nu(x')] \rangle$, obeys a set of Ward identities that follow from gauge invariance and the conservation of the current. One of these identities is the f-sum rule:

$$\int_{-\infty}^{\infty} \frac{d\omega}{2\pi} i\omega D_{00}^{\rm R}(\omega, \vec{Q}) = \frac{\rho_0}{M} \vec{Q}^2 \tag{13.145}$$

with $D_{00}^{\rm R}(\omega, \vec{Q}) = -\Pi_{00}^{\rm R}(\omega, \vec{Q})$, where the label R means the retarded function. Equation (13.145) follows from Eq. (12.106) after taking Fourier transforms and integrating over frequencies.

What is the leading behavior of $\Pi_{00}(\omega, \vec{Q})$ at small momentum, $\vec{Q} \to 0$? This we can determine from Eq. (13.141) for (the time-ordered) $\Pi_{00}(\omega, \vec{Q})$ at small $\vec{Q}$ with ω fixed:

$$\Pi_{00}(\omega, \vec{Q}) \simeq \vec{Q}^2 \Pi_0(\omega, 0) = -\theta^2 \vec{Q}^2 \frac{\Pi_0^{\rm F}(\omega, 0)}{D(\omega, 0)} = -\frac{\rho_0}{M} \frac{\vec{Q}^2}{\omega^2 - \omega_{\rm c}^2 + i\epsilon} \tag{13.146}$$

where $\omega_{\rm c}$ is the "bare" cyclotron frequency (of electrons!)

$$\omega_{\rm c} = \frac{eB}{Mc} \equiv \frac{B}{M} \tag{13.147}$$

As we can see, this result is consistent with the f-sum rule, Eq. (13.145). It also implies that, in the $\vec{Q} \to 0$ limit, it has no corrections since the sum rule is saturated. This is, of course, equivalent to Kohn's theorem, which states that for a 2DEG in a Galilean-invariant system, the cyclotron resonance (the denominator of

Eq. (13.146)) lies exactly at the cyclotron frequency ω_c, without any renormalizations due to particle–particle interactions (Kohn, 1961). Physically this means that in a Galilean-invariant system this resonance is due to the motion of the fluid as a whole, namely of its center of mass, which is not affected by the interactions between the particles.

This result also corrects a serious difficulty of the average-field approximation. Indeed, as Kohn's theorem tells us, in a Galilean-invariant system we can replace the entire 2DEG by its center of mass, which behaves as a particle with the total charge of the fluid, $Q = Ne$, moving in the perpendicular magnetic field. Therefore the total linear momentum of the fluid $\vec{P}$ (where $\vec{P}$ is the total canonical and gauge-invariant momentum operator) should obey exactly the magnetic algebra of a particle of mass NM and charge Ne,

$$\left[P_i, P_j\right] = i\frac{e\hbar}{c}B\epsilon_{ij} \tag{13.148}$$

and see the full external magnetic field. Instead, the composite fermions (in the average-field approximation) see the partially screened magnetic field, $B_{\text{eff}} < B$. Thus, as we see, the quantum fluctuations, already at the Gaussian level, change this result by restoring the correct magnetic algebra, and yield the exact long-distance limit.

13.6.1 The anyon superfluid

We have used a similar approach to describe a theory of anyons at finite density, see Chapter 11. In that case, the value of θ was such that we found an exact cancellation of the effective coupling constant of the Chern–Simons coupling, $\theta_{\text{eff}} = \sigma_{xy}^0 + \theta = 0$. As we can see from the expression of $D(\omega, \vec{Q})$, in the limit $\vec{Q} = 0$, the analog of the Kohn mode is now a linearly dispersing mode with

$$\omega = v|\vec{Q}|, \qquad v = \sqrt{\frac{2\pi\rho_0}{M^2}} \tag{13.149}$$

which we identify with the phase mode of the anyon superfluid, with a velocity consistent with the requirements of Galilean invariance.

13.6.2 Collective modes

In addition to Kohn's mode, which has an energy at the cyclotron frequency and a residue of the order of $|\vec{Q}|^2$, this theory predicts that the lowest-energy collective mode, a magneto-phonon, has an energy at $\vec{Q} = 0$ that is a fraction of the cyclotron frequency ω_c. This mode has a residue proportional to $|\vec{Q}|^4$. In the case of the Laughlin state ($p = 1$) Girvin, MacDonald, and Platzman (Girvin *et al.*, 1986) also found a magneto-phonon mode with the same residue. However, in their theory (which works directly in the lowest Landau level using the single-mode approximation) the gap of the mode is given by the Coulomb energy and does not depend on the bare mass M of the particles. In contrast, the calculation we have just described predicts (incorrectly) that the energy depends explicitly on M and not on the Coulomb interaction. This is one of the difficulties of this approach. Indeed, this mode is expected to be corrected by terms higher in the expansion about the average-field approximation that occur at order $|\vec{Q}|^4$. Thus, although this theory predicts the correct behavior at long distances, the predictions for dimension-full quantities (such as energy gaps) not protected by symmetries cannot be trusted, even if the qualitative "level scheme" is actually correct. A program to eliminate these systematic problems was developed by Murthy and Shankar (2003).

13.7 The Hall conductance and Chern–Simons theory

The effective action of Eq. (13.139) is sufficient to find the Hall conductance, as well as the charge and statistics of the quasiparticles. Notice that, if the electromagnetic fluctuation A_ν is turned off, the action for the statistical gauge field has a Chern–Simons term with a coupling constant equal to the *sum* of the bare (θ) and induced (σ_{xy}) couplings. In the anyon superconductor of Chapter 11, these two contributions *cancelled* each other out, leading to a *compressible* state. In a fractional quantum Hall state, they *add up*, and the state is *incompressible*. It is also worthy of note that, except for the "Maxwell-like" first two terms, this expression is exact and independent of the gradient fluctuation. In particular, it contains the *exact* dependence on the interaction pair potential V.

I will show now how this formalism can be used to compute the Hall conductance σ_{xy} and the statistics of the quasiparticles. Let us first note that the quantity σ_{xy}^0 is the Hall conductance of the effective fermions in mean-field theory and that it is not equal to σ_{xy}. In particular, it is equal to $\sigma_{xy}^0 = p/(2\pi)$, and predicts an *integer* instead of a *fractional* Hall conductance. The full Hall conductance is obtained by calculating the electromagnetic response function.

If we are interested only in the behavior at very low frequency and momentum, we can further approximate S_{eff} by keeping only the terms with the smallest number of derivatives. The Chern–Simons terms have just one derivative, whereas the other terms have at least two. Thus, at long wavelengths and low frequencies, we can use the approximation

$$S_{\text{eff}}[\mathcal{A}, \tilde{A}] \approx (\sigma_{xy}^0 + \theta)\mathcal{S}_{\text{CS}}(\mathcal{A}_\mu) + \theta\mathcal{S}_{\text{CS}}(\tilde{A}_\mu) - \int d^3z\, \frac{\theta}{2}\epsilon_{\mu\nu\lambda}\mathcal{A}^\mu \tilde{F}^{\nu\lambda} \tag{13.150}$$

where only the statistical gauge field $\mathcal{A}_\mu$ is dynamical. This approximation is sufficient for our purposes. We will see below that this approximate form of the effective action is sufficient to determine the charge and statistics of the quasiparticles as well as the Hall conductance.

The electromagnetic response is calculated from the partition function

$$\begin{aligned}\mathcal{Z}[\tilde{A}] &= \int \mathcal{D}\mathcal{A}_\mu \, e^{\,iS_{\text{eff}}[\mathcal{A},\tilde{A}]} \\ &= \exp\left(\frac{i}{2}\int d^3z \int d^3z' \; \tilde{A}_\mu(z)\Pi^{\mu\nu}_{\text{eff}}(z,z')\tilde{A}_\nu(z')\right)\end{aligned} \tag{13.151}$$

where $\Pi^{\mu\nu}_{\text{eff}}(z,z')$ is the effective polarization tensor (i.e. the current–current correlation function for the full system) in the Gaussian (RPA) approximation. The calculation is particularly simple in the infrared limit.

In Chapter 10 we showed that a theory with two gauge fields, $\tilde{A}_\mu$ and $\mathcal{A}_\mu$, with just Chern–Simons terms in the action, with couplings $\theta_1 = \theta$ and $\theta_2 = \sigma^0_{xy}$, respectively, is equivalent, upon integration over $\mathcal{A}_\mu$, to a theory with a Lagrangian $\mathcal{L}_{\text{eff}}[\tilde{A}] \equiv -i \, \ln \mathcal{Z}[\tilde{A}]$ that has the Chern–Simons form

$$\mathcal{L}_{\text{eff}}[\tilde{A}] \approx \theta_{\text{eff}}\mathcal{L}_{\text{CS}}[\tilde{A}] \tag{13.152}$$

The effective Chern–Simons coupling θ_{eff} is given by

$$\frac{1}{\theta_{\text{eff}}} = \frac{1}{\theta_1} + \frac{1}{\theta_2} \tag{13.153}$$

For the values $1/\theta = 2\pi(2s)$ and $\sigma^0_{xy} = p/(2\pi)$, which we found above, we get

$$\frac{1}{\theta_{\text{eff}}} = 2\pi(2s) + \frac{2\pi}{p} \tag{13.154}$$

Since in the effective Lagrangian $\mathcal{L}_{\text{eff}}[\tilde{A}]$ we are keeping only the terms with the smallest number of gradients, we are neglecting the (even) Maxwell terms coming both from electrodynamics and from their renormalization by the charge fluctuations.

The (induced) current $J_\mu(x)$ is computed by using its usual definition:

$$J_\mu(x) = -i\,\frac{\delta \ln \mathcal{Z}[\tilde{A}]}{\delta \tilde{A}_\mu(x)} \equiv \frac{\delta \mathcal{L}_{\text{eff}}[\tilde{A}]}{\delta \tilde{A}_\mu(x)} \tag{13.155}$$

The current $J_\mu(x)$ is determined by the Chern–Simons term alone:

$$J_\mu(x) = \frac{\theta_{\text{eff}}}{2}\epsilon_{\mu\nu\lambda}\tilde{F}^{\nu\lambda}(x) \tag{13.156}$$

For a weak external static electric field $\tilde{E}_j(\vec{x})$, we find that the induced charge vanishes and that there is a non-zero Hall current, i.e.

$$\begin{aligned} \rho_{\text{ind}}(\vec{x}) &\equiv J_0(\vec{x}) = 0 \\ J_k^{\text{ind}}(\vec{x}) &\equiv \theta_{\text{eff}} \epsilon_{kj} \tilde{E}_j(\vec{x}) \end{aligned} \tag{13.157}$$

The form of the Hall current enables us to identify the Hall conductance σ_{xy} with θ_{eff}. Thus, the Hall conductance for this ground state is

$$\sigma_{xy} = \theta_{\text{eff}} = \frac{1}{2\pi} \left(\frac{p}{2sp+1} \right) \tag{13.158}$$

For the *odd* integers m, in the sequence $m = 2sp + 1$, we can write the Hall conductance as the *fraction*

$$\sigma_{xy} = \frac{1}{2\pi} \frac{p}{m} \left(\frac{e^2}{\hbar} \right) \tag{13.159}$$

where we have restored the factor $e^2/\hbar$. Hence, we get a *fractional* quantum Hall effect. The particular choice $p = 1$ yields the family of Laughlin states Ψ_m, with $m = 2s + 1$.

13.8 Quantum numbers of the quasiparticles: fractional charge

Let us now evaluate the quantum numbers of the quasiparticles within the Chern–Simons theory. In particular, we want to compute their charge and statistics. Much of what follows is a rederivation, directly from the path integral, of results that were obtained before using Berry-phase arguments. The path-integral methods have the great advantage that they are very general and widely applicable.

We first need to identify the operators which create the quasiparticles in the Chern–Simons theory. Or, at least, we need to find a set of operators whose correlation functions yield information about the spectrum of the quasiparticles.

We have already identified the collective modes. Let us now identify the *quasi-hole*. From Laughlin's theory we know that the quasihole is an *anyon* that carries fractional charge.

We will now define a gauge-invariant operator that creates an excitation at $\vec{x}$ at time x_0 and destroys it at $\vec{x}'$ at time x_0', and behaves like a quasihole. Let us consider the gauge-invariant "bilinear" operator

$$\psi^\dagger(x) \exp\left(i \int_{\Gamma(x,x')} (A_\mu + \mathcal{A}_\mu) dx_\mu \right) \psi(x') \tag{13.160}$$

where $\Gamma(x, x')$ is a path in space-time going from x to x'. By construction, this operator is invariant under gauge transformations of the statistical gauge field $\mathcal{A}_\mu$.

We will assume for the moment that the fluctuating component $\tilde{A}_\mu$ of the electromagnetic field is switched off and, therefore, this object feels only the uniform magnetic field A_μ (insofar as electromagnetism is concerned). In any event, the line integral in the exponent of the bilinear depends only on the sum of all the vector potentials. According to the procedure we used above, the fields $\tilde{A}_\mu$ and λ have already been shifted away, and do not appear explicitly in this operator. Their effect is felt through their coupling to the vector potential $\mathcal{A}_\mu$.

Let us evaluate the path-dependent correlation function $G_\Gamma(x, x')$ defined by

$$G_\Gamma(x, x') = \left\langle T \left[\psi^\dagger(x) \exp\left(i \int_{\Gamma(x,x')} (A_\mu + \mathcal{A}_\mu) dx_\mu \right) \psi(x') \right] \right\rangle \tag{13.161}$$

in a fractional quantum Hall state, where T is the time-ordering operator. This correlation function is gauge-invariant but depends on the choice of path Γ.

In path-integral language, this Green function is given by an average over the histories of Fermi and statistical fields, weighted with the amplitude $\exp(iS_\theta)$ defined earlier in this section. We now proceed to integrate out the Fermi fields, and find that the Green function is given by the average

$$G_\Gamma(x, x') = \left\langle G(x, x'|\{A_\mu + \mathcal{A}_\mu\}) \exp\left(i \int_{\Gamma(x,x')} (A_\mu + \mathcal{A}_\mu) dx_\mu \right) \right\rangle_{\mathcal{A}} \tag{13.162}$$

The function $G(x, x'|\{A_\mu + \mathcal{A}_\mu\})$ is the one-particle Green function for a problem of fermions in *fixed* statistical and electromagnetic gauge fields at finite particle density, determined by the chemical potential μ. It is straightforward to see that $G(x, x'|\{A_\mu + \mathcal{A}_\mu\})$ is the inverse of the Schrödinger operator, i.e.

$$G(x, x'|\{A_\mu + \mathcal{A}_\mu\}) = \langle x | \frac{1}{iD_0 + \mu + \lambda + (1/(2M))\vec{D}^2} | x' \rangle \tag{13.163}$$

From now on we will not write down explicitly in our formulas the constant part of the electromagnetic field, A_μ. Its presence will be assumed throughout the rest of the discussion.

The average of any operator $\mathcal{O}[\{\mathcal{A}\}]$ over all configurations of the fields $\mathcal{A}_\mu$ is given by the path integral

$$\langle \mathcal{O}[\{\mathcal{A}\}] \rangle = \frac{1}{\mathcal{Z}} \int \mathcal{D}\mathcal{A}_\mu \, \mathcal{O}[\{\mathcal{A}\}] e^{iS_{\text{eff}}[\mathcal{A}]} \tag{13.164}$$

where $\mathcal{Z}$ is the partition function and $S_{\text{eff}}[\mathcal{A}]$ is the effective action, which turns out to be given by

$$S_{\text{eff}} = -i \operatorname{tr} \ln \left[iD_0 + \mu + \lambda + \frac{1}{2M}\vec{D}^2 \right] + \theta S_{\text{CS}}(\mathcal{A}_\mu) + S_{\text{int}}[\mathcal{A}_\mu] \tag{13.165}$$

Let us now represent the one-particle Green function $G(x, x'|\{\mathcal{A}_\mu\})$ in terms of a Feynman path integral (Feynman and Hibbs, 1965; Polyakov, 1987). We first use the representation of the propagator (or Green function) as an integral of a transition-matrix element, namely

$$G(x, x'|\{\mathcal{A}_\mu\}) = -i \int_0^{+\infty} dT \, \langle \vec{x}, 0|\vec{x}', T\rangle e^{i\mu T} \tag{13.166}$$

where the weight $\exp(i\mu T)$ serves to fix the number of particles. Since the saddle-point has p filled Landau levels, the chemical potential has to be set to lie between the levels p and $p+1$. The matrix element $\langle \vec{x}, 0|\vec{x}', T\rangle$ can be written as a sum over histories by means of the Feynman formula

$$\langle \vec{x}, 0|\vec{x}', \tau\rangle = \int \mathcal{D}\vec{z}[t] e^{iS[\vec{z}(t)]} \tag{13.167}$$

with the boundary conditions

$$\lim_{t\to 0} \vec{z}(t) = \vec{x}, \qquad \lim_{t\to T} \vec{z}(t) = \vec{x}\,' \tag{13.168}$$

Thus, as usual, the matrix element $\langle \vec{x}, 0|\vec{x}', T\rangle$ is a sum over all paths $\tilde{\Gamma}$ that go from $\vec{x}$ to $\vec{x}'$ in time T. The action S in the path integral is the standard action for non-relativistic quantum mechanics for particles coupled to a gauge field,

$$S = \int_0^{\tau} dt \left\{ \frac{M}{2} \left(\frac{d\vec{z}}{dt} \right)^2 + \frac{e}{c} \frac{dz^\mu}{dt}(t) A_\mu(\vec{z}(t)) \right\} \tag{13.169}$$

where we have used the notation $z_0 \equiv t$. The second term in the integrand is a shorthand notation for the coupling to the electromagnetic and statistical gauge fields,

$$\begin{aligned} \frac{d\vec{z}}{dt}(t) A_\mu(\vec{z}(t)) \equiv & \frac{e}{c} \frac{d\vec{z}}{dt}(t) \cdot \vec{A}(\vec{z}(t)) + e A_0(\vec{z}(t)) \\ & + \frac{d\vec{z}}{dt}(t) \cdot \vec{\mathcal{A}}(\vec{z}(t)) + \mathcal{A}_0(\vec{z}(t)) + \lambda(\vec{z}(t)) \end{aligned} \tag{13.170}$$

Similar-looking formulas can be derived for the two-particle and other many-particle propagators.

For a problem with an energy gap, the long-distance, long-time limit, $|x - x'| \to \infty$, of the path integral is dominated by paths close to the solution of the classical equations of motion. Thus, in this case, the dominant trajectories are smooth. Thus, it should be a good approximation for our problem to pull the integral over the trajectories $\{\vec{z}(t)\}$ outside of the functional integral over the statistical gauge fields and over all the configurations of these fields for a fixed path γ. The averaging over the trajectories of the particle is done at a later stage. We should keep in mind that these averages are performed around the saddle-point configuration, which has an effective constant uniform magnetic field B_{eff} and a total number p of Landau levels that are completely filled. Formally, we can write the average in the form

$$G_\Gamma(x,x') = \int_0^{+\infty} dT \int \mathcal{D}\vec{z}(t) e^{i\mu T} \exp\left[i \int_0^T dt \, \frac{M}{2} \left(\frac{d\vec{z}}{dt}\right)^2 \right] \times \left\langle \exp\left(i \oint_\gamma A^\mu \, dz_\mu \right) \right\rangle_{\mathcal{A}} \tag{13.171}$$

where the set of *closed curves* $\{\gamma\}$ represents paths that are the oriented sum of the path Γ and the histories of the particle $\tilde{\Gamma}$. It is important to keep in mind that this formula is a sum over all trajectories that go from $\vec{x}$ to $\vec{x}'$ with a *fixed* return path Γ. Notice that the particle does not return to $\vec{x}$; only the gauge fields see the *closed* paths γ.

It is straightforward to find a generalization of this formalism for the calculation of the two-particle Green function. The main difference is that, for the two-particle case, there are two sets of trajectories to be summed over. The Grassmann integral automatically antisymmetrizes the two-particle Green function, which comes in the form of a sum over direct and exchange processes with the gauge fields as a fixed background.

In the semi-classical approximation, the exact average is replaced by an expansion around the solutions of the classical equations of motion. Thus, in this approximation, the particle feels only the average of the sum of the electromagnetic and statistical gauge fields. The effective field felt by the particle is equal to $B_{\text{eff}} = B - \rho_0/\theta$. Thus, for each closed trajectory γ, there is a constant factor that can be factored out from the functional integral. This factor corresponds to an Aharonov–Bohm phase factor for a particle moving in the field B_{eff}, not in the external field B. It is easy to show that, as a result of the screening of the external magnetic field, the Aharonov–Bohm phase factor is that of a particle of charge $1/m$ of the electron charge moving in the unscreened field B.

Indeed, we have that the exponent of the Aharonov–Bohm phase factor is $(2\pi/\phi_0)B_{\text{eff}}A_\perp(\gamma)$, where $A_\perp(\gamma)$ is the (spatial) cross-sectional area bounded by the path γ. Since $B_{\text{eff}} = B - \rho_0/\theta$, we can define the *effective charge* (in units of e) $q_{\text{eff}} \equiv 1 - \rho_0/(\theta B)$ and write $B_{\text{eff}} = q_{\text{eff}} B$. The effective charge q_{eff} can also be written in the more useful form

$$q_{\text{eff}} = 1 - \frac{\rho_0}{\theta B} = 1 - \frac{\rho_0 L^2}{\theta B L^2} = 1 - \frac{N}{2\pi\theta N_\phi} \tag{13.172}$$

where L is the linear size of the system. Thus, we get

$$q_{\text{eff}} = 1 - \frac{\nu}{2\pi\theta} \tag{13.173}$$

For a filling fraction $\nu = p/m \equiv p/(2sp+1)$ and $\theta = 1/(4\pi s)$, we find that the effective charge is

$$q_{\text{eff}} = 1 - \frac{2sp}{2sp+1} = \frac{1}{2sp+1} \equiv \frac{1}{m} \tag{13.174}$$

Hence, the effective charge is $\pm e/m$.

13.9 Quantum numbers of the quasiparticles: fractional statistics

The fractional statistics can be studied by considering the two-particle Green function. Recall that now we have to consider two sets of trajectories, one for each particle, which constitute a half-braid such as the one shown in Fig. 13.2. We now consider two paths γ_1 and γ_2, such as the ones discussed in Section 10.5. Here too, the configurations of paths can be classified according to their *linking number* ν_{L}. The weights of configurations with different linking numbers have different phase factors. Likewise, configurations of paths from direct and exchange processes also have different linking numbers. While the phase factors themselves depend on the trajectories, and thus on the arbitrarily chosen paths for the two particles, the *relative phase* depends only on the topological properties of the configurations of paths, and is determined entirely by the relative linking number $\Delta\nu_{\text{L}}$. In particular, we want to compare two paths that form a linked knot with two paths that do not. In this case, the linking number changes by $\Delta\nu_{\text{L}} = 1$.

If the paths are very long and wide, such as the dominant paths for the low-energy excitations, the average over the statistical gauge fields can be calculated using the effective action in the infrared approximation. This effective action

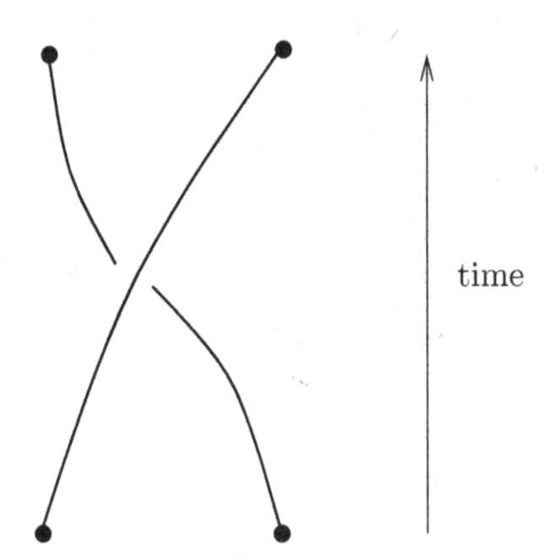

Figure 13.2 A half-braid of the worldlines of two quasiparticles is equivalent to a counterclockwise exchange.

contains only one Chern–Simons term (if $\tilde{A} = 0$), with coupling constant $\bar{\theta}$ equal to

$$\bar{\theta} = \sigma_{xy}^0 + \theta = \frac{p}{2\pi} + \frac{1}{4\pi s} \tag{13.175}$$

The arguments of Section 10.5 show that these two amplitudes differ by a factor W_{ex}, given by

$$W_{\text{ex}} = -\exp\left(i\frac{\Delta\nu_{\text{L}}}{2\bar{\theta}}\right) \tag{13.176}$$

Thus, the statistical angle δ (including the fermion sign) for all fractional quantum Hall states in the Jain sequences, with filling fraction $\nu = p/(2sp + 1)$, is given by

$$\delta = \pm\left(\frac{2s(p-1)+1}{2sp+1}\right)\pi \pmod{2\pi} \tag{13.177}$$

For the special case of the Laughlin states, with $\nu = 1/m$ (and thus $p = 1$), the statistical angle reduces to

$$\delta = \pm\frac{\pi}{m} \tag{13.178}$$

We conclude that the operator we found creates quasiholes (or, conversely, quasiparticles) of charge $\pm(e/m)$ and statistics π/m. This result agrees with the Berry-phase arguments of Arovas, Schrieffer, and Wilczek and with the calculations based on the Laughlin wave function, which we summarized in Section 13.1.4. The power of the derivation that we just gave lies in the fact that it follows directly from the general principles of quantum mechanics (just as the Berry-phase arguments do), but without the need to make any specific ansatz for the wave functions for the ground state and for the quasihole. The adiabatic approximation, which is essential to the Berry-phase argument, is just as important here, since it results from the existence of an energy gap. But the general formula for the path integral is valid even in the absence of a gap.

14

Topological fluids

In this chapter we will develop the effective-field theory of topological fluids, focusing on the fractional quantum Hall states as the prototype.

14.1 Quantum Hall fluids on a torus

Quantum Hall states are topological electron fluids whose properties depend on the topology of the surface on which the electrons are bound. In this section we will discuss the case of a fluid confined to a 2D torus. Although considering a 2D electron gas on a torus is of little experimental value, it is a great conceptual (and numerical) tool.

For the Laughlin states we have discussed there is the implicit assumption that the electrons are confined in a simply connected region of the plane by an external confining potential. Thus, the surface on which the electrons live has the topology of a disk or, which is equivalent, a sphere. In this geometry the ground state is unique.

In Section 12.7 we discussed the state of a free-fermion system on a torus with filling factor $\nu = 1$ (a full Landau level), and we showed that the ground state on a torus is also unique. The reason for this is that the filling factor is 1. Thus, in a translationally invariant system the motion of the center of mass of the electron fluid decouples from the relative motion of the electrons in the fluid. The motion of the center of mass is that of a single particle moving in a magnetic field with one flux quantum. Thus the state is unique.

The situation is different in the case of Laughlin states, since they have a filling fraction $\nu = 1/m$. Thus the center of mass behaves as a charged particle moving in the field of m flux quanta. Hence we expect that the ground-state wave function should be m-fold degenerate. Haldane and Rezayi (1985) gave an explicit construction of the Laughlin state on a torus (essentially the generalization of what we discussed in Section 12.7). These wave functions (in the Landau gauge) have the same factorized form as that of Eq. (12.157) expressed in terms of theta functions. The difference is that in the case of Laughlin states there are m linearly independent states. Later in this chapter we will see that in the case of the non-abelian quantum Hall states the degeneracy on a torus is not determined solely by the motion of the center of mass of the fluid.

In this section we will see how the topology of the torus affects the concept of flux attachment and the construction of the effective-field theory of the quantum Hall states.

14.1.1 Flux attachment on a torus

We will now show how to define flux attachment in a manner compatible with the requirement of quantization of the abelian Chern–Simons coupling constant or, which amounts to the same thing, of invariance under large gauge transformations. Here we follow in detail the work of Fradkin, Nayak, Tsvelik, and Wilczek (Fradkin *et al.*, 1998).

Consider a theory of particles (in first quantization) that interact with each other as they evolve in time. We will assume in what follows that the particles are fermions (in two spatial dimensions) and that their worldlines never cross. The actual choice of statistics is not important in what follows, but the requirement that there is no crossing is important and, for bosons, it implies the assumption that there is a hard-core interaction, whereas for fermions the Pauli principle takes care of this issue automatically. For simplicity, we will assume that the time evolution is periodic, with a very long period.

The worldlines of the particles can be represented by a conserved current j_μ. For a given history of the system, the worldlines form braids with well-defined linking numbers $\nu_{\rm L}[j_\mu]$, which, as we saw before, are given by

$$\nu_{\rm L}[j_\mu] = \int d^3x \; j_\mu(x) B^\mu(x) \tag{14.1}$$

where j_μ and B_μ are related through Ampère's law, cf. Eq. (10.109),

$$\epsilon_{\mu\nu\lambda}\, \partial^\nu B^\lambda(x) = j_\mu(x) \tag{14.2}$$

Under the assumption of the absence of crossing of the worldlines of the particles, the linking $\nu_{\rm L}[j_\mu]$ is a topological invariant. Thus, if $S[j_\mu]$ is the action for a given history, then the quantum-mechanical amplitudes of all physical observables remain unchanged if the action is modified by

$$S[j_\mu] \to S[j_\mu] + 2\pi s \nu_{\rm L}[j_\mu] \tag{14.3}$$

where $s \in \mathbb{Z}$ is an arbitrary integer.

The quantum-mechanical amplitudes are sums over histories of the particles, and take the form

$$W[\{j_\mu\}] = \sum_{[j_\mu]} e^{\,iS[j_\mu]+2\pi i s \nu_{\rm L}[j_\mu]} e^{\,i\phi[j_\mu]} \tag{14.4}$$

where $\phi[j_\mu]$ is a phase factor that accounts for the statistics of the particles (0 for bosons and π for fermions).

However, the amplitudes remain unchanged if in the integrand of Eq. (14.4) we insert the number 1 written as the following expression:

$$\begin{aligned} 1 &\equiv \int \mathcal{D}b_\mu \prod_x \delta(\epsilon_{\mu\nu\lambda}\, \partial^\nu b^\lambda - j_\mu) \\ &= \mathcal{N} \int \mathcal{D}b_\mu\, \mathcal{D}a_\mu\, \exp\left(\frac{i}{2\pi} \int d^3x\, a^\mu \left[\epsilon_{\mu\nu\lambda}\, \partial^\nu b^\lambda - j_\mu \right] \right) \end{aligned} \tag{14.5}$$

where $\mathcal{N}$ is a normalization constant and we have used a representation of the delta function in terms of a Lagrange-multiplier vector field a_μ. Notice that, since j_μ is locally conserved, i.e. $\partial_\mu j^\mu = 0$, these expressions are invariant under the gauge transformations $a_\mu(x) \to a_\mu(x) + \partial_\mu \Lambda(x)$.

After using the constraint $j_\mu = \epsilon_{\mu\nu\lambda}\, \partial^\nu b^\lambda$, the amplitude can also be written in the equivalent form

$$\begin{aligned} W[\{j_\mu\}] = \sum_{[j_\mu]} \int \mathcal{D}b_\mu\, \mathcal{D}a_\mu\, e^{\,iS[j_\mu]+2\pi i s \nu_{\rm L}[j_\mu]}\, e^{i\phi[j_\mu]} \\ \times \exp\left(i \int d^3x\, a^\mu(x) \frac{1}{2\pi} \left[\epsilon_{\mu\nu\lambda}\, \partial^\nu b^\lambda - j_\mu \right] \right) \end{aligned} \tag{14.6}$$

We can then compute this amplitude as a path integral of a theory in which the particles whose worldlines are represented by the currents j_μ interact with the gauge fields a_μ and b_μ. These interactions are encoded in the effective Lagrangian

$$\mathcal{L}_{\rm eff}[a, b, j] = \frac{1}{2\pi} \epsilon_{\mu\nu\lambda} a^\mu\, \partial^\nu b^\lambda - a^\mu j_\mu - \frac{2s}{4\pi} \epsilon_{\mu\nu\lambda} b^\mu\, \partial^\nu b^\lambda \tag{14.7}$$

where we have used the constraint $j_\mu = \epsilon_{\mu\nu\lambda}\, \partial^\nu b^\lambda$ to write the winding number in the form of a Chern–Simons action for the gauge field b_μ (Wilczek and Zee, 1983; Wu and Zee, 1984). Hence, the amplitudes can be written in terms of a path integral over an abelian Chern–Simons gauge field with a correctly quantized coupling constant equal to $2s/(4\pi)$. The first term of the effective Lagrangian of Eq. (14.7), the cross term involving both of the gauge fields, a^μ and b^μ, is called the BF Lagrangian since it couples a vector potential to the field strength of another field.

As we noted earlier in this chapter, the usual form of the flux-attachment transformation is found by integrating out the gauge field b_μ. For vanishing boundary conditions at infinity, which is to say for a system on a surface with the topology of a disk, this leads to an effective action for the field a_μ of the conventional form (López and Fradkin, 1991)

$$S_{\text{eff}}[a] = \frac{\theta}{2} \int d^3x \; \epsilon_{\mu\nu\lambda} a^\mu \, \partial^\nu a^\lambda, \qquad \theta = \frac{1}{2\pi \times 2s} \tag{14.8}$$

which is the expression we have used before for the statistical gauge field. This form of the effective action is not valid for manifolds with non-trivial topology. However, Eq. (14.7) is correct in all cases since it is invariant under both local and large gauge transformations. In addition to being consistent on closed manifolds, the action of Eq. (14.7) treats the statistical gauge field a_μ and its dual b_μ on an equal footing. We will see in the next section that the dual field b_μ arises naturally in a hydrodynamic theory, and plays a central role in Wen's construction of the abelian fractional quantum Hall hierarchy (Wen, 1995).

We can now proceed as before, the only difference being that the (composite) fermions (or bosons, depending on our choice) couple to the gauge field b_μ rather than to the field a_μ (which plays the role of the statistical gauge field). In this fashion the mean-field theory in the composite-fermion language proceeds by first spreading out the field and constructing an effective integer Hall effect of the partially screened magnetic field. The composite-fermions fill up p effective Landau levels. The effective action in the composite-fermion picture is found by integrating out the local particle–hole fluctuations of the fermions about the uniform mean-field state. This leads to an effective-field theory with the following effective Lagrangian (López and Fradkin, 1999):

$$\begin{aligned}\mathcal{L} = {} & \frac{p}{4\pi} \epsilon_{\mu\nu\lambda} a^\mu \, \partial^\nu a^\lambda - \frac{2s}{4\pi} \epsilon_{\mu\nu\lambda} b^\mu \, \partial^\nu b^\lambda + \frac{1}{4\pi} \epsilon_{\mu\nu\lambda} c^\mu \, \partial^\nu c^\lambda \\ & + \frac{1}{2\pi} \epsilon_{\mu\nu\lambda} a^\mu \, \partial^\nu b^\lambda - \frac{e}{2\pi} \epsilon_{\mu\nu\lambda} b^\mu \, \partial^\nu A^\lambda - a_\mu j^\mu_{\text{qp}} - e_\mu j^\mu_{\text{qp}}\end{aligned} \tag{14.9}$$

where we introduced an additional gauge field e_μ to track the fermion sign of the composite fermions, and A_μ is an external electromagnetic perturbation. The currents j^μ_{qp} describe the worldlines of the excitations, i.e. the particles outside the condensate. The result is a description of the states in the generalized Jain hierarchies $\nu_\pm(s, p) = p/(2sp \pm 1)$, where $p, s \in \mathbb{Z}$ and the $\pm$ signs apply for an electron-like and a hole-like FQH fractional quantum Hall state, respectively.

The low-energy effective Lagrangian for the gauge fields can be written in terms of a 3×3 matrix of coupling constants (the K-matrix of Wen and Zee's generalized fractional quantum Hall fluids (Wen and Zee, 1992), see the next section)

$$\mathcal{L}_{\text{eff}} = \frac{1}{4\pi} K_{IJ} \epsilon^{\mu\nu\lambda} a_\mu^I \, \partial_\nu a_\lambda^J - e \frac{1}{2\pi} t_I \epsilon_{\mu\nu\lambda} a_I^\mu \, \partial^\nu A^\lambda - \ell_I j_{\text{qp}}^\mu a_\mu^I \tag{14.10}$$

with $(I, J = 1, 2, 3)$

$$K_{IJ} = \begin{pmatrix} -2s & 1 & 0 \\ 1 & p & 0 \\ 0 & 0 & 1 \end{pmatrix} \tag{14.11}$$

where $a_\mu = a_\mu^1$, $b_\mu = a_\mu^2$, and $c_\mu = a_\mu^3$. The charge vector $t_I = (1, 0, 0)$ indicates which gauge field represents the charge current, and the excitation vector $\vec{\ell} = (0, 1, -1)$ assigns the quantum numbers to the excitations.

We now notice that this effective theory is globally well defined since the Chern–Simons coupling constants are correctly quantized. Indeed, if we integrate out the gauge field $b_\mu = a_\mu^2$, we find the same effective action for a_μ as that of Eq. (13.150). For the Laughlin sequence, with $p = 1$, we can integrate out the gauge field b_μ (for a system on a disk!) and recover the effective action for the statistical field a_μ of Eq. (14.8).

14.1.2 Chern–Simons on a torus

Chern–Simons gauge theories (both one- and multi-component abelian as well as non-abelian) are topological field theories. We have already seen that this means that they do not have local degrees of freedom and that their energy–momentum tensors (and hence their Hamiltonians) are equal to zero. Nevertheless, on topologically non-trivial closed manifolds, such as the torus, they support finite-dimensional Hilbert spaces. The dimension of these Hilbert spaces is their ground-state degeneracy. We will not give here a detailed proof of this statement (Wen, 1989; Witten, 1989; Wen, 1990c; Wen and Niu, 1990; Wesolowski *et al.*, 1994) but present the basic ideas.

Let us consider for concreteness the case of a rectangular torus with coordinates $0 \leq x_i \leq L_i$ ($i = 1, 2$ being the two orthogonal directions on the torus). We will consider an M-component Chern–Simons theory with a Lagrangian of the form of Eq. (14.10). In this case we have M gauge fields a_μ^I with $I = 1, \ldots, M$, and K_{IJ} is an $M \times M$ symmetric matrix of integer coefficients. The gauge group in this case is $\mathrm{U}(1)^M$. The Gauss-law constraint for this theory, in the absence of sources (temporal Wilson loops), simply states that the vacuum states $|\text{vac}\rangle$ satisfy

$$K_{IJ} \epsilon_{ij} \, \partial_i a_j^J |\text{vac}\rangle = 0 \tag{14.12}$$

which is obeyed by pure gauge configurations of the form (with no summation over the spatial label i) (for each $I = 1, \ldots, M$)

$$a_i^I(x) = \partial_i \Phi^I(x) + \frac{\bar{a}_i^I}{L_i} \tag{14.13}$$

where $\Phi^I(x)$ are local ("small") smooth gauge transformations and $\bar{a}_i^I$ are the two holonomies of the torus (for each field),

$$\int_0^{L_i} dx_i \, a_i^I(x) = \bar{a}_i^I \tag{14.14}$$

By plugging this solution into a Lagrangian of the form of Eq. (14.10) we find that the holonomies $\bar{a}_i^I$ obey the effective Lagrangian (not density!) (repeated indices are summed over)

$$L = \frac{1}{4\pi} K_{IJ} \epsilon_{ij} \bar{a}_i^I \, \partial_0 \bar{a}_j^J = \frac{1}{4\pi} K_{IJ} \left(\bar{a}_1^I \, \partial_0 \bar{a}_2^J - \bar{a}_2^I \, \partial_0 \bar{a}_1^J \right) \tag{14.15}$$

This Lagrangian tells us that the x- and y-components of the holonomies form canonical pairs. Upon quantization they become operators acting on the Hilbert space and satisfy the equal-time commutation relations

$$\left[\bar{a}_1^I, \bar{a}_2^J \right] = i 2\pi K_{IJ}^{-1} \tag{14.16}$$

Thus, in the $\bar{a}_1^I$ representation, the operator $\bar{a}_2^I$ is a differential operator (and vice versa)

$$\bar{a}_2^I = -2\pi i K_{IJ}^{-1} \frac{\partial}{\partial \bar{a}_1^J} \tag{14.17}$$

Let us now define the Wilson lines for the holonomies, W_i^I (with no summation over i):

$$W_i^I = \exp\left(i \int_0^{L_i} dx_i \, a_i^I \right) = e^{i \bar{a}_i^I} \tag{14.18}$$

The requirement that the path integral be invariant under both local and large gauge transformations (which is the reason for the quantization of the Chern–Simons coupling constants, as we saw in Chapter 10) now implies the invariance of the Wilson lines, W_i^I, under large gauge transformations that shift $\bar{a}_i^I \to \bar{a}_i^I + 2\pi$. This has the effect of *compactifying* the target space to an M-torus. The unitary operators that induce these large gauge transformations are

$$U_i^I = e^{i \epsilon_{ij} K_{IJ} \bar{a}_j^J} \tag{14.19}$$

and obey the following algebra:

$$U_1^I U_2^J = e^{-2\pi i K_{IJ}} U_2^J U_1^I \tag{14.20}$$

$$W_1^I W_2^J = e^{-2\pi i K_{IJ}^{-1}} W_2^J W_1^I \tag{14.21}$$

$$U_i^I W_j^J = W_j^J U_i^I \tag{14.22}$$

We recognize the close similarity of this structure to that in our discussion of Wilson and 't Hooft loops in Chapter 9 and also to the magnetic algebra discussed in Chapter 12.

A straightforward (but lengthy) line of argument (Wesolowski *et al.*, 1994), similar to what we did for Landau levels in Chapter 12, shows that the dimension of this (topological) Hilbert space is $|\det K|$. For a general surface of genus g (with g handles) the degeneracy is (Wen and Zee, 1992; Wen, 1999)

$$|\det K|^g \tag{14.23}$$

Chern–Simons gauge theories with non-abelian gauge groups have a similar ground-state degeneracy on topologically non-trivial manifolds (Witten, 1989).

We conclude that the ground-state degeneracy of Jain states on a torus is the absolute value of the determinant

$$|\det K| = |2sp \pm 1| \tag{14.24}$$

Thus, we find that the Jain states are $|2np \pm 1|$-fold degenerate on the torus, which is the correct result.

14.2 Hydrodynamic theory

The microscopic description of the FQHE of the preceding sections led us to an effective action for the long-distance and low-energy physics involving a Chern–Simons action for a gauge field. The validity of this description is based on several key observations.

We begin with the fact of the existence of a set of filling fractions at which the fractional quantum Hall fluid is formed. At those precisely defined densities the 2DEG behaves as an incompressible fluid, and responds as such to an external electromagnetic perturbation. In particular, it exhibits a non-dissipative Hall current. If the number of electrons at fixed magnetic field does not precisely correspond to one of these "magic fractions," the excess (or defect) electrons produce a number of excitations. In the bosonic picture we saw that the excess electrons can be viewed as defects or vortices in the fluid, excitations that carry fractional charge and fractional statistics, *anyons*. Since the vortex charge is a fraction of the charge of one electron, adding or removing a full electron is equivalent to adding or removing several quasiparticles (or vortices). Therefore, in a fluid state that exhibits the FQHE, a fractional quantum Hall state, the electron behaves as if it were a composite object, with the quasiparticles being the "fundamental" entities. In other terms, in fractional quantum Hall states the electron fractionalizes. This picture is naturally correct for describing excitations very close to the ground state. At very high energies (or at very short distances) the quasiparticle picture is no longer accurate, and the electron behaves as what it is, an electron.

We will now take a different approach and treat the electron gas in a large magnetic field as an incompressible fluid. The physics of these fluids can be deduced from general basic considerations of symmetries and conservation laws, rather than from a detailed microscopic theory. Since this description is applicable only for the low-energy physics, which describes slowly moving excitations, it is natural to try to reformulate the problem in terms of hydrodynamics. In this section we will develop a theory with this structure. This approach was originally suggested by Fröhlich and Kerler (1991) and Fröhlich and Zee (1991), was developed in full by Wen and Zee (1992), and was further generalized by Wen (1995). Here we will follow in some detail the analysis and notation of Wen (1995).

We begin by considering a system of (for the moment) fully polarized electrons, which we will treat as if they were spinless. The particle-coordinate Heisenberg operators are $\{\vec{x}_i\}$ and the velocity Heisenberg operators are $\{\vec{v}_i = d\vec{x}_i/dt\}$ $(i = 1, \ldots, N)$. We can define the local density $J_0(\vec{x})$ and current $\vec{J}(\vec{x})$ operators by the obvious expressions

$$J_0(\vec{x}) = \sum_{i=1}^{N} \delta(\vec{x} - \vec{x}_i), \qquad \vec{J}(\vec{x}) = \sum_{i=1}^{N} \vec{v}_i \delta(\vec{x} - \vec{x}_i) \tag{14.25}$$

The condition of local charge conservation means that the 3-vector $J_\mu(x) = (J_0, \vec{J})$ (with $x = (t, \vec{x})$) obeys the continuity equation

$$\partial_\mu J^\mu = 0 \quad \Leftrightarrow \quad \partial_t J_0 + \vec{\nabla} \cdot \vec{J} = 0 \tag{14.26}$$

What we want to do is formulate a theory of the FQHE that is based on hydrodynamics, i.e. a theory of locally conserved currents and densities in a large magnetic field. We will postulate an action that depends only on the distribution of currents and densities, $S[J_\mu]$. Incompressibility of the quantum fluid then implies that the effective action must be a local function of the currents and their derivatives, i.e.

$$S[J_\mu] = \int d^3x \, \mathcal{L}[J_\mu] \tag{14.27}$$

where $\mathcal{L}[J_\mu]$ is a local Lagrangian density.

On the other hand, since the current J_μ is locally conserved, and hence obeys Eq. (14.26), it can be expressed as the curl of a vector field $b_\mu(x)$,

$$J_\mu(x) = \frac{1}{2\pi}\epsilon_{\mu\nu\lambda}\,\partial^\nu b^\lambda \tag{14.28}$$

which guarantees that the current is conserved. Here, as before, $\epsilon_{\mu\nu\lambda}$ is the totally antisymmetric Levi-Civita third-rank tensor. The vector field b_μ is actually a gauge field. Indeed, under a local gauge transformation $\Phi(x)$ (where $\Phi(x)$ is a smooth function of the space-time coordinates), the current distribution remains unchanged,

$$b_\mu(x) \to b_\mu(x) + \partial_\mu\Phi(x), \qquad J_\mu(x) \to J_\mu(x) \tag{14.29}$$

Hence, the effective action of the currents, $S[J_\mu]$, must also be invariant under the gauge transformation, Eq. (14.29). In other words, the hydrodynamic theory of an incompressible fluid with a conserved current is a gauge theory. Locality and gauge invariance now require that the action be a local gauge-invariant function of the hydrodynamic gauge field, $S[b_\mu]$.

What is the form of $S[b_\mu]$? A natural guess is to write a Maxwell-type action, which has two derivatives of the gauge field. Since the current and charge densities are dimension-2 operators (since their integrals over finite regions of space must have units of charge), the hydrodynamic gauge field b_μ must be a dimension-1 operator (to be consistent with it being a 1-form). This means that a Maxwell-type Lagrangian density, which has two fields and two derivatives, is a dimension-4 operator. For the action to be dimensionless in $(2+1)$ space-time dimensions, a Maxwell-type term must have a coefficient with the units of length, or, which amounts to the same thing, the inverse of the energy gap of the incompressible fluid. We have already found the same scaling in the theory of the chiral spin liquid, Eq. (10.78).

However, a time-reversal-invariant Maxwell action cannot describe a fluid of charged particles in an external magnetic field, since the latter breaks time-reversal invariance. Only an action with an odd number of time derivatives can do that. There is a unique gauge-invariant action that is odd under both time-reversal invariance and parity (but it is invariant under their product): the Chern–Simons action. The Chern–Simons action not only has the correct transformation properties under time reversal and parity but also has just one derivative. Hence it is more relevant than a Maxwell action. The conclusion of this analysis is that the leading term of the effective action at low energies of the hydrodynamic gauge field must have the Chern–Simons form. However, gauge invariance must also apply to fluids on closed surfaces. We saw in Section 10.4 that the coupling constant of the Chern–Simons action, the level, must obey a quantization condition in order for the path integral to be gauge-invariant on closed surfaces.

Therefore, we are led to postulate that the natural low-energy effective Lagrangian density for an incompressible charged fluid in an external uniform magnetic field is a Chern–Simons Lagrangian for the statistical gauge field b_μ,

$$\mathcal{L}[b_\mu] = -\frac{m}{4\pi}\epsilon^{\mu\nu\lambda} b_\mu \, \partial_\nu b_\lambda - \frac{e}{2\pi} A^\mu \epsilon_{\mu\nu\lambda} \, \partial^\nu b^\lambda \tag{14.30}$$

where $m \in \mathbb{Z}$ must be an integer in order to satisfy the quantization condition. The last term is the $J_\mu A^\mu$ coupling of the external electromagnetic field (the *total* field, not just a probe).

To check whether this hydrodynamic theory is correct, we will now find its predictions. We will first compute the effective action for the electromagnetic field A_μ in order to determine the Hall conductance. Upon integrating out the hydrodynamic gauge field b_μ we find, as expected, that it also has a Chern–Simons form:

$$\mathcal{L}_{\text{eff}}[A_\mu] = \frac{e^2}{4\pi m}\epsilon_{\mu\nu\lambda} A^\mu \, \partial^\nu A^\lambda \tag{14.31}$$

The induced current J_μ^{ind} is

$$-eJ_\mu^{\text{ind}} = \frac{\delta \mathcal{L}_{\text{eff}}}{\delta A^\mu} = \frac{e^2}{2\pi m}\epsilon_{\mu\nu\lambda} \, \partial^\nu A^\lambda \tag{14.32}$$

Therefore the fluid has a Hall conductivity (in units with $\hbar = c = 1$) of

$$\sigma_{xy} = \frac{1}{m}\frac{e^2}{2\pi} \tag{14.33}$$

In other words, for $m \neq 1$ the fluid has an FQHE, corresponding to a filling fraction $\nu = 1/m$.

Notice that these arguments do not require m to be an odd integer, and they work just as well for m even. Thus, they also apply for a system of charged bosons in a magnetic field. Indeed, for a system of charged bosons we can also write the same Laughlin wave function as that of Eq. (13.5). For bosons the wave function must be symmetric under exchange, which requires m to be even. The simplest bosonic Laughlin state has filling factor $\nu = 1/2$.

To check this theory further, we will now compute the statistics of the quasiparticles/quasiholes. We will assume that the quasiparticle/quasihole is the lowest-energy excitation of this incompressible fluid. In the low-energy regime the smooth worldlines of these excitations can be represented by a set of currents, j_μ, that couple in a gauge-invariant way to the hydrodynamic gauge field b_μ. Including the excitations, the total Lagrangian density is (Wen, 1995)

$$\mathcal{L}[b_\mu] = -\frac{m}{4\pi}\epsilon^{\mu\nu\lambda} b_\mu \, \partial_\nu b_\lambda - \frac{e}{2\pi} A^\mu \epsilon_{\mu\nu\lambda} \, \partial^\nu b^\lambda + q j^\mu b_\mu \tag{14.34}$$

where $q = \pm 1$ correspond to the histories of the quasiparticles (+) and quasiholes (−), respectively.

We can now integrate out the hydrodynamic gauge field b_μ to obtain an effective action for the excitations (represented by the quasiparticle current j_μ) and the total external electromagnetic field A_μ:

$$\mathcal{L}_{\text{eff}}[A_\mu, j_\mu] = \frac{e^2}{4\pi m}\epsilon_{\mu\nu\lambda}A^\mu \partial^\nu A^\lambda - \frac{e}{em} j_\mu A^\mu + q^2\frac{\pi}{m} j_\mu B^\mu \tag{14.35}$$

where B_μ is a "magnetostatic" field generated by the quasiparticle current using Ampère's law, $j_\mu = \epsilon_{\mu\nu\lambda}\partial^\nu B^\lambda$.

As in our discussion on fractional statistics in Section 10.5, here too the integral of the last term in Eq. (14.35) is identified with the Hopf invariant, Eq. (10.105):

$$\int d^3x\; j_\mu B^\mu = \int d^3x \int d^3y\; j_\mu(x)\epsilon^{\mu\nu\lambda}\left\langle x\left|\frac{1}{\partial^2}\right|x'\right\rangle \partial^y_\lambda j_\nu(y) \tag{14.36}$$

In fact, the analysis we did in Section 10.5 tells us that the quasiparticles (and quasiholes), represented by their currents j_μ, have fractional statistics with a statistical angle given by the prefactor of the Hopf/Chern–Simons term of Eq. (14.35):

$$\delta = q^2\frac{\pi}{m} = \frac{\pi}{m} \tag{14.37}$$

which agrees with our results earlier in this chapter. Here we used the fact that $q^2 = 1$ both for quasiparticles and for quasiholes.

We can now compute the electromagnetic current, $-J_\mu$, by differentiating the effective action with respect to the external electromagnetic field:

$$-eJ_\mu = \frac{e^2}{2\pi m}\epsilon_{\mu\nu\lambda}\partial^\nu A^\lambda - \frac{e}{m} j_\mu \tag{14.38}$$

where we identify the first term with the current of the ground state (in the presence of external fields) and the second term with a quasiparticle contribution. Let us consider the case in which we have just one static quasiparticle at rest at the origin, which we represent by the quasiparticle density $j_0 = \delta(\vec{x})$ and current $\vec{j} = 0$. Equation (14.38) now becomes (with B being the uniform external magnetic field)

$$-eJ_0 = \frac{e^2}{2\pi m}B - q\frac{e}{m}\delta(\vec{x}) \tag{14.39}$$

Upon integrating this expression over the region occupied by the fluid, we find the total charge $Q = Q_{\text{gnd}} + Q_{\text{excitation}}$, where

$$Q_{\text{gnd}} = \frac{e^2}{2\pi m}BL^2 = \frac{e^2}{2\pi m}N_\phi\frac{2\pi}{e} = e\frac{1}{m}N_\phi = eN \tag{14.40}$$

is the ground-state charge (and tells us that the filling fraction is $\nu = 1/m$), and

$$Q_{\text{excitation}} = q\frac{e}{m} \tag{14.41}$$

is the charge of the excitation. For a *quasiparticle* the charge is $Q_{\text{qp}} = e/m$, whereas for a quasihole it is $Q_{\text{qh}} = -e/m$. Hence, as we found before, the excitations have fractional charge $\pm e/m$ and fractional statistics π/m.

Using the arguments of Wilczek and Zee (1983), we can assign an effective *fractional spin* S to the quasiparticles by demanding consistency with the spin-statistics theorem:

$$2\pi S = \delta = \frac{\pi}{m} \qquad \Rightarrow \qquad S = \frac{1}{2m} \tag{14.42}$$

This result agrees with the detailed analysis of Einarsson *et al.* (1995). In this context, fractional spin means that the quasiparticles have an internal structure with an associated fractional intrinsic angular momentum. Unlike the relativistic concept of spin, fractional spin is not associated with a spinor structure and does not require the existence of more degrees of freedom. In this sense, unlike the spin of the electron, the fractional spin is not an independent quantum number of the quasiparticles. Nevertheless, it is possible to couple to the fractional spin of the quasiparticles in some circumstances. One situation in which this matters is a quantum Hall fluid on a sphere, where the fractional spin couples to the curvature of the surface (Wen, 1995). This is important in numerical simulations, which are often done on the sphere. The concept of fractional spin plays a crucial role in the explanation of the physical properties of fractional quantum Hall fluids, such as the (non-dissipative) Hall viscosity (Avron *et al.*, 1995; Read, 2009; Haldane, 2011; Hoyos and Son, 2012).

Notice that if we pick $q = m$ (i.e. m quasiparticles) the corresponding charge of this object is $Q = e$ and the statistical angle is $\delta = m\pi$. Thus, for m odd, the composite object of m quasiparticles has charge e and is a fermion, i.e. it is the *electron*. Hence, in this theory the electron is fractionalized.

14.3 Hierarchical states

We will now briefly describe a generalized hydrodynamic theory that describes the fractional quantum Hall hierarchical states as well as multi-component fluids. Once again we follow the work of Wen (1995).

In this approach the hierarchical states are constructed from a set of nested Bose condensates of quasiparticles, and this method hence follows closely the Haldane–Halperin construction. An equivalent construction can be used for the equivalent Jain generalized hierarchical states.

We begin by allowing the quasiparticles to be dynamical excitations. To this end, we will add a quasiparticle kinetic-energy term $\mathcal{L}_{\text{qp-KE}}[j_\mu]$ to the effective Lagrangian of Eq. (14.34). We will write the total effective Lagrangian as the sum of a condensate contribution (the first two terms) and a quasiparticle contribution (the last two terms):

$$\mathcal{L}[b_\mu, A_\mu, j_\mu] = -\frac{m}{4\pi}\epsilon^{\mu\nu\lambda} b_\mu\, \partial_\nu b_\lambda - \frac{e}{2\pi} A^\mu \epsilon_{\mu\nu\lambda}\, \partial^\nu b^\lambda + q b_\mu j^\mu + \mathcal{L}_{\text{qp-KE}} \tag{14.43}$$

where $q = \pm 1$ (for quasiparticles and quasiholes).

The quasiparticles will be assumed to be in a fractional quantum Hall fluid of their own. Since the quasiparticle current j_μ is conserved, it can also be described by its own hydrodynamic gauge field, which we will denote by c_μ, defined by

$$j_\mu = \frac{1}{2\pi}\epsilon_{\mu\nu\lambda}\, \partial^\nu c^\lambda \tag{14.44}$$

The same assumptions as those we made for electrons in a magnetic field now apply to the quasiparticle current. Thus, the field c_μ will be governed by a Chern–Simons Lagrangian $\mathcal{L}'$, also with a quantized coupling constant, representing the ideal fractional quantum Hall state of the excitations:

$$\mathcal{L}' = -\frac{n}{4\pi}\epsilon_{\mu\nu\lambda} c^\mu\, \partial^\nu c^\lambda + \frac{1}{2\pi} b^\mu \epsilon_{\mu\nu\lambda}\, \partial^\nu c^\lambda \tag{14.45}$$

where n is an *even* integer. Thus we will postulate that, in the absence of additional excitations, the effective Lagrangian at the second level of this hierarchy is

$$\begin{aligned}\mathcal{L}[b_\mu, A_\mu, c_\mu] = &-\frac{m}{4\pi}\epsilon^{\mu\nu\lambda} b_\mu\, \partial_\nu b_\lambda - \frac{e}{2\pi} A_\mu \epsilon^{\mu\nu\lambda}\, \partial_\nu b_\lambda \\ &- \frac{n}{4\pi}\epsilon^{\mu\nu\lambda} c_\mu\, \partial_\nu c_\lambda + \frac{1}{2\pi} b_\mu \epsilon^{\mu\nu\lambda}\, \partial_\nu c_\lambda\end{aligned} \tag{14.46}$$

What filling fraction does this effective theory describe? To find out, we will derive the Chern–Simons constraint (Gauss's law). Since we now have two gauge fields, b_μ and c_μ, we have two constraints:

$$\begin{aligned}\frac{\delta\mathcal{L}}{\delta b_0} = 0 \;&\Rightarrow\; -eB = m\langle\epsilon_{ij}\, \partial_i b_j\rangle - \langle\epsilon_{ij}\, \partial_i c_j\rangle \\ \frac{\delta\mathcal{L}}{\delta c_0} = 0 \;&\Rightarrow\; n\langle\epsilon_{ij}\, \partial_i c_j\rangle = \langle\epsilon_{ij}\, \partial_i b_j\rangle\end{aligned} \tag{14.47}$$

The filling factor is the ratio of the number of electrons, $N_e = (1/(2\pi)) \times \langle\epsilon_{ij}\, \partial_i b_j\rangle L^2$, to the total number of flux quanta, $N_\phi = -(e/2\pi))BL^2$. Hence, from Eq. (14.47) we find that the filling fraction at the second level of the hierarchy is

$$\nu = \frac{\langle\epsilon_{ij}\, \partial_i b_j\rangle}{-eB} = \frac{1}{m - 1/n} = \frac{n}{nm - 1} \tag{14.48}$$

with the Hall conductance $\sigma_{xy} = \nu e^2/h$. For example, if the "parent" state is the Laughlin state at $\nu = 1/3$ ($m = 3$), the first "daughter" state will have $n = 2$, and the total filling fraction is $\nu = 2/5$ (which we saw is also describable as a Jain state).

Once again, if the system is not precisely at this filling fraction, the fluid will have a number of excitations. They are also represented by a set of quasiparticle currents, $j_\mu^{\rm qp}$, which are minimally coupled to the two hydrodynamic fields and two integer charges.

It will be convenient to change to a more compact notation introduced by Wen and Zee (1992). We will relabel the gauge fields as $b_\mu = b_\mu^1$ and $c_\mu = b_\mu^2$, and the quasiparticle charges as $q = \ell_1$ and $q' = \ell_2$. We will also introduce a 2×2 K-matrix, K_{IJ} (with $I, J = 1, 2$), in this case

$$K = \begin{pmatrix} p_1 & -1 \\ -1 & p_2 \end{pmatrix} \tag{14.49}$$

where $m = p_1$ and $n = p_2$. We will also introduce a charge vector $\vec{t} = (1, 0)$ and a vector $\vec{\ell} = (\ell_1, \ell_2)$ for the quasiparticle couplings (which we will use to label the quasiparticles). In this notation the effective Lagrangian is (as usual repeated indices are summed over)

$$\mathcal{L}[b_\mu^I, A_\mu] = -\frac{m}{4\pi} K_{IJ} \epsilon^{\mu\nu\lambda} b_\mu^I \, \partial_\nu b_\lambda^J - \frac{e}{2\pi} A_\mu t_I \epsilon^{\mu\nu\lambda} \, \partial_\nu b_\lambda^I + j_{\rm qp}^\mu \ell_I b_\mu^I \tag{14.50}$$

In this notation, the filling fraction (cf. Eq. (14.48)) becomes

$$\nu = \sum_{I,J=1,2} K_{IJ}^{-1} t_I t_J \tag{14.51}$$

where K^{-1} is the inverse of the K-matrix of Eq. (14.49).

The results of Section 14.1.2 tell us that the ground-state degeneracy on a torus is

$$|\det K| = |p_1 p_2 - 1| \tag{14.52}$$

The electric charges and statistics of the quasiparticles can be computed with the same methods as above. Thus by integrating out the hydrodynamic gauge fields we find that the quasiparticle (electric) charges are

$$Q = -e \sum_{I,J=1,2} K_{IJ}^{-1} t_I \ell_J = -e \frac{p_2 \ell_1 + \ell_2}{p_1 p_2 - 1} \tag{14.53}$$

and their statistical angles δ are

$$\delta = \pi \sum_{I,J=1,2} K_{IJ}^{-1} \ell_I \ell_J = \pi \frac{p_2 \ell_1^2 + p_1 \ell_2^2 + 2\ell_1 \ell_2}{p_1 p_2 - 1} \tag{14.54}$$

For example, for the $\nu = 2/5$ fractional quantum Hall state, which is regarded as the daughter state of the $\nu = 1/3$ Laughlin state, the K-matrix is

$$K = \begin{pmatrix} 3 & -1 \\ -1 & 2 \end{pmatrix} \tag{14.55}$$

This state thus has a $\det K = 5$-fold-degenerate ground state on a torus. This analysis then predicts that this fractional quantum Hall state has two types of quasiparticles (and quasiholes): (1) a quasihole $\vec{\ell} = (0, 1)$ with fractional charge $Q = e/5$ and statistics $\delta = 3\pi/5$, and (2) a quasihole $\vec{\ell} = (1, 0)$ with fractional charge $Q = 2e/5$ and statistics $\delta = 2\pi/5$. Notice that in this case the quasihole $\vec{\ell} = (1, 0)$ can also be regarded as a composite object made of two quasiholes $\vec{\ell} = (0, 1)$, which will have the same electric charge and the same statistical angle (modulo 2π), hence $(1, 0) \equiv (0, 2)$. Similarly, an electron is regarded as a composite excitation made of five elementary quasiparticles. Indeed, the excitation $\vec{\ell} = (0, -5)$ has charge $-e$ and statistics π (modulo 2π).

Another interesting example is the fractional quantum Hall state with $\nu = 2/3$, which can be regarded as the particle–hole conjugate of the Laughlin state at $\nu = 1/3$. However, unlike the $\nu = 1/3$ Laughlin state, the $\nu = 2/3$ state appears at level 2 in this hierarchical construction, and it is represented by the 2×2 K-matrix

$$K = \begin{pmatrix} 1 & 1 \\ 1 & -2 \end{pmatrix} \tag{14.56}$$

This construction is generalized to arbitrary levels of the hierarchy. For instance, we could now proceed further and consider a state that results from the condensation of a set of quasiparticles of level 2 into a new quantum Hall state to find a new state at level 3, and so on and so forth. Hence the effective Lagrangian $\mathcal{L}_n$ at level n of the hierarchy is constructed from the effective Lagrangian $\mathcal{L}_{n-1}$ at level $n-1$. Let $I, J = 1, \ldots, n-1$ and let $K_{IJ}^{(n-1)}$ be the K-matrix at level $n-1$. We will consider a state resulting from condensing the level-$(n-1)$ quasiparticles labeled by $\vec{\ell}^{(n-1)} = (\ell_1^{(n-1)}, \ldots, \ell_{n-1}^{(n-1)})$ and charge vector $\vec{t}^{\,(n-1)}$. The effective Lagrangian at level n has the same form as before,

$$\mathcal{L} = -\sum_{I,J=1}^{n} \frac{1}{4\pi} K_{IJ}^{(n)} \epsilon^{\mu\nu\lambda} b_\mu^I \, \partial_\nu b_\lambda^J - \frac{e}{2\pi} \sum_{I=1}^{n} A_\mu t_I^{(n)} \epsilon^{\mu\nu\lambda} \, \partial_\nu b_\lambda^I + \sum_{I=1}^{n} \ell_I^{(n)} j^\mu b_\mu^I \tag{14.57}$$

where the level-n K-matrix has the block form

$$K^{(n)} = \begin{pmatrix} K^{(n-1)} & -[\vec{\ell}^{\,(n-1)}]^t \\ -\vec{\ell}^{\,(n-1)} & p_n \end{pmatrix} \tag{14.58}$$

with a ground-state degeneracy on a torus, cf. Eq. (14.23).

In all cases the filling fraction ν of the resulting fractional quantum Hall state and the electric charges and statistical angles of the excitations are given by the obvious generalization of the level-2 expressions of Eq. (14.51), Eq. (14.53), and Eq. (14.54). For example, if we always condense the quasiparticle with smallest charge at level $n-1$ (with vector $\vec{\ell}^{\,(n-1)} = (0, \ldots, 1)$), the K-matrix at level n is the tridiagonal matrix

$$K_{IJ} = p_I \delta_{IJ} - \delta_{I,J-1} - \delta_{I,J+1}, \qquad p_1 \text{ odd and } p_I \text{ even } (I \geq 2) \tag{14.59}$$

with charge vector $t_I = \delta_{I,1}$. In this case the filling fraction has the partial-fraction decomposition (Haldane, 1983b)

$$\nu = \cfrac{1}{p_1 - \cfrac{1}{p_2 - \cfrac{1}{p_3 - \cdots}}} \tag{14.60}$$

However, do all the choices of a K-matrix and charge vector $\vec{t}$ represent physically distinct quantum Hall fluids? In general the answer to this question is no, since under a suitable change of basis for the gauge fields, i.e. a linear transformation of the form

$$b_\mu^I \to W_{IJ} b_\mu^J = b_\mu'^I \tag{14.61}$$

we can seemingly always bring the K-matrix to a diagonal form. However, not all such linear transformations are allowed. The reason is that the set of allowed "independent" quasiparticle vectors, $\{\ell^{(I)}\}$, with $\ell_J^{(I)} = \delta_{IJ}$, constitutes a basis that spans the *charge lattice*, $\ell = \sum_{I=1}^n l_I \ell^{(I)}$, with integer coefficients l_I (required by charge quantization). However, the quasiparticle vectors $\vec{\ell}$ will also transform under a general transformation of Eq. (14.61), namely $\ell_I' = W_{IJ} \ell_J$, and, for a general transformation, the transformed vector is not an element of the (*integer*) charge lattice. Therefore only those transformations W that map the charge lattice into the same charge lattice are allowed. In other terms, the allowed linear transformations are not general linear transformations of $\mathbb{R}^n$ but elements of the group $\mathrm{SL}(n, \mathbb{Z})$, the group of integer-valued matrices with unit determinant. Therefore two quantum Hall fluids characterized by two K-matrices, K_1 and K_2, and two charge vectors, $\vec{t}_1$ and $\vec{t}_2$, are equivalent (that is, the same state) if there exists a linear transformation $W \in \mathrm{SL}(n, \mathbb{Z})$ such that $K_2 = W K_1 W^{\mathrm{T}}$ and $\vec{t}_2 = W t_1$. Here we have neglected the role of the spin vector, which must be considered for a full analysis of the equivalence classes (Wen, 1995).

14.4 Multi-component abelian fluids

A very similar effective theory can be developed for multi-component quantum Hall fluids. For simplicity we will consider a two-component system, which we can think of either as a fully polarized electron gas in a bilayer system or as a single-layer system in which the spin is not fully polarized by the magnetic field. The latter case occurs in many heterostructures at high electron density, in which the g-factor can be made very small. More complicated systems can be (and have been) considered. Here we will follow the hydrodynamic approach of the previous subsection. One can alternatively use a flux-attachment approach, see e.g. López and Fradkin (1995, 2001).

Wave functions with a Laughlin structure for these systems were proposed long ago by Halperin (1983, 1984). Let us label by $\{z_i\}$ and $\{w_i\}$ (with $i = 1, \ldots, N/2$) the complex coordinates of particles of type 1 (say, with spin up) and the complex coordinates of particles of type 2 (say, spin down), respectively. The total number of particles (which I will take to be fermions) is N, and the number of flux quanta is N_ϕ. The total filling fraction is $\nu = N/N_\phi$ (as before), which can be written as the sum of the filling fractions of each layer, $\nu = \nu_1 + \nu_2$. For simplicity we are considering the case in which we have the same number of particles for each type. In the spin interpretation (when possible) this state would be spin-unpolarized.

A simple generalization of the Laughlin state is provided by the Halperin wave functions, which have the following form:

$$\begin{aligned}\Psi_{m_1,m_2,n}(z_1, \ldots, z_{N/2}, w_1, \ldots w_{N/2}) \\ = \prod_{i<j}(z_i - z_j)^{m_1} \prod_{i<j}(w_i - w_j)^{m_2} \prod_{i\leq j}(z_i - w_j)^n \\ \times \exp\left(-\frac{1}{4\ell_0^2} \sum_{i=1}^{N/2} \left(|z_i|^2 + |w_i|^2\right)\right)\end{aligned} \tag{14.62}$$

As in the case of a one-component fluid, we will require the wave function to be completely antisymmetric under exchange. This will require m_1 and m_2 to be odd integers and n to be an integer (or zero). We will refer to these as the (m_1, m_2, n) Halperin states.

In the hydrodynamic approach we will define two separate currents, one for each type of electron, J_μ^I (with $I = 1, 2$). In a bilayer system without inter-layer tunneling each current is separately conserved. Thus we will define two hydrodynamic gauge fields, b_μ^I $(I = 1, 2)$, whose curls are the two currents:

$$j_I^\mu = \frac{1}{2\pi}\epsilon^{\mu\nu\lambda}\,\partial_\nu b_\lambda^I \tag{14.63}$$

Thus, in the bilayer systems the two hydrodynamic currents represent distinct degrees of freedom of the electron gas with the label $I = 1, 2$ indicating the layer (or spin projection) of the electrons. In contrast, in the hierarchical construction the label indicates the order in the hierarchy of a single-layer system of fully polarized electrons. The formal similarities of the effective theories should not obscure the physical differences between the two systems.

The Halperin wave functions also tell us that in addition to intra-Landau-level interactions in each layer (represented in the Halperin wave function by the odd integers m_1 and m_2), there are also inter-layer interactions, represented by the integer n. In our discussion of the fractional Hall effect in terms of flux attachments we saw that the exponents m_1 and m_2 mean that we have attached an even number $m_1 - 1$ of fluxes to the electrons in layer 1 and an even number $m_2 - 1$ of fluxes to the electrons in layer 2. The exponent n is telling us that the electrons of the two layers repel each other (since $n > 0$). It can also be interpreted as saying that the repulsion can be represented by the attachment of n fluxes to an electron in layer 2 due to an electron in layer 1 and vice versa.

There is a natural candidate for the effective Lagrangian of this fractional quantum Hall state that also has the form of a K-matrix, namely

$$\mathcal{L}[b_\mu^I, A_\mu] = -\frac{1}{4\pi} K_{IJ} \epsilon_{\mu\nu\lambda} b_I^\mu \, \partial^\nu b_J^\lambda - \frac{e}{2\pi} A_\mu t_I \epsilon_{\mu\nu\lambda} \, \partial^\nu b_I^\lambda \tag{14.64}$$

where K is a 2×2 symmetric matrix. We will show that the correct matrix is

$$K = \begin{pmatrix} m_1 & n \\ n & m_2 \end{pmatrix} \tag{14.65}$$

The ground-state degeneracy on a torus for this state is $|m_1 m_2 - n^2|$.

The electrons in each layer (or spin projection) must couple in the same way to the external field A_μ. Therefore the charge vector $\vec{t}$ must assign the same electric charge to the electrons in each layer. Hence, we must choose

$$\vec{t} = (1, 1) \tag{14.66}$$

If the K-matrix is non-singular, i.e. $m_1 m_2 - n^2 \neq 0$, we can use, once again, Eq. (14.51) to read off the filling fraction and the quantum numbers of the excitations of this theory. We find that the two layers have the filling fractions

$$\nu_1 = \frac{m_2 - n}{m_1 m_2 - n^2}, \qquad \nu_2 = \frac{m_1 - n}{m_1 m_2 - n^2} \tag{14.67}$$

and hence that the total filling fraction ν is

$$\nu = \frac{m_1 + m_2 - 2n}{m_1 m_2 - n^2} \tag{14.68}$$

As we see, a state with $m_1 \neq m_2$ represents a system with unequal filling fractions in the two layers, with $\nu_1 - \nu_2 = (m - 1 - m_2)/(m_1 m_2 - n^2)$ being the charge imbalance. In the spin language, the electron gas has a net spin polarization.

In this system we have quasiparticles for each layer with currents $j_\mu^{\text{qp}^I}$ coupled to the hydrodynamic gauge fields b_μ^I by the quasiparticle charge vectors $\vec{\ell} = (\ell_1, \ell_2)$. We can determine the fractional electric charges and the statistical angles of the quasiparticles using Eqs. (14.53) and (14.54) for the charge vector $\vec{t}$ of Eq. (14.66). The charges and statistical angles of the excitations are

$$Q(\ell_1, \ell_2) = -e \frac{(m_2 - n)\ell_1 + (m_1 - n)\ell_2}{m_1 m_2 - n^2} \tag{14.69}$$

$$\delta(\ell_1, \ell_2) = \pi \frac{m_2 \ell_1^2 + m_1 \ell_2^2 - 2n\ell_1\ell_2}{m_1 m_2 - n^2} \tag{14.70}$$

We will now discuss the special case of the (m, m, n) symmetric Halperin states. These states have very simple quantum numbers. The filling fractions are $(\nu_1 = \nu_2 = \nu/2)$

$$\nu = \frac{2}{m + n} \tag{14.71}$$

The fundamental quasiparticles (with smallest charge) are $(1, 0)$ and $(0, 1)$. Their charges and statistics are

$$Q = \pm \frac{e}{m + n}, \qquad \delta = \pi \frac{m}{m^2 - n^2} \tag{14.72}$$

In the case of the symmetric states (m, m, n) it is convenient to rotate to a new basis in which the fields decouple. This can be done by the orthogonal transformation to the fields, $b_\pm^\mu$, given by

$$b_\pm^\mu = \frac{1}{\sqrt{2}}(b_1^\mu \pm b_2^\mu) \tag{14.73}$$

in terms of which the action becomes

$$\begin{aligned} \mathcal{L} = & \frac{m + n}{4\pi} \epsilon_{\mu\nu\lambda} b_+^\mu \, \partial^\nu b_+^\lambda - \sqrt{2} \frac{e}{2\pi} \epsilon_{\mu\nu\lambda} A^\mu \, \partial^\nu b_+^\lambda \\ & + \frac{m - n}{4\pi} \epsilon_{\mu\nu\lambda} b_-^\mu \, \partial^\nu b_-^\lambda \\ & + j_\mu^{\text{qp}} \left[\frac{1}{\sqrt{2}}(\ell_1 + \ell_2) b_+^\mu + \frac{1}{\sqrt{2}}(\ell_1 - \ell_2) b_-^\mu \right] \end{aligned} \tag{14.74}$$

In this basis, the effective Lagrangian decouples into a charge mode, b_+^μ, and a neutral mode, b_-^μ. Notice, however, that the quasiparticles carry both quantum numbers.

The most prominent states seen in balanced bilayer systems are the $(3, 3, 1)$ state with filling fraction $\nu = 1/2$, the $(3, 3, 2)$ state with $\nu = 2/5$, and the $(1, 1, 2)$ state with $\nu = 2/3$. The K-matrices, charge vectors $\vec{t}$, and ground-state degeneracies (on a torus) for these Halperin states are

$$\begin{aligned}
(331): \quad \nu &= \frac{1}{2}, \quad & K &= \begin{pmatrix} 3 & 1 \\ 1 & 3 \end{pmatrix}, \quad & \vec{t} &= (1, 1), \quad & 8 \\
(332): \quad \nu &= \frac{2}{5}, \quad & K &= \begin{pmatrix} 3 & 2 \\ 2 & 3 \end{pmatrix}, \quad & \vec{t} &= (1, 1), \quad & 5 \\
(112): \quad \nu &= \frac{2}{3}, \quad & K &= \begin{pmatrix} 1 & 2 \\ 2 & 1 \end{pmatrix}, \quad & \vec{t} &= (1, 1), \quad & 3
\end{aligned} \tag{14.75}$$

The $(3, 3, 1)$ state is a bilayer state with filling fraction $\nu = 1/2$ that occurs when the inter-layer interaction is large enough. It is seen in experiments in wide quantum wells. The elementary quasihole has charge $e/4$ and statistics $\delta = 3\pi/5$. As a function of external electric fields (normal to the 2DEG) the system has a phase transition from two essentially decoupled 2DEGs in a compressible "Fermi-liquid" state (at low bias), in which each 2DEG is in a Halperin–Lee–Read $\nu = 1/2$ compressible state weakly coupled to the other, to a $\nu = 1/2$ incompressible fractional quantum Hall state, namely the $(3, 3, 1)$ bilayer state (Eisenstein *et al.*, 1992).

Similar spin transitions have been seen in experiments in single-layer high-density 2DEGs at filling fraction $\nu = 2/3$, namely from an incompressible fully polarized state to a spin-singlet state at the same filling fraction (Eisenstein *et al.*, 1990), presumably a Halperin $(1, 1, 2)$ state. Similar phase transitions have been seen at $\nu = 2/5$, between a fully polarized 2DEG Jain state and a $(3, 3, 2)$ spin-singlet Halperin state (Cho *et al.*, 1998).

What happens if the K-matrix is singular? Many of our general expressions are invalid in this case. We will discuss here the simple case of the balanced singular (m, m, n) states. They become singular for $m^2 - n^2 = 0$ or, which amounts to the same thing, for the (m, m, m) states, with filling fractions $\nu = 1/m$. In the decoupled basis, the effective theory of the (m, m, m) states has a charge mode with a Chern–Simons coupling $2m$ and a seemingly absent neutral mode. Of course, in this case the terms which we neglected before (since they were subleading) yield the leading behavior and cannot be dropped. At any rate, the absence of a Chern–Simons term in the neutral mode is telling us that it describes a condensed superfluid state.

The behavior of states of this type has been studied in quite some detail, both theoretically and experimentally, in the simplest example, the $(1, 1, 1)$ state with filling fraction $\nu = 1$. Hence, this is an integer quantum Hall state. However, in spite of this, this state cannot be described in terms of free electrons in a magnetic field. It can be realized either as a bilayer system or as a single-layer *ferromagnetic* quantum Hall state with a weak Zeeman coupling. A unique feature of this $\nu = 1$ "integer" quantum Hall state is that the addition (or subtraction) of an electron results not in an excitation in an excited Landau level (as in the case of a non-interacting system) but, instead, in the formation of a bound state of an electron with enough flux to generate a soliton state known as a skyrmion, as introduced in Chapter 7, a topologically non-trivial spatial texture of the spin of the electrons stabilized by the combined effects of the Zeeman energy and the Coulomb interaction (Sondhi *et al.*, 1993). Skyrmions were detected in this $\nu = 1$ fluid in NMR experiments (Barrett *et al.*, 1995). On the other hand, the absence of a Chern–Simons term in the neutral mode was predicted to give rise to a Goldstone mode (Sondhi *et al.*, 1993; Yang *et al.*, 1994), which was observed as a sharp resonance in tunneling into the $\nu = 1$ state (Spielman *et al.*, 2000). Similar effects have been seen in fully polarized bilayer systems, also at $\nu = 1$, where they are interpreted as evidence of a superfluid state and an exciton condensate (Eisenstein and MacDonald, 2004).

14.5 Superconductors as topological fluids

We will make a brief intermission in our analysis of quantum Hall states and examine the properties of a superconductor from a different perspective. We will consider for simplicity an s-wave superconductor. According to the Bardeen–Cooper–Schrieffer (BCS) theory, a superconductor is a ground state of a system of charged spin-1/2 fermions in which the spin-singlet pair-field operator $\Delta(\vec{x}, \vec{y}) = \epsilon_{\sigma\sigma'}\psi_\sigma^\dagger(\vec{x})\psi_{\sigma'}^\dagger(\vec{y})$ has a non-vanishing expectation value in the superconducting ground state (Schrieffer, 1964). In the case of a spin-singlet superconductor, in momentum space the pair-field operator for spin-singlet Cooper pairs is $\Delta(\vec{k}) = \epsilon_{\sigma\sigma'}\psi_\sigma^\dagger(\vec{k})\psi_{\sigma'}^\dagger(-\vec{k})$, where $\vec{k}$ lies on the Fermi surface of the metal whose quasiparticles have paired and condensed. Since the Hamiltonians of all interacting fermionic systems are gauge-invariant, they are also invariant under the global part of the gauge transformation, $\psi_\sigma(\vec{x}) \to e^{i\theta}\psi_\sigma(\vec{x})$. The pair-field operators transform under a global gauge transformation as $\Delta \to e^{i2\theta}\Delta$ and hence are not gauge-invariant.

The superconducting ground state is thus a state in which the global U(1) gauge symmetry is spontaneously broken. For this reason the expectation value of the pair field plays the role of the order parameter for the superconducting state, and it is the basic building block of the Landau–Ginzburg theory (de Gennes, 1966). We will see below that, strictly speaking, the concept of an order parameter is problematic in a superconductor.

The spectrum of the superconducting state has charge-neutral fermionic excitations (Kivelson and Rokhsar, 1990). In an s-wave superconductor, the expectation value of the Cooper-pair condensate is isotropic and hence it is the same for all directions $\vec{k}$. Its fermionic (Bogoliubov) quasiparticles have a finite and isotropic energy gap whose magnitude is the amplitude of the condensate $|\Delta|$. In the case of a fluid of neutral fermions, the phase of the pair-field condensate φ is the Goldstone boson of the broken global U(1) symmetry. On the other hand, if the fermions are charged, e.g. are electrons of charge e, which couple to the electromagnetic field, then in this case the transverse components (the photon) of the electromagnetic gauge field become massive, and static magnetic fields are expelled. This is the Meissner effect. For the same reasons, the phase field φ effectively becomes the longitudinal component of the electromagnetic vector potential and also acquires a finite energy gap, which in a superconductor is the plasma frequency. Hence, in a superconducting state, which is a *condensate of electric charges*, electric fields are screened and magnetic fields are expelled.

In this (very brief!) description of the superconducting state the electromagnetic gauge field is used as a probe of the condensed state. However the quantum dynamics of the gauge field is ignored. In particular, the order-parameter field is constructed without taking the gauge fields into consideration. In Section 9.12 we discussed the definition of a Higgs phase in a theory that has dynamical matter and gauge fields. In that theory the superconductor pair field plays the role of the Higgs field. It was noted there, following the arguments of Fradkin and Shenker (1979), that it is not possible to construct a gauge-invariant operator that plays the role of the order-parameter field if the gauge group is compact. In that case its deconfined phase (when it exists) is characterized by the behavior of the Wilson and 't Hooft loop operators.

One may object that in the case of a superconductor the electromagnetic gauge field, even if it is considered as a quantum field, is non-compact and, hence, this objection should be moot. If we regard the pair field as a *local* field $\Delta(\vec{r}, t)$ that transforms under gauge transformations as a charge-$2e$ field, we can construct a gauge-invariant order-parameter field by creating together with the Cooper pair its static Coulomb field (Dirac, 1955). Thus, the operator (here we are discussing the 2D case) which creates a coherent state of static photons (the Coulomb field) together with the Cooper pair

$$\mathcal{O}(\vec{x}, \vec{y}) = \Delta(\vec{x}, \vec{y}) \exp\left(ie \int d^2z \vec{A}(\vec{z}) \cdot \vec{E}_c(\vec{z}) \right) \tag{14.76}$$

is gauge-invariant if the classical field $\vec{E}_c(\vec{z})$ satisfies the condition

$$\vec{\nabla} \cdot \vec{E}_c(\vec{z}) = e\left(\delta(\vec{z} - \vec{x}) + \delta(\vec{z} - \vec{y}) \right) \tag{14.77}$$

The classical static field $\vec{E}_c$ is the Coulomb field of the Cooper pair. On the other hand, we can also write $\vec{E}_c(\vec{z}) = -\vec{\nabla} U_c(\vec{z})$, where $U(z)$ is the electrostatic potential.

Upon integration by parts, we find that, in the Coulomb gauge, $\vec{\nabla} \cdot \vec{A} = 0$, the exponential operator becomes

$$\exp\left(-ie \int d^2z\, \vec{A}(\vec{z}) \cdot \vec{\nabla} U_c(\vec{z})\right) = \exp\left(ie \int d^2z\, \vec{\nabla} \cdot \vec{A}(\vec{z}) U_c(\vec{z})\right) = 1 \tag{14.78}$$

Therefore, in the Coulomb gauge the exponential operator is equivalent to the identity operator. However, it was shown by Kennedy and King (1985) that this non-local operator has a power-law dependence in time, which is inconsistent with the gauge field being massive. So, even in the case of the non-compact gauge theory, this operator cannot represent an order-parameter field. The reader may recognize here the construction that we used in Section 13.3.6 for the Girvin–MacDonald non-local order parameter for the FQHE.

Moreover, the argument that led to Eq. (14.78) holds only if the Coulomb-gauge condition can be consistently imposed. However, the Coulomb-gauge condition cannot be imposed if the gauge group is compact since, even in the case of an abelian theory, the Coulomb-gauge condition can only be imposed modulo 2π due to the existence of monopole configurations. This fact was discussed in some detail in Section 9.12, where we showed, following a standard result from gauge theories (Fradkin and Shenker, 1979), that the concept of a spontaneously broken symmetry does not truly exist if the (compact) gauge fields are dynamical.

In an insightful and deep paper, Hansson, Oganesyan, and Sondhi (Hansson *et al.*, 2004) found an additional and more important loophole in the construction. We will follow their analysis closely. In a fermionic superconductor the pair field, which plays the role of the scalar Higgs field, is a composite operator of fermions. Although the fermionic excitations of the superconductor are charge-neutral and as such do not couple directly to the gauge field, they have a non-local behavior with respect to the vortices of the superconductor. Indeed, the vortex of the superconductor carries magnetic flux $hc/(2e)$. This flux quantization follows from the requirement that the pair field be local (i.e. single-valued). Hence the phase field of the pair field winds by 2π on large contours about the core of the vortex. On the other hand, the fermionic quasiparticle is essentially half of the pair field and *changes sign* on a large contour that encircles the vortex core.

The origin of this phenomenon is the nature of the superconducting state as a pair condensate. Since it is a pair condensate, the Bogoliubov quasiparticles *are not conserved* since they can emerge from (or be absorbed by) the condensate. However, the *parity* of the fermion number is an exact quantum number even in the superconducting state since the excitations can be created and destroyed in pairs. The fermionic Bogoliubov excitations carry zero electric charge (at low energies) and effectively they are Majorana fermions. Therefore the quasiparticle states are sensitive only to a local change in sign. From the point of view of the Bogoliubov quasiparticles, the vortices carry a $\mathbb{Z}_2$ charge or, equivalently, they carry a flux of π. This sign is all that is left of the Aharonov–Bohm phase factor for the fermionic quasiparticles of a superconductor. On the other hand, the $\mathbb{Z}_2$ nature of the excitation spectrum suggests that there should be a connection between a superconductor and the deconfined phase of a $\mathbb{Z}_2$ gauge theory.

Hansson and coworkers established that the superconductor–$\mathbb{Z}_2$ gauge-theory relation exists using the following line of argument. They first showed that a theory in which quasiparticles and vortices have a topological interaction is a topological BF field theory. We have already encountered the BF Lagrangian at the beginning of this chapter in our discussion of flux attachment on a torus in Section 14.1.1. In particular, see the discussion surrounding Eq. (14.7), which is relevant to the problem we are considering. Let us describe the histories of the quasiparticle excitation by a current j_μ^{qp}. Here we are making the simplifying assumption of ignoring the spin of the quasiparticles, which would require us to add extra currents if we took it into account. For the time being we will ignore the fact that this current is not truly conserved, since the quasiparticles can go in and out of the condensate in pairs, and assume that the quasiparticle current is locally conserved and obeys $\partial_\mu j_{\text{qp}}^\mu = 0$. We will come back and fix this problem later. We will further assume that the Bogoliubov quasiparticles have only short-range interactions, which is generally the case. We will see that, in spite of their being charge-neutral, there still is an interaction of topological origin between quasiparticles and vortices. A very transparent and detailed discussion of this problem can be found in the work of Nayak and coworkers (Nayak *et al.*, 2001).

Likewise we will assume that the vortex excitations are light enough that they can be created (in pairs) by quantum-mechanical processes. Let the locally conserved current j_μ^v denote the vortex histories and also obey a continuity equation, $\partial^\mu j_\mu^v = 0$. We will also make the assumption that the vortices have only short-range interactions. This assumption is correct in a 3D superconductor since it has a complete Meissner state. In 2D this assumption is somewhat problematic since the field lines can escape from the plane wherein the 2D superconductor lies. In practice this problem introduces some degree of non-locality into the interaction, which, as it turns out, is not a serious problem for the argument that we will use here. Thus we will assume that the magnetic interactions are also screened even in 2D (which rigorously should happen only in $(2 + 1)$-dimensional electrodynamics).

The non-local topological interaction between vortices and quasiparticles can be encoded in an effective Lagrangian in which the quasiparticle current j_μ^{qp} and the vortex current $j^{\rm v}_\mu$ couple to two separate gauge fields, a^μ and b^μ, whose only coupling has the BF form,

$$\mathcal{L}_{\rm topo}[a_\mu, b_\mu] = \frac{1}{\pi}\epsilon^{\mu\nu\lambda} a_\mu\, \partial_\nu b_\lambda - a^\mu\, j^{\rm qp}_\mu - j^{\rm v}_\mu b^\mu \tag{14.79}$$

The quantization of electric charge and of the magnetic charge (the vorticity) now implies that the gauge groups for the fields a_μ and b_μ must be *compact*. Hence, gauge transformations are defined modulo an integer multiple of 2π, i.e. $a_\mu \to a_\mu + \partial_\mu \Lambda$ with $\Lambda \equiv \Lambda + 2\pi$ at every point in space-time (and the same for b_μ) (this argument can be made simpler by working on a Euclidean space-time lattice). However, the fact that the gauge fields a_μ and b_μ are compact implies that one cannot ignore the effects of monopoles in these fields, as in the case of Polyakov's compact electrodynamics (Polyakov, 1977) (see Section 9.11).

This effective Lagrangian has the same form as the theories for multi-component (and hierarchical) fractional quantum Hall fluids but with a K-matrix of the form

$$K = \begin{pmatrix} 0 & 2 \\ 2 & 0 \end{pmatrix} \tag{14.80}$$

which is traceless and has determinant -4. Using the results of Section 14.3, and the general arguments of Wen (1995), one readily finds that the quasiparticles and vortices have a mutual statistical angle $\delta = \pi$. This theory has an explicit self-duality that exchanges quasiparticles with vortices and for the same reason it is also time-reversal-invariant. On the other hand, this effective action describes a topological fluid that has degeneracy 4 on the torus. Thus, if the description of the superconductor could be reduced to Eq. (14.79), we would have shown that it is a topological fluid with these two types of topological excitations, quasiparticles and vortices, and a four-fold degeneracy on the torus.

A two-component Chern–Simons theory with a K-matrix with the structure shown in Eq. (14.80) is invariant under time-reversal transformations. This is so since a time reversal is equivalent to the exchange of the two species of gauge fields. This is most apparent on rotating the basis of fields in which the K-matrix is diagonal and traceless. Thus a "doubled" Chern–Simons gauge theory can describe time-reversal-invariant topological phases (Freedman *et al.*, 2004).

The alert reader most likely would now notice that this is the same ground state degeneracy and the same mutual statistics as those we found in the deconfined (topological) phase of the $\mathbb{Z}_2$ gauge theory discussed in Section 9.9 (where the excitations were holons and visons). We will show below that the topological phase of the $\mathbb{Z}_2$ gauge theory is actually the correct effective theory since its excitations are also conserved modulo 2.

However, there is a more important question we need to address. How do we know that this effective action, or the $\mathbb{Z}_2$ gauge theory for this matter, actually has anything to do with a superconductor? It certainly looks very different from the Landau–Ginzburg theory. Hansson *et al.* offered an impressionistic (but physically correct) derivation of the BF theory. They start with the abelian Higgs model, which is a relativistic field theory of a charged scalar field ϕ coupled to the electromagnetic gauge field A_μ whose Lagrangian (in Euclidean (2+1)-continuum space-time dimensions) is

$$\mathcal{L} = (D_\mu \phi)^* D^\mu \phi + \frac{1}{4} F_{\mu\nu}^2 + V(\phi^* \phi) - e A_\mu j_{\mathrm{qp}}^\mu \tag{14.81}$$

where

$$D_\mu = \partial_\mu + i \frac{2e}{\hbar c} A_\mu \tag{14.82}$$

is the covariant derivative for a charge-$2e$ complex scalar field. In Eq. (14.81) we have added a set of (fermionic) quasiparticle currents j_{qp}^μ to describe the slow (adiabatic) dynamics of the gapped Bogoliubov quasiparticles. These are absent from the standard abelian Higgs model. Notice that in this Lagrangian they carry charge $-e$, half of the charge of the charged scalar field ϕ. We will see shortly that, although the charge of the quasiparticle will be completely screened (giving a neutral quasiparticle), there will be a remnant of its magnetic (Aharonov–Bohm) interaction with the vortices of the superconductor.

The potential $V(\phi^* \phi)$ is taken to represent the classical broken-symmetry state: a "Mexican-hat" potential with a deep minimum at $\phi = \sqrt{\rho_{\mathrm{s}}} e^{i\varphi}$, where ρ_{s} is the superfluid density of the superconductor and, hence, proportional to the magnitude of the superconducting gap, $\rho_{\mathrm{s}} \propto |\Delta|$. We will assume that we are sufficiently deep in the superconducting state that the amplitude of the order parameter or, equivalently, the superfluid density ρ_{s}, can be taken to be essentially uniform in space and time. This theory is, qualitatively, a quantum version of the Landau–Ginzburg theory of a superconductor.

In this picture the vortices (in $(2+1)$ dimensions) are point-like singularities of the order-parameter field that evolve in time along a set of worldlines. Thus, we will split the phase of the order parameter φ into a smooth part, which we will denote by η, and a singular piece χ that satisfies

$$\oint_\Gamma dx_\mu \, \partial^\mu \chi = (\Delta \chi)_\Gamma = 2\pi N_{\mathrm{v}}[\Sigma] \tag{14.83}$$

where Γ is a closed contour and $N_{\mathrm{v}}[\Sigma]$ is the number of vortex lines (with their signs) piercing a surface Σ whose boundary is $\Gamma = \partial\Sigma$. It will be convenient to define a gauge field $a_\mu \equiv \frac{1}{2} \partial_\mu \chi$ such that its circulation on the same contour Γ is πN_{v}. Since the phase field has singularities on the worldlines of the vortices, the amplitude of the charged scalar field must vanish along these lines. To make the arguments simpler, we will assume that we are dealing with a superconductor with a short enough coherence length ξ that the vanishing of the amplitude of the charged scalar field occurs on very short length scales, as in the case of a strong type II superconductor. It will also be convenient to define a set of vortex currents $j_\mu^{\rm v}$,

$$j_\mu^{\rm v} = \frac{1}{\pi}\epsilon_{\mu\nu\lambda}\,\partial^\nu a^\lambda \tag{14.84}$$

such that the flux of the vortex currents through the surface Σ is the vorticity $N_{\rm v}[\sigma]$ going through that surface,

$$\int_\Sigma dS_\mu\, j_{\rm v}^\mu = N_{\rm v}[\Sigma] \tag{14.85}$$

On the other hand, the smooth part of the phase field η (the "Goldstone boson") can be "eaten" by the electromagnetic gauge field A_μ or, which amounts to the same thing, we can fix the London (or unitary) gauge $\eta = 0$. Within these assumptions the Lagrangian now becomes

$$\mathcal{L}_{\rm eff} = \frac{1}{4}F_{\mu\nu}^2 + \frac{m_{\rm s}^2}{2}\left(A_\mu - \frac{1}{e}a_\mu\right)^2 - eA_\mu j^\mu + \cdots \tag{14.86}$$

where $m_{\rm s} = \lambda_{\rm L}^{-1} = \rho_{\rm s}(2e/(\hbar c))^2$ and $\lambda_{\rm L}$ is the London penetration depth.

We can now resort to the same trick as that we used in Section 14.1.1 and rewrite the partition function in the background of the quasiparticle and vortex currents as

$$\begin{aligned} Z[j_\mu^{\rm qp}, j_\mu^{\rm v}] &= \int \mathcal{D}A_\mu\,\mathcal{D}a_\mu\,\delta\left(j_\mu^{\rm v} - \frac{1}{\pi}a_\mu\right)\,\exp\!\left(-\int d^3x\,\mathcal{L}_{\rm eff}[A_\mu, a_\mu, b_\mu]\right) \\ &= \int \mathcal{D}a_\mu\,\mathcal{D}b_\mu\,\exp\!\left(-\int d^3x\,\mathcal{L}_{\rm eff}[a_\mu, b_\mu]\right) \end{aligned} \tag{14.87}$$

where in the last line we introduced a representation of the delta function through a Lagrange-multiplier vector field b_μ and subsequently integrated out the fluctuations of the electromagnetic gauge field A_μ. The effective action prior to integrating out the electromagnetic field in Minkowski space-time is

$$\begin{aligned} \mathcal{L}_{\rm eff}[A_\mu, a_\mu, b_\mu] = &-\frac{1}{4}F_{\mu\nu}^2 - \frac{m_{\rm s}^2}{2}\left(A_\mu - \frac{1}{e}a_\mu\right)^2 + eA_\mu j_{\rm qp}^\mu + b_\mu j_{\rm v}^\mu \\ &+ \frac{1}{\pi}\epsilon_{\mu\nu\lambda}a^\mu\,\partial^\nu b^\lambda \end{aligned} \tag{14.88}$$

Upon integrating out the massive electromagnetic gauge field A_μ and keeping only terms with the lowest order in derivatives we find that the effective action

$$\mathcal{L}[a_\mu, b_\mu] = \mathcal{L}_{\rm topo}[a_\mu, b_\mu] + \cdots \tag{14.89}$$

is precisely the topological BF effective action of Eq. (14.79).

To show that it is a superconductor, we will need to show that all local excitations are massive and in particular that in this theory the photon is massive. Within this local theory this is indeed the case, as one readily finds by setting the quasiparticle and vortex currents to zero and then integrating out the auxiliary field b_μ, resulting in the effective action (to second order in derivatives)

$$\mathcal{L}_{\text{eff}}[a_\mu] = -\frac{1}{4e^2}\left(\partial_\mu a_\nu - \partial_\nu a_\mu\right)^2 - \frac{m_{\text{s}}^2}{2e^2} a_\mu a^\mu \tag{14.90}$$

(again in the London unitary gauge). It is straightforward to derive the London equation of a superconductor from this effective action.

Therefore we have shown that the low-energy effective action of a superconductor with a quantum-mechanical electromagnetic gauge field is indeed a topological field theory! As we noted above, this theory has a four-fold degeneracy on the torus. Here "fractionalization" is realized into terms of the fundamental particles that entered into the pairing instability. This seems to complete our argument.

However, we are not quite yet done, since we still have a debt to pay. It remains to remove the unnecessary assumption that the quasiparticle currents are locally conserved. To remove this assumption, we have to recall that quasiparticles can go in and out of the condensate in pairs. Thus, our partition function must contain instanton processes (in Euclidean space-time) that describe pairs of quasiparticles being removed at some locations in space-time and, likewise, pairs of quasiparticles that are created at some locations in space-time. These processes are monopole configurations in the gauge field b_μ. However, these instanton processes do not affect the topological sector of the theory.

Finally, there remains one question to answer: is there an order parameter? The answer to this question depends on whether the electromagnetic gauge field itself is compact and, hence, whether Dirac (or Polyakov) magnetic monopoles play a role. If they do, this problem is mapped exactly into the theory discussed in Sections 9.6 and 9.12, following the results of Fradkin and Shenker (1979). If this were literally the case, the deconfined phase would be strictly topological and no trace of superconductivity would be found. This is so, since vorticity in that $\mathbb{Z}_2$ case is also defined modulo 2. On the other hand, as far as we know, the electrodynamics sector of the Standard Model does not appear to have monopoles (or they are so heavy that they do not play a role in experiments). Thus, in this more realistic scenario, although a superconductor is actually a topological state, it still exhibits all the features that define superconductivity.

14.6 Non-abelian quantum Hall states

The quantum Hall states we have discussed in this section have excitations with fractional statistics. Thus, the amplitude for a clockwise adiabatic exchange of two quasiparticles 1 and 2 (a *braiding* operation of the quasiparticle worldlines; see Fig. 13.2) acquires a phase factor $W(1, 2) \to \exp(i\delta)W(2, 1)$. We assigned the statistical phase δ as a quantum number that characterizes the excitations of the state. A state with several excitations will undergo many such processes under time evolution, and each process will yield an additional phase factor.

In Section 10.8 we discussed briefly the fact that we can regard the sequence of exchange processes as a sequence of concatenated braiding processes during which the linking number of the worldlines of the quasiparticles changes. The successive action of two braiding processes defines a product operation that is closed in the space of braids, i.e. the two successive adiabatic clockwise exchanges yield another braid. Obviously, we can also regard a clockwise exchange as the inverse operation. Thus adiabatic exchanges define a group of transformations known as the *braid group*. The excitations with anyon or fractional statistics characterized by a single phase δ are one-dimensional representations of the braid group.

However, as we also noted in Section 10.8, the braid group also has multi-dimensional representations in which a braid is characterized by a *matrix* of phase factors instead of a single number. In this section we will discuss a set of quantum Hall states whose excitations transform under a braiding process precisely in this fashion. For this reason these quantum Hall states are said to be non-abelian.

Chern–Simons gauge theories with non-abelian gauge groups have non-abelian representations of the braid group (if the level k of the Chern–Simons theory is greater than $k = 1$). Thus it is natural to suspect that there is a relation between non-abelian Chern–Simons gauge theory and non-abelian quantum Hall states. We will see that this is the case. On the other hand, Chern–Simons gauge theory (in $(2+1)$ dimensions) has a close connection with chiral conformal field theory in two dimensions (Witten, 1989, 1992). We will see in this section (and in Chapter 15) that there is also a close connection between fractional quantum Hall states and chiral conformal field theory (CFT). This connection will come in two forms, (a) through the universal structure of the ground-state wave functions and (b) through the excitation spectrum of their edge states.

14.6.1 Conformal field theory and quantum Hall wave functions

We will begin by rewriting the Laughlin (and Halperin) wave functions in a manner that makes the connection with chiral CFT self-evident. The Jastrow prefactor of the Laughlin wave function

$$\Psi_m(z_1, \ldots, z_N) = \prod_{i<j}(z_i - z_j)^m \exp\left(-\sum_{i=1}^{N} \frac{|z_i|^2}{4\ell_0^2}\right) \tag{14.91}$$

has a structure reminiscent of the Coulomb-gas expressions discussed in Section 4.6 except for the important fact that the exponent is positive. Thus, the Jastrow factor vanishes at short distances and diverges at long distances, whereas the Coulomb-gas expressions behave in the opposite fashion. In addition the Jastrow factor is a holomorphic function of the complex coordinates $z_1, \ldots, z_N$, while the Coulomb-gas expressions are a product of holomorphic and anti-holomorphic factors.

The Laughlin wave functions have the peculiar property (for wave functions) of being *universal*, i.e. independent of the details of the Hamiltonian of the interacting electron gas. The only dimension-full parameter they contain is the magnetic length, which enters only in the Gaussian integrating factor. While these "ideal" wave functions are the exact ground states of a class of local interacting Hamiltonians, they are an outstanding ansatz for a large class of Hamiltonians (including the physically relevant case of the Coulomb interaction). The exact ground-state wave functions for specific Hamiltonians exhibit the same asymptotic behaviors as those of the Laughlin states: they (a) have the same behavior as two electron coordinates approach each other (they have zeros of the same order) and (b) also have the same asymptotic power-law behavior at long distances. Thus, each Laughlin wave function is the universal representative of a phase of the 2DEG, a stable fixed point that has a large basin of attraction. The exact wave functions for generic Hamiltonians in the same phase differ from the Laughlin states by complicated functional dependences that do not affect these universal behaviors. In the language of the renormalization group they differ from the Laughlin states by the action of irrelevant operators.

The universal structure of the Laughlin wave functions naturally motivates the use of a description of these states in the language of a 2D CFT. This approach, which was pioneered by Moore and Read (1991), constitutes an extremely powerful tool to generalize and classify these topological states. Moore and Read showed that it is possible to rewrite the full Laughlin wave function (including the integrating exponential factor) in the following Coulomb-gas expression:

$$\Psi_m(z_1, \ldots, z_N) = \left\langle \left(\prod_{i=1}^{N} e^{\,i\sqrt{m}\varphi(z_i)} \right) \exp\left(- \int d^2z' \sqrt{m}\rho_0 \varphi(z') \right) \right\rangle \tag{14.92}$$

where $\varphi(z)$ is a chiral boson in two Euclidean dimensions whose correlation function is (cf. Eq. (4.70))

$$\langle \varphi(z)\varphi(z') \rangle = -\ln(z - z') \tag{14.93}$$

A straightforward inspection of the results discussed in Section 4.6 in the context of the Kosterlitz–Thouless theory tells us that this is the propagator of the holomorphic half of a boson CFT with stiffness $K = 1/(4\pi)$. In Coulomb-gas language the Jastrow factor of Eq. (14.92) represents a set of positive charges, each of strength $\sqrt{m}$, and the exponential factor represents a uniform background neutralizing charge with areal density $\rho_0 = 1/(2\pi m)$ (measured in units in which the magnetic length is $\ell_0 = 1$).

Similarly, in this representation the wave function of a state with a quasihole with complex coordinate w is

$$\begin{aligned}
\Psi_m^+(w, z_1, \ldots, z_N; m) \\
&= \left\langle \exp\left(\frac{i}{\sqrt{m}} \varphi(w)\right) \left(\prod_{i=1}^{N} e^{i\sqrt{m}\varphi(z_i)}\right) \exp\left(-\int d^2z' \sqrt{m}\rho_0 \varphi(z')\right)\right\rangle \\
&= \prod_{i=1}^{N} (z_i - w) \prod_{i<j} (z_i - z_j)^m \exp\left(-\frac{1}{4}\sum_{i=1}^{N} |z_i|^2 - \frac{1}{4m}|w|^2\right)
\end{aligned} \tag{14.94}$$

Thus we conclude that the construction of wave functions of Laughlin states Ψ_m is equivalent to the computation of expectation values of a conformal field theory for a free chiral boson $\varphi(z)$. In this construction the action of removing an electron (a hole of positive charge e) at a complex coordinate z is represented by a chiral vertex operator $V_e(z)$,

$$V_e(z) = e^{i\sqrt{m}\varphi(z)} \tag{14.95}$$

which we will call the *electron operator*. Similarly, we can identify the operator which, when acting on a Laughlin state Ψ_m, creates a quasihole at complex coordinate w with the vertex $V_1(w)$ operator of the CFT of a chiral free boson,

$$V_1(w) = \exp\left(\frac{i}{\sqrt{m}}\varphi(w)\right) \tag{14.96}$$

The Halperin wave function for two quasiholes of a Laughlin state, which was presented in Eq. (13.36), can also be represented as a correlator in this chiral CFT involving now the insertion of an additional vertex operator $V_1(u)$. This wave function has the same structure in its dependence on the electron coordinates as the Laughlin wave function Ψ_m; it has a non-local branch cut dependence on the relative coordinates of the quasiholes, $(u - w)^{1/m}$. Thus, the transport of a quasihole around any of the electron coordinates is trivial. Instead, transporting a quasihole around the other quasihole senses the branch cut. Mathematically this is called a non-trivial monodromy. As we saw, this is a manifestation of fractional statistics.

In other terms, the quasihole operators are not local with respect to each other, but are local with respect to the electron operators. Furthermore, two quasihole operators at coordinates u and w closer to each other than to the other coordinates behave in the same way as the two quasiholes *fused* together into a quasihole of twice the charge with vertex operator V_2. This process has exactly the same structure as the operator-product expansion (OPE) in a CFT (see Sections 4.3 and 7.11). We can continue this fusion process to produce quasiholes of charge p/m (obtained from fusing p fundamental quasiholes).

However, this process cannot be continued indefinitely, since if we fuse m fundamental quasiholes we obtain an electron, or rather a hole, since we are removing an electron from the fluid. Hence, m quasiparticles fuse into the condensate and have essentially the same effect as acting with the identity operator. Only the fusion of quasiholes modulo an electron leads to physically distinct excitations. In the language of an OPE we write

$$\lim_{w \to u} V_p(u) V_q(w) \simeq V_{[p+q]_m}(u) \tag{14.97}$$

where $[p]_m$ is the integer p modulo a multiple of m, i.e. if $p = mr + s$ then $[p]_m = s$ (with $r \geq 0$ and $0 \leq s < m$). This means that the vertex operators which create physically distinct excitations must be truncated to this set.

In other words, there is a correspondence between the spectrum of allowed excitations of quasiholes in a Laughlin state and the spectrum of primary fields of the chiral boson CFT. This correspondence includes the wave function itself, which is written as a correlator in a chiral rational conformal field theory (RCFT). More precisely, the wave function is the holomorphic half of a correlator of the RCFT. These quantities are known as *conformal blocks* of the RCFT since the correlators are built from them (see Moore and Seiberg (1989)). The truncation of the spectrum of primary fields (the allowed vertex operators) is equivalent to the statement that the chiral boson is *compactified* with a compactification radius $R = 1/\sqrt{m}$ determined by the invariance of the electron operator under shifts $\varphi \to \varphi + 2\pi/\sqrt{m}$. This makes this CFT an RCFT (Ginsparg, 1989; Di Francesco *et al.*, 1997). On the other hand, this construction has the additional feature that all the allowed primary fields are local with respect to the electron operator. In the language of Moore and Read (1991), the electron operator plays the role of a "current" that generates an extended symmetry algebra (of the primary fields). We encountered a similar construction in our discussion of Luttinger liquids in Chapter 6 and of the quantum critical points of 2D quantum dimer models in Chapter 9 (although we were not as explicit as we are here). In Chapter 15 we will see that there is a closely related structure of the edge states of the quantum Hall states. We will use the same principles in the rest of this section to construct generalized quantum Hall states.

This structure is also present in a Chern–Simons gauge theory. Consider a Chern–Simons gauge theory with a U(1) gauge group

$$\mathcal{L}[a_\mu] = \frac{m}{4\pi} \epsilon_{\mu\nu\lambda} a^\mu \, \partial^\nu a^\lambda \tag{14.98}$$

which we will refer to as a U(1) level-m Chern–Simons theory and denote by $U(1)_m$. The observables of this theory are Wilson loop operators that carry p units of charge,

$$W_\Gamma^p = \exp\left(ip \oint_\Gamma dx_\mu \, a^\mu \right) \tag{14.99}$$

We have already shown that the Wilson loops represent heavy particles carrying fractional statistics with a statistical angle $\delta_p = \pi p^2/m$. Thus, in this simple abelian system, fusing $p = m$ particles together is equivalent to an electron (since $\delta_m = \pi$). Here too we can define an extended symmetry algebra by defining as being physically distinct the Wilson loop operators modulo the level m of the Chern–Simons theory. Hence, there are only m physically distinct states (including the identity) allowed in this theory. As we saw, this structure is natural when defining the theory on a torus with each physically distinct Wilson loop on a non-contractible path of the torus, labeling each of the m distinct topological sectors of the Hilbert space.

14.6.2 Conformal blocks

Armed with the principles we presented above, we can construct many potentially interesting quantum Hall states. The strategy is to find an RCFT that may be of physical relevance and construct the wave functions by application of these principles. In fact, what we did in the preceding subsection can easily be applied to the wave functions of the Halperin states of multi-component systems, for instance to the spin-singlet Halperin states (Balatsky and Fradkin, 1991). It has also been used to suggest wave functions for the hierarchical states (Blok and Wen, 1990; Moore and Read, 1991). The RCFT approach has been a particularly powerful tool to uncover the existence of a class of non-abelian quantum Hall phases with unexpected and extremely interesting properties.

Before we introduce the non-abelian fractional quantum Hall states, let us discuss what is meant by non-abelian fractional (or braiding) statistics. Let us consider an abstract system in two dimensions. An N-particle state will be denoted by $\psi_{p;i_1,\ldots,i_N}(z_1, \ldots, z_N)$ with complex coordinates $z_1, \ldots, z_N$, where $i_1, \ldots, i_N$ denote the possible quantum numbers of the particles. The integer p labels the basis in a multi-dimensional vector space of states and it is not related to the quantum numbers of the individual particles (which, as usual, are associated with the states of local observables). Let us consider an exchange process of two of the particles, with labels k and l, understood as a quasi-static braiding process under which the particles exchange places. Since these states are assumed to be a basis, the general result of any unitary operation, including braiding processes, must in general be equivalent to a linear combination of the basis states. Therefore (Moore and Read, 1991)

$$\psi_{p;i_1,\ldots,i_s,\ldots,i_r,\ldots,i_N}(z_1,\ldots,z_s,\ldots,z_r,\ldots,z_N)$$
$$= \sum_q B_{pq}[i_1,\ldots,i_N]\psi_{q;i_1,\ldots,i_r,\ldots,i_s,\ldots,i_N}(z_1,\ldots,z_r,\ldots,z_s,\ldots,z_N) \tag{14.100}$$

where q labels the states in the basis. The two-particle exchange process is then represented by the *matrix* B_{pq}. In other terms, the Hilbert spaces spanned by these basis states are multi-dimensional representations of the braid group. Since these matrices in general do not commute, it is said that the particles ("non-abelions") have non-abelian fractional statistics.

It turns out that the conformal blocks that appear in the correlators in CFTs are analytic functions of the coordinates of the operators and satisfy monodromy properties that are consistent with the transformation law of Eq. (14.100). If the coordinates of the operators are regarded as the coordinates of quantum-mechanical particles on the complex plane, the conformal blocks can be interpreted as the wave functions for anyon-type objects that are multi-dimensional representations of the braid group (Moore and Read, 1991). The rules that we will import from conformal field theory can be used to define in a precise way the laws that govern the behavior of non-abelian anyons as quantum objects in two dimensions (Kitaev, 2003; Preskill, 2004). The mathematical term is a modular tensor category (or modular functor) (Moore and Seiberg, 1989; Freedman, 2003).

Although a full discussion of conformal blocks is beyond the scope of this book (see e.g. Di Francesco *et al.* (1997)) we will summarize some of their important properties that will be useful for us. Let $\{\phi_j\}$ be the set of primary fields of some CFT. We will assume that this set is finite and that it is an RCFT. As we have seen, in an RCFT the primary fields satisfy a closed operator algebra known as the operator-product expansion (OPE). In an OPE two primary fields, say ϕ_i and ϕ_j, can be fused together into several possible other operators. The result is a sum over all the possible primaries ϕ_k with singular coefficients. The label k in this sum runs over all possible primary fields, including the identity, with coefficients C_{ijk}. Now, in the fusion of the primary fields ϕ_i and ϕ_j, the operator ϕ_k will appear a number of times, which we will denote by the integers N_{ij}^k. This fact is expressed with the notation

$$\phi_i \star \phi_j = \sum_k N_{ij}^k \phi_k \tag{14.101}$$

For instance, in the case of the compactified boson which we discussed above $N_{ij}^k = \delta_{k,i+j}$ since only operators with the correct charges will mix. We will see below that non-abelian fractional statistics arises if the integers N_{ij}^k do not vanish for two or more fields.

Since the order of fusion is immaterial, the fusion algebra is commutative,

$$N_{ij}^k = N_{ji}^k \tag{14.102}$$

This algebra is also associative,

$$(\phi_i \star \phi_j) \star \phi_k = \phi_i \star (\phi_j \star \phi_k) \tag{14.103}$$

since the order of successive fusions does not matter. This means that the integers N_{ij}^k satisfy

$$\sum_l N_{il}^m N_{jk}^l = \sum_n N_{ij}^n N_{nk}^m \tag{14.104}$$

The integers N_{ij}^k can be regarded as matrix elements in an associative and commutative algebra of matrices known as the Verlinde algebra (Verlinde, 1988).

The requirement of conformal invariance imposes stringent conditions on the analytic structure of the correlation functions. For example the two-point function of the primary field ϕ_i factorizes into a product of a holomorphic factor and an anti-holomorphic factor (see Section 7.11, Eq. (7.173)),

$$\langle \phi_i(z, \bar{z})\phi_i(0, 0)\rangle = \frac{1}{z^{2h_i}\bar{z}^{2\bar{h}_i}} \tag{14.105}$$

where h_i and $\bar{h}_i$ are the conformal dimensions (where $\Delta_i = h_i + \bar{h}_i$ is the scaling dimension and $s_i = h_i - \bar{h}_i$ is the conformal spin). A factorized analytic structure that can be expressed as a sum of the form

$$\left\langle \prod_{s=1}^{N} \phi_i(z_{i_s}, \bar{z}_{i_s}) \right\rangle = \sum_p \left| \mathcal{F}_{p;\{i_j\}}(z_1, \ldots, z_N) \right|^2 \tag{14.106}$$

also applies for more general correlators of primary fields. This expression states that the general correlator of primaries is a sum of terms, each of which can be written as a product of holomorphic and anti-holomorphic functions $\mathcal{F}_{p;\{i_j\}}(z_1, \ldots, z_N)$ known as conformal blocks. The structure of the conformal blocks depends on the number and type of primary fields present in the correlator. For a *given set of coordinates* $z_1, \ldots, z_N$ (and labels $i_1, \ldots, i_N$) the conformal blocks are linearly independent holomorphic functions that span a basis of the vector space whose dimension depends on the number of fusion channels of the primary fields ϕ_i (and hence on the integers N_{ij}^k) (Belavin *et al.*, 1984; Friedan and Shenker, 1987).

If the primary fields ϕ_i and ϕ_j have more than one fusion channel (i.e. $N_{ij}^k \neq 0$ for more than one k for i and j fixed), the number of conformal blocks in

Eq. (14.106) is $p > 1$. The conformal blocks are analytic functions of the coordinates that have isolated singularities. These singularities are such that, as one displaces a coordinate z_i along a closed smooth curve on the complex plane which encloses another coordinate z_j, the conformal blocks transform into linear combinations of each other. This transformation, called a monodromy, is equivalent to two successive braids. In this way non-abelian statistics enters into the structure of the correlators.

In CFT each primary field ϕ_i labels a space of states generated by a tower of irrelevant operators (its descendants). Let us consider a 2D space with the topology of a torus, defined as a parallelogram of vertices 0, 1, τ, and $1+\tau$, where the complex number $\tau = iL_2/L_1$ is the modular parameter of the torus. Let $\vec{a}$ and $\vec{b}$ be the two cycles of the torus and we will pick $\vec{b}$ to denote the direction of time evolution. The partition function restricted to this space of states is the character χ_i of the representation labeled by ϕ_i,

$$\chi_i \equiv \text{tr}_{[\phi_i]} q^{H-c/24}, \qquad q = e^{2\pi i \tau} \tag{14.107}$$

where H is the Hamiltonian for the right-moving states and c is the central charge. The character χ_i is related to the short-distance behavior of the conformal block $\mathcal{F}_j^{ii^*}(z-w)$ (ϕ_i and ϕ_{i^*} are conjugate fields) by (Di Francesco *et al.*, 1997)

$$\chi_i = \lim_{z\to w} (z-w)^{2h_j} \mathcal{F}_i^{jj^*}(z-w) \tag{14.108}$$

where h_i is the conformal dimension of the primary field ϕ_i.

We can define the modular transformations T and S of the torus:

$$T : \tau \to \tau + 1, \qquad S : \tau \to -\frac{1}{\tau} \tag{14.109}$$

Verlinde showed that under these transformations the characters χ_i obey the transformation laws

$$T : \chi_i \to \exp\left[2\pi i \left(h_i - \frac{c}{24}\right)\right] \chi_i, \qquad S : \chi_i \to \sum_j \mathcal{S}_i^j \chi_j \tag{14.110}$$

where the unitary matrix $\mathcal{S}_i^j$ is called the modular $\mathcal{S}$-matrix. Thus, the conformal blocks also transform under an S modular transformation by the action of the modular $\mathcal{S}$-matrix.

Following Verlinde, we now define the action of inserting a primary field ϕ_i into the partition function $Z = \chi_0$ of the system and moving it along the full closed cycle $\vec{b}$. He showed that the result was

$$\phi_i(\vec{b})\chi_0 = \chi_i \tag{14.111}$$

If the same operation is carried on the character χ_i, one obtains

$$\phi_j(\vec{b})\chi_i = \sum_k N_{ij}^k \chi_k \tag{14.112}$$

with N_{ij}^k being the same matrix as that which defines the fusion rules for ϕ_i and ϕ_j. Using these relations, Verlinde further proved that the coefficients N_{ij}^k of the fusion algebra are related to the elements of the modular $\mathcal{S}$-matrix by the Verlinde formula (Verlinde, 1988)

$$N_{ij}^k = \sum_n \frac{\mathcal{S}_j^n \mathcal{S}_i^n (\mathcal{S}^{-1})_n^k}{\mathcal{S}_0^n} \tag{14.113}$$

The same modular $\mathcal{S}$-matrix also appears in Chern–Simons gauge theory. Witten showed that the states of a Chern–Simons theory are accounted for by the conformal blocks of a CFT. As a result, the Chern–Simons states are representations of the braid group, and may be identified with the characters of a CFT. The modular $\mathcal{S}$-matrix of the CFT will then enter into calculations of Wilson loop observables in the Chern–Simons theory, as was illustrated by Witten (1989). He showed that in Chern–Simons gauge theory the expectation value of the Wilson loop operator which carries the representation R_j in a space-time with the topology of the 3-sphere S^3 is given by a matrix element of the modular $\mathcal{S}$-matrix,

$$Z(S^3, R_j) = \mathcal{S}_0^j \tag{14.114}$$

and that the partition function for the vacuum $Z(S^3)$ is given by

$$Z(S^3) = \mathcal{S}_0^0 \tag{14.115}$$

For reasons that will become clear shortly, it is convenient to define the *quantum dimension* d_j of the representation R_j,

$$d_j = \frac{\mathcal{S}_0^j}{\mathcal{S}_0^0} \tag{14.116}$$

The unitarity of the modular $\mathcal{S}$-matrix implies that

$$\left(\mathcal{S}_0^0\right)^{-1} = \sqrt{\sum_j |d_j|^2} \equiv \mathcal{D} \tag{14.117}$$

where the quantity $\mathcal{D}$ defines the *effective quantum dimension*.

The important insight that we will retain is that the conformal blocks are holomorphic functions that span a basis of a finite-dimensional vector space and transform under braiding operations following the law of Eq. (14.100) (Moore and Seiberg, 1989). This observation led Moore and Read to propose the use of conformal blocks of RCFTs to construct holomorphic wave functions (hence those in the lowest Landau level) whose excitations may realize non-abelian statistics (Moore and Read, 1991). In Section 14.6.1 we saw that the Laughlin states can be recast as expectation values of the chiral CFT of a compactified boson. We will now generalize this construction.

14.7 The spin-singlet Halperin states

The spin-singlet Halperin $(n+1, n+1, n)$ states with filling fraction $\nu = 2/(2n+1)$ can also be recast as a CFT correlator. To this end, we factorize these wave functions for N electrons (with N even) (Balatsky and Fradkin, 1991; Moore and Read, 1991),

$$\Psi_{(n+1,n+1,n)}(\{z_i^\uparrow\}, \{z_i^\downarrow\}) = \prod_{i<j} (z_i^\uparrow - z_j^\uparrow)^{n+1} (z_i^\downarrow - z_j^\downarrow)^{n+1} \prod_{i,j} (z_i^\uparrow - z_j^\downarrow)^n \times \exp\left(-\frac{1}{4\ell_0^2} \sum_i \left(|z_i^\uparrow|^2 + |z_i^\downarrow|^2 \right) \right) \tag{14.118}$$

into a product of a wave function for the charge degrees of freedom, which takes the form of a Laughlin state for *semions* (with statistical angle $\delta = \pi/2$),

$$\Psi_m^{(\pi/2)}(\{z_i^\uparrow\}, \{z_i^\downarrow\}) = \prod_{i<j} (z_i^\uparrow - z_j^\uparrow)^{n+1/2} (z_i^\downarrow - z_j^\downarrow)^{n+1/2} (z_i^\uparrow - z_j^\downarrow)^{n+1/2} \times \exp\left(-\frac{1}{4\ell_0^2} \sum_i \left(|z_i^\uparrow|^2 + |z_i^\downarrow|^2 \right) \right) \tag{14.119}$$

and the spin-singlet wave function for the spin degrees of freedom,

$$\Psi_{\text{singlet}}(\{z_i^\uparrow\}, \{z_i^\downarrow\}) = \prod_{i<j} \frac{(z_i^\uparrow - z_j^\uparrow)^{1/2} (z_i^\downarrow - z_j^\downarrow)^{1/2}}{(z_i^\uparrow - z_j^\downarrow)^{1/2}} \tag{14.120}$$

The Laughlin state for semions can be written as the CFT correlator, a conformal block of a compactified boson RCFT at level $k = 2n + 1$,

$$\Psi_m^{(\pi/2)}(\{z_i^\uparrow\}, \{z_i^\downarrow\}) = \left\langle \left(\prod_{i=1}^{N} \exp\left(i\sqrt{n + \frac{1}{2}}\varphi(z_i) \right) \right) \times \exp\left(-\int d^2z' \sqrt{n + \frac{1}{2}} \rho_0 \varphi(z') \right) \right\rangle_{U(1)_k} \tag{14.121}$$

where the label i runs over all N particles, irrespective of their spin polarization.

The wave function for the spin sector, Ψ_{singlet}, is equal to the correlation function of the spin-doublet primary field (with $j = 1/2$) in a chiral $SU(2)_1$ Wess–Zumino–Witten (WZW) model discussed in Section 7.12 (see Eq. (7.199) for $SU(2)_1$). If we denote the spin $j = 1/2$ doublet primary fields of the $\mathrm{SU}(2)_1$ WZW chiral CFT by $V_{\pm 1/2}(z)$, the spin-singlet wave function is

$$\Psi_{\text{singlet}}(\{z_i^\uparrow\}, \{z_i^\downarrow\}) = \left\langle V_{+1/2}(z_1^\uparrow) \dots V_{+1/2}(z_{N/2}^\uparrow) V_{-1/2}(z_1^\downarrow) \dots V_{-1/2}(z_{N/2}^\downarrow) \right\rangle_{\mathrm{SU}(2)_1} \tag{14.122}$$

Interestingly (but not accidentally!), this wave function for the spin sector is the same as the conjectured Kalmeyer–Laughlin wave function for a spin-liquid state of a frustrated quantum antiferromagnet (Kalmeyer and Laughlin, 1987).

Thus, both the charge sector and the spin sector are given by a conformal block of a chiral RCFT, the U(1) chiral boson at level $k = 2n+1$ for the charge sector and the $\mathrm{SU}(2)_1$ chiral WZW CFT for the spin sector. In both cases only one conformal block is involved. This is expected for the U(1) chiral boson, since we showed above that the fusion rules have only one channel.

However, the spin sector requires some discussion. In a sense this result is not surprising since we can picture the spin-singlet states as a special case of a bilayer system that has a $\mathrm{U}(1) \times \mathrm{U}(1)$ symmetry. However, as shown in Balatsky and Fradkin (1991), the spin-singlet fractional quantum Hall states can be constructed as a $\mathrm{U}(1) \times \mathrm{SU}(2)$ Chern–Simons theory whose spin sector has level $k = 1$ that has a very simple structure. In a theory with SU(2) symmetry we expect the states (and also the fields) to be organized into multiplets, namely the irreducible representations of the group SU(2). Thus, if we have states with angular momenta j_1 and j_2, the tensor product decomposes into a sum of states with total angular momentum j ranging from $|j_1 - j_2|$ to $j_1 + j_2$. However, for the primary fields ϕ_j of the Kac–Moody *current algebra* $\mathrm{su}(2)_k$ (associated with the $\mathrm{SU}(2)_k$ WZW model) the corresponding fusion rules are truncated at $j_{\max} = k/2$,

$$\phi_{j_1} \star \phi_{j_2} = \phi_{|j_1 - j_2|} + \dots + \phi_{j_{\max}} \tag{14.123}$$

In particular, in the case of $\mathrm{SU}(2)_1$, the level is $k = 1$ and the fusion is truncated at $j_{\max} = 1/2$. This means that for $\mathrm{SU}(2)_1$ two spin-$1/2$ primary fields can only fuse into the $j = 0$ identity field since $j = 1$ is projected out. Hence, for $\mathrm{SU}(2)_1$ we simply write

$$[1/2] \star [1/2] = [0], \qquad \text{for } \mathrm{SU}(2)_1 \tag{14.124}$$

This means that, for $\mathrm{SU}(2)_1$ (this is also true for $\mathrm{SU}(N)_1$ (Knizhnik and Zamolodchikov, 1984)), there is only one fusion channel and hence only one conformal block in the correlator. This also implies that, from the point of view of the braid group, $\mathrm{SU}(2)_1$ has only one-dimensional representations and the *fractional statistics* is *abelian* (even though the group SU(2) is non-abelian!).

The conclusion is that $\mathrm{SU}(2)_1$ theory (both as a WZW CFT and as a Chern–Simons theory) is secretly an abelian theory. But we already know this. In fact, in our discussion of the Luttinger model in Chapter 6 we found using abelian bosonization that a theory with an $SU(2)_1$ current algebra can be represented in terms of an abelian bosonic theory with a special choice of compactification radius. Thus, the right-moving currents $J_R^\pm$ and J_R^3 can be represented in terms of a single compactified chiral boson $\phi(z)$ (with a suitable choice of normalization)

$$J_R^3(z) \sim i\,\partial_z\phi(z), \qquad J_R^\pm(z) \sim e^{\pm i\sqrt{2}\phi(z)} \tag{14.125}$$

In this representation the spin-1/2 doublet chiral primary fields are

$$V_{\pm 1/2}(z) \sim \exp\left(\pm\frac{i}{\sqrt{2}}\phi(z)\right) \tag{14.126}$$

14.8 Moore–Read states and their generalizations

Moore and Read (1991) generalized this approach to propose several fractional quantum Hall states with strikingly interesting properties. Specifically they proposed the Pfaffian wave functions for a system of fully polarized electrons that has the form of a product of a Pfaffian and a Laughlin wave function,

$$\Psi_{\rm MR}(z_1,\ldots,z_N) = \text{Pf}\left(\frac{1}{z_i - z_j}\right)\prod_{i<j}(z_i - z_j)^n\ \exp\left(-\frac{1}{4\ell_0^2}\sum_i |z_i|^2\right) \tag{14.127}$$

This wave function represents a system of electrons in the lowest Landau level at filling factor $\nu = 1/n$.

The Pfaffian of the antisymmetric matrix $M_{ij} = 1/(z_i - z_j)$ is the fully antisymmetrized product of all possible *pairs* of matrix elements,

$$\text{Pf}(M_{ij}) = \frac{1}{2^{N/2}(N/2)!}\sum_P \text{sgn}(P)\prod_{r=1}^{N/2} M_{P(2r-1)P(2r)} = \mathcal{A}\big(M_{12}\ldots M_{N-1,N}\big) \tag{14.128}$$

where P labels all possible permutations of N elements, and $\mathcal{A}$ denotes the operation of antisymmetrization. Up to an overall sign, the Pfaffian of an antisymmetric matrix is equal to the square root of the determinant of the same matrix,

$$\text{Pf}(M) = \pm\sqrt{\det M} \tag{14.129}$$

For n even this state is antisymmetric and hence is a candidate wave function for a fractional quantum Hall state for electrons. More interestingly, unlike Laughlin states and its generalizations, for a fermionic state the filling fraction $\nu = 1/n$ of the Moore–Read states has an *even denominator*. Conversely, for n odd these wave functions are symmetric under exchange and describe a system of bosons. Since the Pfaffian prefactor is a set of poles it allows the particles to be packed more densely than a Laughlin-type state. For $n \geq 1$ the Laughlin factor insures the integrability of this wave function. It can also be shown that for a fixed number of particles this state is still a polynomial of the complex coordinates of the particles.

We already know that on a torus the Laughlin state at $\nu = 1/n$ has n degenerate ground states due to the motion of the center of mass of the system on the torus. The Pfaffian state on a torus is obtained by the replacement (as in the Laughlin state)

$$\mathrm{Pf}\left(\frac{1}{z_i - z_j}\right) \longmapsto \mathrm{Pf}\left(\frac{\theta_a(z_i - z_j)}{\theta_1(z_i - z_j)}\right) \tag{14.130}$$

where $\theta_a(z)$ (with $a = 1, \ldots, 4$) are the four theta functions. This gives us a total degeneracy of $3n$ for a Pfaffian state with $\nu = 1/n$.

The structure of the Pfaffian factor indicates that this state has strong pairing correlations, and for this reason it is often called a paired Hall state (Greiter *et al.*, 1991). Since the Pfaffian factor is a product of single poles, the pairing correlations are similar to those of a BCS-type superconductor with an order parameter with $p_x + ip_y$ symmetry, similar to the A phase of superfluid ^{3}He: a fully spin-polarized paired state of fermions with orbital angular momentum $l = 1$ and $m = \pm 1$ (Leggett, 1975).

Of particular interest is the Moore–Read state at $\nu = 1/2$ ($n = 2$). It has been known for many years that in the first Landau level there is a fractional quantum Hall state with filling fraction $\nu = 5/2$. We can regard the lowest Landau level (which is completely filled) as having a filling factor of $\nu = 2$. Thus, a $\nu = 5/2$ state can then be pictured as a half-filled first Landau level. Currently available numerical results indicate that the Moore–Read state is the most likely candidate to explain the fractional quantum Hall state at 5/2 (Morf, 1998; Rezayi and Haldane, 2000). Although it is possible to have a paired state even in the lowest Landau level (through some version of the Kohn–Luttinger mechanism (Kohn and Luttinger 1965; Chubukov 1993; Raghu and Kivelson, 2011)), so far there is no evidence (experimental or numerical) of paired states in the lowest Landau level. The bosonic Moore–Read state with $n = 1$ has been conjectured to occur in a system of ultra-cold bosons rotating at very high angular velocity (Cooper *et al.*, 2001).

An insight into the physical origin of the paired states is gained by noticing that the Moore–Read state has an interesting and physically illuminating connection with the Halperin (3, 3, 1) state which has the same filling fraction, $\nu = 1/2$:

$$\Psi_{(3,3,1)}(z_1, \ldots, z_{N/2}, w_1, \ldots w_{N/2}) = \prod_{i<j}(z_i - z_j)^3 \prod_{i<j}(w_i - w_j)^3 \prod_{i \leq j}(z_i - w_j)^1 \times \exp\left(-\frac{1}{4\ell_0^2}\sum_{i=1}^{N/2}\left(|z_i|^2 + |w_i|^2\right)\right) \tag{14.131}$$

We can regard this state as a wave function for a fully spin-polarized system in a bilayer (with two half-filled layers), or as a spin-unpolarized single-layer system with an explicitly broken SU(2) spin rotational invariance. The Halperin (3, 3, 1) wave function is a good state for a bilayer system with weak inter-layer tunneling. This state is not antisymmetric under the exchange of electrons between the two layers (or spin polarizations). As we increase the inter-layer tunneling from weak to strong (or, equivalently, the spin flip rate), the wave function must be antisymmetric in the coordinates of all the electrons.

It is a remarkable fact that the fully antisymmetrized (3, 3, 1) wave function is equal to the Moore–Read state (Ho, 1995). This can be seen as follows. Let us consider a system of N electrons with both spin orientations. Let u_i and v_i be the up- and down-spin spinors. In this form the (3, 3, 1) state can be written as

$$\begin{aligned}\Psi_{(3,3,1)} &= \mathrm{Pf}\left(\frac{u_i v_j + u_j v_i}{z_i - z_j}\right) \prod_{i<j}(z_i - z_j)^2 \exp\left(-\frac{1}{4\ell_0^2}\sum_i |z_i|^2\right) \\ &= \mathcal{A}\left\{\prod_{i>j}(z_{2i-1} - z_{2j-1})^3 (z_{2i} - z_{2j})^3 \prod_{i,j}(z_{2i-1} - z_{2j})^1 \right. \\ &\quad \left. \times \prod_i u_{2i-1} v_{2i}\right\} \exp\left(-\frac{1}{4\ell_0^2}\sum_i |z_i|^2\right)\end{aligned} \tag{14.132}$$

If we now rotate the spin quantization axis to the x axis this wave function becomes

$$\Psi_{(3,3,1)} = \mathrm{Pf}\left(\frac{u_i^x u_j^x - v_i^x v_j^x}{z_i - z_j}\right) \prod_{i<j}(z_i - z_j)^2 \exp\left(-\frac{1}{4\ell_0^2}\sum_i |z_i|^2\right) \tag{14.133}$$

Thus, the Moore–Read (or Pfaffian) state is a (3, 3, 1) state with its down spins (with respect to the x axis) projected out. This suggests that the Moore–Read state arises in the strong-tunneling limit in a bilayer system.

The other insight into the origin of this non-abelian state comes from thinking of the Moore–Read state as a paired state and making the connection with a 2D $p_x + ip_y$ superconductor explicit. Greiter, Wen, and Wilczek suggested the existence of a paired state at $\nu = 1/2$ and showed that the likely pairing channel has angular momentum $l = -1$ (p wave) which must be a spin-triplet state (Greiter *et al.*, 1991, 1992). Numerical calculations showed that a system of composite fermions, namely the quasiparticles of the compressible state at $\nu = 1/2$, has attractive interactions in the $l = 1$ channel (Park *et al.*, 1998). Hence, a state with p-wave pairing of composite fermions is favored. Read and Green (2000) showed that the superconducting analog of the Moore–Read state describes the weak-pairing regime as a BCS superfluid. As the strength of the pairing interactions increases, there is a phase transition to an abelian quantum Hall state of bosonic pairs ("molecules"). The connection of the Moore–Read state with a $p_x + ip_y$ superconductor comes from the Pfaffian factor in the wave function (see Read and Green (2000)).

The Moore–Read state also has the structure of a conformal block of a CFT. As we saw, the Laughlin factor has that structure. But so does the Pfaffian factor. In fact, the Pfaffian can be written as a correlator of a free *chiral Majorana fermion* $\chi(z)$,

$$\langle \chi(z_1)\ldots\chi(z_N)\rangle = \mathrm{Pf}\left(\frac{1}{z_i - z_j}\right) \tag{14.134}$$

This result follows from using Wick's theorem, with the propagator for the (Euclidean) chiral Majorana field being

$$\langle \chi(z)\chi(w)\rangle = \frac{1}{z-w} \tag{14.135}$$

Recall that a Majorana fermion is a real field and satisfies $\chi^\dagger = \chi$. In contrast, for charged (or complex) (Dirac) fermions the correlators are determinants, which is consistent since the square of a Pfaffian is a determinant. Thus the Moore–Read states can be written as

$$\begin{aligned}\Psi_{\mathrm{MR}}[\{z_i\}] &= \langle \chi(z_1)\ldots\chi(z_N)\rangle \\ &\times \left\langle \left(\prod_{i=1}^{N} e^{i\sqrt{n}\,\phi(z_i)}\right) \exp\left(-\int d^2z'\sqrt{n}\rho_0\phi(z')\right)\right\rangle_{\mathrm{U}(1)_n}\end{aligned} \tag{14.136}$$

This wave function is the exact ground state of a three-body Hamiltonian of the form (Greiter *et al.*, 1991)

$$H = \sum_{i;j\neq i;k\neq i,j} P_{\mathrm{LLL}}\delta'(z_i - z_j)\delta'(z_i - z_k)P_{\mathrm{LLL}} \tag{14.137}$$

where P_{LLL} denotes the projection onto the lowest Landau level, and it is an excellent wave function for a system of electrons projected onto the first Landau level with Coulomb interactions (Morf, 1998; Rezayi and Haldane, 2000).

The structure of this wave function suggests that the states are constructed by gluing together (with some rules that we will specify shortly) a *charge sector* represented by a $\mathrm{U}(1)_n$ compactified chiral boson $\phi(z)$ and a *neutral sector* represented in the wave function by the Pfaffian factor. In Chapter 15 we will discuss the edge states and we will find precisely the same construction. The charge of the excitations will be determined by the charge sector, which for this state with filling fraction $\nu = 1/n$ has a chiral current operator

$$J(z) \sim \frac{i}{\sqrt{n}} \partial_z \phi(z) \tag{14.138}$$

Our next task is to find an RCFT that has a Majorana fermion as a primary field. But we already know the answer: it is the CFT of the 2D Ising model (or $d = 1$ quantum)! In fact, in Section 5.3 we discussed the 1D quantum Ising model and saw that it is equivalent to a theory of free Majorana fermions. It has been known since Onsager's solution of the classical Ising model in two dimensions that the partition function is equal to the Pfaffian of a matrix (McCoy and Wu, 1973).

The CFT of the critical point of the 2D Ising model has the following primary fields (Belavin *et al.*, 1984; Friedan *et al.*, 1984; Ginsparg, 1989): (a) the identity I, (b) the energy density ε (the mass term of the Majorana fermion), (c) the Majorana fermion χ, and (d) the Ising field (order parameter) σ, which have the following values of the scaling dimension Δ and conformal spin s: $(0, 0)$, $(1, 0)$, $(1/2, 1/2)$, and $(1/8, 0)$, respectively. Here we are interested not in the operators of the full critical Ising model but only in its chiral (right-moving) piece represented by the chiral primary fields I, χ, and σ. Here we have used the fact that the right-moving piece of the energy density ε is the chiral Majorana fermion χ. These chiral primary fields obey the following chiral fusion algebra (Di Francesco *et al.*, 1997):

$$\chi \star \chi = I, \qquad \sigma \star \sigma = I + \chi, \qquad \sigma \star \chi = \chi \tag{14.139}$$

Thus, we now have a case in which the fusion of two σ fields has *two* fusion channels. In turn the $U(1)_n$ CFT of the boson ϕ that describes the Laughlin sector has compactification radius $1/\sqrt{n}$ and has n allowed primary fields.

The Moore–Read (or Pfaffian) quantum Hall state is a special case of a class of non-abelian states. In these generalized states, which were initially proposed by Read and Rezayi (1999), the electrons are not paired but have strong cluster correlations involving three (or more) particles at a time. In the Read–Rezayi states the Pfaffian factor is replaced by a correlator of *parafermions* (that we discussed in Section 10.7) in CFT. In these states the Laughlin–Jastrow factor is replaced by the wave function with an exponent of $m + 2/k$, i.e. a fractional quantum Hall state of anyons.

The parafermion wave functions are exact zero-energy ground states of a class of local Hamiltonians of the form of Eq. (14.137) but involving $k + 1$ delta functions for a system of N electrons, with N being divisible by k. Read and Rezayi showed that these states have filling fraction $\nu = k/(mk + 2)$ (with $m \geq 0$) with a ground-state degeneracy on a torus of $(k + 1)(mk + 2)/2$. For m even the wave functions are symmetric (and describe bosons), whereas for m odd they are antisymmetric and describe fermions.

The (particle–hole-conjugate) $k = 3$ parafermion state is a candidate to represent the quantum Hall plateau that is observed at filling fraction $\nu = 12/5$ (i.e. $2 + 2/5$). However, at that filling fraction there is a competing (abelian) Jain state. In contrast, at $\nu = 5/2$ there is essentially no competing abelian state (since this filling fraction has an even denominator) except for a possible trivial paired abelian state, the bosonic Laughlin state of tightly bound pairs. This trivial state is unlikely to be relevant because it requires strong attractive interactions, instead of the weak suppression of repulsion naturally present in the first Landau level.

Parafermions arise in CFT in the context of the critical phenomena of the $\mathbb{Z}_n$ clock models (of which the three-state Potts model is a special case). In CFT these operators appear in a subclass of minimal models (Belavin *et al.*, 1984; Friedan *et al.*, 1984). The simplest example is the CFT of the $\mathbb{Z}_3$ Potts model (Dotsenko, 1984). Although, unlike Majorana fermions, parafermions cannot be realized as free fields, correlators of parafermions are known from CFT and, in particular, their conformal blocks are also known explicitly.

Returning to the Moore–Read states, we need to determine the states for the quasiparticles/quasiholes. The allowed states must be such that their operators are local with respect to the electron operator (since the fluid is made of electrons!). By inspection of the Moore–Read wave function we see that the operator

$$\psi_e(z) \sim \chi(z) e^{i\sqrt{n}\phi(z)} \tag{14.140}$$

plays the role of the electron. The condition that the allowed operators be *local* (or single-valued) with respect to the electron operator leads to the following set of allowed states.

1. The identity I, which represents the quiescent fluid.
2. The "σ particle" (the "non-abelion" or "half-vortex")

$$\sigma(z) \exp\left(\frac{i}{2\sqrt{n}}\phi(z)\right) \tag{14.141}$$

where $\sigma(z)$ is the "chiral piece" of the order-parameter field of the Ising model. This state has charge $Q = e/(2n)$ and (as we will see below) non-abelian braiding statistics. Notice that $\sigma(z)$ is non-local with respect to the Majorana fermion $\chi(z)$ since it changes the fermion boundary conditions from periodic to anti-periodic, and hence it is double-valued. Similarly, the vertex operator from the charge sector is also double-valued. Nevertheless, their product is single-valued (and hence local) with respect to the electron operator ψ_e.

3. The Majorana fermion, a fermionic $Q = 0$ neutral excitation,

$$\chi(z) \tag{14.142}$$

4. The Laughlin quasihole (a vortex)

$$\exp\left(\frac{i}{\sqrt{n}}\phi(z)\right) \tag{14.143}$$

with charge $Q = e/n$ and *abelian* fractional statistics $\delta = \pi/n$.

The wave function for two quasiholes located with complex coordinates η_1 and η_2 is obtained from the Moore–Read wave function, Eq. (14.127), by the following replacement inside the Pfaffian factor (Greiter *et al.*, 1992; Nayak and Wilczek, 1996):

$$\operatorname{Pf}\left(\frac{1}{z_i - z_j}\right) \longmapsto \operatorname{Pf}\left(\frac{(z_i - \eta_1)(z_j - \eta_2) + (i \leftrightarrow j)}{z_i - z_j}\right) \tag{14.144}$$

The wave function for two quasiholes in the Moore–Read state can be expressed as the following CFT expectation value:

$$\begin{aligned} \Psi_{\text{MR}}^{\text{2qh}}[\{z_i\}] = &\langle \sigma(\eta_1)\sigma(\eta_2)\chi(z_1)\ldots\chi(z_N)\rangle_{\text{Ising CFT}} \\ &\times \left\langle \left(\exp\left(\frac{i}{2\sqrt{n}}\phi(\eta_1)\right) \exp\left(\frac{i}{2\sqrt{n}}\phi(\eta_2)\right) \prod_{i=1}^{N} e^{i\sqrt{n}\phi(z_i)}\right) \right. \\ &\left. \times \exp\left(-\int d^2z' \sqrt{n}\rho_0\phi(z')\right)\right\rangle_{\text{U}(1)_n} \end{aligned} \tag{14.145}$$

which is then also a conformal block.

Similarly, for four quasiholes with coordinates η_i $(i = 1, \ldots, 4)$ we can write

$$\begin{aligned} \operatorname{Pf}\left(\frac{1}{z_i - z_j}\right) &\mapsto \operatorname{Pf}\left(\frac{(z_i - \eta_1)(z_i - \eta_2)(z_j - \eta_3)(z_j - \eta_4) + (i \leftrightarrow j)}{z_i - z_j}\right) \\ &\equiv \operatorname{Pf}_{(12)(34)} \end{aligned} \tag{14.146}$$

However, this is not the only possible wave function of this type, since the two following pairings, $\operatorname{Pf}_{(13)(24)}$ and $\operatorname{Pf}_{(14)(23)}$, are equally good. In fact these three wave functions are not linearly independent, since the following algebraic identity holds:

$$\operatorname{Pf}_{(12)(34)} - \operatorname{Pf}_{(14)(23)} = \frac{\eta_{14}\eta_{23}}{\eta_{13}\eta_{24}}\left(\operatorname{Pf}_{(12)(34)} - \operatorname{Pf}_{(13)(24)}\right) \tag{14.147}$$

where $\eta_{ij} = \eta_i - \eta_j$. Therefore if we specify the coordinates of four quasiholes there are two linearly independent wave functions that span a two-dimensional

Hilbert space. Nayak and Wilczek further showed that there are 2^{p-1} linearly independent states with $2p$ quasiholes.

These Hilbert spaces of states of quasiholes have a topological origin. To see this we will write the states of four quasiholes as the following CFT conformal block:

$$\begin{aligned}\Psi^{4\text{qh}}_{\text{MR}}[\{z_i\}] = &\left\langle \left(\prod_{r=1}^{4} \sigma(\eta_r)\right) \chi(z_1)\dots\chi(z_N)\right\rangle_{\text{Ising CFT}} \\ &\times \left\langle \left(\prod_{r=1}^{4} \exp\left(\frac{i}{2\sqrt{n}}\phi(\eta_r)\right)\prod_{i=1}^{N} e^{i\sqrt{n}\phi(z_i)}\right)\right. \\ &\left.\times \exp\left(-\int d^2z' \sqrt{n}\rho_0\phi(z')\right)\right\rangle_{\text{U}(1)_n}\end{aligned} \qquad (14.148)$$

The origin of the two linearly independent states is the existence of two fusion channels, [0] and [1/2], for the σ fields in the Ising model CFT. For the expectation value to be non-vanishing all the operators in the Ising CFT must fuse (together) into the identity, operator I. This in turn implies that the four σ fields themselves must fuse into the identity, and there are two ways for them to do this. These two possible expectation values are (essentially) the two chiral conformal blocks of the Ising CFT. By means of an explicit (but lengthy) calculation, Nayak and Wilczek obtained the following wave functions for four quasiholes, $\Psi^{4\text{qh}}_{[0]}$ and $\Psi^{4\text{qh}}_{[1/2]}$ (including the contributions from the charge sector):

$$\begin{aligned}\Psi^{4\text{qh}}_{[0]} &= \frac{(\eta_{13}\eta_{24})^{1/4}}{(1+\sqrt{1-x})^{1/2}}\left(\Psi_{(13)(24)} + \sqrt{1-x}\;\Psi_{(14)(23)}\right) \\ \Psi^{4\text{qh}}_{[1/2]} &= \frac{(\eta_{13}\eta_{24})^{1/4}}{(1-\sqrt{1-x})^{1/2}}\left(\Psi_{(13)(24)} - \sqrt{1-x}\;\Psi_{(14)(23)}\right)\end{aligned} \qquad (14.149)$$

where x is the cross ratio

$$x = \frac{\eta_{12}\eta_{34}}{\eta_{13}\eta_{24}} \qquad (14.150)$$

The wave functions for four quasiholes $\Psi^{4\text{qh}}_{[0]}$ and $\Psi^{4\text{qh}}_{[1/2]}$, Eq. (14.149), have a branch-cut structure that is intimately related to non-abelian statistics. Given the coordinates of the four quasiholes, $\eta_1, \dots, \eta_4$, we can associate with each one of the two linearly independent wave functions a prescription regarding how to pair the quasiholes by running branch cuts between them (as shown in Fig. 14.1).

Owing to the branch cuts present in the wave functions, under a braiding operation B (a unitary transformation representing a half monodromy) the conformal block wave functions transform into linear combinations of each other. For

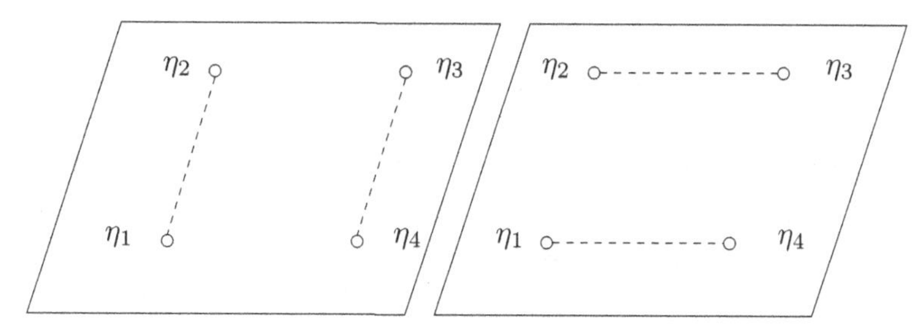

Figure 14.1 The two linearly independent wave functions for four quasiholes in the Moore–Read state have associated with them two inequivalent sets of branch cuts.

instance, an exchange of η_1 and η_3 induces in the degenerate Hilbert space the unitary transformation

$$U = \frac{1}{\sqrt{2}} \exp\left[i\pi \left(\frac{1}{8} + \frac{1}{4n} \right) \right] \begin{pmatrix} 1 & 1 \\ -1 & 1 \end{pmatrix} \tag{14.151}$$

Thus, the degenerate Hilbert space of quasiholes provides a two-dimensional representation of the braid group: the exchange (a half-braid) of two quasiholes in a four-quasihole state is represented by the 2×2 unitary matrix (of statistical angles) of Eq. (14.151). This feature of these states is called non-abelian (fractional) braid statistics.

It is important to stress that what we have done does not mean that each quasihole has an internal Hilbert space. This degeneracy has a topological origin, and it is a non-local shared property of the four quasiholes. In fact, since the dimension of this Hilbert space of $2p$ quasiholes is 2^{p-1} (instead of 2^{2p} as would be the case for an internal degree of freedom with two states), we see that the degeneracy is $\sqrt{2}$ per quasihole! In other words, these degenerate states are not localized (or localizable) on any of the quasiholes which are collectively in a state of this Hilbert space.

The same results arise in $\mathrm{SU}(2)_k$ Chern–Simons theory, a non-abelian generalization of the theory we discussed earlier in this book (see Section 10.4). Here we will follow in detail the work of Fradkin *et al.* (1998) on the connection between braiding in Chern–Simons theory and non-abelian quantum Hall states. The action of a non-abelian Chern–Simons gauge theory is

$$S_{\mathrm{CS}}[A_\mu] = \frac{k}{4\pi} \int_{\Sigma \times \mathbb{R}} d^3x \, \epsilon_{\mu\nu\lambda} \left[A_a^\mu \partial^\nu A_a^\lambda + \frac{2}{3} f_{abc} A_a^\mu A_b^\nu A_c^\lambda \right] \tag{14.152}$$

where the gauge field A_μ takes values in the algebra of $\mathrm{SU}(N)$, and the integer k is known as the level. In Chern–Simons theory (or in any field theory) the wave function $\Psi[a]$ of the ground state has the path-integral representation

$$\Psi[a] = \int_{A_\mu|_\Sigma = a_\mu} \mathcal{D}A_\mu \, \exp\left[i\frac{k}{4\pi} \int_{\Sigma\times\mathbb{R}} d^3x \, \epsilon^{\mu\nu\lambda} \left(A_\mu^a \, \partial_\nu A_\lambda^a + \frac{2}{3} f_{abc} A_\mu^a A_\nu^b A_\lambda^c \right) \right] \tag{14.153}$$

where Σ is a space-like surface (a sphere S^2, a disk D, a torus T^2, etc).

The wave functions of Chern–Simons theory on a torus T^2 correspond to the insertion of Wilson loops that carry the representations j on a non-contractible loop of the torus T^2. For $SU(2)_2$ there are only three allowed representations: (a) the identity ($j = 0$), (b) the $j = 1/2$ (doublet) representation, and (c) the $j = 1$ (triplet) representation. Thus, the $SU(2)_2$ Chern–Simons theory has three inequivalent states on a torus. The Read–Rezayi state for bosons with $k = 3$ is related to the Chern–Simons gauge theory with $SU(2)_3$ (Fradkin *et al.*, 1999) and has four ground states on the torus (since the allowed representations are now $j = 0, 1/2, 1, 3/2$). The simplest fermionic Read–Rezayi state has ten sectors.

We can picture the four quasiholes by considering their worldlines $\gamma_1, \ldots, \gamma_4$. Initially the quasiholes are located at the "punctures" with coordinates $\eta_1, \ldots, \eta_4$ on the surface Σ, see Fig. 14.2. Each quasihole carries a representation $j_1, \ldots, j_4$. In the case of $SU(2)_2$, since the level is $k = 2$, there are only three allowed representations, $j = 0, 1/2, 1$. Chern–Simons theory provides a beautiful way to understand the braiding properties of the quasiparticles, which are regarded as Wilson lines along a set of worldlines that we will collectively denote by γ.

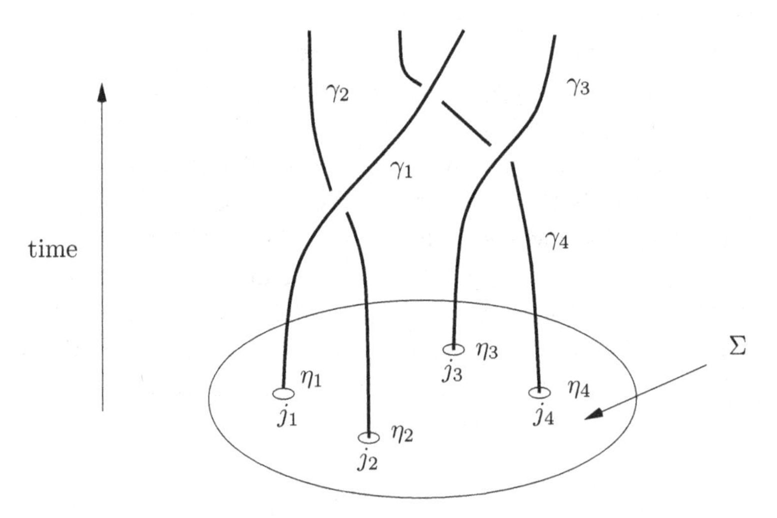

Figure 14.2 Four Wilson worldlines in Chern–Simons theory with representations $j_1, \ldots, j_4$ along the worldlines $\gamma_1, \ldots, \gamma_4$ puncturing the surface Σ at $\eta_1, \ldots, \eta_4$.

Witten showed that the expectation value of the Wilson lines carrying the fundamental, $j = 1/2$, representation is given by a topological invariant known as the Jones polynomial $V_\gamma(q)$ (with $q = \exp(i\pi/4)$) of the loops γ:

$$\int \mathcal{D}A_\mu \, \mathrm{tr}_{1/2} \left\{ P \exp\left(i \oint dx_i \, A_i^a t^a \right) \right\}$$
$$\times \exp\left[i \frac{k}{4\pi} \int d^3x \, \epsilon^{\mu\nu\lambda} \left(A_\mu^a \, \partial_\nu A_\lambda^a + \frac{2}{3} f_{abc} A_\mu^a A_\nu^b A_\lambda^c \right) \right] = V_\gamma(e^{i\pi/4}) \tag{14.154}$$

The Jones polynomial is a Laurent series in a variable q, which is a topological invariant of a knot, γ. In Chern–Simons theory the Jones polynomial defines a hierarchy of topological invariants whose first member is the linking number which, as we saw in Chapter 10, determines the phase factor for abelian fractional statistics. The Jones polynomial is defined to be $V_\gamma(q) = 1$ if γ is the unknot and by the skein relation

$$q^{-1} V_\gamma(q) - q V_{\gamma''}(q) = \left(q^{1/2} - q^{-1/2} \right) V_{\gamma'}(q) \tag{14.155}$$

where γ' and γ'' are obtained by performing successive counterclockwise half-braids of any two worldlines in γ, as shown in Fig. 14.3.

Equation (14.155) tells us how the quantum-mechanical amplitudes of the quasiholes (represented by the expectation values of the Wilson lines) are modified by braiding operations that take place during the time evolution. Let us consider a state $|\Psi\rangle$ in the two-dimensional Hilbert space of the four quasiholes, and let us denote by B the braiding operator of two quasiholes. In this context the skein relation means

$$q^{-1} |\Psi\rangle - q B^2 |\Psi\rangle = \left(q^{1/2} - q^{-1/2} \right) B |\Psi\rangle \tag{14.156}$$

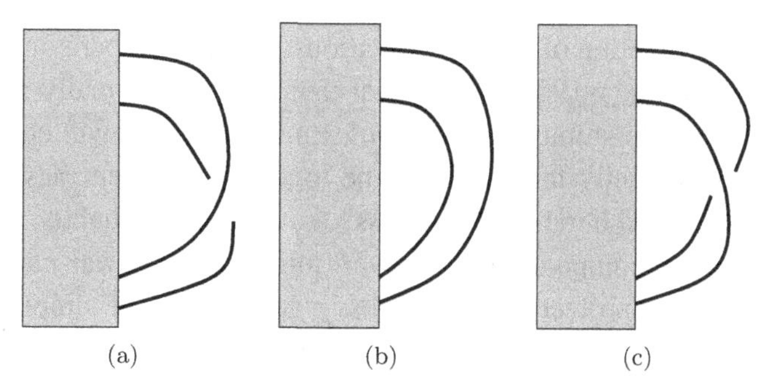

Figure 14.3 The loops (a) γ, (b) γ', and (c) γ'' that enter into the skein relation. The three loops differ only by the braiding shown here. The shaded area is arbitrary.

This equation implies a quadratic equation for the eigenvalues of the braiding operator B. Its eigenvalues are $\exp(-3\pi i/8)$ and $\exp(i\pi/8)$, which is consistent with the unitary transformation defined by Eq. (14.151).

These results imply that non-abelian quantum Hall states have excitations (in the case at hand, a σ quasihole or half-vortex) whose braiding properties depend on how many such quasiholes are present. Thus, in a state with $2p$ quasiholes, successive braiding operations involving different pairs of σ particles induce a sequence of different unitary transformations in their degenerate Hilbert space. As we saw, this Hilbert space has a topological origin. This means that local processes that would normally lead to decoherence cannot couple to these finite-dimensional Hilbert spaces of non-abelian quasiholes, which are thereby protected. Similarly, while local disorder may lead to the localization of the quasiholes, it cannot affect their topological Hilbert spaces. These observations led Kitaev to propose in 1997 (Kitaev, 2003) that the states of these topological Hilbert spaces may be used to make a *topological* quantum computer. This work was subsequently expanded by Freedman *et al.* (2002b). In this framework the unitary transformations induced by successive braiding operations can be regarded as unitary transformations in a space of *topological qubits*. In this scheme, a computation is equivalent to a sequence of braiding operations.

We will not discuss here the details of this proposal, since it is beyond the scope of this book and there are excellent reviews on this subject (e.g. Das Sarma *et al.*, 2008). Nevertheless, it is worthwhile to note that it raises several important questions, including in particular that of the conditions under which topological protection works. It should be clear from our discussion that non-abelian statistics (just as much as its abelian cousin) is a *long-distance* property of the excitations of these systems. Thus, the quasiholes must be far apart from each other (i.e. further apart than all microscopic scales such as the pairing length and the magnetic length) and move slowly enough that the processes of interest are in the adiabatic limit and do not involve the creation of further excitations, and the temperature must be low enough (lower than the quasihole gap) that there are no thermally excited quasiholes. In particular, the number of quasiholes must also be small enough that the quasiholes are sufficiently far apart that the topological degeneracy is not lifted. In addition, one should be able to "address" the topological qubits, that is to find out the result of the computation. However, this is not a trivial matter precisely due to the topological protection which these states enjoy. The simplest (but by no means trivial!) way to interact with these non-abelian excitations is through devices known as quantum interferometers (that are discussed in Chapter 15). This is a subject of intense current research both theoretically and experimentally. We will see that quasihole interferometers in non-abelian quantum Hall states measure directly the Jones polynomial of their worldlines.

14.9 Topological superconductors

As part of our discussion of non-abelian states we will now summarize the construction of a topological superconductor, in this case a $p_x + ip_y$ state, in the BCS theory of superconductivity (Schrieffer, 1964), which is a problem of wider interest. Our main interest here is to understand how non-abelian fractional statistics arises in this system. The beauty of the $p_x + ip_y$ superconductor is that, as we will see, the concept of non-abelian statistics can be presented explicitly using only standard methods of fermionic systems. The downside is that this approach is very specific to this system and cannot be generalized. In contrast, the more formal but more powerful approach of CFT provides for a more general setting.

The defining property of a superconductor (and for that matter of any superfluid) is that it is a state with a spontaneously broken symmetry, the U(1) phase symmetry associated with local gauge invariance. In the case of all superconductors, and also for fermionic superfluids such as superfluid ^{3}He and in superfluid phases of fermionic cold atoms, the spontaneously broken symmetry is characterized by the existence of an order-parameter field, the pair field (see also Section 2.5)

$$\Delta_{\sigma_1,\sigma_2}(\vec{r}_1, \vec{r}_2) = \langle c_{\sigma_1}(\vec{r}_1) c_{\sigma_2}(\vec{r}_2) \rangle \tag{14.157}$$

which acquires a non-vanishing expectation value in the superfluid (or superconducting) state. Here we have allowed for both spin-singlet and spin-triplet superfluid states. Alternatively, the superconducting state is defined by the existence of the limit (known as off-diagonal long-range order)

$$\lim_{\vec{R}\to\infty} \langle c_{\sigma_1}(\vec{r}_1) c_{\sigma_2}(\vec{r}_2) c^\dagger_{\sigma_1}(\vec{r}_1 + \vec{R}) c^\dagger_{\sigma_2}(\vec{r}_2 + \vec{R}) \rangle = |\Delta_{\sigma_1,\sigma_2}(\vec{r}_1, \vec{r}_2)|^2 \neq 0 \tag{14.158}$$

This definition is independent of the choice of basis or representation, and holds equally in a system with a fixed number of fermions (the canonical ensemble) and in a system in which the number of particles is not fixed (the grand canonical ensemble). The direct consequence of a ground state with a spontaneously broken global symmetry is the existence of a finite stiffness for the phase field in the superfluid state, i.e. the Nambu–Goldstone boson of the spontaneously broken symmetry. The quantum Hamiltonian (and action) of a physical system is invariant under a global phase transformation of the Fermi fields, $c_\sigma(\vec{r}) \to e^{i\phi} c_\sigma(\vec{r})$ (where ϕ is constant and spin-independent). However, the pair-field operator is not invariant under a global phase transformation and changes as $\Delta_{\sigma_1,\sigma_2}(\vec{r}_1, \vec{r}_2) \to e^{i2\phi} \Delta_{\sigma_1,\sigma_2}(\vec{r}_1, \vec{r}_2)$. This transformation law implies that the pair field couples to an electromagnetic gauge field as a charge-$2e$ scalar field.

We should keep in mind, as we discussed extensively in Section 13.2, that a superconducting state (topological or not) is not and cannot be equivalent to any quantum Hall state, abelian or non-abelian. In the case of the fermionic

Moore–Read quantum Hall states, the pairing instability involves the composite fermions, which, as we have seen, are strongly coupled to a dynamically generated Chern–Simons gauge field. Thus, the would-be Goldstone boson of the putative superconducting state is already eaten by the Chern–Simons gauge field. This means that, although the Moore–Read states have pairing correlations, it is not a superfluid state.

The construction we are following here relies crucially on the concept of pairing and on regarding the non-abelian state as a paired state. However, it is not obvious that this is a necessary mechanism. In fact, other non-abelian states that are known, such as the Read–Rezayi state, use instead a clustering property that generalizes the concept of pairing.

14.9.1 BCS mean-field theory of the $p_x + ip_y$ state

In the $p_x + ip_y$ state the fermions are in a spin triplet fully spin-polarized state. It is a 2D analog of the A phase of superfluid ^{3}He (see e.g. Vollhardt and Wölfle (1990)). This will allow us to ignore the spin degree of freedom. Let $c_{\vec{k}}^\dagger$ be the fermion operator that creates a fermion with momentum $\vec{k}$ and let $c_{\vec{k}}$ be the adjoint operator (which in the quantum Hall context are interpreted as composite-fermion operators). These operators obey the standard fermionic algebra,

$$\{c_{\vec{k}}^\dagger, c_{\vec{q}}\} = \delta_{\vec{k},\vec{q}}, \qquad \{c_{\vec{k}}^\dagger, c_{\vec{q}}^\dagger\} = 0 \tag{14.159}$$

The kinetic energy of the effective fermionic (mean-field) BCS Hamiltonian for the quasiparticles of the superconducting state is

$$\begin{aligned} H_{\mathrm{F}} = &\int d^2x\, c^\dagger(\vec{x})\hat{h}_0\, c(\vec{x}) \\ &+ \int d^2x\, d^2x' \left[\Delta^*(\vec{x},\vec{x}\,')c(\vec{x})c(\vec{x}\,') + \Delta(\vec{x},\vec{x}\,')c^\dagger(\vec{x})c^\dagger(\vec{x}\,')\right] \end{aligned} \tag{14.160}$$

where $\hat{h}_0$ is the one-particle kinetic-energy operator. For a complex p-wave condensate the complex pair field $\Delta(\vec{x},\vec{x}\,')$ is given by

$$\Delta(\vec{x},\vec{x}\,') \equiv \Delta\left(\frac{\vec{x}+\vec{x}\,'}{2}\right)\left(i\,\partial_{x'} - \partial_{y'}\right)\delta(\vec{x}-\vec{x}\,') \tag{14.161}$$

In momentum space the fermionic Hamiltonian becomes

$$H_{\mathrm{F}} = \int d^2k \left[(\varepsilon_{\vec{k}} - \mu)c_{\vec{k}}^\dagger c_{\vec{k}} + \frac{1}{2}\left(\Delta_{\vec{k}}^* c_{-\vec{k}} c_{\vec{k}} + \Delta_{\vec{k}} c_{\vec{k}}^\dagger c_{-\vec{k}}^\dagger\right)\right] \tag{14.162}$$

where $\varepsilon_{\vec{k}}$ is the quasiparticle kinetic energy and μ is an effective chemical potential. At small momenta $\varepsilon_{\vec{k}} \simeq \vec{k}^2/(2m^*)$, where m^* is an effective mass. $\Delta_{\vec{k}}$ is the gap function, which is proportional to the order parameter of the superconducting state. Since we will not discuss the superconducting instability itself, we will focus only on the kinetic-energy part of the Hamiltonian, Eq. (14.162), and we will not include the terms that involve the fluctuations of the superconducting state (see e.g. Schrieffer (1964)).

For a spatially isotropic system the Fourier transform of the pair field $\Delta_{\vec{k}}$ (the "pair wave function") is an eigenstate of angular momentum. For a $p_x + ip_y$ state the complex gap function $\Delta_{\vec{k}}$ transforms as the $l = -1$ representation of the group of rotations in two dimensions. In the limit of small momenta, $\vec{k} \to 0$, $\Delta_{\vec{k}}$ has the asymptotic behavior

$$\Delta_{\vec{k}} = (k_x - ik_y)\Delta \qquad \text{as } \vec{k} \to 0 \tag{14.163}$$

(where Δ is a constant pairing amplitude) and vanishes at large momenta.

The mean-field ground state $|G\rangle$ has the standard BCS form

$$|G\rangle = \prod_{\vec{k}}{}' \left(u_{\vec{k}} + v_{\vec{k}} c^\dagger_{\vec{k}} c^\dagger_{-\vec{k}} \right) |0\rangle \tag{14.164}$$

where $|0\rangle$ is the state without fermions. The prime on the product symbol in Eq. (14.164) indicates that each pair $(\vec{k}, -\vec{k})$ enters only once. The complex amplitudes $u_{\vec{k}}$ and $v_{\vec{k}}$ are determined by a self-consistency condition, which is equivalent to a variational argument, and obey the condition

$$|u_{\vec{k}}|^2 + |v_{\vec{k}}|^2 = 1 \tag{14.165}$$

which follows from the condition that the ground state $|G\rangle$ is normalized to unity.

In order to diagonalize the kinetic energy of the mean-field Hamiltonian of Eq. (14.162), we will proceed in the same way as we did in the case of the quantum Ising chain in Chapter 5. Let $\eta_{\vec{k}}$ and $\eta^\dagger_{\vec{k}}$ be a set of new fermion operators that obey standard fermionic anticommutation relations, and are related to $c_{\vec{k}}$ and $c^\dagger_{\vec{k}}$ by the Bogoliubov transformation

$$\eta_{\vec{k}} = u_{\vec{k}} c_{\vec{k}} - v_{\vec{k}} c^\dagger_{-\vec{k}}, \qquad \eta^\dagger_{\vec{k}} = u^*_{\vec{k}} c^\dagger_{\vec{k}} - v^*_{\vec{k}} c_{-\vec{k}} \tag{14.166}$$

and annihilate the BCS ground state, $\eta_{\vec{k}}|G\rangle = 0$. The requirement that the new fermions create the actual eigenstates of the (full) mean-field Hamiltonian,

$$H = \sum_{\vec{k}} E_{\vec{k}} \eta^\dagger_{\vec{k}} \eta_{\vec{k}} + E_{\text{gnd}} \tag{14.167}$$

(where $E_{\vec{k}} \geq 0$ are the quasiparticle excitation energies and E_{gnd} is the ground-state energy) is met by the condition that the new fermions be eigenoperators of the mean-field Hamiltonian,

$$\left[\eta_{\vec{k}}, H\right] = E_{\vec{k}}\, \eta_{\vec{k}} \tag{14.168}$$

As a result the amplitudes $(u_{\vec{k}}, v_{\vec{k}})$ (written in a spinor form) obey the Bogoliubov–de Gennes (BdG) equation

$$\begin{pmatrix} \xi_{\vec{k}} & -\Delta^*_{\vec{k}} \\ -\Delta_{\vec{k}} & -\xi_{\vec{k}} \end{pmatrix} \begin{pmatrix} u_{\vec{k}} \\ v_{\vec{k}} \end{pmatrix} \equiv E_{\vec{k}}\, \vec{n}_{\vec{k}} \cdot \vec{\sigma} \begin{pmatrix} u_{\vec{k}} \\ v_{\vec{k}} \end{pmatrix} = E_{\vec{k}} \begin{pmatrix} u_{\vec{k}} \\ v_{\vec{k}} \end{pmatrix} \tag{14.169}$$

where $\xi_{\vec{k}} = \varepsilon_{\vec{k}} - \mu$, $\vec{\sigma} = (\sigma_x, \sigma_y, \sigma_z)$ is a three-component vector made of the three Pauli matrices, and $\vec{n}_k$ is the unit vector

$$\vec{n}_k = \begin{pmatrix} u^*_{\vec{k}}, & v^*_{\vec{k}} \end{pmatrix} \vec{\sigma} \begin{pmatrix} u_{\vec{k}} \\ v_{\vec{k}} \end{pmatrix} = \frac{1}{E_{\vec{k}}} (-\text{Re}\, \Delta_{\vec{k}}, \quad \text{Im}\, \Delta_{\vec{k}}, \quad \xi_{\vec{k}}) \tag{14.170}$$

The eigenvalues $E_{\vec{k}}$ and eigenvectors $(u_{\vec{k}}, \ v_{\vec{k}})$ are

$$E_{\vec{k}} = \sqrt{\xi^2_{\vec{k}} + |\Delta_{\vec{k}}|^2} \tag{14.171}$$

$$\frac{v_{\vec{k}}}{u_{\vec{k}}} = -\frac{E_{\vec{k}} - \xi_{\vec{k}}}{\Delta^*_{\vec{k}}} \tag{14.172}$$

Up to a momentum-dependent phase, the spinor amplitudes are given by

$$|u_{\vec{k}}|^2 = \frac{1}{2}\left(1 + \frac{\xi_{\vec{k}}}{E_{\vec{k}}}\right), \qquad |v_{\vec{k}}|^2 = \frac{1}{2}\left(1 - \frac{\xi_{\vec{k}}}{E_{\vec{k}}}\right) \tag{14.173}$$

In the low-momentum regime, $\vec{k} \to 0$, $\xi_{\vec{k}} \to -\mu$ and $\Delta_{\vec{k}}$ has the form of Eq. (14.163). Thus at low momenta the BdG equation takes the form

$$\begin{pmatrix} -\mu & -(k_x + ik_y)\Delta^* \\ -(k_x - ik_y)\Delta & \mu \end{pmatrix} \begin{pmatrix} u_{\vec{k}} \\ v_{\vec{k}} \end{pmatrix} = E_{\vec{k}} \begin{pmatrix} u_{\vec{k}} \\ v_{\vec{k}} \end{pmatrix} \tag{14.174}$$

which in real space becomes

$$\begin{aligned} i\,\partial_t u &= -\mu u + \Delta^* i(\partial_x + i\,\partial_y) v \\ i\,\partial_t v &= \mu v + \Delta i(\partial_x - i\,\partial_y) u \end{aligned} \tag{14.175}$$

We recognize this result as the Dirac equation in $(2+1)$ dimensions, with the constraint that the spinor (u, v) obeys the Majorana condition

$$(u, v) \begin{pmatrix} 0 & 1 \\ 1 & 0 \end{pmatrix} = \begin{pmatrix} u \\ v \end{pmatrix}^\dagger \tag{14.176}$$

In other terms, the quasiparticles of the superconductor are Majorana fermions, a result that we also encountered in the solution of the quantum Ising chain in Chapter 5. Notice that in this language the chemical potential μ became the mass of the Majorana fermion. Furthermore, the BdG equation, Eq. (14.169) (and hence also the Dirac approximation) has the symmetry (with σ_1 being the Pauli matrix)

$$\sigma_1 \begin{pmatrix} \xi_{\vec{k}} & -\Delta^*_{\vec{k}} \\ -\Delta_{\vec{k}} & -\xi_{\vec{k}} \end{pmatrix} \sigma_1 = -\begin{pmatrix} \xi_{\vec{k}} & -\Delta_{\vec{k}} \\ -\Delta^*_{\vec{k}} & -\xi_{\vec{k}} \end{pmatrix} \tag{14.177}$$

This implies that, if the spinor $(u_{\vec{k}}, v_{\vec{k}})$ is a solution with energy $E_{\vec{k}}$, then the spinor $(u^*_{\vec{k}}, v^*_{\vec{k}})\sigma_1$ has energy $-E_{\vec{k}}$ and the spectrum is symmetric. However, the Majorana condition tells us that these two states are the same state. Hence the quasiparticle is its own anti-particle.

Following this line of argument, Read and Green showed that the BCS wave function $|G\rangle$ can be written in the suggestive form of a coherent state of Cooper pairs:

$$|G\rangle = \left(\prod_{\vec{k}} |u_{\vec{k}}|^{1/2}\right) \exp\left(\frac{1}{2} \sum_{\vec{k}} g(\vec{k}) c^\dagger_{\vec{k}} c^\dagger_{-\vec{k}}\right) |0\rangle \tag{14.178}$$

where

$$g(\vec{k}) = \frac{v_{\vec{k}}}{u_{\vec{k}}} \tag{14.179}$$

Furthermore, when projected onto a state with N fermions with real-space coordinates $\vec{x}_i$ ($i = 1, \ldots, N$, with N even) this state has the form of a Pfaffian wave function (Read and Green, 2000):

$$\Psi(\vec{x}_1, \ldots, \vec{x}_N) = \langle \vec{x}_1, \ldots, \vec{x}_N | G\rangle = \text{Pf}\left(g(\vec{x}_i - \vec{x}_j)\right) \tag{14.180}$$

where $g(\vec{x})$ is the Fourier transform of the function $g(\vec{k})$ defined by Eq. (14.179).

The function $g(\vec{k})$ has different possible behaviors at small momenta. The behavior of the amplitudes $u_{\vec{k}}$ and $v_{\vec{k}}$ in this regime depends on how $E_{\vec{k}} \pm \xi_{\vec{k}}$ behaves in this regime. Since, as $\vec{k} \to 0$, $\xi_{\vec{k}} \to -\mu$ and $\Delta_{\vec{k}} \simeq (k_x - ik_y)\Delta$, we find

$$\lim_{\vec{k}\to 0} \left(E_{\vec{k}} \pm \xi_{\vec{k}}\right) = \lim_{\vec{k}\to 0} \left(E_{\vec{k}} \pm \text{sgn}(\xi_{\vec{k}})\right) = |\mu|(1 \mp \text{sgn}(\mu)) \tag{14.181}$$

Hence, we have three different behaviors depending on whether $\mu > 0$ (the *weak-pairing* regime), $\mu < 0$ (the *strong-pairing* regime), or $\mu = 0$.

For $\mu > 0$, the small-momentum behavior of $g(\vec{k})$ is

$$g(k) \simeq -\frac{2|\mu|}{(k_x + ik_y)\Delta^*}, \qquad \text{as } \vec{k} \to 0 \tag{14.182}$$

The Fourier transform to real space of $g(\vec{k})$ has the long-distance behavior

$$g(z) = \left(\frac{i\mu}{\pi\Delta^*}\right) \frac{1}{z} \tag{14.183}$$

where $z = x + iy$. Thus, in the weak-pairing regime, in which $\mu > 0$, the function $g(\vec{x})$ has a power-law behavior and is an analytic function of the complex coordinates. In particular, in this regime the wave function of this superconducting state is a Pfaffian with the same analytic dependence in the fermion coordinates as in the Moore–Read fractional quantum Hall state, Eq. (14.127).

On the other hand, for $\mu < 0$, the small-momentum behavior of $g(\vec{k})$ is instead

$$g(k) \simeq -\frac{(k_x - ik_y)A}{a_0^{-2} + \vec{k}^{\,2}} \tag{14.184}$$

where

$$A = \frac{2|\mu| m^* \Delta}{2|\mu| + m^*|\Delta|^2}, \qquad a_0 = \frac{1}{2|\mu|}\sqrt{\frac{2|\mu|}{m^*} + |\Delta|^2} \tag{14.185}$$

The form of $g(\vec{k})$ in this regime tells us that its Fourier transform in real space exhibits an exponential decay at separations long compared with the length scale a_0. In the BCS theory of a superconductor the regime with $\mu < 0$ means that the chemical potential lies below the band of single-particle states. In this regime the Cooper pairs behave effectively as bosonic "molecules," and the superconducting state can be legitimately regarded as a Bose–Einstein condensate of these bosons.

14.9.2 Topology and the superconducting state

Finally, if $\mu = 0$, the amplitudes reach constant values at small momenta, $|u_{\vec{k}}| \to 1/2$ and $|v_{\vec{k}}| \to 1/2$. In this regime the quasiparticle excitation energy vanishes linearly with the momentum, $E_{\vec{k}} \simeq |\vec{k}||\Delta|$, and hence there is no energy gap! Furthermore, it is easy to see that at $\mu = 0$ the function $g(\vec{k})$ becomes

$$g(\vec{k}) \simeq -\frac{|\vec{k}|}{k_x + ik_y}\frac{|\Delta|}{\Delta^*}, \qquad \mu = 0 \tag{14.186}$$

which, in real space, has the non-analytic behavior

$$g(z) \simeq \left(\frac{i|\Delta|}{2\pi\Delta^*}\right)\frac{1}{z|z|} \tag{14.187}$$

The scenario we have presented does not occur for a weak-coupling system and, indeed, a large attractive interaction is needed in order for this phase transition to be reached, which is clearly outside the regime of validity of the BCS theory of superconductivity. The BCS theory describes the superconducting state as an instability of the Fermi surface of the quasiparticles of a Fermi liquid that happens for arbitrarily weak attractive interactions in the Fermi sea (Schrieffer, 1964), and it is accurate only in this weak-coupling regime. In this regime the minimum of the quasiparticle excitation energy occurs at the Fermi surface whose wave vector is $|\vec{k}| = k_{\rm F}$ (i.e. at the chemical potential or Fermi energy), where the $p_x + ip_y$ state has a full and isotropic energy gap $\sim|\Delta|$. On the other hand, as the interactions grow stronger the value of the chemical potential μ (the Fermi energy) begins to decrease (from positive values) and the minimum excitation energy progressively moves to $\vec{k} = 0$. Since the spectrum is gapped both for $\mu > 0$ and for $\mu < 0$, we must conclude that $\mu = 0$ represents a (quantum) phase transition *inside* the superconducting state.

The weak- and strong-pairing phases, with $\mu > 0$ and $\mu < 0$, respectively, have different topological properties (Volovik, 1988). The solutions of the BdG equation, Eq. (14.169), are the complex spinors $(u_{\vec{k}}, v_{\vec{k}})$ and obey the normalization $|u_{\vec{k}}|^2 + |v_{\vec{k}}|^2 = 1$. In addition, a smooth change in the phase of both components of the spinor does not change the state. Thus, the solutions to the BdG equation are effectively labeled by two real parameters and can thus be regarded as points on a 2-sphere S_2. Therefore, the solutions to the BdG equation are mappings of the momentum space (labeled by $\vec{k}$) to the unit sphere S_2. The three-component real unit vector $\vec{n}_{\vec{k}}$ defined in Eq. (14.170) can be used to parametrize the 2-sphere S_2. Since $v_{\vec{k}} \to 0$ for $|\vec{k}| \to \infty$ (and, hence, $u_{\vec{k}} \to 1$ as $|\vec{k}| \to \infty$), we can add the point of infinity to the momentum space, by which means it also becomes topologically equivalent to a 2-sphere, S_2.

Therefore, we find that the solutions of the BdG equation are smooth mappings of $S_2 \mapsto S_2$. In Section 7.9 we showed that the mappings of the S_2 base space (momentum space) to the S_2 target space are classified by homotopy classes associated with the homotopy group $\pi_2(S_2) \cong \mathbb{Z}$. Each topological (homotopy) class is labeled by a topological invariant, the Chern number $\mathcal{Q} \in \mathbb{Z}$:

$$\mathcal{Q} = \frac{1}{8\pi} \int d^2k \; \epsilon_{ij} \epsilon_{abc} n^a \, \partial^i n_b \, \partial^j n_c = \frac{1}{4\pi} \int d^2k \; \vec{n} \cdot (\partial_{k_x} \vec{n} \times \partial_{k_y} \vec{n}) \qquad (14.188)$$

Therefore, the solutions to the BdG equation admit the same classification. For this connection with the Chern number, the $p_x + ip_y$ superconductor is (at the level of the BdG equation) closely related to the theory of *topological insulators* (even though it is a superconductor!) discussed in Chapter 16.

In the strong-pairing phase, $\mu < 0$ and $\xi_{\vec{k}} = \vec{k}^2/(2m^*) - \mu > 0$ for all $\vec{k}$. Hence, in this phase the vector $\vec{n}_{\vec{k}}$ takes values only on the northern hemisphere of the sphere S_2. Therefore, the solutions of the BdG equation with $\mu < 0$ can be smoothly deformed to their value at the North Pole, $\vec{n}_{\vec{k}} = (0, 0, 1)$ (corresponding to the spinor $(u_{\vec{k}}, v_{\vec{k}}) = (1, 0)$), and belong to the topologically trivial homotopy class (with $\mathcal{Q} = 0$).

On the other hand, in the weak-pairing phase $\mu > 0$ and $\xi_{\vec{k}}$ can take all possible real values, both positive and negative. Hence, for $\mu > 0$ the solutions of the BdG equation are non-trivial maps of $S_2 \mapsto S_2$ and have a non-vanishing winding number $\mathcal{Q} = 1$ (or -1). Since the integer-valued topological invariant $\mathcal{Q}$ cannot be smoothly deformed from 0 to ± 1, the strong- and weak-pairing regimes must correspond to separate phases. For this reason, the weak-pairing phase of the $p_x + ip_y$ state is identified as a topological superconductor.

14.9.3 The half-vortex

The vortex excitations of the $p_x + ip_y$ state have very interesting properties. The vortices of the strong-pairing phase have similar properties to those of conventional superconductors and will not be discussed here. We will focus instead on the vortices of the weak-pairing phase, which, as we will see, have non-abelian statistics. A half-vortex has been observed experimentally in the superconducting phase of Sr_2RuO_4, a quasi-2D system (Jang *et al.*, 2011). For an extensive review of the superconducting properties of Sr_2RuO_4 see the work of Mackenzie and Maeno (2003).

In the condensed state the amplitude of the superconducting order parameter Δ is essentially constant in space. However, the pair field couples to a gauge field as a charge-$2e$ scalar field. For fields stronger than a critical field, usually called H_{c1}, the uniform Meissner state is destroyed and the system enters the vortex (or mixed) phase, see e.g. de Gennes (1966). Here we will be interested in the long-distance properties of a superconducting vortex. The p-wave superconducting state has to be in a spin-triplet state. In addition to the orbital dependence of Eq. (14.163), the pair field has also a spin dependence. For a p-wave state the spin state must be symmetric (and hence a triplet). If we retain both the orbital components and the spin components, the pair field has the form

$$\begin{aligned}\Delta = e^{i\varphi}\big[&d_x(|\uparrow\uparrow\rangle + |\downarrow\downarrow\rangle) - id_y(|\uparrow\uparrow\rangle - |\downarrow\downarrow\rangle)\\ &+ d_z(|\uparrow\downarrow\rangle + |\downarrow\uparrow\rangle)\big](k_x - ik_y)\end{aligned} \tag{14.189}$$

where we introduced the three-component vector $\vec{d} = (d_x, d_y, d_z)$ (as in the A phase of superfluid ^{3}He, see Leggett (1975)). This order parameter is invariant under a shift of the phase by π, $\varphi \mapsto \varphi + \pi$, and a simultaneous inversion of the $\vec{d}$ vector, $\vec{d} \mapsto -\vec{d}$. In other terms, the order parameter involves not a vector but a *director* (as in the description of nematic liquid crystals (Chaikin and Lubensky, 1995)). Thus, the half-vortex is a topological soliton of this condensed state.

We are interested in a 2D system and we will assume a superconducting state in which the $\vec{d}$ vector lies on the plane and is hence normal to the angular momentum of the pair (this is the A phase). Provided that the $\vec{d}$ vector can rotate in the plane, this symmetry allows this superconducting state to support *half-vortex* excitations. In a half-vortex, depicted in Fig. 14.4, the superconducting phase $\varphi(\vec{x})$ varies slowly on large circle by π, provided that the $\vec{d}$ vector also rotates by 180°. This state is equivalent to a fully polarized state in which only the phase of the up component of the spin winds by 2π. Hence, the half-vortex of this state is equivalent to a full (2π) vortex of a $p_x + ip_y$ condensate of spinless fermions (Ivanov, 2001; Chung *et al.*, 2007; Vakaryuk and Leggett, 2009).

Let us now discuss the spectrum of quasiparticles in the half-vortex state. Here we will use the equivalent description in terms of spinless fermions in the $p_x + ip_y$

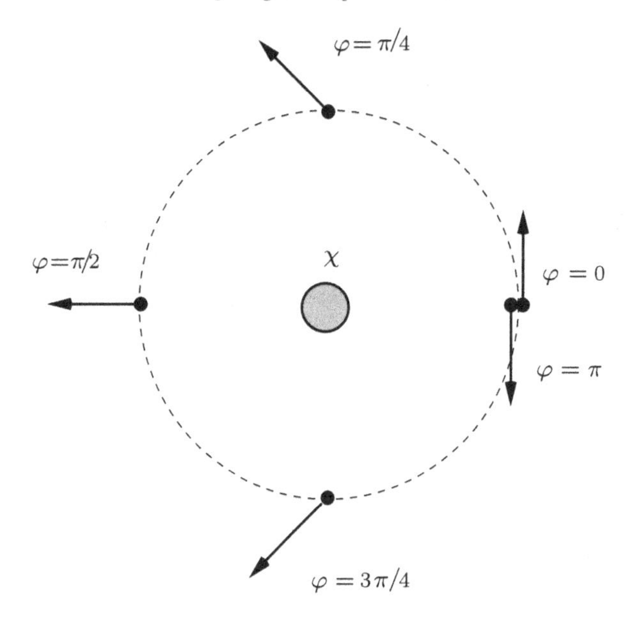

Figure 14.4 A half-vortex in a $p_x + ip_y$ superconductor. The arrows represent the local direction of the $\vec{d}$ vector. The shaded area represents the core of the half-vortex with a Majorana zero mode χ.

state with a full 2π vortex. Since we are interested only in the properties far from the core of the vortex, where the order parameter must vanish, we will assume that the absolute value of the pair field is essentially constant in space. The core of the vortex will be represented by a small region in which there are no quasiparticles, and hence we take the chemical potential μ to be large and negative in this region, while everywhere else μ is positive (since we are in the weak-pairing phase). In the presence of a vortex, at long distances the phase of the order parameter winds by 2π. In addition to propagating quasiparticle states, whose spectrum obeys the symmetry of Eq. (14.177), in the background of a vortex the BdG equation now has a state with exactly zero energy (a "zero mode"), which in polar coordinates (r, ϕ) is the solution of (Read and Green, 2000)

$$\begin{aligned} \Delta i e^{i\phi}\left(\partial_r + \frac{i}{r}\,\partial_\phi\right) v &= \mu u \\ \Delta i e^{-i\phi}\left(\partial_r - \frac{i}{r}\,\partial_\phi\right) u &= -\mu v \end{aligned} \tag{14.190}$$

The explicit form of the (normalizable) zero-mode spinor is

$$\begin{pmatrix} u(r,\phi) \\ v(r,\phi) \end{pmatrix} = \begin{pmatrix} \dfrac{1}{\sqrt{i}} e^{i\phi/2} \\ \dfrac{1}{\sqrt{-i}} e^{-i\phi/2} \end{pmatrix} \frac{f(r)}{\sqrt{r}} \tag{14.191}$$

where $f(r)$ is given by

$$f(r) \propto \exp\left(-\int^r dr' \frac{\mu(r')}{|\Delta|} \right) \sim \exp\left(-\frac{\mu}{|\Delta|} r \right) \tag{14.192}$$

In particular, we see that the spinor solution is double-valued, $(u(r, \phi + 2\pi), v(r, \phi + 2\pi)) = -(u(r, \phi), v(r, \phi))$. This property follows from global phase invariance. Indeed, under a global transformation of the pair field by a uniform phase ϕ, the fermions must transform with half of that phase,

$$\Delta(\vec{r}) \to e^{i\phi} \Delta(\vec{r}), \qquad c(x) \to e^{i\phi/2} c(\vec{x}), \qquad c^\dagger(x) \to e^{-i\phi/2} c^\dagger(\vec{x}) \tag{14.193}$$

Hence, under a change by $\phi = 2\pi$, the fermions must change sign. Consequently, in a 2π vortex the spinor solution must be double-valued.

Thus, in the background of a vortex the quasiparticle has a bound state with exactly zero energy that decays exponentially fast away from the vortex location. While states of fermionic quasiparticles bound to vortices are a common occurrence, such states typically have finite energy. What is special about the topological superconductors is that these fermionic bound states have exactly zero energy and that this is a robust topologically protected feature of this superconducting state (Roy, 2010).

In conclusion, the half-vortex is a topological soliton, a non-trivial collective excitation of the $p_x + ip_y$ condensate that, as we saw, has a zero-energy fermionic bound state. Topological solitons with fermionic zero modes have been discussed in high-energy physics (Jackiw and Rebbi, 1976; Jackiw and Rossi, 1981; Rajaraman, 1985) and in condensed matter physics in the context of 1D conductors (Heeger *et al.*, 1988). In both cases the topological solitons acquire fractional quantum numbers through the occurrence of fermionic zero modes. We will discuss this problem in Chapter 16.

However, in the case of topological superconductors, these topological solitons, the half-vortices, exhibit more unusual properties, one of them being non-abelian statistics, which we will discuss now. At the root of these differences is the nature of the broken-symmetry state. For example, in the case of the 1D conductors such as polyacetylene (Su and Schrieffer, 1981), the excitation that carries a zero-energy mode is a soliton of the spontaneously broken $\mathbb{Z}_2$ symmetry. The existence of a zero mode then implies that the soliton carries fractional charge $\pm e/2$. In polyacetylene, a system in which charge is locally and globally conserved, the zero mode is associated with a charged fermion state, which can be either unoccupied (corresponding to a positively charged soliton of charge $+e/2$) or occupied (corresponding to a negatively charged soliton of charge $-e/2$) (Heeger *et al.*, 1988).

In contrast, in the case of superconductors, the quasiparticles are charge-neutral (Majorana) fermions (Kivelson and Rokhsar, 1990). Thus, in a system such as a

$p_x + ip_y$ superconductor, the zero mode of a vortex cannot be occupied or empty since the excitations have no charge. To see how this works, we will follow the construction of Ivanov (2001). We first observe that, since the quasiparticles of a superconductor are Majorana fermions, the vortex with its fermion bound state must also have a Majorana fermion character. In the low-energy limit, the operator that creates a Majorana quasiparticle in the vortex background is reduced to a self-adjoint (Majorana) fermion operator $\gamma_i = \gamma_i^\dagger$, which is the zero mode of each vortex of a $p_x + ip_y$ superconductor. Let us consider a more general case in which we have $2n$ vortices with fixed coordinates, sufficiently far apart from each other, compared with the zero-temperature coherence length $\xi_0 \sim v_F/|\Delta|$ of the superconducting state, that their fermionic zero modes do not mix with each other and hence with exactly degenerate states. In this limit we have a set of $2n$ Majorana fermions, γ_i (with $i = 1, \dots, 2n$), with $\{\gamma_i, \gamma_j\} = 2\delta_{ij}$ and zero energy.

We can group the $2n$ Majorana fermions into n pairs. For each pair of Majorana operators we can define a complex (Dirac) fermion operator satisfying the standard anti-commuting algebra,

$$\psi_j = \frac{1}{2}(\gamma_{2j} + i\gamma_{2j+1}), \qquad \psi_j^\dagger = \frac{1}{2}(\gamma_{2j} - i\gamma_{2j+1}), \qquad \{\psi_j, \psi_k^\dagger\} = \delta_{jk} \quad (14.194)$$

Each complex fermion has two states, $|0_i\rangle$ (empty) and $|1_i\rangle$ (occupied), which span a two-dimensional Hilbert space for each pair of vortices. Notice that the state that is either empty or occupied is shared by two vortices, which can be very far apart from each other; it is not associated with each independent vortex, as is the case in the soliton in polyacetylene. This also means that, associated with each configuration of $2n$ vortex coordinates, there is a Hilbert space of states of dimension 2^{n-1} (the degeneracy) associated with the Majorana fermions. As we saw, the existence of this "topological" Hilbert space is the key ingredient of non-abelian statistics. These are the reasons why topological superconductors as well as non-abelian quantum Hall states are primary candidates to realize schemes of topological quantum computing. In this context, the states $|0\rangle_i$ and $|1\rangle_i$ can be regarded as topological quantum qubits that are immune to the effects of local perturbations such as disorder, phonons, etc.

14.10 Braiding and fusion

We will now discuss the braiding properties of non-abelian quasiparticles. For concreteness we will focus on the half-vortices of the $p_x + ip_y$ 2D superconductor and of the Moore–Read state. These ideas can be extended to the more general cases. We will see that the braiding properties of the non-abelian quasiparticles are intimately related to the fusion rules they obey.

14.10.1 Braiding of half-vortices

The way in which we grouped the $2n$ Majorana fermions into pairs is clearly arbitrary. Different groupings must correspond to physically identical states. To rearrange one grouping into another grouping of pairs, we must swap vortices around slowly enough that no additional quasiparticle states are created. This process amounts to an adiabatic *braiding* of the vortex worldlines. Since all groupings must describe the same Hilbert space, the process of swapping Majorana fermions (braids) must be equivalent to a set of unitary operators acting on the 2^n-dimensional Hilbert space. In other words, we can construct a representation of the braid group B_{2n} (of $2n$ "particles") in this space of states.

To this end, let $i = 1, \ldots, 2n$ label the set of vortices (and hence of Majorana fermions) on various locations on the plane. For $2n$ particles we can define a set of $2n - 1$ elementary particle exchanges σ_i ($i = 1, \ldots, 2n - 1$). Each operation represents the braiding of a pair of particles. Successive braidings define a natural product of these operations. The braid group is generated by the elementary exchange operators σ_i (not to be confused with a Pauli matrix!) that satisfy the following algebra (shown in Fig. 14.5):

$$\sigma_i \sigma_j = \sigma_j \sigma_i, \qquad |i - j| > 1 \tag{14.195}$$

$$\sigma_i \sigma_j \sigma_i = \sigma_j \sigma_i \sigma_j, \qquad |i - j| = 1 \tag{14.196}$$

Since the Majorana fermions anti-commute with each other, we will attach a branch cut to each vortex that will indicate how the different vortices are ordered on the plane (see Fig. 14.6). The phase of the superconducting order parameter is single-valued and jumps by 2π as the cut is crossed. This construction is very similar to the Jordan–Wigner transformation discussed in Chapter 8. In this picture, an elementary braiding operation amounts to a rearrangement of the branch cuts,

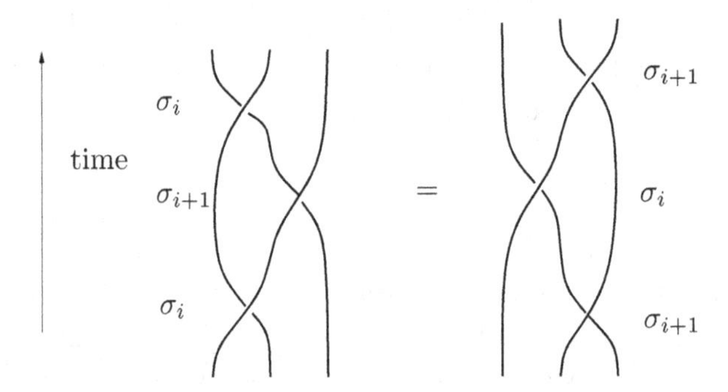

Figure 14.5 A schematic depiction of the braid-group relation Eq. (14.196) on the worldlines of three vortices. See the text.

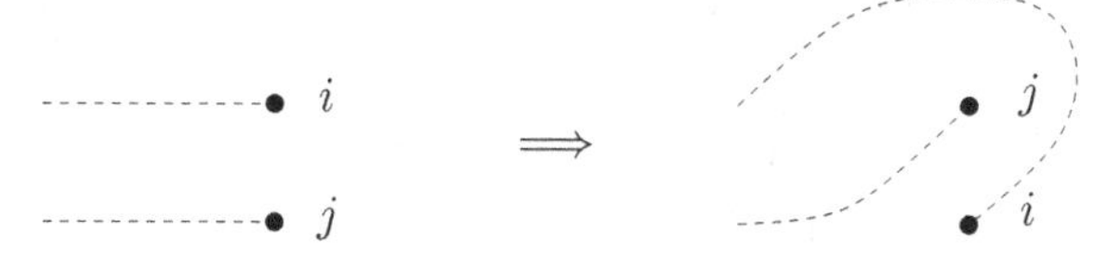

Figure 14.6 An elementary braid exchange of two vortices. See the text.

as shown in Fig. 14.6, which leads to a change in the phase of the order parameter of one vortex by 2π and of the fermion zero mode by π. This property tells us that the braid operators σ_i act on the Majorana zero modes as follows:

$$\begin{aligned} \sigma_i &: \gamma_i \mapsto \gamma_{i+1} \\ \sigma_i &: \gamma_{i+1} \mapsto -\gamma_i \\ \sigma_i &: \gamma_j \mapsto \gamma_j, \qquad j \neq i, i+1 \end{aligned} \tag{14.197}$$

We will now find a representation of the braid operations, σ_i, in the 2^n-dimensional Hilbert space $\mathcal{M}_{2n}$ of the $2n$ Majorana zero modes. Using an obvious notation, we will denote by $\tau(\sigma_i) \equiv \tau_i$ the operators that act on the Hilbert space $\mathcal{M}_{2n}$. The operators τ_i will be constructed as a unitary representation of the braid group. Hence the braid operator τ is required to obey

$$\tau(\sigma_i)\gamma_j\tau(\sigma_i)^{-1} = \tau(\sigma_j) \tag{14.198}$$

for $\tau(\gamma)$ defined by Eq. (14.197). It is easy to check that the following operators satisfy the requirements of the Braid group, Eq. (14.196) (Ivanov, 2001):

$$\tau(\sigma_i) = e^{\frac{\pi}{4}\gamma_{i+1}\gamma_i} = \frac{1}{\sqrt{2}}(1 + \gamma_{i+1}\gamma_i) \tag{14.199}$$

This result holds also in the case of the Moore–Read state (Nayak and Wilczek, 1996).

On the other hand, since the braid operators of Eq. (14.199) are bilinear functions of the $2n$ Majorana fermions, they commute with the fermion parity operator,

$$P = (-1)^{N_{\rm F}}, \qquad N_{\rm F} = \sum_j \psi_j^\dagger \psi_j = \sum_j \frac{1}{2}(1 + \gamma_{2j}\gamma_{2j+1}) \tag{14.200}$$

Hence, we can consistently restrict the representation to either the sector with $N_{\rm F}$ even or the sector with $N_{\rm F}$ odd. We conclude that the total degeneracy of the states supported by $2n$ vortices is actually 2^{n-1} (instead of 2^n). This is also the degeneracy found in the case of the Moore–Read state (Nayak and Wilczek, 1996).

As we have seen, a set of $2n$ half-vortices of a $p_x + ip_y$ superconductor supports a finite-dimensional Hilbert space of dimension 2^{n-1}. This Hilbert space has a topological origin. Indeed, provided that the vortices are far from each other, this degeneracy is not lifted. Furthermore, all other states are separated from this Hilbert space by a finite energy gap. Thus, physical processes in which the vortices are being braided adiabatically do not mix this Hilbert space with the rest of the states of the system. In addition, this degeneracy is also robust against the effects of disorder, which primarily will lead to the localization of the vortices themselves.

14.10.2 Fusion of half-vortices

There is another way to think about this degeneracy, which makes contact with the CFT construction of the states. In the CFT approach we identified the Pfaffian factor with a correlator of Majorana chiral fields in a critical classical 2D Ising model. We also noted that the wave functions for the quasiparticles amount to an insertion of the chiral component of the Ising primary field into the fermion correlators. For historical reasons, and at the risk of confusing the reader with notation, we will denote by χ the Majorana fermion of the Ising model and by σ the Ising primary field. In Eq. (14.139) we gave their fusion rules,

$$\chi \star \chi = I, \qquad \sigma \star \sigma = I + \chi, \qquad \sigma \star \chi = \chi$$

We can see immediately that there is a correspondence between the Ising fusion rules and the properties of the Majorana zero modes of the half-vortices. Indeed, if we identify the Ising primary field σ with the insertion of a half-vortex, we see that fusing two half-vortices leads either to a state with an occupied fermion state or to a state with an empty fermion state. Thus we can identify the state $|0\rangle$ with the identity field in the Ising CFT and the state $|1\rangle$ with the Majorana fermion χ in the Ising CFT. The degeneracy then arises from counting how many ways we can fuse a given set of half-vortices into the identity (so it can have an expectation value). For example, if we have $2n = 8$ vortices our analysis predicts a degeneracy of $2^{4-1} = 8$. This is also the number of ways in which eight Ising σ primary fields can fuse into the identity field I.

This result is part of a general rule. We will see in Chapter 15 that the chiral CFT associated with the chiral sector of the Ising CFT is not precisely the level $k = 2$ chiral SU(2) Wess–Zumino–Witten model, but what is known as the (chiral) coset $\mathrm{SU}(2)_2/\mathrm{U}(1)_2$. We will also see there that the chiral $\mathrm{SU}(2)_2$ WZW CFT has a one-to-one correspondence with the level $k = 2$ SU(2) Chern–Simons gauge theory. On the other hand, we also know that the observables of the $\mathrm{SU}(2)_k$ Chern–Simons theory are Wilson loops with quantum number $j = 0, 1/2, \ldots, k/2$ (where $k/2$ stands for the integer part of $k/2$). Thus, for $k = 2$ only three representations are allowed: (a) the singlet $j = 0$ representation denoted by [0], (b) the doublet representation $j = 1/2$, denoted by [1/2]; and (c) the triplet representation $j = 1$, denoted by [1]. The Wilson loop operators with representations $[j_1]$ and $[j_2]$ in the $SU(2)_k$ theory obey the fusion rule

$$[j_1] \star [j_2] = [|j_1 - j_2|] + \cdots + [\min(j_1 + j_2, k - j_1 - j_2)] \tag{14.201}$$

Thus, in contrast to the case of the SU(2) group, for which the upper end of the range would have been $j_1 + j_2$, in $SU(2)_k$ the representations are truncated by the level k. Mathematical structures that obey these modified rules are called quantum groups (for a detailed treatment of quantum groups see Fuchs (1992)).

For the case of interest, $SU(2)_2$, the non-trivial fusion rules reduce to

$$[1/2] \star [1/2] = [0] + [1], \qquad [1/2] \star [1] = [1/2], \qquad [1] \star [1] = [0] \tag{14.202}$$

As we can see, the fusion rules are the same as those of the Ising CFT if we identify the identity primary field I with the singlet representation [0], the Ising primary field σ with the doublet representation [1/2], and the Majorana primary field with the representation [1]. However, these two CFTs are not identical since, in addition to the scaling dimensions being generally different, in the Ising CFT each primary field appears once, whereas in $SU(2)_2$ they appear in multiplets (Di Francesco *et al.*, 1997). Below we will discuss a quantum Hall state with $SU(2)_2$ symmetry.

Of course, we want not just the dimensions of the Hilbert spaces but also the wave functions themselves. This requires the computation of the conformal blocks, which do depend on other information such as the scaling dimensions, etc. Nevertheless, the conformal blocks of $SU(2)_2$ are part of this construction. We saw a similar example in our discussion of non-abelian bosonization in Section 7.10, where we considered the case of $SU(2)_1$.

The fusion rules provide a simple pictorial way to compute the dimensions of the topological Hilbert spaces $\mathcal{M}_{2n}$. For instance, let us suppose that we have a state with $2n$ insertions of the σ field. We can begin to fuse pairs of σ fields, then fuse the result of their fusion, and so on until we get to the point at which all fields have fused into the identity field I. Only the contribution of such a fusion

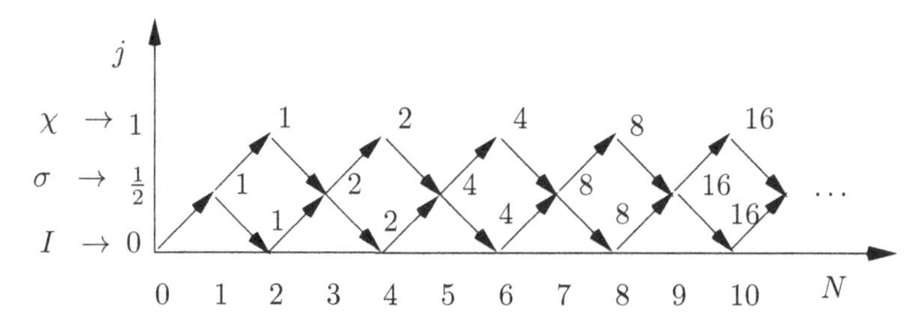

Figure 14.7 The Bratteli diagram for $SU(2)_2$. The numbers at the vertices are the numbers of ways to fuse into this channel (see the text).

process can yield a non-vanishing expectation value resulting in a non-vanishing wave function. The degeneracy of the Hilbert space is the number of ways in which this fusion can be done. This is depicted in the Bratteli diagram shown in Fig. 14.7 for the case of $SU(2)_2$. The vertical axis of the Bratteli diagram has the list of allowed representations, which for $SU(2)_2$ are [0], $[\frac{1}{2}]$, and [1] (for the Ising CFT they are I, σ, and χ). If we want to know the dimension of the Hilbert space for, say, ten σ fields, we must count the number of paths on the Bratteli diagram (16) that will reach (10, 0) starting from (0, 0). For $SU(2)_2$ this is just the 2^{n-1} result (with $N = 2n$).

15

Physics at the edge

In an incompressible quantum fluid, such as the Laughlin state, the fluctuations in the bulk induce fluctuations at the boundary. While the local fluctuations in the bulk are associated with local changes in the density, the fluctuations of the states at the boundary are associated with changes in the shape of the "droplet" of the electron fluid, shown in Fig. 15.1. These "edge waves" are the only gapless excitations of the system. It may seem surprising that an incompressible fluid may have gapless modes at the surface, although this is quite common in conventional fluids such as water! In the FQHE the gaplessness arises from the fact that the geometric edge of the fluid coincides with the locus of points in which the Fermi energy crosses the external potential which confines the fluid. Thus, the boundary of the fluid behaves like a "Fermi surface" and, as we move from the edge and into the bulk, we get deeper and deeper in the Fermi sea of occupied states. Because of the presence of the magnetic field, the edge waves are *chiral* excitations that move at the drift velocity of the particles at that point. Thus, edge states move only in one direction, which is specified by the magnetic field. The importance of the edge states to the observability of the quantum Hall effect was first emphasized by Halperin (1982). The description of the chiral quantum dynamics of the edge states is due to X. G. Wen (Wen, 1990a, 1990b, 1991b) and M. Stone (Stone, 1991).

15.1 Edge states of integer quantum Hall fluids

Let us consider the physics of the edge waves in the context of the simplest system: non-interacting electrons filling up the lowest Landau level ($\nu = 1$). Strictly speaking, we are discussing the behavior of the edge states in a system with an *integer* quantum Hall effect (IQHE). However, at least within a mean (or average)-field approximation, the *fractional* quantum Hall effect can also be regarded as an IQHE of an equivalent system of fermions. We will discuss this point of view (originally

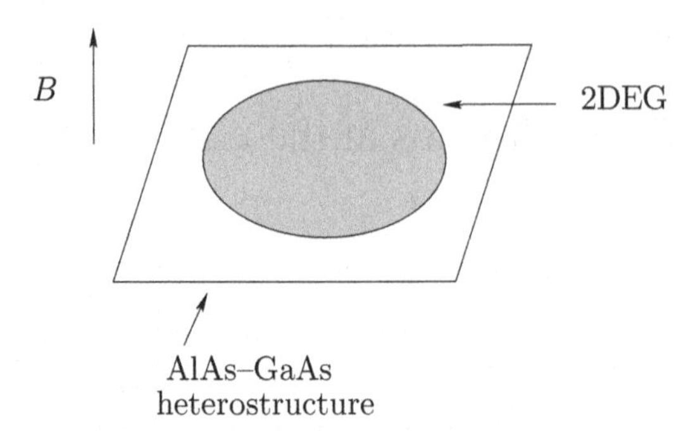

Figure 15.1 The two-dimensional electron-gas (2DEG) droplet in an AlAs–GaAs heterostructure in a perpendicular magnetic field B.

due to Jain) in this chapter, where we present the Chern–Simons approach to the FQHE. In this section we follow the methods of Stone (1991).

Let us, once again, consider a set of N electrons that are filling up the lowest Landau level of a system with $N_\phi = N$ flux quanta piercing the surface. In the absence of any other forces, the system has uniform density $\langle \rho \rangle = B/(2\pi)$, if the units are such that $\hbar = c = e = 1$. But, if no external forces are present, a system with N electrons in an *infinite* plane cannot have a fixed density. Furthermore, since we are interested in the physics at the edge, we must assume that the N electrons are constrained to remain within some region of the sample by the action of some external force. In the experimental setting, even in the purest samples there are forces as we examine the system close to the edge. So, we should assume that, in addition to experiencing the uniform magnetic field B, the electrons also feel an *electrostatic* potential $V(\vec{x})$ that keeps them inside the sample. We will consider the simple geometry of an infinite strip. The system has finite width L_1 along the axis x_1 and length L_2 along the axis x_2, with $L_2 \gg L_1$. We can also assume *periodic boundary conditions* along x_2. I will also assume that the potential V varies only along x_1 and that its variation is so slow that locally it can always be approximated by a linear function of x_1, $V(x_1) \approx E x_1$. In this geometry, it is natural to use the axial-Landau gauge $A_1 = 0$, $A_2 = B x_1$.

Let us now expand the second-quantized electron field operators $\psi(\vec{x})$ as a sum over states of the lowest Landau level, namely

$$\psi(x_1, x_2) = \sqrt{\frac{B}{\pi L_2}} \sum_{n=-\infty}^{+\infty} a_n e^{i k_n x_2} e^{-\frac{B}{2}(x_1 - k_n/B)^2} \tag{15.1}$$

which satisfies the boundary conditions. The allowed momenta k_n are $k_n = 2\pi n/L_2$. The creation and destruction operators, $a_n^\dagger$ and a_n, obey the anticommutation relations

$$\left\{a_n, a_m^\dagger\right\} = \delta_{nm} \tag{15.2}$$

In the presence of an external potential, the degeneracy of the Landau level is lifted. Thus, in perturbation theory the energy of the lowest, $N = 0$, Landau level has a first-order shift

$$E_0(k) = \frac{1}{2}\hbar\omega_c + \langle 0, k|V(x)|0, k\rangle + \cdots \tag{15.3}$$

This shift effectively "lifts" the Landau level in the vicinity of the edge where the potential is acting (see Fig. 15.2). For the particular case of the linear potential, the wave functions are the same as the wave functions in the absence of the potential, but the single-particle energies $E_0(k)$ become

$$E_0(k) = \frac{E}{B}k \tag{15.4}$$

with a sign determined by the sign of B (for $V(x_1)$ fixed). Thus, the states near the Fermi energy E_F have momentum k along the edge. The Fermi velocity v_F of these states is

$$v_F = \frac{\partial E_0(k)}{\partial k} = \frac{e}{c}\frac{|\vec{E}|}{|B|} \tag{15.5}$$

which is the drift velocity of a charged particle moving in an external magnetic field that is perpendicular to the plane and in an electric field $\vec{E} = -(\partial V/\partial x)e_x$ pointing inwards towards the droplet of electron fluid. Here I have assumed that $B > 0$. These states are chiral and move with the drift velocity. Semi-classically we can picture the edge states as electrons that move along skipping orbits along the edge (Halperin, 1982).

This expression is accurate for those states whose energies are close to the Fermi energy, which I have set to zero. Away from the boundaries, the potential is

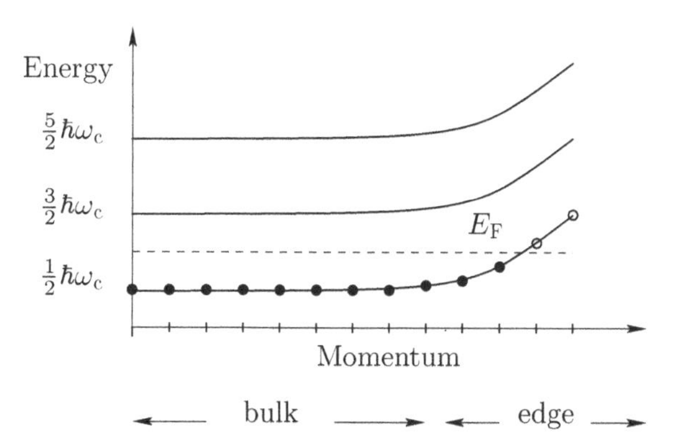

Figure 15.2 Lifting of the Landau levels by a confining potential $V(x_1)$ (see the text). For a cylindrical geometry, in the Landau gauge the levels are labeled by their momentum. E_F is the Fermi energy and $\hbar\omega_c$ is the cyclotron energy. Bulk and edge regions of the system are shown.

essentially constant and the Landau level effectively has a degeneracy. The origin of my coordinate system is at the point where the potential crosses the Fermi energy. Thus, far to the left of the crossing point, the density is constant, and to the right of the crossing point there are no particles. It is clear from this picture that it takes a negligible amount of energy to add a particle to the system, but the particle is added to the surface, not to the bulk. The low-energy excitations of the system are local changes of density at the surface, the edge waves, as shown in Fig. 15.3. Notice that, since the number of particles is fixed and since the next Landau level is separated from the ground state by a very large energy gap, a lower density at a point on the surface means that there should be an excess density at some other point of the same surface.

As usual, we are interested only in the excitations with low energy. Here, close to the Fermi energy means close to the surface. Let $j(x_2)$ be the operator which measures the amount of charge localized within some region of size Λ of the edge,

$$j(x_2) = \int_{-\infty}^{+\infty} dx_1\; f_\Lambda(x_1)\psi^\dagger(x_1, x_2)\psi(x_1, x_2) \tag{15.6}$$

The cutoff function $f_\Lambda(x_1)$ must be chosen in such a way that it is vanishingly small in the region $|x_1| \gg \Lambda$, and the cutoff Λ must be larger than the typical amplitude fluctuation of the low-energy states. We will choose the cutoff function to be a Gaussian, $f_\Lambda(x_1) = (1/(\sqrt{2\pi}\Lambda))\exp(-x_1^2/(2\Lambda^2))$.

Since we are using periodic boundary conditions in x_2, it is convenient to consider the Fourier transform of the operator $j(x_2)$, i.e.

$$j(x_2) = \sum_n e^{-ik_n x_2} j_n \tag{15.7}$$

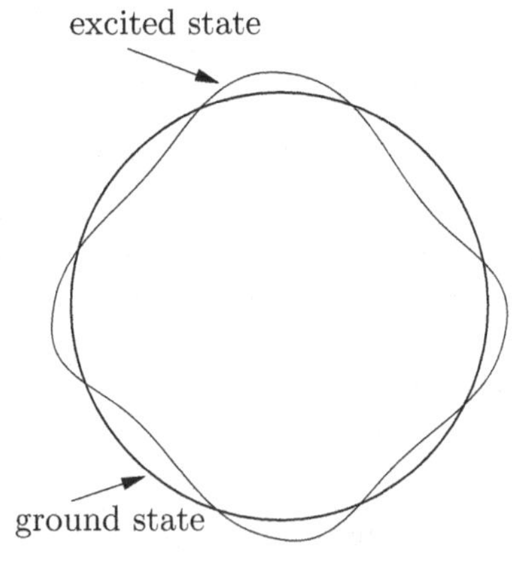

Figure 15.3 An excitation of the edge states is a deformation of the incompressible charge fluid resulting in a chiral wave propagating along the edge with fixed direction.

Conversely, we can write

$$j_n = \sum_{m=-\infty}^{+\infty} a_{m+n}^{\dagger} a_m e^{-\frac{B}{4} k_n^2} \tag{15.8}$$

It is apparent that the Gaussian factor $\exp(-Bk_n^2/4)$ is negligibly small away from the Fermi surface.

If we ignore the Gaussian factor, the density operator j_n coincides with the density operator for a system of fermions in one space dimension that are allowed to move only in one direction. In this case the direction is specified by the sign of the magnetic field B. Indeed, the dispersion law $\epsilon(k) = v_{\mathrm{F}} k$, with a Fermi velocity $v_{\mathrm{F}} = E/B$, follows from the Hamiltonian

$$H = \int dx_2 \; \psi_{\mathrm{R}}^{\dagger}(x_2)(-i v_{\mathrm{F}} \, \partial_2) \psi_{\mathrm{R}}(x_2) \tag{15.9}$$

which governs the dynamics of right-moving chiral fermions in one dimension. Notice that this is precisely the same Hamiltonian as that we found in Chapter 5 when we discussed bosonization.

The results of Chapter 5 enable us to write down the commutation relations obeyed by the operators j_n. There we found that the commutator of the Fourier-transformed density operators is different from zero due to the presence of a Schwinger term,

$$\left[j_n, j_m \right] = -n \delta_{n+m,0} \tag{15.10}$$

Alternatively, in position space we can write

$$\left[j(x_2), j(x_2') \right] = -\frac{i}{2\pi} \, \partial_2 \delta(x_2 - x_2') \tag{15.11}$$

This algebra is known as the level-1 U(1) chiral Kac–Moody algebra. Wen has shown that the spectrum of the edge states is always determined by an appropriate Kac–Moody algebra. For instance, if the fermions were not fully polarized, spin would have to be included in the dynamics. In that case the relevant algebra is the (level-1) SU(2) Kac–Moody algebra. These results are similar to what we already found in Luttinger liquids (except that the states here are chiral), see Chapter 6.

Stone has also given the following explicit construction of the edge density waves. Let $|0\rangle$ denote the ground state, which corresponds to an undisturbed droplet. Let us define the family of coherent states $\{|\theta(x_2)\rangle\}$,

$$|\theta(x_2)\rangle = e^{i \int dx_2 \, \theta(x_2) j(x_2)} |0\rangle \tag{15.12}$$

which represent *coherent excitations* of the edge states. Throughout it is assumed that the density operator has been *normal-ordered* relative to the undisturbed state, namely $j(x_2)|0\rangle \equiv 0$.

We now show that the states $|\theta(x_2)\rangle$ are eigenstates of the (normal-ordered) density operator $j(x_2)$:

$$j(x_2)|\theta(x_2)\rangle = \frac{1}{2\pi}\,\partial_2\theta(x_2)|\theta(x_2)\rangle \tag{15.13}$$

This property can be derived by using the identity

$$e^{-i\int dx_2'\,\theta(x_2')j(x_2')}\,j(x_2)e^{+i\int dx_2'\,\theta(x_2')j(x_2')} = j(x_2) + \frac{1}{2\pi}\,\partial_2\theta(x_2) \tag{15.14}$$

These states represent local changes in the density. This can be seen from the following argument: the state $|\theta(x_2)\rangle$ has a local excess of charge equal to $(1/(2\pi))\partial_2\theta(x_2)$. From the linearity of the energy–momentum relation we know that an extra number of particles means that the local position of the Fermi level has gone from zero to $\partial_2\theta$, which is still much less than the Landau gap. Likewise, the momentum k has changed by the same amount. Since we also saw that, for these states, there is a precise relation between the energy of the state and its location on the axis x_1, we conclude that this state is in fact a local change of the *shape* of the droplet. Moreover, at least within the accuracy of the linear approximation for the dispersion relation, these states propagate without deformation, since all the excitations propagate at the same speed $v_{\rm F}$.

Throughout this discussion we have focused on the states close to the edge. But, as we have already warned the reader, the bulk cannot be decoupled from the edge. In fact, theories of chiral fermions, such as the one we are discussing here, are intrinsically sick. The reason is that, if the linear spectrum is taken literally, this system would not be able to keep track of the conservation of charge once it is coupled to a fluctuating vector potential. Indeed, in one dimension, all the components of the vector potential are longitudinal, since there is no way to "enclose flux inside a line." But it is possible to do it if the line closes on itself, forming a closed curve. This is precisely the case of interest to us. For example, in the gauge $A_0 = 0$, the only component we are left with is the component $A_\parallel$ tangent to the curve (the edge). By general arguments of gauge invariance we know that the Hamiltonian for the chiral fermions coupled to the gauge field is obtained by the minimal-coupling procedure, which replaces the derivative ∂_2 by the covariant derivative $D_2 = \partial_2 - ieA_\parallel(x_2)$.

Thus, the Hamiltonian picks up an extra term $H_{\rm gauge}$ of the form

$$H_{\rm gauge} = \int dx_2\; eA_\parallel(x_2)\psi_{\rm R}^\dagger(x_2)\psi(x_2) \tag{15.15}$$

This term shows that the local fluctuations of $A_\parallel(x_2)$ will cause the Fermi level to move up and down. Thus, charge has to "leak in" or "leak out" through the bottom of the Fermi sea. For a theory "without a bottom," such as a relativistic field theory, this is a disaster. The chiral theories are then said to be sick and to break gauge invariance and to have a gauge anomaly. But, in the problem we are considering, the Fermi sea does have a bottom. It is determined by the Landau level, which acts like a reservoir of particles and redistributes the particles from one point of the edge to another.

The configurations with a non-zero circulation of $A_{\parallel}$ have a very interesting meaning: the circulation of $A_{\parallel}$ on a closed curve such as the edge is just the amount of flux enclosed inside the curve. Thus, the uniform field causes the electrons on edge states to move around the system. A change in the circulation means that flux has been added to or removed from the system. Thus, the addition of one quasihole should cause a jump in the circulation by exactly one flux quantum. The edge states see this extra flux as a change of the position of the Fermi level. This is then interpreted as the generation of a net charge at the edge. For a non-interacting problem, the net charge is equal to e. But, for a Laughlin state, it is equal to e/m. The extra charge accumulated at the edge is interpreted as a lack of charge conservation; that is, as a gauge anomaly of the theory of the edge states. The precise cancellation of the gauge anomaly of the bulk with the gauge anomaly of the edge, which was first discussed by Wen, is a consequence of the gauge invariance of the system as a whole (Wen, 1991b).

15.2 Hydrodynamic theory of the edge states

We will now turn to the more interesting case of the edge states of the fractional quantum Hall states on an open geometry, which we will take to be a disk. The fractional quantum Hall ground states on a disk are unique and have a gap to all excited states. As we saw in the non-interacting case, in general we expect that these states will have a gapless spectrum of excitations localized near the edges of the system. We will call all of these states the edge states.

The necessity for the existence of edge states can be seen by invoking an elegant argument due to Wen that is based on gauge invariance (Wen, 1990b; Wen and Zee, 1992). It goes as follows. Let us consider a 2DEG confined to a finite (but large) region of a sample by a confining potential. Such a system in a quantum Hall state (integer or fractional) is an incompressible fluid since all states in the bulk have a gap that can be made arbitrarily large by turning up the external magnetic field (while keeping the filling fraction of the Landau level fixed). In this situation the action of a weak external electromagnetic perturbation on this charge fluid can only have a net effect on its boundary, leading to slow and long-wavelength changes in its shape such as those shown in Fig. 15.3.

Owing to the incompressibility of the fluid, adiabatically adding or removing some amount of charge from the bulk of the fluid is equivalent to adding or removing the same amount of charge from the edge. In other words, the whole fluid (bulk plus edge) must conserve charge. We will see below that the local conservation of charge leads to a simple and elegant hydrodynamic theory. More importantly, a fluid with local conservation of charge obeys locally a continuity equation, which in turn means that its electromagnetic response must be gauge-invariant (see Chapter 12).

Let us now imagine that we want to define some region of the fluid as the bulk and the rest as the edge region, as in Fig. 15.2. However, the arguments we gave above tell us that charge cannot be conserved in the bulk or the edge separately, but only in the system as a whole. In other terms, the electromagnetic response of the bulk must violate gauge invariance at the boundary of this region. Similarly, the edge region also violates electromagnetic gauge invariance. However, since the system as a whole must be gauge-invariant, the violations of gauge invariance in the bulk and at the edge must exactly cancel each other out. In the language of quantum field theory, the effective theory of the edge states will turn out to be a chiral system that has a gauge anomaly. We will see that the bulk system will be generally described by an effective gauge theory, namely the Chern–Simons gauge theory, which is not gauge-invariant on systems with an open manifold. However, the anomaly of the edge degrees of freedom is (and must be) equal and opposite to the anomaly of the Chern–Simons theory in the bulk. This requirement is the physical basis of the bulk–edge correspondence in quantum Hall fluids and will be discussed in detail in Section 15.4.

In Section 14.2 we showed that it is possible to give a purely hydrodynamic description of the bulk physics of the fractional quantum Hall states. There is a similar and quite powerful hydrodynamic picture of the edge states. This theory, which is largely due to the work of Wen (1995), whose work we will follow here, gives an essentially universal description of the edge states. A key ingredient of this theory is that the incompressibility of the electron fluid in the bulk forces the existence of a one-to-one correspondence with the physics of the edge states, and that the bulk physics is encoded at the edge. A system with these properties is often referred to as being "holographic" ('t Hooft, 1993; Susskind, 1995).

We will assume that the 2DEG is in a fractional quantum Hall fluid state. This state has a gap to all local excitations and hence it is incompressible. We will further assume that the fluid is uniform and hence that the electron density is constant in the bulk, and that it falls to zero smoothly across the edge over a length scale of the order of the magnetic length, ℓ_0. The latter assumption is not trivial since Coulomb interactions can give rise to a modulation of the density of the 2DEG near the edge. This phenomenon is known as edge reconstruction, and when it happens it can alter the properties of the electron gas near the edge substantially from the properties in the bulk of the system. In what follows we will consider a simplified situation

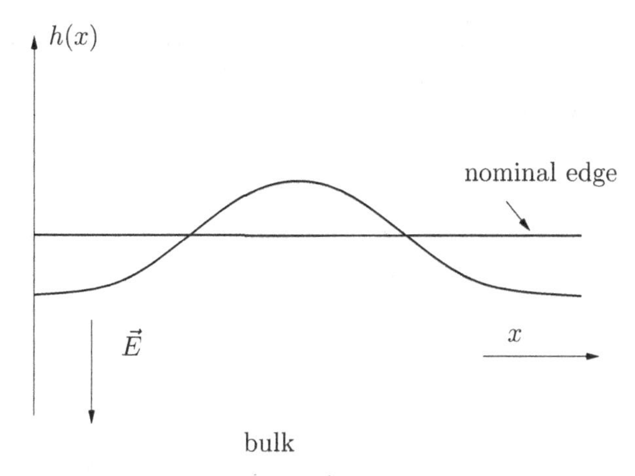

Figure 15.4 The straight edge. Here $\vec{E}$ is the electric field which keeps the electrons in the droplet and x is the coordinate along the edge.

and, for the sake of clarity, ignore the possible effects of edge reconstruction. An excellent and extensive review of the physics of edge states, particularly on the experiments, is given in Chang (2003).

Since the bulk fluid is incompressible, the low-energy fluctuations at the edge which do not change the total number of particles of the system are equivalent to changes in the shape of the edge. In the thermodynamic limit, namely the regime in which the radius R of the electron droplet is much larger than the magnetic length ℓ_0, there are many closely spaced electronic states close to the edge. As in the non-interacting case, these states are closely spaced and their single-particle energies rise smoothly, with an average slope largely determined by the potential that confines the electrons within the interior of the system. Let $\vec{E}$ be the local electrostatic field created by the confining potential (see Fig. 15.4). Since the electrons are moving in an external magnetic field of magnitude B perpendicular to the sample, there is a drift current $\vec{j} = \sigma_{xy}\hat{z} \times \vec{E}$ flowing (with velocity $v = |\vec{E}|c/B$) along the edge, where $\sigma_{xy} = (\nu/(2\pi))e^2/\hbar$ is the Hall conductance and ν is the filling fraction.

Regardless of the nature of the excitations, fermionic for filling factor $\nu = 1$ or generally anyonic for $\nu \neq 1$, the fluctuations of the edge are made up of large numbers of particle–hole excitations. These coherent states are bosonic and behave similarly to the density fluctuations in a 1D electron fluid, except for the important difference that they are chiral: the direction of propagation is determined by the sign of the perpendicular magnetic field. Thus, the edge excitations are chiral waves that propagate only in one direction at the drift velocity. Let n_0 be the 2D uniform particle density

$$n_0 = \frac{\nu}{2\pi \ell_0^2}, \qquad \ell_0 = \sqrt{\frac{\hbar c}{eB}} \tag{15.16}$$

The 1D density along the straight edge is $\rho(x) = n_0 h(x)$, where $h(x)$ is the local displacement of the edge (as shown in Fig. 15.4). The density wave is chiral if it obeys the classical equation of motion

$$\partial_t \rho(x,t) - v\, \partial_x \rho(x,t) = 0 \tag{15.17}$$

where v is the drift velocity. For small deformations $h(x)$ (compared with the radius R of the system), the external electrostatic field E is uniform, and the total electrostatic energy H (the classical Hamiltonian) stored in this edge distortion is

$$H = \int dx\, \frac{1}{2} eh\rho(x)E = \int dx\, \frac{\pi\hbar v}{\nu}\rho(x)^2 \tag{15.18}$$

Let $L = 2\pi R$ be the total length of the edge of the electron droplet. For an isolated system the edge is a simply connected closed curve that acts as a boundary. Hence, the density fluctuation must obey periodic boundary conditions

$$\rho(x) = \rho(x+L) \tag{15.19}$$

We can now consider the Fourier modes of the edge fluctuation and write

$$\begin{aligned} \rho(x) &= \frac{1}{\sqrt{L}} \sum_{n=-\infty}^{\infty} \exp\left(i\frac{2\pi n}{L}x\right)_{\rho_n} \\ \rho_n &= \frac{1}{\sqrt{L}} \int_0^L dx\, \exp\left(-i\frac{2\pi n}{L}x\right) \end{aligned} \tag{15.20}$$

In terms of the Fourier modes the classical Hamiltonian is simply given by

$$H = \frac{\pi}{\nu}\hbar v \sum_{n=-\infty}^{\infty} \rho_n\rho_{-n} \equiv \frac{2\pi}{\nu}\hbar v \sum_{k>0} \rho_k\rho_{-k} \tag{15.21}$$

where we introduced the edge momentum labels $k_n = 2\pi n/L$. The classical equation of motion of the Fourier density modes is

$$\partial_t \rho_k = i v k \rho_k \tag{15.22}$$

If we compare this equation with Hamilton's equations of a system with coordinate q and momentum p,

$$\dot{q} = \partial_p H, \qquad \dot{p} = -\partial_q H \tag{15.23}$$

we see that we can identify generalized coordinates Q_k and generalized canonical momenta P_k by writing

$$Q_k \equiv \rho_k, \qquad P_k \equiv -i\frac{2\pi}{\nu k}\rho_{-k} \tag{15.24}$$

such that

$$\begin{aligned} \dot{Q}_k &= \frac{\partial H}{\partial P_k} \\ \dot{P}_k &= -\frac{\partial H}{\partial Q_k} \end{aligned} \quad \Rightarrow \quad \begin{aligned} \dot{\rho}_k &= i v k \rho_k \\ \dot{\rho}_{-k} &= -i v k \rho_{-k} \end{aligned} \tag{15.25}$$

which implies that the classical Hamiltonian is

$$H = i v \sum_{k>0} Q_k P_k \tag{15.26}$$

To quantize this system we promote the coordinates and momenta to operators in a Hilbert space, and satisfy the canonical equal-time commutation relations

$$[Q_k, P_{k'}] = i\hbar \delta_{k,k'} \tag{15.27}$$

with $k = 2\pi n/L$. Hence, the Fourier modes of the density operators satisfy the commutation relation

$$[\rho_k, \rho_{k'}] = \frac{\nu}{2\pi} k \delta_{k+k',0} \tag{15.28}$$

Hence, even in this more general hydrodynamic theory, the edge-mode operators ρ_k still satisfy a chiral U(1) Kac–Moody algebra. We will see shortly that the "level" in general is not equal to 1 except for $\nu = 1$. Similarly, the classical Hamiltonian of Eq. (15.21) is promoted to an operator acting on the Hilbert space of states of the edge modes. The (normal-ordered) quantum Hamiltonian generates the time evolution of the density mode operators ρ_k which satisfy the Heisenberg equation

$$[H, \rho_k] = \hbar v k \rho_k \tag{15.29}$$

How do we describe in this picture the excitations that change the total charge of the system? For instance, let's say we want an operator that describes the action of adding an electron at a location x of the edge. We will denote the electron-creation operator by $\psi_e^\dagger(x)$ and demand that it satisfies the following commutation relation with the (normal-ordered) local number-density operator $\rho(x)$:

$$\left[\rho(x), \psi_e^\dagger(x')\right] = \delta(x - x')\psi_e^\dagger(x') \tag{15.30}$$

But we already know how to do this using (abelian) bosonization! (See Chapters 5 and 6.) We represent the Kac–Moody current (and density) field $\rho(x)$ in terms of a Bose (scalar) field $\phi(x)$,

$$\rho(x) = \frac{1}{2\pi} \partial_x \phi(x) \tag{15.31}$$

where we have assumed a normalization for future convenience. The density operator $\rho(x)$ has been assumed to be normal-ordered with respect to an edge state without excitations (the ground state) of an isolated 2DEG. Thus, if the bulk state of a 2D electron gas has the precise number of electrons to satisfy exactly the condition that the filling factor is $\nu = 1/m$, then the normal-ordered total charge of its edge states, measured in units of the electron charge e, is exactly zero:

$$Q = \int_0^L dx \; \rho(x) = 0 \tag{15.32}$$

This also implies that, since the normal-ordered density of the edge states of an isolated 2DEG obeys periodic boundary conditions, Eq. (15.19), then the chiral boson $\phi(x)$ of an isolated 2DEG also obeys periodic boundary conditions,

$$\phi(x + L) = \phi(x) \tag{15.33}$$

With these assumptions, the electron operator $\psi_{\mathrm{e}}(x)$ has the boson representation

$$\psi_{\mathrm{e}}(x) \sim e^{\frac{i}{\nu}\phi(x)} \tag{15.34}$$

Using the methods we discussed in Chapters 5 and 6, we find that the operators of Eq. (15.34) satisfy the algebra

$$\psi_{\mathrm{e}}(x)\psi_{\mathrm{e}}(x') = e^{i\frac{\pi}{\nu}}\psi_{\mathrm{e}}(x')\psi_{\mathrm{e}}(x) \tag{15.35}$$

Thus, the operator $\psi_{\mathrm{e}}(x)$ obeys fermionic anticommutation relations only for the filling factor $\nu = 1/m$ where m is an *odd* integer:

$$\nu = \frac{1}{m} \qquad \Rightarrow \qquad \left\{\psi_{\mathrm{e}}(x), \psi_{\mathrm{e}}(x')\right\} = 0 \tag{15.36}$$

In contrast, for m even the operator $\psi_{\mathrm{e}}(x)$ represents a boson. We will see that this corresponds to the theory of the edge states of bosonic Laughlin states at filling factor $\nu = 1/m$.

We can now relate the scalar field $\phi(x)$ to the edge modes and to the generalized coordinates Q_k and canonical momenta P_k. After a Fourier transform we find

$$Q_k = i\frac{k}{2\pi}\phi_k, \qquad P_k = -\frac{1}{\nu}\phi_k \tag{15.37}$$

Thus, we find that the Fourier modes of the scalar field satisfy the equal-time commutators

$$\left[\phi_k, \phi_{-k'}\right] = i\nu\delta_{k,k'} \tag{15.38}$$

and that the Fourier modes ϕ_k obey chiral equations of motion (as expected) of the form

$$\dot{\phi}_k = i v k \phi_k \tag{15.39}$$

Therefore $\phi(x)$ is a free chiral scalar field whose Hamiltonian (in position space) is (after we set $\hbar = 1$, as we will do from now on)

$$H = \int dx \, \frac{v}{4\pi \nu} (\partial_x \phi)^2 \tag{15.40}$$

and the associated Lagrangian density $\mathcal{L}$ of the chiral boson is

$$\mathcal{L} = \frac{m}{4\pi} \left[\partial_t \phi \, \partial_x \phi - v (\partial_x \phi)^2 \right] \tag{15.41}$$

where we used the fact that $\nu = 1/m$.

So far we have ignored the effects of interactions on the edge states. Since these states are chiral, and as such can propagate only in a direction fixed by the magnetic field, backscattering processes arising from either electron–electron interactions or localized impurities are forbidden. Thus, the incompressible fluid simply skirts about the location of the impurity without changing the direction of propagation of the edge excitations. This leaves forward-scattering processes as the only allowed interactions. However, as we saw in our discussion of the conventional Luttinger liquid in Chapter 6, forward-scattering processes merely change the velocity of propagation and do not change the scaling dimensions of the operators. In the present context, an interaction term for the edge states becomes

$$\begin{aligned} H_{\text{int}} &= \frac{1}{2} \int dx \int dx' \, \rho(x) V(x - x') \rho(x') \\ &= \frac{1}{8\pi^2} \int dx \int dx' \, \partial_x \phi(x) V(x - x') \partial_{x'} \phi(x') \end{aligned} \tag{15.42}$$

For short-range interactions, with forward-scattering coupling constant g, the only effect is a finite renormalization of the velocity,

$$v_{\text{eff}} = v + \frac{g}{2\pi} \tag{15.43}$$

Thus, for short-range interactions the excitations of the edge modes have an energy $\omega(k) = v_{\text{eff}} k$.

On the other hand, in the case of Coulomb interactions, the electrons are coupled by the singular potential $V(x - x') = e^2/|x - x'|$. The Fourier transform at small k of this interaction potential is $V(k) = -e^2 \ln(|k|\ell_0) + \cdots$. As in the case of short-range interactions, the only effect of Coulomb interactions for chiral fermions is also a renormalization of the excitation energy, albeit with the singular form $\omega(k) = -(e^2/(2\pi)) k \ln(|k|\ell_0)$. At any rate, even these interactions cannot open a gap in a chiral system, since backscattering processes are not allowed. This feature is the key to the robustness of the edge states. We will see below that, while edge reconstructions may complicate the picture, they do not alter this basic central fact.

Hence, the edge states of a fractional quantum Hall fluid constitute a *chiral* Luttinger liquid. The scalar field $\phi(x)$ is the chiral (right-moving) half, $\phi_{\rm R}$, of the conventional scalar field (see Chapter 6). In what follows we will denote by $\phi(x)$ the right-moving (chiral) field. We showed in Eq. (6.121) that the propagator of the chiral boson is (with a different normalization of the scalar field)

$$\langle T\phi(x,t)\phi(0,0)\rangle = -\nu \ \ln\left(\frac{x - vt + i\epsilon}{a_0}\right) \tag{15.44}$$

where T denotes time ordering, a_0 is a short-distance cutoff, and $\epsilon \to 0^+$. Similarly, the electron propagator is

$$\begin{aligned} G_{\rm F}(x,t) &= \langle T\psi_{\rm e}^\dagger(x,t)\psi_{\rm e}(0,0)\rangle \\ &= \left\langle Te^{-\frac{i}{\nu}\phi(x,t)}e^{\frac{i}{\nu}\phi(0,0)}\right\rangle \\ &= e^{\frac{1}{\nu^2}\langle T\phi(x,t)\phi(0,0)\rangle} \\ &\propto \frac{\text{constant}}{(x - vt)^{1/\nu}} \end{aligned} \tag{15.45}$$

(where the $i\epsilon$ has been omitted) up to a prefactor that oscillates with the "Fermi wave vector" of the edge, namely the characteristic average momentum of the states near the edge determined by the location of the edge and the electrostatic confining potential.

Thus, for the edge states of the Laughlin states we find that the electron propagator is

$$G_{\rm F}(x,t) = \frac{\text{constant}}{(x - vt)^m} \tag{15.46}$$

This propagator is only a function of $x - vt$ and hence it is explicitly chiral. It clearly obeys the condition of being odd (antisymmetric) under the exchange of the coordinates of the electron operators, $x \leftrightarrow -x$ and $t \leftrightarrow -t$,

$$G_{\rm F}(x,t) = -G_{\rm F}(-x,-t) \tag{15.47}$$

as required by the Pauli principle, only if m is an odd integer. On the other hand, the analytic structure is not that of a free-fermion system. Indeed, this propagator reduces to a simple pole of the argument only for the case of the integer Hall effect, for which it reduces to the free-field chiral fermion propagator. This analytic structure with a multiple pole (of order m) is a direct manifestation of the strongly correlated nature of the 2D electron fluid in a large magnetic field.

In this hydrodynamic theory we have assumed all along that the incompressible fluid has a *unique edge* with natural properties. The results of this quantized theory are telling us that, without assuming any additional structure, a fractional quantum

Hall state with a single edge can exist only for the Laughlin states at $\nu = 1/m$. We conclude that for the Laughlin states the electron operator at the edge is given by (up to a normalization)

$$\psi_{\mathrm{e}}(x) = e^{im\phi(x)}, \qquad \text{for } \nu = \frac{1}{m} \tag{15.48}$$

Let us consider now a process by which we either add an electron to the 2DEG or remove an electron from the 2DEG. Clearly, in this process the total number of electrons has been changed. Since the bulk is incompressible and uniform, the removal (or addition) of an electron from the bulk must become the same as removal (or addition) of an electron from the edge of the fluid. Since the local edge density operator $\rho(x)$ was assumed to be normal-ordered with respect to the ground state of the 2DEG with a number of electrons fixed precisely at $\nu = 1/m$, this implies that, if n_{e} electrons are added to the bulk (or removed), the total charge at the edge must now be changed by n_{e} units,

$$n_{\mathrm{e}} = \int_0^L dx\, \rho(x) = \frac{1}{2\pi}(\phi(L) - \phi(0)) \tag{15.49}$$

This result tells us that a change in the number of electrons in the bulk (in the same fractional quantum Hall state) leads to a change in the boundary conditions of the chiral boson, which now become "twisted,"

$$\phi(x + L) - \phi(x) = 2\pi n_{\mathrm{e}} \tag{15.50}$$

This means that the theory of the edge states is a chiral *compactified* boson with compactification radius $R = 1$ (not to be confused with the radius of the fluid droplet!).

As we saw in Chapter 13, the elementary excitations of a fractional quantum Hall fluid in a Laughlin state are not electrons but vortices of the charged fluid with fractional charge $\pm e/m$ and fractional braiding statistics $\delta = \pi/m$. Since we can interact with the 2DEG only in a manner that either does not change the total number of electrons or changes it by an integer number, we can produce or destroy these vortices only in groups of m of them. Nonetheless, since the vortices have very-short-range interactions with each other, the m vortices in each of these groups are not bound to each other but are (qualitatively speaking) essentially free. Since the bulk is gapped, these vortex excitations can only become "light" (gapless) at the edge where the energy gap collapses. Therefore, there should be gapless edge excitations. However, these edge excitations are not arbitrary but the projection at the edge of the bulk state.

This line of argument tells us that m vortex excitations must at the edge coalesce (or fuse) into an electron operator. The unique choice for the quasiparticle (vortex)-creation operator at the edge of a Laughlin state that satisfies this condition is

$$\psi_{\text{qp}}(x) \propto e^{i\phi(x)} \tag{15.51}$$

Indeed, m such quasiparticles *fuse* into an electron operator! Moreover, the commutator of the quasiparticle operator with the local density $\rho(x)$ is (for $\nu = 1/m$)

$$\left[\rho(x), \psi^{\dagger}_{\text{qp}}(x')\right] = \frac{1}{m}\delta(x - x')\psi^{\dagger}_{\text{qp}}(x') \tag{15.52}$$

Thus, the quasiparticle charge operator creates an excitation of charge $1/m$ (in units of the electron charge e).

To determine the statistics of this excitation, we compute the propagator of the quasiparticle operator:

$$\left\langle T\psi^{\dagger}_{\text{qp}}(x, t)\psi_{\text{qp}}(0, 0)\right\rangle = e^{\langle T\phi(x,t)\phi(0,0)\rangle} = \frac{\text{constant}}{(x - vt)^{1/m}} \tag{15.53}$$

In contrast with the case of the electron propagator, which has a pole of order m (see Eq. (15.46)), the quasiparticle propagator of Eq. (15.53) has a branch-cut singularity with exponent $1/m$. Consequently, under an exchange of the quasiparticle coordinates, which amounts to setting $x \to -x$ and $t \to -t$, the quasiparticle propagator changes by a *phase factor*,

$$\left\langle T\psi^{\dagger}_{\text{qp}}(-x, -t)\psi_{\text{qp}}(0, 0)\right\rangle \to e^{\pm i\frac{\pi}{m}}\left\langle T\psi^{\dagger}_{\text{qp}}(x, t)\psi_{\text{qp}}(0, 0)\right\rangle \tag{15.54}$$

Thus, the quasiparticles (and quasiholes) are anyons with statistical angle $\delta = \pi/m$, as expected.

These results imply that we should regard the edge states as a rational chiral conformal field theory (RCFT). In Sections 7.12 and 14.6 we defined an RCFT as a CFT with a finite number of primary fields. The compactified chiral boson is an example of a rational CFT. Why do we have a finite number of primary fields in this case? The reason is that the compactification condition implies that the only admissible operators in this theory must be invariant under the global shift $\phi \mapsto \phi + 2\pi$ (since $R = 1$). The operators that satisfy this condition are the chiral vertex operators $V_n(x)$,

$$V_n(x) = e^{in\phi(x)} \tag{15.55}$$

where $\phi(x)$ is the chiral boson, since they are all invariant under a phase shift by 2π,

$$V_n(x) \mapsto V_n(x), \qquad \text{as } \phi(x) \mapsto \phi(x) + 2\pi \tag{15.56}$$

The electron operator, which, as we saw, satisfies this condition, is just the operator $V_m(x)$.

The chiral vertex operators $V_n(x)$ with $1 \leq n < m$ have the special property that, in addition to satisfying the compactification condition, they are *local* with respect to the electron operator, i.e. they commute with the electron operator. However, they are non-local with respect to each other. Indeed, a correlator in which the operators V_n and $V_{n'}$ are present has a branch cut attached to each of their coordinates. Under a monodromy, namely a smooth displacement of the coordinate of one vertex operator around that of the other one along the closed contour shown in Fig. 15.5, which as we saw is equivalent to a double braid, the correlator changes by the phase factor $\exp(i2\pi nn'/m)$. Hence V_n is single-valued with respect to $V_m = \psi_e$ but not with respect to the other operators.

On the other hand, all operators with $n = lm$ can be regarded as being equivalent to the creation (or removal) of l electrons (with their associated fluxes), which does not change the state, i.e. the filling fraction. Therefore we must consider *only* chiral vertex operators with "charge" n modulo an integer (or zero) number of electrons. Thus, these conditions amount to restricting the set of distinct primaries to being the set $\{V_n\}$ with $0 \leq n < m$, which has m primary fields, m being the number of linearly independent ground states of the bulk 2DEG on a torus. In the language of CFT what we have done is define an extended or chiral algebra with respect to the electron operator (Moore and Read, 1991).

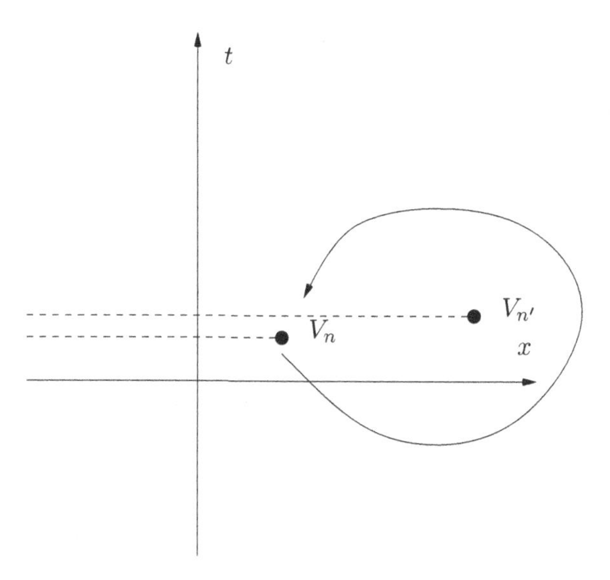

Figure 15.5 The chiral vertex operators, which create the excitations at the edge, transform non-trivially under a monodromy. The broken lines are branch cuts.

Therefore, the integer m can be identified with the "level" of the Kac–Moody algebra which restricts the tower of allowed primaries (representations). We will say that the edge state of the $\nu = 1/m$ Laughlin state is a $\mathrm{U}(1)_m$ compactified chiral boson CFT. The m primary fields of this chiral CFT, the chiral vertex operators V_n, have scaling dimensions

$$\Delta_n = \frac{n^2}{2m} \tag{15.57}$$

with a conformal spin equal to the scaling dimension. Thus, the electron operator has scaling dimension $\Delta_e = \Delta_m = m/2$, while the "fundamental" quasiparticle, V_1, has scaling dimension $\Delta_{\text{qp}} = \Delta_1 = 1/(2m)$. We will see below that there is duality transformation relating the electron and the quasiparticle.

15.3 Edges of general abelian quantum Hall states

The hydrodynamic theory of the preceding section describes only the edge states of the Laughlin fractional quantum Hall states. A theory of wider applicability has to take into account the different character of these more general states. We have already discussed the construction of these bulk states in Chapter 14. We will see below that more general fractional quantum Hall states require that more degrees of freedom be included in the description of the edge. Thus, in addition to the edge degree of freedom we used to describe the Laughlin states, which we will call the charge mode, we will need to include in the description one or more neutral modes. While the physical origin of the neutral modes is transparent in the simple edges of multi-component fluids, its appearance in the case of one-component fluids is connected with the hierarchical descriptions of the bulk states. With variants, this is true in all the theories of the edge states.

Two simple and interesting examples of states for which the simple edge description fails are the Jain states at filling fractions $\nu = 2/5$ and $\nu = 2/3$. As in the preceding section, here we will follow Wen's approach (Wen, 1995), which is based on the Haldane–Halperin hierarchy. An alternative picture has been proposed by López and Fradkin (1999, 2001).

We will begin with the $\nu = 2/5$ case. In the Haldane–Halperin hierarchy construction, the $\nu = 2/5$ state is described as a fractional quantum Hall condensate of quasiparticles of the $\nu = 1/3$ Laughlin state. The $\nu = 2/5$ state has a larger electronic density than that of the "parent" $\nu = 1/3$ Laughlin state. Suppose we have a 2DEG with an overall density close to what is needed for the $\nu = 2/5$ state to be the ground state. Although deep in the bulk the density is constant, as the edge is approached the effects of the rising electrostatic potential that keeps the 2DEG inside the sample become more pronounced. In particular, the potential gives rise

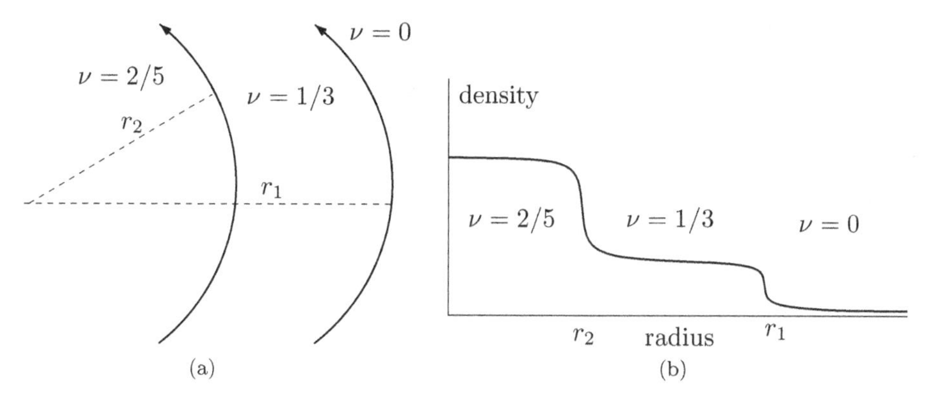

Figure 15.6 (a) Edge and bulk structure of the $\nu = 2/5$ state with its co-propagating chiral edge states. (b) The density profile of the $\nu = 2/5$ state.

to a lowering (as gradual as possible) of the electron density, so much so that in some outlying areas of the sample the density is that of the parent $\nu = 1/3$ state. Thus, the system tends to "phase separate" into an interior region with a filling fraction of $\nu = 2/5$ and an exterior region with filling fraction $\nu = 1/3$.

For a rotationally invariant system we thus get a situation such as the one depicted in Fig. 15.6. The 2DEG acquires a layered structure, with an inner droplet (of radius r_2) containing the $\nu = 2/5$ state, an annular region (between r_2 and r_1) with a $\nu = 1/3$ state, and the exterior of the sample ($r > r_1$) labeled by $\nu = 0$. We will picture the $\nu = 2/5$ state as a $\nu = 1/15$ condensate on top of a $\nu = 1/3$ state. For $|r_1 - r_2| \gg \ell_0$ the regions of the sample occupied by the bulk $\nu = 1/3$ and $\nu = 2/5$ states are macroscopic in size and can be regarded as being in the thermodynamic limit. The "interfaces" between the $\nu = 2/5$ and $\mu = 1/3$ regions and between the $\nu = 1/3$ region and the exterior of the sample ("$\nu = 0$") are the edge states. We will assume that the radial width of each interface (edge) is of the order of the magnetic length, ℓ_0, and hence infinitesimally small compared with the radii r_1 and r_2 as well as with $|r_1 - r_2|$. Since the 2DEG in each region is in an incompressible state, only the fluctuations of the charge densities in the edges (interfaces) have low-energy excitations (as in the Laughlin states).

The $\nu = 2/3$ fractional quantum Hall state also has a "layered" structure. In the Haldane–Halperin hierarchy the $\nu = 2/3$ state is a $\nu = 1/3$ Laughlin condensate of holes of the integer quantum Hall state at $\nu = 1$, i.e. it is the particle–hole version of the $\nu = 1/3$ Laughlin state. For a fixed orientation of the perpendicular magnetic field B, this picture is possible if the outer region is the "parent" $\nu = 1$ state and the inner region is viewed as a $\nu = -1/3$ state, a Laughlin state of holes. This leads to a density profile and edge structure for $\nu = 2/3$ as shown

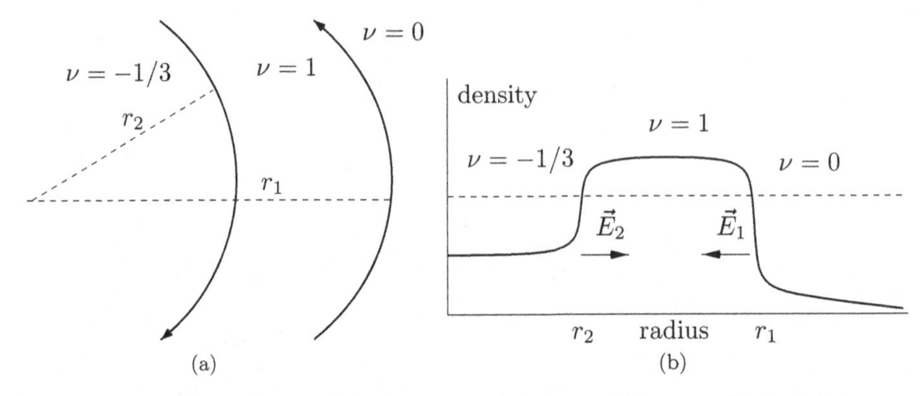

Figure 15.7 (a) Edge and bulk structure of the $\nu = 2/3$ state with its counter-propagating chiral edge states. (b) The density profile of the $\nu = 2/3$ state. The bulk $\nu = 2/3$ state is being described as a $\nu = -1/3$ Laughlin state of holes of the full Landau level.

in Fig. 15.7, with a density that has a non-monotonic dependence on the radial coordinate.

The main (and essentially only) difference between the $\nu = 2/3$ case and the $\nu = 2/5$ state is that we now must have two counter-propagating modes. Thus, in the $\nu = 2/3$ case, the effective electric fields have opposite signs, $\text{sgn}\, E_1^* = -\text{sgn}\, E_2^*$ (to accommodate the non-monotonic radial dependence of the density), and the edge modes have also velocities with opposite signs, $\text{sgn}\, v_1 = -\text{sgn}\, v_2$. Keeping in mind this important difference, we can give a similar description for both (and many other) states.

Let $\rho_{I,k}$ (with $I = 1, 2$) describe the Fourier modes of momentum k of the density fluctuations of the two edges,

$$\rho_{I,k} = \frac{\nu_I}{2\pi \ell_0^2} h_I \tag{15.58}$$

where h_I is the local displacement of each edge. For each edge these Fourier modes obey a U(1) Kac–Moody algebra of the form (with $I, J = 1, 2$)

$$\left[\rho_{I,k}, \rho_{J,k'}\right] = \frac{\nu_I}{2\pi} k \delta_{IJ} \delta_{k+k',0} \tag{15.59}$$

If the edges are far apart no inter-edge interactions are possible. In this limit, the Hamiltonian of each edge is

$$H = 2\pi \sum_{I=1,2} \sum_{k>0} \frac{v_I}{\nu_I} \rho_{I,k} \rho_{I,-k} \tag{15.60}$$

where v_I are the velocities of the two edge modes. For the case of $\nu = 2/5$, the effective electrostatic field E_I^* that develops at each interface that separates two incompressible bulk states has the same direction (inwards) as the external electric field which keeps the 2DEG inside the sample. Hence, the two edge modes propagate in the same direction, and the signs of the two velocities are the same, $v_1 > 0$ and $v_2 > 0$, with $v_I = E_I^* c/B$. The spectrum of the Hamiltonian Eq. (15.60) is positive (bounded from below) if $v_I/\nu_I > 0$, which requires that $v_I > 0$. For $\nu = 2/3$ the edge velocities have opposite signs. The Hamiltonian is nevertheless positive, since in this description the inner region has a negative filling factor, $\nu = -1/3$. In what follows we will assume that all intra-edge interactions have been taken into account in the magnitude of the renormalized velocities.

We can now use abelian bosonization and introduce two chiral bosons (scalar fields) ϕ_I (again with $I = 1, 2$) to parametrize the edge fluctuations,

$$\rho_I(x) = \frac{1}{2\pi}\, \partial_x \phi_I(x) \tag{15.61}$$

and to define an electron operator $\psi_{e,I}(x)$ for *each* edge I,

$$\psi_{e,I}(x) \sim e^{\frac{i}{\nu_I}\phi_I(x)} \tag{15.62}$$

The electron propagators of these so-far decoupled edges are

$$\left\langle T\psi_{e,I}(x,t)\psi^\dagger_{e,I}(0,0)\right\rangle = \frac{e^{ik_I x}}{(x - v_I t)^{1/|\nu_I|}} \tag{15.63}$$

The electron states on each edge have a characteristic momentum, the Fermi wave vector $k_I = r_I/(2\ell_0^2)$.

Let us now consider the effects of possible inter-edge interactions. As the edges approach each other, while keeping the inter-edge distance large enough for it to make sense for the intervening region to be regarded as a bulk state, density–density interactions (Coulomb for long-range forces) become increasingly important. Here we will consider for simplicity the case of short-range translation-invariant interactions whose strength is parametrized by the positive-definite and symmetric matrix of coupling constants V_{IJ},

$$H_{\text{inter-edge}} = 2\pi \sum_{I,J} \sum_{k>0} V_{IJ}\, \rho_{I,k}\, \rho_{J,-k} \tag{15.64}$$

which, using the language of the Luttinger model discussed in Chapter 6, describes only forward-scattering processes. It turns out that for a translationally invariant system no other inter-edge interactions are allowed. We may wonder whether this system may have processes that open energy gaps. In the case of co-propagating edges backscattering processes are simply forbidden, since the states have the same chirality. On the other hand, in the case of counter-propagating states, although backscattering processes can now exist, processes that can open energy gaps are forbidden by momentum conservation. For the same reason electron-tunneling (and quasiparticle-tunneling) processes are not allowed either. The situation becomes more involved (and richer) in the presence of disorder (Kane *et al.*, 1994). So, up to irrelevant operators, the only allowed interactions have the Luttinger form of Eq. (15.64). An exception to this rule is the case of two 2DEGs separated by a barrier. In this case there is a value of the electron density (the Fermi energy) for which the two oppositely propagating edge states have zero momentum and a gap may open up.

We have thus reduced the problem to a system of coupled chiral Luttinger liquids. In Chapter 6 we discussed the theory of the 1D Luttinger fluids. The problem of coupled edge states is a version of that problem and is solved by the same methods. We will not reproduce these calculations here because they are similar to the Bogoliubov transformations that we used in Chapter 6 (details can be found in Wen (1995)). What matters to us is the main result of these calculations, namely that the only effect of inter-edge interactions is a renormalization of the velocities of the edge modes. However, neither the number of edge states nor their chiralities can be changed by the interactions. We will see in the next section that the chirality of the edge modes has a topological origin. Clearly, this description is generic and can be applied to more diverse systems such as bilayers and partially polarized states as well (among many others).

15.4 The bulk–edge correspondence

We will now discuss in detail the connection between the edge states and the properties of the bulk quantum Hall fluid. We will see that there is a one-to-one correspondence between the universal properties of the bulk incompressible fluid and the physics of its edge states (Wen, 1990a, 1991a). In Chapters 13 and 14 we presented in some detail a theory of the bulk fractional quantum Hall states. There we saw that the robust properties (i.e. those which are independent of microscopic details) of the incompressible fluid are describable in terms of an effective-field theory that is a Chern–Simons gauge theory. Furthermore, the Chern–Simons gauge theory is a topological field theory. Its observables are Wilson loop operators carrying the quantum numbers of the representations of the gauge group, whose expectation values are topological invariants. Thus the bulk fractional quantum Hall states are topological fluids. On the other hand, in the preceding section we saw that the edge states of these incompressible topological fluids in open geometries are chiral and scale-invariant 1D systems. In other terms, the edge states are 1D chiral quantum critical systems. As such they are not only scale-invariant but also conformally invariant. We will now see that these universal descriptions of the bulk and the edge are two sides of the same coin.

We will begin by considering a generic effective theory of an abelian fractional quantum Hall state of a possibly multi-component fluid. Let $\mathcal{A}_{\mu I}$ be a set of n

U(1) fields of a hydrodynamic description of a fractional quantum Hall state (as discussed in Chapter 13) with the effective Lagrangian density (Wen and Zee, 1992)

$$\mathcal{L} = -\frac{1}{4\pi} K_{IJ} \epsilon^{\mu\nu\lambda} \mathcal{A}_{\mu I}\, \partial_\nu \mathcal{A}_{\lambda J} - \frac{e}{2\pi} t_I \epsilon^{\mu\nu\lambda} A_{\mu I}\, \partial_\nu \mathcal{A}_{\lambda J} \tag{15.65}$$

where $\vec{t} = (1, \ldots, 1)$ is the charge vector, A_μ is an external electromagnetic field, and K_{IJ} is the symmetric K-matrix. Invariance under large gauge transformations for systems on closed manifolds (such as a 2-torus) requires that all the entries of the K-matrix be integers. However, for a fermionic fluid the diagonal matrix elements of the K-matrix are odd integers. The number of condensates p in the fluid is equal to the rank of the K-matrix, which is equal to the number of quasiparticle excitations.

The quasiparticles are vortices of the condensates. A generic quasiparticle is labeled by k integers l_I (with $I = 1, \ldots, k$) and enters into the effective field theory through a source term

$$\mathcal{L}_{\text{qp}} = l_I \mathcal{A}_{\mu I} j^\mu_{\text{qp}} \tag{15.66}$$

where the currents $j^\mu_{\text{qp}} = (j^0_{\text{qp}}, \vec{j}_{\text{qp}})$ represent the worldlines of the gapped quasiparticles,

$$j^0_{\text{qp}} = \delta(\vec{x} - \vec{x}_0(t)), \qquad \vec{j}_{\text{qp}} = \hat{v}\delta(\vec{x} - \vec{x}_0(t)) \tag{15.67}$$

where $\hat{v}$ is the unit vector tangent to the quasiparticle trajectory.

The density and current of the Ith condensate are, by definition,

$$J_{\mu I} = \frac{1}{2\pi} \epsilon_{\mu\nu\lambda}\, \partial^\nu \mathcal{A}^\lambda_I \tag{15.68}$$

As the quasiparticle $\psi_{\vec{l}}$ is created it induces a change in the charge and currents of the condensates. The total induced charge is

$$Q[\vec{l}] = -e t_I \int d^2x\, \delta J^0_I = \int d^2x\, l_J K^{-1}_{JI} j^0_{\text{qp}}(x) = -e l_I K^{-1}_{IJ} t_J \tag{15.69}$$

Similarly, the statistical angle $\delta_{\vec{l}}$ is

$$\delta[\vec{l}] = \pi l_I K^{-1}_{IJ} l_J \tag{15.70}$$

In this effective theory there is always at least one quasiparticle $\psi_{\vec{l}_e}$ (but often several) whose quantum numbers are

$$l_{eI} = K_{IJ} L_J, \qquad \sum_I L_I = 1 \tag{15.71}$$

(where L_I are integers) that can be identified as an electron. Such quasiparticles are electrons since (a) they have charge $-e$, (b) they are fermions, (c) they are local with respect to all other quasiparticles (have trivial phase factors with them), and (d) these are the only excitations which satisfy these conditions.

15.4.1 The correspondence for the $U(1)_m$ Chern–Simons theory

In order to make a connection between the bulk effective-field theory and the edge states, we need to consider the effects of edges (or boundaries). We will discuss this problem first for the Laughlin states. Thus we will consider a Laughlin fractional quantum Hall fluid represented by an effective U(1) Chern–Simons gauge theory on a region Ω of the plane with a boundary (see Fig. 15.8). The total 3-manifold is then $\Omega \times \mathbb{R}$, where $\mathbb{R}$ is the time evolution.

The action for a $\nu = 1/m$ Laughlin fractional quantum Hall state with a boundary is

$$S = \frac{m}{4\pi} \int_{\Omega\times\mathbb{R}} d^3x\ \epsilon_{\mu\nu\lambda} \mathcal{A}^{\mu}\ \partial^{\nu} \mathcal{A}^{\lambda} \tag{15.72}$$

Under a gauge transformation $\mathcal{A}_{\mu} \mapsto \mathcal{A}_{\mu} + \partial_{\mu} f$ the Chern–Simons action changes by $S \mapsto S + \Delta S$, where

$$\begin{aligned} \Delta S &= \frac{m}{4\pi} \int_{\Omega} d^2x\ \epsilon_{\mu\nu\lambda}\ \partial^{\mu} f\ \partial^{\nu} \mathcal{A}^{\lambda} \\ &= \frac{m}{4\pi} \int dx\, dt\ f(x, y, t)(\partial_t \mathcal{A}_1 - \partial_x \mathcal{A}_0)\Big|_{y=0} \end{aligned} \tag{15.73}$$

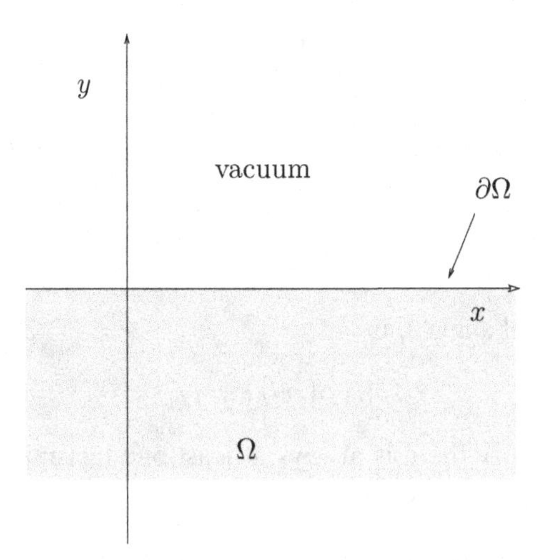

Figure 15.8 The Chern–Simons theory is defined on a 3-manifold $\Omega \times \mathbb{R}$, where Ω has the topology of a disk and has a boundary $\partial\Omega$. The shaded area Ω represents the region occupied by the fractional quantum Hall fluid.

To maintain gauge invariance we must require the gauge transformations to vanish at the boundary,

$$f(x, y = 0, t) = 0 \tag{15.74}$$

We know that the states at the edge propagate at a velocity that is fixed by the properties of the edge itself. However, the Chern–Simons theory is an effective long-distance theory and as such has no knowledge of the properties of the edge. There are several ways to incorporate this physics. One way is to construct an effective theory that takes into account the fact that the excitations are actually gapless along the edge (López and Fradkin, 1999). There is another, perhaps more formal, but "more economical," approach to this problem that is due to Wen (1995). It consists in realizing that a topological field theory such as the Chern–Simons theory, in addition to being locally gauge-invariant, is also invariant under local coordinate transformations and is independent of the metric of the manifold on which it is defined. Thus, in order for the theory to be completely defined on a manifold Ω with a boundary, in addition to imposing a gauge-fixing condition (as in all gauge theories) it is also necessary to specify the metric along the boundary. In fact, many gauge-fixing conditions also break general coordinate invariance and serve to specify a metric as well. The same problem arises in defining the wave function in Chern–Simons theory. It is not a gauge-invariant object (although the overlaps are), and depends on the choice of metric (Witten, 1992).

We will follow Wen's approach and impose the generalized axial gauge-fixing condition (everywhere, not just along the edge)

$$\mathcal{A}_t + v\mathcal{A}_x = 0 \tag{15.75}$$

where $\mathcal{A}_x$ is the component of the gauge field $\mathcal{A}_\mu$ tangent to the edge (the boundary of Ω), and v is an arbitrary parameter with the dimensions of a velocity. Under a change of coordinates

$$x + vt = \bar{x}, \qquad t = \bar{t}, \qquad y = \bar{y} \tag{15.76}$$

the gauge fields transform to

$$\bar{\mathcal{A}}_{\bar{t}} = \mathcal{A}_t - v\mathcal{A}_x, \qquad \bar{\mathcal{A}}_{\bar{x}} = \mathcal{A}_x, \qquad \bar{\mathcal{A}}_{\bar{y}} = \mathcal{A}_y \tag{15.77}$$

Under this coordinate transformation the gauge-fixing condition simply becomes the temporal gauge condition in the new coordinates,

$$\bar{\mathcal{A}}_{\bar{t}} = 0 \tag{15.78}$$

However, the Chern–Simons action does not change under the coordinate transformation (since it satisfies general coordinate invariance),

$$S = \frac{m}{4\pi} \int_{\Omega\times\mathbb{R}} d^3x\, \epsilon_{\mu\nu\lambda} \mathcal{A}^\mu\, \partial^\nu \mathcal{A}^\lambda = \frac{m}{4\pi} \int_{\Omega\times\mathbb{R}} d^3x\, \epsilon_{\mu\nu\lambda} \bar{\mathcal{A}}^\mu\, \partial^\nu \bar{\mathcal{A}}^\lambda \tag{15.79}$$

The quantization of Chern–Simons theory in the temporal gauge, $\bar{\mathcal{A}}_{\bar{t}} = 0$, is very simple (see Section 10.8). In this gauge the Gauss-law condition becomes the constraint that the allowed states have no gauge flux,

$$\partial_{\bar{x}} \bar{\mathcal{A}}_{\bar{y}} - \partial_{\bar{y}} \bar{\mathcal{A}}_{\bar{x}} = 0 \tag{15.80}$$

which is solved by the flat (pure-gauge) configurations

$$\bar{\mathcal{A}}_{\bar{x}} = \partial_{\bar{x}} \phi, \qquad \bar{\mathcal{A}}_{\bar{y}} = \partial_{\bar{y}} \phi \tag{15.81}$$

On a manifold with a boundary the Chern–Simons action of a pure gauge configuration, Eq. (15.81), does not vanish but is a total derivative and integrates to the boundary,

$$S = \frac{m}{4\pi} \int d\bar{x}\, d\bar{t}\, \partial_{\bar{x}}\phi\, \partial_{\bar{t}}\phi = \frac{m}{4\pi} \int dx\, dt \big(\partial_t \phi\, \partial_x \phi - v (\partial_x \phi)^2\big) \tag{15.82}$$

which is precisely the action of the chiral boson theory. Thus, the degrees of freedom of the gauge theory at the edge became the physical degrees of freedom.

We can now apply canonical quantization to this system and find that the canonical momentum $\Pi(x)$ of the chiral boson is

$$\Pi(x) = \frac{\delta S}{\partial_t \phi} = \frac{m}{4\pi}\, \partial_x \phi \tag{15.83}$$

Then, after demanding that the field and the canonical momentum obey canonical equal-time commutation relations, which for this chiral system are

$$\big[\phi(x), \Pi(y)\big] = \frac{i}{2} \delta(x - y) \tag{15.84}$$

we find that the chiral boson field does not commute with itself at equal times,

$$\big[\phi(x), \phi(y)\big] = i \frac{\pi}{m} \operatorname{sgn}(x - y) \tag{15.85}$$

With $L = \int dx\, \mathcal{L}$ being the Lagrangian, the Hamiltonian of this system is

$$H = \int \Pi\, \partial_t \phi - L = \frac{m}{4\pi} v \int dx (\partial_x \phi)^2 \tag{15.86}$$

which is positive definite for $mv > 0$. Hence, the sign of the velocity, the chirality, is determined by the sign of the Chern–Simons term in the bulk.

To find a connection with physical observables such as the density, we need to couple this theory to the electromagnetic field. This is done through the (gauge-invariant) source term

$$\int_{\Omega\times\mathbb{R}} d^3x\, A_\mu J^\mu = \frac{1}{2\pi}\int_{\Omega\times\mathbb{R}} d^3x\, \epsilon_{\mu\nu\lambda} A^\mu\, \partial^\nu \mathcal{A}^\lambda = \frac{1}{2\pi}\int_{\Omega\times\mathbb{R}} d^3x\, \epsilon_{\mu\nu\lambda} \mathcal{A}^\mu\, \partial^\nu A^\lambda \tag{15.87}$$

Let A_μ be independent of y and also set $A_y = A_{\bar{y}} = 0$. Since $\bar{\mathcal{A}}_\mu$ is a pure gauge, the source term becomes

$$\begin{aligned}\frac{1}{2\pi}\int_{\Omega\times\mathbb{R}} d^3x\, \epsilon_{\mu\nu\lambda}\mathcal{A}^\mu\, \partial^\nu A^\lambda &= -\int d\bar{x}\, d\bar{y}\, d\bar{t}\, \frac{1}{2\pi}\, \partial_{\bar{y}}\phi(\partial_{\bar{x}}A_{\bar{t}} - \partial_{\bar{t}}A_{\bar{x}}) \\ &= \int dx\, dt\, \frac{1}{2\pi}(A_t - vA_x)\partial_x\phi)\bigg|_{y=0}\end{aligned} \tag{15.88}$$

This result allows us to identify the physical edge charge density $\rho(x)$ and current $j(x)$ as

$$\rho(x) = \frac{\delta S}{\delta A_t} = \frac{1}{2\pi}\,\partial_x\phi, \qquad j(x) = \frac{\delta S}{\delta A_x} = -\frac{v}{2\pi}\,\partial_x\phi \tag{15.89}$$

Hence, as expected, for the chiral edge the density and the current are (essentially) the same physical observable.

The observables of the bulk Chern–Simons theory on a manifold without boundaries are Wilson loops on closed curves. On a manifold with a boundary, such as $\Omega\times\mathbb{R}$ that we are discussing here, Wilson loops on open paths are allowed, provided that they are defined on arcs $\Gamma(x, y)$ that begin and end on points (x and y) of the boundary (with $n \in \mathbb{Z}$)

$$W_n[\Gamma(x, y)] = \left\langle P \exp\left(in \int_{\Gamma(x,y)} dz_\mu\, \mathcal{A}^\mu \right)\right\rangle \tag{15.90}$$

With the definitions on gauge fixing that we used above, the Wilson arc operators become gauge-invariant and are physical observables. Using the condition that the gauge fields are pure gauge configurations, $\mathcal{A}_\mu = \partial_\mu\phi$, we obtain the result that the Wilson arc is the correlator for the chiral vertex operator $V_n = e^{-in\phi}$,

$$W_n[\Gamma(x, y)] = \left\langle T e^{-in\phi(x)} e^{\,in\phi(y)}\right\rangle \tag{15.91}$$

Thus, the correlators of quasiparticle operators of the edge states are expectation values of Wilson arcs. This result also makes manifest the relation between the compactification radius of the chiral boson and the topological properties of the bulk Chern–Simons gauge theory. See Fig. 15.9.

In summary, what we showed is that there is a one-to-one correspondence between the abelian $U(1)_m$ Chern–Simons gauge theory in the bulk and the RCFT

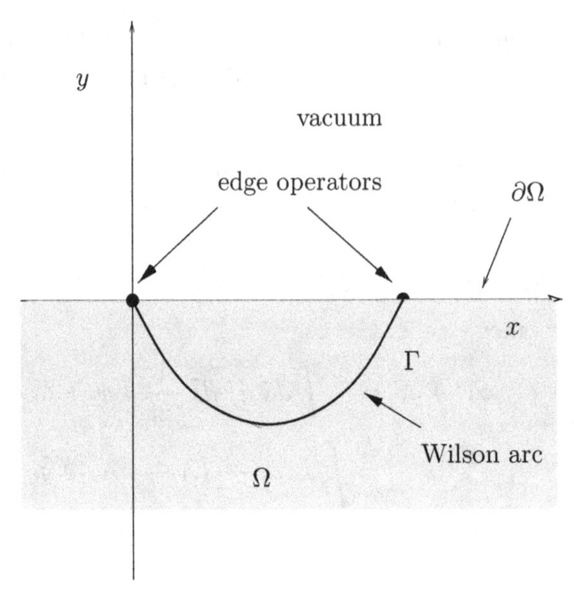

Figure 15.9 The Wilson arc in $\Omega \times \mathbb{R}$ with endpoints, on the boundary $\partial\Omega$, where the edge operators act (see the text).

of a compactified boson also at level m. Here too, by a rescaling of the field $\phi \mapsto \phi/\sqrt{m}$, we can trade the level m of the chiral boson for a change in the compactification radius from $R = 1$ to $R = \sqrt{m}$.

15.4.2 The general abelian Chern–Simons theory

Let us now describe how the bulk–edge correspondence works for a general abelian case. The effective Lagrangian for the general case is a multi-component Chern–Simons gauge theory, given in Eq. (15.65). It has a $\mathrm{U}(1)^p$ gauge invariance, where p is the rank of the K-matrix. The only difference is that we now need to impose p gauge-fixing conditions. Once again, on a manifold such as $\Omega \times \mathbb{R}$ this system is equivalent to a theory of k chiral bosons defined at the edge with the effective action

$$S_{\text{edge}} = \frac{1}{4\pi} \int dx\, dt [K_{IJ}\, \partial_t \phi_I\, \partial_x \phi_J - V_{IJ}\, \partial_x \phi_I\, \partial_x \phi_J] \tag{15.92}$$

with a Hamiltonian for the edge states

$$H = \frac{1}{4\pi} \int dx\; V_{IJ}\, \partial_x \phi_I\, \partial_x \phi_J \tag{15.93}$$

where the matrix V_{IJ} is positive definite. In this effective theory the eigenmodes of the K-matrix with positive eigenvalue are left-moving states and the eigenmodes with negative eigenvalue are right-moving states.

For example, the $\nu = 2/5$ state has a K-matrix and charge vector $\vec{t}$

$$K_{2/5} = \begin{pmatrix} 3 & 2 \\ 2 & 3 \end{pmatrix}, \qquad \vec{t} = \begin{pmatrix} 1 \\ 1 \end{pmatrix} \tag{15.94}$$

This K-matrix has two positive eigenvalues, 5 for the charge mode and 1 for the neutral mode. Thus both the charge and the neutral mode are left-moving.

On the other hand, the $\nu = 2/3$ state has

$$K_{2/3} = \begin{pmatrix} 1 & 0 \\ 0 & -3 \end{pmatrix}, \qquad \vec{t} = \begin{pmatrix} 1 \\ 1 \end{pmatrix} \tag{15.95}$$

Thus, in this case there are two oppositely moving branches of edge modes.

Armed with these results we can now give a full description of the universal properties of the observables of the edge states in the multi-component case. The Hilbert space of the edge excitations consists of representations of the Kac–Moody algebra for a theory with p components,

$$\left[\rho_{I,k}, \rho_{J,k'}\right] = i K^{-1}_{IJ} \frac{1}{2\pi} k \delta_{k+k',0} \tag{15.96}$$

where $k, k' = 2\pi/L \times$ integer, $I, J = 1, \ldots, p$ (the rank of the K-matrix), and

$$\rho_I = \frac{1}{2\pi} \partial_x \phi_I \tag{15.97}$$

Here we are ignoring several interesting cases, such as the (n, n, n) bilayer states, in which the K-matrix is singular and hence is not invertible. After a linear transformation that brings the K-matrix into a block-diagonal form, these cases can be treated similarly.

The total electronic density $\rho_{\mathrm{e}} = -e \sum_I \rho_I$. The Hamiltonian takes the chiral Luttinger form

$$H = 2\pi \sum_{I,J} \sum_{k>0} V_{IJ} \rho_{I,k} \rho_{J,-k} \tag{15.98}$$

The quasiparticle operators for this case can also be constructed from Wilson arcs. The only difference is that each Wilson arc will now carry a quantum number l_I telling us how this quasiparticle couples with the Ith gauge field. Let us denote by $\psi_{\vec{l}}$, with

$$\psi_{\vec{l}} \sim \exp\left(i \sum_I l_I \phi_I \right) \tag{15.99}$$

the quasiparticle operators that we obtained in this way. Using the Kac–Moody algebra, we find that

$$\left[\rho_I(x), \psi_{\vec{l}}(x')\right] = l_J K^{-1}_{JI} \psi_{\vec{l}}(x') \delta(x - x') \tag{15.100}$$

which says that $\psi_{\vec{l}}$ is a local operator that creates a quasiparticle of charge

$$Q_{\vec{l}} = -e \sum_J l_J K_{JI}^{-1} \tag{15.101}$$

For a theory with a K-matrix whose diagonal matrix elements are odd integers, the electron operator is obtained in this way if $l_I = \sum_J K_{IJ} L_J$, where L_J are integers such that $\sum_I L_I = 1$. Clearly, more than one such electron operator can be constructed in this way and they are all equally physical.

Let us finally compute the propagators of these quasiparticles. The K-matrix and the V-matrix can be diagonalized simultaneously. To this end, we will perform a unitary transformation, U_{IJ}, to bring the Hamiltonian to a diagonal form. Under this transformation the densities mix $\tilde{\rho}_I = \sum_J U_{IJ} \rho_J$. The new density operators satisfy the commutation relations

$$\left[\tilde{\rho}_{I,k}, \tilde{\rho}_{J,k'}\right] = i \frac{s_I}{2\pi} \delta_{IJ} k \delta_{k+k',0} \tag{15.102}$$

where s_I is the sign of the Ith eigenvalue of the K-matrix. The diagonalized Hamiltonian reads

$$H = 2\pi \sum_I |v_I| \tilde{\rho}_{I,k} \tilde{\rho}_{I,-k} \tag{15.103}$$

where $v_I = s_I |v_I|$ are the (renormalized) velocities of the modes. The chiral bosons have also rotated to the fields $\tilde{\phi}_I$, and we write the quasiparticle operators in the new basis

$$\psi_{\vec{l}} = \exp\left(i \sum_I \tilde{l}_I \tilde{\phi}_I \right), \qquad \tilde{l}_I = \sum_J l_J U_{JI}^{-1} \tag{15.104}$$

We find that the propagators of the quasiparticles are

$$\left\langle T \psi_{\vec{l}}^{\dagger}(x,t) \psi_{\vec{l}}(0,0) \right\rangle \sim \exp\left(i \sum_I k_I l_I x \right) \prod_I (x - v_I t + i s_I \epsilon)^{-\tilde{l}_I^2} \tag{15.105}$$

which leads to the result that the statistical angles are, as before,

$$\delta[\vec{l}] = \pi \sum_{I,J} l_I K_{IJ}^{-1} l_J \tag{15.106}$$

15.4.3 The non-abelian Chern–Simons theory

In Chapter 14 we discussed the non-abelian quantum Hall states. There we explained that there is a close connection between their wave functions and RCFTs associated with the chiral version of the Wess–Zumino–Witten (WZW) model discussed in Chapter 7. We will see here that there is a one-to-one correspondence between the chiral WZW model, a (1 + 1)-dimensional RCFT, and the (2 + 1)-dimensional non-abelian Chern–Simons gauge theory (Elitzur *et al.*, 1989). In another section we will use this connection to guess the form of the effective-field theory in the bulk for the non-abelian fractional quantum Hall states.

The action of the non-abelian Chern–Simons theory with gauge group $SU(N)$ at level k on a 3-manifold $\Sigma \times \mathbb{R}$ is (see Section 10.4)

$$kS_{CS}[A_\mu] = \frac{k}{4\pi} \int_{\Sigma \times \mathbb{R}} d^3x \, \epsilon_{\mu\nu\lambda} \left[A_a^\mu \, \partial^\nu A_a^\lambda + \frac{2}{3} f_{abc} A_a^\mu A_b^\nu A_c^\lambda \right] \tag{15.107}$$

As in the case of its abelian counterpart, provided that the level k is quantized to be an integer, this theory is gauge-invariant on a closed 3-manifold (see the discussion at the end of Section 10.4).

Also as in the abelian case, this theory is not gauge-invariant if the manifold has a boundary. This problem can be solved using the same procedure as that which we just used in the abelian theory. Thus, we will fix the gauge in the bulk by imposing Eq. (15.75) on the non-abelian gauge fields, which are matrices in the algebra of the gauge group, and requiring the gauge transformations to approach the identity element in the gauge group G on the boundary (Cabra *et al.*, 2000). In the non-abelian theory, as was the case in the abelian theory, the Gauss-law condition requires the vacuum states to be flat, i.e. to have zero non-abelian flux, $F_{ij} = 0$. This means that the gauge fields must be just gauge transformations and thus have the form (with $i = 1, 2$ being the two spatial components)

$$A_i = -(\partial_i g) g^{-1} \tag{15.108}$$

where $g(x)$ is a gauge transformation. Upon substitution into the Chern–Simons action in the gauge of Eq. (15.75), we find that the Chern–Simons action is once again a total derivative that integrates to the boundary of the manifold $\Omega \times \mathbb{R}$, where it becomes the action for the $SU(N)_k$ chiral WZW model in (1 + 1) dimensions,

$$\begin{aligned} S = {} & \frac{k}{16\pi} \int_{\partial\Omega \times \mathbb{R}} d^2x \, \mathrm{tr} \left[g^{-1} \, \partial_x g \; g^{-1} (\partial_t - v \, \partial_x) g \right] \\ & - \frac{k}{24\pi} \int_{\Omega \times \mathbb{R}} d^3x \, \epsilon_{\mu\nu\lambda} \, \mathrm{tr} \left(g^{-1} \, \partial_\mu g \; g^{-1} \, \partial_\nu g \; g^{-1} \, \partial_\lambda g \right) \end{aligned} \tag{15.109}$$

Here I use the same normalization of the trace as in Section 7.10, to be consistent with the conventions of Witten (1984, 1989).

The arguments in Section 15.4.1 on the bulk–edge correspondence for the $U(1)_m$ abelian Chern–Simons theory apply also, without any essential formal change, to the non-abelian Chern–Simons theory. Thus, a Wilson arc operator, carrying some representation λ of the gauge group, with endpoints on the edge manifold $\partial\Omega \times \mathbb{R}$, maps onto the correlator of two primary fields of the chiral WZW model on the edge carrying the same quantum numbers.

We discussed the properties of the CFT of the WZW model in Section 7.12. Although we will not repeat that analysis here, it is useful to recall the main results for the case of $SU(2)_k$, which, as we will now see, is of interest in the theory of fractional non-abelian quantum Hall states (Wen 1999; Read and Rezayi 1999; Fradkin *et al.* 1998, 1999).

The primary fields of $SU(2)_k$, $\Phi_{(j,m)}$, with $0 \leq j \leq k/2$ and with the same restrictions as in Section 7.12, can be factorized into holomorphic (or right-moving) and anti-holomorphic (left-moving) components. For a bulk Chern–Simons theory with $k > 0$, the edge theory is the chiral $SU(2)_k$ WZW CFT which is built from the holomorphic components of the fields. This theory has a chiral $SU(2)_k$ Kac–Moody algebra of three chiral (holomorphic) currents $J^a(z)$ that generate the spectrum physical states. With this caveat in mind, we can then use the result that the scaling dimensions for the representations (j, m) (in the allowed range) are $\Delta_{(j,m)} = j(j+1)/(k+2)$ (see Eq. (7.193)), and are independent of m (due to the SU(2) symmetry) (for more details see Section 7.12).

We will now discuss two simple cases, with $k = 1$ and $k = 2$, respectively, and see how they are related to bulk quantum Hall states.

15.4.4 The $SU(2)_1$ correspondence

The $SU(2)_1$ case is very simple. We have already encountered this case in our discussion of the spin-singlet Halperin states (in Section 14.7). This theory has three currents, all with scaling dimension 1. Since the level is $k = 1$, this theory has two representations: [0] the field $\Phi_{(0,0)}$ (the identity field I) and [1/2] with primary fields $\Phi_{(1/2,\pm 1/2)}$, each with scaling dimension 1/4. In Section 14.7 we showed that there is a simple abelian theory that is equivalent, namely the chiral boson ϕ with a special choice of compactification radius. Indeed the chiral current of the boson $(1/(2\pi))\partial_x\phi$ and the operators $\exp(\pm i\sqrt{2}\phi)$ have dimension 1 and form an $SU(2)_1$ Kac–Moody algebra. Thus the propagators of all three currents are

$$\langle T J^a(x,t) J^b(0,0)\rangle \sim \frac{1}{(x - vt)^2} \tag{15.110}$$

The only vertex operator which is local with respect to the currents is $V_{\pm 1/2} \sim \exp(\pm i\phi/\sqrt{2})$, which has scaling dimension 1/4 and carries spin $j = 1/2$. The propagator of this vertex operator is

$$\langle T V_{1/2}(x,t) V^{\dagger}_{1/2}(0,0)\rangle \sim \frac{1}{(x - vt)^{1/2}} \tag{15.111}$$

This is precisely the same result as we found for the quasiparticle propagator in the edge states of the *bosonic* Laughlin state at $\nu = 1/2$ with bulk wave function

$$\Psi_2(z_1, \ldots, z_N) \sim \prod_{i<j}(z_i - z_j)^2 \exp\left(-\frac{1}{4\ell_0^2}\sum_j |z_j|^2\right) \tag{15.112}$$

So we see that in this language the currents $J^\pm \sim \exp(\pm i\sqrt{2}\phi)$ are the *boson* (instead of fermion) operators and that the current J^3 is the charge current of the bosonic edge state.

From this analysis we see that the effective theory of the bulk $\nu = 1/2$ Laughlin fractional quantum Hall state (of bosons!) can be equivalently described by the abelian $U(1)_2$ Chern–Simons theory at level $m = 2$, or by the $SU(2)_1$ Chern–Simons theory. At the level of the edge theory, this states the equivalence of the theory of the compactified chiral boson with compactification radius $R = 1/\sqrt{2}$, the "SU(2) radius" (Ginsparg, 1989), and the $SU(2)_1$ chiral WZW CFT.

However, while these two theories have the same universal content, their equivalence in the bulk does not necessarily hold microscopically. Indeed, as we saw in a system with a one-component fluid, the $U(1)_2$ abelian Chern–Simons theory arises naturally as a description of the fractional quantum Hall state of bosons. On the other hand, although the SU(2) symmetry of the $SU(2)_1$ Chern–Simons theory could appear naturally in a system of "bosons" with spin 1/2, the equivalence to $U(1)_2$ has to be viewed as a dynamical symmetry (like the accidental O(4) symmetry in the quantum mechanics of the Kepler problem in the hydrogen atom). Thus, in general, in the bulk this seemingly larger symmetry would require some fine-tuning. In contrast, the equivalence at the level of the edge states is more robust (although it may require the fine-tuning of the velocities of different modes).

15.4.5 The $SU(2)_2$ correspondence and the Moore–Read edge states

We will now discuss the $SU(2)_2$ case. The $SU(2)_2$ chiral WZW CFT has central charge $c = 3/2$ (see Eq. (7.189)). This theory has three representations, [0], [1/2], and [1], with scaling dimensions 0, 3/16, and 1/2, respectively.

Let us discuss first the adjoint representation [1], which, as such, is a triplet of hermitian fields. The propagators are (with $m = 0, \pm 1$)

$$\langle T\Phi_{(1,m)}(x,t)\Phi_{(1,m)}(0,0)\rangle \sim \frac{1}{x - vt} \tag{15.113}$$

These three propagators are *antisymmetric* under $(x,t) \mapsto (-x,-t)$. Therefore the spin-1 primary fields $\Phi_{(1,m)}$ can be identified as three chiral Majorana fermions $\chi_a(x,t)$ $(a = 1,2,3)$. Indeed, a theory of three Majorana fermions generates the $SU(2)_2$ Kac–Moody current algebra with the currents

$$J^a \sim \chi_b T^a_{bc} \chi_c \tag{15.114}$$

where $T^a_{bc} = i\epsilon_{abc}$ are the SU(2) generators in the adjoint (spin-1) representation and ϵ_{abc} is the third-rank Levi-Civita tensor. In this language we can think of $SU(2)_2$ as being $SO(3)_1$, see Di Francesco *et al.* (1997).

The primary field $\Phi_{(1/2,\pm 1/2)}$ is the WZW field itself. It is the primary field in the representation [1/2]. It has scaling dimension $\Delta_{(1/2,\pm)} = 3/16$ and its propagator is given by

$$\langle T\Phi_{(1/2,\pm 1/2)}(x,t)\Phi^{\dagger}_{(1/2,\pm 1/2)}(0,0)\rangle \sim \frac{1}{(x-vt)^{3/8}} \tag{15.115}$$

which is not single-valued. In fact, under an exchange process this propagator acquires a phase factor $\delta = 3\pi/8$. In other terms, the primary fields are associated with particles (solitons) with statistical angle $3\pi/8$ that propagate along the edge. The fusion properties of these quasiparticles can be determined from the OPE that these fields satisfy,

$$\left[1/2\right]\star\left[1/2\right] = [0]+[1], \qquad \Leftrightarrow \qquad \Phi_{1/2}\star\Phi_{1/2} = \Phi_0+\Phi_1 \tag{15.116}$$

This is the edge version of the fact that there are two fusion channels in this case and hence that there are two conformal blocks, as we discussed in Section 14.8.

Is there a fractional quantum Hall bulk state with the same properties? The answer is yes, the $n=1$ Moore–Read state whose wave function for general n is

$$\Psi_{\rm MR}(z_1,\ldots,z_N) = \text{Pf}\left(\frac{1}{z_i-z_j}\right)\prod_{i<j}(z_i-z_j)^n \exp\left(-\frac{1}{4\ell_0^2}\sum_i |z_i|^2\right) \tag{15.117}$$

This wave function describes a fractional quantum Hall state of bosons at filling factor $\nu = 1/n$. In Section 14.8 we discussed the fact that the Pfaffian factor can be represented by a Majorana fermion, which here we will denote by χ_3. For $n=1$ the "Laughlin factor" is just a Vandermonde determinant, the wave function for a filled Landau level of fermions (with $\nu = 1$). We saw at the beginning of this chapter that the edge states of a $\nu = 1$ state are described by a theory of a free chiral Dirac (charged) fermion, which we will denote by $\psi(x)$. This suggests that in the case of the bosonic $n=1$ Moore–Read state the effective field theory in the bulk in the topological limit is the $SU(2)_2$ Chern–Simons gauge theory (Fradkin *et al.*, 1998). This also suggests that for $n>1$ the non-abelian properties of the Moore–Read states may also be connected in some way with the $SU(2)_2$ Chern–Simons in the bulk and WZW on the edge. We will now examine how (and whether) this guess is correct.

There is a simple way to describe the edge states for all Moore–Read states, which follows from the structure of the wave function itself (Milovanović and Read, 1996), and for most purposes it is the most efficient representation. Let us first bosonize the chiral Dirac fermion and use its representation by the vertex operator $V_1 \sim \exp(i\phi)$ of a compactified chiral boson ϕ with compactification radius $R = 1$ (the "fermionic radius" (Ginsparg, 1989)). Hence in this picture, which we saw in Eq. (14.136) is natural from the wave-function viewpoint, at the edge we have a charge mode (the charge boson ϕ, in this case with compactification radius $R = 1$) and a neutral mode represented by a chiral Majorana fermion χ. In fact, this argument holds for all $\nu = 1/n$ Moore–Read states, with the only change being that the level of the chiral boson in the general case is n (or that the compactification radius of the chiral boson is rescaled to $R = \sqrt{n}$). Therefore, in this representation the effective Lagrangian of the chiral edge states is the sum of two decoupled terms,

$$\mathcal{L} = \chi i(\partial_t - v_n\,\partial_x)\chi + \frac{n}{4\pi}\,\partial_x\phi(\partial_t - v_{\rm c}\,\partial_x)\phi \tag{15.118}$$

In Eq. (15.118) we have allowed the charge and the neutral modes to have different velocities. In this normalization of the charge boson, the charge current is $j = (1/(2\pi\sqrt{n}))\partial_x\phi$. Since the chiral Majorana fermion is the chiral half of the CFT of the critical Ising model, we say that $\mathbb{Z}_2 \times \mathrm{U}(1)_n$ is the CFT of the edge states of the $\nu = 1/n$ Moore–Read state.

The effective Lagrangian of Eq. (15.118) is a sum of two apparently decoupled terms. This Lagrangian suggests that the allowed primary fields of these edge states are products of the two sectors. This structure is reminiscent of the phenomenon of spin–charge separation which we discussed in the context of the theory of 1D Luttinger liquids (see Chapter 6). However, this similarity is only superficial. Indeed, while Eq. (15.118) is the correct Lagrangian for these edge states, we still have to impose the condition that the allowed primary fields are local with respect to the "electron" (i.e. the particles the quantum Hall fluid is made of). This selection rule, which reduces the number and type of allowed operators, tells us how to glue the charge and neutral sectors together. In this sense there is no separation between the $\mathrm{U}(1)_n$ charge sector and the $\mathbb{Z}_2$ neutral sector.

To see how this works, we first identify the operator $\psi_{\rm e} \sim \chi\exp(i\sqrt{n}\phi)$ with the "electron" operator. This operator has scaling dimension $\Delta_{\rm e} = n/2$ and charge $Q = e$. Clearly the Majorana fermion χ, with scaling dimension $\Delta_\chi = 1/2$ and charge $Q = 0$, is local with respect to the electron operator, and is an allowed primary field. Similarly the vertex operators $V_p \sim \exp(ip\phi/\sqrt{n})$ (with $p = 0, \ldots, n-1$), with scaling dimension $\Delta_p = p^2/(2n)$, statistical angle $\delta_p = \pi p/n$, and charge $Q = p/n$, satisfy all the requirements and are also allowed primary fields.

However, there are more allowed operators in addition to the ones we considered. Consider the primary field σ of the chiral sector of the Ising model. This operator has scaling dimension $1/16$ and charge $Q = 0$. However, this operator twists the Majorana fermion, thereby changing its boundary conditions from periodic to anti-periodic (and vice versa), and, for this reason, it is also called the twist field (Dixon *et al.*, 1987). This means that it is double-valued and is not local with respect to the electron (since it is not local with respect to the Majorana fermion). Similarly, the vertex operator $V_{1/2} \sim \exp(i\phi/2\sqrt{n})$ is not local with respect to the electron either; it is also a "branch-cut operator" and is double-valued. However, the composite operator $\psi_{\text{qp}} \sim \sigma V_{1/2}$ is local with respect to the electron operator, and it is an allowed primary field. This operator has scaling dimension $\Delta_{\text{qp}} = 1/16 + 1/(8n)$, statistical angle $\delta_{\text{qp}} = \pi(n+2)/(8n)$, and charge $Q = 1/(2n)$. In addition, given the Ising fusion rule $\sigma \star \sigma = I + \chi$, this primary field has two fusion channels, and thus it has non-abelian braiding statistics. Clearly, any primary field obtained from fusing any pair of operators of the types we considered will lead to another operator that should be on the list. This tower, however, truncates when the operators end up fusing into the electron operator. As relevant examples, in Tables 15.1 and 15.2 we give the list of primary fields (and their scaling dimensions) for the edge states of the $n = 1$ bosonic and $n = 2$ fermionic Moore–Read states, respectively.

The description of the edge states in terms of a $\mathbb{Z}_2 \times \text{U}(1)_n$ chiral CFT gives a simple and economical way to describe the edge states. However, while the U(1) charge sector has a simple representation in terms of an effective-field theory of the bulk state, in the form of the $\text{U}(1)_n$ Chern–Simons abelian gauge theory, it is far

Table 15.1 *The $SU(2)_2$ quantum numbers, scaling dimension Δ, and $U(1)$ charge Q of the primary fields of the edge states of the bosonic $n = 1$ Moore–Read fractional quantum Hall state*

	I	$\sigma e^{\frac{i}{2}\phi}$	$e^{\pm i\phi}$	χ	$\chi e^{\pm i\phi}$
(j, m)	$(0, 0)$	$(1/2, \pm 1/2)$	$(1, \pm 1)$	$(1, 0)$	current
Δ	0	3/16	1/2	1/2	1
Q	0	1/2	0	1	± 1

Table 15.2 *The scaling dimension Δ and charge Q (in units of e) of the primary fields of the edge states of the fermionic $n = 2$ Moore–Read fractional quantum Hall state*

	I	$\sigma e^{\frac{i}{2\sqrt{2}}\phi}$	χ	$e^{\frac{i}{\sqrt{2}}\phi}$	$\chi e^{i\sqrt{2}\phi}$
Δ	0	1/8	1/2	1/4	3/2
Q	0	1/4	0	1/2	1

from obvious what topological field theory in the bulk has a chiral edge Majorana fermion state and how the rules for gluing different representations at the edge manifest themselves in the bulk.

The bulk–edge correspondence between the non-abelian Chern–Simons theory and the chiral WZW model offers a hint on how to do this. This raises the question of how $SU(2)_2$ is related to this description of the edge states. To make this connection explicit, we return to the case of the $n = 1$ bosonic Moore–Read state. In this case, the "electron" operators $\chi \exp(\pm i\phi)$, which create and destroy a boson, have scaling dimension 1 and hence have the same dimension as a chiral current. Indeed, the three operators

$$J^{\pm} \sim \chi \exp(\pm i\phi), \qquad J^3 \sim \partial_x \phi \tag{15.119}$$

form an $SU(2)_2$ Kac–Moody current algebra. Thus, the "diagonal" (Cartan) generator J^3 is the charge density, and its integral is the total charge operator (in units of e)

$$Q = \int dx\ J^3(x) \propto \int dx\ \partial_x \phi \tag{15.120}$$

This means that the coupling to an external electromagnetic perturbation breaks the SU(2) symmetry down to U(1), since it couples only to the J^3 generator. In this sense SU(2) is an "accidental symmetry," however useful it may be.

We now recall that a Dirac fermion can be decomposed into two Majorana fermions, $\psi \sim \chi_2 + i\chi_3$, its "real" and "imaginary" parts, and we see that these are just the currents of Eq. (15.114). Hence, the edge states of the $\nu = 1$ bosonic Moore–Read fractional quantum Hall state constitute a theory with three chiral Majorana fermions χ_a (with $a = 1, 2, 3$) with Lagrangian

$$\mathcal{L} = \sum_{a=1}^{3} \chi_a i(\partial_t - v\,\partial_x)\chi_a \tag{15.121}$$

Since each free chiral Majorana fermion is a CFT with central charge $c = 1/2$ (as in the chiral sector of the Ising model), the central charge of this theory should be $c = 3/2$. We also know that the non-abelian quasiparticles of the $n = 1$ Moore–Read state have statistical angle $\delta = 3\pi/8$, see Eq. (14.151). This is consistent with what we just found from the analytic properties of the propagator of the WZW field in $SU(2)_2$. More significantly, the four-point function of four WZW primary fields has two conformal blocks (Knizhnik and Zamolodchikov, 1984), which is also consistent with these fields representing particles with non-abelian braiding statistics. Similarly, the Laughlin quasihole (and quasiparticle) operators, the vertex operators $V_{\pm 1} = \exp(\pm i\phi)$, in the $\nu = 1$ bosonic Moore–Read state, are Dirac *fermions* (i.e. made of two Majorana fermions) and have statistical angle $\delta = \pi$.

Thus we conclude that the universal properties of the $n = 1$ bosonic Moore–Read state are reproduced by an effective-field theory, the $SU(2)_2$ Chern–Simons theory, in the bulk and the $SU(2)_2$ chiral WZW model at the edge.

This line of argument does not precisely work for $n > 1$. The problem is that for $n > 1$ the compactification radius of the boson is no longer $R = 1$, and the electron operator is no longer an $SU(2)_2$ current (as it is for $n = 1$). Thus, the general Moore–Read state can be thought of as an $SU(2)_2$ theory in which the compactification radius of the diagonal (Cartan) U(1) subgroup has been changed. Naturally, this deformation breaks the SU(2) symmetry explicitly. There is, however, a way to make this deformation of $SU(2)_2$ explicit. The answer is to consider a theory with a coset CFT, which is constructed as follows (Gepner, 1987; Di Francesco *et al.*, 1997).

Let us consider a CFT with a Kac–Moody current algebra for a Lie group G at some level k, and let H be a subgroup of G, $H \subset G$, which is also a Lie group. Let J_G^a be the chiral currents that generate the Kac–Moody current algebra of G_k and let J_H^i be the currents that generate the Kac–Moody current algebra for the subgroup H. Let us now gauge the subgroup H by coupling its currents to a dynamical gauge field A_μ that takes values in the algebra of the subgroup H. In the strong-coupling limit of this $(1+1)$-dimensional gauge theory, the gauge fields do not have an action of their own, and act as Lagrange multiplier fields that set the currents that act on H to zero. In other words, a coset is a theory with a current algebra in which we have made a projection onto the subset of states of the Hilbert space which is annihilated by the currents J_H^i of the subgroup H,

$$J_H^i|\text{Phys}\rangle = 0 \tag{15.122}$$

Since the theory with current algebra in the group G at level k is the G_k WZW model, the coset is a gauged (chiral) WZW model (in which only the subgroup H has been gauged).

In the case of $SU(2)_k$, we can consider the U(1) diagonal subgroup, and the resulting coset is $SU(2)_k/U(1)$. This case was solved in great detail by Gepner (1987), who showed that this CFT is equivalent to the $\mathbb{Z}_k$ *parafermion* CFT. For $k = 2$ this is the Ising CFT and for $k = 3$ it is the CFT of the critical three-state Potts model. The solution is quite technical for the scope of this book. We will state the main results without giving the details of their derivation.

Since the coset is effectively a constrained version of the original system, the central charge of the CFT (which counts the number of critical degrees of freedom) of the coset G/H is smaller, $c(G/H) = c(G) - c(H)$. Since $H = U(1)$ and its central charge is $c(U(1)) = 1$, we have

$$c(SU(2)_k/U(1)) = \frac{2(k-1)}{k+2} \tag{15.123}$$

Furthermore, the scaling dimensions of the primary fields $\Phi_{(j,m)}$ are also reduced in $\mathrm{SU}(2)_k/\mathrm{U}(1)$, to

$$\Delta_{(j,m)} = \frac{j(j+1)}{k+2} - \frac{m^2}{k} \tag{15.124}$$

and now depend not only on j but also on m since the symmetry SU(2) has been broken explicitly by gauging its diagonal (Cartan) U(1) subgroup.

The simplest example of a coset theory is $\mathrm{SU}(2)_2/\mathrm{U}(1)$. We see that for $k = 2$ the central charge is $c(\mathrm{SU}(2)_2/\mathrm{U}(1)) = 1/2$. Also for $k = 2$ the only non-trivial primary fields of this theory are $\Phi_{(1/2,\pm 1/2)}$ and $\Phi_{(1,0)}$, with scaling dimensions $\Delta_{(1/2,1/2)} = 1/16$ and $\Delta_{(1,0)} = 1/2$. Thus we can identify the fields of the coset $\mathrm{SU}(2)_2/\mathrm{U}(1)$ theory with the fields of the Ising CFT, $\mathbb{Z}_2$, $\Phi_{(1/2,1/2)} \equiv \sigma$ (the twist field of the Ising model), and $\Phi_{(1,0)} \equiv \chi$ (the Majorana fermion). From these results, we conclude that the chiral edge states of the Moore–Read fractional quantum Hall states are described by the chiral CFTs

$$\mathbb{Z}_2 \times \mathrm{U}(1)_n \simeq (\mathrm{SU}(2)_2/\mathrm{U}(1)) \times \mathrm{U}(1)_n \tag{15.125}$$

where the coset $\mathrm{SU}(2)_2/\mathrm{U}(1)$ is a chiral gauged $\mathrm{SU}(2)_2$ WZW. The coset $\mathrm{SU}(2)_3/\mathrm{U}(1) \simeq \mathbb{Z}_3$, the CFT of the three-state Potts model, enters into the description of the edge states of the Read–Rezayi states (Read, 1998). Instead of a Majorana fermion, the theory of the Read–Rezayi states involves fields called parafermions, which we defined in Eq. (10.127). The properties of parafermion theories are understood from CFT, and they are never free fields (Dotsenko, 1984). One interesting feature of the Read–Rezayi states is that they are leading candidates to realize a universal quantum computer (Das Sarma *et al.*, 2008).

15.5 Effective-field theory of non-abelian states

The use of the bulk–edge correspondence combined with CFT methods enabled us to construct generalized fractional quantum Hall states and to investigate their properties. In the last section we constructed the theory of the edge states using the structure of the wave functions of the bulk states. We will see in the next sections that the CFT description of the edge states is sufficient (and efficient) to study the properties of these states. This is important, since many of the properties of the edge states are directly accessible to (very challenging!) experiments. While the bulk–edge correspondence suggests that there should be an effective topological field theory for all of these states, there is no straightforward non-abelian version of the hydrodynamic arguments that we used to formulate the abelian fractional quantum Hall states in terms of abelian Chern–Simons gauge theory in Chapter 14.

What is missing is an effective-field-theory approach to this problem that includes the non-abelian states as well. Such an approach was discussed in detail for the case of the abelian states, for which it provides a general construction, which amounts to a classification of such states (Wen, 1995). So far no general classification of non-abelian quantum Hall states has been established and, for this reason, there is no generic effective-field theory of non-abelian states either. Nevertheless, there has been some significant progress on this question (Fradkin *et al.*, 1998, 1999; Wen, 1999; Cabra *et al.*, 2000; Barkeshli and Wen, 2010a, b).

The structure of the CFTs of their edge states naturally suggests that the effective-field theory for the bulk must somehow involve an $SU(2)_k$ Chern–Simons gauge theory at some non-trivial level k. However, we are now left to explain the physical origin of the SU(2) symmetry in the bulk, how (and whether) it is related to the concept of pairing, and what the bulk counterpart of the concept of "gauging a subgroup" at the edge is. Since the theory of the bulk state is already a gauge theory, one cannot gauge what has already been gauged!

Here we will outline the construction of the effective-field theory of the bulk for the simplest case, the bosonic Moore–Read state at filling factor $\nu = 1$. We will see that the construction can be generalized to the bosonic Read–Rezayi states with filling factor $\nu = k/2$. The effective-field theory of the edge states for the $\nu = 1$ bosonic Moore–Read state has $SU(2)_2$ chiral Kac–Moody symmetry, namely the $SU(2)_2$ chiral WZW model. We saw in the preceding section that there is a one-to-one correspondence between the bulk $SU(2)_2$ non-abelian Chern–Simons gauge theory and the chiral $SU(2)_2$ WZW model on the edge.

So in the case of the $\nu = 1$ Moore–Read state the effective-field theory of the bulk is the $SU(2)_2$ Chern–Simons theory. Is this related in any way to the concept of pairing? In Chapter 14 we compared the structure of the wave function of the fermionic Moore–Read state at $\nu = 1/2$ with the behavior of the pair field in the BCS theory of a $p_x + ip_y$ superconductor. There we saw that the Pfaffian factor embodies the physics of this superconducting state. We now want to ask a similar question regarding the bosonic state.

To this end, following the results of Fradkin *et al.* (1999), we will consider a quantum Hall system with two species of charged bosons in two dimensions. We will assume that each species of charged bosons has sufficiently strong short-range repulsive interactions that each Landau level state can be occupied by only one boson and that each species of bosons is in a Laughlin state with filling fractions $\nu = 1/2$. Thus we have a (2, 2, 0) Halperin state with wave function

$$\Psi_{(2,2,0)} = \prod_{i<j}(z_i - z_j)^2 \prod_{i<j}(w_i - w_j)^2 e^{-\frac{1}{4\ell^2}\sum_i\left(|z_i|^2+|w_i|^2\right)} \tag{15.126}$$

where z_i are the complex coordinates of type 1 bosons and w_i are the complex coordinates of type 2 bosons. Since the two species are decoupled, the effective-field theory involves two Chern–Simons gauge fields, $\mathcal{A}_\mu^l$ (with $l = 1, 2$), for the currents of each species of bosons, and has a $U(1)_2 \times U(1)_2$ gauge symmetry. The effective action is

$$S_{(2,2,0)}^{\text{bulk}} = \frac{2}{4\pi} \sum_{l=1}^{2} \int_{\Sigma\times\mathbb{R}} d^3x \, \epsilon^{\nu\nu\lambda} \mathcal{A}_\mu^l \, \partial_\nu \mathcal{A}_\lambda^l \tag{15.127}$$

where the level of each Chern–Simons theory is $m = 2$ since the filling factor is $\nu = 1/2$ for each species of bosons. The effective action of the edge states of the (2, 2, 0) state is then a theory of two chiral bosons ϕ_a ($l = 1, 2$) with the same velocity v, also at level $m = 2$,

$$S_{(2,2,0)}^{\text{edge}} = \frac{m}{4\pi} \sum_{l=1}^{2} \int dx \, dt \; \partial_x \phi_l (\partial_t \phi_l - v \, \partial_x \phi_l) \tag{15.128}$$

In Section 15.4.4 we showed that the $U(1)_2$ Chern–Simons theory is equivalent to the $SU(2)_1$ Chern–Simons theory in the sense that they have the same topological invariants. Similarly, the theory of the $U(1)_2$ chiral boson is equivalent to the $SU(2)_1$ chiral WZW model. So the universal (topological) properties of the (2, 2, 0) bosonic Halperin state are equally described by a sum of two $SU(2)_1$ Chern–Simons gauge theories, each with its own SU(2) gauge field, $\mathcal{A}_\mu^{a,l}$, where $a = 1, 2, 3$ runs over the su(2) algebra. In this language the action becomes

$$S_{(2,2,0)}^{\text{bulk}} = \frac{1}{4\pi} \sum_{l=1,2} \int_{\Sigma\times\mathbb{R}} d^3x \, \epsilon^{\mu\nu\lambda} \left[\mathcal{A}_\mu^{l,a}, \partial_\nu \mathcal{A}_\lambda^{l,a} + \frac{2}{3} f_{abc} \mathcal{A}_\mu^{l,a} \mathcal{A}_\nu^{l,b} \mathcal{A}_\lambda^{l,c} \right] \tag{15.129}$$

We return momentarily to the theory of the edge and ask how an $SU(2)_2$ WZW model arises from two decoupled $SU(2)_1$ WZW models. Each $SU(2)_1$ WZW model has an $SU(2)_1$ Kac–Moody algebra (at level $k = 1$) of the currents $J^{a,l}$. We now consider perturbing this system by a marginally relevant off-diagonal current–current interaction whose coupling constant flows to a strong-coupling fixed point at which the relative current vanishes

$$\left(J^{a,1} - J^{a,2}\right) |\text{Phys}\rangle = 0 \tag{15.130}$$

We saw at the end of Chapter 7 that this limit is equivalent to coupling the system to a gauge field that projects out all states that violate the constraint of Eq. (15.130), i.e. it is a gauged WZW model. However, this condition is equivalent to imposing that the WZW currents are identified, and hence the WZW fields themselves are also identified, $g^1 = g^2 \equiv g$. Since the original action was the sum of two decoupled $SU(2)_1$ WZW models (with fields g^1 and g^2), the action of the theory at this fixed point is simply an $SU(2)_2$ WZW model for the field g. Thus, under the action of this perturbation, the theory flows to the $SU(2)_2$ fixed point, with a single dynamical field g, and a Kac–Moody algebra with level raised to $k = 2$,

$$S^{\mathrm{WZW}}_{\mathrm{SU}(2)_1}[g^1] + S^{\mathrm{WZW}}_{\mathrm{SU}(2)_1}[g^2] \mapsto S^{\mathrm{WZW}}_{\mathrm{SU}(2)_2}[g] \tag{15.131}$$

On the other hand, the $\mathrm{SU}(2)_1$ currents are just the "electron" (here the boson) operators $V_\pm = \exp(\pm i\sqrt{2}\phi)$ and the edge current $j = \partial_x \phi$. We also saw that, for a Laughlin state of bosons at $\nu = 1/2$, these three operators are the generators of the $\mathrm{SU}(2)_1$ Kac–Moody algebra, and the boson (and its hole) are part of the $\mathrm{SU}(2)_1$ Kac–Moody algebra. Hence, if the $\mathrm{SU}(2)_1$ Kac–Moody currents are identified, this means that the bosons themselves are glued together. This picture suggests that the bulk theory must describe a paired state of the bosons!

We can describe this physics from the point of view of the bulk as follows. Let B_l^a be two three-component real fields ($a = 1, 2, 3$) representing the boson excitations in the bulk for each species of bosons ($l = 1, 2$), each made of the fields that create and destroy each type of boson, $B_l^\dagger = B_{1,l} + i B_{2,l}$ and its complex conjugate, and the local boson density fluctuation, which we denote by $B_{3,l}$. We will assume that the bosons themselves do not undergo Bose condensation but that they can pair. A paired state of "spin-1" bosons is naturally represented by the (p-wave) triplet pair field (Fradkin *et al.*, 1999)

$$\Psi_a(z_1 - z_2) = \epsilon_{abc} B_1^b(z_1) B_2^c(z_2) f(z_1 - z_2) \tag{15.132}$$

where the kernel has the long-distance behavior $f(z) \sim 1/z$. This pair field has p-wave symmetry and maximally violates time-reversal invariance. This pair field transforms in the spin-1 (adjoint) representation of each SU(2) and in the spin-1 representation of the diagonal SU(2) subgroup of SU(2) × SU(2).

A pair field of two triplet fields can be best understood as a matrix field that transforms under the action of each group. Let $O \sim \exp(i T_l \Psi_l)$ (where T_a are the three SU(2) generators in the spin-1 representation) be the (3×3) matrix field. Since it is in the adjoint representation of SU(2), it is blind under the action of the $\mathbb{Z}_2$ center of SU(2). Hence the field O effectively is an element of SO(3), and transforms under the action of the groups G_1 and G_2 (here both are SO(3)) as

$$O \mapsto G_1 O G_2^{-1} \tag{15.133}$$

We will postulate that the boson pair field has a Landau–Ginzburg action of the form (Fradkin *et al.*, 1999)

$$\begin{aligned} S = & \int d^3x \left\{ \kappa \operatorname{tr}\left(O^{-1} D_\mu O O^{-1} D^\mu O\right) + \lambda B_1^{\mathrm{T}} O B_2 \right\} \\ & + \int d^3x \sum_{l=1,2} \left| (\partial_\mu B_{a,l} + i\epsilon_{abc} \mathcal{A}_\mu^b B_{c,l}) \right|^2 \\ & + \frac{1}{4\pi} \sum_{l=1,2} \int_{\Sigma\times\mathbb{R}} d^3x\, \epsilon^{\mu\nu\lambda} \left[\mathcal{A}_\mu^{l,a}\, \partial_\nu \mathcal{A}_\lambda^{l,a} + \frac{2}{3} f_{abc}\, \mathcal{A}_\mu^{l,a} \mathcal{A}_\nu^{l,b} \mathcal{A}_\lambda^{l,c} \right] \end{aligned} \tag{15.134}$$

where $D_\mu = \partial_\mu + i\left[\mathcal{A}_{\mu,1} - \mathcal{A}_{\mu,2}\right]$ is the covariant derivative in the adjoint (spin-1) representation of the diagonal SU(2) subgroup, and κ and λ are two coupling constants.

In the phase in which the pair field O acquires an expectation value, $\langle O \rangle = I$, the symmetry is broken spontaneously as

$$\mathrm{SO}(3) \times \mathrm{SO}(3) \mapsto \mathrm{SO}(3) \tag{15.135}$$

and the symmetry is broken down to the diagonal SO(3) subgroup. Therefore, in this broken-symmetry phase, the order-parameter manifold is the coset $(\mathrm{SO}(3) \times \mathrm{SO}(3))/\mathrm{SO}(3)$.

In this phase we have a Higgs mechanism and there is a Meissner effect. Indeed, deep in this phase the first term of the action of Eq. (15.134) reduces to a mass term for the relative gauge field $\propto \kappa\left(\mathcal{A}^a_{\mu,1} - \mathcal{A}^a_{\mu,2}\right)^2$. Therefore, in the low-energy limit of the paired phase, the massive relative gauge field is frozen out and the two gauge fields become identified with each other. The direct consequence of this result is that the effective action of the average gauge field $\mathcal{A}^a_\mu = (\mathcal{A}^a_{\mu,1} + \mathcal{A}^a_{\mu,2})$ is an $\mathrm{SU}(2)_2$ Chern–Simons gauge theory,

$$S_{\mathrm{CS}}[\mathcal{A}^a_{\mu,1}] + S_{\mathrm{CS}}[\mathcal{A}^a_{\mu,2}] \mapsto 2S_{\mathrm{CS}}[\mathcal{A}^a_\mu] \tag{15.136}$$

This construction is easily generalized to the case of a system with k species of bosons. We start again with a system in which the ground state of each species is a $\nu = 1/2$ Laughlin state (of bosons) with symmetry $\mathrm{SU}(2)_1$ (as before) and the total filling fraction is $\nu = k/2$. We now have k spin-1 Bose fields, $B_{a,l}$, now with $l = 1, \ldots, k$. If the bosons pair up again in a spin-triplet p-wave state with the pattern that B_1 pairs with B_2, B_2 with $B_3, \ldots,$ and B_{k-1} pairs with B_k, we will now have a theory with the following symmetry breaking:

$$\mathrm{SU}(2)_1 \times \cdots \times \mathrm{SU}(2)_1 \mapsto \mathrm{SU}(2)_k \tag{15.137}$$

Hence this "clustered" state has an effective-field theory that is the $\mathrm{SU}(2)_k$ Chern–Simons gauge theory. For $k > 2$ the resulting effective-field theory describes the Read–Rezayi parafermionic states (Fradkin *et al.*, 1999; Read and Rezayi, 1999).

The order-parameter field has a non-trivial pattern of spontaneous symmetry breaking, given in Eq. (15.135). This effective non-linear sigma model admits topological skyrmion-like excitations. We can now apply the ideas of homotopy theory that were discussed in Section 7.6 to analyze the nature of the topological defects of the $\nu = 1$ Moore–Read state. However, unlike the conventional non-linear sigma model discussed in Chapter 7, which had an order-parameter-field manifold $\mathrm{SO}(3)/\mathrm{U}(1) \simeq S_2$, the order-parameter manifold for the problem we are now discussing is the non-trivial coset $\mathrm{SO}(3) \times \mathrm{SO}(3)/\mathrm{SO}(3)$. In this case, the skyrmions are similar to vortices and the topologically non-trivial configurations are classified by the homotopy group (Mermin, 1979)

$$\pi_1(\mathrm{SO}(3) \times \mathrm{SO}(3)/\mathrm{SO}(3)) \simeq \mathbb{Z}_2 \tag{15.138}$$

The origin of this result can be traced back to the fact the SO(3) is blind to the center $\mathbb{Z}_2$ of SU(2), or, equivalently, this order parameter is a *nematic* and it is invariant under a rotation by π. In contrast, if we had a pair condensate of the fundamental (Laughlin) quasiparticles of the $\nu = 1/2$ fluids, the order-parameter manifold would have been instead $\mathrm{SU}(2)\times\mathrm{SU}(2)/\mathrm{SU}(2)$, which has a trivial homotopy class, $\pi_1(\mathrm{SU}(2) \times \mathrm{SU}(2)/\mathrm{SU}(2)) \simeq 0$, and hence does not have stable half-vortices.

We then conclude that the pair condensate field O supports skyrmions with a $\mathbb{Z}_2$ topological charge. The skyrmions of this theory are vortex solutions of the Landau–Ginzburg theory. On points (r, θ) of a very large circumference S_1 of radius $r \to \infty$, the field O approaches asymptotically a rotation matrix

$$\lim_{r\to\infty} O(r, \theta) = \mathcal{R}(\hat{n}, \theta) \tag{15.139}$$

where $\hat{n}$ is a unit vector and $0 \leq \theta < 2\pi$. For the energy of this soliton to be finite the relative gauge field $\mathcal{A}^a_{1,i} - \mathcal{A}^a_{2,i}$ must have the asymptotic behavior

$$\lim_{r\to\infty} \left(\mathcal{A}_{1,i}(r, \theta) - \mathcal{A}_{2,i}(r, \theta)\right) = \mathcal{R}(\hat{n}, \theta)^{-1}\, \partial_\theta \mathcal{R}(\hat{n}, \theta) \tag{15.140}$$

There is a different vortex solution for each unit vector $\hat{n}$. These solutions rotate into each other under the action of the unbroken SO(3). Thus, at the quantum level the vortices carry an SO(3) quantum number. However, the $\mathbb{Z}_2$ topological charge makes the vortices double-valued and hence they carry the spin-$1/2$ representation of SO(3). This double-valuedness is compensated for by a rotation in real space by $\theta \to \theta + \pi$. Thus, this vortex solution is actually the same as the half-vortex of the $p_x + ip_y$ superconductor which we discussed in Section 14.9.3. On the other hand, due to the coupling to the $\mathrm{SU}(2)_2$ Chern–Simons gauge field, these soliton states give rise to a representation of a quantum group symmetry and exhibit non-abelian braiding statistics. Thus we identity these half-vortices with the Ising twist field (denoted by σ) of the theory of the edge states.

Similarly, the triplet of unpaired bosons will carry the spin-1 representation of SO(3). However, due to the coupling of the triplet field to the $\mathrm{SU}(2)_2$ Chern–Simons gauge theory, these states undergo a statistics transmutation and become a triplet of Majorana fermions, again in agreement with our analysis of the edge states. The boson operator (the "electron") of the $\nu = 1$ Moore–Read state, which at the edge is a current, is obtained by fusing the neutral component of the triplet, B_3 (a "fermionic dipole"), with one of the other components, say $B^\dagger$ (which carries the unit of flux of the U(1) Cartan subgroup of SU(2) which we identified with the charge sector).

15.6 Tunneling conductance at point contacts

In spite of all the beautiful properties of the fractional quantum Hall fluids, it is experimentally very hard to probe them in the bulk. The reason is that, due to the incompressibility of the fluid, i.e. the large energy gap in the bulk, only two types of experimental probes can access the physics of the bulk. One of these types is transport experiments that measure the Hall and longitudinal conductances. These experiments are used to establish the existence of a quantum Hall fluid by observing a plateau in the Hall conductance (quantized to integer or fractional multiples of e^2/h) and the energy gap through the temperature dependence of the longitudinal conductance. The other type consists of resonance experiments, microwave resonance, nuclear magnetic resonance, and Raman light scattering, which probe the gapped density fluctuations in the bulk. In practice, experiments of the second type can be done only in the fractional quantum Hall states with the largest energy gaps. These experiments have confirmed that incompressible fractional quantum Hall fluids have density fluctuations, known as magneto-phonons, and that their behavior can be predicted by theories that are based on the Laughlin wave function (Girvin *et al.*, 1986) (see Das Sarma and Pinczuk (1997)), on the existence of skyrmion excitations in quantum Hall ferromagnets (Sondhi *et al.*, 1993; Barrett *et al.*, 1995), and on an exciton Bose–Einstein condensate in bilayer quantum Hall systems (Eisenstein and MacDonald, 2004).

On the other hand, since the edge states are gapless, they offer the opportunity to test the more subtle predictions of the theory. The catch is that these experiments are technically quite challenging. We will now discuss several important experimentally testable (and tested) consequences of the theory of the edge states of quantum Hall fluids (integer and fractional).

The ideal way to test the gapless edge states is to tunnel electrons either into these systems or from one edge to the other. However, the edge states on both integer and fractional quantum Hall fluids have a Fermi momentum, which is determined by the Fermi energy of the quantum Hall fluid subject to the confining potential. Thus momentum conservation allows tunneling of electrons from an external reservoir only if translation invariance is broken, for instance by a defect. Such a tunneling center is called a quantum point contact. Similarly, the edge states at the opposite ends of a fluid have opposite momentum and it is not possible to tunnel (electrons or quasiparticles) unless translation invariance somehow is also broken. In this case translation invariance is typically broken by an external gate at some voltage. Since the fluid is incompressible, the gate creates a constriction in the fluid, as shown schematically in Fig. 15.10(a). This point contact is a quantum Hall junction. Here the tunneling process is internal and occurs across the fractional quantum Hall fluid. Two types of experiments have been done with this setup, concerning transport across the junction in Laughlin states (Milliken *et al.*, 1996;

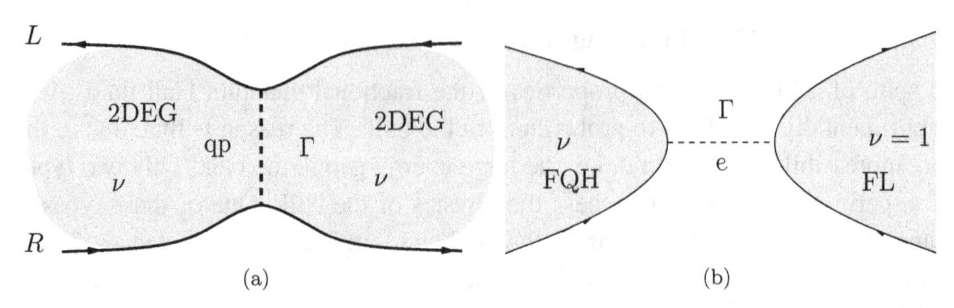

Figure 15.10 (a) Internal quasiparticle tunneling between two edges of a fractional quantum Hall fluid at a junction created by a constriction of the 2DEG. (b) Tunneling of electrons from a Fermi liquid (FL) (equivalent to a 2DEG with $\nu = 1$ on the right) to a fractional quantum Hall (FQH) Laughlin fluid with filling fraction $\nu = 1/m$ (on the left) at a point contact with tunneling amplitude Γ.

Roddaro *et al.*, 2004; Stefano *et al.*, 2004) and in non-abelian states (at $\nu = 5/2$) (Miller *et al.*, 2007; Radu *et al.*, 2008), and noise in the tunneling current both in Laughlin states (de Picciotto *et al.*, 1997; Saminadayar *et al.*, 1997) and in the non-abelian state at $\nu = 5/2$ (Dolev *et al.*, 2008). We will discuss these experiments and the theory below.

The other typical experimental setup involves tunneling from some external reservoir of electrons into the edge of a quantum Hall system. A highly idealized version of this junction is shown in Fig. 15.10(b). A more realistic setup involves a quantum Hall fluid separated by a barrier from a usually 3D electron gas. In this case, tunneling takes place at isolated defects along the barrier, which act as tunneling centers. To the extent that these tunneling centers can be regarded as acting independently of each other, this setup reduces to our idealized case. Since the external electron reservoir is 3D, in most cases of experimental relevance in practice it can be regarded as a Fermi liquid, which is a weakly interacting system. For this reason this tunnel junction is equivalent to a junction between the fractional quantum Hall fluid and an integer quantum Hall state, since the quantum numbers of the latter state are the same as in a Fermi liquid. Transport experiments in Laughlin states have been done in this setup (Chang *et al.*, 1996; Chang, 2003).

15.6.1 Tunneling Hamiltonians

Let us begin with a theoretical picture of the problem of tunneling of electrons into a fractional quantum Hall edge (Wen, 1991b). As we saw before, in the general case we have several edge states, which can be regarded as a charged edge state and possibly several neutral edge states. We will discuss the simpler case of the Laughlin states, since they have a unique edge. In all of these cases the edge states are chiral Luttinger liquids. Thus, the quantum Hall junction is just the chiral version of the setup of a scanning tunneling microscope discussed in Section 6.8.2. The results we derived there for a Luttinger liquid apply to the chiral case with only minor changes.

In the case of a quantum Hall junction with a Fermi liquid (see Fig. 15.10(b)), the tunneling Hamiltonian describes processes of electron tunneling from the external reservoir (the "tip," which we regard as a Fermi liquid) and the edge state of the fractional quantum Hall fluid. This system is formally equivalent to the problem of a spinless (non-chiral) Luttinger liquid coupled to an impurity. In this interpretation, the top and bottom edges are the two chiral components of a spinless Luttinger liquid with Luttinger parameter $K = \nu = 1/m$ and the constriction represents backscattering at the impurity (Kane and Fisher, 1992).

The tunneling Hamiltonian is (with the point contact at $x = 0$)

$$H = H_{\text{edge}} + H_{\text{FL}} + \Gamma e^{i\omega_0 t}\psi^{\dagger}_{\text{e,edge}}(0,t)\psi_{\text{e,FL}}(0,t) + \text{h.c.} \tag{15.141}$$

where the "Josephson frequency" is $\omega_0 = eV/\hbar$, and Γ is the tunneling matrix element.

The electron propagator of a Laughlin edge state is

$$G_{\text{e}}(x,t) = \text{constant} \times \frac{1}{(i(x - vt) + \epsilon)^m} \tag{15.142}$$

in the limit $\epsilon \to 0^+$. This exponent leads to a tunneling density of states $N(\omega)$,

$$N(\omega) \sim \text{Im} \lim_{x \to 0^+} \int_{-\infty}^{\infty} dt\; G_{\text{e}}(x,t) e^{i\omega t} \sim \text{constant} \times |\omega|^{m-1} \tag{15.143}$$

where we have used that

$$\int_{-\infty}^{\infty} dt\; \frac{e^{i\omega t}}{(\beta \pm it)^{\alpha}} = \frac{2\pi}{\Gamma(\alpha)}(\pm\omega)^{\alpha-1} e^{\mp\beta\omega}\theta(\pm\omega) \tag{15.144}$$

where $\beta = \epsilon + ix$, α is real and positive, $\Gamma(x)$ is the Euler gamma function, and $\theta(x)$ is the step function.

This result, combined with Fermi's Golden Rule at a tunnel junction with bias V, predicts a tunneling current I for electrons tunneling into a chiral Luttinger liquid (CLL) from a Fermi liquid (FL) (see Eq. (6.166))

$$I_{\text{e}}(V) = 2\pi \frac{e}{\hbar} |\Gamma|^2 \int_{-eV}^{0} dE\; N_{\text{CLL}}(E,T) N_{\text{FL}}(E + eV, T) \propto V^m \tag{15.145}$$

and a differential tunneling conductance $G(V)$ (see Eq. (6.168))

$$G_{\text{e}}(V,T) = \frac{dI}{dV} \simeq \frac{2\pi e}{\hbar} |\Gamma|^2 N_{\text{FL}}(0) N_{\text{CLL}}(V,T) \propto V^{m-1} \tag{15.146}$$

Hence, we conclude that tunneling conductance of electrons from an external tip (a Fermi liquid) to the edge of a fractional quantum Hall state has a power-law form with an exponent $\alpha = m - 1$. Thus, tunneling is suppressed at low bias and vanishes as $V \to 0$.

The case of a constriction of a quantum Hall fluid created by an external gate, shown in Fig. 15.10(a), can be understood using a similar approach. The difference is that the tunneling is now internal and takes place between two identical edges across the fractional quantum Hall fluid. More importantly, the "particles" which tunnel are the quasiparticles of the quantum Hall state. The tunneling Hamiltonian for Laughlin quasiparticles is

$$H = H_{\rm R} + H_{\rm L} + \Gamma e^{i\omega_0^* t} \psi^\dagger_{\rm qp,L}(0,t)\psi_{\rm qp,R}(0,t) + {\rm h.c.} \tag{15.147}$$

where we denoted by $H_{\rm R}$ and $H_{\rm L}$ the Hamiltonians of the bottom (right-moving, R) and top edges (left-moving, L), respectively, which have opposite chirality. In this case the "Josephson frequency" is $\omega_0^* = q^* V/\hbar$, where $q^* = e/m$ is the (fractional) charge of the Laughlin quasiparticle (or quasihole). Similarly the quasiparticle tunneling current $J_{\rm qp}$ is

$$J_{\rm qp} = i\Gamma \frac{e}{m} \left(\psi^\dagger_{\rm qp,L}(0,t)\psi_{\rm qp,R}(0,t) - {\rm h.c.} \right) \tag{15.148}$$

We can now repeat the same steps to obtain a formula for the tunneling current for Laughlin quasiparticles $G_{\rm qp}$ with the same form as the Golden Rule expression of Eq. (15.145), with the important difference that the densities of states are now those for Laughlin quasiparticles at each edge. The density of states for Laughlin quasiparticles scales as

$$N_{\rm qp}(\omega) \sim |\omega|^{1/m-1} \tag{15.149}$$

which diverges at low frequencies. On putting it all together, we find that the tunneling current of Laughlin quasiparticles has the voltage dependence

$$I_{\rm qp}(V) = 2\pi \frac{e}{m\hbar} |\Gamma|^2 \int_{-eV}^{0} dE\, N_{\rm CLL}(E,T) N_{\rm CLL}(E+eV,T) \propto V^{2/m-1} \tag{15.150}$$

Thus, the quasiparticle differential conductance $G_{\rm qp}(V)$ has the scaling form

$$G_{\rm qp}(V) = \frac{dI_{\rm qp}}{dV} \propto V^{2(1/m-1)} \tag{15.151}$$

For Laughlin states $\nu = 1/m < 1$. Hence the exponent of the power law of the differential conductance is negative, which seemingly implies that the conductance becomes *large* (and divergent) as the bias $V \to 0$. This indicates a breakdown of perturbation theory in the tunneling matrix element Γ that we will address below.

15.6.2 Scaling behavior

We can understand these results using a simple perturbative RG analysis. Indeed, in all the cases we discussed in the limit of no tunneling, $\Gamma = 0$, the theory is scale-invariant, and hence it is a fixed point. The question is whether this fixed point is stable or unstable or, equivalently, whether the tunneling operator is relevant or irrelevant. The chiral electron and quasiparticle operators have scaling dimensions $\Delta_{\text{e}} = m/2$ and $\Delta_{\text{qp}} = 1/(2m)$, respectively.

The scaling dimension of the tunneling operator depends on what is tunneling and between which type of states. In the case of electron tunneling between a Fermi liquid and a Laughlin state with $\nu = 1/m$, the tunneling operator involves removing an electron from one system and adding it to the other. Since the scaling dimension of the tunneling operator of electrons is the sum of the scaling dimensions of the electron operators of each system, we find that in this case the scaling dimension of the electron tunneling operator is $\Delta_{\text{e,tunnel}} = (m+1)/2$. In the case of electron tunneling between the edge states of two $\nu = 1/m$ Laughlin states the scaling dimension of the electron tunneling operator is $\Delta_{\text{e,tunnel}}[\nu = 1/m] = m$. Instead, the scaling dimension of the quasiparticle tunneling operator is $\Delta_{\text{qp,tunnel}}[\nu = 1/m] = 1/m$.

By following the same analysis as we did in Chapter 4, we can write down the RG beta function for the tunneling amplitude Γ. The only difference here is that the tunneling operator acts at only one point, say $x = 0$, which means that the term in the action of the tunneling process has a delta function $\delta(x)$ that has scaling dimension 1. Thus, if we denote by Δ the scaling dimension of the tunneling operator, we can define a dimensionless tunnel amplitude g by

$$\Gamma = a^{\Delta - 1} g \tag{15.152}$$

where a is a short-distance cutoff. We then find that the beta function, to zeroth ("tree-level") order, is

$$\beta(g) = a\frac{\partial g}{\partial a} = (1 - \Delta) g + O(g^2) \tag{15.153}$$

Notice that the coefficient of the (leading) linear term is not $(2 - \Delta)$ (where 2 is the dimension of the space-time) but $1 - \Delta$, since the operator acts at only one point. This is characteristic of all quantum impurity problems, the prototype of which is the Kondo problem, which describes the coupling of a magnetic impurity embedded in a metal. We will see below that we are actually discussing a boundary CFT.

The form of the beta function, Eq. (15.153), indicates that a tunneling operator is relevant if its scaling dimension is $\Delta < 1$, irrelevant if $\Delta > 1$, and marginal if $\Delta = 1$. Thus, both in the case of *electron tunneling* from a Fermi liquid to the edge state of a Laughlin state and in that of electron tunneling between the edge states of Laughlin states, the tunneling operators are *irrelevant* since in both cases their scaling dimensions are $\Delta = (m+1)/2 > 1$ and $\Delta = m > 1$. Hence, in the case of electron tunneling the decoupled fixed point is stable. The exception is the case of $\nu = 1$, for which $\Delta = 1$ and the tunneling operator is marginal. In contrast, the tunneling operator of quasiparticles between the edges of Laughlin states has scaling dimension $\Delta = 1/m < 1$, the tunneling operator is always *relevant*, and hence the decoupled fixed point is *unstable*.

These results can be summarized in the following appealing picture, which is due to Kane and Fisher (1992). Let us consider the case of a fractional quantum Hall state with a constriction, with a quasiparticle tunneling amplitude Γ. The scaling analysis shows that the decoupled fixed point, $\Gamma = 0$, is unstable and hence that the effective tunneling amplitude grows at low energies or, equivalently, that there is a growth in the backscattered current. Since the growth of the tunneling amplitude can be interpreted as a narrowing of the constriction, this leads to the natural assumption that the end result of this process is a state in which the quantum Hall fluid is split in two, as shown in Figs. 15.11(a) and (b). In the limit implied by Fig. 15.11(b) the only tunneling process allowed is that of electrons across vacuum between the two edge states, which is irrelevant.

Hence, the main effect of the constriction is a crossover between an unstable weak-coupling fixed point (dominated by quasiparticle tunneling) and a stable strong-coupling fixed point (dominated by electron tunneling). From the point of view of the bulk quantum Hall fluid, this crossover is a change in the topology of the region occupied by the quantum Hall fluid, which is split in two by the constriction.

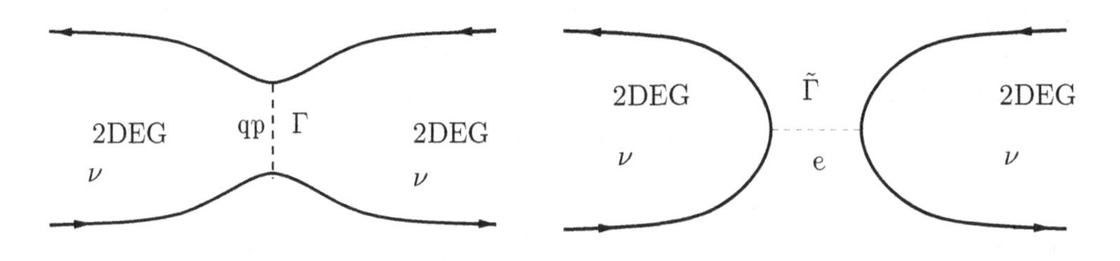

Figure 15.11 Two limiting regimes of a fractional quantum Hall tunneling junction: (a) in the weak-tunneling regime, Γ small, quasiparticles tunnel across the fractional quantum Hall fluid at a constriction; and (b) in the strong-tunneling regime, Γ large (or dual $\tilde{\Gamma}$ small), the fractional quantum Hall fluid is split and we get electron tunneling between the two fractional quantum Hall fluids across the "vacuum."

The unstable fixed point of a constriction can be regarded as a (quantum) critical fixed point, following the same logic as in Chapter 4. In particular, in this case there is a characteristic crossover *energy scale* that characterizes the crossover. In the case of the Kondo problem the two fixed points represent the magnetic impurity decoupled from the metal (the unstable weak-coupling fixed point) and the magnetic impurity being screened (or over-screened, depending on the number of channels in the metal) (the stable strong-coupling fixed point). By analogy with the Kondo problem, in the case of the quantum Hall constriction (and junction) we will refer to the crossover energy scale as the "Kondo scale" and denote it by $T_{\mathrm{K}}(g)$ (even though the scaling is different). A simple RG analysis now tells us that the crossover scale depends on the (dimensionless) tunneling amplitude g through a universal scaling law of the form

$$T_{\mathrm{K}}(g) \sim g^{1/(1-\Delta)} \tag{15.154}$$

where Δ is the scaling dimension of the tunneling operator at the unstable fixed point. The exponent of this scaling law is, as in all cases with a relevant perturbation, given by the reciprocal of the slope of the beta function at the unstable fixed point. Thus, if the junction (or constriction) is probed at an energy scale (voltage or temperature) that is large compared with $T_{\mathrm{K}}(g)$ we access the physics of the weak-coupling fixed point. Conversely, if we probe the junction at energy scales low that are compared with $T_{\mathrm{K}}(g)$ we see the physics of the strong-coupling fixed point. In the case of the Kondo problem of a magnetic impurity coupled by exchange to a metal, the exchange interaction is marginally relevant, and the Kondo scale (conventionally denoted by T_{K}) depends on the coupling constant J through an essential singularity $T_{\mathrm{K}} \sim \sqrt{J}\exp(-\text{constant}/J)$ (Anderson, 1970; Wilson, 1975).

15.6.3 Effective-field theory and boundary conformal field theory

We will now use the chiral-boson description of the edge states of the Laughlin states to develop a non-perturbative description of all three cases: (a) electron tunneling from a Fermi liquid ("$\nu = 1$") to a $\nu = 1/m$ Laughlin state, (b) electron tunneling between the edges of two identical Laughlin states, and (c) quasiparticle tunneling between the edges of a single Laughlin quantum Hall fluid. In all cases we will denote by ϕ_1 and ϕ_2 the right- and left-moving chiral bosons associated with the two edges. The first two cases, (a) and (b), can be treated in a single unified approach (Chamon and Fradkin, 1997), while case (c) will be treated in a simply related fashion (Kane and Fisher, 1992; Fendley *et al.*, 1995a). We will also see that all of these problems are described by a perturbed boundary CFT (Cardy, 1986).

Extensions of this theory have been made in order to describe constrictions in the $\nu = 5/2$ non-abelian quantum Hall state in which the tunneling quasiparticle has non-abelian fractional statistics (Fendley *et al.*, 2006). In this case, as we saw in Section 15.4.5, each edge has a charge sector, which is described by a chiral boson that carries the charge, and a neutral sector, which is described by a chiral Majorana fermion. The non-abelian quasiparticle σ is a composite object made of the Majorana fermion and a vertex operator of the chiral boson. We will not discuss this problem here.

To develop this theory we will work with the Lagrangian density rather than the Hamiltonian. For the cases (a) and (b) of electron tunneling between two edges with filling fractions ν_1 and ν_2 (with $\nu_1 = 1/m$ and $\nu_2 = 1$ in case (a), and $\nu_1 = \nu_2 = 1/m$ in case (b)) at a point contact (or constriction) the Lagrangian density is given by

$$\mathcal{L} = \frac{1}{4\pi}\,\partial_x\phi_1(\partial_t\phi_1 - \partial_x\phi_1) + \frac{1}{4\pi}\,\partial_x\phi_2(\partial_t\phi_2 - \partial_x\phi_2) + \Gamma_e\delta(x)\left(e^{i\omega_0 t}e^{i\left(\frac{1}{\sqrt{\nu_1}}\phi_1 - \frac{1}{\sqrt{\nu_2}}\phi_2\right)} + \text{h.c.}\right) \tag{15.155}$$

where Γ_e is the amplitude for electron tunneling at the point contact. For quasiparticle tunneling at a constriction with amplitude Γ_{qp} (case (c)) the Lagrangian is instead given by

$$\mathcal{L} = \frac{1}{4\pi}\,\partial_x\phi_1\,(\partial_t\phi_1 - \partial_x\phi_1) + \frac{1}{4\pi}\,\partial_x\phi_2(\partial_t\phi_2 - \partial_x\phi_2) + \Gamma_{qp}\delta(x)\left(e^{i\omega_0^* t}e^{i\sqrt{\nu}(\phi_1 - \phi_2)} + \text{h.c.}\right) \tag{15.156}$$

where, as before, $\omega_0 = eV/\hbar$ and $\omega_0^* = q^*V/\hbar = eV/(m\hbar)$ are the Josephson frequencies.

In Eqs. (15.155) and (15.156) we have omitted the Klein factors that insure that the electron operators of different edges anti-commute with each other and that the quasiparticle operators of different edges obey the anyon algebra, since their product depends on the total combined charge of both edges and it is a constant of motion. Hence these factors can be absorbed (in this case) into a simple redefinition of the tunneling matrix element Γ. In addition, in both cases we have set the edge velocities to $v = 1$, which we can do without loss of generality since we have a single point contact. For the same reason we can also use the same coordinate x to denote the arc lengths along each edge. Also, and for the same reason, both in Eq. (15.155) and in Eq. (15.156) we carried out a parity transformation $x \leftrightarrow -x$ on the left-moving edge (which amounts to flipping the direction of the top edge) and so both chiral bosons are "right-moving." For reasons of clarity we still denote the chiral bosons by ϕ_1 and ϕ_2.

Next, in Eq. (15.155) we perform an orthogonal transformation to new fields ϕ_1' and ϕ_2' to map the problem of electron tunneling between inequivalent edges into a problem of "electron" tunneling between the (equivalent) edges of two fluids with the same effective filling fraction $\bar{\nu}$. The transformation is

$$\begin{pmatrix} \phi_1' \\ \phi_2' \end{pmatrix} = \begin{pmatrix} \cos\theta & \sin\theta \\ -\sin\theta & \cos\theta \end{pmatrix} \begin{pmatrix} \phi_a \\ \phi_b \end{pmatrix} \tag{15.157}$$

where

$$\cos\theta = \frac{1}{\sqrt{2}} \frac{\sqrt{\nu_1^{-1}} + \sqrt{\nu_2^{-1}}}{\sqrt{\nu_1^{-1} + \nu_2^{-1}}}, \qquad \sin\theta = \frac{1}{\sqrt{2}} \frac{\sqrt{\nu_1^{-1}} - \sqrt{\nu_2^{-1}}}{\sqrt{\nu_1^{-1} + \nu_2^{-1}}} \tag{15.158}$$

The Lagrangian of Eq. (15.155) in terms of the transformed fields becomes

$$\begin{aligned} \mathcal{L} = & \frac{1}{4\pi} \partial_x \phi_1' \left(\partial_t \phi_1' - \partial_x \phi_1' \right) + \frac{1}{4\pi} \partial_x \phi_2' \left(\partial_t \phi_2' - \partial_x \phi_2' \right) \\ & + \Gamma_e \delta(x) \left(e^{i\omega_0 t} e^{i \frac{1}{\sqrt{\bar{\nu}}} (\phi_1' - \phi_2')} + \text{h.c.} \right) \end{aligned} \tag{15.159}$$

where $\bar{\nu}$ is given by

$$\bar{\nu}^{-1} = \frac{1}{2} \left(\nu_1^{-1} + \nu_2^{-1} \right) \tag{15.160}$$

Thus, the rotated fields describe a problem of a quantum point contact for electron tunneling between two identical edges of filling fraction $\bar{\nu}$. In particular, the problem of electron tunneling between a Fermi liquid (which we represent as the edge of a quantum Hall state with $\nu_1 = 1$) and a Laughlin state with $\nu_2 = \nu = 1/m$ is equivalent to the tunneling of electrons between two identical quantum Hall fluids with a filling fraction $\bar{\nu}$ given by

$$\bar{\nu}^{-1} = \frac{1 + \nu^{-1}}{2} = \frac{m+1}{2} \tag{15.161}$$

For instance, for $\nu = 1/3$ the equivalent system has an effective filling fraction $\bar{\nu} = 1/2$, which is equivalent to a problem of tunneling between identical bosonic Laughlin states, and hence the tunneling "electron" becomes a boson of charge 1. For $\nu = 1/5$ the equivalent problem has $\bar{\nu} = 1/3$, and the tunneling "electron" is a fermion of charge 1.

The mapping to a theory of two identical edges with effective filling fraction $\bar{\nu}$ suggests that there should exist a quasiparticle of charge $\bar{\nu}$. However, such a quasiparticle does not exist in each decoupled edge! This is telling us that this innocent-looking orthogonal transformation is gluing together the Hilbert spaces of the individual edges (two macroscopic systems) in a non-trivial way, and that this quasiparticle has to be interpreted as a soliton of the tunneling process. Actually, it is necessary to enlarge the Hilbert space in order to describe the decoupled edges and the system at finite tunneling amplitude (Hsu *et al.*, 2009a).

In this formulation the problems of electron tunneling (cases (a) and (b)) and of quasiparticle tunneling (case (c)) have very similar forms. In what follows we will drop the primes on the fields for the electron-tunneling cases and set $\phi_1' \to \phi_1$ and $\phi_2' \to \phi_2$. Similarly, we will not distinguish the filling factor ν of case (a) from the effective filling factor $\bar{\nu}$ of case (b), and specify which case it is as needed.

We will take advantage now of the form of Eq. (15.159) and Eq. (15.156) to make another orthogonal transformation, to even and odd combinations of the fields, $\phi_{\rm e}$ and $\phi_{\rm o}$,

$$\phi_1 = \frac{1}{\sqrt{2}}(\phi_{\rm e} + \phi_{\rm o}), \qquad \phi_2 = \frac{1}{\sqrt{2}}(\phi_{\rm e} - \phi_{\rm o}) \tag{15.162}$$

In this new basis, the Lagrangian for electron tunneling becomes

$$\begin{aligned}\mathcal{L} = {} & \frac{1}{4\pi}\,\partial_x\phi_{\rm e}(\partial_t\phi_{\rm e} - \partial_x\phi_{\rm e}) + \frac{1}{4\pi}\,\partial_x\phi_{\rm o}(\partial_t\phi_{\rm o} - \partial_x\phi_{\rm o}) \\ & + \Gamma_{\rm e}\delta(x)\left(e^{i\omega_0 t}e^{i\sqrt{\frac{2}{\nu}}\phi_{\rm o}} + \text{h.c.}\right)\end{aligned} \tag{15.163}$$

while the Lagrangian for quasiparticle tunneling is

$$\begin{aligned}\mathcal{L} = {} & \frac{1}{4\pi}\,\partial_x\phi_{\rm e}(\partial_t\phi_{\rm e} - \partial_x\phi_{\rm e}) + \frac{1}{4\pi}\,\partial_x\phi_{\rm o}(\partial_t\phi_{\rm o} - \partial_x\phi_{\rm o}) \\ & + \Gamma_{\rm qp}\delta(x)\left(e^{i\omega_0^* t}e^{i\sqrt{2\nu}\phi_{\rm o}} + \text{h.c.}\right)\end{aligned} \tag{15.164}$$

We see that in both electron and quasiparticle tunneling the even field $\phi_{\rm e}$ is insensitive to the tunneling process and decouples. This is not an accident since the charge defined by the even field is the total charge of both edges, $Q_{\rm e} = Q_1 + Q_2 = Q$, which is conserved by the tunneling process, which involves only the field $\phi_{\rm o}$. Naturally, the tunneling currents both for electrons, $I_{\rm e}$, and for quasiparticles, $I_{\rm qp}$, involve only the field $\phi_{\rm o}$,

$$I_{\rm e} \sim 2e\Gamma_{\rm e}\sin\left(\sqrt{\frac{2}{\nu}}\,\phi_{\rm o} + \frac{eV}{\hbar}t\right), \quad I_{\rm qp} \sim 2\frac{e}{m}\Gamma_{\rm qp}\sin\left(\sqrt{2\nu}\,\phi_{\rm o} + \frac{e\nu V}{\hbar}t\right) \tag{15.165}$$

In what follows we will focus on the field $\phi_{\rm o}$, which describes the tunneling process, and drop the decoupled total charge field $\phi_{\rm e}$.

The odd field $\phi_{\rm o}$ is a chiral boson that is coupled to a vertex operator (the tunneling operator) at $x = 0$. We can map this problem to a system described by a non-chiral boson defined (for all times) only on the half-line $x \geq 0$. Let $\varphi(x, t)$ be a non-chiral boson defined on the half-line. The field $\phi(x, t)$ in turn can be decomposed into two chiral bosons, namely its right- and left-moving components φ_R and φ_L (as we did in Chapter 6), which are related to the odd field ϕ_o. These fields are now defined on the entire line as follows:

$$\varphi_R(x, t) = \phi_o(x, t), \qquad \varphi_L(x, t) = \phi_o(-x, t) \tag{15.166}$$

We will refer to this transformation as the "folding" of the x axis on the positive half-line. In terms of the folded field φ the Euclidean action for the odd field becomes, for the two electron tunneling cases we discussed,

$$S = \frac{1}{8\pi}\int_{-\infty}^{\infty} dt \int_0^{\infty} dx (\partial_\mu \varphi)^2 + \Gamma_e \int_{-\infty}^{\infty} dt \cos\left(\frac{1}{\sqrt{2\nu}}\varphi(0, t)\right) \tag{15.167}$$

and, for quasiparticle tunneling,

$$S = \frac{1}{8\pi}\int_{-\infty}^{\infty} dt \int_0^{\infty} dx (\partial_\mu \varphi)^2 + \Gamma_{qp} \int_{-\infty}^{\infty} dt \cos\left(\sqrt{\frac{\nu}{2}}\varphi(0, t)\right) \tag{15.168}$$

The action(s) of Eqs. (15.167) and (15.168) look very similar to the action of the sine–Gordon theory, which was discussed in Chapters 4 and 5, except for the fact that (a) the space is restricted to the half-line $x \geq 0$ and (b) the vertex operator acts only on the boundary of the space-time, $x = 0$. For this reason this system is known as the boundary sine–Gordon theory.

The free compactified boson φ in two (or 1 + 1) dimensions is a CFT (see Chapters 4 and 7). In the case at hand this CFT lives on a space that is half the line, $x > 0$, and for all times. In its Euclidean version the space-time is the right half-plane. The boundary vertex operators that enter in Eq. (15.167) and Eq. (15.168) can affect only the boundary conditions of the field φ. Thus, if the vertex operators representing tunneling are absent, the field φ obeys *Neumann boundary conditions*, $\partial_x \varphi = 0$. More physically, this means that, in the absence of tunneling, the tunneling current must vanish. On the other hand, in the limit of strong tunneling, the field φ is pinned at the classical minima of the cosine operators, given by $\varphi = \sqrt{2\nu}2\pi n$ (with $n \in \mathbb{Z}$) for electron tunneling and $\varphi = \sqrt{2/\nu}2\pi n$ for quasiparticle tunneling, and hence obeys *Dirichlet boundary conditions*, $\varphi =$ constant.

Let us consider the case of quasiparticle tunneling at a constriction (case (c)). This situation is described by a theory of a free compactified boson on the half-plane $x > 0$, with Neumann boundary conditions, perturbed by the boundary vertex operator $\cos(\sqrt{\nu/2}\varphi)$, which represents quasiparticle-tunneling processes. We saw above that the scaling dimension of the quasiparticle-tunneling operator is $\nu = 1/m < 1$. Hence it is a relevant perturbation and this fixed point is unstable. To see how this comes about in the boundary sine–Gordon picture, we note first that the Neumann boundary condition changes the correlators of the free boson. We will work in the Euclidean theory. If we denote by $G_0(\vec{x}, \vec{x}') = G_0(\vec{x}-\vec{x}')$ the correlator of the free boson on the entire plane, we obtain the correlator on the half-plane with Neumann boundary conditions $G_N(\vec{x}, \vec{x}')$ using the method of image charges,

$$G_N(\vec{x}, \vec{x}') = G_0(x - x', \tau - \tau') + G_0(x + x', \tau - \tau') \tag{15.169}$$

which satisfies the Neumann boundary condition. Notice that this propagator is not invariant under translations along the x axis but is invariant under translations along the τ axis, as it should be. As a result of the boundary (and of the boundary condition), the scaling dimension of the vertex operator on the boundary is twice the scaling dimension in the bulk, $\Delta = 2(\nu/2) = \nu$, which is consistent with our earlier analysis.

Since this fixed point is unstable, we guess that it must flow to a regime in which the tunneling amplitude Γ_{qp} becomes large, $\Gamma_{\text{qp}} \to \infty$, and the boson φ now obeys Dirichlet boundary conditions. In other terms, the effect of the boundary vertex operator is to induce a flow in the boundary conditions from Neumann (the unstable fixed point) to Dirichlet (the stable fixed point). On the other hand, in terms of the *dual field* ϑ, which we introduced in Section 5.6.2, Eqs. (5.278), the Neumann boundary condition maps onto a Dirichlet boundary condition since φ and the dual field ϑ satisfy the dual (Cauchy–Riemann) relations Eq. (5.280) which imply that

$$\partial_x \varphi = -\partial_\tau \vartheta, \qquad \partial_\tau \varphi = \partial_x \vartheta \tag{15.170}$$

Hence, Dirichlet boundary conditions for the field φ, $\partial_\tau \varphi = 0$, i.e. $\varphi =$ constant, maps onto Neumann boundary conditions for the dual field, $\partial_x \vartheta = 0$. This mapping is known as T duality. On the other hand, in our discussion of the Luttinger model in Chapter 6, we saw that under T duality the Luttinger parameter $K = \nu \mapsto 1/K = 1/\nu$. However, in the boundary sine–Gordon system, Eq. (15.167) and Eq. (15.168), we see that we can identify the Luttinger parameter with the filling factor, $K = \nu$. Thus, T duality maps the quasiparticle operator onto the electron operator and vice versa. Hence, T duality maps the problem of Eq. (15.167) at strong coupling to Eq. (15.168) at weak coupling. The crossover between these two fixed points takes place at the crossover scale (the "Kondo scale") $T_K \sim \Gamma_{\text{qp}}^{1/(1-\nu)}$, since the boundary scaling dimension of the vertex operator is $\nu = 1/m$.

Affleck and Ludwig (1991) considered the effects of boundary conditions on the entropy of CFTs. Using the methods of boundary CFT (Cardy, 1986, 1989), they examined the behavior of the thermodynamic entropy of a CFT as a function of temperature T and system size L. They showed that in a 1D quantum critical system, i.e. a CFT, in the thermodynamic limit $L \to \infty$ the entropy has a finite non-extensive limit as $T \to 0$ given by

$$S = \ln g \tag{15.171}$$

where g is known as the Affleck–Ludwig degeneracy. Furthermore, Affleck and Ludwig also showed that, if the CFT has a perturbation that acts only at the boundary, it induces a flow in the degeneracy g that has universal values for conformal boundary conditions. This is precisely the case in the boundary sine–Gordon theory we are interested in here. In particular, the Affleck–Ludwig degeneracy vanishes for Dirichlet boundary conditions but is non-zero for Neumann boundary conditions, under which it takes the value $g = \ln\sqrt{m}$. In general the Affleck–Ludwig entropy is determined by the fusion rules of the primary fields associated with the conformal boundary conditions and can be computed from the modular S-matrix of the conformal field theory using the Verlinde formula.

It is instructive to construct the perturbation series in powers of the coupling constant $\Gamma_{\rm qp}$ of the Euclidean form of the path integral for the boundary sine–Gordon field φ obeying Neumann boundary conditions at $x = 0$, whose action is given in Eq. (15.168). The correlator of the field φ along the imaginary time axis is

$$\langle N|\varphi(\tau, 0)\varphi(0, 0)|N\rangle = -\ln \tau^2 \tag{15.172}$$

where $|N\rangle$ denotes the ground state of the field φ with Neumann boundary conditions. Let $T^{\pm}$ denote the vertex operator for quasiparticle tunneling, $T^{\pm} = \exp(\pm i\sqrt{\nu/2}\varphi)$. The tunneling term of the action now has the form

$$S_{\rm tun} = \int d\tau \left[\Gamma T^{+}(\tau) + \Gamma^{*} T^{-}(\tau)\right] \tag{15.173}$$

The nth-order term of the expansion in powers of the tunneling amplitude $\Gamma_{\rm qp}$ (where we absorb the oscillatory factor) involves the computation of an expectation value of the n vertex operators $T^{\pm}$ at n (imaginary) times τ_j ($j = 1, \ldots, n$) for a field φ with Neumann boundary conditions that has the form of a logarithmic gas along the imaginary-time axis (see Kane and Fisher (1992) and Chamon *et al.* (2007)),

$$\langle N|T^{q_n} \ldots T_1^{q}|N\rangle = \delta\left(\sum_j q_j\right) \exp\left(\nu \sum_{j>k} q_j q_k \ln|\tau_j - \tau_k|^2\right) \tag{15.174}$$

where we defined the charges $q_j = \pm 1$ to represent the insertions of the vertex operators. The delta-function factor enforces charge neutrality.

We can also construct a similar expansion about the strong-coupling fixed point at which the field φ is pinned (at the boundary) and obeys Dirichlet boundary conditions, which we will represent as the state $|D\rangle$. This expansion now involves a series of instanton and anti-instanton processes that represent the tunneling between successive vacua of the field φ at the boundary. However, we can use T duality to map the limit of strong quasiparticle tunneling to the weak-coupling regime for tunneling of electrons in terms of the dual field ϑ obeying Neumann boundary conditions. The propagator of the field now is

$$\langle D|\varphi(\tau)\varphi(0)|D\rangle = \langle N|\vartheta(\tau)\vartheta(0)|N\rangle = -\ln \tau^2 \tag{15.175}$$

Hence the new perturbation series now involves the computation of the expectation value of the insertion of n vertex operators of the dual field, $\tilde{T}^{\pm} = \exp(\pm i\varphi/\sqrt{2\nu})$, which takes the form

$$\langle D|\tilde{T}^{q_1}(\tau_1)\ldots\tilde{T}^{q_n}(\tau_n)|D\rangle = \delta\left(\sum_j q_j\right)\exp\left(\frac{1}{\nu}\sum_{j>k} q_j q_k \ln|\tau_j - \tau_k|^2\right) \tag{15.176}$$

which is the same expression as that which we would have obtained for electron tunneling. Thus, up to a redefinition of the coupling constant, we see that T duality maps quasiparticle tunneling to electron tunneling and vice versa.

It turns out that the boundary sine–Gordon theory is actually an integrable system solvable by the (thermodynamic) Bethe ansatz (Fendley *et al.*, 1994). The exact solution of the boundary sine–Gordon theory is explicitly self-dual. Using this approach, whose details we will not go into here, Fendley, Ludwig, and Saleur (Fendley *et al.*, 1995a, b) showed that at zero temperature, $T = 0$, the quasiparticle tunneling current I_t obeys explicitly the duality relation

$$I_t(T_K, V, \nu) = \frac{e^2}{h}\nu V - \nu^2 I_t(T_K, V, \nu^{-1}) \tag{15.177}$$

Furthermore, the differential tunneling conductance $G_t(V)$ at voltage V can be expressed in terms of two series (related to each other by T duality):

$$G_t(V) = \begin{cases} \dfrac{e^2}{h}\displaystyle\sum_{n=1}^{\infty} c_n(\nu^{-1})\left(\frac{eV}{T_K}\right)^{2n(\nu^{-1}-1)}, & \dfrac{eV}{T_K} < e^{\delta} \\ \dfrac{e^2}{h}\nu\left[1 - \displaystyle\sum_{n=1}^{\infty} c_n(\nu)\left(\frac{eV}{T_K}\right)^{2n(\nu-1)}\right], & \dfrac{eV}{T_K} > e^{\delta} \end{cases} \tag{15.178}$$

where, as before, the crossover scale is $T_K = \text{constant} \times \Gamma_{qp}^{1/(1-\nu)}$. The coefficients c_n are

$$c_n(\nu) = (-1)^{n+1}\frac{\Gamma(n\nu)\sqrt{\pi}}{\Gamma(n)\Gamma(n(\nu-1)+1/2)} \tag{15.179}$$

where $\Gamma(z)$ is the Euler gamma function, and in this case $\delta = \frac{1}{2}\ln(\nu^{-1} - 1) - (1/(2(1-\nu)))\ln \nu$.

The results of Eq. (15.178) show that in the low-voltage regime the differential conductance is suppressed, and vanishes as $V^{2(m-1)}$. This is consistent with the low-energy limit being governed by the stable fixed point at which the fluid is split into two parts that are weakly coupled by electron tunneling, which is irrelevant. Conversely, in the large-voltage regime the differential conductance approaches $\nu e^2/h$, the Hall conductance of the bulk fluid. This is consistent with this limit being controlled by the unstable fixed point at which the constriction is open and for which quasiparticle tunneling is relevant. This non-trivial behavior, including the crossover, which was originally proposed by Kane and Fisher (1992), has been verified experimentally by Roddaro and coworkers (Roddaro *et al.*, 2004; Stefano *et al.*, 2004).

The problem of tunneling into a quantum Hall edge state from a Fermi liquid at a point contact has been solved by a similar approach (Chamon and Fradkin, 1997). The crossover behavior for this case was in general terms verified by experiment (Chang *et al.*, 1996; Chang, 2003), although the observed tunneling exponent does not quite agree with the theoretical prediction. Presumably this discrepancy may be due to the fact that in this experiment the setup is closer to a line junction (or, rather, an array of point contacts) instead of a single point contact, which is what the theory actually describes.

15.7 Noise and fractional charge

We will now apply the formalism of the preceding section to the problem of shot noise in the tunneling current. The interest of this question is that it gives a direct way to measure the fractional charge of the quasiparticles. Here we will discuss the simplest case, namely the tunneling current in a quantum Hall constriction of a Laughlin state.

The noise spectrum $S(\omega)$ of the quasiparticle-tunneling current I_{qp} is obtained from the tunneling-current correlation function by the expression (Kane and Fisher, 1994; Chamon *et al.*, 1996)

$$S(\omega) = \int_{-\infty}^{\infty} dt \left\langle \left\{ I_{\text{qp}}(t), I_{\text{qp}}(0) \right\} \right\rangle e^{i\omega t} \tag{15.180}$$

where $I_{\text{qp}} = 2e\nu\Gamma_{\text{qp}} \sin\left(\sqrt{\nu/2}\varphi + \omega_0^* t\right)$ is the quasiparticle-tunneling current defined in Eq. (15.165) and $\omega_0^* = e\nu V/\hbar$.

The quasiparticle-tunneling-current correlator is

$$\begin{aligned} \langle I_{\text{qp}}(t) I_{\text{qp}}(0)\rangle = {} & \left(\frac{e}{m}\right)^2 |\Gamma_{\text{qp}}|^2 e^{i\omega_0^* t} \\ & \times \left\langle \exp\left(i\sqrt{\frac{\nu}{2}}\varphi(t)\right) \exp\left(-i\sqrt{\frac{\nu}{2}}\varphi(0)\right)\right\rangle + \text{c.c.} \end{aligned} \tag{15.181}$$

To lowest order in the matrix element $\Gamma_{\rm qp}$ it is given by

$$\langle I_{\rm qp}(t) I_{\rm qp}(0)\rangle = (e\nu)^2 |\Gamma_{\rm qp}|^2 \frac{2\cos(\omega_0^* t)}{(\epsilon + it)^{2\nu}} \tag{15.182}$$

Thus, the spectral function $S(\omega)$ is found to be (Chamon *et al.*, 1996)

$$S(\omega) = (e\nu)^2 \Gamma_{\rm qp}^2 \left[f_+(\omega_0^* + \omega) + f_-(\omega_0^* + \omega) + f_+(\omega_0^* - \omega) + f_+(\omega_0^* - \omega) \right] \tag{15.183}$$

where

$$f_\pm(\omega) = \int_{-\infty}^{\infty} dp \, \frac{e^{i\omega p}}{(\epsilon \mp ip)^{2\nu}} = \frac{2\pi}{\Gamma(2\nu)} |\omega|^{2\nu - 1} e^{-|\omega|\epsilon} \theta(\pm\omega) \tag{15.184}$$

where $\Gamma(x)$ is the Euler gamma function and $\theta(x)$ is the step function. Using the result that the expectation value of the tunneling current at voltage V, to lowest order in $\Gamma_{\rm qp}$, is given by

$$\langle I_{\rm qp}\rangle = \frac{2\pi}{\Gamma(2\nu)} e\nu |\Gamma_{\rm qp}|^2 \omega_0^{*2\nu - 1} \tag{15.185}$$

we find that the noise spectrum is

$$S(\omega) = e\nu \langle I_{\rm qp}\rangle \left[\left(1 - \frac{\omega}{\omega_0^*}\right)^{2\nu - 1} + \left(1 + \frac{\omega}{\omega_0^*}\right)^{2\nu - 1} \right] \tag{15.186}$$

In the limit of zero frequency, the noise spectrum approaches the shot-noise form

$$\lim_{\omega \to 0} S(\omega) = 2e^* \langle I_{\rm qp}\rangle \tag{15.187}$$

with $e^* = e\nu = e/m$. Therefore a measurement of the shot noise of the tunneling current measures directly the fractional charge of the quasiparticles (Kane and Fisher, 1994).

The result of Eq. (15.187) has been used to measure the fractional charge of the quasiparticles in tunneling-current-noise experiments (de Picciotto *et al.*, 1997; Saminadayar *et al.*, 1997) in Laughlin states, and in the non-abelian state with filling fraction 5/2 (Dolev *et al.*, 2008). The experimental results are generally consistent with the quasiparticles having a fractional charge, although there are unresolved questions regarding the experiments.

15.8 Quantum interferometers

Constrictions and point-contact tunneling offer a way to measure the fractional charge of the quasiparticles. However, to measure the fractional statistics it is necessary for the quasiparticles to have more than one interference pathway. In principle the simplest way to measure the fractional statistics is a quantum interferometer of the type shown in Fig. 15.12, which is an idealized description of such a device, a "Fabry–Pérot" interferometer.

The theory of the quantum Hall interferometer for Laughlin states was developed by Chamon *et al.* (1997). This theory was extended to the case of non-abelian states (Fradkin *et al.*, 1998; Bonderson *et al.*, 2006; Stern and Halperin, 2006; Bishara *et al.*, 2009). Several experimental groups have tested these theoretical results. The predictions of Chamon *et al.* (1997) for the Laughlin state at $\nu = 1/3$ have been tested by Camino *et al.* (2005), who, up to some interpretation issues, have qualitatively confirmed the theoretical results. The interferometer for the non-abelian quantum Hall state at $\nu = 5/2$ has been the focus of challenging experiments that at least qualitatively have produced results agreeing with the theoretical predictions (Willett *et al.*, 2010).

We will consider a Laughlin quantum Hall state with two constrictions, as shown in Fig. 15.12, located at coordinates x_j $(j = 1, 2)$, with the x axis running parallel to the edges. The tunneling-matrix elements are, respectively, Γ_1 and Γ_2. The flux Φ enclosed in the region $x_1 \leq x \leq x_2$, which enters into the theory through the phase of the tunneling amplitudes, creates a number $N_{\rm q}$ of Laughlin quasiparticles/quasiholes, and thus controls the deviation of the flux–charge relation for the ideal fractional quantum Hall state. The tunneling currents in the two constrictions are labeled by I_1 and I_2.

In the absence of constrictions, the transmitted current along (say) the bottom edge is simply the Hall current $I_{\rm H}$, with a Hall conductance $\nu e^2/h$. In the presence of the constrictions, the transmitted current is reduced to $I_{\rm trans} = I_{\rm H} - (I_1 + I_2)$. This transmitted current will exhibit oscillations as a function of the flux Φ due to (a) the Aharonov–Bohm processes affecting the tunneling quasiparticles and (b) an

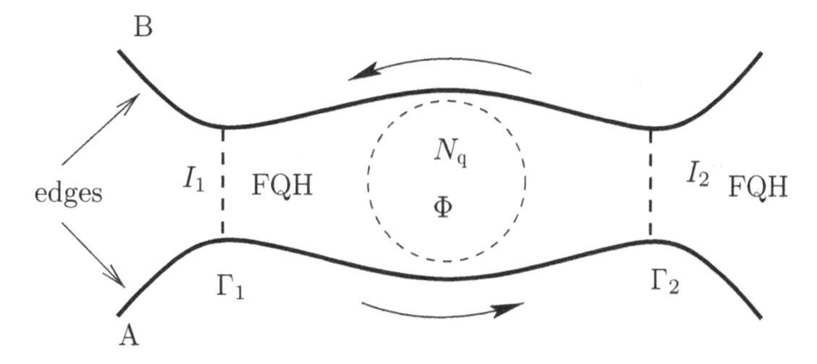

Figure 15.12 A fractional quantum Hall (FQH) interferometer.

additional interference due to the fractional statistics of the tunneling quasiparticles with localized quasiparticles in the region between the two constrictions. Which interference process is seen depends on how the interferometer is operated. If we keep the filling fraction fixed as the external magnetic field is varied, the oscillation will be due entirely to the Aharonov–Bohm effect, since no quasiparticles are added to (or subtracted from) the bulk. If, on the other hand, the number of particles in the interference region is held fixed, the oscillations are due to both the Aharonov–Bohm effect and the fractional statistics of the quasiparticles. Since the Aharonov–Bohm oscillation depends on the charge of the tunneling quasiparticle, in this mode the interferometer can also be used to measure the fractional charge. It is in practice hard to control the total charge enclosed in the interference region, and this makes the operation of the interferometer difficult, leading to subtle but revealing effects (Halperin *et al.*, 2011). For simplicity, we will ignore these problems and assume that the ideal conditions can be reproduced in the experiment.

15.8.1 Interferometers for abelian quantum Hall states

The theory of the quantum Hall interferometer for abelian (Laughlin) states developed by Chamon *et al.* (1997), whose work we will follow in this section, is constructed as an extension of the theory of tunneling at quantum Hall constrictions which we presented in the preceding section. Let $\phi_{\rm L}$ and $\phi_{\rm R}$ be, respectively, the chiral bosons for the top and bottom edges of the quantum Hall fluid as depicted in Fig. 15.12. We will assume that the tunneling-matrix elements Γ_1 and Γ_2 are weak enough that we can use perturbation theory to describe the interference. However, unlike the case of a single point contact, we will not be able to simplify the discussion by "flipping" the direction of flow along one of the edges. Instead we will regard $\phi_{\rm R}$ and $\phi_{\rm L}$ as the two chiral components of a single non-chiral boson $\phi = \phi_{\rm R} + \phi_{\rm L}$ as in the bosonized theory of the Luttinger liquid discussed in Chapter 6. In this language, the quasiparticle operators $\Psi_{\rm qp,R}$ and $\Psi_{\rm qp,L}$ of each edge are

$$\Psi_{\rm qp,R} \propto e^{i\sqrt{\nu}\phi_{\rm R}} e^{ik_{\rm F}x}, \qquad \Psi_{\rm qp,L} \propto e^{-i\sqrt{\nu}\phi_{\rm L}} e^{-ik_{\rm F}x} \tag{15.188}$$

where $k_{\rm F}$ is the momentum of the quasiparticles (which can be neglected for a single point contact). If A is the area of the quantum Hall fluid enclosed between the two point contacts, then the momentum difference of electrons in the top and bottom edges is $2k_{\rm F,e} = 2\pi BA/\phi_0$ (where $BA = \Phi$ and $\phi_0 = hc/e$ is the flux quantum), whereas for quasiparticles it is instead $2k_{\rm F} = \nu 2k_{\rm F,e}$. These phase factors can be absorbed into the definition of the tunneling amplitudes,

$$\Gamma_{1,2} = \bar{\Gamma}_{1,2} e^{\pm i\pi\nu\Phi/\phi_0} \tag{15.189}$$

With these definitions the Lagrangian (density) of the edge states of a Laughlin quantum Hall fluid with two constrictions is

$$\mathcal{L} = \frac{1}{8\pi}\left[(\partial_t\phi)^2 - v^2(\partial_x\phi)^2\right] - \sum_{j=1,2} \Gamma_j e^{-i\omega_0^* t}\delta(x-x_j)e^{i\sqrt{\nu}\phi(t,x_j)} + \text{h.c.} \quad (15.190)$$

which is also the Lagrangian of a (spinless) Luttinger liquid with two back-scattering impurities. Here, as before, the Josephson frequencies are $\omega^* = e^*V/\hbar$, where V is the voltage across both constrictions and v is the velocity of the edge states.

A straightforward calculation to second order in the tunneling-matrix elements yields the result that the tunneling current $I_t = I_1 + I_2$ at zero temperature is (Chamon *et al.*, 2007)

$$I_t = e^*|\Gamma_{\text{eff}}|^2 \frac{2\pi}{\Gamma(2\nu)} |\omega_0^*|^{2\nu-1}\,\text{sgn}(\omega_0^*) \quad (15.191)$$

where the "effective" matrix element Γ_{eff} is given by

$$|\Gamma_{\text{eff}}|^2 = |\Gamma_1|^2 + |\Gamma_2|^2 + \left(\Gamma_1\Gamma_2^* + \Gamma_1^*\Gamma_2\right) F_\nu\left(\frac{\omega_0^* a}{v}\right) \quad (15.192)$$

where a is the distance between the two constrictions and

$$F_\nu(x) = \sqrt{\pi}\frac{\Gamma(2\nu)}{\Gamma(\nu)}\frac{J_{\nu-1/2}(x)}{(2x)^{\nu-1/2}} \quad (15.193)$$

Here $\Gamma(x)$ is the Euler gamma function and $J_{\nu-1/2}(x)$ is the Bessel function of the first kind.

In the presence of N_q quasiholes in the area between the two constrictions, the contribution of the phases of the tunneling-matrix elements gets shifted to

$$\Gamma_1^*\Gamma_2 = \bar{\Gamma}_1^*\bar{\Gamma}_2 \exp\left[-i2\pi\left(\nu\frac{\Phi}{\phi_0} - N_q\nu\right)\right] \quad (15.194)$$

where the phase shift $2\pi\nu N_q$ is the contribution of the fractional statistics of the tunneling quasiparticle as its worldline braids with the N_q localized quasi-particles. Thus, there is an interference contribution to the tunneling current (and hence also to the transmitted current), which is sensitive both to the charge of the quasiparticles and to their fractional statistics.

15.8.2 Interferometers for non-abelian quantum Hall states

This analysis has been extended to the case of non-abelian quantum Hall states. Let us consider for concreteness the case of the Moore–Read state at filling factor $\nu = 5/2$, which was discussed in Sections 14.8 and 15.4.5. The Moore–Read state has a composite edge with a charge sector described by a chiral boson ϕ and a chiral Majorana fermion χ, each with its own propagation velocity. The fundamental quasiparticle is a composite object of the charge and neutral sectors,

$$\sigma \sim \chi \exp\left(\frac{i}{2\sqrt{2}}\phi\right) \tag{15.195}$$

Thus, the main difference is that the quasiparticle which takes part in the tunneling processes, the σ quasiparticle, obeys non-abelian braiding statistics.

The tunneling current has an interference component that is directly related to the measurement of expectation values of Wilson loop operators in the effective Chern–Simons gauge theory. To see this we follow the work of Fradkin and coworkers (Fradkin *et al.*, 1998) and consider a quasihole that is injected at point A in the lower edge and then tunnels at the first contact, and arrives at point B at the left end of the top edge in state $|\psi\rangle$ (see Fig. 15.12). Let us now consider a second quasihole also injected at point A, but which now tunnels to the top edge at the second constriction, arriving at point B in state $e^{i\alpha} B_{N_q}|\psi\rangle$, where α is the Aharonov–Bohm phase determined by the flux Φ piercing the interferometer and B_{N_q} is the braiding operator for the second quasihole to encircle the N_q quasiholes in the region between the constrictions.

Then, the longitudinal tunneling conductance measured at point B of Fig. 15.12 is proportional to (Fradkin *et al.*, 1998)

$$\sigma_{xx} \propto |\Gamma_1|^2 + |\Gamma_2|^2 + \text{Re}\left[\Gamma_1^* \Gamma_2 e^{i\alpha} \langle\psi|B_{N_q}|\psi\rangle\right] \tag{15.196}$$

The matrix element $\langle\psi|B_{N_q}|\psi\rangle$ is the expectation value of the Wilson loop operators of the tunneling quasiholes braided with the Wilson loops of the static quasiholes in the enclosed region, which is equal to the Jones polynomial $V_{N_q}(e^{i\pi/4})$ (see Section 14.8). Therefore, the oscillatory component of the tunneling current (and conductance) measures a topological invariant!

In the simplest non-abelian quantum Hall state, the Moore–Read state, we saw that the effective-field theory is closely related to the $SU(2)_2$ Chern–Simons gauge theory. In this case the general result of Eq. (15.196) takes the much simpler form. Using the skein relation, Eq. (14.156), the matrix element $\langle\psi|B_{N_q}|\psi\rangle$ can be computed explicitly, with the result (Bonderson *et al.*, 2006; Stern and Halperin, 2006)

$$\sigma_{xx} \propto |\Gamma_1|^2 + |\Gamma_2|^2 \tag{15.197}$$

for N_q odd, and

$$\sigma_{xx} \propto |\Gamma_1|^2 + |\Gamma_2|^2 + 2|\Gamma_1||\Gamma_2|(-1)^{N_\psi} \cos\left(\alpha + \arg\left(\frac{\Gamma_2}{\Gamma_1}\right) + N_q\frac{\pi}{4}\right) \tag{15.198}$$

for N_q even. Here $N_\psi = 1$ when the N_q quasiholes are (or fuse into) the state ψ and $N_\psi = 0$ otherwise. The interference term for N_q odd vanishes since an odd number of σ particles cannot fuse into the identity I. Hence the expectation value vanishes for N_q odd, and the interference disappears if there is an odd number of σ quasiholes in the enclosed region.

Thus, in addition to the oscillations due to the Aharonov–Bohm effect and to fractional statistics, in the non-abelian case there is an extra oscillation, which for the Moore–Read state takes different values depending on whether there is an odd or even number of quasiparticles trapped inside the interferometer. The origin of this even–odd effect lies in the fusion rules of the non-abelian excitations. In the case of the Moore–Read states the fundamental quasiparticle, which we denoted by σ, obeys the Ising fusion algebra: $\sigma \star \sigma = 1 + \psi$, $\sigma \star \psi = \psi$, and $\psi \star \psi = 1$ (here ψ is a chiral Majorana fermion and 1 is the identity operator). Since σ particles can fuse into two different channels, 1 and ψ, the expectation value of a set of σ operators will vanish if the resulting state is not the identity, 1. Hence, the even–odd oscillation is a signature of the collective state of these non-abelian quasiparticles. Therefore, the interferometer can also detect non-abelian statistics.

15.9 Topological quantum computation

Classical computers (i.e. all the computers we know so far) operate by making sequential binary operations of bits. Bits are physical systems with two possible states, usually denoted by 0 and 1. There are many quantum-mechanical physical systems that also have two states. We call them two-level systems. One example of such systems is the electron spin, which can assume two possible states, $|\uparrow\rangle$ and $|\downarrow\rangle$. In contrast to the classical bits, these quantum bits or *qubits*, can be in *any* linear combination of these two states. In other words, in quantum mechanics we operate on a Hilbert space of states spanned by the basis states of the qubit. This linear feature of quantum mechanics makes the notion of using the quantum evolution of states for a computation a very appealing prospect. There is, however, a drawback (there always is!): decoherence. All physical systems are coupled in one way or another to their physical environment, and in most cases this leads to a loss of coherence, which in this context implies a loss of information. This problem is the main obstacle to most schemes of quantum computation. A beautiful introduction to quantum computing, including topological quantum computing, can be seen in Preskill's lectures (Preskill, 2004).

It was realized first by Kitaev (2003), in a paper that has circulated since 1997, and was later expanded and developed by Freedman and coworkers (Freedman, 2001; Freedman *et al.*, 2002a, b), that topological field theories offer, in principle, a pathway for quantum computation without decoherence. The essence of this proposal is that topological field theories quite generally have finite-dimensional topologically protected Hilbert spaces. We saw elsewhere in this book that the effective low-energy theory of physical systems in a topological phase is a topological field theory. Thus one is led to the notion of regarding the topologically protected degeneracies of topological phases as the qubits themselves.

For instance, in the case of the Moore–Read state we can consider a system with four σ quasiparticles, which, as we saw, supports a two-dimensional topologically protected Hilbert space. A braiding operation by which one σ particle is adiabatically transported leads to a state that is a linear combination of the two basis states. In this context, adiabatic means a sufficiently slow process that does not generate states outside this Hilbert space. Hence a braiding operation is represented as a unitary transformation in this Hilbert space. Topological protection here means that local excitations and/or disorder are decoupled from this Hilbert space and, hence, that there is no available mechanism for decoherence.

Since we know that non-abelian quantum Hall states support such non-abelian excitations with topologically protected Hilbert spaces, it is natural to regard these degenerate states as topologically protected qubits. Then the next question is that of how to manipulate these states and how to read off the result of a computation. One possible way is to use variants of the quantum interferometers discussed in the preceding section as devices that manipulate qubits. There is an on-going effort, both experimental and theoretical, to generate devices that allow the controlled manipulation of these degenerate Hilbert spaces (Das Sarma *et al.*, 2008). An idealized qubit based on the $\nu = 5/2$ Moore–Read state consists in adding two islands (quantum dots) to the central region of the interferometer of Fig. 15.12 with a control gate in between them. The purpose of the two islands is to trap the σ particles, and the control gate monitors the state (Das Sarma *et al.*, 2005).

From the physics point of view this is a fascinating prospect. For this scheme (and its variants) to work, several formidable problems need to be solved. One is the issue that quantum Hall states only occur at very low temperatures and high magnetic fields. It may be possible to circumvent these very practical issues by using devices made of topological insulators, which is currently actively being explored. This is an appealing possibility but with problems of its own (such as making these insulators insulate!). At any rate, at the time of writing this field is still in its infancy.

16

Topological insulators

16.1 Topological insulators and topological band structures

The term topological insulator refers to a novel (in 2011) class of solid-state systems that have quantized transport properties due to topological properties of their band structures. In this chapter I will provide a description of the salient ideas behind this new and rapidly growing field. I will certainly not attempt to be exhaustive in the presentation. Several specialized reviews have recently become available and the reader is referred to them for more details (including an extensive list of references) (Hasan and Kane, 2010; Hasan and Moore, 2011; Maciejko *et al.*, 2011; Qi and Zhang, 2011).

What is a topological insulator? It is an electronic system that is an insulator but whose band structure is characterized by a topological invariant, i.e. a number that in general is quantized to be an integer. As such, states of this type are robust in the sense that their physical properties are stable (unchanged) under the action of local perturbations of finite size. From this definition it follows that the properties of topological insulators can be characterized at the level of free-fermion systems and are not necessarily the result of strong-correlation physics. They are a generalization of the conceptual framework behind the integer quantum Hall states. However, in spite of the topological properties of their band structures, the ground states of topological insulators are essentially unique and, even when degeneracies may be present, they do not depend on the topology of the space. Hence topological insulators are not topological fluids in the sense of the fractional quantum Hall states or of the deconfined gauge theories (and spin liquids). It is quite likely that, at least in two dimensions, there may exist topological fluids that are generalizations of topological insulators. At the time of writing this is an open area of research whose future is difficult to predict.

As we will see, one important consequence of the topological character of these insulators is the existence of edge states, which are gapless excitations with support at the edges of the physical system, and cannot be affected in an essential way by the effects of disorder and interactions. Quite surprisingly, in the simplest cases these non-trivial insulators can be characterized at the level of a one-electron theory, that is, by a property of their band structure. In some special, but very interesting, cases, these states arise as the low-lying excitations (or quasiparticles) of condensates, such as certain superfluids and superconductors, as well as in special condensates in the particle–hole channel.

In two dimensions, systems with topologically non-trivial band structures turn out to exhibit a variety of unusual quantized transport properties such as the *anomalous Hall effect*, i.e. a quantum Hall effect in the absence of an external magnetic field, and the quantum *spin Hall effect*, which is a similar phenomenon involving instead the spin current rather than the charge current. Experimental evidence for the quantized spin Hall effect has been found in 2D electron gases in HgTe–CdTe quantum wells (König *et al.*, 2008). An anomalous quantum Hall effect has been predicted to exist in graphene bilayers, but so far has not been seen experimentally. In three dimensions topological insulators are more subtle, and are characterized by having topologically protected surface (or edge) states whose excitations are *gapless chiral Dirac fermions*, i.e. the spin of the excitation points along the direction of the momentum in much the same way as the spins of neutrinos were supposed to do (if they were massless, which we now know they are not). Materials that are predicted to be 3D topological insulators include Bi_xSb_{1-x}, Bi_2Se_3, and Bi_2Te_3. Although so far these materials are not insulating in the bulk, the chiral Dirac surface fermions have been detected in angle-resolved photo-emission experiments (Hasan and Kane, 2010; Hasan and Moore, 2011).

In retrospect, the first example of a topological insulator was actually found by Thouless, Kohmoto, Nightingale, and den Nijs (TKNN) (Thouless *et al.*, 1982) in their theory of the integer quantum Hall effect. We discussed this theory in Chapter 12. In addition, the mean-field theory of the chiral spin liquid, which was discussed in Chapter 10, can also be interpreted in hindsight as describing a system with a quantum anomalous Hall effect. The theory of topological insulators is, in essence, a generalization of the TKNN theory to band structures with non-trivial topology in time-reversal-invariant electronic systems. For this reason we will begin with a discussion of the TKNN theory but from a more general perspective.

16.2 The integer quantum Hall effect as a topological insulator

The role of topology in band structure is simple to formulate. Here we will revisit the theory of Thouless and coworkers (TKNN) of the integer Hall effect on lattices, the Hofstadter problem, since it is the prototype of the topological insulator. Let us consider a system of spinless fermions on a 2D lattice. Let us imagine a case in which the system has M electronic bands with eigenvalues $\{E_m(\vec{k})\}$ (with $m = 1, \ldots, M$) and the eigenvectors are the Bloch states $\{|u_m(\vec{k})\rangle\}$ such that the wave functions are

$$\psi_m(\vec{x}) = u_m(\vec{k}) e^{i\vec{k}\cdot\vec{x}} \tag{16.1}$$

where $\vec{k}$ is a (quasi-)momentum in the first Brillouin zone of the lattice. The Bloch states will be assumed to be non-degenerate and hence the eigenvalues obey the strict inequality

$$|E_m(\vec{k}) - E_n(\vec{k})| > 0 \tag{16.2}$$

for all momenta in the first Brillouin zone. Here we will consider the case in which $N < M$ bands are fully occupied, and hence the gap between the Nth and $(N+1)$th bands does not close in the Brillouin zone.

Next we follow TKNN and define the *Berry connection* $\vec{\mathcal{A}}_m(\vec{k})$, the two-component vector field

$$\mathcal{A}_i^{(m)}(\vec{k}) = i\langle u_m(\vec{k})|\nabla_{k_i}|u_m(\vec{k})\rangle \tag{16.3}$$

where $i = 1, 2$ are two orthogonal directions in momentum space. A redefinition of the basis of Bloch states $\{|u_m(\vec{k})\rangle\}$ induces a unitary transformation on the vector $\{u_m(\vec{k})\}$ of occupied bands. In general this is a transformation with gauge group $\mathrm{U}(1)^N$. Occasionally degeneracies in the band structure may lead to non-abelian symmetries. A $\mathrm{U}(1)^N$ gauge transformation is induced by a change in the local (on the Brillouin zone) phase of the Bloch state,

$$\begin{aligned} |u_m(\vec{k})\rangle &\mapsto e^{if_m(\vec{k})}|u_m(\vec{k})\rangle \\ \mathcal{A}_i^{(m)}(\vec{k}) &\mapsto \mathcal{A}_i^{(m)}(\vec{k}) + \nabla_{k_i} f_m(\vec{k}) \end{aligned} \tag{16.4}$$

where we assumed that the functions $f_m(\vec{k})$ are continuous and differentiable. Since the physics cannot depend on the choice of basis (or, rather, its redefinition), we are led to the conclusion that only gauge-invariant quantities that are invariant under these smooth redefinitions are physically meaningful. Thus, the physical content must be encoded in the curl (or curvature) of the Berry connection, which, in two dimensions, is the pseudo-scalar quantity

$$\mathcal{F}_m(\vec{k}) = \epsilon_{ij}\,\partial_{k_i}\mathcal{A}_j^{(m)}(\vec{k}) \tag{16.5}$$

The flux of the Berry curvature $\mathcal{F}_m$ over the Brillouin zone (BZ) is

$$\int_{\mathrm{BZ}} d^2k\,\mathcal{F}_m(\vec{k}) = \oint_\Gamma d\vec{k}\cdot\vec{\mathcal{A}}^{(m)}(\vec{k}) \tag{16.6}$$

where Γ is the boundary of the Brillouin zone. However, the circulation of the Berry connection on the boundary Γ of the Brillouin zone must obey the (Dirac) quantization condition

$$\oint_{\Gamma} d\vec{k} \cdot \vec{\mathcal{A}}^{(m)}(\vec{k}) = 2\pi N_m \tag{16.7}$$

where N_m are integers. This condition is required in order for the Bloch states to be single-valued over the Brillouin zone.

The integers N_m are the topological invariants known as the first *Chern number*. We recognize that these integers are the same as the topological invariants of the TKNN construction for the Hofstadter problem. They are topological in the following sense. The Bloch states are eigenstates of the band Hamiltonian $\mathcal{H}(\vec{k})$,

$$H = \sum_{n,m} \int_{\mathrm{BZ}} d^2k \, c^{\dagger}(\vec{k}) \mathcal{H}_{n,m}(\vec{k}) c_m(\vec{k}) \tag{16.8}$$

where BZ denotes the 2D Brillouin zone.

Let us assume that for a particular band m the Chern number does not vanish, $N_m \neq 0$. Then, a smooth change of the parameters of the Hamiltonian $\mathcal{H}_{n,m}(\vec{k})$ cannot change the value of the Chern number (since it is an integer!). The only way to change the Chern number by a smooth deformation of the Hamiltonian is for the gap to close (under the deformation) at some point $\vec{k}_0$ of the Brillouin zone. In fact, we have already seen in Section 12.8 that in the TKNN theory of the integer quantum Hall effect the Chern number yields the value of the quantized Hall conductance.

We will now apply these ideas to insulators whose band structures are also characterized by the value of the Chern number, even though these physical 2D systems do not have an applied external magnetic field. Nevertheless, depending on whether they are spin-polarized or not, these insulators exhibit either an anomalous Hall effect, i.e. a quantum Hall effect at zero magnetic field, or a quantum spin Hall effect. Some of these ideas extend to 3D systems. However, in three dimensions the topological invariant is not an integer but instead can take only two possible values. The Berry-phase concept is also useful to describe phases of Fermi fluids with broken time-reversal invariance, with an unquantized anomalous Hall effect (Sun and Fradkin, 2008).

16.3 The quantum anomalous Hall effect

Let us consider a 2D non-interacting electronic system with the Fermi energy located at a gap between two of its energy bands. This system has all its bands below the Fermi energy fully occupied and all the bands above the Fermi energy empty. This is the prototype of a band insulator. From a macroscopic point of view, a band insulator is characterized by having a finite dielectric constant and, at $T = 0$, a vanishing conductivity tensor. However, if the insulator breaks time-reversal invariance, the Hall conductivity may be non-vanishing even in the absence of an external magnetic field. The existence of a finite Hall conductivity in the absence of an external magnetic field is known as the *anomalous Hall effect* regardless of whether the system is a metal or an insulator. However, as we will see below, if the system is an insulator, the anomalous Hall conductivity is quantized and we have a *quantum anomalous Hall effect.*

16.3.1 A square lattice with flux π per plaquette

There are two simple model systems that have an anomalous quantum Hall effect. One example is a system of spinless fermions on a square lattice with flux $\Phi = \pi$ (half the flux quantum) per plaquette with the following free-fermion Hamiltonian:

$$\begin{aligned} H = &-t \sum_{\vec{r},j=1,2} \left(c^\dagger(\vec{r})e^{iA_j(\vec{r})}c(\vec{r}+\vec{e}_j) + \text{h.c.}\right) \\ &- t'\sum_{\vec{r}} \left(c^\dagger(\vec{r})e^{i\chi_+}c(\vec{r}+\vec{e}_1+\vec{e}_2) + \; c(\vec{r}+\vec{e}_2)^\dagger e^{i\chi_-}c(\vec{r}+\vec{e}_1) + \text{h.c}\right) \end{aligned} \tag{16.9}$$

where $\vec{r} = (x_1, x_2)$ runs over the sites of the square lattice, and the vector field A_i is chosen to have flux $\Phi = \pi$ on each plaquette of the square lattice. In the Landau gauge we adopted in Chapter 12, $A_2 = 0$, $A_1 = \pi$ for x_1 even, and $A_1 = 0$ for x_1 odd, the magnetic unit cell has two inequivalent lattice sites. The t' terms open a gap in the spectrum. Here, as in the case of the chiral spin liquid, we will choose the phases along the diagonals of the plaquette to take the values $\chi_\pm = \pm\pi/2$ in such a way that the flux on every elementary triangle of the square lattice is $+\pi/2$.

For $t' = 0$ this is a special case of the Hofstadter problem discussed in Sections 12.2 and 12.8. As we saw in Chapter 8, and in the theory of the chiral spin liquid in Chapter 10, the first two terms of the Hamiltonian of Eq. (16.9) represent a theory of gapless lattice Dirac (or Kogut–Susskind) fermions. Except for the site-potential term in Eq. (16.9), the spectrum of this system was derived in Section 10.2, and it was given in Eq. (10.50). The spectrum of this Hamiltonian is particle–hole-symmetric and, in the gauge we chose above, it is given by

$$E_\pm(\vec{k}) = \pm\sqrt{(2t\cos k_1)^2 + (2t\cos k_2)^2 + (4t'\sin k_1 \sin k_2)^2} \tag{16.10}$$

At half-filling the Fermi energy is at $E_\text{F} = 0$ and the lower band (labeled $-$) is the filled valence band, while the upper band (labeled $+$) is the empty conduction band. For small t' the spectrum has a small energy gap at the Fermi energy located in momentum space near the "nodal" point $(\pi/2, \pi/2)$ and its reflections across the three other quadrants of the first Brillouin zone. For $t' = 0$ the valence and conduction bands cross at these four points. For excitation energies that are small on the scale of the Fermi energy the excitation energies have a linear, relativistic-like, spectrum,

$$E_{\pm}(\vec{q}) = \pm 2t|\vec{q}| + O(q^2) \tag{16.11}$$

where the momentum $\vec{q}$ is measured from one of the crossing points.

The low-energy physics of this system is described by a system of two species (or valleys) of two-component Dirac fermions, $u_a(x)$ and $v_a(x)$ (with $a = 1, 2$ being the two-spinor index), which for $t' \neq 0$ obey a Dirac equation

$$\begin{aligned} \left(i\gamma_0\,\partial_0 - i v_{\mathrm{F}}\vec{\gamma}\cdot\vec{\nabla} + m\right)_{ab} u_b(\vec{x}) = 0 \\ \left(i\gamma_0\,\partial_0 - i v_{\mathrm{F}}\vec{\gamma}\cdot\vec{\nabla} + m\right)_{ab} v_b(\vec{x}) = 0 \end{aligned} \tag{16.12}$$

where the 2D Dirac gamma matrices are given in terms of the three 2×2 Pauli matrices

$$\gamma_0 = -\sigma_2, \qquad \gamma_1 = -i\sigma_1, \qquad \gamma_2 = -i\sigma_3 \tag{16.13}$$

and obey the Dirac algebra,

$$\{\gamma^{\mu}, \gamma^{\nu}\} = 2g^{\mu\nu} \tag{16.14}$$

where $g^{\mu\nu} = \mathrm{diag}(1, -1, -1)$ is the metric tensor of the $(2+1)$-dimensional (Minkowski) space-time. Here we have used the fact that the Fermi velocity is $v_{\mathrm{F}} = 2ta_0$ (with a_0 being the lattice spacing). The mass terms in Eq. (16.12) have *the same sign* for both species of fermions. While the magnitude of the mass is set by the next-nearest-neighbor hopping amplitude, $m \propto t'$, the sign of the mass term is determined by the sign of the $\pi/2$ flux threading each elementary triangle of the lattice.

A mass term for the two species of fermions can also be generated by a site-potential energy that alternates between the two sublattices of the square lattice, or by a Peierls unidirectional distortion of the bonds, as shown in Section 10.2. However, the resulting mass terms have *opposite* signs for the two species. We will see below that the relative sign of the mass terms is related to the role of time-reversal invariance.

16.3.2 Graphene

Another simple system with a similar spectrum is graphene. Graphene is a system of carbon atoms arranged into a 2D honeycomb lattice. Although as a conceptual model 2D carbon had "existed" for many years, the 2D form of carbon known as graphene was only discovered quite recently (Novoselov *et al.*, 2004). Most of the observed transport properties of graphene, both without a magnetic field and with a strong magnetic field, can be explained in terms of its low-energy theory, a system of gapless Dirac fermions (Castro Neto *et al.*, 2009).

In charge-neutral graphene, only one orbital of the carbon atom, the π orbital, is partially occupied, while the other orbitals are either empty or full, and hence separated by a large energy gap. The simplest description of the electronic states of graphene is a tight-binding model on the honeycomb lattice with only one orbital (or state) per site. The honeycomb lattice, shown in Fig. 16.1(a), has two sites in each unit cell and hence can be regarded as two interpenetrating triangular lattices, which we label by A and B. Let $\vec{r}_A$ denote the A site of the unit cell. Each A site is separated from its neighboring B sites by the vectors (in units of the spacing between two nearest-neighboring atoms on the same sublattice if $a = 1$)

$$\vec{d}_1 = \left(\frac{1}{2\sqrt{3}}, \frac{1}{2}\right), \quad \vec{d}_2 = \left(-\frac{1}{\sqrt{3}}, 0\right), \quad \vec{d}_3 = \left(\frac{1}{2\sqrt{3}}, -\frac{1}{2}\right) \tag{16.15}$$

For future use we will also define the six next-nearest-neighbor displacement vectors $\pm\vec{a}_i$ (with $i = 1, 2, 3$) by

$$\vec{a}_1 = \vec{d}_2 - \vec{d}_3, \quad \vec{a}_2 = \vec{d}_3 - \vec{d}_1, \quad \vec{a}_3 = \vec{d}_1 - \vec{d}_2 \tag{16.16}$$

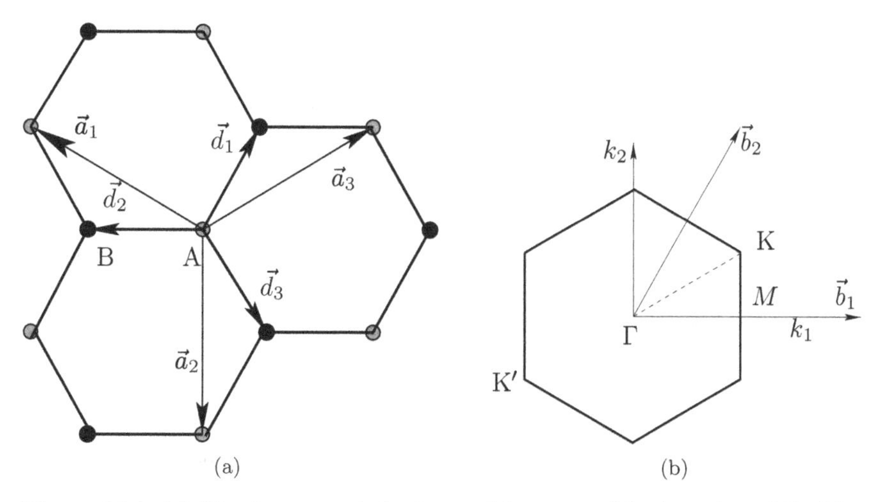

Figure 16.1 (a) The honeycomb lattice and its two sublattices A and B. The rhombus is the unit cell. The two triangular sublattices are connected by the vectors $\vec{d}_1$, $\vec{d}_2$, and $\vec{d}_3$, and the three next-nearest-neighbor displacement vectors are $\vec{a}_1$, $\vec{a}_2$, and $\vec{a}_3$ (see the text). (b) The first Brillouin zone; $\vec{b}_1$ and $\vec{b}_2$ are the two fundamental reciprocal-lattice vectors (see the text).

Let us denote the fermion operator on the A sites by $\psi(\vec{r}_A)$ and the fermion operators on the neighboring B sites by $\chi(\vec{r}_A + \vec{d}_i)$ (with $i = 1, 2, 3$).

With this notation, the Hamiltonian for this tight-binding model of non-interacting (spinless) electrons with hopping amplitude t_1 between nearest-neighbor sites is (Semenoff, 1984)

$$H_0 = t_1 \sum_{\vec{r}_A, i=1,2,3} \left[\psi^\dagger(\vec{r}_A)\chi(\vec{r}_A + \vec{d}_i) + \text{h.c.} \right] \tag{16.17}$$

For the moment we will consider the case in which the fermions are spinless, or that the spin degree of freedom does not affect the physics beyond the requirements of the Pauli principle. We will shortly consider the effects of the electron spin.

In Fourier space we write

$$\psi(\vec{r}_A) = \int_{\text{BZ}} \frac{d^2k}{(2\pi)^2} \psi(\vec{k}) e^{i\vec{k}\cdot\vec{r}_A}, \qquad \chi(\vec{r}_B) = \int_{\text{BZ}} \frac{d^2k}{(2\pi)^2} \psi(\vec{k}) e^{i\vec{k}\cdot\vec{r}_B} \tag{16.18}$$

where the momentum integrals run over the first Brillouin zone of the honeycomb lattice, shown in Fig. 16.1(b). In Fourier space this free fermion system has the Hamiltonian

$$H_0 = \int_{\text{BZ}} \frac{d^2k}{(2\pi)^2} \left(\psi^\dagger(k), \chi^\dagger(\vec{k}) \right) \times \begin{pmatrix} 0 & t_1 \sum_{i=1,2,3} e^{i\vec{k}\cdot\vec{d}_i} \\ t_1 \sum_{i=1,2,3} e^{-i\vec{k}\cdot\vec{d}_i} & 0 \end{pmatrix} \begin{pmatrix} \psi(\vec{k}) \\ \chi(\vec{k}) \end{pmatrix} \tag{16.19}$$

The single-particle energy eigenvalues of this Hamiltonian are

$$E_\pm(\vec{k}) = \pm t_1 \sqrt{\left| e^{i\vec{k}\cdot\vec{d}_1} + e^{i\vec{k}\cdot\vec{d}_2} + e^{i\vec{k}\cdot\vec{d}_3} \right|^2} \tag{16.20}$$

Hence we have two bands, a valence band of negative-energy states $E_-(\vec{k})$, and a conduction band of positive-energy states $E_+(\vec{k})$. The energy gap between these two bands vanishes at the zeros of the function $\sum_{i=1,2,3} e^{i\vec{k}\cdot\vec{d}_i}$, which occur at the corners of the Brillouin zone, labeled by K and K′ in Fig. 16.1(b). Their wave vectors are $\vec{q}_K = \left(2\pi/\sqrt{3}, 2\pi/3\right)$ and $\vec{q}_{K'} = -\vec{q}_K$ such that $\vec{q}_K \cdot \vec{d}_1 = 2\pi/3$, $\vec{q}_K \cdot \vec{d}_2 = -2\pi/3$, and $\vec{q}_K \cdot \vec{d}_3 = 0$, and similarly for the other equivalent corners of the Brillouin zone obtained by successive rotations of $\vec{q}_K$ (and $\vec{q}_{K'}$) by $2\pi/3$ (see Fig. 16.1(b)).

For charge-neutral graphene the Fermi energy is at $E_F = 0$. In this case the valence band, with dispersion $E_-(\vec{k})$, is full, and the conduction band, with dispersion $E_+(\vec{k})$, is empty. However, the only states at the Fermi energy are the crossing points K and K′ (and their symmetry-related points at the corners of the first Brillouin zone) of the two bands. Thus graphene is not a semiconductor, since its energy gap is zero, but it is not a metal either, since the Fermi surface reduces to the crossing points. Hence graphene is an example of a (direct) semi-metal.

Near the crossing points K and K′ the energy–momentum relations can be linearized and become $E_\pm = \pm t_1|\vec{q}|$, where $\vec{q}$ is a small deviation from $\vec{q}_{\rm K}$ or $\vec{q}_{\rm K'}$. Thus, in the low-energy limit, only the single-particle states with wave vectors close to K and K′ (the two "valleys") contribute to the physics of this system. Thus, we are led to define two species (or flavors) of two-component Dirac fermions (or, more properly, Weyl fermions), $\psi_1(\vec{k})$ and $\psi_2(\vec{k})$,

$$\psi_1(\vec{q}) = \begin{pmatrix} e^{-i\frac{\pi}{6}}\psi_{\rm K}(\vec{q}) \\ e^{i\frac{\pi}{6}}\chi(\vec{q}) \end{pmatrix}, \qquad \psi_2(\vec{q}) = \begin{pmatrix} e^{-i\frac{\pi}{6}}\chi_{\rm K'}(\vec{q}) \\ e^{i\frac{\pi}{6}}\psi_{\rm K'}(\vec{q}) \end{pmatrix} \tag{16.21}$$

where we have used the notation for the components of the Fermi fields near the crossing points at K and K′ to be, respectively, $\psi_{\rm K}(\vec{q}) = \psi(\vec{q}_{\rm K}+\vec{q})$, $\chi_{\rm K}(\vec{q}) = \chi(\vec{q}_{\rm K}+\vec{q})$, $\psi_{\rm K'}(\vec{q}) = \psi(\vec{q}_{\rm K'}+\vec{q})$, and $\chi_{\rm K'}(\vec{q}) = \chi(\vec{q}_{\rm K'}+\vec{q})$.

With this notation the effective low-energy graphene Hamiltonian becomes

$$\begin{aligned} H_0 &= \int \frac{d^2q}{(2\pi)^2} \sum_{a=1,2} \psi_a^\dagger(\vec{q}) v_{\rm F}\Big(\sigma_1 q_1 + \sigma_2 q_2\Big)\psi_a(\vec{q}) \\ &= \int d^2x \sum_{a=1,2} \psi_a^\dagger(x) v_{\rm F}\Big(i\sigma_1\,\partial_1 + i\sigma_2\,\partial_2\Big)\psi_a(x) \end{aligned} \tag{16.22}$$

where $a = 1, 2$ labels the two species (or flavors) of two-component Dirac (Weyl) fermions, and $v_{\rm F} = (\sqrt{3}/2)t_1$ is the Fermi velocity. Here, as before, σ_1 and σ_2 are the two off-diagonal 2×2 Pauli matrices.

Let us consider two simple extensions of the simple graphene electronic structure. Thus we will consider adding to the Hamiltonian a site-potential-energy term that assumes two values, ε on the sites of the A sublattice and $-\varepsilon$ on the sites of the B sublattice. This does not happen in graphene but does happen in graphene grown on boron nitride, which has the same lattice structure. The other case that we will consider is adding a next-nearest-neighbor hopping term that connects A sites with each other and B sites with each other. The amplitude for the next-nearest-neighbor hopping is $t_2 e^{\pm i\phi}$ and represents a staggered magnetic flux (as shown in Fig. 16.2). However, differently from the case in the Hofstadter problem we discussed before, the flux through each hexagon is zero. Nevertheless, this flux breaks time-reversal invariance. These time-reversal-symmetry-breaking terms are absent in the case of graphene.

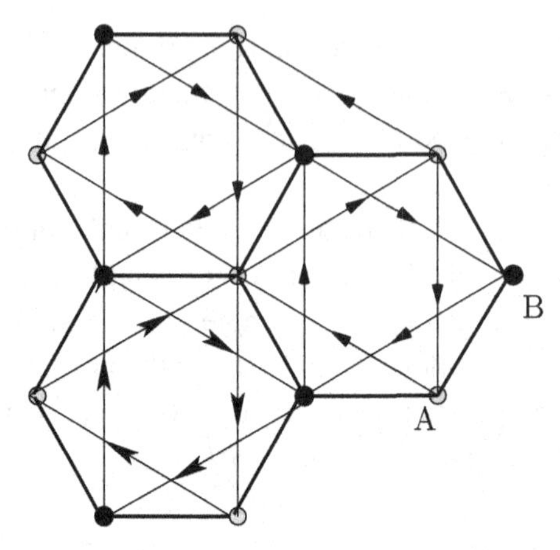

Figure 16.2 Next-nearest-neighbor hopping amplitudes on the honeycomb lattice. The arrows represent the orientation of the next-nearest-neighbor links with hopping processes with amplitude $t_2 e^{i\phi}$. On one hexagon, the flux on the triangles on the A and B sublattices is -3ϕ, whereas on all the adjacent triangles of the same sublattice the flux is $+3\phi$. The total flux on each hexagon is zero.

With these additional terms the free-fermion Hamiltonian in Fourier space has the form (known as the Haldane model (Haldane, 1988a))

$$H_0 = \int_{\text{BZ}} \frac{d^2k}{(2\pi)^2} \left(\psi^\dagger(\vec{k}), \chi^\dagger(\vec{k})\right) h(\vec{k}) \begin{pmatrix} \psi(\vec{k}) \\ \chi(\vec{k}) \end{pmatrix} \tag{16.23}$$

where we have defined the one-particle Hamiltonian $h(\vec{k})$, which is a hermitian 2×2 matrix for each wave vector $\vec{k}$ of the Brillouin zone. As such it can always be expressed as a linear combination of the three Pauli matrices and of the 2×2 identity matrix $\mathbb{I}$ of the form

$$h(\vec{k}) = h_0(\vec{k})\mathbb{I} + \vec{h}(\vec{k}) \cdot \vec{\sigma} \tag{16.24}$$

which is generic for any two-band system. The same considerations apply for the π flux model we discussed before. The one-particle states have energy eigenvalues

$$E_\pm(\vec{k}) = h_0(\vec{k}) \pm ||\vec{h}(\vec{k})|| \tag{16.25}$$

where

$$||\vec{h}(\vec{k})|| = (\vec{h}(\vec{k}) \cdot \vec{h}(\vec{k}))^{1/2} \tag{16.26}$$

is the norm of the vector $\vec{h}(\vec{k})$. The term proportional to the identity matrix $\mathbb{I}$ in Eq. (16.24) reflects the fact that the lattice model with a flux is not invariant under particle–hole conjugation. In what follows we will neglect this term, since it turns out to be unimportant to the physics of the quantum anomalous Hall effect.

In the particular case of the honeycomb lattice the scalar $h_0(\vec{k})$ and the three-component vector $\vec{h}(\vec{k})$ are given by

$$\begin{aligned} h_0(\vec{k}) &= 2t_2 \cos\phi \sum_{i=1}^{3} \cos(\vec{k}\cdot\vec{a}_i), & h_1(\vec{k}) &= t_1 \sum_{i=1}^{3} \cos(\vec{k}\cdot\vec{d}_i) \\ h_2(\vec{k}) &= t_1 \sum_{i=1}^{3} \sin(\vec{k}\cdot\vec{d}_i), & h_3(\vec{k}) &= \varepsilon + 2t_2 \sin\phi \sum_{i=1}^{3} \sin(\vec{k}\cdot\vec{a}_i) \end{aligned} \tag{16.27}$$

Since all three Pauli matrices are present, it is not possible to find a coordinate system for the spinors in which the Hamiltonian is real and symmetric. Hence in this system, as expected, time-reversal invariance is broken explicitly, whereas in the absence of the flux it is possible to rotate the spinors to a basis in which the Hamiltonian is real.

Following now the same steps as those which led to the effective low-energy theory for graphene, Eq. (16.22), the effective low-energy Hamiltonian for the Haldane model, after rescaling energies by the Fermi velocity $v_F = (\sqrt{3}/2)t_1$, takes the standard Dirac form

$$H_0 = \int d^2x \sum_{a=1,2} \psi_a^\dagger(x)\Big(i\alpha_1\,\partial_1 + i\alpha_2\,\partial_2 + m_a\beta\Big)\psi_a(x) \tag{16.28}$$

Here we have defined the 2×2 Dirac matrices $\alpha_1 = \sigma_1$, $\alpha_2 = \sigma_2$ and $\beta = \sigma_3$. In what follows we will use the set of 2×2 Dirac gamma matrices,

$$\gamma_0 = \beta = \sigma_3, \qquad \gamma_1 = \beta\alpha_1 = i\sigma_2, \qquad \gamma_2 = \beta\alpha_2 = -i\sigma_1 \tag{16.29}$$

which satisfy the Dirac algebra, $\{\gamma_\mu, \gamma_\nu\} = 2g_{\mu\nu}$, where $g_{\mu\nu}$ is the standard (Bjorken and Drell) Minkowski metric in $(2+1)$-dimensional space-time. In this notation the low-energy action for this system is that of two Weyl–Dirac fields (with different masses)

$$S = \int d^3x \sum_{a=1,2} \bar{\psi}_a\big(i\gamma^\mu\,\partial_\mu - m_a\big)\psi_a \tag{16.30}$$

where, as usual, $\bar{\psi}_a = \psi_a^\dagger\gamma_0$.

In the language of the Dirac spinors, time reversal is the operation that flips the spin of the fermion and complex-conjugates the state. Thus, under time reversal, which we will denote by the anti-linear operator Θ, a Dirac fermion in real space transforms as

$$\Theta\psi(x,y,t) = -\gamma_1\psi(-x,-y,t) = -i\sigma_2\psi(-x,-y,t) \tag{16.31}$$

whereas under parity the Dirac spinor transforms as

$$\mathcal{P}\psi(x,y,t) = i\gamma_2\psi(x,-y,t) = \sigma_1\psi(x,-y,t) \tag{16.32}$$

The (single-particle) Dirac Hamiltonian transforms under time reversal as follows:

$$\Theta h(\vec{p}, m)\Theta^{-1} = -i\sigma_2 h^*(-\vec{p}, m) i\sigma_2 = p_1\sigma_1 + p_2\sigma_2 - m\sigma_3 = h(\vec{p}, -m) \tag{16.33}$$

which amounts to saying that the mass term breaks time-reversal invariance in two dimensions. On the other hand, under 2D parity the Dirac Hamiltonian transforms as

$$\mathcal{P} h(\vec{p}, m)\mathcal{P}^{-1} = \sigma_1\left[p_1\sigma_1 - p_2\sigma_2 + m\sigma_3\right] = p_1\sigma_1 + p_2\sigma_2 - m\sigma_3 \tag{16.34}$$

which is equivalent to a time-reversal transformation.

In Eq. (16.28) we denoted the Dirac masses of the two flavors by

$$\begin{aligned} m_1 &= \frac{3}{2}\frac{t_2}{t_1}\sin\phi - \frac{2}{\sqrt{3}}\frac{\varepsilon}{t_1} \\ m_2 &= \frac{3}{2}\frac{t_2}{t_1}\sin\phi + \frac{2}{\sqrt{3}}\frac{\varepsilon}{t_1} \end{aligned} \tag{16.35}$$

If $\varepsilon = 0$ the two masses of the two flavors of Dirac fermions have *the same sign*, whereas in the absence of a flux, $\phi = 0$, the two flavors have masses with *opposite signs*. In general $m_1 \neq m_2$ will have different magnitudes and/or signs.

We should note that much of the condensed matter literature, e.g. Haldane (1988a) and Kane and Mele (2005a), uses a convention in which either the α_2 matrix or the α_1 matrix, but not both, has opposite signs for the two flavors. In that convention time-reversal symmetry is broken when the Dirac mass terms have opposite signs, which is the opposite convention to that which we have adopted here. In terms of the Dirac gamma matrices, the reason for this difference is that in odd space-time dimensions there is no γ_5 matrix and instead there are two inequivalent frames for the Dirac spinors, which are distinguished from each other by a handedness or helicity. At any rate, it is always possible, and consistent, to define the frames of the Dirac spinors to have the same handedness, as we have done here, and to follow the criterion that time reversal amounts to changing the sign of the mass term. This criterion is also intuitive insofar as, when a mass term is present, the one-particle Dirac Hamiltonian is hermitian (as it should be) but cannot be real and symmetric.

16.3.3 Quantization of the anomalous Hall effect

To see that we get an anomalous quantum Hall effect, we will couple the system to a weak electromagnetic field in order to compute the conductance. We will first discuss this calculation at the level of the effective-field theory of Dirac fermions. Thus, we will consider the coupling of the Dirac fermions (either for the π flux model or for the case of Haldane's honeycomb model) to a weak electromagnetic field $A_\mu(x)$. This interaction has the standard form of minimal coupling (dictated by gauge invariance). We will consider first the case of one species of two-component Dirac (or Weyl) fermions of mass m for which the Lagrangian is (setting $v_{\rm F} = 1$)

$$\mathcal{L} = \bar{\psi}\left(i\gamma^\mu \partial_\mu - e\gamma^\mu A_\mu - m\right)\psi \tag{16.36}$$

To compute the conductivity, we first need to compute the polarization tensor $\Pi_{\mu\nu}$, i.e. the current–current correlation functions. However, we have already done this calculation in Section 10.4. There we found that the effective low-energy (compared with the energy gap m) Lagrangian for the gauge fields is (Redlich, 1984)

$$\mathcal{L}[A_\mu] = -\frac{1}{4g^2}F_{\mu\nu}F^{\mu\nu} + \frac{\sigma_{xy}}{4}\epsilon_{\mu\nu\lambda}A^\mu F^{\nu\lambda} \tag{16.37}$$

where (in standard units) $g^2 = \pi$ and $\sigma_{xy} = (e^2/(4\pi))\text{sgn}(m)$. The presence in the effective Lagrangian of a Chern–Simons term, which is odd under 2D parity and time reversal, is known as the *parity anomaly*. After restoring standard units, a single two-component Dirac fermion in two space dimensions has a Hall conductivity

$$\sigma_{xy} = \frac{1}{2}\frac{e^2}{h}\,\text{sgn}(m) \tag{16.38}$$

which is half of the minimum integer Hall conductivity, e^2/h.

This result seemingly violates the quantization of the Hall conductivity of free fermions that we discussed in the context of the integer quantum Hall effect. Indeed, in Chapter 12 we showed that for systems with a full band of one-particle states the Hall conductivity is quantized and, moreover, that this quantization has a topological meaning insofar as it can be expressed in terms of a topological invariant, the first Chern number, C_1. We expect that these general arguments should also apply to the π flux model and the honeycomb-lattice Haldane model since both are free-fermion lattice models with a filled band. Hence the Hall conductivity should be an integer, not a half-integer.

The loophole in our arguments is that in both models, and, indeed, in all lattice models with band crossings (or models that are close to having band crossings), the number of such crossings must be an *even* integer (even neglecting other degrees of freedom such as spin), in accordance with the Nielsen–Ninomiya theorem (Nielsen and Ninomiya, 1981; Friedan, 1982). Thus, in both systems, we encountered two flavors of Dirac fermions, and found that each flavor contributes to the total Hall conductivity an amount equal to $(e^2/(2h))\text{sgn}(m)$, where m is the mass for that fermion flavor. In both models we found two situations. In one case the mass terms of the Dirac Lagrangian have the same sign, $\text{sgn}(m_1) = \text{sgn}(m_2)$, and the *total* Hall conductivity is an integer (in accordance with our expectations)

$$\sigma_{xy} = \pm\frac{e^2}{h} \tag{16.39}$$

We encountered a closely related problem in the theory of the chiral spin liquid discussed in Chapter 10.

In this case we have an *anomalous quantum Hall effect* since the system is an insulator with a vanishing *net* magnetic field (as in Haldane's honeycomb model) or with half the quantum of flux (as in the π flux model). Although the total flux is either zero or π, in both systems the Hamiltonian has terms that violate time-reversal invariance $\mathcal{T}$, as well as 2D parity $\mathcal{P}$, although the product $\mathcal{TP}$ is an unbroken symmetry of the Hamiltonian. A system of this type is often called a Chern insulator. Recall that 2D parity is a mirror symmetry, such as $x \to -x$ and $y \to y$; it is not inversion symmetry, $\vec{r} \to -\vec{r}$.

However, in the second case we found that the signs of the mass terms were opposite, $\text{sgn}(m_1) = -\text{sgn}(m_2)$. Hence, in this case the two contributions to the Hall conductivity cancel each other out, and we find that the total Hall conductivity vanishes, $\sigma_{xy} = 0$. In this case both time reversal $\mathcal{T}$ and 2D parity $\mathcal{P}$ are unbroken. In other terms, this system is a conventional insulator.

In what follows we will adopt the physical criterion that a 2D system is time-reversal-invariant if its *physical response* to an external electromagnetic field A_μ, expressed in terms of the effective action $S[A_\mu]$, is invariant under time reversal, $t \to -t$. Thus a system that exhibits the anomalous quantum Hall effect has a broken time-reversal invariance since the effective action of the external electromagnetic field has a Chern–Simons term that is odd under time reversal $\mathcal{T}$ and parity $\mathcal{P}$ (but invariant under $\mathcal{PT}$). We will see below that this definition can be extended to systems in three dimensions, where the time-reversal-symmetry-breaking term is more subtle.

It is important to keep in mind that this field-theoretic, or, more properly, *macroscopic*, definition of time-reversal symmetry is different from what is meant by time reversal microscopically at the level of the one-particle theory (expressed in the band structure). At the level of the single-particle theory time reversal is an anti-linear and anti-unitary operator that relates a single-particle state with momentum $\vec{k}$ and spin $\uparrow$ ($\downarrow$) to a single-particle state with momentum $-\vec{k}$ and spin $\downarrow$ ($\uparrow$). Thus, both in the π flux model and in graphene, the one-particle definition of time reversal is equivalent to the exchange of the handedness of the two flavors (or valleys) of fermions.

16.3.4 A two-band topological invariant

Since our discussion has been based largely on results derived from the effective low-energy theory of Dirac fermions, one might suspect that the approximations we have made may in some subtle way invalidate our analysis. We will now see that there is a way to reach the same conclusions without taking the continuum (or low-energy) limit by means of a generalization to a two-band insulating system of the topological arguments of Thouless, Kohmoto, Nightingale, and den Nijs (Thouless *et al.*, 1982).

This generalization is based on analysis of the Kubo formula for the Hall conductivity in the two-band case. This result is due to Qi, Wu, and Zhang (Qi *et al.*, 2006b). We will see in the next section that this result plays a key role in the theory of the spin quantum Hall effect. In Chapter 12, see Eq. (12.118), we saw that the Kubo formula for the Hall conductivity implied that the latter is determined as the zero-frequency limit, $\omega \to 0$, of the xy component of the current-correlation function at $\vec{Q} = 0$,

$$\sigma_{xy} = \lim_{\omega \to 0} \frac{i}{\omega} \Pi_{xy}(\omega, \vec{Q} = 0) \tag{16.40}$$

For a free-fermion system the current correlator $\Pi_{xy}(\omega, \vec{Q} = 0)$ is

$$\Pi_{xy}(\omega, \vec{Q} = 0) = \int \frac{d^2k}{(2\pi)^2} \int \frac{d\Omega}{2\pi} \operatorname{tr}\Big[J_x(\vec{k}) G(\vec{k}, \omega + \Omega) J_y(\vec{k}) G(\vec{k}, \Omega) \Big] \tag{16.41}$$

which can be calculated from the expressions for the current operators, which in momentum space are given in terms of the one-particle two-band Hamiltonian $h(\vec{k})$ of Eq. (16.24),

$$J_l(\vec{k}) = \frac{\partial h(\vec{k})}{\partial \vec{k}} = \frac{\partial h_0(\vec{k})}{\partial \vec{k}} \mathbb{I} + \frac{\partial h_a(\vec{k})}{\partial \vec{k}} \sigma^a \tag{16.42}$$

(with $a = 1, 2, 3$), and the free-fermion propagator of the two-band system $G(\vec{k}, \omega)$, which is also a 2×2 matrix in the band indices,

$$G(\vec{k}, \omega) = \Big(\omega \mathbb{I} - h(\vec{k}) + i\epsilon \Big)^{-1} = \frac{P_+(\vec{k})}{\omega - E_+(\vec{k}) + i\epsilon} + \frac{P_-(\vec{k})}{\omega - E_-(\vec{k}) + i\epsilon} \tag{16.43}$$

where we have used the projection operators $P_\pm(\vec{k})$,

$$P_\pm(\vec{k}) = \frac{1}{2} \Big(\mathbb{I} \pm \hat{h}_a(\vec{k}) \sigma^a \Big) \tag{16.44}$$

where $\hat{h}_a(\vec{k})$ (with $a = 1, 2, 3$) is a unit vector with components

$$\hat{h}_a(\vec{k}) = \frac{h_a(\vec{k})}{||\vec{h}(\vec{k})||} \tag{16.45}$$

After some straightforward algebra it is found that the Hall conductivity is given by the expression

$$\sigma_{xy} = \frac{e^2}{2} \int_{\text{BZ}} \frac{d^2k}{(2\pi)^2} \epsilon_{abc} \frac{\partial \hat{h}_a(\vec{k})}{\partial k_x} \frac{\partial \hat{h}_b(\vec{k})}{\partial k_y} \hat{h}_c(\vec{k}) \left(n_+(\vec{k}) - n_-(\vec{k}) \right) \tag{16.46}$$

where $n_{\pm}(\vec{k})$ are the Fermi functions (at $T = 0$ in this case) for the two bands. Since $E_+(\vec{k}) - E_-(\vec{k}) = 2||\vec{h}(\vec{k})|| > 0$, there is a finite energy gap for the entire Brillouin zone. In the case of an insulating state the Fermi energy is in the gap between the minimum energy of the conduction band $\min_{\text{BZ}}\{E_+(\vec{k})\}$ and the maximum energy of the valence band $\max_{\text{BZ}}\{E_-(\vec{k})\}$, the valence band will be fully occupied, $n_-(\vec{k}) = 1$ (for all $\vec{k}$ in the Brillouin zone), and the conduction band will be empty, $n_+(\vec{k}) = 0$ (again for all $\vec{k}$ in the Brillouin zone). Hence, the Hall conductivity of the insulating state is given by the much simpler expression (Qi *et al.*, 2006b; Yakovenko, 1990)

$$\sigma_{xy} = -\frac{e^2}{8\pi^2} \int_{\text{BZ}} d^2k \;\; \epsilon_{abc} \hat{h}_a(\vec{k}) \partial_{k_x} \hat{h}_b(\vec{k}) \partial_{k_y} \hat{h}_c(\vec{k}) \tag{16.47}$$

We now recall that we encountered essentially the same expression in our discussion of topological terms in 1D quantum antiferromagnets in Section 7.6, where we showed that the topological charge $\mathcal{Q}$ of the mappings of a smooth closed 2-manifold (which in that case was the sphere S_2) to the target space S_2 of a nonlinear sigma model with global symmetry O(3) is given by a topological invariant known as the Pontryagin index (or winding number) of Eq. (7.75). The expression for the Pontryagin index $\mathcal{Q}$

$$\mathcal{Q} = \frac{1}{4\pi} \int_{\text{BZ}} d^2k \;\; \epsilon_{abc} \hat{h}_a(\vec{k}) \partial_{k_x} \hat{h}_b(\vec{k}) \partial_{k_y} \hat{h}_c(\vec{k}) \tag{16.48}$$

is indeed essentially the same as our result for the Hall conductivity of Eq. (16.47) (up to a prefactor of $e^2/(2\pi)$). This is also the same topological invariant as that which we encountered in the theory of the path integral for spin in Chapter 7, where it entered into the Berry phase for a two-level system.

The main difference in the case at hand is that the topological charge $\mathcal{Q}$ now is the integer that classifies the homotopy classes of maps of the first Brillouin zone, the torus T^2_{BZ} (instead of the sphere S_2), onto the target space S_2 of the unit vector $\hat{h}(\vec{k})$ which parametrizes the one-particle Hamiltonians, i.e. the homotopy class $\pi_2(S^2) = \mathbb{Z}$,

$$\hat{h}(\vec{k}) : T^2_{\text{BZ}} \mapsto S^2 \tag{16.49}$$

(Fig. 16.3). Nevertheless, these maps are still classified by a topological charge $\mathcal{Q}$ that can take only integer values. We thus conclude that the Hall conductivity

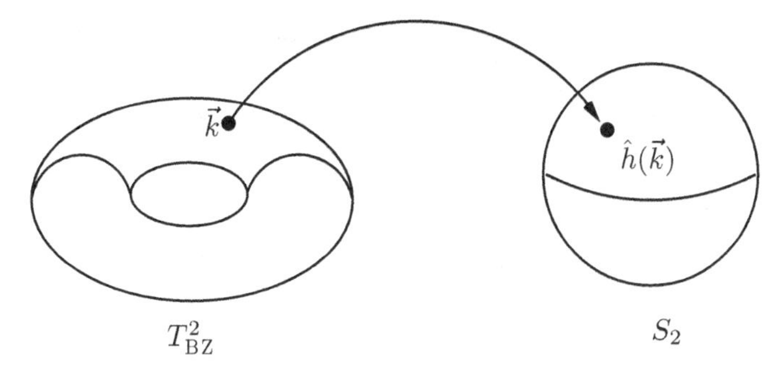

Figure 16.3 The maps of the Brillouin zone, the torus T^2_{BZ}, onto the target space S^2 that parametrize the one-particle Dirac Hamiltonian, Eq. (16.49).

(in standard units) is given now in terms of the Pontryagin index $\mathcal{Q}$,

$$\sigma_{xy} = -\frac{e^2}{2\pi\hbar}\mathcal{Q} \tag{16.50}$$

Thus, if the topological charge $\mathcal{Q} \neq 0$ we have a topological insulator with a quantized anomalous Hall effect. Moreover, and more subtly, the topological charge $\mathcal{Q}$ actually computes the TKNN integers of the total curvature of the (abelian) Berry connections of the spinors over the Brillouin zone. Thus, in this case the first Chern number C_1 is computed by the Pontryagin index $\mathcal{Q}$, i.e. $C_1 = \mathcal{Q}$.

In the theory of the non-linear sigma model of 1D quantum antiferromagnets the Pontryagin index $\mathcal{Q}$ was used to classify instanton processes, whereas in 2D quantum antiferromagnets it classified soliton states known as skyrmions. In the case of the topological insulators the index $\mathcal{Q}$ classifies one-particle Hamiltonians labeled by the parameters $\hat{h}(\vec{k})$. Thus the vector $\hat{h}(\vec{k})$ is associated with skyrmion-like configurations on the first Brillouin zone.

In particular, this important result implies that the Hall conductivity has to be an integer (in units of e^2/h) and cannot be a half-integer. Naturally, this is consistent with the requirement of the Nielsen–Ninomiya theorem that the number of Dirac flavors must be an even integer. It is interesting to see in this language the origin of the half-integer value of the Hall conductivity for a single Dirac fermion. Let us consider the one-particle (Dirac) Hamiltonian for a two-component Dirac spinor. In momentum space it is (in units with $v_{\text{F}} = 1$ and $\hbar = 1$)

$$h(\vec{p}) = \vec{\alpha} \cdot \vec{p} + \beta m = \vec{h} \cdot \sigma \tag{16.51}$$

where we defined the three-component vector $\vec{h}$ by

$$\vec{h} = (p_x, p_y, m) \tag{16.52}$$

The norm of this vector is, of course, the one-particle energy $E(\vec{p})$,

$$||\vec{h}(\vec{p})|| = E(\vec{p}) = \sqrt{\vec{p}^{\,2} + m^2} \tag{16.53}$$

Let $\hat{h}(\vec{p})$ be the unit vector

$$\hat{h}(\vec{p}) = \frac{\vec{h}(\vec{p})}{||\vec{h}(\vec{p})||} = \frac{1}{E(\vec{p})}(p_x, p_y, m) \tag{16.54}$$

This unit vector has the limiting behaviors

$$\lim_{|\vec{p}|\to\infty} \hat{h}(\vec{p}) = \frac{1}{|\vec{p}|}(p_x, p_y, 0), \qquad \lim_{|\vec{p}|\to 0} \hat{h}(\vec{p}) = \mathrm{sgn}(m)(0, 0, 1) \tag{16.55}$$

Therefore, we see that $\hat{h}(\vec{k})$ has the form of the meron configuration shown in Fig. 16.4, and corresponds to the case of $m > 0$. It is half a skyrmion and as such it sweeps half of the area of the unit sphere, i.e. 2π. Hence the meron has half of the topological charge, $\mathcal{Q} = -\frac{1}{2}$. Similarly, the anti-meron, which corresponds to the case of $m < 0$, also has half of the topological charge but with opposite sign, $\mathcal{Q} = +\frac{1}{2}$. Thus the topological-charge contribution is $\mathcal{Q} = -\frac{1}{2}\,\mathrm{sgn}(m)$. The general result of Eq. (16.50) then tells us that each two-component Dirac fermion contributes to the Hall conductivity (or, which amounts to the same thing, to the coupling constant of the effective Chern–Simons action) with $\sigma_{xy} = (e^2/(2h))\mathrm{sgn}(m)$, which is the result we found above.

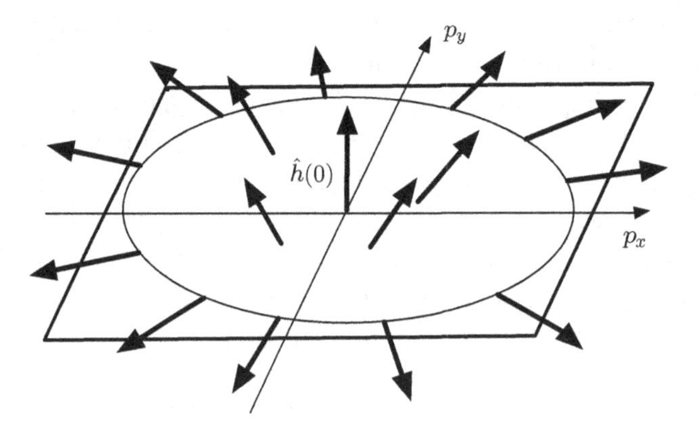

Figure 16.4 A meron configuration for the unit vector $\hat{h}(\vec{k})$ in momentum space for a two-component Dirac fermion with $m > 0$.

16.4 The quantum spin Hall effect

The quantum spin Hall effect is a close relative of the quantum anomalous Hall effect. However, unlike the quantum Hall effects (both "normal" and anomalous), the quantum spin Hall effect involves the spin current instead of the charge current. Although the spin Hall effect had been predicted to exist in metallic systems (Hirsch, 1999), we will focus here on the case of insulators, where this effect may be quantized. Indeed, a quantized spin Hall effect was discovered recently in 2DEGs that exist, under suitable circumstances that we will discuss shortly, in HgTe/CdTe quantum wells (König *et al.*, 2008). We will see that this is an example of a time-reversal-invariant topological insulator.

However, before we get started we need to clarify what is meant by a spin conductivity and under what circumstances the spin Hall effect may occur. Spin transport is defined by analogy with charge transport, i.e. as a spin accumulation at one end of the sample as a consequence of electronic motion. However, a fundamental difference between spin and charge is that they are related to very different symmetries. Thus, charge conservation is related to the (gauge-invariant) current J_μ which is the generator of the $U(1)$ electromagnetic (gauge) symmetry, which is abelian. In contrast, a system with an unbroken $SU(2)$ spin invariance has as many components of the spin currents as generators in the $SU(2)$ algebra, J_μ^a, where $a = 1, 2, 3$ runs in the algebra of $SU(2)$ and $\mu = 0, 1, \ldots, d$ is the space-time index in d space dimensions. Thus, the spin symmetry is non-abelian and the spin currents generate the (non-abelian) $SU(2)$ algebra.

In this sense a spin conductivity can be defined only if the spin symmetry is somehow broken down to a $U(1)$ subgroup so that the J^3 current is conserved. This is often the case in solid-state systems due to the effects of the (atomic scale) spin–orbit interaction which ties the spin of the conduction electrons to the lattice. However, we will see that it is also possible to have a quantum spin Hall effect even if the spin symmetry is broken down to a $\mathbb{Z}_2$ subgroup of the spin symmetry group $SU(2)$. Although in this case a bulk spin Hall conductance can no longer be defined, this state will still be characterized by having gapless edge states that transport spin.

16.4.1 The Kane–Mele model

Kane and Mele (2005a) suggested that the quantum spin Hall effect could be observed in graphene. Their proposal relied on an estimate of the magnitude of the spin–orbit interaction in graphene, which, unfortunately, turns out to be much smaller than their estimate – too small, in fact, to matter. Nevertheless, their analysis provides a simple model to study the quantum spin Hall effect.

The Kane–Mele tight-binding (free-fermion) Hamiltonian is a modification of the honeycomb model for graphene, Eq. (16.17), of the form

$$\begin{aligned} H_{\mathrm{KM}} = t_1 & \sum_{\vec{r}_{\mathrm{A}};i=1,2,3;\sigma=\uparrow,\downarrow} \left[\psi_\sigma^\dagger(\vec{r}_{\mathrm{A}})\chi_\sigma(\vec{r}_{\mathrm{A}}+\vec{d}_i)+\text{h.c.}\right] \\ & + \sum_{\langle\vec{r}_{\mathrm{A}};\vec{r}_{\mathrm{A}}'\rangle;\sigma,\sigma'=\uparrow,\downarrow} \left[it_2\psi_\sigma^\dagger(\vec{r}_{\mathrm{A}})\nu[\vec{r}_{\mathrm{A}},\vec{r}_{\mathrm{A}}']s^z_{\sigma\sigma'}\psi_{\sigma'}(\vec{r}_{\mathrm{A}}')+\text{h.c.}\right] \\ & + \sum_{\langle\vec{r}_{\mathrm{B}};\vec{r}_{\mathrm{B}}'\rangle;\sigma,\sigma'=\uparrow,\downarrow} \left[it_2\chi_\sigma^\dagger(\vec{r}_{\mathrm{B}})\nu[\vec{r}_{\mathrm{B}},\vec{r}_{\mathrm{B}}']s^z_{\sigma\sigma'}\chi_{\sigma'}(\vec{r}_{\mathrm{B}}')+\text{h.c.}\right] \end{aligned} \tag{16.56}$$

where s^z is the diagonal Pauli matrix σ^3 and acts on the spin labels. Here $\langle\vec{r}_{\mathrm{A}};\vec{r}_{\mathrm{A}}'\rangle$ denotes nearest-neighbor sites on the A sublattice (which are second-nearest neighbors for the honeycomb lattice), and similarly for the B sublattice. The amplitude $\nu[\vec{r}_{\mathrm{A}},\vec{r}_{\mathrm{A}}']$ takes the values ± 1 depending on the orientation of the two nearest-neighbor bonds $\vec{d}_i$ and $\vec{d}_j$ which the electron traverses in going from site $\vec{r}_{\mathrm{A}}$ to $\vec{r}_{\mathrm{A}}'$ (and analogously for sublattice B): $\nu[\vec{r}_{\mathrm{A}},\vec{r}_{\mathrm{A}}'] = +1$ (-1) if the electron makes a left (right) turn on the second traversed bond (also with the same rule for sublattice B). In other words, this spin-dependent amplitude is $i\vec{d}_i \times \vec{d}_j \cdot \vec{s}$ (where $\vec{s}$ is the electron spin), which has also been derived from a microscopic model of graphene that includes the spin–orbit interaction.

We can easily recognize that this tight-binding free-fermion model is simply two copies of a Haldane honeycomb model with flux $\phi = \pi/2$ for electrons with spin $\uparrow$ and flux $-\pi/2$ for electrons with spin $\downarrow$. This system is time-reversal-invariant since the time-reversal-violating term, t_2, has opposite signs for the two spin components.

It is now straightforward to find the effective low-energy theory in terms of Dirac fermions by mimicking our earlier analysis for the Haldane honeycomb model. It is essentially the same as for the Haldane honeycomb model (and also the π flux model), but doubled to account for spin. Thus the effective theory now involves Fermi fields that have not only a Dirac (or sublattice) index $a = 1, 2$ and a flavor (or valley) index $I = 1, 2$, but also a spin index $\sigma = \uparrow, \downarrow$. The effective Dirac Hamiltonian is simply

$$\mathcal{H} = -i\hbar v_{\mathrm{F}}\psi^\dagger(x)\big(\alpha_1\,\partial_x + \alpha_2\,\partial_y\big)\psi(x) + \Delta_{\mathrm{so}}\psi^\dagger(x)\beta s^z\psi(x) \tag{16.57}$$

where $\Delta_{\mathrm{so}} = 3\sqrt{3}t_2$ measures the strength of the spin–orbit coupling (Kane and Mele, 2005a). Here we have, once again, used the notation for the two-component Dirac spinors

$$\psi_{1,\sigma} = \begin{pmatrix}\psi_{\mathrm{K},\sigma}\\ \chi_{\mathrm{K},\sigma}\end{pmatrix}, \qquad \psi_{2,\sigma} = \begin{pmatrix}-i\chi_{\mathrm{K}',\sigma}\\ i\psi_{\mathrm{K}',\sigma}\end{pmatrix} \tag{16.58}$$

which amounts to a change of basis for the spinor at valley K′ relative to the spinor at valley K without changing the spin.

At the level of the effective Dirac theory the energy gap generated by the spin–orbit coupling is a mass term, which we denoted by $\Delta_{\rm so}$ in Eq. (16.57), which has opposite signs for the two spin components of the Dirac fermions. Since the mass term for each spin component signals a violation of time-reversal invariance, we conclude that in this system time reversal is not broken. In particular, if no other terms are included in the Hamiltonian, our earlier analysis implies that the Hall conductivity of this system is zero. This happens since the two spin orientations make equal and opposite contributions to $\sigma_{xy}^{\uparrow;\downarrow} = \pm e^2/h$, and cancel each other out,

$$\sigma_{xy} = \sigma_{xy}^{\uparrow} + \sigma_{xy}^{\downarrow} = \frac{e^2}{h} - \frac{e^2}{h} = 0 \tag{16.59}$$

which is required by time-reversal invariance.

Although the Kane–Mele Hamiltonian is time-reversal-invariant, it has an explicitly broken SU(2) spin-rotational invariance by virtue of spin–orbit effects. However, unless further terms are included, this Hamiltonian has a conserved z component of the total spin, S^z, which generates an unbroken U(1) subgroup of SU(2). Consequently, in addition to the charge current, the Kane–Mele model also has a conserved S^z spin current,

$$\vec{J}_{\rm spin} = \frac{\hbar}{2e}(\vec{J}_{\uparrow} - \vec{J}_{\downarrow}) = \frac{\hbar}{2e}\sum_{I=1,2} \psi_I^{\dagger} s^z \vec{\alpha} \psi_I = \frac{\hbar}{2e}\sum_{I=1,2} \bar{\psi}_I s^z \vec{\gamma} \psi_I \tag{16.60}$$

This means that an external electric field will induce a vanishing Hall current (since $\sigma_{xy} = 0$), and, at the same time, equal (and opposite) currents of electrons carrying opposite spins are also induced. Hence, at this level it is possible to define a bulk spin Hall conductivity that is the *difference* of the Hall conductivities for electrons with ↑ and ↓ spins. However, since the two spin components have Hall conductivities that are equal but have opposite signs, the spin Hall conductivity is non-vanishing and quantized,

$$\sigma_{xy}^{\rm spin} = \frac{\hbar}{2e}\left(\sigma_{xy}^{\uparrow} - \sigma_{xy}^{\downarrow}\right) = \frac{e}{2\pi} \tag{16.61}$$

This is the *quantum spin Hall effect.*

This analysis of the bulk currents induced by electric fields relies on the conservation of the z-component of the spin. As noted by Kane and Mele, in the presence of a coupling to an external perpendicular electric field (or to a substrate) graphene admits an additional (Rashba) coupling, which at the level of the graphene model is

$$H_{\rm R} = i\lambda_{\rm R} \sum_{\vec{r}_{\rm A},i} \psi^{\dagger}(\vec{r}_{\rm A}) \left(\vec{\sigma} \times \vec{d}_i\right)_z \chi(\vec{r}_{\rm A} + \vec{d}_i) + {\rm h.c.} \tag{16.62}$$

where λ_R is the Rashba coupling. This term breaks the mirror symmetry of the plane, $z \to -z$, and it is also due to spin–orbit interactions. Furthermore, while this term is still consistent with time-reversal invariance, it breaks the spin-conservation law. Although this term is very small in graphene, it means that, strictly speaking, a conserved bulk spin current no longer exists and that the spin Hall effect cannot be defined as we have done above. In terms of the Dirac fermions the Rashba coupling contributes to the effective Hamiltonian a term of the form

$$\mathcal{H}_R = \lambda_R \sum_{I=1,2} \psi_I^\dagger \, (\vec{\sigma} \times \vec{s})_z \, \psi_I \tag{16.63}$$

which is a linear combination of the spatial components of the x and y spin currents. However, provided that $\lambda_R < \Delta_{so}$, the Rashba term cannot close the gap and hence the system remains a topological insulator. We will see below that the more precise way to characterize the quantum spin Hall effect is in terms of its edge states.

16.4.2 The quantum spin Hall effect in HgTe quantum wells

The analysis of the Kane–Mele model shows that the natural place to look for the spin quantum Hall effect is in systems that naturally have large spin–orbit effects. Bernevig, Hughes, and Zhang (Bernevig *et al.*, 2006) showed that the quantum spin Hall effect should be observable in a CdTe/HgTe/CdTe quantum well, involving two narrow-gap semiconductors with large spin–orbit coupling. The main predictions of their theory, which we will now discuss, were confirmed in the experiments of Molenkamp and coworkers (König *et al.*, 2007, 2008).

HgTe and CdTe are semiconductors with a narrow gap at the Γ point (the center) of their Brillouin zone. In both materials the bands that nearly cross at the Γ point are the s-type band Γ_6 and the p-type band which is split by spin–orbit coupling into a $J = 3/2$ band, Γ_8, and a higher-energy $J = 1/2$ band, Γ_7. While the details will not be important to us, it will matter that in HgTe the Γ_8 band lies above the Γ_6 band in energy, while in CdTe the order of the bands is reversed.

For the geometry of a quantum well along the z axis, centered at $z = 0$ and with width d (the thickness of the HgTe region), Bernevig, Hughes, and Zhang derived an effective Hamiltonian for a CdTe/HgTe/CdTe quantum well that is accurate near the Γ point, $\vec{k} = 0$, of the Brillouin zone, for the 2D bands $|\mathrm{E1}, m_J = 1/2\rangle$, $|\mathrm{H1}, m_J = 3/2\rangle$, $|\mathrm{E1}, m_J = -1/2\rangle$, and $|\mathrm{H1}, m_J = -3/2\rangle$ (which are linear combinations of the Γ_6 and the Γ_8 states). In this basis, the effective one-particle Hamiltonian near the Γ point, $\vec{k} = 0$, is a 4×4 block-diagonal matrix of the form (Bernevig *et al.*, 2006)

$$H_{\text{eff}}(k_x, k_y) = \begin{pmatrix} h(\vec{k}) & 0 \\ 0 & h^*(-\vec{k}) \end{pmatrix} \tag{16.64}$$

Here $h(\vec{k})$ is a 2×2 hermitian matrix

$$h(\vec{k}) = \varepsilon(\vec{k})\mathbb{I} + \vec{d}(\vec{k}) \cdot \vec{\sigma} \tag{16.65}$$

where $\mathbb{I} = \text{diag}(1, 1)$ is the 2×2 identity matrix. The components of the vector $\vec{d}(\vec{k})$ near the Γ point, $\vec{k} = 0$, are

$$\begin{aligned} d_1 \pm i d_2 &= A\big(k_x \pm i k_y\big) + \cdots \\ d_3 &= M - B\vec{k}^2 + \cdots \\ \varepsilon(\vec{k}) &= C - D\vec{k}^2 + \cdots \end{aligned} \tag{16.66}$$

where A, B, C, and D are real and positive coefficients. The parameter M, which opens a gap in the one-particle spectrum, is real (as it should be), and changes sign as the thickness d of the HgTe in the quantum well increases. The effective one-particle Hamiltonian of Eq. (16.64) is manifestly time-reversal-invariant, with the states in the two 2×2 blocks related by time reversal. By inspection we see that, as expected, these two pairs of states have opposite parity. Thus the system is time-reversal- and parity-invariant.

This effective Hamiltonian has the same structure as that we encountered in the Kane–Mele model. However, there are important differences between the two models. One important difference is that we now have four bands that have near crossings at the Γ point, whereas the Kane–Mele model has two bands with two valleys at the vertices K and K′ of the Brillouin zone of the honeycomb lattice where they have near crossings. In addition, due to the effects of spin–orbit coupling, spin and orbital angular momentum are not good quantum numbers to label the states. In this context, by "spin" we mean the direction of the projection of the total (atomic) angular momentum J^z since the states in the upper block have $m_J = 1/2, 3/2$ while the states in the lower block have $m_J = -1/2, -3/2$.

In the Kane–Mele model the two flavors of Dirac fermions (at K and K′) have opposite parity. This led us to the condition (after a redefinition of the spinor at K′) that the quantized Hall conductivity is $(e^2/(2h))\big[\text{sgn}(m_1) + \text{sgn}(m_2)\big] = 0, e^2/h$, depending on whether the mass terms have opposite signs, and we have an unbroken time-reversal symmetry and a normal insulator, or they have the same sign, and we have a broken time-reversal symmetry and an anomalous quantum Hall effect.

Here we can proceed in the same fashion. The effective Hamiltonian for the upper block, for states with spin $\uparrow$, is parametrized by a vector $\vec{d}_\uparrow(\vec{k})$,

$$\vec{d}_\uparrow(\vec{k}) = \frac{1}{\sqrt{\vec{k}^2 + M^2}}\big(k_x, k_y, M\big) \tag{16.67}$$

where we have rescaled all the momenta by $1/A > 0$. Its unit vector $\vec{m}_\uparrow(\vec{k})$ has the asymptotic behaviors

$$\lim_{\vec{k}\to 0} \vec{m}_\uparrow(\vec{k}) = \text{sgn}(M)\big(0, 0, 1\big), \qquad \lim_{|\vec{k}|\to\infty} \vec{m}_\uparrow(\vec{k}) = \frac{1}{|\vec{k}|}\big(k_x, k_y, 0\big) \tag{16.68}$$

which is a meron with topological charge $\mathcal{Q}_\uparrow = -\frac{1}{2}\,\text{sgn}(M)$.

For the states with spin $\downarrow$, the Hamiltonian is parametrized by $\vec{d}_\downarrow(\vec{k})$,

$$\vec{d}_\downarrow(\vec{k}) = \frac{1}{\sqrt{\vec{k}^{\,2} + M^2}}\big(-k_x, k_y, M\big) \tag{16.69}$$

which is the transform of $\vec{d}_\uparrow$ by parity, $k_x \to -k_x$ and $k_y \to k_y$. A rotation of the $\downarrow$ spinor by $\exp(i(\pi/2)\sigma_2)$ rotates the $\vec{d}_\downarrow$ vector about the k_y axis to

$$\vec{d}_\downarrow \mapsto \frac{A}{\sqrt{A^2\vec{k}^2 + M^2}}\big(k_x, k_y, -M\big) \tag{16.70}$$

Its unit vector $\vec{m}_\downarrow(\vec{k})$ has the asymptotic behaviors

$$\lim_{\vec{k}\to 0} \vec{m}_\downarrow(\vec{k}) = -\text{sgn}(M)\big(0, 0, 1\big), \qquad \lim_{|\vec{k}|\to\infty} \vec{m}_\downarrow(\vec{k}) = \frac{1}{|\vec{k}|}\big(k_x, k_y, 0\big) \tag{16.71}$$

which is a meron with topological charge $\mathcal{Q}_\downarrow = +\frac{1}{2}\,\text{sgn}(M)$.

Although it would be tempting to proceed as we did in the Kane–Mele model and use the topological charges to compute the conductivities, in this case we would be led to an incorrect conclusion. The reason is that we actually started with a four-band model, instead of a two-band model as in the Kane–Mele model, and we need to determine the contribution (if any) of the rest of the bands away from the Γ point to the topological charges. They clearly must contribute, since each band cannot have a fractional Chern number.

Bernevig, Hughes, and Zhang (BHZ) solved this problem by writing down a tight-binding model that, near the Γ point, reduces to the Hamiltonian of Eq. (16.64). They proposed to replace each 2×2 block of the effective band model by the following lattice model with two flavors, $\uparrow$ and $\downarrow$, of two-component fermion spinors on the square lattice. In real space the Hamiltonian for each block of the BHZ model becomes (where we have dropped the spinor indices)

$$\begin{aligned} H = \sum_{\vec{r}} \Big\{ &\Big[\mp i c^\dagger(\vec{r} + \hat{e}_x)\sigma^x c(\vec{r}) - i c^\dagger(\vec{r} + \hat{e}_y)\sigma^y c(\vec{r}) + \text{h.c.}\Big] \\ &+ \Big[c^\dagger(\vec{r} + \hat{e}_x)\sigma^z c(\vec{r}) + c^\dagger(\vec{r} + \hat{e}_y)\sigma^z c(\vec{r}) + \text{h.c.}\Big] \\ &+ (M - 2)c^\dagger(\vec{r})\sigma^z c(\vec{r})\Big\} \end{aligned} \tag{16.72}$$

where the $+$ $(-)$ denotes the $\uparrow$ $(\downarrow)$ spinors.

The (second-quantized) Hamiltonian (in momentum space) is

$$H = \int_{\rm BZ} \frac{d^2k}{(2\pi)^2} \sum_{\alpha,\beta=1,2} \Big[c^\dagger_{\alpha,\uparrow}(\vec{k}) \vec{d}_\uparrow(\vec{k}) \cdot \vec{\sigma}_{\alpha\beta}\, c_{\beta,\uparrow}(\vec{k}) + c^\dagger_{\alpha,\downarrow}(\vec{k}) \vec{d}_\downarrow(\vec{k}) \cdot \vec{\sigma}_{\alpha\beta}\, c_{\beta,\downarrow}(\vec{k}) \Big] \tag{16.73}$$

where the integral runs over the first Brillouin zone of the square lattice, $|k_x| \leq \pi$ and $|k_y| \leq \pi$, and $\vec{\sigma}$ is a three-component vector of the three Pauli matrices. Here we used the notation $\vec{d}_\pm(\vec{k}) = (\pm \sin k_x, \sin k_y, M + \cos k_x + \cos k_y - 2)$, where $+$ stands for $\uparrow$ and $-$ for $\downarrow$. This form, which makes manifest the fact that the two bands have opposite 2D parities, follows from the requirement that $\vec{d}_\downarrow(\vec{k}) \cdot \vec{\sigma} = \vec{d}_\uparrow(-\vec{k}) \cdot \vec{\sigma}^*$ so that the one-particle Hamiltonian has the form of Eq. (16.64), as is demanded by time-reversal invariance.

This theory looks like a lattice version of the theory of Dirac fermions. There are two ways to discretize the Dirac theory in such a way that it reproduces the standard field theory of relativistic Dirac fermions at low energies. One way is to use Kogut–Susskind (also known as "staggered") fermions, in which half of the components of the Dirac fermion are assigned to one sublattice (of a bipartite lattice) and the other half to the other sublattice (Kogut and Susskind, 1975; Susskind, 1977; Kogut, 1983). The Hamiltonians of flux phases, and hence of fermions in a lattice with flux $\Phi = \pi$ per plaquette, are examples of Kogut–Susskind fermions.

The other approach is to use Wilson fermions, in which all components of the Dirac fermion are defined at each lattice site (Wilson, 1974; Creutz, 2001). The Hamiltonian of Eq. (16.72) is identical to that of Wilson fermions. However, in both approaches some symmetries of the continuum field theory of massless Dirac fermions are broken in the lattice versions. In $(1+1)$ and $(3+1)$ dimensions, the continuous chiral symmetry, $\psi \to \exp(i\theta\gamma_5)\psi$, is in general broken to a discrete subgroup by lattice effects. We have discussed the 1D version of this in Chapters 5 and 6, where we saw that the continuous chiral symmetry is equivalent to a uniform shift of the charge-density profile. In both cases the spontaneous breaking of the discrete chiral symmetry led to the fermions acquiring a gap, a dynamical mass generation as in the example of the Gross–Neveu model.

In $(2+1)$ dimensions these questions have a certain subtlety, since there is no γ_5 Dirac matrix. In fact, as we saw, for two-component fermions the mass term, $\bar{\psi}\psi$, breaks parity and time-reversal invariance. Since the lattice models have "doublers" in the Kogut–Susskind version, the different flavors may also acquire a mass. For instance, this happens in the case of the graphene model with sufficiently strong (but much too large for real graphene) repulsive interactions. In this case the mass term which is generated breaks the sublattice symmetry of graphene and hence breaks its point-group symmetry. If a mass term has Haldane's form, the broken symmetry is time-reversal-invariance. Likewise, in the case of the Wilson-fermion model, the fermion "doublers," i.e. the massive fermions at the corners of the Brillouin zone, break time-reversal invariance separately. In more formal terms, lattice regularizations generally break symmetries of the continuum field theory of Dirac fermions, while keeping gauge invariance intact. In $(2+1)$ dimensions these considerations lead to the parity anomaly, i.e. the breaking of parity (and time reversal) in a gauge-invariant regularization. In $(1+1)$ and $(3+1)$ dimensions, these gauge-invariant regularizations lead to the breaking of the continuum chiral symmetries.

In order to compute the Hall (and spin Hall) conductivities we need to compute the topological charge for the $\uparrow$ bands and for the $\downarrow$ bands. We can do this by expanding the vectors $\vec{d}_\pm(\vec{k})$ in patches of the Brillouin zone centered at the special time-reversal-invariant points of the Brillouin zone $\vec{Q}=(0,0)$, $(\pi,0)$, $(0,\pi)$, (π,π), and then evaluate the contribution of each patch to the topological charge. After setting $\vec{k}=\vec{Q}+\vec{q}$, with $|\vec{q}|$ small, we find both for $\uparrow$ bands and for $\downarrow$ bands

$$\begin{aligned}
\vec{d}_{\uparrow,(0,0)}(\vec{q}) &\simeq \big(q_x, q_y, M\big), & \vec{d}_{\downarrow,(0,0)}(\vec{q}) &\simeq \big(-q_x, q_y, M\big),\\
\vec{d}_{\uparrow,(\pi,0)}(\vec{q}) &\simeq \big(-q_x, q_y, M-2\big), & \vec{d}_{\downarrow,(\pi,0)}(\vec{q}) &\simeq \big(q_x, q_y, M-2\big),\\
\vec{d}_{\uparrow,(0,\pi)}(\vec{q}) &\simeq \big(q_x, -q_y, M-2\big), & \vec{d}_{\downarrow,(0,\pi)}(\vec{q}) &\simeq \big(-q_x, -q_y, M-2\big),\\
\vec{d}_{\uparrow,(\pi,\pi)}(\vec{q}) &\simeq \big(-q_x, -q_y, M-4\big), & \vec{d}_{\downarrow,(\pi,\pi)}(\vec{q}) &\simeq \big(q_x, -q_y, M-4\big)
\end{aligned} \tag{16.74}$$

Notice, for instance, that the vectors $\vec{d}_{\uparrow,(0,0)}$ and $\vec{d}_{\downarrow,(0,0)}$ indicate that near the Γ point the two blocks have opposite 2D parities and hence are also time-reversed. The same feature applies to the three other points.

We can now read off their contributions to the topological charges of the $\uparrow$ and $\downarrow$ bands to be

$$\begin{aligned}
\mathcal{Q}^\pm_{(0,0)} &= \mp\frac{1}{2}\,\mathrm{sgn}(M), & \mathcal{Q}^\pm_{(\pi,0)} &= \pm\frac{1}{2}\,\mathrm{sgn}(M-2),\\
\mathcal{Q}^\pm_{(0,\pi)} &= \pm\frac{1}{2}\,\mathrm{sgn}(M-2), & \mathcal{Q}^\pm_{(\pi,\pi)} &= \mp\frac{1}{2}\,\mathrm{sgn}(M-4)
\end{aligned} \tag{16.75}$$

The total topological charge of the $\pm$ bands is $\mathcal{Q}^\pm_{\mathrm{T}} = \mathcal{Q}^\pm_{(0,0)} + \mathcal{Q}^\pm_{(\pi,0)} + \mathcal{Q}^\pm_{(0,\pi)} + \mathcal{Q}^\pm_{(\pi,\pi)}$, which can now be computed for each regime. The result is

$$\mathcal{Q}^\pm_{\mathrm{T}} = \begin{cases} 0, & \text{for } M<0\\ \mp 1, & \text{for } 0<M<2\\ \pm 1, & \text{for } 2<M<4\\ 0, & \text{for } 4<M \end{cases} \tag{16.76}$$

Therefore, in the regimes $M < 0$ and $M > 4$ both $\uparrow$ bands and $\downarrow$ bands have zero (first) Chern number $\mathcal{Q}^{\pm}$. This is the trivial (or normal) insulator regime for each band. On the other hand, in the two remaining regimes, $0 < M < 2$ and $2 < M < 4$, their Chern numbers $\mathcal{Q}^{\pm}$ are ± 1. Hence, both regimes correspond to a quantum anomalous Hall insulator for each band.

Thus, as M increases *continuously* from $M < 0$ to $M > 0$, and the gap closes and opens up again at $(0, 0)$, the Chern number of each band jumps *discontinuously* from 0 to ± 1. The two subsequent jumps predicted by Eq. (16.76) correspond to the gap of the tight-binding model closing at $(\pi, 0)$ (and $(0, \pi)$) at $M = 2$ and at (π, π) at $M = 4$. Only the first gap closing is relevant to the physics of the quantum well we are discussing here.

By inspection of Eq. (16.76) we see that the total topological charge of the full four-band model is zero,

$$\mathcal{Q} = \mathcal{Q}_{\mathrm{T}}^{\uparrow} + \mathcal{Q}_{\mathrm{T}}^{\downarrow} = 0 \tag{16.77}$$

Therefore, this system has vanishing Hall conductivity, $\sigma_{xy} = 0$, as it should, since it is time-reversal-invariant.

On the other hand, the *spin Hall conductivity*, the difference of the Hall conductivities of the $\uparrow$ and $\downarrow$ bands,

$$\sigma_{xy}^{\mathrm{QSH}} = \sigma_{xy}^{\uparrow} - \sigma_{xy}^{\downarrow} = -\frac{e^2}{h}\left(\mathcal{Q}_{\mathrm{T}}^{\uparrow} - \mathcal{Q}_{\mathrm{T}}^{\downarrow}\right) \tag{16.78}$$

takes the quantized values

$$\sigma_{xy}^{\mathrm{QSH}} = \begin{cases} 0, & \text{for } M < 0 \\ +\dfrac{2e^2}{h}, & \text{for } 0 < M < 2 \\ -\dfrac{2e^2}{h}, & \text{for } 2 < M < 4 \\ 0, & \text{for } 4 < M \end{cases} \tag{16.79}$$

Thus, this simple lattice model exhibits the quantum spin Hall effect.

As we mentioned above, band-structure calculations (Bernevig *et al.*, 2006) show that the sign of the gap M changes as a function of the quantum-well thickness d from negative, $M < 0$, where the E1 and H1 bands of the quantum well are ordered (in energy) as in bulk CdTe, to positive, $M > 0$, where the order of the bands is *inverted*. Thus, this theory predicts that in the inverted-band regime there is a range of values of the gap within which the quantum well should display a quantum spin Hall effect.

16.5 $\mathbb{Z}_2$ topological invariants

The Chern number is the topological invariant that classifies the integer quantum Hall states and the quantum anomalous Hall states. In the case of the quantum spin Hall states the Chern number of the spin current cannot be defined since the current in general is not conserved. However, as we saw, a form of parity can be associated with the quantum spin Hall states. We will now consider a more general quantum spin Hall system with $N > 1$ right-moving edges with spin $\uparrow$ and N left-moving edges with spin $\downarrow$. The above considerations require that N be an *odd* integer for a quantum spin Hall insulator, whereas N is *even* for a normal insulator. In particular, in both cases the only way in which the edge states can become insulating (either by acquiring a gap or by localization due to disorder) requires that the states that become gapped have opposite chiralities. Thus edge states become gapped (and thus "disappear") in pairs.

From this perspective a normal insulator is equivalent, modulo 2, to a state without edge states, whereas the edge states of a generic quantum Hall insulator are equivalent to those of a quantum Hall insulator with only one chiral edge state for a given spin orientation. Hence, only the *parity* of the number of edge states is well defined. This observation leads to the concept that the quantum spin Hall insulators have $\mathbb{Z}_2$ topological invariants. In contrast, the quantum anomalous Hall insulators are classified by an integer, the Chern invariant, which coincides with the number of chiral edge states (as in the integer quantum Hall state).

What we are interested in is the topological classification of time-reversal-invariant insulators. We will follow here the work and notation of Kane, Fu, and Mele (Kane and Mele, 2005b; Fu and Kane, 2006; Fu *et al.*, 2007; Fu and Kane, 2007) as well as the work of Roy (2009), and Moore and Balents (2007). It turns out that these concepts, with some caveats, apply both to 2D and to 3D systems, and hence we will discuss them together. To this end, let us consider a time-reversal-invariant periodic (one-particle) Hamiltonian $\mathcal{H}$ with $2N$ occupied bands. $\mathcal{H}$ has Bloch wave functions $|\psi_n(\vec{k})\rangle = \exp(i\vec{k} \cdot \vec{r})|u_n(\vec{k})\rangle$, where the states $|u_n(\vec{k})\rangle$ are periodic in the unit cell and are eigenstates of the (reduced) Bloch Hamiltonian $H(\vec{k})$,

$$H(\vec{k}) = e^{-i\vec{k}\cdot\vec{r}}\mathcal{H}e^{i\vec{k}\cdot r} \tag{16.80}$$

If we denote by $\vec{G}$ the reciprocal-lattice vectors, then the eigenstates are periodic, $|\psi_n(\vec{k}+\vec{G})\rangle = |\psi_n(\vec{k})\rangle$, and the Brillouin zone is a torus. Hence $|u_n(\vec{k}+\vec{G})\rangle = \exp(-i\vec{G}\cdot\vec{r})|u_n(\vec{k})\rangle$.

Time reversal is the operation that complex-conjugates the state and reverses the spin of the particle. Thus, when acting on one-particle states with spin S, it is represented by the operator $\Theta = \exp(i\pi S_y)\mathcal{K}$, where S_y is the y-component of the spin and $\mathcal{K}$ is the complex conjugation. For spin $S = 1/2$ particles time reversal satisfies $\Theta^2 = -1$. If the one-particle Hamiltonian $\mathcal{H}$ is time-reversal-invariant, $[\mathcal{H}, \Theta] = 0$, then the Bloch Hamiltonian $H(\vec{k})$ satisfies

$$\Theta H(\vec{k})\Theta^{-1} = H(-\vec{k}) \tag{16.81}$$

The time-reversal operation induces a transformation in the Hilbert space of Bloch states. Let us assume that the system has two occupied Bloch bands, $|u_{i=1,2}(\vec{k})\rangle$, for each $\vec{k}$ in the Brillouin zone. Hence the states of the occupied bands form a rank-2 vector bundle over the Brillouin-zone torus. The time-reversal transformation $\mathcal{T}$ induces an involution in the Brillouin zone that identifies the points $\vec{k}$ and $-\vec{k}$. The states at these two points are related by an anti-unitary operator Θ, $|u_i(-\vec{k})\rangle = \Theta|u_i(\vec{k})\rangle$, which implies that the bundle is real. From the condition $\Theta^2 = -1$, the bundle is found to be twisted. In algebraic topology these bundles are classified by an integer (here the number of occupied bands) and a $\mathbb{Z}_2$ index that will allow us to classify the quantum spin Hall insulators, which for this reason are called $\mathbb{Z}_2$ topological insulators.

In a periodic lattice there is a set of points of the Brillouin zone that we will denote by $\vec{Q}_i$ with the property that they differ from their images under the action of time reversal by a vector $\vec{G}$ of the reciprocal lattice, $-\vec{Q}_i = \vec{Q}_i + \vec{G}$. There are four such points in 2D and eight in 3D. These time-reversal-invariant points of the Brillouin zone can be labeled by two (three) integers (mod 2) $n_i = 0, 1$ in 2D (3D), such that $\vec{Q}_i = \frac{1}{2}\sum_j n_j \vec{b}_j$, where $\{\vec{b}_j\}$ are the primitive reciprocal-lattice vectors. Kane and Mele defined the unitary $2N \times 2N$ antisymmetric matrix $w_{m,n}(\vec{k})$,

$$w_{m,n}(\vec{k}) = \langle u_m(-\vec{k})\big|\Theta\big|u_n(\vec{k})\rangle \tag{16.82}$$

and showed that the quantities δ_i are given by

$$\delta_i = \frac{\sqrt{\det[w(\vec{Q}_i)]}}{\mathrm{Pf}[w(\vec{Q}_i)]} = \pm 1 \tag{16.83}$$

where $\det[w]$ and $\mathrm{Pf}[w]$ are the determinant and the Pfaffian of the matrix w. Recall that $\det[w] = \mathrm{Pf}[w]^2$.

Because of the square root, the sign of the quantities δ_i is ambiguous. However, by requiring the states $|u_n(\vec{k})\rangle$ to be continuous, $\sqrt{\det[w(\vec{k})]}$ is defined globally in the Brillouin zone since $\det[w(\vec{k})]$ is single-valued on closed loops C and the square root has no branch cuts. For contractible loops this follows from the continuity of the states $|u_n(\vec{k})\rangle$. For non-contractible loops, which can be deformed from C to $-C$, this follows from $\det[w(\vec{k})] = \det[w(-\vec{k})]$. The quantities δ_i are also gauge-dependent, i.e. changes of the phases of the Bloch states at the invariant points. However, suitable products of them are gauge and are also topological invariants. In 2D the product of these quantities for the four time-reversal-invariant points is gauge-invariant. This defines the $\mathbb{Z}_2$ topological invariant ν in 2D,

$$(-1)^{\nu} = \prod_{i=1}^{4} \delta_i \tag{16.84}$$

In 3D one can use the quantities δ_i computed from the eight time-reversal-invariant points Q_i to construct four $\mathbb{Z}_2$ topological invariants, namely ν_0 and ν_k, $k = 1, 2, 3$. They are given by (Fu and Kane, 2007)

$$(-1)^{\nu_0} = \prod_{i=1}^{8} \delta_i \tag{16.85}$$

and

$$(-1)^{\nu_k} = \prod_{n_k=1, n_{j\neq k}=0,1} \delta_{i=(n_1,n_2,n_3)} \tag{16.86}$$

Clearly, the three invariants ν_k treat the three orthogonal planes which include four time-reversal-invariant points at a time as 2D projections, and the system as if it were layered. It turns out that these three invariants are weak in the sense that they are not robust in the presence of disorder. In contrast, ν_0 is robust and hence "more fundamental." Insulators characterized by $\nu_0 = 1$ are said to be *strong topological insulators*.

Fu and Kane showed that the computation of the $\mathbb{Z}_2$ invariant is simpler in the case of systems with inversion symmetry. If the one-particle Hamiltonian $\mathcal{H}$ is invariant under inversion P, $[\mathcal{H}, P] = 0$, where the parity operator P is defined as $P|\vec{r}, s_z\rangle = |-\vec{r}, s_z\rangle$, then the Bloch Hamiltonian satisfies $PH(\vec{k})P^{-1} = H(-\vec{k})$. Let $\mathcal{A}(\vec{k})$ be the Berry connection

$$\mathcal{A}(\vec{k}) = -i \sum_{i=1}^{2N} \langle u_n(\vec{k})|\nabla_{\vec{k}}|u_n(\vec{k})\rangle \tag{16.87}$$

and let $\mathcal{F}(\vec{k})$ be the Berry curvature,

$$\mathcal{F}(\vec{k}) = \nabla_{\vec{k}} \times \mathcal{A}(\vec{k}) \tag{16.88}$$

However, the Berry curvature $\mathcal{F}(\vec{k})$ is simultaneously odd under time reversal, $\mathcal{F}(-\vec{k}) = -\mathcal{F}(\vec{k})$, and even under inversion, $\mathcal{F}(-\vec{k}) = +\mathcal{F}(\vec{k})$. Hence the Berry curvature must vanish, $\mathcal{F}(\vec{k}) = 0$, in systems that are both time-reversal- and inversion-invariant. In this case, it is possible to choose the phases of the Bloch states, i.e. to make a choice of gauge, so that the Berry connection vanishes as well, $\mathcal{A}(\vec{k}) = 0$. Fu and Kane found that in this gauge the quantities δ_i are given by (Fu and Kane, 2007)

$$\delta_i = \prod_{m=1}^{N} \xi_{2m}(\vec{Q}_i) \tag{16.89}$$

where $\xi_n(\vec{Q}_i) = \pm 1$ are the parity eigenvalues of the occupied parity eigenstates $|\psi_n(\vec{Q}_i)\rangle$. They further showed that the strong $\mathbb{Z}_2$ invariant ν_0, which is the product of the eight (four) δ_i in 3D (2D), does not rely on the existence of inversion symmetry.

We will now focus on the case of a system with two bands (plus spin), for which the Hilbert space at each $\vec{k}$ is four-dimensional. The one-particle Hamiltonian $h(\vec{k})$ can always be expanded in a basis of the space of hermitian 4×4 matrices. This basis can be chosen to be the 4×4 identity matrix I, the five Dirac hermitian gamma matrices, Γ^a (with $a = 1, \ldots, 5$), which satisfy the Clifford algebra $\{\Gamma_a, \Gamma_b\} = \delta_{ab} I$, and their ten commutators $\Gamma^{ab} = (1/(2i))[\Gamma^a, \Gamma^b]$. For example, we can take the Γ^a matrices to be the standard Dirac matrices

$$\Gamma_i \equiv \alpha_i = \sigma_i \otimes \tau_1, \quad \Gamma_4 \equiv \beta = \gamma_0 = I \otimes \tau_3, \quad \Gamma_5 \equiv i\gamma_0\gamma_5 = -I \otimes \tau_2 \tag{16.90}$$

of the Dirac equation in $(3+1)$ dimensions. As usual, in this notation the first factor acts on the two spin components, while the second factor acts on the two bands (the positive- and negative-energy states of the Dirac equation). In the Dirac basis, the time-reversal operation is represented by $\Theta = (i\sigma_2 \otimes I)\mathcal{K}$, where $\mathcal{K}$ is the operation of complex conjugation, and parity by $P = \mathcal{I} \otimes \tau_3 = \beta$. It is straightforward to see that the five gamma matrices commute with $P\Theta$, while the commutators anti-commute with it.

For general time-reversal- and parity-invariant systems, the two-band Bloch Hamiltonian has the form (both in 2D and in 3D)

$$H(\vec{k}) = d_0(\vec{k})\mathcal{I} + \sum_{a=1}^{5} d_a(\vec{k})\Gamma_a \tag{16.91}$$

By symmetry, at the time-reversal- and parity-invariant momenta Q_i the Bloch Hamiltonian $H(\vec{Q}_i)$ can depend only on the identity matrix $\mathcal{I}$ and on the gamma matrix $\Gamma_4 = \beta$,

$$H(\vec{Q}_i) = d_0(Q_i)\mathcal{I} + d_4(\vec{Q}_i)\beta \tag{16.92}$$

From this result, and recalling that the parities of the spinors are the eigenstates of the gamma matrix β, we can read off the parities of the occupied states at each time-reversal-invariant point $\vec{Q}_i$, leading to the result that, in this simpler case, the quantities δ_i are simply given by (Fu and Kane, 2007)

$$\delta_i = -\text{sgn}[d_4(\vec{Q}_i)] \tag{16.93}$$

The fact that the $\mathbb{Z}_2$ topological invariants are given by the parities of the eigenstates at the time-reversal-invariant points of the Brillouin zone suggests that in the regime in which the two-band model has a small gap, and hence can be approximated by the continuum Dirac equation, there may be a relation with the ground-state expectation value of the Dirac fermion bilinear $\langle\bar{\psi}\psi\rangle$, which is non-zero if there is a mass gap. Indeed, if the two-band model has local gap minima at the invariant points, $\vec{Q}_i$, as in the case of the Wilson-fermion model, the computation of $\langle\bar{\psi}\psi\rangle$ *for each fermionic species* involves the matrix element of the Dirac gamma matrix β for the occupied eigenvectors (spinors) whose values are precisely the parities.

It may seem surprising that a global topological property of the band structure, such as the $\mathbb{Z}_2$ invariants, can be expressed in terms of the parities at certain invariant points of the Brillouin zone, or equivalently in terms of the expectation values of local operators such as $\bar{\psi}\psi$. This is consistent since the band structure is continuous over the Brillouin zone. This also tells us that it is not possible to determine whether a system is a topological insulator solely by a local analysis near a *single* band crossing. However, suppose we know that for some range of parameters the system is topologically trivial, and then we find that for some other range of parameters there is a band crossing (and inversion) near some invariant point of the Brillouin zone. Then we can assert that in the second range the system is topologically non-trivial and became a $\mathbb{Z}_2$ topological insulator.

On the other hand, if the gaps are essentially of $O(1)$ everywhere in the Brillouin zone, a natural approximation to compute the $\mathbb{Z}_2$ invariant is to take a "flat-band" limit of the Hamiltonian, in which its momentum dependence essentially disappears. The computation of the invariants in the flat-band limit is considerably simpler, and it is often used to classify the Hamiltonians (Qi *et al.*, 2008; Kitaev, 2009).

As an example, let us compute the $\mathbb{Z}_2$ invariant ν_0 for the BHZ model for the quantum spin Hall effect in 2D. By inspection of Eq. (16.74) and Eq. (16.75) we see that $\delta(0,0) = -\text{sgn}(M)$, $\delta(\pi,0) = -\text{sgn}(M-2) = \delta(0,\pi)$, and $\delta(\pi,\pi) = -\text{sgn}(M-4)$. Hence the $\mathbb{Z}_2$ topological invariant ν_0 is

$$(-1)^{\nu_0} = \delta(0,0)\delta(\pi,0)\delta(0,\pi)\delta(\pi,\pi) = +\text{sgn}(M)\text{sgn}(M-4) \tag{16.94}$$

In other terms, this system has $\nu_0 = 0 \pmod 2$ for $M < 0$ and for $M > 4$ (since the two signs are equal in these regimes), where it is a trivial insulator. Conversely, it has $\nu_0 = 1 \pmod 2$ for $0 < M < 4$ (where the signs are opposite), which we identify as a $\mathbb{Z}_2$ topological insulator. This is the quantum spin Hall regime. Clearly, in this regime the parity of the occupied bands at the Γ point is opposite to the parity of the occupied bands at (π,π). Hence, as expected, we have a case of a band inversion.

While the analysis of the previous section predicted a quantized spin Hall conductance, and in fact two different regimes in which the spin Hall conductance was found to have the quantized value $\pm 2e^2/h$, the $\mathbb{Z}_2$ topological invariant has the same value, $\nu_0 = 1$, regardless of the sign of the spin Hall conductance. The reason is that the characterization of the quantum spin Hall state in terms of a Chern number requires the conservation of the component S_z of the spin (and of the associated spin current). In contrast, the $\mathbb{Z}_2$ topological insulator is well characterized by the topological invariant ν_0 even in the absence of a conservation law for the S_z component of the spin. We will see below that a system with a non-trivial value of the $\mathbb{Z}_2$ topological invariant, $\nu_0 = 1 \pmod 2$, has protected edge states, and a quantum but not quantized spin Hall effect, whereas such edge states are generally absent if $\nu_0 = 0 \pmod 2$.

16.6 Three-dimensional topological insulators

The construction that we have used in 2D can be extended to 3D topological insulators. Three-dimensional semiconductors with near band crossings, and hence small gaps, at certain symmetry points of their Brillouin zones have been known for a long time. In all cases the systems have strong spin–orbit couplings and hence the continuous SU(2) spin invariance is broken to a discrete subgroup. One of the first examples investigated with this approach was PbTe, which is not a topological insulator (although it is close to being one). Its bands have near crossings at the L points of the cubic lattice, $(\pm\pi/2, \pm\pi/2, \pm\pi/2)$, which suggests that it can be described by a Kogut–Susskind version of 3D lattice Dirac fermions (Fradkin *et al.*, 1986; Boyanovsky *et al.*, 1987). It was suggested (by the same authors) that the 2D parity anomaly may occur in these systems. It turns out that this does not happen in PbTe, since it is a standard insulator, not a topological insulator. This can be checked by a direct computation of the $\mathbb{Z}_2$ topological invariant (Fu and Kane, 2007).

Several 3D narrow-gap semiconductors have been proposed to be topological insulators, notably $Bi_{1-x}Sb_x$ and HgTe (under uniaxial stress) (Fu and Kane, 2007; Fu *et al.*, 2007), Bi_2Se_3 and Bi_2Te_3 (Zhang *et al.*, 2009). We will see that all these systems exhibit an odd number of surface states that behave as 2D chiral fermions (different odd numbers depending on the material). These surface states have been detected by angle-resolved photoemission spectroscopy (Hasan and Kane, 2010; Hasan and Moore, 2011; Qi and Zhang, 2011). Unfortunately, at the time of writing, the available materials are conducting, not insulating.

The simplest examples of 3D topological insulators are Bi_2Se_3 and Bi_2Te_3. In both cases spin–orbit coupling is strong and there is a near band crossing at the Γ point, the center $(0, 0, 0)$ of the Brillouin zone. We will not go through the complications of the band structure. It will be sufficient for us to use the effective two-band model derived by Zhang and coworkers (Zhang *et al.*, 2009), who showed that the minimal model involves two orbitals, $|\mathrm{P1}_z^+, \sigma\rangle$ and $|\mathrm{P2}_z^-, \sigma\rangle$, where $\pm$ denotes the parity of the state, and $\sigma = \uparrow, \downarrow$ are the two spin (actually J_z) components. Thus the states can be represented by a four-component spinor, where the two upper components have $+$ parity and the two lower components have $-$ parity. In this basis the one-particle Hamiltonian near the Γ point has the standard Dirac form

$$H = \varepsilon_0(\vec{p})\mathbb{I} + \vec{A}(\vec{p}) \cdot \vec{\alpha} + \mathcal{M}(\vec{p})\beta \tag{16.95}$$

where $\vec{\alpha}$ and β are the standard Dirac matrices (in the Dirac basis used earlier), $\mathbb{I}$ is the 4×4 identity matrix, and $\mathcal{M}(\vec{p}) = M - Bp_z^2 - B'(p_x^2 + p_y^2)$ and $\vec{A}(p) = (Ap_x, Ap_y, A'p_z)$. The parameters B, B', A, and A' are positive.

It will be important to our analysis that, reflecting the inversion symmetry and time-reversal invariance of these materials, the effective Hamiltonian does not include the fifth Dirac matrix $\Gamma_5 = i\beta\gamma_5$. One consequence of this is that the spectrum is particle–hole (or charge-conjugation)-invariant. This fact plays an important role in the physics of the "edge states" of these systems and is a necessary condition in order for these states to be gapless Weyl fermions.

The ("Dirac mass") M, the gap at the Γ point, is *positive* for Bi_2Se_3 (and Bi_2Te_3), but it is *negative* for Sb_2Se_3. Hence there is a band inversion at the Γ point. By analogy with the 2D quantum spin-Hall-system cousin that we just discussed, these systems are candidates to be $\mathbb{Z}_2$ topological insulators, and the order of the bands with opposite band parity is switched in one material relative to the others. This result suggests that Bi_2Se_3 and Bi_2Te_3 are topological insulators, while Sb_2Se_3 is not. That this guess was correct was verified by Zhang *et al.* (2009) by computating the topological invariant parities introduced by Fu *et al.* (2007) (which we discussed in the preceding section) at the four inequivalent time-reversal-invariant points of the Brillouin zone, $\Gamma(0, 0, 0)$, $L(\pi, 0, 0)$, $F(\pi, \pi, 0)$, and $Z(\pi, \pi, \pi)$ and verifying that their product is negative for Bi_2Se_3 and Bi_2Te_3, but positive for Sb_2Se_3.

We can give a simpler description by using a Wilson-fermion lattice Hamiltonian (which we have already used in the discussion of the 2D quantum spin Hall effect) on a 3D cubic lattice of the form (with cubic symmetry for simplicity)

$$H = \sin \vec{p} \cdot \vec{\alpha} + M(\vec{p})\beta \tag{16.96}$$

where $M(\vec{p}) = M + \cos p_x + \cos p_y + \cos p_z - 3$. This simpler Hamiltonian has the same qualitative behavior and has the same time-reversal-invariant points as the one derived from band-structure calculations. Thus, e.g. at the Γ point, $\vec{p} = (0, 0, 0)$, if the Dirac mass $M > 0$ the positive-energy states have $+$ parity and the negative-energy states have − parity. For $M < 0$ the order of the states is reversed. This construction shows that the 3D $\mathbb{Z}_2$ topological insulators are a generalization of the 2D quantum spin Hall effect.

It is straightforward to compute the $\mathbb{Z}_2$ topological invariants for the Wilson-fermion model. The parities at the eight invariant points are ($k = 1, 2, 3$)

$$\begin{aligned} \delta(0,0,0) &= -\mathrm{sgn}(M) \\ \delta(\pi,0,0) = \delta(0,\pi,0) = \delta(0,0,\pi) &= -\mathrm{sgn}(M-2) \\ \delta(\pi,\pi,0) = \delta(0,\pi,\pi) = \delta(\pi,0,\pi) &= -\mathrm{sgn}(M-4) \\ \delta(\pi,\pi,\pi) &= -\mathrm{sgn}(M-6) \end{aligned} \tag{16.97}$$

and the invariants are

$$(-1)^{\nu_0} = \mathrm{sgn}(M)\mathrm{sgn}(M-2)\mathrm{sgn}(M-4)\mathrm{sgn}(M-6) \tag{16.98}$$

$$(-1)^{\nu_k} = \mathrm{sgn}(M-2)\mathrm{sgn}(M-6) \tag{16.99}$$

from which we find that this model describes a strong topological insulator for $0 < M < 2$ (with $\mathbb{Z}_2$ invariants $\nu_0 = 1 \pmod 2$ and $(\nu_1, \nu_2, \nu_3) = (0,0,0) \pmod 2$) and $2 < M < 4$ (with $\mathbb{Z}_2$ invariants $\nu_0 = 1 \pmod 2$ and $(\nu_1, \nu_2, \nu_3) = (1,1,1) \pmod 2$), a weak topological insulator for $2 < M < 4$ (with $\mathbb{Z}_2$ invariants $\nu_0 = 0 \pmod 2$ and $(\nu_1, \nu_2, \nu_3) = (1,1,1) \pmod 2$), and a trivial insulator (with $\mathbb{Z}_2$ invariants $\nu_0 = 0 \pmod 2$ and $(\nu_1, \nu_2, \nu_3) = (0,0,0) \pmod 2$) for the other regimes.

It is interesting and useful to write the free Dirac field theory of this model. Thus, following the standard Wilson construction at each lattice site $\vec{r}$ of the cubic lattice, and hence at every momentum $\vec{p}$ of the cubic Brillouin zone, we introduce a set of four component fermions, $\psi_\alpha(\vec{r})$, with $\alpha = 1, \ldots, 4$. Here too, the two upper (lower) components represent the spin ↑ and ↓ components of the fermion. However the parities of the valence-band (negative-energy) and conduction-band (positive-energy) states may be different near the Γ point and near the corners of the Brillouin zone.

The Hamiltonian of (free) Wilson fermions on a cubic lattice is

$$H = \int_{\mathrm{BZ}} \frac{d^3p}{(2\pi)^3} \sum_{\alpha,\beta} \psi_\alpha^\dagger(\vec{p}) \left(\sin\vec{p}\cdot\vec{\alpha} + M(\vec{p})\beta \right)_{\alpha\beta} \psi_\beta(\vec{p}) \tag{16.100}$$

We recall that the covariant definition of the Dirac gamma matrices is

$$\gamma_0 = \beta, \qquad \gamma^i = \beta\alpha^i, \qquad \gamma_5 = i\gamma_0\gamma_1\gamma_2\gamma_3 \tag{16.101}$$

and that they obey the covariant Dirac algebra

$$\{\gamma_\mu, \gamma_\nu\} = 2g_{\mu\nu}\mathcal{I}, \qquad \{\gamma_5, \gamma_\mu\} = 0 \tag{16.102}$$

where $g_{\mu\nu} = \text{diag}(1, -1, -1, -1)$ is the (Bjorken and Drell) Minkowski metric in $(3+1)$ dimensions.

We now observe that the *one-particle* Dirac Hamiltonian H of Eq. (16.96) (and Eq. (16.95)) (not to be confused with the Hamiltonian of the field theory, Eq. (16.100)) does not involve the Dirac matrix γ_5. Hence γ_5 *anti-commutes* with the one-particle Dirac Hamiltonian, $\{\gamma_5, H\} = 0$. If $u_\alpha^\pm(\vec{p}, \sigma)$ (with $\sigma = \uparrow, \downarrow$) are the four linearly independent spinors with energies $E_\pm = \pm\sqrt{\sin^2 \vec{p} + M(\vec{p})^2}$, it is easy to see that the spinors $\gamma_5 u^\pm(\vec{p}, \sigma)$ have energies $\mp E(\vec{p}, \sigma)$. In other terms, we conclude that $\gamma_5 u^\pm(\vec{p}, \sigma) = u^\mp(\vec{p}, \sigma)$, since they have the same spin projection. Thus, γ_5 maps spinors with positive energy to spinors with negative energy (and vice versa), and hence it also maps spinors with opposite parities into each other.

Wilson introduced the discrete version of Dirac fermions as a regularization of the continuum field theory. From that perspective, only the region of momentum space near $\vec{p} = 0$ is physically relevant. In this regime one simply approximates the dispersion by setting $\sin \vec{p} \cdot \vec{\alpha} \simeq \vec{p} \cdot \vec{\alpha}$. The Hamiltonian of the continuum free Dirac theory is

$$H = \int d^3x \; \psi_\alpha^\dagger(\vec{p})\big(\vec{p} \cdot \vec{\alpha} + M\beta\big) \tag{16.103}$$

and its Lagrangian density has the Lorentz-invariant form

$$\mathcal{L} = \bar{\psi}(x)\big(i\gamma^\mu \partial_\mu - M\big)\psi(x) \tag{16.104}$$

where we dropped the Dirac indices and, as usual, $\bar{\psi} = \psi^\dagger \gamma_0$. In the massless limit, $M = 0$, the Dirac Lagrangian is invariant under global continuous chiral transformations,

$$\psi(x) \to e^{i\theta\gamma_5}\psi(x), \qquad \bar{\psi}(x) \to \bar{\psi}(x)e^{i\theta\gamma_5} \tag{16.105}$$

where $0 \leq \theta < 2\pi$.

The Dirac theory admits two types of mass terms. One is the $\bar{\psi}\psi$ operator we already have. The other possible mass term is $i\bar{\psi}\gamma_5\psi$. At the level of the one-particle Dirac Hamiltonian, the γ_5 mass term enters with the matrix $\Gamma_5 = i\gamma_0\gamma_5$ (see Eq. (16.90)). If both mass terms are present, charge-conjugation symmetry C is broken and so is parity P. Hence CP is broken, which is equivalent to breaking time-reversal invariance, T. The global continuous chiral symmetry is broken explicitly by both mass terms, which transform as

$$\begin{pmatrix} \bar{\psi}(x)\psi(x) \\ i\bar{\psi}(x)\gamma_5\psi(x) \end{pmatrix} \to \begin{pmatrix} \cos(2\theta) & \sin(2\theta) \\ -\sin(2\theta) & \cos(2\theta) \end{pmatrix} \begin{pmatrix} \bar{\psi}(x)\psi(x) \\ i\bar{\psi}(x)\gamma_5\psi(x) \end{pmatrix} \tag{16.106}$$

This transformation has the same form as the one we used in Section 6.3 to represent a translation of a charge-density wave in $(1+1)$ dimensions.

The lattice Hamiltonian, Eq. (16.100), has a broken chiral symmetry even if we fine-tune $\lim_{\vec{p}\to 0} M(\vec{p}) = 0$. With this choice there are terms that break the continuous chiral symmetry, but which vanish as $\vec{p} \to 0$. These operators are irrelevant. However the "doublers," i.e. the states near $(\pi, 0, 0)$ and so on, have $O(1)$ mass terms that break the global continuous chiral symmetry. In fact the signs of the operator $\langle \bar{\psi}\psi \rangle$ for the fermion "doublers" are needed, as we saw, in order to determine the $\mathbb{Z}_2$ topological class.

16.7 Solitons in polyacetylene

In Section 15.1 we discussed the theory of the edge states of the 2DEG in a uniform perpendicular magnetic field in the regime in which the integer quantum Hall effect is observed. There we saw that, due to the incompressibility of the bulk 2DEG in an integer (and fractional) quantum Hall state, its only gapless excitations are edge states (residing, naturally, at the boundary of the 2DEG!). Moreover, due to the explicitly broken time-reversal invariance caused by the external magnetic field, these edge states are chiral and hence propagate in just one direction. We will now see that a similar set of states generically exists in free-fermion systems that display the quantum anomalous Hall effect and the quantum spin Hall effect.

We will also see how this theory explains the chiral surface states of 3D $\mathbb{Z}_2$ topological insulators. The basic conceptual explanation of all these phenomena is based on the concept of anomalies in the Dirac field theory, which has different manifestations in different dimensions. The oldest application of these ideas was in $(1+1)$ dimensions, where it led to a theory of fractional quantum numbers of solitons (Jackiw and Rebbi, 1976; Goldstone and Wilczek, 1981), and was famously applied in a condensed matter context to the theory of solitons in polyacetylene (Su *et al.*, 1979; Jackiw and Schrieffer, 1981; Heeger *et al.*, 1988). The extension of these ideas to higher dimensions is known as the Callan–Harvey effect in fermionic domain walls (Callan and Harvey, 1985), which will also play a key role in our discussion of edge states in topological insulators (with or without time-reversal symmetry).

Domain walls in 1D systems have been studied extensively in the theory of soliton states both in quantum field theory (Dashen *et al.*, 1975; Jackiw and Rebbi, 1976; Rajaraman, 1985) and in condensed matter physics, mostly in the context of 1D conductors such as polyacetylene (Su *et al.*, 1979; Jackiw and Schrieffer, 1981; Heeger *et al.*, 1988). Polyacetylene is a polymer chain of carbon and hydrogen atoms with one hydrogen per carbon. This is denoted as $(CH)_n$. In the stable *trans* configuration, the chain has a zig-zag pattern with the carbon atoms at the vertices and the hydrogen atoms being placed in a staggered fashion to the right and to the left of the carbon atoms. As in the example of graphene, only the electrons in the π orbitals of the carbon atoms are effectively mobile, and the other σ bands are occupied and separated by a large energy gap.

Su, Schrieffer, and Heeger (SSH) proposed the following simple 1D lattice model for a single *trans* polyacetylene chain (Su *et al.*, 1979). Let the integer n denote the position of the carbon atoms along the chain. If we denote by $\psi^\dagger_{n,\sigma}$ the fermion operator that creates a π electron at site n with spin projection $\sigma = \uparrow, \downarrow$ and by u_n the displacement of the CH unit at site n (relative to the equilibrium position), the SSH Hamiltonian for a polyacetylene chain with N sites is

$$H = -\sum_{n,\sigma} \left(t_{n,n+1} \psi^\dagger_{n,\sigma} \psi_{n+1,\sigma} + \text{h.c.}\right) + \sum_n \frac{D}{2}(u_{n+1} - u_n)^2 + \sum_n \frac{P_n^2}{2M} \tag{16.107}$$

where P_n is the momentum of the CH group labeled by n, and M and D are the mass of the CH unit and the elastic constant of the chain (due to the σ bonding). As usual, the displacements u_n and the momenta P_n obey equal-time commutation relations, $[u_n, P_{n'}] = i\delta_{n,n'}$, and the fermion creation and annihilation operators obey the standard anticommutator algebra, $\{\psi_{n,\sigma}, \psi^\dagger_{n',\sigma'}\} = \delta_{n,n'}\delta_{\sigma,\sigma'}$. The position-dependent hopping term reflects the electron–phonon coupling between the electron hopping and the local displacements. For small relative distortions, $|u_{n+1} - u_n| \ll a$ (where a is the lattice spacing), it can be written as

$$t_{n,n+1} = t - \alpha(u_{n+1} - u_n) \tag{16.108}$$

where α is the electron–phonon-coupling constant.

Undoped polyacetylene has one π electron for each CH group. Hence, the chain is at half-filling and the Fermi momenta are at $p_{\text{F}} = \pm\pi/2$ (in units in which the lattice spacing is $a = 1$). As is well known, polyacetylene is the prototype of the systems that exhibit the Peierls instability, by which means it lowers its ground-state energy through a lattice distortion that breaks the symmetry of translations by one lattice spacing. Since polyacetylene is half-filled, it can lower the energy by dimerizing the chain, i.e. by the development of an expectation value of the displacement field $\langle u_n \rangle = (-1)^n \Delta$ with a staggered pattern. In other words, the effective hopping amplitude has a dimerized pattern, $t_{n,n+1} = t + (-1)^n 2\alpha\Delta$. Since there are two possible patterns (that differ by a rigid displacement of the state by one lattice spacing) this ground state is doubly degenerate. In this ground state the chain looks like a sequence of "single" and "double" bonds (as it is commonly depicted in chemistry) or, in the physicist's language, it is a period-2 commensurate charge-density wave on the bonds of the chain. In other terms, the $\mathbb{Z}_2$

symmetry of rigid displacements of the dimerization pattern in *trans* polyacetylene is spontaneously broken.

We will not present here the detailed theory of polyacetylene, which can be found in excellent reviews (Heeger *et al.*, 1988). We will rather write down an effective-field theory that bears out the physics we just described, which is accurate in the weak-coupling regime but also gives a correct qualitative description of the physics at substantial values of the coupling constant. We will follow the same approach as we used in Chapters 5 and 6 and expand the Fermi fields in fast and slow components (here $x = na$)

$$\psi_{n,\sigma} = e^{i\pi n/2}\psi_{\mathrm{R},\sigma}(x) + e^{-i\pi n/2}\psi_{\mathrm{L},\sigma}(x) \tag{16.109}$$

where we set $p_{\mathrm{F}} = \pi/2$. We will use the standard Dirac notation and represent the right- and left-moving fermions by a Dirac doublet, $\psi_{a,\sigma}(x)$ (with $a = \mathrm{R, L}$). For later convenience we will work in the Dirac basis (rather than with chiral components) of the fermions,

$$\begin{aligned}\psi_{1,\sigma} &= \frac{1}{\sqrt{2}}\left(-\psi_{\mathrm{R},\sigma} + \psi_{\mathrm{L},\sigma}\right)\\ \psi_{2,\sigma} &= \frac{1}{\sqrt{2}}\left(\psi_{\mathrm{R},\sigma} + \psi_{\mathrm{L},\sigma}\right)\end{aligned} \tag{16.110}$$

and denote the doublet Fermi field by $\psi_{a,\sigma}(x)$, with $a = 1, 2$.

The displacement fields admit an expansion of the form

$$u_n = u_0(x) + (-1)^n\Delta(x) \tag{16.111}$$

and similarly for the canonical momenta. Here $u_0(x)$ represents the smooth long-wavelength fluctuations of "acoustic" phonons, while $\Delta(x)$ represents the fluctuations of "optical" phonons with wave vector $Q = 2p_{\mathrm{F}} = \pi$. Much as in the weak-coupling theory of antiferromagnetism of Chapter 3, the 1D chain at half-filling obeys a nesting condition and is unstable with respect to a Peierls distortion, a backscattering process that is here due to the exchange of phonons with wave vector $Q = 2p_{\mathrm{F}} = \pi$ and hence represented by the fluctuations $\Delta(x)$. In what follows we will focus on the coupling of the fermions to the dimerization field $\Delta(x)$ and neglect the "acoustic" component u_0 since it essentially decouples.

The Hamiltonian density of the effective continuum theory is (summation over repeated indices is understood) (Fradkin and Hirsch, 1983)

$$\begin{aligned}\mathcal{H} = {} & \psi^{\dagger}_{a,\sigma}(x)(-iv_{\mathrm{F}}\sigma_1\,\partial_x)\psi_{a,\sigma}(x) + g\Delta(x)\bar{\psi}_{\sigma}(x)\psi_{\sigma}(x) + \frac{1}{8Ma_0^2}\Pi^2(x)\\ & + \frac{1}{2}\Delta^2(x)\end{aligned} \tag{16.112}$$

where $\Pi(x)$ is the canonical momentum for $\Delta(x)$, and obeys $[\Delta(x), \Pi(y)] = i\delta(x-y)$. This effective Hamiltonian is invariant under the discrete chiral symmetry $\psi \to \gamma_5 \psi$ and $\Delta \to -\Delta$, which on the lattice is a shift by one lattice constant.

In Eq. (16.112) we introduced the effective coupling constant $g \sim \alpha/\sqrt{Dt}$. M is the mass of the CH group and a_0 is the lattice spacing, t is the hopping amplitude, and the Fermi velocity is $v_{\text{F}} = 2t$. We have used the 2×2 Dirac matrices in the Dirac basis, $\alpha = \gamma_5 = \sigma_1$, $\gamma_0 = \sigma_3$, $\gamma_1 = i\sigma_2$. In this notation the dimerization field $\Delta(x)$ couples to the fermion bilinear $\bar{\psi}(x)\psi(x)$. In the chiral basis, the fermion bilinear is $\bar{\psi}(x)\psi(x) = i\psi^\dagger_{\text{R},\sigma}\psi_{\text{L},\sigma} + \text{h.c.}$ On changing to the Dirac basis (after a subsequent chiral rotation by $\theta = \pi/4$) the fermion bilinear becomes $\bar{\psi}\psi = \psi^\dagger_{1,\sigma}\psi_{2,\sigma} + \text{h.c.}$

Two limits of the Hamiltonian of Eq. (16.112) are worth considering. One is the regime in which the mass M of the CH group is taken to be so large that the quantum fluctuations of the dimerization field $\Delta(x)$, i.e. its kinetic-energy term, can be neglected. In this *adiabatic* limit the dimerization field becomes classical. The ground state of the system is found by finding the value $\Delta(x) = \Delta_0$ which minimizes the total energy density $\mathcal{E}$,

$$\mathcal{E} = \frac{\Delta_0^2}{2} - \int_{-\Lambda}^{\Lambda} \frac{dp}{2\pi} \sqrt{p^2 v_{\text{F}}^2 + g^2 \Delta_0^2} \tag{16.113}$$

where $\Lambda \simeq \pi/a_0$ is the momentum cutoff. By requiring that Δ_0 be a local extremum of the energy density $\mathcal{E}$ we find the gap equation

$$\frac{\partial \mathcal{E}}{\partial \Delta_0} = 0 \Rightarrow \Delta_0 = \int_{-\Lambda}^{\Lambda} \frac{dp}{2\pi} \frac{g^2 \Delta_0}{\sqrt{p^2 v_{\text{F}}^2 + g^2 \Delta_0^2}} \tag{16.114}$$

which has the non-trivial solution

$$\Delta_0 = \frac{2\Lambda v_{\text{F}}}{g} \exp\left(-\frac{\pi v_{\text{F}}}{g^2}\right) \tag{16.115}$$

which implies that there is a dynamically generated exponentially small gap in the fermionic spectrum $\sqrt{2} g \Delta_0 = 2\sqrt{2}\Lambda v_{\text{F}} \exp(-\pi v_{\text{F}}/g^2)$.

The opposite limit of interest is the regime in which the mass of the CH group is taken to be very small, $M \to 0$. In this regime the quantum fluctuations of the dimerization field are as large as they can be. In fact, in this limit the dimerization field is not only not classical but also can be integrated out (in the path integral). The effective Lagrangian density of the fermions is found to be (Fradkin and Hirsch, 1983) (on setting $v_{\text{F}} = 1$)

$$\mathcal{L} = \bar{\psi}_\sigma(x) i\gamma^\mu \partial_\mu \psi_\sigma(x) + g^2 \left(\bar{\psi}_\sigma(x)\psi_\sigma(x)\right)^2 \tag{16.116}$$

which is the Gross–Neveu model which we encountered in Chapter 5. It turns out that the corrections for taking the $M \to 0$ limit are irrelevant operators. Nevertheless, the RG beta function of the Gross–Neveu model is

$$\beta(g) = a_0 \frac{\partial g}{\partial a_0} = \frac{n-1}{\pi} g^3 + \cdots \tag{16.117}$$

where $n = 2$ is the number of spin components. This means that this system is asymptotically free, and that the effective coupling constant grows at low energies, flowing to a strong-coupling fixed point with a spontaneously broken discrete chiral symmetry (i.e. dimerization) and a dynamically generated energy gap (Gross and Neveu, 1974). From semi-classical analyses it is known that, in addition to massive fermions, the Gross–Neveu model has massive solitons. Notice, however, that the spinless case, $n = 1$, is different. It turns out that in this case the fermions remain massless until some critical value of the coupling constant at which there is a (Kosterlitz–Thouless) quantum phase transition to massive phase.

We will now discuss the solitons that appear in polyacetylene. The domain walls in polyacetylene are topological solitons since they interpolate between two non-trivial boundary conditions of the manifold of broken-symmetry ground states. In this sense they are close relatives of the skyrmions we discussed elsewhere in this book and of the solitons in sine–Gordon theory. Since the adiabatic limit and the ultra-quantum limit lead to essentially the same physics, we will consider this problem in the simpler adiabatic regime (Su *et al.*, 1979; Jackiw and Schrieffer, 1981). The soliton can be easily constructed in a lattice model. Here we will use a continuum description that is simpler to apply and is accurate if the energy gap is small compared with the Fermi energy.

The soliton can be regarded as a domain wall between two different condensates of the dimerization field $\Delta(x) = \pm\Delta_0$. For instance, the soliton may take a profile of the form $\Delta(x) = \Delta_0 \tanh[(x - x_0)/\xi]$, where ξ is the correlation length (typically the inverse of the mass gap) and plays the role here of the size of the soliton, while x_0 is the soliton coordinate. In our discussion we will take x_0 as being fixed, $x_0 = 0$. However, a soliton is an actual quantum eigenstate of the quantum field theory. Thus the soliton is a topological excitation that has a momentum and an energy. A full computation of these properties in the semi-classical regime is beyond the scope of this book. A thorough discussion can be found in the classic papers of Dashen, Hasslacher, and Neveu (Dashen *et al.*, 1975) or in Rajaraman's book (Rajaraman, 1985).

In what follows we will focus only on the behavior of the single-particle states of the fermionic flavor (or valley) whose mass term is changing sign at $x = 0$. Let H

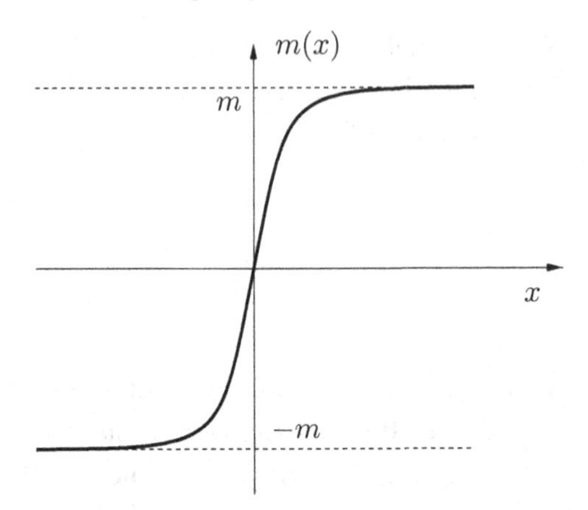

Figure 16.5 A soliton in a 1D system is a domain wall that interpolates between two ground states with opposite expectation values of the fermion bilinear $\langle\bar{\psi}\psi\rangle = \pm m$, where m is the dynamically generated mass of the Dirac fermion. In polyacetylene the soliton interpolates between the two ground states of the commensurate (period-2) charge-density wave (see the text).

be the one-particle (Dirac) Hamiltonian for this two-component Dirac fermion with a position-dependent mass term in one space dimension and $m(x) = \sqrt{2}g\Delta(x)$,

$$H = -i\sigma_1\,\partial_x + m(x)\sigma_3 = \begin{pmatrix} m(x) & -i\,\partial_x \\ -i\,\partial_x & -m(x) \end{pmatrix} \tag{16.118}$$

which is hermitian and real. We will take $m(x)$ to have qualitatively the behavior shown in Fig. 16.5.

It is straightforward to show that the spectrum of the one-particle Hamiltonian H of Eq. (16.118) is particle–hole-symmetric. What is more interesting (and relevant to us) is that H has a normalizable zero mode, a one-particle state with exactly zero energy and wave function $\psi_0(x)$,

$$\psi_0(x) = \frac{1}{\sqrt{2}}\begin{pmatrix} -i \\ 1 \end{pmatrix} \exp\left(-\text{sgn}(m)\int_0^x dx'\, m(x')\right) \tag{16.119}$$

In addition, there is a non-normalizable solution, also with $E = 0$, which is not part of the spectrum. Another way to think about this problem is to imagine that a soliton–anti-soliton pair is created in some region of space away from the boundaries. The two normalizable zero modes of the isolated solitons now mix, their degeneracy is lifted, and they no longer have zero energy. It is easy to see that, as the two solitons are slowly separated, the zero modes are recovered. It is also easy to see that the un-normalizable solution at the soliton can be regarded as the normalizable solution of the anti-soliton. In what follows we will assume that the domain walls are always sufficiently far apart for their zero modes to remain unmixed.

Likewise, if we add an electron to a polyacetylene chain, it becomes fractionalized into two solitons, each carrying half of the charge of the electron.

In addition to this zero mode the Dirac Hamiltonian has a continuous spectrum of positive-energy states with eigenspinors $u_p(x)$ and of negative-energy states with eigenspinors $v_p(x)$. (We are neglecting the spin label of the electrons of polyacetylene.) Charge-conjugation invariance (or, rather, *CP* invariance) implies that for each positive-energy state with momentum p and energy $E(p)$ there is a negative-energy state with momentum p and energy $-E$. However, the existence of the zero mode changes the completeness relation of the eigenstates to

$$\int \frac{dp}{2\pi}[u_p^*(x)u_p(y) + v_p^*(x)v_p(y)] + \psi_0^*(x)\psi_0(y) = \delta(x-y) \tag{16.120}$$

The existence of the zero mode also changes the mode expansion of the Dirac field. In the standard notation of the Dirac theory the mode expansion is (Jackiw and Rebbi, 1976)

$$\psi = a\psi_0 + \sum_p (b_p u_p + d_p^\dagger v_p^*) \tag{16.121}$$

where b_p and d_p are fermion-annihilation operators, and a is a fermion-annihilation operator for the zero mode. Since the zero mode is a discrete state, the operator a and its adjoint satisfy the standard algebra, $\{a, a\} = \{a^\dagger, a^\dagger\} = 0$ and $\{a, a^\dagger\} = 1$.

We will now compute the charge of the soliton using the method of Jackiw and Rebbi (1976). The charge in the presence of the soliton has to be computed relative to the charge of the uniform ground state which has been normal-ordered to be zero. The normal-ordered charge-density operator is $j_0(x) = \frac{1}{2}\big(\psi^\dagger(x)\psi(x) - \psi(x)\psi^\dagger(x)\big)$ (where we have not written down the Dirac indices). The charge of the soliton is the (integral of the) expectation value of the local charge density $j_0(x)$ in the state with one soliton, which we will take to be at rest. Let us denote by $\pm$ the soliton and anti-soliton states, where the sign is the sign of the mass at $+\infty$, i.e. the sign of the asymptotic value of $\Delta_0(x)$ at $x \to +\infty$. The total (integrated) soliton charge (in units of the electron charge $-e$) is

$$\begin{aligned} Q_\pm &= -\frac{e}{2}\left\{\pm\int dx\, \psi_0(x)^*\psi_0(x) + \int dx \int \frac{dp}{2\pi}\big[v_p^*(x)v_p(x) - u_p^*(x)u_p(x)\big]\right\} \\ &= \mp\frac{e}{2} \end{aligned} \tag{16.122}$$

where we have used the completeness relation and the fact that due to charge conjugation symmetry the contributions of the continuum states cancel out exactly, i.e. they have equal and opposite charges. Hence, a soliton has charge $Q_+ = -e/2$ and an anti-soliton has charge $Q_- = +e/2$. Notice that charge is not a label but a quantum number of a state. We have encountered this concept in other places in this book, e.g. in the theory of the fractional quantum Hall states, but charge fractionalization was first discussed in the problem we are now looking at.

The fractional charge result of Eq. (16.122) can be expressed in terms of the *spectral asymmetry* of the Dirac operator (Jackiw and Schrieffer, 1981). If we denote by $\rho_0(E)$ and $\rho_S(E)$ the density of single-particle states in the absence and in the presence of the soliton, the change in the charge of the ground state can be written as

$$Q = \int_{-\infty}^{0^-} dE\big(\rho_S(E) - \rho_0(E)\big) \tag{16.123}$$

which, using completeness, becomes

$$Q = -\frac{1}{2}\int_0^\infty \big(\rho_S(E) - \rho_S(-E)\big)dE \tag{16.124}$$

Then, as before, the symmetry of the spectrum implies that only the zero mode contributes to the spectral asymmetry, Eq. (16.124), and hence to the charge, Eq. (16.122).

An alternative way to think about fractional quantum numbers is due to Goldstone and Wilczek (1981) and is based on the use of (in this case) the chiral anomaly. Goldstone and Wilczek considered a theory of massless Dirac fermions in (1 + 1) dimensions coupled to two scalar fields, ϕ_1 and ϕ_2, with Lagrangian (with the speed $v_F = 1$)

$$\mathcal{L} = \bar{\psi} i\gamma^\mu \partial_\mu \psi + g\bar{\psi}(\phi_1 + i\gamma_5\phi_2)\psi \tag{16.125}$$

Notice that the one-particle (Dirac) Hamiltonian now becomes

$$H = i\sigma_1 \partial_x + g\phi_1\sigma_3 + g\phi_2\sigma_2 \tag{16.126}$$

which is no longer a real symmetric matrix but is complex and hermitian. Thus, in the presence of both mass terms CP (or, equivalently, time-reversal) invariance is broken. We will take ϕ_1 and ϕ_2 to be slowly varying and in general everywhere non-zero. For polyacetylene, in the notation we have been using, ϕ_1 is our dimerization field (a charge-density wave on the bonds) and ϕ_2 is a charge-density wave on the sites. By setting $\phi_1 = |\phi|\cos\theta$ and $\phi_2 = |\phi|\sin\theta$ we can write the Lagrangian in the suggestive form

$$\mathcal{L} = \bar{\psi} i\gamma^\mu \partial_\mu \psi + g|\phi|\bar{\psi} e^{i\gamma_5\theta}\psi \tag{16.127}$$

Let us now imagine that we start with some constant value of ϕ_1 and ϕ_2, which opens a mass gap in the fermion spectrum of $g\sqrt{\phi_1^2 + \phi_2^2}$, and that we make an infinitesimal space- and time-dependent smooth local change in the values of both fields. Goldstone and Wilczek (1981) computed the current j_μ which such a smooth adiabatic change induces, and found the result ($a, b = 1, 2$)

$$\langle j_\mu \rangle = \frac{1}{2\pi} \epsilon_{\mu\nu} \epsilon_{ab} \frac{\phi_a \, \partial^\nu \phi_b}{|\phi|^2} = \frac{1}{2\pi} \epsilon_{\mu\nu} \, \partial^\nu \theta \tag{16.128}$$

where $|\phi|^2 = \phi_1^2 + \phi_2^2$ and $\theta(x) = \tan^{-1}(\phi_2/\phi_1)$. Notice that the induced current is locally conserved, $\partial_\mu \langle j_\mu \rangle = 0$, as required by charge conservation and gauge invariance.

Let us suppose now that we begin with the system in the sector without a soliton, say with $\phi_1 = \text{constant}$ and $\phi_2 = 0$, and that, through a sequence of slow changes, after a long time we arrive at a system with one soliton. To compute the charge of the soliton we can now just integrate the induced charge density $\langle j_0 \rangle$,

$$Q = \int_{-\infty}^{+\infty} dx_1 \langle j_0(x_1) \rangle = \frac{1}{2\pi} (\theta(+\infty) - \theta(-\infty)) \equiv \frac{\Delta\theta}{2\pi} \tag{16.129}$$

Let us now fix the fermion mass to be $m = \phi_1/g$ and represent the soliton by a twist in which ϕ_2 slowly changes from $\phi_2(-\infty) = -\phi_0$ to $\phi_2(+\infty) = +\phi_0$. The charge now is

$$Q = \frac{1}{\pi} \tan^{-1}\left(\frac{g\phi_0}{m}\right) \tag{16.130}$$

If we now let $m \to 0$ (which restores time-reversal invariance) we find the Jackiw–Rebbi result that the charge is $Q = 1/2$ (in units of the electron charge $-e$). Notice, however, that if time-reversal (or CP) invariance is broken, the Goldstone–Wilczek formula, Eq. (16.129), allows any fractional value of the charge, depending on the value of the twist of the chiral angle θ. We will see shortly below that this result can be extended to higher dimensions.

Finally, let us note that in the special case of (1 + 1) dimensions we can reach the same conclusion using abelian bosonization (which was discussed in Chapters 5 and 6). In terms of the real scalar field φ the Lagrangian of Eq. (16.127) takes the bosonized form

$$\mathcal{L} = \frac{1}{2}(\partial_\mu \varphi)^2 + \frac{g|\phi|}{2\pi a_0} \cos(\sqrt{4\pi}\, \varphi - \theta) \tag{16.131}$$

Thus, in the presence of a non-trivial chiral twist $\theta(x_1)$, the Bose field takes the value $\varphi(x_1) = \theta(x_1)/\sqrt{4\pi}$. Using the bosonization identity for the current, cf. Eq. (5.253), we find

$$j_\mu = \frac{1}{\sqrt{\pi}} \epsilon_{\mu\nu} \, \partial^\nu \varphi = \frac{1}{2\pi} \epsilon_{\mu\nu} \, \partial^\nu \theta \tag{16.132}$$

which agrees with the Goldstone–Wilczek result, Eq. (16.128), which was derived using an adiabatic argument in the massive theory. This result leads again to the

Goldstone–Wilczek formula for the fractional charge in terms of the chiral twist. In some sense this result is not surprising since bosonization is a consequence of the chiral anomaly in the massless theory.

16.8 Edge states in the quantum anomalous Hall effect

Let us now consider a 2D system that on the $x > 0$ right-hand half-plane is a topological insulator with a broken time-reversal symmetry (TRS) and displays the quantized anomalous Hall effect, while for the $x < 0$ left-hand half-plane it is either a conventional insulator with an unbroken TRS or, more simply, vacuum. This domain wall will be taken to run parallel to the y axis. While this calculation can be done numerically quite efficiently, it is simpler, and conceptually more instructive, to use the effective-field theory of Dirac fermions in two space dimensions to analyze this problem. Indeed, in Section 16.3.3 we saw that the quantum anomalous-Hall-state insulator can be represented by a system of two flavors of massive (two-component) Dirac fermions with the same sign of the mass, while the trivial insulator can also be represented as two flavors of massive Dirac fermions, but with mass terms with opposite signs. In this language, if we want to represent "vacuum" we will simply send the mass gap of the trivial insulator to infinity. For simplicity, we have represented the edge as a domain wall created by changing the sign of the mass m_1 of one of the Dirac flavors, which is a smooth monotonically increasing function $m(x)$ that varies only along the x coordinate. The mass term of the other flavor, m_2, is kept fixed and positive.

We will see below that this procedure yields a set of chiral edge states residing at the interface between the two systems. Microscopically, we can picture a domain wall in Haldane's honeycomb model (for example) as being obtained by changing smoothly across the wall the strength of the site potential ε relative to the next-nearest-neighbor hopping amplitude t_2 (see Eq. (16.35)). In this case, in general, both mass terms will vary with position but only one will change sign. Provided that the position dependence is sufficiently smooth, this will not lead to additional bound states (that is, aside from the edge states themselves). It should be stressed that this description of the trivial insulator does not affect the low-energy behavior of these edge states. At any rate the same results are found in numerical simulations that use a conventional description of the trivial insulator.

In two space dimensions the coupling of the two-component spinors to the domain wall is described by the one-particle Dirac Hamiltonian

$$H = -i\sigma_1\,\partial_x - i\sigma_2\,\partial_y - m(x)\sigma_3 = \begin{pmatrix} m(x) & -i\,\partial_x + \partial_y \\ -i\,\partial_x + \partial_y & -m(x) \end{pmatrix} \tag{16.133}$$

which is hermitian and complex. Since the domain wall is parallel to the y axis and, hence, the mass term is a function of x alone, the component p_y of the momentum parallel to the wall is conserved. The one-particle states are thus plane waves along the y direction. We can then take the two-component spinors to be of the form

$$\psi(x, y) = e^{ip_y y} \begin{pmatrix} u_{p_y}(x) \\ v_{p_y}(x) \end{pmatrix} \tag{16.134}$$

which we will require to be an eigenstate of the one-particle Dirac Hamiltonian, Eq. (16.133), with energy E.

We will now consider two cases. In the first case, shown in Fig. 16.6(a), the mass of the Dirac fermion $m(x)$ changes from being negative to the left of the wall, $m(x) < 0$ for $x < 0$, to being positive to the right of the wall, $m(x) > 0$ for $x > 0$. In this case the quantum anomalous Hall system is on the right half-plane. In the second case, shown in Fig. 16.6(b), the situation is reversed.

In addition to massive bulk states, it is straightforward to see that the states

$$\psi(x, y) = \frac{1}{\sqrt{2}} e^{ip_y y} \begin{pmatrix} -i \\ 1 \end{pmatrix} \exp\left(-\text{sgn}(m) \int_0^x m(x')dx' \right) \tag{16.135}$$

are eigenstates of the Hamiltonian of the Dirac equation with a domain wall, Eq. (16.133). Here we have set $m = \lim_{x\to\infty} m(x)$ to be the bulk value of the

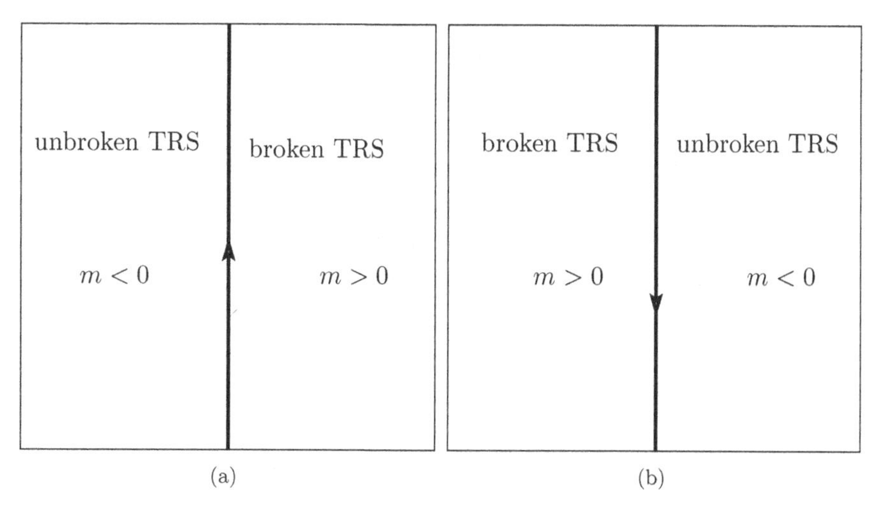

Figure 16.6 Chiral edge states propagating along the domain wall upwards (case (a)) and downwards (case (b)). In case (a) $x < 0$ is an insulator with unbroken TRS and $x > 0$ is a topological (quantum anomalous-Hall) insulator with broken TRS (see the text).

mass, far away from the domain wall. These states are normalizable eigenstates (along the x direction) of the Dirac Hamiltonian with energy

$$E(p_y) = \text{sgn}(m) v_F p_y \tag{16.136}$$

where we have restored the Fermi velocity. Just as in the case of polyacetylene, there are also non-normalizable states of this form, but with the sign of the exponent of the exponential reversed.

These results tell us that there are states inside the energy gap and that they have support only near the location of the domain wall (or physical edge of the system). The result that the energy–momentum relationship is strictly linear, Eq. (16.136), is exact for the continuum Dirac theory. For the lattice models there are (expected) corrections to the linear dispersion. In addition, at high energies, the edge modes merge with the bulk spectrum.

Using the same line of reasoning as that we used in the theory of the edge states of the integer quantum Hall fluids, we can now deduce that the effective Lagrangian for the chiral Dirac fermions of the 1D edge states bound to the wall for the quantum anomalous Hall state is

$$\mathcal{L}_{\text{edge}} = \psi(y,t) i(\partial_t - v_F \,\text{sgn}(m)\partial_y)\psi(y,t) \tag{16.137}$$

where $\psi(y,t)$ is a chiral right (left)-moving Dirac fermion in one space dimension for $m > 0$ ($m < 0$). As we can see, this is identical to the theory of the chiral edge states for the $\nu = 1$ integer quantum Hall state. Here we have neglected the spin degree of freedom which, if it were included, would lead to two branches of chiral Dirac fermions, one for each spin orientation.

The same line of reasoning can be applied to domain walls of the chiral spin liquid discussed in Chapter 10 and to a domain wall between two states with broken TRS (but with opposite signs). The domain wall will now have two chiral edge Dirac fermions in the case of the quantum anomalous Hall domain wall (obtained by flipping the sign of the flux ϕ in half of the system in a Haldane honeycomb model with $\varepsilon = 0$), and four chiral Dirac edge modes for the case of the chiral spin liquid.

The chiral edge states of topological insulators, such as the quantum anomalous Hall systems we are discussing here, are robust for the same reason as that which explains why they are robust in the case of the integer quantum Hall states: being chiral, no backscattering processes can be induced by impurities, and hence they are immune to localization. Of course, this is not the case for the edge states of systems that do not break time-reversal invariance. In that case the edge states come with both chiralities, and backscattering processes are allowed and localization effects render these states insulating. This difference is completely natural since there is no topological invariant associated with the insulators with an unbroken time-reversal invariance.

We see that there is a close analogy between the edge states of the quantum Hall effect, discussed in Section 15.1, and the quantum anomalous Hall effect. We can make the connection more apparent by computing the currents induced by an external electromagnetic field A_μ on both sides of the domain wall and on the domain wall itself. Let us begin by looking at the effective action of the gauge field. For the sake of definiteness we will consider the case shown in Fig. 16.6(a), which is a conventional insulator for $x < 0$ and a quantum anomalous Hall insulator for $x > 0$. Away from the wall the effective action is just the Chern–Simons action

$$S_{\text{eff}}[A] = \frac{N_+ - N_-}{4\pi} \int d^3x \, \epsilon_{\mu\nu\lambda} A^\mu \, \partial^\nu A^\lambda \tag{16.138}$$

where $N_\pm$ is the number of fermionic species with topological charge $\pm 1/2$. The only difference between the two regions is that the sign of the mass term of one species changes from positive for $x > 0$ to negative for $x < 0$ and hence $N_+ = N_-$ for $x < 0$, but $N_+ = N_- + 2$ for $x > 0$. Hence, the effective action can be written as

$$S_{\text{eff}} = \frac{2}{4\pi} \int_\Omega d^3x \, \text{sgn}(x_1) \int d^3x \, \epsilon_{\mu\nu\lambda} A^\mu \, \partial^\nu A^\lambda \tag{16.139}$$

where Ω is the 3D space-time excluding the edge-state, namely the 2D manifold spanned by the x_2 spatial axis and time x_0, in order to avoid the singularity of the sign function. However, it is easy to see that this action is not gauge-invariant. Indeed, under a gauge transformation $A_\mu \to A_\mu + \partial_\mu \Phi$ it changes by (here $\mu, \nu = 0, 2$ only)

$$S_{\text{eff}}[A + \partial\Phi] - S_{\text{eff}}[A] = -\frac{1}{\pi} \int_{-\infty}^{\infty} dx_2 \int dx_0 \; \Phi(x_1 = 0, x_2, x_0) \epsilon_{\mu\nu} \, \partial^\mu A^\nu \tag{16.140}$$

and we get, as in the quantum Hall effect, a gauge anomaly. However, here too, we must also include the contribution of the edge states to the effective action. Since the edge states are chiral, they also have a gauge anomaly, which is equal and opposite to the gauge anomaly of the bulk. Hence, they cancel each other out. It is now easy to compute the currents. In the bulk the currents are just those of the anomalous Hall state for $x > 0$ and 0 for $x < 0$,

$$j_\mu = \frac{\delta S_{\text{eff}}[A]}{\delta a_\mu(x)} = \frac{1}{\pi} \epsilon_{\mu\nu\lambda} \, \partial^\nu A^\lambda \tag{16.141}$$

For a configuration of gauge fields with zero magnetic field, $B = 0$, and electric field parallel to the length of the wall, $E_2 = E$, the bulk current is the Hall current of the anomalous Hall state and flows *towards the wall* along the x_1 axis.

Instead, along the wall we get the edge current, which flows upwards along the x_2 axis. In this context this argument, which is formally analogous to Wen's anomaly-cancellation argument for the integer and fractional quantum Hall states, was first formulated by Callan and Harvey (1985), and it is also known as the Callan–Harvey effect.

16.9 Edge states and the quantum spin Hall effect

We now turn to a discussion of the edge states in the quantum spin Hall effect. It was stressed by Kane and Mele (2005b) that, since spin is not generally conserved in systems with spin–orbit couplings, the bulk quantum spin Hall effect cannot possibly be a robust feature of these systems. In particular, they argued that, while spin accumulation may take place as a result of applying an electric field, the associated spin Hall conductivity will generally be lower than the idealized calculation predicts. Nevertheless, Kane and Mele found that the edge states still have robust properties.

To see how this works, let us consider a system that is a time-reversal-invariant topological insulator in 2D (and hence one that ideally would display the quantum spin Hall effect) with a boundary. The arguments apply both to the Kane–Mele model and to the CdTe/HgTe/CdTE quantum-well model. As before, the outside vacuum will be described as a normal insulator with a very large and negative energy gap, while the quantum spin Hall insulator will have a positive energy gap. Thus, this system can then be pictured as a domain wall created by a sign change of the mass term. This picture is analogous to the edge of the quantum anomalous Hall insulator shown in Fig. 16.6. The main, and important, difference is that we now have twice as many degrees of freedom (due to the spin), and it is time-reversal invariant.

Thus, there will be one chiral edge state for the $\uparrow$ fermions and one chiral edge state for the $\downarrow$ fermions. However, since these fermions are related by time reversal (and hence parity), the $\uparrow$ fermions have positive chirality, since they have $\mathcal{Q}_\uparrow = +1/2$ and obey a dispersion $E_\uparrow(p) = +v_{\mathrm{F}}p$, while the $\downarrow$ fermions have negative chirality, since they have $\mathcal{Q}_\downarrow = +1/2$ and obey a dispersion $E_\downarrow(p) = -v_{\mathrm{F}}p$.

We conclude that a topological insulator that exhibits the quantum spin Hall effect has a *pair* of edge states that are chiral Dirac (or Weyl) fermions and have opposite chiralities for $\uparrow$ and $\downarrow$ spins. In other terms, the $\uparrow$ spins are right movers $R_\uparrow$ while the $\downarrow$ spins are left movers $L_\downarrow$, as shown in Fig. 16.7. A system with this feature is called a "spin-split" metal (Hirsch, 1990) or, more generally, a "helical metal." In contrast, a normal time-reversal-invariant insulator either has no edge states (and hence also the edge is insulating) or it has edges states with *both* chiralities for *each* spin orientation. We will see shortly that these "normal" edge states either become localized by disorder or become gapped by interaction effects.

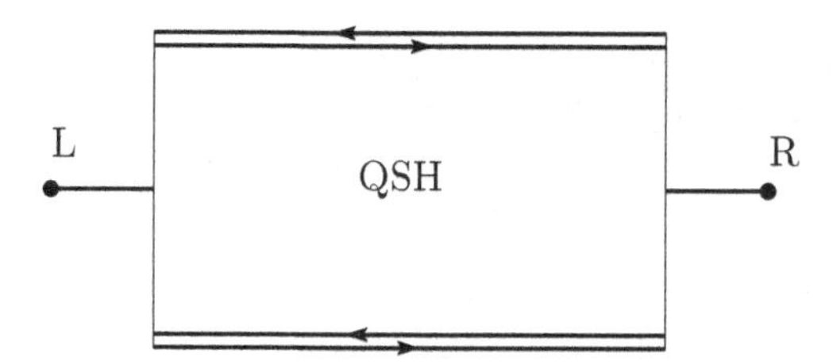

Figure 16.7 A two-terminal (L and R) setup used to detect the quantum spin Hall (QSH) effect (König *et al.*, 2007). Counter-propagating chiral edge states propagate along the boundary of a quantum spin Hall topological insulator (see the text). The ↑ spins move to the right and the ↓ spins move to the left.

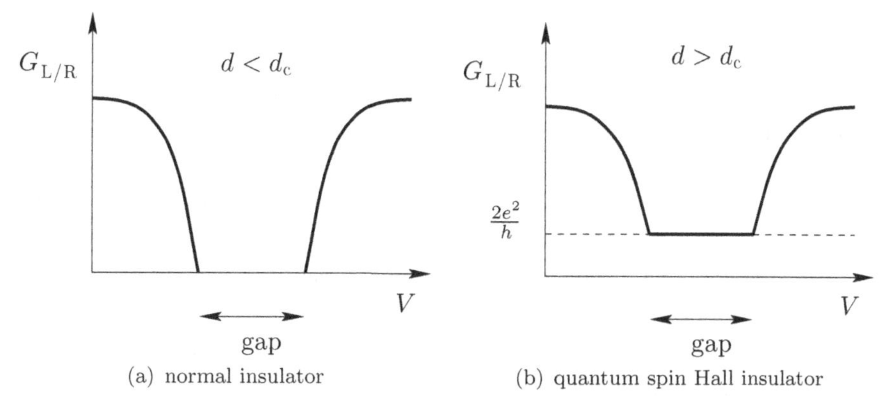

Figure 16.8 A schematic plot of the two-terminal conductance $G_{L/R}$ as a function of the bias V between the terminals L and R of the device sketched in Fig. 16.7, (a) in the normal regime of the quantum well, where it shows insulating behavior in the energy gap, $d < d_c$, and (b) in the quantum spin Hall regime, $d > d_c$, where it shows quantized edge spin Hall conduction. In the spin Hall regime Molenkamp *et al.* (König *et al.*, 2007, 2008) measured in CdTe/HgTe/CdTe quantum wells a quantum spin Hall conductance of $2e^2/h$ in agreement with the spin Hall current being carried by the edge states (see the text).

The effective Hamiltonian for the edge states of a quantum spin Hall insulator is

$$\mathcal{H}_{\text{QSH edge}} = R_\uparrow^\dagger(x) i v_F \hbar\ \partial_x R_\uparrow(x) - L_\downarrow^\dagger(x) i v_F \hbar\ \partial_x L_\downarrow(x) \tag{16.142}$$

Therefore the edge states of a time-reversal-invariant topological insulator constitute a non-chiral 1D system.

Given that the edge states of a quantum spin Hall insulator have both chiralities, one may suspect that they might not have topological protection. This turns out to be a subtle problem in which time-reversal invariance plays a key role. In a time-reversal-invariant *free-fermion* system the exact degeneracy between two time-reversed states, $|\uparrow, p\rangle$ and $|\downarrow, -p\rangle$, is guaranteed by Kramers' theorem. It insures that this degeneracy cannot be lifted unless time-reversal symmetry is broken explicitly since, in this case, there is no symmetry distinction between a normal insulator and a quantum spin Hall insulator. For example, a magnetic field parallel to the edge (say, along the x direction) will induce a spin-flip process. However, in this case a spin-flip process requires one to flip the momentum and hence to exchange a right with a left mover. Thus, this term induces a term at the edge of the form

$$\mathcal{H}_{\text{flip}} = gB\left(R_{\uparrow}^{\dagger}L_{\downarrow} + L_{\downarrow}^{\dagger}R_{\uparrow}\right) \tag{16.143}$$

With this term, by virtue of mixing right and left movers, the energy spectrum now is $E(p) = \sqrt{p^2 v_{\text{F}}^2 + (gB)^2}$. Therefore now there is a gap $g|B|$ in the spectrum of edge states and the edge is insulating.

Can interactions open a gap in the spectrum of edge states? We suspect that this may be possible since the edge states in this case are not chiral. This question is equivalent to asking whether time reversal may be broken spontaneously. It is easy to see that in this system a backscattering *interaction*, which is a four-fermion process, is a marginal operator. In fact, the edge states of this system are identical to those of a *spinless* Luttinger liquid, which we discussed in depth in Chapters 5 and 6. This connection tells us that the edge states in general have anomalous dimensions and hence are not the simple free-fermion states we discussed. From this line of argument we can also deduce that the only possible operators that could open an energy gap are Umklapp processes. However, Umklapp processes are allowed only for a half-filled edge and become relevant only for a sufficiently large value of their coupling constant.

What are the effects of disorder on the edge states of the quantum spin Hall insulator? The chiral edge states of the quantum Hall fluid, and of the anomalous Hall insulator, are protected from localization effects since backscattering processes are forbidden. The quantum spin Hall edge states are also stable against localization except in the case of backscattering processes with a spin flip, i.e. magnetic impurities. Thus, magnetic impurities make the quantum spin Hall edge states insulating. In the case of the edge states of a normal insulator, if it has edge states at all, they are not protected, and impurities will generally induce backscattering, turning it into an Anderson insulator due to localization effects.

16.10 $\mathbb{Z}_2$ topological insulators and the parity anomaly

The last example we will discuss concerns the edge states of 3D $\mathbb{Z}_2$ topological insulators. We will use a continuum-field-theory notation while keeping in mind the central importance of lattice effects associated with fermion doubling. We have already discussed these subjects for the bulk states. An alternative and very elegant approach that unifies all these effects and phenomena can be found in the beautiful work of Qi, Hughes, and Zhang (Qi *et al.*, 2008), which will be discussed below.

We will imagine now that we have a simply connected region Ω of 3D space occupied by a $\mathbb{Z}_2$ topological insulator, e.g. Bi_2Se_3. The boundary of the region, that we will denote by Σ, is a simply connected closed 2D manifold with the topology of the sphere S_2. Outside this region we have vacuum or, which amounts to the samething, a trivial insulator with a very large energy gap. To simplify the analysis further, we will take the region Ω to occupy the entire half-space $x_3 < 0$ of 3D space, and its boundary Σ becomes $\mathbb{R}^2$, the $x_1 x_2$ plane. We will begin by constructing the zero modes of this system, the 2D edge states of the 3D $\mathbb{Z}_2$ topological insulator. We will see that the edge states are chiral (Weyl) fermions that move in 2D space, the boundary Σ. We will follow the work of Callan and Harvey. This problem was considered before (Fradkin *et al.*, 1986; Boyanovsky *et al.*, 1987), except that in that work it was applied to PbTe, which, as it happens, is not a topological insulator. Consequently, it has edge states with both chiralities, and as such does not enjoy any form of topological protection. Nevertheless, their analysis does apply to the case of $\mathbb{Z}_2$ topological insulators, which had not yet been discovered at that time.

Once again, we will assume that inside region Ω we have an even number of species of massive Dirac fermions, as required by the Nielsen–Ninomiya theorem. For the sake of concreteness we will keep in mind the 3D Wilson-fermion model as a simple example of a $\mathbb{Z}_2$ topological insulator. Thus we will work with four component spinors that represent states in the conduction and valence bands with both spin projections. We will assume that inside region Ω the mass parameter of the Wilson-fermion model lies in the regime in which the mass is negative at the Γ point and positive at all other time-reversal-invariant points (see Section 16.6), but that it is negative outside region Ω. This can be described in terms of a single four-component Dirac spinor in $(3+1)$ dimensions with a mass term that changes sign across the $x_1 x_2$ plane. Although we will use a continuum description, which is accurate for the states near the Γ point, the results are consistent with lattice calculations of Dirac fermions with domain walls.

There is an extensive literature in lattice gauge theory on the subject of "domain-wall fermions," which were proposed in connection with the realization of chiral symmetry in lattice systems (see e.g. Kaplan (1992), Jansen (1996), and Creutz (2001)), which has explicitly reproduced the results with Wilson fermions we discuss here. However, while in lattice gauge theory the bulk is a system in $(4+1)$-dimensional space-time and the boundary is $(3+1)$-dimensional space-time, here we will work with one spatial dimension fewer. In the next section we will see that it is also convenient to consider the problem beginning in $(4+1)$ dimensions (Qi *et al.*, 2008).

Let H be the Dirac Hamiltonian in three space dimensions with a domain wall normal to the x_3 axis. Thus we will assume that the Dirac mass term is a smooth and monotonically decreasing function $m(x_3)$ that changes sign at $x_3 = 0$ and takes the asymptotic values $\lim_{x_3 \to \pm\infty} m(x_3) = \mp m$, with $m > 0$. The (one-particle) Dirac Hamiltonian is

$$H = -i\vec{\alpha} \cdot \vec{\nabla} + m(x_3)\beta \tag{16.144}$$

which can be written as

$$H = H_\perp + H_3 \tag{16.145}$$

where

$$H_\perp = -i\alpha_1 \partial_1 - i\alpha_2 \partial_2 \tag{16.146}$$

and

$$H_3 = -i\alpha_3 \partial_3 + m(x_3)\beta \tag{16.147}$$

Let $\psi_0^\pm$ be an eigenstate of the (anti-hermitian) Dirac matrix $\gamma_3 = \beta\alpha_3$ with eigenvalue $\pm i$,

$$\gamma_3 \psi_0^\pm = \pm i \psi_0^\pm \tag{16.148}$$

We will demand that $\psi_0^\pm$ be a zero mode of H_3,

$$H_3 \psi_0^\pm = 0 \quad \Rightarrow \quad \pm\partial_3 \psi_0^\pm + m(x_3)\psi_0^\pm = 0 \tag{16.149}$$

and a solution of the Dirac equation in $(2+1)$ dimensions,

$$i\gamma_0 \, \partial_0 \psi_0^\pm - i\left(\vec{\gamma} \cdot \vec{\nabla}\right)_\perp \psi_0^\pm = 0 \tag{16.150}$$

where the label $\perp$ indicates that the space derivatives act only on the domain-wall coordinates x_1 and x_2.

We can satisfy these conditions by writing

$$\psi_0^\pm = \eta_\pm(x_0, x_1, x_2) f(x_3) \tag{16.151}$$

where $f(x_3)$ is the solution of

$$\pm\partial_3 f_\pm(x_3) = -m(x_3) f_\pm(x_3) \tag{16.152}$$

which has the form

$$f_\pm(x_3) = f(0)\exp\left(\mp \int_0^{x_3} dx_3' \, m(x_3')\right) \tag{16.153}$$

Since $m(x_3) < 0$ for $x_3 > 0$, it is monotonically decreasing and $\lim_{x_3 \to \infty} m(x_3) = m < 0$, the requirement that the solution be normalizable implies that we must choose the solution $f_+(x_3)$, which decays exponentially fast at long distances from the wall, $|x_3| \gg m^{-1}$. This also implies that the spinor $\eta_+(x_0, x_1, x_2)$ must be chosen to be an eigenspinor of γ_3 with eigenvalue $-i$. Since in the Dirac basis $\gamma_3 = i\sigma_3 \otimes \tau_1$, it follows that the spinor $\eta_+(x_0, x_1, x_2)$ is a superposition of positive-energy states (the conduction band of the insulator) with spin $\uparrow$ and negative-energy states (the valence band of the insulator) with spin $\downarrow$, which should also satisfy the *massless* Dirac equation in $(2+1)$ dimensions, Eq. (16.150).

The nature of the edge states is more easily seen by using a basis of the Dirac gamma matrices in which γ_3 is diagonal. In this basis we have the 4×4 Dirac gamma matrices are $\gamma_0 = -\mathcal{I} \otimes \tau_2$, $\gamma_5 = \mathcal{I} \otimes \tau_1$, and $\vec{\gamma} = i\vec{\sigma} \otimes \tau_3$. The subspace of spinors with γ_3 eigenvalue $+i$ is spanned by $(1, 0, 0, 0)^{\mathrm{T}}$ and $(0, 0, 0, 1)^{\mathrm{T}}$ (here the label T means transposed), while the subspace in which γ_3 has eigenvalue $-i$ is spanned by the spinors $(0, 1, 0, 0)^{\mathrm{T}}$ and $(0, 0, 1, 0)^{\mathrm{T}}$. We can find the effective Dirac Hamiltonian in each of these two-dimensional subspaces. In the $\gamma_3 = i$ subspace the matrices $\alpha_1 = \gamma_0\gamma_1$ and $\alpha_2 = \gamma_0\gamma_2$ become $\alpha_1 = \sigma_1$ and $\alpha_2 = -\sigma_2$. On the other hand, in $\gamma_3 = -i$ subspace they are instead $\alpha_1 = \sigma_1$ and $\alpha_2 = +\sigma_2$. Thus, in both subspaces the effective Hamiltonian has the form of a Weyl fermion in $(2+1)$ dimensions,

$$H_{2\mathrm{D}} = -i\alpha_1\,\partial_1 - i\alpha_2\,\partial_2 \tag{16.154}$$

and the states have energy $E(\vec{p}) = \pm|\vec{p}|$. However, since in the two subspaces the term that contains the matrix α_2 has opposite signs, the subspaces have states that are related by a parity transformation. Hence the edge states with $\gamma_3 = +i$ have positive chirality and those with $\gamma_3 = -i$ have negative chirality.

We conclude that there are states bound to the wall whose wave functions have the form

$$\psi_0^+(x_0, x_1, x_2, x_3) = \eta_+(x_0, x_1, x_2)\exp\left(-\int_0^{x_3} dx_3'|m(x_3')|\right) \tag{16.155}$$

where the spinor η_+ satisfies $\gamma_3\eta_+ = i\eta_+$ and obeys a Dirac equation for bispinors in $(2+1)$ dimensions, $\left(i\partial\!\!\!/\right)_\perp\eta_+ = 0$. Therefore, a 3D $\mathbb{Z}_2$ topological insulator has chiral edge states, bound states localized at its open surface, which are massless two-component Weyl fermions.

However, since the surface of the insulator is simply connected, this result also means that the *opposite* surface also has massless two-component Weyl fermions but with the *opposite* chirality. Hence, in this system the fermion-doubling theorem is satisfied by spatially separating the chiral partners, with the top surface having positive-chirality Weyl fermions and the bottom surface having the negative-chirality Weyl fermions.

In the simple Wilson model, which is qualitatively adequate to describe topological insulators such as Bi_2Se_3, there is just *one* species of Weyl fermion on each surface. Other topological insulators, such as $Bi_{1-x}Sb_x$, have a diamond lattice structure and more complex band structure, leading to five Weyl modes on each surface. Nevertheless, what matters is that the number of surface Weyl fermions is *odd*. Indeed, if there were an even number of edge Weyl modes (as in the case of PbTe) a mass term would be allowed on the surface even in a system with time-reversal invariance. Hence *pairs* of Weyl fermions can acquire a gap. Thus insulators with an *even* number of surface Weyl fermions are $\mathbb{Z}_2$-trivial. On the other hand, if the number of surface Weyl modes is *odd* there will be at least one gapless mode left gapless. This is what happens in the $\mathbb{Z}_2$ topologically non-trivial class. The Wilson-fermion model is in this class and can be used as its representative.

Let us explore the consequences of these results by considering the effects of external electromagnetic fields. The coupling to the external electromagnetic field is given by the usual minimal coupling which is dictated by gauge invariance. However, we will have to be careful to include also the Zeeman coupling of the "microscopic" electrons, which in this language is a "non-minimal" coupling, although it is consistent with gauge invariance. Such a term breaks time-reversal invariance since it couples directly to the spin. For simplicity we will take the magnetic field to be normal to the surface, $\vec{B} = B\vec{e}_3$. We will work in the gauge $A_3 = 0$.

The one-particle Dirac Hamiltonian in the presence of a domain wall and coupled to the external magnetic field can also be written in the split form of Eq. (16.145), with $H_\perp$ and H_3 now being given by

$$\begin{aligned} H_\perp &= \left[-i\vec{\alpha} \cdot \left(\vec{\nabla} - i\frac{e}{\hbar c}\vec{A} \right) \right]_\perp - gB\Sigma_3 \\ H_3 &= -\, i\alpha_3\, \partial_3 + m(x_3)\beta \end{aligned} \tag{16.156}$$

where g is the Zeeman coupling and $\vec{\Sigma} = \mathrm{diag}(\vec{\sigma}, \vec{\sigma})$ is the Dirac spin matrix. Notice that the Zeeman term is equivalent to a mass term for the Weyl fermion with mass $m = -gB$. So the sign of the mass, which sets the sign of the time-reversal symmetry breaking, is the opposite of the sign of B, the component of the magnetic field perpendicular to the wall.

Since $\Sigma_3 = \mathrm{diag}(\sigma_3, \sigma_3)$, it commutes with γ_3. Thus in the subspace spanned by the normalizable zero modes of H_3, which above was denoted by ψ_0^+, the effective Hamiltonian in $(2+1)$ dimensions now is

$$H_{2D} = \vec{\alpha} \cdot \left(\vec{p} + \frac{e}{\hbar c}\vec{A} \right) - gB\sigma_3 \tag{16.157}$$

where $\alpha_1 = \sigma_1$, $\alpha_2 = \sigma_2$ (given that our spinors satisfy $\gamma_3 = +i$), and $\vec{A}$ denotes the components of the electromagnetic vector potential tangent to the wall, the x_1x_2 plane.

We see that the Zeeman coupling for a field normal to the wall opens a gap in the energy spectrum. Moreover, the Hamiltonian now has all three Pauli matrices and breaks time-reversal invariance. On the other hand, for the same reason an in-plane magnetic field does not open a gap in the spectrum and amounts to a shift of the momentum. Thus, an in-plane magnetic field can be thought of as a large gauge transformation or, equivalently, as a twist in the boundary conditions of the spinors. The other way to open a gap in the spectrum of Weyl surface states is to have a term in the bulk Hamiltonian that breaks parity and charge-conjugation symmetry. As we will see below, this term involves the matrix $\beta\gamma_5$.

The relativistic energy levels $\varepsilon_{n,\sigma}$ in $(2+1)$ dimensions are (setting $m = -gB$)

$$\begin{aligned} \varepsilon_{n,\sigma} &= \pm\big((2n+1)B - \sigma B + m^2\big)^{1/2}, \qquad n = 0, 1, 2, \ldots, \qquad \sigma_3 = \pm 1 \\ \varepsilon_0 &= m, \qquad n = 0, \qquad \sigma = +1 \end{aligned} \tag{16.158}$$

It is easy to see that the $n = 0, \sigma = 1$ states are N_ϕ-fold degenerate, while all other states are $2N_\phi$-fold degenerate. Therefore, in the ground state, which is found by filling up all negative-energy states, the Landau level with $n = 0$ and spin $\uparrow$ will be empty if $m > 0$ or full if $m < 0$. Thus, we expect a charge (and spin) accumulation at the surface by an amount Q.

How much charge accumulates? Since the Hamiltonian H_3 is nothing but the Hamiltonian of a soliton in 1D, we easily find that the induced charge is

$$Q = \frac{e}{2}\,\mathrm{sgn}(m) N_\phi = \frac{e}{2}\frac{BL^2}{\phi_0}\,\mathrm{sgn}(m) = \frac{e^2}{\hbar c}\frac{1}{4\pi}\,\mathrm{sgn}(m) BL^2 \tag{16.159}$$

Where does this charge come from? It necessarily has to come from the bulk of the system. However, since the system is isolated and charge is conserved, an equal and opposite amount of charge, $-Q$, has to be somewhere else. Indeed, the "missing charge" is at the opposite surface! As we saw above, at the opposite surface we also get 2D Weyl fermions, but with the opposite "chirality" $\gamma_3 = -i$. It is straightforward to see that the charge that accumulates at the "anti-domain wall" is indeed $-Q$ and has opposite spin projection. Hence the magnetic field induces a *charge polarization* in the system. In $\mathbb{Z}_2$ topological insulators this effect is known as the topological magneto-electric effect (TME) (Qi *et al.*, 2008; Essin *et al.*, 2009).

If we now impose an additional in-plane electric field $\vec{E}$, we expect that the charge Q (or rather its center of mass) will move at the drift velocity $v = c|\vec{E}|/|B|$, and that the system will have a *Hall current* $\vec{J}$, perpendicular to both $\vec{B}$ and $\vec{E}$,

$$J_i = \sigma_{xy}\epsilon_{ij}E_j \tag{16.160}$$

with a Hall conductance

$$\sigma_{xy} = \frac{1}{4\pi}\frac{e^2}{\hbar}\,\mathrm{sgn}(m) \tag{16.161}$$

Hence the Weyl fermions have a Hall conductance $\sigma_{xy} = \pm 1/2$ (in units of e^2/h)! Since the electric field acts throughout the system (the bulk is an insulator and cannot screen the electric field!), the Hall current has the opposite sign at the opposite surface. Thus, there is no total Hall current, since the currents cancel out in the system as a whole (but not locally!), but there is instead a *spin Hall current*. We see that this effect is the 3D analog of the quantum spin Hall effect. However, the significant difference is that the chiral partners are spatially separated. This is another manifestation of the Callan–Harvey effect. We see that the system exhibits a *parity anomaly* at each surface, but that this cancels out in the system as a whole. Equivalently, we have shown that the surface Weyl fermions of a 3D $\mathbb{Z}_2$ topological insulator exhibit an *anomalous Hall effect* with a Hall conductance that is equal to $e^2/(2h)$ (up to a sign), for this case in which there is a single Weyl fermion. For a more general topological insulator the surface Hall conductance is $(N_+ - N_-)e^2/(2h)$, where $N_\pm$ is the number of Weyl fermions of each chirality. The opposite surface has the opposite Hall conductance.

We can recast these conclusions in terms of an effective action for the external electromagnetic field (Qi *et al.*, 2008). From the point of view of the surface states their response to the external electromagnetic field is given by a Chern–Simons action

$$
\begin{aligned}
S_{\text{eff}}[A] = & \int_{\partial\Omega^+ \times \mathbb{R}} d^3x \left[\frac{1}{4\pi} \operatorname{sgn}(m) \epsilon^{\mu\nu\lambda} A_\mu \, \partial_\nu A_\lambda \right] \\
& - \int_{\partial\Omega^- \times \mathbb{R}} d^3x \left[\frac{1}{4\pi} \operatorname{sgn}(m) \epsilon^{\mu\nu\lambda} A_\mu \, \partial_\nu A_\lambda \right]
\end{aligned} \tag{16.162}
$$

where $\partial\Omega^\pm$ are the top $(+)$ and bottom $(-)$ surfaces of a 3D region Ω occupied by the $\mathbb{Z}_2$ topological insulator. Here $\mathbb{R}$ represents the time coordinate and $\mu, \nu, \lambda = 0, 1, 2$. We can express this result in the form of a *volume* integral (and time)

$$
S_{\text{eff}}[A] = \int_{\Omega \times \mathbb{R}} d^4x \, \frac{\theta}{8\pi^2} \epsilon_{\mu\nu\lambda\rho} \, \partial^\mu A^\nu \, \partial^\rho A^\lambda \tag{16.163}
$$

with $\theta = \operatorname{sgn}(m)\pi$.

It is straightforward to see that the integrand of Eq. (16.163) is a total derivative, which therefore integrates to the boundary (which is what we wanted) and that the effective action at the boundary has the Chern–Simons form. Furthermore, on a 4-manifold without boundaries (topologically equivalent to the 4-sphere S_4) the *Pontryagin index*

$$
\mathcal{Q} = \int_{S_4} d^4x \, \frac{1}{32\pi^2} F^{\mu\nu} \widetilde{F}_{\mu\nu} = \int_{S_4} d^4x \, \frac{1}{32\pi^2} \epsilon_{\mu\nu\lambda\rho} F^{\mu\nu} F^{\lambda\rho} \tag{16.164}
$$

is a topological invariant and is quantized to be an integer. Here $F^{\mu\nu} = \partial^\mu A^\nu - \partial^\nu A^\mu$ is the field tensor and $\widetilde{F}_{\mu\nu} = \frac{1}{2}\epsilon_{\mu\nu\lambda\rho} F^{\lambda\rho}$ is the dual tensor; $\mu, \nu, \lambda, \rho =$

0, 1, 2, 3. It can be shown that in the case of a 4D gauge theory, the Pontryagin index classifies the smooth maps $S_4 \mapsto S_4$, and thus with homotopy group $\pi_4(S_4) = \mathbb{Z}$, which counts the instanton number, as discussed beautifully in Coleman's book (Coleman, 1985) (see also Eguchi *et al.* (1980)). We see that this topological invariant is the analog of the Pontryagin index of the 2D non-linear sigma model, see Eq. (7.75).

The parameter θ is called the θ angle or axion field, and in the case at hand $\theta = \pm\pi$, which (aside from 0) is the only value (mod 2π) compatible with time-reversal invariance. Thus a trivial insulator, which does not have protected surface Weyl fermions, has $\theta = 0$ (mod 2π), whereas a $\mathbb{Z}_2$ 3D topological insulator has $\theta = \pi$ (mod 2π). We will see shortly that in the context of topological insulators it is closely related to the induced charge polarization. These observations led Qi, Hughes, and Zhang to the conclusion that the effective action of 3D insulators is (Qi *et al.*, 2008)

$$
\begin{aligned}
S_{\text{eff}}[A] &= \int d^4x \left[-\frac{1}{4g^2} F^{\mu\nu} F_{\mu\nu} + \frac{\theta}{32\pi^2} F^{\mu\nu} \widetilde{F}_{\mu\nu} \right] \\
&= \int d^4x \left[\frac{1}{2g^2} \left(\frac{1}{v^2} \vec{E}^{\,2} - \vec{B}^{\,2} \right) + \frac{\theta}{8\pi^2} \vec{E} \cdot \vec{B} \right]
\end{aligned}
\tag{16.165}
$$

with $\theta = 0, \pi$. The first term in Eq. (16.165) is a Maxwell term, which, for an isotropic system, is parametrized by the dielectric constant and the magnetic susceptibility of the insulator, which we represented in terms of the coupling constant $g \propto e^2$ and the speed of light v in the insulator.

Callan and Harvey gave a general descriptions of problems of this type (Callan and Harvey, 1985) by adapting the ideas of Goldstone and Wilczek we discussed in the polyacetylene case. In the present context the Callan–Harvey approach requires one to add a γ_5 mass term to the action or, equivalently, a $\beta\gamma_5$ term to the Dirac Hamiltonian. In terms of the topological insulator a bulk term of this type breaks time-reversal invariance (or CP) explicitly. Thus, we imagine that our Dirac fermion has both a Dirac mass term and a γ_5 mass term, each coupled to two scalar fields, ϕ_1 and ϕ_2, which we will take to be slowly varying. In the 3D case at hand, the Lagrangian is

$$
\mathcal{L} = \bar{\psi} i \not{\partial} \psi + g\phi_1 \bar{\psi}\psi + i g \phi_2 \bar{\psi}\gamma_5\psi = \bar{\psi} i \not{\partial} \psi + g|\phi| \bar{\psi} e^{i\gamma_5\theta} \psi \tag{16.166}
$$

where $\not{\partial} = \gamma^\mu \partial_\mu$, $\gamma_5 = i\gamma_0\gamma_1\gamma_2\gamma_3$, and γ_μ are the four 4×4 Dirac gamma matrices that satisfy the Dirac algebra, $\{\gamma_\mu, \gamma_\nu\} = 2g_{\mu\nu}\mathcal{I}$, with $g_{\mu\nu} = \text{diag}(1, -1, -1, -1)$, and $\{\gamma_\mu, \gamma_5\} = 0$. Here we have used the fact that $\gamma_5^2 = \mathbb{I}$ and expressed the complex field $\phi = \phi_1 + i\phi_2 = |\phi| \exp(i\theta)$ in terms of an amplitude field $|\phi|$ and a phase field, the axion field θ.

Using the same line of argument as that which led Goldstone and Wilczek to the induced current of Eq. (16.128), Callan and Harvey showed that in (3 + 1) dimensions an electromagnetic field A^μ, whose field tensor is $F^{\mu\nu} = \partial^\mu A^\nu - \partial^\nu A^\mu$, in the background of a complex scalar field $\phi = \phi_1 + i\phi_2$ induces the charge current $\langle j_\mu \rangle$ given by

$$\begin{aligned}\langle j_\mu \rangle &= -i\frac{e}{16\pi^2}\epsilon_{\mu\nu\lambda\rho}\frac{\phi^*\,\partial^\nu\phi - \phi\,\partial^\nu\phi^*}{|\phi|^2}F^{\lambda\rho} \\ &= \frac{e}{8\pi^2}\epsilon_{\mu\nu\lambda\rho}\,\partial^\nu\theta\,F^{\lambda\rho}\end{aligned} \tag{16.167}$$

which is a consequence of the axial anomaly in (3 + 1) dimensions. Callan and Harvey considered a rather different problem in which the complex scalar field ϕ has a vortex in the 3D space, called an axion string. The vortex has a 1D core and the Callan–Harvey analysis predicts that it has chiral Dirac–Weyl fermions that move along the vortex with the direction determined by the sign of the vorticity.

Here we are interested in the case of a domain wall that is closely related to the soliton that we discussed in Section 16.7. We will proceed similarly. Thus, we will assume that $g\phi_1 = m$ is the Dirac mass and that ϕ_2 varies slowly (and smoothly) from the value ϕ_0 as $x_3 \to -\infty$ to $-\phi_0$ as $x_3 \to +\infty$. We can compute the total charge accumulated on the wall using the Callan–Harvey result, Eq. (16.167), to find (after restoring units)

$$Q = e\int d^3x\,\langle j_0 \rangle = \frac{e^2}{\hbar c}\frac{\Delta\theta}{2\pi}\frac{B_3 L^2}{2\pi} \tag{16.168}$$

where we assumed that there is a non-vanishing magnetic field B_3 perpendicular to the wall. Since the chiral angle $\theta = \tan^{-1}(\phi_2/\phi_1)$ its total variation is

$$\Delta\theta = \theta(x_3 \to +\infty) - \theta(x_3 \to -\infty) = -2\tan^{-1}\left(\frac{g\phi_0}{m}\right) \tag{16.169}$$

Then in the time-reversal invariant limit, in which $\Delta\Theta \to -1/2$, the charge Q becomes

$$\lim_{m\to 0} Q = -\frac{e^2}{\hbar c}\frac{B_3 L^2}{4\pi} \tag{16.170}$$

which agrees with Eq. (16.159). Similarly, we can easily see that the Callan–Harvey current predicts that, if an electric field is applied parallel to the wall, there is an electric current on the wall perpendicular to it. This Hall current is peculiar, since no magnetic field is applied. However, we must be careful to see that it appears in the limit of vanishingly small breaking of time-reversal symmetry. The result is discontinuous, and its sign depends on how the limit is taken.

By now the reader may have wondered whether we have not lost something important using the continuum approximations. In particular, the topological nature of the effects has been to some extent hidden by going to the continuum limit. In fact, the only place where we used the topological nature of the insulator was at the beginning of the section, where we assumed that we would be working close to the point in which a band inversion first appears at the Γ point, and relied on the fact that the invariants at the other time-reversal-invariant points of the Brillouin zone were unaffected. Therefore, if we are just "inside" the $\mathbb{Z}_2$ topological insulator phase, our assumption that only the 3D Dirac fermion near the Γ point has a domain wall (and that the other "species" do not) yields a consistent description of the changes in the electronic structure. On the other hand, it would be desirable to have a more general framework that does not rely on special arguments and, for that matter, on a special form of the Hamiltonian. This was accomplished in the general and elegant work of Qi, Hughes, and Zhang (QHZ) (Qi *et al.*, 2008), whose main ideas we will now describe.

They began their analysis by reexamining the 1D problem of fractionally charged solitons, see Section 16.7, but with a somewhat different perspective. They took the point of view that fractional charge in one dimension is an expression of the problem of quantized charge transport and polarization in 1D insulators (Thouless, 1983; King-Smith and Vanderbilt, 1993; Ortiz and Martin, 1994).

Following the QHZ construction, we will first consider a simple Wilson-fermion model for the anomalous quantum Hall state in $d = 2$ space dimensions whose Hamiltonian in real space is

$$
\begin{aligned}
H = & \sum_{\vec{r}, j=1,2} \psi^\dagger(\vec{r}) i \Gamma_j \left(\psi(\vec{r} + e_j) - \psi(\vec{r} - e_j)\right) + m \sum_{\vec{r}} \psi^\dagger(\vec{r}) \Gamma_3 \psi(\vec{r}) \\
& + \sum_{\vec{r}, j=1,2} \psi^\dagger(\vec{r}) \Gamma_3 \left(\psi(\vec{r} + e_j) + \psi(\vec{r} - e_j) - 2\psi(\vec{r})\right)
\end{aligned}
\tag{16.171}
$$

where Γ_j (with $j = 1, 2, 3$) are the three 2×2 hermitian matrices, $\Gamma_1 = \alpha_1 = \sigma_1$, $\Gamma_2 = \alpha_2 = \sigma_2$, and $\Gamma_3 = \beta = \sigma_3$, and satisfy the Clifford algebra $\{\Gamma_j, \Gamma_k\} = \delta_{jk} \mathcal{I}$, where $\mathcal{I}$ is the 2×2 identity matrix. The last term in Eq. (16.171) is known as the Wilson mass term, and its role is to give a mass gap to all time-reversal-invariant points in the Brillouin zone except the origin (the Γ point).

In momentum space this Hamiltonian is

$$
H = \int_{\text{BZ}} \frac{d^2k}{(2\pi)^2} \psi^\dagger(\vec{k}) \vec{h}(\vec{k}) \cdot \vec{\Gamma}\ \psi(\vec{k})
\tag{16.172}
$$

It is characterized by the vector $\vec{h}(\vec{k}) = (\sin k_1, \sin k_2, m + \cos k_1 + \cos k_2 - 2)$ which defines a mapping of the Brillouin zone onto the 2-sphere S_2. In Section 16.3.4 we showed that these mappings are classified by the Pontryagin index (or topological charge) $\mathcal{Q}$ of Eq. (16.48), which in this case computes the (first) Chern number C_1 of the Berry connection of the spinors on the Brillouin-zone torus. This result is a general property of this class of systems rather than of the specific form of the Wilson Hamiltonian. In this example, the Chern number C_1 takes the values 1 and -1 for $m < 0$ and $0 < m < 2$, respectively, and 0 otherwise (this result can be derived from Eq. (16.75)). We will focus on the regime $0 < m < 2$, for which this system displays the anomalous quantum Hall effect with a quantized Hall conductance $\sigma_{xy} = +e^2/h$ (Qi *et al.*, 2008).

We also know that the physical response to an external electromagnetic field A_μ of a system with a quantized anomalous quantum Hall effect is given by the Maxwell–Chern–Simons effective action of the form of Eq. (16.37). In this case, however, the coefficient of the Chern–Simons term is *quantized* and equal to the first Chern number C_1 of the system. This effective action is gauge-invariant for a system without boundaries, e.g. with periodic boundary conditions. If the system has boundaries the full gauge-invariant effective action also includes the electromagnetic response of the (topologically protected) edge states.

Let us consider the idealized problem of a 2D system with a quantum anomalous Hall state *on a cylinder* of length L_1 and circumference L_2. We will picture this situation, as in Fig. 16.9, as a *compactification* of the plane, which has been wrapped into a cylinder. The effective action of the $(2+1)$-dimensional quantum anomalous Hall insulator in the asymptotic low-energy limit is the Chern–Simons action on the cylinder,

$$S_{2+1}[A] = \frac{C_1}{4\pi} \int_{S_1 \times S_1 \times \mathbb{R}} d^3x \; \epsilon_{\mu\nu\lambda} A^\mu \, \partial^\nu A^\lambda \qquad (16.173)$$

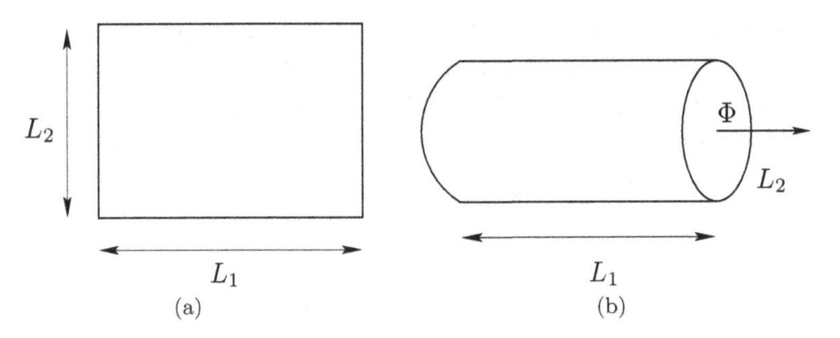

Figure 16.9 (a) The rectangle of area $L_1 L_2$. (b) The compactified rectangle is a cylinder of length L_1 and circumference (or period) L_2. The magnetic flux Φ going through the cylinder in (b) is a twist in the boundary conditions for the states on the plane in (a).

where we denoted the cylinder as $S_1 \times S_1 \times \mathbb{R}$, where the first S_1 represents time x_0 (with periodic boundary conditions), the second S_1 represents the compact (periodic) direction x_2 (with $0 \leq x_2 < L_2$), and $\mathbb{R}$ is the x_1 direction along the cylinder of length L_1. Here we have not included the edge-state contribution.

In the "thin-cylinder" limit, in which formally $L_2 \to 0$, the Chern–Simons term reduces to the $(1 + 1)$-dimensional action for the projected gauge field $A_\mu = (A_0, A_1)$

$$S_{1+1}[A] = \frac{C_1}{2\pi} \int_{S_1 \times \mathbb{R}} dx\, dt\; \Phi(x, t) \epsilon_{\mu\nu}\, \partial^\mu A^\nu \tag{16.174}$$

where the local field $\Phi(x, t) = \oint dx_2\, A_2$ is the flux through the cylinder at location x at time t. We will now see that there is a $(1 + 1)$-dimensional interpretation of the flux Φ. Indeed, the action $S_{1+1}[A]$ predicts that the $(1+1)$-dimensional current charge density j_0 and charge current j_1 are given by

$$j_0 = \frac{\delta S_{1+1}[A]}{\delta A_0} = \frac{C_1}{2\pi}\, \partial_1 \Phi, \qquad j_1 = \frac{\delta S_{1+1}[A]}{\delta A_1} = -\frac{C_1}{2\pi}\, \partial_0 \Phi \tag{16.175}$$

In a more compact form, the current $j^\mu = (j_0, j_1)$ is given by

$$j^\mu = \frac{\delta S_{1+1}[A]}{\delta A_\mu} = \frac{C_1}{2\pi} \epsilon^{\mu\nu}\, \partial_\nu \Phi \tag{16.176}$$

We now see that this result is essentially the same as the Goldstone–Wilczek formula, Eq. (16.128), provided that we identify the flux through the cylinder Φ with the (local) value of the chiral angle θ. Qi, Hughes, and Zhang, who derived this result by means of a dimensional-reduction argument of the fermion Hamiltonian, argued (quite convincingly) that this result implies that we should identify the θ angle with the *polarization* of the 1D insulator. In this language the quantization of the charge in 1D is the same as (or descends from) the quantization of flux in 2D. Since the only values of the flux compatible with time-reversal invariance are $\Phi = 0, \pi \pmod{2\pi}$ (in units of the flux quantum $\phi_0 = hc/e$), the only allowed values of charge in 1D are 0 and $1/2 \pmod 1$, which is consistent with our previous conclusions.

Qi, Hughes, and Zhang generalized these results to higher-dimensional insulators. They began by considering a topological insulator in $d = 4$ space dimensions. For the sake of argument, we can imagine a Wilson-fermion Hamiltonian for four-component spinors on a 4D hypercubic lattice. The simplest model of this type has a one-particle Dirac Hamiltonian $h(\vec{k}) = \sum_{i=a}^{5} h_a(\vec{k}) \cdot \Gamma_a$, which involves all five Dirac Γ matrices, reflecting the broken time-reversal invariance of this insulator. Qi, Hughes, and Zhang noted that in systems of this type one can define a non-abelian Berry connection $\mathcal{A}_i^{\alpha\beta}(\vec{k})$ on the 4-torus T^4 of the 4D Brillouin zone, where α and β run over the labels of the occupied bands (there are two occupied bands in the present example), and $i = 1, \ldots, 4$ are the four orthogonal coordinates of the 4D Brillouin zone. This non-abelian Berry connection is

$$\mathcal{A}_i^{\alpha\beta}(\vec{k}) = -i\langle \alpha, \vec{k}|\partial_{k_i}|\beta, \vec{k}\rangle \tag{16.177}$$

which for two occupied bands takes values in the algebra of the gauge group $SU(2) \times U(1)$. We can define the (non-abelian) field strength $\mathcal{F}_{ij}^{\alpha\beta}$,

$$\mathcal{F}_{ij}^{\alpha\beta} = -i\left[D_i, D_j\right]^{\alpha\beta} = \partial_i \mathcal{A}_j^{\alpha\beta} - \partial_j \mathcal{A}_i^{\alpha\beta} + i\left[\mathcal{A}_i, \mathcal{A}_j\right]^{\alpha\beta} \tag{16.178}$$

where $\partial_i = \partial_{k_i}$ and D_i is the covariant derivative in the adjoint representation of the gauge group ($SU(2) \times U(1)$ in the case at hand). In this system we can define a topological invariant called the second Chern number, C_2, which is given by

$$C_2 = \frac{1}{32\pi^2}\int_{BZ} d^4k\, \epsilon^{ijkl}\, \text{tr}[\mathcal{F}_{ij}\mathcal{F}_{kl}] \tag{16.179}$$

On the other hand, given the five-component vector $\vec{h}(\vec{k})$ which defines the Hamiltonian, we can construct the five-component unit vector $\hat{h}(\vec{k}) = \vec{h}(\vec{k})/||\vec{h}(\vec{k})||$. This unit vector defines a mapping of the 4-torus T^4 (the Brillouin zone) into the 4-sphere S_4. These mappings can be classified by the homotopy group $\pi_4(S_4) = \mathbb{Z}$ whose topological invariant is the Pontryagin index $\mathcal{Q}$,

$$\mathcal{Q} = \frac{3}{8\pi^2}\int_{BZ} d^4k\, \epsilon^{ijlmn}\hat{h}_i(\vec{k})\partial_{k_1}\hat{h}_j(\vec{k})\partial_{k_2}\hat{h}_l(\vec{k})\partial_{k_3}\hat{h}_m(\vec{k})\partial_{k_4}\hat{h}_m(\vec{k}) \tag{16.180}$$

Exactly as in the 2D example we discussed before, the second Chern number C_2 is computed by the Pontryagin index $\mathcal{Q}$ of Eq. (16.180), i.e. $C_2 = \mathcal{Q}$.

These results determine the form of the effective action for an external electromagnetic field A_μ (in $(4+1)$ dimensions) for this 4D insulator. The effective action is a Chern–Simons term in $(4+1)$ dimensions,

$$S_{4+1}[A] = \frac{C_2}{24\pi^2}\int d^5x\, \epsilon^{\mu\nu\lambda\rho\sigma} A_\mu\, \partial_\nu A_\lambda\, \partial_\rho A_\sigma \tag{16.181}$$

This result was derived by QHZ (Qi *et al.*, 2008) by computing the non-linear response to the external electromagnetic field, and is a generalization to 4D of the TKNN result that in 2D the Hall conductance is given by the first Chern number. This result was derived earlier on in the high-energy literature by Golterman and coworkers (Golterman *et al.*, 1993), and in the reviews by Jansen (1996) and by Creutz (2001).

We can now proceed to reduce the dimensionality by one by compactifying one of the four directions of 4D space, say x_4 (the fifth dimension of the space-time), into a circle S_1 of circumference L_4. The manifold now is $S_1 \times S_1 \times \mathbb{R}^3$. The effective action in the reduced dimension can be deduced by analogy with the arguments that led to Eq. (16.174). By taking the compact direction to be very small, $L_4 \to 0$, we find that the effective action in (3 + 1) dimensions is

$$S_{3+1}[A] = \frac{C_2}{8\pi^2} \int d^4x \; \Phi(x) \epsilon^{\mu\nu\lambda\rho} \, \partial_\mu A_\nu \, \partial_\lambda A_\rho \tag{16.182}$$

where $\Phi = \oint dx_4 \, A_4$ is the flux through the compactified fifth dimension. Naturally, this result can also be derived by direct calculation in (3 + 1) dimensions.

We have thus reproduced the result of QHZ, Eq. (16.165), who further identified the θ angle, given here by the flux Φ, with the local *polarization* (using the notation of QHZ)

$$P_3(\vec{x}, t) = \frac{1}{2\pi} \Phi(\vec{x}, t) \tag{16.183}$$

It is now straightforward to show that the 4-current j_μ is given by

$$j_\mu = \frac{\delta S_{3+1}[A]}{\delta A^\mu} = \frac{C_2}{2\pi} \epsilon_{\mu\nu\lambda\rho} \, \partial^\nu P_3 \, \partial^\lambda A^\rho \tag{16.184}$$

which is the same as the Callan–Harvey current, Eq. (16.167). Thus, this analysis also predicts the existence of the topological magneto-electric effect.

16.11 Topological insulators and interactions

In this chapter we have focused on the properties of insulating systems whose band structures have non-trivial topology. Since our discussion has been done entirely at the level of free-fermion systems, it is worthwhile to raise some important questions.

(a) The first question is whether, and to what extent, these topological properties, and their consequences, are stable if interactions between the electrons are included.
(b) A second important question is whether these topological properties can be the result of a phase transition in a trivial system into a state with non-trivial topology. In other words, can we have a *topological Mott insulator*?
(c) Finally, a third (but related) question is what is the relation, if any, between topological insulators and the topological *phases* that we discussed in depth in other chapters of this book?

We will now address the first of these questions: stability. The stability of topological insulators, and for that matter of all insulators, is guaranteed by the energy gap in the spectrum. We have already encountered this issue in other chapters of this book, e.g. the stability of the integer and fractional quantum Hall states in Chapters 12 and 13. Thus, the effects of interaction terms in the Hamiltonian are suppressed at low energies since perturbation theory is convergent when the system has a gap, provided that the interactions are sufficiently local and the coupling constants are smaller than a non-universal critical value. This fact has been checked by exact diagonalization studies (Varney *et al.*, 2010) in the quantum anomalous Hall state of the Haldane model on the honeycomb lattice for spinless fermions with a repulsive interaction V between fermions on nearest-neighboring sites. Varney *et al.* found that the repulsive interaction tends to close the gap of the topological insulators and to favor a non-topological insulating Mott state in which the charge density is different on the two sublattices of the honeycomb lattice. This uniform state, which breaks the point-group symmetry of the honeycomb lattice spontaneously, is also the strong-coupling ground state if the time-reversal-symmetry-breaking terms are absent. In other words, in the strong-coupling regime the physics is local and insensitive with regard to the subtleties of the band structure which are relevant at weak coupling.

A similar effect was found in the quantum spin Hall regime of the Kane–Mele–Hubbard model. Quantum Monte Carlo simulations (Hohenadler *et al.*, 2011, 2012; Zheng *et al.*, 2011) found that the on-site Hubbard interaction with coupling constant U does not affect the quantum spin Hall state, provided that it is smaller than a critical value set by the insulating gap. If the insulating gap of the quantum spin Hall state is increased, the system becomes more stable with respect to Hubbard-type perturbations. When the interactions begin to dominate, these simulations find quantum phase transitions to one of two Mott states, either a $\mathbb{Z}_2$ topological spin-liquid phase at intermediate couplings, similar to the ones we discussed in Chapter 9, or a quantum antiferromagnet in a Néel phase (typically at stronger coupling). However, the existence of these Mott states is unrelated to the physics of the topological insulator, since they are also found even if the non-interacting limit is not a topological insulator, see e.g. Meng *et al.* (2010). Nevertheless, the $\mathbb{Z}_2$ topological state seems to compete with the topological insulator since the size of the phase diagram it occupies shrinks (and disappears rapidly) as the strength of the Kane–Mele coupling increases. In contrast, although the threshold to the Néel antiferromagnetic phase is monotonically pushed to stronger couplings with increasing Kane–Mele coupling, it never disappears from the phase diagram.

In other words, the topological insulators that arise from free-fermion systems are stable. In fact, they are *as stable* as any other *band insulator* regardless of whether their bands are topological or trivial. This does not mean that topology does not matter. Indeed, unlike trivial insulators, topological insulators have gapless fermionic edge states that are protected.

Since the edge states are gapless, one may wonder whether they are stable with respect to interactions. In the case of 2D topological insulators in the quantum anomalous Hall state, the stability of their edge states is guaranteed by the fact that these states are chiral and interactions can lead only to finite renormalizations of the Fermi velocity, exactly as in the case of the integer and fractional quantum Hall states. In the case of the quantum spin Hall states, their stability is guaranteed by time-reversal invariance, which does not allow for mixing between spin-reversed states. Thus, in the quantum spin Hall state the edge states may become Luttinger liquids but will remain gapless, and are always present, provided that the bulk remains in a topological insulator phase. This is consistent with the general arguments on the topological protection and stability of the edge states (Qi *et al.*, 2006a).

Let us now consider the stability of the edge states of 3D $\mathbb{Z}_2$ topological insulators. As we saw earlier in this chapter, their edge states are 2D Weyl fermions that are confined to the surface of the insulator and are also protected, provided that the bulk has a gap of topological origin. Protection of the 2D Weyl fermions works in two ways. On the one hand, they live in $D = 2 + 1$ dimensions. As we saw in Section 4.2.5, the action of relativistic free Dirac (or Weyl) fermions in $D > 2$ space-time dimensions is a *stable fixed point* of the RG in a theory with only local four-fermion interactions (Wilson, 1973). This in fact would also be true if the bulk were not topological but had gapless edge states by virtue of an accidental fine-tuning of the Hamiltonian. Recall that the scaling dimension of a Dirac (and Weyl) fermion is $\Delta_\psi = (D - 1)/2$, where D is the dimension of space-time. The scaling dimension of a local four-fermion interaction at the free Dirac fixed point is $\Delta_4 = 2(D-1)$. Hence, for all $D > 2$ (space-time) dimensions, the scaling dimension $\Delta_4 > D$. Hence the four-fermion operator is irrelevant for $D > 2$ and the free (Dirac)-fermion fixed point is stable. This scaling argument also protects the Dirac semi-metal phase of single-layer graphene. Coulomb interactions, which are marginally irrelevant here too (as in graphene), cannot open a gap for similar reasons (unless the coupling is too strong).

However, the 2D Weyl-fermion surface states of 3D $\mathbb{Z}_2$ insulators are protected also by the topological nature of the bulk. As we saw in Section 16.10, 3D $\mathbb{Z}_2$ topological insulators have a bulk–edge correspondence in the form of the Callan–Harvey effect, which guarantees that the anomaly of the bulk is exactly compensated by the anomaly of the edge. Since the edge states have an anomaly, many possible interactions are not allowed, since they would violate the bulk–edge correspondence. Thus, exactly as in the case of the quantum spin Hall state, time-reversal-invariant interactions can only renormalize the Fermi velocity of the Weyl fermions. Even if time-reversal invariance were broken explicitly, the irrelevance of local interactions would kick in and protect the Weyl surface states.

16.12 Topological Mott insulators and nematic phases

Our second question was whether a topological insulator could arise from a non-topological system as a result of a phase transition to a *topological Mott insulator*. This question was first addressed by Raghu *et al.* (2008), who considered two related problems. They (mainly) used mean-field theory (Hartree–Fock) in a system of spinless fermions on the honeycomb lattice with repulsive interactions V_1 and V_2 between nearest-neighbor and next-nearest-neighbor sites. In the weak-coupling limit this reduces to two species of Dirac fermions with local four-fermion interactions. Since, as we just saw, this interaction is irrelevant at weak coupling, all phase transitions are pushed to finite values of V_1 and V_2, presumably of the order of the bandwidth of the free fermions.

The main result of this work was that Raghu *et al.* found a competition between two phases. For V_1 larger than a critical value they found a state that they called a "charge-density-wave" (CDW) phase. It is actually a uniform state with intra-unit-cell charge order that breaks spontaneously the point-group symmetry of the honeycomb lattice by an unequal occupation of the two sublattices. This is a non-topological Mott insulator and can be pictured as two Dirac (Weyl) fermions with mass gaps of opposite sign or, equivalently, with opposite 2D parities.

On the other hand, for V_2 larger than a critical value they found a phase with a dynamically generated quantum anomalous Hall phase. This phase is characterized by the development of a *complex* expectation value of a next-nearest-neighbor fermion bilinear $\langle c_A(\vec{r})^\dagger c_A(\vec{r}+\vec{a}_i)\rangle$. In other words, the phase with a spontaneous quantum anomalous Hall phase appears when a Haldane term is generated dynamically. This state can be pictured as two Dirac fermions with masses with the same sign and hence with a spontaneously broken time-reversal invariance. In other words, this state has spontaneous circulating currents.

Raghu *et al.* also found a direct (first-order) transition between the non-topological charge-ordered Mott insulator and the spontaneous quantum anomalous Hall state merging at a bicritical point where $V_1 \sim V_2$. They checked this result using an RG analysis, which is reliable only in the weak-coupling regime. They also considered a graphene system with spin-$1/2$ fermions. In this case, in addition to the nearest- and next-nearest-neighbor interactions, V_1 and V_2, they also included the on-site repulsive Hubbard U term. In addition to the CDW, spontaneous quantum anomalous Hall, and semi-metal phases of the spinless case, they now also found a Néel phase, or spin-density wave (SDW), and a phase with a spontaneous quantum spin Hall state. The latter state is a time-reversal-invariant phase characterized by an expectation value of the Kane–Mele term. This is a time-reversal-invariant insulating uniform phase that breaks spontaneously the SU(2) spin symmetry. It should be noted that their mean-field theory (not surprisingly!) did not find a $\mathbb{Z}_2$ spin liquid that is seen in Monte Carlo simulations for $V_1 = V_2 = 0$ (Meng *et al.*, 2010).

The interesting results of Raghu and coworkers (Raghu *et al.*, 2008) have the drawback of relying on approximations that can be trusted only in the weak-coupling regime. In fact, while the conventional Mott states, namely the CDW phase and three Néel phases, are easily obtained in the strong-coupling limits of large V_1 or large U, the existence of topological phases is far from clear in the regime of very large V_2. One may wonder whether there are perhaps other systems in which the topological phases appear in the weak-coupling regime and the phases with obvious types of broken symmetry appear in the strong-coupling regime.

The way around the problem we just discussed is to change the band structure so that the four-fermion operators become marginal. This approach was taken by Sun and coworkers (Sun *et al.*, 2009), whose work we will follow here. Let us suppose that we have a system in which the single-particle kinetic energy scales as L^{-2}, instead of as L^{-1} in the Dirac theory. This would happen if instead of a linear band crossing we had a system with a quadratic band crossing. In such a system time will have to scale as L^z, where z is the dynamic critical exponent. Since the fermion operators must still satisfy canonical anticommutation relations, the fermion action must be linear in time derivatives. Since it is also quadratic in the space derivatives, this fixes the dynamic exponent $z = 2$. On the other hand, the free-fermion action is a bilinear in the fermionic fields. Hence, if the system is in $d = 2$ space dimensions, the scaling dimension of the fermionic field is $\Delta = z/2 = 1$. From here it follows that the scaling dimension of all four-fermion operators is $\Delta_4 = 4\Delta = 4$, which happens to be equal to $d + z = 4$. Hence this naive scaling, which is correct for a free field, implies that in a system with these scaling properties the four-fermion interactions are marginal, and one has a chance to obtain non-trivial condensates that are accessible by means of a perturbative RG. Or, equivalently, it implies that the new phases result from infinitesimal instabilities.

The catch in this argument is that, unlike linear band crossings, which are generic and are described by a theory of relativistic Dirac fermions, in general quadratic band crossings are accidental degeneracies that require two parameters to be simultaneously tuned to zero. Such accidental degeneracies can be lifted by the addition of an arbitrarily small parameter without violating any symmetries. The way out of this problem is to find a symmetry-protected quadratic band crossing. The key to this question is for the crossing to have a non-trivial Berry phase. Let $|\psi(\vec{k})\rangle$ be a two-component spinor state defined on each point $\vec{k}$ of the Brillouin zone. The Berry connection is $\mathcal{A}_i[\vec{k}] = -i\langle\psi(\vec{k})|\partial_{k_i}|\psi(\vec{k})\rangle$. Let Γ be a closed path on the Brillouin zone. The Berry flux is (see Chapter 12 and Section 16.2)

$$\Phi_\Gamma = -i \oint_\Gamma d\vec{k} \cdot \langle\psi(\vec{k})|\partial_{\vec{k}}|\psi(\vec{k})\rangle \tag{16.185}$$

The Berry flux can be non-zero only if the spinors have a singularity at some point $\vec{k}_0$ of the Brillouin zone. If the closed path Γ does not enclose a singularity, the Berry flux must vanish. If the spinors have a singularity at $\vec{k}_0$ then the Berry flux will not vanish for all paths Γ that enclose $\vec{k}_0$. If the system is further assumed to be time-reversal-invariant then the Berry flux can only be $\Phi_\Gamma = n\pi$. We saw before that lattice systems have a special set of momenta for which the Hamiltonian is invariant under time reversal. We called them the time-reversal-invariant momenta. Let us consider a case in which the band crossing occurs at a time-reversal-invariant momentum $\vec{k}_0$.

In a system of massless two-component Dirac (Weyl) fermions, the Berry flux is $\pm\pi$. In such a system we cannot open a gap unless time-reversal invariance is broken, since with the extra operator all three Pauli matrices enter into the Hamiltonian. This is the path we followed to discuss the quantum anomalous Hall state. A time-reversal-invariant system with a quadratic band crossing at a time-reversal-invariant point $\vec{k}_0$ has a Berry flux that is twice as large, $\Phi_\Gamma = 2n\pi$. If the integer is $n = 0$ there is no singularity and the quadratic band crossing can be removed without breaking any symmetry. However, if the Berry flux is $\pm 2\pi$ the quadratic band crossing is protected by time-reversal invariance and cannot be removed without breaking time-reversal invariance.

On the other hand, one can imagine adding operators that "split" (in momentum space) the quadratic band crossing with Berry phase $+2\pi$ into two Dirac "cones," each with Berry phase $+\pi$. However, such a splitting of the quadratic crossing breaks rotational invariance and makes the system anisotropic. We will see below an example on a square lattice, whose point-group symmetry is C_4, the symmetry group of the square. In that case, the splitting would amount to the spontaneous breaking of the four-fold C_4 rotational symmetry down to a two-fold symmetry C_2 (of rotations by π). We will say that a system with a spontaneous breaking of rotational invariance is in an *electronic nematic state* (or quantum nematic state), as in the case of nematic liquid crystals (de Gennes and Prost, 1993; Chaikin and Lubensky, 1995). This state would be an example of an electronic liquid-crystal phase (Kivelson *et al.*, 1998). Since the broken symmetry is $\mathbb{Z}_2$, this state should be called an "Ising nematic." Below we will give a specific example in which the topological Mott insulator competes with an electronic nematic phase (which here is a semi-metal).

There are many examples of electronic systems that have nematic phases of purely quantum-mechanical origin, such as several high-T_c superconductors, including $YBa_2Cu_3O_{6+x}$ in much of its "pseudogap" regime, 2D electron gases in the second Landau level (Lilly *et al.*, 1999) (near the center of the Landau level, where the quantum Hall effect is not seen), the bilayer material $Sr_3Ru_2O_7$ in magnetic fields (Borzi *et al.*, 2007), and the "hidden-order" phase of the heavy-fermion superconductor URu_2Si_2. For reasons of space the theory of quantum electronic liquid crystals will not be discussed in this book. A review of the experimental evidence for electronic nematic states and of the theory can be found in Fradkin *et al.* (2010).

The main difference between these quantum nematic states and their classical cousins is that, while in the latter the constituent degrees of freedom are rod-like molecules ("nematogens"), in the electronic system the constituent degrees of freedom are point particles, electrons! In the electronic case the formation of an anisotropic phase is a result of several possible quantum-mechanical mechanisms, such as the lifting of a high-order degeneracy (as in the quadratic band crossing), the spontaneous distortion of the Fermi surface of a Fermi liquid (the Pomeranchuk instability) (Oganesyan *et al.*, 2001), the spontaneous ordering of "orbital" degrees of freedom, or the quantum-mechanical melting of a stripe phase (a unidirectional CDW) (Kivelson *et al.*, 1998).

Another example with a quadratic band crossing is bilayer graphene. Bilayer graphene is a system of two layers of carbon atoms, with each layer arranged on a honeycomb lattice and stacked in such a way that one sublattice is right on top of the other (Castro Neto *et al.*, 2009). This arrangement is called a Bernal stacking. There is a small tunneling-matrix element for the electrons in one layer to tunnel to the electronic states in the other layer (and vice versa).

The particular features of the Bernal stacking cause the Dirac spectrum of single-layer graphene to be replaced by a quadratic band crossing at the same points of the Brillouin zone, K and K^*. In this case this is an accidental degeneracy, since in principle it is possible to have a term in the Hamiltonian that lifts the quadratic crossing. In bilayer graphene with Bernal stacking this term is caused by a trigonal warping of the crystal structure, which is compatible with the symmetries of bilayer graphene. The trigonal warping term does indeed split each of the quadratic crossings (with Berry flux 2π) into three Dirac cones (each with Berry flux $+\pi$) away from the K and K^* points and another Dirac cone with Berry flux $-\pi$. However, although the quadratic crossing in bilayer graphene is not protected by symmetry, it is protected by accident (or by carbon!) since the trigonal warping term is three orders of magnitude smaller than all other terms in the Hamiltonian. So, for all practical purposes it can be ignored. Consequently, many aspects of the theory that we will now discuss also apply in practice for the case of bilayer graphene and, not surprisingly, many of these ideas have also been discussed in that context (Nandkishore and Levitov, 2010; Vafek, 2010; Vafek and Yang, 2010).

Sun and coworkers (Sun *et al.*, 2009) showed that topological insulators (as well as nematic phases) can arise dynamically within a weak-coupling theory by considering a system of fermions (with and without spin) with a symmetry-protected quadratic band crossing. An example of such a system is a half-filled checkerboard

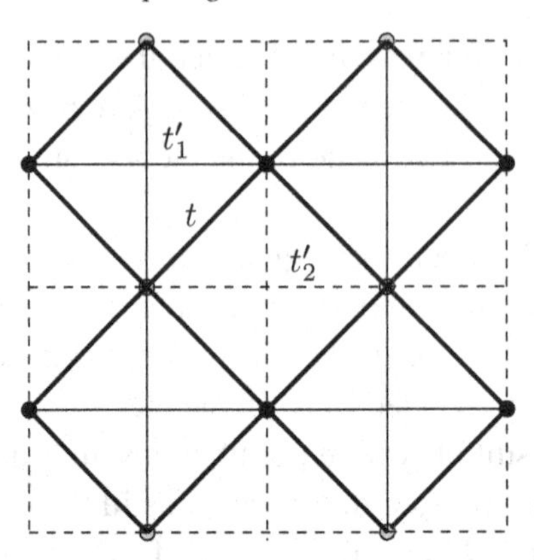

Figure 16.10 The checkerboard lattice is a 2D lattice with corner-sharing tetrahedra, a 2D version of the pyrochlore lattice. The hopping-matrix elements are t along the diagonals, t'_1 along the horizontal and vertical full lines, and t'_2 along the horizontal and vertical broken lines.

lattice, as shown in Fig. 16.10. This lattice can be regarded as a 2D version of the 3D pyrochlore lattice, a lattice of corner-sharing tetrahedra. This is also the lattice of the oxygen sites in the strong-coupling limit of the Emery model for the copper-oxide plane of the high-temperature superconductors (Kivelson *et al.*, 2004; Sun and Fradkin, 2008). (The half-filled case which we will discuss below is not physically meaningful for the copper oxides.)

Before discussing the specific aspects of the lattice model, it is useful to write down a theory of interacting fermions with symmetry-protected quadratic band crossing. Let $\Psi(\vec{r}) = (\psi_1(\vec{r}), \psi_2(\vec{r}))^{\mathrm{T}}$ be a two-component Fermi field, with or without spin (below the two components will correspond to the sublattice labels), whose low-energy Hamiltonian is

$$H = \int d\vec{r} \left[\Psi^\dagger(\vec{r}) \mathcal{H}_0 \Psi(\vec{r}) + u \psi_1^\dagger(\mathbf{r}) \psi_1(\mathbf{r}) \psi_2^\dagger(\mathbf{r}) \psi_2(\mathbf{r}) \right] \tag{16.186}$$

where $\mathcal{H}_0$ is the one-particle Hamiltonian for a quadratic band crossing and u is a coupling constant. In momentum space $\mathcal{H}_0(\vec{k})$ is

$$\mathcal{H}_0(\mathbf{k}) = d_{\mathrm{I}} \mathbb{I} + d_x \sigma_1 + d_z \sigma_3 \tag{16.187}$$

where, as before, $\mathbb{I}$ is the 2×2 identity matrix and σ_x and σ_z are the corresponding Pauli matrices. The coefficients are

$$d_{\mathrm{I}} = t_{\mathrm{I}}(k_x^2 + k_y^2), \qquad d_x = 2t_x k_x k_y, \qquad d_z = t_z(k_x^2 - k_y^2) \tag{16.188}$$

With this "d-wave" (quadrupolar) structure the Berry flux associated with the quadratic band crossing is 2π (or -2π depending on the relative signs). If $|t_x| =$

$|t_z|$ the system is rotationally invariant in this continuum approximation and has C_6 point-group symmetry in the case of a lattice system. This effective Hamiltonian manifestly exhibits the $z = 2$ scaling we discussed above. The term proportional to the identity matrix breaks the particle–hole symmetry of this theory. This term is not very important, provided that the parameters of the theory are such that the upper band lies entirely above the lower band, and never crosses the Fermi energy, which we will set to be at zero. Below we will see that in specific lattice models this condition can easily be satisfied. Unless we state the contrary, in what follows we will work with a particle–hole-symmetric theory.

Let us discuss briefly some aspects of the free-fermion system. For simplicity we will take the system to have maximal symmetry and set $|t_x| = |t_y| = t$. The free-fermion theory has the same matrix structure as the relativistic Dirac fermion except that it has dynamic critical exponent $z = 2$, and is not relativistically invariant. In Chapter 9 we discussed the quantum Lifshitz model, which is a free scalar field with $z = 2$. We will now see that the quadratic band-crossing system is closely related to the quantum Lifshitz model. Let us write the $(2+1)$-dimensional massless Dirac operator in its covariant form $i\partial\!\!\!/ = i\gamma^\mu \partial_\mu$, where we will take the Dirac gamma matrices to be $\gamma_0 = \sigma_2$, $\gamma_1 = -i\sigma_3$, and $\gamma_2 = i\sigma_1$. From the algebra of the gamma matrices it follows that the square of the Dirac operator is simply the d'Alembertian,

$$(i\partial\!\!\!/)^2 = -\partial^2 \tag{16.189}$$

Likewise, from the form of the Hamiltonian of the quadratic band crossing, Eq. (16.187), we see that we can write a $z = 2$ version of the Dirac operator in the "covariant" form $i\gamma^\mu D_\mu$ (here D_μ is not the covariant derivative!) with the same gamma matrices but with $D_0 = \partial_0$, $D_1 = t(\partial_x^2 - \partial_y^2)$, and $D_2 = 2t\,\partial_x\partial_y$. The square of the $z = 2$ "Dirac" operator is

$$(i\gamma^\mu D_\mu)^2 = -\partial_0^2 + t^2(\nabla^2)^2 \tag{16.190}$$

which we recognize as the $z = 2$ version of the d'Alembertian, the differential operator that enters in the action of the quantum Lifshitz model, Eq. (9.161), after setting $t = \kappa$.

Let us now return to the interacting Hamiltonian, Eq. (16.186), and use the RG to see what the effects of the interactions are. The parameter $g = u/|t_x|$ plays the role of the dimensionless coupling constant of this problem, and at the tree-level (i.e. without computing quantum corrections) the four-fermion operator is marginal. Hence, at this level, this theory seems entirely analogous to the theory of the Luttinger liquid of Chapter 6, and the four-fermion interaction looks like the backscattering term in the 1DEG. So one is tempted to guess that here too this superficially marginal operator may remain exactly marginal to all orders in perturbation theory. However, this guess is incorrect. In $(1+1)$ dimensions the exact marginality of the backscattering interaction is due to the exact cancellation of the bubble diagram against the Cooper channel. This is true at one-loop level and to all orders in perturbation theory. However, this cancellation does not happen in the case of the Hamiltonian of Eq. (16.186). As a result, not only is the beta function for the coupling constant g not zero but already at one-loop level it is given by (Sun *et al.*, 2009)

$$\beta(g) = \frac{\partial g}{\partial l} = Ag^2 + O(g^3) \tag{16.191}$$

where $l = \ln a$ and a is the short-distance cutoff. The coefficient A is a finite *positive* function of the ratio $\lambda = t_z/t_x$ and is given by

$$A(\lambda) = \frac{1}{2\pi^2} K(\sqrt{1-\lambda^2}) \tag{16.192}$$

where $K(x)$ is the complete elliptic integral

$$K(x) = \int_0^{\pi/2} \frac{d\theta}{\left(1 - x^2 \sin^2\theta\right)^{1/2}} \tag{16.193}$$

where $x = \sqrt{1-\lambda^2}$. For the special case of a system with C_6 symmetry, $|t_x| = |t_z|$ (and hence $\lambda = 1$), the coefficient is $A(1) = 1/(4\pi)$. This result for the beta function means that the four-fermion interaction is a marginally relevant operator and the effective coupling constant $g > 0$ flows to strong coupling at low energies. This means that the free-fermion ground state of a theory of gapless fermions with a quadratic band crossing is unstable in two space dimensions. To find out what results from this instability one has to know the nature of the stable fixed point to which the RG flow is converging. This cannot be determined in perturbation theory.

To explore the consequences of this instability, we need to investigate using mean-field theory the possible condensates of order parameters expressed in terms of fermion bilinears. In this system they are

$$\Phi = \langle \Psi^\dagger(\vec{r})\sigma_y \Psi(\vec{r})\rangle, \quad Q_1 = \langle \Psi^\dagger(\vec{r})\sigma_z \Psi(\vec{r})\rangle, \quad Q_2 = \langle \Psi^\dagger(\vec{r})\sigma_x \Psi(\vec{r})\rangle \tag{16.194}$$

Here Φ is the order parameter of a time-reversal-symmetry-breaking gapped phase with a *spontaneous quantum anomalous-Hall-state* phase. This phase has a zero-field quantized Hall conductivity $\sigma_{xy} = e^2/h$. Notice that this is an *integer* quantum Hall state. This phase is a time-reversal-symmetry-breaking topological Mott insulator similar to the one proposed by Raghu and collaborators (Raghu *et al.*, 2008). The order parameters Q_1 and Q_2 describe the *nematic phases* in which the C_4 or C_6 rotational symmetry is broken down to C_2 by splitting the quadratic band crossing into two types of Dirac points located along the direction of one of the main axes

(Q_1), or along a diagonal (Q_2). The nematic phase is an anisotropic semi-metal. However, unlike in graphene, where the two Dirac points have Berry fluxes π and $-\pi$, in this nematic phase the two types of Dirac points have the same Berry flux. In addition, there is also a phase in which the nematic ($Q_1 \neq 0$ or $Q_2 \neq 0$) and the quantum anomalous Hall orders ($\Phi \neq 0$) *coexist*. This phase is a nematic integer quantum Hall state.

Since there is only one coupling constant, denoted by u in Eq. (16.186), the weak-coupling ordering tendencies are determined by the logarithmically divergent normal-state susceptibilities χ_Φ, for broken time-reversal invariance, and χ_{Q_1} and χ_{Q_2}, for the two nematic order parameters. For general t_x and t_z, they can be shown to satisfy the identity $\chi_\Phi = \chi_{Q_1} + \chi_{Q_2}$. Therefore, $\chi_\Phi > \chi_{Q_i}$ ($i = 1, 2$), so the leading weak-coupling instability is with respect to the gapped quantum anomalous Hall state (here we are assuming that all processes have the same coupling constant).

The mean-field Hamiltonian is

$$H_{\text{MF}} = \int d\vec{r}\;\; \Psi^\dagger(\vec{r}) \left[\mathcal{H}_0 - \frac{u}{2}\left(Q_1\sigma_z + Q_2\sigma_x + \Phi\sigma_y\right)\right]\Psi(\vec{r}) + \frac{V}{4}\int d\vec{r}\left(Q_1^2 + Q_2^2 + \Phi^2\right) \tag{16.195}$$

By minimizing the ground-state energy of H_{MF} we find that at weak coupling the ground state is indeed the spontaneous quantum anomalous Hall phase, with a gap $\Delta \sim \Lambda \exp(-2/(\alpha g))$ (where Λ is a cutoff) and a mean-field critical temperature $T_c \sim \Delta$, which is consistent with the scaling predicted by the RG. This mean-field theory also predicts nematic phases, provided that (marginally) irrelevant operators, such as $\int d\vec{r}\, d\vec{r}\,' \sum_{i=1,2} U(\mathbf{r}-\mathbf{r}')\psi_i^\dagger(\vec{r})\psi_i(\vec{r})\psi_i^\dagger(\vec{r}\,')\psi_i(\vec{r}\,')$, are also included. The nematic phase Q_1 is energetically favored at small $V > 0$ and $U < 0$ if $|U/V|$ is large enough. As $|U/V|$ is reduced, the nematic phase gives way to the quantum anomalous Hall phase (and to a mixed phase).

Sun and coworkers (Sun *et al.*, 2009) found a relatively simple lattice model that exhibits these behaviors. They considered a system of fermions at half-filling (both spinless and spin-1/2) on the checkerboard lattice shown in Fig. 16.10, with Hamiltonian

$$H = -\sum_{\vec{r},\vec{r}\,'} t_{\vec{r},\vec{r}\,'} c^\dagger(\vec{r})c(\vec{r}\,') + \sum_{\vec{r},\vec{r}\,'} V_{\vec{r},\vec{r}\,'} c^\dagger(\vec{r})c(\vec{r})c^\dagger(\vec{r}\,')c(\vec{r}\,') \tag{16.196}$$

It is a square lattice (rotated by 45°) with hopping-matrix element t for hopping between nearest neighbors, t_1' for hopping between next-nearest neighbors on half of the plaquettes that share only vertices, and t_2' for hopping between the next-nearest neighbors on the remaining plaquettes. We will always assume that

$t_1' \gg t_2'$. The half-filled system has special properties. Many of the same properties are found on the kagome lattice at one-third filling. The interactions will be taken to be repulsive for nearest neighbors, $V > 0$. A next-nearest-neighbor attraction $V_{\rm nnn} < 0$ generates the marginally irrelevant operator discussed above.

Let A and B be the two sublattices of the square lattice and let $\psi_{\rm A}(\vec{r})$ and $\psi_{\rm B}(\vec{r})$ be the Fermi fields on sites of the two sublattices. We will consider the spinless case first. The free-fermion term of the Hamiltonian in momentum space has the usual two-band form

$$H_0 = \int_{\rm BZ} \frac{d^2p}{(2\pi)^2} \Psi^\dagger(\vec{p}) \Big[h_{\rm I}(\vec{p}) \mathcal{I} + \vec{h}(\vec{p}) \cdot \vec{\sigma} \Big] \Psi(\vec{p}) \tag{16.197}$$

where $\Psi(\vec{p}) = (\psi_{\rm A}(\vec{p}), \psi_{\rm B}(\vec{p}))^{\rm T}$ is a two-component spinor to account for the two sublattices, $\vec{\sigma} = (\sigma_1, \sigma_2, \sigma_3)$ is a three-component vector of the three Pauli matrices, and $\mathcal{I}$ is the identity matrix. The momentum $\vec{p}$ takes values on the first Brillouin zone of the square lattice. For the problem at hand the components of the vector $\vec{h}(\vec{p})$ are

$$\begin{aligned} h_1(\vec{p}) &= -4t \cos\left(\frac{p_x}{2}\right) \cos\left(\frac{p_y}{2}\right) \\ h_2(\vec{p}) &= 0 \\ h_3(\vec{p}) &= -(t_1' - t_2')(\cos p_x - \cos p_y) \end{aligned} \tag{16.198}$$

and $h_{\rm I}(\vec{p}) = -(t_1' + t_2')(\cos p_x + \cos p_y)$. The condition $h_2(\vec{p}) = 0$ follows from time-reversal invariance. The energy eigenvalues $E_\pm(\vec{p})$ are

$$\begin{aligned} E_\pm(\vec{p}) = {} & -(t_1' + t_2')(\cos p_x + \cos p_y) \\ & \pm \sqrt{\left(4t \cos\left(\frac{p_x}{2}\right) \cos\left(\frac{p_y}{2}\right)\right)^2 + \left((t_1' - t_2')(\cos p_x - \cos p_y)\right)^2} \end{aligned} \tag{16.199}$$

Provided that, on the first Brillouin zone, the minimum of $E_+(\vec{p})$ is greater than the maximum of $E_-(\vec{p})$, the ground state at half-filling of this free-fermion system has the states of the $-$ band occupied and the states of the $+$ band empty. We can choose the parameters for this to be the case.

This system has a band crossing at the zeros of the square-root term in Eq. (16.199), $\vec{p}_0 = (\pi, \pi)$ (up to reciprocal-lattice vectors). It will be important to what follows that (π, π) is the only crossing point, which is also a time-reversal-invariant momentum, and that the bands are analytic functions of the momenta everywhere else on the first Brillouin zone. However, unlike the cases we discussed before in this chapter, the dispersion near (π, π) has a *quadratic* dependence on the momentum (measured from (π, π)), instead of a linear dependence. Thus these spinors are not Dirac fermions. If we put $\vec{q} = \vec{p} - (\pi, \pi)$, the effective one-particle

Hamiltonian near the crossing point is a 2×2 matrix of the form of Eq. (16.187). The parameters of the continuum Hamiltonian (near the quadratic band crossing) are $t_I = (t' + t'')/2$, $t_x = t/2$, and $t_z = (t' - t'')/2$.

The order parameters for the lattice model are

$$
\begin{aligned}
Q_1 &= \frac{1}{4} \sum_{\vec{\delta}} \langle c_{\rm A}^\dagger(\vec{r}) c_{\rm A}(\vec{r}) - c_{\rm B}^\dagger(\vec{r}+\vec{\delta}) c_{\rm B}(\vec{r}+\vec{\delta}) \rangle \\
Q_2 &= \frac{1}{2} \sum_{\vec{\delta}} D_{\vec{\delta}} \, \mathrm{Re} \langle c_{\rm A}^\dagger(\vec{r}) c_{\rm B}(\vec{r}+\vec{\delta}) \rangle \\
\Phi &= \frac{1}{2} \sum_{\vec{\delta}} D_{\vec{\delta}} \, \mathrm{Im} \langle c_{\rm A}^\dagger(\vec{r}) c_{\rm B}(\vec{r}+\vec{\delta}) \rangle
\end{aligned}
\tag{16.200}
$$

where Q_1 is the "site-nematic" order parameter, Q_2 is the "bond-nematic" order parameter, and Φ is the quantum anomalous Hall order parameter. Here $\vec{\delta} = \pm\hat{x}/2 \pm \hat{y}/2$ connects nearest neighbors, $D_{\vec{\delta}} = \pm 1$, $D_{\pm(\hat{x}/2+\hat{y}/2)} = 1$, and $D_{\pm(\hat{x}/2-\hat{y}/2)} = -1$.

The details of the phase diagram found in mean-field theory depend on the parameters of this model (which has many!). A quantum anomalous Hall phase is found for V small and below a critical temperature. This phase has a zero-field quantized Hall conductivity e^2/h, and the quasiparticle spectrum has topologically protected chiral edge states, as predicted from general considerations (Qi *et al.*, 2006a). A site nematic is found for $V \sim |t_1' - t_2'|$, while the bond nematic is not favored. For $|t_1'| \geq |t_2'|$ and $|t''|/|t'| \ll 1$, there is a direct nematic–quantum-anomalous-Hall first-order phase transition. If $|t''|/|t'| \sim 1$, there is also a phase with coexisting quantum anomalous Hall and nematic orders. For other values, one has a direct first-order transition and a coexisting phase.

In summary, if a system has a quadratic band crossing, interaction effects lead to a phase diagram in which the topological insulator appears as a result of repulsive interactions and is a topological Mott insulator. In the spinless case we discussed above, the topological insulator is a state with a spontaneous quantum Hall effect and the other phases are two types of nematic states. If spin is included, in addition to the spin-polarized version of the phases we just discussed, one has also a phase with a spontaneous quantum spin Hall state and phases with spin-triplet versions of the nematic order parameters (Sun *et al.*, 2009). Analogs of these phases can also exist in metallic states (i.e. systems with a Fermi surface) (Wu *et al.*, 2007; Sun and Fradkin, 2008).

16.13 Topological insulators and topological phases

We will now come to the third, and most subtle, of our three questions: what is the relation, if any, between topological insulators and topological phases? As we noted at the beginning of this chapter, topological insulators are states with topologically protected properties but are not topological phases. However, they may be a "springboard" from which to access a topological phase. This can happen in several possible ways.

One possibility is that the topological insulator may become a topological phase, for instance by moving the Fermi energy away from the gap, and that this system may become unstable. For example, the doped topological insulator (which is now a metal with an "interesting" band structure) can become a superconductor. In Section 14.9 we showed that, at least in two dimensions, a superconductor with an order parameter with symmetry $p_x + ip_y$ is topological, which is a state in which time-reversal invariance is spontaneously broken. In this sense this is a superconducting cousin of a quantum anomalous Hall state. This state is believed to occur in Sr_2RuO_4. We have already discussed this state in Chapter 14, where we also discussed its connection with the non-abelian Moore–Read fractional quantum Hall state. There we saw that both condensed states have half-vortices with non-abelian braiding statistics whose origin was traced to the existence of Majorana zero modes in the vortex cores.

This observation motivated Fu and Kane (2008) to ask whether something similar could happen in a superconducting state of a topological insulator. They considered a strong 3D time-reversal-invariant topological insulator in contact with a 3D conventional s-wave superconductor. Since the topological insulator has a gap, the effects of the s-wave superconductor will matter only near their interface. On the other hand, the surface of the strong topological insulator supports gapless Weyl fermions. These surface excitations are charged and hence will couple to the nearby superconductor by a proximity effect.

Here we will follow the work of Fu and Kane (2008). Their argument goes as follows. We will consider only the Weyl fermions at the top surface of a time-reversal-invariant $\mathbb{Z}_2$ (strong) topological insulator. The Weyl fermions are described by a two-component spinor field $\psi = (\psi_\uparrow, \psi_\downarrow)^{\mathrm{T}}$ with a free gapless fermion 2D Hamiltonian density

$$\mathcal{H}_0 = \psi^\dagger(-iv_{\mathrm{F}}\vec{\sigma}\cdot\vec{\nabla} - \mu)\psi \tag{16.201}$$

where $\vec{\sigma} = (\sigma_x, \sigma_y)$ and μ is the chemical potential. Let us imagine now that a standard s-wave superconductor is deposited on the top surface. The superconductor is a ground state of an electronic system with a spontaneously broken U(1) global symmetry. The superconducting state exists if the pair-field operator $c_\uparrow^\dagger(\vec{k})c_\downarrow^\dagger(-\vec{k})$ has a non-vanishing expectation value or, equivalently, if its real-space counterpart exhibits long-range order. The expectation value of the pair field is the local order parameter of the superconducting state and, if the superconducting state is uniform, it is a condensate of spin-singlet Cooper pairs with total momentum zero. As in Chapter 14, we will denote the superconducting order parameter by $\Delta = \Delta_0 e^{i\phi}$. In the case of an s-wave superconductor, the fermionic excitations, which are charge-neutral, have an isotropic energy gap of magnitude Δ_0.

Since the superconductor has a finite gap Δ_0, electrons from another system cannot tunnel into the superconductor unless their energy is larger than Δ_0. On the other hand, pairs of electrons can tunnel into the superconductor since they have a finite amplitude to become part of the superconducting pair condensate. Thus, when a superconductor is placed next to a nearby electronic system, spin-singlet pairs of electrons with total momentum zero can tunnel into the superconductor and, similarly, spin-singlet pairs of electrons of the superconductor can tunnel into the Weyl-fermion surface states. By this mechanism, known as the superconducting proximity effect, the electrons of the nearby system themselves become superconducting. This process cannot affect the bulk states of the nearby system, but affects only those states which are sufficiently close to the superconductor, a distance that typically is of the order of the superconducting coherence length ξ_0. If the nearby system is a topological insulator, its bulk states cannot become superconducting. However, its surface states can become superconducting since they are gapless and confined to the surface.

The proximity-effect coupling we have just discussed is described by an extra term of the Hamiltonian that describes the spin-singlet Cooper-pair tunneling process (Schrieffer, 1964; de Gennes, 1966), which has the form

$$\mathcal{H}_{\text{tunnel}} = \Delta \psi_\uparrow^\dagger(x) \psi_\downarrow^\dagger(x) + \text{h.c.} \tag{16.202}$$

where $\psi_\uparrow(x)$ and $\psi_\downarrow(x)$ are the two components of the Weyl spinor. Here we assume that the s-wave superconducting condensate Δ is fixed by the bulk superconductor and is unaffected by the surface states, although in reality there may be a suppression of Δ near the surface of the superconductor. Nevertheless, what matters in what follows is that there is a finite value of Δ at the surface even if it is smaller than in the bulk.

We will use the Nambu four-component spinor notation,

$$\Psi = \left((\psi_\uparrow, \psi_\downarrow), (\psi_\downarrow^\dagger, -\psi_\uparrow^\dagger) \right)^{\text{T}} \tag{16.203}$$

and write the full Hamiltonian density of the surface states in the form (Fu and Kane, 2008)

$$\mathcal{H} = \Psi^\dagger \left[-i v_{\text{F}} \tau_3 \vec{\sigma} \cdot \vec{\nabla} - \mu \tau_3 + \Delta_0 (\tau_1 \cos\phi + \tau_2 \sin\phi) \right] \Psi \tag{16.204}$$

where τ_1, τ_2, and τ_3 are (as before) the three Pauli matrices that act on the upper, ψ, and lower, $\psi^\dagger$, Weyl components of the Nambu 4-spinor. It is easy to see that this is the chiral basis for 4×4 gamma matrices with $\gamma_0 = \mathbb{I} \otimes \tau_1$, $\gamma_1 = -i\sigma_1 \otimes \tau_2$,

$\gamma_2 = -i\sigma_2 \otimes \tau_2$, $\gamma_3 = -i\sigma_3 \otimes \tau_2$, and $\gamma_5 = \mathbb{I} \otimes \tau_3$. Time-reversal is represented by the operator $\Theta = i\sigma_2 \mathcal{K}$, where $\mathcal{K}$ is complex conjugation, and commutes with the Hamiltonian, $[\Theta, \mathcal{H}] = 0$.

When Δ is constant in space, the excitation spectrum in

$$E(\vec{k}) = \pm\sqrt{(\pm v_{\rm F}|\vec{k}| - \mu)^2 + \Delta_0^2} \tag{16.205}$$

For $\mu \gg \Delta_0$ the low-energy spectrum is the same as that of a $p_x + ip_y$ superconductor, see Section 14.9. This can also be seen by defining the field operators

$$c(\vec{k}) = \frac{1}{\sqrt{2}}\left(\psi_\uparrow(\vec{k}) + e^{i\,\arg(\vec{k})}\psi_\downarrow(\vec{k})\right) \tag{16.206}$$

where $\arg(\vec{k}) = \tan^{-1}(k_y/k_x)$. If we project the Hamiltonian of Eq. (16.204) to act only on the subspace defined by the operators $c(\vec{k})$, we readily find that the projected Hamiltonian becomes

$$H = \int \frac{d^2k}{(2\pi)^2}\Big[(v_{\rm F}|\vec{k}| - \mu)c^\dagger(\vec{k})c(\vec{k}) + \frac{1}{2}(\Delta e^{i\,\arg(\vec{k})}c^\dagger(\vec{k})c^\dagger(-\vec{k}) + \text{h.c.})\Big] \tag{16.207}$$

which is the Hamiltonian of a $p_x + ip_y$ superconductor for spinless fermions. Fu and Kane further noted that the full Hamiltonian of Eq. (16.204) is invariant under time reversal, whereas the projected Hamiltonian is not.

In Section 14.9.3 we saw that an $hc/(2e)$ vortex in a $p_x + ip_y$ superconductor has an exact Majorana zero mode in its core. We will now see that the junction of an s-wave superconductor with the Weyl surface states of a 3D $\mathbb{Z}_2$ topological insulator also has vortices that have an exact Majorana zero mode in their cores. This problem turns out to be essentially equivalent to the problem of a vortex of a relativistic charged scalar field coupled to Weyl fermions studied long ago by Jackiw and Rossi (1981). The precise mapping between the two problems can be found in the work of Chamon and collaborators (Chamon *et al.*, 2010; Nishida *et al.*, 2010). Let us consider a configuration of the order parameter $\Delta(\vec{r})$ that corresponds to a vortex. In polar coordinates $\vec{r} = (r, \varphi)$ (with $0 \leq \varphi < 2\pi$), the vortex configuration has the form

$$\Delta_\pm(r, \varphi) = \Delta(r)e^{\pm i\varphi} \tag{16.208}$$

where $\lim_{r\to\infty} \Delta(r) = \Delta_0$ in the far field, and $\lim_{r\to 0} \Delta(r) = 0$ deep in the core. Here $\Delta_\pm(r, \theta)$ corresponds to the vortex (+) and to the anti-vortex (−). It is easy to see that the vortex has the exact zero mode (on setting $v_{\rm F} = 1$)

$$\eta_0^+(r,\theta) \sim \exp\left(-\int_0^r dr'\ \Delta_0(r')\right)\begin{pmatrix}0\\ i\\ 1\\ 0\end{pmatrix} \tag{16.209}$$

whereas for the anti-vortex the zero mode has the same radial dependence but the spinor now is $(1, 0, 0, -i)^{\mathrm{T}}$.

These results led Fu and Kane to state, following the arguments of Read and Green (2000) and of Ivanov (2001), that the vortices in this system are non-abelian anyons that exhibit "Ising" fusion rules. Hence, these vortices have the same behavior as that of the half-vortices that we discussed extensively in Section 14.9. This result showed that "hybrid" structures of superconductors and topological insulators offer an alternative approach by means of which to construct topological qubits!

Fu and Kane also considered other geometries. A particularly interesting case was a hybrid Josephson line junction of two superconductors deposited on top of a 3D topological insulator, which also supplied the material for the barrier in the line junction. They showed that, if the phases of the order parameters of the two superconductors differ by π, a gapless Majorana *field* propagates along the line junction. In a recent experiment on a device of this type, made with a Bi_2Se_3 topological insulator as a "substrate," it was found that the current–phase relation of the Josephson line junction has a sharp resonance when the phase difference between the superconducting leads is exactly equal to π (Williams *et al.*, 2012). This feature is suppressed by a low magnetic field, in contradiction with the behavior of conventional Josephson junctions, but in agreement with the predictions of Fu and Kane (2008).

The Fu–Kane result led to an explosion of work whose goal is to find simple condensed matter systems that may support Majorana zero modes. Some time earlier Kitaev (2001) had considered a simple model of a mean-field theory of spinless fermions in one dimension with p-wave superconductivity. In one dimension p-wave simply means that the order parameter is odd under parity. The model considered by Kitaev is a chain of N sites with a spinless fermion operator c_j and its adjoint $c_j^\dagger$ defined at every site. The chain has open boundary conditions. The Hamiltonian of this problem is

$$H = -\sum_{j=1}^{N}\mu c(j)^\dagger c(j) - \sum_{j=1}^{N-1}\left(c(j)^\dagger c(j+1) + |\Delta| e^{i\phi} c(j)c(j+1) + \text{h.c.}\right) \tag{16.210}$$

Here μ is the chemical potential, t is the hopping term, $|\Delta|$ is the amplitude of the superconducting order parameter, and ϕ is its phase. Since the superconducting order parameter is defined on the bonds of the 1D chain, it is odd under parity, defined as inversion with respect to a site of the chain.

The alert reader should at once recognize that this problem is essentially the same as the fermion representation of the quantum Ising chain discussed in Section 5.5. Much as in the case of the 1D quantum Ising model, in this system (as in all pairing Hamiltonians) fermion number is not conserved, but it is conserved modulo 2, i.e. the parity of the fermion number is conserved. Here too this system is equivalent to a system of Majorana fermions. In the simpler case of $\mu = 0$ and $t = |\Delta|$ the (Dirac) fermion operator can be decomposed into its two Majorana components. Using the same notation as in Section 5.5, we define two Majorana fermions $\chi_1(j)$ and $\chi_2(j)$, which are self-adjoint fermions, $\chi_a(j)^\dagger = \chi_a(j)$ ($a = 1, 2$), and satisfy the Majorana anticommutation relations $\{\chi_a(j), \chi_b(l)\} = 2\delta_{ab}\delta_{jl}$. The Dirac fermion is expressed in terms of the Majorana fermions as

$$c(j) = \frac{1}{2}e^{-i\phi/2}\Big(\chi_1(j) + i\chi_2(j)\Big), \qquad c^\dagger(j) = \frac{1}{2}e^{+i\phi/2}\Big(\chi_1(j) - i\chi_2(j)\Big) \tag{16.211}$$

In terms of the Majorana fermions the Hamiltonian of Eq. (16.210) (for $\mu = 0$ and $t = |\Delta|$) becomes

$$H = -it\sum_{j=1}^{N-1} \chi_2(j)\chi_1(j+1) = 2t\sum_{j=1}^{N-1} d^\dagger(j)d(j) \tag{16.212}$$

where we have defined a new Dirac fermion $d(j) = \frac{1}{2}(\chi_1(j+1) + i\chi_2(j))$ by combining one Majorana component on one site with the other Majorana component on the next site. Clearly, this system has local excitations of energy $2t$. Hence the bulk states are gapped. However, in this construction the Majorana fermions $\chi_1(1)$ and $\chi_2(N)$ are absent. We can now combine these two Majorana fermions into a single boundary Dirac fermion $d_0 = \frac{1}{2}(\chi_1(1) + i\chi_2(N))$. The states $|0\rangle$ and $|1\rangle = d^\dagger|0\rangle$ are zero-energy states, *zero modes*, of the Hamiltonian. Hence this system has two degenerate ground states with exactly zero energy, i.e. a *qubit*. The reader can check that what we have described is the Ising chain with open (instead of periodic) boundary conditions deep in the broken-symmetry states, $\lambda \gg 1$. These zero modes are analogous to the zero modes in polyacetylene except that here they are Majorana fermions and do not carry charge. Alicea and his collaborators showed that this simple model proposed by Kitaev can be realized physically in a 1D semiconductor with strong spin–orbit interactions coupled to an s-wave superconductor. This system has Majorana zero modes at the endpoints of the wire (Alicea *et al.*, 2011).

We close this chapter with a brief discussion of topological fluids emerging from topological insulators. These are open problems and we will merely touch on some important questions without attempting to give a thorough description. This transition can happen in several possible ways. One possibility is to consider changing the chemical potential in a topological insulator so that it becomes a metal. The question is whether the topological properties of the electronic bands of the single particle states imply that the resulting superconductor should be topological. By topological here we mean that the single-particle states of the Bogoliubov–de Gennes equations of the superconductor have topological properties similar to those of the "parent" insulator. Consideration of this problem lead to the realization that the quasiparticle states of the superconductors can also have non-trivial topology (Schnyder *et al.*, 2008; Kitaev, 2009; Roy, 2010; Ryu *et al.*, 2010). However, it turns out that, while this scenario is certainly possible, it is by no means necessary. Thus, the metal obtained by doping a topological insulator can become either a standard superconductor or a topological one.

A very exciting pathway from a topological insulator to a topological phase is via a fractional topological insulator, the fractionalized version of the quantum anomalous Hall state (and of the quantum spin Hall state). If such a state were to exist, it would be a topological fluid in the same sense as the fractional quantum Hall states. Although we do not know (yet) of any experimental examples of this fractionalized phase, there are models with local interactions that have been shown to do the job. Several authors (Neupert *et al.*, 2011; Sheng *et al.*, 2011; Tang *et al.*, 2011a) have looked at models of topological insulators in a regime in which one of the bands with non-trivial topology becomes essentially "flat." They showed that if this flat band with non-trivial Chern number is partially filled the system can develop a spontaneous anomalous quantum Hall state with a fractional Hall conductance. In other words, this system has a fractional quantum Hall state in the absence of a magnetic field. In this case time-reversal invariance is broken spontaneously. In these states, and for the same reasons as in the fractional quantum Hall effect, the electron becomes fractionalized. The new feature here is that the emerging fractional quantum Hall state is the result of spontaneous breaking of time-reversal invariance in the absence of an external magnetic field. Clearly, something very similar can happen in the case of the quantum spin Hall state.

In two dimensions these fractionalized phases are close relatives of the fractional quantum Hall states. Thus, it is possible to write down an effective hydrodynamic field theory of the form of a multi-component Chern–Simons gauge theory as discussed in detail in Section 14.2 (Levin and Stern, 2009; Cho and Moore, 2011; Santos *et al.*, 2011). Time-reversal invariance requires that there should not be a Hall current, a condition that is met by any real symmetric and traceless K-matrix. For instance, in the case of the time-reversal-invariant fractional quantum spin Hall states the effective theory must have at least two components and its K-matrix must be traceless. In its simpler form the K-matrix has the form

$$K = \begin{pmatrix} 0 & k \\ k & 0 \end{pmatrix} \tag{16.213}$$

which is both traceless and real. With a matrix of this form the Chern–Simons action implies that the (hydrodynamic) flux of one component couples to the hydrodynamic gauge field of the other component and hence has the form of a BF theory of the type we discussed in Sections 14.1.1 and 14.5. This theory is equivalent to a state of a system of two types of fermions with opposite "charges" (actually spin projections) each at filling fraction $1/k$. Since the absolute value of the determinant of the K-matrix is k^2, this topological fluid has degeneracy k^2 on a torus.

What is still missing is a general microscopic theory of these fractionalized phases. Nonetheless, some simple microscopic lattice models that exhibit some of these key features have been constructed (Levin *et al.*, 2011), and many properties of the wave functions for the fractionalized states in 2D have been investigated (Qi, 2011). The extension of these ideas to 3D topological insulators is far less obvious, since the notion of fractional statistics for particles itself cannot be extended to three spatial dimensions because in three dimensions the braid group is trivial (although it may be applicable for extended objects such as strings and domain walls). Nevertheless, the notion of statistical transmutation of bosons into fermions, and vice versa, is still meaningful. A famous example is the case of a spinless particle of charge e that, bound to a Wu–Yang magnetic monopole, becomes a spin-$1/2$ particle, a problem we have already discussed in Section 13.1.2. From general principles of local quantum field theory, we expect that if the bound state is described by a local field theory then this particle should be a fermion and should be described by a Dirac field (Wu and Yang, 1975, 1976; Witten, 1979).

17
Quantum entanglement

17.1 Classical and quantum criticality

In most cases the phases of quantum field theories, in particular those of interest in condensed matter physics, can be described in terms of the behavior of local observables, such as order parameters or currents that transform properly under the symmetries of the theory. Quantum and thermal phase transitions are characterized by the behavior of these observables as a function of temperature and of the coupling constants of the theory. The phase transitions themselves, quantum or thermal, are classified into universality classes, which are represented by the critical exponents which specify the scaling laws of the expectation values of the observables. Historically, the development of this approach to critical behavior goes back to the Landau theory of critical behavior. It acquired its most complete form with the development of the renormalization group (RG) in the late 1960s and early 1970s. It is the centerpiece of Wilson's approach to quantum field theory, in which all local quantum field theories are defined by the scaling regime of a physical system near a continuous phase transition. From this point of view there is no fundamental difference between classical (or thermal) phase transitions, which are described by the theory of classical critical behavior, and quantum phase transitions.

For example, the expectation value of a local order parameter $\mathcal{M}$ as the thermal phase transition is approached from below behaves as $\mathcal{M} \sim (T_c - T)^\beta$. Here T_c is the critical temperature and β is a critical exponent that depends on the universality class of the thermal phase transition and on the dimensionality of space. While quantum mechanics can play a key role in the existence of the ordered phase, e.g. superfluidity and superconductivity are macroscopic manifestations of essentially quantum-mechanical phenomena, the thermal transition itself is governed entirely by classical statistical mechanics, and quantum mechanics plays a role in setting the value of non-universal quantities such as the critical temperature, etc. On the other hand, in the case of a quantum phase transition, the order parameter $\mathcal{M}$ has a similar scaling behavior as a function of the coupling constant, $\mathcal{M} \sim (g_c - g)^{\tilde{\beta}}$, where g is the coupling constant, g_c is the critical coupling constant, and $\tilde{\beta}$ is a critical exponent that depends on the universality class of the quantum phase transition. Here we assume that $\mathcal{M}$ has a non-vanishing expectation value only for $g < g_c$.

However, as is apparent in the description of a quantum phase transition in terms of a path integral in imaginary time, quantum phase transitions look like classical phase transitions in a space with an extra dimension, imaginary time. The main difference between the two types of transitions is the existence of a dynamical critical exponent z that specifies how space and time scale in the quantum case. This scaling law is absent in the theory of equilibrium classical critical phenomena. In the case of quantum systems that obey relativistic invariance the dynamical exponent $z = 1$, and, in these cases, the quantum transition is literally a classical transition in a space with one additional dimension. Thus, near a continuous quantum phase transition the correlation length diverges with a universal exponent ν as $\xi \sim |g - g_c|^{-\nu}$, while the energy gap G vanishes at the same quantum phase transition as $G \sim |g - g_c|^{\nu z}$.

Gauge theories, which we discussed in other chapters in connection with the physics of the topological fluids and spin liquids, have a similar description. The crucial difference is that in the case of gauge theory the local nature of the gauge symmetry requires the consideration of non-local gauge-invariant observables, such as the Wilson loop and the 't Hooft loop. As we saw in Chapter 9, the Wilson loop operators have distinct behaviors in different phases of the theory, and obey an area law in a confining phase and a perimeter law in a deconfined phase. Hence, near a continuous phase transition the energy gap of the gauge theory, which is set by the string tension in the confining phase and enters into the definition of the area law, vanishes as a function of the coupling constant with a universal critical exponent.

In Chapter 9 we saw that gauge theories with a compact gauge group have non-trivial topological properties in their deconfined phases. Likewise, Chern–Simons gauge theory is a topological field theory, see Chapter 10, and we used it extensively in the field-theoretic description of the fractional quantum Hall fluids. In topological field theories the expectation values of the observables depend only on the topological properties of the observables, such as the linking number of the Wilson loops, and on the global topology of the space. In condensed matter physics topological field theories describe the effective low-energy and long-distance behavior in a topological phase. The best-understood example of a system in a topological phase is a fractional quantum Hall fluid, see Chapters 13 and 14. The consideration of topological phases, and of topological field theories, led us to describe the excitations by their transformation properties under the braid group, and led to the concept of abelian and non-abelian fractional statistics. More generally, we also had to consider the quantum dimensions of the excitations and their fusion rules.

Although in all cases of interest the quantum phase transition itself can be described in terms of the behavior of local operators, the existence of topological phases tells us that some important features of the theory cannot be reduced to local physics. This raises the question of whether perhaps even at the quantum phase transition some global non-local properties of the quantum field theory may be important. The topological properties of the topological phases have a quantum origin and do not have a counterpart in classical systems. This raises the need to consider other tools to describe topological phases and their phase transitions.

The information on all the static properties of a physical system is stored in its ground-state wave function. This naturally includes the behavior of the equal-time correlators of local operators. Since we can compute their properties directly, we did not need to investigate the properties of the ground-state wave functions. In general, the wave functions do not by themselves play a great role since the wave functions of thermodynamically large systems contain all kinds of information, much of which is highly non-universal and depends on all the microscopic details. In almost all cases very little of the nature of the physical system would change if these microscopic details were to be modified. In addition, it is often more difficult to understand the physical properties of a system by staring at an expression of the wave function than to look at the correlators of the local observables. Of course, if we know the behavior of the observables, we know the information that is encoded in the wave functions. One may nevertheless ask whether there is important physical information encoded in the wave functions that might not be readily accessible via properties gleaned from the behavior of local operators. For instance, the wave function for the ground state of the spin-$1/2$ quantum antiferromagnetic Heisenberg chain obtained using the Bethe ansatz (see Chapter 5) has been known since Hans Bethe derived it in 1931. However, the physics that was encoded in this wave function could not be understood until the properties of correlators of local observables were found many years later, beginning in the 1970s.

On the other hand, the wave functions that describe systems in topological phases, e.g. the Laughlin wave function of the fractional quantum Hall effect, have manifestly universal properties that are closely related to the properties and characterization of these phases. We saw in Chapters 13 and 14 that the wave functions of topological phases encode universal non-local properties of these systems such as the braiding fractional statistics of their excitations. Therefore, at least in these cases, we need an understanding of non-local properties of the ground-state wave functions.

17.2 Quantum entanglement

The natural tool to investigate the non-local properties stored in the ground-state wave functions is the concept of quantum entanglement and the associated entanglement entropy. In this chapter we will define what is meant by quantum entanglement and its measures, focusing on the entanglement entropy. The concept of quantum entanglement was born with quantum mechanics itself, and predates our motivations by many decades (Einstein *et al.*, 1935).

Historically the concept of quantum entanglement was formulated to describe how a finite quantum system is coupled to its essentially infinite environment. The prime example is the non-local information stored in the wave functions of systems of identical particles. Consider, for instance, a system of two identical spin-$1/2$ particles, A and B. The wave function of a spin-singlet state, with total spin $S = 0$ and $S_z = 0$, is $|0, 0\rangle = (1/\sqrt{2})(|\uparrow, \downarrow\rangle - |\downarrow, \uparrow\rangle)$. Since this wave function does not contain any length scale, the physical size of the spin singlet can be as large as we want. If we *measure* the spin of particle A, say on Earth, and find that it is $\uparrow$ then, provided that *we know* a priori that the state is indeed a spin singlet, we *know* that the spin of particle B is $\downarrow$ even if particle B is, let us say, on the Moon. On the other hand, if the two spins are in a product state, such as $|1, \pm 1\rangle$, the measurement of the spin of particle A to be $\uparrow$ implies that the spin of particle B is $\uparrow$ only if we know both that $S = 1$ and that $S_z = 1$. For this reason we say that the spins are entangled in the singlet state, while in the product state they are not.

A measure of the degree of entanglement of a quantum state is the von Neumann entanglement entropy, which is defined as follows. Let us consider a partition of a physical system Σ into two disjoint subsystems that we will label by A and B. Hence, $\Sigma = A \bigcup B$ and the two subsystems have a vanishing intersection, $A \bigcap B = \emptyset$. Let $\mathcal{H}_A$ and $\mathcal{H}_B$ be the Hilbert spaces of the states with separate support in system A and in system B such that the Hilbert space of the states on Σ is $\mathcal{H}_\Sigma = \mathcal{H}_A \oplus \mathcal{H}_B$. Let $|\Psi\rangle$ be a pure quantum state of the system on $A \bigcup B$. As such it can be decomposed as

$$|\Psi\rangle = \sum_{m,n} M_{m,n} |\psi_n^A\rangle |\psi_m^B\rangle \tag{17.1}$$

where $\{|\psi_n^A\rangle\}$ and $\{|\psi_m^B\rangle\}$ are orthonormal basis states of $\mathcal{H}_A$ and $\mathcal{H}_B$, respectively, and $M_{n,m}$ are the matrix elements of an (in general) rectangular matrix $\mathbf{M}$. However, using the singular-value-decomposition theorem, a rectangular matrix can always be written as a product of a unitary matrix $\mathbf{U}$, a diagonal matrix $\mathbf{D} = \text{diag}(\lambda_1, \ldots, \lambda_n, \ldots)$, and a rectangular matrix $\mathbf{V}$ (whose rows are orthonormal vectors). Then, after going to the new bases, $|\psi_n^A\rangle \to \mathbf{U}|\psi_n^A\rangle$ and $|\psi_m^B\rangle \to \mathbf{V}|\psi_m^B\rangle$, we find the Schmidt decomposition of the state vector $|\Psi\rangle$,

$$|\Psi\rangle = \sum_{n}^{D} \lambda_n |\psi_n^A\rangle |\psi_n^B\rangle \tag{17.2}$$

where $D = \min\{d_A, d_B\}$, with d_A and d_B being the dimensions of the Hilbert spaces $\mathcal{H}_A$ abd $\mathcal{H}_B$. Also, if the state vector $|\Psi\rangle$ is normalized to unity, $||\Psi|| = 1$, then the set of (generally complex) numbers $\{\lambda_n\}$ must satisfy the sum rule

$$\sum_{n}^{D} |\lambda_n|^2 = 1 \tag{17.3}$$

Given the pure state $|\Psi\rangle$ of the total system $A \bigcup B$, its (trivial) density matrix is

$$\rho_{A \bigcup B} = |\Psi\rangle\langle\Psi| \tag{17.4}$$

We can now define the reduced density matrix for subsystem A to be the partial trace of $\rho_{A \bigcup B}$ over the degrees of freedom in B,

$$\rho_A = \mathrm{tr}_B \rho_{A \bigcup B} \tag{17.5}$$

and similarly for the reduced density matrix ρ_B. Therefore, if we observe only the subsystem A, it is in a mixed state defined by the reduced density matrix ρ_A, and similarly for B.

The von Neumann entanglement entropy S_A for subsystem A, when the total system is in state $|\Psi\rangle$, is defined to be the entropy of the reduced density matrix,

$$S_A \equiv -\mathrm{tr}_A(\rho_A \ln \rho_A) \tag{17.6}$$

(For historical reasons in the quantum-information literature the von Neumann entropy uses $\log_2$ instead of the standard natural logarithm used in statistical mechanics.)

Using the Schmidt decomposition, Eq. (17.2), we can write the reduced density matrix ρ_A as

$$\rho_A = \sum_{n}^{D} |\lambda_n|^2 |\psi_n^A\rangle\langle\psi_n^A| \tag{17.7}$$

and similarly for B. Hence, the quantities $p_n \equiv |\lambda_n|^2$ represent the probability of observing the subsystem A in the state $|\psi_n^A\rangle$. From these expressions it follows that the reduced density matrices ρ_A and ρ_B have the same non-zero eigenvalues, that they are given by $p_n = |\lambda_n|^2$, and that both reduced density matrices have unit trace, $\mathrm{tr}_A \rho_A = \mathrm{tr}_B \rho_B = 1$.

It also follows that the von Neumann entanglement entropy can be written as

$$S_A = -\mathrm{tr}_A(\rho_A \ln \rho_A) = -\sum_{n}^{D} |\lambda_n|^2 \ln|\lambda_n|^2 = -\mathrm{tr}_B(\rho_B \ln \rho_B) = S_B \tag{17.8}$$

In other words, the entanglement entropy is symmetric in the two (entangled) subsystems. This symmetry property is a consequence of our assumption that the total system $A \bigcup B$ is in a pure state $|\Psi\rangle$. Conversely, the symmetry property of the entanglement entropy does not hold if the total system is in a mixed state, e.g. in a thermal state defined by a Gibbs density matrix. The expression of the von Neumann entanglement entropy in terms of the probabilities $\{|\lambda_n|^2\}$ also tells us that the entanglement entropy can vanish only if the reduced density matrix ρ_A (and hence also ρ_B) itself represents a pure state. This can hold only if the state Ψ of the system on $A \bigcup B$ is a *product state*. In this case the reduced density matrix is diagonal and equal to $\rho_A = \text{diag}(1, 0, \ldots, 0)$, which has a vanishing entropy. Therefore the von Neumann entropy is a measure of the entanglement of the two subsystems A and B in state $|\Psi\rangle$. The spectrum of (Schmidt) eigenvalues $\{|\lambda_n|^2\}$ of the reduced density matrix is known as the entanglement spectrum.

Finally, the von Neumann entanglement entropy satisfies the following properties. For two regions, A and B, the entanglement entropy is subadditive, i.e. $S_{A\cup B} \leq S_A + S_B$. In addition, for three regions, A, B, and C, the von Neumann entanglement entropy satisfies the condition of strong subadditivity (of Lieb and Ruskai (1973)), i.e. $S_{A\cup B\cup C} \leq S_{A\cup B} + S_{B\cup C} - S_B$.

17.3 Entanglement in quantum field theory

For the rest of this chapter we will focus on entanglement entropy, how it is computed, and how it behaves in different systems of interest. We will see that the entanglement entropy is a very efficient and powerful tool to characterize the physics of topological phases in condensed matter and in quantum field theory. We will also discuss in detail the scaling laws obeyed by the entanglement entropy.

How do we compute the entanglement entropy in an extended, macroscopic, system such as a quantum field theory? In systems with few degrees of freedom it is relatively straightforward to construct the reduced density matrices directly from their wave functions. However, except for a few exceptional cases, the wave functions of systems with an infinite number of degrees of freedom are in almost all circumstances forbiddingly complex. There are a few exceptional systems in which the wave functions are known explicitly and have universal properties. However, even for these cases the computation of the entanglement entropy is non-trivial.

For definiteness let us consider a field theory with a Euclidean action $S[\phi] = \int_\Omega d^D x \ \mathcal{L}[\phi]$, where ϕ is some field and $\mathcal{L}$ is the Lagrangian density. Here Ω is some D-dimensional space-time manifold. In the Euclidean theory the imaginary-time coordinate τ is periodic (compactified) with period $\beta = 1/T$, while the space directions are for now arbitrary. Thus, we will generally refer to the space-time Ω

as a cylinder of circumference $\beta = 1/T$. What follows does not depend on whether the theory is relativistic or not.

The Gibbs density matrix of the system is $\rho = \exp(-\beta H)$, where H is the quantum Hamiltonian and $\beta = 1/T$. The partition function is

$$Z = \text{tr}\,\rho = \text{tr}\,e^{-\beta H} = \int \mathcal{D}\phi\, e^{-\int_\Omega d^D x\, \mathcal{L}[\phi]} \tag{17.9}$$

In the Euclidean metric the imaginary-time τ evolution operator is $U(\tau) = \exp(-\tau H)$. For large τ the Euclidean evolution operator $U(\tau)$ projects any initial state onto the ground-state $|0\rangle$. Therefore, the ground-state wave function $\Psi_0[\phi]$ is expressed in terms of a path integral by a similar-looking expression (see Fig. 17.1),

$$\Psi_0[\phi] = \langle\phi|0\rangle = \frac{1}{Z}\int \mathcal{D}\phi\, e^{-\int_\Sigma d^D x\, \mathcal{L}[\phi]} \tag{17.10}$$

where the manifold Σ has a boundary at some (imaginary) time $\tau = 0$ where we specify that the state is given by the field configuration, $|\phi\rangle$, and where we have taken the zero-temperature limit $T \to 0$ or, which amounts to the same thing, made the time dimension infinite in size, $\beta \to \infty$. This last step projects onto the ground state $|0\rangle$, assuming that it is unique. If the ground state is not unique, the evolution will project onto a state that is the linear superposition of the degenerate

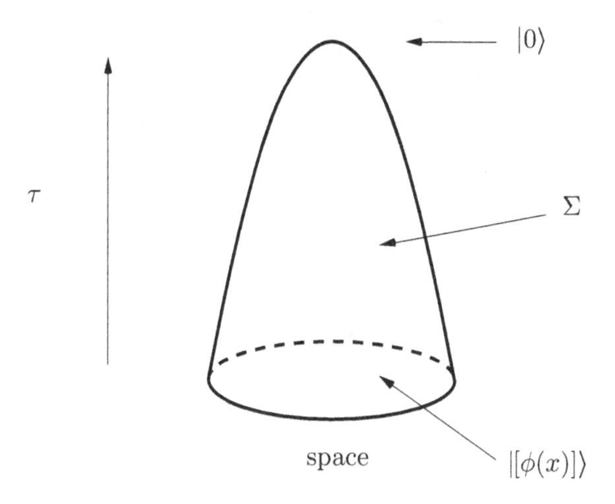

Figure 17.1 The path-integral picture of the wave function as the amplitude for an evolution from some initial state $|[\phi(x)]\rangle$ to the vacuum state $|0\rangle$ of the system. The initial state is a field configuration defined on the boundary of the open space-time manifold Σ. The boundary is a spatial manifold at the initial imaginary time slice $\tau = 0$ of the (Euclidean) evolution.

vacua (ground states), unless of course the degeneracy is explicitly lifted. Similarly, the matrix element of the density matrix ρ between states $|\phi(x)\rangle$ and $|\phi'(x)\rangle$ is

$$\langle\phi(x)|\rho|\phi'(x)\rangle$$
$$= \frac{1}{Z}\int \mathcal{D}\phi \;\prod_x \delta(\phi(x,\tau=0)-\phi(x))\prod_x \delta(\phi(x,\tau=\beta)-\phi'(x))e^{-S_{\mathrm{E}}[\phi]} \tag{17.11}$$

where $S_{\mathrm{E}}[\phi]$ is the Euclidean action of the field theory on a manifold that is a strip of width β with the specified boundary conditions.

We can use the same approach to, at least formally, find a path-integral expression for the reduced density matrix. So, once again we will consider a partition of the spatial manifold into two disjoint sets A and B. The reduced thermal density matrix ρ_A is then obtained by tracing over the degrees of freedom in its complement, region B. Hence we will require that in region B the initial and final states are the same and summed over, and the imaginary-time coordinate is periodic with period β. Thus, in region B the strip is wrapped into a cylinder of circumference β. Let $|[\phi_A(x)]\rangle$ and $|[\phi'_A(x)]\rangle$ be two field configurations with support in region A. The matrix elements of the reduced density matrix for region A are obtained by computing a trace of the full density matrix restricted to the states in region B. However, in region A the states evolve from the state $|[\phi_A(x)]\rangle$ to $|[\phi'_A(x)]\rangle$. In other words, the new manifold is smooth and periodic on region B but has a cut in region A with a discontinuity expressed in terms of the two configurations specified by the states $|[\phi_A(x)]\rangle$ and $|[\phi'_A(x)]\rangle$ (see Fig. 17.2),

$$\langle[\phi_A(x)]|\rho_A|[\phi'_A(x)]\rangle = \langle[\phi_A(x)]|\mathrm{tr}_B\rho|[\phi'_A(x)]\rangle \tag{17.12}$$

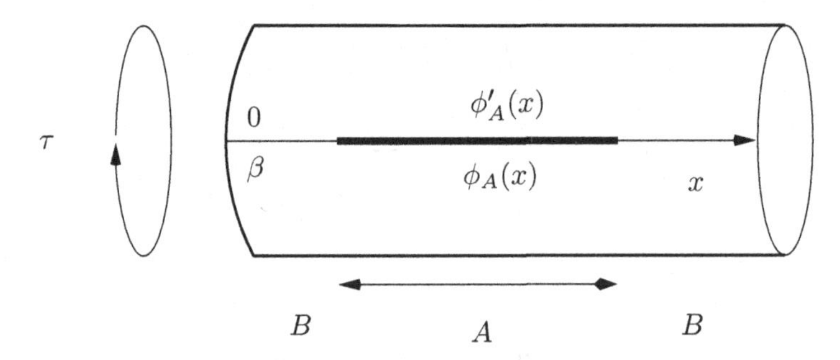

Figure 17.2 The path integral for the reduced density matrix ρ_A is defined on a manifold with the topology of a cylinder with a cut, the dark line shown in the figure. Here $\phi_A(x)$ and $\phi'_A(x)$ are two field configurations on region A. The origin and end of imaginary time τ are at $\tau = 0$ and $\tau = \beta$.

To compute the entanglement entropy we will use the "replica trick." To this end, we compute first

$$Z_n[A] = \mathrm{tr}_A \rho_A^n \tag{17.13}$$

which is defined for all positive integers n. In terms of the spectrum of eigenvalues of the reduced density matrix, $\{\lambda_k\}$ (the entanglement spectrum), it becomes

$$Z_n[A] = \sum_k^D \lambda_k^n \tag{17.14}$$

Since the spectrum of eigenvalues lies within the interval $[0, 1]$ and obeys the normalization condition $\sum_k^D \lambda_k = 1$, the trace (defined by $Z_n[A]$) is absolutely convergent and analytic for $\mathrm{Re}\, n > 1$. Therefore, $Z_n[A]$ can be extended by analytic continuation to the complex-n plane as $Z[n, A]$.

The von Neumann entanglement entropy S_A is computed by taking the limit (Callan and Wilczek, 1994; Calabrese and Cardy, 2004)

$$S_A = -\mathrm{tr}(\rho_A \ln \rho_A) = -\lim_{n\to 1^+} \frac{\partial}{\partial n} \mathrm{tr}_A \rho_A^n \tag{17.15}$$

We will also define the Rényi entropies S_A^n,

$$S_A^n = \frac{1}{1-n} \ln \mathrm{tr}\, \rho_A^n \tag{17.16}$$

Since we are taking traces of matrix products, it is clear that the final state in region A of one factor is the initial state of the next factor (also in region A). The overall trace means that the final state of the last factor is identified with the initial state of the first factor. This means that the manifold on which the path integral is computed can be regarded as an n-sheeted Riemann surface in region A with overall imaginary-time span $n\beta$ glued to cylinders defined for region B, each with imaginary-time span β.

Alternatively, we can think of this calculation as involving n identical copies of the field theory, and hence having n copies or replicas of the fields ϕ_i ($i = 1, \ldots, n$), each with the same action such that in region B the replicas obey periodic boundary conditions in imaginary time separately, whereas in region A they are identified with each other sequentially and cyclically. More specifically, the replicated fields obey the boundary conditions

$$\begin{aligned} \phi_j(0^+, x) &= \phi_{j+1}(0^-, x), \quad \text{for } x \in A \\ \phi_j(0^+, x) &= \phi_j(0^-, x), \quad \text{for } x \in B \end{aligned} \tag{17.17}$$

A consequence of the boundary conditions that specify how the n replicas are glued to each other is that there exists a set of twist operators that act on the replicated theory and identify the different copies at the cuts. At the common boundary between regions A and B, the fields have a *conical singularity*. This formulation is originally due to Holzhey, Larsen, and Wilczek (Holzhey *et al.*, 1994), and was developed and extended by Calabrese and Cardy (2004, 2009), whose work we follow here.

Following Calabrese and Cardy, we define a partition function with n copies of the field, obeying the boundary conditions of Eq. (17.17). The action of the replicated fields is simply the sum of the actions for each copy,

$$S[\phi_1, \ldots, \phi_n] = \sum_{i=1}^{n} S[\phi_i] \tag{17.18}$$

The path integral, the partition function, is then restricted by the condition that the local fields $\{\phi_i\}$ are glued according to the boundary conditions of Eq. (17.17). The replicated theory thus defined now has a global symmetry since it is invariant under the exchange of the replicas. Local fields that obey these conditions are called *twist fields*. With the particular structure of the boundary conditions we are using, they are called branch-point twist fields, and are associated with two cyclic permutation symmetries, $\sigma : j \to j+1 \pmod n$ and $\sigma^{-1} : j+1 \to j \pmod n$ (with $j = 1, \ldots, n$). The branch-cut twist fields that map the different copies across the branch cuts are denoted by $\mathcal{T}_n \equiv \mathcal{T}_\sigma$, associated with the permutation $\sigma : j \to j+1 \pmod n$, and $\tilde{\mathcal{T}}_n \equiv \mathcal{T}_{\sigma^{-1}}$, associated with the (inverse) permutation $\sigma^{-1} : j+1 \to j \pmod n$. Thus, the reduced density matrix, which we defined as the partition function with a branch cut on every point along the boundary between regions A and B, amounts to a partition function in the replicated theory with insertions of the branch-cut twist fields along the boundary between the two regions. A detailed treatment of this approach can be found in the work of Cardy and coworkers (Cardy *et al.*, 2007). This formulation is particularly powerful in $(1+1)$ dimensions.

17.4 The area law

So far, much of the discussion in this chapter has been on the formal aspects of how entanglement and entanglement entropy are defined. As the reader can easily see, the expressions we have presented are rather formal. They are also very non-local. Although there are many reasons for being interested in entanglement, the complexity and non-locality of these measures have prevented the development of a comprehensive theory. However, in recent years there has been a sustained effort devoted to understanding the properties of entanglement measures (and of entanglement entropy in particular) in quantum field theory, in gravity, in string theory, and more recently in condensed matter physics. Different motivations have converged on this problem.

One of the main motivations for work on entanglement originates from the physics of black holes. In spite of their origin in a classical theory, the general theory of relativity, black holes behave as if they were thermodynamic objects that have entropy and temperature. A fundamental result in black-hole physics is the expression for the black-hole entropy (Bekenstein, 1973; Hawking, 1975)

$$S_{\mathrm{BH}} = \frac{1}{4\ell_{\mathrm{P}}^2}\mathcal{A} \tag{17.19}$$

where $\mathcal{A}$ is the *area* of the event horizon of the black hole, $\ell_{\mathrm{P}} = \sqrt{\hbar G_{\mathrm{N}}/c^3}$ is the Planck length, and G_{N} is the Newton constant. This formula, known as the Bekenstein–Hawking entropy of the black hole, is most intriguing. Since black holes are some of the most classical objects in the Universe (or universe?), it is unclear what this entropy actually means.

The way the Bekenstein–Hawking formula was originally derived involved the idea that near the strong gravitational fields of a black hole pair production would exist. The existence of these processes then implies that some form of radiation, Hawking radiation, should be produced in the vicinity of the black hole, and hence these objects would acquire thermal properties such as a temperature. In addition, there are many examples in general relativity of so-called extremal black holes, which have entropy but no temperature. In particular, if a particle–anti-particle pair is created by the gravitational field of the black hole, the pair will be in an entangled state. One is then led to assume that the black-hole entropy must have a relation with entanglement entropy.

So the following question arises: if they have entropy, what degrees of freedom are being counted by this entropy? One of the triumphs of modern string theory is that it offers an explanation for the black-hole entropy that is consistent with statistical mechanics (Maldacena and Strominger, 1996). On the other hand, since some fraction of the particles produced by pair production would fall into the black hole and hence disappear for ever, this led to the idea that, once a black hole has formed, the system somehow on its own evolves into a mixed quantum state, since some information seems to be eaten by the black hole. This notion is in conflict with quantum mechanics since it violates unitary time evolution. This problem has led to a profound examination of the relation among gravitation, quantum mechanics, and information (Susskind, 2008).

If one assumes that the black-hole entropy is entanglement entropy, the Bekenstein–Hawking area law motivates the question of whether quantum field theory itself is consistent with this law. This problem was first studied by Bombelli, Koul, and Sorkin (Bombelli *et al.*, 1986) and by Srednicki (1993), who found that in the case of a free massive relativistic scalar field of mass m the von Neumann entanglement entropy of a region A of linear size ℓ in the limit $\ell \gg m^{-1}$ behaves as

$$S[\ell] = \alpha \ell^{d-1} \tag{17.20}$$

where d is the dimensionality of *space* and α is a dimension-full, and hence non-universal, constant. Here, by non-universality we mean that it depends explicitly on the short-distance cutoff, e.g. the lattice spacing a or the momentum cutoff $\Lambda \sim a^{-1}$. The non-universal character of the coefficient of the area law in Eq. (17.20) indicates that it is governed by the short-distance correlations encoded in the wave function. However, in contrast with the Bekenstein–Hawking formula, the area law of the entanglement entropy of local field theories cannot be expressed in terms of a fundamental length and depends on the explicit way in which the theory is cut off at short distances.

For example, in the case of a free scalar field in $(d+1)$ space-time dimensions, the entanglement entropy for a spherical region in d space dimensions can be expressed in terms of the two-point function of the free massive scalar field in the n-sheeted geometry, see e.g. Calabrese and Cardy (2004), Casini *et al.* (2005), and Casini and Huerta (2005, 2009), resulting in the expression

$$\ln \operatorname{tr} \rho_A^n = \mathcal{A}\frac{1}{24}\left(n - \frac{1}{n}\right) \int \frac{d^{d-1}k_\perp}{(2\pi)^{d-1}} \ln\left(k_\perp^2 + m^2\right) \tag{17.21}$$

where $k_\perp$ acts on the $d-1$ non-radial coordinates. From here it follows that the von Neumann entanglement entropy for the spherical region has an area law

$$S = -\frac{\mathcal{A}}{12} \int \frac{d^{d-1}k_\perp}{(2\pi)^{d-1}} \ln\left(\frac{k_\perp^2 + m^2}{k_\perp^2 + a^{-2}}\right) \tag{17.22}$$

where $\mathcal{A}$ is the area of the d-dimensional hypersphere and a is the short-distance cutoff. This integral diverges as $a^{-(d-1)}$.

In conclusion, the entanglement entropy of free massive scalar fields is sub-extensive and scales as the *area* of the observed region, instead of the *volume* of a region as in the case of the *thermodynamic* entropy. This result holds essentially for all local theories. Since the area law follows from short-distance physics, it should apply equally to systems in the non-critical regime and to quantum critical systems. The only known exception to the strict area-law scaling is the case of systems of fermions at finite density, i.e. with a Fermi surface, where it has been shown that the entanglement entropy scales instead as $S_A \sim L^{d-1} \ln L$ (Gioev and Klich, 2006; Wolf, 2006), which is still sub-extensive. The sub-extensive scaling of the entanglement entropy is a necessary condition for the success of quantum-information-based approaches to the simulation of quantum critical systems such as the projected entangled-pair-state representation of quantum states on 2D lattices (Verstraete *et al.*, 2006).

17.5 Entanglement entropy in conformal field theory

The generic existence of the area-law result naturally poses the following question: what scaling law does the entanglement entropy obey in a quantum critical system? In particular, what is the manifestation of the expected universal behavior of a quantum critical system in the entanglement entropy? One expects to find universal contributions to the entanglement entropy. But then the next logical question is to ask what determines these universal contributions and how they are related to the scaling of local observables at a quantum critical point.

To this date there isn't a generally known answer to these questions. However, these issues have been investigated in several important cases: (a) CFTs in $(1+1)$ dimensions, (b) the $(2+1)$-dimensional quantum Lifshitz universality class (see Section 9.15), and (c) ϕ^4 field theory in the $4-\epsilon$ expansion. In this section we will describe the $(1+1)$-dimensional case.

The scaling behavior of the entanglement von Neumann entropy can be determined by general arguments of CFT. We will follow here in detail the arguments and results of Calabrese and Cardy (Calabrese and Cardy, 2004; Calabrese *et al.*, 2009), which in turn are an extension of the early work of Holzhey, Larsen, and Wilczek (Holzhey *et al.*, 1994). We will use the representation of $Z_n[A] = \text{tr}\, \rho_A^n$ in terms of a path integral with a replicated target space.

For simplicity region A will be a segment of length ℓ, and region B is the complement. The entire system has length $L \gg \ell$. We will work in the long distance limit and hence assume that $\ell \gg a$, where a is a microscopic spatial cutoff, e.g. the lattice spacing. We will compute Z_n and from there we will determine the von Neumann entanglement entropy. This problem was also considered in the context of specific lattice systems, such as quantum spin chains. The main result that we will prove is that in this limit, that provided the system is quantum critical, the entanglement entropy has the behavior

$$S = \frac{c}{3} \ln\left(\frac{\ell}{a}\right) + \text{finite terms} + O(\ell^{-1}) \tag{17.23}$$

where c is the central charge of the Virasoro algebra associated with the specific CFT that describes the 1D quantum critical point. From the point of view of the area law this result is expected, since in one dimension the boundaries are sets of points. Hence, as $d \to 1$, the area law can become a logarithm in one dimension. In this case, a redefinition of the microscopic cutoff will change the $O(1)$ terms in the expansion of Eq. (17.23) but cannot change the prefactor of the logarithm. Hence the prefactor must be a universal quantity, which, we will see, is related to the central charge c discussed in Section 7.11. However, the finite terms have non-universal contributions. On the other hand, if the logarithmic contributions were absent, the finite terms should be universal. We will see below an example of this case in two space dimensions.

The existence of universal terms in the finite-size scaling of the entanglement entropy is important since they signal the existence of *large-scale entanglement*. Indeed, although the area-law term is the most singular term as a function of the size of the observed region, the (generally subleading) universal term indicates that it measures the contributions to the entanglement entropy from all length scales. We will see that the entanglement entropy generally has a universal term in systems at quantum criticality, i.e. whose effective field theory displays scale and conformal invariance, and in topological phases.

In the replica formulation the replicated fields become twisted by the non-trivial boundary conditions. The fields are twisted by the exchange symmetries, which are represented by the local twist fields $\mathcal{T}_n$ and $\mathcal{T}_n^{-1}$, respectively. The local twist fields $\mathcal{T}_n$ and $\mathcal{T}_n^{-1}$ act at the boundary between the two regions A and B, and effectively serve to link the different copies of the CFT with each other. In the case of a single interval with endpoints at the spatial coordinates $x = u$ and $x = v$, with $\ell = |u - v|$, the partition function is to be calculated on fields on a Riemann surface with n sheets. In two dimensions we can expect to be able to compute this object as a path integral for replicated fields on the complex plane $\mathbb{C}$ where the Riemann surface is specified by a set of boundary conditions, Eq. (17.17). For the n-sheeted Riemann surface along interval A the partition function $Z_n[A]$ becomes

$$Z_n[A] \sim \langle \mathcal{T}_n(u, 0)\mathcal{T}_n^{-1}(v, 0)\rangle \tag{17.24}$$

where the expectation value is computed from the replica theory on the complex plane. The role of the operators $\mathcal{T}_n(u, 0)$ and $\mathcal{T}_n^{-1}(v, 0)$ is to enforce the boundary conditions of Eq. (17.17). Similarly, the expectation value of an operator in sheet $i = 1, \ldots, n$, $\mathcal{O}(x, \tau; i)$, on the n-sheeted Riemann surface is

$$\langle \mathcal{O}(x, \tau; i)\rangle = \frac{\langle \mathcal{O}(x, \tau; i)\mathcal{T}_n(u, 0)\mathcal{T}_n^{-1}(v, 0)\rangle}{\langle \mathcal{T}_n(u, 0)\mathcal{T}_n^{-1}(v, 0)\rangle} \tag{17.25}$$

In the thermodynamic limit in which the length of the system $L \to \infty$, the conformal mapping

$$\zeta = \frac{w - u}{w - v} \tag{17.26}$$

(where $w = x + i\tau$) maps the branch points to $(0, \infty)$. The conformal mapping

$$z = \zeta^{1/n} = \left(\frac{w - u}{w - v}\right)^{1/n} \tag{17.27}$$

maps the n-sheeted Riemann surface onto the complex plane $\mathbb{C}$. In the case of a CFT the energy–momentum tensor (or stress–energy tensor) decomposes into holomorphic and anti-holomorphic components (see Section 7.11). The holomorphic component of the stress–energy tensor $T(w)$ under a conformal mapping transforms as (Belavin *et al.*, 1984; Di Francesco *et al.*, 1997)

$$T(w) = \left(\frac{dz}{dw}\right)^2 T(z) + \frac{c}{12}\{z, w\} \tag{17.28}$$

where $\{z, w\} = (z'''z' - \frac{3}{2}z''^2)/z'^2$ is the Schwartzian derivative. However, in the plane $\langle T(z)\rangle_{\mathbb{C}} = 0$, since the system is rotational and translation-invariant, it follows that the expectation value of the stress–energy tensor on the n-sheeted Riemann surface is

$$\langle T(w)\rangle = \frac{c}{12}\{z, w\} = \frac{c}{24}\left(1 - \frac{1}{n^2}\right)\frac{(v-u)^2}{(w-u)^2(w-v)^2} \tag{17.29}$$

But this should be the same as the computation of the expectation value of the stress–energy tensor in the n-sheeted Riemann surface. In particular, if the stress–energy tensor for the replicated Lagrangian is n times this answer, this leads to the result

$$\frac{\langle T(w)^{(n)}\mathcal{T}_n(u,0)\mathcal{T}_n^{-1}(v,0)\rangle}{\langle \mathcal{T}_n(u,0)\mathcal{T}_n^{-1}(v,0)\rangle} = \frac{c}{24n}(n^2-1)\frac{(v-u)^2}{(w-u)^2(w-v)^2} \tag{17.30}$$

In CFT the fields obey a set of conformal Ward identities (Belavin *et al.*, 1984), which in this case take the form (Calabrese and Cardy, 2004; Calabrese *et al.*, 2009)

$$\begin{aligned}\langle T(w)^{(n)}&\mathcal{T}_n(u,0)\mathcal{T}_n^{-1}(v,0)\rangle \\ &= \left(\frac{1}{w-u}\partial_u + \frac{d_{\mathcal{T}_n}}{(w-u)^2} + \frac{1}{w-v}\partial_v + \frac{d_{\mathcal{T}_n^{-1}}}{(w-v)^2}\right)\langle \mathcal{T}_n(u,0)\mathcal{T}_n^{-1}(v,0)\rangle\end{aligned} \tag{17.31}$$

where $d_{\mathcal{T}_n} = d_{\mathcal{T}_n^{-1}}$ is the scaling dimension of the primary field $\mathcal{T}_n$ (and of $\mathcal{T}_n^{-1}$). Using that, by definition of the scaling dimension

$$\langle \mathcal{T}_n(u,0)\mathcal{T}_n^{-1}(v,0)\rangle = \frac{1}{|u-v|^{2d_n}} \tag{17.32}$$

we can now identify the scaling dimension of the twist fields d_n to be

$$d_n = \frac{c}{12}\left(n - \frac{1}{n}\right) \tag{17.33}$$

Since the partition function

$$\mathrm{tr}\,\rho_A^n = \frac{Z_n[A]}{Z^n} \tag{17.34}$$

is equivalent to the two-point function of the twist fields, Eq. (17.24), it should behave in the same way under conformal transformations. Hence, up to a non-universal constant C_n, we can make the identification

$$\operatorname{tr} \rho_A^n = C_n \left(\frac{v-u}{a} \right)^{-(c/6)(n-1/n)} \tag{17.35}$$

where a is a short-distance cutoff. It then follows that the Rényi and von Neumann entanglement entropies obey the scaling (Holzhey *et al.*, 1994; Calabrese and Cardy, 2004)

$$\begin{aligned} S_A^{(n)} &= \frac{c}{6}\left(1+\frac{1}{n}\right) \ln\left(\frac{\ell}{a}\right) + \text{constant} \\ S_A &= \frac{c}{3} \ln\left(\frac{\ell}{a}\right) + \text{constant} \end{aligned} \tag{17.36}$$

This is the main result.

Calabrese and Cardy also derived expressions for the entanglement entropy for a system with finite size L. In the case of periodic boundary conditions they found the result for a system of length L and an interval A of length ℓ:

$$S_A[L,\ell] = \frac{c}{3} \ln\left(\frac{L}{\pi a} \sin\left(\frac{\pi \ell}{L}\right)\right) + \text{constant} \tag{17.37}$$

Using a conformal mapping $w \to z = (\beta/(2\pi)) \ln w$ that maps each sheet on the w plane onto an infinitely long cylinder of circumference β, which plays the role of the inverse temperature, Calabrese and Cardy derived a formula for $\operatorname{tr} \rho_A^n$ in a thermally mixed state at temperature β^{-1} resulting in an entropy

$$S_A(T,\ell) = \frac{c}{3} \ln\left(\frac{\beta}{\pi a} \sinh\left(\frac{\pi \ell}{\beta}\right)\right) + \text{constant} \tag{17.38}$$

At $T = 0$ this result reproduces the result for the entanglement entropy in a 1D conformal field theory, Eq. (17.36). It also recovers the (expected) extensive local thermodynamic entropy in the high-temperature limit $T \gg \ell^{-1}$,

$$S = \frac{\pi c}{3} T\ell + \cdots \tag{17.39}$$

The most powerful numerical method to simulate 1D systems is, at present, the density-matrix renormalization group (DMRG) (White, 1992; Schollwöck, 2005; Hallberg, 2006). The geometry used in DMRG calculations is a half-line with some specified boundary condition at the endpoint. This method involves the computation of the reduced density matrix of some region A, which we will take to be the interval $[0, \ell)$. The rest of the system is region B, and has length $L - \ell$. In the case of this geometry Calabrese and Cardy found the following result for the entanglement entropy of region A (Calabrese *et al.*, 2009):

$$S_A[\ell, L] = \frac{c}{6} \ln\left(\frac{2L}{\pi a} \sin\left(\frac{\pi \ell}{L}\right)\right) + \ln g + \text{constant} \tag{17.40}$$

This result provides the most direct and efficient way to compute the central charge c using numerical DMRG calculations. It shows that the entanglement entropy is a scaling function of ℓ/L with a form that allows the determination of the finite-size corrections. The constant term, $\ln g$, is the so-called boundary entropy of Affleck and Ludwig, as discussed in Section 15.6.3, which is a universal quantity that depends only on the boundary conditions. One may object to keeping the constant term, since it is in principle non-universal. However, if two different entanglement entropies are computed using the same regularization, their difference is finite and universal. Thus, the finite universal terms are actually meaningful and universal (in this sense) up to a choice of which conformally invariant boundary condition is considered. We will see in the next sections that in space dimensions $D > 1$ the entanglement entropy is characterized by universal constant corrections to the non-universal area-law term.

17.6 Entanglement entropy in the quantum Lifshitz universality class

Very few results are known for quantum critical systems in space dimensions $D > 1$. A special case in which the scaling behavior of the entanglement entropy in quantum critical systems has been studied in some detail is the quantum Lifshitz universality class, discussed in Section 9.15, which includes the quantum dimer model (QDM) and its generalizations, see Chapter 9. The scaling of the entanglement entropy in the quantum Lifshitz universality class was discussed first by Fradkin and Moore (2006), whose work we follow here in some detail, and was further refined in Fradkin (2009), Hsu and Fradkin (2010), Hsu *et al.* (2009b), and Oshikawa (2010). The scaling of the entanglement entropy in the quantum critical dimer-model wave function was determined (mostly numerically) by Stéphan and collaborators (Stéphan *et al.*, 2009, 2011).

We will discuss this problem in terms of the ground-state wave function for the quantum Lifshitz model, which is given by (see Eq. (9.179))

$$\Psi_0[\varphi(x)] = \frac{1}{\sqrt{Z_0}} \exp\left(-\int d^2x \, \frac{\kappa}{2} \left(\vec{\nabla}\varphi(x)\right)^2\right) \tag{17.41}$$

where κ is a parameter of the quantum Lifshitz model, which is specified in Eq. (9.161), and Z_0 is the norm of this wave function (see Eq. (9.180)),

$$Z_0 = ||\Psi_0||^2 = \int \mathcal{D}\varphi \, \exp\Big(-\int d^2x \, \kappa \, \left(\vec{\nabla}\varphi(\vec{x})\right)^2\Big) \tag{17.42}$$

Recall that the field φ is compactified with compactification radius $R = 1$. The quantum critical point of the Rokhsar–Kivelson QDM is represented by the choice of the parameter $\kappa = 1/(8\pi)$.

The wave functions of the quantum Lifshitz universality class (and its generalization) have the special property that the weight of a field configuration $[\varphi]$ in the wave function is local and has the form of the Gibbs weights in a conformally invariant classical system in two space dimensions. Consequently, the *norm* Z_0 of these wave functions is equivalent to a conformally invariant classical partition function in two dimensions (Ardonne *et al.*, 2004).

Let us first consider the case in which the system is defined on a large disk of diameter L with some boundary conditions at infinity. We will consider the plane geometry shown in Fig. 17.3. In this case the "entangling region" will be a simply connected region A with the topology of a disk of diameter ℓ with boundary Γ. The complement of region A is region B. The wave function has the form $\Psi_0[\varphi_A, \varphi_B]$, where $[\varphi_A]$ and $[\varphi_B]$ label the degrees of freedom on regions A and B, respectively.

For conformal quantum critical points, the Hilbert space has an orthonormal basis of states $|\{\varphi\}\rangle$ indexed by classical configurations $\{\varphi\}$, and the ground state $|\Psi_0\rangle$ of the bipartite system is determined by a local CFT action $S[\varphi]$:

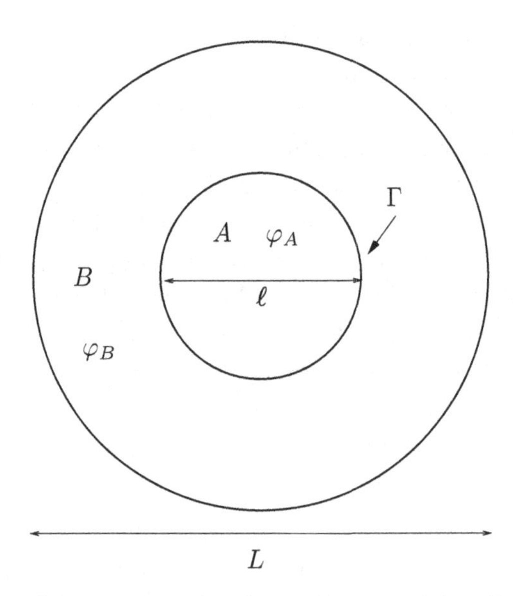

Figure 17.3 The disk geometry: the observed region A has diameter ℓ and the entire system is also a disk of diameter L. B is the annular region and is the complement of the disk A.

$$|\Psi_0\rangle = \frac{1}{\sqrt{Z_0}} \int \mathcal{D}\varphi \, e^{-S(\{\varphi\})/2} |\{\varphi\}\rangle \tag{17.43}$$

Here

$$Z_0 = \int \mathcal{D}\varphi \, e^{-S(\{\varphi\})} \tag{17.44}$$

and expectation values in this state reproduce CFT correlators.

Fradkin and Moore (2006) used the "replica trick" to compute the entanglement entropy (Holzhey *et al.*, 1994; Calabrese and Cardy, 2004) for conformally invariant wave functions, and showed that the trace of the nth power of the reduced density matrix, $\operatorname{tr} \rho_A^n$, where ρ_A is the reduced density matrix of a region A (where A and B form a *partition* of the entire system $A \cup B$, and are separated by their common boundary Γ) for the ground state Ψ_0 on $A \cup B$, is given by

$$\operatorname{tr} \rho_A^n = \frac{Z_n}{Z^n} \tag{17.45}$$

Here Z_n is the partition function of n copies of the equivalent 2D classical statistical-mechanical system satisfying the constraint that their degrees of freedom are identified on the boundary Γ, and Z^n is the partition function for n decoupled systems.

In order to construct $\operatorname{tr} \rho_A^n$, we need an expression for the matrix elements of the reduced density matrix $\langle \varphi^A | \rho_A | \varphi'^A \rangle$. Since the ground-state wave function is a local function of the field $\varphi(x)$, a general matrix element of the reduced density matrix is a trace of the density matrix of the pure state $\Psi_{\rm GS}[\varphi]$ over the degrees of freedom of the "unobserved" region B, denoted by $\varphi^B(x)$. Hence the matrix elements of ρ_A take the form

$$\langle \varphi^A | \hat{\rho}_A | \varphi'^A \rangle = \frac{1}{Z} \int [D\varphi^B] \exp\left[-\left(\frac{1}{2} S^A(\varphi^A) + \frac{1}{2} S^A(\varphi'^A) + S^B(\varphi^B) \right) \right] \tag{17.46}$$

where the degrees of freedom satisfy the *boundary condition* at the common boundary Γ:

$$BC_\Gamma : \quad \varphi^B|_\Gamma = \varphi^A|_\Gamma = \varphi'^A|_\Gamma \tag{17.47}$$

Proceeding with the computation of $\operatorname{tr} \rho_A^n$, one immediately sees that the matrix product requires the condition $\varphi_i^A = \varphi'^A_{i-1}$ for $i = 1, \ldots, n$, and $\varphi'^A_n = \varphi_1^A$ from the trace condition. Hence, $\operatorname{tr} \rho_A^n$ takes the form

$$\begin{aligned} \operatorname{tr} \rho_A^n &\equiv \frac{Z_n}{Z^n} \\ &= \frac{1}{Z^n} \int_{BC_\Gamma} \prod_i D\varphi_i^A \, D\varphi_i^B \, e^{-\sum_{i=1}^n \left(S(\varphi_i^A)+S(\varphi_i^B)\right)} \end{aligned} \tag{17.48}$$

subject to the boundary condition BC_Γ of Eq. (17.47).

This result shows that the numerator of Eq. (17.48), Z_n, is the partition function of n 2D *classical* systems, each with the same energy functional $S[\varphi]$, whose degrees of freedom are identified with each other on the boundary Γ that separates regions A and B, but which are otherwise independent. Hence, Z_n is the partition function of a classical 2D critical system on the "book" geometry shown in Fig. 17.4. In contrast, the denominator, Z^n, is the partition functor of the same n decoupled systems. Since the systems are conformally invariant away from their common boundary and since the boundary condition is consistent with the condition of conformal invariance, we have effectively mapped the computation of the entanglement entropies to a problem in 2D boundary CFT.

The expression for $\operatorname{tr} \rho_A^n$ given in Eq. (17.48) implies that the replicated partition function Z_n is invariant under the permutations of the n replicas. Actually, it is invariant under the action of the permutation group S_n independently on regions A and B. Also, since the fields have to agree on the boundary Γ, Z_n is also invariant under a simultaneous global shift of all the replicated fields in both regions. This condition implies that nothing physical happens at the common boundary Γ, which, after all, is an arbitrary device.

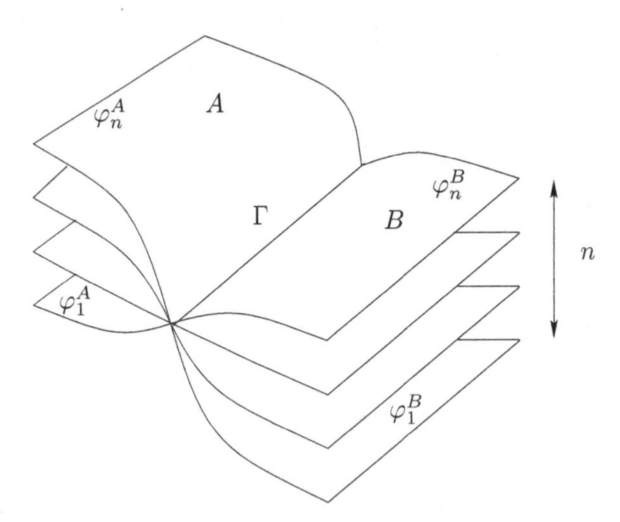

Figure 17.4 The book geometry of the replicated system. The fields are identified with each other at the common boundary Γ between regions A and B.

The specific problem we want to address here is the scaling behavior of the von Neumann entanglement entropy (and of the associated Rényi entropies). To this end, it will be useful to introduce the following notation. Let $F_1 = -\ln Z$ be the classical partition function of one of the n decoupled copies (or replicas) and $F_n = -\ln Z_n$ the classical partition function of the system of n replicas identified along Γ. Then we can formally relate the von Neumann entanglement entropy S_A to the expression

$$S_A = \lim_{n \to 1} \frac{F_n - nF_1}{n-1} \tag{17.49}$$

where the dependence on n has to be understood as an analytic continuation to the complex-n plane. The free energy of a physical system in two dimensions with a well-defined thermodynamic limit has a leading extensive term, $F = L^2 f + \ldots$, where L is the linear size of the system and f is the free-energy density. Here the ellipsis indicates the finite-size corrections. These are generally expected to include a perimeter term, $O(L)$, constant terms $O(L^0)$, and possibly a term that depends logarithmically on L (Privman and Fisher, 1984; Privman, 1988),

$$F = L^2 f + \sigma L + a \ln L + b + O(L^{-1}) \tag{17.50}$$

Since the copies in the replicated system are identical away from the contour Γ, which can be viewed as a "defect" in the replicated system, it is clear that the extensive terms (proportional to the free-energy density) must be the same for F_n and for nF_1 and hence must cancel out exactly. Therefore, in the limit $L \gg \ell \gg a$, we expect the von Neumann entanglement entropy to have the following form:

$$S_A = \alpha\ell + C \ln\left(\frac{\ell}{a}\right) + \Phi\left(\frac{\ell}{L}\right) + O(\ell^{-1}) \tag{17.51}$$

where the first term is the non-universal "area law" (in 2D it is a perimeter), C is a constant, and $\Phi(x)$ is a dimensionless function of the aspect ratio ℓ/L. We will see below that this general argument is correct.

As a first step let us consider the case in which the field φ is free, which is the case in the quantum Lifshitz model. In this case it is obvious that the field $\varphi = \sum_{j=1}^{n} \varphi_n$ decouples and does not see the boundary Γ. It further obeys the same boundary conditions at spatial infinity as the individual copies. So one expects that it should be possible to write the partition function Z_n as a product of two terms, one of which is the partition function of just one copy (the field φ) and the other is a partition function for the remaining $n-1$ copies, $\tilde{Z}_{n-1}$. However, in the partition function $\tilde{Z}_{n-1}$ the replica fields can enter only as the $n-1$ differences, say $\tilde{\varphi}_j = \varphi_{j+1} - \varphi_j$ (for $j = 1, \ldots, n-1$), and these fields must now vanish at the common boundary Γ, which is to say they must obey Dirichlet boundary conditions on Γ. This assumption implies that the trace of the nth power of the reduced density matrix is a simple power, $\mathrm{tr}\,\rho_A^n \propto \text{constant}^n$. We will see below that this is not quite correct.

If this argument were literally correct, we would be able to write a simpler expression for the entanglement entropy. The partition functions on the r.h.s. of Eq. (17.45) are $Z_A = ||\Psi_0^A||^2$ with support in region A and $||\Psi_0^B||^2$ with support in region B, both satisfying generalized Dirichlet (i.e. fixed) boundary conditions on Γ of A and B, and $Z_{A\cup B} = ||\Psi_0||^2$ is the norm squared for the full system. The entanglement entropy S is then obtained by an analytic continuation in n (Fradkin and Moore, 2006; Hsu *et al.*, 2009b),

$$\begin{aligned} S_A = -\mathrm{tr}(\rho_A \ln \rho_A) &= -\lim_{n\to 1} \frac{\partial}{\partial n} \mathrm{tr}\,\rho_A^n = -\log\left(\frac{Z_A Z_B}{Z_{A\cup B}}\right) \\ &= F_A + F_B - F_{A\cup B} \end{aligned} \tag{17.52}$$

where F_A and F_B are the free energies of a free field in regions A and B with Dirichlet boundary conditions on Γ, and $F_{A\cup B}$ is the free energy for the whole system. Hence, the computation of the entanglement entropy is reduced to the computation of a ratio of partition functions in a 2D classical statistical-mechanical problem, a Euclidean CFT in the case of a critical wave function, each satisfying specific boundary conditions. Notice that if these arguments are correct then the ratio Z_n/Z^n can be written as a simple power of n with no other dependence on n left. We will see below that, while this argument is almost right, there are small but conceptually significant corrections. We will first proceed with the assumption that Eq. (17.52) is an identity.

The dependence of the free energy on the size of the system for different boundary conditions is a problem in boundary CFT. This problem was first discussed by the mathematician Mark Kac (Kac, 1966), who was interested in the problem of spectral geometry. In physics terms Kac was interested in the spectrum of the Laplacian operator in two dimensions in regions of various shapes. He posed this problem as the following question: can you hear the shape of a drum? In the present context we can rephrase the question as "can you hear the shape of a quantum drum?" (or of Schrödinger's cat?). This problem has since been reanalyzed by people working in CFT (and in string theory). The most complete expression of the asymptotic form of the free energy, and the one that will be useful to us, is due to Cardy and Peschel (1988), who showed that the free energy of a 2D Euclidean CFT of central (Virasoro) charge c has the following finite-size dependence in a region of linear size L and smooth boundary Γ:

$$F = fL^2 + \sigma L - \frac{c}{6} \ln\left(\frac{L}{a}\right) + O(L^0) \tag{17.53}$$

where f is the (non-universal) free-energy density, σ is the (non-universal) surface tension, c is the central charge of the conformal field theory, and χ is the Euler characteristic of the region under consideration,

$$\chi = 2 - 2h - b \tag{17.54}$$

where h is the number of handles of the region and b is the number of boundaries.

By direct application of Eq. (17.53), the entanglement entropy S_A becomes (using Eq. (17.52))

$$S_A = \alpha\ell - \frac{c}{6}\Delta\chi \ln\left(\frac{\ell}{a}\right) + O(1) \tag{17.55}$$

where the Euler characteristics of the disk A, the annular region B, and the large disk $A \cup B$ enter through the expression

$$\Delta\chi = \chi_A + \chi_B - \chi_{A\cup B} \tag{17.56}$$

However, for a region $A \subset A \cup B$ with a smooth boundary Γ, $\Delta\chi = 0$ since in this case $\chi_A + \chi_B = \chi_{A\cup B}$. Hence, this argument predicts that in this case there is no term that scales as $\ln(\ell/a)$. This raises the possibility that the $O(1)$ term may actually be universal.

On the other hand, if the boundary Γ is not smooth, logarithmic terms do exist. Imagine, for instance, a simply connected region A whose boundary Γ is piecewise smooth but has cusp singularities (corners) at isolated points, with interior angle γ, $0 < \gamma < 2\pi$. In this case, Cardy and Peschel found a logarithmic contribution to the free energy of the form

$$\Delta F = \frac{c\gamma}{24\pi}\left(1 - \frac{\pi^2}{\gamma^2}\right) \tag{17.57}$$

Since all other contributions to the entanglement entropy vanish, aside from the "area-law term," we conclude that a corner will give a logarithmic contribution of the form of Eq. (17.57). A similar logarithmic term is found if region A is not fully contained inside region B, but its boundary now disconnects the system in two disjoint and simply connected regions. The logarithmic term originates from the conical singularities at the intersection of the boundary γ with the outer boundaries of the system (Fradkin and Moore, 2006).

Let us now ask whether the assumptions that led to Eq. (17.52) are actually correct. Let us consider for simplicity the case of the cylinder geometry shown in Fig. 17.5. This problem was discussed by Hsu and collaborators (Hsu *et al.*, 2009b; Hsu and Fradkin, 2010) and by Oshikawa (2010). It was also investigated numerically and analytically by Stéphan and coworkers in the QDM wave functions (Stéphan *et al.*, 2009). A direct application of the result of Eq. (17.52) leads to

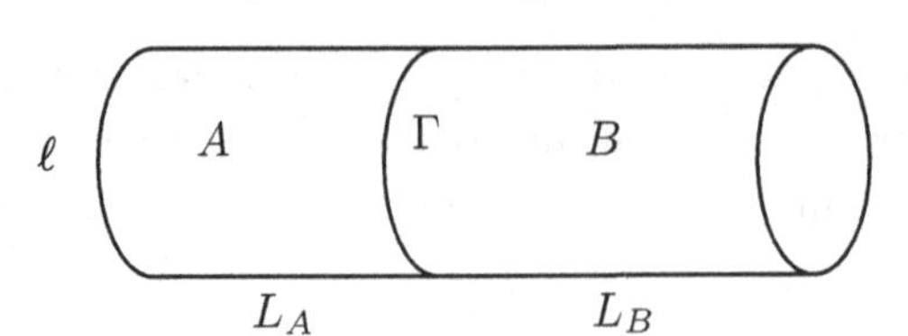

Figure 17.5 The cylinder geometry.

the computation of the partition function of a 2D Euclidean scalar field with compactification radius $R = \sqrt{8\pi\kappa}$, where κ is the parameter of the quantum Lifshitz model. Recall that for the QDM $\kappa = 1/(8\pi)$, and hence $R = 1$. For the generalized QDMs (Papanikolaou *et al.*, 2007b) and for the quantum eight-vertex model (Ardonne *et al.*, 2004), the parameter κ varies continuously from $\kappa = 1/(8\pi)$ to $\kappa = 1/(4\pi)$ at a Kosterlitz–Thouless transition.

I will assume that the system obeys Dirichlet boundary conditions on the circles (each with circumference ℓ) at both ends of the cylinder. Thus in this case Z_A, Z_B, and $Z_{A \cup B}$ obey Dirichlet boundary conditions at both ends. The partition function $Z_{\rm DD}$ for a free boson on a cylinder with compactification radius R with Dirichlet boundary conditions at both ends of the cylinder is (Fendley *et al.*, 1994)

$$Z_{\rm DD} = \frac{\text{constant}}{R} \times \frac{\vartheta_3(2\tau/R^2)}{\eta(q^2)} \tag{17.58}$$

where $\tau = iL/\ell$ is the modular parameter and $q = e^{2\pi i \tau}$, and where we have introduced the elliptic theta function $\vartheta_3(\tau)$ and the Dedekind eta function $\eta(q)$, which are, respectively, given by

$$\vartheta_3(\tau) = \sum_{n=-\infty}^{\infty} q^{n^2/2}, \qquad \eta(q) = q^{1/24} \prod_{n=1}^{\infty} (1 - q^n) \tag{17.59}$$

In general the resulting expression for the entanglement entropy depends on the various aspect ratios ℓ/L. However, in the limit of long cylinders, $L \gg \ell$, we obtain a simple expression for the $O(1)$ term of the entanglement entropy (Hsu *et al.*, 2009b),

$$S_A^{\rm cylinder} = \alpha\ell + \ln R \tag{17.60}$$

which is a continuous function of the compactification radius R.

A similar computation for the case of the torus (shown in Fig. 17.6) also leads to a finite and universal $O(1)$ term in the entanglement entropy,

$$S_A^{\rm torus} = \alpha\ell + 2\ \ln\left(\frac{R^2}{2}\right) \tag{17.61}$$

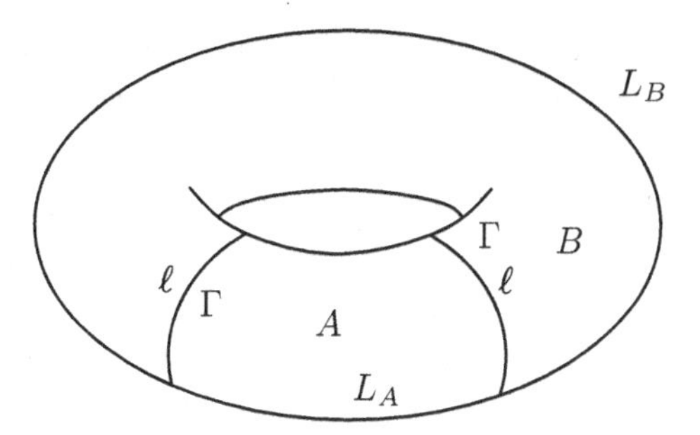

Figure 17.6 The torus geometry.

On the other hand, for the case of the disk geometry of Fig. 17.3, in addition to a universal constant term, one also finds a term that depends explicitly on the aspect ratio ℓ/L:

$$S_A^{\text{disk}} = \alpha\ell + \frac{1}{2}\ln\left[\frac{1}{\pi}\ln\left(\frac{L}{\ell}\right)\right] + \ln R \tag{17.62}$$

The origin of the dependence on the aspect ratio L/ℓ is the fact that on the disk the boundary has a finite radius of curvature.

These results, which are based on Eq. (17.52), lead to the conclusion that, except for possible functions of the aspect ratio, the entanglement entropy of 2D scale-invariant wave functions has a universal $O(1)$ term. These results were based on the factorization assumption which implied that the dependence on n of $\operatorname{tr}\rho_A^n$ is a simple power. It turns out that the correct dependence on n has a $\sqrt{n}$ prefactor. This result was derived by direct computation with the compactified theory by constructing the boundary state for the compactified boson explicitly (Hsu and Fradkin, 2010; Oshikawa, 2010) and leads to the result (for the cylinder geometry)

$$S_A^{\text{cylinder}} = \alpha\ell + \ln R - \frac{1}{2} \tag{17.63}$$

which agrees with the numerical results of Stéphan *et al.* (2009). The origin of this term is the structure of the compactification lattice of the difference fields, $\tilde{\varphi}_j$. Notice that, since the wave function of the quantum Lifshitz model maps onto a 2D Euclidean CFT with central charge $c = 1$, this quantity does not appear in our formulas. Nevertheless, they do depend on the compactification radius R which determines the operator content of the theory.

The generalization of this universal condition for generic conformally invariant wave functions is not known at present. Hsu and coworkers (Hsu *et al.*, 2009b) used the (most likely generally incorrect) formula of Eq. (17.48) which relates the universal constant term to the modular S-matrix of the 2D CFT of the wave function. It is curious that for the special case of the 2D Ising wave function this prediction is consistent with the numerical results (Stéphan *et al.*, 2010) (except for a subtle behavior as $n \to 1$). At any rate, these results confirm the expectation that in 2D quantum critical systems the entanglement entropy has a universal $O(1)$ correction to the area-law term. We will see next that conventional relativistic ϕ^4 theory has an entanglement entropy with the same scaling behavior at its quantum critical point. Thus, we expect this to be the generic universal scaling behavior for space dimensions $d > 1$, regardless of the type of dynamics the system has.

17.7 Entanglement entropy in ϕ^4 theory

The behavior of the entanglement entropy in a generic quantum critical, i.e. relativistically invariant, quantum field theory in $D = d + 1 > 2$ dimensions, where d is the number of space dimensions, is much less well understood than the two previous examples. Aside from the omnipresent area law, in general logarithmic terms are absent unless the region being observed is not smooth (Fursaev, 2006; Casini and Huerta, 2007, 2009). A logarithmic dependence on the linear size of the observed region is also present in systems with a spontaneously broken continuous symmetry, e.g. in the non-linear sigma model for $d > 1$ and the related quantum Heisenberg antiferromagnet (discussed in Chapter 7) (Metlitski and Grover, 2011; Ju *et al.*, 2012).

However, we are interested here in the scaling at the quantum critical point of the entanglement entropy. Metlitski, Fuertes, and Sachdev (Metlitski *et al.*, 2009) considered the scaling of the entanglement entropy in the relativistically invariant $O(N)$ ϕ^4 theory in $d = 3-\epsilon$ space dimensions. The Euclidean action in $D = d+1$ dimensions, with imaginary time τ, for an N-component real field with $O(N)$ global symmetry is given by ($\mu = 1, \ldots, D$)

$$S = \int d^d x \, d\tau \left[\frac{1}{2} \left(\partial_\mu \vec{\phi} \right)^2 + \frac{t}{2} \vec{\phi}^{\,2} + \frac{\lambda}{4} \left(\vec{\phi}^{\,2} \right)^2 \right] \tag{17.64}$$

Here one is interested in the quantum critical theory, which means that we will need to set the renormalized mass (squared) $t_{\mathrm{R}} = 0$. Since it is a relativistically invariant theory it has dynamical exponent $z = 1$. They considered the cylinder geometry shown in Fig. 17.5. The calculation is technically complex and I will not give the details here.

As in the example of the last section, we divide space (but not time!) into two disjoint but complementary regions A and B, with the boundary being a $(d-1)$-dimensional region $\mathcal{B}$. Here we will take $\mathcal{B}$ to be a plane located at the spatial coordinate $x = 0$. The coordinates along the boundary directions are denoted by $x_\perp$. To compute the entanglement entropies (von Neumann and Rényi) we will use again the replica trick. This means that we will once again compute the partition function Z_n for a theory with n copies that are stitched together at the boundary

$\mathcal{B}$. The result is again an n-sheeted Riemann surface that lies in the plane spanned by $x_{\parallel} = (\tau, x)$ and has a conical singularity at $(\tau, x) = (0, 0)$. This surface is invariant under translations and rotations along the perpendicular directions $x_{\perp}$. In this geometry it is natural to use polar coordinates in the $x_{\parallel}$ plane, (r, θ). The metric in these coordinates is simply $ds^2 = dr^2 + r^2 d\theta^2 + dx_{\perp}^2$. This is just the usual Euclidean metric except that the angular variable has a modified period $\theta \to \theta + 2\pi n$ to reflect the conical singularity.

As a function of the ultraviolet (UV) cutoff a the entanglement entropy has the general scaling form

$$S_A = g_{d-1}(\mathcal{B})a^{-(d-1)} + g_{d-2}(\mathcal{B})a^{-(d-2)} + \cdots + g_0(\mathcal{B})\ln\left(\frac{\ell}{a}\right) \tag{17.65}$$

where ℓ is the linear size of the region A. The first term is the area law discussed before. For $d = 3 - \epsilon$ Metlitski and collaborators found that the entanglement entropy for the cylinder geometry at the Wilson–Fisher fixed point has the same scaling behavior as what we found in the quantum Lifshitz model,

$$S_A = C\left(\frac{\ell}{a}\right)^{d-1} + \gamma \tag{17.66}$$

To leading order in the ϵ expansion at the Wilson–Fisher fixed point the universal contribution γ is given by

$$\gamma = -\frac{N\epsilon}{6(N+8)}\left[\ln\left|\vartheta_1\left(\frac{\phi}{2\pi}(1+i), i\right)\right| - \frac{\phi^2}{4\pi} - \ln\eta(i)\right] \tag{17.67}$$

where $i = \sqrt{-1}$, and ϑ_1 and η are the Jacobi elliptic and Dedekind eta functions. Here ϕ is a twist in the boundary conditions along the directions labeled by $x_{\perp}$, which was introduced by Metlitski and coworkers in order to remove a zero mode. They also suggested that for zero twist the result is non-analytic in ϵ,

$$\gamma = -\frac{N\epsilon}{12(N+8)}\ln\epsilon \tag{17.68}$$

On the other hand, at the infrared (IR)-unstable free-field fixed point in $d = 3 - \epsilon$ dimensions γ is instead given by

$$\gamma = -\frac{N}{6}\left[\ln\left|\vartheta_1\left(\frac{\phi}{2\pi}(1+i), i\right)\right| - \frac{\phi^2}{4\pi} - \ln\eta(i)\right] \tag{17.69}$$

This result suggests that the universal term γ decreases under the RG flow. The decrease of the universal term of the entanglement entropy under the action of the RG flow is reminiscent of Zamolodchikov's c theorem in 2D perturbed CFT (Zamolodchikov, 1986). This behavior of the universal term of the entanglement entropy was also found by Myers and Singh (2012) using the AdS/CFT program that we will discuss in the next section.

17.8 Entanglement entropy and holography

A different perspective on the scaling and universal properties of quantum entanglement in quantum field theories, and hence also in strongly coupled systems in condensed matter physics, is related to the concept of holography. Holography here means that the quantum field theories which describe our world are actually holographic images of a theory of gravity in higher dimensions ('t Hooft, 1993; Susskind, 1995). As a concept, holography originally took shape as a way to understand the physics of black holes. It has since become central to the understanding of strongly coupled quantum field theories. In 1997 Maldacena realized that the classical (weak-coupling) limit of a superstring theory in a 5D space-time with a background anti-de Sitter (AdS) metric is equivalent, or "dual," to the strong-coupling limit of a super-Yang–Mills gauge theory in four (flat) Minkowski space-time dimensions. For this reason the Maldacena conjecture is known as the gauge/gravity duality (Gubser *et al.*, 1998; Maldacena, 1998; Witten, 1998; Aharony *et al.*, 2000; Maldacena, 2012). In this section we will briefly explain the main ideas of holography and then use them to discuss the problem of the scaling of entanglement entropy in general CFTs (Ryu and Takayanagi, 2006a, 2006b; Nishioka *et al.*, 2009). A particularly insightful introduction to this problem is given in McGreevy (2010).

17.8.1 The CFT/AdS correspondence

Anti-de Sitter (AdS) space is a space-time with negative curvature in $(d+1)$ dimensions with the metric

$$ds^2 = \frac{R^2}{u^2}\left(du^2 - dx_0^2 + \sum_{i=1}^{d-1} dx_i^2\right) \tag{17.70}$$

In the AdS geometry the limit $u \to 0$ can be viewed as the boundary of the space-time, and $u \to \infty$ corresponds to space-time points deep in this geometry, with a horizon at $u = \infty$. The parameter R is the radius of curvature of the AdS space-time, measured in units in which the Planck length $\ell_P = 1$. The AdS metric has the special property that a scale transformation (dilatations) of the Minkowski part of the metric of the form $x_\mu \to \lambda x_\mu$ can be absorbed in a rescaling of the extra dimension on the AdS space, $u \to \lambda u$. Hence dilatations are isometries of the AdS geometry. This also means that the limit of $u \to 0$ can be viewed as the short-distance, UV, limit in Minkowski space-time. Similarly, the $u \to \infty$ limit corresponds to space-time points deep inside the AdS geometry and represents the long-distance, IR, limit in Minkowski space-time.

The gauge/gravity duality or, more generally, the CFT/gravity duality, is the statement that all the physics in an asymptotically AdS space-time can be described by a local conformal quantum field theory that resides at the boundary, which looks like flat Minkowski space-time in d space-time dimensions. In particular, the isometries of the AdS geometry act as space-time symmetries on the boundary. These isometries are equivalent to the conformal group in d dimensions, SO(2, d), which includes the Poincaré group of the flat Minkowski space-time $\mathcal{M}$, the dilatation, and the special conformal transformations. Hence, the quantum field theory on the boundary is a CFT. In particular, the dilatation isometry of the AdS geometry becomes scale invariance of the quantum field theory at the boundary. In this holographic picture the short-distance (UV) behavior of a scale-invariant, conformal, field theory on the boundary of the AdS space maps to the long-distance (IR) behavior of a classical theory of gravity defined deep in the bulk of AdS space-time.

Let us consider the classical field theory of a field ϕ in the AdS space-time, and let ϕ_0 be the value of the classical field on the boundary of AdS space-time. We will denote by $Z(\phi_0)$ the partition function of the gravity theory coupled to the field ϕ, with boundary condition ϕ_0, and by $\mathcal{O}$ a local operator of the CFT defined on the boundary of AdS space-time whose source is ϕ_0. Then, the CFT/gravity duality is the identification (Gubser *et al.*, 1998; Witten, 1998)

$$
\begin{aligned}
Z_{\text{CFT}} &= Z_{\text{AdS gravity}} \\
\left\langle \exp\left(\int_{\mathcal{M}} d^d x \; \phi_0(x) \mathcal{O}(x) \right) \right\rangle_{\text{CFT}} &= Z(\phi_0)_{\text{AdS gravity}}
\end{aligned}
\tag{17.71}
$$

The picture of the 4D Minkowski space-time as the boundary or "edge" of the 5D AdS space is somewhat (but not completely) analogous to the bulk–edge correspondence in quantum Hall fluids. In this context, duality is understood, as in other sections of this book, as a mapping relating a weakly coupled theory to another strongly coupled but generally different theory. Here the two theories are so different that they even live in different dimensions!

This CFT/gravity duality further suggests that the AdS coordinate should be regarded as a scale transformation in quantum field theory (Susskind and Witten, 1998; Witten, 1998). In other words, the action of successive RG transformations on a quantum field theory, which, as we saw, makes the coupling constants scale-dependent quantities, can be regarded geometrically as defining an extra dimension (the local scale) in addition to the usual space-time dimensions. Hence the family of theories related by RG transformations can be viewed as a single theory defined on a higher-dimensional space-time with an AdS geometry (Heemskerk and Polchinski, 2011). This interpretation of the RG flow is shown in Fig. 17.7. Heemskerk and Polchinski (2011) have shown that there is a precise connection between the Wilsonian construction of the RG in quantum field theory and the holographic RG.

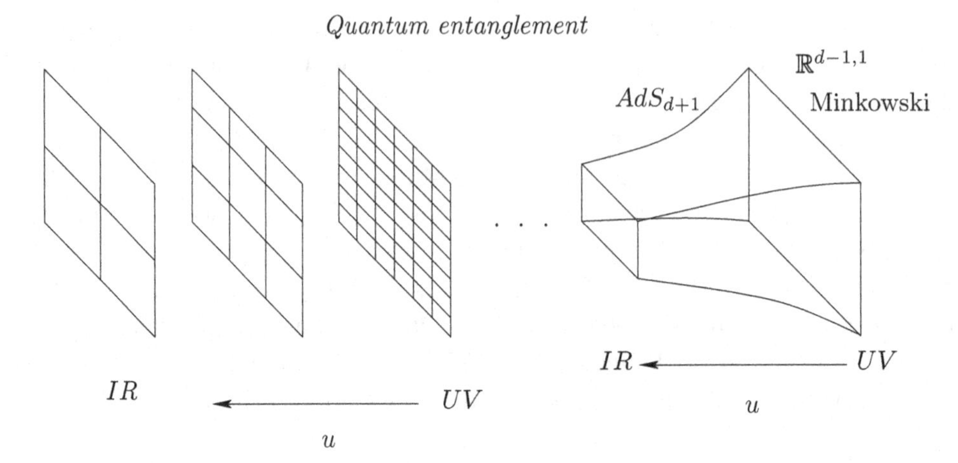

Figure 17.7 The AdS/CFT correspondence: a sequence of RG transformations (left) is equivalent to the AdS geometry (right) (after J. McGreevy (McGreevy, 2010)).

Also, an interesting, explicit, construction of a gravity dual as a theory of a family of RG transformations has been proposed by Lee (2010, 2011).

The CFT/gravity duality (or CFT/AdS correspondence) has been extended from its original formulation relating string theory (or supergravity) to the strong-coupling limit of the large-N limit of super-Yang–Mills theories to describe more general strongly coupled theories. For instance a CFT at finite temperature is described by this duality as a theory of gravity on AdS space-time with a black hole deep in the $(d+1)$-dimensional geometry. This description relates the Bekenstein–Hawking temperature of the black hole to the temperature of the CFT. It also provides an explanation for the Bekenstein–Hawking entropy that is consistent with the requirements of unitarity and the conservation of information. This relation resolved the information paradox posed by Hawking in 1980 (Susskind, 2008). The CFT/AdS correspondence also has important applications to poorly understood problems in condensed matter physics such as the behavior of Fermi fluids near quantum critical points and in so-called "non-Fermi-liquid" phases. We will not discuss here this important problem (since we have not discussed Fermi-liquid theory in the first place!), work on which is rapidly developing (Hartnoll *et al.*, 2010; McGreevy, 2010; Liu *et al.*, 2011; Hartnoll, 2012).

17.8.2 Holography and entanglement entropy

We will now apply the ideas of holography to the problem of the scaling of entanglement entropy in CFTs. This problem was first considered by Ryu and

Takayanagi, who used this approach to find a startling result (Ryu and Takayanagi, 2006b).

Let us consider first the CFT/gravity duality for the case of a bulk AdS_3. The boundary theory is then a conformal field theory in $(1+1)$ dimensions. In principle, to apply the CFT/AdS gravity correspondence to this problem, using Eq. (17.71), requires one to find the AdS_3 geometry that corresponds to the n-sheeted Riemann surface on which the CFT_2 is defined (as discussed in Section 17.5). Since this is a complex problem, Ryu and Takayanagi opted to conjecture that the following, much simpler, relation applies. It is worth emphasizing that there is strong evidence that this conjecture predicts correct results. For instance, its predictions are consistent with the physical argument that the entanglement entropy should obey strong subadditivity.

For concreteness we will consider the situation shown in Fig. 17.8, in which region A is a segment of length L (on the boundary) and region B is the complement of a CFT defined on a circle, i.e. we have periodic boundary conditions in space. Ryu and Takayanagi conjectured that this geometry is the n-sheeted AdS_3 and its generalization to higher dimensions, which is defined by putting the deficit angle $\delta = 2\pi(1-n)$ localized on a co-dimension-2 surface γ_A (including time). With this assumption they showed that the action for Einstein's gravity on the n-sheeted AdS_{d+1} geometry in general dimension d is the generalization of the Einstein–Hilbert action (this assumption was shown to be correct by Casini *et al.* (2011) for the case in which the entangling surface is S^{d-1} that is bipartitioning the total space S^d),

$$S_{\text{AdS}} = -\frac{1}{16\pi G_{\text{N}}^{(d+1)}} \int_M d^{d+1}x \sqrt{g}(R+\Lambda) + \cdots \tag{17.72}$$

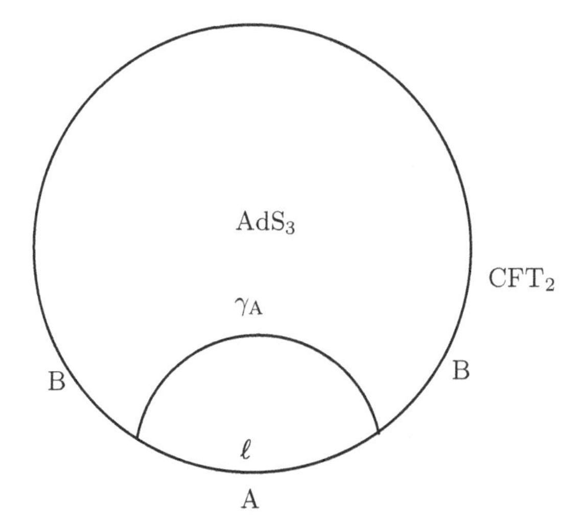

Figure 17.8 Entanglement entropy and the CFT/AdS correspondence.

where $G_N^{(d+1)}$ is the Newton constant of the AdS_{d+1} gravity, and R is the (Ricci scalar) curvature

$$R = 4\pi(1-n)\delta(\gamma_A) + R^0 \tag{17.73}$$

where R^0 is the scalar curvature of the AdS_{d+1} space-time.

We will now equate the partition function of the CFT_d on the n-sheeted Riemann surface with the partition function for gravity on the modified AdS_{d+1} geometry we just described. This allows us to find an expression for the entanglement entropy for region A. That is, S_A is given, in the gravity dual, defined on AdS_{d+1} (a $(d+1)$-dimensional AdS space-time), by the Bekenstein–Hawking entropy of AdS_{d+1}, which involves the *area* $\mathcal{A}(\gamma_A)$ of the *minimal surface* γ_A. Explicitly, they found the result (Nishioka *et al.*, 2009)

$$S_A = -\frac{\partial}{\partial n}\ln \operatorname{tr} \rho_A^n\Big|_{n\to 1} = -\frac{\partial}{\partial n}\left[\frac{(1-n)\mathcal{A}(\gamma_A)}{4G_N^{(d+1)}}\right]_{n\to 1} = \frac{\mathrm{A}(\gamma_A)}{4G_N^{(d+1)}} \tag{17.74}$$

which is the Bekenstein–Hawking formula for the entropy! (cf. Eq. (17.19)).

In the particular case in which the boundary is a 2D space-time, Eq. (17.74) predicts the entanglement entropy for CFT_2 in terms of the Bekenstein–Hawking entropy. To this end, we consider an interval of length ℓ on the boundary. The endpoints of this interval will also be the endpoints of a minimal surface in AdS_3, which is to say a geodesic between the points $(x, u) = (-\ell/2, a)$ and $(x, u) = (\ell/2, a)$, where $a \to 0$ plays the role of a UV cutoff. This geodesic is given by the half-circle

$$(x, u) = \frac{\ell}{2}(\cos\varphi, \sin\varphi) \tag{17.75}$$

with $\epsilon \leq \varphi \leq \pi - \epsilon$; here $\epsilon = 2a/\ell \to 0^+$ plays the role of the cutoff a. The length $\mathcal{A}(\gamma_A)$ of the circle γ_A is

$$\mathcal{A}(\gamma_A) = 2R\int_\epsilon^{\pi/2}\frac{d\varphi}{\sin\varphi} = 2R\ln\left(\frac{\ell}{a}\right) \tag{17.76}$$

This result then implies that the entropy is given by

$$S_A = \frac{R}{2G_N^{(3)}}\ln\left(\frac{\ell}{a}\right) \tag{17.77}$$

This result has the same scaling as the general expression for the entanglement entropy of an interval in a $(1+1)$-dimensional CFT of Eq. (17.36) (Holzhey *et al.*, 1994; Calabrese and Cardy, 2004). This suggests that we identify the central charge of the CFT as being related to its gravity dual by the relation

$$c = \frac{3R}{2G_N^{(3)}} \tag{17.78}$$

However, this identification makes sense only if the AdS radius R is large compared with the Planck length so that the classical description of gravity makes sense. In other words, the CFT_2/AdS_3 correspondence makes sense only if the central charge of the CFT is very large. In contrast, the result of Eq. (17.36) is nevertheless valid *for all* values of the central charge, even for values small enough that the AdS/CFT correspondence does not hold.

The CFT/gravity duality has been used by Ryu and Takayanagi to derive expressions for the entanglement entropy at finite temperatures, in which case there is a black hole deep in the AdS_3 geometry, and they found a result that agrees exactly with the expression of Eq. (17.38) obtained by Calabrese and Cardy using CFT_2. They also used this approach to obtain results for the entanglement entropy in higher-dimensional CFTs, where they found universal corrections to the area law. These corrections generally depend on scale-invariants of the geometry of the observed region, such as its aspect ratio.

17.9 Quantum entanglement and topological phases

We end this chapter with a discussion of the characterization of topological phases in terms of the behavior of the entanglement entropy. We will now see that the defining properties of topological phases are strongly apparent in entanglement-entropy measurements. Topological phases of matter are fully gapped states whose low-energy physics is described by a topological quantum field theory. A consequence of this feature is that the entanglement entropy has, in addition to the non-universal area law which is governed by short-distance physics, universal terms that are determined entirely by topological invariants. Here we will focus exclusively on the properties of topological phases in two space dimensions.

Before discussing the behavior of entanglement entropy in topological phases, let us summarize the defining properties of topological phases of matter discussed in Chapters 9 and 14 (see in particular Section 14.6).

1. Topological phases of matter are translationally and rotationally invariant states. Since they do not break any symmetries, neither of space-time nor internal, these fluid states cannot be characterized by a local order parameter. Hence, the ground states of topological phases respect all the symmetries of the system. However, on long length scales, that is, long compared with the lattice constant, the low-energy effective action of a topological phase is given by a topological field theory. The prototype topological field theories are the Chern–Simons gauge theories discussed extensively in Chapter 14 and the discrete gauge theories discussed in Chapter 9. In both cases the effective action has the key feature

that it does not depend on the local metric of the 2D surface on which the system is defined.

2. The ground state of a topological phase is generally not unique and, on a 2D surface of genus g, the number of handles of the surface, the degeneracy of the ground states grows exponentially with the genus as k^g, where k is an integer.
3. Topological phases come in two types, namely even and odd under time-reversal transformations. Time-reversal-odd topological phases occur in electron fluids in large magnetic fields, as discussed in Chapters 13 and 14, such as the fractional quantum Hall states, and in spin liquids with a spontaneously broken time-reversal invariance, such as the chiral spin liquid discussed in Chapter 10. Examples of time-reversal-invariant topological phases are the deconfined ("Coulomb") phases of discrete gauge theories, topological phases of QDMs, doubled Chern–Simons gauge theories (and BF gauge theories), and Kitaev's toric code (all discussed in Chapter 9) and its generalizations (Freedman *et al.*, 2004; Fendley and Fradkin, 2005; Fidkowski *et al.*, 2009).
4. In a topological phase all the local excitations have a large energy gap and have non-trivial quantum numbers. If the topological phase occurs in a charge fluid, the excitations are charged and their charge is generally fractional. The excitations in a topological phase carry the quantum numbers of non-trivial representations under the braid group, which are determined by their gauge charges in the topological field theory. If these representations are one-dimensional, the excitations are abelian anyons, and if they are multi-dimensional the excitations are non-abelian anyons.
5. The states of the topological field theory are conformal blocks of a 2D Euclidean CFT. The number of non-trivial representations, i.e. the number of non-trivial quasiparticles (including the identity), is equal to the ground-state degeneracy on a torus. The transformation laws of conformal blocks under the action of the modular group are given by the modular S-matrix, $\mathcal{S}$, of the CFT. The quasiparticles obey a fusion algebra with the same structure as in conformal field theory. Each quasiparticle state (conformal block) has a *quantum dimension* d_j, which is determined by the matrix elements of the modular S-matrix (see Eq. (14.116)). The quantum dimension d_j governs the rate of growth of the topologically protected Hilbert spaces of multi-quasiparticle states.

Let us consider now the problem of the scaling of the entanglement entropy in a topological phase. Let us consider first the simpler case of a region A of linear size ℓ that is simply connected and surrounded by its complement, region B. For simplicity, in Fig. 17.9 I show the case of the sphere. We will assume for now that the boundary Γ, of perimeter $L(\Gamma)$, is a smooth closed curve. Kitaev and Preskill (2006) and Levin and Wen (2006) showed that in the case of this geometry the

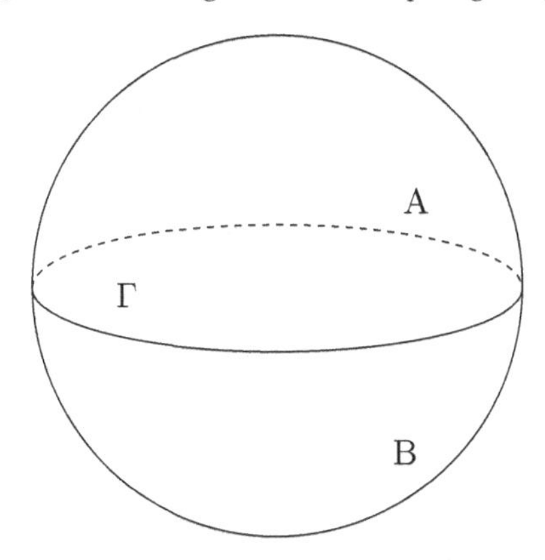

Figure 17.9 The entangling region A is simply connected and has a smooth boundary Γ.

entanglement entropy of region A has a universal finite correction to the "area-law" term:

$$S_{\mathrm{A}} = \alpha L(\Gamma) - \gamma_{\mathrm{topo}} \tag{17.79}$$

where α is non-universal (and hence is not predicted by the effective topological field theory). The quantity γ_{topo} is independent of the size of the region. It is not only universal but also a topological invariant. More specifically, we will see that γ_{topo} is given by certain matrix elements of the modular S-matrix of the topological field theory which describes the low-energy physics of the system of interest.

The modular S-matrix was introduced in Section 14.6.2. There we recalled the result of Witten's work on Chern–Simons theory (Witten, 1989) that relates the matrix elements of the modular S-matrix to the expectation value of Wilson loops in Chern–Simons theory. The same modular S-matrix determines the fusion rules of the Wilson loops in Chern–Simons theory and in its associated 2D Euclidean CFT, namely the WZW model. In particular, the quantum dimensions of the primary fields of the CFT, and of the excitations of the topological field theory, are given by the matrix elements of the modular S-matrix through Eq. (14.116).

For the case of a singly connected region A, with a smooth boundary Γ, γ_{topo} is given by

$$\gamma_{\mathrm{topo}} = \ln \mathcal{D} \tag{17.80}$$

where $\mathcal{D}$ is the *effective quantum dimension* of the topological field theory, and it is given in terms of the matrix element $\mathcal{S}_{00}$ of the modular S-matrix by (see Eq. (14.117))

$$\mathcal{D} = (\mathcal{S}_{00})^{-1} = \sqrt{\sum_j |d_j|^2} \tag{17.81}$$

where

$$d_j = \frac{\mathcal{S}_{0j}}{\mathcal{S}_{00}} \tag{17.82}$$

are the quantum dimensions of the states labeled by the representation j, and measure the rate of growth of the degenerate Hilbert spaces of particles with that representation. Thus, similarly to the case of quantum criticality which we discussed earlier in this chapter, there is a *finite universal term* that is a topological invariant (or expressed in terms of invariants), although in this case its contribution to the entanglement entropy is negative.

What will matter to the present discussion is that in his work on the relation between the theory of knots and Chern–Simons gauge theory Witten showed that the computation of the expectation value of a Wilson loop in the gauge theory reduces to the computation of a matrix element of the modular S-matrix in CFT (Witten, 1989) (see the discussion following Eq. (14.113) in Section 14.6.2). The following seminal results from Witten's work will be important to our discussion. For concreteness we will consider a Chern–Simons gauge theory with gauge group $G_k = \mathrm{SU}(2)_k$, where k is the level. The Chern–Simons action is

$$S(A) = \frac{k}{4\pi} \int \mathrm{tr}\left(A \wedge dA + \frac{2}{3} A \wedge A \wedge A \right) \tag{17.83}$$

where, as usual, A_μ is a vector field taking values in the algebra of a (compact) gauge group G. Here we will be primarily interested in the case of $G = \mathrm{SU}(2)$.

We will need a few important results on the structure of the Chern–Simons theory and its solution. First, following Witten (1989), we realize the states on a closed 2D surface as a path integral over a 3D volume. Witten showed that the Chern–Simons states on a spatial manifold Σ (which we will take to be closed) are in one-to-one correspondence with the conformal blocks of a Wess–Zumino–Witten (WZW) CFT. Furthermore, the ground-state degeneracy depends on the level k and on the topology of the surface Σ. The partition functions, i.e. the values of the path integral, depend on the matrix elements of the modular $\mathcal{S}$-matrix, e.g. the partition function on a space-time with the topology of a sphere S^3 with a Wilson loop in the representation ρ_j is

$$Z(S^3, \rho_j) = \mathcal{S}_{0j} \tag{17.84}$$

We have already encountered the modular S-matrix in Section 14.6.2, where we discussed the concept of non-abelian statistics. There we noted that the degenerate states of the topological fluids on a torus have a one-to-one correspondence with the conformal blocks of an associated 2D CFT. We also discussed the fact that the short-distance behavior of the conformal blocks is equivalent to the characters χ_j of the representations that label the conformal blocks. The modular S-matrix is the transformation matrix of the characters under modular transformations of the torus, Eq. (14.109).

We will also need the modular S-matrices of the conformal blocks for the theories we are interested in. For the gauge group $\mathrm{U}(1)_m$, $n = 0, \ldots, m-1$, the modular S-matrix is

$$\mathcal{S}_{n,n'} = \frac{1}{\sqrt{m}} e^{2\pi i n n'/m} \tag{17.85}$$

For the gauge group $\mathrm{SU}(2)_k$, $j, j' = 0, 1/2, \ldots, k/2$, the modular S-matrix is

$$\mathcal{S}^{(k)}_{j,j'} = \sqrt{\frac{2}{k+2}} \sin\left(\frac{(2j+1)(2j'+1)\pi}{k+2}\right) \tag{17.86}$$

The Chern–Simons path integral, the partition function on various manifolds, can be reduced to its computation on a sphere S^3 using the method of (Chern–Simons) surgeries (Witten, 1989). Using surgeries, it can be shown that, if a 3-manifold M is the connected sum of two 3-manifolds M_1 and M_2 joined along an S^2, then the Chern–Simons partition functions on these manifolds are related by

$$Z(M)Z(S^3) = Z(M_1)Z(M_2) \tag{17.87}$$

In particular, if M is M_1 and M_2 joined along n S^2s, the resulting partition function is

$$Z(M) = \frac{Z(M_1)Z(M_2)}{Z(S^3)^n} \tag{17.88}$$

Witten's result can be used to compute the entanglement entropy in various cases of interest, and was used by Kitaev and Preskill (2006) to derive the result of Eq. (17.80) for a simply connected region. The Kitaev–Preskill results were extended and generalized by Dong and coworkers (Dong *et al.*, 2008), who used the replica approach to compute the entanglement entropies for topologically nontrivial regions on a torus and for states with quasiparticles, represented by punctures carrying specific quantum numbers (representation labels).

The replica calculation computes the Chern–Simons partition function on an n-sheeted Riemann-surface space-time (as in the previously discussed cases). Although the space-time manifold needed for the replica approach is rather involved, explicit results for the entanglement entropies can nevertheless be obtained using Witten's method of surgeries (Dong *et al.*, 2008).

Let us consider first the simplest case, in which the spatial manifold is a sphere, $\Sigma = S^2$, and hence the space-time manifold is just a 3-sphere, $\Sigma \times S^1 \cong S^3$. The

Hilbert space on S^3 is one-dimensional. Using the method of surgeries, Dong and coworkers (Dong *et al.*, 2008) considered the case of S^2 with a single boundary between regions A and B, as shown in Fig. 17.9, i.e. the observed region is a hemisphere, and the two regions A and B are two hemispheres (disks). The 3-geometry is a ball.

To construct $\operatorname{tr} \rho_{\rm A}^n$ we glue $2n$ such pieces together. When they are glued together to form $\operatorname{tr} \rho_{\rm A}^n$, one finds a manifold with the topology of S^3 for all n, and therefore has the same partition function. Thus, we find that the (normalized) trace of $\rho_{\rm A}^n$ is

$$\frac{\operatorname{tr} \rho^n_{{\rm A}(S^2,1)}}{\left(\operatorname{tr} \rho_{{\rm A}(S^2,1)}\right)^n} = \frac{Z(S^3)}{\left(Z(S^3)\right)^n} = \left(Z(S^3)\right)^{1-n} = \mathcal{S}_{00}^{1-n} \tag{17.89}$$

In the replica limit, $n \to 1$, we obtain for the entanglement entropy

$$S_{\rm A}^{(S^2,1)} = \ln \mathcal{S}_{00} = -\ln \mathcal{D} \tag{17.90}$$

which is the result of Kitaev and Preskill (2006) and Levin and Wen (2006) for the universal topological entanglement entropy. Notice that the topological-field-theory calculation computes only the universal contribution. The non-universal, and hence cutoff-dependent, area-law term has been regularized to zero by the methods used by Witten. This is not surprising, since the Chern–Simons gauge theory does not depend on the metric. In other words, the area-law term is given by the non-topological short-distance corrections to the topological field theory.

The result of Eq. (17.90) also holds for surfaces with arbitrary topology, provided that the region A being observed is topologically trivial, regardless of the pure state labeled by the representations ρ_j. For the case of a sphere S^2 and a disconnected region A with p boundaries, we trivially find that they are additive, $S_{\rm A}^{(S^2,p)} = p \ln \mathcal{S}_{00} = -p \ln \mathcal{D}$.

Let us compute the entanglement entropy for a Chern–Simons theory on the torus T^2. For a torus T^2 split into two regions we have two cases, shown in Fig. 17.10. If the torus is in the trivial state, that is, without any Wilson loop threading the torus, the entropy for Fig. 17.10(a) is the same as for the sphere, cf. Eq. (17.90), whereas for the case of Fig. 17.10(b) it is twice as large, $S_{\rm A}(T^2, 2) = 2 \ln \mathcal{S}_{00}$. However, if there is a Wilson loop with a non-trivial representation ρ_j threading the torus, we obtain the same result for the case depicted in Fig. 17.10(a), but, for the case of Fig. 17.10(b), we obtain instead

$$S_{\rm A}(T^2, 2, \rho_j) = 2 \ln \mathcal{S}_{0\rho_j} \tag{17.91}$$

In other words in this case, the entanglement entropy is different for the different degenerate states on the torus, each labeled by a representation ρ_j of the Wilson

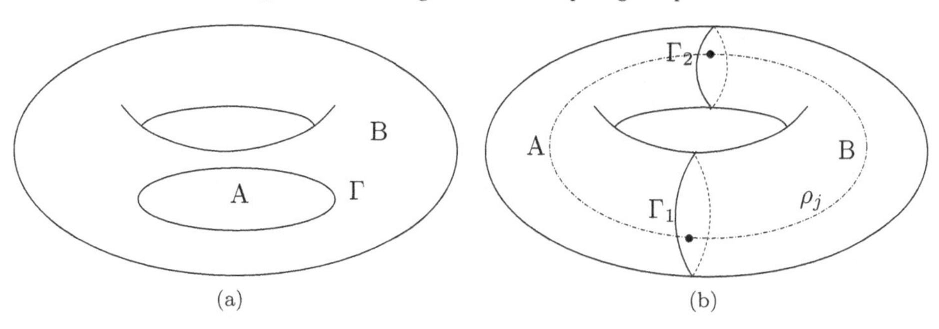

Figure 17.10 Entanglement on a torus geometry. (a) Region A is simply connected and the entanglement entropy is independent of the state of the torus. (b) Region A is topologically non-trivial, and its entanglement entropy depends on the ground state of the system on the torus, which is labeled by the representation ρ_j of the Wilson loop (the dot–dashed loop).

loop. In addition, if the torus is in a state that is a linear superposition, $|\psi\rangle = \sum_\rho \psi_\rho |\rho\rangle$, we further find

$$S_{\mathrm{A}}(T^2, 2, \psi) = 2 \ln \mathcal{S}_{00} - \sum_\rho d_\rho^2 \left(\frac{|\psi_\rho|^2}{d_\rho^2} \ln \left(\frac{|\psi_\rho|^2}{d_\rho^2} \right) \right) \tag{17.92}$$

Clearly, the entanglement entropy now depends not only on the effective quantum dimension $\mathcal{D} = \mathcal{S}_{00}^{-1}$ but also on the quantum dimension of the excitation labeled by the representation ρ, as well as on the particular linear combination of the degenerate ground states on the torus in which the system is prepared.

Following the same line of argument, one can consider other situations of interest. For example, in Fig. 17.11 we consider a simply connected entangling region A on a sphere with four quasiparticles represented here by four punctures. A temporal Wilson loop pierces the sphere S^2 at each puncture and carries a representation label, γ_1, γ_2, γ_3, and γ_4. Dong and coworkers (Dong *et al.*, 2008) showed that the entanglement entropy now depends on the fusion rules of the quasiparticles in the case of Fig. 17.11(b) since the quasiparticles need to fuse across the boundary Γ, but does not depend on these properties for the case of Fig. 17.11(a), where they must fuse into the identity. The reason for this difference is that the fusion of the quasiparticles amounts to changing the topology of the surface with Wilson loops running around the glued circles, with each fusion channel adding a handle to the surface. In this way the entanglement entropy can detect in which state the set of quasiparticles ("qubits") is. This property is important to the concept of topological quantum computation and is the key to the topological robustness of the state.

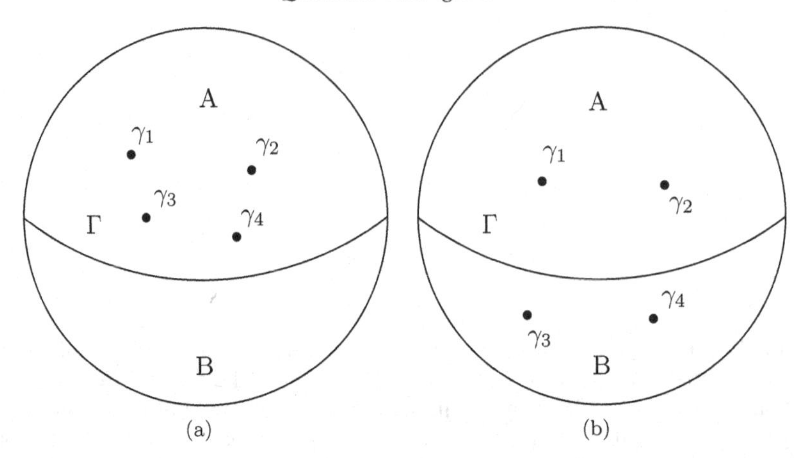

Figure 17.11 States on the sphere with four quasiparticles (punctures) labeled by γ_i (with $i = 1, \ldots, 4$). (a) All the quasiparticles are inside the entangling region. (b) Two of the quasiparticles are inside the entangling region and two are outside.

The upshot of this discussion is that the entanglement entropy provides a way to determine all the universal properties of the topological phases, including their degeneracy, their quantum numbers, and their fusion rules.

We close this section with an application of these ideas and results to several cases of physical interest.

The deconfined phase of the $\mathbb{Z}_2$ gauge theory. The simplest topological phase is the deconfined phase of the $\mathbb{Z}_2$ gauge theory, a.k.a. the $\mathbb{Z}_2$ spin liquid. In $(2 + 1)$ dimensions this phase is essentially equivalent to Kitaev's toric code and to the topological phase of the quantum dimer model on the triangular lattice. This theory has a four-fold-degenerate ground state on a torus (see Section 9.8). The four states are obtained by acting with the 't Hooft magnetic loops along the two non-contractible loops of the torus. In this case each operator creates a one-dimensional representation of the braid group (hence it is abelian). Therefore this theory has four sectors, each with quantum dimension $d = 1$, and the total effective dimension is $\mathcal{D}[\mathbb{Z}_2] = 2$. The topological term in the entanglement entropy is

$$\gamma_{\text{topo}}[\mathbb{Z}_2] = \ln 2 \tag{17.93}$$

This result was obtained by an explicit calculation with the Kitaev state on the lattice by Hamma and coworkers (Hamma *et al.*, 2005a, 2005b) and by Levin and Wen (2006). The validity of this result for the case of the $\mathbb{Z}_2$ spin liquid of the QDM on the triangular lattice was shown by Furukawa and Misguich (2007). Papanikolaou and coworkers (Papanikolaou *et al.*, 2007c) extended this result by computing the entanglement entropy for the eight-vertex wave function discussed in Section 9.8, Eq. (9.67). This is a topological phase that includes the Kitaev state as a particular case. They showed that the topological contribution to the entanglement entropy is given by Eq. (17.93) without correction in the entire topological phase. In contrast, the non-universal area-law term varies continuously throughout the phase.

The Laughlin states of the fractional quantum Hall effect. The Laughlin states of the fractional quantum Hall effect at filling fraction $\nu = 1/m$ have an m-fold degenerate ground state on the torus. As we showed in Chapters 13 and 14, the effective low-energy action for the Laughlin states is a $U(1)_m$ abelian Chern–Simons gauge theory. The modular S-matrix for this theory is given by Eq. (17.85). Each one-dimensional subspace is assigned to one of the m distinct quasiparticles of the Laughlin states: the identity I (no quasiparticle) and the $m-1$ quasihole states $V_n = \exp(-in\phi(x)/\sqrt{m})$, with $n = 1, \ldots, m-1$. These excitations are abelian anyons, with statistical angle $\delta_n = n\pi/m$ (and charge ne/m) and with quantum dimension $d_n = 1$. Each ground state is created by a Wilson loop operator (in the effective $U(1)_m$ Chern–Simons gauge theory) with gauge charge n. Since all the sectors are associated with states with quantum dimension $d = 1$ or, equivalently, since $\mathcal{S}_{00} = 1/\sqrt{m}$, the effective quantum dimension $\mathcal{D}$ for the Laughlin states is $\mathcal{D}[U(1)_m] = \sqrt{m}$, and the topological contribution to the entanglement entropy is

$$\gamma_{\text{topo}}[U(1)_m] = \ln \sqrt{m} \tag{17.94}$$

This result was confirmed numerically by Haque and coworkers (Zozulya *et al.*, 2009) using both the Laughlin wave function and the wave function for the Coulomb interaction (which has a large overlap with the Laughlin wave function). This result is trivially extended to all abelian fractional quantum Hall states, such as the hierarchical and multi-component states discussed in Sections 14.3 and 14.4. Indeed, for a generic abelian fractional quantum Hall fluid defined by a K-matrix, the number of sectors, and hence the ground-state degeneracy on a torus, is $|\det K|$. Since each sector has quantum dimension 1 (being abelian) the topological contribution to the entanglement entropy is

$$\gamma_{\text{topo}}[\text{abelian}] = \ln \sqrt{|\det K|} \tag{17.95}$$

The result for the Laughlin states is a particular case.

Abelian double Chern–Simons theories. Abelian double Chern–Simons theories represent time-reversal-invariant fractionalized phases. We encountered this example in our discussion of superconductors as topological fluids in Section 14.5 and in our discussion of fractionalized (abelian) topological insulators in Section 16.13, where it took the form of a "BF theory." The $\mathbb{Z}_2$ spin liquid is equivalent to this theory for level $k = 2$.

The non-abelian fractional quantum Hall states. We have discussed several non-abelian fractional quantum Hall states and related systems, see Section 14.8. The simplest non-abelian state is the Moore–Read fractional quantum Hall state for bosons at filling fraction $\nu = 1$. We showed in Sections 14.8 and 15.4.5 that the effective topological field theory for this state is the $SU(2)_2$ non-abelian Chern–Simons gauge theory. This system has a three-fold-degenerate ground state on the torus, corresponding to the conformal blocks labeled by the identity I, the non-abelion $\sigma e^{i\phi}$, and the Majorana fermion ψ. The modular S-matrix, $\mathcal{S}$, for this system, in the basis of the conformal blocks listed above, is

$$\mathcal{S}[SU(2)_2] = \frac{1}{2}\begin{pmatrix} 1 & \sqrt{2} & 1 \\ \sqrt{2} & 0 & -\sqrt{2} \\ 1 & -\sqrt{2} & 1 \end{pmatrix} \tag{17.96}$$

Thus, we see that the effective quantum dimension is $\mathcal{D}[SU(2)_2] = \mathcal{S}_{00}^{-1} = \ln 2$, and the topological contribution to the entanglement entropy is

$$\gamma_{\text{topo}}[SU(2)_2] = \ln 2 \tag{17.97}$$

Another relevant case is the $p_x + ip_y$ superconductor, which was discussed in Section 14.9. The CFT associated with this problem is the chiral critical 2D Ising model, which is represented by the $SU(2)_2/U(1)_2$ coset. This chiral CFT has three primaries: the identity I, the spin field σ, and the Majorana fermion ψ. The modular S-matrix turns out to be the same as in the $SU(2)_2$ case we just discussed (see Dong *et al.* (2008)). Hence the quantum dimensions are the same, and we conclude that the entanglement entropy of Eq. (17.97) also applies to the $p_x + ip_y$ chiral superconductor.

The non-abelian fractional quantum Hall Moore–Read fermionic state (with total filling fraction $\nu = 5/2$) has a six-fold-degenerate ground state on the torus. These six ground states correspond to the conformal blocks of the identity I, the non-abelian quasiparticle and quasihole $\sigma e^{\pm i\phi/2\sqrt{2}}$, the Majorana fermion ψ, and the abelian (Laughlin) quasiparticle and quasihole $e^{\pm i\sqrt{2}\phi}$. Except for the non-abelian quasiparticle and quasihole, which have quantum dimension $d_\sigma = \sqrt{2}$, all other states are abelian and have quantum dimension $d = 1$. The fermionic Moore–Read state is represented by the chiral coset Chern–Simons gauge theory (and chiral CFT) $(SU(2)_2/U(1)_2) \times U(1)_8 \simeq \mathbb{Z}_2 \times U(1)_2$. Its modular S-matrix, in the basis of the conformal block listed above, is given by

$$\mathcal{S}[\text{MR}] = \frac{1}{2\sqrt{2}}\begin{pmatrix} 1 & 1 & \sqrt{2} & \sqrt{2} & 1 & 1 \\ 1 & 1 & -\sqrt{2} & -\sqrt{2} & 1 & 1 \\ \sqrt{2} & -\sqrt{2} & 0 & 0 & i\sqrt{2} & -i\sqrt{2} \\ \sqrt{2} & -\sqrt{2} & 0 & 0 & -i\sqrt{2} & i\sqrt{2} \\ 1 & 1 & i\sqrt{2} & -i\sqrt{2} & -1 & -1 \\ 1 & 1 & -i\sqrt{2} & i\sqrt{2} & -1 & -1 \end{pmatrix} \tag{17.98}$$

Thus, the total effective dimension of the Moore–Read fermionic state is $\mathcal{S}_{00}^{-1} = \mathcal{D}[\mathrm{MR}] = 2\sqrt{2}$, and the topological contribution to the entanglement entropy now is

$$\gamma_{\mathrm{topo}}[\mathrm{MR}] = \ln(2\sqrt{2}) \tag{17.99}$$

Notice that for this non-abelian state the effective quantum dimension is $\mathcal{D}[\mathrm{MR}] = \sqrt{8}$ and it is not equal to the square root of the number of degenerate ground states on a torus (six in this case). This is a generic feature of all non-abelian states. Similar, but more complex, expressions can be obtained for the $SU(2)_3$ Read–Rezayi states (both fermionic and bosonic) (Dong *et al.*, 2008). These non-abelian states are candidates for universal topological qubits (Freedman *et al.*, 2002a; Das Sarma *et al.*, 2008).

17.10 Outlook

In this chapter we have discussed the role of quantum entanglement in condensed matter physics. We have focused primarily on the scale-dependence of the entanglement entropy as a way to characterize phases and quantum critical points where large-scale entanglement is realized. For this reason we have centered our attention on the behavior of the von Neumann entropy at quantum critical points and in topological phases. It is fair to say that, while the latter case is by now fairly well understood (although, as we will see below, not completely), the understanding of its behavior at quantum criticality is still in its initial stages, except possibly in one dimension.

We have left out many important problems and questions. In our discussion of the scaling of the entanglement entropy in quantum critical systems we have not discussed the effects of perturbations away from the conformal limit (i.e. away from the fixed point). This problem has so far been studied only in the 1D case (Pollmann *et al.*, 2009; Calabrese and Essler, 2010; Cardy and Calabrese, 2010). We have also not discussed the problem of the entanglement of quantum impurities within critical systems. This question was studied in one dimension in detail by Affleck and coworkers (Laflorencie *et al.*, 2006; Sørensen *et al.*, 2007a, 2007b; Affleck, 2010), and independently by Kopp and Chakravarty (Kopp *et al.*, 2007), who showed explicitly how different boundary conditions enter into the universal behavior of the entanglement entropy.

A problem in which the ideas of quantum entanglement will very likely play an important role is the behavior of disordered quantum systems. By disorder here we do not mean uniform systems without long-range order but rather systems that are physically disordered and are best regarded as random systems. This is a notoriously difficult problem, which we have not discussed in this book. It has remained an open problem even in classical statistical mechanics, which in spite of decades of effort still has a host of so far poorly understood problems such as spin glasses and random field systems.

The quantum version of these problems is certainly no less difficult and the study of the behavior (and role) of quantum entanglement in these systems is in its beginnings. Chakravarty has shown that the scaling of the entanglement entropy can be used to study the Anderson localization–delocalization transition in disordered systems (Jia *et al.*, 2008; Chakravarty, 2010). In a pioneering series of papers Refael and Moore showed that, in the case of random spin chains at their infinite-disorder fixed point (Fisher, 1994, 1995), the (ensemble-average) entanglement entropy of 1D random critical spin chains has a logarithmic dependence on the linear size ℓ of the entangling region, $S_{\mathrm{A}} \sim A \ln \ell$ (Refael and Moore, 2004, 2009). The constant A turns out to be universal and to be different only for different universality classes of random fixed points. Although this result has the same form as that in the case of 1D CFT, the random fixed points are not conformal and there is no natural definition of a central charge. So the connection between the universal prefactor A and the critical behavior of local operators at random fixed points is not so far understood.

One question that we have not discussed is that of the so-called entanglement spectrum. The entanglement spectrum is the spectrum of the reduced density matrix of the entangling region. It is typically presented by writing the reduced density matrix in the suggestive exponential form $\rho_{\mathrm{A}} = \exp(-\mathcal{H}_{\mathrm{A}})$. Since the eigenvalues of the reduced density matrix, $\{\lambda_i\}$, by definition are real numbers between 0 and 1, the eigenvalues $\{\varepsilon_i\}$ of the "pseudo-Hamiltonian" $\mathcal{H}_{\mathrm{A}}$ are real positive numbers (or zero). The numerical study of the entanglement spectrum of fractional quantum Hall states has revealed that it contains a wealth of information on the nature of these states. Li and Haldane (2008), who pioneered those studies, found that the degeneracies of the low-energy entanglement spectra of the fractional quantum Hall states agree with the partitions of states encoded in the characters of the associated CFT. This relation extends to much of the spectrum for the "ideal states," namely wave functions constructed using conformal blocks such as the Laughlin and Moore–Read states. However, this relation holds also for the low-energy entanglement spectrum for the "realistic" wave functions obtained by exact diagonalization with Coulomb interactions. For the same reason the Rényi entropies S_n have the same behavior in the large-n limit, which selects the low-pseudo-energy portion of the entanglement spectrum.

At the time of writing there isn't a general theory of the entanglement spectrum, at least not a theory developed to the same extent as what we have discussed in this chapter. Nevertheless, what is clear is that the interesting universal features which have been found in numerical studies, such as in Li and Haldane (2008), can also be extracted by studying finite-size effects in the Rényi entropies in the

$n \to \infty$ regime. Earlier in this chapter we saw that the topological field theory of topological phases computes only the finite, universal, terms in the entanglement entropy. This also follows from the fact that the result we found for $\operatorname{tr} \rho_A^n$ is a simple power of n, such as $\mathcal{S}_{00}^{n-1}$. Hence, in the topological limit the spectrum of the density matrix ought to consist only of zeros or ones.

On the other hand, this should hardly be surprising since any finite-size effect requires the definition of a distance in the system which the topological field theory does not have (since by definition it is independent of the metric!). To compute these effects requires the computation of (non-topological) corrections to the topological-field-theory results. The empirical evidence that a "ghost" of the topological structure survives in the finite-size corrections is very interesting. This is an important but challenging open problem. To date there isn't a scaling theory of the entanglement spectra, although some very interesting results have been obtained in 1D spin chains (Calabrese and Lefevre, 2008; Pollmann *et al.*, 2010; Pollmann and Moore, 2010), in quantum disordered systems (Fagotti *et al.*, 2011), and in the quantum Lifshitz model (Hsu and Fradkin, 2010).

Finally, we will need to address the most important omission we have made in this chapter, namely, can the entanglement entropy be measured? This is the "elephant in the room" question of this problem. Although we showed that the behavior of large-scale entanglement can be used to characterize phases (and quantum phase transitions), it is far from obvious whether as a matter of principle this quantity can be measured in any reasonable experiment. In contrast to what happens in a system with a small number of degrees of freedom, in which the reduced density matrix can be tested directly in experiment, the reduced density matrix of a large extended system is a very non-local object. It is this non-local nature of the reduced density matrix which makes it useful to characterize large-scale entanglement.

However, most experiments in physics consist in the measurement of local observables, including their time evolution. This is true even for thermodynamic measurements, since these test the behavior of the spatial average of local observables, such as the charge density, current, magnetization, energy density, etc. The entanglement entropy (and, for the same reason, the entanglement spectra) cannot be reduced to measurements of this type. Nevertheless, a few proposals for measuring the entanglement entropy have been put forth.

So far the most practical suggestion has been that made by Klich and Levitov (2009). They proposed the use of an externally driven tunneling point contact between two electronic reservoirs. In their scheme the point contact would be open and closed suddenly and sequentially. They then showed that, at least for a free-fermion system, the noise in the tunneling current can detect the entanglement between the two free-fermion reservoirs. This problem was reexamined in detail by Hsu and coworkers (Hsu *et al.*, 2009a), who considered the problem of a quantum point contact between the edge states in a Laughlin quantum Hall state. This work considered the simpler case of a single sudden opening of the point contact, i.e. a local *quantum quench*. They showed that the entanglement and the quantum noise are two essentially unrelated quantities that have the same scaling in time for dimensional reasons. In addition, during a quantum quench it is the dynamical entanglement which is being tested, which in the case of a local quench in 1D critical systems has a universal logarithmic growth in *time* (Calabrese and Cardy, 2007; Hsu *et al.*, 2009a). On the other hand, for systems that do not have a conserved charge, such as the quantum Ising chain, the noise in the *energy* current has a different time dependence as the entanglement entropy (as required by dimensional analysis). The general problem of the dynamical behavior of entanglement under a quench (both local and global) is of great interest in cold-atomic systems, in which an external manipulation of the effective Hamiltonians is possible.

References

Abrahams, E., Anderson, P. W., Licciardello, D. C., and Ramakrishnan, T. V. 1979. Scaling theory of localization: Absence of quantum diffusion in two dimensions. *Phys. Rev. Lett.*, **42**, 673.

Abrikosov, A. A., Gor'kov, L. P., and Dzyaloshinskii, I. E. 1963. *Methods of Quantum Field Theory in Statistical Physics*. Englewood Cliffs, NJ: Prentice-Hall.

Affleck, I. 1985. Large-N limit of SU(N) quantum "spin" chains. *Phys. Rev. Lett.*, **54**, 966.

Affleck, I. 1986a. Exact critical exponents for quantum spin chains, non-linear σ-models at $\theta = \pi$ and the quantum Hall effect. *Nucl. Phys. B*, **265**, 409.

Affleck, I. 1986b. Universal term in the free energy at a critical point and the conformal anomaly. *Phys. Rev. Lett.*, **56**, 746.

Affleck, I. 1990. Field theory methods and quantum critical phenomena, in *Fields, Strings and Critical Phenomena. Proceedings of the Les Houches Summer School 1988, Session XLIX*, E. Brézin and J. Zinn-Justin (eds.). Amsterdam: North-Holland, p. 563.

Affleck, I. 1998. Exact correlation amplitude for the $S = \frac{1}{2}$ Heisenberg antiferromagnetic chain. *J. Phys. A: Math. Gen.*, **31**, 4573.

Affleck, I. 2010. Quantum impurity problems in condensed matter physics, in *Exact Methods in Low-Dimensional Statistical Physics and Quantum Computing. Proceedings of the Les Houches Summer School 2008, Session LXXXIX*, J. Jacobson, S. Ouvry, V. Pasquier, D. Serban, and L. F. Cugliandolo (eds.). Oxford: Oxford University Press.

Affleck, I., and Haldane, F. D. M. 1987. Critical theory of quantum spin chains. *Phys. Rev. B*, **36**, 5291.

Affleck, I., and Ludwig, A. W. W. 1991. Universal noninteger "ground-state degeneracy" in critical quantum systems. *Phys. Rev. Lett.*, **67**, 161.

Affleck, I., and Marston, J. B. 1988. Large-N limit of the Heisenberg–Hubbard model: Implications for high-T_c superconductors. *Phys. Rev. B*, **37**, 3774.

Affleck, I., Zou, Z., Hsu, T., and Anderson, P. W. 1988a. SU(2) gauge symmetry of the large-U limit of the Hubbard model. *Phys. Rev. B*, **38**, 745.

Affleck, I., Kennedy, T., Lieb, E. H., and Tasaki, H. 1988b. Valence bond ground states in isotropic quantum antiferromagnets. *Commun. Math. Phys.*, **115**, 477.

Affleck, I., Harvey, J., Palla, L., and Semenoff, G. W. 1989. The Chern–Simons term versus the monopole. *Nucl. Phys. B*, **328**, 575.

Aharony, O., Gubser, S. S., Maldacena, J. M., Ooguri, H., and Oz, Y. 2000. Large N field theories, string theory and gravity. *Phys. Rep.*, **323**, 183.

Albuquerque, A. F., and Alet, F. 2010. Critical correlations for short-range valence-bond wave functions on the square lattice. *Phys. Rev. B*, **82**, 180408(R).

Alet, F., Jacobsen, J. L., Misguich, G. *et al.* 2005. Interacting classical dimers on the square lattice. *Phys. Rev. Lett.*, **94**, 235702.

Alet, F., Ikhlef, Y., Jacobsen, J. L., Misguich, G., and Pasquier, V. 2006. Classical dimers with aligning interactions on the square lattice. *Phys. Rev. E*, **74**, 041124.

Alicea, J., Oreg, Y., Refael, G., von Oppen, F., and Fisher, M. P. A. 2011. Non-Abelian statistics and topological quantum information processing in 1D wire networks. *Nature Phys.*, **7**, 412.

Amit, D. J. 1980. *Field Theory, the Renormalization Group and Critical Phenomena.* New York, NY: McGraw-Hill.

Amit, D. J., Goldschmidt, Y., and Grinstein, G. 1980. Renormalisation group analysis of the phase transition in the 2D Coulomb gas, sine–Gordon theory and XY-model. *J. Phys. A: Math. Gen.*, **13**, 585.

Anderson, P. W. 1958. Absence of diffusion in certain random lattices. *Phys. Rev.*, **109**, 1492.

Anderson, P. W. 1970. A poor man's derivation of the scaling laws of the Kondo problem. *J. Phys. C: Solid State Phys.*, **3**, 2436.

Anderson, P. W. 1973. Resonating valence bonds: A new kind of insulator? *Mater. Res. Bull.*, **8**, 153.

Anderson, P. W. 1987. The resonating valence bond state in La_2CuO_4 and superconductivity. *Science*, **235**, 1196.

Anderson, P. W., Yuval, G., and Hamann, D. R. 1970. Exact results in the Kondo problem. II. Scaling theory, qualitatively correct solution, and some new results on one-dimensional classical statistical models. *Phys. Rev. B*, **1**, 4464.

Ardonne, E., Fendley, P., and Fradkin, E. 2004. Topological order and conformal quantum critical points. *Ann. Phys.*, **310**, 493.

Armour, W., Hands, S., Kogut, J. B. *et al.* 2011. Magnetic monopole plasma phase in $(2+1)$d compact quantum electrodynamics with fermionic matter. *Phys. Rev. D*, **84**, 014502.

Arovas, D., Schrieffer, J. R., and Wilczek, F. 1984. Fractional statistics and the quantum Hall effect. *Phys. Rev. Lett.*, **53**, 722.

Arovas, D. P., and Auerbach, A. 1988. Functional integral theories of low-dimensional quantum Heisenberg models. *Phys. Rev. B*, **38**, 316.

Arovas, D. P., Schrieffer, J. R., Wilczek, F., and Zee, A. 1985. Statistical mechanics of anyons. *Nucl. Phys. B*, **251**, 117.

Assaad, F. F. 2005. Phase diagram of the half-filled two-dimensional SU(N) Hubbard–Heisenberg model: A quantum Monte Carlo study. *Phys. Rev. B*, **71**, 075103.

Auerbach, A. 1994. *Interacting Electrons and Quantum Magnetism*, 2nd edn. Berlin: Springer-Verlag.

Avron, J. E., Seiler, R., and Simon, B. 1983. Homotopy and quantization in condensed matter physics. *Phys. Rev. Lett.*, **51**, 51.

Avron, J. E., Seiler, R. and Zograf, P. G. 1995. Viscosity in quantum Hall fluids. *Phys. Rev. Lett.*, **75**, 697.

Babujian, H. M., and Tsvelik, A. M. 1986. Heisenberg magnet with an arbitrary spin and anisotropic chiral field. *Nucl. Phys. B*, **265**, 24.

Bais, F. A., van Driel, P., and de Wild Propitius, M. 1992. Quantum symmetries in discrete gauge theories. *Phys. Lett. B*, **280**, 63.

Balatsky, A., and Fradkin, E. 1991. Singlet quantum Hall effect and Chern–Simons theories. *Phys. Rev. B*, **43**, 10622.

Balian, R., Drouffe, J. M., and Itzykson, C. 1975. Gauge fields on a lattice. II. Gauge-invariant Ising model. *Phys. Rev. D*, **11**, 2098.

Bander, M., and Itzykson, C. 1977. Quantum-field-theory calculation of the two-dimensional Ising model correlation function. *Phys. Rev. D*, **15**, 463.

Banks, T., and Lykken, J. D. 1990. Landau–Ginzburg description of anyonic superconductors. *Nucl. Phys. B*, **336**, 500.

Banks, T., Myerson, R., and Kogut, J. 1977. Phase transitions in abelian lattice gauge theories. *Nucl. Phys. B*, **129**, 493.

Barkeshli, M., and Wen, X. G. 2010a. Classification of Abelian and non-Abelian multi-layer fractional quantum Hall states through the pattern of zeros. *Phys. Rev. B*, **82**, 245301.

Barkeshli, M., and Wen, X. G. 2010b. Effective field theory and projective construction for $\mathbb{Z}_k$ parafermion fractional quantum Hall states. *Phys. Rev. B*, **81**, 155302.

Barrett, S. E., Dabbagh, G., Pfeiffer, L. N., West, K. W., and Tycko, R. 1995. Optically pumped NMR evidence for finite-size skyrmions in GaAs quantum wells near Landau level filling $\nu = 1$. *Phys. Rev. Lett.*, **74**, 5112.

Baskaran, G., and Anderson, P. W. 1988. Gauge theory of high-temperature superconductors and strongly correlated Fermi systems. *Phys. Rev. B*, **37**, 580.

Baskaran, G., Zou, Z., and Anderson, P. W. 1987. The resonating valence bond state and high-T_c superconductivity – a mean field theory. *Solid State Commun.*, **63**, 973.

Baxter, R.J. 1982. *Exactly Solved Models in Statistical Mechanics*. New York, NY: Academic Press.

Baym, G. 1974. *Lectures on Quantum Mechanics*. New York, NY: Benjamin.

Baym, G., and Pethick, C. J. 1991. *Landau Fermi Liquid Theory*. New York, NY: John Wiley & Sons.

Bekenstein, J. D. 1973. Black holes and entropy. *Phys. Rev. D*, **7**, 2333.

Belavin, A. A., Polyakov, A. M., and Zamolodchikov, A. B. 1984. Infinite conformal symmetry in two-dimensional quantum field theory. *Nucl. Phys. B*, **241**, 333.

Bergknoff, H., and Thacker, H. B. 1979. Structure and solution of the massive Thirring model. *Phys. Rev. D*, **19**, 3666.

Bernevig, B. A., Hughes, T. L., and Zhang, S. C. 2006. Quantum spin Hall effect and topological phase transition in HgTe quantum wells. *Science*, **314**, 1757.

Berry, M. V. 1984. Quantal phase factors accompanying adiabatic changes. *Proc. Roy. Soc. London A*, **392**, 45.

Bethe, H. 1931. Theory of metals. I. Eigenvalues and eigenfunctions of the linear atomic chain. *Z. Phys.*, **71**, 205.

Bishara, W., Bonderson, P., Nayak, C., Shtengel, K., and Slingerland, J. K. 2009. Interferometric signature of non-abelian anyons. *Phys. Rev. B*, **80**, 155303.

Bloch, F. 1930. Theory of ferromagnetism. *Z. Phys.*, **61**, 206.

Bloch, F. 1933. Stopping power of atoms with several electrons. *Z. Phys.*, **81**, 363.

Blok, B., and Wen, X. G. 1990. Effective theories of the fractional quantum Hall effect: Hierarchy construction. *Phys. Rev. B*, **42**, 8145.

Blöte, H. W. J., Cardy, J. L., and Nightingale, M. P. 1986. Conformal invariance, the central charge, and universal finite-size amplitudes at criticality. *Phys. Rev. Lett.*, **56**, 742.

Bombelli, L., Koul, R. K., Lee, J., and Sorkin, R. D. 1986. Quantum source of entropy for black holes. *Phys. Rev. D*, **34**, 373.

Bonderson, P., Kitaev, A., and Shtengel, K. 2006. Detecting non-abelian statistics in the $\nu = 5/2$ fractional quantum Hall state. *Phys. Rev. Lett.*, **96**, 016803.

Borzi, R. A., Grigera, S. A., Farrell, J. *et al.* 2007. Formation of a nematic fluid at high fields in $Sr_3Ru_2O_7$. *Science*, **315**, 214.

Boyanovsky, D., Dagotto, E., and Fradkin, E. 1987. Anomalous currents, induced charge and bound states on a domain wall of a semiconductor. *Nucl. Phys. B*, **285**, 340.

Cabra, D. C., Fradkin, E., Rossini, G. L., and Schaposnik, F. A. 2000. Non-abelian fractional quantum Hall states and chiral coset conformal field theories. *Int. J. Mod. Phys. A*, **30**, 4857.

Calabrese, P., and Cardy, J. 2004. Entanglement entropy and quantum field theory. *JSTAT J. Statist. Mech.: Theor. Exp.* **04**, P06002.

Calabrese, P., and Cardy, J. 2007. Entanglement and correlation functions following a local quench: A conformal field theory approach. *JSTAT J. Statist. Mech.: Theor. Exp.*, **2007**, P10004.

Calabrese, P., and Cardy, J. 2009. Entanglement entropy and conformal field theory. *J. Phys. A: Math. Theor.*, **42**, 504005.

Calabrese, P., and Essler, F. H. L. 2010. Universal corrections to scaling for block entanglement in spin-1/2 XX chains. *JSTAT J. Statist. Mech.: Theor. Exp.*, **2010**, P08029.

Calabrese, P., and Lefevre, A. 2008. Entanglement spectrum in one-dimensional systems. *Phys. Rev. A*, **78**, 032329.

Calabrese, P., Cardy, J., and Tonni, E. 2009. Entanglement entropy of two disjoint intervals in conformal field theory. *JSTAT J. Statist. Mech.: Theor. Exp.*, **2009**, P11001.

Callan, C. G., and Harvey, J. A. 1985. Anomalies and fermion zero modes on strings and domain walls. *Nucl. Phys. B*, **250**, 427.

Callan, C. G., and Wilczek, F. 1994. On geometric entropy. *Phys. Lett. B*, **333**, 55.

Camino, F. E., Zhou, W., and Goldman, V. J. 2005. Realization of a Laughlin quasiparticle interferometer: Observation of fractional statistics. *Phys. Rev. B*, **72**, 075342.

Cano, J., and Fendley, P. 2010. Spin Hamiltonians with resonating-valence-bond ground states. *Phys. Rev. Lett.*, **105**, 067205.

Canright, G. S., Girvin, S. M., and Brass, A. 1989. Statistics and flux in two dimensions. *Phys. Rev. Lett.*, **63**, 2291.

Cardy, J. 1996. *Scaling and Renormalization in Statistical Physics*. Cambridge: Cambridge University Press.

Cardy, J., and Peschel, I. 1988. Finite-size dependence of the free energy in two-dimensional critical systems. *Nucl. Phys. B*, **300**, 377.

Cardy, J. L. 1984. Conformal invariance and universality in finite-size scaling. *J. Phys. A: Math. Gen.*, **17**, L385.

Cardy, J. L. 1986. Effect of boundary conditions on the operator content of two-dimensional conformally invariant theories. *Nucl. Phys. B*, **275**, 200.

Cardy, J. L. 1989. Boundary conditions, fusion rules, and the Verlinde formula. *Nucl. Phys. B*, **324**, 581.

Cardy, J. L., and Calabrese, P. 2010. Unusual corrections to scaling in entanglement entropy. *JSTAT J. Statist. Mech.: Theor. Exp.*, **P04023**.

Cardy, J. L., Castro-Alvaredo, O. A., and Doyon, B. 2007. Form factors of branch-point twist fields in quantum integrable models and entanglement entropy. *J. Stat. Phys.*, **130**, 129.

Carlson, E. W., Emery, V. J., Kivelson, S. A., and Orgad, D. 2004. Concepts in high temperature superconductivity, in *The Physics of Conventional and Unconventional Superconductors*, K. H. Bennemann and J. B. Ketterson (eds.). Berlin: Springer-Verlag. arXiv:cond-mat/0206217.

Casini, H., and Huerta, M. 2005. Entanglement and alpha entropies for a massive scalar field in two dimensions. *JSTAT J. Statist. Mech.: Theor. Exp.*, **2005**, P12012.

Casini, H., and Huerta, M. 2007. Universal terms for the entanglement entropy in $2+1$ dimensions. *Nucl. Phys. B*, **764**, 183.

Casini, H., and Huerta, M. 2009. Entanglement entropy in free quantum field theory. *J. Phys. A: Math. Theor.*, **42**, 504007.

Casini, H., Fosco, C. D., and Huerta, M. 2005. Entanglement and alpha entropies for a massive Dirac field in two dimensions. *JSTAT J. Statist. Mech.: Theor. Exp.*, **2005**, P07007.

Casini, H., Huerta, M., and Myers, R. C. 2011. Towards a derivation of holographic entanglement entropy. *JHEP J. High Energy Phys.*, **2011**, 036.

Castelnovo, C., Chamon, C., Mudry, C., and Pujol, P. 2004. From quantum mechanics to classical statistical physics: Generalized Rokhsar–Kivelson Hamiltonians and the "stochastic matrix form" decomposition. *Ann. Phys.*, **318**, 316.

Castro Neto, A. H., Guinea, F., Peres, N. M. R., Novoselov, K. S., and Geim, A. K. 2009. The electronic properties of graphene. *Rev. Mod. Phys.*, **81**, 109.

Chaikin, P. M., and Lubensky, T. C. 1995. *Principles of Condensed Matter Physics*. Cambridge: Cambridge University Press.

Chakravarty, S. 2010. Scaling of von Neumann entropy at the Anderson transition. *Int. J. Mod. Phys. B*, **24**, 1823 (Special volume on Fifty Years of Anderson Localization).

Chakravarty, S., Halperin, B. I., and Nelson, D. R. 1988. Low-temperature behavior of two-dimensional quantum antiferromagnets. *Phys. Rev. Lett.*, **60**, 1057.

Chamon, C., Fradkin, E., and López, A. 2007. Fractional statistics and duality: Strong tunneling behavior of edge states of quantum Hall liquids in the Jain sequence. *Phys. Rev. Lett.*, **98**, 176801.

Chamon, C., Jackiw, R., Nishida, Y., Pi, S. Y., and Santos, L. 2010. Quantizing Majorana fermions in a superconductor. *Phys. Rev. B*, **81**, 224515.

Chamon, C. de C., and Fradkin, E. 1997. Distinct universal conductances in tunneling to quantum Hall states: The role of contacts. *Phys. Rev. B*, **56**, 2012.

Chamon, C. de C., Freed, D. E., and Wen, X. G. 1996. Nonequilibrium quantum noise in chiral Luttinger liquids. *Phys. Rev. B*, **53**, 4033.

Chamon, C. de C., Freed, D. E., Kivelson, S. A., Sondhi, S. L., and Wen, X. G. 1997. Two-point-contact interferometer for quantum Hall systems. *Phys. Rev. B*, **55**, 2331.

Chandra, P., Coleman, P., and Larkin, A. I. 1990. Ising transition in frustrated Heisenberg models. *Phys. Rev. Lett.*, **64**, 88.

Chang, A. M. 2003. Chiral Luttinger liquids at the fractional quantum Hall edge. *Rev. Mod. Phys.*, **75**, 1449.

Chang, A. M., Pfeiffer, L. N., and West, K. W. 1996. Observation of chiral Luttinger behavior in electron tunneling into fractional quantum Hall edges. *Phys. Rev. Lett.*, **77**, 2538.

Chen, Y.-H., Wilczek, F., Witten, E., and Halperin, B. I. 1989. On anyon superconductivity. *Int. J. Mod. Phys. B*, **3**, 1001.

Cho, G. Y., and Moore, J. E. 2011. Topological BF field theory description of topological insulators. *Ann. Phys.*, **326**, 1515.

Cho, H., Young, J. B., Kang, W. *et al.* 1998. Hysteresis and spin transitions in the fractional quantum Hall effect. *Phys. Rev. Lett.*, **81**, 2522.

Chubukov, A. V. 1993. Kohn–Luttinger effect and the instability of a two-dimensional repulsive Fermi liquid at $T = 0$. *Phys. Rev. B*, **48**, 1097.

Chung, S. B., Bluhm, H., and Kim, E.-A. 2007. Stability of half-quantum vortices in $p_x + ip_y$ superconductors. *Phys. Rev. Lett.*, **99**, 197002.

Coleman, P. 1984. New approach to the mixed-valence problem. *Phys. Rev. B*, **29**, 3035.

Coleman, S. 1975. Quantum sine–Gordon equation as the massive Thirring model. *Phys. Rev. D*, **11**, 2088.

Coleman, S. 1985. *Aspects of Symmetry*. Cambridge: Cambridge University Press.

Cooper, N. R., Wilkin, N. K., and Gunn, J. M. F. 2001. Quantum phase of vortices in rotating Bose–Einstein condensates. *Phys. Rev. Lett.*, **87**, 120405.

Creutz, M. 2001. Aspects of chiral symmetry and the lattice. *Rev. Mod. Phys.*, **73**, 119.

Dagotto, E., and Moreo, A. 1989. Phase diagram of the frustrated spin-1/2 Heisenberg antiferromagnet in 2 dimensions. *Phys. Rev. Lett.*, **63**, 2148.

Dagotto, E., Fradkin, E., and Moreo, A. 1988. SU(2) gauge invariance and order parameters in strongly coupled electronic systems. *Phys. Rev. B*, **38**, 2926.

Das Sarma, S., and Pinczuk, A. (eds.). 1997. *Perspectives in Quantum Hall Effects: Novel Quantum Liquids in Two-Dimensional Semiconductor Structures*. New York, NY: Wiley.

Das Sarma, S., Freedman, M., and Nayak, C. 2005. Topologically protected qubits from a possible non-abelian fractional quantum Hall state. *Phys. Rev. Lett.*, **94**, 166802.

Das Sarma, S., Freedman, M., Nayak, C., Simon, S. H., and Stern, A. 2008. Non-abelian anyons and topological quantum computation. *Rev. Mod. Phys.*, **80**, 1083.

Dashen, R., and Frishman, Y. 1975. Four-fermion interactions and scale invariance. *Phys. Rev. D*, **11**, 2781.

Dashen, R., Hasslacher, B., and Neveu, A. 1975. Particle spectrum in model field theories from semiclassical functional integral techniques. *Phys. Rev. D*, **12**, 2443.

de Gennes, P. G. 1966. *Superconductivity of Metals and Alloys*. New York, NY: W. A. Benjamin.

de Gennes, P. G., and Prost, J. 1993. *The Physics of Liquid Crystals*. Oxford: Oxford Science/Clarendon.

de Picciotto, R., Reznikov, M., Heiblum, M. *et al.* 1997. Direct observation of a fractional charge. *Nature*, **389**, 162.

den Nijs, M. P. M. 1981. Derivation of extended scaling relations between critical exponents in two-dimensional models from the one-dimensional Luttinger model. *Phys. Rev. B*, **23**, 6111.

Deser, S., Jackiw, R., and Templeton, S. 1982. Three-dimensional massive gauge theories. *Phys. Rev. Lett.*, **48**, 975.

Di Francesco, P., Mathieu, P., and Sénéchal, D. 1997. *Conformal Field Theory*. Berlin: Springer-Verlag.

Dirac, P. A. M. 1931. Quantised singularities in the electromagnetic field. *Proc. Roy. Soc. London*, **133**, 60.

Dirac, P. A. M. 1955. Gauge invariant formulation of quantum electrodynamics. *Can. J. Phys.*, **33**, 650.

Dixon, L., Friedan, D., Martinec, E., and Shenker, S. 1987. The conformal field theory of orbifolds. *Nucl. Phys. B*, **282**, 13.

Dolev, M., Heiblum, M., Umansky, V., Stern, A., and Mahalu, D. 2008. Observation of a quarter of an electron charge at the $\nu = 5/2$ quantum Hall state. *Nature*, **452**, 829.

Dombre, T., and Kotliar, G. 1989. Instability of the long-range resonating-valence-bond state in the mean-field approach. *Phys. Rev. B*, **39**, 855.

Dombre, T., and Read, N. 1988. Absence of the Hopf invariant in the long-wavelength action of two-dimensional antiferromagnets. *Phys. Rev. B*, **38**, 7181.

Dong, S., Fradkin, E., Leigh, R. G., and Nowling, S. 2008. Topological entanglement entropy in Chern–Simons theories and quantum Hall fluids. *JHEP J. High Energy Phys.*, **05**, 016.

Doniach, S., and Sondheimer, E. H. 1974. *Green's Functions for Solid State Physicists*. New York, NY: Benjamin.

Dotsenko, V. S. 1984. Critical behaviour and associated conformal algebra of the $\mathbb{Z}_3$ Potts model. *Nucl. Phys. B*, **235**, 54.

Dyson, F. J. 1956a. General theory of spin-wave interactions. *Phys. Rev.*, **102**, 1217.
Dyson, F. J. 1956b. Thermodynamic behavior of an ideal ferromagnet. *Phys. Rev.*, **102**, 1230.
Dzyaloshinskii, I., Polyakov, A. M., and Wiegmann, P. B. 1988. Neutral fermions in paramagnetic insulators. *Phys. Lett. A*, **127**, 112.
Eguchi, T., Gilkey, P. B., and Hanson, A. J. 1980. Gravitation, gauge theories and differential geometry. *Phys. Rep.*, **66**, 213.
Einarsson, T., Sondhi, S. L., Girvin, S. M., and Arovas, D. P. 1995. Fractional spin for quantum Hall effect quasiparticles. *Nucl. Phys. B*, **441**, 515.
Einstein, A., Podolsky, P., and Rosen, N. 1935. Can quantum-mechanical description of physical reality be considered complete? *Phys. Rev.*, **47**, 777.
Eisenstein, J. P., and MacDonald, A. H. 2004. Bose–Einstein condensation of excitons in bilayer electron systems. *Nature*, **432**, 691.
Eisenstein, J. P., Störmer, H. L., Pfeiffer, L. N., and West, K. W. 1990. Evidence for a spin transition in the $\nu = 2/3$ fractional quantum Hall effect. *Phys. Rev. B*, **41**, 7910.
Eisenstein, J. P., Boebinger, G. S., Pfeiffer, L. N., West, K. W., and He, S. 1992. New fractional quantum Hall state in double-layer two-dimensional electron systems. *Phys. Rev. Lett.*, **68**.
Eliezer, D., and Semenoff, G. W. 1992a. Anyonization of lattice Chern–Simons theory. *Ann. Phys.*, **217**, 66.
Eliezer, D., and Semenoff, G. W. 1992b. Intersection forms and the geometry of lattice Chern–Simons theory. *Phys. Lett. B*, **286**, 118.
Elitzur, S. 1975. Impossibility of spontaneous breaking of local symmetries. *Phys. Rev. D*, **12**, 3978.
Elitzur, S., Moore, G., Schwimmer, A., and Seiberg, N. 1989. Remarks on the canonical quantization of the Chern–Simons–Witten theory. *Nucl. Phys. B*, **326**, 108.
Elstner, N., Singh, R. P. P., and Young, A. P. 1993. Finite temperature properties of the spin-1/2 Heisenberg antiferromagnet on the triangular lattice. *Phys. Rev. Lett.*, **71**, 1629.
Emery, V. J. 1979. Theory of the one-dimensional electron gas, in *Highly Conducting One-Dimensional Solids*. J. T. Devreese, R. P. Evrard, and V. E. van Doren (eds.). New York, NY: Plenum.
Erdélyi, A. (ed). 1953. *Higher Transcendental Functions*. New York, NY: McGraw-Hill.
Essin, A. M., Moore, J. E., and Vanderbilt, D. 2009. Magnetoelectric polarizability and axion electrodynamics in crystalline insulators. *Phys. Rev. Lett.*, **102**, 146805.
Essler, F. H. L., Frahm, H., Gohmann, F., Klumper, A., and Korepin, V. E. 2005. *The One-Dimensional Hubbard Model*. Cambridge: Cambridge University Press.
Evenbly, G., and Vidal, G. 2009. Entanglement renormalization in two spatial dimensions. *Phys. Rev. Lett.*, **102**, 180406.
Faddeev, L. D. 1976. Introduction to functional methods, in *Methods of Field Theory. Proceedings of the Les Houches Summer School 1975, Session XXVIII*, R. Stora and J. Zinn-Justin (eds.). Amsterdam: North-Holland.
Faddeev, L. D. 1984. Integrable models in (1 + 1)-dimensional quantum field theory, in *Recent Advances in Field Theory and Statistical Mechanics. Proceedings of the 1982 Les Houches Summer School, Session XXXIX*, J.-B. Zuber and R. Stora (eds.). Amsterdam: North-Holland.
Fagotti, M., Calabrese, P., and Moore, J. E. 2011. Entanglement spectrum of random-singlet quantum critical points. *Phys. Rev. B*, **83**, 045110.
Feenberg, E. 1969. *Theory of Quantum Fluids*. New York, NY: Academic Press.

Fendley, P., and Fradkin, E. 2005. Realizing non-abelian statistics in time-reversal-invariant systems. *Phys. Rev. B*, **72**, 024412.

Fendley, P., Saleur, H., and Warner, N.P. 1994. Exact solution of a massless scalar field with a relevant boundary interaction. *Nucl. Phys. B*, **430**, 577.

Fendley, P., Ludwig, A. W. W., and Saleur, H. 1995a. Exact conductance through point contacts in the $\nu = 1/3$ fractional quantum Hall effect. *Phys. Rev. Lett.*, **74**, 3005.

Fendley, P., Ludwig, A. W. W., and Saleur, H. 1995b. Exact nonequilibrium transport through point contacts in quantum wires and fractional quantum Hall devices. *Phys. Rev. B*, **52**, 8934.

Fendley, P., Moessner, R., and Sondhi, S. L. 2002. Classical dimers on the triangular lattice. *Phys. Rev. B*, **66**, 214513.

Fendley, P., Fisher, M. P. A., and Nayak, C. 2006. Dynamical disentanglement across a point contact in a non-Abelian quantum Hall state. *Phys. Rev. Lett.*, **97**, 036801.

Fetter, A. L., and Walecka, J. D. 1971. *Quantum Theory of Many-Particle Systems*. New York, NY: McGraw-Hill.

Fetter, A. L., Hanna, C. B., and Laughlin, R. B. 1989. Random-phase approximation in the fractional-statistics gas. *Phys. Rev. B*, **39**, 9679.

Feynman, R. P. 1972. *Statistical Mechanics, A Set of Lectures*. Reading, MA: W. A. Benjamin Inc.

Feynman, R. P., and Hibbs, A. R. 1965. *Path Integrals and Quantum Mechanics*. New York, NY: McGraw-Hill.

Fidkowski, L., Freedman, M., Nayak, C., Walker, K., and Wang, Z. 2009. From string nets to nonabelions. *Commun. Math. Phys.*, **287**, 805.

Fisher, D. S. 1994. Random antiferromagnetic quantum spin chains. *Phys. Rev. B*, **50**, 3799.

Fisher, D. S. 1995. Critical behavior of random transverse-field Ising spin chains. *Phys. Rev. B*, **51**, 6411.

Fisher, M. E., and Stephenson, J. 1963. Statistical mechanics of dimers on a plane lattice. II. Dimer correlations and monomers. *Phys. Rev*, **132**, 1411.

Fradkin, E. 1989. Jordan–Wigner transformation for quantum-spin systems in two dimensions and fractional statistics. *Phys. Rev. Lett.*, **63**, 322.

Fradkin, E. 1990a. Superfluidity of the lattice anyon gas and topological invariance. *Phys. Rev. B*, **42**, 570.

Fradkin, E. 1990b. The spectrum of short range resonating valence bond theories, in *Field Theories in Condensed Matter Physics, A Workshop. Proceedings of the Johns Hopkins Workshop on Field Theories in Condensed Matter Physics, Baltimore 1988*, Z. Tešanović (ed.). Redwood City, CA: Addison-Wesley, p. 73.

Fradkin, E. 2009. Scaling of entanglement entropy at 2D quantum Lifshitz fixed points and topological fluids. *J. Phys. A: Math. Theor.*, **42**, 504011.

Fradkin, E., and Hirsch, J. E. 1983. Phase diagram of one-dimensional electron–phonon systems. I. The Su–Schrieffer–Heeger model. *Phys. Rev. B*, **27**, 1680.

Fradkin, E., and Kadanoff, L. P. 1980. Disorder variables and para-fermions in two-dimensional statistical mechanics. *Nucl. Phys. B*, **170**, 1.

Fradkin, E., and Kivelson, S. A. 1990. Short range resonating valence bond theories and superconductivity. *Mod. Phys. Lett. B*, **4**, 225.

Fradkin, E., and Kohmoto, M. 1987. Quantized Hall effect and geometric localization of electrons on lattices. *Phys. Rev. B*, **35**, 6017.

Fradkin, E., and Moore, J. E. 2006. Entanglement entropy of 2D conformal quantum critical points: Hearing the shape of a quantum drum. *Phys. Rev. Lett.*, **97**, 050404.

Fradkin, E., and Schaposnik, F. A. 1991. Chern–Simons gauge theories, confinement, and the chiral spin liquid. *Phys. Rev. Lett.*, **66**, 276.

Fradkin, E., and Shenker, S. H. 1979. Phase diagrams of lattice gauge theories with Higgs fields. *Phys. Rev. D*, **19**, 3682.
Fradkin, E., and Stone, M. 1988. Topological terms in one- and two-dimensional quantum Heisenberg antiferromagnets. *Phys. Rev. B*, **38**, 7215(R).
Fradkin, E., and Susskind, L. 1978. Order and disorder in gauge systems and magnets. *Phys. Rev. D*, **17**, 2637.
Fradkin, E., Dagotto, E., and Boyanovsky, D. 1986. Physical realization of the parity anomaly in condensed matter physics. *Phys. Rev. Lett.*, **57**, 2967. Erratum: *Ibid.* **58**, 961 (1987).
Fradkin, E., Moreno, E., and Schaposnik, F. A. 1993. Ground state wave functions for $1+1$ dimensional fermion field theories. *Nucl. Phys. B*, **392**, 667.
Fradkin, E., Nayak, C., Tsvelik, A., and Wilczek, F. 1998. A Chern–Simons effective field theory for the Pfaffian quantum Hall state. *Nucl. Phys. B*, **516**, 704.
Fradkin, E., Nayak, C., and Schoutens, K. 1999. Landau–Ginzburg theories for non-Abelian quantum Hall states. *Nucl. Phys. B*, **546**, 711.
Fradkin, E., Huse, D., Moessner, R., Oganesyan, V., and Sondhi, S. L. 2004. On bipartite Rokhsar–Kivelson points and Cantor deconfinement. *Phys. Rev. B*, **69**, 224415.
Fradkin, E., Kivelson, S. A., Lawler, M. J., Eisenstein, J. P., and Mackenzie, A. P. 2010. Nematic Fermi fluids in condensed matter physics. *Annu. Rev. Condens. Matter Phys.*, **1**, 7.1.
Freedman, M., Nayak, C., Shtengel, K., and Walker, K. 2004. A class of P, T-invariant topological phases of interacting electrons. *Ann. Phys.*, **310**, 428.
Freedman, M. H. 2001. Quantum computation and the localization of modular functors. *Found. Comput. Math.*, **1**, 183.
Freedman, M. H. 2003. A magnetic model with a possible Chern–Simons phase. *Commun. Math. Phys.*, **234**, 129.
Freedman, M. H., Kitaev, A., and Wang, Z. 2002a. Simulation of topological field theories by quantum computers. *Commun. Math. Phys.*, **227**, 587.
Freedman, M. H., Kitaev, A., Larsen, M. J., and Wang, Z. 2002b. Topological quantum computation. *Commun. Math. Phys.*, **227**, 605.
Friedan, D. 1982. A proof of the Nielsen–Ninomiya theorem. *Commun. Math. Phys.*, **85**, 481.
Friedan, D., and Shenker, S. 1987. The analytic geometry of two-dimensional conformal field theory. *Nucl. Phys. B*, **281**, 509.
Friedan, D. H. 1984. Introduction to Polyakov's string theory, in *Recent Advances in Field Theory and Statistical Mechanics. Proceedings of the Les Houches Summer School 1982, Session XXXIX*, J.-B. Zuber and R. Stora (eds.). Amsterdam: North-Holland.
Friedan, D. H. 1985. Nonlinear models in $2+\epsilon$ dimensions. *Ann. Phys.*, **163**, 318.
Friedan, D. H., Qiu, Z., and Shenker, S. H. 1984. Conformal invariance, unitarity, and critical exponents in two dimensions. *Phys. Rev. Lett.*, **52**, 1575.
Fröhlich, J., and Kerler, T. 1991. Universality in quantum Hall systems. *Nucl. Phys. B*, **354**, 369.
Fröhlich, J., and Marchetti, P. A. 1988. Quantum field theory of anyons. *Lett. Math. Phys.*, **16**, 347.
Fröhlich, J., and Zee, A. 1991. Large-scale physics of the quantum Hall fluid. *Nucl. Phys. B*, **364**, 517.
Fu, L., and Kane, C. L. 2006. Time reversal polarization and a Z_2 adiabatic spin pump. *Phys. Rev. B*, **74**, 195312.
Fu, L., and Kane, C. L. 2007. Topological insulators with inversion symmetry. *Phys. Rev. B*, **76**, 045302.

Fu, L., and Kane, C. L. 2008. Superconducting proximity effect and Majorana fermions at the surface of a topological insulator. *Phys. Rev. Lett.*, **100**, 096407.

Fu, L., Kane, C. L., and Mele, E. J. 2007. Topological insulators in three dimensions. *Phys. Rev. Lett.*, **98**, 106803.

Fuchs, J. 1992. *Affine Lie Algebras and Quantum Groups*. Cambridge: Cambridge University Press.

Fursaev, D. V. 2006. Entanglement entropy in critical phenomena and analog models of quantum gravity. *Phys. Rev. D*, **73**, 124025.

Furukawa, S., and Misguich, G. 2007. Topological entanglement entropy in the quantum dimer model on the triangular lattice. *Phys. Rev. B*, **75**, 214407.

Georgi, H. 1982. *Lie Algebras in Particle Physics*. New York, NY: Benjamin/Cummings.

Gepner, D. 1987. New conformal field theories associated with Lie algebras and their partition functions. *Nucl. Phys. B*, **290**, 10.

Ginsparg, P. 1989. Applied conformal field theory, in *Fields, Strings and Critical Phenomena. Proceedings of the Les Houches Summer School 1988, Session XLIX*, E. Brézin and J. Zinn-Justin (eds.). Amsterdam: North-Holland.

Gioev, D., and Klich, I. 2006. Entanglement entropy of fermions in any dimension and the Widom conjecture. *Phys. Rev. Lett.*, **96**, 100503.

Girvin, S. M., and Jach, T. 1984. Formalism for the quantum Hall effect: Hilbert space of analytic functions. *Phys. Rev. B*, **29**, 5617.

Girvin, S. M., and MacDonald, A. H. 1987. Off-diagonal long-range order, oblique confinement, and the fractional quantum Hall effect. *Phys. Rev. Lett.*, **58**, 1252.

Girvin, S. M., MacDonald, A. H., and Platzman, P. M. 1986. Magneto-roton theory of collective excitations in the fractional quantum Hall effect. *Phys. Rev. B*, **33**, 2481.

Gogolin, A. O., Nersesyan, A. A., and Tsvelik, A. M. 1998. *Bosonization and Strongly Correlated Systems*. Cambridge: Cambridge University Press.

Goldenfeld, N. 1992. *Lectures on Phase Transitions and the Renormalization Group*. Reading, MA: Addison-Wesley.

Goldhaber, A. S. 1998. Hairs on the unicorn: Fine structure of monopoles and other solitons, in *Proceedings of the CRM-FIELDS-CAP Workshop "Solitons," Queen's University, Kingston (Ontario, Canada), July 1997*. New York: Springer-Verlag. arXiv:9712190.

Goldstone, J., and Wilczek, F. 1981. Fractional quantum numbers on solitons. *Phys. Rev. Lett.*, **47**, 986.

Golterman, M. F. L., Jansen, K., and Kaplan, D. B. 1993. Chern–Simons currents and chiral fermions on the lattice. *Phys. Lett. B*, **301**, 219.

Greiter, M., Wen, X. G., and Wilczek, F. 1991. Paired Hall state at half-filling. *Phys. Rev. Lett.*, **66**, 3205.

Greiter, M., Wen, X. G., and Wilczek, F. 1992. Paired Hall states. *Nucl. Phys. B*, **374**, 567.

Grinstein, G. 1981. Anisotropic sine–Gordon model and infinite-order phase transitions in three dimensions. *Phys. Rev. B*, **23**, 4615.

Grinstein, G., and Pelcovits, R. A. 1982. Nonlinear elastic theory of smectic liquid crystals. *Phys. Rev. A*, **26**, 915.

Gross, D. J., and Neveu, A. 1974. Dynamical symmetry breaking in asymptotically free field theories. *Phys. Rev. D*, **10**, 3235.

Gubser, S. S., Klebanov, I. R., and Polyakov, A. M. 1998. Gauge theory correlators from non-critical string theory. *Phys. Lett. B*, **428**, 105.

Haldane, F. D. M. 1981. Luttinger liquid theory of one-dimensional quantum fluids. I. Properties of the Luttinger model and their extension to the general 1D interacting spinless Fermi gas. *J. Phys. C: Solid State Phys.*, **14**, 2585.

Haldane, F. D. M. 1982. Spontaneous dimerization in the $S = 1/2$ Heisenberg antiferromagnetic chain with competing interactions. *Phys. Rev. B*, **25**, 4925.
Haldane, F. D. M. 1983a. Continuum dynamics of the 1-D Heisenberg antiferromagnet: Identification with the O(3) nonlinear sigma model. *Phys. Lett. A*, **93**, 464.
Haldane, F. D. M. 1983b. Fractional quantization of the Hall effect: A hierarchy of incompressible quantum fluid states. *Phys. Rev. Lett.*, **51**, 605.
Haldane, F. D. M. 1983c. Nonlinear field theory of large-spin Heisenberg antiferromagnets: Semiclassically quantized solitons of the one-dimensional easy-axis Néel state. *Phys. Rev. Lett.*, **50**, 1153.
Haldane, F. D. M. 1985a. Many-particle translational symmetries of two-dimensional electrons at rational Landau-level filling. *Phys. Rev. Lett.*, **55**, 2095.
Haldane, F. D. M. 1985b. "θ Physics" and quantum spin chains. *J. Appl. Phys.*, **57**, 3359.
Haldane, F. D. M. 1988a. Model for a quantum Hall effect without Landau levels: Condensed-matter realization of the "parity anomaly." *Phys. Rev. Lett.*, **61**, 2015.
Haldane, F. D. M. 1988b. O(3) non-linear σ model and the topological distinction between integer- and half-integer-spin antiferromagnets in two dimensions. *Phys. Rev. Lett.*, **61**, 1029.
Haldane, F. D. M. 2011. Geometrical description of the fractional quantum Hall effect. *Phys. Rev. Lett.*, **107**, 116801.
Haldane, F. D. M., and Rezayi, E. H. 1985. Periodic Laughlin–Jastrow wave functions for the fractional quantized Hall effect. *Phys. Rev. B*, **31**, 2529.
Hallberg, K. 2006. New trends in density matrix renormalization. *Adv. Phys.*, **55**, 477.
Halperin, B. I. 1982. Quantized Hall conductance, current-carrying edge states, and the existence of extended states in a two-dimensional disordered potential. *Phys. Rev. B*, **25**, 2185.
Halperin, B. I. 1983. Theory of the quantized Hall conductance. *Helv. Phys. Acta*, **56**, 75.
Halperin, B. I. 1984. Statistics of quasiparticles and the hierarchy of fractional quantized Hall states. *Phys. Rev. Lett.*, **52**, 1583.
Halperin, B. I., Lee, P. A., and Read, N. 1993. Theory of the half-filled Landau level. *Phys. Rev. B*, **47**, 7312.
Halperin, B. I., Stern, A., Neder, I., and Rosenow, B. 2011. Theory of the Fabry–Pérot quantum Hall interferometer. *Phys. Rev. B*, **83**, 155440.
Hamma, A., Ionicioiu, R., and Zanardi, P. 2005a. Bipartite entanglement and entropic boundary law in lattice spin systems. *Phys. Rev. A*, **71**, 022315.
Hamma, A., Ionicioiu, R., and Zanardi, P. 2005b. Ground state entanglement and geometric entropy in the Kitaev model. *Phys. Lett. A*, **337**, 22.
Hansson, T. H., Oganesyan, V., and Sondhi, S. L. 2004. Superconductors are topologically ordered. *Ann. Phys.*, **313**, 497.
Hartnoll, S. A. 2012. Horizons, holography and condensed matter, in *Black Holes in Higher Dimensions*, G. Horowitz (ed.). Cambridge: Cambridge University Press. pp. 387–419.
Hartnoll, S. A., Polchinski, J., Silverstein, E., and Tong, D. 2010. Towards strange metal holography. *J. High Energy Phys. JHEP*, **2010**, 120.
Hasan, M. Z., and Kane, C. L. 2010. Colloquium: Topological insulators. *Rev. Mod. Phys.*, **82**, 3045.
Hasan, M. Z., and Moore, J. E. 2011. Three-dimensional topological insulators. *Annu. Rev. Condens. Matter Phys.*, **2**, 55.
Hastings, M., and Koma, T. 2006. Spectral gap and exponential decay of correlations. *Commun. Math. Phys.*, **265**, 781.
Hawking, S. W. 1975. Particle creation by black holes. *Commun. Math. Phys.*, **43**, 199.

Heeger, A. J., Kivelson, S., Schrieffer, J. R., and Su, W. P. 1988. Solitons in conducting polymers. *Rev. Mod. Phys.*, **60**, 781.

Heemskerk, I., and Polchinski, J. 2011. Holographic and Wilsonian renormalization groups. *J. High Energy Phys. JHEP*, **2011**, 031.

Heinonen, O. (ed.). 1998. *Composite Fermions: A Unified View of the Quantum Hall Regime*. Singapore: World-Scientific Publishing Co.

Hirsch, J. E. 1990. Spin-split states in metals. *Phys. Rev. B*, **41**, 6820.

Hirsch, J. E. 1999. Spin Hall effect. *Phys. Rev. Lett.*, **83**, 1834.

Ho, T. L. 1995. Broken symmetry of two-component $\nu = 1/2$ quantum Hall states. *Phys. Rev. Lett.*, **75**, 1186.

Hofstadter, D. R. 1976. Energy levels and wave functions of Bloch electrons in rational and irrational magnetic fields. *Phys. Rev. B*, **14**, 2239.

Hohenadler, M., Lang, T. C., and Assaad, F. F. 2011. Correlation effects in quantum spin-Hall insulators: A quantum Monte Carlo study. *Phys. Rev. Lett.*, **106**, 100403.

Hohenadler, M., Meng, Z. Y., Lang, T. C. *et al.* 2012. Quantum phase transitions in the Kane–Mele–Hubbard model. *Phys. Rev. B*, **85**, 115132.

Holstein, T., and Primakoff, H. 1940. Field dependence of the intrinsic domain magnetization of a ferromagnet. *Phys. Rev.*, **58**, 1098.

Holzhey, C., Larsen, F., and Wilczek, F. 1994. Geometric and renormalized entropy in conformal field theory. *Nucl. Phys. B*, **424**, 443.

Hosotani, Y., and Chakravarty, S. 1990. Superconductivity in the anyon model. *Phys. Rev. B*, **42**, 342.

Hoyos, C., and Son, D. T. 2012. Hall viscosity and electromagnetic response. *Phys. Rev. Lett.*, **108**, 066805.

Hsu, B., and Fradkin, E. 2010. Universal behavior of entanglement in 2D quantum critical dimer models. *JSTAT J. Statist. Mech.: Theor. Exp.*, **2010**, P09004.

Hsu, B., Grosfeld, E., and Fradkin, E. 2009a. Quantum noise and entanglement generated by a local quantum quench. *Phys. Rev. B*, **80**, 235412.

Hsu, B., Mulligan, M., Fradkin, E., and Kim, E.-A. 2009b. Universal behavior of the entanglement entropy in 2D conformal quantum critical points. *Phys. Rev. B*, **79**, 115421.

Ioffe, L. B., and Larkin, A. I. 1988. Effective action of a two-dimensional antiferromagnet. *Int. J. Mod. Phys. B*, **2**, 203.

Itzykson, C., and Zuber, J. B. 1980. *Quantum Field Theory*, 1st edn. New York, NY: McGraw-Hill.

Ivanov, D. A. 2001. Non-Abelian statistics of half-quantum vortices in p-wave superconductors. *Phys. Rev. Lett.*, **86**, 268.

Jackiw, R., and Rebbi, C. 1976. Solitons with fermion number 1/2. *Phys. Rev. D*, **13**, 3398.

Jackiw, R., and Rossi, P. 1981. Zero modes of the vortex–fermion system. *Nucl. Phys. B*, **190**, 681.

Jackiw, R., and Schrieffer, J. R. 1981. Solitons with fermion number 1/2 in condensed matter and relativistic field theories. *Nucl. Phys. B*, **190**, 253.

Jain, J. K. 1989a. Composite-fermion approach for the fractional quantum Hall effect. *Phys. Rev. Lett.*, **63**, 199.

Jain, J. K. 1989b. Incompressible quantum Hall states. *Phys. Rev. B*, **40**, 8079.

Jain, J. K. 1990. Theory of the fractional quantum Hall effect. *Phys. Rev. B*, **41**, 7653.

Jain, J. K. 2007. *Composite Fermions*, 1st edn. Cambridge: Cambridge University Press.

Jalabert, R. A., and Sachdev, S. 1991. Spontaneous alignment of frustrated bonds in an anisotropic, three-dimensional Ising model. *Phys. Rev. B*, **44**, 686.

Jang, J., Ferguson, D. G., Vakaryuk, V. *et al.* 2011. Observation of half-height magnetization steps in Sr_2RuO_4. *Science*, **331**, 186.
Jansen, K. 1996. Domain wall fermions and chiral gauge theories. *Phys. Rep.*, **273**, 1.
Jia, X., Subramanian, A. R., Gruzberg, I. A., and Chakravarty, S. 2008. Entanglement entropy and multifractality at the localization transition. *Phys. Rev. B*, **77**, 014208.
Jiang, H. C., Yao, H., and Balents, L. 2012. Spin liquid ground state of the spin-$1/2$ square J_1–J_2 Heisenberg model. *Phys. Rev. B*, **86**, 024424.
Jongeward, G. A., Stack, J. D., and Jayaprakash, C. 1980. Monte Carlo calculations on Z_2 gauge-Higgs theories. *Phys. Rev. D*, **21**, 3360.
Jordan, P., and Wigner, E. P. 1928. Pauli's equivalence prohibition. *Z. Phys.*, **47**, 631.
José, J. V., Kadanoff, L. P., Kirkpatrick, S., and Nelson, D. R. 1977. Renormalization, vortices, and symmetry-breaking perturbations in the two-dimensional planar model. *Phys. Rev. B*, **16**, 1217.
Ju, H., Kallin, A. B., Fendley, P., Hastings, M. B., and Melko, R. G. 2012. Universal large-scale entanglement in two-dimensional gapless systems. *Phys. Rev. B*, **85**, 165121.
Kac, M. 1966. Can you hear the shape of a drum? *Amer. Math. Monthly*, **73**, 1.
Kadanoff, L. P. 1969. Operator algebra and the determination of critical indices. *Phys. Rev. Lett.*, **23**, 1430.
Kadanoff, L. P. 1977. The application of renormalization group techniques to quarks and strings. *Rev. Mod. Phys.*, **49**, 267.
Kadanoff, L. P. 1979. Multicritical behavior at the Kosterlitz–Thouless critical point. *Ann. Phys.*, **120**, 39.
Kadanoff, L. P., and Baym, G. 1962. *Quantum Statistical Mechanics: Green's Function Methods in Equilibrium and Non-Equilibrium Problems*. New York, NY: Benjamin.
Kadanoff, L. P., and Brown, A. C. 1979. Correlation functions on the critical lines of the Baxter and Ashkin–Teller models. *Ann. Phys.*, **121**, 318.
Kadanoff, L. P., and Ceva, H. 1971. Determination of an operator algebra for the two-dimensional Ising model. *Phys. Rev. B*, **3**, 3918.
Kadanoff, L. P., and Martin, P. C. 1961. Theory of many-particle systems. II. Superconductivity. *Phys. Rev.*, **124**, 670.
Kalmeyer, V., and Laughlin, R. B. 1987. Equivalence of the resonating-valence-bond and fractional quantum Hall states. *Phys. Rev. Lett.*, **59**, 2095.
Kane, C. L., and Fisher, M. P. A. 1992. Transmission through barriers and resonant tunneling in an interacting one-dimensional electron gas. *Phys. Rev. B*, **46**, 15233.
Kane, C. L., and Fisher, M. P. A. 1994. Nonequilibrium noise and fractional charge in the quantum Hall effect. *Phys. Rev. Lett.*, **72**, 724.
Kane, C. L., and Mele, E. J. 2005a. Quantum spin Hall effect in graphene. *Phys. Rev. Lett.*, **95**, 226801.
Kane, C. L., and Mele, E. J. 2005b. Z_2 Topological order and the quantum spin Hall effect. *Phys. Rev. Lett.*, **95**, 146802.
Kane, C. L., Lee, P. A., Ng, T. K., Chakraborty, B., and Read, N. 1990. Mean-field theory of the spiral phases of a doped antiferromagnet. *Phys. Rev. B*, **41**, 2653.
Kane, C. L., Fisher, M. P. A., and Polchinski, J. 1994. Randomness at the edge: Theory of quantum Hall transport at filling $\nu = 2/3$. *Phys. Rev. Lett.*, **72**, 4129.
Kaplan, D. B. 1992. A method for simulating chiral fermions on the lattice. *Phys. Lett. B*, **288**, 342.
Kennedy, T., and King, C. 1985. Symmetry breaking in the lattice Abelian Higgs model. *Phys. Rev. Lett.*, **55**, 776.
Kennedy, T., Lieb, E., and Shastri, S. 1988. The XY model has long-range order for all spins and all dimensions greater than one. *Phys. Rev. Lett.*, **61**, 2582.

King-Smith, R. D., and Vanderbilt, D. 1993. Theory of polarization of crystalline solids. *Phys. Rev. B*, **47**, 1651.

Kitaev, A. 2009. Periodic table for topological insulators and superconductors, in *Advances in Theoretical Physics: Landau Memorial Conference*, M. Feigelman (ed.). College Park, MA: AIP Conference Proceedings, for the American Institute of Physics, p. 22.

Kitaev, A., and Preskill, J. 2006. Topological entanglement entropy. *Phys. Rev. Lett.*, **96**, 110404.

Kitaev, A. Yu. 2001. Unpaired Majorana fermions in quantum wires. *Physics – Uspekhi*, **44**, 131. (*Proceedings of the Mesoscopic and Strongly Correlated Electron Systems Conference* (9–16 July 2000, Chernogolovka, Moscow Oblast).)

Kitaev, A. Yu. 2003. Fault-tolerant quantum computation by anyons. *Ann. Phys.*, **303**, 2. arXiv:quant-ph/9707021.

Kitazawa, Y., and Murayama, H. 1990. Topological phase transition of anyon systems. *Nucl. Phys. B*, **338**, 777.

Kivelson, S., and Roček, M. 1985. Consequences of gauge invariance for fractionally charged quasi-particles. *Phys. Lett. B*, **156**, 85.

Kivelson, S. A., and Rokhsar, D. S. 1990. Bogoliubov quasiparticles, spinons, and spin–charge decoupling in superconductors. *Phys. Rev. B*, **41**, 11693(R).

Kivelson, S. A., Kallin, C., Arovas, D. P., and Schrieffer, J. R. 1986. Cooperative ring exchange theory of the fractional quantized Hall effect. *Phys. Rev. Lett.*, **56**, 873.

Kivelson, S. A., Rokhsar, D., and Sethna, J. P. 1987. Topology of the resonating valence-bond state: Solitons and high T_c superconductivity. *Phys. Rev. B*, **35**, 865.

Kivelson, S. A., Fradkin, E., and Emery, V. J. 1998. Electronic liquid-crystal phases of a doped Mott insulator. *Nature*, **393**, 550.

Kivelson, S. A., Fradkin, E., and Geballe, T. H. 2004. Quasi-1D dynamics and the nematic phase of the 2D Emery model. *Phys. Rev. B*, **69**, 144505.

Klauder, J. 1979. Path integrals and stationary phase approximations. *Phys. Rev. D*, **19**, 2349.

Klich, I., and Levitov, L. 2009. Quantum noise and an entanglement meter. *Phys. Rev. Lett.*, **102**, 100502.

Knizhnik, V. G., and Zamolodchikov, A. B. 1984. Current algebra and Wess–Zumino model in two dimensions. *Nucl. Phys. B*, **247**, 83.

Kogut, J., and Susskind, L. 1975. Hamiltonian formulation of Wilson's lattice gauge theories. *Phys. Rev. D*, **11**, 395.

Kogut, J. B. 1979. An introduction to lattice gauge theory and spin systems. *Rev. Mod. Phys.*, **51**, 659.

Kogut, J. B. 1983. The lattice gauge theory approach to quantum chromodynamics. *Rev. Mod. Phys.*, **55**, 775.

Kogut, J. B. 1984. A review of the lattice gauge theory approach to quantum chromodynamics, 1982, in *Recent Advances in Field Theory and Statistical Mechanics. Proceedings of the Les Houches Summer School in Theoretical Physics, 1982, Session XXXIX*, L. B. Zuber and R. Stora (eds.). Amsterdam: North-Holland.

Kohmoto, M. 1983. Metal–insulator transition and scaling for incommensurate systems. *Phys. Rev. Lett.*, **51**, 1198.

Kohmoto, M. 1985. Topological invariant and the quantization of the Hall conductance. *Ann. Phys.*, **160**, 343.

Kohmoto, M., and Shapir, Y. 1988. Antiferromagnetic correlations of the resonating-valence-bond state. *Phys. Rev. B*, **37**, 9439.

Kohn, W. 1961. Cyclotron resonance and de Haas–van Alphen oscillations of an interacting electron gas. *Phys. Rev.*, **123**, 1242.

Kohn, W., and Luttinger, J. M. 1965. New mechanism for superconductivity. *Phys. Rev. Lett.*, **15**, 524.

Kondev, J. 1997. Liouville field theory of fluctuating loops. *Phys. Rev. Lett.*, **78**, 4320.

Kondev, J., and Henley, C. L. 1996. Kac–Moody symmetries of critical ground states. *Nucl. Phys. B*, **464**, 540.

König, M., Steffen, S., Brüne, C. *et al.* 2007. Quantum spin Hall insulator state in HgTe quantum wells. *Science*, **318**, 766.

König, M., Buhmann, H., Molenkamp, L. W. *et al.* 2008. The quantum spin Hall effect: Theory and experiment. *J. Phys. Soc. Japan*, **77**, 031007.

Kopp, A., Jia, X., and Chakravarty, S. 2007. Non-analyticity of von Neumann entropy as a criterion for quantum phase transitions. *Ann. Phys. (N.Y.)*, **322**, 1466.

Kosterlitz, J. M. 1974. The critical properties of the two-dimensional XY model. *J. Phys. C: Solid State Phys.*, **7**, 1046.

Kosterlitz, J. M. 1977. The d-dimensional Coulomb gas and the roughening transition. *J. Phys. C: Solid State Phys.*, **10**, 3753.

Kosterlitz, J. M., and Thouless, D. J. 1973. Order, metastability and phase transitions in two-dimensional systems. *J. Phys. C: Solid State Phys.*, **6**, 1181.

Kotliar, G. 1988. Resonating valence bonds and d-wave superconductivity. *Phys. Rev. B*, **37**, 3664.

Kramers, H. A., and Wannier, G. H. 1941. Statistics of the two-dimensional ferromagnet. Part I. *Phys. Rev.*, **60**, 252.

Krauss, L. M., and Wilczek, F. 1989. Discrete gauge symmetry in continuum theories. *Phys. Rev. Lett.*, **62**, 1221.

Kuklov, A. B., Matsumoto, M., Prokof'ev, N. V., Svistunov, B. V., and Troyer, M. 2008. Deconfined criticality: Generic first-order transition in the SU(2) symmetry case. *Phys. Rev. Lett.*, **101**, 050405.

Kwon, H. J., Houghton, A., and Marston, J. B. 1994. Gauge interactions and bosonized fermion liquids. *Phys. Rev. Lett.*, **73**, 284.

Laflorencie, N., Sørensen, E. S., Chang, M.-S., and Affleck, I. 2006. Boundary effects in the critical scaling of entanglement entropy in 1D systems. *Phys. Rev. Lett.*, **96**, 100603.

Landau, L. D., and Lifshitz, E. M. 1975a. *Statistical Physics*, 3rd edn. Oxford: Pergamon Press.

Landau, L. D., and Lifshitz, E. M. 1975b. *The Classical Theory of Fields*, 3rd edn. Oxford: Pergamon Press.

Laughlin, R. B. 1983. Anomalous quantum Hall effect: An incompressible quantum fluid with fractionally charged excitations. *Phys. Rev. Lett.*, **50**, 1395.

Laughlin, R. B. 1987. Elementary theory: The incompressible quantum fluid, in *The Quantum Hall Effect*, R. Prange and S. M. Girvin (eds.). New York, NY: Springer-Verlag, p. 233.

Laughlin, R. B. 1988a. Superconducting ground state of noninteracting particles obeying fractional statistics. *Phys. Rev. Lett.*, **60**, 2677.

Laughlin, R. B. 1988b. The relationship between high-temperature superconductivity and the fractional quantum Hall effect. *Science*, **242**, 525.

Lee, D.-H., and Fisher, M. P. A. 1989. Anyon superconductivity and the fractional quantum Hall effect. *Phys. Rev. Lett.*, **63**, 903.

Lee, S. S. 2008. Stability of the U(1) spin liquid with a spinon Fermi surface in $2+1$ dimensions. *Phys. Rev. B*, **78**, 085129.

Lee, S. S. 2010. Holographic description of quantum field theory. *Nucl. Phys. B*, **832**, 567.

Lee, S. S. 2011. Holographic description of large N gauge theory. *Nucl. Phys. B*, **851**, 143.

Leggett, A. J. 1975. A theoretical description of the new phases of liquid ^{3}He. *Rev. Mod. Phys.*, **47**, 331.

Leinaas, J. M., and Myrheim, J. 1977. On the theory of identical particles. *Il Nuovo Cimento*, **37B**, 1.

Leung, P. W., and Elser, V. 1993. Numerical studies of a 36-site kagome antiferromagnet. *Phys. Rev. B*, **47**, 5459.

Leung, P. W., Chiu, K. C., and Runge, K. J. 1996. Columnar dimer and plaquette resonating-valence-bond orders in the quantum dimer model. *Phys. Rev. B*, **54**, 12938.

Levin, M., and Stern, A. 2009. Fractional topological insulators. *Phys. Rev. Lett.*, **103**, 196803.

Levin, M., and Wen, X.-G. 2005. String-net condensation: A physical mechanism for topological phases. *Phys. Rev. B*, **71**, 045110.

Levin, M., and Wen, X.-G. 2006. Detecting topological order in a ground state wave function. *Phys. Rev. Lett.*, **96**, 110405.

Levin, M., Burnell, F. J., Koch-Janusz, M., and Stern, A. 2011. Exactly soluble models for fractional topological insulators in two and three dimensions. *Phys. Rev. B*, **84**, 235145.

Levine, H., Libby, S. B., and Pruisken, A. M. M. 1983. Electron delocalization by a magnetic field in two dimensions. *Phys. Rev. Lett.*, **51**, 1915.

Li, H., and Haldane, F. D. M. 2008. Entanglement spectrum as a generalization of entanglement entropy: Identification of topological order in non-Abelian fractional quantum Hall effect states. *Phys. Rev. Lett.*, **101**, 010504.

Liang, S. 1990a. Existence of Néel order at $T = 0$ in the spin-1/2 antiferromagnetic Heisenberg model on a square lattice. *Phys. Rev. B*, **42**, 6555.

Liang, S. 1990b. Monte Carlo simulations of the correlation functions for Heisenberg spin chains at $T = 0$. *Phys. Rev. Lett.*, **64**, 1597.

Liang, S. D., Douçot, B., and Anderson, P. W. 1988. Some new variational resonating-valence-bond-type wave functions for the spin-1/2 antiferromagnetic Heisenberg model on a square lattice. *Phys. Rev. B*, **61**, 365.

Lieb, E., and Mattis, D. C. 1965. Exact solution of a many fermion system and its associated boson field. *J. Math. Phys.*, **6**, 304.

Lieb, E., Schultz, T., and Mattis, D. C. 1961. Two soluble models of an antiferromagnetic chain. *Ann. Phys. (N.Y.)*, **16**, 407.

Lieb, E. H., and Ruskai, M. B. 1973. Proof of the strong subadditivity of quantum mechanical entropy. *J. Math. Phys.*, **14**, 1938.

Lieb, E. H., and Wu, F. Y. 1968. Absence of Mott transition in an exact solution of the short-range, one-band model in one dimension. *Phys. Rev. Lett.*, **20**, 1445.

Lilly, M. P., Cooper, K. B., Eisenstein, J. P., Pfeiffer, L. N., and West, K. W. 1999. Evidence for an anisotropic state of two-dimensional electrons in high Landau levels. *Phys. Rev. Lett.*, **82**, 394.

Liu, H., McGreevy, J., and Vegh, D. 2011. Non-Fermi liquids from holography. *Phys. Rev. D*, **83**, 065029.

López, A., and Fradkin, E. 1991. Fractional quantum Hall effect and Chern–Simons gauge theories. *Phys. Rev. B*, **44**, 5246.

López, A., and Fradkin, E. 1993. Response functions and spectrum of collective excitations of fractional quantum Hall effect systems. *Phys. Rev. B*, **47**, 7080.

López, A., and Fradkin, E. 1995. Fermionic Chern–Simons theory for the fractional quantum Hall effect in bilayers. *Phys. Rev. B*, **51**, 4347.

López, A., and Fradkin, E. 1999. Universal structure of the edge states of the fractional quantum Hall states. *Phys. Rev. B*, **59**, 15323.

López, A., and Fradkin, E. 2001. Effective field theory for the bulk and edge states of quantum Hall states in unpolarized single layer and bilayer systems. *Phys. Rev. B*, **63**, 085306.

López, A., Rojo, A. G., and Fradkin, E. 1994. Chern–Simons theory of the anisotropic quantum Heisenberg antiferromagnet on a square lattice. *Phys. Rev. B*, **49**, 15139.

Lowenstein, J. 1984. Introduction to the Bethe-Ansatz approach in $(1 + 1)$-dimensional models, in *Recent Advances in Field Theory and Statistical Mechanics. Proceedings of the 1982 Les Houches Summer School, Session XXXIX*, J.-B. Zuber and R. Stora (eds.). Amsterdam: North Holland.

Luther, A., and Peschel, I. 1975. Calculation of critical exponents in two dimensions from quantum field theory in one dimension. *Phys. Rev. B*, **12**, 3908.

Maciejko, J., Hughes, T. L., and Zhang, S. C. 2011. The quantum spin Hall effect. *Annu. Rev. Condens. Matter Phys.*, **2**, 31.

Mackenzie, A. P., and Maeno, Y. 2003. The superconductivity of Sr_2RuO_4 and the physics of spin-triplet pairing. *Rev. Mod. Phys.*, **75**, 657.

Mahan, G. 1990. *Many-Particle Physics*, 2nd edn. New York, NY: Plenum Press.

Maldacena, J. M. 1998. The large N limit of superconformal field theories and supergravity. *Adv. Theor. Math. Phys.*, **2**, 231.

Maldacena, J. M. 2012. The gauge/gravity duality, in *Black Holes in Higher Dimensions*, G. Horowitz (ed.). Cambridge: Cambridge University Press.

Maldacena, J. M., and Strominger, A. 1996. Statistical entropy of four-dimensional extremal black holes. *Phys. Rev. Lett.*, **77**, 428.

Maleev, S. V. 1957. Scattering of slow neutrons in ferromagnetics (in Russian). *Zh. Éksp. Teor. Fiz.*, **33**, 1010; English translation *JETP* **6**, 776 (1958).

Mandelstam, S. 1975. Soliton operators for the quantized sine–Gordon equation. *Phys. Rev. D*, **11**, 3026.

Manousakis, E. 1991. The spin-1/2 Heisenberg antiferromagnet on a square lattice and its application to the cuprous oxides. *Rev. Mod. Phys.*, **63**, 1.

Marshall, W. 1955. Antiferromagnetism. *Proc. Roy. Soc. A*, **232**, 48.

Marston, J. B., and Affleck, I. 1989. Large-N limit of the Hubbard–Heisenberg model. *Phys. Rev. B*, **39**, 11538.

Martin, P. C. 1967. *Measurements and Correlation Functions*. New York, NY: Gordon & Breach.

Mattis, D. C. 1965. *The Theory of Magnetism*. New York, NY: Harper & Row.

McCoy, B., and Wu, T. T. 1973. *The Two-Dimensional Ising Model*, 1st edn. Cambridge, MA: Harvard University Press.

McGreevy, J. 2010. Holographic duality with a view toward many-body physics. *Adv. High Energy Phys.*, **2010**, 723105.

Melik-Alaverdian, K. Park V., Bonesteel, N. E., and Jain, J. K. 1998. Possibility of p-wave pairing of composite fermions at $\nu = 12$. *Phys. Rev. B*, **58**, R10167.

Meng, Z. Y., Lang, T. C., Wessel, S., Assaad, F. F., and Muramatsu, A. 2010. Quantum spin liquid emerging in two-dimensional correlated Dirac fermions. *Nature*, **464**, 847.

Mermin, N. D. 1979. The topological theory of defects in ordered media. *Rev. Mod. Phys.*, **51**, 591.

Metlitski, M., Hermele, M., Senthil, T., and M. P. A. Fisher. 2008. Monopoles in CP^{N-1} model via the state-operator correspondence. *Phys. Rev. B*, **78**, 214418.

Metlitski, M. A., and Grover, T. 2011. Entanglement entropy in systems with spontaneously broken continuous symmetry. arXiv:1112.5166 (unpublished).

Metlitski, M. A., and Sachdev, S. 2010. Quantum phase transitions of metals in two spatial dimensions. I. Ising-nematic order. *Phys. Rev. B*, **82**, 075127.

Metlitski, M. A., Fuertes, C. A., and Sachdev, S. 2009. Entanglement entropy in the $O(N)$ model. *Phys. Rev. B*, **80**, 115122.

Miller, J. B., Radu, I. P., Zumbühl, D. Z. *et al.* 2007. Fractional quantum Hall effect in a quantum point contact at filling fraction 5/2. *Nature Phys.*, **3**, 561.

Milliken, F. P., Umbach, C. P., and Webb, R. A. 1996. Indications of a Luttinger liquid in the fractional quantum Hall regime. *Solid State Commun.*, **97**, 309.

Milovanović, M., and Read, N. 1996. Edge excitations of paired fractional quantum Hall states. *Phys. Rev. B*, **53**, 13559.

Misguich, G., Jolicoeur, Th., and Girvin, S. M. 2001. Magnetization plateaus of $SrCu_2(BO_3)_2$ from a Chern–Simons theory. *Phys. Rev. Lett.*, **87**, 097203.

Moessner, R., and Sondhi, S. L. 2001a. Ising models of quantum frustration. *Phys. Rev. B*, **63**, 224401.

Moessner, R., and Sondhi, S. L. 2001b. Resonating valence bond phase in the triangular lattice quantum dimer model. *Phys. Rev. Lett.*, **86**, 1881.

Moessner, R., Sondhi, S. L., and Chandra, P. 2000. Two-dimensional periodic frustrated Ising models in a transverse field. *Phys. Rev. Lett.*, **84**, 4457.

Moessner, R., Sondhi, S. L., and Fradkin, E. 2001. Short-ranged resonating valence bond physics, quantum dimer models, and Ising gauge theories. *Phys. Rev. B*, **65**, 024504.

Moore, G., and Read, N. 1991. Non-abelions in the fractional quantum Hall effect. *Nucl. Phys. B*, **360**, 362.

Moore, G., and Seiberg, N. 1989. Classical and quantum conformal field theory. *Commun. Math. Phys.*, **123**, 177.

Moore, J. E., and Balents, L. 2007. Topological invariants of time-reversal-invariant band structures. *Phys. Rev. B*, **75**, 121306.

Moreo, A. 1987. Conformal anomaly and critical exponents of Heisenberg spin models with half-integer spin. *Phys. Rev. B*, **36**, 8582.

Morf, R. H. 1998. Transition from quantum Hall to compressible states in the second Landau level: New light on the $\nu = 5/2$ enigma. *Phys. Rev. Lett.*, **80**, 1505.

Mudry, C., and Fradkin, E. 1994. Separation of spin and charge quantum numbers in strongly correlated systems. *Phys. Rev. B*, **49**, 5200.

Murthy, G., and Shankar, R. 2003. Hamiltonian theories of the fractional quantum Hall effect. *Rev. Mod. Phys.*, **75**, 1101.

Myers, R. C., and Singh, A. 2012. Comments on holographic entanglement entropy and RG flows. *JHEP J. High Energy Phys.*, **2012**, 122.

Nandkishore, R., and Levitov, L. 2010. Quantum anomalous Hall state in bilayer graphene. *Phys. Rev. B*, **82**, 115124.

Nash, C., and Sen, S. 1983. *Topology and Geometry for Physicists*. New York, NY: Academic Press.

Nayak, C., and Wilczek, F. 1994. Non-Fermi liquid fixed point in 2 + 1 dimensions. *Nucl. Phys. B*, **417**, 359.

Nayak, C., and Wilczek, F. 1996. $2n$ Quasihole states realize 2^{n-1}-dimensional spinor braiding statistics in paired quantum Hall states. *Nucl. Phys. B*, **479**, 529.

Nayak, C., Shtengel, K., Orgad, D., Fisher, M. P. A., and Girvin, S. M. 2001. Electrical current carried by neutral quasiparticles. *Phys. Rev. B*, **64**, 235113.

Negele, J. W., and Orland, H. 1988. *Quantum Many-Particle Systems*. New York, NY: Addison-Wesley.

Neupert, T., Santos, L., Chamon, C., and Mudry, C. 2011. Fractional quantum Hall states at zero magnetic field. *Phys. Rev. Lett.*, **106**, 236804.

Nielsen, H. B., and Ninomiya, M. 1981. Absence of neutrinos on a lattice: (I). Proof by homotopy theory. *Nucl. Phys. B*, **185**, 20.

Nienhuis, B. 1987. Two dimensional critical phenomena and the Coulomb gas, in *Phase Transitions and Critical Phenomena*, vol. 11, C. Domb, M. Green, and J. L. Lebowitz (eds.). London: Academic Press.

Nishida, Y., Santos, L., and Chamon, C. 2010. Topological superconductors as non-relativistic limits of Jackiw–Rossi and Jackiw–Rebbi models. *Phys. Rev. B*, **82**, 144513.

Nishioka, T., Ryu, S., and Takayanagi, T. 2009. Holographic entanglement entropy: An overview. *J. Phys. A: Math. Theor.*, **42**, 504008.

Niu, Q., Thouless, D. J., and Wu, Y.-S. 1985. Quantized Hall conductance as a topological invariant. *Phys. Rev. B*, **31**, 3372.

Novoselov, K. S., Geim, A. K., Morozov, S. V. *et al.* 2004. Electric field effect in atomically thin carbon films. *Science*, **306**, 666.

Oganesyan, V., Kivelson, S. A., and Fradkin, E. 2001. Quantum theory of a nematic Fermi fluid. *Phys. Rev. B*, **64**, 195109.

Onsager, L. 1944. Crystal statistics. I. A two-dimensional model with an order–disorder transition. *Phys. Rev.*, **65**, 117.

Ortiz, G., and Martin, R. M. 1994. Macroscopic polarization as a geometric quantum phase: Many-body formulation. *Phys. Rev. B*, **49**, 14202.

Oshikawa, M. 2010. Boundary conformal field theory and entanglement entropy in two-dimensional quantum Lifshitz critical point. arXiv:1007.3739v1.

Pan, W., Xia, J. S., Shvarts, V. *et al.* 1999. Exact quantization of the even-denominator fractional quantum Hall state at $\nu = 5/2$ Landau level filling factor. *Phys. Rev. Lett.*, **83**, 3530.

Pan, W., Störmer, H. L., Tsui, D. C. *et al.* 2003. Fractional quantum Hall effect of composite fermions. *Phys. Rev. Lett.*, **90**, 016801.

Pan, W., Xia, J. S., Störmer, H. L. *et al.* 2008. Experimental studies of the fractional quantum Hall effect in the first excited Landau level. *Phys. Rev. B*, **77**, 075307.

Papanikolaou, S., Raman, K. S., and Fradkin, E. 2007a. Devil's staircases, quantum dimer models, and stripe formation in strong coupling models of quantum frustration. *Phys. Rev. B*, **75**, 094406.

Papanikolaou, S., Luijten, E., and Fradkin, E. 2007b. Quantum criticality, lines of fixed points, and phase separation in doped two-dimensional quantum dimer models. *Phys. Rev. B*, **76**, 134514.

Papanikolaou, S., Raman, K. S., and Fradkin, E. 2007c. Topological phases and topological entropy of two-dimensional systems with finite correlation length. *Phys. Rev. B*, **76**, 224421.

Pasquier, V., and Haldane, F. D. M. 1998. A dipole interpretation of the $\nu = 1/2$ state. *Nucl. Phys. B*, **516**, 719.

Perelomov, A. 1986. *Generalized Coherent States and Their Applications*. Berlin: Springer-Verlag.

Peskin, M. E. 1980. Critical point behavior of the Wilson loop. *Phys. Lett. B*, **94**, 161.

Pines, D., and Nozières, P. 1966. *The Theory of Quantum Liquids*. New York, NY: Benjamin.

Polchinski, J. 1984. Renormalization and effective Lagrangians. *Nucl. Phys. B*, **231**, 269.

Polchinski, J. 1993. Effective field theory and the Fermi surface, in *Recent Directions in Particle Theory: From Superstrings and Black Holes to the Standard Model*

(TASI - 92), J. Harvey and J. Polchinski (eds.). Singapore: World Scientific, for the Theoretical Advanced Study Institute in High Elementary Particle Physics, Boulder, CO.

Polchinski, J. 1994. Low-energy dynamics of the spinon-gauge system. *Nucl. Phys. B*, **422**, 617.

Polchinski, J. 1998. *String Theory*. Cambridge: Cambridge University Press.

Pollmann, F., and Moore, J. E. 2010. Entanglement spectra of critical and near-critical systems in one dimension. *New J. Phys.*, **12**, 025006.

Pollmann, F., Mukerjee, S., Turner, A. M., and Moore, J. E. 2009. Theory of finite-entanglement scaling at one-dimensional quantum critical points. *Phys. Rev. Lett.*, **102**, 255701.

Pollmann, F., Turner, A. M., Berg, E., and Oshikawa, M. 2010. Entanglement spectrum of a topological phase in one dimension. *Phys. Rev. B*, **81**, 064439.

Polyakov, A. M. 1975. Interaction of Goldstone particles in two dimensions. Applications to ferromagnets and massive Yang–Mills fields. *Phys. Lett. B*, **59**, 79.

Polyakov, A. M. 1977. Quark confinement and topology of gauge theories. *Nucl. Phys. B*, **120**, 429.

Polyakov, A. M. 1981. Quantum geometry of bosonic strings. *Phys. Lett. B*, **103**, 211.

Polyakov, A. M. 1987. *Gauge Fields and Strings*. London: Harwood Academic Publishers.

Polyakov, A. M. 1988. Fermi–Bose transmutations induced by gauge fields. *Mod. Phys. Lett. A*, **3**, 325.

Polyakov, A. M., and Wiegmann, P. B. 1983. Theory of non-Abelian Goldstone bosons in 2 dimensions. *Phys. Lett. B*, **131**, 121.

Polyakov, A. M., and Wiegmann, P. B. 1984. Goldstone fields in two dimensions with multivalued actions. *Phys. Lett. B*, **141**, 223.

Prange, R., and Girvin, S. M. 1990. *The Quantum Hall Effect*, 2nd edn. Berlin: Springer-Verlag.

Preskill, J. 2004. Topological quantum computation, in Lecture Notes for Physics 219: Quantum Computation, Chapter 9; Caltech (unpublished).

Preskill, J., and Krauss, P. 1990. Local discrete symmetry and quantum-mechanical hair. *Nucl. Phys. B*, **341**, 50.

Privman, V. 1988. Universal size dependence of the free energy of finite systems near criticality. *Phys. Rev. B*, **38**, 9261.

Privman, V., and Fisher, M. E. 1984. Universal critical amplitudes in finite-size scaling. *Phys. Rev. B*, **30**, 322.

Pruisken, A. M. M. 1984. On localization in the theory of the quantized Hall effect: A two-dimensional realization of the θ-vacuum. *Nucl. Phys. B*, **235**, 277.

Qi, X. L. 2011. Generic wave-function description of fractional quantum anomalous Hall states and fractional topological insulators. *Phys. Rev. Lett.*, **107**, 126803.

Qi, X. L., and Zhang, S. C. 2011. Topological insulators and superconductors. *Rev. Mod. Phys.*, **83**, 1057.

Qi, X. L., Wu, Y. S., and Zhang, S. C. 2006a. General theorem relating the bulk topological number to edge states in two-dimensional insulators. *Phys. Rev. B*, **74**, 045125.

Qi, X. L., Wu, Y. S., and Zhang, S. C. 2006b. Topological quantization of the spin Hall effect in two-dimensional paramagnetic semiconductors. *Phys. Rev. B*, **74**, 085308.

Qi, X. L., Hughes, T. L., and Zhang, S. C. 2008. Topological field theory of time-reversal invariant insulators. *Phys. Rev. B*, **78**, 195424.

Radu, I., Miller, J. B., Marcus, C. M. *et al.* 2008. Quasi-particle properties from tunneling in the $\nu = 5/2$ fractional quantum Hall state. *Science*, **320**, 899.

Raghu, S., and Kivelson, S. A. 2011. Superconductivity from repulsive interactions in the two-dimensional electron gas. *Phys. Rev. B*, **83**, 094518.
Raghu, S., Qi, X.-L., Honerkamp, C., and Zhang, S. C. 2008. Topological Mott insulators. *Phys. Rev. Lett.*, **100**, 156401.
Rajaraman, R. 1985. *Solitons and Instantons*. Amsterdam: North-Holland.
Randjbar-Daemi, S., Salam, A., and Strathdee, J. 1990. Chern–Simons superconductivity at finite temperature. *Nucl. Phys. B*, **340**, 403.
Read, N. 1989. Order parameter and Ginzburg–Landau theory for the fractional quantum Hall effect. *Phys. Rev. Lett.*, **62**, 86.
Read, N. 1998. Lowest-Landau-level theory of the quantum Hall effect: The Fermi-liquid-like state of bosons at filling factor one. *Phys. Rev. B*, **58**, 16262.
Read, N. 2009. Non-abelian adiabatic statistics and Hall viscosity in quantum Hall states and $p_x + ip_y$ paired superfluids. *Phys. Rev. B*, **79**, 045308.
Read, N., and Green, D. 2000. Paired states of fermions in two dimensions with breaking of parity and time-reversal symmetries and the fractional quantum Hall effect. *Phys. Rev. B*, **61**, 10267.
Read, N., and Newns, D. M. 1983. On the solution of the Coqblin–Schreiffer Hamiltonian by the large-N expansion technique. *J. Phys. C: Solid State Phys.*, **16**, 3273.
Read, N., and Rezayi, E. 1999. Beyond paired quantum Hall states: Parafermions and incompressible states in the first excited Landau level. *Phys. Rev. B*, **59**, 8084.
Read, N., and Sachdev, S. 1989. Some features of the phase diagram of the square lattice SU(N) antiferromagnet. *Nucl. Phys. B*, **316**, 609.
Read, N., and Sachdev, S. 1991. Large-N expansion for frustrated quantum antiferromagnets. *Phys. Rev. Lett.*, **66**, 1773.
Redlich, A. N. 1984. Parity violation and gauge noninvariance of the effective gauge field action in three dimensions. *Phys. Rev. D*, **29**, 2366.
Refael, G., and Moore, J. E. 2004. Entanglement entropy of random quantum critical points in one dimension. *Phys. Rev. Lett.*, **93**, 260602.
Refael, G., and Moore, J. E. 2009. Criticality and entanglement in random quantum systems. *J. Phys. A: Math. Theor.*, **42**, 504010.
Rezayi, E. H., and Haldane, F. D. M. 1988. Off-diagonal long-range order in fractional quantum-Hall-effect states. *Phys. Rev. Lett.*, **61**, 1985.
Rezayi, E. H., and Haldane, F. D. M. 2000. Incompressible paired Hall state, stripe order, and the composite fermion liquid phase in half-filled Landau levels. *Phys. Rev. Lett.*, **84**, 4685.
Rieger, H., and Kawashima, N. 1999. Application of a continuous time cluster algorithm to the two-dimensional random quantum Ising ferromagnet. *Eur. Phys. J. B – Condens. Matter Complex Systems*, **9**, 233.
Roddaro, S., Pellegrini, V., Beltram, F., Biasiol, G., and Sorba, L. 2004. Interedge strong-to-weak scattering evolution at a constriction in the fractional quantum Hall regime. *Phys. Rev. Lett.*, **93**, 046801.
Rokhsar, D., and Kivelson, S. A. 1988. Superconductivity and the quantum hard-core dimer gas. *Phys. Rev. Lett.*, **61**, 2376.
Roy, R. 2009. Z_2 classification of quantum spin Hall systems: An approach using time-reversal invariance. *Phys. Rev. B*, **79**, 195321.
Roy, R. 2010. Topological Majorana and Dirac zero modes in superconducting vortex cores. *Phys. Rev. Lett.*, **105**, 186401.
Ruckenstein, A. E., Hirschfeld, P. J., and Appel, J. 1987. Mean-field theory of high-T_c superconductivity: The superexchange mechanism. *Phys. Rev. B*, **36**, 857.

Ryu, S., and Takayanagi, T. 2006a. Aspects of holographic entanglement entropy. *JHEP J. High Energy Phys.*, **08**, 045.
Ryu, S., and Takayanagi, T. 2006b. Holographic derivation of entanglement entropy from AdS/CFT. *Phys. Rev. Lett.*, **96**, 181602.
Ryu, S., Schnyder, A. P., Furusaki, A., and Ludwig, A. W. W. 2010. Topological insulators and superconductors: Tenfold way and dimensional hierarchy. *New J. Phys.*, **12**, 065010.
Sachdev, S. 1999. *Quantum Phase Transitions*. Cambridge: Cambridge University Press.
Sachdev, S., and Read, N. 1991. Large N expansion for frustrated and doped quantum antiferromagnets. *Int. J. Mod. Phys. B*, **5**, 219.
Saminadayar, L., Glattli, D. C., Jin, Y., and Etienne, B. 1997. Observation of the $e/3$ fractionally charged Laughlin quasiparticle. *Phys. Rev. Lett.*, **79**, 2526.
Sandvik, A. W. 2007. Evidence for deconfined quantum criticality in a two-dimensional Heisenberg model with four-spin interactions. *Phys. Rev. Lett.*, **98**, 227202.
Sandvik, A. W. 2010. Continuous quantum phase transition between an antiferromagnet and a valence-bond solid in two dimensions: Evidence for logarithmic corrections to scaling. *Phys. Rev. Lett.*, **104**, 177201.
Santos, L., Neupert, T., Ryu, S., Chamon, C., and Mudry, C. 2011. Time-reversal symmetric hierarchy of fractional incompressible liquids. *Phys. Rev. B*, **84**, 165138.
Schnyder, A. P., Ryu, S., Furusaki, A., and Ludwig, A. W. W. 2008. Classification of topological insulators and superconductors in three spatial dimensions. *Phys. Rev. B*, **78**, 195125.
Schollwöck, U. 2005. The density-matrix renormalization group. *Rev. Mod. Phys.*, **77**, 259.
Schonfeld, J. F. 1981. A mass term for three-dimensional gauge fields. *Nucl. Phys. B*, **185**, 157.
Schrieffer, J. R. 1964. *Theory of Superconductivity*. New York, NY: Addison-Wesley.
Schulman, L. S. 1981. *Techniques and Applications of Path Integration*. New York, NY: Wiley & Sons.
Schultz, T. D., Mattis., D. C., and Lieb, E. H. 1964. Two-dimensional Ising model as a soluble problem of many fermions. *Rev. Mod. Phys.*, **36**, 856.
Semenoff, G. W. 1984. Condensed-matter simulation of a three-dimensional anomaly. *Phys. Rev. Lett.*, **53**, 2449.
Semenoff, G. W. 1988. Canonical quantum field theory with exotic statistics. *Phys. Rev. Lett.*, **61**, 517.
Semenoff, G. W., Sodano, P., and Wu, Y. S. 1989. Renormalization of the statistics parameter in three-dimensional electrodynamics. *Phys. Rev. Lett.*, **62**, 715.
Senthil, T., Vishwanath, A., Balents, L., Sachdev, S., and M. P. A. Fisher. 2004a. Deconfined quantum critical points. *Science*, **303**, 1490.
Senthil, T., Balents, L., Sachdev, S., Vishwanath, A., and M. P. A. Fisher. 2004b. Quantum criticality beyond the Landau–Ginzburg–Wilson paradigm. *Phys. Rev. B*, **70**, 144407.
Shankar, R. 1994. Renormalization-group approach to interacting fermions. *Rev. Mod. Phys.*, **66**, 129.
Sheng, D. N., Gu, Z. C., Sun, K., and Sheng, L. 2011. Fractional quantum Hall effect in the absence of Landau levels. *Nature Commun.*, **2**, 389.
Shenker, S. H., and Tobochnik, J. 1980. Monte Carlo renormalization group analysis of the classical Heisenberg model in two dimensions. *Phys. Rev. B*, **22**, 4462.
Shirane, G., Endoh, Y., Birgeneau, R., and Kastner, M. 1987. Two-dimensional antiferromagnetic quantum spin-fluid state in La_2CuO_4. *Phys. Rev. Lett.*, **59**, 1613.
Shraiman, B. I., and Siggia, E. D. 1989. Spiral phase of a doped quantum antiferromagnet. *Phys. Rev. Lett.*, **62**, 1564.

Simon, B. 1983. Holonomy, the quantum adiabatic theorem, and Berry's phase. *Phys. Rev. Lett.*, **51**, 2167.

Singh, R. R. P., and Huse, D. A. 1992. Three-sublattice order in triangular- and kagomé-lattice spin-half antiferromagnets. *Phys. Rev. Lett.*, **68**, 1766.

Sondhi, S. L., Karlhede, A., Kivelson, S. A., and Rezayi, E. H. 1993. Skyrmions and the crossover from the integer to fractional quantum Hall effect at small Zeeman energies. *Phys. Rev. B*, **47**, 16419.

Sørensen, E. S., Chang, M.-S., Laflorencie, N., and Affleck, I. 2007a. Impurity entanglement entropy and the Kondo screening cloud. *JSTAT J. Statist. Mech.: Theor. Exp.*, **07**, L01001.

Sørensen, E. S., Chang, M.-S., Laflorencie, N., and Affleck, I. 2007b. Quantum impurity entanglement. *JSTAT J. Statist. Mech.: Theor. Exp.*, **07**, P08003.

Spielman, I. B., Eisenstein, J. P., Pfeiffer, L. N., and West, K. W. 2000. Resonantly enhanced tunneling in a double layer quantum Hall ferromagnet. *Phys. Rev. Lett.*, **84**, 5808.

Srednicki, M. 1993. Entropy and area. *Phys. Rev. Lett.*, **71**, 666.

Stefano, S., Pellegrini, V., and Beltram, F. 2004. Quasi-particle tunneling at a constriction in a fractional quantum Hall state. *Solid State Commun.*, **131**, 565.

Stell, G. 1964. Cluster expansions for classical systems in equilibrium, in *The Equilibrium Theory of Classical Fluids*, H. L. Frisch and J. L. Lebowitz (eds.). New York, NY: W. A. Benjamin Inc, pp. 171–267.

Stéphan, J. M., Furukawa, S., Misguich, G., and Pasquier, V. 2009. Shannon and entanglement entropies of one- and two-dimensional critical wave functions. *Phys. Rev. B*, **80**, 184421.

Stéphan, J. M., Misguich, G., and Pasquier, V. 2010. Rényi entropy of a line in two-dimensional Ising models. *Phys. Rev. B*, **82**, 125455.

Stéphan, J. M., Misguich, G., and Pasquier, V. 2011. Phase transition in the Rényi–Shannon entropy of Luttinger liquids. *Phys. Rev. B*, **84**, 195128.

Stern, A., and Halperin, B. I. 2006. Proposed experiments to probe the non-Abelian $\nu = 5/2$ quantum Hall state. *Phys. Rev. Lett.*, **96**, 016802.

Stone, M. 1986. Born–Oppenheimer approximation and the origin of Wess–Zumino terms: Some quantum-mechanical examples. *Phys. Rev. D*, **33**, 1191.

Stone, M. 1991. Vertex operators in the quantum Hall effect. *Int. J. Mod. Phys. B*, **5**, 509.

Su, W. P., and Schrieffer, J. R. 1981. Fractionally charged excitations in charge-density-wave systems with commensurability 3. *Phys. Rev. Lett.*, **46**, 738.

Su, W. P., Schrieffer, J. R., and Heeger, A. J. 1979. Solitons in polyacetylene. *Phys. Rev. Lett.*, **42**, 1698.

Sun, K., and Fradkin, E. 2008. Time-reversal symmetry breaking and spontaneous anomalous Hall effect in Fermi fluids. *Phys. Rev. B*, **78**, 245122.

Sun, K., Yao, H., Fradkin, E., and Kivelson, S. A. 2009. Topological insulators and nematic phases from spontaneous symmetry breaking in 2D Fermi systems with a quadratic band crossing. *Phys. Rev. Lett.*, **103**, 046811.

Susskind, L. 1977. Lattice fermions. *Phys. Rev. D*, **16**, 3031.

Susskind, L. 1995. The world as a hologram. *J. Math. Phys.*, **36**, 6377.

Susskind, L. 2008. *The Black Hole War*. New York, NY: Back Bay Books, Little Brown & Co.

Susskind, L., and Witten, E. 1998. The holographic bound in anti-de Sitter space. arXiv:hep-th/9805114 (unpublished).

Sutherland, B. 1988. Systems with resonating-valence-bond ground states: Correlations and excitations. *Phys. Rev. B*, **37**, 3786.

't Hooft, G. 1978. On the phase transition towards permanent quark confinement. *Nucl. Phys. B*, **138**, 1.

't Hooft, G. 1979. A property of electric and magnetic flux in non-Abelian gauge theories. *Nucl. Phys. B*, **153**, 141.

't Hooft, G. 1993. Dimensional reduction in quantum gravity. arXiv:gr-gc/9310026v2 (unpublished).

Tang, E., Mei, J. W., and Wen, X. G. 2011a. High-temperature fractional quantum Hall states. *Phys. Rev. Lett.*, **106**, 236802.

Tang, Y., Sandvik, A. W., and Henley, C. L. 2011b. Properties of resonating-valence-bond spin liquids and critical dimer models. *Phys. Rev. B*, **84**, 174427.

Thouless, D. J. 1983. Quantization of particle transport. *Phys. Rev. B*, **27**, 6083.

Thouless, D. J., Kohmoto, M., Nightingale, M. P., and den Nijs, M. P. M. 1982. Quantized Hall conductance in a two-dimensional periodic potential. *Phys. Rev. Lett.*, **49**, 405.

Tomonaga, S. 1950. Remarks on Bloch's method of sound waves applied to many-fermion problems. *Prog. Theor. Phys.*, **5**, 544.

Trebst, S., Werner, P., Troyer, M., Shtengel, K., and Nayak, C. 2007. Breakdown of a topological phase: Quantum phase transition in a loop gas model with tension. *Phys. Rev. Lett.*, **98**, 070602.

Trugman, S. A. 1983. Localization, percolation, and the quantum Hall effect. *Phys. Rev. B*, **27**, 7539.

Trugman, S. A., and Kivelson, S. 1985. Exact results for the fractional quantum Hall effect with general interactions. *Phys. Rev. B*, **31**, 5280.

Tsui, D. C., Stormer, H. L., and Gossard, A. C. 1982. Two-dimensional magnetotransport in the extreme quantum limit. *Phys. Rev. Lett.*, **48**, 1559.

Tupitsyn, I. S., Kitaev, A., Prokof'ev, N. V., and Stamp, P. C. E. 2010. Topological multicritical point in the phase diagram of the toric code model and three-dimensional lattice gauge Higgs model. *Phys. Rev. B*, **82**, 085114.

Vafek, O. 2010. Interacting fermions on the honeycomb bilayer: From weak to strong coupling. *Phys. Rev. B*, **82**, 205106.

Vafek, O., and Yang, K. 2010. Many-body instability of Coulomb interacting bilayer graphene: Renormalization group approach. *Phys. Rev. B*, **81**, 041401(R).

Vakaryuk, V., and Leggett, A. J. 2009. Spin polarization of half-quantum vortex in systems with equal spin pairing. *Phys. Rev. Lett.*, **103**, 057003.

Varney, C. N., Sun, K., Rigol, M., and Galitski, V. 2010. Interaction effects and quantum phase transitions in topological insulators. *Phys. Rev. B*, **82**, 115125.

Verlinde, E. 1988. Fusion rules and modular transformations in 2D conformal field theory. *Nucl. Phys. B*, **300**, 360.

Verstraete, F., Wolf, M. M., Perez-García, D., and Cirac, J. I. 2006. Criticality, the area law, and the computational power of PEPS. *Phys. Rev. Lett.*, **96**, 220601.

Vishwanath, A., Balents, L., and Senthil, T. 2004. Quantum criticality and deconfinement in phase transitions between valence bond solids. *Phys. Rev. B*, **69**, 224416.

Vollhardt, D., and Wölfle, P. 1990. *The Superfluid Phases of Helium 3*. London: Taylor & Francis.

Volovik, G. E. 1988. An analog of the quantum Hall effect in a superfluid ^{3}He film. *Sov. Phys. JETP*, **67**, 1804.

von Klitzing, K., Dorda, G., and Pepper, M. 1980. New method for high-accuracy determination of the fine-structure constant based on quantized Hall resistance. *Phys. Rev. Lett.*, **45**, 494.

Wang, L., Gu, Z. C., Wen, X. G., and Verstraete, F. 2011. Possible spin liquid state in the spin 1/2 J_1–J_2 antiferromagnetic Heisenberg model on square lattice: A tensor product state approach. arXiv:1112.3331.

Wegner, F. 1979. The mobility edge problem: Continuous symmetry and a conjecture. *Z. Phys. B Condens. Matter*, **35**, 207.

Wegner, F. J. 1971. Duality in generalized Ising models and phase transitions without local order parameters. *J. Math. Phys.*, **12**, 2259.

Wen, X. G. 1989. Vacuum degeneracy of chiral spin states in compactified space. *Phys. Rev. B*, **40**, 7387.

Wen, X. G. 1990a. Chiral Luttinger liquid and the edge excitations in the fractional quantum Hall states. *Phys. Rev. B*, **41**, 12838.

Wen, X. G. 1990b. Electrodynamical properties of gapless edge excitations in the fractional quantum Hall states. *Phys. Rev. Lett.*, **64**, 2206.

Wen, X. G. 1990c. Topological orders in rigid states. *Int. J. Mod. Phys. B*, **4**, 239.

Wen, X. G. 1991a. Edge excitations in the fractional quantum Hall states at general filling fractions. *Mod. Phys. Lett. B*, **5**, 39.

Wen, X. G. 1991b. Gapless boundary excitations in the quantum Hall states and in the chiral spin states. *Phys. Rev. B*, **43**, 11025.

Wen, X. G. 1991c. Mean-field theory of spin-liquid states with finite energy gap and topological orders. *Phys. Rev. B*, **44**, 2664.

Wen, X. G. 1995. Topological orders and edge excitations in fractional quantum Hall states. *Adv. Phys.*, **44**, 405.

Wen, X. G. 1999. Projective construction of non-Abelian quantum Hall liquids. *Phys. Rev. B*, **60**, 8827.

Wen, X.-G. 2002. Quantum orders and symmetric spin liquids. *Phys. Rev. B*, **65**, 165113.

Wen, X. G., and Niu, Q. 1990. Ground-state degeneracy of the fractional quantum Hall states in the presence of a random potential and on high-genus Riemann surfaces. *Phys. Rev. B*, **41**, 9377.

Wen, X. G., and Zee, A. 1988. Spin waves and topological terms in the mean-field theory of two-dimensional ferromagnets and antiferromagnets. *Phys. Rev. Lett.*, **61**(1025).

Wen, X. G., and Zee, A. 1989. Winding number, family index theorem, and electron hopping in a magnetic field. *Nucl. Phys. B*, **316**, 641.

Wen, X. G., and Zee, A. 1990. Compressibility and superfluidity in the fractional-statistics liquid. *Phys. Rev. B*, **41**, 240.

Wen, X.-G., and Zee, A. 1992. Classification of Abelian quantum Hall states and matrix formulation of topological fluids. *Phys. Rev. B*, **46**, 2290.

Wen, X. G., Wilczek, F., and Zee, A. 1989. Chiral spin states and superconductivity. *Phys. Rev. B*, **39**, 11413.

Wen, X. G., Dagotto, E., and Fradkin, E. 1990. Anyons on a torus. *Phys. Rev. B*, **42**, 6110.

Wesolowski, D., Hosotani, Y., and Ho, C. L. 1994. Multiple Chern–Simons fields on a torus. *Int. J. Mod. Phys.*, **A9**, 969.

White, S. R. 1992. Density matrix formulation for quantum renormalization groups. *Phys. Rev. Lett.*, **69**, 2863.

Wiegmann, P. B. 1988. Superconductivity in strongly correlated electronic systems and confinement versus deconfinement phenomenon. *Phys. Rev. Lett.*, **60**, 821.

Wiegmann, P. B. 1989. Multivalued functionals and geometrical approach for quantization of relativistic particles and strings. *Nucl. Phys. B*, **323**, 311.

Wilczek, F. 1982. Magnetic flux, angular momentum, and statistics. *Phys. Rev. Lett.*, **48**, 1144.

Wilczek, F., and Zee, A. 1983. Linking numbers, spin, and statistics of solitons. *Phys. Rev. Lett.*, **51**, 2250.
Willett, R., Eisenstein, J. P., Störmer, H. L. *et al.* 1987. Observation of an even-denominator quantum number in the fractional quantum Hall effect. *Phys. Rev. Lett.*, **59**, 1776.
Willett, R. L., l. N. Pfeiffer, and West, K. W. 2010. Alternation and interchange of $e/4$ and $e/2$ period interference oscillations consistent with filling factor $5/2$ non-Abelian quasiparticles. *Phys. Rev. B*, **82**, 205301.
Williams, J. R., Bestwick, A. J., Gallagher, P. *et al.* 2012. Signatures of Majorana fermions in hybrid superconductor–topological insulator devices. *Phys. Rev. Lett.*, **109**, 056803.
Wilson, K. G. 1969. Non-Lagrangian models of current algebra. *Phys. Rev.*, **179**, 1499.
Wilson, K. G. 1973. Quantum field theory models in less than 4 dimensions. *Phys. Rev. D*, **7**, 2911.
Wilson, K. G. 1974. Confinement of quarks. *Phys. Rev. D*, **10**, 2445.
Wilson, K. G. 1975. The renormalization group: Critical phenomena and the Kondo problem. *Rev. Mod. Phys.*, **47**, 773.
Wilson, K. G. 1983. The renormalization group and critical phenomena. *Rev. Mod. Phys.*, **55**, 583.
Wilson, K. G., and Kogut, J. B. 1974. The renormalization group and the ϵ expansion. *Physics Reports C*, **12**, 75.
Witten, E. 1979. Dyons of charge $e\theta/2\pi$. *Phys. Lett. B*, **86**, 283.
Witten, E. 1983. Current algebra, baryons, and quark confinement. *Nucl. Phys. B*, **223**, 422.
Witten, E. 1984. Non-Abelian bosonization in two dimensions. *Commun. Math. Phys.*, **92**, 455.
Witten, E. 1989. Quantum field theory and the Jones polynomial. *Commun. Math. Phys.*, **121**, 351.
Witten, E. 1992. On holomorphic factorization of WZW and coset models. *Commun. Math. Phys.*, **144**, 189.
Witten, E. 1998. Anti de Sitter space and holography. *Adv. Theor. Math. Phys.*, **2**, 253.
Wolf, M. M. 2006. Violation of the entropic area law for fermions. *Phys. Rev. Lett.*, **96**, 010404.
Wu, C., Sun, K., Fradkin, E., and Zhang, S. C. 2007. Fermi liquid instabilities in the spin channel. *Phys. Rev. B*, **75**, 115103.
Wu, T. T., and Yang, C. N. 1975. Concept of nonintegrable phase factors and global formulation of gauge fields. *Phys. Rev. D*, **12**, 3845.
Wu, T. T., and Yang, C. N. 1976. Dirac monopole without strings: Monopole harmonics. *Nucl. Phys. B*, **107**, 365.
Wu, Y. S., and Zee, A. 1984. Comments on the Hopf Lagrangian and fractional statistics of solitons. *Phys. Lett. B*, **147**, 325.
Xia, J. S., Pan, W., Vicente, C. L. *et al.* 2004. Electron correlation in the second Landau level: A competition between many nearly degenerate quantum phases. *Phys. Rev. Lett.*, **93**, 176809.
Yakovenko, V. M. 1990. Chern–Simons terms and *n* field in Haldane's model for the quantum Hall effect without Landau levels. *Phys. Rev. Lett.*, **65**(Jul), 251.
Yang, C. N., and Yang, C. P. 1969. Thermodynamics of a one-dimensional system of bosons with repulsive delta-function interaction. *J. Math. Phys.*, **10**, 1115.
Yang, K., Warman, L. K., and Girvin, S. M. 1993. Possible spin-liquid states on the triangular and kagomé lattices. *Phys. Rev. Lett.*, **70**, 2641.
Yang, K., Moon, K., Zheng, L. *et al.* 1994. Quantum ferromagnetism and phase transitions in double-layer quantum Hall systems. *Phys. Rev. Lett.*, **72**, 732.

Youngblood, R., Axe, J., and McCoy, B. M. 1980. Correlations in ice-rule ferroelectrics. *Phys. Rev. B*, **21**, 5212.
Zak, J. 1964. Magnetic translation group. *Phys. Rev.*, **134**, A1602.
Zamolodchikov, A. B. 1986. "Irreversibility" of the flux of the renormalization group in a 2D field theory. *Pis'ma Zh. Éksp. Teor. Fiz.*, **43**, 565; *JETP Lett.* **43**, 730 (1986).
Zamolodchikov, A. B., and Zamolodchikov, A. B. 1979. Factorized S-matrices in two dimensions as the exact solutions of certain relativistic quantum field theory models. *Ann. Phys.*, **120**, 253.
Zhang, H., Liu, C. X., Qi, X. L. *et al.* 2009. Topological insulators in Bi_2Se_3, Bi_2Te_3 and Sb_2Te_3 with a single Dirac cone on the surface. *Nature Phys.*, **5**, 438.
Zhang, S. C., Hansson, T. H., and Kivelson, S. 1989. Effective-field-theory model for the fractional quantum Hall effect. *Phys. Rev. Lett.*, **62**, 82.
Zheng, D., Zhang, G. M., and Wu, C. 2011. Particle–hole symmetry and interaction effects in the Kane–Mele–Hubbard model. *Phys. Rev. B*, **84**, 205121.
Ziman, T., and Schulz, H. J. 1987. Are antiferromagnetic spin chains representations of the higher Wess–Zumino models? *Phys. Rev. Lett.*, **59**, 140.
Zinn-Justin, J. 2002. *Quantum Field Theory and Critical Phenomena*, 4th edn. Oxford: Oxford University Press.
Zozulya, O. S., Haque, M., and Regnault, N. 2009. Entanglement signatures of quantum Hall phase transitions. *Phys. Rev. B*, **79**, 045409.
Zuber, J. B., and Itzykson, C. 1977. Quantum field theory and the two-dimensional Ising model. *Phys. Rev. D*, **15**, 2875.

Index

Printed in the United States
By Bookmasters